AF797955

Fat Transfer in Plastic Surgery

Alberto Di Giuseppe • Franco Bassetto
Foad Nahai
Editors

Fat Transfer in Plastic Surgery

Techniques, Technology and Safety

Springer

Editors
Alberto Di Giuseppe
Department of Plastic surgery
Plastic Surgery - University of Padova
Padova, Italy

Franco Bassetto
Univ of Padova, Plastic Surgery Clinic
Padova, Italy

Foad Nahai
The Center for Plastic Surgery at Metroderm
Atlanta, GA, USA

ISBN 978-3-031-10880-8 ISBN 978-3-031-10881-5 (eBook)
https://doi.org/10.1007/978-3-031-10881-5

© The Editor(s) (if applicable) and The Author(s), under exclusive license to Springer Nature Switzerland AG 2023
This work is subject to copyright. All rights are solely and exclusively licensed by the Publisher, whether the whole or part of the material is concerned, specifically the rights of translation, reprinting, reuse of illustrations, recitation, broadcasting, reproduction on microfilms or in any other physical way, and transmission or information storage and retrieval, electronic adaptation, computer software, or by similar or dissimilar methodology now known or hereafter developed.
The use of general descriptive names, registered names, trademarks, service marks, etc. in this publication does not imply, even in the absence of a specific statement, that such names are exempt from the relevant protective laws and regulations and therefore free for general use.
The publisher, the authors, and the editors are safe to assume that the advice and information in this book are believed to be true and accurate at the date of publication. Neither the publisher nor the authors or the editors give a warranty, expressed or implied, with respect to the material contained herein or for any errors or omissions that may have been made. The publisher remains neutral with regard to jurisdictional claims in published maps and institutional affiliations.

This Springer imprint is published by the registered company Springer Nature Switzerland AG
The registered company address is: Gewerbestrasse 11, 6330 Cham, Switzerland

The pandemic needlessly claimed millions of victims, disrupted and forever changed even more lives. This book was conceived and completed during that disruptive time. As we emerge from the nightmare of the pandemic, we recognize our front-line healthcare workers who placed their own health and lives at risk whilst selflessly caring for the victims of COVID. We dedicate this book to healthcare workers worldwide: from the front-line medics, paramedics, and nurses to the doctors who worked tirelessly to save lives. We thank you for your dedication, hard work, and service to humanity. We are indebted to you all.

Foreword

The use of adipose tissue in aesthetic and reconstructive procedures has gained tremendous importance during the last three decades and has nowadays become state-of-the-art.

In the last 30 years, when *Sydney Coleman* developed his defined "Lipostructuring" protocol medical doctors from various disciplines, especially plastic surgeons and basic researchers worldwide have intensely worked on the optimization of lipoaspirate processing. The techniques of harvesting, processing, and injecting fat grafts have been further developed.

In addition to the macro-fat graft technique developed by *Sydney Coleman* [1], the milli-fat graft technique described by *Stephen Cohen* [2] and the micro-fat graft technique introduced by *Guy Magalon* [3] have been used for different indications. *Patrick Tonnard's* development of the nano-fat technique has not only taken Regenerative Aesthetic Surgery a big step further but has also extended its use to various medical indications [4]. *Norbert Pallua* [5] with his lipoconcentrate optimized the use of emulsified fat grafts for regenerative indications by a mechanical technique getting a significantly higher number of stromal vascular fraction (SVF) cells, adipose-derived stem cells (ASCs), and endothelial progenitor cells (EPCs) when compared to the native fat and the nano fat.

Fat grafting became an important tool in Plastic and Aesthetic Surgery to improve the appearance of scars and aged skin as well as to correct facial and body contour defects. Traditional body contouring methods with silicone implants on the breasts and buttocks have been supplemented and partially replaced by an alternative of grafting with autologous fat tissue.

An immense number of experimental and clinical studies have been focusing on an increased retention rate as well as on an optimized regenerative effect of fat grafts and ASCs [6]. In his chapter in this book, the working group of *Ramon Llull* discusses the volumetric and regenerative components of fat grafts.

For increased retention—in addition to the improvement of the fat harvesting and grafting technique—the high significance of angiogenic growth factors contained in fat cells has been demonstrated [7]. High-quality research projects, such as Maxim Geeroms, aimed to reduce fibrosis and to improve vascularization by an enrichment of the transplanted fat grafts with EPCs—he addresses this approach in his chapter [8].

Since Patricia Zuk discovered ASCs, the regenerative potential of adipose tissue has become an important area in Plastic and Aesthetic Surgery [9]. Over the past 20 years, the regenerative effect of fat grafts has been scientifically studied and mainly attributed to the SVF cells with the ASCs to the angiogenic growth factors and to the proliferation of the keratinocytes [10]. This progress in the field of regenerative capacities of fat grafts and especially ASCs has opened new indications and applications. In this way, the spectrum of clinical use of fat grafts in many medical fields has expanded enormously.

Fat grafting as a minimal-invasive procedure has become more and more established in many specialties that Plastic Surgeons share with other medical fields, for example, in Dermatology for the treatment of scleroderma; in Orthopedics for the regeneration of articular cartilage in the hip and knee joints; in Gynecology for augmentation of the female genitalia; in Hand Surgery for the treatment of rhizarthrosis or Dupuytren's disease. Thus, Plastic Surgery has also become a link between different medical disciplines through its interdisciplinary use in many areas of the body.

In this book *Fat Transfer in Plastic Surgery: Techniques, Technology and Safety* edited by Alberto Di Giuseppe, Franco Bassetto, and Foad Nahai published by Springer, 46 chapters cover important clinical and scientific aspects of fat grafting and Regenerative Plastic Surgery.

In the three parts of the book, more than 30 invited authors and their co-authors describe general concepts, technologies, as well as new and established surgical procedures of fat grafting. The editors have invited national and international authors according to their expertise, which guarantees this book's high clinical and scientific level. The articles of these international experts encompass the current concepts, techniques, and technologies and provide an outlook on the future of fat grafting as well as on Regenerative Plastic Surgery. It is particularly noteworthy that several chapters in this book deal with the safety of fat grafting, which is of great importance as a guide especially for younger, but also for established colleagues.

The publication of this book which reflects the current standard of research and clinical use, and the safety of adipose tissue transfer contributes significantly to the establishment of fat grafting and regenerative medicine. It will serve as a textbook and guideline for fat grafting for young plastic surgeons and for established colleagues from different medical specialties.

References

1. Coleman SR. Facial recontouring with lipostructure. Clin Plast Surg 1997;24:347–67.
2. Cohen SR, Hewett S, Ross L, Delaunay F, Goodacre A, Ramos C, Leong T, Saad A, Regenerative cells for facial surgery: biofilling and biocontouring. Aesthetic Surg J. 2017;37(Issue suppl 3):S16–32.
3. Nguyen PS, Desouches C, Gay AM, Hautier A, Magalon G. Development of microinjection as an innovative autologous fat graft technique: the use of adipose tissue as dermal filler. J Plast Reconstr Aesthet Surg. 2012;65(12):1692–99.
4. Tonnard P, Verpaele A, Peeters G, Hamdi M, Cornelissen M, Declercq H. Nanofat grafting: basic research and clinical applications. Plast Reconstr Surg. 2013;132(4):1017–26.
5. Pallua N, Grasys J, Kim BS. Enhancement of progenitor cells by two-step centrifugation of emulsified lipoaspirates. Plast Reconstr Surg. 2018;142:99–109.
6. Coleman SR. Long-term survival of fat transplants: controlled demonstrations. Aesthetic Plast Surg. 1995;19:421–5.
7. Pallua N, Pulsfort AK, Suschek C, Wolter TP. Content of the growth factors bFGF, IGF-1, VEGF, and PDGF-BB in freshly harvested lipoaspirate after centrifugation and incubation. Plast Reconstr Surg. 2009;123(3):826–33.
8. Geeroms M, Hamdi M, Hirano R, Hagiwara H, Fujimura S, Mizuno H, Tanaka R. Quality and quantity-cultured murine endothelial progenitor cells increase vascularization and decrease fibrosis in the fat graft. Plast Reconstr Surg. 2019;143(4):744e–55e.
9. Zuk PA, Zhu M, Mizuno H, Huang J, Futrell JW, Katz AJ, Benhaim AL, Lorenz HP, Hedrick MH. Multilineage cells from human adipose tissue: implications for cell-based therapies. Tissue Eng. 2001;7(2):211–18.
10. Kim BS, Gaul C, Paul NE, Dewor M, Stromps JP, Hwang SS, Nourbakhsh M, Bernhagen J, Rennekampff HO, Pallua N. The effect of lipoaspirates on human keratinocytes. Aesthet Surg J. 2016;36(8):941–51.

Düsseldorf, Germany Norbert Pallua

Preface

Over the past decade or so, fat transfer has become one of the most effective "weapons" or addition to the practice of Aesthetic and Reconstructive Surgery.

We undertook the task of editing this book in an attempt to acquaint the medical profession with the latest studies, research, techniques, trends, and technologies on fat transfer to breast, buttocks, and face for aesthetic and reconstructive purposes.

Fat affords unique opportunity to enhance aesthetic outcomes in:

Breast augmentation, on its own or combined with breast implants (hybrid technique).

Body contouring procedures, where simultaneous fat removal and fat grafting are now state-of-the-art to enhance body contouring outcomes.

High-definition techniques in body shaping to enhance abdominal contour and etching, augmentation of pectoral muscles, deltoid, biceps and triceps reshaping with additional fat offer new options for total body contour enhancements.

New technologies, such as laser liposuction, vaser ultrasound liposuction, and expanded vibro liposuction, have been successfully applied to fat transfer facilitating the technique making it easier and faster whilst retaining.

A high percentage of viable fat effecting fat survival rates.

New technologies for fat processing, concentration, and rinsing (revolve canister) have been instrumental in improving fat quality for transfer.

The book includes the latest advancement in research and clinical studies to enhance the safety of Buttock Fat Grafting (Brazilian Butt Lift) BBL.

The numerous fatal events connected with the procedure have been extensively researched and studied in the last 5 years.

The research has led to guidelines including limiting the fat grafting to the subcutaneous layer and avoiding intra- or submuscular deposition of fat.

Technologies as portable ultrasound to guide fat deposition have been added to standard practice of fat buttock transfer, in order to continuously check the level of fat infiltrations.

I want to thank Franco Bassetto and Foad Nahai, two great friends beyond their international reputation in aesthetic plastic and reconstructive surgery, for the continuous support and help in editing such a book.

Finally, I must mention that editing a book in the time of the pandemic has been challenging and more time consuming than normal times.

Reasons for the challenges are psychological, such as less enthusiasm during the darkest days of the pandemic. Finally, despite a bit delay, the "baby," this new book is here.

The book is complimented by videos of the latest techniques and technologies.

I trust our readers will find the book interesting and valuable.

Finally, a heartfelt thanks to Juliette Kleemann of Springer for the great support from the start with encouragement and patience leading to the realization of the book.

Thank you Rakesh Jotheeswaran, book manager, for your great patience and perseverance over these months and your meticulous attention to the details of the book.

Enjoy the book!

Padova, Italy — Alberto Di Giuseppe
Padova, Italy — Franco Bassetto
Atlanta, GA, USA — Foad Nahai

Contents

Part I

General Concepts

Safety in Body Contouring

1

Andrew L. Weinstein and Foad Nahai

Contents

1.1 Introduction

Body contouring encompasses a diverse group of surgical procedures, including brachioplasty, transverse upper body lift, panniculectomy, abdominoplasty, belt lipectomy, lower body lift, and medial thighplasty, with the common aim of removing excess skin and fat. Additionally, liposuction, fat grafting, and their combination fall into this category. While certain body contouring procedures are undertaken to ameliorate medical problems such as intertriginous dermatitis, they all ultimately strive to improve body appearance, quality of life, and self-esteem. Regardless of the indication, these procedures carry perioperative risks ranging from unsightly or misplaced scars to life-threatening venous thromboembolism (VTE) and fat embolism. In outpatient plastic surgery, abdominoplasty has the highest risk of postoperative VTE and gluteal fat grafting, otherwise known as the Brazilian Butt Lift (BBL), the highest risk of fatal fat embolism [1, 2]. As body contouring surgery is always elective, safety is paramount and must be prioritized in every way possible.

A. L. Weinstein
Division of Plastic and Reconstructive Surgery, Emory University School of Medicine, Atlanta, GA, USA

Division of Plastic and Reconstructive Surgery, Weill Cornell Medicine, New York, NY, USA

F. Nahai (✉)
Division of Plastic and Reconstructive Surgery, Emory University School of Medicine, Atlanta, GA, USA
e-mail: fnaha02@emory.edu

The goal of this chapter is to review safety in body contouring as pertaining to the patient, procedure, surgeon, and facility, four critical parameters represented visually as facets of a "Safety Diamond," a concept first introduced by the senior author (F.N.) in 2009 (Fig. 1.1) and approved by The *International* Society of Aesthetic Plastic Surgery (ISAPS) [3].

Fig. 1.1 Four facets of the ISAPS Safety Diamond

© The Author(s), under exclusive license to Springer Nature Switzerland AG 2023
A. Di Giuseppe et al. (eds.), *Fat Transfer in Plastic Surgery*, https://doi.org/10.1007/978-3-031-10881-5_1

1.2 The Patient

As with all plastic surgery procedures, the goal of body contouring is to achieve the most aesthetic result safely, with minimal or no perioperative complications. The more common complications of body contouring are similar to those of other plastic surgery procedures, namely hematoma, seroma, wound dehiscence, surgical site infection, neuropathy, and VTE. Differentiating body contouring, however, is the greater amount of soft tissue undermining for surgical manipulation and the above average number of patient medical comorbidities, especially in massive weight loss patients, that heighten the risk of complications compared with other outpatient plastic surgery procedures [4].

1.2.1 Comorbidities

Patients presenting for body contouring typically have excess skin and fat from one or more heterogenous etiologies: weight loss, weight gain, postpartum, and aging. Subsequently, the overall health of body contouring patients can range widely from a young multiparous woman with no medical comorbidities desiring an improved appearance of her stretched abdominal skin to an older post-bariatric man with residual adiposity, diabetes, and hypertension who has failed to achieve his goal weight and continues to struggle exercising due to a cumbersome overhanging pannus and intertriginous dermatitis. Needless to say, the risk to these two kinds of patients is not similar: while the former is low risk, the latter carries significant risks. As such, patient selection and surgical timing requires thoughtful evaluation and postponing surgery until medical optimization has been achieved.

By the nature of body contouring surgery, many patients at the time of their initial consultation fall outside the normal range body mass index (BMI) with over one-third of all cosmetic surgical procedures performed on patients with a BMI $\geq$25 [5]. Nevertheless, studies have shown that a BMI as low as 25 to be an independent risk factor for postoperative complications, including surgical site infection and VTE, with the magnitude of risk increasing with a BMI $\geq$30 [6–10]. Additionally, diffuse adiposity present in obese patients markedly limits the amount of excess skin that can be safely removed, leading to suboptimal aesthetic results.

Obesity is also strongly associated with diabetes, hypertension, and many other obesity-related comorbidities known to increase the risk of perioperative complications [7, 11]. Diabetes, especially with uncontrolled hyperglycemia, increases the risk of both surgical site infection and wound dehiscence [12]. Similarly, inadequately treated hypertension increases the risk of cerebral hemorrhage and hematoma. Obstructive sleep apnea increases the risk of respiratory depression, episodic sleep-associated oxygen desaturation, and cardiovascular dysfunction and may need longer postoperative monitoring to minimize the risk of perioperative morbidity or mortality [13]. Consequently, losing weight to attain a normal BMI, or as close to a normal BMI as possible, improves the safety profile body contouring surgery on multiple levels as well as overall patient health and well-being.

However, many comorbidities may not be improved by weight loss and other issues may be related to, or even caused by, the loss of weight. Malnutrition and protein deficiency are common developments among post-bariatric patients who have undergone surgical manipulation of their gastrointestinal tract and major alterations in their diet [14, 15]. For example, undiagnosed or untreated iron-deficiency anemia increases the risk of perioperative cardiovascular events and protein deficiency the risk of wound dehiscence. Tobacco smoking, the factor perhaps most preventable by lifestyle modification, increases the risk of numerous perioperative complications including wound dehiscence, surgical site infection, cardiovascular events, and VTE [16].

Although the success of bariatric surgery in treating obesity and obesity-related comorbidities is well-documented, post-bariatric patients often remain overweight with incompletely resolved comorbidities. Additional weight loss may not be feasible due to their maxing out dieting or body habitus with extensive lipodystrophy that is not conducive to exercising. In any case, when it becomes apparent that such a patient has reached his or her lowest attainable BMI and further weight loss may only be achievable after body contouring surgery, attention must be focused on preoperative optimization of medical comorbidities. All patient medical problems and active medications should be assessed on an individual basis and association with perioperative complications determined for risk-stratification and surgical planning.

1.2.2 Preoperative Medical Optimization

For successful body contouring surgery, patient risk factors must be identified at the initial consultation and interdisciplinary collaboration together with the patient and his or her medical doctors pursued in order to optimize them. Equally important is that patients have a clear and detailed understanding of the risks associated with their comorbidities through informed consent. For patients of suitable BMI for body contouring, medical clearance by the patient's primary care physician and, if indicated, other medical specialists and an anesthesiologist is advised to ensure there are no contraindications to surgery and general anesthesia, respectively.

Although there is variation in practice, ideal surgical candidates should have a BMI <25 and attain weight stability, defined as no more than an average change of 5 lbs/month over 3 months, and, for post-bariatric patients, be at least 1 year out from their bariatric surgery [17, 18]. Patients with a BMI >25 should maximize weight loss through lifestyle modification of diet and exercise and, if meeting criteria, be evaluated for bariatric surgery.

Nutritional deficiencies diagnosed on laboratory testing should be appropriately treated and labs rechecked until normalized. Among post-bariatric patients, for example, up to half may be iron deficient and develop iron-deficiency anemia, and many also have low levels of calcium, zinc, selenium, folic acid, thiamine, and vitamins A, B12, D, E, K, C [9, 19, 20]. Similarly, at least one-quarter of post-bariatric patients may be protein deficient, reflected by pre-albumin levels below 25 mg/dL on laboratory testing [21]. Given that surgery can increase protein demand from baseline by 25%, preoperative protein supplementation should be considered with goal protein intake of 70–100 g/day in order to support collagen synthesis and wound healing [17, 20, 22].

1.3 The Procedure

While patients seeking body contouring may be interested in only one procedure, others such as those who have undergone massive weight loss may require multiple procedures to achieve their surgical goals. For the latter group, body contouring can be planned as one or multiple staged procedures [9]. While combining procedures into a single stage reduces the total number of surgical events for the patient and can be done relatively safely, studies have shown that combined procedures are associated with a higher risk of several complications including postoperative VTE [1, 23, 24]. Still, certain procedures such as abdominoplasty, BBL, and large volume liposuction carry higher risks about which the patient should be clearly informed. Therefore, patient selection is critical to safe surgical planning and discussing safety considerations with patients preoperatively will help set reasonable expectations for their body contouring journey [9, 25].

Generally, strong consideration should be given to staging in three clinical situations: (1) procedures involving opposite vectors of pull that may compromise blood supply to skin flaps and place undue tension for the closure; (2) operative time exceeding 6 h; and (3) patient-centric factors that are high risk for general anesthesia or surgical complications [8, 9, 17, 26]. In such cases, staging procedures with a recommended minimum interval of 3 months may reduce perioperative complications as well as the need for revision surgery due to recurrent skin laxity [17].

When combining procedures into a single stage, the following opposing-vector combinations should be avoided: brachioplasty and transverse upper body lift, lower body lift and vertical thigh lift, and lower body lift and transverse upper body lift [17]. Studies have shown operative time as low as 3 h to be an independent risk factor for complications from anesthesia, blood loss, and fluid shifts, and cases above 6 h to be an indication for performing the surgery in a hospital setting where closer monitoring and postoperative management is available [14, 27]. Taken together, these and other surgical principles may serve as useful guidelines for determining the combination of procedures and number of stages that is safest and most appropriate for each patient.

1.4 The Surgeon

A major component of patient safety for any procedure rests on the training, qualifications, judgment, and experience of the surgeon. In order to minimize surgical risk and maximize aesthetic result, the surgeon must make a number of multifactorial decisions about on whom to operate, where to operate, and which procedures to perform [3]. In turn, body contouring surgery is best performed by board-certified plastic surgeons who have completed rigorous training in these procedures as well as being well-versed in both preventing and managing the following complications.

1.4.1 Hematoma

The incidence of hematoma after body contouring surgery is estimated to be 0–6% depending on risk factors that include male gender, uncontrolled hypertension, and combined procedures [24, 28–31]. Preoperatively, patient comorbidities should be optimized and medications with blood-thinning effects, including certain supplements such as fish oils (omega-3 fatty acids), should be held as they are well-known to increase the risk of postoperative hematoma [32].

Intraoperatively, several measures may be taken to reduces the risk of postoperative hematoma. First, careful surgical technique should involve securely ligating larger vessels and meticulous hemostasis. Clear and open communication with the anesthesia team is also important to avoid clinically significant hypo- or hypertension and preventing coughing during extubation that could acutely increase intravascular pressure and lead to bleeding [18]. As postoperative nausea and vomiting is very common, experienced by up to 70% of surgical patients, antiemetics should be administered prophylactically and therapeutically when indicated [11]. Additionally, effective pain management strategies, which may include enhanced recovery after surgery pain protocols and nerve blocks (e.g., transversus abdominis plane block) with long-active anesthetic agents such as liposomal bupivacaine, are vital to minimizing pain-induced tachycardia and hypertension [8].

Although the majority of hematomas develop within the first 48 h after surgery, late-presenting hematomas do occur [28]. In either case, prompt evaluation of both the patient's hemodynamic stability and surgical site is warranted. While small hematomas in hemodynamically stable patients may be treated with needle aspiration and compression, larger hematomas necessitate urgent operative evacuation to avoid compromising skin flap viability and the potential development of skin necrosis and wound dehiscence. In turn, patients should be educated about hematoma risk reduction by complying with pre- and postoperative activity instructions as well as hematoma symptoms in order to allow for early identification and management.

1.4.2 Seroma

The incidence of seroma after body contouring may exceed 20%, varying widely based on risk factors that include obesity and extensive dissections with wide undermining [33, 34]. Preoperatively, obese patients should undergo weight loss to lower the amount of excess adipose tissue and accompanying perfusing blood vessel capillaries, which when transected may contribute to increased serous drainage postoperatively [35]. Intraoperatively, care should be taken to preserve lymph node basins, which may be injured in the axilla during brachioplasty or groin during abdominoplasty, particularly if the lower abdominal incision is marked below the inguinal ligament [10]. The benefit of leaving a layer of sub-Scarpa's fascia on the abdominal wall in abdominoplasty to minimize seroma formation has also been supported by clinical studies [36, 37]. Quilting sutures can be used at the time of surgical site closure to collapse dead space and percutaneous drains placed for continuous evacuation of any serous fluid that may accumulate. Postoperatively, compression garments may be applied and drains left in place until low serous output (<30 cc/day) is observed.

When they occur, seromas can typically be managed by percutaneous aspiration and compression, but if left untreated may lead to surgical site infection and contour irregularities. In rare cases, seromas may become chronic, and treatment with sclerotherapy needed [38].

1.4.3 Surgical Site Infection

The incidence of surgical site infection after body contouring may be as high as 16% depending on risk factors that include obesity, active smoking, diabetes, procedure performed, operative time, and combined procedures [39]. Preoperatively, all medical comorbidities should be optimized. Preoperative skin decolonization with chlorhexidine body washes and mupirocin nasal ointment has been advocated, but largely unsupported by study results [40, 41]. Nevertheless, all active fungal or bacterial skin infections, which have a propensity to occur within skin folds, should be treated and resolved prior to surgery.

Although there is a paucity of high-level evidence on perioperative prophylactic antibiotics in plastic surgery, a consensus statement from the American Association of Plastic Surgeons concluded there to be no significant benefit in such body contouring procedures as abdominoplasty [42, 43]. If prophylactic antibiotics are to be administered, a one-time preoperative dose of first-generation cephalosporin (e.g., cefazolin) is recommended [44]. After sterile skin prepping, intraoperative measures to reduce surgical site infection include avoiding patient hypothermia, defined as core body temperature of ≤36.0 °C, by increasing the operating room temperature, applying an air-forced warming blanket, limiting skin exposure to the surgical site, using warmed fluids, and operating efficiently to minimizing operating time [45]. Due to the presence of bacteria in the axilla and groin regions, brachioplasty and thighplasty are associated with the highest surgical site infection rates among body contouring procedures [46, 47].

While superficial skin infections such as cellulitis can be safely managed non-operatively with oral antibiotics, more serious infections with systemic symptoms should be treated with incision and drainage and may also require hospital admission for intravenous antibiotic therapy. Rarely, necrotizing soft tissue infections can occur, with immunocompromised and diabetic patients most at risk, and must be managed emergently with broad-spectrum intravenous antibiotics and serial operative debridement. Therefore, educating patients about promptly reporting signs and symptoms of infection and having a high index of suspicion for rare, but dangerous, soft tissue infections are key to early diagnosis and treatment.

1.4.4 Wound Dehiscence

With incidence rates as high as 60%, wound dehiscence is the most common complication after body contouring surgery and dependent on risk factors that include obesity, diabetes, malnutrition, smoking, and procedure [18, 48]. Preoperatively, weight loss and all medical comorbidities should be optimized. Nutritional supplements may be started preoperatively and continued throughout the wound healing process. Active smokers should cease smoking at least 1 month prior to surgery and may undergo urine nicotine testing for confirmation [8, 9].

Intraoperatively, appropriate surgical technique should involve maximizing flap perfusion by minimizing soft tissue dissection and avoiding any undermining of subcostal scars

[18]. Final skin tension on the reapproximated wound edges should be minimized by not over-resecting tissue and closing the wound in multiple layers. Particularly high-risk areas for wound breakdown are the axilla and groin where movement is essential for activities of daily life and moisture and bacteria accumulate. As a result, anchoring the superficial fascial system (Scarpa's fascia) to the deep fascia (Colles' fascia in the groin and axillary fascia in the axilla) provides important reinforcement during brachioplasty and medial thighplasty procedures, respectively. Although concerns surround combining liposuction with body contouring procedures, studies assessing postoperative complications have not found liposuction to increase the risk of wound dehiscence in brachioplasty and abdominoplasty [49–51]. Nevertheless, care should always be taken to avoid aggressive liposuction of surgically undermined areas in order to prevent compromising flap perfusion.

Postoperatively, surgical sites should be kept clean and dry, especially if in the axilla and groin areas. While abdominal binders and other supportive garments are often applied, excessive compression and pressure on the healing skin flaps may lead to ischemic changes, skin necrosis, and wound breakdown. Patients should be counseled about appropriate range of motion and limiting activity as directed by their surgeon. While most cases of small wound dehiscence can be treated successfully with local wound care and dressing changes, larger wound dehiscence involving the superficial fascial system warrants surgical repair [18].

1.4.5 Venous Thromboembolism

The leading cause of death after outpatient plastic surgery, VTE is estimated to occur with an incidence as high as 3.4% depending on risk factors that contribute to Virchow's triad of hypercoagulability, venous stasis, and endothelial injury [52]. VTE, which is comprised of deep vein thrombosis (DVT) and pulmonary embolism (PE), is also potentially preventable and major efforts have focused on developing a valid VTE screening tool as well as safe and effective VTE prophylaxis. Currently, the screening tool approved by the American Society of Plastic Surgeons (ASPS) and utilized by the majority of US plastic surgeons to identify patients who would most benefit from chemoprophylaxis is the 2005 Caprini Risk Assessment Model (RAM), wherein a score of 3–4 corresponds to low risk, 5–6 moderate risk, 7–8 high risk, and ≥8 super high risk [53–55].

According to the validation study by Pannucci et al., [56] 11.3% of reconstructive plastic surgery patients with a Caprini score >8 experienced a postoperative VTE when chemoprophylaxis was not provided, supporting the use of VTE chemoprophylaxis in this super high-risk group. However, in a study of aesthetic surgery patients by Keyes et al., [1] over 95% of VTEs occurred in patients with a Caprini score between 2 and 8, with an average Caprini score of 5, suggesting that the Caprini RAM alone may not be adequate for risk-stratifying body contouring patients. Such findings have led the ASPS VTE Task Force to recommend postoperative low molecular weight heparin (LMWH) or unfractionated heparin (UH) for a Caprini score of 3–6 and an extended duration of LMWH for a Caprini score ≥7 for patients undergoing body contouring under general anesthesia with procedure time >60 min [54].

Moreover, studies of outpatient surgery have found that procedures involving abdominoplasty may be responsible for nearly 60% of all VTEs and more than 90% of deaths caused by PE [1, 57]. Given that the Caprini RAM does not directly factor in specific procedures, many plastic surgeons have subsequently modified their VTE prophylaxis regimens to account for "procedural risk" [58]. In the so-called "procedure-specific prophylaxis," studies have supported the empiric use of heparin as well novel oral anticoagulants (NOAC) such as fondaparinux and rivaroxaban [59]. Studies have supported the administration of LMWH for effective VTE prophylaxis in the highest risk patients, but the associated risk of postoperative hematoma is less clear with positive and negative study results, leading to reluctance among surgeons to use chemoprophylaxis unless absolutely indicated [60, 61]. Regarding NOACs, Sarhaddi and colleagues prescribe a 7-day course of fondaparinux beginning 8 h postoperatively for all abdominal body contouring surgeries involving abdominoplasty and found a significant reduction in postoperative VTE with no increase in hematoma [62]. Vasilakis and colleagues prescribe a 7-day course of rivaroxaban, FDA-approved for DVT prophylaxis in orthopedic surgery, beginning on the first postoperative day for all abdominal body contouring surgeries involving rectus abdominis plication, unless contraindicated. Taken together, determining the ideal VTE screening tool and VTE prophylaxis regimen for body contouring continue to be areas of active research.

As with minimizing the risk of all perioperative complications, comorbidities should be optimized preoperatively. Patients with a personal or family history of VTE benefit from laboratory testing for an inherited thrombophilia such as Factor V Leiden, which is present in 5% of the population as heterozygous and 0.02% as homozygous, and in up to 20% of VTE cases [53]. The decision of whether and for how long to administer chemoprophylaxis for VTE prophylaxis should be based on objective clinical assessment using the Caprini RAM together with risk based on the procedure to be performed. Intraoperatively, sequential compression devices should be used routinely as mechanical VTE prophylaxis and normothermia and euvolemia maintained. Non-narcotic-based pain regimens and nerve blocks should be employed in order to minimize discomfort and promote ambulation postoperatively [63, 64]. Compression body garments, such as abdominal binders, should be applied only when necessary and judiciously to prevent increases in intraabdominal pressure and slowing of venous return. Postoperatively, VTE risk reduction measures include ambulation, avoiding dehydration, and wearing compression stockings.

If DVT is suspected, due to pain or edema in the lower extremities, duplex ultrasound should be used for diagnosis. If concern for PE develops, diagnosis requires emergent CT scan. In both cases, prompt medical evaluation and treatment with anticoagulation or, if contraindicated, IVC filter is indicated. Further management of VTE is beyond the scope of this chapter.

1.4.6 Neuropathy

Although less common, neuropathies in body contouring can occur from nerve compression or traction and surgical transection. In the operating room, the patient's arms, elbows, wrists, and all bony prominences should be sufficiently padded and shoulder abduction remain below 90 degrees when in the supine position [8, 12]. In the lateral decubitus position, the axilla should be supported with an axillary roll to avoid traction on the brachial plexus [65]. A second cause of neuropathy in body contouring is inadvertent sensory nerve transection by sharp dissection. During the abdominoplasty, dissection should proceed superficially in the area 2 cm medial to the anterior superior iliac spine to avoid injury to the lateral femoral cutaneous nerve and development of painful meralgia paresthetica. Similarly, care should be taken during brachioplasty dissection to avoid injury to the medial brachial cutaneous nerve, which emerges superficially in the distal third of the arm.

If compression or traction neuropathy does occur, the nerve injury usually constitutes neuropraxia and full spontaneous recovery should be expected. Sensory nerve transection may lead to formation of a neuroma, which can be painful and require excision for treatment. Additionally, patients with nerve pain such as meralgia paresthetica may benefit from treatment with neuropathic analgesics such as gabapentin.

1.5 The Facility

Body contouring surgery may be safely performed in both outpatient ambulatory facility and hospital settings, but never in the "back room" of the doctor's office [3]. To meet the highest safety guidelines set forth by the American Society of Plastic Surgeons, outpatient facilities should be accredited by the American Association for the Accreditation of Ambulatory Surgical Facilities (AAAASF). In general, procedures performed in the hospital allow for the highest level of monitoring, potential emergent intervention, and postoperative care, but come at the cost of an increased financial burden and resource utilization. Nevertheless, operating in the hospital setting with planned overnight observation should be strongly considered for patients at higher risks for perioperative complications due to medical comorbidities or prolonged duration of surgery in an effort to always prioritize patient safety in body contouring.

References

1. Keyes GR, Singer R, Iverson RE, Nahai F. Incidence and predictors of venous thromboembolism in abdominoplasty. Aesthet Surg J. 2017;38:162–73.
2. Mofid MM, Teitelbaum S, Suissa D, Ramirez-Montañana A, Astarita DC, Mendieta C, Singer R. Report on mortality from gluteal fat grafting: recommendations from the ASERF task force. Aesthet Surg J. 2017;37:sjx004.
3. Nahai F. Minimizing risk in aesthetic surgery. Clin Risk. 2009;15:232–6.
4. Colwell AS, Borud LJ. Optimization of patient safety in post-bariatric body contouring: a current review. Aesthet Surg J. 2008;28:437–42.
5. Cosmetic Surgery National Data Bank Statistics. Aesthet Surg J. 2015;35:1–24.
6. Au K, Hazard SW, Dyer A-M, Boustred AM, Mackay DR, Miraliakbari R. Correlation of complications of body contouring surgery with increasing body mass index. Aesthet Surg J. 2008;28:425–9.
7. Gupta V, Winocour J, Rodriguez-Feo C, Bamba R, Shack RB, Grotting JC, Higdon KK. Safety of aesthetic surgery in the overweight patient: analysis of 127,961 patients. Aesthet Surg J. 2016;36:718–29.
8. Kokosis G, Coon D. Safety in body contouring to avoid complications. Clin Plast Surg. 2018;46:25–32.
9. Capla J, Shikowitz-Behr L. Patient evaluation and surgical staging. Clin Plast Surg. 2019;46:9–14.
10. Almutairi K, Gusenoff JA, Rubin JP. Body contouring. Plast Reconstr Surg. 2016;137:586e–602e.
11. Tokin CA, Kaoutzanis C, Higdon KK, Grotting J. Patient safety in aesthetic surgery. In: Nahai F, Nahai F, editors. The art of aesthetic surgery: principles & techniques. 3rd ed. New York: Thieme Medical Publishers, Inc.; 2020. p. 39–50.
12. Davison SP, Clemens MW. Safety first: precautions for the massive weight loss patient. Clin Plast Surg. 2008;35:173–83.
13. Apnea AS of ATF on PM of Patients with Obstructive Sleep. Practice guidelines for the perioperative Management of Patients with obstructive sleep apnea. Anesthesiology. 2014;120:268–86.
14. Small KH, Constantine R, Eaves FF, Kenkel JM. Lessons learned after 15 years of circumferential bodylift surgery. Aesthet Surg J. 2016;36:681–92.
15. Sebastian JL. Bariatric surgery and work-up of the massive weight loss patient. Clin Plast Surg. 2008;35:11–26.
16. Bloom JA, Rashad R, Chatterjee A. The impact on mortality and societal costs from smoking cessation in aesthetic plastic surgery in the United States. Aesthet Surg J. 2019;39:439–44.
17. Naghshineh N, Rubin JP. Preoperative evaluation of the body contouring patient the cornerstone of patient safety. Clin Plast Surg. 2014;41:637–43.
18. Gusenoff JA. Prevention and management of complications in body contouring surgery. Clin Plast Surg. 2014;41:805–18.

19. Heber D, Greenway FL, Kaplan LM, Livingston E, Salvador J, Still C, Society E. Endocrine and nutritional management of the post-bariatric surgery patient: an endocrine society clinical practice guideline. J Clin Endocrinol Metabolism. 2010;95:4823–43.
20. Agha-Mohammadi S, Hurwitz DJ. Enhanced recovery after body-contouring surgery: reducing surgical complication rates by optimizing nutrition. Aesthet Plast Surg. 2010;34:617–25.
21. Naghshineh N, Coon DO, McTigue K, Courcoulas AP, Fernstrom M, Rubin JP. Nutritional assessment of bariatric surgery patients presenting for plastic surgery; a prospective analysis. Plast Reconstr Surg. 2010;126:602–10.
22. Austin RE, Lista F, Khan A, Ahmad J. The impact of protein nutritional supplementation for massive weight loss patients undergoing abdominoplasty. Aesthet Surg J. 2016;36:204–10.
23. Winocour J, Gupta V, Kaoutzanis C, Shi H, Shack RB, Grotting JC, Higdon KK. Venous thromboembolism in the cosmetic patient: analysis of 129,007 patients. Aesthet Surg J. 2016;37:337–49.
24. Kaoutzanis C, Winocour J, Yeslev M, Gupta V, Asokan I, Roostaeian J, Grotting JC, Higdon KK. Aesthetic surgical procedures in men: major complications and associated risk factors. Aesthet Surg J. 2017;38:429–41.
25. Gusenoff JA, Rubin JP. Plastic surgery after weight loss: current concepts in massive weight loss surgery. Aesthet Surg J. 2008;28:452–5.
26. Procter LD, Davenport DL, Bernard AC, Zwischenberger JB. General surgical operative duration is associated with increased risk-adjusted infectious complication rates and length of hospital stay. J Am Coll Surgeons. 2010;210:60–65.e2.
27. Hardy KL, Davis KE, Constantine RS, Chen M, Hein R, Jewell JL, Dirisala K, Lysikowski J, Reed G, Kenkel JM. The impact of operative time on complications after plastic surgery: a multivariate regression analysis of 1753 cases. Aesthet Surg J. 2014;34:614–22.
28. Hood K, Kumar NG, Kaoutzanis C, Higdon KK. Hematomas in aesthetic surgery. Aesthet Surg J. 2018;38:1013–25.
29. Chong T, Coon D, Toy J, Purnell C, Michaels J, Rubin JP. Body contouring in the male weight loss population. Plast Reconstr Surg. 2012;130:325e–30e.
30. Kaoutzanis C, Winocour J, Gupta V, Kumar NG, Sarosiek K, Wormer B, Tokin C, Grotting JC, Higdon KK. Incidence and risk factors for major hematomas in aesthetic surgery: analysis of 129,007 patients. Aesthet Surg J. 2017;37:1175–85.
31. Farkas JP, Kenkel JM, Hatef DA, Davis G, Truong T, Rohrich RJ, Brown SA. The effect of blood pressure on hematoma formation with perioperative Lovenox in excisional body contouring surgery. Aesthet Surg J. 2007;27:589–93.
32. Rowe DJ, Baker AC. Perioperative risks and benefits of herbal supplements in aesthetic surgery. Aesthet Surg J. 2009;29:150–7.
33. Rosenfield LK, Davis CR. Evidence-based abdominoplasty review with body contouring algorithm. Aesthet Surg J. 2019;39:643–61.
34. Pajula S, Jyränki J, Tukiainen E, Koljonen V. Complications after lower body contouring surgery due to massive weight loss unaffected by weight loss method. J Plastic Reconstr Aesthetic Surg. 2019;72:649–55.
35. Lemoine AY, Ledoux S, Larger E. Adipose tissue angiogenesis in obesity. Thromb Haemost. 2013;110:661–9.
36. Costa-Ferreira A, Rebelo M, Vásconez LO, Amarante J. Scarpa fascia preservation during abdominoplasty: a prospective study. Plast Reconstr Surg. 2010;125:1232–9.
37. Fang RC, Lin SJ, Mustoe TA. Abdominoplasty flap elevation in a more superficial plane: decreasing the need for drains. Plast Reconstr Surg. 2010;125:677–82.
38. Shermak MA, Rotellini-Coltvet LA, Chang D. Seroma development following body contouring surgery for massive weight loss: patient risk factors and treatment strategies. Plast Reconstr Surg. 2008;122:280–8.
39. Kaoutzanis C, Kumar NG, Winocour J, Hood K, Higdon KK. Surgical site infections in aesthetic surgery. Aesthet Surg J. 2019;39:1118–38.
40. Chlebicki MP, Safdar N, O'Horo JC, Maki DG. Preoperative chlorhexidine shower or bath for prevention of surgical site infection: a meta-analysis. Am J Infect Control. 2013;41:167–73.
41. Perl TM, Cullen JJ, Wenzel RP, Zimmerman MB, Pfaller MA, Sheppard D, Twombley J, French PP, Herwaldt LA, Team MATROSAS. Intranasal mupirocin to prevent postoperative Staphylococcus aureus infections. New Engl J Med. 2002;346:1871–7.
42. Ariyan S, Martin J, Lal A, et al. Antibiotic prophylaxis for preventing surgical-site infection in plastic surgery. Plast Reconstr Surg. 2015;135:1723–39.
43. Lane MA, Young VL, Camins BC. Prophylactic antibiotics in aesthetic surgery. Aesthet Surg J. 2010;30:859–71.
44. Sevin A, Senen D, Sevin K, Erdogan B, Orhan E. Antibiotic use in abdominoplasty: prospective analysis of 207 cases. J Plast Reconstr Aesthet Surg. 2007;60:379–82.
45. Lista F, Doherty CD, Backstein RM, Ahmad J. The impact of perioperative warming in an outpatient aesthetic surgery setting. Aesthet Surg J. 2012;32:613–20.
46. Afshari A, Gupta V, Nguyen L, Shack RB, Grotting JC, Higdon KK. Preoperative risk factors and complication rates of thighplasty: analysis of 1,493 patients. Aesthet Surg J. 2016;36:897–907.
47. Nguyen L, Gupta V, Afshari A, Shack RB, Grotting JC, Higdon KK. Incidence and risk factors of major complications in brachioplasty: analysis of 2,294 patients. Aesthet Surg J. 2016;36:792–803.
48. Michaels J, Coon D, Rubin JP. Complications in postbariatric body contouring: postoperative management and treatment. Plast Reconstr Surg. 2011;127:1693–700.
49. Xia Y, Zhao J, Cao DS. Safety of lipoabdominoplasty versus abdominoplasty: a systematic review and meta-analysis. Aesthet Plast Surg. 2019;43:167–74.
50. Bossert RP, Dreifuss S, Coon D, Wollstein A, Clavijo-Alvarez J, Gusenoff JA, Rubin JP. Liposuction of the arm concurrent with brachioplasty in the massive weight loss patient. Plast Reconstr Surg. 2013;131:357–65.
51. Stevens WG, Cohen R, Vath SD, Stoker DA, Hirsch EM. Does lipoplasty really add morbidity to abdominoplasty? Revisiting the controversy with a series of 406 cases. Aesthet Surg J. 2005;25:353–8.
52. Pannucci CJ. Venous thromboembolism in aesthetic surgery: risk optimization in the preoperative, intraoperative, and postoperative settings. Aesthet Surg J. 2019;39:209–19.
53. Aimé VL, Neville MR, Thornburg DA, Noland SS, Mahabir RC, Bernard RW. Venous thromboembolism prophylaxis in aesthetic surgery: a survey of plastic surgeons' practices. Aesthet Surg J. 2020;40:1351–69.
54. Murphy RX, Alderman A, Gutowski K, Kerrigan C, Rosolowski K, Schechter L, Schmitz D, Wilkins E. Evidence-based practices for thromboembolism prevention. Plast Reconstr Surg. 2012;130:168e–75e.
55. Pannucci CJ, Barta RJ, Portschy PR, Dreszer G, Hoxworth RE, Kalliainen LK, Wilkins EG. Assessment of postoperative venous thromboembolism risk in plastic surgery patients using the 2005 and 2010 Caprini risk score. Plast Reconstr Surg. 2012;130:343–53.
56. Pannucci CJ, Bailey SH, Dreszer G, et al. Validation of the caprini risk assessment model in plastic and reconstructive surgery patients. J Am Coll Surgeons. 2011;212:105–12.
57. Keyes GR, Singer R, Iverson RE, McGuire M, Yates J, Gold A, Reed L, Pollack H, Thompson D. Mortality in outpatient surgery. Plast Reconstr Surg. 2008;122:245–50.

58. Swanson E. Why risk assessment models are ineffective in predicting venous thromboembolism in plastic surgery patients. Aesthet Surg J. 2016;36:NP233–4.
59. Hatef DA, Trussler AP, Kenkel JM. Procedural risk for venous thromboembolism in abdominal contouring surgery: a systematic review of the literature. Plast Reconstr Surg. 2010;125:352–62.
60. Hatef DA, Kenkel JM, Nguyen MQ, Farkas JP, Abtahi F, Rohrich RJ, Brown SA. Thromboembolic risk assessment and the efficacy of enoxaparin prophylaxis in excisional body contouring surgery. Plast Reconstr Surg. 2008;122:269–79.
61. Vasilakis V, Klein GM, Trostler M, Mukit M, Marquez JE, Dagum AB, Pannucci CJ, Khan SU. Postoperative venous thromboembolism prophylaxis utilizing enoxaparin does not increase bleeding complications after abdominal body contouring surgery. Aesthet Surg J. 2019;40:989–95.
62. Sarhaddi D, Xu K, Wisbeck A, Deigni O, Kaswan S, Prada C, Lund H. Fondaparinux significantly reduces postoperative venous thromboembolism after body contouring procedures without an increase in bleeding complications. Aesthet Surg J. 2019;39:1214–21.
63. Parsa AA, Sprouse-Blum AS, Jackowe DJ, Lee M, Oyama J, Parsa FD. Combined preoperative use of celecoxib and gabapentin in the management of postoperative pain. Aesthet Plast Surg. 2008;33:98.
64. Norwich A, Narayan D. Pain management and body contouring. Clin Plast Surg. 2019;46:33–9.
65. Shermak MA. Pearls and perils of caring for the postbariatric body contouring patient. Plast Reconstr Surg. 2012;130:585e–96e.

Gluteal Fat Transfer: A Scientific Validation

2

Deniz Sarhaddi, Caitlin Francoisse, and Foad Nahai

Contents

2.1 Introduction

Gluteal augmentation via autologous fat grafting, colloquially termed the "Brazilian Butt Lift" (BBL), represents one of the fastest growing areas of plastic surgery in the United States. The American Society of Plastic Surgery estimates over 28,000 procedures were performed in 2019, nearly double the number performed just 5 years prior [1, 2]. This growth has continued despite its nefarious reputation as one of the deadliest surgical cosmetic procedures. Early estimates from the Aesthetic Surgery Education Research Foundation (ASERF) described the mortality rate as between 1 in 2250 to 1 in 6214, at least triple that of then next riskiest procedure, abdominoplasty [3, 4]. This reputation, however, may be misleading. New safety recommendations have changed the safety profile of this once infamous procedure, dropping the mortality by over 75% [5]. Evidence continues to emerge suggesting that with proper precautions, gluteal fat augmentation provides a safe and effective strategy to body contouring.

D. Sarhaddi
Saint Louis Cosmetic Surgery, Inc., Chesterfield, USA

C. Francoisse
Saint Louis University Hospital, Saint Louis, USA
e-mail: caitlin.francoisse@health.slu.edu

F. Nahai (✉)
Emory University, Atlanta, USA

2.2 History and Controversy

The quest for gluteal rejuvenation traces back over 50 years, with Pitanguy's reports of improved gluteal aesthetics through resection of the trochanteric regions and the gluteal fold [6]. However, it was not until the 1980s that surgeons turned to the relatively new technology of liposuction to restore volume and improve contour. In 1986, Gonzalez et al. published the first report of autologous gluteal fat grafting [7]. Even in these early reports, authors emphasized the importance of avoiding large depots and injecting fat at different levels, themes that would return later as safety concerns would later arise. This body of work continued to grow with Toledo in 1980s, followed by others including Cardenas-Camerena, Roberts, and Mendieta in the decades to follow which demonstrated efficacy of gluteal augmentation with this technique [8–11].

However, early pioneers of this technique noted that resorption of fat grafts can impact long-term results, prompting interest on how to maximize graft longevity. One early study by Guerroerosantos et al. investigated long-term survival of fat grafts placed into different planes in rat models [12]. Experimental results demonstrated improved survival with muscular injection instead of subcutaneous deposition. Although this study was experimental in nature, it quickly translated to clinical applications, namely small volume fat grafting for facial contouring [13–15]. Recognizing its suc-

© The Author(s), under exclusive license to Springer Nature Switzerland AG 2023

A. Di Giuseppe et al. (eds.), *Fat Transfer in Plastic Surgery*, https://doi.org/10.1007/978-3-031-10881-5_2

cess in this context, some surgeons began to scale up this application to large volume, muscular injection for gluteal augmentation. Unfortunately, this technique would have two major implications for patient morbidity and mortality.

As the popularity of this procedure rose, so did concerns over its reported safety. In 2015, Cárdenas-Camarena detailed 22 deaths in Mexico and Colombia due to pulmonary fat embolism after gluteal fat during a 15-year period, sending shockwaves through the aesthetic surgery community [16]. Prior to this, there was only an isolated case report of fat embolism mortality in Los Angeles, published in the year prior [17]. Cardenas-Camarena reported that macroscopic fat emboli caused medium and large vessel obstruction, causing immediate cardiovascular collapse at the time of surgery. The volumes of fat injected were relatively small, just over 200 cc on average, but the impact was profound. Autopsy results showed fat in the deep muscle tissue surrounding the vessels and the rupture of gluteal veins, implicating deep plane placement of fat as a key factor in mortality [16].

2.3 Fat Embolism Versus Fat Embolism Syndrome

Fat embolism syndrome (FES) was a known complication of liposuction, prior to Cárdenas Camerena's work [18]. FES results from small lipid globules entering the vascular system and obstructing blood vessels leading to a systemic inflammatory response and ultimately tissue and organ damage. It usually manifests 12–72 h postoperatively with respiratory distress, altered mental status, and petechial rash [19, 20]. The results can be deadly with an up to 20% reported mortality rate [21].

However, Cardenas-Camarena's landmark study suggested an entirely different mechanism was at play. Fat embolism is distinct from fat embolism syndrome (FES). Fat enters the blood stream through the gluteal veins. These are either cannulated directly or transected traumatically with the grafting cannula. If the vein is transected, the fat enters the blood stream via a pressure gradient. Lipoinjection creates an area of higher pressure relative to the venous system, facilitating movement of the fat into the transected vein [22]. Fat can enter the blood stream in either macroscopic form, causing cardiopulmonary obstruction, or microscopic form, leading to reactive airway compromise, hemorrhage, and edema [19].

Whereas fat embolism syndrome appears up to 3 days postoperatively, macroscopic fat emboli can occur intraoperatively. This was demonstrated by Bayter-Mari et al. who reviewed the medical records and autopsy reports of 16 patients who died during gluteal augmentation with fat grafting [23]. They found 75% of patients developed issues intraoperatively with the remaining patients developing symptoms in the recovery unit. On average, patients died within 7 h of the start of surgery. All patients had microemboli in their lungs; the majority (62.5%) also had macroscopic emboli in their lungs, pulmonary vessels, and cardiac vessels.

2.4 ASERF Task Force Recommendations and Response

Recognizing safety concerns surrounding the procedure, the Aesthetic Surgery Education Research Foundation (ASERF) assembled a dedicated Gluteal Fat Grafting Task Force in 2015 to profile this procedure's risk profile. Their results were astonishing: they estimated the mortality rate between 1 in 2250 to 1 in 6214, a risk 10–20 times greater than the average for aesthetic surgery procedures performed in facilities accredited by the American Association for Accreditation of Ambulatory Surgery Facilities (AAAASF) [3]. The task force outlined the common autopsy findings of deceased BBL patients which included fat in and beneath the gluteal muscles, damage to the superior and inferior gluteal veins, and massive fat emboli in the heart and/or lungs [24]. To help mitigate risk, ASERF also established recommendations which included using a large (>4 mm), single hole cannula, injecting only into the subcutaneous plane only while the cannula is in motion, and vigilantly avoiding a downward cannula angle [3].

In a 2018 statement, the task force announced a multisociety launch of focused research efforts into gluteal fat grafting safety. However, due to the continued death toll from the BBL, a second "Urgent Warning to Surgeons" was sent just 5 months after the first [25]. This time, with writing in all capital letters, ASERF called for improved safety, avoidance of intramuscular injection, surgeon vigilance, and possible reconsideration of whether the procedure should still be offered.

This work sent ripples through the aesthetic surgery community. Leaders across the international aesthetic surgery community questioned whether the procedure posed an acceptable risk to patients [26]. In October 2018, at the British Association of Aesthetic Plastic Surgery (BAAPS) Annual Meeting, all members were advised to stop performing BBL procedures until more data were collected [27]. The British Association of Plastic Reconstructive and Aesthetic Surgeons (BAPRAS) was "fully supportive of the BAAPS decision." [28] In June 2019, the state of Florida went so far as to pass an emergency rule prohibiting surgeons from injecting fat into or below a patient's gluteal muscles [29].

2.5 Enhanced Safety Efforts

Spurred by the Aesthetic Surgery Education Research Foundation (ASERF) recommendations, there has been a boom in literature regarding anatomy, safety, and BBL practice, which has given surgeons more in-depth information to guide techniques when performing this procedure.

Numerous anatomical studies have elucidated ways to fat graft safely in the gluteal region. It has been clearly demonstrated that there is no "safe zone" in the intramuscular or submuscular planes [30]. The subcutaneous "danger zone" for gluteal fat grafting has been defined as a triangle demarcated by the posterior superior iliac spine, the greater trochanter, and the ischial tuberosity [31]. Ordenada et al. described the vascular anatomy of the gluteal region on fresh latex-injected cadavers, reinforcing the danger of injecting in the deeper and more medial planes of the gluteal region [31]. Ghavami et al. defined the ligamentous anatomy of the gluteal region; they shared surgical techniques to strategically release these ligaments to allow expansion of the subcutaneous plane while leaving certain ligaments intact to prevent fat injection in the region of the sciatic nerve and large gluteal veins [32]. Alvarez et al. continued to build on this work by investigating optimal cannula positioning. In a fresh cadaver study, the incidence of "complications," defined as fat found in contact with a neurovascular bundle and/or within the gluteus muscle, was investigated with various cannula angles and entry locations for gluteal fat injection: the highest rate of complication was associated with infiltration at an angulation of −30°, 0°, and +15° through the middle lower gluteal sulcus [33].

There have been numerous reports and innovations to support safe and effective injection techniques. Delvecchio et al. described their expansion vibration lipofilling technique (roller pump-propelled fat) for gluteal fat grafting in the subcutaneous plane [34]. Their series demonstrated safety in over 2419 consecutive cases with an average fat injection volume of 1000 cc and reported no cases of fat embolism or death. Wall et al. further demonstrated the safety of power-assisted lipofilling in a cadaveric study while measuring changes in pressure during injection; they were able to demonstrate that fat does not simply migrate into the subfascial or muscular plane when the fascia is intact [35]. Ghavami et al. described their technique using manual fat injection into the subcutaneous plane using 60 cc Toomey syringes, asserting the safety benefits of manual feedback/resistance to injection, and warning that the use of automated systems requires extensive surgical experience to avoid inadvertent injection of large volumes of fat into incorrect planes [32]. Cansancao et al. have proposed the use of real-time ultrasound assistance when grafting to ensure that they remain in the subcutaneous plane during grafting; they found that this only added 25 min to the surgical time [36]. A separate prospective study performed by Cansancao et al. demonstrated that gluteal fat augmentation in the subcutaneous plane is effective with long-term fat retention when compared to similar studies reporting intramuscular fat injections [37].

Anesthesia modifications for this procedure have also been suggested to increase safety. One study suggests that BBL under oral sedation with tumescent infiltration may offer improved safety, because patients are unlikely to tolerate penetration of the muscle and fascia with the cannula while awake [38].

Positioning during surgery has also been optimized. To minimize the risk of either direct cannulation or vein injury due to traction/tear, studies have shown that the jackknife or lateral decubitus positions can decrease vein caliber by up to 27% [30].

2.6 BBL Today

Thanks to the expanding body of literature on gluteal fat grafting and the widely disseminated recommendations of the ASERF Gluteal Fat Grafting Task Force, the safety profile of gluteal fat grafting has improved greatly. In 2019, a follow-up survey was sent to members of American Society for Aesthetic Plastic Surgery (ASAPS) and the International Society of Aesthetic Plastic Surgery (IASPS) to assess changes to techniques and mortality in the wake of the ASERF recommendations. The results showed a dramatic change in technique. 93.5% of respondents reported they were aware of the ASERF recommendations. Over 85% of surgeons inject in subcutaneous fat only, compared to just 39.8% in 2017. Only 4% of surveyed physicians angled their cannulas down compared to 27.2% just 2 years prior. These changes lead to a profound impact on the mortality rate; it dropped from an estimated 1 in 3448 in 2017 to 1 in 14,952 in 2019, a mortality rate less than abdominoplasty [4, 5].

As some leaders have pointed out, this survey is very promising yet needs to be taken in context [39]. Surveys are inherently flawed by retrospective bias and cannot be the basis for definitive conclusions about the safety profile of this procedure; this makes efforts for objective reporting databases even more crucial [39]. The American Association for Accreditation of Ambulatory Surgery Facilities now mandates the reporting of all gluteal fat grafting cases and complications [40]. In addition, the General Registry for Autologous Fat transfer maintains a web-accessible database which is prospectively gathering outcomes data on this procedure. Survey work by Rios et al. suggests changes are being made; additional database reporting will allow us to optimize safety in the future [5].

The Brazilian Butt Lift appears to be a success story of quality improvement, similar to that of liposuction decades prior. Initially, liposuction raised safety concerns after its advent in the 1980s. This prompted the creation of a task force which led to recommendations which improved the safety of the procedure [41]. We see a similar phenomenon now with gluteal fat grafting. Through research and concerted effort by professional societies, and continued efforts by objective reporting agencies, Brazilian Butt Lift Surgery is safer now than it has ever been. Similar to liposuction, we can expect this surgery to become a mainstay of aesthetic surgery and continue to be performed safely for decades to come.

References

1. The American Society for Aesthetic Plastic Surgery's cosmetic surgery National Data Bank: statistics 2018. Aesthet Surg J. 2019;39(Issue Supplement 4):1–27.
2. American Society for Aesthetic Plastic Surgery. The American Society for Aesthetic Plastic Surgery's Cosmetic Surgery National Data Bank: statistics 2013. Sage; 2014.
3. Mofid MM, et al. Report on mortality from gluteal fat grafting: recommendations from the ASERF task force. Aesthet Surg J. 2017;37(7):796–806.
4. Keyes GR, et al. Mortality in outpatient surgery. Plast Reconstr Surg. 2008;122(1):245–50; discussion 251–3.
5. Rios L, Gupta V. Improvement in Brazilian Butt Lift (BBL) safety with the current recommendations from ASERF, ASAPS, and ISAPS. Aesthet Surg J. 2020;40:864.
6. Pitanguy I. Trochanteric lipodystrophy. Plast Reconstr Surg. 1964;34:280–6.
7. Gonzalez R, Spina L. Grafting of fat obtained by liposuction: technique and instruments. Rev Bras Cir. 1986;76:243–50.
8. Matsudo PKR, Toledo LS. Experience of injected fat grafting. Aesthet Plast Surg. 1988;12(1):35–8.
9. Cárdenas-Camarena L, Lacouture AM, Tobar-Losada A. Combined gluteoplasty: liposuction and lipoinjection. Plast Reconstr Surg. 1999;104(5):1524–31; discussion 1532–3.
10. Roberts TL III, Toledo LS, Badin AZ. Augmentation of the buttocks by micro fat grafting. Aesthet Surg J. 2001;21(4):311–9.
11. Mendieta CG. Gluteal reshaping. Aesthet Surg J. 2007;27(6):641–55.
12. Guerrerosantos J, et al. Long-term survival of free fat grafts in muscle: an experimental study in rats. Aesthet Plast Surg. 1996;20(5):403–8.
13. Guerrerosantos J. Long-term outcome of autologous fat transplantation in aesthetic facial recontouring: sixteen years of experience with 1936 cases. Clin Plast Surg. 2000;27(4):515–43.
14. Guerrerosantos J. Evolution of technique: face and neck lifting and fat injections. Clin Plast Surg. 2008;35(4):663–76.
15. Guerrerosantos J, Haidar F, Paillet JC. Aesthetic facial contour augmentation with microlipofilling. Aesthet Surg J. 2003;23(4):239–47.
16. Cárdenas-Camarena L, et al. Deaths caused by gluteal lipoinjection: what are we doing wrong? Plast Reconstr Surg. 2015;136(1):58.
17. Astarita DC, Scheinin LA, Sathyavagiswaran L. Fat transfer and fatal macroembolization. J Forensic Sci. 2015;60(2):509–10.
18. Teimourian B, Rogers BW III. A national survey of complications associated with suction lipectomy: a comparative study. Plast Reconstr Surg. 1989;84(4):628.
19. Che DH, Xiao ZB. Gluteal augmentation with fat grafting: literature review. Aesthet Plast Surg. 2020;45:1633.
20. Gurd AR, Wilson RI. The fat embolism syndrome. J Bone Joint Surg Br. 1974;56b(3):408–16.
21. Abi-Rafeh J, et al. Comments on "commentary on: the potential role of corticosteroid prophylaxis for the prevention of microscopic fat embolism syndrome in gluteal augmentations". Aesthet Surg J. 2020;40(2):Np77–9.
22. Shah B. Complications in gluteal augmentation. Clin Plast Surg. 2018;45(2):179–86.
23. Bayter-Marin JE, et al. Understanding fatal fat embolism in gluteal lipoinjection: a review of the medical records and autopsy reports of 16 patients. Plast Reconstr Surg. 2018;142(5):1198–208.
24. Multi-society gluteal fat grafting task force issues safety advisory urging practitioners to reevaluate technique [letter]. American Society of Plastic Surgeons (ASPS), American Society for Aesthetic Plastic Surgery (ASAPS), International Society of Plastic Surgery (ISPS), International Society of Plastic Regenerative Surgeons (ISPRES), International Federation for Adipose Therapeutics and Science (IFATS).
25. Urgent warning to surgeons performing fat grafting to the buttocks (Brazilian Butt Lift or "BBL") [letter]. American Society of Plastic Surgeons (ASPS), American Society for Aesthetic Plastic Surgery (ASAPS), International Society of Plastic Surgery (ISPS), International Society of Plastic Regenerative Surgeons (ISPRES), International Federation for Adipose Therapeutics and Science (IFATS).
26. Nahai F. Acceptable risk: who decides? Aesthet Surg J. 2017;37(7):852–3.
27. The Bottom Line [Press release]. British Association of Aesthetic Plastic Surgeons.
28. Bapras statement on 'Brazillian butt lift' surgery [news and views]. British Association of Plastic Reconstructive and Aesthetic Surgeons.
29. Gluteal Fat Grafting Safety Advisory [letter]. American Society of Plastic Surgeons (ASPS), American Society for Aesthetic Plastic Surgery (ASAPS), International Society of Plastic Surgery (ISPS), International Society of Plastic Regenerative Surgeons (ISPRES), International Federation for Adipose Therapeutics and Science (IFATS).
30. Turin SY, et al. Gluteal vein anatomy: location, caliber, impact of patient positioning, and implications for fat grafting. Aesthet Surg J. 2020;40(6):642–9.
31. Ordenana C, et al. Objectifying the risk of vascular complications in gluteal augmentation with fat grafting: a latex casted cadaveric study. Aesthet Surg J. 2020;40(4):402–9.
32. Ghavami A, Villanueva NL, Amirlak B. Gluteal ligamentous anatomy and its implication in safe buttock augmentation. Plast Reconstr Surg. 2018;142(2):363–71.
33. Alvarez-Alvarez FA, González-Gutiérrez HO, Ploneda-Valencia CF. Safe gluteal fat graft avoiding a vascular or nervous injury: an anatomical study in cadavers. Aesthet Surg J. 2019;39(2):174–84.
34. Del Vecchio D, Wall S Jr. Expansion vibration lipofilling: a new technique in large-volume fat transplantation. Plast Reconstr Surg. 2018;141(5):639e–49e.
35. Wall S Jr, et al. Subcutaneous migration: a dynamic anatomical study of gluteal fat grafting. Plast Reconstr Surg. 2019;143(5):1343–51.

36. Cansancao AL, et al. Real-time ultrasound-assisted gluteal fat grafting. Plast Reconstr Surg. 2018;142(2):372–6.
37. Cansancao AL, et al. Subcutaneous-only gluteal fat grafting: a prospective study of the long-term results with ultrasound analysis. Plast Reconstr Surg. 2019;143(2):447–51.
38. Chia CT, et al. "Brazilian Butt Lift" under local anesthesia: a novel technique addressing safety concerns. Plast Reconstr Surg. 2018;142(6):1468–75.
39. Nahai F. No "quick fix" for this: an update on the Brazilian Butt Lift. Aesthet Surg J. 2020;40(8):928–30.
40. Ambulatory, A.A.f.A.o. and S. Facilities What is patient safety data reporting.
41. Iverson, R.E., D.J. Lynch, and A.C.o.P.S. the, Practice advisory on liposuction. Plast Reconstr Surg, 2004. 113(5), 1478.

Large Volume and Combined Fat Grafting Surgery in Postbariatric Patients: Safety Profile

3

Franco Bassetto, Laura Pandis, and Carlotta Scarpa

Contents

Nowadays, fat grafting has become one of the most required treatments for improving body contouring, and there are many possible applications either alone or in combined surgery, for example, we can combine it with Brachioplasty, Facial Lifting, and Buttock Lift.

In the last two decades, the improvement in fat grafting storage and technique brought to the fore this treatment; the soft and natural texture of the adipose tissue plus its versatility and the presence of adult-derived stem cells (ADSCs) have made lipofilling the ideal filler especially in patients who are characterized by a "weak tissue" as postbariatric ones.

Adipose tissue has been considered inert for years, but recently scientists have noticed its potential in terms of regenerative effect, due to the presence of ADSCs which have shown the capability to self-renew and the same multilineage differentiation potential of other MSCs, and in releasing adipokines (such as leptin), cytokines (such as IL6), and many growth factors (such as vascular endothelial growth factor or transforming growth factor β), which can contribute, for example, to cell proliferation and communication, motility, and epithelial to mesenchymal transition (EMT).

Indeed, if on one side fat grafting can be used to improve volume, on the other side the adipose-derived stem cells can have a regenerative effect stimulating both the neoangiogenesis and the differentiation and proliferation of cells ameliorating the tissue's quality.

F. Bassetto · L. Pandis · C. Scarpa (✉)
Clinic of Plastic and Reconstructive Surgery, Department of Neurosciences, University of Padova, Padova, Italy
e-mail: franco.bassetto@unipd.it; carlotta.scarpa@unipd.it

Even if it's considered a low invasiveness surgery, lipofilling can present complications; some are in common with other treatments, others specific. Here we'll analyze the most important ones.

3.1 Autologous Fat Transfer Main Complications (1, 9–13, 17, 18)

- *Infection* (Fig. 3.1)*:* rare event, this complication can be due to a severe contamination of the non-vascularized fat during the harvesting or grafting procedure. In order to avoid this complication, a sterile technique is required and preoperative antibiotic treatment is mandatory. Postoperative antibiotic treatment can be proposed in particular to patients who can be predisposed to infection, as diabetic patients, or in patients treated with steroids, or in case of overt infection.
- *Absorption* (Fig. 3.2)*:* ranged from 0 to 70%, this complication can be due to a wrong technique in which damaged or bloody fat tissue or fat derived from fibrous areas (e.g., upper abdomen) has been used. In order to avoid this complication, both an overcorrection of 30–50% and a gentle manipulation of the fat especially during the grafting step can be useful. Small parts and a subcutaneous infiltration are preferable.
- *Excessive augmentation:* different from the "recommended standard" overcorrection, this event can provoke important compression of a vein and/or nerves especially when performed on face/neck or extremities.

© The Author(s), under exclusive license to Springer Nature Switzerland AG 2023

A. Di Giuseppe et al. (eds.), *Fat Transfer in Plastic Surgery*, https://doi.org/10.1007/978-3-031-10881-5_3

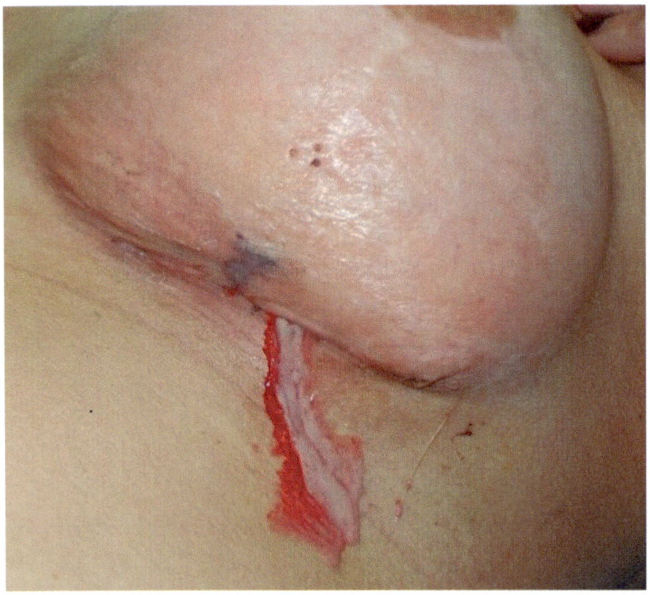

Fig. 3.1 Infection after lipofilling and breast implant

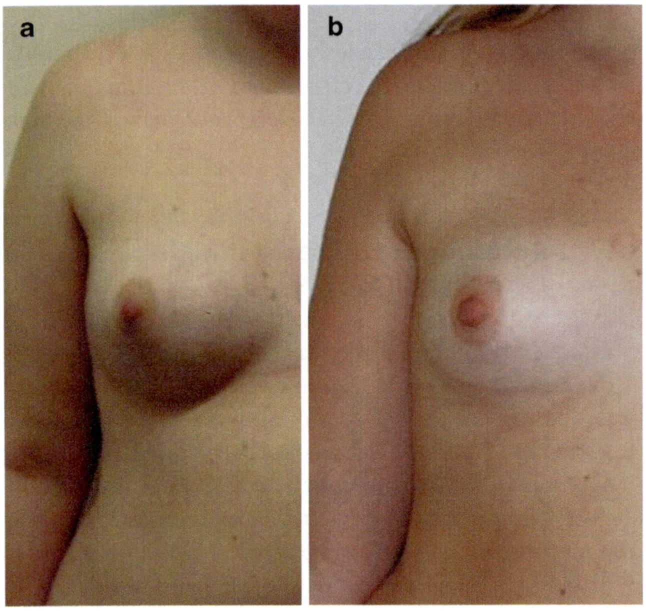

Fig. 3.2 (**a**) Breast lipofilling; (**b**) fat absorption at 3 months

- *Fat necrosis with or without skin necrosis and oil cysts/calcification* (Fig. 3.3)*:* due to infection and/or overcorrection, this complication can lead to oil cysts and/or calcification requiring surgical treatment to remove them. If associated with skin necrosis, it could be possible to associate advanced therapies (e.g., negative pressure therapy) to promote skin healing. If cysts are small, it can also be possible to aspirate them or treating them with triamcinolone injection. As we'll see later, particular attention is required to avoid misdiagnosis, especially in breast lipofilling. Indeed, small calcifications could be misdiagnosed as malignant calcification to an "inexperienced eye."

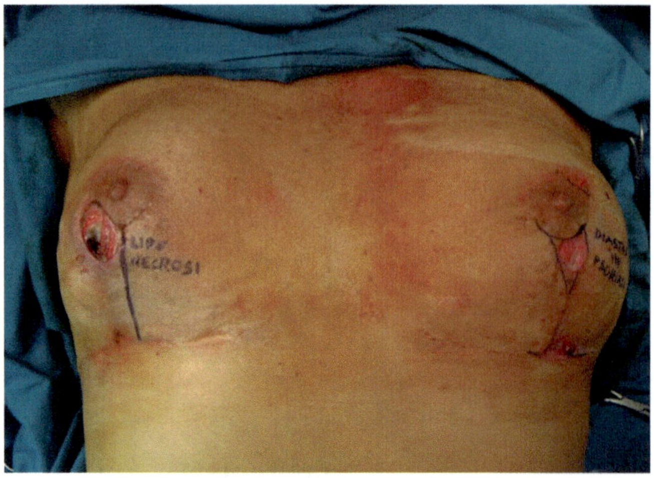

Fig. 3.3 Skin necrosis followed by dehiscence after liponecrosis

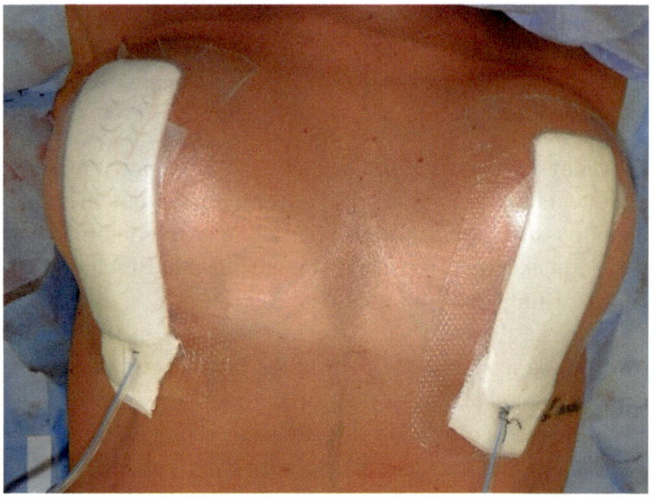

Fig. 3.4 Closed incision negative pressure therapy

- *Asymmetry:* due to different fat absorption and/or incorrect technique, this complication can require surgical treatment as lipoaspiration or new fat transfer. Postoperative external garments are recommended to prevent it.
- *Embolism:* extremely rare event, this complication is due to an incorrect technique that brings to a direct injection of fat in vessels. If not recognized or treated, it can bring patients to death. In order to avoid it, sharp needles are not allowed and it is recommended to graft small fat particles (2 mm or less in diameter).
- *Other complications:* if lymphatic circulation is compromised due to a fat compression in overcorrected patients, persistent edema can also occur. To avoid it, external garments, closed incision negative pressure therapy (Fig. 3.4), and an antiedemigen treatment can be recommended.

3.2 Lipofilling in Breast Surgery

As said earlier, lipofilling in breast surgery has been one the most debated issues of the last decades.

Mammary ptosis means progessive breast drop following lost of volume as in post bariatric condition breast often requesting a mastopexy and breast implant with following cases of capsular contracture and reintervention due to the presence of a foreign body that can chronically stimulate immunity system.

Currently, to avoid this complication, surgeons can perform:

1. A mastopexy plus auto-augmentation: in this case, the mammary glandula works as a prosthesis. In order to obtain this result, many are the reported techniques, such as the "Rubin technique or DSPRM mastopexy" during which the surgeon performs a combination of parenchymal reshaping, dermal suspension, and auto-augmentation.
2. Fat transfer: in this case, large volume of fat can substitute prosthesis.

Proposed in 1987 by Bircoll, this last technique brought many doubts regarding safety profile and different oncological issues have been raised.

There are three main topics: (1) the possible clinical and instrumental misdiagnosis of cancer due to grafted adipose tissue calcification and/or oil cysts [2–8]; (2) the influence on cancer cells' proliferation due to the stimulation of neoangiogenesis, and the reported cross-talking between the cancer cells and the adipose stem cells which can be "educated" and brought to the "dark side"; (3) the aromatases present in adipose tissue which could stimulate cancer onset or differentiation.

3.2.1 The Misdiagnosis

Since the 2000s many authors have faced this issue but only recent systematic reviews and the Expert Consensus Panel have started to dispel any doubts [2–7, 14–16]. These papers report radiological changes in a total of 12% of the patients treated with lipofilling, and only 3.2% of the patients need a biopsy to exclude malignancy [8]. They also underlined that these changes are easily distinguishable from malignant ones due to their morphology and localization and they have to be referred to as fat necrosis, benign calcification, and oil cysts. Furthermore, if radiological changes after fat transfer are compared with the ones after breast reduction, it's possible to note that the second ones are more frequent.

The authors concluded that a regular follow-up is enough for these patients and that FNAC can be done if in doubt, and no significant risk of misdiagnosis is present.

3.2.2 The Influence on Cancer's Cells and the Possibility of a "Bad Education" of the ADSCs

This field has been more debated [3–6] because of recent researches that notice the capability of fat to influence angiogenesis promoting cancer onset or evolution and an in vitro possibility to stimulate cancer cells or being stimulated by cancer cells to differentiate in malignant cells. None of the above sentences have been demonstrated in vivo. Many papers reported no increased cancer in patients treated with lipofilling compared to healthy controls. If the patient has a positive familiar or personal history of breast cancer or there is a suspicious calcification on preoperative exams, for example, mammography and/or ultrasounds, it's better to refuse fat transfer treatment.

3.2.3 The Aromatases

Aromatases are known to synthesize estrogen and to be overexpressed in breast cancer tissue. It's also known that aromatases are present in adipose tissue and that can be overexpressed during inflammation. Based on these claims, scientists wondered whether adipose tissue could arise or stimulate the evolution of estrogen-dependent breast cancer.

To answer this question, it's necessary to consider where the adipose tissue is injected. As Rigotti, Fraser [20, 21], and other authors already noted, breast lipofilling is featured by grafting both in subcutaneous tissue and the layers under the glandula rather than into the parenchyma, thus reducing the risk of rising cancer. Finally, recently a new hypothesis has been published by Zocchi et al. [2] about the possibility of wasting aromatases during fat grafting procedure. The authors indeed suggest that aromatases could be mainly contained in the oily fraction of the adipocyte. Considering that after completing centrifugation, decantation, or mechanical lipocondensation, the superfluous parts, such as the oil one, are discarded before infiltrating, it should therefore be possible to also discard most of the aromatases contained in the fat tissue, avoiding or drastically reducing the oncological risks. Further studies are required to verify this hypothesis.

A final question was raised a few years ago about a direct relationship between a large amount of fat transfer and cancer danger [19]. Again, the clinical studies have shown no relationship [4–6].

In light of the considerations mentioned above, nowadays it's fair to state that fat transfer is a safe and less invasive procedure in postbariatric breast surgery. However, when breast lipofilling is performed, a regular follow-up with mammography and/or ultrasounds is recommended both before and after surgical procedures.

References

1. Ørholt M, et al. Complications after breast augmentation with fat grafting: a systematic review. Plast Reconstr Surg. 2020;145:530e–7e.
2. Zocchi ML, Zocchi L. Large-volume breast fat transfer: technical evolutions and safety aspects based on over 800 cases and 26 years of follow-up. Eur J Plast Surg. 2017;40:367–82.
3. Parrish JN, Metzinger SE. Autogenous fat grafting and breast augmentation: a review of the literature. Aesthet Surg J. 2010;30(4):549–56.
4. Bayram Y, et al. The use of autologous fat grafts in breast surgery: a literature review. Arch Plast Surg. 2019;46(6):498–510.
5. Kronowitz SJ, et al. Lipofilling of the breast does not increase the risk of recurrence of breast cancer: a matched controlled study. Plast Reconstr Surg. 2016;137:385–93.
6. Piccotti F, et al. Lipofilling in breast oncological surgery: a safe opportunity or risk for cancer recurrence? Int J Mol Sci. 2021;22(7):3737.
7. Nava MB, et al. International expert panel consensus on fat grafting of the breast. Plast Reconstr Surg Glob Open. 2019;7:e2426–36.
8. Rubin JP, et al. Mammographic changes after fat transfer to the breast compared with changes after breast reduction: a blinded study. Plast Reconstr Surg. 2012;129:1029–41.
9. Largo RD, et al. Efficacy, safety and complications of autologous fat grafting to healthy breast tissue: a systematic review. J Plast Reconstr Aesthet Surg. 2014;67(4):437–48.
10. Groen JW, et al. Autologous fat grafting in cosmetic breast augmentation: a systematic review on radiological safety, complications, volume retention, and patient/surgeon satisfaction. Aesthet Surg J. 2016;36(9):993–1007.
11. Waked K, et al. Systematic review: the oncological safety of adipose fat transfer after breast cancer surgery. Breast. 2017;3:128–36.
12. Delay E, et al. Fat injection to the breast: technique, results, and indications based on 880 procedures over 10 years. Aesthetic Surg J. 2009;28(5):360–78.
13. Raj S, et al. Safety and regulation of fat grafting. Semin Plast Surg. 2020;34:59–64.
14. Wu Y, et al. Autologous fat transplantation for aesthetic breast augmentation: a systematic review and meta-analysis. Aesthetic Surg J. 2021:1–29.
15. Charvet HJ, et al. The oncologic safety of breast fat grafting and contradictions between basic science and clinical studies. A systematic review of the recent literature. Ann Plast Surg. 2015;75:471–9.
16. Sorrentino L, et al. Autologous fat transfer after breast cancer surgery: an exact-matching study on the long-term oncological safety. Eur J Surg Oncol. 2019;45:1827–34.
17. Khawaja HA, et al. Complications of fat transfer. In: Shiffman MA, editor. Autologous fat transfer 417. Berlin Heidelberg: Springer-Verlag; 2010.
18. Yoshimura K, Coleman S. Complications of fat grafting how they occur and how to find, avoid, and treat them. Clin Plast Surg. 2015;42(3):383–8.
19. Oe B, Rubin JP. Breast reshaping after massive weight loss. Clin Plast Surg. 2019;46(1):71–6.
20. Rigotti G, et al. Determining the oncological risk of autologous lipoaspirate grafting for post-mastectomy breast reconstruction. Aesthet Plast Surg. 2010;34:475–48.
21. Fraser JK, et al. Oncologic risks of autologous fat grafting to the breast. Aesthet Surg J. 2011;31(1):68–75.

Fat Transfer During the Pandemic COVID-19 Time

4

Franco Bassetto and Facchin Federico

Contents

At the end of 2019, a novel coronavirus was recognized as the agent causing the outbreak of pneumonia in Wuhan, Hubei Province, China. The infectious agent was named severe acute respiratory syndrome coronavirus 2 (SARS-CoV-2), while the disease was defined coronavirus disease-19 (COVID-19) [1, 2]. On March 11, 2020, WHO declared the spread of the infection as a pandemic. On May 11, 2021, WHO reported 3,277,834 cumulative deaths and 157,362,408 cumulative cases worldwide [3].

Patients affected by Sars-CoV-2 can be asymptomatic or show the clinical manifestation of the disease. Almost all patients develop symptoms within 12.5 days from virus exposure, while the incubation period reaches up to 14 days with a mean duration of 5.2 days [4–6].

Symptoms range from flu-like disease with fever, dry cough, myalgia, fatigue, and dyspnea to respiratory or multi-organ failure. Uncommon manifestations include diarrhea, abdominal pain, dizziness, productive cough, pleuritic chest pain, and hemoptysis. Skin manifestations have been reported in association to other most common symptoms or alone [4–6].

Furthermore, viral transmission has been demonstrated among asymptomatic patients [7].

The most common, efficient, and safe diagnostic tool used for viral RNA detection is real-time reverse transcriptase-polymerase chain reaction (rRT-PCR) performed on the nasopharyngeal swab. It has been recommended by the Center for Disease Control and Prevention (CDC) instead of most sensitive tools based on lower respiratory samples (i.e., bronchoalveolar lavage fluid specimens, brush biopsy, and sputum) for the lower risk of infection for healthcare workers, avoiding the creation of aerosol droplets [8–11].

During the first pandemic peak, almost all surgical procedures have been limited or postponed including fat transfer. The healthcare systems worldwide focused all their efforts on facing the pandemic limiting elective procedures like fat grafting [12].

In particular, during all surges in COVID-19 cases, plastic surgery departments were forced to limit their activity to allow healthcare workers' redistribution in COVID-19 treating units, to increase room availability for COVID-19 patients and to create intensive care units beds in the operating rooms [13].

The effective and forward-looking management of an epidemic outbreak provides, as a priority, the adoption of strategies aimed at the protection of health personnel and at the maximum containment of the transmission of the contagion between operators and patients.

The physical distancing in the workplace, the limitation of the duration of meetings, the use of personal protective equipment (PPE), and periodic screening of all personnel were effective in protecting operators in many departments worldwide [14].

In addition, hospital reorganization strategies showed positive results in allowing to perform urgent and oncological elective procedures minimizing the risk of infection [15].

Deferrable procedures for SARS-CoV-2-positive patients should be suspended for at least 2 weeks from the resolution

F. Bassetto
University of Padova, Clinic of Plastic and Reconstructive Surgery, Padua, Italy
e-mail: franco.bassetto@unipd.it

F. Federico (✉)
University of Padova, Clinic of Plastic and Reconstructive Surgery, Padua, Italy

Plastic Surgery Unit, Azienda ULSS 8 Berica, Vicenza, Italy

© The Author(s), under exclusive license to Springer Nature Switzerland AG 2023
A. Di Giuseppe et al. (eds.), *Fat Transfer in Plastic Surgery*, https://doi.org/10.1007/978-3-031-10881-5_4

of the infection, while patients at risk by symptomatology or history of close contact should be postponed for at least 2 weeks. Prolonged viral RNA shedding has been reported in the literature and it should be considered when programming the admission of a patient healed from the virus infection [16, 17]. On the contrary, positive patients requiring urgent procedures need to be treated in dedicated operative rooms.

During plateau phases of the pandemic, thanks to recent improvement of healthcare resources, elective surgery, including reconstructive procedures, should restart to address patients suffering from non-urgent diseases requiring treatment. Nonetheless, all patients undergoing operative intervention should be considered potentially infected asymptomatic carriers.

For this reason, specific preoperative screening is desirable. The history of the patient's general condition during the 14 days before the admission including respiratory or gastrointestinal symptoms, anosmia, or risk of exposure to virus need to be collected.

In addition, according to their availability, nasopharyngeal swab tests should be performed for a more sensitive screening of patients undergoing hospital admission.

In this way, maintaining a high suspicious index, medical staff could perform daily clinical tasks with basic personal protective equipment limiting the use of goggles, face shield, gowns, double-layered gloves, and protective footwear only for positive patients [14].

Fat transfer is included among deferrable procedures and it has been postponed during pandemic peaks in the majority of plastic surgery unit. However, giving its role in improving body shape, treating scars, and its regenerative scope to treat many pathological conditions, it should be restored as soon as other elective surgeries. In cases requiring aerosol-generating procedures, such as intubation and non-invasive/manual ventilation, PPE and transmission prevention should be carried out also in patients who screened negative [18]. On the other hand, the risk of infection is lower when aerosol-generating procedures are not required, such as in an outpatient setting.

As the vaccination campaign is proceeding, the risk of infection and virus transmission progressively reduces [19]. No surgical procedures have been associated with an increased risk for vaccine adverse event. However, Food and Drug Administration (FDA) reported adverse events in patients with dermal fillers receiving the SARS-CoV-2 mRNA vaccine [20]. Nonetheless, to limit the risk of in-hospital virus transmission, also vaccinated patients should be screened. In addition, it seems reasonable to program surgeries some days after the vaccination [21].

Indeed, a recent study suggests prioritizing patients needing elective surgery in vaccine administration. In fact, preoperative vaccination of patients needing elective surgical procedures over the general population seems able to prevent additional thousands of COVID-19-related deaths in 1 year [22].

References

1. World Health Organization Clinical management of severe acute respiratory infection when Novel coronavirus (nCoV) infection is suspected http://www.who.int/internalpublications-detail/clinical-management-of-severe-acute-respiratoryinfection-when-novel-coronavirus-(ncov)-infection-is-suspected (2020-01-11) [2020-02-11].
2. Zhu N, Zhang D, Wang W, Li X, Yang B, Song J, et al. A novel coronavirus from patients with pneumonia in China, 2019. N Engl J Med. 2020;382(8):727–33.
3. http://www.who.int/publications/m/item/weekly-epidemiological-update-on-covid-19%2D%2D-11-may-2021.
4. Bai Y, Yao L, Wei T, Tian F, Jin DY, Chen L, et al. Presumed asymptomatic carrier transmission of COVID-19. JAMA. 2020;323(14):1406–7.
5. Li Q, Guan X, Wu P, Wang X, Zhou L, Tong Y, et al. Early transmission dynamics in Wuhan, China, of novel coronavirus-infected pneumonia. N Engl J Med. 2020;382(13):1199–207.
6. Wang D, Hu B, Hu C, Zhu F, Liu X, Zhang J, et al. Clinical characteristics of 138 hospitalized patients with 2019 novel coronavirus-infected pneumonia in Wuhan, China. JAMA. 2020;323(11):1061–9.
7. Lavezzo E, Franchin E, Ciavarella C, Cuomo-Dannenburg G, Barzon L, Del Vecchio C, et al. Suppression of a SARS-CoV-2 outbreak in the Italian municipality of Vo. Nature. 2020;584(7821):425–9.
8. Basso D, Aita A, Navaglia F, Franchin E, Fioretto P, Moz S, et al. SARS-CoV-2 RNA identification in nasopharyngeal swabs: issues in pre-analytics. Clin Chem Lab Med. 2020;58(9):1579–86.
9. Cheng PK, Wong DA, Tong LK, Ip SM, Lo AC, Lau CS, et al. Viral shedding patterns of coronavirus in patients with probable severe acute respiratory syndrome. Lancet. 2004;363(9422):1699–700.
10. Loeffelholz MJ, Tang YW. Laboratory diagnosis of emerging human coronavirus infections—the state of the art. Emerg Microbes Infect. 2020;9(1):747–56.
11. Wang W, Xu Y, Gao R, Lu R, Han K, Wu G, et al. Detection of SARS-CoV-2 in different types of clinical specimens. JAMA. 2020;323(18):1843–4.
12. Bassetto F, Marchica P, Azzena GP, Brambullo T, Facchin F, Masciopinto G, et al. Brief history in the time of SARS-CoV-2 pandemic in Italy. A close look on a plastic surgery unit and plastic surgeons efforts during the COVID-19 outbreak. Ann Ital Chir. 2021:10.
13. Wu M, Wang J, Panayi AC. Plastic surgery during the COVID-19 pandemic: the space, equipment, expertise approach. Aesthet Surg J. 2020;40(9):NP574–7.
14. Facchin F, Scarpa C, Vindigni V, Bassetto F. Effectiveness of preventive measures against coronavirus disease of 2019 in a plastic surgery unit at the epicenter of the pandemic in Italy. Plast Reconstr Surg. 2020;146(1):112e–3e.
15. Facchin F, Messana F, Sonda R, Faccio D, Tiengo C, Bassetto F. COVID-19: initial experience of hand surgeons in northern Italy. Hand Surg Rehabil. 2020;39(4):332–3.
16. Ling Y, Xu SB, Lin YX, Tian D, Zhu ZQ, Dai FH, et al. Persistence and clearance of viral RNA in 2019 novel coronavirus disease rehabilitation patients. Chin Med J. 2020;133(9):1039–43.

17. Xu K, Chen Y, Yuan J, Yi P, Ding C, Wu W, et al. Factors associated with prolonged viral RNA shedding in patients with coronavirus disease 2019 (COVID-19). Clin Infect Dis. 2020;71(15):799–806.
18. Ti LK, Ang LS, Foong TW, Ng BSW. What we do when a COVID-19 patient needs an operation: operating room preparation and guidance. Can J Anaesth. 2020;67(6):756–8.
19. Polack FP, Thomas SJ, Kitchin N, Absalon J, Gurtman A, Lockhart S, et al. Safety and efficacy of the BNT162b2 mRNA Covid-19 vaccine. N Engl J Med. 2020;383(27):2603–15.
20. http://www.plasticsurgery.org/for-medical-professionals/covid19-member-resources/covid19-vaccine-dermal-fillers.
21. Oliver SE, Gargano JW, Marin M, Wallace M, Curran KG, Chamberland M, et al. The advisory committee on immunization Practices' interim recommendation for use of Pfizer-BioNTech COVID-19 vaccine—United States, December 2020. MMWR Morb Mortal Wkly Rep. 2020;69(50):1922–4.
22. SARS-CoV-2 vaccination modelling for safe surgery to save lives: data from an international prospective cohort study. Br J Surg; 2021.

Fat Grafting as an Ancillary Treatment for Burns, Other Complex Wounds, and Their Sequelae

5

Nelson Sarto Piccolo, Mônica Sarto Piccolo, Nelson de Paula Pìccolo, and Paulo de Paula Piccolo

Contents

Key Points

- The use of fat grafting has been incorporated on our everyday routine, changing our practice dramatically.
- In relation to burns and other wounds, there are several changes in our routine acute treatment, as well as after healing occurs, when fat grafting has greatly influenced the way we treat these patients´ acute wounds.
- These changes are even more noticeable in the way we take care of the resulting scar, since fat grafting has also greatly influenced the way we treat hypertrophic scars as a consequence of burn and other wounds.
- One of the most pleasant surprises in using fat grafts is the minimal incidence (or none) of hypertrophic scarring upon the healing of wounds treated with one or more sessions of fat grafting.
- In cases where no fat grafting was used to treat the acute wound, it can be used to treat the sequela and one can note improvement of the scar appearance as well as in volume, as early as 1–2 weeks after fat injection/fat delivery.
- We believe that the effect of fat grafting in treating wounds and/or scars is cumulative, meaning that repeat fat grafting will yield continuous, "overlapping" gain.

N. S. Piccolo (✉) · M. S. Piccolo · N. de Paula Pìccolo
P. de Paula Piccolo
Division of Plastic Surgery, Pronto Socorro para Queimaduras, Goiânia, Goiás, Brazil
e-mail: nelsonpiccolo@grupopiccolo.com.br; monicapiccolo@grupopiccolo.com.br; nelsonppiccolo@grupopiccolo.com.br; paulopiccolo@grupopiccolo.com.br

5.1 Introduction

Fat grafting has become a common procedure in wounds originated from trauma and/or other causes. Fat contains Adipose-Derived Stem Cells and a great variety of growth factors which may have a direct effect in wound healing. Fat grafting has also been used successfully for the management of scars and post-trauma healing fibrosis, scarring, and pain. ADSCs may differentiate into fibroblasts, keratinocytes, and many other cells; they may also secrete mediators with neo-angiogenic and anti-inflammatory properties. This would allow for it to act in all phases of the wound healing process as we understand it today. Fat on the lipoaspirate can be isolated and/or treated by physical or chemical methods, in the OR or in a laboratory setup [1–8].

As it was used more than a century ago to treat facial deformities, fat grafting was originally (re)introduced in the cephalic segment aiming at improvement in aesthetic aspects of the face and peri-orbit by Coleman in the early 1990s, and soon became one of the main options for disease, trauma, or post-surgery-related deformities [9, 10].

The use of fat grafting as an adjuvant treatment in acute and sub-acute burn and other wounds and in (chronic) vascu-

© The Author(s), under exclusive license to Springer Nature Switzerland AG 2023

A. Di Giuseppe et al. (eds.), *Fat Transfer in Plastic Surgery*, https://doi.org/10.1007/978-3-031-10881-5_5

lar wounds (venous insufficiency or diabetic arterial disease) attempts to take advantage of fat's benefits, when a great variety of metabolic and regenerative properties act increasing vascularization and enhancing the tissue regeneration process. When these wounds are treated with (repeated) fat grafting, healing is the planned outcome [11, 12].

When treating scars, the objective is to decrease the amount of hypertrophy (fibrosis), diminishing the scar thickness and increasing scar malleability. We also use this technique aiming at enhancing bone fracture healing, at decreasing fibrosis around bone joints, and at releasing tendon adhesions [13–15].

Additionally, when fat grafting is performed after the wound is healed, it will provide a compounded benefit, in decreasing fibrosis (on the surface, in the skin scar, and deep, around tendons and joints). This will occur regardless of the patient being previously treated with fat grafting or not [16].

Fat can be grafted (fat grafting) under the wound or scar, and fat can also be delivered (fat delivery) directly over the surface of a debrided wound or to scar or to skin surface, after microneedling or laser treatment [17].

The fat intended to be used in this technique of fat grafting and fat delivery is harvested from the patient him- or herself, via common liposuction techniques. The lipoaspirate may be treated by a variety of physical and chemical methods, or a combination of both.

In our institutions, we prepare and inject the fat via the Coleman Technique. We added fat delivery as a routine part of the procedure, aiming at an additional benefit, when we thought that the wound surface (or the substance of the scar) could also be influenced by direct contact with cells and factors concentrated in the centrifuged fat. The findings were remarkable, and we now use both fat grafting and fat delivery in the same surgical setting routinely.

The objective is to obtain full and prompt wound healing, while aiming at lesser fibrosis, and when treating scars, to obtain increased malleability and a progressive reduction of hypertrophic scarring.

5.2 Preoperative Evaluation and Special Considerations

5.2.1 Patient Selection

Patients with wounds or scars who are candidates for fat grafting procedure at our Service are those with:

1. Burn and other wounds with 3 weeks or more with no apparent progression to healing
2. Sub-acute burn wounds or other wounds who are transferred to us within more than 2 weeks after the accident or wound
3. Venous or diabetic ulcers
4. Decubitus ulcers
5. Wound cavities of any origin (avulsion, drained hematomas, tumor resection, etc.)
6. Shoulder, wrist, knee, and ankle tendinitis; post-fracture "bone pain"; major joint arthrosis, fracture lines on long bones on external fixation
7. Hypertrophic scars that are not improving or not being controlled by pressure garments at six or more weeks after healing
8. Patients with scar retractions over noble areas (tendons, nerves, or vessels) when fat is (pre)injected in these areas, aiming at "covering" the deep tissues when they could be exposed upon the incision release of the retraction

Patients with sub-acute burn wounds (more than 3 weeks in our Service) without apparent progression to healing and patients with hypertrophic scarring after healing of a burn or other type of trauma or keloids of any origin are also selected for treatment with fat injection/delivery. Repeat injections (up to four injections total) are performed at 7–10 days intervals for wounds or at 6–8 weeks intervals for scars. Fat is re-harvested each time a new fat grafting is performed.

The use of fat grafting as an adjuvant treatment in acute and sub-acute burn wounds aims at taking advantage of fat's benefits—a variety of metabolic and regenerative properties, increasing vascularization, and enhancing the tissue regeneration process. When these wounds are treated with (repeated) fat grafting, healing (with minimal fibrosis) is the planned outcome. When treating burn and other scars, the objective is to decrease the amount of hypertrophy (fibrosis), diminishing the scar thickness and increasing scar malleability [12, 18, 19].

In patients with chronic wounds, as part of the general preoperative evaluation, fragments of the wound are obtained for culture and sensitivity. These patients very frequently will carry multi-resistant organisms on their wounds, since the vast majority has received treatment in one or more institutions prior to our evaluation. We know the inhabitant flora will aid in the perioperative antibiotic selection, while not precluding performing the procedure.

The actual surgical procedure is performed in the Operating Room, following all rigors and care for sterile procedures.

5.3 Surgical Technique

Fat harvesting and fat injection are sterile surgical procedures and should be performed only in accredited operation rooms under rigorous, completely sterile technique. Patients are submitted to general anesthesia or regional block.

5.3.1 Donor Areas

Donor areas are "rotated" as needed and fat most frequently is obtained from the thighs or lateral upper buttocks—less frequently from the abdomen (when we obtain fat from the abdomen, we first order an ultrasound of the abdominal to verify the presence or not of wall defects and/or hernias, which could preclude the use of this area as a donor area).

When necessary, shaving of the pubic area or proximal thigh is performed in the OR, immediately before the procedure. Puncture incisions for introduction of the liposuction cannula are placed on the midline, at the suprapubic crease, or medial to the femoral pulse, at the inguinal crease, or in the middle axillary line, at the upper border of the iliac bone.

5.3.2 Fat Graft Harvest

The actual volume of harvested lipoaspirate should be at least twice the anticipated volume planned to be injected, and at least four times this volume, if one is also planning to have fat delivered over the wound or the scar.

5.3.3 Patient Positioning

Patients are supine when using the abdomen or thighs as donor areas or on lateral decubitus when obtaining fat from the lateral upper thighs. Fat is usually injected and delivered while the patient is supine. In paraplegic patients with decubitus pressure orders or patients with wounds in the back, fat is usually harvested with the patient supine and then the patient is repositioned into a lateral or prone position as needed.

5.3.4 Recipient Site Preparation

In patients with open wounds, the donor area is initially prepped and draped and fat is then obtained by liposuction; only after the planned amount of fat is obtained, the recipient area is then prepped and draped, while the obtained fat is being centrifuged and distributed into various 1-cc syringes.

In patients with scars (healed wounds), the donor area and recipient area are individually prepped and draped in the usual manner.

5.3.5 Fat Harvesting

Fat is harvested from the patient him- or herself, using a 10-cc Luer Lok syringe, attached to a 3-mm canula, with multiple (8–12) distal side openings, with 10, 15, or 20 cm length, according to the harvesting site. In children weighing less than 25 kg, as well as in ladies with relatively thin thighs, we prefer 20-cc syringes and a 2.5-mm cannula, also multi-perforated distally. In these smaller patients, these cannulas will enforce a higher negative pressure assuring a more even and efficient fat harvesting, respectively. Occasionally in very small patients (our smallest patient weighted 8.145 kg) or in the elderly (our oldest patient was 89 years old), it may be necessary to harvest fat from more than one donor site.

As recommended by Coleman, one or more distally plugged 10-cc syringe containing the obtained fat is/are centrifuged at 3000 rpm for 3 min on a 30-degree angle centrifuge (1200 G's). The obtained compound has a top layer of oil, a middle layer of fat (with the SVF within, at its lower portion), and an aqueous inferior layer. The top layer of oil is discarded while the plug still is on the syringe. The plug is then removed and the aqueous layer drains out per gravity. The remaining compound is sequentially injected anterogradely into "insulin" syringes without the plunger, which is then replaced (Fig. 5.1).

5.3.6 Fat Grafting/Fat Delivery

A perforation is made at an acute angle in healthy skin in the periphery of the wound or the scar using a 16-gauge needle. A 1.8 mm outside diameter 70-mm-long cannula already connected to a Luer Lok 1-cc syringe is inserted through the needle puncture hole and (forcefully, if needed) driven immediately under (or through) the wound bed or the scar. Fat is then deposited in a retrograde manner, in several "passes" until the entire area is grafted (via as many puncture sites as needed around the periphery of the scar or wound). On average, 1.8–2.5 cc of centrifuged fat are injected per each 10 cm^2 area and it is necessary to make 25 to 30 "passes" to inject 1 cc (Fig. 5.2).

Occasionally, when there are fracture lines of bone loss "voids" or exposed bone, fat will be injected through the wound into the fracture line. After the wound area has been completely (under)grafted, the surface of the wound is thoroughly debrided and fat is deposited in enough quantity to cover the entire wound. We usually debride the wound only after the undersurface has been grafted because, by doing so, we avoid having to do multiple punctures around a bleeding wound or running the risk of moving debris along with the injection cannula under the wound. In deeper burn wounds, the dead tissue must be excised, even if tendons or nerves will be exposed – centrifuged fat is then delivered locally covering these noble structures.

An average of 3.5 cc of centrifuged fat is delivered per each 10 cm^2 wound area and the fat is delivered in a zig-zag manner directly over the entire surface of the wound, using a 1-cc or 10-cc syringe connected to a 1.8-mm/70-mm cannula.

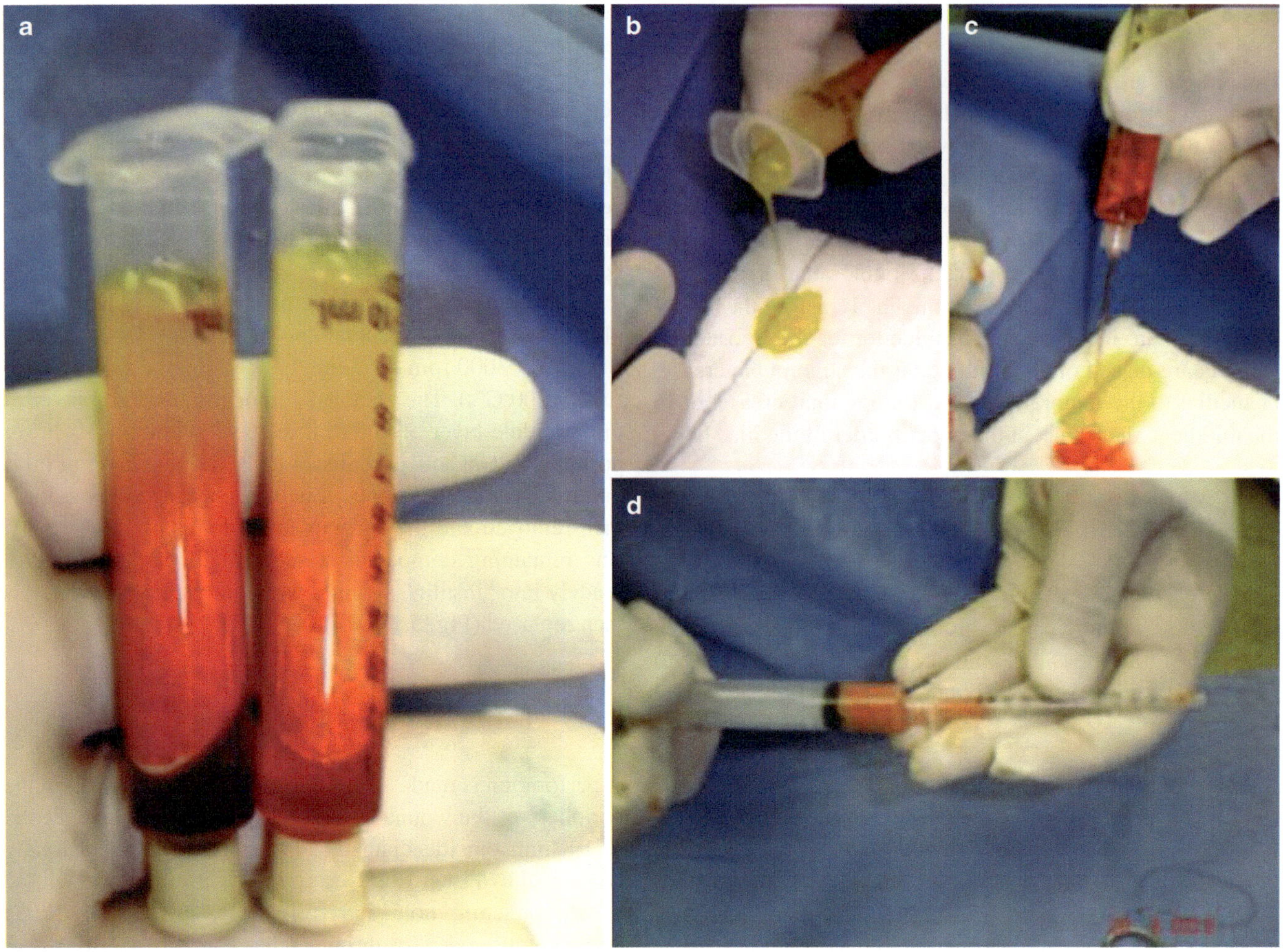

Fig. 5.1 (**a–d**) – Centrifuged lipoaspirate, discarding oil and aqueous layers, and filling "insulin" syringe

After the wound area has been completely (under)grafted, the surface of the wound is thoroughly debrided and fat is deposited in enough quantity to cover the entire wound. If bone (with or without periosteum), tendons, or nerves are already exposed or exposed after debridement, fat is delivered directly over any or all of these structures (Figs. 5.3 and 5.4).

When treating scars, centrifuged fat also treated by the Coleman Technique is injected immediately under the substance of the scar. Fat is injected through as many puncture holes and directions necessary to "cover" the entire undersurface of the wound or scar. The cannula will run immediately under the scar tissue. After fat grafting is complete, the scar surface area is treated with a dermaroller (usually with 0.5–1.5 mm needle length) or the scar surface may be treated with a fractional CO_2 laser, opening "pores" or holes, through the epidermis, into the substance of the scar. Centrifuged fat is then delivered directly to the treated surface (in average 3 cc/10 cm^2) (Figs. 5.5 and 5.6).

Very frequently, in scar cases, we will combine one or more partial intralesional scar resections and primary suturing with fat grafting immediately under the suture line, as well as under the entire surface of the scar. In these cases, one may also run the dermaroller over the sutured area and the surface of the scar and also provide centrifuged fat delivery.

Partial scar removal and fat grafting/fat delivery are associated very frequently. In these cases, we perform the partial scar resection first, keeping the resection within the scar substance (trying not to go into subcutaneous tissue). A running nylon suture closes the surgical wound. Fat is injected under the entire scar surface, including under the suture line. Surprisingly, all these patients who are originally healed from a facial burn with variable amounts of hypertrophic

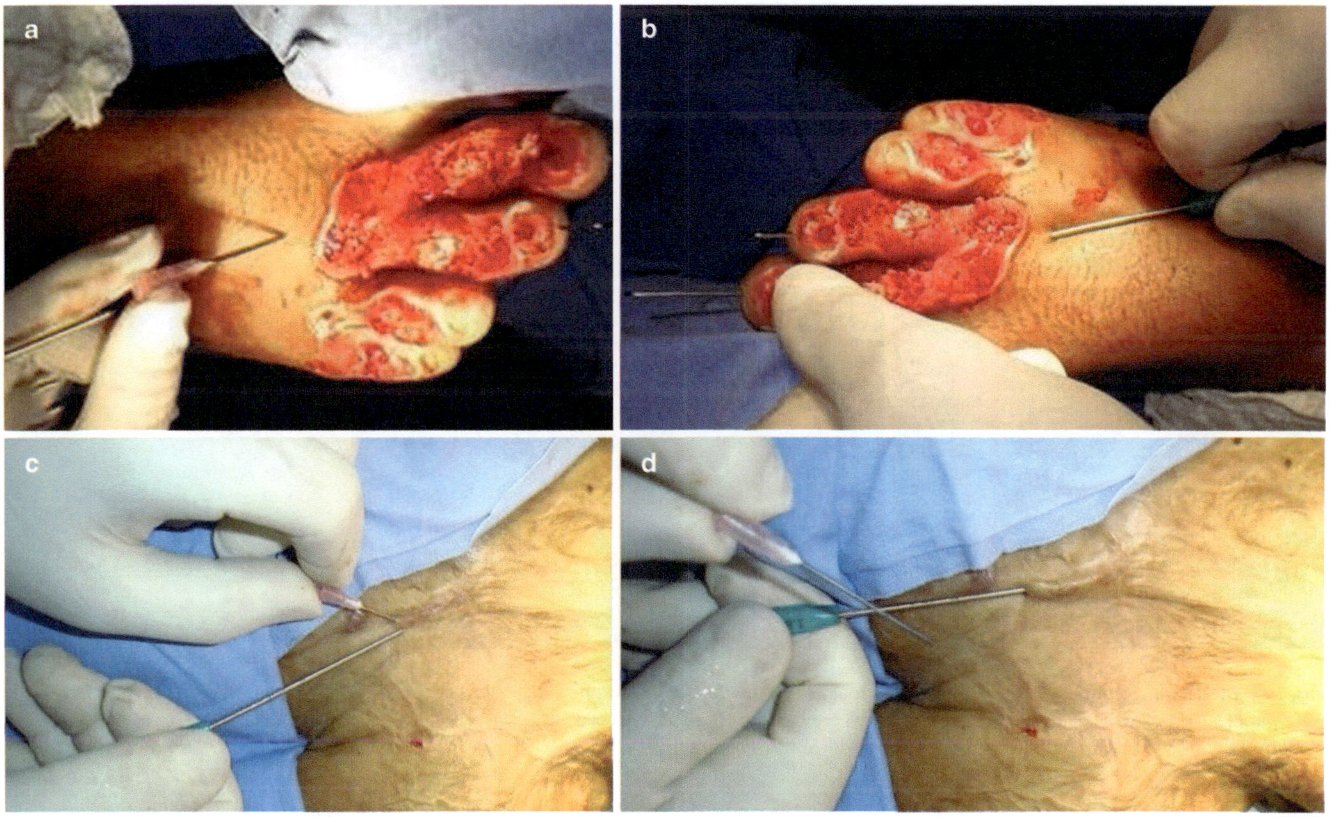

Fig. 5.2 (**a**, **c**) A puncture on the skin peripherally to the wound or scar is made with a pink (18 gauge) needle; **b**, **d**—the injection cannula is then inserted via the needle puncture site

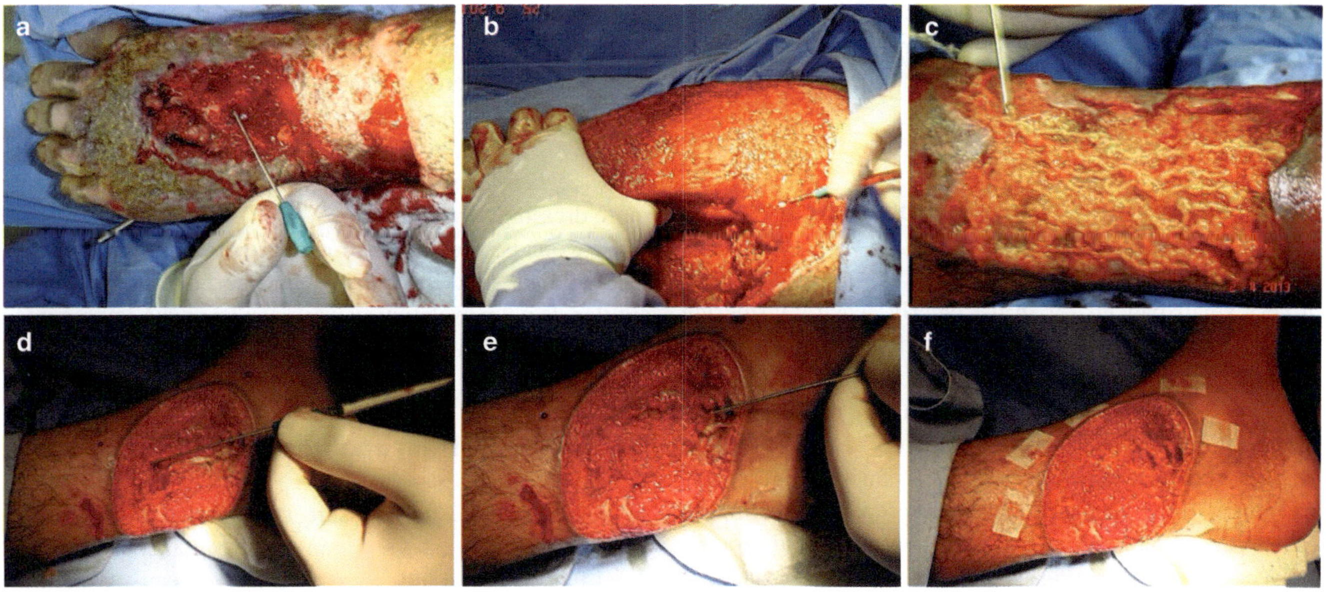

Fig. 5.3 (**a**) Cannula penetrating through the wound into bone fracture line; (**b**) cannula penetrating though the wound (and muscle) delivering fat to wound cavity; (**c**) fat being delivered to wound surface after debridement; (**d**–**f**) fat being delivered to debrided wound covering exposed nerve and tendons

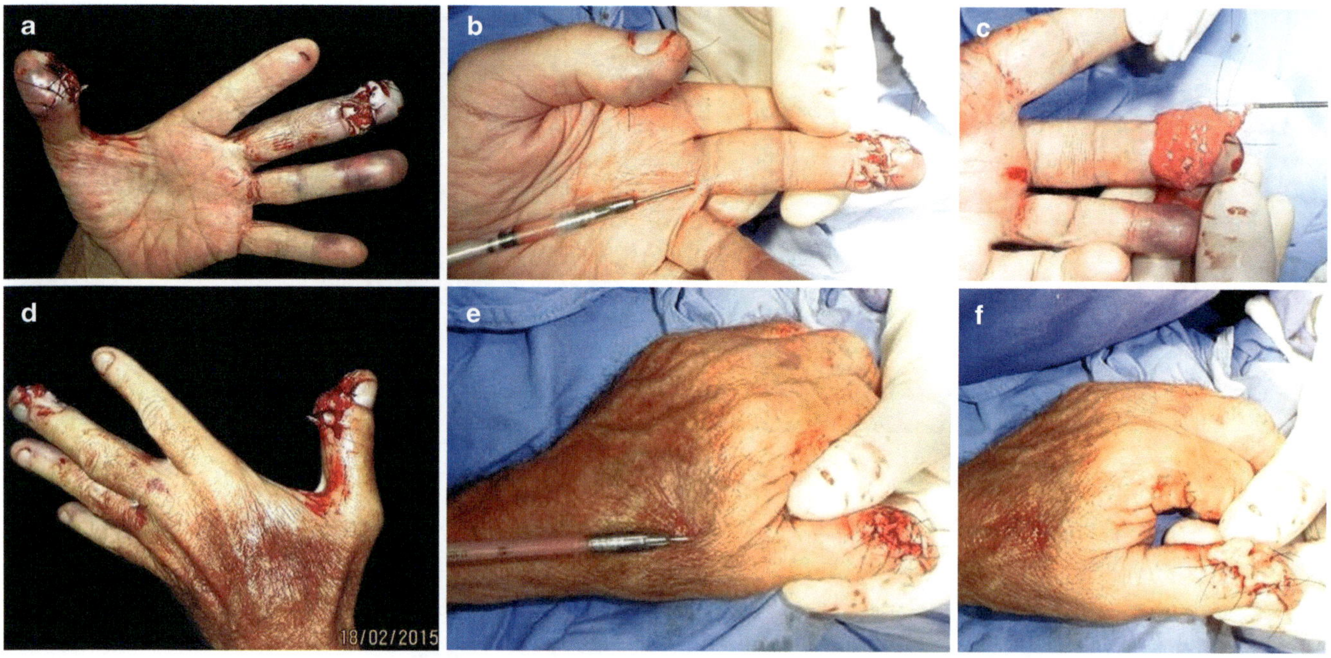

Fig. 5.4 (**a**, **d**) Blast wound to a 68-year-old post-renal transplant diabetic patient—previously "sutured" elsewhere 3 days previously; (**b**, **e**) wounds were opened, debrided, and loosely closed with sutures and fat grafting was performed; (**c**, **f**) fat delivery was also performed to the surface of the wounds

scarring will NOT hypertrophy at the scar resection suture line within the hypertrophic scar. This is attributed to the local effect of the fat grafting and the delivered fat previously treated by the Coleman Technique (Fig. 5.7).

This technique may be also particularly rewarding on massive hypertrophy or keloids of ear scars, when centrifuged fat grafting is performed under the scar resection suture line (Fig. 5.8).

5.3.7 Fat Grafting in Preparation for Scar Release

In very small patients, or in the elderly, when considering severe retraction release in the palm of the hand or dorsum of the foot, fat grafting may be indicated previously to the actual release procedure to warrant a more irrigated and "protected" wound bed when the release is performed and the skin graft placed. In this way, fat is injected 3–6 weeks previously to the planned scar release procedure, adding fat volume to the future skin graft bed, "covering" and protecting deeper noble structures, such as vessels, nerves, and tendons, such as in a small child hand or foot retractions (Figs. 5.9 and 5.10).

5.3.8 Postoperative Care

After the centrifuged fat is delivered to the microneedled area, dressings consist of petrolatum gauze applied directly over the delivered fat and regular gauze impregnated with Dakin's solution, applied directly over the petrolatum gauze, which is then covered with a bandage. Dressings are changed every 2 days, when all gauze is removed as well as all loose fat—if the delivered fat is adherent, it will stay for another 2 days, when the scar surface is cleansed with saline and no more dressings are applied.

In wounds, dressings are changed every 2 days and the delivered fat (and evolving wound surface) is covered in similar way—petrolatum gauze, regular gauze impregnated on Dakin's solution, and regular bandage.

5.3.9 Fat Grafting and Decubitus Ulcers

In this technique, the pressure sore wound is thoroughly debrided and closed via simple fasciocutaneous advancement flaps, sutured in several progressive layers, while centrifuged fat is delivered to each suture line space, layer by layer. The idea is to completely obliterate dead space and as

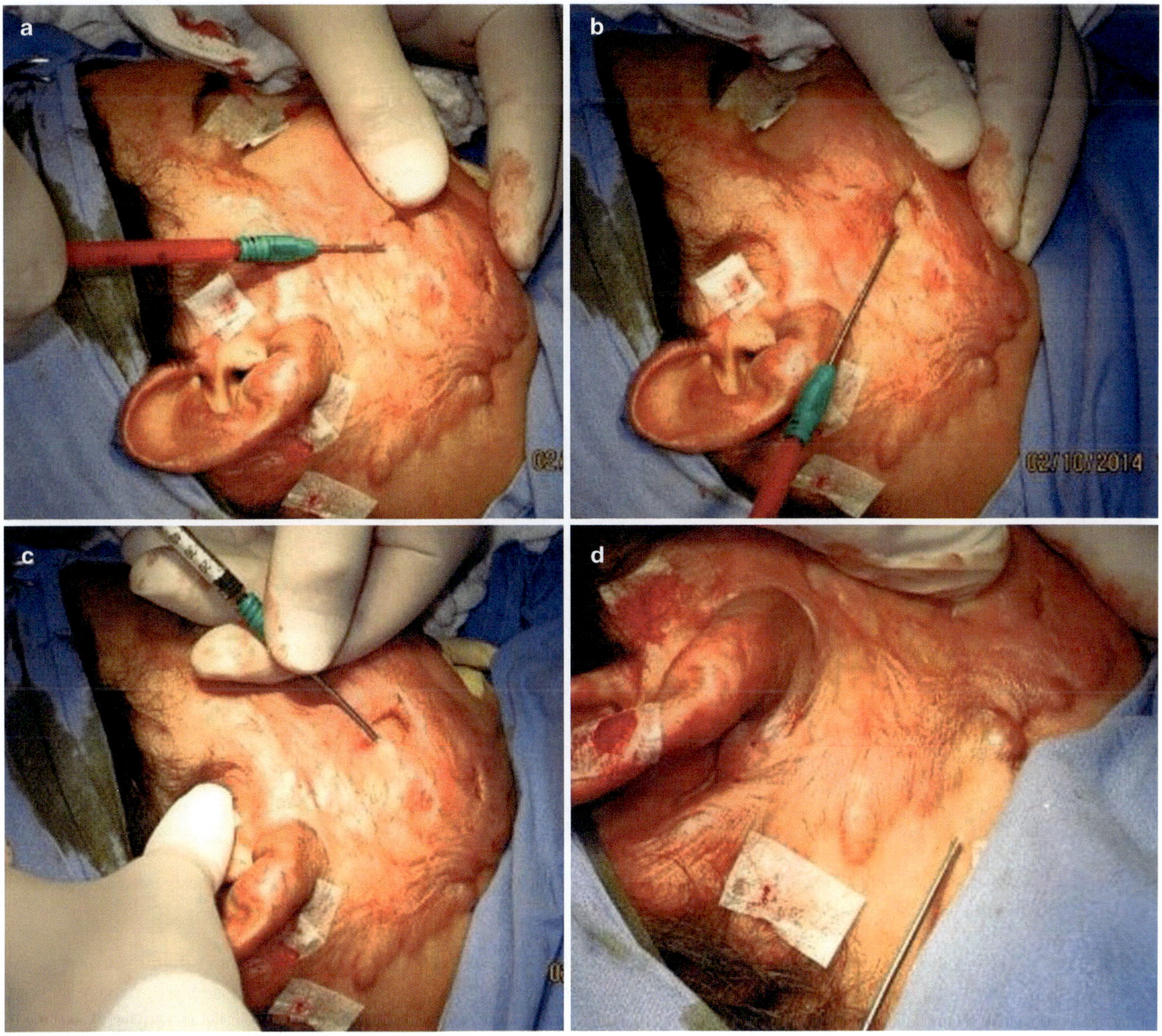

Fig. 5.5 (**a**–**d**) Fat is injected in a retrograde manner, in several directions, via as many puncture sites as needed. Entrance sites are covered with micropore tape

the sutures are tightened, fat remains within the suture line compounding the obliteration of the dead space, while "extra" fat is extruded outward and removed—subsequent suture lines are placed in a similar manner, until the wound is completely closed.

These wounds are treated in such a way as to simply close them with advancement flaps or to compoundly compress the outer flaps against the deep tissues. No drains are placed and the wound is usually healed in 12–15 days. In very large wounds, there may be a need to repeat the procedure, as there may be reoccurrence of a smaller wound, and this (second or even third) closure is done in the exact same way, until complete closure is obtained.

The surgical procedure has a double objective—to obtain a completely debrided wound and to obtain its closure. So, the wound is debrided thoroughly, using sharp cutting instruments, or the electrocautery, progressively removing the entire "lining" of the pressure wound to minimize at most the risk of further contamination and infection after the wound is closed. This tissue can be removed in large plaques or on

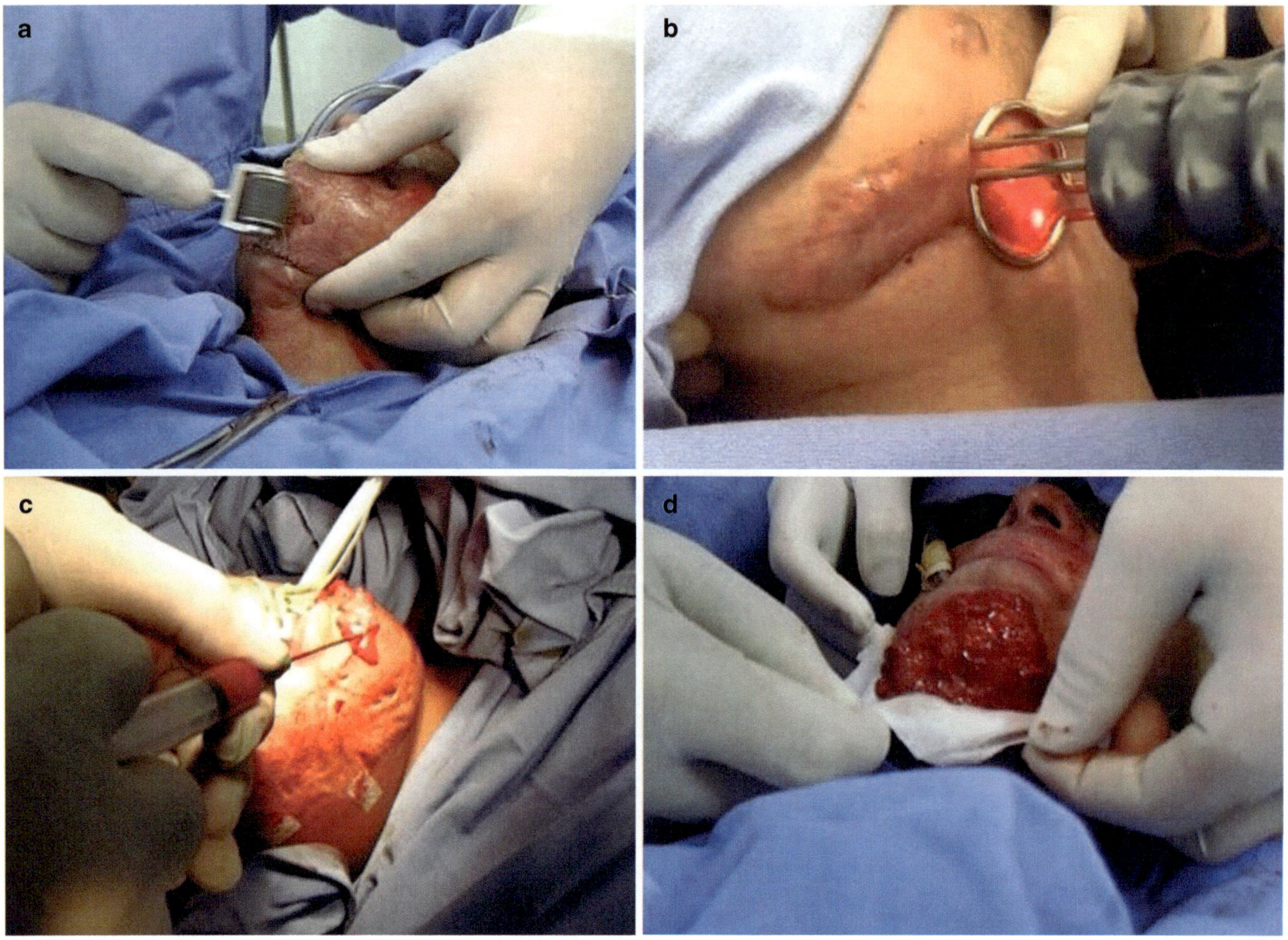

Fig. 5.6 (**a**, **b**) Scar being treated with dermaroller and laser, respectively; (**c**) centrifuged fat being delivered to the previously treated scar surface; (**d**) petrolatum gauze is placed over the delivered fat and is then covered with dry gauze and a bandage

"piece meal" fashion and with whichever fashion the surgeon prefers, all of the lining tissue must be removed. Frequent irrigation with saline may be necessary to rid the wound cavity of debris or fragments of contaminated tissue.

After creating a practically "new" wound, free of debris and undesired tissue, the most geometrically accepted flaps will be created, aiming at simple advancement—no rotation flaps are usually created, as these may yield larger cavities and longer suture lines (both unnecessary for this technique). In very large wounds, when there are cavities extending away from the open wound, it may be necessary to open some of the skin and subcutaneous tissue immediately "over" the cavity, to be able to completely clean the wound—this, however, is uncommon, and when preformed, flaps are also closed in an opposed advancement fashion.

After creating the flaps, the plan is to place enough "suture lines" (simple suture rows) oriented perpendicularly to the flaps as needed. Starting form the most distal portion of the cavity, 0 nylon simple sutures are passed progressively using sutures with 4-cm needles—the needle is removed and both suture ends are temporarily held with straight Kelly clamps, until the complete planned suture line is placed.

Centrifuged fat is then laid in deep to the suture line, in the deeper portion of the cavity and knots are progressively tightened. As the cavity is closed, additional suture lines are placed and centrifuged fat is delivered as needed. No Penrose or suction drains are placed (Fig. 5.11).

One should not worry with how much fat should be delivered since as the suture lines are tightened, considering that they are placed in such a way to leave a minimum of dead space within the previous and the next suture line, the tying of the knots will also provide a very constrict, practically a "virtual" space, while the "extra" fat is extruded—in this way warranting the survival of the delivered fat, since it will

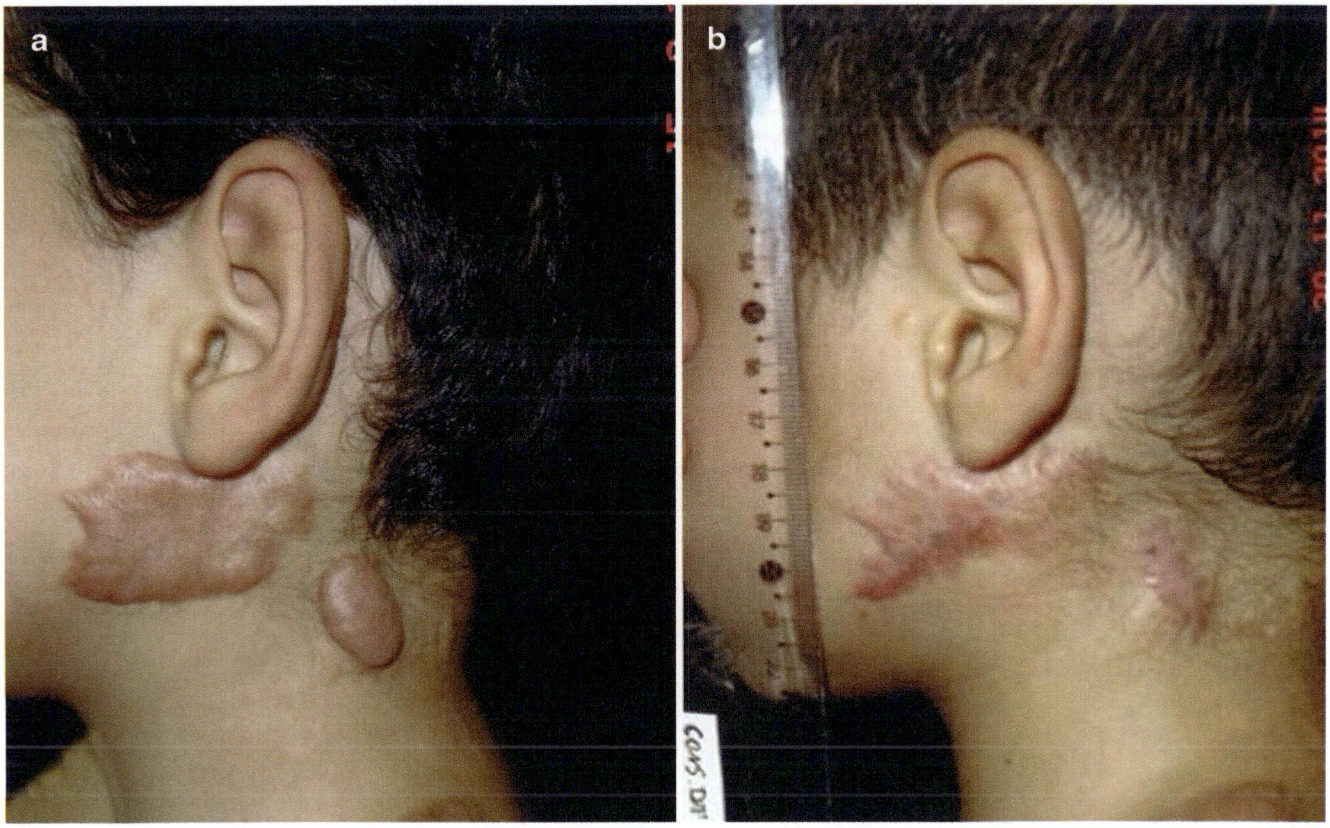

Fig. 5.7 (**a**) A 4-year-old patient previously treated elsewhere after flame burn; (**b**) Result after two intralesional resections and fat grafting/fat delivery—please note the absence of intralesional suture line scars

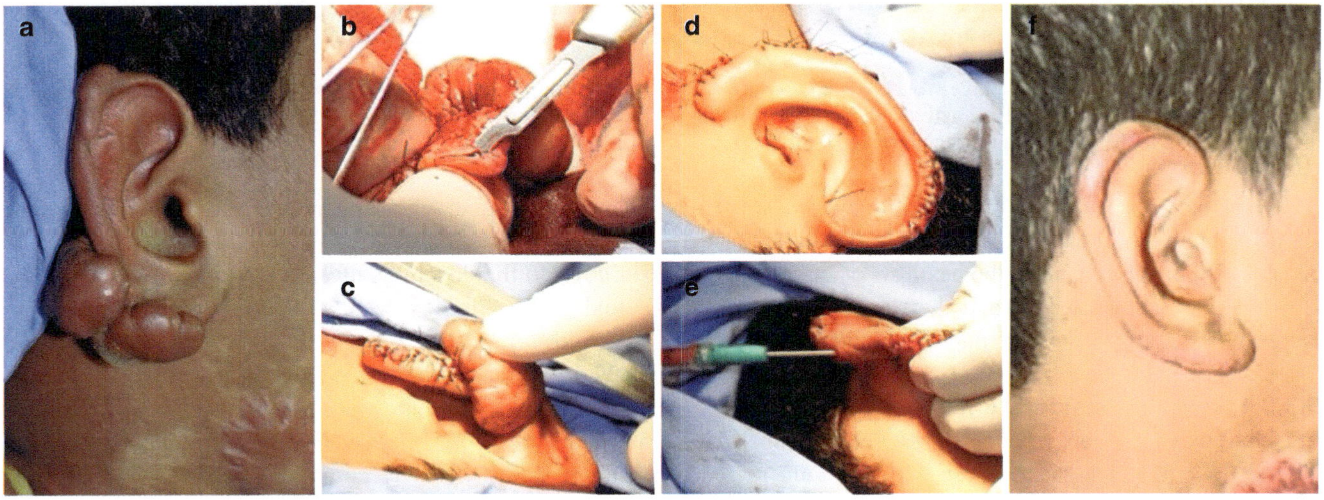

Fig. 5.8 (**a**) A teenager boy 3 months after ethanol burn and severe hypertrophy on ear lobe; (**b**–**d**) almost complete intralesional resection of both lesions; (**e**) fat grafting to the undersurface of the suture line; (**f**) result at 6 months

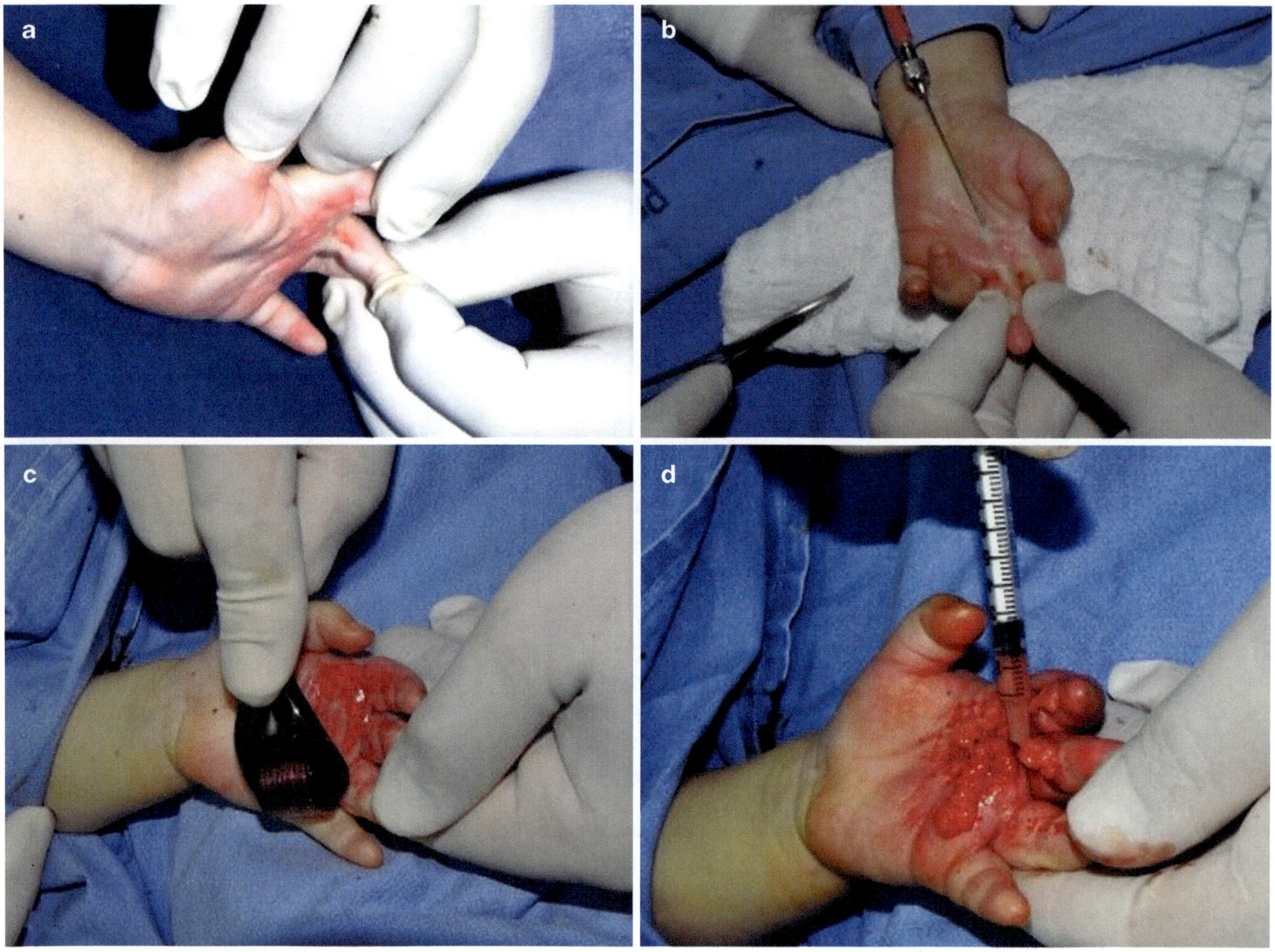

Fig. 5.9 (**a**) Preoperative view of a 2-year-old boy with severe retraction of the II and III right fingers after an oven contact burn; (**b**) centrifuged fat grafting to the undersurface of the scar; (**c**, **d**) microneedling and centrifuged fat delivery to the surface of the scar—this procedure aiming at "preparing" the future scar release area to "better receive" the skin graft

be in contact with live tissue. Extruded fat after each line is completed may be left in place or cleansed, depending on the need of visualization to place the next row of sutures and fat is delivered again as needed and the sutures tied until the complete approximation of the flaps is obtained.

5.3.9.1 Special Considerations

In ischial wounds, there frequently is a need to "open" the wound tract to reach the deepest portion of the wound, which, of course, must be completely seen, for one to be able to get rid of all the contaminated lining and to create a clean, new wound. A helpful method is to inject a Methylene blue solution (1 cc of Methylene blue added to 9 cc of saline) into the decubitus ulcer cavity helping to define the actual ulcer lining—whatever become blue must be removed.

Very frequently in these patients, the wound will be several years old, since the patient only sees a small opening, and thinks that it will eventually heal since apparently it is a small wound—this may complicate the wound, including with ischial bone colonization, infection, and even possibly partial bone loss (Fig. 5.12).

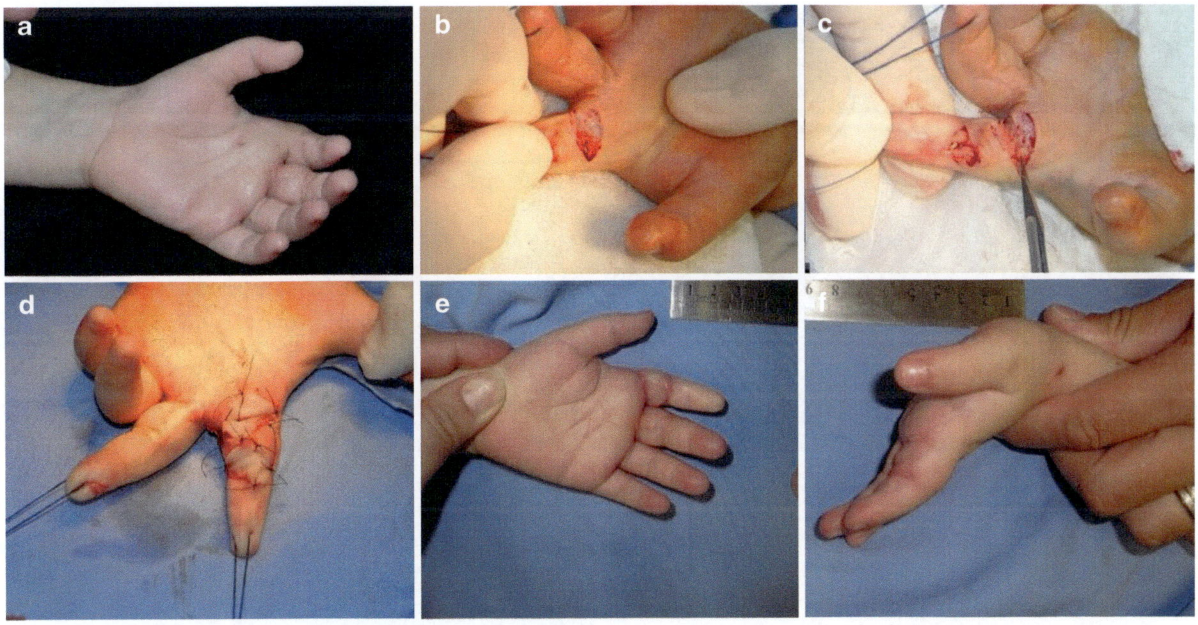

Fig. 5.10 (**a**) Postoperative view after two fat grafting/fat delivery procedures; (**b**, **c**) incision through the scar at the retraction site immediately shows previously grafted fat (with abundant irrigation); fat delivery to the surface of the scar; (**d**) immediate aspect after cavum plantar skin graft; (**e**, **f**) results at 3 months

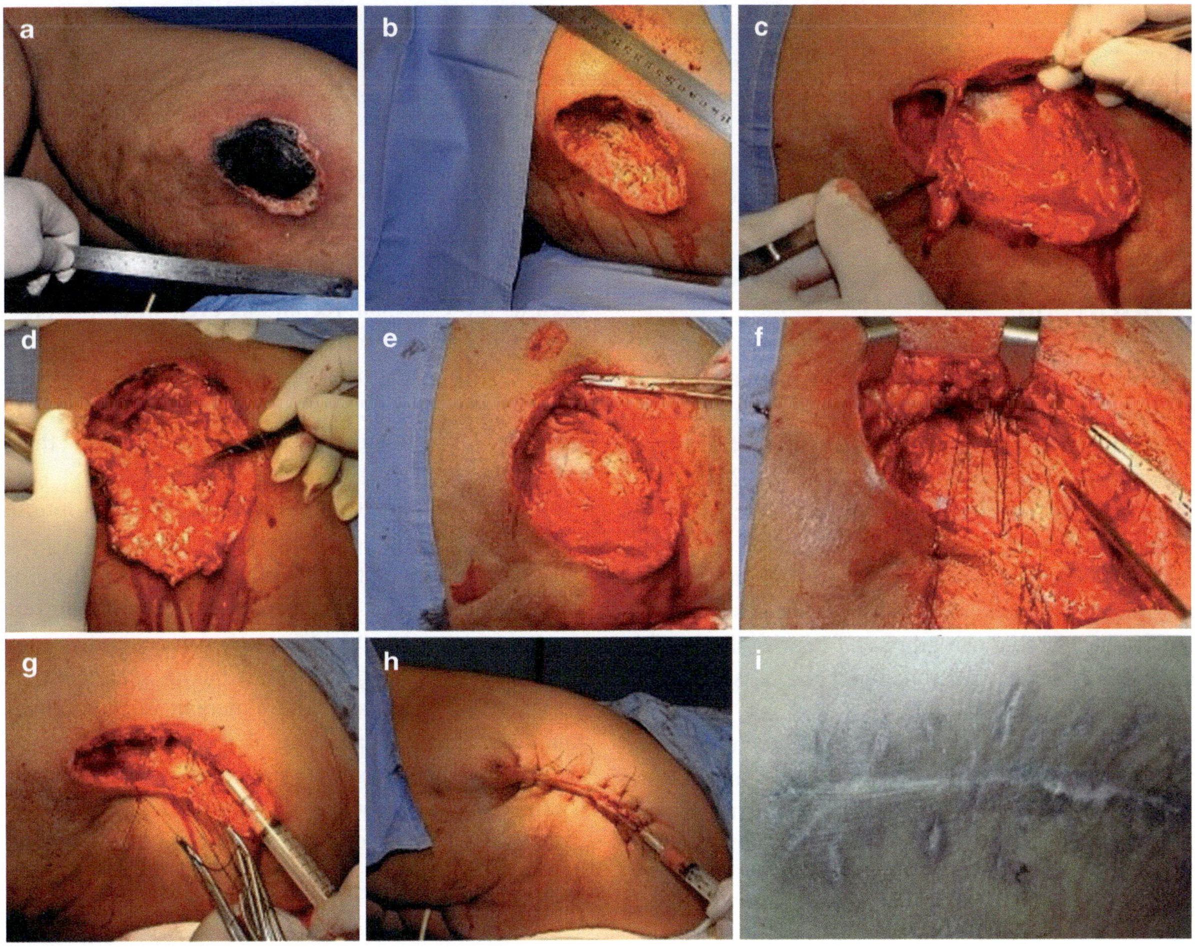

Fig. 5.11 (**a**) Left trochanteric ulcer on a 36-year-old lady with acute spinal cord injury (42 days after the accident); (**b**, **c**) thorough debridement with complete removal of the wound bed; (**d**) flaps are created bilaterally; (**e**–**h**) 0 nylon reverted sutures are placed longitudinally; centrifuged fat is deposited along each suture line, which are tied sequentially approximating the entire width of the wound; (**i**) healed wound aspect, 6 weeks post healing

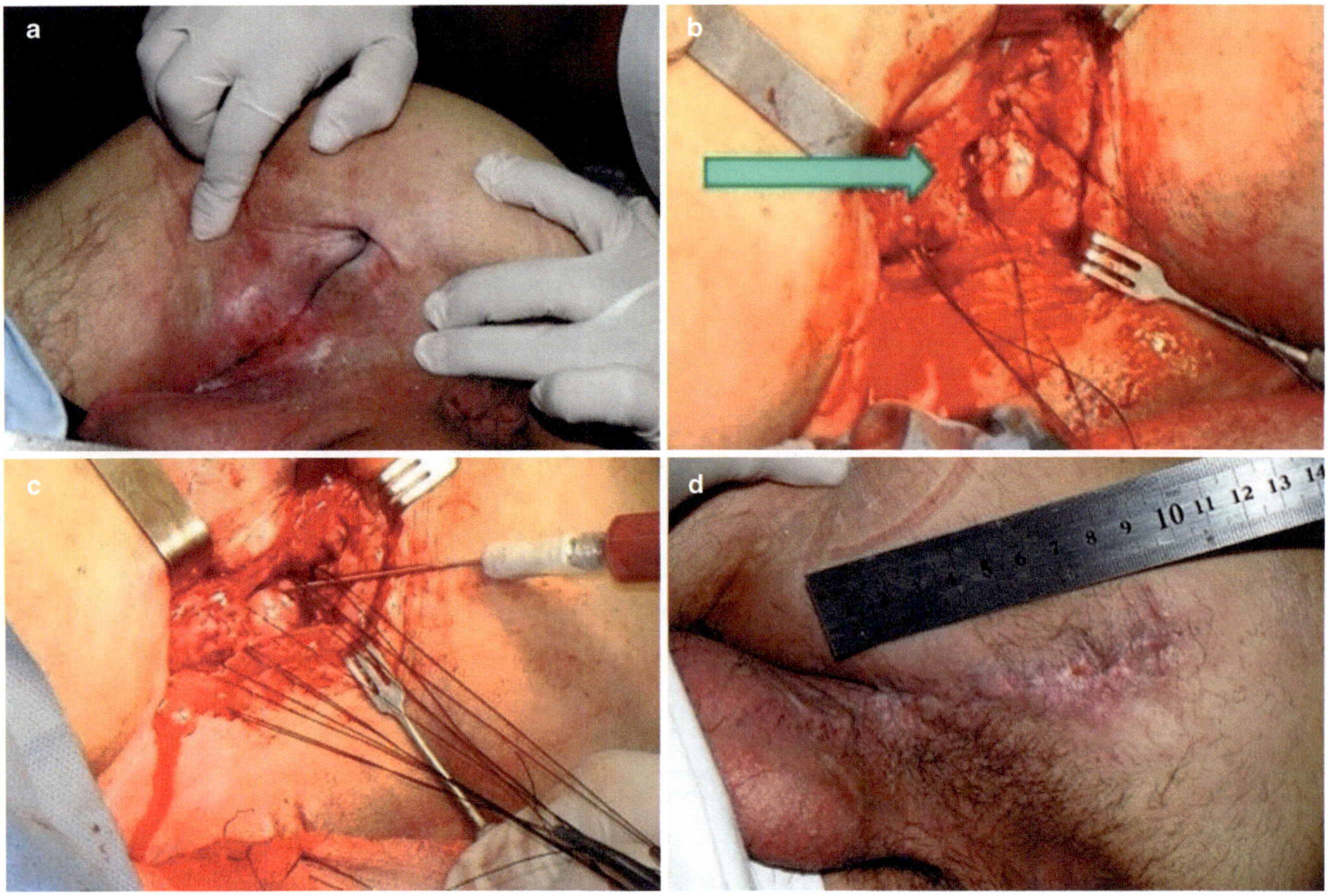

Fig. 5.12 (**a**) Please note the apparent shallow aspect of this 7-year-old ischial ulcer; (**b**) the wound is as deep as the bony surface which has been debrided (please note arrow showing ischial tuberosity); (**c**) fat delivered to the wound and several suture layers provide progressive closure; (**d**) aspect of the wound 3 months post-cure

5.4 Discussion

Fat can easily be grafted practically in any body area. As we have demonstrated, it can also be generously applied over microneedled or laser-treated skin, delivering fat directly within the substance of the skin or the scar. [20–23].

Current (and past) fat grafting literature is rich with different authors indicating different methods of harvesting and different ways to handle, or enrich, the lipoaspirate, or even how it is injected, which could influence the result of each one specific fat injection procedure. Of course, these scientists and practitioners have also had great success in treating similar patients.

However, we believe that the Coleman Technique provides a rather standardized method, with a very short learning curve for the surgeon and with an easy to learn routine by the entire surgical team. When one uses a recognized routine such as the Coleman Technique, even when the procedures are practiced and results presented by different surgeons, this routine will allow to make these results easier to compare, since the harvesting and preparation of the fat was done through this standardized technique and this well-consolidated method [23, 24].

As the technique of blunt cannula insertion optimizes the release of scar retraction, this may also play a part in the analgesic effect of this treatment method. This finding is related in published evidence which supports current theories of mesenchymal stem cell's regenerative and anti-inflammatory properties responsible for scar healing.

We have been using the Coleman Technique for more than a decade now and one of the most pleasant surprises with fat grafting, when we started to use this procedure as a routine, was the presence of very little hypertrophy as a routine result of the procedure. Minimal or practically no fibrosis is the routine outcome and these such results are very frequent. This a common finding in wounds treated with fat grafting, regardless of intermediate coverage with skin grafting or if left to heal spontaneously (Fig. 5.13).

This will be also true when treating the patient that comes to us already with a scar, or was treated acutely by us and developed significant hypertrophic scarring. Fat grafting on itself, when used in scars, will similarly (as when treating acute wounds) yield a most satisfactory result in relation to diminishing fibrosis, be it on the surface (skin scar) or in the deep tissues such as post-traumatic fibrosis in injuries to the hand and distal forearm, like post-electrical burn scarring around tendons and nerves. The effect appears to be cumulative, as proportionally more improvement is perceived after the fat grafting procedure is repeated, at 6–8 weeks intervals and so on. This improvement could be related to the number of injections and/or the time elapsed post-injection, since there are definite, progressive, and cumulative changes after repeat injections (Figs. 5.14 and 5.15).

Fat injection aimed at improving scars most likely brings improvements through mesenchymal cells and numerous growth factors contained in the lipoaspirate which contribute to the skin and scar remodeling. In the patients with scars which were treated by this technique, one of the main related improvements was the increase in elasticity and malleability of the scar tissue, as well as for its significant decrease in volume and thickness. This could be partially due to the marked increase in the number of elastic fibers, easily perceived microscopically in post-injection scar samples, in consequence of these injections.

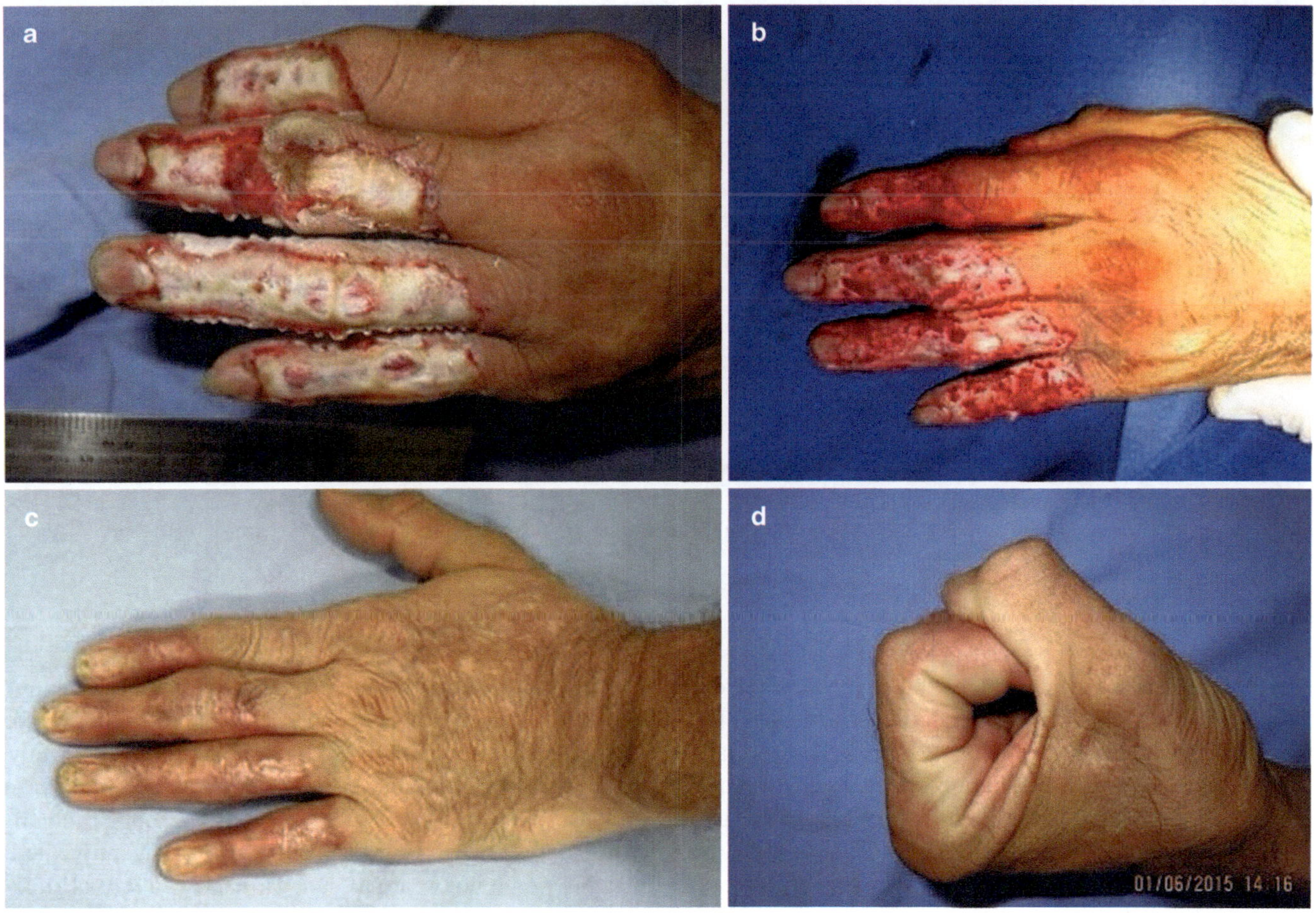

Fig. 5.13 (**a**) Areas of deep second- and third-degree burns on a hot clothing press working accident; (**b**) aspect after tangential excision; (**c**, **d**) aspect 6 months after "spontaneous" cure (no skin grafting)—fully functional hand—no tendon adhesions

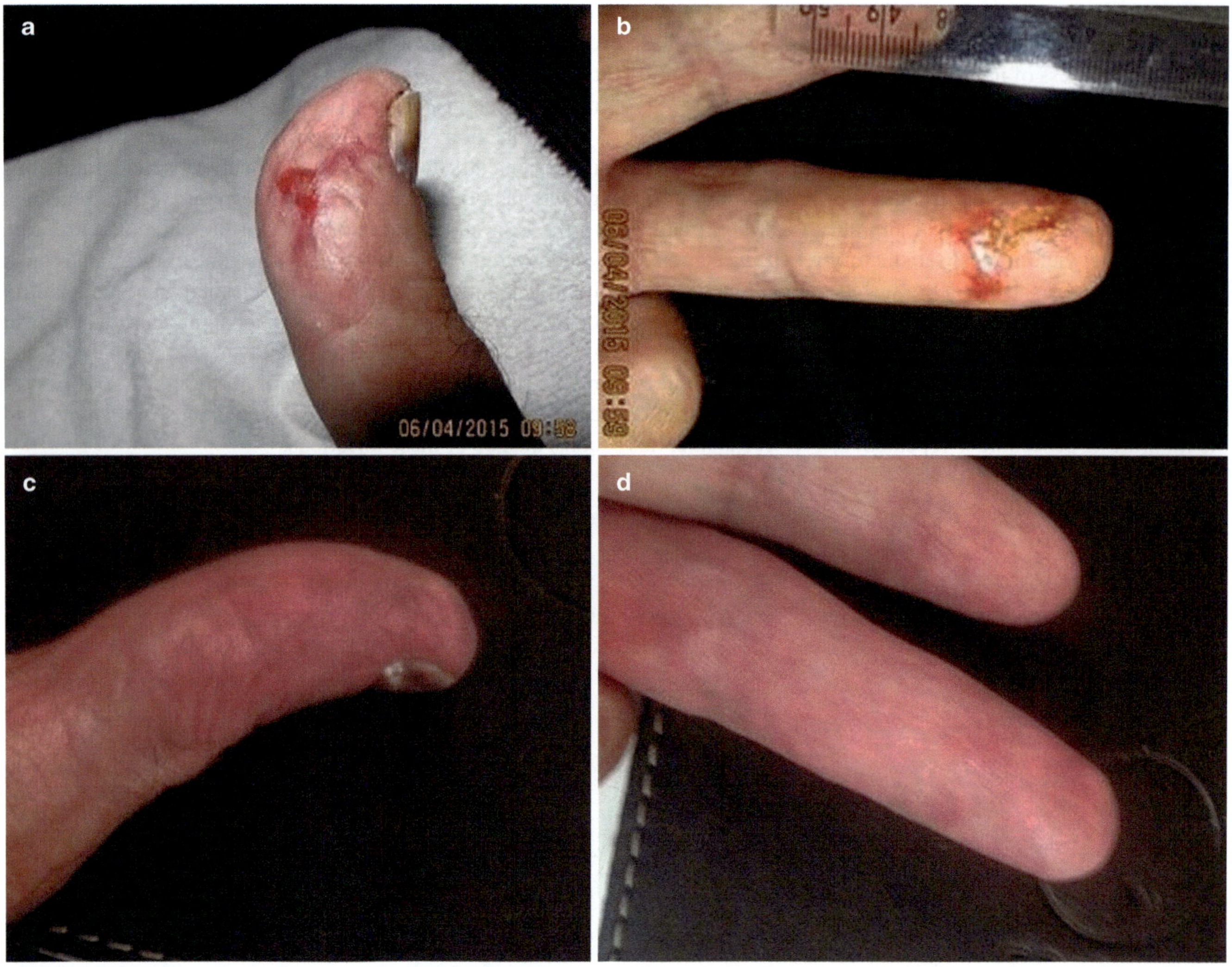

Fig. 5.14 (**a**, **b**) The same patient as in Fig. 5.4, 3 weeks after healing, when a repeat fat grafting procedure was performed; (**c**, **d**) aspect 3 months after healing

Fat grafting/fat delivery has become a routine treatment in our Service for these patients. As surgeons we notice almost immediate changes, but a moste rewarding situation is that the patients also noticed the same. These findings also appear very obviously across the time range of scar evolution and maturation, even at post-matured scar phases, when fat grafting and fat delivery continue to act on diminishing fibrosis (completely matured hypertrophic scarring). As mentioned before, these effects are even more impressive (and appear to be cumulative) (Fig. 5.16).

Another fact to consider is the psychological burden of a sequela, or a chronic wound, regardless of the patient age. And consequently, the emotional difficulties that a patient or his or her family may have to go through to decide when and how to take care of the deformity.

The results of fat grafting/grafting are even more impressive on keloid-like hypertrophic scars which may appear after (flame or other) burns to the ears. Most unfortunately, these are relatively common accidents in our community. In the past, we would make these children go through months and months of clinical treatment. Very frequently, these patients would not wear pressure garments and inserts effectively, usually resulting in a frustrated attempt to control hypertrophy. We have now been able to remove the scar completely in one single surgical act, when centrifuged fat is grafted under the suture line as well as delivered over it. Figure 5.17 shows a typical long-term result.

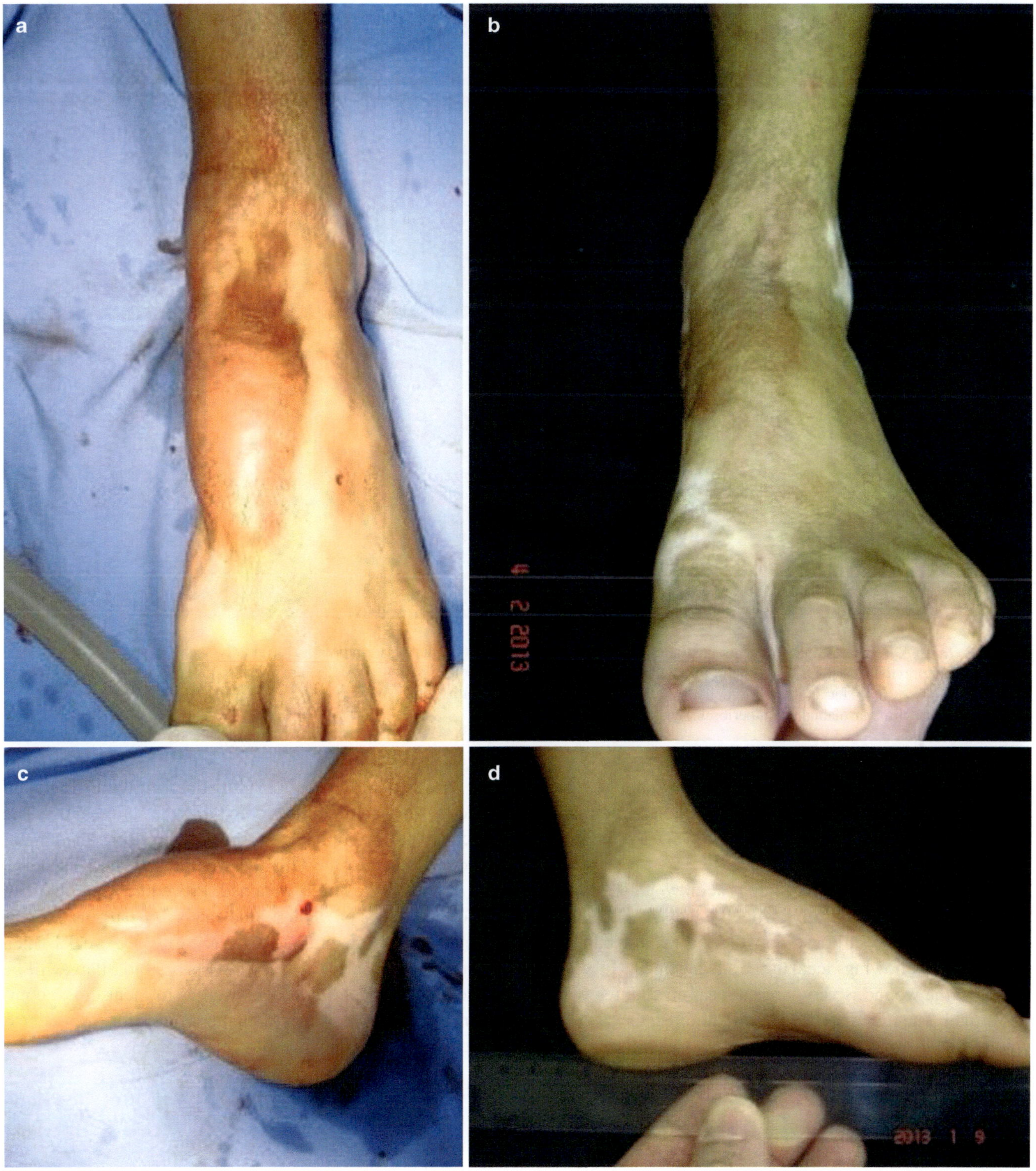

Fig. 5.15 (**a**, **c**) A 3-year-old boy 6 months after an oil burn to the left foot; (**b**, **d**) aspect 4 weeks after the third fat grafting procedure

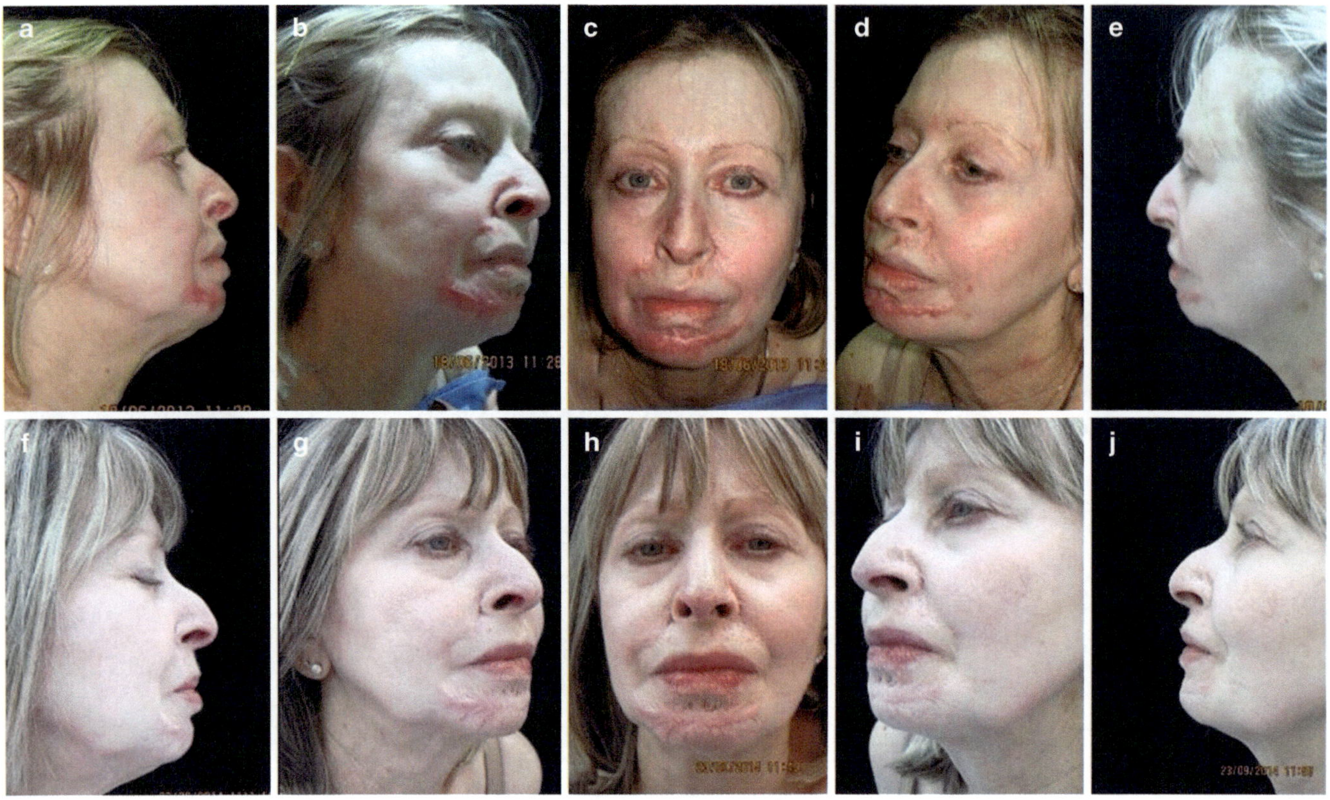

Fig. 5.16 (**a–e**) A 53-year-old lady 4 years post-acid (unknown) burn to the face, on previous "aesthetic" treatment by a local dermatologist; (**f–j**) results after three fat grafting/fat delivery procedures to the affected areas on chin, nasolabial sulci, perioral area, and dorsum of the nose (results 15 months after originally seen, 3 months after the last procedure; patient lives 200 km away, so procedures were performed at 3 months intervals)

Chronic wounds may cause even more damage due to the difficulties in wound care which will then perpetrate the disease, causing severe physical as well as emotional damage.

In decubiti, mostly on very large sacral wounds, although one may consider technically difficult to advance local flaps, one usually will be surprised how "forgiving" the buttock is in relation to "allowing" local tissue advancement (Fig. 5.18).

When the wound is very large, repeat procedures will be performed. The initial procedure will diminish the size of the wound, advancing a flap on each side, and anchoring it to the deep tissues, suture line after suture line, with centrifuged fat added in every suture line, before tying the suture knots. The objective is the complete closure, and every time the patient is brought back to the OR, the wound is treated similarly, until complete closure is obtained. Procedures can be a one side or both sides procedures and can be repeated as early as 1 week on contralateral sides or as early as 2 weeks when advancing both sides. On the "last" procedure, flaps will be anchored to the deep tissues as well as to themselves, completing the closure, always tying the suture knots over centrifuged fat deposited layer by layer (Fig. 5.19).

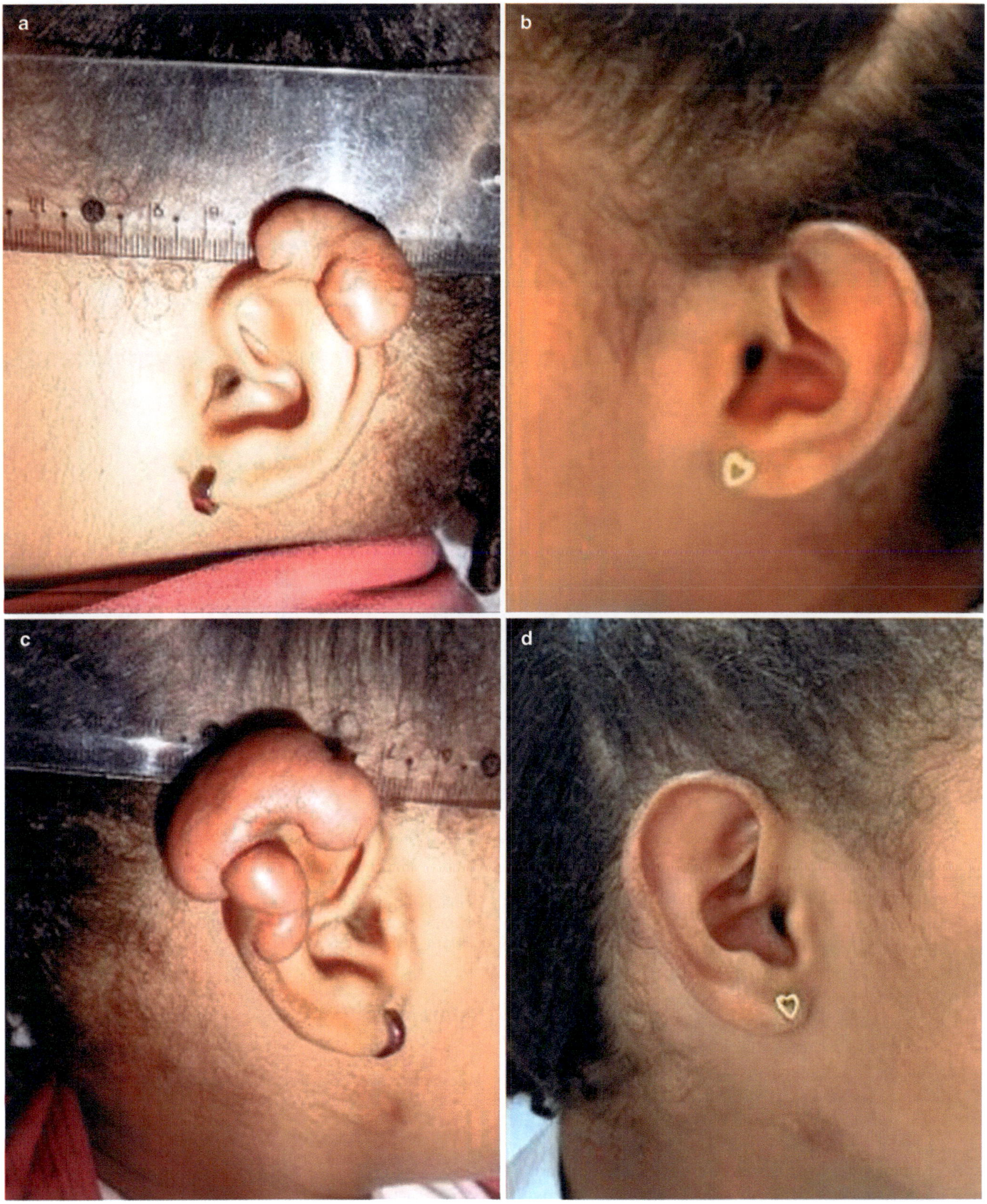

Fig. 5.17 (**a**, **c**) A 3-year-old girl post-flame burn with severe bilateral ear keloid-like hypertrophy; (**b**–**d**) – 3 years post almost complete scar resection and centrifuged fat injection under the suture line—single procedure—please note the complete absence of hypertrophy

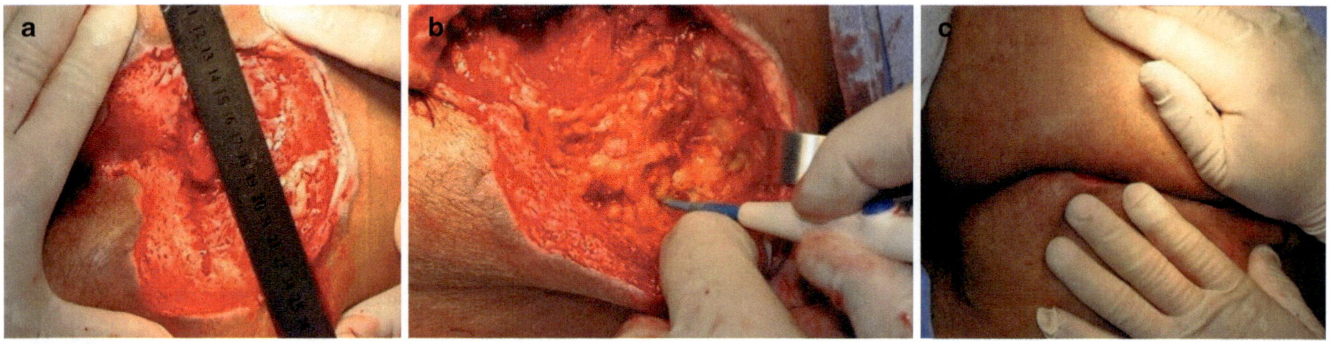

Fig. 5.18 (**a**) Large decubitus sacral ulcer immediately after comprehensive debridement; (**b**) creation of local advancement flap; (**c**) demonstration of flap mobility

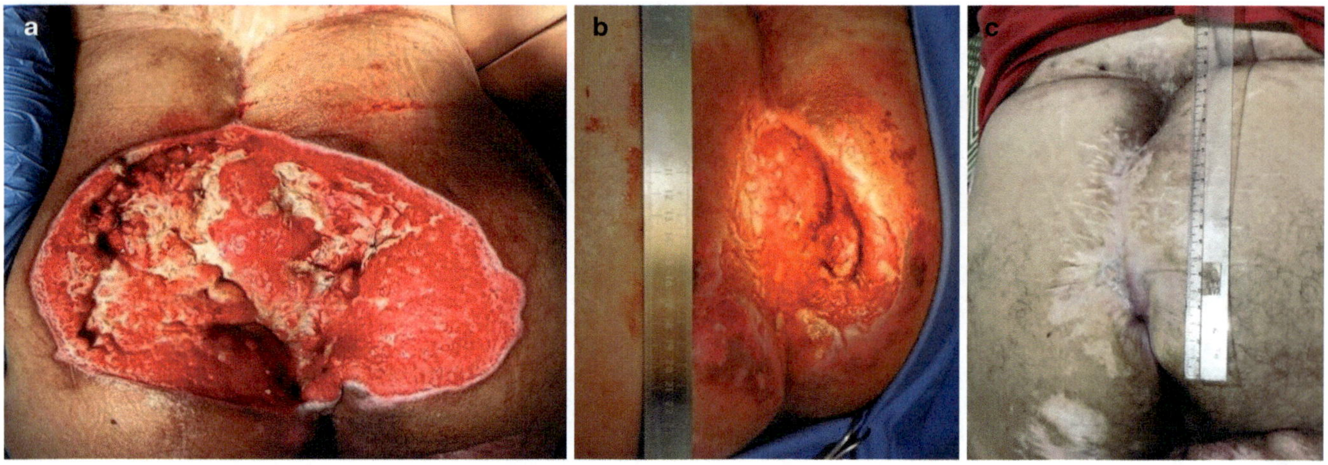

Fig. 5.19 (**a**) A 26-year-old paraplegic patient post several debridement sessions on chronic (4 years), infected, sacral wound; (**b**) appearance 22 days after "**a**," after one bilateral procedure and one more on the right; (**c**) appearance 2 years after one more combined procedure which healed completely in 26 days

5.5 Conclusion

The range of utilization of fat grafting is, almost certainly, still underestimated. There are, and certainly will be, many different entities which could benefit by the deposition of centrifuged fat directly under, over, and around the affected anatomic structure or wound or scar surface.

As for the practitioner in charge of taking care of patients with scars or wounds, one can apply this treatment method with the certainty that it will definitely promote a prompter healing, with relatively little or no hypertrophy. This is also true upon returning on the scarred wound, or treating a scar in consequence of a wound not previously treated by fat grafting/fat delivery, when the result already noticed upon the primary procedure appears to be cumulative upon repeat procedures, and it will usually provide suppleness and functionality, regardless of the timing it was used.

Acknowledgments The authors would like to thank Dr. Altamiro Vieira, MD, for his most valuable help in the photographic documentation of these patients' cases.

Conflicts of Interest The authors would like to declare that there were no conflicts of interest while treating these patients, choosing their cases, or preparing this manuscript.

References

1. Zuk PA, et al. Multilineage cells from human adipose tissue: implications for cell based therapies. Tissue Eng. 2001;7:211.
2. Zuk PA, et al. Human adipose tissue is a source of multipotent stem cells. Mol Biol Cell. 2002;13:4279–95.
3. Fujimura J, et al. Neural differentiation of adipose-derived stem cells isolated from GFP transgenic mice. Biochem Biophys Res Commun. 2005;333(1):116–21.
4. Dominici M, et al. Minimal criteria for defining multipotent mesenchymal stromal cells, the International Society for Cellular Therapy position statement. Cryotherapy. 2006;8:315–7.

5. Rigotti G, et al. Clinical treatment of radiotherapy tissue damage by lipoaspirate transplant: a healing process mediated by adipose derived adult stem cells. Plast Reconstr Surg. 2007;119:1409–22.
6. Gimble JM, Katz AJ, Foster SJ. Adipose-derived stem cells for regenerative medicine. Circ Res. 2007;100:1249–60.
7. Akita S, et al. Non-cultured autologous adipose-derived stem cells therapy for chronic radiation injury. Stem Cells Int. 2010;2010:532704.
8. Brown SA, et al. Basic science review on adipose tissue for clinicians. Plast Reconstr Surg. 2010;126:1936–46.
9. Coleman SR. The technique of periorbital lipoinfiltration. Oper Tech Plast Reconstr Surg. 1994;1:120–6.
10. Coleman SR. Long term survival of fat transplants: controlled demonstrations. Aesth Plast Surg. 1995;19:421–5.
11. Kim W, et al. Wound healing effect of adipose-derived stem cells: a critical role of secretory factors on human dermal fibroblasts. J Derm Sci. 2007;48:15–24.
12. Lolli P, Malleo G, Rigotti G. Treatment of chronic anal fissures and associated stenosis by autologous adipose tissue transplant: a pilot study. Dis Colon Rectum. 2010;53:460–6.
13. Klinger M, et al. Fat injection for cases of severe burn outcomes: a new perspective of scar remodeling and reduction. Aesthet Plast Surg. 2008;32:465–9.
14. Viard R, et al. La lipostructure dans les sequelles de brulures facials. Ann Chir Plast Esthet. 2012;57:217–29.
15. Carpaneda CA, Ribeiro MT. Study of histologic alterations and viability of adipose grafts in humans. Aesthet Plast Surg. 1993;17:43–7.
16. Sultan SM, Barr JS, Butala P, Davidson EH, Weinstein AL, Knobel D, Hazen A. Fat grafting accelerates revascularisation and decreases fibrosis following thermal injury. J Plast Reconstr Aesthet Surg. 2012;65(2):219–27. https://doi.org/10.1016/j.bjps.2011.08.046.
17. Piccolo NS, Piccolo MS, de Paula PN, de Paula PP, de Paula PN, Daher RP, Lobo RP, Daher SP, Sarto Piccolo MT. Fat grafting for treatment of facial burns and burn scars. Clin Plast Surg. 2020;47(1):119–30. https://doi.org/10.1016/j.cps.2019.08.015.
18. Yoshimura K, et al. Cell-assisted lipotransfer for facial lipoatrophy: efficacy of clinical use of adipose-derived stem cells. Dermatol Surg. 2008;34:1178–85.
19. Bene MD, et al. Autologous fat grafting for scleroderma-induced digital ulcers. An effective technique in patients with systemic sclerosis. Handchir Mikrochir Plast Chir. 2014;46:242–7.
20. Chen H, et al. Autologous fat graft for the treatment of sighted post-traumatic enophthalmos and sunken upper eyelid. Ophthalmic Plast Reconstr Surg. 2018;34:381–6.
21. Marten TJ, Elyassnia D. Fat grafting in facial rejuvenation. Clin Plast Surg. 2015;42:219–52.
22. Shukla L, Yuan Y, Shayan R, Greening DW, Karnezis T. Fat therapeutics: the clinical capacity of adipose-derived stem cells and exosomes for human disease and tissue regeneration. Front Pharmacol. 2020;11:158.
23. Coleman SR. Structural fat grafts: the ideal filler? Clin Plast Surg. 2001;28:111–9.
24. Piccolo NS, Piccolo MS, Piccolo MT. Fat grafting for treatment of burns, burn scars, and other difficult wounds. Clin Plast Surg. 2015;42(2):263–83.

Efficacy and Safety of Cell-Enriched Fat Grafting in the Breast

6

Valerio Cervelli, Gabriele Storti, and Andrea A. Pierro

Contents

6.1 Introduction

Fat transfer is a very useful technique in both cosmetic and reconstructive plastic surgery. A recent survey showed that approximately 80 percent of plastic surgeons have used fat grafting in their practice [1].

Since its first introduction in the late 1990s, its popularity has spread rapidly and the original technique has undergone several modifications to improve its efficacy.

Fat grafting has been used for facial contouring, post-traumatic deformities, radiation damage, congenital anomalies, and burn injuries.

Breast surgery is probably the best application for fat grafting (Table 6.1). It can be used for correction of imperfections in breast reconstruction with implants or in hybrid breast reconstruction, treatment of irradiated breast, correction of scar retraction after quartectomy, and breast augmentation for cosmetic purposes or capsular contraction [2–5].

One of the most important problems linked to fat grafting is its resorption rate that can be variable and unpredictable. References note that the resorption rate can range from 20 to 80 percent of the grafted tissue which necessitates multiple repeat procedures [6–9].

V. Cervelli (✉) · G. Storti · A. A. Pierro
Plastic Surgery Division, University of Rome "Tor Vergata", Rome, Italy

Today, we know that the regenerative function of fat transfer is mainly mediated by adipose-derived stem cells (ASCs). ASCs have been identified as a well-defined cellular line with the ability to differentiate in mature mesenchymal cells. This has led to fat grafting becoming not only a filler but also a regenerative tissue for a lot of indications where it is able to improve both the volume and the skin quality, tissue trophism, and vascularity [10].

The interaction between the fat graft and the recipient site has been the object of several studies. In particular, most of these have identified three zones from the periphery to the center of the graft [11].

- *The surviving area:* adipocytes survive thanks to the best oxygen intake.
- *The regenerating area:* adipocytes die while adipose-derived stromal cells survive, and dead adipocytes are replaced with new ones thanks to the ability of ASCs to survive in lower oxygen conditions compared to mature adipocytes.
- *The necrotic area:* both adipocytes and adipose-derived stromal cells die.

It is clear that fat graft, thanks to this variety in terms of cellular composition, can act as a very complex tissue compared to other grafts, with different actions depending on the recipient site conditions. In addition, not only is the

© The Author(s), under exclusive license to Springer Nature Switzerland AG 2023
A. Di Giuseppe et al. (eds.), *Fat Transfer in Plastic Surgery*, https://doi.org/10.1007/978-3-031-10881-5_6

Table 6.1 Fat grafting indications in breast surgery

Breast reconstruction	Aesthetic breast surgery
• Breast Implant Reconstruction – *Filling contour irregolarities* – *Improvement of implant coverage* • Breast Conservation Therapy – *Correction as asymmetries, scar retraction and other contour deficits* • Total Breast Reconstruction – *Brava system* – *Reverse expansion* • Treatment of Irradiated breast • Miscellaneous – *Correction of asymmetries, scar retractions and* – *Other contour deficits in both autologous and implant breast reconstruction*	• Primary Breast Augmentation • Primary Implant Breast Augmentation – *Filling contour irregolarities* – *Improvement of implant coverage*
	Congenital anomalies – Poland Syndrome – Pectus Excavatus – Thoracic Hypoplasia

composition of fat graft in terms of regenerative cells important for obtaining the best grafting, but also the surgical technique must obtain the maximal contact surface between the graft and the recipient site (lipostructure) to ensure the best oxygen intake throughout the graft. Fat harvesting technique has a crucial role for a successful grafting considering that liposuction has a traumatic effect on adipocytes that can affect their vitality. Each step of this process must be carried out effectively in order to maximize cellular vitality.

The first large contribution to this methodology was from Coleman who proposed in 1997 his technique, the current standard, that low-pressure lipoaspiration, low-speed centrifugation, and other standardized steps allow for decreased cellular damage for adipocytes which increased graft vitality [12].

The fat resorption rate, however, continued to be inconsistent and unpredictable which led to researchers looking for methods to increase the number of ASCs in the fat graft.

Initially, some surgeons experimented with different types of growth factors such as insulin, VEGF, or PRP in combination with fat but none of these techniques showed a clear utility [13–16].

In 2006, Matsumoto et al. described for the first time the CAL (Cellular-Assisted Lipotransfer). In particular, they enriched the fat with SVF (Stromal Vascular Fraction) extracted from the lipoaspirate (which contains ASCs, endothelial cells, and other mesenchymal cells such as pericytes) in order to improve the survival rate of the grafted tissue [17].

Since then, several studies have demonstrated CAL efficacy [18–24]. The clinical application, however, is still not widespread due to few randomized clinical trials and the low level of evidence and safety data in the literature. In this chapter, we analyze the CAL in terms of techniques in breast surgery, its efficacy, and safety in a clinical application.

6.2 Cell-Assisted Lipotransfer

CAL (Cell-Assisted Lipotransfer) is a procedure through which the lipoaspirate is enriched with ASC or SVF in order to improve the regenerative function of the fat.

The resorption rate of the fat has always been a challenging problem for many reasons, especially the need to repeat the procedure several times.

Microscopical analysis of the fat, as demonstrated by some studies, showed that the suctioned fat has approximately half of the ASCs compared to the excised fat [25]. This can be explained by the fact that ASCs generally are more represented close to vascular structure and are, for this reason, generally preserved by the cannula. Moreover, a substantial number of ASCs go into the liquid part of the lipoaspirate.

The CAL stems from the need to address this problem using a part of the lipoaspiration as a reservoir of ASCs to enrich the material which acts like a living scaffold for ASCs (Fig. 6.1).

In particular, selective ASC extraction can only be performed through enzymatic digestion of the lipoaspirate and subsequent expansion in culture in a GMP (Good Manufacturing Practice) laboratory. This time-consuming process makes it impractical for clinical use and necessitates two separate surgical procedures (one for fat harvesting and one for fat grafting).

Intraoperative techniques (most commonly non-enzymatic) allow to extract the whole SVF which includes ASCs, MSC, HSC, Treg Cells, Pericytes, mast-cells, complex microvascular beds (fibroblasts, WBC, dendritic cells, intra-adventitial smooth muscular-like cells, etc.) and are more suitable for clinical applications due to their simplicity and rapidity of execution.

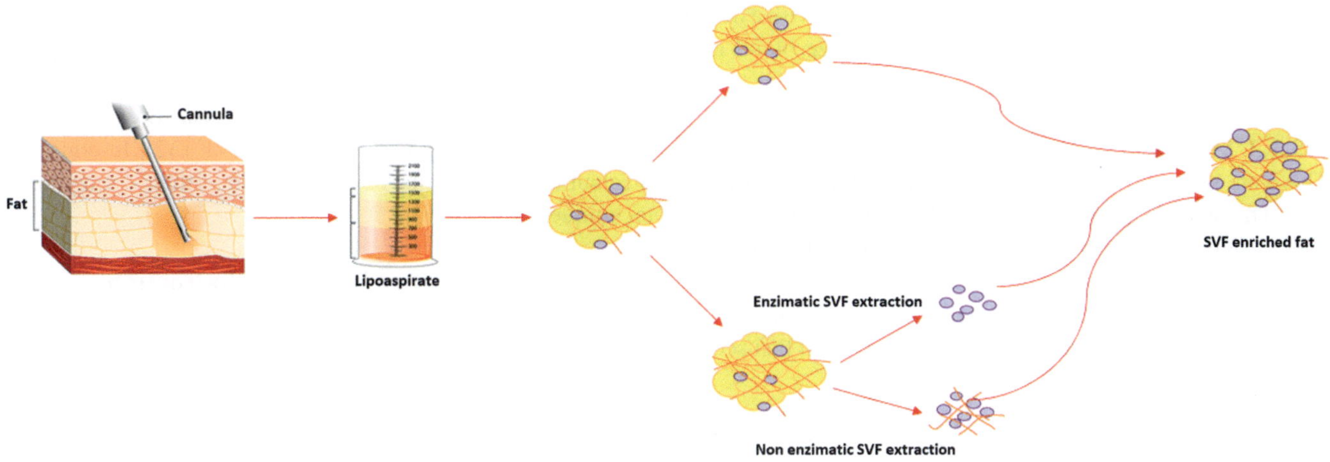

Fig. 6.1 Cell-assisted lipotransfer procedure

6.3 Stromal Vascular Fraction (SVF)

Zuk et al. in 2001 and 2002 were the first to identify the complexity of the cellular population found in adipose tissues [26].

Since the mid-late 2000s, many researchers and clinicians began to understand the complexity, contents, and paracrine functions of the adipose-derived stromal vascular fraction. We now know that it is composed of a highly heterogeneous population of cells: MSC, HSC, Treg Cells, Pericytes, mast-cells, complex microvascular beds (fibroblasts, WBC, dendritic cells, intra-adventitial smooth muscular-like cells, etc.). These cells are linked to each other to form a complex extracellular matrix [27].

If ASCs can be defined as the most important element for the regenerative function of fat graft, their vitality and their regenerative function are closely linked to this complex system of interaction. Fat harvested via en bloc excision delivers mature adipocytes and their attached progenitor cells (pre-adipocytes of presumed unipotent abilities as near terminally differentiated cells) which represent another group of secretory-capable cells within the grafted microenvironment. Appreciation of the importance of the complex, three-dimensional matrix found in adipose tissues has led many to consider the intact, non-manipulated transplantation of such tissues and their accompanying cellular elements may be advantageous.

6.4 Adipose-Derived Stem Cells (ASCs)

MSCs (mesenchymal stromal cells) were described for the first time by Friedenstein in 1970 in the bone marrow [28]. Decades later, we know there are several types of MSCs in many organs and tissues, including adipose tissue.

ASCs, first characterized in 2001, show a higher yield compared to bone marrow stem cells and their extraction is very easy thanks to the low invasiveness of liposuction compared to bone marrow extraction. They can differentiate in several cellular types of the mesenchymal line as adipocytes, osteoblasts, and chondrocytes; express cellular markers similar to other mesenchymal cells like CD105, CD73, and CD90; and are negative for CD31 and CD45. They have regenerative function in many aspects thanks to their angiogenetic, antifibrotic, and immunomodulatory effect. In addition, thanks to the release of paracrine modulators, they can improve the resistance to hypoxic conditions [10].

All these factors put ASCs at the center of scientific interest in regenerative surgery in the last 20 years and research in this field is still very active.

6.5 Cell-Enriched Fat Grafting in the Breast

6.5.1 Surgical Technique

Breast fat grafting procedure starts from a precise analysis of the defect we want to correct or of the breast volume we want to reach depending on the indication.

The patient is marked in the standing position where areas of contour irregularity can be most clearly seen due to shadows created by overhead lighting.

The surgeon estimates the amount of fat needed for grafting and selects one or more donor sites. Donor site selection includes any area that has sufficient fat to donate, typically the abdomen, hips, and thighs. The site selected can be based on ease of harvest and patient preference, since no studies have demonstrated superiority of different sites in graft survival [29].

Standard liposuction infiltration is performed with a dilute epinephrine and anesthetic solution to minimize blood loss (Klein solution).

6.5.1.1 Lipoaspiration

There are several different techniques to perform lipoaspiration. The author's preference is the Coleman technique. The main difference in this step when performing CAL compared to a standard lipofilling is in the amount of lipoaspiration. The volume must be approximately double the amount of final enriched fat required since half of the fat will be used to extract the SVF for enrichment.

The Coleman technique involves the use of a cannula which has a blunt tip and two adjacent holes. This is connected to a 10-mL syringe in order to apply a lowest suction compared to a larger syringe. As the cannula is pushed through the harvest site, a combination of the curetting action of the cannula and the negative pressure pulls fat into the syringe [30].

At this point, standard Coleman technique would proceed with centrifugation of all the lipoaspirate at 3000 rpm for 3 min. When performing CAL, however, the lipoaspirate is divided in two equal parts: one for SVF/ASC extraction and the other to be enriched, typically after centrifugation.

After centrifugation, the lipoaspirate components are separated by density. The upper level of oil (ruptured adipocytes from aspiration) is removed by pouring it off and then removing any remainder with absorbent wicks of Telfa (Kendall). The bottom, the densest level, contains blood and infiltration solution and is removed by briefly removing the cap and allowing it to drain. After this, the fat, ready to be enriched, is placed in ice [30].

6.5.1.2 SVF and ASC Extraction

There are several techniques for extraction of ASCs or SVF [31].

ASCs can be extracted from lipoaspirate and expanded in culture by submitting the fat to several steps of enzymatic digestion, washing, and manipulations in a GMP laboratory. It is a time-consuming process that requires a two-step surgery which makes it impractical for clinical use but is considered, despite this, the gold standard technique for ASC extraction.

There are several kinds of intraoperative extraction techniques that make the process easier and faster and can be divided into enzymatic and non-enzymatic techniques [32].

Intraoperative techniques allow to extract the whole SVF, not only ASCs. In particular, they allow to obtain a cellular SVF (cSVF) or a tissue SVF (tSVF) depending on the technique. Enzymatic techniques acting as digestion of the extracellular matrix generally allow one to obtain a cellular SVF. Non-enzymatic techniques act as a mechanical separation of the lipoaspirate components and generally do not affect the extracellular matrix, giving a tissue SVF that has a greater volume compared to cellular SVF [32]. In addition, mechanical techniques are faster than enzymatic techniques.

6.5.1.3 Intraoperative Enzymatic Extraction

Intraoperative enzymatic extraction can be performed in several ways. Typically, after lipoaspiration the fat for SVF extraction is washed several times with PBS solution (phosphate-buffered saline) and is put under 0.075% collagenase solution at 37.8 °C in a shaker bath for 30 min. It is then washed several times with PBS solution [33].

The collagenase action is inactivated with an equal volume of 10% fetal bovine serum (FBS) and the infranatant is centrifuged for 5 min at 1200 × *g*. The cellular pellet is then resuspended in 10% FBS and passed through a 100-μm mesh filter to remove debris.

At this point, the SVF extracted and the fat are gently mixed and after 10–15 min the SVF-enriched lipoaspirate is placed in the injection syringe.

6.5.1.4 Intraoperative Non-enzymatic Extraction

There are several protocols for non-enzymatic extraction of SVF from lipoaspirate (FAT, FAST, FEF, LIPOGEMS, etc.). Most of these procedures require a mechanical manipulation of the fat with different steps of filtration or centrifugation depending on the type of technique.

For example, Van Dongen et al. proposed in 2016 an SVF extraction performed through 3000 rpm centrifugation, passing through a 1.4-mm opening and then a second centrifugation at 3000 rpm [32].

Domenis et al. performed a filtration through a filter bag with a 120-μm filter and then a 400-*g* centrifugation [34].

The nanofat technique described by Tonnard et al. has been used as a SVF extraction technique (passing adipose tissue through a female-to-female Luer-Lok 30 times and filtering) [35].

As mentioned above, mechanical extraction of SVF leads to a tSVF that has a greater volume compared to cSVF. After the extraction, the SVF is used to enrich the lipoaspirate as mentioned previously.

6.5.1.5 Fat Injection Technique

Different cannulas are used for injection than for harvesting. Typically, the injection cannula is a blunt cannula with a single hole for precise fat deposition and is available in different lengths and flexibilities.

Droplets of fat grafts are deposited with each pass as the cannula is withdrawn and in multiple tissue planes after multiple passes. Regular distribution of the fat is essential to successful fat grafting since a wider surface of contact between the graft and the donor allows a better vascular supply. The total injected amount is calculated to not over graft the areas to treat, since it does not improve grafting and increases the formation of oil cysts.

The fat grafting technique can be different depending on the surgical indication.

Depressions are generally treated with multiple tunnels in a crosshatch fashion, while ridges and rippling respond best to a few long tunnels below the longitudinal depression. Depressed areas or those with scar retraction may benefit from scar release with specialized cannulas that have cutting edges.

For cosmetic breast augmentation, fat can be placed in different planes in order to enhance the volume, generally in the fatty layers on, around, and under the mammary gland. Some authors also infiltrate into the gland and in the pectoralis muscle but is this author's opinion that these procedures can lead to significant complications such as hematoma or fat embolism when performed in the pectoralis muscle or damage to gland ducts after intraglandular injection with risk of mastitis, infection, and excessive post-operative swelling. In fact, we generally use fat grafting alone only for mild cosmetic breast augmentation while very often in combination with breast implants to improve the cosmetic result of breast augmentation and to enhance implant coverage.

In post-oncological patients, fat grafting can be used as mentioned above to improve implant coverage as well as prevention and treatment of implant rippling (Figs. 6.2 and 6.3).

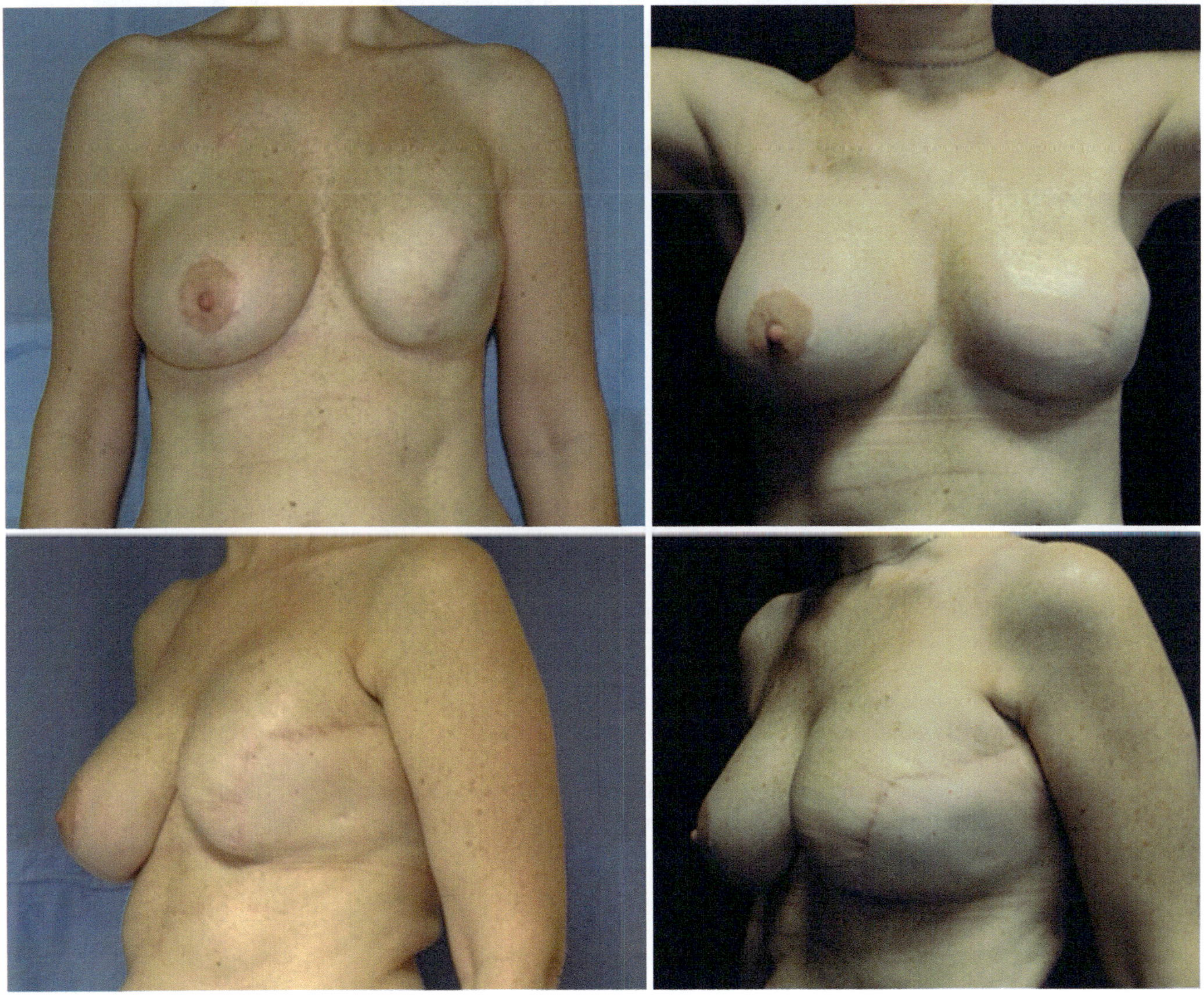

Fig. 6.2 Clinical case: fat grafting on breast implant reconstruction. One year post-op

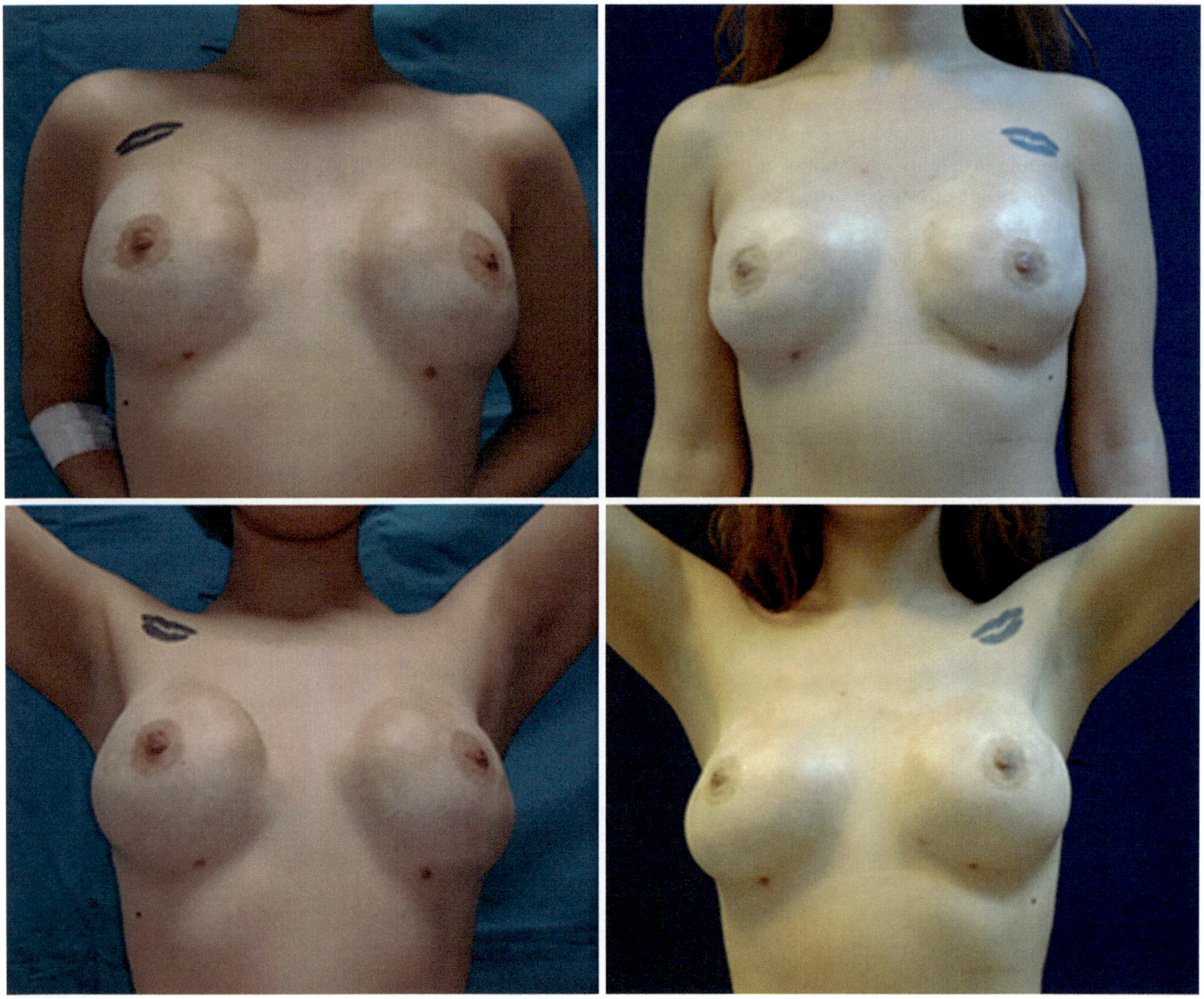

Fig. 6.3 Clinical case: bilateral breast implant reconstruction after preventive mastectomy. One year post-op after breast implant revision and fat grafting

Another common indication is to treat specifical deformity consequently to breast-conserving surgery treatments (BCS) as quartectomy or nodulectomy. These cases illustrate the dual benefits of fat graft on both the volume defect and on the scar retractions that are very common after BCS thanks to its antifibrotic effect (Figs. 6.4 and 6.5).

Finally, fat grafting can be a useful technique in all clinical conditions that require the correction of small asymmetries of the breast in both cosmetic and reconstructive context (Figs. 6.6 and 6.7).

6.5.2 Efficacy

Several studies demonstrated that CAL is a useful technique which can significantly reduce the fat resorption rate.

Two studies selectively investigated the efficacy of fat enriched with cultivated ASCs comparing it to traditional fat grafting. Koh et al. observed a significant reduction of fat resorption rate in the CAL group compared to the control group (20.59% versus 46.81%) [36].

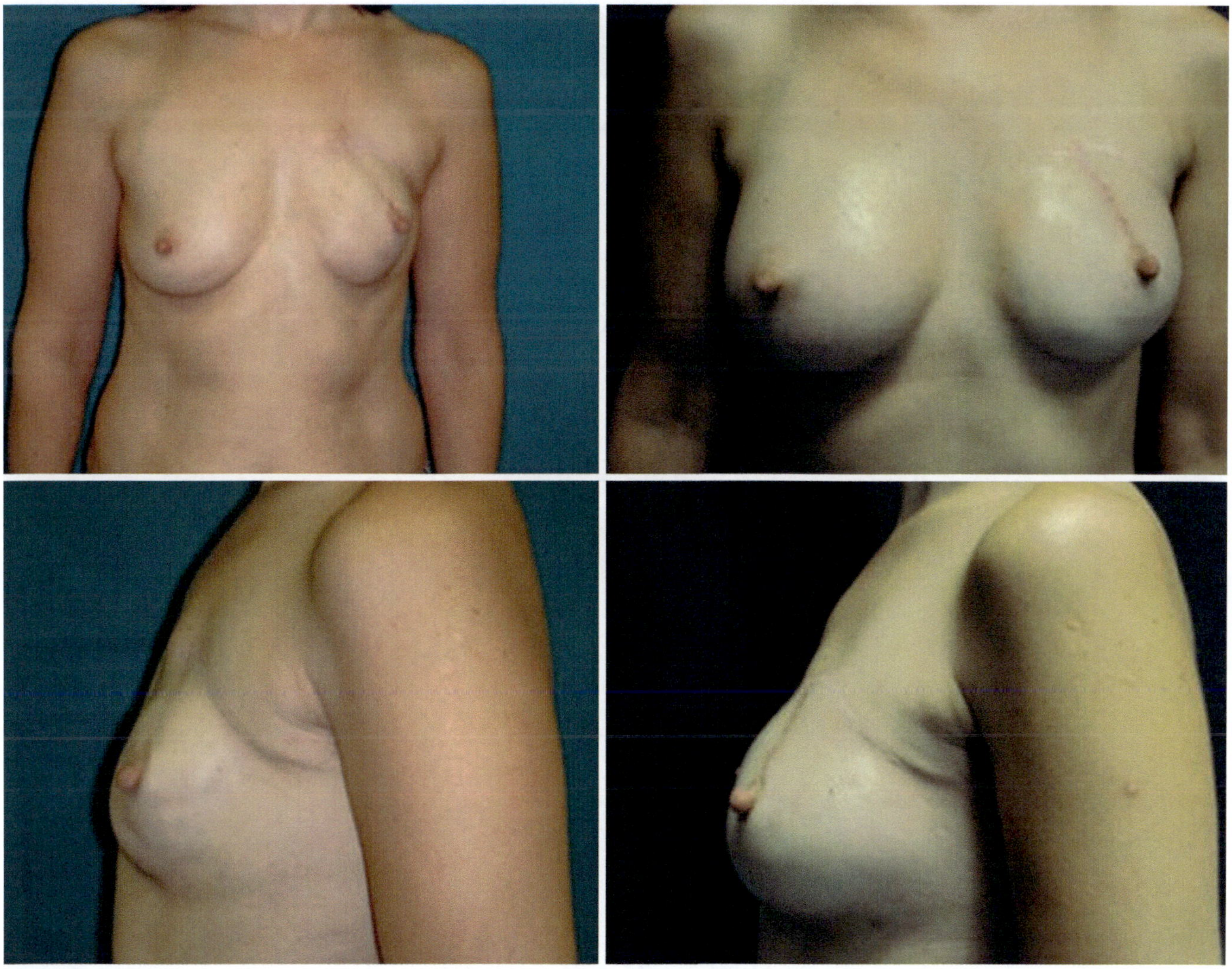

Fig. 6.4 Clinal case: fat grafting for scar retraction and asymmetry correction after BCS. Six months post-op

Similarly, Kolle et al. showed a fat survival rate of 80.9% for ASC-enriched graft compared to the 16.3% of traditional fat grafting, yielding a relative survival improvement of 5.0 [19].

Several studies investigated the general efficacy of CAL (in terms of both ASCs and SVF enrichment) compared to traditional fat grafting.

Zhou et al. recently performed a systematic review with meta-analysis on 17 previous studies, a global case series of 387 cases, and showed an improvement from 45% to 60% of fat survival rate with CAL compared to traditional fat grafting.

Furthermore, this improvement appeared to be even more significant for fat grafting in the face area and able to reduce the number of surgical operations, compared to breast fat grafting [18].

Laloze et al. performed a similar review expanding the case series but obtained slightly different results. They showed a fat survival rate of 64% for CAL vs 44% of traditional fat grafting with no differences in terms of efficacy for different injection sites. However, this improvement appeared to be significant only for small injection volumes (<100 mL) and, in addition, CAL showed a higher rate of complications compared to traditional fat grafting and did not reduce the number of surgical procedures [37].

Li et al. performed a similar review with meta-analysis comparing the efficacy of CAL with traditional fat grafting in the breast and showed a fat survival rate significantly higher in the CAL group. However, analyses of the subgroups comparing the SVF-enhanced fat graft with the conventional fat grafting noted no significant differences in fat survival rate, suggesting that the fat survival rate improvement observed might regard only ASC enrichment [38].

This apparent uncertainty can be partially explained by the lack of a standardized procedure that is universally

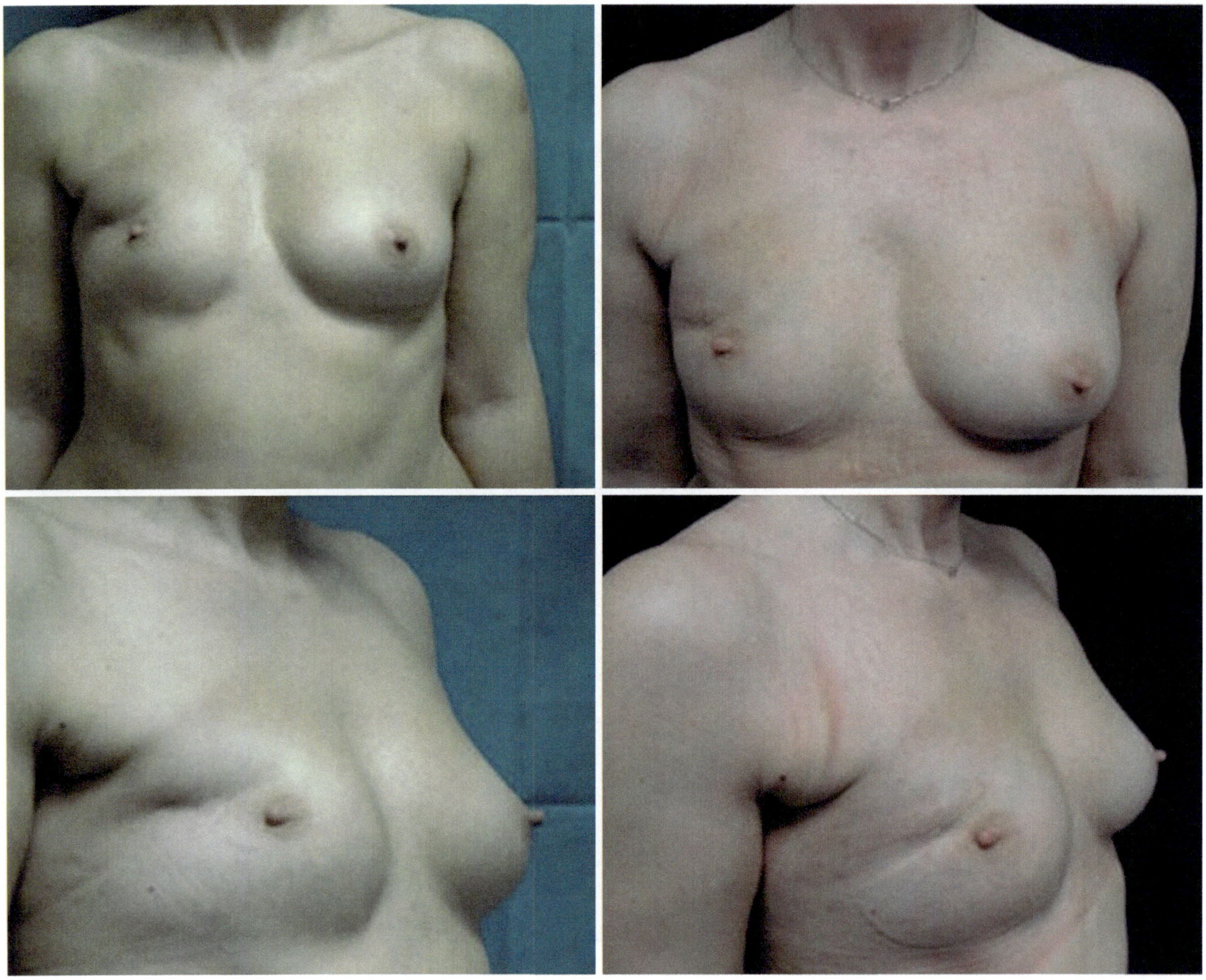

Fig. 6.5 Clinical Case: fat grafting for scar retraction and asymmetry correction after BCS. Eight months post-op

accepted by the scientific community. At the moment, several different processing techniques are described, each one with different outcomes [20, 22, 36].

In addition, some of these trials evaluated the fat survival rate through imaging techniques as MRI or Ecography, while other trials through patient satisfaction or clinical evaluation, creating even more the variability of the results [23].

By now, there is no evidence that any intraoperative isolation procedure could be designated as preferred procedure for isolating SVF [32].

Enzymatic and non-enzymatic procedures had comparable results as it comes to cell yield, viability, and SVF composition. Non-enzymatic isolation procedures produced end products that had greater volumes (tSVF) than the pellets (cSVF) of the enzymatic isolation procedures. The results of intraoperative isolation procedures are comparable with those of the gold standard, the collagenase-based non-intraoperative isolation protocol. Since intraoperative isolation procedures are less time-consuming, but as efficient as the non-intraoperative isolation protocol, the use of intraoperative isolation procedures seems to be more suitable for clinical purposes [32].

Finally, most of the countries have legislative problems with the manipulation of stem cells for a collagenase-based non-intraoperative isolation protocol, with the necessity of a GMP (Good Manufacturing Practice) laboratory, which is generally not present in most clinical centers.

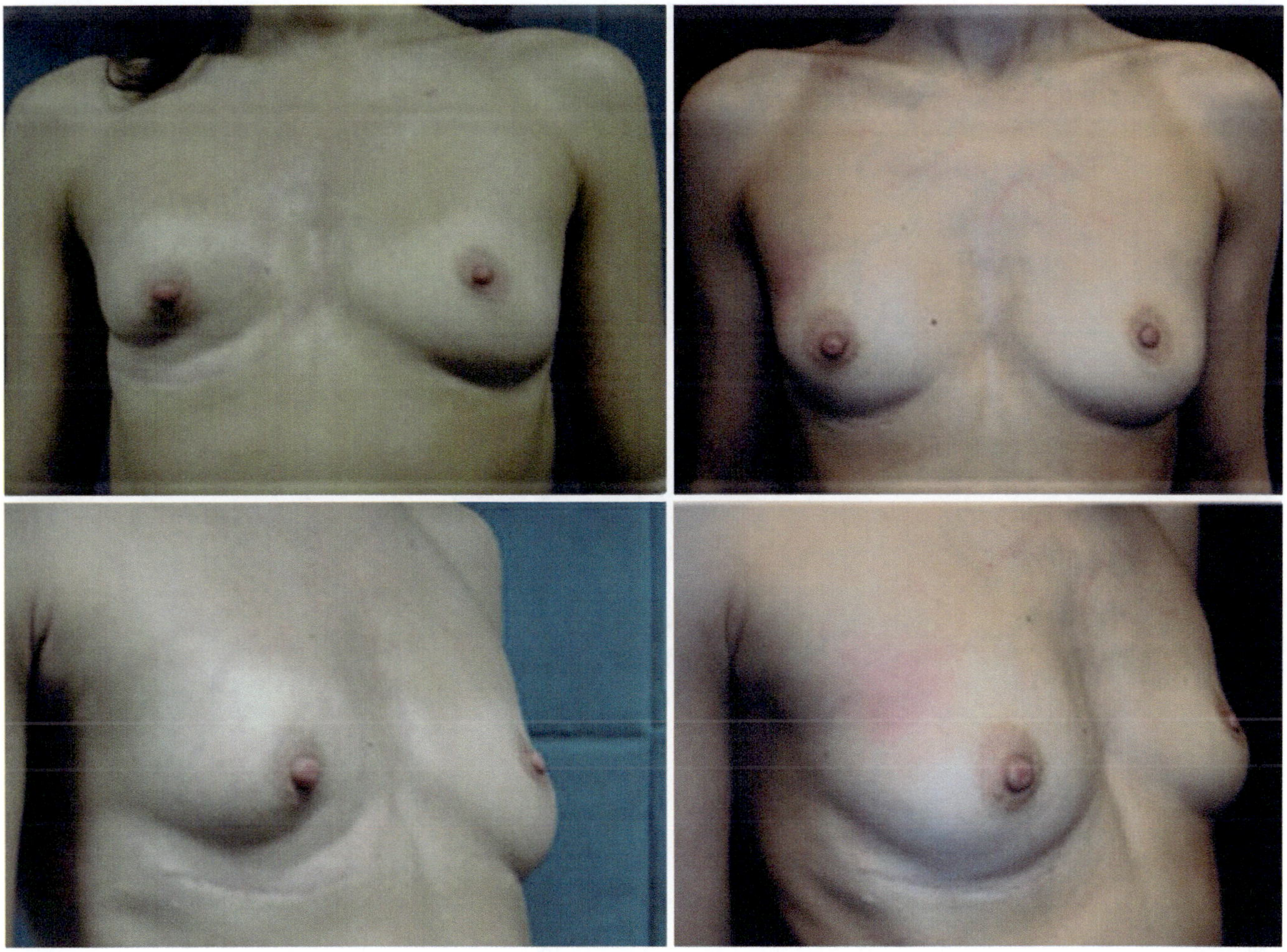

Fig. 6.6 Clinical Case: fat grafting for correction of breast contour irregularity. One year post-op after two procedures

6.5.3 Complications

Some studies showed a higher complication incidence with CAL compared with traditional lipotransfer (8.4% vs 1.5%), while others showed that this increased complication rate appeared to be significant only for breast fat grafting [18, 37].

Currently, the majority of the studies comparing CAL with traditional fat grafting did not show any significant difference in terms of classical complications like oil cysts, infections, fat necrosis, calcifications, nodules, and fibrosis.

6.5.4 Oncological Safety

Today, it is generally accepted by the scientific community that fat grafting is a safe technique when cautiously performed and, most importantly, when performed after complete cancer remission [39].

The subject of CAL is obviously still much debated due to the enhancement of its regenerative effect and the greater rate of ASCs.

Several in vitro trials demonstrate that ASCs can promote tumor growth and increase the risk of cancer recurrence [40–42].

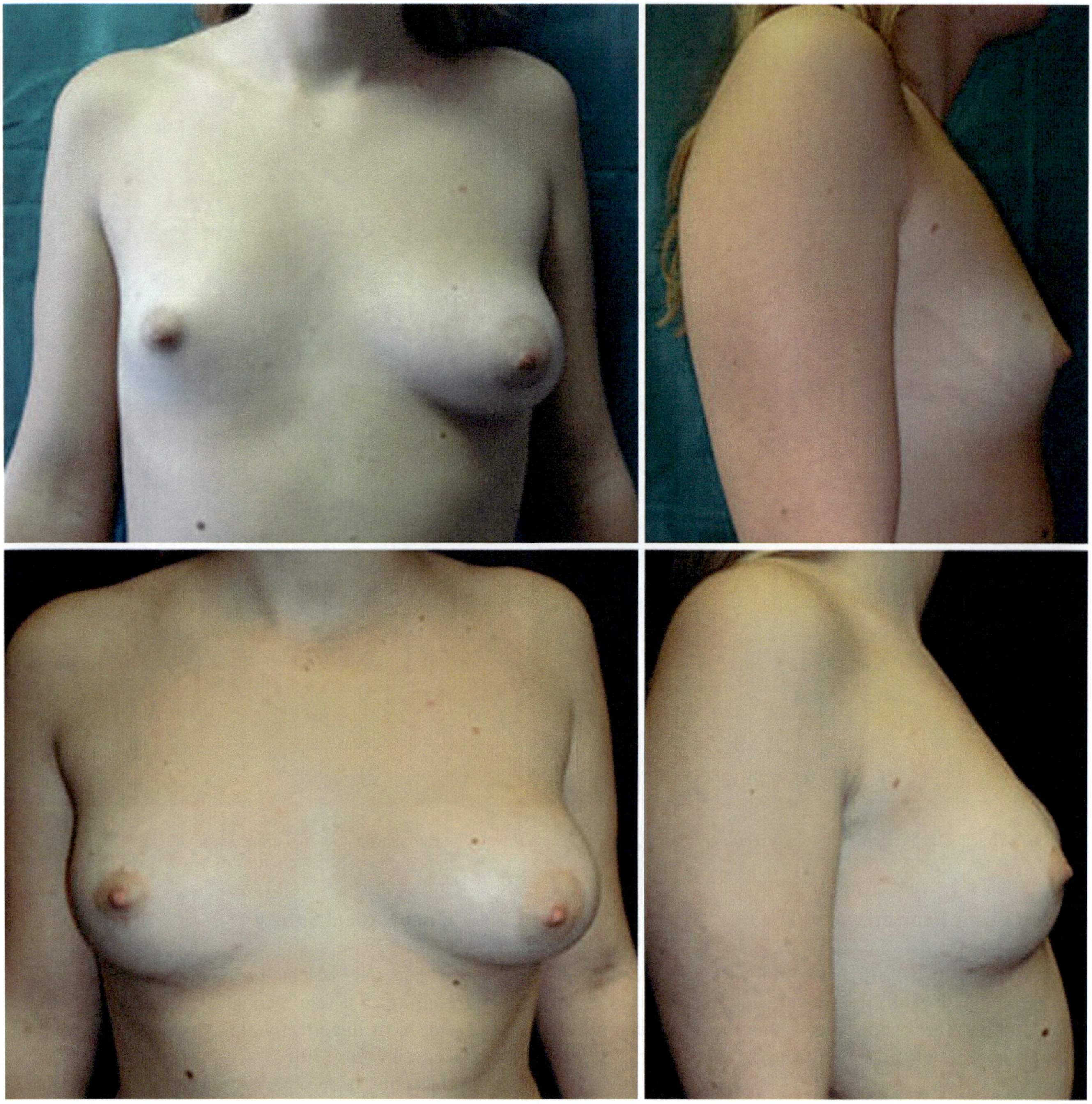

Fig. 6.7 Clinical Case: fat grafting for correction of congenital breast asymmetry. One year Post-op after three procedures

On the other hand, in vivo trials so far have not shown that CAL can increase the risk of breast cancer incidence or recurrence. Only one trial demonstrated the incidence of a single local malignancy, but further characterization showed that to be a secondary localization of pelvic metastasis which was independent from the treatment [43].

Since cancer occurrence is a long-term complication, there is still a lack of evidence about the oncological potential of CAL as those studies are still ongoing.

6.6 Conclusion

The identification of ASCs and their clinical application are quite recent concepts and it is very likely that in the coming years there will be a significant increase of clinical randomized trials on this subject that can hopefully bring a standardization of the technique. Currently, there is evidence that CAL is a useful and promising technique that in the future could lay the foundation for several clinical applications in regenerative surgery and breast surgery.

References

1. Kling RE, Mehrara BJ, Pusic AL, Young VL, Hume KM, Crotty CA, Rubin JP. Trends in autologous fat grafting to the breast: a national survey of the american society of plastic surgeons. Plast Reconstr Surg. 2013;132:35.
2. Coleman SR, Saboeiro AP. Fat grafting to the breast revisited: safety and efficacy. Plast Reconstr Surg. 2007;119:775.
3. Missana MD, Laurent I, Barreau L, et al. Autologous fat transfer in reconstructive breast surgery: indications, technique and results. Eur J Surg Oncol. 2007;33:685.
4. Spear S. Fat for breast: where are we? Plast Reconstr Surg. 2008;122:983.
5. Gutowski K. Current applications and safety of autologous fat grafts: a report of the asps fat graft task force. Plast Reconstr Surg. 2009;124:272.
6. Parrish JN, Metzinger SE. Autogenous fat grafting and breast augmentation: a review of the literature. Aesthet Plast Surg. 2010;30:549.
7. Fournier PF. Fat grafting: my technique. Dermatol Surg Off Publ Am Soc Dermatol Surg Al. 2000;26:1117.
8. Coleman SR. Long-term survival of fat transplants: controlled demonstrations. Aesthet Plast Surg. 1995;19:421.
9. Ersek RA. Transplantation of purified autologous fat: a 3-year follow-up is disappointing. Plast Reconstr Surg. 117:1045.
10. Minteer D, Marra KG, Rubin JP. Adipose-derived mesenchymal stem cells: biology and potential applications. Adv Biochem Eng Biotechnol. 2013;129:59–71.
11. Eto H, Kato H, Suga H, Aoi N, Doi K, Kuno S, Yoshimura K. The fate of adipocytes after nonvascularized fat grafting: evidence of early death and replacement of adipocytes. Plast Reconstr Surg. 2012;129:1081.
12. Coleman SR. Facial recontouring with lipostructure. Clin Plast Surg. 1997;24:347.
13. Gentile P, De Angelis B, Pasin M, Cervelli G, Curcio CB, Floris M, Di Pasquali C, Bocchini I, Balzani A, Nicoli F, Insalaco C, Tati E, Lucarini L, Palla L, Pascali M, De Logu P, Di Segni C, Bottini DJ, Cervelli V. Adipose-derived stromal vascular fraction cells and platelet-rich plasma: basic and clinical evaluation for cell-based therapies in patients with scars on the face. J Craniofac Surg. 2014;25:267.
14. Gentile P, Orlandi A, Scioli MG, Di Pasquali C, Bocchini I, Curcio CB, Floris M, Fiaschetti V, Floris R, Cervell V. A comparative translational study: the combined use of enhanced stromal vascular fraction and platelet-rich plasma improves fat grafting maintenance in breast reconstruction. Stem Cells. 2012;1:341.
15. Yuksel E, Weinfeld AB, Cleek R, Wamsley S, Jensen J, Boutros S, Waugh JM, Shenaq SM, Spira M. Increased free fat-graft survival with the long-term, local delivery of insulin, insulin-like growth factor-I, and basic fibroblast growth factor by PLGA/PEG microspheres. Plast Reconstr Surg. 2000;105:1712.
16. Kim YW, Min HJ, Choi RJ, Lee DH, Cheon YW. Insulin promotes adipose-derived stem cell differentiation after fat grafting. Plast Reconstr Surg. 2018;142:927.
17. Matsumoto D, Sato K, Gonda K, Takaki Y, Shigeura T, Sato T, Aiba-Kojima E, Iizuka F, Inoue K, Suga H, Yoshimura K. Cell-assisted lipotransfer: supportive use of human adipose-derived cells for soft tissue augmentation with lipoinjection. Tissue Eng. 2006:061122053519001.
18. Zhou Y, Wang J, Li H, Liang X, Bae J, Huang X, et al. Efficacy and safety of cell-assisted lipotransfer: a systematic review and meta-analysis. Plast Reconstr Surg. 2016;137:44e.
19. Kølle S-FT, Fischer-Nielsen A, Mathiasen AB, Elberg JJ, Oliveri RS, Glovinski PV, et al. Enrichment of autologous fat grafts with ex-vivo expanded adipose tissue-derived stem cells for graft survival: a randomised placebo-controlled trial. Lancet. 2013;382:1113.
20. Sterodimas A, de Faria J, Nicaretta B, Boriani F. Autologous fat transplantation versus adipose-derived stem cell-enriched lipografts: a study. Aesthet Plast Surg. 2011;31:682.
21. Herold C, Ueberreiter K, Busche MN, Vogt PM. Autologous fat transplantation: volumetric tools for estimation of volume survival. A systematic review. Aesthet Plast Surg. 2013;37:380.
22. Peltoniemi HH, Salmi A, Miettinen S, Mannerström B, Saariniemi K, Mikkonen R, et al. Stem cell enrichment does not warrant a higher graft survival in lipofilling of the breast: a prospective comparative study. J Plast Reconstr Aesthet Surg. 2013;66:1494.
23. Lee SK, Kim D-W, Dhong E-S, Park S-H, Yoon E-S. Facial soft tissue augmentation using autologous fat mixed with stromal vascular fraction. Arch Plast Surg. 2012;39:534.
24. Tiryaki T, Findikli N, Tiryaki D. Staged stem cell-enriched tissue (SET) injections for soft tissue augmentation in hostile recipient areas: a preliminary report. Aesthet Plast Surg. 2011;35:965.
25. Yoshimura K, Sato K, Aoi N, Kurita M, Inoue K, Suga H, et al. Cell-assisted lipotransfer for facial lipoatrophy: efficacy of clinical use of adipose-derived stem cells. Dermatol Surg. 2008;34:1178.
26. Zuk P, et al. Multilineage cells from human adipose tissue: implications for cell-based therapies. J Tiss Eng. 2001;7:211.
27. Doi K, Tanaka S, Iida H, Eto H, Kato H, Aoi N, et al. Stromal vascular fraction isolated from lipo-aspirates using an automated processing system: bench and bed analysis. J Tissue Eng Regen Med. 2013;7:864.
28. Friedenstein AJ, Chailakhjan RK, Lalykina KS. The development of fibroblast colonies in monolayer cultures of Guinea-pig bone marrow and spleen cells. Cell Tissue Kinet. 1970;3:393.
29. Rohrich RJ, Sorokin ES, Brown SA. In search of improved fat transfer viability: a quantitative analysis of the role of centrifugation and harvest site. Plast Reconstr Surg. 2004;114:1945.
30. Pu LL, Coleman SR, Cui X, et al. Autologous fat grafts harvested and refined by the Coleman technique: a comparative study. Plast Reconstr Surg. 2008;122:932.
31. Laloze J, Varin A, Bertheuil N, Grolleau JL, Vaysse C, Chaput B. Cell-assisted lipotransfer: current concepts. Ann Chir Plast Esthet. 2017;62:609.
32. Van Dongen JA, Tuin AJ, Spiekman M, Jansma J, van der Lei B, Harmsen MC. Comparison of intraoperative procedures for isolation of clinical grade stromal vascular fraction for regenerative purposes: a systematic review. J Tissue Eng Regen Med. 2018;12:e261.
33. Yoshimura K, Sato K, Aoi N, Kurita M, Hirohi T, Harii K. Cell-assisted lipotransfer for cosmetic breast augmentation: supportive use of adipose-derived stem/stromal cells. Aesthet Plast Surg. 2008;32:48.

34. Domenis R, Lazzaro L, Calabrese S, et al. Adipose tissue derived stem cells: in vitro and in vivo analysis of a standard and three commercially available cell-assisted lipotransfer techniques, vol. 6. Stem Cell Res Ther; 2015. p. 2.
35. Tonnard P, Verpaele A, Peeters G, et al. Nanofat grafting: basic research and clinical applications. Plast Reconstr Surg. 2013;132:1017.
36. Koh KS, Oh TS, Kim H, Chung IW, Lee KW, Lee HB, et al. Clinical application of human adipose tissue-derived mesenchymal stem cells in progressive hemifacial atrophy (parry-Romberg disease) with microfat grafting techniques using 3-dimensional computed tomography and 3-dimensional camera. Ann Plast Surg. 2012;69:331.
37. Laloze J, Varin A, Gilhodes J, Bertheuil N, Grolleau JL, Brie J, Usseglio J, Sensebe L, Filleron T, Chaput B. Cell-assisted lipotransfer: friend or foe in fat grafting? Systematic review and meta-analysis. J Tissue Eng Regen Med. 2018;12:e1237–50.
38. Li M, Chen C. The efficacy of cell-assisted lipotransfer versus conventional lipotransfer in breast augmentation: a systematic review and meta-analysis. Aesthet Plast Surg. 2021;45:1478.
39. Raj S, Abu-Ghname A, Davis MJ, Izaddoost SA, Winocour SJ. Safety and regulation of fat grafting. Semin Plast Surg. 2020;34:059.
40. Chaput B, Foucras L, Le Guellec S, Grolleau JL, Garrido I. Recurrence of an invasive ductal breast carcinoma 4 months after autologous fat grafting. Plast Reconstr Surg. 2013;131:123e.
41. Chaput B, Grolleau JL, Bertheuil N, Eburdery H, Chavoin JP, Garrido I. Another suspected case of breast cancer recurrence after lipofilling? Remain cautious. J Plast Reconstr Aesthet Surg. 2014;67:1156.
42. Alharbi M, Garrido I, Vaysse C, Chavoin JP, Grolleau JL, Chaput B. Latissimus dorsi flap invasion by ductal breast carcinoma after lipofilling. Plast Reconstr Surg Glob Open. 2013;1:e68.
43. Pérez-Cano R, Vranckx JJ, Lasso JM, Calabrese C, Merck B, Milstein AM, et al. Prospective trial of adipose-derived regenerative cell (ADRC)-enriched fat grafting for partial mastectomy defects: the RESTORE-2 trial. Eur J Surg Oncol. 2012;38:382.

Enrichment of the Fat Graft with Vascular Stem Cells

7

Maxim Geeroms, Moustapha Hamdi, and Rica Tanaka

Contents

7.1 Introduction

Despite its clinical application since 1893 [1], we still struggle with the fact that part of the fat graft perishes postoperatively. Ischemia is known to be the limiting factor in graft survival. Several strategies [2] have been investigated to create an early vessel network in order to avoid this tissue loss [3–5]. In this chapter, we analyze the vascular state of the fat graft which has led to an experimental enrichment modality. Specifically, we describe the development of a highly vasculogenic cell population containing endothelial progenitor cells (EPCs), which can be added to the fat grafts, followed by its advantages, limitations, and future perspectives.

M. Geeroms (✉)
Department of Plastic and Reconstructive Surgery, Vrije Universiteit Brussel—Universitair Ziekenhuis Brussel, Brussels, Belgium

Department of Plastic and Reconstructive Surgery, Juntendo University School of Medicine, Tokyo, Japan

M. Hamdi
Department of Plastic and Reconstructive Surgery, Vrije Universiteit Brussel—Universitair Ziekenhuis Brussel, Brussels, Belgium

R. Tanaka
Department of Plastic and Reconstructive Surgery, Juntendo University School of Medicine, Tokyo, Japan

Department of Regenerative Therapy, Juntendo University Graduate School of Medicine, Tokyo, Japan

7.2 Evolution of Fat Grafting

7.2.1 From 1920 to 2020

> *"Regardless of the technique used, it does not matter what ultimately happens to the transplanted fat as long as it fulfils its goal of permanently filling a space."* (Passot, 1920)

We are one century past this remarkable quote [6, 7], which highlights two facts. On the one hand, a lot happened since the early twentieth century. The main goal still is *filling a space* with a stable product. But we have understood the composition of the adipose tissue that we harvest, we are aware of its potential regenerative properties, we know the common principles to follow when we perform the injection, and we have learned its limitations. This evolution is based on a better scientific understanding, and it has led to clinical progress. Nowadays, fat grafting is part of our *armamentarium* and a plastic surgery practice without fat grafting is outdated.

© The Author(s), under exclusive license to Springer Nature Switzerland AG 2023

A. Di Giuseppe et al. (eds.), *Fat Transfer in Plastic Surgery*, https://doi.org/10.1007/978-3-031-10881-5_7

On the other hand, it is confronting how little we have advanced. The fat remains in the first place a *filler*. But a fat graft which ultimately consists of a high percentage of calcifications, thick fibrosis, or oil cysts is not a desirable outcome, even if it maintains its volume. Despite the rapidly growing number of research experiments, inconsistencies in clinical technique and outcome are the norm, and at conferences we often have to filter the truth out of unsubstantiated theses and claims, which are not rarely company-influenced. Therefore, Aristotle's popular quote "*the more we know, the more we realize we do not know*" is applicable to fat grafting research as well.

7.2.2 Historical Milestones

Neuber is a recurring name in manuscripts on fat grafting, as he was the first to transfer adipose tissue [1]. In 1893, he corrected an osteomyelitis-induced depressed scar in the face. Czerny did the first autologous breast reconstruction by grafting a fist-sized lipoma from the buttock *en bloc* to the breast after a partial mastectomy [8]. In 1909, fat transfer by injection was born thanks to Holländer, who injected fat to contour deformities [9]. Halfway through the twentieth century, fat grafting became in disuse due to its tissue loss, transformation to oil cysts, and unpredictability [10]. It was demonstrated by Peer that half of the transferred adipocytes did not survive the process, and that a lack of vascularization caused cyst formation [11].

Three key revolutions led to the revival of fat grafting. First, Illouz developed the technique of liposuction in the late seventies as a tool to reduce the subcutaneous tissue layer, which caught the attention of American plastic surgeons, as well as the press [12, 13]. He injected this semi-liquid aspirate of adipocytes again to fill dimples, depressions, hollows, atrophied areas, the breast and wrinkles [13].

Second, in the late eighties, Chajchir pointed out the importance of the delicate manipulation of the adipose tissue prior to re-injection, in order to not destruct fat globules; the relevance of the elimination of broken-down fat cells, oil and blood, in order to obtain a pure fat graft; and the value of the vascularity of the recipient site which correlates inversely with the resorption rate of the grafted fat [14]. Coleman was further responsible for making the procedure of harvesting (gentle extraction by means of a blunt liposuction cannula), processing (centrifugation, followed by removal of the aqueous and oily fraction), and injection (placement of the graft in tiny aliquots) standardized for facial recontouring [15]. Nowadays, we still refer to the Coleman method, even though numerous modifications of his technique are being applied. The combined work of Chajchir and Coleman lowered the complication rate and renewed interest in fat grafting, allowing the procedure to regain popularity.

Third, at the beginning of the twenty-first century, and in an effort to better understand the cellular basis of adipose tissue, Zuk [16, 17] and Mizuno [18] described the abundant presence of adult mesenchymal stem cells in the adipose tissue. These adipose-derived stem cells (ASCs) are an undifferentiated self-renewing cell population, capable of multilineage differentiation. Since the discovery of ASCs, the regenerative potential of adipose tissue became a hot topic in our field, opening doors for new research projects (Fig. 7.1).

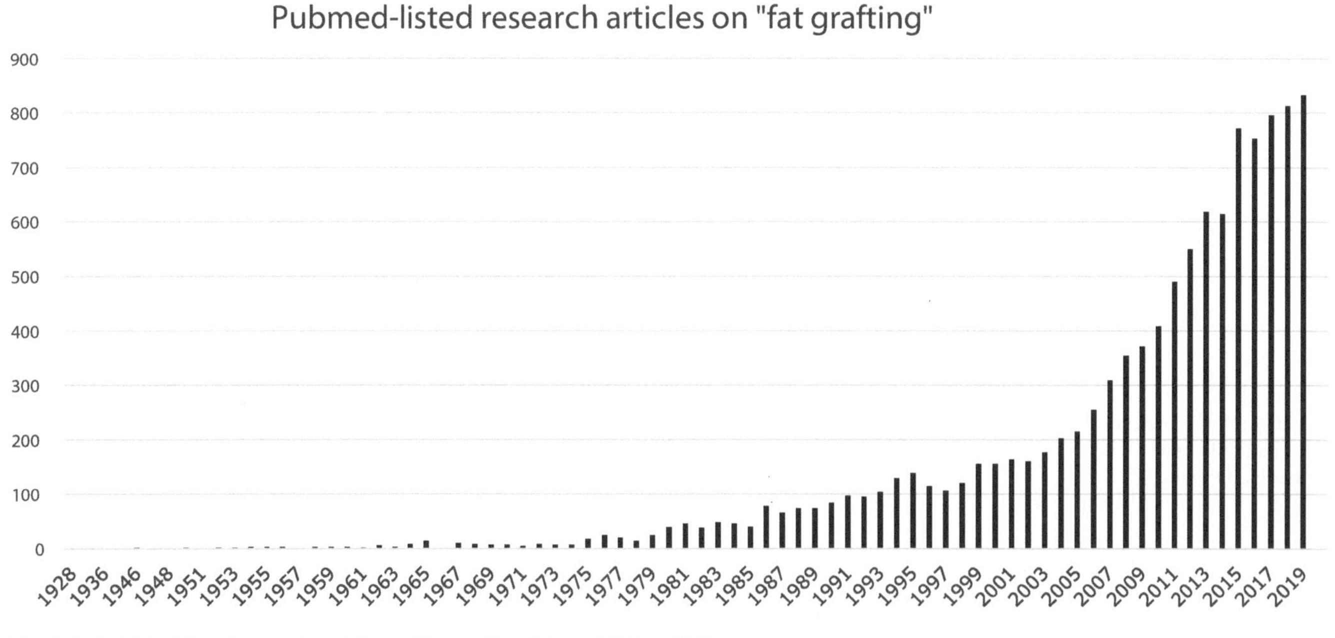

Fig. 7.1 PubMed-listed research articles on "fat grafting" from 1928 to 2019

To date, fat grafting has permeated many aspects of plastic and reconstructive surgery, as well as other medical specialties. Due to its rich composition, it has regenerative potential beyond simple volume augmentation [19–28]. Beneficial effects of fat grafting have been reported in wound healing [29, 30], peripheral nerve regeneration [31], scarring [32, 33], scleroderma [34, 35], radiotherapy damage [36], ulcerations [37], rheumatoid arthritis [38], osteoarthritis [39], dermal atrophy [40], and fibrosis [41].

7.2.3 The Limitations of Fat Grafting

Independent from the technique, and regardless of the indication, the major drawback of fat grafting remains its volume loss due to fat necrosis [42]. Small areas of non-viable adipocytes are reabsorbed or can lead to calcification, and larger zones of fat necrosis can turn into oil cysts or hard fibrosis, which can cause confusion in breast imaging for inexperienced radiologists [43–47]. Patients are informed that fat graft success is unpredictable and never equal to 100%, and that subsequent procedures are often required, but these operations lead to additional costs, donor site morbidity, hospital admissions, anesthesia, patient discomfort, and psychological burden.

One of the main indications for fat grafting is breast reconstruction. Approximately 13% of women will be diagnosed with invasive breast cancer in their lifetime according to the currently available data. The average 5-year and 10-year survival rates for female patients with invasive breast cancer have increased to 90% and 83%, respectively [48]. Thus, patients live a longer time with the consequences after the breast cancer treatment and, therefore, the quality of life of these women should not be neglected. Whether the patient requires a partial or complete reconstruction, with an implant or autologous tissue, reconstructive surgeons can choose a breast reconstruction modality which is tailored to each patient, and fat grafting often plays a role, as it offers several advantages to the outcome and patient satisfaction. By creating a stable fat graft, we can eventually find a solution for its shortcomings, i.e., the unpredictable graft resorption and multiple procedures.

The engraftment of the fat graft depends on various parameters which can all be studied separately and be clinically influenced. Technical variables consist of the tumescent solution being infiltrated prior to liposuction, the harvesting method, the manipulation of the fat, and the injection technique. Different approaches lead to mainly non-significant differences in outcomes, and there is no consensus for a universal clinical protocol [49–59]. No high level of evidence exists that support a preferred technique of infiltration, harvesting, or processing [60]. Patient-related variables that may influence graft yield and viability are recipient site properties [61], age [62, 63], smoking [64], and comorbidities.

To overcome the tissue loss and volume reduction due to fat graft resorption, overcorrection has been applied and advocated by clinicians. Overcorrection means the injection of a volume of fat that exceeds the requirement for correcting a volume depletion or contour defect. However, based on scientific evidence [65] and clinical experience [66], we know now that overgrafting does not lead to an improved survival of the grafted tissue. On the contrary, because of increased interstitial pressure, capillary perfusion can be restricted to the adipocytes, and the individual cells will be more prone to apoptosis. By consequence, overcorrecting causes complications such as necrosis and oil cysts. We cannot expect more adipose tissue to survive than what the recipient site can accommodate, and, therefore, grafting beyond the capacity of the recipient site is avoided.

7.3 The Vascular State of the Fat Graft

7.3.1 Grafted Tissues Depend on Revascularization

The most essential requirement of fat graft survival is blood flow, because we need circulation for all tissues to live. Adipose tissue has an inherent rich vascular network in which adipocytes are connected to capillary vessels. By consequence, the tissue oxygen tension is very high in the subcutaneous adipose tissue [67]. When liposuction is performed, the fragile parts of adipose tissue are aspirated, but the fibrous honeycomb-like network remains intact in the donor area. This network consists of connective tissue as well as the relatively large vessels. Thus, individual adipocytes are disrupted from their original blood supply once they enter the suction cannula. Upon injection, the adipose tissue will lack a vascular network.

Harold Gillies already mentioned in 1957 "*a (skin) graft is a piece of detached skin which is dead when you put it on and comes to life later*" [68]. By "life" and "death," Gillies meant the presence or absence of a blood vessel network. Similarly, because a fat graft lacks a blood supply at the moment of injection, it goes through ischemia. The fat graft depends initially on plasmatic imbibition from the surrounding recipient tissue to obtain its nutrients and oxygen, until adequate revascularization is established to supply metabolically active cells. Adipocytes in the graft are prone to cell death within 24 hours in this hypoxic situation, and, inevitably, a percentage of the adipocytes will die [65]. Other cells in the adipose tissue, such as bone marrow-derived mesenchymal stem cells, are less susceptible to cell death under hypoxic circumstances, which makes them a preferable cellular source for cell-based therapy in ischemic tissues [69].

7.3.2 Vascularization Determines Fat Graft Survival

The adequacy of blood perfusion early after the grafting appears to be key to the fate of the grafted fat. Yamaguchi proved in a murine experiment that the administration of an angiogenesis inhibitor led to less blood flow from the recipient tissue toward the graft, resulting in a lower graft weight and impaired adipocyte function [4]. Lu demonstrated a link between vessel density and graft retention in a murine experiment with VEGF-transduced ASCs [70, 71]. In rabbits, Karacaoglu had the best outcome when the graft is placed in the supramuscular plane, presuming this is a consequence of being a more appropriate and highly vascular recipient site, due to the musculocutaneous perforators emerging from the outer muscular surface [3]. Also Lee has shown that by preconditioning and increasing the recipient site's vascularity, an increased graft weight can be obtained [5]. Taking into account these findings, the establishment of an early blood supply is of great importance for the fat graft.

7.3.3 Vascularization Determines Adipogenesis

The vascular supply of oxygen and nutrients is essential to adipose function, including their contribution to metabolism, and their secretion of soluble factors for adipocyte survival and adipocyte regulation [72]. Not only the survival of the grafted fat but also the de novo adipogenesis is regulated by factors that drive angiogenesis. Evidence indicates that blood vessel development influences adipocytes and adipogenesis [73]. During fetal development, arteriolar differentiation precedes adipocyte development, and blood vessel extracellular matrix (ECM) differentiation precedes differentiation of adipocyte ECM [74, 75]. Administration of anti-angiogenic factors to mice decreased fat pad weights by 12–22% and decreased body weights in a dose-dependent manner [76]. Kawaguchi injected bFGF and Matrigel subcutaneously in mice for neovascularization which resulted in migration of adipose precursor cells and their differentiation into adipocytes. Their results demonstrated that enhancement of *de novo* adipogenesis was due to increased neovascularization [77]. Rophael showed how combinations of angiogenic cytokines enhanced early angiogenesis and cell infiltration which led to synergistically increased adipose tissue growth [78]. Fukumura described a reciprocal regulation of adipogenesis and angiogenesis in which inhibition of angiogenesis by blocking VEGFR-2 reduced angiogenesis but also blocked pre-adipocyte differentiation [79]. Based on these reports, an adequate blood vessel network in the fat graft can drive adipogenesis after adipocyte apoptosis.

7.3.4 The Graft's Three Layers and the Farmer's Four "S"

The importance of the vascularity has been reflected in publications by Yoshimura, who provided us with great insight on how the fat graft can be subdivided in three areas from the periphery toward the graft center: surviving zone, regenerative layer, necrotic core [80]. The fate of the individual adipocytes depends on their location in the graft:

1. Outer (surviving) layer: Cells survive the grafting process because of the proximity of the surrounding ingrowing capillaries.
2. Intermediate (regenerative) layer: Adipocytes suffer from ischemia and will be phagocytized. ASCs repair the ischemic and injured tissue, together with stem and progenitor cells, that originate from the bone marrow and migrate in response to ischemia-related paracrine signaling. ASCs are activated by adjacent adipocytes undergoing apoptosis to differentiate to adipocytes, and contribute to the cell replacement, if the microenvironment, including the capillary network and oxygen tension, is improved within 3 days after the grafting. If vessel ingrowth from surrounding host tissue does not reach this area within 3 days, the ASCs will also degenerate, thereby aborting the regenerative process. In that case, this regenerative layer becomes part of the necrotic area [80].
3. Innermost (necrotic) layer: The vasculogenic delay impairs the survival of all cell types which undergo destruction. No adipogenesis or regeneration occurs in this necrotic core. The inner layer will undergo resorption, fibrogenesis, and cicatrization, or can develop into sand-like calcifications or oil cysts, which can lead to lumps with never-ending inflammation and tenderness [66, 67, 81]. Thus, this model emphasizes the early and abundant vascularization for successful retention of fat grafts.

A demarcation in the fat graft, between the surviving outer layer and the regenerating middle layer, becomes rapidly apparent. Based on immunohistochemistry experiments, viable adipocytes can be distinguished from the intermediate regenerative zone at 1 week [67, 80]. Two to four weeks after the grafting process, the demarcation between the regenerative zone and the necrotic center becomes clear [67, 80]. At 2 weeks post-grafting, immature adipocytes appear between moribund adipocytes [67, 80]. After the inflammatory and regenerative phase, a stable graft results, with capillaries consisting of endothelial cells, which are derived equally from the graft and the host bone marrow [82].

Based on experimental work, the regenerative zone has a 600- to 1 200-μm thickness, whereas the radius of the superficial surviving zone measures maximum 300 μm, which became approximate numbers to remember for clinicians [67, 80]. This indicates that a close relationship should be respected between adipocytes and surrounding host tissue in order to avoid necrosis, and stresses the biological violation of injecting fat particles larger than 2 mm. Khouri had launched the analogy with how a farmer plants seeds in a field and has to respect the four "S": *soil*, *sowing technique*, *seeds*, and *support* [83, 84]. Surgical craftsmanship aims for a close contact between fat particles and the recipient tissue. Through the diffuse fan-like pattern injection of small noodles of adipose tissue without overgrafting, we optimize the contact and potential vascular ingrowth in the graft. However, even when the fat grafting is properly executed (*sowing technique*), when the recipient site is well-vascularized (*soil*: absence of radiotherapy sequelae, peripheral vascular disease, diabetes mellitus-induced micro-angiopathy, tobacco use, etc.), and when the postoperative care is optimal (*support*: avoidance of trauma, shear force, high pressure, or tight compression), there remains room for technical improvement. Maximizing the potential of the fat graft can be done through cell enrichment (*seeds*). On the other hand, overloading a fat graft with stem cells while neglecting the other three "S" would add little benefit to the outcome, similar to adding horsepower to a car stuck in traffic.

7.4 Endothelial Progenitor Cells (EPCs)

7.4.1 EPCs Target Ischemic Repair

Endothelial progenitor cells (EPCs) are precursors of endothelial cells, the squamous cells which line our circulatory system, from heart and aorta to peripheral capillaries. EPCs were first discovered by Asahara [85]. After isolating $CD34^+$ mononuclear cells (MNCs) from peripheral blood, he noticed their fast proliferation, ability to form tubular structures and CD31 expression. He demonstrated how these cells homed to ischemic limbs after distant injection in the murine tail vein, as they "sensed" the requirement of the ischemic tissues. The integration of the labeled $CD34^+$ MNCs in vascular structures was confirmed histologically. EPCs were shown to contribute to postnatal physiological and pathological endothelial repair and neovascularization. These findings opened the door to the concept of therapeutic vasculogenesis.

7.4.2 Role of EPCs in the Fat Graft

In conventional fat grafting, EPCs can originate from the graft itself, from the surrounding tissues, or from the bone marrow [86]. Doi described the cellular origin of regenerated adipose tissue and vessels in the graft. Based on histologic and flow cytometric analyses, nearly half of vascular endothelial cells in the grafts originated from host bone marrow, and a chimeric capillary network was seen constructed with graft-derived and host-derived cells [82]. This has two interesting consequences:

- Graft-derived vascular structures are involved in revascularization, but EPCs mobilized from bone marrow make an equal contribution. As in situations of ischemia, EPCs will be released from the bone marrow and migrate toward ischemic sites where it can differentiate into endothelial cells and incorporate in newly formed vessels [87].
- A rapidly established vascular network toward and in the fat graft is necessary to deliver these EPCs as well as ASCs.

The creation of a vascular network takes 3–7 days [88–92], and even longer to reach the innermost core of the graft, where they might not even reach. The addition of EPCs to the fat graft is a legitimate manipulation because we skip the transportation step, and we deliver building blocks for vessels. This is especially useful since conventional fat grafts have relatively low stem cell numbers compared to excised fat [93, 94], because the viable endothelial cell component in lipoaspirates decreases quickly as opposed to ASCs [65], and because a positive significant correlation exists between the innate stem cell percentage in adipose tissue and graft retention [95].

7.4.3 Previous Experiments on EPC-Enriched Fat Grafting

In 2006, Yi conducted an animal experiment, in which human EPCs were mixed with murine adipose tissue and grafted in mice. A dual pro-vasculogenic and anti-fibrotic effect was attributed to the EPCs. The EPCs originated from umbilical cord blood $CD133^+$ cells from a different human donor than the adipose tissue donor, and there was no proof of their integration into vascular structures. A clinical application based on this study would be impractical because human umbilical cord blood is not readily available, and their application would not be autologous [89].

Hamed also added EPCs to fat grafts and showed how EPCs are required for fat graft vascularization, graft retention, and reduced adipocyte apoptosis [86]. In another experiment, Hamed has demonstrated how the induction of EPCs in the bone marrow and their migration toward the graft contribute to graft retention [96].

Xu indirectly used EPCs as an enrichment modality in fat grafting by adding CXCR4-transfected ASCs to the fat graft. CXCR4 is a membrane receptor involved in the recruitment of stem cells such as EPCs, neovascularization, and cell proliferation. In his animal experiment, Xu demonstrated a higher capillary density in nude mice that received the grafts mixed with CXCR4-transfected ASCs [97].

More recently, Pallua has described *lipoconcentrate*, which is a processed lipoaspirate and obtained through a modification of Tonnard and Verpaele's *nanofat* protocol [98, 99]. By including an extra centrifugation step, the *lipoconcentrate* contains a higher amount of EPCs and it is currently being used clinically [33].

7.5 Fat Grafting Enriched with QQ-cultured MNCs

7.5.1 Development of the QQ Culture

The limitation of using EPCs hinges on its sparse presence in peripheral blood [100]. With 70–210 EPCs per mL, EPCs account for only 0.002% of the peripheral blood mononuclear cells (PBMNC), or for 0.4% of the total peripheral blood $CD34^{+}$ cell population [101]. EPCs are characterized by even lower numbers, and impaired proliferation, adhesion, and vascular incorporation, in elderly patients, patients with diabetes, hyperlipidemia, peripheral vascular disease and ischemia, cardiac and lung disease, endocrine, gastrointestinal, and hematologic disorders, musculoskeletal problems, alcohol or drug abuse, tobacco use, the use of immunosuppressive drugs and chemotherapy [102–110]. This quantitative and qualitative impairment of EPCs results in a lack of clinical value for autologous EPC therapy, and indicates the need to pre-treat these cells.

The Quality and Quantity culture (QQ) overcomes this problem, by increasing cell counts and ameliorating the functional deficits of EPCs to a highly vasculogenic and regenerative level. The QQ culture is a 1-week culture which is based on an optimal cytokine and growth factor mix consisting of stem cell factor (SCF), thrombopoietin, vascular endothelial growth factor (VEGF), interleukin-6 (IL-6), and Flt-3 ligand.

In *in vitro* experiments in 2012, Masuda discovered both a quantitative increase as well as a qualitative enhancement (differentiation of EPCs into endothelial cell lineage as shown by an increase in definitive EPC colony-forming units; endothelial cell surface marker expression of VEGFR-2, CD146, and vWF; and VEGF and HGF secretion) of $CD133^{+}$ EPCs that underwent QQ culture. *In vivo* was demonstrated how the intramyocardially injected QQ-cultured cells were able to preserve more left ventricle function and contractility in a myocardial infarction model, and more endothelial cells and cardiomyocytes were discovered on immunohistochemistry [106, 111].

In 2014, Masuda obtained QQ-cultured mononuclear cells (MNC-QQ) enriched in EPCs from unfractionated mononuclear cells (MNC) [112]. The MNC-QQ formed blood vessels in an ischemic hind limb mouse model. Anti-inflammatory cytokines (IL-10) and genes related to angiogenesis and tissue regeneration (VEGF, IGF-1, MMP-2, MMP-9) were up-regulated in MNC-QQ.

To date, the QQ culture is the only *ex vivo* culture system with the ability to expand and improve the vasculogenic potential of EPCs for clinical usage [106]. The *in vitro* tendency of MNC-QQ to develop into an EPC population with pro-vasculogenic characteristics as cell adhesion, tube-like structure formation, and endothelial lineage differentiation was demonstrated through EPC colony-forming assay [113–118]. In EPC culture assays, MNC-QQ were shown to be adherent cells that fulfill the three requirements of being viable early EPCs in a more differentiated phase compared to the colony-forming EPCs (DiI-acLDL uptake, UEA-1 binding, and DAPI staining) [116, 119, 120]. However, describing the QQ-cultured MNCs as a pure EPC population would be incomplete: the QQ culture also induces a phenotypical switch from classically activated pro-inflammatory M1 macrophages ($CCR2^{+}$) to the alternatively activated anti-inflammatory M2 macrophages ($CD206^{+}$), and activates helper T-lymphocytes [107, 113, 120–122]. The higher angiogenic potential of M2 macrophages can be explained by their FGF and PlGF signaling [123], because of their PDGF-BB and MMP9 signaling and their effect on anastomosing sprouting endothelial cells [124]. Phipps grafted adipose tissue with M2 macrophages and noticed an improved graft retention which was attributable to vascular development [125]. Early depletion of macrophages resulted in incompetent angiogenesis and eventually a poor retention rate, whereas up-regulated macrophages allowed better angiogenesis and survival [126].

Therapeutic vasculogenesis with QQ-cultured cells has been shown more efficient than conventional modalities for wound healing [107, 114], ischemic stroke [127], myocardial infarction [122], and acute kidney injury [121].

7.5.2 Current Scientific Findings on Fat Grafting Enriched with QQ-cultured MNCs

In 2019, the addition of a homogenous QQ-cultured EPC-population to fat grafts was studied for the first time in a murine recipient model [128]. Our main finding was the significantly increased vessel density in these fat grafts compared to conventional (non-enriched) fat grafts, fat grafts enriched with non-cultured EPC, and ASC-enriched fat grafts. Five weeks post-grafting, a significantly denser vessel network was created in grafts mixed with QQ-cultured cells. Another 5 weeks later, it was seen saw how the amount of vessels had further increased, which suggested that the previously formed vessels were being maintained while new vessels were added. Additionally, the rate of fibrotic tissue in these fat grafts was significantly lower, underlining the antifibrotic effects of the QQ-culture. This was attributed to a dual function. Due to capillary ingrowth, more adipocytes survive the grafting process, and, by consequence, less transformation to fibrotic tissue was observed. We also demonstrated a down-regulated TGF-β/Smad3 pathway with a lower expression of the pro-fibrotic COL1A1 gene in the QQ-cultured cells. Earlier, Masuda had described that MNC-QQ potentiated protective effects against fibrosis [112]. Similarly, MNC-QQ decreased fibrosis in acute kidney injury [121] and myocardial infarction [122].

The pro-vasculogenic properties of QQ-cultured cells on fat grafting were confirmed in a second study involving the use of human MNC-QQ, as a source of EPCs and M2 macrophages [113]. Human lipoaspirates were enriched with the MNC-QQ, or with freshly harvested non-cultured MNC, or with Stromal Vascular Fraction (SVF), and grafted in nude mice (Fig. 7.2). In this study, it was demonstrated that the MNC-QQ contribute to vessel formation because of the histologically determined quantitative increase in vessel density. Grafts supplemented with MNC-QQ had a significantly higher vessel density (96.6 ± 6.5/mm^2, $p < 0.05$) than the control (70.4 ± 5.6/mm^2). In contrast, adding PBMNC or SVF to the fat grafts did not lead to an improved vessel network (74.4 ± 7.0/mm^2, $p > 0.05$; and 75.0 ± 7.0/mm^2, $p > 0.05$, respectively). Immunohistochemistry revealed that MNC-QQ were directly involved in vessel build-up by differentiating into CD31$^+$ cells and integrating into the vessel lining (Fig. 7.3). An additional in vivo paracrine effect of the QQ-cultured cells was assumed: grafts enriched with MNC-QQ secreted significantly more VEGF and VEGFR-1 [129–131]. An average 90% weight persistence after 7 weeks was found in the MNC-QQ group and SVF group, which was significantly higher than grafts in the control group (70.4 ± 6.3%). The high graft survival was considered a consequence of the rich vascular network that is able to supply a larger part of the adipocytes. Furthermore, through these vascular interconnections, host-derived ASCs and EPCs can migrate more efficiently to injured or ischemic tissues inside the fat graft [72, 77–79]. A dense vessel network also creates

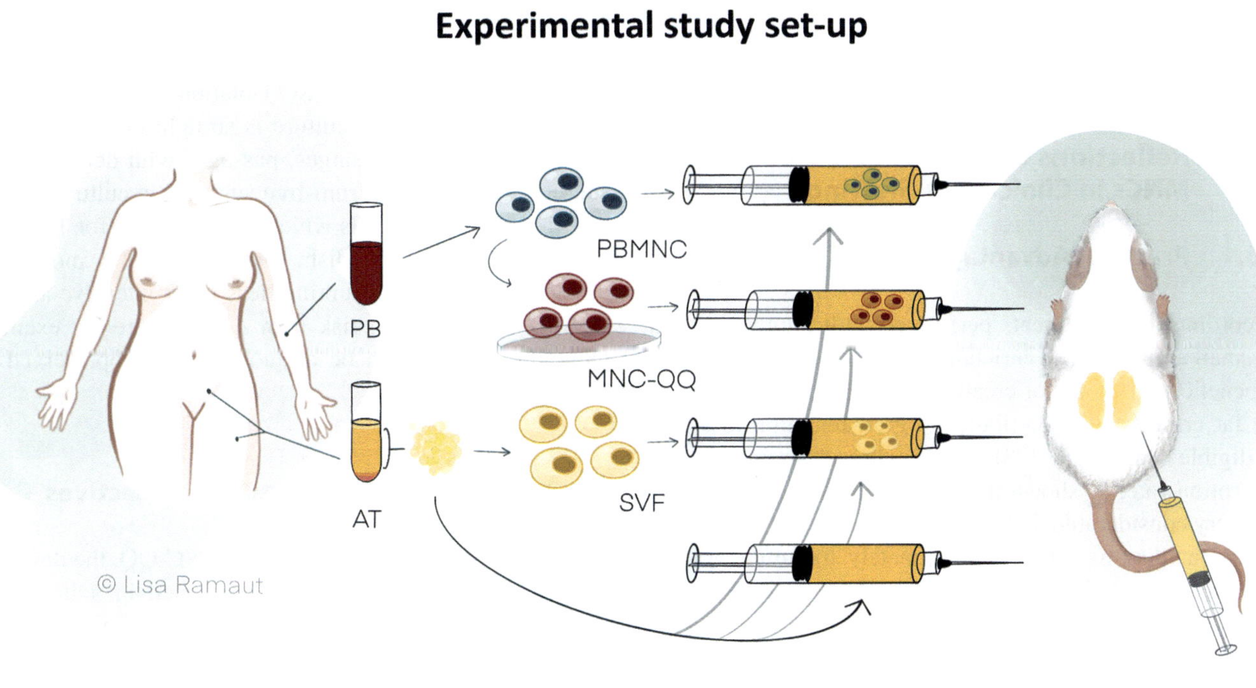

Fig. 7.2 Healthy female patients underwent a peripheral blood draw for PBMNC isolation and subsequent QQ culture in order to obtain MNC-QQ. Liposuction was performed to collect adipose tissue and to isolate SVF. Recipient nude mice (BALB/cAJcl-nu/nu) were randomly divided in four groups. Each recipient mouse received two dorsal fat grafts and each graft consisted of 0.25 g of adipose tissue. Grafts were supplemented with 1 × 10^6 freshly harvested (non-cultured) MNCs (blue), with 1 × 10^6 QQ-cultured MNCs (red), with 1 × 10^6 SVF cells (yellow), or with PBS-lacking cells (control group). 16 grafts were created in each group. Abbreviations: *PB* Peripheral blood, *PBMNC* Peripheral blood-derived mononuclear cells, *MNC-QQ* Quality and Quantity-cultured mononuclear cells, *AT* Adipose tissue, *SVF* Stromal vascular fraction, *PBS* Phosphate-buffered saline. Original drawing by Lisa Ramaut, MD. Reprinted with permission from *Geeroms M, Fujimura S, Aiba E, Orgun D, Arita K, Kitamura R, et al. Quality and Quantity-Cultured Human Mononuclear Cells Improve the Human Fat Graft Vascularization and Survival in an In Vivo Murine Experimental Model. Plast Reconstr Surg. 2020 Nov 2. doi: 10.1097/PRS.0000000000007580* [113]

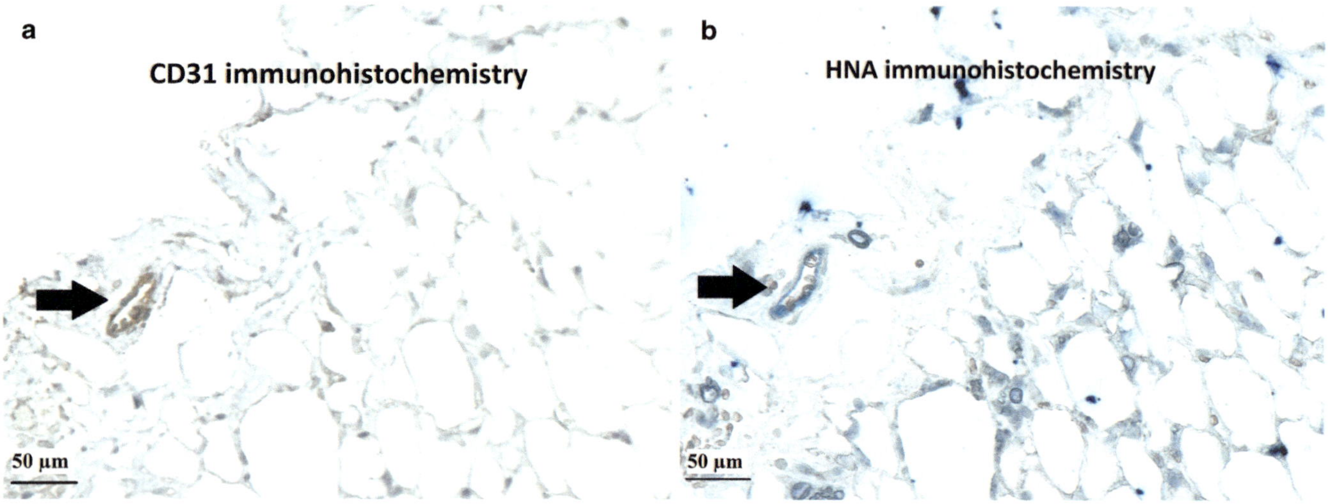

Fig. 7.3 In vivo vasculogenic effect of MNC-QQ. The QQ-cultured MNCs are able to survive in the grafted fat and contribute to the vascular build-up. This was demonstrated by the presence of HNA$^+$ cells (i.e., human-derived material) in a CD31$^+$ luminal structure (i.e., vessel with endothelial lining) in a fat graft consisting of murine adipose tissue and human QQ-cultured MNCs. (**a**) CD31 immunohistochemistry identifies vascular structures (black arrow) in the explanted fat graft. (**b**) HNA immunohistochemistry specifically detects structures derived from human origin (black arrow). Together, these two consecutive sections demonstrate the CD31$^+$ phenotype of the grafted MNC-QQ. Scale bar = 50 µm. *Reprinted with permission from Geeroms M, Fujimura S, Aiba E, Orgun D, Arita K, Kitamura R, et al. Quality and Quantity-Cultured Human Mononuclear Cells Improve the Human Fat Graft Vascularization and Survival in an In Vivo Murine Experimental Model. Plast Reconstr Surg. 2020 Nov 2. doi: 10.1097/PRS.0000000000007580* [113]

a favorable environment for adipogenesis in which ASCs develop into next-generation adipocytes and replace moribund fat cells [72, 77–79].

7.6 Reflections on the Use of QQ-cultured MNCs in Clinical Fat Grafting

7.6.1 Practical Advantages

According to experiments performed with human material on small animals, graft enrichment with QQ-cultured MNCs is beneficial in terms of creating a vascular network inside the fat graft, which contributes to graft retention in a non-negligible way [113, 128]. In the clinical setting, a 30% resorption rate, as shown in control samples, is a common and very considerable tissue loss which leads very frequently to patient dissatisfaction and ultimately to future grafting sessions. In a clinical situation, an approximately 20% gain in graft retention (90.0% compared to 70.4%) can make a tremendous difference for a patient and can reduce the need for secondary corrections [113].

The QQ culture has the additional benefit of not containing animal products and being a serum-free culture, which carries less risk of transmission of bovine spongiform encephalopathy, less risk of antibody-based responses which would lead to serum sickness, and it does not contain xenoproteins.

Patient admission is not necessary for someone who requires treatment with QQ-cultured cells. The outpatient peripheral blood draw can be routinely done during the preoperative anesthesia consultation, and is less invasive and more practical than harvesting bone marrow content, or lipoaspirates for SVF or ASC isolation. The production of MNC-QQ through QQ culture is straightforward and does not require medium changes, passages with detachment or sub-culturing. It is a serum-free suspension culture consisting of non-adherent cells which are incubated for 1 week in the same cell culture dish, and is therefore more cost-efficient, less time-consuming, less labor-intensive and carries a lower infection risk than ASC culture for example. QQ-cultured cells can be aliquoted and cryopreserved for future use [106].

7.6.2 Limitations and Future Perspectives

To validate the clinical potential of MNC-QQ, the described findings [113, 128] should be confirmed in a larger and immunocompetent animal model with a range of medium-sized to large fat particles, before the results can be translated to human subjects. Even though the described immunodeficient animal model is considered a standard recipient for the study of fat grafting, the method is different from the clinically recommended technique of injecting small noodles of adipose tissue in an adipose tissue matrix, with multiple passes in multiple tunnels. The investigation in a pig recipient model would be useful given its similarity to human physiology. Porcine tissue structures and turn-over cycles are similar to those of humans, and the swine itself is an appropriate mass to model that of an adult human [107]. The subcutaneous adipose tissue in pigs consists of two fibrofatty layers, comparable to the superficial and deep fat separated by a superficial fascia in humans, making it a better research model [132].

A fat graft consists of regenerating and metabolically active tissue. Suggested by the in vivo immunohistochemistry experiments by Kato, replacement of dying adipocytes in the regenerative layer by differentiating ASCs peaks at 4 weeks after grafting, and is finishing between the 8th and 12th week after grafting [80]. This underlines the importance of studying whether MNC-QQ-induced neovascularization, fat graft retention, and tissue quality are stable over a longer period of time than 10 weeks [128] and 7 weeks [113]. For example, Luo managed to augment the graft vessel density with endothelial cells in his experiments, but this decreased again afterward, indicating that a fat graft is a dynamic biological entity [133].

Apart from exploring the benefit of MNC-QQ-enriched fat grafting on pigs with lengthy follow-up, another important study topic is the dosing. Dose efficacy is essential in medicine and cell therapy. Regarding the cell number for fat graft enrichment, a high variation exists among the available literature, from relatively low to supra-physiologic cell numbers. The question of the ideal cell dose of MNC-QQ that needs to be added to a milliliter of fat for optimal effect remains unanswered. At this moment, it is known which concentration of QQ-cultured MNC has an effect on vascularization and on graft retention, i.e., 1×10^6 MNC-QQ per 0.25 g of adipose tissue [113]. It remains to be determined whether a lower dose is equally effective, whether a dose-dependent response of the fat graft to the cell enrichment exists, and whether a higher dose is potentially counterproductive. It cannot be automatically assumed that overloading the fat graft with QQ-cultured MNC would correlate with a higher vascularization, more retention, and less fibrosis. In enhancing the fat graft with certain cell types or cytokines, a linear relationship is not always observed between the concentration of the enriching substance and the results. Kakudo discovered in his SVF-enriched fat grafts that higher concentrations of the enriching cell source led to more cysts and fibrosis [134]. Cai saw how granulocyte colony-stimulating factor improved fat graft survival and induced angiogenesis. However, in a higher dose, it had an adverse effect [135].

At this stage, MNC-QQ, or other cell-enrichment methods, cannot guarantee immediate perfusion to the individual adipocytes in the center of large grafts [65], which should discourage clinicians from injecting large-particle grafts. We always have to keep in mind the analogy with how the farmer plants the seeds in his field [83, 84]. Even though results indicate the beneficial effect of enhancing the fat graft by addition of QQ-cultured MNCs (*seeds*), the recipient site (*soil*), the surgical craftsmanship (*sowing technique*), and postoperative care (*support*) remain equally important, and the weakest link will influence the outcome.

7.6.3 Oncological Safety

Adipocytes work like glandular cells and secrete cytokines (adipokines). These cytokines have local, metabolic, and cardiovascular effects. Because of the high global prevalence of breast cancer, the adipokines, and abundant mesenchymal stem cells in adipose tissue, fat grafting to the breast was fraught with fear that it might stimulate *dormant* or potentially (pre)malignant cells. Our goal, whether we are oncologists or plastic surgeons, remains *primum non nocere*, first do not harm. Promoting tumorigenesis or cancerogenesis in a cancer patient, who underwent surgery, chemotherapy, and/or radiotherapy, would be the worst outcome possible to fat grafting.

In 1987, the American Society of Plastic Surgeons (ASPS) issued a report that condemned fat grafting to the breast, because of concerns that the fat graft could cause detectable breast tumors to go undiscovered [136]. Later, and based on the confirmation that fat grafting does not interfere with the radiographic follow-up of the breast [44–47, 137], the ASPS built a Fat Graft Task Force [138]. The Task Force concluded that fat grafting could be considered for breast augmentation and corrections, and that there was no association between fat grafting and higher rates of malignancy, even though there was a paucity of studies in this field and further investigations were demanded.

The interactions between adipocytes and adipokines have been studied and their tumor promoting effect have been determined *in vitro* or in small-animal experiments [139–149]. Conclusions based on experimental studies differ, however, from clinical studies in the susceptible breast after an oncologic treatment. Strong evidence from large case-controlled studies exists nowadays that (non-enriched) fat grafting in treated breast cancer patients is not associated with an increased risk of locoregional recurrence of breast cancer, systemic recurrence, or a second breast cancer [150–160]. However, some studies conclude confusingly by stating that future studies are still recommended to investigate its safety. Therefore, the safety of fat grafting should be an ongoing area of research by multiple independent research groups, in long-term follow-up prospective randomized controlled trials.

While stem cell enrichment is exciting and promising, concerns were also raised surrounding the oncological potential of the cells. By enriching fat grafts with cellular sources, the theoretical risk of promoting the proliferation of residual cancer cells can be increased, and controversy remained. A possible disadvantage of MNC-QQ in fat grafting for reconstruction after cancer treatment follows from the actual advantage of this cell population. Although the MNC-QQ's vasculogenic and regenerative qualities make it valuable in reconstructive procedures, neoplastic processes also depend on these mechanisms. This potentially tumorigenic effect of the highly vasculogenic QQ-cultured cells on adipose tissue, breast glandular tissue, malignant cells, and tumor stroma should be individually analyzed. It will be critical to exclude a role of QQ-cultured cells in tumor biology and breast cancer recurrence, before grafting into humans.

7.7 Comparison to Other Enrichment Techniques

7.7.1 Cell Enrichment of the Fat Graft with Stromal Vascular Fraction (SVF)

Subcutaneous adipose tissue consists of several cell types, and adipocytes account for only 20% of the total cell number [161]. The SVF consists of the cellular components of adipose tissue without the mature adipocytes. It is a heterogeneous group of cells such as EPCs, endothelial cells, vascular smooth cells, pericytes, macrophages, neutrophils, fibroblasts, and mesenchymal stem cells or ASCs [161, 162]. To obtain SVF, the freshly harvested lipoaspirate has to be processed enzymatically through collagenase digestion or mechanically through disruption. Cells expressing a CD45$^-$ CD235a$^-$ CD31$^-$ CD34$^+$ surface marker profile respond to the IFATS criteria for SVF and are considered the stromal cells of interest [163].

The main advantage of SVF compared to ASCs is that their harvesting and clinical use in the patient is possible in one surgical procedure. Compared to ASCs, there is a lower risk for contamination and less associated costs. SVF does not require cell culture, which makes it more practical from a legal perspective than ASCs. Additionally, to use ASCs, we would need our patients to come to the clinic 2 to 3 weeks before their operation, so we could harvest adipose tissue from their body in order to start the cell culture. Recent research also pointed out that there is no preference between ASCs and SVF for cell-enriched fat grafting in terms of graft retention [164].

With the aim to improve fat grafting results, cell-assisted lipotransfer (CAL) has been developed by the Yoshimura lab [94, 165]. In his technique, adipose tissue is enriched with SVF prior to the grafting. The initial results with SVF-enriched fat grafting were promising. Matsumoto demonstrated a higher survival of SVF-enriched fat grafts compared to conventional fat grafts, as well as advantageous histological changes, and a prominent microvasculature, although these data were not quantified [94]. In other studies and in a recent review, however, no convincing evidence could be attributed to SVF-enriched fat grafting in human subjects [166–168]. In general, enriched fat grafting is relevant if it decreases the number of necessary additional procedures to achieve the desired result. Based on reviews by Zhou and Laloze, SVF enrichment does not prevent multiple operations, and, secondly, does not reverse the ischemic state in the fat graft [168, 169].

However, reviews about cell-enriched fat grafting often have low value as they compare studies with different modalities in different settings. Therefore, graft enrichment with QQ-cultured cells was immediately compared with SVF in identical situations. Regarding graft retention, SVF-enriched grafts showed the same pattern over time as the MNC-QQ group [113]. However, supplementation with SVF did not increase vessel density, and no SVF cells were able to differentiate to CD31$^+$ endothelial cells or able to integrate in the vascular structures. Also, no increase in the expression of VEGF or other angiogenic cytokines was observed. These data suggest that SVF works through a different mechanism compared to MNC-QQ, or multiple mechanisms since it consists of a mix of cells. At the moment, it is unclear what the role of each cell type in SVF is. SVF contains leukocytes which might have unfavorable effects on the fat survival or quality. The responsible cell population in the heterogeneous SVF sample might be ASCs, which account for 1 out of 30 SVF cells [170]. A standard or optimized SVF isolation procedure does not exist, even though there is a plethora of devices promoting the mechanical SVF isolation.

A practical disadvantage to the use of SVF, compared to MNC-QQ, is the additional intra-operative time for the mechanical (or enzymatic) isolation of SVF from the lipoaspirate. In addition, a high volume of the available adipose tissue is sacrificed for the isolation of SVF, leading to more donor site morbidity, and this fat cannot be "recycled" for injection. Especially in thin patients requiring large volumes, this can lead to a reduced volume available for grafting. To further determine possible differences in outcome, SVF and MNC-QQ can be compared after a longer follow-up time or can be mixed and added to the fat graft to determine a potential synergistic effect: because they appear to be working through different mechanisms, a better result could be achieved.

7.7.2 Cell Enrichment of the Fat Graft with Adipose-derived Stem Cells (ASCs)

A widely studied cell therapy for graft enhancement are adipose-derived stem cells (ASCs) [17]. ASCs are present surrounding blood vessels and embedded in the connective tissue of fat. The reported percentage of ASCs in SVF is approximately 3% [170]. However, the abundance of SVF within the adipose tissue and the ease in terms of harvesting the SVF make adipose tissue a reliable source for ASCs.

In order to transform a heterogeneous SVF population into a homogeneous ASCs population, cell culture has to be done starting from freshly harvested adipose tissue and isolated SVF. The SVF are plated into culture, and the ASCs can expand and become purified after each passage based on selection by adherence to the plastic culture dishes. Other cells are washed out of the culture at each culture passage. ASCs can be further characterized through their fibroblast-like phenotype, their mesenchymal stem cell surface marker profile ($CD44^+$, $CD73^+$, $CD90^+$, $CD105^+$, $CD36^+$, $CD31^-$, $CD45^-$, $CD106^-$), and their potential to differentiate into adipocytes, osteoblasts, chondrocytes, skeletal and cardiac muscle cells [171].

ASC enrichment is based on the host replacement theory which involves a dynamic remodeling of the fat graft. Specifically, the regenerative zone can become repopulated with ASCs differentiating into adipocytes, if adequate vascularization is established in a timely manner [65, 88].

Compared to SVF applications, ASCs are subject to more regulatory hurdles. A GMP-approved (good manufacturing practice) set-up for clinical cell isolation has to be established. The culturing of ASCs is time-consuming which is cumbersome for the patient, as he or she has to undergo two operations. A minor liposuction 2 to 3 weeks before the main operation is necessary, to isolate the SVF, and expand the ASCs. During the second operation, liposuction will be performed for fat harvesting, followed by ASC-enriched fat grafting. Because of these reasons, cell enrichment of the fat graft with ASCs is expensive.

In 2013, Lancet published a study by Kølle who conducted a clinical trial in which patients underwent bolus fat injection in the arm, with or without culture-expanded ASCs, followed by excision [172]. A remarkable increase in fat graft retention was noted in favor of the ASC-enriched fat grafts: 81% vs 16%. However, 7 years later, this effect of ASCs on fat grafting has not been confirmed in other studies [166] and patient satisfaction is not different between ASC-enriched and conventional fat grafting [173]. Interestingly, in Kølle's experiment, no increase in capillary density was noted, despite a fivefold increase in graft retention, which suggests the main contribution of ASCs involves adipogenesis [172]. ASC-enriched fat grafting is promising but continued development is necessary to obtain consistent results [166, 174]. Despite their release of pro-angiogenic cytokines [175–177] and in vitro endothelial differentiation [178], *in vivo* experiments with ASCs emphasize their mainly paracrine effect on vasculogenesis [177, 179], or suggest non-vasculogenic mechanisms by which ASCs enhance grafting [172]. This indicates that improvements can still be achieved through promoting the graft vascularization.

In the described experiment, ASCs were tested using the same model as QQ-cultured EPCs. A closer look teaches us that QQ-cultured EPCs were superior to ASCs regarding vasculogenic and anti-fibrotic properties. Together, QQ-cultured EPCs and ASCs had a stronger effect on vascularization [128]. QQ-cultured EPC and ASC had similar influence on graft retention, adipose tissue integrity, and inflammation. When adding ASCs to the fat graft, the anti-fibrotic effect of QQ-cultured cells was diminished, explaining the increased fibrosis in these grafts despite the induced vascular response. This is in accordance with a previous study in which adipose-derived regenerative cells increased fibrosis development in a dose-dependent manner [134].

7.8 Conclusion

Fat grafting has become an essential asset to our daily practice in plastic surgery. The combined effort of researchers and clinicians in this rapidly expanding field and stem cell research, especially over the last 20 years, has led to a refinement of existing techniques and has resulted in considerable progress. Due to its broad range of clinical applications and the interest in the regenerative potential of adipose tissue, a large number of commercialized products have been developed to facilitate the harvesting of adipocytes, to enhance their yield or viability, to isolate progenitor cells for enrichment, to deliver the fat graft, etc. Manufacturers claim to provide the best results but evidence and unbiased experiments are often lacking. This underlines the importance of independent basic research in fat grafting.

The goal of this chapter is to further enhance the outcome of the fat grafting process by understanding the vascular network inside the grafted adipose tissue. A pre-clinical innovative tool is described with the aim to increase the vascularity of the fat graft. The Quality and Quantity (QQ) culture has been developed to convert heterogeneous MNCs from the peripheral blood into a highly vasculogenic EPC-containing cell solution for therapeutic vasculogenesis. The addition of QQ-cultured human MNCs has a beneficial effect on human fat grafting in nude mice. Through VEGF signaling and through differentiation into $CD31^+$ luminal structures, MNC-QQ actively contribute to the revascularization of the fat graft, which leads to higher grafting rate. These pre-clinical data are important for future applications in different fields.

References

1. Neuber G. Über die Wiederanheilung vollständig vom Körper getrennter, die ganze Fettschicht enthaltender Hautstücke. Zbl f Chir. 1893;30:16–7.
2. Oranges CM, Striebel J, Tremp M, Madduri S, Kalbermatten DF, Harder Y, et al. The preparation of the recipient site in fat grafting: a comprehensive review of the preclinical evidence. Plast Reconstr Surg. 2019;143(4):1099–107.
3. Karacaoglu E, Kizilkaya E, Cermik H, Zienowicz R. The role of recipient sites in fat-graft survival: experimental study. Ann Plast Surg. 2005;55(1):63–8; Discussion 8.
4. Yamaguchi M, Matsumoto F, Bujo H, Shibasaki M, Takahashi K, Yoshimoto S, et al. Revascularization determines volume reten-

tion and gene expression by fat grafts in mice. Exp Biol Med (Maywood). 2005;230(10):742–8.
5. Lee JW, Han YS, Kim SR, Kim HK, Kim H, Park JH. A rabbit model of fat graft recipient site preconditioning using external negative pressure. Arch Plast Surg. 2015;42(2):150–8.
6. Passot R. Chirurgie Esthétique. Paris: Doin; 1931.
7. Asken S. Microliposuction and autologous fat transplantation for aesthetic enhancement of the aging face. J Dermatol Surg Oncol. 1990;16(10):965–72.
8. Czerny V. Drei plastische Operationen. III. Plastischer Ersatz der Brustdrüse durch ein Lipom. Arch F Klin Chir. 1895;50:544–50.
9. Holländer E. Über einen Fall von fortschreitenden Schwund des Fettgewebes und seinen kosmetischen Ersatz durch Menschenfett. Münch Med Wochenschr. 1910;57:1794–5.
10. Mazzola RF, Mazzola IC. History of fat grafting: from ram fat to stem cells. Clin Plast Surg. 2015;42(2):147–53.
11. Peer LA. Transplantation of tissues. Baltimore (MD): Williams & Wilkins; 1955.
12. Field LM. Lipo-suction surgery: a review. Dermatol Surg. 1984;10(7):530–8.
13. Illouz Y-G. The fat cell "graft": a new technique to fill depressions. Plast Reconstr Surg. 1986;78(1):122.
14. Chajchir A, Benzaquen I. Fat-grafting injection for soft-tissue augmentation. Plast Reconstr Surg. 1989;84(6):921–34; Discussion 35.
15. Coleman SR. Facial recontouring with lipostructure. Clin Plast Surg. 1997;24(2):347–67.
16. Zuk PA, Zhu M, Mizuno H, Huang J, Futrell JW, Katz AJ, et al. Multilineage cells from human adipose tissue: implications for cell-based therapies. Tissue Eng. 2001;7(2):211–28.
17. Zuk PA, Zhu M, Ashjian P, De Ugarte DA, Huang JI, Mizuno H, et al. Human adipose tissue is a source of multipotent stem cells. Mol Biol Cell. 2002;13(12):4279–95.
18. Mizuno H, Zuk PA, Zhu M, Lorenz HP, Benhaim P, Hedrick MH. Myogenic differentiation by human processed lipoaspirate cells. Plast Reconstr Surg. 2002;109(1):199–209; discussion 10-1.
19. The American Society for Aesthetic Plastic Surgery 2018 Procedural Statistics. http://www.surgery.org/sites/default/files/ASAPS-Stats2018.pdf.
20. Coleman SR. Hand rejuvenation with structural fat grafting. Plast Reconstr Surg. 2002;110(7):1731–44; discussion 45–7.
21. Coleman SR. Facial augmentation with structural fat grafting. Clin Plast Surg. 2006;33(4):567–77.
22. Coleman SR, Saboeiro AP. Fat grafting to the breast revisited: safety and efficacy. Plast Reconstr Surg. 2007;119(3):775–85; discussion 86–7.
23. Conlon CJ, Abu-Ghname A, Davis MJ, Ali K, Winocour SJ, Eisemann ML, et al. Fat grafting for hand rejuvenation. Semin Plast Surg. 2020;34(1):47–52.
24. Davis MJ, Perdanasari AT, Abu-Ghname A, Gonzalez SR, Chamata E, Rammos CK, et al. Application of fat grafting in cosmetic breast surgery. Semin Plast Surg. 2020;34(1):24–9.
25. Hersant B, Jabbour S, Noel W, Benadiba L, La Padula S, SidAhmed-Mezi M, et al. Labia majora augmentation combined with minimal labia minora resection: a safe and global approach to the external female genitalia. Ann Plast Surg. 2018;80(4):323–7.
26. Shiffman MA, Kaminski MV. Fat transfer to the face: technique and new concepts. Facial Plast Surg Clin North Am. 2001;9(2):229–37, viii.
27. Turner A, Abu-Ghname A, Davis MJ, Winocour SJ, Hanson SE, Chu CK. Fat grafting in breast reconstruction. Semin Plast Surg. 2020;34(1):17–23.
28. Abboud M, Geeroms M, El Hajj H, Abboud N. Improving the female silhouette and gluteal projection: an anatomy-based, safe, and harmonious approach through liposuction, suspension loops, and moderate lipofilling. Aesthet Surg J. 2020;
29. Sukho P, Cohen A, Hesselink JW, Kirpensteijn J, Verseijden F, Bastiaansen-Jenniskens YM. Adipose tissue-derived stem cell sheet application for tissue healing in vivo: a systematic review. Tissue Eng Part B Rev. 2018;24(1):37–52.
30. Kim BS, Gaul C, Paul NE, Dewor M, Stromps JP, Hwang SS, et al. The effect of lipoaspirates on human keratinocytes. Aesthet Surg J. 2016;36(8):941–51.
31. Walocko FM, Khouri RK Jr, Urbanchek MG, Levi B, Cederna PS. The potential roles for adipose tissue in peripheral nerve regeneration. Microsurgery. 2016;36(1):81–8.
32. Klinger M, Marazzi M, Vigo D, Torre M. Fat injection for cases of severe burn outcomes: a new perspective of scar remodeling and reduction. Aesthetic Plast Surg. 2008;32(3):465–9.
33. Pallua N, Kim BS. Microfat and lipoconcentrate for the treatment of facial scars. Clin Plast Surg. 2020;47(1):139–45.
34. Strong AL, Rubin JP, Kozlow JH, Cederna PS. Fat grafting for the treatment of scleroderma. Plast Reconstr Surg. 2019;144(6):1498–507.
35. Magalon G, Daumas A, Sautereau N, Magalon J, Sabatier F, Granel B. Regenerative approach to scleroderma with fat grafting. Clin Plast Surg. 2015;42(3):353–64, viii–ix.
36. Rigotti G, Marchi A, Galie M, Baroni G, Benati D, Krampera M, et al. Clinical treatment of radiotherapy tissue damage by lipoaspirate transplant: a healing process mediated by adipose-derived adult stem cells. Plast Reconstr Surg. 2007;119(5):1409–22; discussion 23–4.
37. Klinger M, Caviggioli F, Vinci V, Salval A, Villani F. Treatment of chronic posttraumatic ulcers using autologous fat graft. Plast Reconstr Surg. 2010;126(3):154e–5e.
38. Lopez-Santalla M, Menta R, Mancheno-Corvo P, Lopez-Belmonte J, DelaRosa O, Bueren JA, et al. Adipose-derived mesenchymal stromal cells modulate experimental autoimmune arthritis by inducing an early regulatory innate cell signature. Immun Inflamm Dis. 2016;4(2):213–24.
39. Wu L, Cai X, Zhang S, Karperien M, Lin Y. Regeneration of articular cartilage by adipose tissue derived mesenchymal stem cells: perspectives from stem cell biology and molecular medicine. J Cell Physiol. 2013;228(5):938–44.
40. Charles-de-Sa L, Gontijo-de-Amorim NF, Maeda Takiya C, Borojevic R, Benati D, Bernardi P, et al. Antiaging treatment of the facial skin by fat graft and adipose-derived stem cells. Plast Reconstr Surg. 2015;135(4):999–1009.
41. Sun W, Ni X, Sun S, Cai L, Yu J, Wang J, et al. Adipose-derived stem cells alleviate radiation-induced muscular fibrosis by suppressing the expression of TGF-beta1. Stem Cells Int. 2016;2016:5638204.
42. Lv Q, Li X, Qi Y, Gu Y, Liu Z, Ma GE. Volume retention after facial fat grafting and relevant factors: a systematic review and meta-analysis. Aesthetic Plast Surg. 2020;
43. Parikh RP, Doren EL, Mooney B, Sun WV, Laronga C, Smith PD. Differentiating fat necrosis from recurrent malignancy in fat-grafted breasts: an imaging classification system to guide management. Plast Reconstr Surg. 2012;130(4):761–72.
44. Rubin JP, Coon D, Zuley M, Toy J, Asano Y, Kurita M, et al. Mammographic changes after fat transfer to the breast compared with changes after breast reduction: a blinded study. Plast Reconstr Surg. 2012;129(5):1029–38.
45. Alvarez S. Natural breast augmentation or fat transfer: the mammographic and sonographic correlation. J Diagn Med Sonogr. 2012;28(1):26–32.
46. Veber M, Tourasse C, Toussoun G, Moutran M, Mojallal A, Delay E. Radiographic findings after breast augmentation by autologous fat transfer. Plast Reconstr Surg. 2011;127(3):1289–99.
47. Gosset J, Guerin N, Toussoun G, Delaporte T, Delay E. Radiological evaluation after lipomodelling for correction of breast conservative treatment sequelae. Ann Chir Plast Esthet. 2008;53(2):178–89.

48. American Society of Clinical Oncology. Breast Cancer: Statistics. 2019, January.
49. Sinno S, Wilson S, Brownstone N, Levine SM. Current thoughts on fat grafting: using the evidence to determine fact or fiction. Plast Reconstr Surg. 2016;137(3):818–24.
50. Strong AL, Cederna PS, Rubin JP, Coleman SR, Levi B. The current state of fat grafting: a review of harvesting, processing, and injection techniques. Plast Reconstr Surg. 2015;136(4):897–912.
51. Ross RJ, Shayan R, Mutimer KL, Ashton MW. Autologous fat grafting: current state of the art and critical review. Ann Plast Surg. 2014;73(3):352–7.
52. Li K, Gao J, Zhang Z, Li J, Cha P, Liao Y, et al. Selection of donor site for fat grafting and cell isolation. Aesthetic Plast Surg. 2013;37(1):153–8.
53. Xue EY, Narvaez L, Chu CK, Hanson SE. Fat processing techniques. Semin Plast Surg. 2020;34(1):11–6.
54. Lee JH, Kirkham JC, McCormack MC, Nicholls AM, Randolph MA, Austen WG Jr. The effect of pressure and shear on autologous fat grafting. Plast Reconstr Surg. 2013;131(5):1125–36.
55. Livaoglu M, Buruk CK, Uraloglu M, Ersoz S, Livaogglu A, Sozen E, et al. Effects of lidocaine plus epinephrine and prilocaine on autologous fat graft survival. J Craniofac Surg. 2012;23(4):1015–8.
56. Moore JH Jr, Kolaczynski JW, Morales LM, Considine RV, Pietrzkowski Z, Noto PF, et al. Viability of fat obtained by syringe suction lipectomy: effects of local anesthesia with lidocaine. Aesthetic Plast Surg. 1995;19(4):335–9.
57. Nguyen A, Pasyk KA, Bouvier TN, Hassett CA, Argenta LC. Comparative study of survival of autologous adipose tissue taken and transplanted by different techniques. Plast Reconstr Surg. 1990;85(3):378–86; discussion 87-9.
58. Rohrich RJ, Sorokin ES, Brown SA. In search of improved fat transfer viability: a quantitative analysis of the role of centrifugation and harvest site. Plast Reconstr Surg. 2004;113(1):391–5; discussion 6–7.
59. Gir P, Brown SA, Oni G, Kashefi N, Mojallal A, Rohrich RJ. Fat grafting: evidence-based review on autologous fat harvesting, processing, reinjection, and storage. Plast Reconstr Surg. 2012;130(1):249–58.
60. Nava MB, Blondeel P, Botti G, Casabona F, Catanuto G, Clemens MW, et al. International expert panel consensus on fat grafting of the breast. Plast Reconstr Surg Glob Open. 2019;7(10):e2426.
61. Herly M, Orholt M, Larsen A, Pipper CB, Bredgaard R, Gramkow CS, et al. Efficacy of breast reconstruction with fat grafting: a systematic review and meta-analysis. J Plast Reconstr Aesthet Surg. 2018;71(12):1740–50.
62. Chung NN, Ransom RC, Blackshear CP, Irizarry DM, Yen D, Momeni A, et al. Fat grafting into younger recipients improves volume retention in an animal model. Plast Reconstr Surg. 2019;143(4):1067–75.
63. Geissler PJ, Davis K, Roostaeian J, Unger J, Huang J, Rohrich RJ. Improving fat transfer viability: the role of aging, body mass index, and harvest site. Plast Reconstr Surg. 2014;134(2):227–32.
64. Ercan A, Baghaki S, Suleymanov S, Aydin O, Konukoglu D, Cetinkale O. Effects of cigarette smoke on fat graft survival in an experimental rat model. Aesthetic Plast Surg. 2019;43(3):815–25.
65. Eto H, Kato H, Suga H, Aoi N, Doi K, Kuno S, et al. The fate of adipocytes after nonvascularized fat grafting: evidence of early death and replacement of adipocytes. Plast Reconstr Surg. 2012;129(5):1081–92.
66. Mineda K, Kuno S, Kato H, Kinoshita K, Doi K, Hashimoto I, et al. Chronic inflammation and progressive calcification as a result of fat necrosis: the worst outcome in fat grafting. Plast Reconstr Surg. 2014;133(5):1064–72.
67. Mashiko T, Yoshimura K. How does fat survive and remodel after grafting? Clin Plast Surg. 2015;42(2):181–90.
68. Gillies HM, Ralph D. The principles and art of plastic surgery. Boston: Little, Brown & Company; 1957.
69. Hoffmann J, Glassford AJ, Doyle TC, Robbins RC, Schrepfer S, Pelletier MP. Angiogenic effects despite limited cell survival of bone marrow-derived mesenchymal stem cells under ischemia. Thorac Cardiovasc Surg. 2010;58(3):136–42.
70. Pu LL. Discussion. Improvement of the survival of human autologous fat transplantation by using VEGF-transfected adipose-derived stem cells. Plast Reconstr Surg. 2009;124(5):1447–9.
71. Lu F, Li J, Gao J, Ogawa R, Ou C, Yang B, et al. Improvement of the survival of human autologous fat transplantation by using VEGF-transfected adipose-derived stem cells. Plast Reconstr Surg. 2009;124(5):1437–46.
72. Chiu YC, Cheng MH, Uriel S, Brey EM. Materials for engineering vascularized adipose tissue. J Tissue Viability. 2011;20(2):37–48.
73. Hausman GJ, Dodson MV. Stromal vascular cells and adipogenesis: cells within adipose depots regulate adipogenesis. J Genomics. 2013;1:56–66.
74. Hausman GJ, Richardson RL. Adipose tissue angiogenesis. J Anim Sci. 2004;82(3):925–34.
75. Crandall DL, Hausman GJ, Kral JG. A review of the microcirculation of adipose tissue: anatomic, metabolic, and angiogenic perspectives. Microcirculation. 1997;4(2):211–32.
76. Rupnick MA, Panigrahy D, Zhang CY, Dallabrida SM, Lowell BB, Langer R, et al. Adipose tissue mass can be regulated through the vasculature. Proc Natl Acad Sci U S A. 2002;99(16):10,730–5.
77. Kawaguchi N, Toriyama K, Nicodemou-Lena E, Inou K, Torii S, Kitagawa Y. De novo adipogenesis in mice at the site of injection of basement membrane and basic fibroblast growth factor. Proc Natl Acad Sci U S A. 1998;95(3):1062–6.
78. Rophael JA, Craft RO, Palmer JA, Hussey AJ, Thomas GP, Morrison WA, et al. Angiogenic growth factor synergism in a murine tissue engineering model of angiogenesis and adipogenesis. Am J Pathol. 2007;171(6):2048–57.
79. Fukumura D, Ushiyama A, Duda DG, Xu L, Tam J, Krishna V, et al. Paracrine regulation of angiogenesis and adipocyte differentiation during in vivo adipogenesis. Circ Res. 2003;93(9):e88–97.
80. Kato H, Mineda K, Eto H, Doi K, Kuno S, Kinoshita K, et al. Degeneration, regeneration, and cicatrization after fat grafting: dynamic total tissue remodeling during the first 3 months. Plast Reconstr Surg. 2014;133(3):303e–13e.
81. Yoshimura K, Coleman SR. Complications of fat grafting: how they occur and how to find, avoid, and treat them. Clin Plast Surg. 2015;42(3):383–8. ix
82. Doi K, Ogata F, Eto H, Kato H, Kuno S, Kinoshita K, et al. Differential contributions of graft-derived and host-derived cells in tissue regeneration/remodeling after fat grafting. Plast Reconstr Surg. 2015;135(6):1607–17.
83. Khouri RK, Rigotti G, Cardoso E, Khouri RK Jr, Biggs TM. Megavolume autologous fat transfer: part II. Practice and techniques. Plast Reconstr Surg. 2014;133(6):1369–77.
84. Hyun JS, Montoro DT, Lo DD, Flynn RA, Wong V, Chung MT, et al. The seed and the soil: optimizing stem cells and their environment for tissue regeneration. Ann Plast Surg. 2013;70(2):235–9.
85. Asahara T, Murohara T, Sullivan A, Silver M, van der Zee R, Li T, et al. Isolation of putative progenitor endothelial cells for angiogenesis. Science. 1997;275(5302):964–7.
86. Hamed S, Ben-Nun O, Egozi D, Keren A, Malyarova N, Kruchevsky D, et al. Treating fat grafts with human endothelial progenitor cells promotes their vascularization and improves their survival in diabetes mellitus. Plast Reconstr Surg. 2012;130(4):801–11.
87. Dimarino AM, Caplan AI, Bonfield TL. Mesenchymal stem cells in tissue repair. Front Immunol. 2013;4:201.

88. Sunaga A, Sugawara Y, Katsuragi-Tomioka Y, Kobayashi E. The fate of nonvascularized fat grafts: histological and bioluminescent study. Plast Reconstr Surg Glob Open. 2013;1(6):e40.
89. Yi C, Pan Y, Zhen Y, Zhang L, Zhang X, Shu M, et al. Enhancement of viability of fat grafts in nude mice by endothelial progenitor cells. Dermatol Surg. 2006;32(12):1437–43.
90. Bartynski J, Marion MS, Wang TD. Histopathologic evaluation of adipose autografts in a rabbit ear model. Otolaryngol Head Neck Surg. 1990;102(4):314–21.
91. Saunders MC, Keller JT, Dunsker SB, Mayfield FH. Survival of autologous fat grafts in humans and in mice. Connect Tissue Res. 1981;8(2):85–91.
92. Rossatti B. Revascularisation and phagocytosis in free fat autografts: an experimental study. Br J Plast Surg. 1960;13:35–41.
93. Kuno S, Yoshimura K. Condensation of tissue and stem cells for fat grafting. Clin Plast Surg. 2015;42(2):191–7.
94. Matsumoto D, Sato K, Gonda K, Takaki Y, Shigeura T, Sato T, et al. Cell-assisted lipotransfer: supportive use of human adipose-derived cells for soft tissue augmentation with lipoinjection. Tissue Eng. 2006;12(12):3375–82.
95. Philips BJ, Grahovac TL, Valentin JE, Chung CW, Bliley JM, Pfeifer ME, et al. Prevalence of endogenous CD34+ adipose stem cells predicts human fat graft retention in a xenograft model. Plast Reconstr Surg. 2013;132(4):845–58.
96. Hamed S, Egozi D, Dawood H, Keren A, Kruchevsky D, Ben-Nun O, et al. The chemokine stromal cell-derived factor-1alpha promotes endothelial progenitor cell-mediated neovascularization of human transplanted fat tissue in diabetic immunocompromised mice. Plast Reconstr Surg. 2013;132(2):239e–50e.
97. Xu FT, Li HM, Yin QS, Liu DL, Nan H, Zhao PR, et al. Human breast adipose-derived stem cells transfected with the stromal cell-derived factor-1 receptor CXCR4 exhibit enhanced viability in human autologous free fat grafts. Cell Physiol Biochem. 2014;34(6):2091–104.
98. Pallua N, Grasys J, Kim BS. Enhancement of progenitor cells by two-step centrifugation of emulsified lipoaspirates. Plast Reconstr Surg. 2018;142(1):99–109.
99. Tonnard P, Verpaele A, Peeters G, Hamdi M, Cornelissen M, Declercq H. Nanofat grafting: basic research and clinical applications. Plast Reconstr Surg. 2013;132(4):1017–26.
100. Ingram DA, Caplice NM, Yoder MC. Unresolved questions, changing definitions, and novel paradigms for defining endothelial progenitor cells. Blood. 2005;106(5):1525–31.
101. Peichev M, Naiyer AJ, Pereira D, Zhu Z, Lane WJ, Williams M, et al. Expression of VEGFR-2 and AC133 by circulating human CD34(+) cells identifies a population of functional endothelial precursors. Blood. 2000;95(3):952–8.
102. Tepper OM, Galiano RD, Capla JM, Kalka C, Gagne PJ, Jacobowitz GR, et al. Human endothelial progenitor cells from type II diabetics exhibit impaired proliferation, adhesion, and incorporation into vascular structures. Circulation. 2002;106(22):2781–6.
103. Loomans CJ, De Koning EJ, Staal FJ, Rabelink TJ, Zonneveld AJ. Endothelial progenitor cell dysfunction in type 1 diabetes: another consequence of oxidative stress? Antioxid Redox Signal. 2005;7(11-12):1468–75.
104. Fadini GP, Miorin M, Facco M, Bonamico S, Baesso I, Grego F, et al. Circulating endothelial progenitor cells are reduced in peripheral vascular complications of type 2 diabetes mellitus. J Am Coll Cardiol. 2005;45(9):1449–57.
105. Fadini GP, Agostini C, Avogaro A. Endothelial progenitor cells and vascular biology in diabetes mellitus: current knowledge and future perspectives. Curr Diabetes Rev. 2005;1(1):41–58.
106. Tanaka R, Masuda H, Fujimura S, Ito-Hirano R, Arita K, Kakinuma Y, et al. Quality-quantity control culture enhances vasculogenesis and wound healing efficacy of human diabetic peripheral blood CD34+ cells. Stem Cells Transl Med. 2018;7(5):428–38.
107. Kado M, Tanaka R, Arita K, Okada K, Ito-Hirano R, Fujimura S, et al. Human peripheral blood mononuclear cells enriched in endothelial progenitor cells via quality and quantity controlled culture accelerate vascularization and wound healing in a porcine wound model. Cell Transplant. 2018;27(7):1068–79.
108. Scheubel RJ, Zorn H, Silber RE, Kuss O, Morawietz H, Holtz J, et al. Age-dependent depression in circulating endothelial progenitor cells in patients undergoing coronary artery bypass grafting. J Am Coll Cardiol. 2003;42(12):2073–80.
109. Ma FX, Zhou B, Chen Z, Ren Q, Lu SH, Sawamura T, et al. Oxidized low density lipoprotein impairs endothelial progenitor cells by regulation of endothelial nitric oxide synthase. J Lipid Res. 2006;47(6):1227–37.
110. Heiss C, Keymel S, Niesler U, Ziemann J, Kelm M, Kalka C. Impaired progenitor cell activity in age-related endothelial dysfunction. J Am Coll Cardiol. 2005;45(9):1441–8.
111. Masuda H, Iwasaki H, Kawamoto A, Akimaru H, Ishikawa M, Ii M, et al. Development of serum-free quality and quantity control culture of colony-forming endothelial progenitor cell for vasculogenesis. Stem Cells Transl Med. 2012;1(2):160–71.
112. Masuda H, Tanaka R, Fujimura S, Ishikawa M, Akimaru H, Shizuno T, et al. Vasculogenic conditioning of peripheral blood mononuclear cells promotes endothelial progenitor cell expansion and phenotype transition of anti-inflammatory macrophage and T lymphocyte to cells with regenerative potential. J Am Heart Assoc. 2014;3(3):e000743.
113. Geeroms M, Fujimura S, Aiba E, Orgun D, Arita K, Kitamura R, et al. Quality and quantity-cultured human mononuclear cells improve the human fat graft vascularization and survival in an in vivo murine experimental model. Plast Reconstr Surg. 2020; https://doi.org/10.1097/PRS.0000000000007580. Online ahead of print.
114. Tanaka R, Vaynrub M, Masuda H, Ito R, Kobori M, Miyasaka M, et al. Quality-control culture system restores diabetic endothelial progenitor cell vasculogenesis and accelerates wound closure. Diabetes. 2013;62(9):3207–17.
115. Kwon SM, Lee YK, Yokoyama A, Jung SY, Masuda H, Kawamoto A, et al. Differential activity of bone marrow hematopoietic stem cell subpopulations for EPC development and ischemic neovascularization. J Mol Cell Cardiol. 2011;51(3):308–17.
116. Masuda H, Alev C, Akimaru H, Ito R, Shizuno T, Kobori M, et al. Methodological development of a clonogenic assay to determine endothelial progenitor cell potential. Circ Res. 2011;109(1):20–37.
117. Tsukada S, Kwon SM, Matsuda T, Jung SY, Lee JH, Lee SH, et al. Identification of mouse colony-forming endothelial progenitor cells for postnatal neovascularization: a novel insight highlighted by new mouse colony-forming assay. Stem Cell Res Ther. 2013;4(1):20.
118. Masuda H, Asahara T. Clonogenic assay of endothelial progenitor cells. Trends Cardiovasc Med. 2013;23(4):99–103.
119. Tanaka R, Wada M, Kwon SM, Masuda H, Carr J, Ito R, et al. The effects of flap ischemia on normal and diabetic progenitor cell function. Plast Reconstr Surg. 2008;121(6):1929–42.
120. Hagiwara H, Higashibata A, Ogawa S, Kanazawa S, Mizuno H, Tanaka R. Effectiveness of endothelial progenitor cell culture under microgravity for improved angiogenic potential. Sci Rep. 2018;8(1):14,239.
121. Ohtake T, Kobayashi S, Slavin S, Mochida Y, Ishioka K, Moriya H, et al. Human peripheral blood mononuclear cells incubated in vasculogenic conditioning medium dramatically improve ischemia/reperfusion acute kidney injury in mice. Cell Transplant. 2018;27(3):520–30.
122. Salybekov AA, Kawaguchi AT, Masuda H, Vorateera K, Okada C, Asahara T. Regeneration-associated cells improve recovery from myocardial infarction through enhanced vasculo-

genesis, anti-inflammation, and cardiomyogenesis. PloS One. 2018;13(11):e0203244.
123. Jetten N, Verbruggen S, Gijbels MJ, Post MJ, De Winther MP, Donners MM. Anti-inflammatory M2, but not pro-inflammatory M1 macrophages promote angiogenesis in vivo. Angiogenesis. 2014;17(1):109–18.
124. Spiller KL, Anfang RR, Spiller KJ, Ng J, Nakazawa KR, Daulton JW, et al. The role of macrophage phenotype in vascularization of tissue engineering scaffolds. Biomaterials. 2014;35(15):4477–88.
125. Phipps KD, Gebremeskel S, Gillis J, Hong P, Johnston B, Bezuhly M. Alternatively activated M2 macrophages improve autologous Fat Graft survival in a mouse model through induction of angiogenesis. Plast Reconstr Surg. 2015;135(1):140–9.
126. Cai J, Feng J, Liu K, Zhou S, Lu F. Early macrophage infiltration improves fat graft survival by inducing angiogenesis and hematopoietic stem cell recruitment. Plast Reconstr Surg. 2018;141(2):376–86.
127. Nakayama T, Nagata E, Masuda H, Asahara T, Takizawa S. Regeneration-associated cell transplantation contributes to tissue recovery in mice with acute ischemic stroke. PloS One. 2019;14(1):e0210198.
128. Geeroms M, Hamdi M, Hirano R, Hagiwara H, Fujimura S, Mizuno H, et al. Quality and quantity-cultured murine endothelial progenitor cells increase vascularization and decrease fibrosis in the fat graft. Plast Reconstr Surg. 2019;143(4):744e–55e.
129. Fong GH, Rossant J, Gertsenstein M, Breitman ML. Role of the Flt-1 receptor tyrosine kinase in regulating the assembly of vascular endothelium. Nature. 1995;376(6535):66–70.
130. Autiero M, Waltenberger J, Communi D, Kranz A, Moons L, Lambrechts D, et al. Role of PlGF in the intra- and intermolecular cross talk between the VEGF receptors Flt1 and Flk1. Nat Med. 2003;9(7):936–43.
131. Rahimi N. VEGFR-1 and VEGFR-2: two non-identical twins with a unique physiognomy. Front Biosci. 2006;11:818–29.
132. Wang Q, Wang J, Wang T. Pigs can be used as a large animal model for autologous fat grafting. Ophthal Plast Reconstr Surg. 2016;32(1):73–4.
133. Luo X, Cao W, Xu H, Wang L, Zhang Z, Lu Y, et al. Coimplanted endothelial cells improve adipose tissue grafts' survival by increasing vascularization. J Craniofac Surg. 2015;26(2):358–64.
134. Kakudo N, Tanaka Y, Morimoto N, Ogawa T, Kushida S, Hara T, et al. Adipose-derived regenerative cell (ADRC)-enriched fat grafting: optimal cell concentration and effects on grafted fat characteristics. J Transl Med. 2013;11:254.
135. Cai J, Li B, Liu K, Feng J, Gao K, Lu F. Low-dose G-CSF improves fat graft retention by mobilizing endogenous stem cells and inducing angiogenesis, whereas high-dose G-CSF inhibits adipogenesis with prolonged inflammation and severe fibrosis. Biochem Biophys Res Commun. 2017;491(3):662–7.
136. Coleman SR, Saboeiro AP. Primary breast augmentation with fat grafting. Clin Plast Surg. 2015;42(3):301–6, vii.
137. Raj S, Abu-Ghname A, Davis MJ, Izaddoost SA, Winocour SJ. Safety and regulation of fat grafting. Semin Plast Surg. 2020;34(1):59–64.
138. Gutowski KA, Force AFGT. Current applications and safety of autologous fat grafts: a report of the ASPS fat graft task force. Plast Reconstr Surg. 2009;124(1):272–80.
139. Lohsiriwat V, Curigliano G, Rietjens M, Goldhirsch A, Petit JY. Autologous fat transplantation in patients with breast cancer: "silencing" or "fueling" cancer recurrence? Breast. 2011;20(4):351–7.
140. Vona-Davis L, Rose DP. Adipokines as endocrine, paracrine, and autocrine factors in breast cancer risk and progression. Endocr Relat Cancer. 2007;14(2):189–206.
141. Massa M, Gasparini S, Baldelli I, Scarabelli L, Santi P, Quarto R, et al. Interaction between breast cancer cells and adipose tissue cells derived from fat grafting. Aesthet Surg J. 2016;36(3):358–63.
142. Hou WK, Xu YX, Yu T, Zhang L, Zhang WW, Fu CL, et al. Adipocytokines and breast cancer risk. Chin Med J (Engl). 2007;120(18):1592–6.
143. Gennari R, Griguolo G, Dieci MV, Guarneri V, Tavaniello B, Sibilio A, et al. Fat grafting for breast cancer patients: from basic science to clinical studies. Eur J Surg Oncol. 2016;42(8):1088–102.
144. Fraser JK, Hedrick MH, Cohen SR. Oncologic risks of autologous fat grafting to the breast. Aesthet Surg J. 2011;31(1):68–75.
145. Zhang Y, Daquinag A, Traktuev DO, Amaya-Manzanares F, Simmons PJ, March KL, et al. White adipose tissue cells are recruited by experimental tumors and promote cancer progression in mouse models. Cancer Res. 2009;69(12):5259–66.
146. Rowan BG, Gimble JM, Sheng M, Anbalagan M, Jones RK, Frazier TP, et al. Human adipose tissue-derived stromal/stem cells promote migration and early metastasis of triple negative breast cancer xenografts. PloS One. 2014;9(2):e89595.
147. Orecchioni S, Gregato G, Martin-Padura I, Reggiani F, Braidotti P, Mancuso P, et al. Complementary populations of human adipose CD34+ progenitor cells promote growth, angiogenesis, and metastasis of breast cancer. Cancer Res. 2013;73(19):5880–91.
148. Orbay H, Hinchcliff KM, Charvet HJ, Sahar DE. Fat graft safety after oncologic surgery: addressing the contradiction between in vitro and clinical studies. Plast Reconstr Surg. 2018;142(6):1489–99.
149. Charvet HJ, Orbay H, Harrison L, Devi K, Sahar DE. In vitro effects of adipose-derived stem cells on breast cancer cells harvested from the same patient. Ann Plast Surg. 2016;76(Suppl 3):S241–5.
150. Wang K, Dai Y, Pan Y, Cheng P, Jin X. Local-regional recurrence risk after autologous fat grafting in breast cancer patients: a systematic review and meta-analysis. J Surg Oncol. 2020;121(3):435–40.
151. Petit JY, Maisonneuve P, Rotmensz N, Bertolini F, Rietjens M. Fat grafting after invasive breast cancer: a matched case-control study. Plast Reconstr Surg. 2017;139(6):1292–6.
152. Petit JY, Maisonneuve P. Lipofilling of the breast does not increase the risk of recurrence of breast cancer: a matched controlled study. Plast Reconstr Surg. 2016;138(5):937e–8e.
153. Kronowitz SJ, Mandujano CC, Liu J, Kuerer HM, Smith B, Garvey P, et al. Lipofilling of the breast does not increase the risk of recurrence of breast cancer: a matched controlled study. Plast Reconstr Surg. 2016;137(2):385–93.
154. Kim JYS. Discussion: fat grafting after invasive breast cancer: a matched case-control study. Plast Reconstr Surg. 2017;139(6):1297–9.
155. Cohen O, Lam G, Karp N, Choi M. Determining the oncologic safety of autologous fat grafting as a reconstructive modality: an institutional review of breast cancer recurrence rates and surgical outcomes. Plast Reconstr Surg. 2017;140(3):382e–92e.
156. Hamidian JA. Determining the oncologic safety of autologous fat grafting as a reconstructive modality: an institutional review of breast cancer recurrence rates and surgical outcomes. Plast Reconstr Surg. 2018;142(4):579e–80e.
157. Cohen O, Karp N, Choi M. Reply: determining the oncologic safety of autologous fat grafting as a reconstructive modality: an institutional review of breast cancer recurrence rates and surgical outcomes. Plast Reconstr Surg. 2018;142(4):580e–1e.
158. Delay E, Garson S, Tousson G, Sinna R. Fat injection to the breast: technique, results, and indications based on 880 procedures over 10 years. Aesthet Surg J. 2009;29(5):360–76.
159. Petit JY, Maisonneuve P, Rotmensz N, Bertolini F, Clough KB, Sarfati I, et al. Safety of lipofilling in patients with breast cancer. Clin Plast Surg. 2015;42(3):339–44, viii.

160. Rigotti G, Marchi A, Stringhini P, Baroni G, Galie M, Molino AM, et al. Determining the oncological risk of autologous lipoaspirate grafting for post-mastectomy breast reconstruction. Aesthetic Plast Surg. 2010;34(4):475–80.
161. Eto H, Suga H, Matsumoto D, Inoue K, Aoi N, Kato H, et al. Characterization of structure and cellular components of aspirated and excised adipose tissue. Plast Reconstr Surg. 2009;124(4):1087–97.
162. Brown JC, Katz AJ. Stem cells derived from fat. Principles of regenerative medicine. Elsevier; 2019. p. 295–305.
163. Bourin P, Bunnell BA, Casteilla L, Dominici M, Katz AJ, March KL, et al. Stromal cells from the adipose tissue-derived stromal vascular fraction and culture expanded adipose tissue-derived stromal/stem cells: a joint statement of the International Federation for Adipose Therapeutics and Science (IFATS) and the International Society for Cellular Therapy (ISCT). Cytotherapy. 2013;15(6):641–8.
164. Rasmussen BS, Sorensen CL, Kurbegovic S, Orholt M, Talman MM, Herly M, et al. Cell-enriched fat grafting improves graft retention in a porcine model: a dose-response study of adipose-derived stem cells versus stromal vascular fraction. Plast Reconstr Surg. 2019;144(3):397e–408e.
165. Yoshimura K, Sato K, Aoi N, Kurita M, Inoue K, Suga H, et al. Cell-assisted lipotransfer for facial lipoatrophy: efficacy of clinical use of adipose-derived stem cells. Dermatol Surg. 2008;34(9):1178–85.
166. Rasmussen BS, Lykke Sorensen C, Vester-Glowinski PV, Herly M, Trojahn Kolle SF, Fischer-Nielsen A, et al. Effect, feasibility, and clinical relevance of cell enrichment in large volume fat grafting: a systematic review. Aesthet Surg J. 2017;37(suppl 3):S46–58.
167. Wang L, Luo X, Lu Y, Fan ZH, Hu X. Is the resorption of grafted fat reduced in cell-assisted lipotransfer for breast augmentation? Ann Plast Surg. 2015;75(2):128–34.
168. Laloze J, Varin A, Gilhodes J, Bertheuil N, Grolleau JL, Brie J, et al. Cell-assisted lipotransfer: friend or foe in fat grafting? Systematic review and meta-analysis. J Tissue Eng Regen Med. 2018;12(2):e1237–e50.
169. Zhou Y, Wang J, Li H, Liang X, Bae J, Huang X, et al. Efficacy and safety of cell-assisted lipotransfer: a systematic review and meta-analysis. Plast Reconstr Surg. 2016;137(1):44e–57e.
170. Mitchell JB, McIntosh K, Zvonic S, Garrett S, Floyd ZE, Kloster A, et al. Immunophenotype of human adipose-derived cells: temporal changes in stromal-associated and stem cell-associated markers. Stem Cells. 2006;24(2):376–85.
171. Kapur SK, Dos-Anjos Vilaboa S, Llull R, Katz AJ. Adipose tissue and stem/progenitor cells: discovery and development. Clin Plast Surg. 2015;42(2):155–67.
172. Kolle SF, Fischer-Nielsen A, Mathiasen AB, Elberg JJ, Oliveri RS, Glovinski PV, et al. Enrichment of autologous fat grafts with ex-vivo expanded adipose tissue-derived stem cells for graft survival: a randomised placebo-controlled trial. Lancet. 2013;382(9898):1113–20.
173. Sterodimas A, de Faria J, Nicaretta B, Boriani F. Autologous fat transplantation versus adipose-derived stem cell-enriched lipografts: a study. Aesthet Surg J. 2011;31(6):682–93.
174. Peltoniemi HH, Salmi A, Miettinen S, Mannerstrom B, Saariniemi K, Mikkonen R, et al. Stem cell enrichment does not warrant a higher graft survival in lipofilling of the breast: a prospective comparative study. J Plast Reconstr Aesthet Surg. 2013;66(11):1494–503.
175. Rehman J, Traktuev D, Li J, Merfeld-Clauss S, Temm-Grove CJ, Bovenkerk JE, et al. Secretion of angiogenic and anti-apoptotic factors by human adipose stromal cells. Circulation. 2004;109(10):1292–8.
176. Thangarajah H, Vial IN, Chang E, El-Ftesi S, Januszyk M, Chang EI, et al. IFATS collection: adipose stromal cells adopt a proangiogenic phenotype under the influence of hypoxia. Stem Cells. 2009;27(1):266–74.
177. Garza RM, Rennert RC, Paik KJ, Atashroo D, Chung MT, Duscher D, et al. Studies in fat grafting: Part IV. Adipose-derived stromal cell gene expression in cell-assisted lipotransfer. Plast Reconstr Surg. 2015;135(4):1045–55.
178. Planat-Benard V, Silvestre JS, Cousin B, Andre M, Nibbelink M, Tamarat R, et al. Plasticity of human adipose lineage cells toward endothelial cells: physiological and therapeutic perspectives. Circulation. 2004;109(5):656–63.
179. Fan W, Sun D, Liu J, Liang D, Wang Y, Narsinh KH, et al. Adipose stromal cells amplify angiogenic signaling via the VEGF/mTOR/Akt pathway in a murine hindlimb ischemia model: a 3D multimodality imaging study. PloS One. 2012;7(9):e45621.

8 Volumetric and Regenerative Components of Fat Graft: Positioning in the Fat-Nanofat Spectrum

Marion W. Tapp, Kelsey M. Lloyd, Adam J. Katz, and Ramon Llull

Contents

8.1 Introduction

Adipose tissue was believed to be a relatively simple organ with a primary role to protect more "valuable structures" as it stored fuel. Over the past few decades, advances in the field have revealed that adipocytes and their supportive cells constitute a dynamic organ which we propose to be an organ of healing [1].

Volumetric interest in adipose tissue began more than a century ago when the first adipose tissue transfer was performed by Gustav Neuber in 1893 for periorbital scar correction [2]. Years later, Viktor Czerny would transplant a lipoma for breast reconstruction and Erich Lexer began fat grafting of disfigured soldiers. These efforts represented en-bloc transfer of fat tissue, which fell out of favor due to its high resorption rate and tendency to cause oily cysts. The first exogenous injectable used to correct contour deformities utilized paraffin wax, which was historically employed to correct syphilitic saddle nose deformities. Paraffin, too, was found to be problematic in that it migrated, created firm nodules, and also caused pulmonary emboli. In 1909, Eugene Hollander pioneered the first injectable fat transfer using ram fat mixed with minced fat [3]. Nevertheless, due to unreliable volumetric retention over time, injection of fat was questioned until the advent of liposuction by Fournier and Illouz in the 1980s and its adoption in the 1990s, when Sydney Coleman would harness the use of liposuction for fat injection [2].

The regenerative interest in adipose tissue is relatively still in its youth. In the 1960s, adipose tissue was enzymatically dissociated into a single cell suspension, which allowed the ability to study its individual cell components [4, 5]. The therapeutic, regenerative potential of these cells was first conceived by members of our lab and later popularized by Zuk et al. [6–10]. Prior research showed that connective tissue matrices of other animals contained uncommitted mesenchymal stem cells [11]. Mesenchymal stem cells have the potential to differentiate into adipocytes, chondrocytes, myoblasts, or osteoblasts [10–13]. Additionally, adipose tissue is known to derive from embryonic mesoderm and have a heterogeneous stromal cell population much like bone marrow, where mesenchymal stem cells had been previously harvested [10, 14–18]. Therefore, it was proposed that adipose tissue may represent a source of stem cells that would be easy to obtain, and capable of yielding high cell numbers. Their results supported their hypothesis and paved the way for future research utilizing these adipocyte-derived stromal cells (ADSC).

M. W. Tapp · K. M. Lloyd · A. J. Katz · R. Llull (✉)
Department of Plastic and Reconstructive Surgery, Wake Forest School of Medicine, Winston-Salem, NC, USA
e-mail: mtapp@wakehealth.edu; kmlloyd@wakehealth.edu; akatz@wakehealth.edu; rllull@wakehealth.edu

© The Author(s), under exclusive license to Springer Nature Switzerland AG 2023
A. Di Giuseppe et al. (eds.), *Fat Transfer in Plastic Surgery*, https://doi.org/10.1007/978-3-031-10881-5_8

As illustrated in other chapters, the use of adipose tissue and fat grafting began as an attempt at satisfying the simple goal of adding volume and has now expanded to nearly endless possibilities. Therefore, much like our understanding of adipose tissue has evolved, so has the responsibility of plastic surgeons. We must deeply understand the components of our fat graft as we are ultimately ushering new generations of cell surgeons: those who, wearing the hat of a cell biologist and a surgical gown, uncover the volumetric and regenerative properties within the adipose-related biospace. The foundation to fully harness the potential of fat grafting begins with our belief that you must "know your graft."

8.2 The Graft

The quest to refine a lipoaspirated fat tissue parcel into a graft has driven the entire evolution of fat fragmentation. While not a novel idea, we conceptualize successful fat grafting into optimization of five crucial areas: graft content, graft procurement, graft processing, the recipient bed, and the engraftment process. A complete review of all these components is outside the scope of the chapter; however, a few fundamentals are necessary to establish a knowledge foundation for a post-fragmentation fat graft spectrum and allow further discussion.

The graft itself begins with understanding the components of the lipoaspirate. The lipoaspirate sample is a complex coarse dispersion in which tissue fragments, composed of parenchymal (adipocyte) and stromal (mesenchymal) elements, are suspended in a liquid medium [19]. Liposuction transforms the adipose tissue into a flowable medium of small particles in liquid amenable to injection. This very transformation from solid tissue to liquid injectable allows volumetric augmentation with adipose tissue to be a minimally invasive procedure. However, in doing so, the slurry composition becomes a highly unpredictable graft; oil and solid components dramatically change from sample to sample, surgeon to surgeon, case after case. We have established the goal of maintaining viable adipocytes and their stromal components within a reproducible ratio in respect to their containing liquid medium, without violating their integrity. There are numerous ways to do this, but one of the most popular is centrifugation. Following the Coleman technique and his described use of the centrifuge, we find ourselves with three distinct layers when viewing the syringe [20]. An oil layer resides on the surface with fatty tissue components just beneath, and a sero-sanguinous heterogeneous mixture on the bottom. Systematic removal of oil and sero-sanguinous layers is paramount to reproducible volumetric retention. Once removed, we are left with our fatty tissue comprised of adipocytes and stromal vascular cells. A histological analysis of the fatty tissue discloses a fragment stratification in which adipocyte-predominant fragments float in the superficial layer, while stromal cell-rich fragments are found in greater proportion in the lower level [21, 22] (Fig. 8.1).

Adipose tissue is composed of mature adipocytes and the stromal vascular fraction (SVF). The mature adipocyte consists of a large lipid droplet surrounded by cytoplasm and a plasma membrane. During times of caloric abundance, energy storage is promoted and adipocytes undergo hypertrophy and hyperplasia [23, 24]. The stromal vascular fraction is composed of stromal reparative cells including fibroblasts, pericytes, vascular endothelial cells, and immune cells. SVF also contains a pool of progenitor cells capable of proliferation and differentiation [1, 21, 25].

Due to their constitutive reparative properties and representing a substitutive pool to parenchymal adipocytes, the SVF is a target for further research in tissue regeneration. SVF may be isolated from the mature adipocytes by collagenase and more recently by mechanical means and then can be further modified or expanded to isolate those descendant cells from the precursor fraction in SVF [10, 26–29]. The constitutive properties and heterogeneous nature of SVF have been thought to add clinical benefit, as opposed to the stem pool alone, in outcomes such as immunomodulation and angiogenesis which are critical mechanisms to secure engraftment when and if the recipient bed is ready [26, 30–33].

The ideal recipient bed should be compatible, adjacent, perfused, and competent in order to allow successful engraftment. The engraftment process itself returns vascular perfusion to a temporarily parasitic and ischemic fat graft [34]. There are two general theories as to how a fat graft survives: Graft Replacement vs. Graft Survival. The graft replacement theory is based on the thought of pre-adipocytes and other undifferentiated cells being more resistant and less fragile than adipocytes. Work by Eto and Yoshimura et al. has shown in their models that adipose progenitor cells undergo activation and differentiation into new adipocytes and thus propose the final amount of fat graft retention is based on the graft replacement by these progenitor cells [35–37]. This is argued by Peer et al. that believe adipocytes undergo a similar process of engraftment as would a skin graft beginning with imbibition [38]. This theory was further strengthened by studies such as Carpaneda and Ribeiro that found adipocytes within 2 mm of vascularized tissue after transplant survive via imbibition, and by Zhao et al. who demonstrated histologically the survival of graft adipocytes in mice after transplantation [35, 39, 40]. We dare to advance that a more precise description of engraftment entails an interplay of these two theories during a four-step process involving the graft's implantation, imbibition, revascularization, and integration.

That being said, as a community we still argue over a standardized method of fat grafting and boast variable ranges

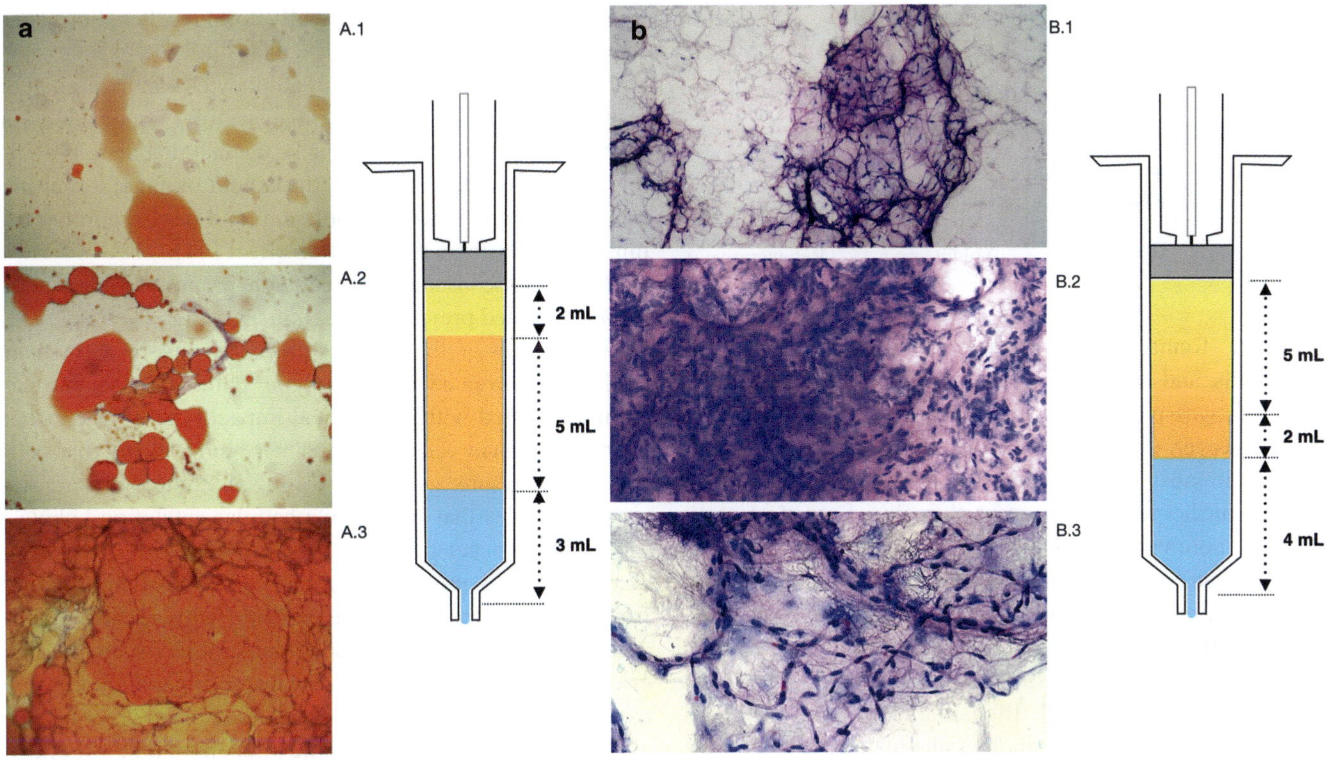

Fig. 8.1 Physical composition of a lipoaspirated fat graft (macrofat, panels **A**) versus a fragmented nanofat (panels **B**). Panels **A** (H&E). Upon centrifugation (see syringe graphic LEFT), a typical yet highly variable composition of a standardized fat graft (macrofat) is depicted from a 10-cc sample: 2 cc of oil (yellow, **A.1**), 5 cc of tissue (orange), and 3 cc of water (blue), containing minimal number of loose cells. Within the tissue layer, adipose tissue particles stratify immediately below the oil micelles according to their cell densities and pack themselves in the mid-tissue layer: those rich in buoyant adipocytes on top (panel **A.2**), those rich in stroma on the bottom (panel **A.3**). Panels **B** (Oil red O). In contrast (compare above to syringe graphic RIGHT), nanofat fragmentation renders a significantly higher volume (5 cc) of emulsified oil on top in which 500 μ tissue fragments (panel **B.1**), a densely populated tissue fraction with progressively decreasing number of adipocytes (2 cc, **B.2**). The tissue fragments immediately over the water phase are constituted by heavy fiber and vascular structures (**B.3**). Of note, centrifugation deposits a pellet of scant fibers and isolated cells (50,000 cells/ml of tissue) with low viability (not shown)

of graft take [41]. It is our firm conviction that the knowledge of the graft's physical (water and oil contents) and biological (graft) composition is key to optimize outcomes for all patients. Once the graft composition is truly reproducible, then a greater scientific appreciation of adipose tissue engraftment and its components will usher in a new world where fat grafting lends itself to not only volumetric but also regenerative capabilities.

8.3 The Spectrum of Fat Fragmentation

Fat grafting has evolved to meet specific demands of both the surgeon and the scientist, demanding reproducibility and understanding, while minimizing donor volume requirements and implantation morbidity. This has been done through mechanical fragmentation of the graft, thus reducing its particle size while optimizing its implantation. Quite unexpectedly, the apparent cell injury caused by mechanical fragmentation did not result in an increment of morbidity, but rather to a trend of reports confirming both volumetric enhancement in addition to trophic changes in the overlying skin and adnexa.

Our sequence of studies on the fat fragmentation byproducts and their histological structure, cell quantity and viability, allows us to present the conceptualization of fat grafting on a spectrum in which particle size correlates directly with the volumetric potential of mature adipocytes and inversely with the regenerative potential of stromal cells.

We will focus specifically on the spectrum of fat particles achieved from liposuction as produced through harvesting, with liposuction cannula size thought to correlate with particle size [42]. This spectrum is loosely divided into the unscientific, inconsistent terminology such as macrofat, microfat (or millifat), and nanofat (Table 8.1).

By popular convention, macrofat is the fat particle size correlated with use of large bore cannulas, typically 2–3 mm in size, with larger side ports [43]. In the landmark paper by Tonnard et al., he used a 3-mm Mercedes-type liposuction cannula with side ports measuring 2 × 7 mm. High-negative-pressure liposuction was used and the lipoaspirate was rinsed and filtered using a nylon cloth [44]. Upon analysis of the

Table 8.1 The spectrum of fat fragmentation

	Macrofat	Microfat	Nanofat
Process	Collection	Collection	Post collection
Size	2–3 mm	1 mm	<1 mm
Composition	Adipocytes and SVF	Reduced adipocytes SVF	SVF
Application	Volumetric	Volumetric	Regenerative

macrofat, Tonnard et al. found viable adipocytes with few dead cells and overall normal histological structure [44]. Macrofat was placed in small aliquots to maximize diffusion; however, the limiting factor upon injection is the cannula as it must allow for the larger fat particles to pass without further fragmentation or clumping. Larger fat particles and more viable adipocytes make macrofat the preferred source when considering volumetric correction, in particular when implanted in deep planes.

We define microfat as the fat particle size correlated with use of liposuction cannulas with smaller side ports, typically 1 mm in diameter [43, 44]. Tonnard and his team harvested and processed the microfat in the same manner as that of the macrofat. Despite the smaller side ports on the cannula, qualitative histological analysis of the adipocytes was found to be similar both in structure and viability to that of the macrofat [44]. Since microfat grafting is often utilized in the facial region and uses smaller blunt-tipped cannulas often ranging as small as 0.7 mm, microfat is preferred to macrofat as it allows smooth injection which would likely result in less irregular fat deposition [44]. Microfat grafting offers the volumetric benefits of macrofat grafting in areas that have more shallow implantation sites and lower volumetric requirement.

Both macrofat and microfat are adipose tissue fragments generated by suction-assisted mechanical force. This is in contrast to nanofat which is a fat particle generated when the above macro- or microfat grafts are further fragmented post collection [43, 44]. The fragmentation is classically described by transferring fat grafts between two syringes connected by a Luer-lock connector 30 times resulting in a milky (Tyndall effect) emulsification of the fat. The liquid is then filtered over a nylon cloth, leaving an effluent first termed nanofat by the Belgian group [44].

Nanofat grafting is typically combined with microfat grafting and is injected using a 27-gauge needle. Analysis of nanofat reveals complete disruption of adipose tissue structure with no viable adipocytes; however, the SVF and CD34+ cell count resembled that of the microfat [44]. SVF and the CD34+ subset are obtained via enzymatic dissociation of only collagenase-sensitive cells, which remain unchanged in micro versus nanofat samples. The issue is in the enzymatic digestion itself which results in an unrepresentatively small fraction of cell species that actually colonize the adipose tissue [45].

In fact, and due to the lack of adipocytes which ultimately results in low volumetric potential, nanofat is considered a misnomer by many, as critics voice the fact that there is nothing nano in nanofat [43]. Still, the benefit of nanofat grafting remains in its regenerative potential and case reports accumulate supporting a rejuvenation effect [44].

The adipose tissue fragmentation process has not been fully elucidated previously, but a recent study by Kang et al. clarified some of the cellular processes involved. In their study, they revealed the underlying cellular injury mechanism associated with mechanical impact is more dependent on pressure than acceleration. Their study shows that cell-bursting pressure generates localized compressive/tensile pressure cycles that induce intracellular cavitation bubbles, more so than acceleration-induced pressure, resulting in cell death [46]. This injury mechanism can be applied to the fat fragmentation spectrum and explains why mature adipocytes, larger in size and loaded with a highly viscous triglyceride droplet, are the most fragile cell population when exposed to compressive-tensile pressure cycles. Since mature adipocytes account for nearly the total amount of volume in an aspirated fat particle, preferential lysis of adipocytes during processing compromises the clinical potential of processed tissue fragments when grafted to restore volumetric defects. Yet, by the same reason, adipocyte lysis significantly concentrates the stromal cell component, vastly retained in the tissue layer, responsible for regenerative applications [47]. In other words, higher degrees of fragmentation result in an increasing number of viable stromal cells relative to adipocytes in the sample.

As mentioned before, adipocyte stroma consists of cell lineages including fibrocytes, angiocytes, and histiocytes. They represent the stromal support to the adipocyte parenchyma and remain attached to vascular axles after adipocyte stripping. These critical cells account for the upkeep of the tissue matrix in addition to comprising the vascular network and immunomodulatory properties: the precise mechanisms sought after in site-specific regenerative therapies. This is the reason highly fragmented fat grafts, such as nanofat grafts, garner so much attention.

In short, the fat grafting spectrum ranges from macrofat to nanofat based on the amount of disaggregation. As fat particle size is reduced so is the volumetric potential, yet the regenerative potential remains due to the presence of the smaller SVF cellular components. Perhaps inspired by French Surgeon, Le Fort, who published his iconic study reviewing facial bone fracture patterns following repeated trauma using cannonballs and bats; we believe that further studies of fat disaggregation will reveal that as fat particle size is reduced, that it fragments in predictable way [48].

8.4 Volumetric Potential

Volumetric potential is the ability of a fat graft to provide volume. In a fat graft sample, the greatest volumetric factor is that of the viable adipocyte and therefore volumetric potential is greatest in a macrofat graft with lesser potential in microfat graft. Volume has historically been a subjective measurement of the perceived alteration in the contour of the overlying skin although objective measurements may now be obtained through the use of imaging including magnetic resonance imaging (MRI) and 3-dimensional (3D) surface imaging [49–51].

The volumetric potential in fat grafting often seems more of an art than a science as retention rates may range from 25 to 70% [41]. Additionally reviews of grafting for volume restoration present variable results [52, 53]. Beyond the lack of standardization in physical and biological graft composition, a compelling factor underlying dubious reproducibility and reliability is the lack of outcome metrics.

For long-term success, we expect that volume retention is directly related to fat viability. Therefore, we must strive to achieve reproducibility via optimizing the physical and physiological conditions of the fat graft, which are based on our optimization of the five essential steps of fat grafting outlined earlier.

An issue with the above statements is the assumption that injecting 1 cc of fat graft would equal a direct volume change of 1 cc in the recipient bed. Dos Anjos et al. directly evaluated this using two measurements: Intraoperative Volume Restoration Index (IVRI) and Postoperative Volume Retention Index (POVRI) [54] (Fig. 8.2). IVRI is conceptualized as the comparison of measured volume change versus volume injected. Using 3D scanning models, these authors objectively measured their intraoperative changes in volume following fat grafting to the breast. They found that the measured changes in volume were never equal to that of the graft placed. In fact, the majority of scenarios resulted in an Intraoperative Volume Restoration Index of 85–92% of the injected volume. Upon further exploration, they attempted to measure the IVRI with placement of silicone implants. Silicone implants are a more cohesive structure and therefore one might expect an IVRI of 1; nevertheless, the authors found values of only 0.84 (84%). The authors proposed that this phenomenon is likely due to either the viscoelastic behavior of both the breast and graft or possibly due to packing density [54]. In addition to the challenge posed by a graft that is subject to wide ranges in viability, we now are faced with the predicament that if even 100% of our graft were viable, it would not correlate to a 1:1 volume restoration in the body. The authors propose this is more reason to strive for reproducible fat grafting results so that the surgeon may accurately predict surgical outcomes, once again lending to the thought of fat grafting being a combination of art and science.

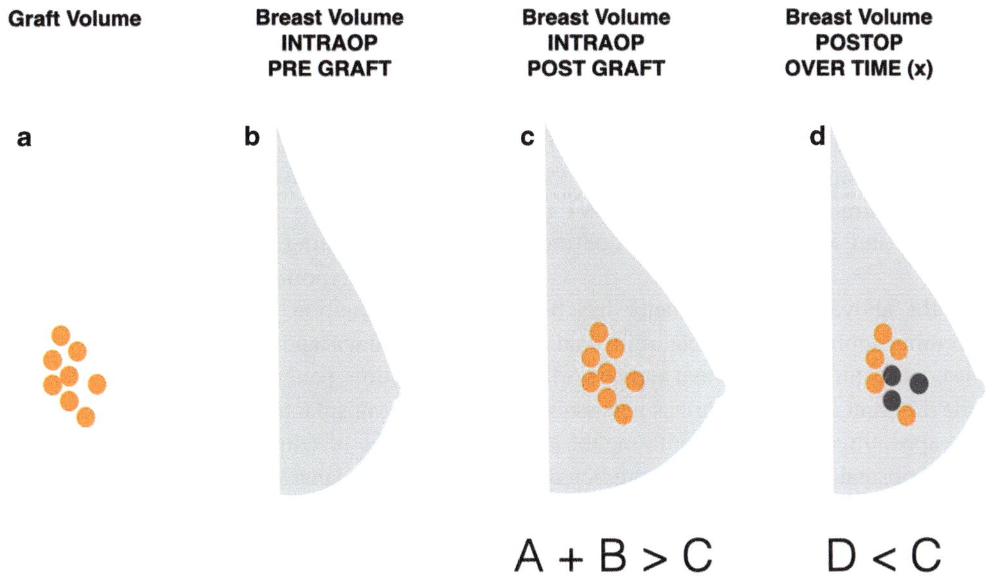

Fig. 8.2 The concept of volumetric restoration and retention
The Intraoperative Volume Restoration Index (IVRI) provides the practitioner with a quantification of how much graft is needed to satisfy (restore) a volumetric deficit. It is counterintuitive to realize that the graft volume (A) plus the initial breast volume (B) does not equal the immediately post-grafted breast (C), but it is actually consistently inferior to that addition. The Postoperative Volume Retention Index (POVRI) is the ratio between the breast volume at a given postoperative time (D) over the volume achieved immediately post graft (C). POVRI allows the practitioner to quantify the fraction of volumetric resorption resulting from non-viable (gray) fat grafts and determine when the reabsorption subsides. Both parameters will aid the practitioner to consequently predict the graft volume (A) requirements to reach the necessary overcorrection at a given postoperative time

In the age of extensive research in SVF and ADSC, clinicians have explored the volumetric potential of augmenting fat graft with these constituents. The same authors that proposed IVRI and POVRI explored this potential [54]. They studied the "dose" effect of adding SVF to adipose graft to observe long-term volume retention rates. In doing so, they injected 77 fat grafts with 21 enriched with low dose of SVF cells and 56 enriched with high SVF cell doses. The low doses were defined as <50,000 SVF cells, considered to be inconsequential and thus the control group. The high dose was defined as SVF cells >200,000. They measured outcomes with POVRI defined as the change in postoperative volume compared to pre-operative values using the 3D imaging software. In their study, they found that initial volumes decreased, but over time the high-dose SVF samples had significantly higher volumes compared to the low-dose group with long-term retention at 1.5 years of 75% of the initial injected volume [54]. The results of their study are not isolated; in fact, other studies have utilized SVF and ADSCs and shown increased retention rates as well [55, 56]. These results are conflicting compared to other studies which did not reproduce these same results [49, 57]. Chiu et al. argue against use of SVF as they did not find significant retention rate differences in their fat grafting. They felt patients are better served by avoiding the excess liposuction required for the SVF and saving those donor areas to be utilized at a later time for additional grafting if needed [49].

The principle of volumetric potential with the added SVF is based on the hypothesis that the constitutive cells of the stroma provide cellular supportive functions including cellular repair via fibroblasts, improved vascularity via endothelial and perivascular elements, and immunomodulatory control via immune cells [54, 58–61].

While the foundation of volumetric potential is based on proper harvesting, processing, and engraftment, there appears to be a role for optimization of grafts with SVF and adipose progenitor cells and will be the focus of continued research in the future.

In relation with the above, a new parameter has been described with intriguing applications in facial rejuvenation: the surface-volume coefficient [62]. Cotofana and coworkers demonstrate that the different injection volumes necessary to deliver aesthetically appealing results when utilizing the subdermal vs. the supraperiosteal technique can be predicted by the region-specific surface volume ratio.

8.5 Regenerative Potential

Graft standardization and outcome metrics are currently advancing to improve reproducible and reliable surgical performance for large, minimally fragmented grafts. When it comes to smaller, highly fragmented grafts, standardization and clinical outcomes remain highly elusive especially in regard to their beseeched regenerative properties. In lack of precise characterization of highly fragmented graft and quantification of clinical effects, a raising body of experimental evidence allows one to speculate on nanofat's mechanisms of action.

The regenerative potential of fat grafting is based on the interaction of the SVF and its resident progenitor. The effect can be thought of by two key mechanisms that are likely synergistic in nature: the substitutive effect and the supportive effect. The substitutive effect is based on the thought that parenchymal cells (adipocytes) may be replaced or substituted by precursors found in the stroma. This includes both pre-adipocytes (adipocyte-committed precursors) that have already been differentiated and more primitive progenitor cells that still retain the potential to differentiate into other non-adipocytic cell lineages [1, 10–13, 21, 25]. The supportive effect mentioned previously is based on the blastic potential that fibroblasts, endothelial cells, immune competent cells, and others express in the SVF.

Nanofat is composed predominantly of stromal elements and non-viable adipocytes dispersed in the emulsification process. This renders a graft with no volumetric potential, but apparently with high regenerative potential: Lipofilling has been known to produce skin regeneration in the setting of radiation and scarring [44, 63–65]. The mechanism is thought to be secondary to the potential of the SVF ultimately resulting in increased production and remodeling of collagen and elastin [44]. Since Tonnard's original paper, research has shown nanofat possesses the ability to improve hypertrophic scars, atrophic scarring, skin pliability, rhytid attenuation, and rejuvenate skin [28, 66–70]. The effects of nanofat grafting is typically delayed with results in 4 weeks up to 3 months [44].

The regenerative potential of fat grating continues to be studied and as of 2016 there were over 130 active clinical trials investigating the potential of ADSCs [71]. Some of the most highly reported applications include bone regeneration, fat reconstruction, cardiovascular and myocardial regeneration, cartilage and intervertebral disc regeneration, hepatic regeneration, pancreatic preservation, and vascular augmentation in transplanted tissue [29, 72–83]. Further regenerative capabilities of adipose tissue continue to evolve due to the overall abundance and ease of accessibility to this tissue.

8.6 A Position on Fat Fragmentation Byproducts

Successful use of fat grafting, whether for volumetric or regenerative indications, relies on the principle of "knowing your graft." Adherence to standardizing and optimizing each step of the fat grafting process will ensure reproducible results and better outcomes for our patients. Modulating the fraction of volumetric over regenerative components of adipose tissue via mechanical fragmentation may well satisfy both volumetric and regenerative clinical requirements. Conceptualizing the grafting options over a spectrum of these different fragmentation byproducts (macrofat to "nanofat") is already contributing to stimulate discussions and innovation on fundamental aspects of structural and functional aspects of tissue repair.

References

1. Kapur SK, Vilaboa SD, Llull R, Katz, AJ. Adipose tissue and stem/progenitor cells: discovery and Development. Clinics in plastic surgery. 2015;42(2):155–167.
2. Mazzola RF, Mazzola IC. History of fat grafting: from ram fat to stem cells. Clin Plast Surg [Internet]. 2015;42(2):147–53. https://doi.org/10.1016/j.cps.2014.12.002.
3. Mazzola RF, Mazzola IC. The fascinating history of fat grafting. J Craniofac Surg. 2013;24(4):1069–71.
4. Rodbell M. Localization of lipoprotein lipase in fat cells of rat adipose tissue. J Biol Chem [Internet]. 1964;239:753–5. http://www.ncbi.nlm.nih.gov/pubmed/14154450
5. Rodbell M. Metabolism of isolated fat cells. i. Effects of hormones on glucose metabolism and lipolysis. J Biol Chem [Internet]. 1964;239:375–80. http://www.ncbi.nlm.nih.gov/pubmed/14169133
6. Tholpady SS, Aojanepong C, Llull R, Jeong JH, Mason AC, Futrell JW, et al. The cellular plasticity of human adipocytes. Ann Plast Surg. 2005;54(6):651–6.
7. Tholpady SS, Llull R, Ogle RC, Rubin JP, Futrell JW, Katz AJ. Adipose tissue: stem cells and beyond. Clin Plast Surg [Internet]. 2006;33(1):55–62. vi. http://www.ncbi.nlm.nih.gov/pubmed/16427974
8. Marshall CB. University of Pittsburgh of the Commonweath system of Education, (Plaintiff) versus Marc H. Hedrick, Prosper Benhaim, Herman Peter Lorenz, and Minh Zhu (Defendants) Findings of fact and conclusion of law [Internet]; 2008. https://caselaw.findlaw.com/us-federal-circuit/1155097.html
9. Nayer R. University of Pittsburgh of the Commonwealth System of Higher Education v. Hedrick, 573 F.3d 1290; 2009
10. Zuk PA, Zhu M, Mizuno H, Huang J, Futrell JW, Katz AJ, et al. Multilineage cells from human adipose tissue: implications for cell-based therapies. Tissue Eng. 2001;7:211.
11. Caplan AI. Mesenchymal stem cells. J Orthop Res [Internet]. 1991;9(5):641–50. http://www.ncbi.nlm.nih.gov/pubmed/1870029
12. Pittenger MF, Mackay AM, Beck SC, Jaiswal RK, Douglas R, Mosca JD, et al. Multilineage potential of adult human mesenchymal stem cells. Science [Internet]. 1999;284(5411):143–7. http://www.ncbi.nlm.nih.gov/pubmed/10102814
13. Wakitani S, Saito T, Caplan AI. Myogenic cells derived from rat bone marrow mesenchymal stem cells exposed to 5-azacytidine. Muscle Nerve [Internet]. 1995;18(12):1417–26. http://www.ncbi.nlm.nih.gov/pubmed/7477065
14. Hausman GJ. Techniques for studying adipocytes. Stain Technol [Internet]. 1981;56(3):149–54. http://www.ncbi.nlm.nih.gov/pubmed/6168034
15. Hausman GJ, Campion DR. Histology of the stroma in developing rat subcutaneous adipose tissue. J Anim Sci [Internet]. 1982;55(6):1336–42. http://www.ncbi.nlm.nih.gov/pubmed/7161207
16. Pettersson P, Cigolini M, Sjöström L, Smith U, Björntorp P. Cells in human adipose tissue developing into adipocytes. Acta Med Scand [Internet]. 1984;215(5):447–51. http://www.ncbi.nlm.nih.gov/pubmed/6741607
17. Pettersson P, Van R, Karlsson M, Björntorp P. Adipocyte precursor cells in obese and nonobese humans. Metabolism [Internet]. 1985;34(9):808–12. http://www.ncbi.nlm.nih.gov/pubmed/4033422
18. Löffler G, Hauner H. Adipose tissue development: the role of precursor cells and adipogenic factors. Part II: the regulation of the adipogenic conversion by hormones and serum factors. Klin Wochenschr [Internet]. 1987;65(17):812–7. http://www.ncbi.nlm.nih.gov/pubmed/3309457
19. Everett DH. Manual of symbols and terminology for physicochemical quantities and units, appendix II: definitions, terminology and symbols in colloid and surface chemistry. Pure Appl Chem. 2008;
20. Coleman SR. Facial recontouring with lipostructure. Clin Plast Surg [Internet]. 1997;24(2):347–67. http://www.ncbi.nlm.nih.gov/pubmed/9142473
21. Strong AL, Cederna PS, Rubin JP, Coleman SR, Levi B. The current state of fat grafting: a review of harvesting, processing, and injection techniques. Plast Reconstr Surg. 2015;136(4):897–912.
22. Allen RJ, Canizares O, Scharf C, Nguyen PD, Thanik V, Saadeh PB, et al. Grading lipoaspirate: is there an optimal density for fat grafting? Plast Reconstr Surg. 2013;131:38.
23. Shepherd PR, Gnudi L, Tozzo E, Yang H, Leach F, Kahn BB. Adipose cell hyperplasia and enhanced glucose disposal in transgenic mice overexpressing GLUT4 selectively in adipose tissue. J Biol Chem [Internet]. 1993;268(30):22243–6. http://www.ncbi.nlm.nih.gov/pubmed/8226728
24. Tang QQ, Lane MD. Adipogenesis: from stem cell to adipocyte. Annu Rev Biochem [Internet]. 2012;81:715–36. http://www.ncbi.nlm.nih.gov/pubmed/22463691
25. Bourin P, Bunnell BA, Casteilla L, Dominici M, Katz AJ, March KL, et al. Stromal cells from the adipose tissue-derived stromal vascular fraction and culture expanded adipose tissue-derived stromal/stem cells: a joint statement of the International Federation for Adipose Therapeutics and Science (IFATS) and the international so. Cytotherapy [Internet]. 2013;15(6):641–8. https://www.cambridge.org/core/product/identifier/CBO9781107360631A083/type/book_part
26. Bora P, Majumdar AS. Adipose tissue-derived stromal vascular fraction in regenerative medicine: a brief review on biology and translation. Stem Cell Res Ther. 2017;8(1):1–10.
27. Matsumoto D, Sato K, Gonda K, Takaki Y, Shigeura T, Sato T, et al. Cell-assisted lipotransfer: supportive use of human adipose-derived cells for soft tissue augmentation with lipoinjection. Tissue Eng [Internet]. 2006;12(12):3375–82. http://www.ncbi.nlm.nih.gov/pubmed/17518674
28. Kamat P, Frueh FS, McLuckie M, Sanchez-Macedo N, Wolint P, Lindenblatt N, et al. Adipose tissue and the vascularization of biomaterials: stem cells, microvascular fragments and nanofat—a review. Cytotherapy. Elsevier B.V. 2020;22:400–11.
29. Jones VM, Suarez-Martinez AD, Hodges NA, Murfee WL, Llull R, Katz AJ. A clinical perspective on adipose-derived cell therapy for enhancing microvascular health and function: implications and applications for reconstructive surgery. Microcirculation. 2021;28(3):1–16.
30. van Dijk A, Naaijkens BA, Jurgens WJFM, Nalliah K, Sairras S, van der Pijl RJ, et al. Reduction of infarct size by intravenous injection of uncultured adipose derived stromal cells in a rat model is dependent on the time point of application. Stem Cell Res [Internet]. 2011;7(3):219–29. http://www.ncbi.nlm.nih.gov/pubmed/21907165
31. Charles-de-Sá L, Gontijo-de-Amorim NF, Maeda Takiya C, Borojevic R, Benati D, Bernardi P, et al. Antiaging treatment of the facial skin by fat graft and adipose-derived stem cells. Plast

Reconstr Surg [Internet]. 2015;135(4):999–1009. http://www.ncbi.nlm.nih.gov/pubmed/25811565
32. Semon JA, Zhang X, Pandey AC, Alandete SM, Maness C, Zhang S, et al. Administration of murine stromal vascular fraction ameliorates chronic experimental autoimmune encephalomyelitis. Stem Cells Transl Med [Internet]. 2013;2(10):789–96. http://www.ncbi.nlm.nih.gov/pubmed/23981726
33. You D, Jang MJ, Kim BH, Song G, Lee C, Suh N, et al. Comparative study of autologous stromal vascular fraction and adipose-derived stem cells for erectile function recovery in a rat model of cavernous nerve injury. Stem Cells Transl Med [Internet]. 2015;4(4):351–8. http://www.ncbi.nlm.nih.gov/pubmed/25792486
34. Thornton J, Gosman A. Skin grafts and skin substitutes and principles of flaps. Sel Readings Plast Surg. 2004;
35. Gal S, Xue Y, Pu LLQ. What do we know now about autologous fat grafting? Ann Plast Surg. 2019;83(4S Suppl 1):S17–20.
36. Yoshimura K, Eto H, Kato H, Doi K, Aoi N. In vivo manipulation of stem cells for adipose tissue repair/reconstruction. Regen Med. 2011;6:33.
37. Eto H, Kato H, Suga H, Aoi N, Doi K, Kuno S, et al. The fate of adipocytes after nonvascularized fat grafting: evidence of early death and replacement of adipocytes. Plast Reconstr Surg. 2012;129:1081.
38. Peer LA. Cell survival theory versus replacement theory. Plast Reconstr Surg. 1955;16:161.
39. Carpaneda CA, Ribeiro MT. Percentage of graft viability versus injected volume in adipose autotransplants. Aesthet Plast Surg. 1994;18:17.
40. Zhao J, Yi C, Li L, Zheng Y, Wu K, Liang L, et al. Observations on the survival and neovascularization of fat grafts interchanged between C57BL/6-gfp and C57BL/6 mice. Plast Reconstr Surg. 2012;130:398e.
41. Bellini E, Grieco MP, Raposio E. The science behind autologous fat grafting. Ann Med Surg. 2017;24:65.
42. Gause TM, Kling RE, Sivak WN, Marra KG, Rubin JP, Kokai LE. Particle size in fat graft retention: a review on the impact of harvesting technique in lipofilling surgical outcomes. Adipocytes. 2014;3:273.
43. Rihani J. Microfat and nanofat: when and where these treatments work. Vol. 27, Facial plastic surgery clinics of North America. W.B. Saunders; 2019. p. 321–30.
44. Tonnard P, Verpaele A, Peeters G, Hamdi M, Cornelissen M, Declercq H. Nanofat grafting: basic research and clinical applications. Plast Reconstr Surg. 2013;132(4):1017–26.
45. Sesé B, Sanmartín JM, Ortega B, Matas-Palau A, Llull R. Nanofat cell aggregates: a nearly constitutive stromal cell inoculum for regenerative site-specific therapies. Plast Reconstr Surg. 2019;144(5):1079–88.
46. Kang W, Robitaille MC, Merrill M, Teferra K, Kim C, Raphael MP. Mechanisms of cell damage due to mechanical impact: an in vitro investigation. Sci Rep [Internet]. 2020;10(1):12009. http://www.ncbi.nlm.nih.gov/pubmed/32686715
47. Sesé B, Sanmartín JM, Ortega B, Llull R. Human stromal cell aggregates concentrate adipose tissue constitutive cell population by in vitro DNA quantification analysis. Plast Reconstr Surg [Internet]. 2020;146(6):1285–93.
48. Tessier P. The classic reprint. Experimental study of fractures of the upper jaw. I and II. René Le Fort, M.D. Plast Reconstr Surg [Internet]. 1972;50(5):497–506. contd. http://www.ncbi.nlm.nih.gov/pubmed/4563382
49. Chiu CH. Does stromal vascular fraction ensure a higher survival in autologous fat grafting for breast augmentation? A volumetric study using 3-dimensional laser scanning. Aesthetic Surg J. 2019;39(1):41–52.
50. Chen YE, Gerstle TL, Liang F, Lee BT. Use of a novel laser projection grid to assess symmetry in breast surgery. Plast Reconstr Surg [Internet]. 2012;130(1):231e–3e. http://www.ncbi.nlm.nih.gov/pubmed/22743944
51. Isogai N, Sai K, Kamiishi H, Watatani M, Inui H, Shiozaki H. Quantitative analysis of the reconstructed breast using a 3-dimensional laser light scanner. Ann Plast Surg [Internet]. 2006;56(3):237–42. http://www.ncbi.nlm.nih.gov/pubmed/16508350
52. Delay E, Garson S, Tousson G, Sinna R. Fat injection to the breast: technique, results, and indications based on 880 procedures over 10 years. Aesthetic Surg J [Internet]. 2009;29(5):360–76. https://doi.org/10.1016/j.asj.2009.08.010.
53. Dos Anjos S, Matas-Palau A, Mercader J, Katz AJ, Llull R. Reproducible volume restoration and efficient long-term volume retention after point-of-care standardized cell-enhanced fat grafting in breast surgery. Plast Reconstr Surg - Glob Open. 2015;3(10):1–11.
54. Dos Anjos S, Matas-Palau A, Mercader J, Katz AJ, Llull R. Reproducible volume restoration and efficient long-term volume retention after point-of-care standardized cell-enhanced fat grafting in breast surgery. Plast Reconstr Surg Glob Open. 2015;3:e547.
55. Kakudo N, Tanaka Y, Morimoto N, Ogawa T, Kushida S, Hara T, et al. Adipose-derived regenerative cell (ADRC)-enriched fat grafting: optimal cell concentration and effects on grafted fat characteristics. J Transl Med [Internet]. 2013;11:254. http://www.ncbi.nlm.nih.gov/pubmed/24107489
56. Kølle SFT, Fischer-Nielsen A, Mathiasen AB, Elberg JJ, Oliveri RS, Glovinski PV, et al. Enrichment of autologous fat grafts with ex-vivo expanded adipose tissue-derived stem cells for graft survival: a randomised placebo-controlled trial. Lancet. 2013;382:1113.
57. Peltoniemi HH, Salmi A, Miettinen S, Mannerström B, Saariniemi K, Mikkonen R, et al. Stem cell enrichment does not warrant a higher graft survival in lipofilling of the breast: a prospective comparative study. J Plast Reconstr Aesthet Surg [Internet]. 2013;66(11):1494–503. http://www.ncbi.nlm.nih.gov/pubmed/23845909
58. Matsuda K, Falkenberg KJ, Woods AA, Choi YS, Morrison WA, Dilley RJ. Adipose-derived stem cells promote angiogenesis and tissue formation for in vivo tissue engineering. Tissue Eng Part A [Internet]. 2013;19(11–12):1327–35. http://www.ncbi.nlm.nih.gov/pubmed/23394225
59. Traktuev DO, Prater DN, Merfeld-Clauss S, Sanjeevaiah AR, Saadatzadeh MR, Murphy M, et al. Robust functional vascular network formation in vivo by cooperation of adipose progenitor and endothelial cells. Circ Res [Internet]. 2009;104(12):1410–20. http://www.ncbi.nlm.nih.gov/pubmed/19443841
60. Melief SM, Zwaginga JJ, Fibbe WE, Roelofs H. Adipose tissue-derived multipotent stromal cells have a higher immunomodulatory capacity than their bone marrow-derived counterparts. Stem Cells Transl Med [Internet]. 2013;2(6):455–63. http://www.ncbi.nlm.nih.gov/pubmed/23694810
61. Dong Z, Peng Z, Chang Q, Zhan W, Zeng Z, Zhang S, et al. The angiogenic and adipogenic modes of adipose tissue after free fat grafting. Plast Reconstr Surg. 2015;135:556e.
62. Cotofana S, Koban K, Pavicic T, Yankonva M, Frank K, Green JB, et al. Clinical validation of the surface volume coefficient for minimally invasive treatment of the Temple. J Drugs Dermatol [Internet]. 2019;18(6):533. http://www.ncbi.nlm.nih.gov/pubmed/31251545
63. Rigotti G, Marchi A, Galiè M, Baroni G, Benati D, Krampera M, et al. Clinical treatment of radiotherapy tissue damage by lipoaspirate transplant: a healing process mediated by adipose-derived adult stem cells. Plast Reconstr Surg. 2007;119:1409.
64. Sardesai MG, Moore CC. Quantitative and qualitative dermal change with microfat grafting of facial scars. Otolaryngol Head Neck Surg [Internet]. 2007;137(6):868–72. http://www.ncbi.nlm.nih.gov/pubmed/18036412
65. Akita S, Yoshimoto H, Ohtsuru A, Hirano A, Yamashita S. Autologous adipose-derived regenerative cells are effective for chronic intractable radiation injuries. Radiat Prot Dosimetry

[Internet]. 2012;151(4):656–60. http://www.ncbi.nlm.nih.gov/pubmed/22914335
66. Uyulmaz S, Sanchez Macedo N, Rezaeian F, Giovanoli P, Lindenblatt N. Nanofat grafting for scar treatment and skin quality improvement. Aesthetic Surg J [Internet]. 2018;38(4):421–8. http://www.ncbi.nlm.nih.gov/pubmed/29365061
67. Bhooshan LS, Devi MG, Aniraj R, Binod P, Lekshmi M. Autologous emulsified fat injection for rejuvenation of scars: a prospective observational study. Indian J Plast Surg [Internet]. 51(1):77–83. http://www.ncbi.nlm.nih.gov/pubmed/29928084
68. Martin A, Maladry D, Esmaeli A, Gaucher S, Philippe H-J. Fat grafting of hairy areas of head and neck—comparison between lipofilling and nanofat grafting procedures in a cadaveric study. J Stomatol oral Maxillofac Surg [Internet]. 2018;119(4):274–8. http://www.ncbi.nlm.nih.gov/pubmed/29964199
69. Gentile P, Scioli MG, Bielli A, Orlandi A, Cervelli V. Comparing different nanofat procedures on scars: role of the stromal vascular fraction and its clinical implications. Regen Med [Internet]. 2017;12(8):939–52. http://www.ncbi.nlm.nih.gov/pubmed/29236575
70. Gu Z, Li Y, Li H. Use of condensed Nanofat combined with fat grafts to treat atrophic scars. JAMA Facial Plast Surg [Internet]. 2018;20(2):128–35. http://www.ncbi.nlm.nih.gov/pubmed/28975248
71. Frese L, Dijkman PE, Hoerstrup SP. Adipose tissue-derived stem cells in regenerative medicine. Transfus Med Hemother. 2016;43:268.
72. Godoy Zanicotti D, Coates DE, Duncan WJ. In vivo bone regeneration on titanium devices using serum-free grown adipose-derived stem cells, in a sheep femur model. Clin Oral Implants Res [Internet]. 2017;28(1):64–75. http://www.ncbi.nlm.nih.gov/pubmed/26853552
73. Lu W, Ji K, Kirkham J, Yan Y, Boccaccini AR, Kellett M, et al. Bone tissue engineering by using a combination of polymer/bioglass composites with human adipose-derived stem cells. Cell Tissue Res [Internet]. 2014;356(1):97–107. http://www.ncbi.nlm.nih.gov/pubmed/24408074
74. Weichman KE, Broer PN, Tanna N, Wilson SC, Allan A, Levine JP, et al. The role of autologous fat grafting in secondary microsurgical breast reconstruction. Ann Plast Surg [Internet]. 2013;71(1):24–30. http://www.ncbi.nlm.nih.gov/pubmed/23788122
75. Li G, Fu N, Xie J, Fu Y, Deng S, Cun X, et al. Poly(3-hydroxybutyrate-co-4-hydroxybutyrate) based electrospun 3D scaffolds for delivery of autogeneic chondrocytes and adipose-derived stem cells: evaluation of cartilage defects in rabbit. J Biomed Nanotechnol [Internet]. 2015;11(1):105–16. http://www.ncbi.nlm.nih.gov/pubmed/26301304
76. Latief N, Raza FA, Bhatti F-U-R, Tarar MN, Khan SN, Riazuddin S. Adipose stem cells differentiated chondrocytes regenerate damaged cartilage in rat model of osteoarthritis. Cell Biol Int [Internet]. 2016;40(5):579–88. http://www.ncbi.nlm.nih.gov/pubmed/26888708
77. Valina C, Pinkernell K, Song Y-H, Bai X, Sadat S, Campeau RJ, et al. Intracoronary administration of autologous adipose tissue-derived stem cells improves left ventricular function, perfusion, and remodelling after acute myocardial infarction. Eur Heart J [Internet]. 2007;28(21):2667–77. http://www.ncbi.nlm.nih.gov/pubmed/17933755
78. Nagata H, Ii M, Kohbayashi E, Hoshiga M, Hanafusa T, Asahi M. Cardiac adipose-derived stem cells exhibit high differentiation potential to cardiovascular cells in C57BL/6 mice. Stem Cells Transl Med. 2016;5:141.
79. Seo MJ, Suh SY, Bae YC, Jung JS. Differentiation of human adipose stromal cells into hepatic lineage in vitro and in vivo. Biochem Biophys Res Commun [Internet]. 2005;328(1):258–64. http://www.ncbi.nlm.nih.gov/pubmed/15670778
80. Lee J, Han D-J, Kim S-C. In vitro differentiation of human adipose tissue-derived stem cells into cells with pancreatic phenotype by regenerating pancreas extract. Biochem Biophys Res Commun [Internet]. 2008;375(4):547–51. http://www.ncbi.nlm.nih.gov/pubmed/18725201
81. Suga H, Eto H, Shigeura T, Inoue K, Aoi N, Kato H, et al. IFATS collection: fibroblast growth factor-2-induced hepatocyte growth factor secretion by adipose-derived stromal cells inhibits postinjury fibrogenesis through a c-Jun N-terminal kinase-dependent mechanism. Stem Cells [Internet]. 2009;27(1):238–49. http://www.ncbi.nlm.nih.gov/pubmed/18772314
82. Hefei W, Yu R, Haiqing W, Xiao W, Jingyuan W, Dongjun L. Morphological characteristics and identification of islet-like cells derived from rat adipose-derived stem cells cocultured with pancreas adult stem cells. Cell Biol Int [Internet]. 2015;39(3):253–63. http://www.ncbi.nlm.nih.gov/pubmed/25262665
83. Missana MC, Laurent I, Barreau L, Balleyguier C. Autologous fat transfer in reconstructive breast surgery: indications, technique and results. Eur J Surg Oncol [Internet]. 2007;33(6):685–90. http://www.ncbi.nlm.nih.gov/pubmed/17241760

Fat Transfer to Improve Results after Breast Surgery and in Breast Abnormalities

9

M. W. Payne and J. M. M. Nijboer

Contents

Breast abnormalities, i.e., tubular breast deformity, Poland's syndrome, significant asymmetries (between breasts or thorax halves), after trauma or surgery, are difficult to deal with for both patient and surgeon. Expectations are high, surgical procedures are plenty, and disappointments are frequent. Equally challenging are certain unfavorable results after breast surgery, no matter how experienced the surgeon.

Often the results after corrective procedures are acceptable but not completely satisfactory, for all parties involved. An improvement is achievable for an experienced surgeon. However, often a degree of "imperfection" remains postoperatively. If the patient is unsatisfied with the final result, often the only solution seems to reoperate but unfortunately without guarantee of success. The key is, and always will be, to create realistic expectations before any procedure, especially in these complex cases. Fat transfer does not change that.

Fat transfer gives us, besides volume increase, the possibility to even out irregularities in a subtle way, not achievable with other surgical procedures. The particulars and pitfalls of fat transfer are discussed in previous chapters. This chapter will illustrate its usability to improve aesthetics in patients with breast abnormalities and/or unfavorable results after breast surgery.

9.1 Asymmetries

Naturally asymmetric breasts are the rule rather than the exception. In case of a volume discrepancy, the correction involves making a larger breast smaller, a smaller breast larger or a combination of the two. Often correcting the volume is not enough and a shape modification, with changing the nipple position and/or breast shape, is necessary. Without fat transfer, adequate volume augmentation may result in a volume balance but with an unnatural appearance of the smaller breast, with the larger implant aspect being relatively evident. A solution is to increase the soft tissue cover with fat transfer.

Beware of abnormalities due to boney causes, i.e. scoliosis, asymmetries between ribs, pectus carinatum and excavatum. Although the latter are more common in males, regularly we see light forms in women, resulting in postoperative asymmetry. Often the patient is unaware of the abnormality before the operation, but will notice the asymmetry afterwards. A combination of the right implants and fat transfer may limit its manifestation but addressing it preoperatively (documenting in informed consent) remains of vital importance.

Supplementary Information The online version contains supplementary material available at https://doi.org/10.1007/978-3-031-10881-5_9.

M. W. Payne (✉)
Private practice for plastic and cosmetic surgery, General Guisan - Quai 30, Zürich, Switzerland

J. M. M. Nijboer
M1 Med Beauty Swiss, Zürich, Switzerland

© The Author(s), under exclusive license to Springer Nature Switzerland AG 2023
A. Di Giuseppe et al. (eds.), *Fat Transfer in Plastic Surgery*, https://doi.org/10.1007/978-3-031-10881-5_9

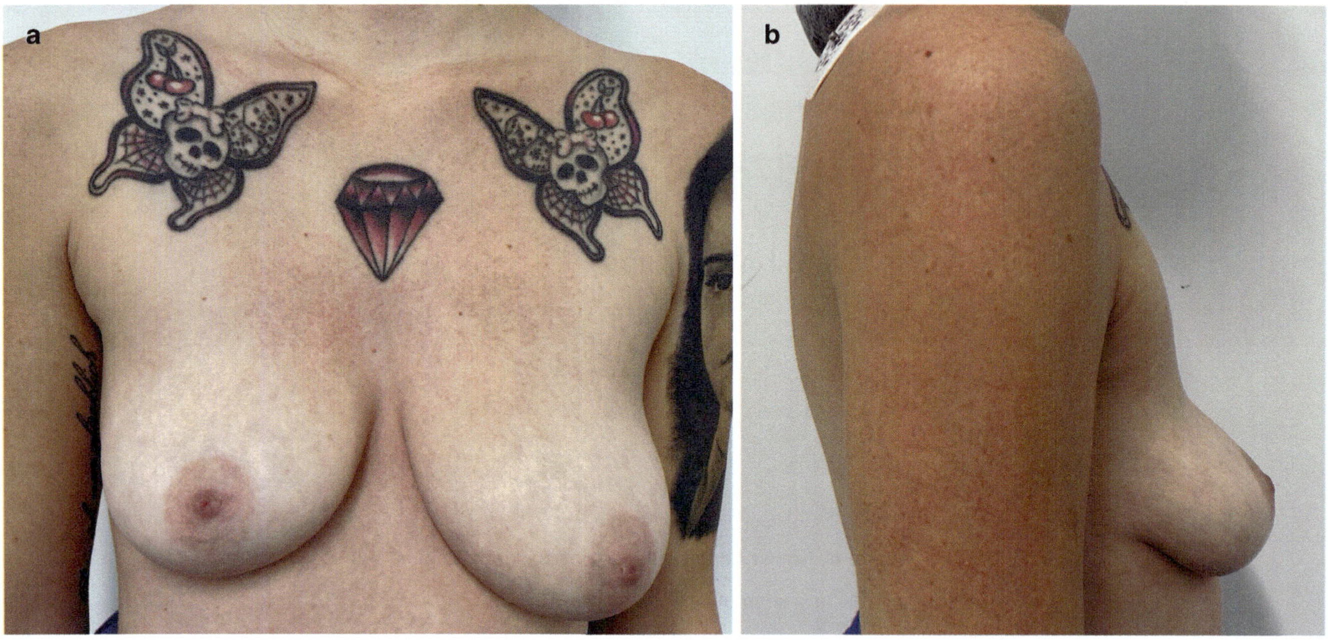

Image 9.1 (**a**) Case I before surgical procedure. (**b**) Case I before surgical procedure, right lateral view

Case I
Our first case is a 29-year-old woman, who presented after a voluntary 56Kg weight loss, expressing the wish for "large and balanced" breasts. Upon examination, we observed an asymmetry in breast volume and shape. Combined with a chest wall asymmetry due to a mild pectus excavatum; ribs 3–5 on the right side are concave shaped and a boney defect lateral of the distal sternum/xyphoid process exists (Image 9.1a, b).

After simulation, we decided on round Motiva® SilkSurface implants, right 400 mL Full profile, left 265 mL Demi profile. The authors recommended round implants in this case to help conceal the cranial boney defect. It was clear that the volume asymmetry exceeded 135 mL, but we decided 400 mL to be the maximum implant size on the right side because of the pectus excavatum and subsequent possibly abnormal origin of the pectoralis major and the dual plane positioning of the implants. Meaning, we had to plan for a left-sided volume reduction and uplift. Transfer of 100–150 mL fat was estimated to increase the natural look of the right breast, and reduce both boney defects. The patient signed an extensive informed consent, including remaining asymmetry and possible indication for a repetition of fat transfer.

9.1.1 The Operation

Step
0. Preoperative photos and mark-up with patient standing.
1. Patient under general anesthesia, first dose of antibiotics (amoxicillin/clavulanic acid, 7 days in total) at induction, supine position with both arms out, prepped and draped.
2. Start on the smaller, **right side.** Approach through IMF and submuscular positioning of the implant (round Motiva® SilkSurface 400 mL Full profile), dual plane. For maximum sterility, use of MotivaImagine® Insertion Sleeve and immediate wound closure in three layers, with special attention to IMF.
3. On the **left side** positioning of round 265 mL Demi sizer, submuscular, dual plane. Temporary wound closure. With the patient in upright position, adjustment of preoperative marking.
4a. With sizer in situ, mastopexy with reduction of 80 g of breast tissue. After temporary wound closure and again upright positioning, check for symmetry. Satisfied with result, exchange of sizer for definitive implant (round Motiva® SilkSurface 265 mL Demi profile), wound closure in three layers, including repositioning the nipple.
4b. Infiltration of marked area of lower abdomen with liposuction solution.
5. Liposuction of 300 mL from the abdomen, using body-jet® and MicroAire PAL® machines. Patient is very slim and after losing 56 kg, the remaining fat tissue appeared very resistant. Preparation of fat, using the MotivaImagine® Puregraft 250 System. 120 mL of pure fat is available for transfer.
6a. Marking the areas for fat transfer; 40 mL in lateral breast, 50 mL in upper breast/boney defect, 30 mL around Xyphoid.

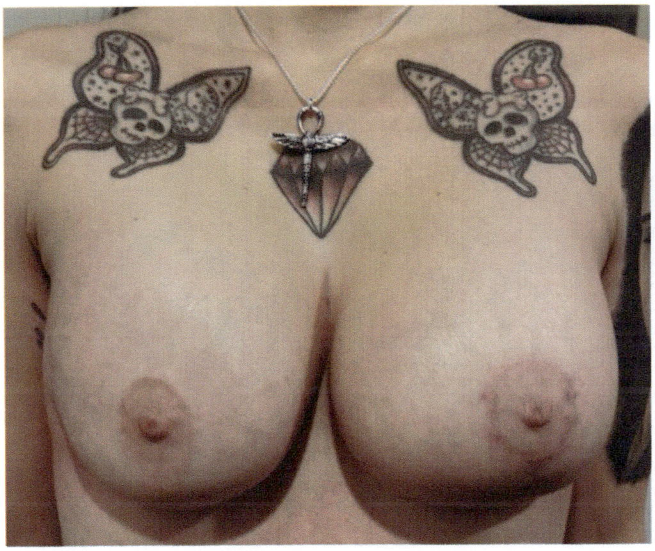

Image 9.2 Case I 1 year after surgical procedure

6b. Injection of fat using the MotivaImagine® Bulb-Canula. Theoretically, one can perforate the implant during fat transfer! Without a Bulb-Canula, consider transfer of fat with sizer in situ.
7. Application of dressings, support-bra, support-body. End of the procedure. Duration 3 h.

No complications occurred during or after surgery. The patient was fit and comfortably discharged the next day. One year after surgery, she is happy with the result; she no longer finds her breasts awkward, feels more feminine and her self-esteem has increased (Image 9.2).

The boney defects are no longer visible, and the asymmetry in volume and shape between both breasts have improved significantly, without an "implanty" look of the previously smaller side.

9.2 Correction of Breast Augmentation Complications/Imperfections

Case II

This case describes the use of fat transfer to increase the soft tissue covering of an implant after unilateral skin sparing mastectomy because of breast cancer. Cosmetically, this is a tricky situation, as per definition only skin and muscle is left to cover the implant.

The concept of using autologous fat grafting after breast cancer treatment has received a lukewarm response for fears of oncological safety and complicating radiological follow-up. Although understandable, to date, there is no clinical evidence confirming these fears. Currently, fat grafting is even used in the early stages of breast cancer (as part of a direct reconstruction) [1–3].

Our patient is a 44-year-old woman, 7 years after successful unilateral cancer treatment with a right-sided breast ablation, chemotherapy and hormonal therapy. A direct reconstruction with a submuscular silicone implant and nipple reconstruction was performed. Four years later, a reoperation followed to correct asymmetry in both volume and shape, by means of a larger anatomical implant (480 mL medium height, moderate profile) on the ablated side, and second mastopexy on the healthy, left side. Currently, she presented to us, requesting to improve the appearance of the ablated breast and an improvement of the still present asymmetry (fluctuating body weight being a factor). Upon examination, we observed an asymmetry based on volume and shape, with extensive rippling and skin denting on the ablated side (Image 9.3a, b). She practices boxing, and a well-developed pectoralis major probably does not improve her breast's appearance.

After simulation, we opted for a right-sided volume increase to a 625 mL Motiva® Ergonomix implant with Demi profile + fat transfer, combined with another mastopexy and reduction of the left side. Besides signing an extensive informed consent, including remaining asymmetry and skin denting, we informed the patient of the possible need for a repeat fat transfer 3–6 months postoperatively.

9.2.1 The Operation

The procedure started on the ablated, **right side**. For our approach, we chose the new "IMF" as we wanted to avoid reopening the scar through the reconstructed nipple. After removal of implant, we found the pectoralis major to be relatively thick, 2.5 cm, and the covering tissue, mainly skin, to be very thin, 3–5 mm (Image 9.4). After expansion of the implant pocket caudally, a sizer was positioned, followed by temporary closure.

On the **left side**, we performed a mastopexy with 133 g reduction. Liposuction was performed on the lower abdomen. With the sizer in situ, 100 mL of fat graft was used to thicken the soft tissue covering the pectoralis major, using the MotivaImagine® Bulb-Cannula, followed by exchanging sizer for definitive Implant (Motiva® Ergonomix 625 mL Demi profile) with submuscular position and closure of incision. The operation took 2.5 h and the patient was discharged the next day. No complications occurred.

One year after surgery, the result is good, as expected a mild asymmetry still exists (Image 9.5). The patient is delighted with her cleavage and the feminine shape of the right breast. She does not find the remaining asymmetry bothersome.

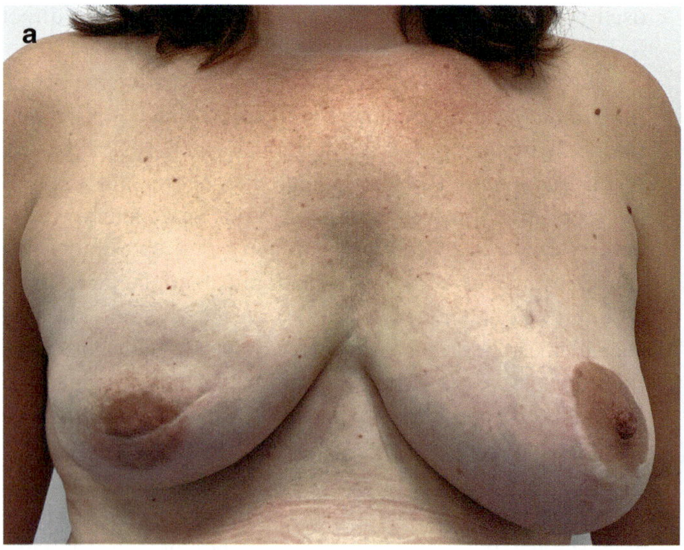

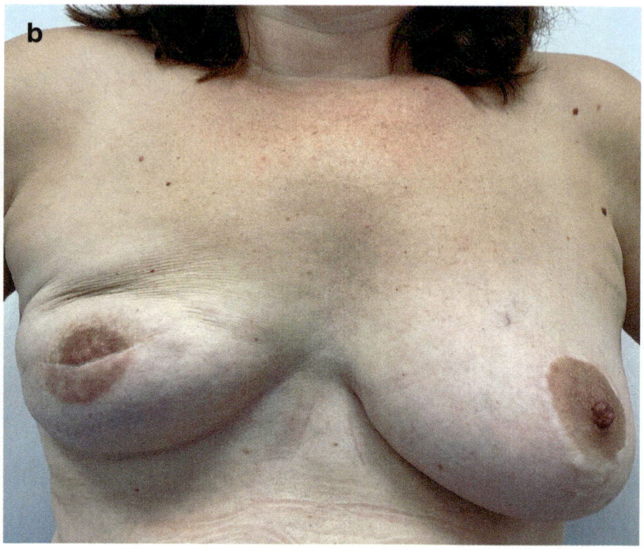

Image 9.3 (**a**) Case II before surgical procedure. (**b**) Case II before surgical procedure, right pectoralis major activated

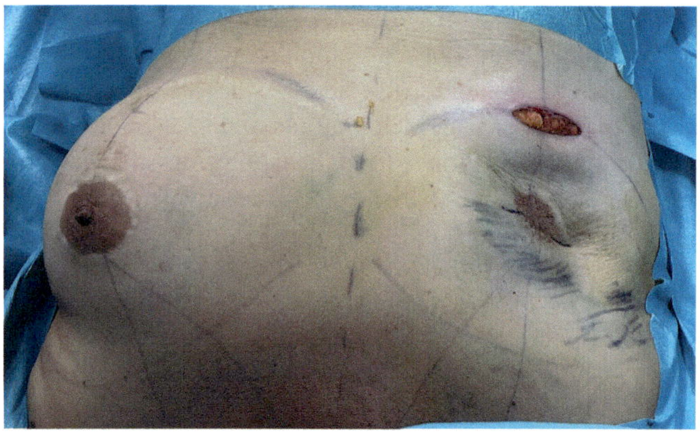

Image 9.4 Case II during surgery, notice thinness of soft tissue covering after removal of implant

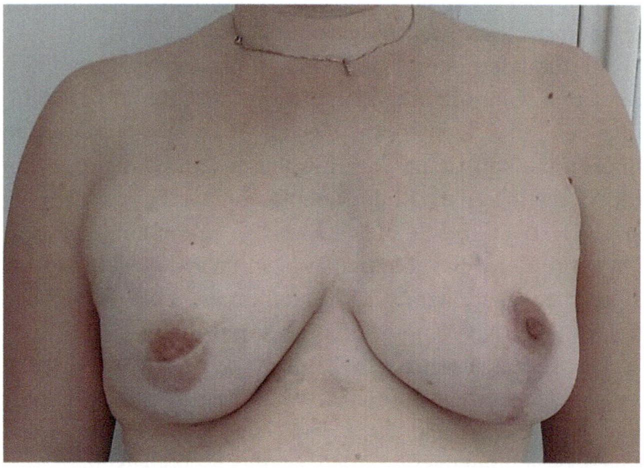

Image 9.5 Case II 1 year after surgical procedure

9.3 As Part of Correction of Breast Abnormalities

Case III This case describes the use of fat transfer in addition to mastopexy, to improve shape and symmetry in tubular breast deformity [4].

Our patient is a 27-year-old woman presenting with the wish for a better breast shape, after two pregnancies and breast feeding. Upon examination, a right-sided hypoplasia of the lower medial and lateral quadrant but with sufficient skin in the subareolar region was observed. Her left breast showed a larger volume, mainly of the upper quadrants, combined with a mild hypoplasia of the lower quadrants (Image 9.6). Implants were not desired.

9.3.1 The Operation

We corrected the shape of her **right** breast by a periareolar mastopexy, of her **left** side with a vertical mastopexy. The right-sided lack of volume of the lower quadrants was corrected by lipofilling using 100 mL of fat graft, another 80 mL was used to increase volume of the upper quadrants. The left breast received 100 mL fat graft, mainly in the lower quadrants (Video 9.1). The operation took 2 h and the patient was discharged later that day.

No postoperative complications occurred and 1 year after surgery both parties are satisfied with the result (Image 9.7). Both breasts have a good, feminine shape with only a minor asymmetry in size.

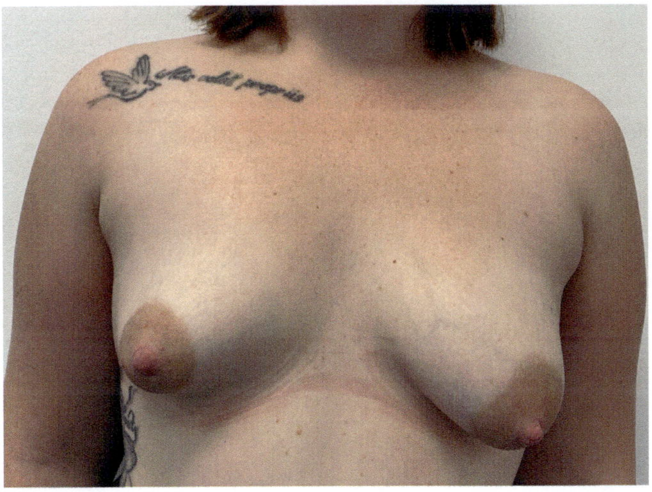

Image 9.6 Case III before surgical procedure

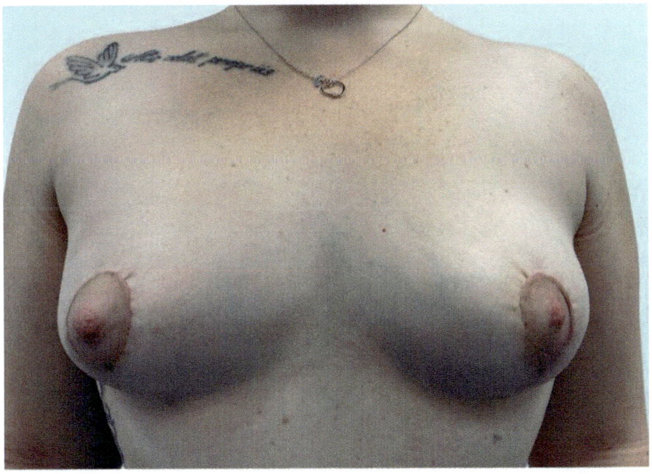

Image 9.7 Case III 1 year after surgical procedure

References

1. Krastev T, van Turnhout A, Vriens E, et al. Long-term follow-up of autologous fat transfer vs. conventional breast reconstruction and association with cancer relapse in patients with breast cancer. JAMA Surg. 2019;154(1):56–63.
2. Groen J, Negenborn V, Twisk D, et al. Autologous fat grafting in onco-plastic breast reconstruction: a systematic review on oncological and radiological safety, complications, volume retention and patient/surgeon satisfaction. J Plast Reconstr Aesthet Surg. 2016;69(6):742–64.
3. Stumpf C, Zucatto A, Cavalheiro J, et al. Oncologic safety of immediate autologous fat grafting for reconstruction in breast-conserving surgery. Breast Cancer Res Treat. 2020;180(2):301–9.
4. Walters J, Bourn L, Tessler O, et al. Single staged mastopexy with autologous fat grafting; an alternative to augmentation mastopexy with implants. Aesthet Surg J. 2020;40(4):152–8.

N.I.L.—Nutational Infrasonic Liposculpture. Lipo and Lipofilling

10

Ângelo Rebelo

Contents

Safety, Precision, Much Less Complications, Much Less Morbidity, Quicker Recovery, Less Pain, No Mortality, Less Revisions. This is the author's experience from 23 years with vibroliposuction/N.I.L.®

What's new?

- Device up date
- Vibro local tumescent anesthesia
- Lipofilling with fat harvested with N.I.L.®

The **N.I.L.®**—Nutational Infrasonic Liposculpture—*is one of the "recent" techniques of liposuction/liposculpture* [1–4], *an evolution of the liposculpture with compressed air, after the name vibroliposuction and now* **N.I.L.®**. *It was invented by Malak, a Belgium medical doctor, and presented for the first time in October of 1997 in Belgium by Dr. Ângelo Rebelo. He and the producer invited the author to make that presentation and to collaborate in the further developments needed to transform the prototype in a much better machine. From the first device Lipomatic, we are nowadays with the fifth prototype—the* EVAmatic®5.

After 23 years and more than 26.000 cases, the author presents the technique and results of **N.I.L.®**—Nutational Infrasonic Liposculpture and the new lipofilling with fat harvested with this device and the innovate vibro local tumescent anesthesia.

10.1 Materials and Methods

From December of 1997, he has performed over 26,000 vibroliposuctions/N.I.L.® (Table 10.1) with this technique. The **LIPOMATIC®/EVAmatic®**5 (Fig. 10.1) system was used; it is a small machine that works with compressed air, is easy to handle, weighing less than 700 g, easy to clean and wash, and may be sterilized in autoclave or other systems. Specific cannulas (Fig. 10.2) of different lengths and diameters (3–5 mm) can be connected. The passage of air induces movements of "come and go" with a frequency of 10 Hz and a course of 6 mm. The cannulas vibrate and have a rotation and translation, which causes a unique movement called **Nutation**. Ten hertz corresponds to 600 times of back-and-forth movement in each minute. The goal is to break down the fat so that it is emulsified and aspirated at the same time. It is of the utmost importance that the frequency stays closer of that value. With higher frequency, the course becomes shorter, influencing the safety and increasing the operating time. The option for the **N.I.L.®** Nutational Infrasonic Liposculpture technique is due to the fact that in the technical characteristics of **LIPOMATIC** we find a low frequency and a long course.

In terms of technologies, there is an important difference between nutational infrasonic liposculpture ("N.I.L.®") and ultrasonic-assisted liposuction ("UAL"):

Â. Rebelo (✉)
Plastic, Reconstructive and Aesthetic Surgery, Clínica Milénio, Lisboa, Portugal

© The Author(s), under exclusive license to Springer Nature Switzerland AG 2023
A. Di Giuseppe et al. (eds.), *Fat Transfer in Plastic Surgery*, https://doi.org/10.1007/978-3-031-10881-5_10

Table 10.1 Difference between ultrasonic-assisted liposuction ("UAL") and nutational infrasonic liposculpture ("N.I.L.®")

Ultrasound vs N.I.L.

Criteria	*UAL*	*NIL*
Principle	**Ultrasound to burn the fat**	**Nutational Infrasonic Liposculpture**
Surgery process	Slow (25h procces)	Fast and not tiring
Traumatism	Heating up of the cannula Risk of burning and necrosis	Respect the tissues
Anaesthesia	Local anaesthesia	Local anaesthesia
Recovery	3-6 months Painful	Short and comfortable
Skin aspect	Poor results- Needs additional treatment to tighten the skin	Excellent skin retraction
Security	Risk of burning and necrosis	Active security system
Servicing	Expensive	Affordable
Material	Expensive	Affordable
Brands	Vaser (int UAL), BodyTite (ext UAL)	Lipomatic
Infiltration System	Vaser : peristaltic pump BodyTite : no infiltration system	Control box for compressed air and power of the infiltration system regulated by compressed air

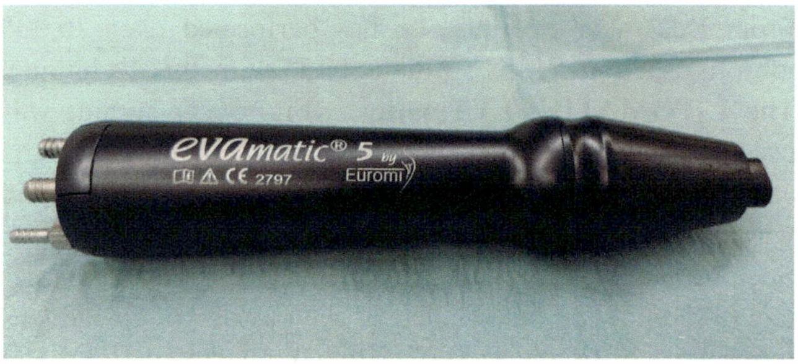

Fig. 10.1 LIPOMATIC®/EVAmatic®5

10.1.1 Technique

The **N.I.L.**® Nutational Infrasonic Liposculpture is performed with patients monitored under vibro local tumescent anesthesia and IV sedation with Propofol and Midazolam.

Infiltration with EVAsp®6: it's the most recent technique to perform the tumescent anesthesia with a better and easier diffusion of the liquid, especially in difficult and/or fibrotic areas. The EVAsp® is composed of a control box (with an entirely digitalized display) regulating the compressed air supply to the Lipomatic/EVAmatic® which allows the Nutational Infrasonic Liposculpture, a high-speed aspirator and a quantitative infiltration system. The EVA sp®6 is a complete system. The quantitative infiltration system allows to control precisely the infiltrated quantity in different predefined zones. The system of pistons ensures a homogeneous infiltration at a constant pressure, whatever the type of fat, even fibrous and allows a faster infiltration with a better and easier diffusion of the liquid, especially in difficult and/or fibrotic areas.

The tumescent solution is based in the Klein [5–10] formula. In general and epidural anesthesia, the patient bleeds much more than with local tumescent anesthesia. In the latter, we achieve a very good vasoconstriction resulting in an almost null loss of blood. The control is more effective cutting the risks down and patient can leave the clinic some hours after the surgery.

The association of more than one anatomical region depends on the total amount of tumescence and Lidocaine to be used. According to Klein, the amount of Lidocaine we have been using is far below their limits. We have confirmed that it is safer and more comfortable for the patient to perform the **N.I.L.**® Nutational Infrasonic Liposculpture in more than one session when indicated and necessary. Apart from the recovery being quicker, they are less limited in their everyday life. Thanks to **N.I.L.**® Nutational Infrasonic Liposculpture, it is rare to spend more than two hours on a surgery even in bigger or associated ones. The mean time between administrating the anesthesia and completing the **N.I.L.**® Nutational Infrasonic Liposculpture is between 45 min and 2 h.

Oral medication is administered 60 min before the surgery with Lysine Clonixinate—250 mg + Hydroxyzine—50 mg and all surgeries are performed with patients monitored.

Every patient follows a protocol with photos, measurements, and weight before and after surgery.

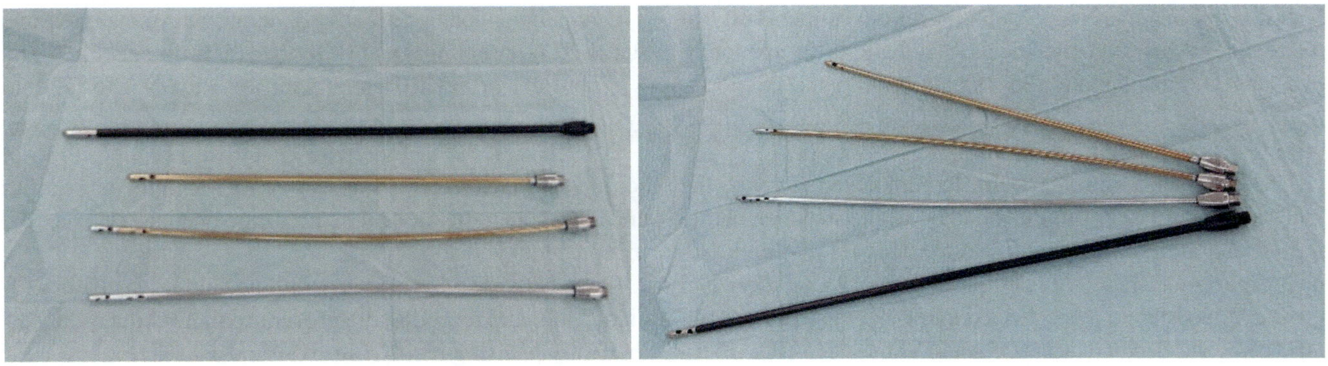

Fig. 10.2 Specific cannulas of different lengths and diameters (3–5 mm)

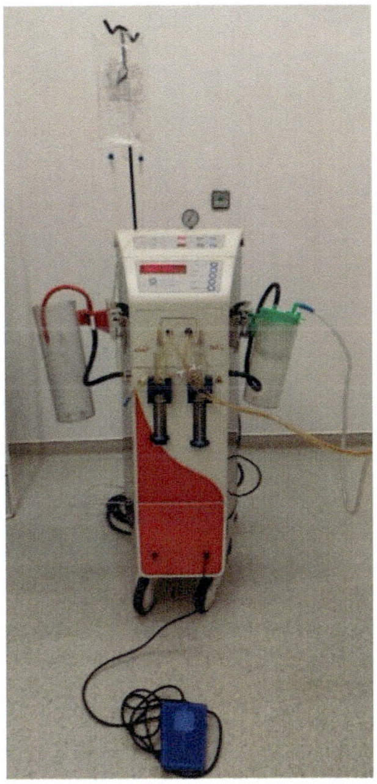

Fig. 10.3 The EVAsp®6 Infiltrator from Euromi

10.1.2 The Vibro Local Tumescent Anesthesia

To each 1.000 cc of 9% saline solution ambient temperature add:

1. 1 or 2 ampoule of adrenaline 1 mg/mL
2. 30 cc of Lidocaine 2% WITHOUT adrenaline

For the tumescence, we use the **EVAsp®6** Infiltrator from Euromi (Fig. 10.3) with special cannulas of 2.5 mm in diameter (Fig. 10.4). It's not common to use more than 3.000 cc per surgery. For a very good vasoconstriction, one should wait at least 30 min, especially in the abdominal area. In our experience, with this technique, there is no relationship between the infiltrated volume and the fat removed.

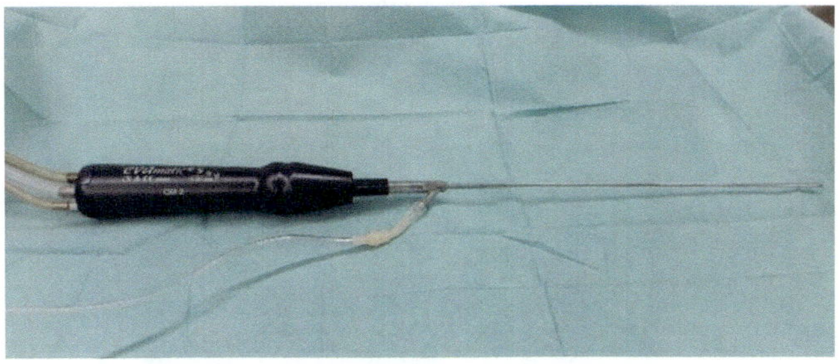

Fig. 10.4 Special cannulas of 2.5 mm in diameter

10.1.3 The N.I.L.®: Nutational Infrasonic Liposculpture

To perform the **N.I.L.®**—Nutational Infrasonic Liposculpture, different types of cannulas may be used. A short or long cannula, with a small or large diameter and specific holes, can be chosen, depending on the local, the area to be treated and the volume to be aspirated.

The 3.5-mm cannula, with different models, are more frequently used.

In the post-op period, patients can, from the day after, take a daily shower and they use moderate compression during 2–4 weeks. All patients have a manual lymphatic drainage program during one month, starting the day after the surgery. The medication after surgery consists of an oral antibiotic, anti-inflammatory and analgesic (rarely used). We recommend massage with an anti-inflammatory cream and heparinoid cream twice a day, until the bruising disappears. Sun block protection is recommended on the small scars and ecchymosis.

10.1.4 The Lipofilling with Fat Harvested with N.I.L.®

The authors describe a new approach to *large-volume lipofilling* that combines liposculpting and fat harvesting that were performed in the same session with Lipomatic/N.I.L.® system.

N.I.L.® transfer of autologous fat improves the ability of the recipient site to receive the graft and allows for explantation and fat transplantation in a single session and the possibility of large-volume fat grafting. This surgical approach minimizes operating time.

The Lipomatic/N.I.L.® system was employed to make the vibro local anesthesia with the modified Klein's solution and to harvest and graft fat. Fat was harvested with a 3-mm multiple-hole cannula attached to a hand-piece and set to *4,2 bars and 0.9 atm.* Lipoaspirate was collected into a closed sterile system, after being washed and decanted with Dextrose 5% (Fig. 10.5).

The remaining solution of adipose tissue was transferred into sterile 20, 60 cc syringes and 100 cc bags to be frozen (Fig. 10.6).

There are no relationships with the Revolve™ or similar system.

The harvest fat is frozen to be utilized in lipofilling of some regions like buttocks, correction of lipo deformities and all others indications for lipofilling, and can also be used to store stem cells.

Power-assisted lipofilling with autologous fat is an efficient, safe, and reproducible procedure that produces an aesthetically pleasing result (Fig. 10.7 and 10.8).

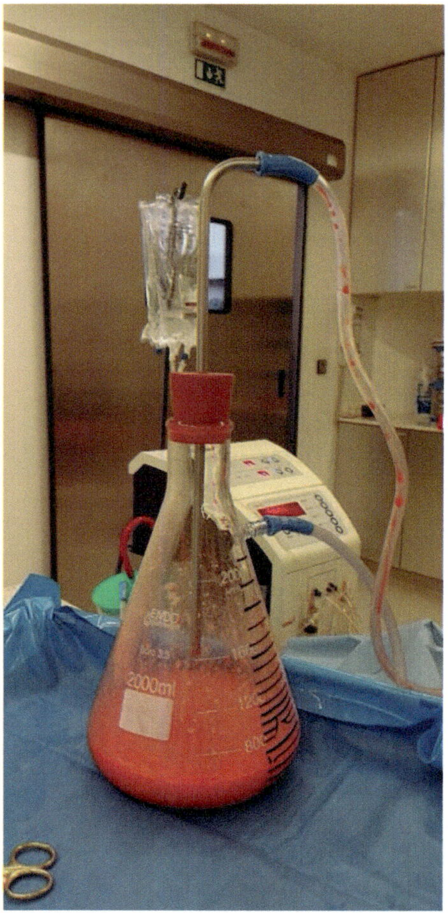

Fig. 10.5 Lipoaspirate collected into a closed sterile system, and after washed and decanted

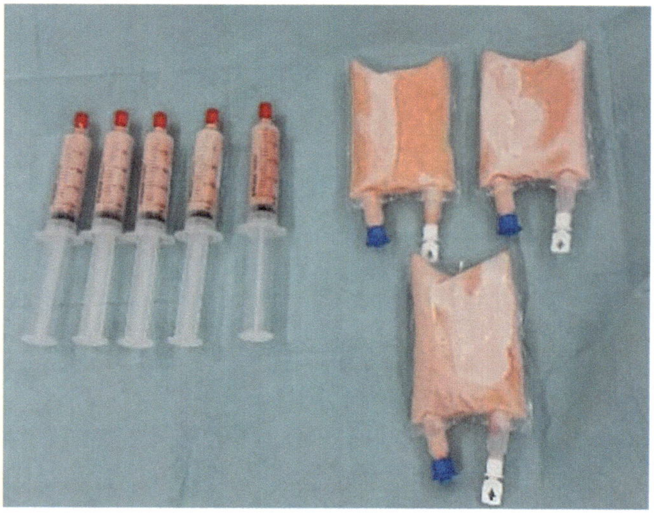

Fig. 10.6 Solution of adipose tissue transferred into sterile 20, 60 cc syringes and 100 cc bags to frozen

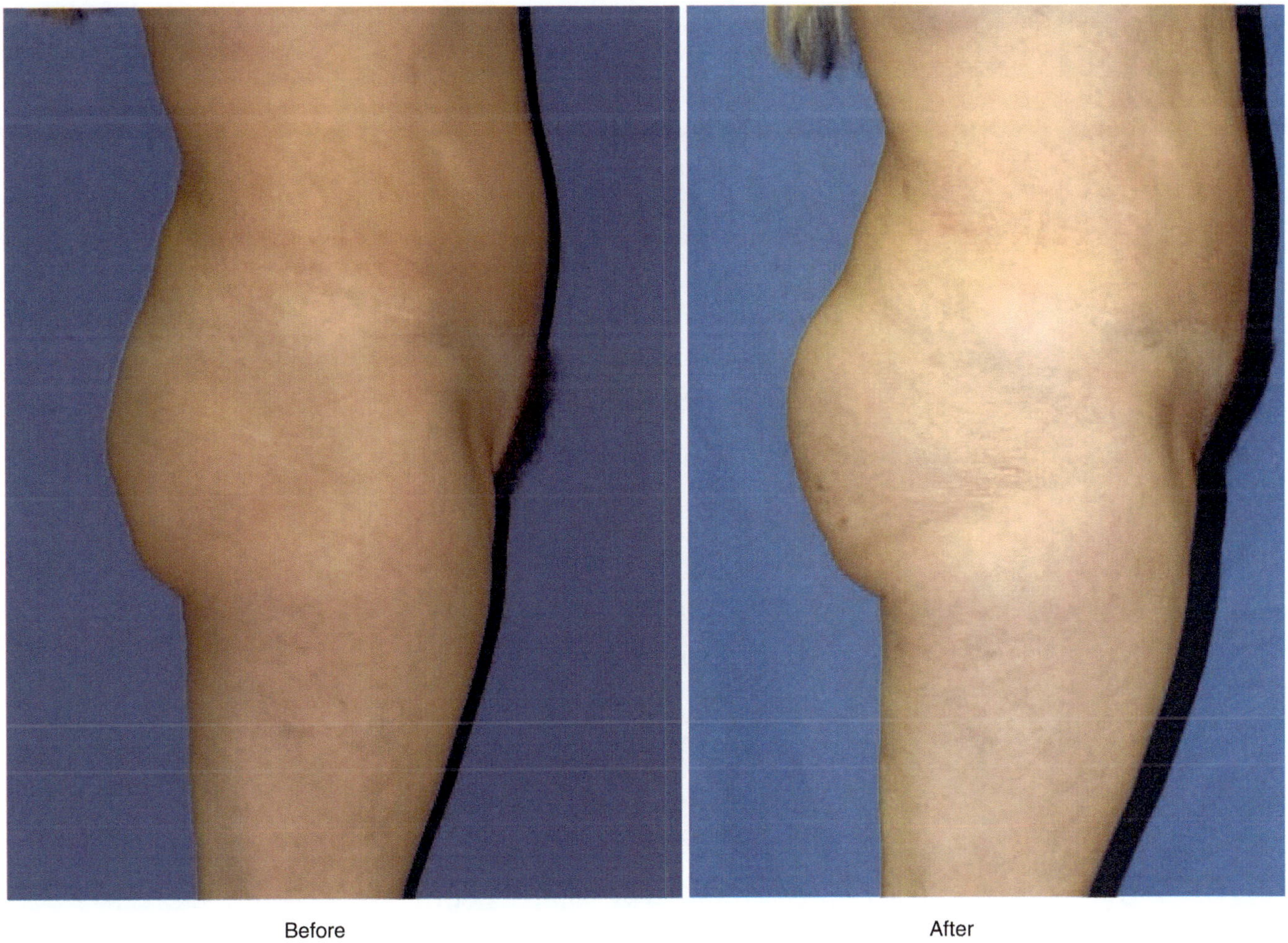

Fig. 10.7 Buttocks (55 years old; five sessions; 920 cc)

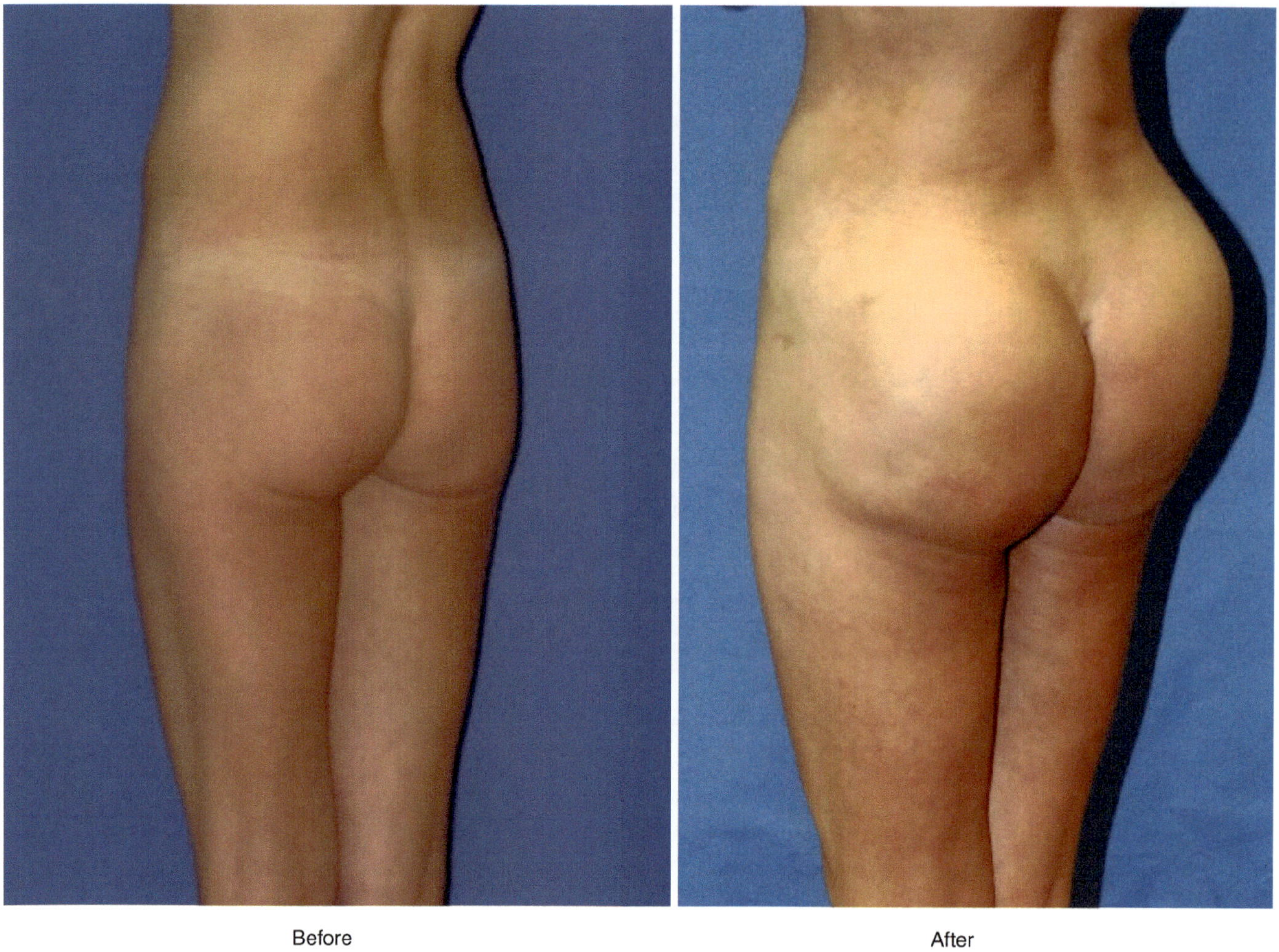

Fig. 10.8 Buttocks (45 years old; six sessions; 860 cc)

10.1.5 Complications and Management

We can say that **N.I.L.**®—Nutational Infrasonic Liposculpture is with no doubt a great and important development and it is proven that it is safe and efficient and has no counter indications. No special maintenance is needed, with good results and less problems and risks. It is very easy to execute especially where the other techniques fail (difficult areas, secondary lipos, etc.) with no technical problems. Much less tiring, this allows the surgeon to perform several surgeries daily. It is more convenient for the patient, with less swelling and bruising, with faster and better recovery as a result.

10.1.6 Results

All regions (Fig. 10.9, 10.10, 10.11, 10.12) can be treated by **N.I.L.**®—Nutational Infrasonic Liposculpture, from small to big areas, from one to several areas at once, but in author's opinions and experience, the best indicator is for localized fat and localized lipo-dystrophies, which means there is a genetic cause, besides that it works very well in difficult areas with fibrotic fat and in secondary liposuctions. In a specific program, it can give a very good help to treat obesity.

The author's experience to treat obese patients is by following an important protocol, a non-predominant of visceral fat. During the consultation and according to a previous evaluation, decisions are made about how many sessions are needed, 2 or 3, to remove fat of different body regions with N.I.L.®. With this technique, we can obtain a good skin retraction, and between 6 and 12 months the patients are revaluated to remove more fat and/or skin.

All of us performing liposuction over several years know very well that the physical effort expended by the surgeon during this surgery is considerable and it reflects on the number of patients operated in a working day. The surgery itself, being long and tiring in the ultimate analysis, can jeopardize the final results.

We cannot ignore the fact that many surgeons and medical doctors refuse to perform liposuction/liposculpture because of the sheer physical effort involved, age and physical being the disability factors involved. With the **N.I.L.**® —Nutational Infrasonic Liposculpture, this don't happen (Table 10.2).

Clinical Cases

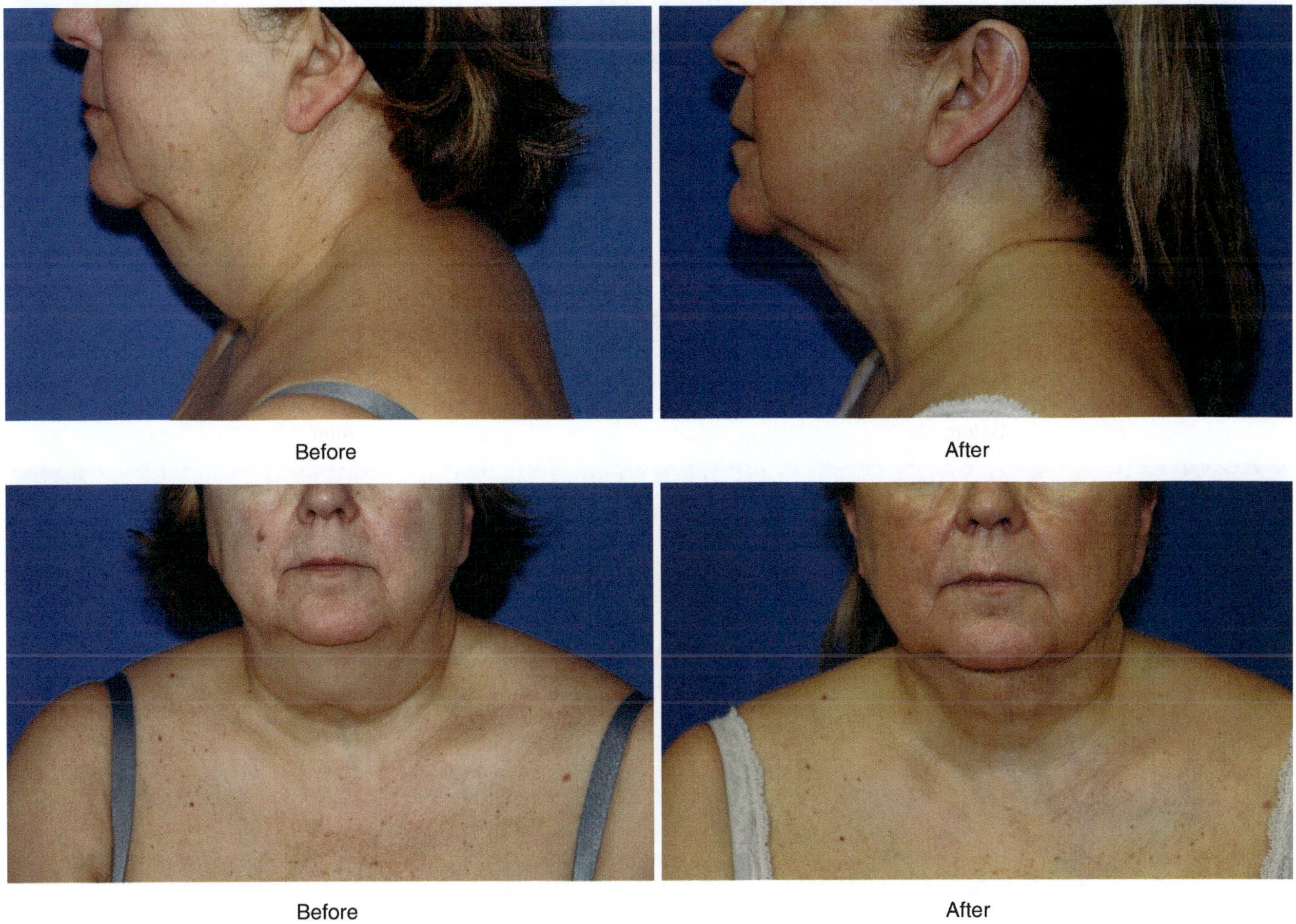

Fig. 10.9 Neck (62 years old; 4 months after; 120 cc)

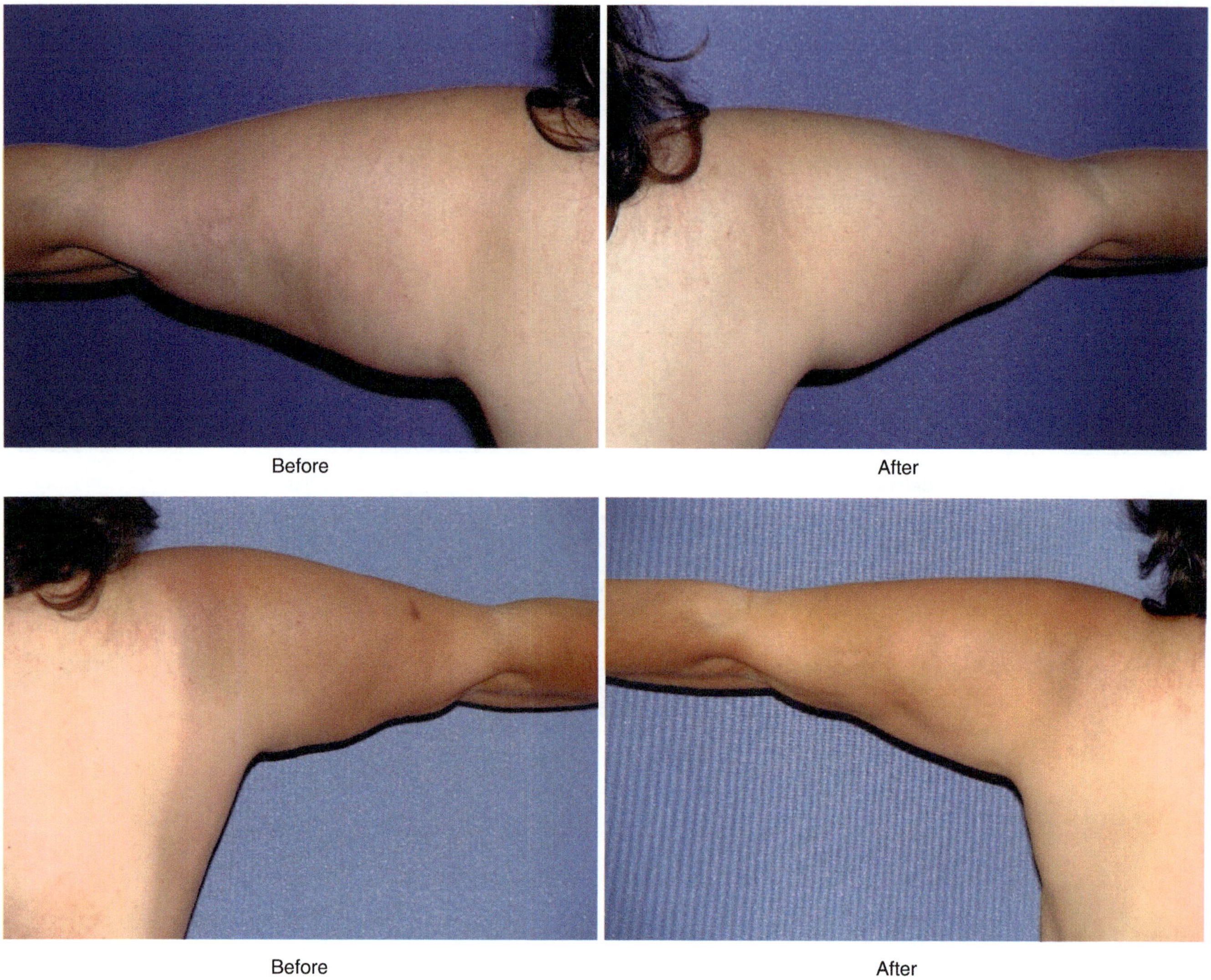

Fig. 10.10 Arms (53 years old; 6 months after; 300 cc)

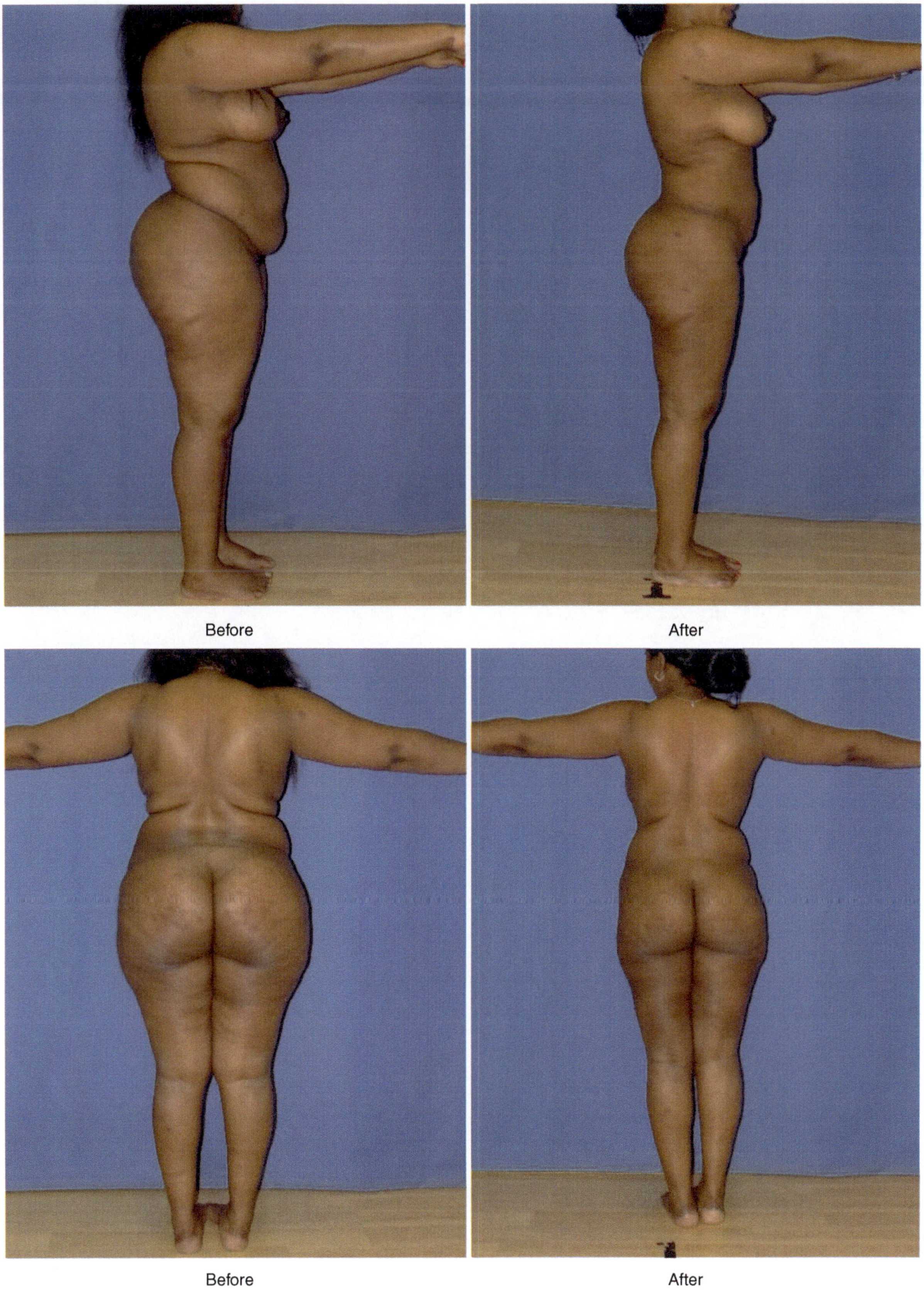

Fig. 10.11 Body (43 years old; 7 months, after two sessions N.I.L.® only; 16.500 cc)

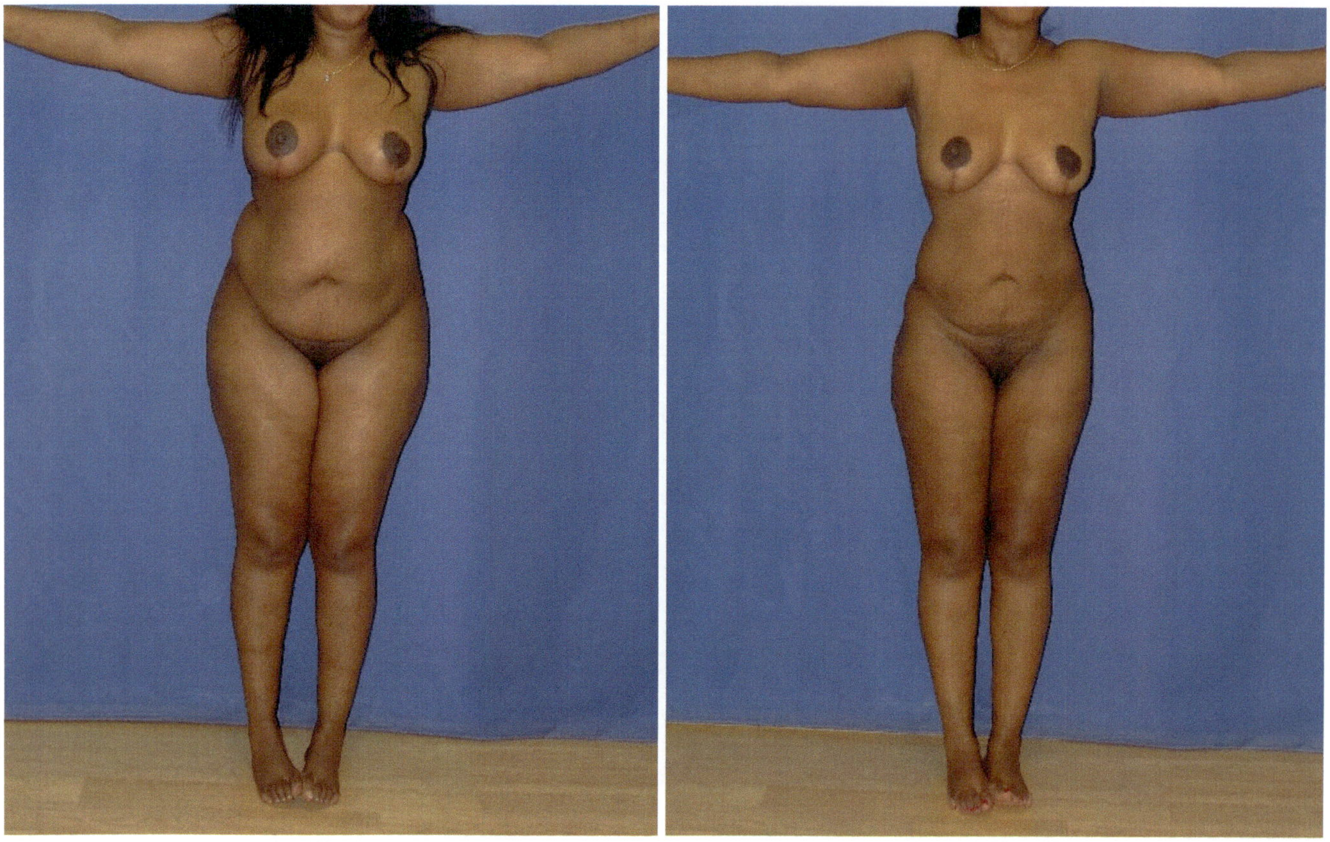

Before After

Fig. 10.11 (continued)

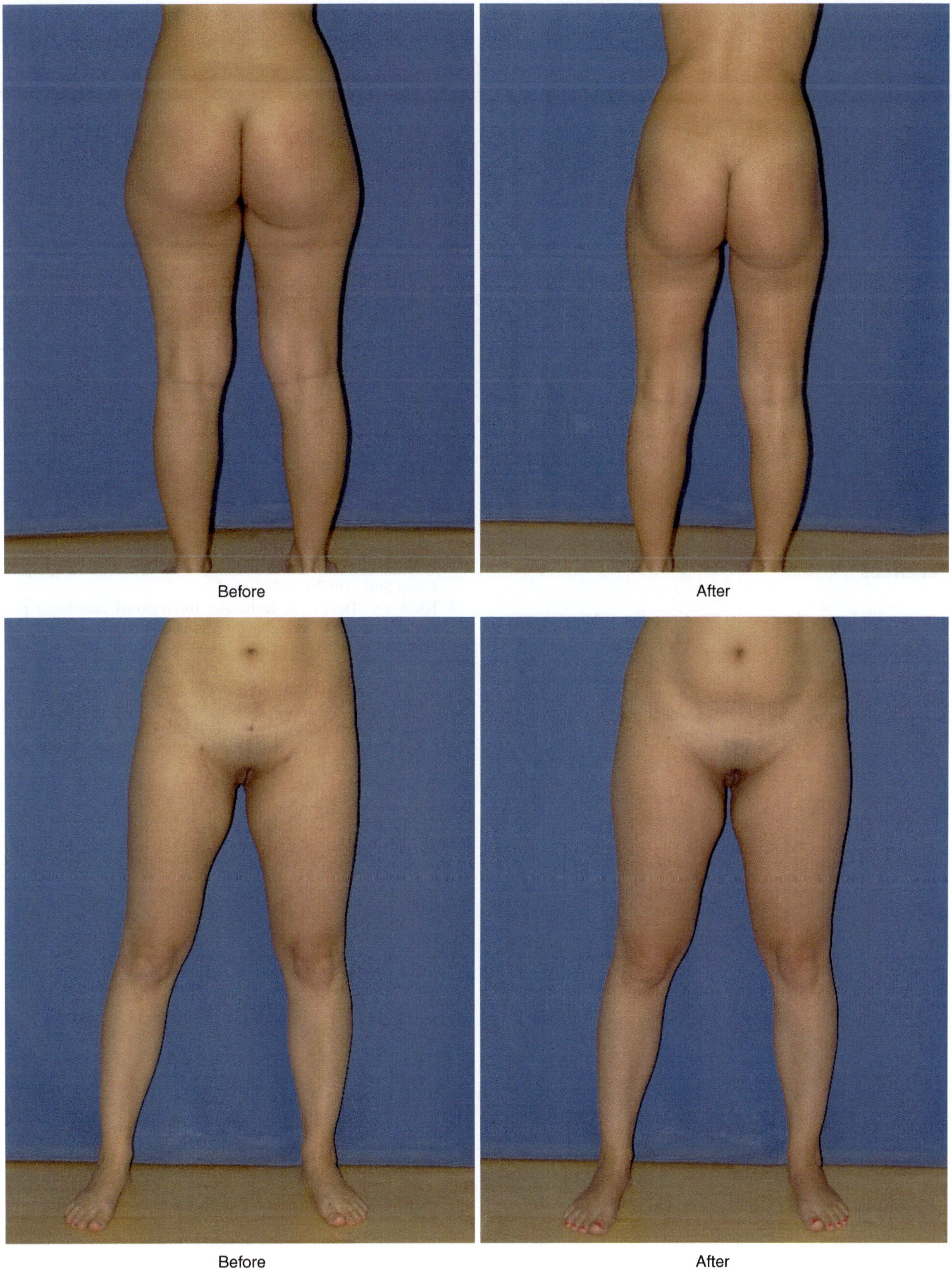

Fig. 10.12 Thighs (22 years old; 3 months after; 3000 cc)

Table 10.2 Body parts where N.I.L.® is performed

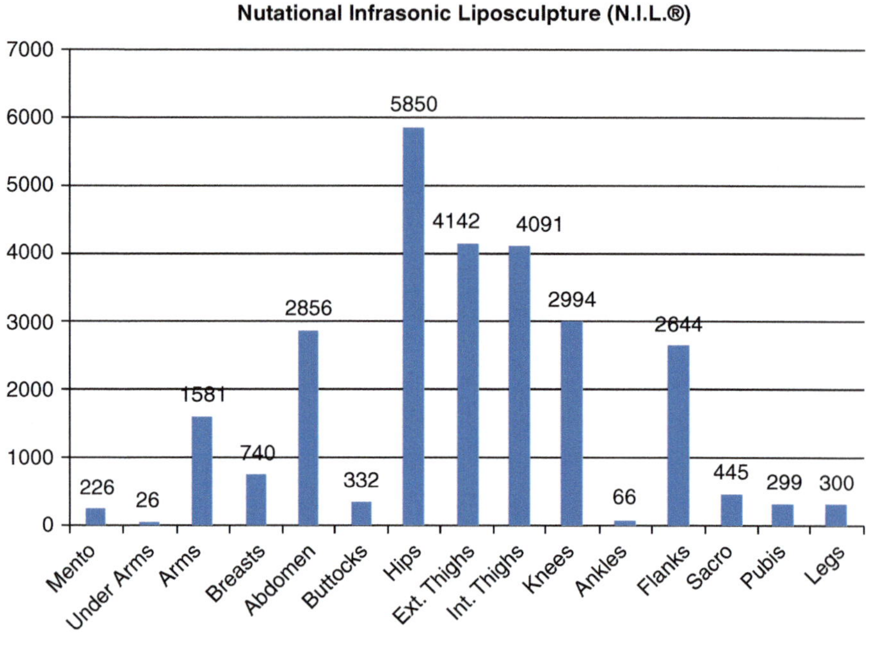

References

1. Fournier PF. Liposculpture – ma technique. Paris: Arnette; 1989.
2. Pierre F. Fournier Liposculpture – ma technique. 2nd ed. Paris: Arnette; 1996.
3. Illouz YG. Body conturing by lipolysis: a 5-year experience with over 3000 cases. Plast Reconstr Surg. 1983;72:591.
4. Zochi M. Ultrasound-assisted lipoplasty. Adv Plast Reconstr Surg. 1998;11:197–221.
5. Klein JA. Anesthesia for liposuction in dermatologic surgery. J Dermatol Surg Oncol. 1988;14:1124–32.
6. Klein JA. The tumescent technique. Anesthesia and modified liposuction technique. Derm Clin. 1990;8(3):425–37.
7. Klein JA. The tumescent technique for liposuction surgery. Am J Cosm Surg. 1987;4:263–7.
8. Klein JA. Tumescent technique for regional anesthesia permits lidocaine doses of 35mg/kg for liposuction. J Dermatol Surg Oncol. 1990;16:3.
9. Klein JA. Tumescent technique for local anesthesia improves safety in large-volume liposuction. Plast Reconstr Surg. 1993;92:1085–98.
10. Klein JA. Anesthesia for dermatologic cosmetic surgery: principles and techniques. In: Coleman WP, Hanke CW, Alt TH, et al., editors. Cosmetic surgery of the skin. Philadelphia: BC Decker Inc.; 1991. p. 39–45.

Part II

Technologies

Stromal-Enriched Lipograft™: Combining Chemical Automatic Cell Station and Mechanical Lipocube for Supercharged Fat Graft

11

Aris Sterodimas

Contents

11.1 Introduction

The improvement of surgical techniques in autologous fat transplantation has increased the survival rate of the grafted fat and has reduced the side effects such as fat necrosis. Several mechanisms may contribute to the variability of outcomes for fat-grafting procedures. The harvesting process is traumatic to adipocytes, which can lead to apoptosis. Additionally, results for recipient wound beds with different degrees of blood supply and fluctuations in oxygen delivery may range from adequate revascularization and good "take" to insufficient revascularization and ischemia, apoptosis, and dedifferentiation of central adipocytes [1]. A significant proportion of the engrafted fat undergoes resorption and necrosis due to the non-vascularized nature of the transplant. Resorption rates after fat grafting are generally reported to be between 55 and 80% [2]. Therefore, in order to maximize survival, multiple passes in different tissue planes are required in order to optimize plasmatic imbibition and neovascularization of the transplanted fat grafts. Injection of the fat graft should be met with minimal resistance, and a small volume should be administered upon entry, with the remaining volume given during withdrawal to minimize trauma. Adipose tissue is composed of mature adipocytes constituting about 90% of the tissue volume and a stromal vascular fraction (SVF), including fibroblasts, endothelial cells, preadipocytes, vascular smooth muscle cells, lymphocytes, and resident monocytes/macrophages. It has become apparent over the years that white adipose tissue (WAT) is the most suitable autologous injectable filler for correcting soft tissue defects [3]. Adipose tissue is considered as a source of mesenchymal stem cells (MSCs), termed adipose-derived stem cells (ADSCs). They are ubiquitous and easily obtained in large quantities with little donor site morbidity or patient discomfort, making the use of autologous ADSCs an appropriate cellular therapy. The use of ADSCs may enhance angiogenesis, improve the survival of grafts, and thus reduce the atrophy of the fat grafting [4]. ADSCs have been evaluated in clinical studies for soft tissue augmentation and represent a novel approach to cell-based therapies, such as autologous fat transplantation. Nowadays, autologous fat transplantation or adipose cell grafting incorporates adipocytes, ADSCs, and growth factors already present in the lipoaspirate. In 2005, the research team led by the senior author Dr. Sterodimas described a method of supplementing the lipoaspirate used for fat grafting with the stromal vascular fraction found in adipose. This process was named Stromal-Enriched Lipograft™. The Stromal-Enriched Lipograft™ (SEL™) is a surgical technique of autologous fat grafting for face and body contouring that converts a stem cell-poor fat graft to a

A. Sterodimas (✉)
Plastic & Reconstructive Surgery Department, Metropolitan General Hospital, Athens, Greece
e-mail: aris@sterodimas.com; https://www.sterodimas.com

© The Author(s), under exclusive license to Springer Nature Switzerland AG 2023

A. Di Giuseppe et al. (eds.), *Fat Transfer in Plastic Surgery*, https://doi.org/10.1007/978-3-031-10881-5_11

stem cell-rich fat graft [5]. The clinical outcomes of scientific research on the last decade have shown that SEL is one of the core elements of regenerative medicine.

11.2 Patient Selection

Contouring of the face, breast, trunk, and extremities combine liposuction, excisional surgical techniques, and autologous fat grafting [6]. These patients should be carefully evaluated for their expectations and the actual results that can be achieved by the proposed procedures. Planning each procedure and indicating it properly is just as important as the surgical procedure itself. The ideal patient has a small amount of adipose tissue to be suctioned out in order to improve the face and body contour, has small skin laxity, and is dedicated to a healthy lifestyle, including daily physical exercise and a healthy diet. It is the responsibility of the surgeon to address all concerns, risks, goals and expectations of the procedure chosen to the patient in order to avoid false expectations and future frustrations. In this chapter, we will address special attention to fat grafting of the face, breast, upper and lower abdomen, upper and lower back, flanks, pubis and sacrum V-zone, gluteal area, inner thighs, outer thighs, inner knees, and hands.

11.3 Technique

Marking of the areas to be liposucted is made while the patient is in a standing position. Preoperative sedation in the surgical suite is administered. Anesthesia consists of an epidural block and intravenous sedation. The patient is placed in a prone position. After the injection of normal saline wetting solution containing 1:500,000 of adrenaline by a small bore cannula and waiting 15 min, a 60-cc syringe attached to a 4 mm blunt cannula is inserted through small incisions. Fat is aspirated by using the syringe method. Two-third of the aspirated fat is used in order to isolate the SVF. Digestion is done with 0.075% collagenase (Sigma, St. Louis, MO) in buffered saline and agitated for 30 minutes at 37°C in Automatic Cell Station (ACS) manufactured by BSL in Seoul, S. Korea. Separation of the SVF containing ADSCs is then done by using centrifugation at 1200 × *g* for 5 min using the ACS. The SVF is located in the pellet derived from the centrifuged fat at the bottom of the lipoaspirate. In Stromal-Enriched Lipograft (SEL), freshly isolated SVF is attached to the aspirated fat, with the fat tissue acting as a living bioscaffold before transplantation [7]. The remaining 1/3 of the aspirated fat is treated as follows: with the syringe held vertically with the open end down, the fat and fluid are separated. Isotonic saline is added to the syringe, the fat and saline are separated, and the exudate is discarded. The procedure is repeated until the fat becomes yellow in color, free of blood, and other contaminants [8]. Mixing of the SVF containing ADSCs and the purified fat is then done (Fig. 11.1). This whole procedure is done inside the operating theater by two tissue engineers, manually, and the time required is about 90 min (Video 11.1). The Lipocube™ is a single-use mechanical device for processing lipoaspirate, the Stromal-Enriched Lipograft™, into milli, micro, and nano fat graft according to the depth of transfer (Fig. 11.2). The device processes fat tissue in a closed sterile system with specific structural geometry blades coupled with a flow pattern enabling partial filtration of fibrous material. Tissue planes are created by using specific cannulas in different trajectories, always from the deeper aspect to more superficial areas using the MAFT-GUN produced by Dermato Plastica Beauty Co, Ltd. in Taiwan. Successful fat application is performed using a blunt cannula that creates a tunnel at insertion, and the fat is injected as the cannula is withdrawn in order to avoid intravascular fat injection. Multiple passes are used to fan across the deficient region. Antibiotics, analgesics, and anti-inflammatory medications are prescribed during the following 7 postoperative days.

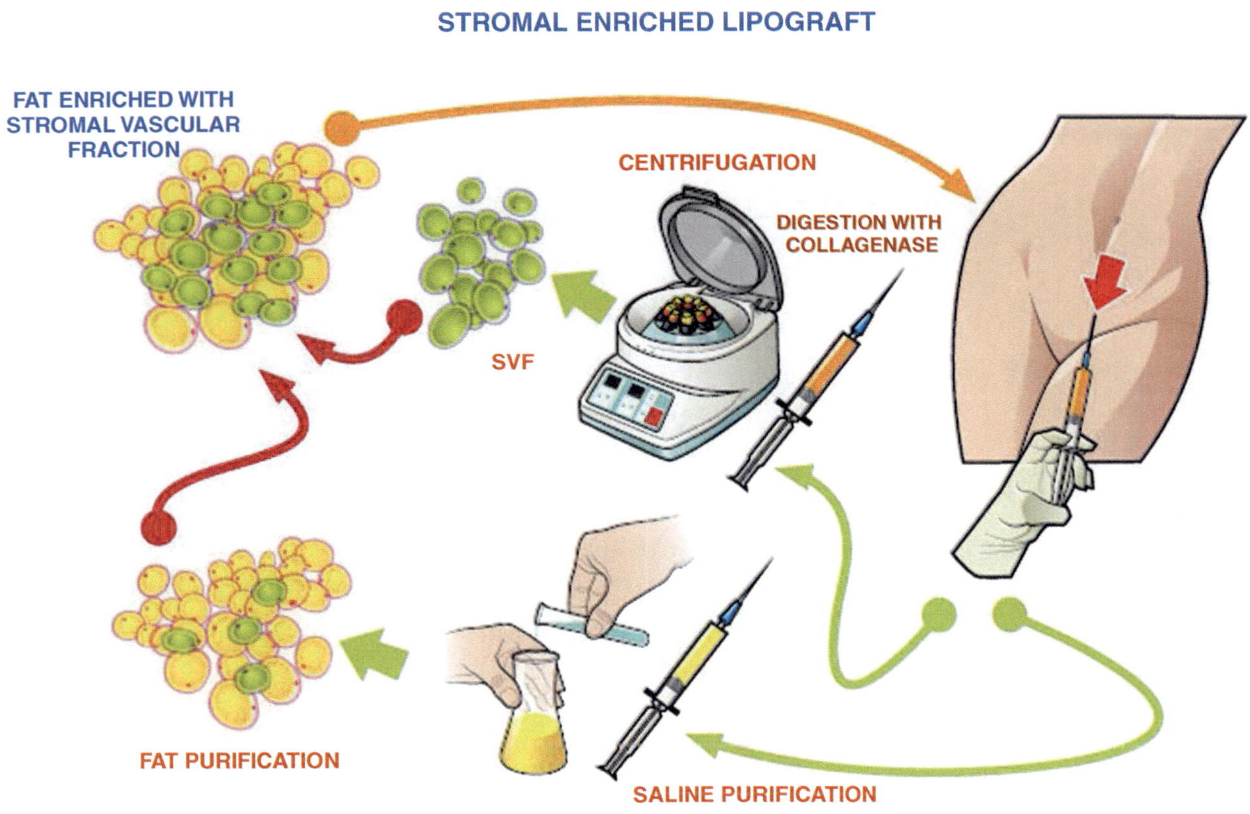

Fig. 11.1 Schematic representation of Stromal Enriched Lipograft technique

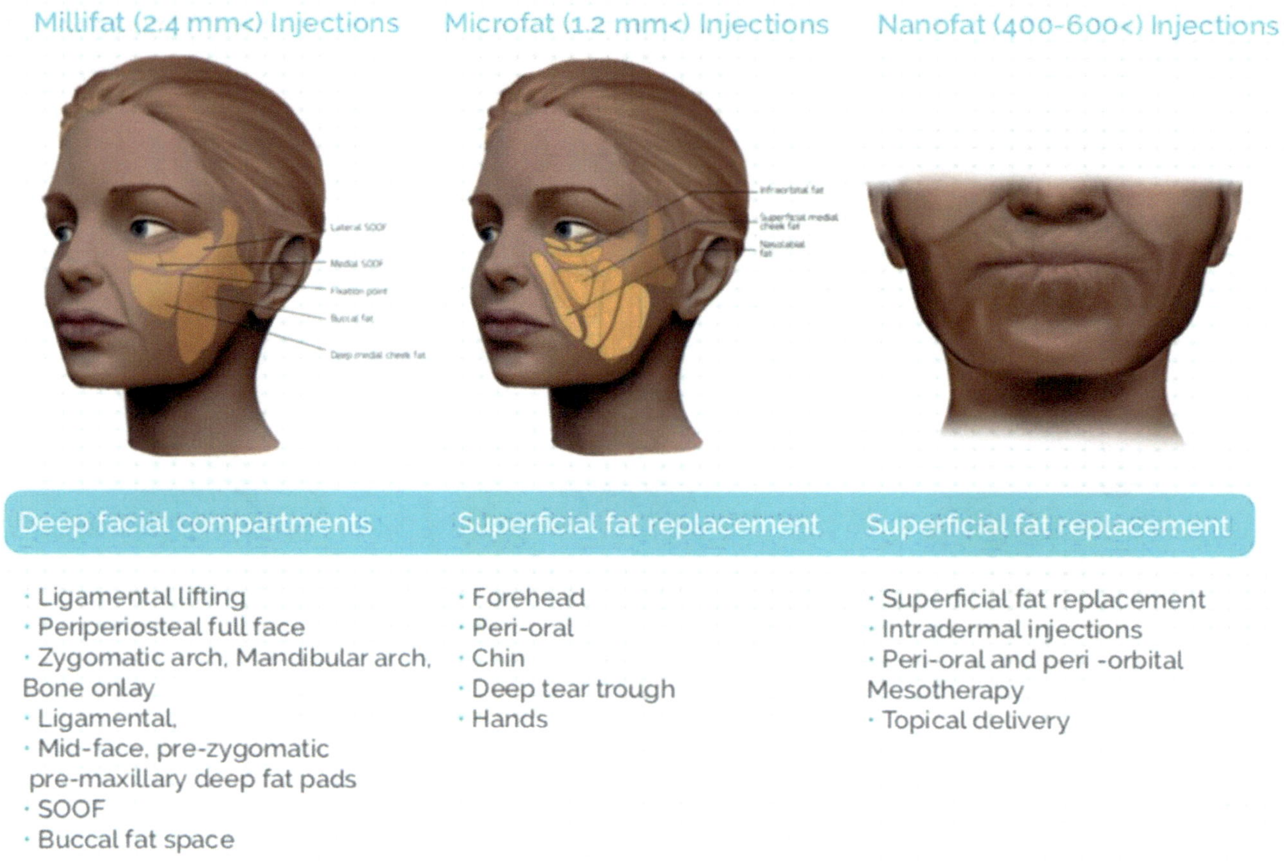

Fig. 11.2 A Schematic representation of the different fat products (millifat, microfat and nanofat) for facial rejuvenation

11.4 Clinical Applications

Stromal-Enriched Lipograft technique aims at filling preoperatively marked deficiencies with adipocytes that will survive and become incorporated into the recipient bed. The combination of circumferential liposuction, SEL of breasts, buttocks, and lower limbs in a single surgical procedure has been performed successfully in the last 10 years, emphasizing the low rate of complications and the high overall patient satisfaction [9]. The procedure is named All-in-One™. SEL™ should be thought as part of body contouring in order to give the balanced and properly proportioned anatomy of the body. The aspirated fat may be utilized to augment, shape, and to correct irregularities or asymmetry that may be detected preoperatively [10]. The use of Stromal-Enriched Lipograft for body contouring has gained attention due to further improvements in fat preparation and processing. Recently the concept of composite body contouring has been introduced, combining lipoabdominoplasty and Stromal-Enriched Lipograft to the thighs and gluteal area has been done in 575 patients with aesthetically favorable results.

11.4.1 Patient 1

This 49-year-old lady presented complaining of facial sagging, which made her "unattractive." Preoperative views of the patient are shown in Fig. 11.3a–c. Face lifting assisted by SEL was performed. The total fat transfer is 30 mL. Postoperative views were taken 2 years after the procedure (Fig. 11.3d–f).

11.4.2 Patient 2

This 63-year-old lady presented complaining of facial sagging, which made her "unattractive." Preoperative views of the patient are shown in Fig. 11.4a–c. Face lifting assisted by SEL was performed. The total fat transfer is 30 mL. Postoperative views were taken 2 years after the procedure. Postoperative views were taken 1 year after the procedure (Fig. 11.4d–f).

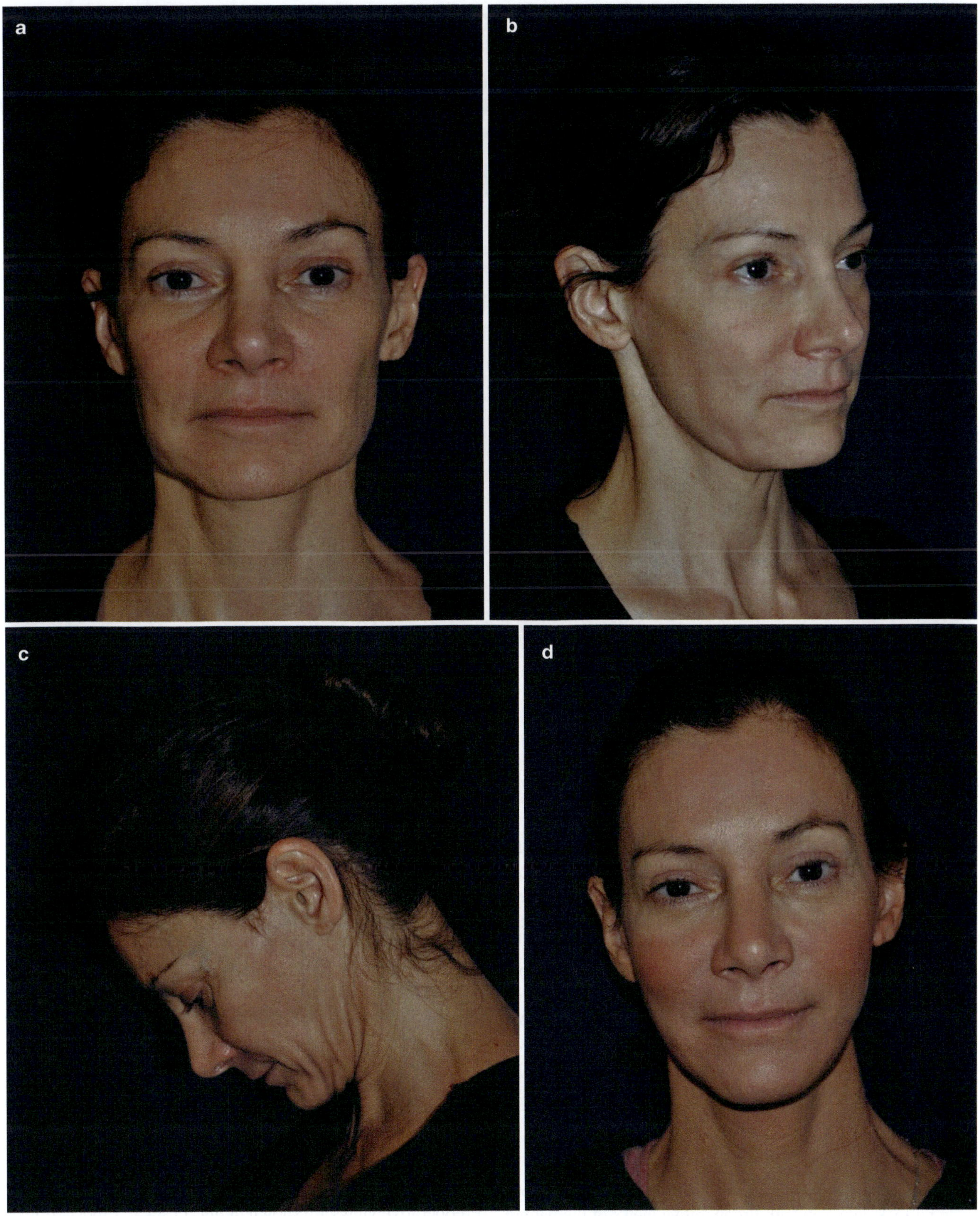

Fig. 11.3 (**a–c**) Preoperative views of a 47-year-old lady presented complaining of facial atrophy and flaccidity. (**d–f**) Postoperative views of a 47-year-old lady 2 years after face lifting assisted by SEL was performed

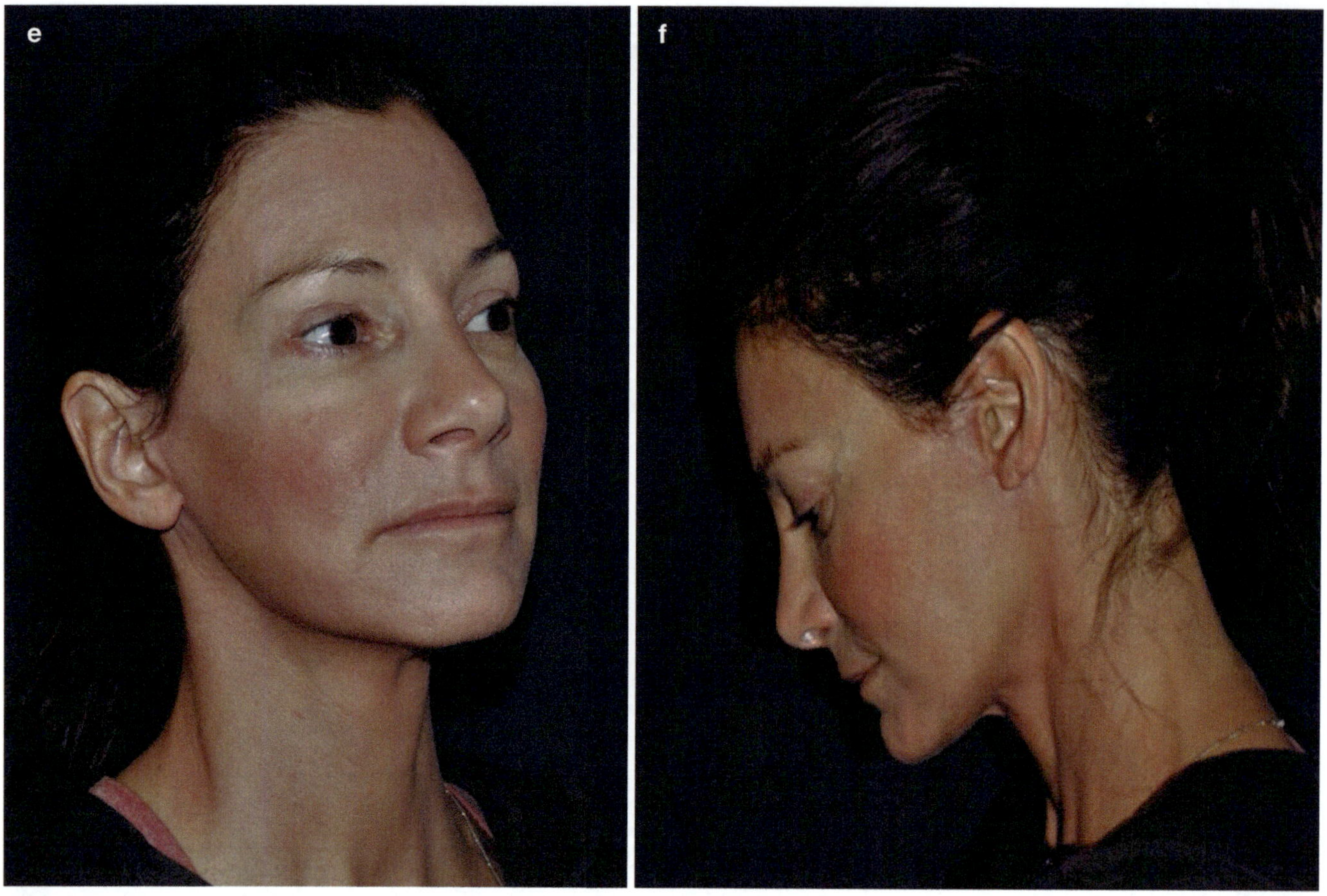

Fig. 11.3 (continued)

11.4.3 Patient 3

This 22-year-old lady presented complaining of breast hypomastia, which made her "unattractive". Preoperative views of the patient are shown in Fig. 11.5a–c. Liposuction of the back, flanks, and abdomen and SEL was performed in both breasts. The fat transfer per breast was 190 mL in the first session. Postoperative views were taken 2 years after the procedure (Fig. 11.5d–f). The patient decided that she wanted to have another SEL session 3 years after the initial operation. She underwent liposuction of the back, flanks, and abdomen and SEL was performed in both breasts. The fat transfer per breast was 250 mL in the second session. Postoperative views were taken 1 year after the second procedure (Fig. 11.5g–i).

11.4.4 Patient 4

This 28-year-old woman presented for liposuction and moderate buttock enhancement. Preoperative views of the patient are shown in Fig. 11.6a, b. Liposuction of the back, flanks, and abdomen was done. Stromal-Enriched Lipograft was performed. The total buttocks fat transfer was 350 mL per side. Postoperative views were taken 3 years after the procedure (Fig. 11.6c, d).

11.4.5 Patient 5

This 35-year-old man complained about his body shape. Preoperative views of the patient are shown in Fig. 11.7a–c. Liposuction of the back, flanks, chest, and abdomen and SEL was performed. The total gluteal fat transfer was 270 mL per side using the SEL technique. Postoperative views were taken 2 years after the procedure (Fig. 11.7d–f).

11.4.6 Patient 6

This 51-year-old lady presented, complaining of having excess skin and fat deposits, which made her "unattractive." Preoperative views are shown in Fig. 11.8a–c. Liposuction of the back, flanks, lipoabdominoplasty, and SEL was performed. The total gluteal fat transfer was 420 mL per side

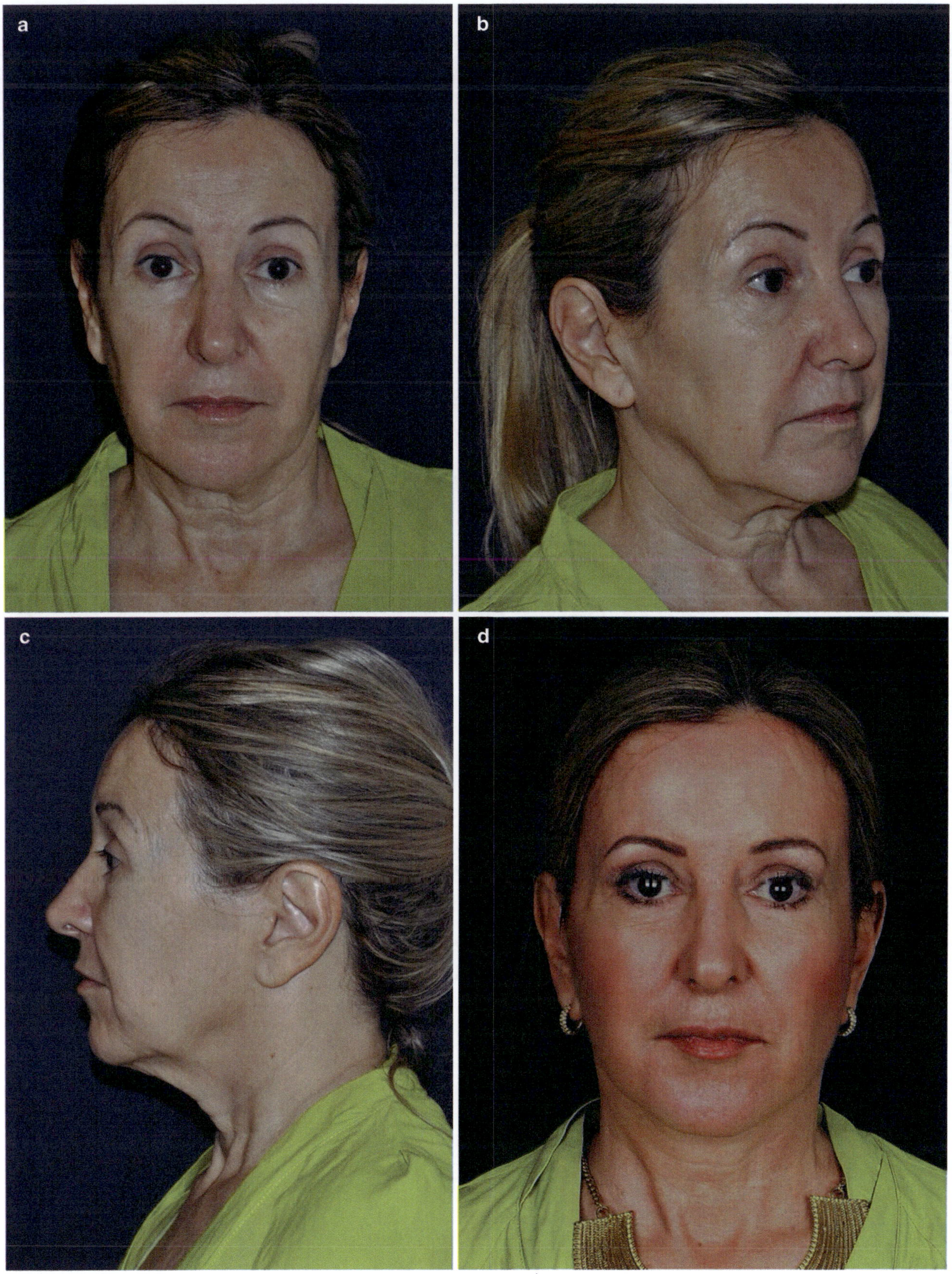

Fig. 11.4 (**a**–**c**) Preoperative views of a 63-year-old woman 2 years after face lifting assisted by SEL was performed. (**d**–**f**) Postoperative views of a 47-year-old lady 2 years after face lifting assisted by SEL was performed

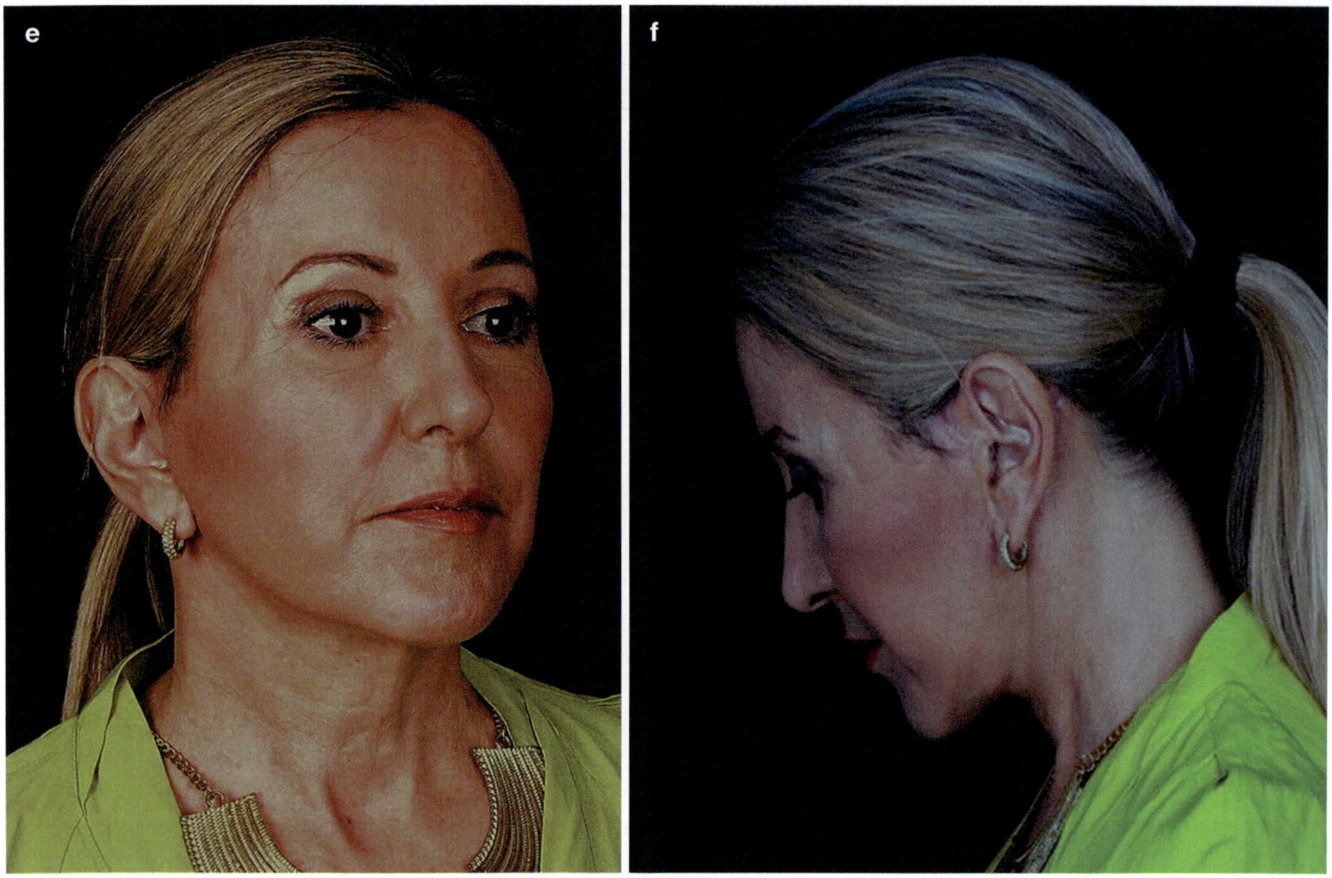

Fig. 11.4 (continued)

using the SEL technique. Postoperative views were taken 4 years after the procedure (Fig. 11.8e–g).

11.4.7 Patient 7

This 73-year-old woman presented, complaining of her "unattractive" hands. Preoperative views are shown in Fig. 11.9a. Liposuction of the back, flanks, and SEL of the hands was performed. The total fat transfer to each hand was 12 mL using the SEL technique. Postoperative views taken 2 years after the procedure (Fig. 11.9b).

11.4.8 Patient 8

This 42-year-old woman presented requesting rejuvenation of her external genitalia. Preoperative views are shown in Fig. 11.10a. Liposuction of the back, flanks, and SEL of the external genitalia was performed. The total fat transfer to the external genitalia was 32 mL using the SEL technique. Postoperative views were taken 1 year after the procedure (Fig. 11.10b).

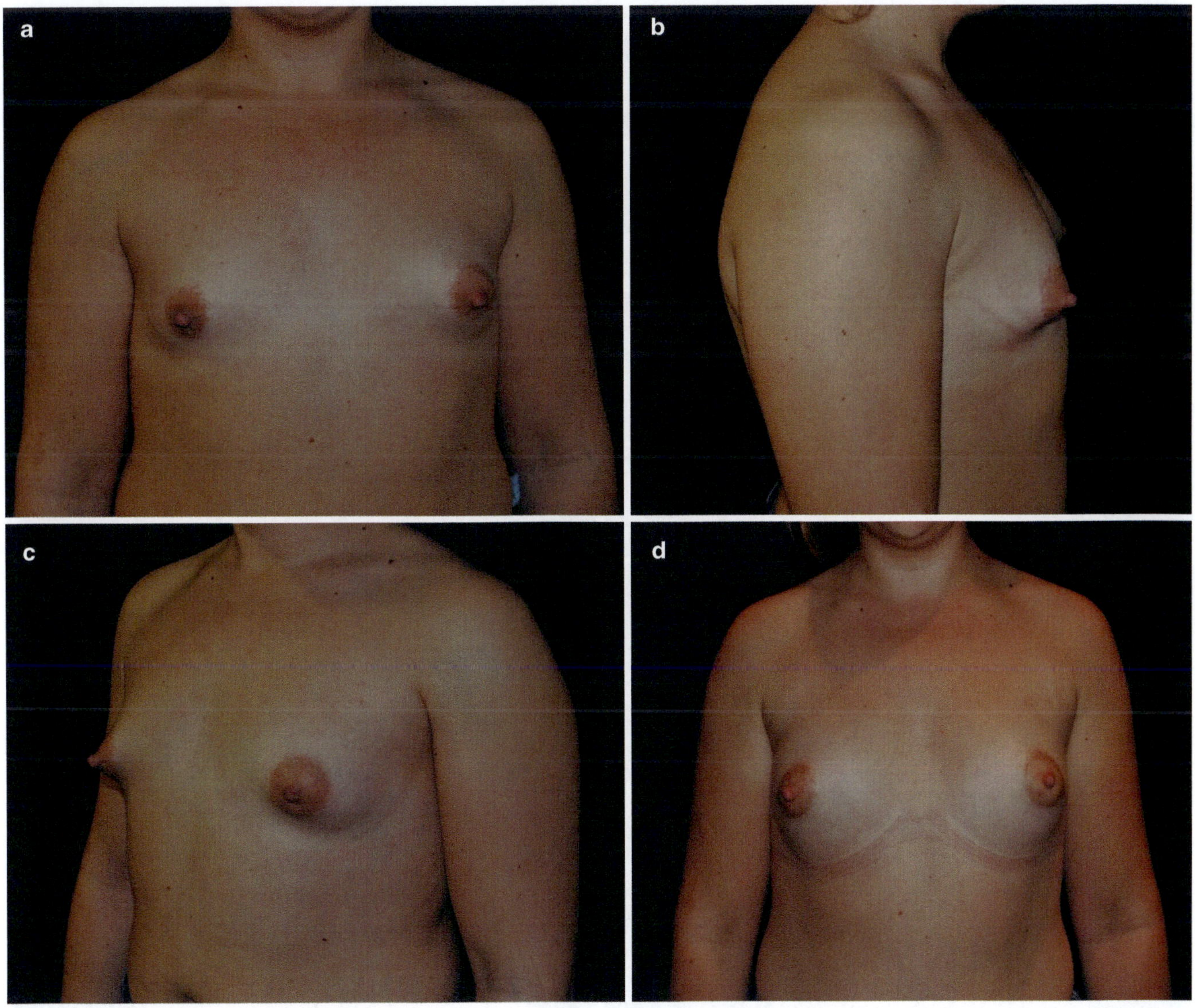

Fig. 11.5 (**a–c**) Preoperative views of a 22-year old female patient with hypomastia. (**d–f**) Postoperative views of a 22-year old female patient after breast SEL 2 years after the first surgical intervention. (**g–i**) Postoperative views of a 22-year old female patient 1 year after the second surgical intervention for breast SEL

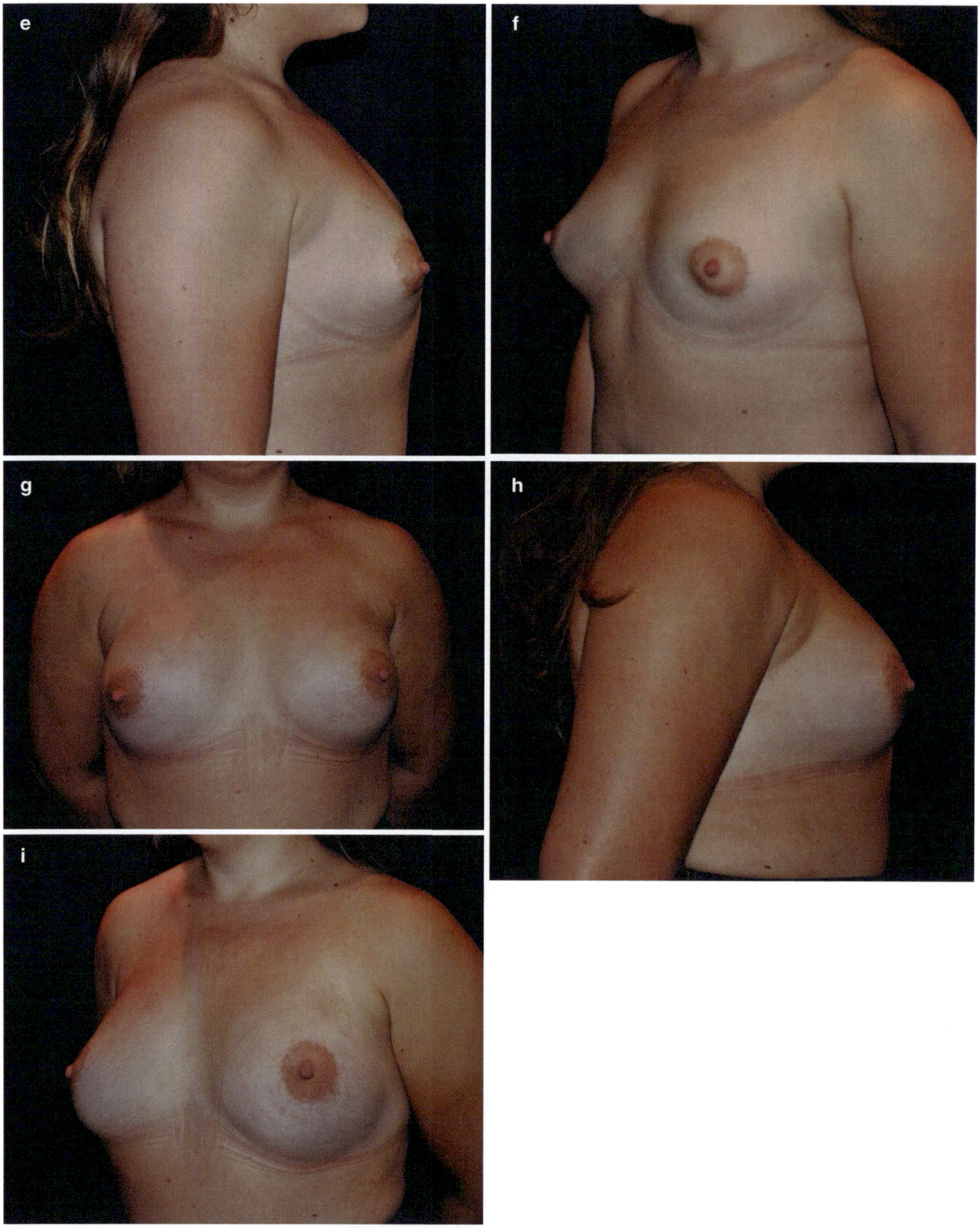

Fig. 11.5 (continued)

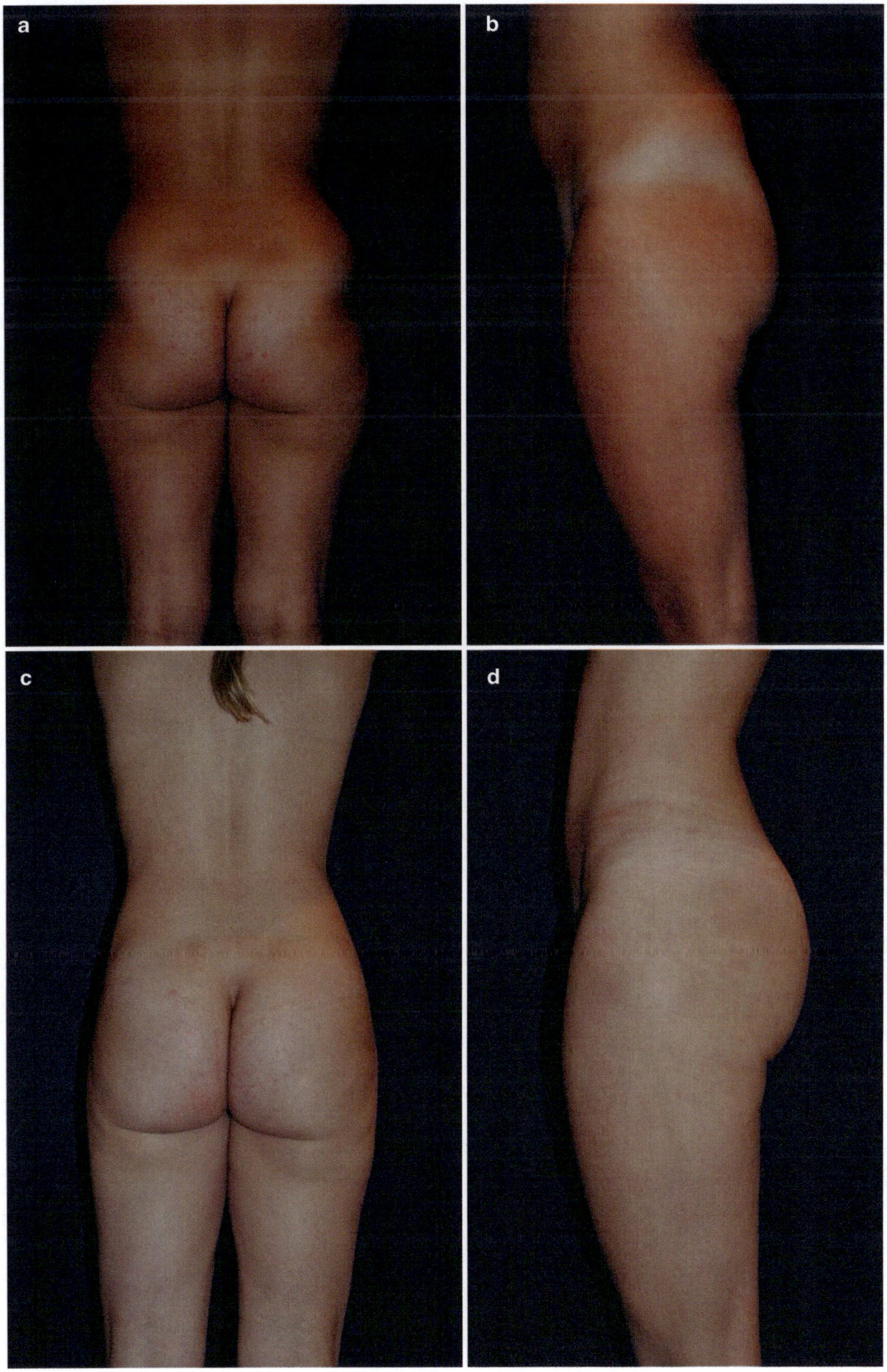

Fig. 11.6 (**a**, **b**) Preoperative vies of a 28-year old female patient requesting buttocks augmentation. (**c**, **d**) Postoperative views of a 28- year old female patient 3 years after undergoing SEL buttocks augmentation

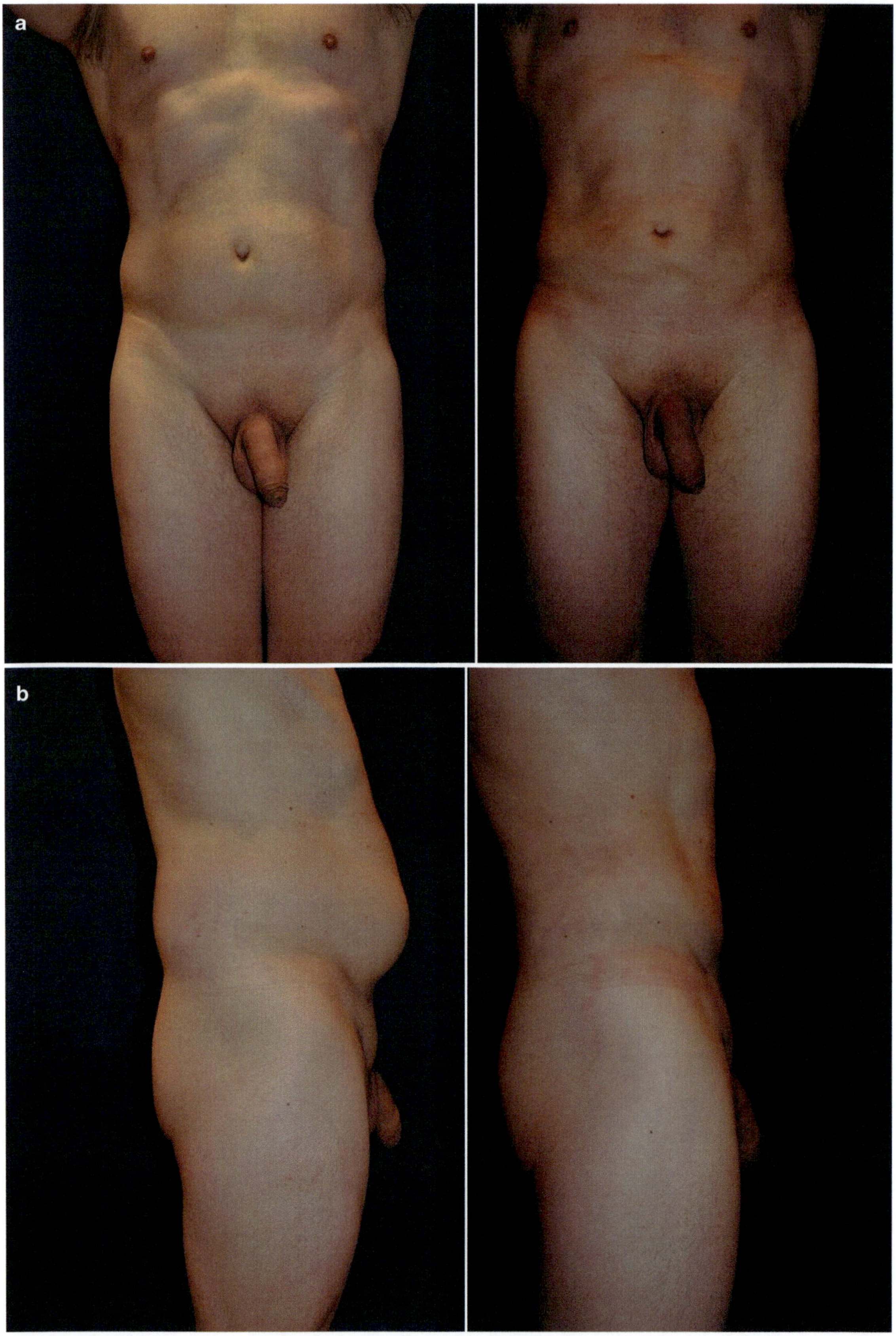

Fig. 11.7 (**a**–**c**) Preoperative views of a 35-years old male patient complaining about his body shape. (**d**–**f**) Postoperative views of a 35-years old male patient 2 years after undergoing liposuction and gluteal augmentation by SEL

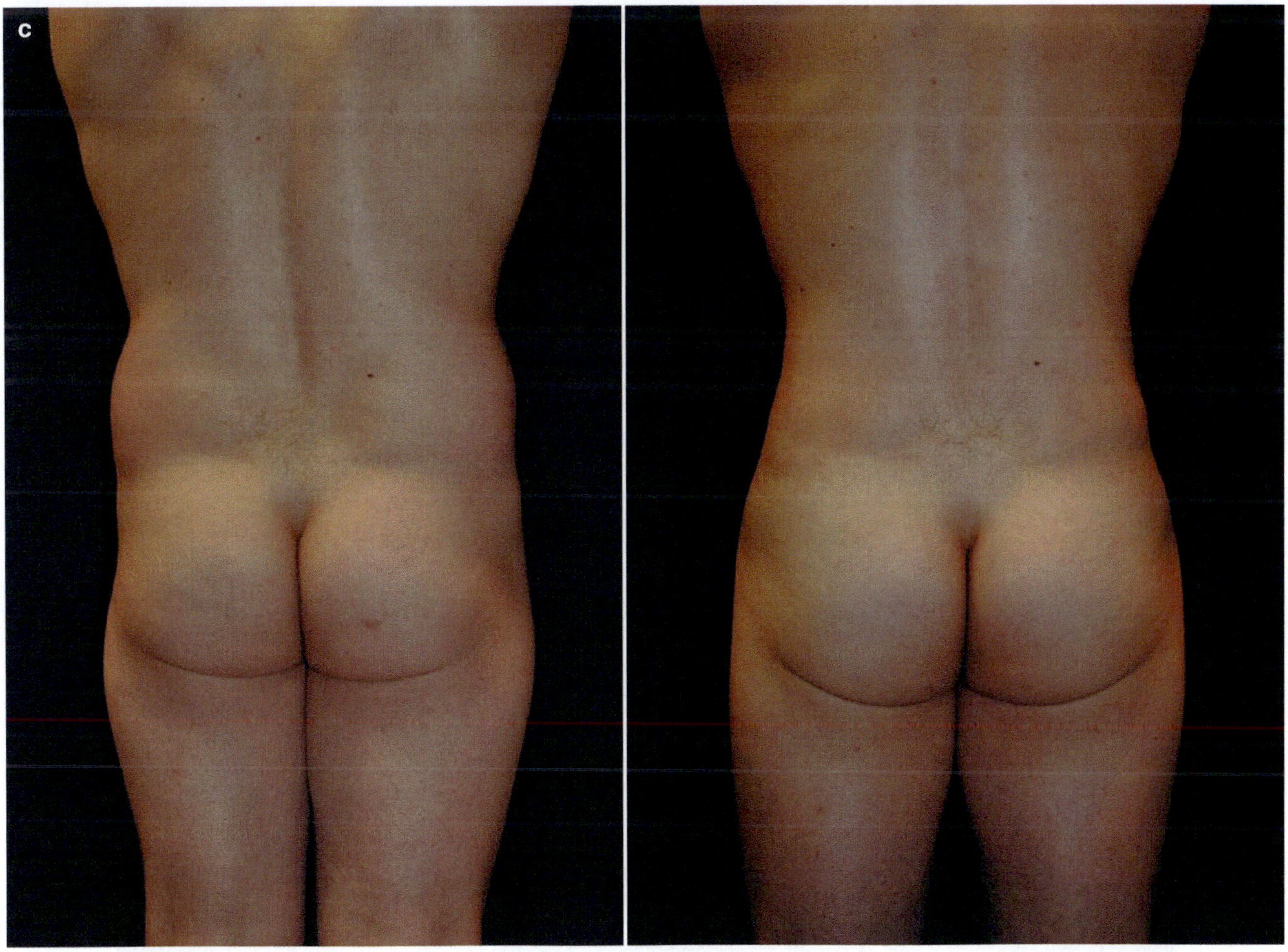

Fig. 11.7 (continued)

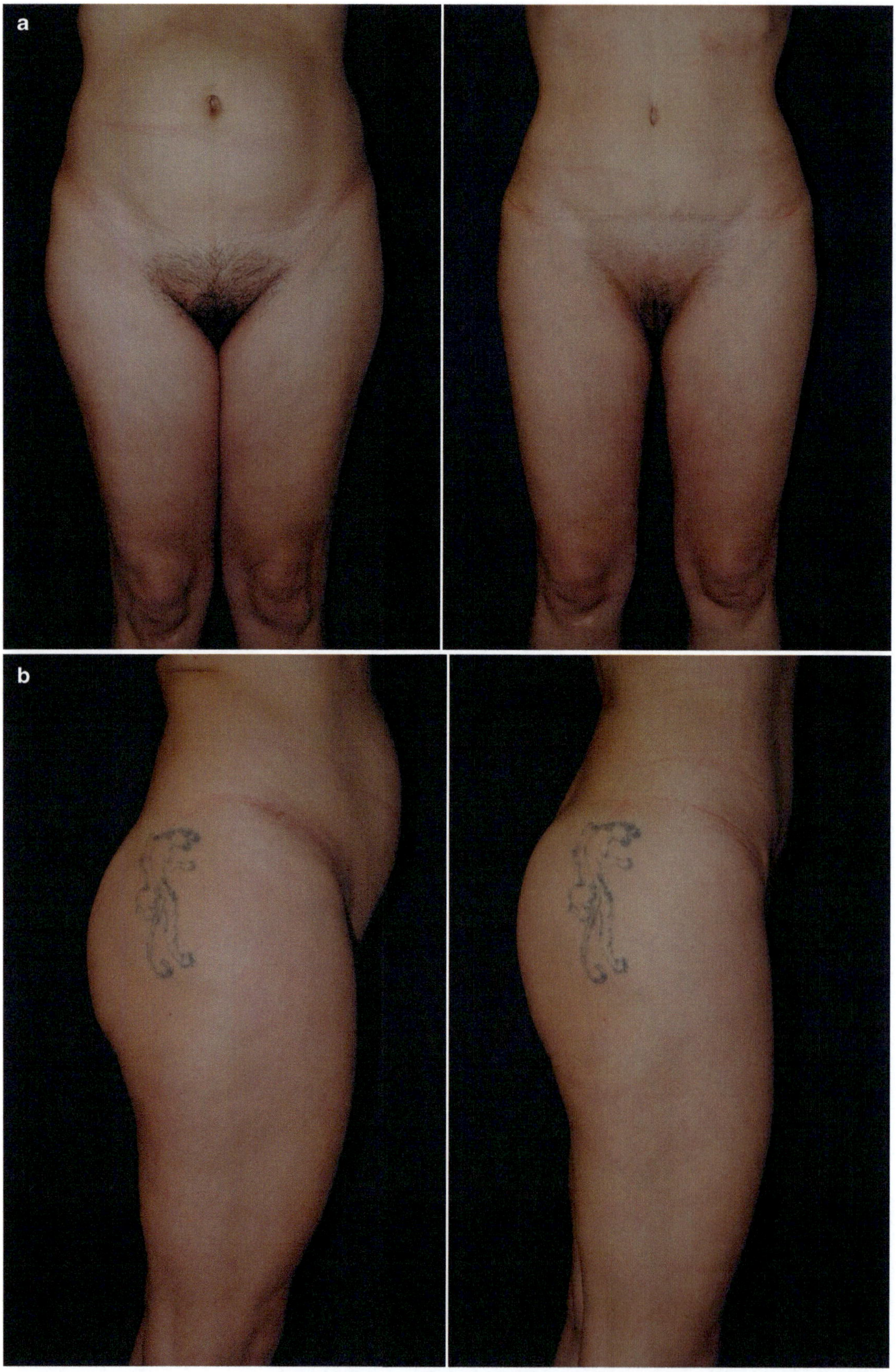

Fig. 11.8 (**a**–**c**) Preoperative views of a 51-year old female patient complaining about her body. (**d**–**f**) Postoperative views of a 51-year old female patient 4 years after undergoing lipoabdominoplasty and gluteal augmentation by SEL

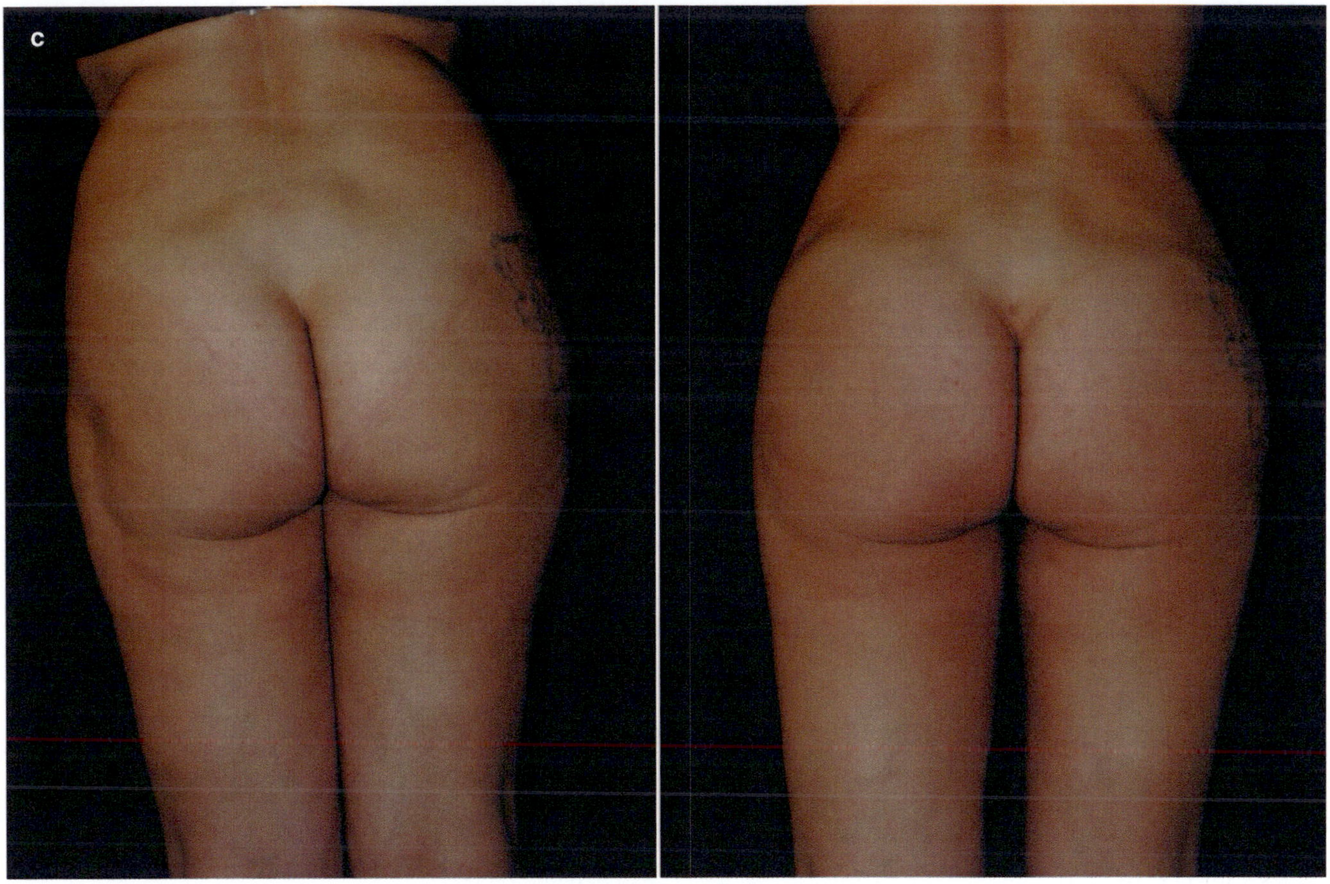

Fig. 11.8 (continued)

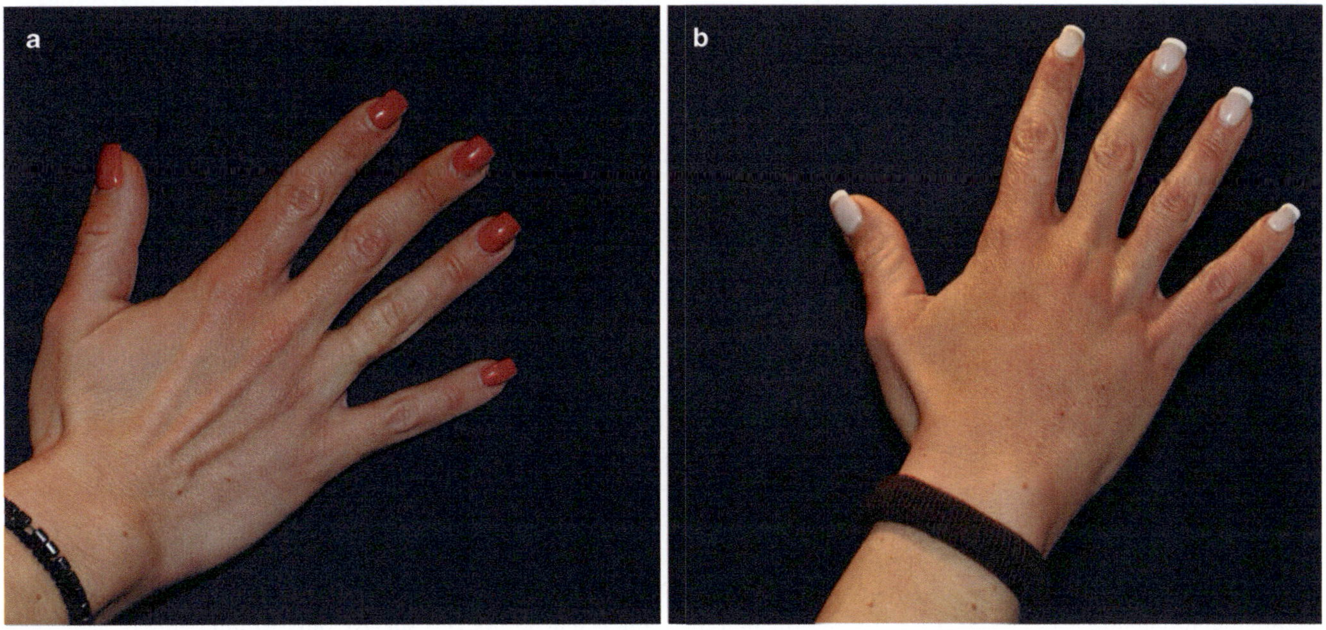

Fig. 11.9 (**a**) Preoperative view of a 73-year-old woman presented complaining of her "unattractive" hands. (**b**) Postoperative view of a 73-year old woman 2 year after undergoing SEL transfer to the hands

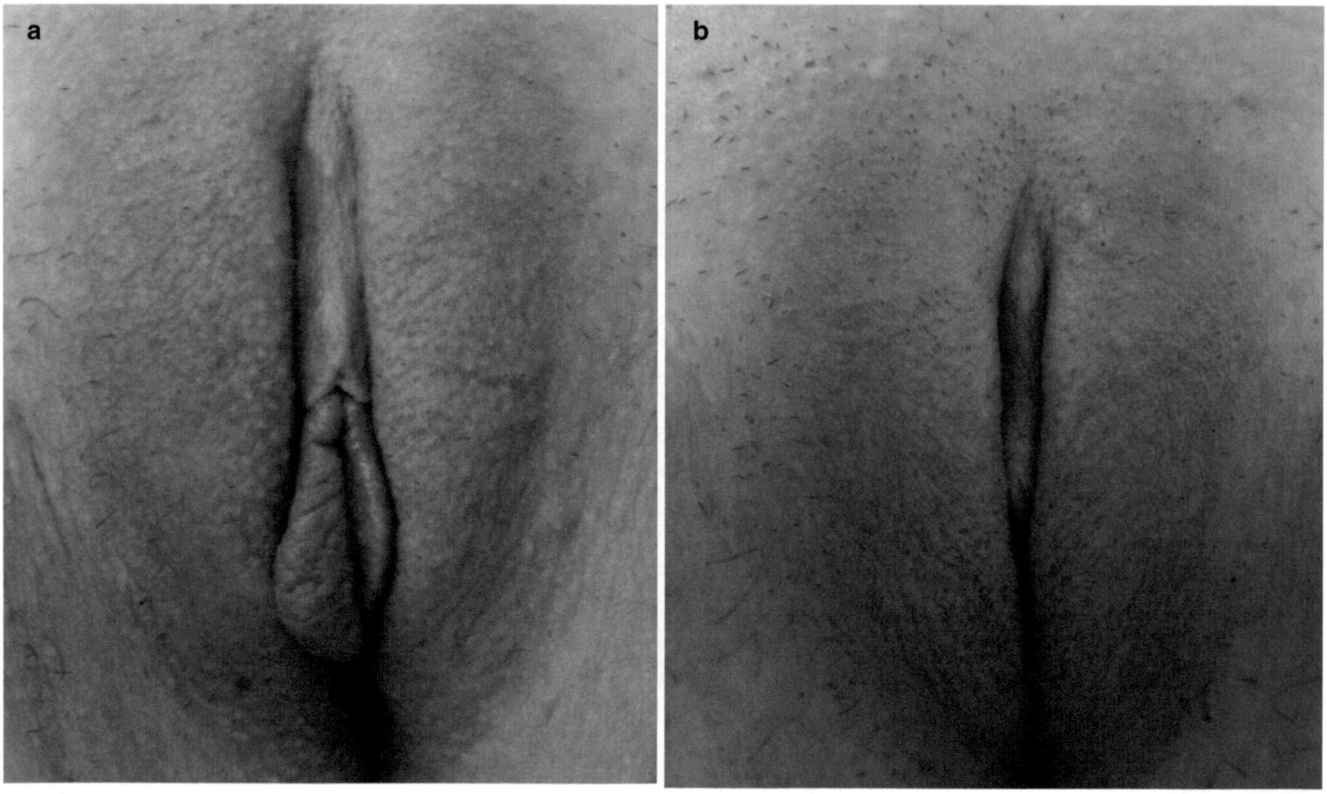

Fig. 11.10 (**a**) Preoperative view of 42-year old female patient requesting rejuvenation of her external genitalia. (**b**) Postoperative view of a 42-year old female patient 1 year after SEL of the external genitalia

11.5 Discussion

Fat grafting for face and body contouring remains shrouded in the stigma of variable results experienced by most plastic surgeons when they first graft fat. The need of standardization of the autologous fat-grafting technique needs to be done. Numerous clinical reports have been published regarding techniques of fat graft harvesting, preparation, and injection. Techniques are still selected according to a surgeon's individual preference, since quantitative evidence of clinical fat survivability and predictability of volume restoration does not exist. ADSCs can be distinguished from other adipocyte progenitor populations based on their expression of a variety of surface markers. The regenerative capacity of ADSCs during graft setting and their contribution to fat regeneration has been proven [9]. Stromal vascular fraction (SVF) is a heterogeneous population of cells that results from the processing of adipose tissue and is composed mainly of various blood cells, pericytes, macrophages, smooth muscle cells, and both adipose-derived and vascular endothelial progenitor cells. The factor in successful engraftment is the presence of ADSCs. These cells are pluripotent mesenchymal stem cells that reside in large numbers in adipose tissue. These small stellate-shaped cells are identified by surface antigens such as CD134 and their ability to form colonies in vitro. It is estimated that 1–3 million of these small stellate-shaped cells typically reside in proximity to small vessels of adipose tissue. They are known to tolerate the conditions associated with harvest and graft injection more successfully than mature adipocytes, participate in the tissue response to these stresses, and direct adipose tissue regeneration. ADSCs are able to differentiate into new adipocytes, replacing a portion of the adipocytes, which succumb to apoptosis due to hypoxic or physical stress and have been shown to actively promote angiogenesis via growth factor secretion and through neovascular differentiation. Clinical studies have demonstrated that the resident ADSCs within fat-grafted tissues can differentiate into adipocytes and add structure to fill the implanted tissue defect; secrete growth factors, cytokines, and chemo-attractants that can enhance angiogenesis and increase local vascularization and blood supply; and inhibit innate immune responses after tissue transplantation [11]. Recent studies have indicated that ADSCs can promote angiogenesis in addition to suppressing inflammation. An accepted principle of autologous fat grafting is that adipocytes survive only within 2 mm of an arterial blood supply. Fat cells outside this boundary may undergo necrosis leading to scar tissue. The adipose tissue graft enriched with SVF is woven into the targeted tissues, injecting only 5–10 mL of fat with each pass as in order to obtain the most reliable clinical outcome. The process of fat regeneration is progressed by ADSCs between 3 and 7 days, so the role of ADSCs is important in fat grafting. ADSCs are also involved with establishing fat homeostasis. These properties support successful tissue regeneration and the long-term survival of the fat graft. It has been shown that ADSCs harvested from superficial abdominal regions are significantly more resistant to apoptosis than other parts [12].

A series of clinical cases have shown that fat graft enriched with adipose-derived stem cells can improve the survival rate of the traditional fat graftting. In 2005, the research team led by the author Dr. Sterodimas described a method of supplementing the lipoaspirate used for fat grafting with the stromal vascular fraction found in adipose. This process was named Stromal-Enriched Lipograft [13]. The rationale behind this technique is that aspirated adipose tissue is poor in progenitor cells, growth factors, and cytokines, which are contributing factors to poor survival in vivo of the fat graft. While certain aspiration sites were initially thought to produce better graft take, these theories have not been supported by recent studies. In the SEL technique, the donor sites are chosen together with the patient, and care is taken in order to avoid reharvesting fat from a previous lipoaspiration site and not to create contour deformities. A stromal vascular fraction (SVF) containing ADSCs is freshly isolated from half of the aspirated fat and recombined with the other half. This process converts relatively ADSC-poor aspirated fat to ADSC-rich fat. SEL is based on the use of adipose-derived stem cells combined with a biomaterial that is the adipose tissue that has been processed to be used as a natural scaffold and biomolecules, cytokines, and growth factors, which are secreted by the stem cells and the adipose tissue [14]. A recent study has confirmed that SEL fat can survive better than non-SEL fat, and microvasculature can be detected more prominently in SEL fat, especially in the outer layers of the fat transfer [15].

11.5.1 Face SEL

The use of SEL can augment certain areas of the breast, for example, in the midline to build cleavage, adding upper pole fullness, or under the intended nipple position to provide additional projection, which cannot be achieved by the insertion of silicone implant alone. SEL can effectively improve implant animation deformity, rippling, and visibility of the implant in dècolletage. With aging, soft tissue loss is inevitable, and the skin of the dorsal surface of the hand becomes thinner, thereby revealing extensor tendons and dorsal veins. Successful, three-dimensional sculpting of the hand requires attention to patient preparation, meticulous planning, and optimizing the harvesting and transplantation of adipose tissue. Hand rejuvenation by SEL™ aims at successfully reversing this three-dimensional process. Recent reports showed ADSCs as a powerful source of skin regeneration because of their capability to provide not only cellular elements, but also numerous cytokines. A standardized technique of combining liposuction of lumbosacral, subgluteal, and trochanteric areas with lipoinjection in the upper middle buttock to improve the gluteal contour together with the projection of the gluteus is paramount for a successful gluteal fat grafting. Deep knowledge of the topography of the gluteal region and familiarity with the technique are essential in order to safely reach a pleasant result.

11.6 Conclusion

SEL is a safe and acceptable method for face, breast, and body fat-grafting surgeries due to its high patient satisfaction and low complication rate. Future research will hopefully refine our understanding of the effect of fat grafting on the local tissue micro-environment and provide clues toward its optimization. In addition, large-scale, controlled studies are needed to advance our ability to tailor the SEL technique for further applications in the field of plastic & reconstructive surgery.

References

1. Sterodimas A, Boriani F, Magarakis E, Nicaretta B, Pereira LH, Illouz YG. Thirty-four years of liposuction: past, present and future. Eur Rev Med Pharmacol Sci. 2012;16(3):393–406.
2. Sterodimas A, De Faria J, Correa WE, Pitanguy I. Tissue engineering in plastic surgery: an up-to-date review of the current literature. Ann Plast Surg. 2009;62(1):97–103.
3. Sterodimas A, de Faria J, Nicaretta B, Pitanguy I. Tissue engineering with adipose-derived stem cells (ADSCs): current and future applications. J Plast Reconstr Aesthet Surg. 2010;63(11):1886–92.
4. Sterodimas A, de Faria J, Nicaretta B, Papadopoulos O, Papalambros E, Illouz YG. Cell-assisted lipotransfer. Aesthet Surg J. 2010;30(1):78–81.
5. Sterodimas A, Illouz YG. Conclusions and future directions. In: Illouz YG, Sterodimas A, editors. Adipose derived stem cells and regenerative medicine. Berlin Heidelberg: Springer-Verlag; 2011. p. 273–6.
6. Pereira LH, Sterodimas A. Composite body contouring. Aesthetic Plast Surg. 2009;33(4):616–24.
7. Sterodimas A. Stromal enriched lipograft for rhinoplasty refinement. Aesthet Surg J. 2013;33(4):612–4.
8. Sterodimas A, Huanquipaco JC, de Souza FS, Bornia FA, Pitanguy I. Autologous fat transplantation for the treatment of Parry-Romberg syndrome. J Plast Reconstr Aesthet Surg. 2009;62(11):e424–6.
9. Sterodimas A, de Faria J, Nicaretta B, Boriani F. Autologous fat transplantation versus adipose-derived stem cell-enriched lipografts: a study. Aesthet Surg J. 2011;31(6):682–93.
10. Nicareta B, Pereira LH, Sterodimas A, Illouz YG. Autologous gluteal lipograft. Aesthetic Plast Surg. 2011;35(2):216–24.
11. Sterodimas A, Pereira LH. Liposuction of the abdomen and trunk. In: Rubin JP, Jewell ML, Richter D, Uebel CO, editors. Body con-

touring & liposuction. New York, NY: W.B. Saunders Elsevier; 2012. p. 311–20.
12. Sterodimas A. Adipose Stem cell engineering: clinical applications in plastic and reconstructive surgery. In: Illouz YG, Sterodimas A, editors. Adipose derived stem cells and regenerative medicine. Berlin Heidelberg: Springer-Verlag; 2011. p. 165–80.
13. Sterodimas A. Tissue engineering with adipose derived stem cells (ADSCs) in plastic & reconstructive surgery: current and future applications. In: Di Giuseppe A, Shiffman MA, editors. New frontiers in plastic and cosmetic surgery. Jaypee; 2015.
14. Sterodimas A, Nicaretta B. Filippo Boriani composite face lifting: the combination of stromal enriched lipograft with face minilift and upper and lower blepharoplasty: a review of 210 cases. Ann Plast Surg. 2020;85(6):e20–3.
15. Gentile P, Sterodimas A, Calabrese C, De Angelis B, Trivisonno A, Pizzicannella J, Dionisi L, De Fazio D, Garcovich S. Regenerative application of stromal vascular fraction cells enhanced fat graft maintenance: clinical assessment in face rejuvenation. Expert Opin Biol Ther. 2020;20(12):1503–13.

The Legacy of Micro-Autologous Fat Transplantation: A Reality from Evidence-Based Medicine

12

Tsai-Ming Lin, Hidenobu Takahashi, and Chih-Kong Chou

Contents

12.1 Brief History of Fat Grafting

Fat grafting was first addressed in 1893 by the German surgeon, Neuber [1]. Since then, techniques, methods, and protocols for fat grafting have been enhanced and developed frequently to improve the fat survival/retention rate. Numerous surgical mentors in this field have demonstrated innovative strategies for improving the results of fat grafting [2–7].

Financial Disclosure Statement: Dr. Tsai-Ming Lin owns the patent rights to the MAFT-GUN and is a scientific adviser for Dermato Plastica Beauty Co., the manufacturer of the MAFT-GUN device. None of the other authors has any financial disclosures or conflicts of interest.

T.-M. Lin (✉) · C.-K. Chou
Charming Institute of Aesthetic and Regenerative Surgery (CIARS), Kaohsiung City, Taiwan

H. Takahashi
Department of Surgery, Kaohsiung Medical University Hospital, Kaohsiung City, Taiwan

12.2 Evidence-Based Medicine of the Volume of Each Delivered Fat Parcel

Fat grafting has been indispensably regarded as an important plastic surgery innovation as hand/face transplantation of the modern era in a survey conducted by the American Council of Academic Plastic Surgeons (ACAPS) and the Southeastern Society of Plastic and Reconstructive Surgeons (SESPRS) [8]. Furthermore, Sydney R. Coleman has been ranked as one of the top 10 most influential plastic surgeons currently in practice. In the past decade, fat grafting has been used for facial rejuvenation, craniofacial deformity, cosmetic and reconstructive breast surgeries, gluteal augmentation, hand rejuvenation, and regenerative surgeries [9]. In the literature, a few studies illustrate the critical issues of the size of each fat parcel and the consistent control of fat parcels during transplantation. The evolution of the volume of each fat parcel has been suggested from a volume between 1/8 [10] and 1/30 mL [11] to between 1/50 and 1/30 mL [12] by Coleman;

© The Author(s), under exclusive license to Springer Nature Switzerland AG 2023
A. Di Giuseppe et al. (eds.), *Fat Transfer in Plastic Surgery*, https://doi.org/10.1007/978-3-031-10881-5_12

however, in terms of the consistent control of grafting parcels, no conclusive methods or maneuvers have been agreed, especially using the surgeon's hands. Most commercial instruments, such as the Rachet gun, reach a volume per parcel of only 1/20 mL, considerably large than the recommended volume of 1/50 to 1/30 mL [12].

12.3 Introduction of Micro-autologous Fat Transplantation (MAFT) and MAFT-GUN

During the initial 24–48 h after fat grafting, the inflow of nutrients and outflow of metabolites from adipocytes depend on physiologic diffusion and plasmatic imbibition in the recipient niche. The neo-vascularization of the grafted fat is reformed by regaining blood circulation after 48 h. Carpaneda et al. [13, 14] demonstrated that the central area eventually necrotized, and only the marginal zone survived at a rate of approximately 40% at 1.5 ± 0.5 mm from the grafted margin, regardless of the shape of the fat graft (spherical or cylindrical). Moreover, previous studies have reported several postoperative complications associated with the large size of an implanted fat parcel, including absorption, cyst formation, fibrosis, nodulation, calcification, ossification, and asymmetry [15]. Inevitably, the phenomenon of "central necrosis" at any large fat parcel results in unpredictable graft survival/retention rate and induces a vicious cycle of postoperative complications. Therefore, the size of the implanted fat droplets should be as small as possible [16]. Based on the theory postulated by Carpaneda, the mathematical formula for determining the optimal volume of fat parcels (here, the fat graft is presumed to be spherical in shape) is calculated as follows: the volume of a globe is (4/3) πr^3, where r is the global radius and π is the ratio of any circle's circumference to its diameter in Euclidean space. Table 12.1 presents information regarding the volume of each fat parcel at radii of 1, 1.5, and 2 mm. We concluded that the injection frequency of 1 mL fat parcels for a spherical graft with a 2 mm radius can be calculated by dividing 1000 mm^3 (1 mL) by (4/3) π (2 mm^3). Therefore, a minimal injection frequency of 30 was set for each 1 mL of fat-graft parcel to achieve superior graft survival rates. A patented instrument, the MAFT-GUN [17], was invented to achieve precise, accurate, and consistent control of each fat parcel from 1/60, 1/90, 1/120, 1/150, 1/180, and 1/240 mL [18].

Table 12.1 The volume of a spherical at a radius of 1.0, 1.5, and 2.0 mm is calculated, and the total injection frequency of 1 mL (1000 mm^3) is concluded accordingly

Radius	Volume of Spherical	Total Injection frequency for each 1 mL (1000 mm^3) fat graft
1.0 mm	4.2 mm^3	1000÷4.2 = 240
1.5 mm	14.1 mm^3	1000÷14.1 = 70
2.0 mm	33.5 mm^3	1000÷33.5 = 30

Therefore, the total injection frequency of 1 mL (1000 mm^3) of fat is 240, 70, and 30 in sequence.

12.4 Basic Procedures for MAFT

Since 2010, more than 2800 cases of fat grafting involving the face, neck, hand, breast, buttock, and other sites with the assistance of MAFT-GUN have been performed by the author (Tsai-Ming Lin) at the Charming Institute of Aesthetic and Regenerative Surgery, Kaohsiung, Taiwan. The basic procedures for MAFT are briefly described here.

12.4.1 Fat Harvesting

The donor area (lower abdomen, flank, or thigh) was pre-infiltrated with a tumescent solution prepared at a ratio of 10 mL of 2% lidocaine (20 mg/mL): 30 mL of Ringer's lactate solution: 0.2 mL of epinephrine (1:1000) after the incision site was anesthetized. Approximately 10–15 min after the tumescent solution was administered, a blunt-tip liposuction cannula (2.5 mm in diameter with one-sided-hole) was used to harvest the fat, and the lipoaspirate volume was the same as that of the infiltrated tumescent solution to achieve a high proportion of purified fat after centrifugation. Damage to the lipoaspirate was minimized by pulling back the plunger of a 10 mL Luer lock syringe for 2–3 mL and was sustained so that while it was connected to the liposuction cannula, the reactive aspirating negative pressure between 270- and 330-mmHg was maintained (Fig. 12.1a) [19].

12.4.2 Fat Processing and Refinement

Various methods, such as the sieve method, multiple-layer gauze filtration, and centrifugation, have been proposed for fat processing [20, 21]. The internationally accepted Coleman's technique was used for processing the lipoaspirate through centrifugation because of its advantages of less environmental exposure and lower manual manipulation in the aseptic procedure [22]. Standard centrifugation at 3000 rpm, which was approximately 1200 × g for 3 min (Fig. 12.1b), was applied to process (purify) the fat. Thereafter, the purified (condensed) fat was in the middle layer of each 10 mL syringe and meticulously collected after removing the top oil layer and the bottom bloody tumescent solution (Fig. 12.1c).

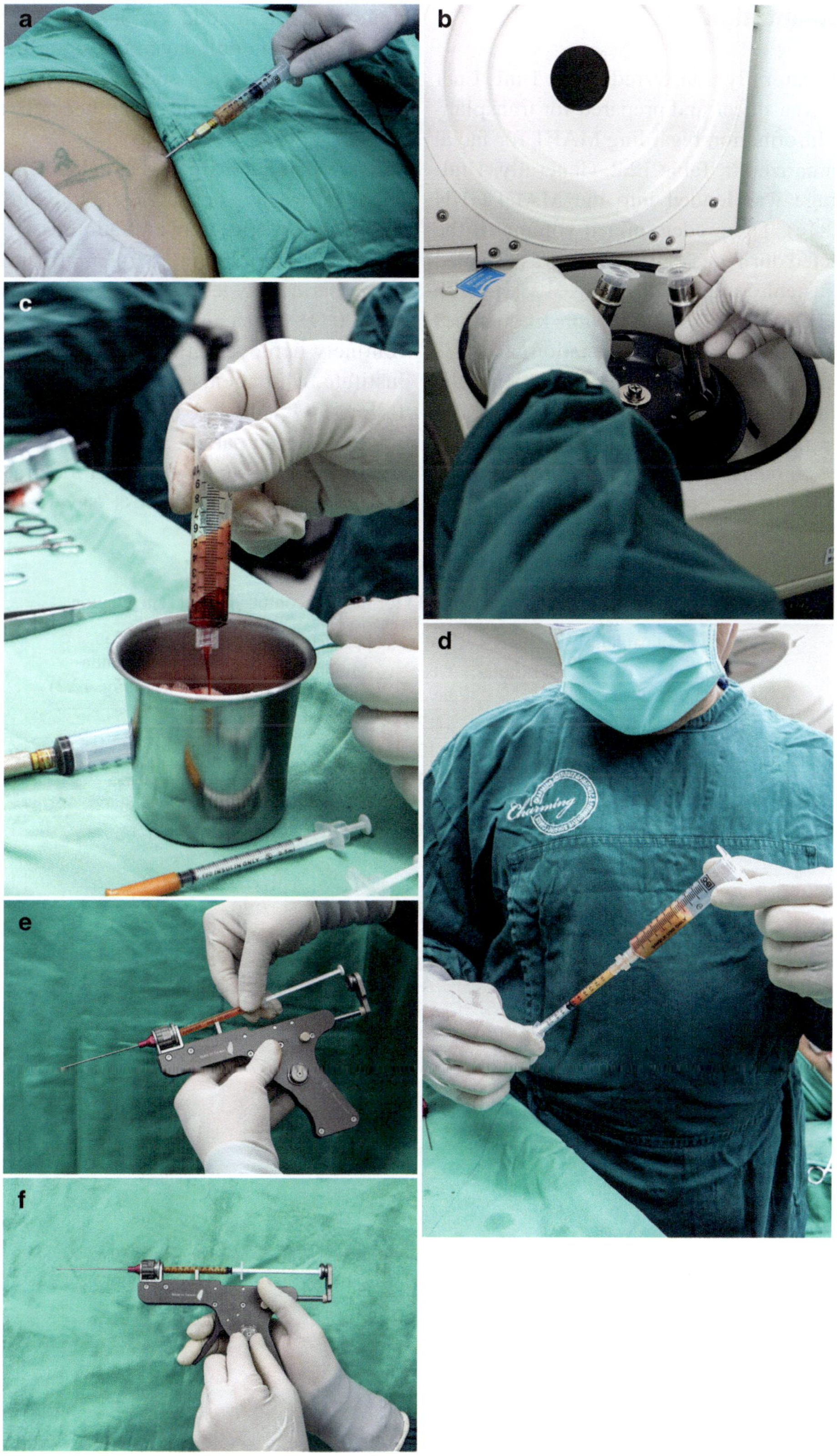

Fig. 12.1 (**a**) Fat aspiration is performed by back-pulling the plunger of a 10 mL syringe to approximately 2–3 mL to maintain an appropriate negative pressure. (**b**) Processing by centrifugation is maintained at 3000 rpm (~1200 × *g*) for 3 min to purify the fat. (**c**) The lower part of the centrifuged lipoaspirate (bloody content) is leaked out, and the oil in the upper part is wiped off using gauze. (**d**) The purified, condensed fat is transferred to a 1 mL Luer slip syringe for transplantation. (**e**) The fat-filled syringe is loaded into a MAFT-GUN (Dermato Plastica Beauty Co., Ltd. Taiwan). (**f**) The six-graded dial is set at 60 for neck and dorsal hand rejuvenation, 120 for most facial areas, and 240 for periorbital areas, and the delivered volumes of the fat parcel are 1/60 mL (0.017 mL), 1/120 mL (0.008 mL), and 1/240 mL (0.004 mL), respectively

12.4.3 Fat Transplantation

The purified fat was carefully transferred into a 1 mL Luer slip syringe using a transducer and prepared for transplantation (Fig. 12.1d). Information regarding MAFT for facial rejuvenation is summarized in Table 12.2. Generally, 1 mL of a fat-filled syringe was loaded into the MAFT-GUN instrument (Fig. 12.1e). The predetermined volume of the fat parcel to be injected during each triggering was selected by rotating the adjustable volume dial (Fig. 12.1f), with labeled numbers depicting the total injection frequencies per 1 mL of the fat graft. For facial rejuvenation, an 18-G blunt cannula was used to inject fat while withdrawing the MAFT-GUN. Each delivered fat volume was set at 1/120 mL (however, the periorbital areas were preferred at 1/240 mL) and meticulously transplanted in three to four layers according to the characteristics and anatomy of the recipient tissue. For example, in sunken eyelid rejuvenation, there are three layers: a deep layer above the inferior orbital rim, a middle layer, the suborbicularis oculi muscle (deep in the muscle), and a superficial layer, the supraorbicularis oculi muscle (just beneath the dermis of the eyelid).

Postoperative care was provided regularly without any special dressings or massage. Oral and non-steroidal anti-inflammatory drugs were administered for 3 days, as required. Hyperbaric oxygen treatment ("HearMEC" Hyperbaric Oxygen Chamber 02 Prime Cabin Model: CA200 OX-II, Health Way Biomedical Co, Ltd. Taiwan) is routinely deployed after MAFT procedure in the author's institute. All patients were regularly photographed at each follow-up visit, and the preoperative and postoperative images of each patient were compared. Touchup MAFT is recommended 4–6 months after the first session to enhance the appearance or address severe depression or scar tissue.

Table 12.2 Information of micro-autologous fat transplantation (MAFT) for facial recontouring and rejuvenation

	Type of cannula			Fat parcel per triggering (mL)			Insertion site of cannula (see Fig. 12.2)	Average volume (mL)
Facial area	Size (gauge)	Tip	Side hole	Deep layer	Middle layer	Superficial layer		
Forehead [23, 24]	18	Blunt	One	1/120	1/120	1/120	A	10–25
Sunken temporal fossa [23, 25]	18	Blunt	One	1/120	1/120	1/120	B_1, B_2	6.0–12/unilateral
Sunken upper eyelid [23, 26, 27]	18	Blunt	One	1/240	1/240	1/240	C_1, C_2	0.5–2.0/unilateral
Nasojugal groove and Lid-cheek junction [23, 28]	18	Blunt	One	1/240	1/240	1/240	D_1, D_2	1.0–4.0/unilateral
Nasal dorsum [23, 29]	18	Blunt	One	1/120	1/120	1/120	E	1.0–4.0
Cheek [23]	18	Blunt	One	1/120	1/120	1/120	D_1, D_2	3.0–8.0/unilateral
Nasolabial groove [23, 30, 31]	18	Blunt	One	1/120	1/120	1/120	D_1'G_1, D_2'G_2	4.0–10/unilateral
Upper/ lower lip [23]	18	Blunt	One	1/120	1/120	1/120	F_1, F_2	1.0–2.0/upper or lower
Chin [23, 32]	18	Blunt	One	1/120	1/120	1/120	F_1, F_2	4.0–8.0

12.5 Clinical Applications of MAFT

12.5.1 Facial Contouring

Table 12.2 and Fig. 12.2 summarize the fundamental information on the MAFT technique for facial contouring and rejuvenation, including the authors' preferred cannula pattern, volume per delivered parcel, insertion site, and suggestive average volume (mL) with insertion sites for each recipient area. Briefly, an 18-G (1.2 mm in diameter) blunt-tip one-sided hole injection cannula (disposable cannula preferred) was selected for transplantation. The ideal volume for grafting was 1/120 mL per fat parcel with the adjustable dial switched to 120 and 240 (to reflect each delivered parcel 1/120 and 1/240 mL) for periorbital areas, sunken upper eyelid, tear trough, and lid-cheek junction, where the skin is thin to avoid irregularity. In more than 2500 cases performed by the senior author (TM Lin) for facial contouring, the average volume for each area was suggested (Table 12.2) but subjective to change due to individual variation. Figures 12.3, 12.4, 12.5, 12.6 demonstrate the clinical results of facial recontouring and rejuvenation by MAFT [23–30].

12.5.2 Neck Recontouring

Using a 16-G (1.6 mm in diameter) blunt tip with a one-sided hole injection cannula, the predetermined volume was set at 1/60 mL per parcel, and the procedure was performed for neck recontouring in the elderly group for alleviating neck lines in the younger group. Satisfactory results demonstrated the long-term effects (Figs. 12.7 and 12.8). The total transplanted volume varies from 4 to 8 mL for each neck line to 6–12 mL for recontouring per depression individually. Touchup might be necessary 4–6 months after the first session to enhance the effect of the procedure.

12.5.3 Dorsal Hand Rejuvenation

In the aging of the dorsal hand surface, soft tissue loss is inevitable, and the skin becomes thinner, thereby revealing the dorsal veins and extensor tendons that could hinder social activities in some individuals. Using a 16-G injection cannula, with a delivered volume of 1/60 mL per parcel, an average volume of 15–25 mL for each dorsal hand, including digits (up to the proximal interphalangeal joint), was transplanted in three layers [33]. In our earlier study, favorable outcomes (very satisfied: 58.8% and satisfied: 39.7%) were demonstrated in patients who underwent a single session of MAFT. The volume of the aging dorsal hand surface was restored, and skin texture improved with sustainable long-term effectiveness, confirming that the technique is a reliable and simple strategy for rejuvenating the dorsal surface of aging hands (Fig. 12.9).

12.5.4 Scar and Related Skin Rejuvenation Issues

For old traumatic scars, there are not only fibrotic tissue formations that cause subcutaneous hardness and/or tissue deficiency but also widening or hypertrophic scar forma-

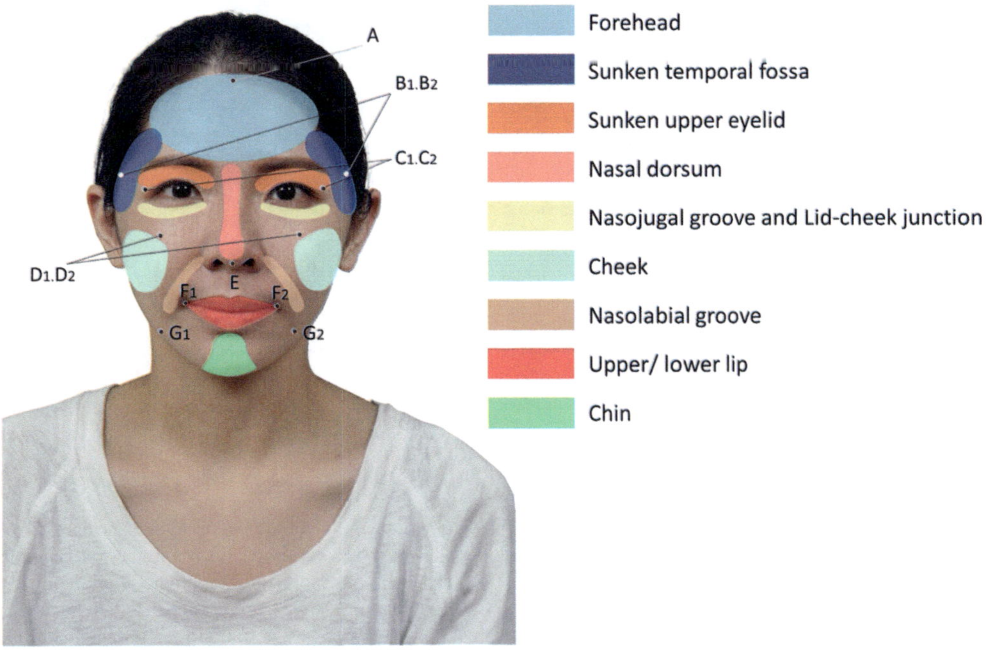

Fig. 12.2 Recommended cannula insertion sites for facial rejuvenation in the MAFT procedure

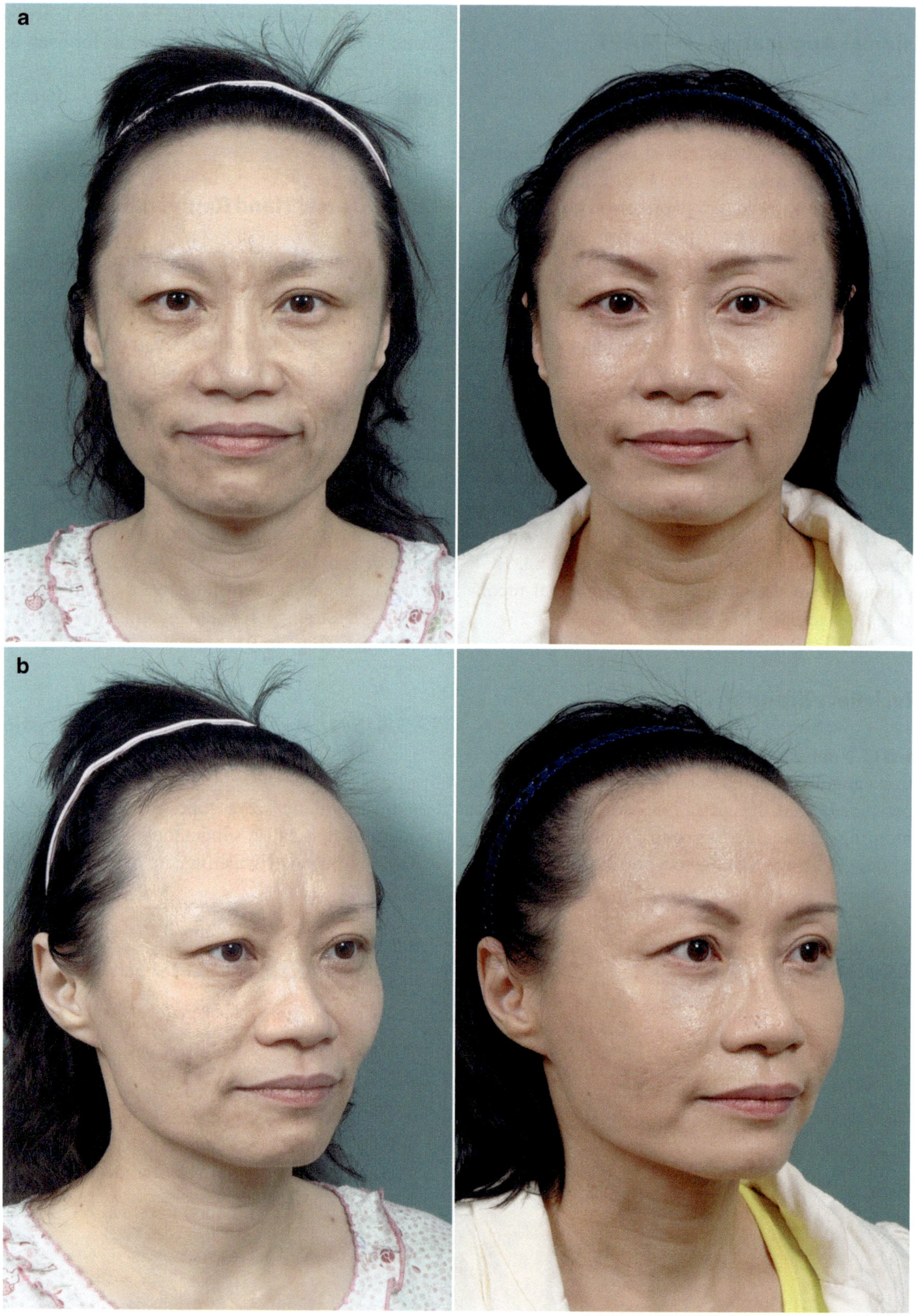

Fig. 12.3 A 48-year-old woman underwent trans-subcutaneously lower blepharoplasty with the MAFT procedure on her temple (R't/L't, 5.0/5.0 mL), tear trough/lid-cheek junction (R't/L't, 3.0/3.0 mL), nasal dorsum (3.0 mL), nasolabial groove (R't/L't, 4.0/4.0 mL), upper/lower lip (1.5/2.0 mL), cheek (R't/L't, 6.0/6.0 mL), and chin (4.0 mL) under intravenous general anesthesia. After 1 year of a single MAFT session, the volume restoration of all recipient areas was maintained with fullness appearance. No nodulation or irregularity was found, and the skin texture was rejuvenated (left/right and upper/lower: pre-/post-MAFT) (**a**: AP view, **b**: oblique view, **c**: lateral view, **d**: downward-looking view)

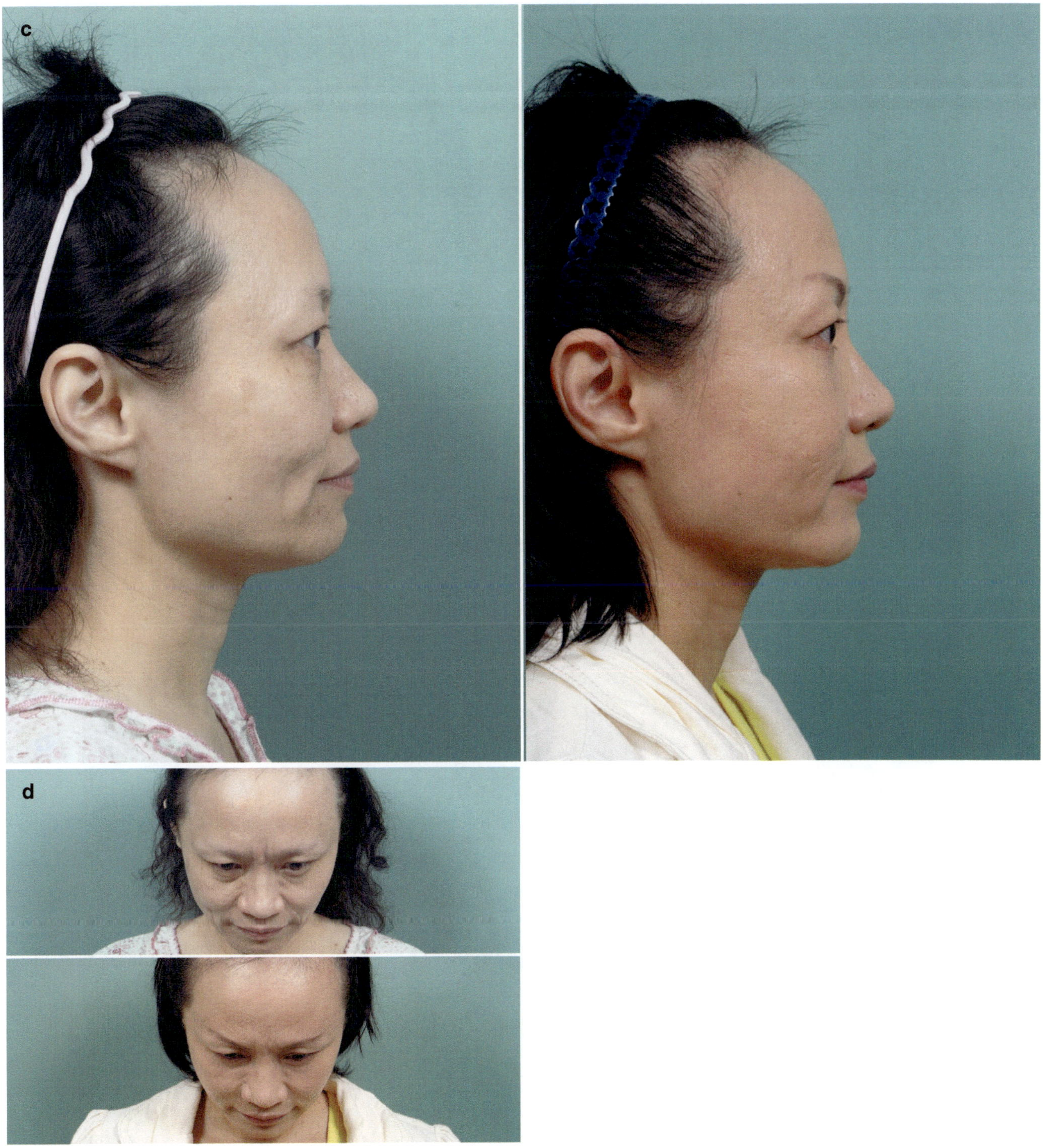

Fig. 12.3 (continued)

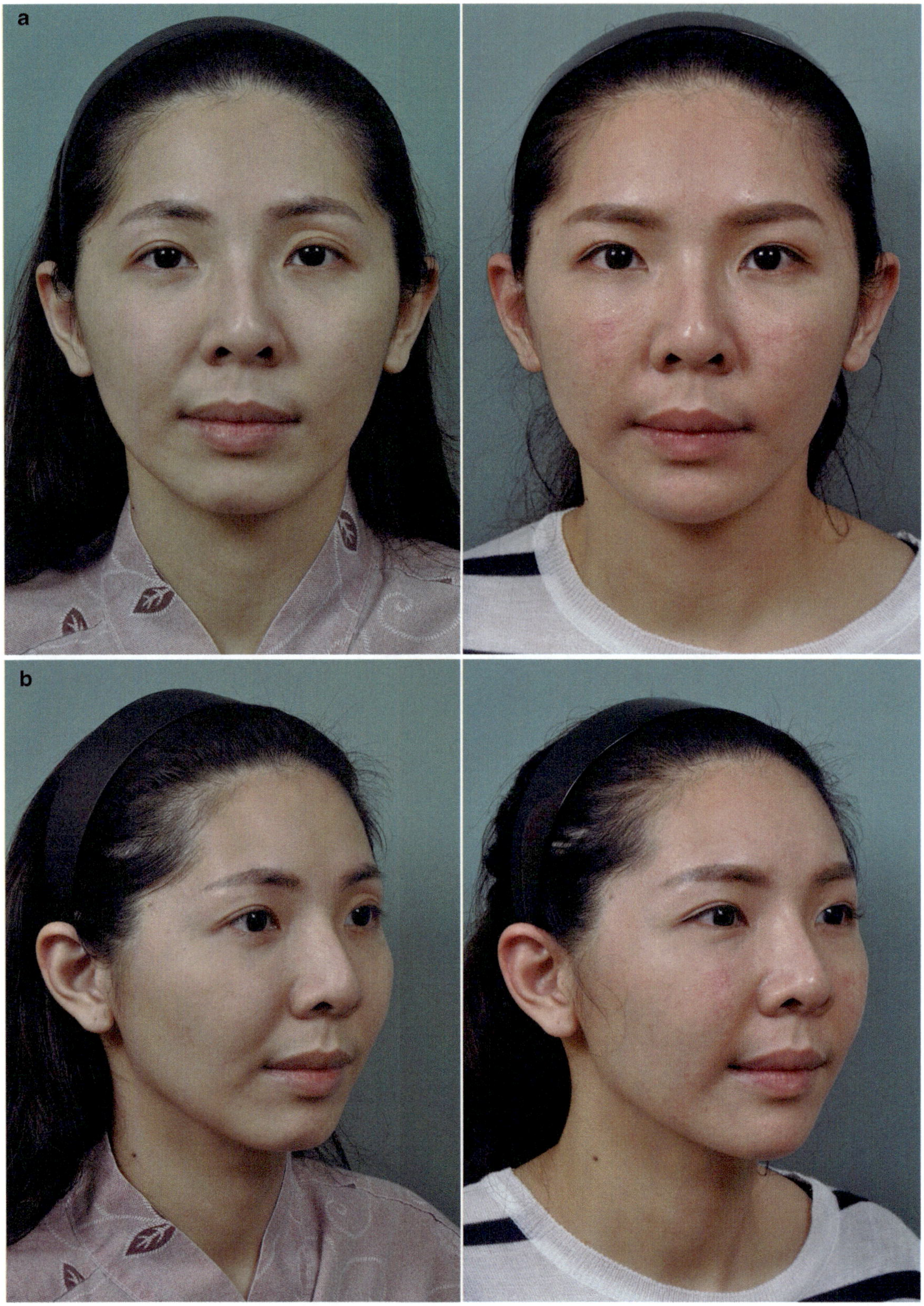

Fig. 12.4 A 32-year-old woman underwent total facial rejuvenation with the MAFT procedure on her forehead (16 mL), temple (R't/L't, 10/10 mL), left sunken upper eyelid (0.8 mL), tear trough/lid-cheek junction (R't/L't, 3.0/3.0 mL), nasolabial groove (R't/L't, 6.0/6.0 mL), ergotrid (R't/L't, 1.5/1.5 mL), upper lip (1.5 mL), and chin (5.0 mL) under intravenous general anesthesia. After 6 months of a single MAFT session, the volume restoration of all recipient areas was maintained with a youthful appearance. The gummy smile appearance was also satisfactorily corrected (left/right and upper/lower: pre-/post-MAFT) (**a**: AP view, **b**: oblique view, **c**: downward-looking view, **d**: full-smile view)

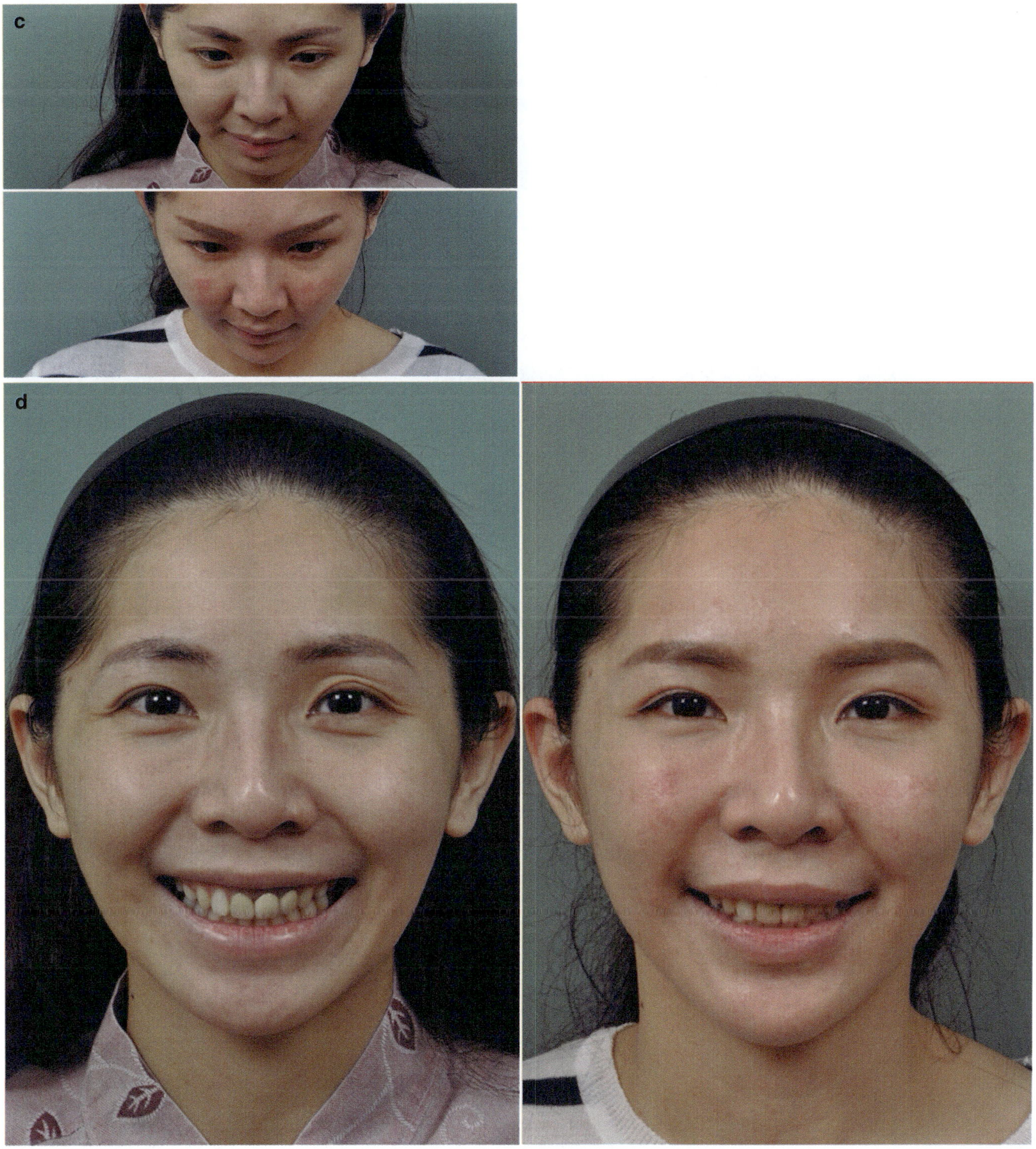

Fig. 12.4 (continued)

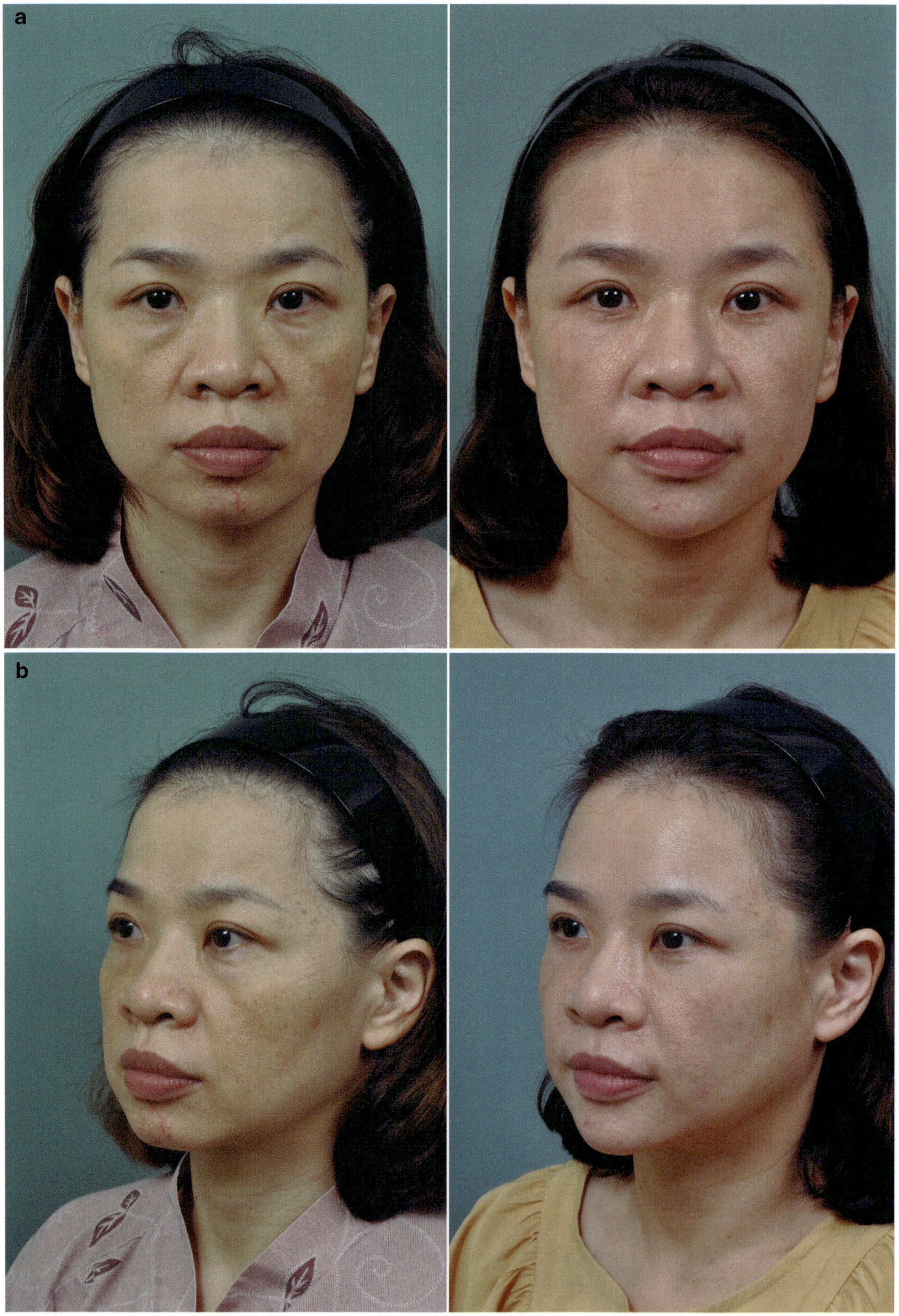

Fig. 12.5 A 45-year-old woman underwent trans-subcutaneously lower blepharoplasty with the MAFT procedure on her forehead (14 mL), temple (R't/L't, 10/10 mL), tear trough/lid-cheek junction (R't/L't, 3.5/3.5 mL), nasal dorsum (3.0 mL), nasolabial groove (R't/L't, 6.0/7.0 mL), and chin (4.0 mL) under intravenous general anesthesia. After 8 months of a single MAFT session, the volume restoration of all recipient areas was maintained with fullness. No complication was found, and the skin texture over the crow's feet areas was well rejuvenated (left/right and upper/lower: pre-/post-MAFT) (**a**: AP view, **b**: oblique view, **c**: lateral view, **d**: smiling view)

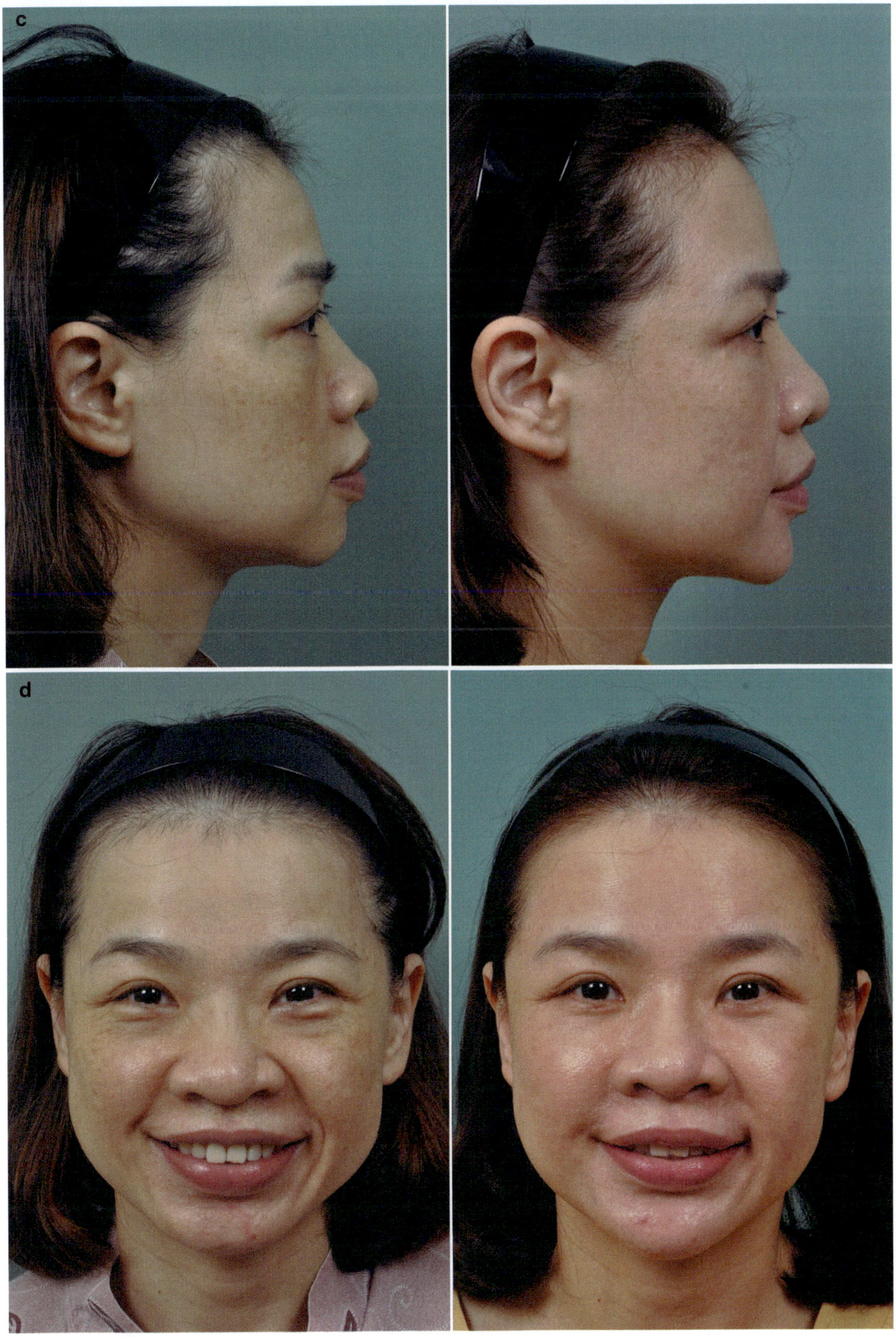

Fig. 12.5 (continued)

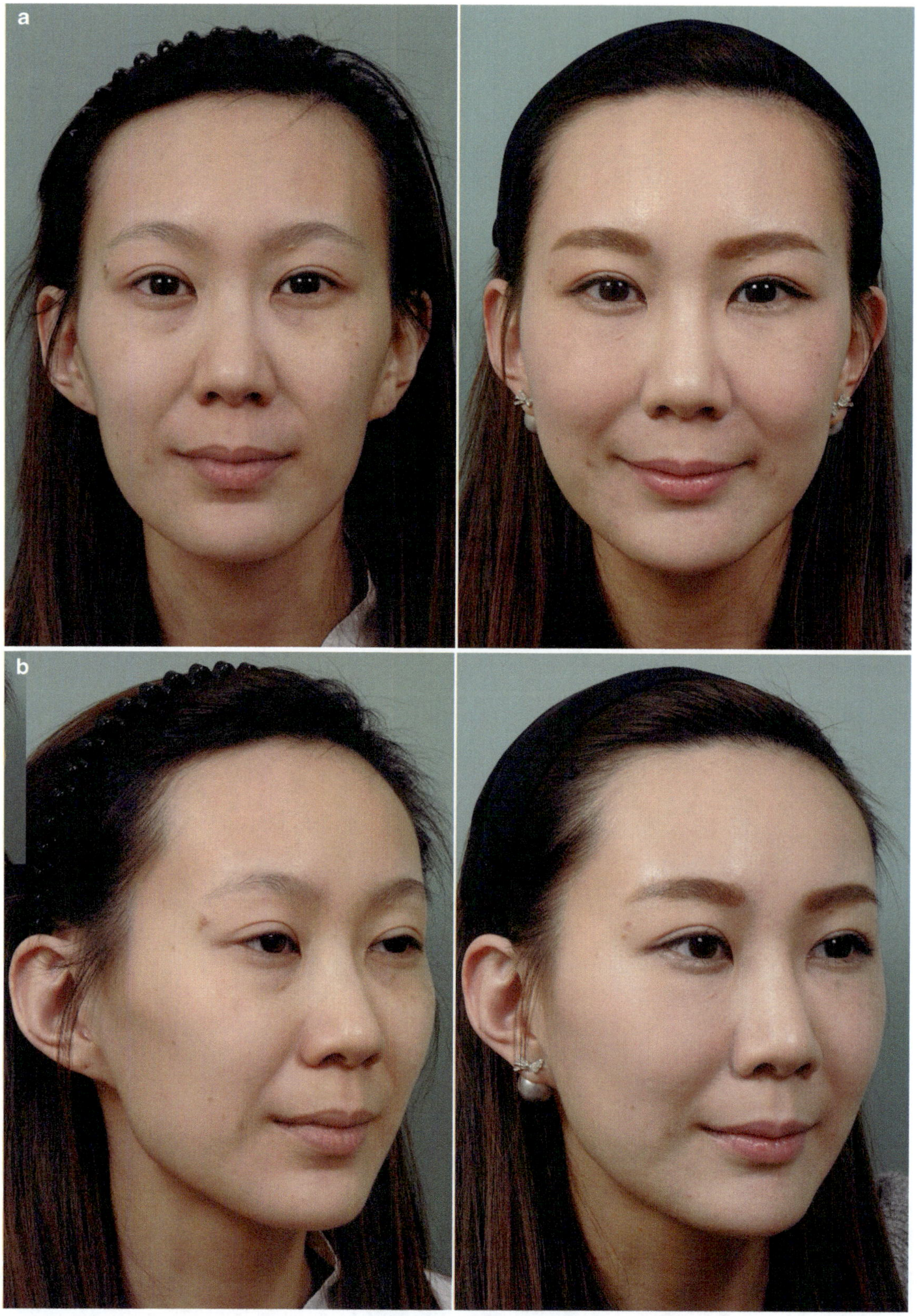

Fig. 12.6 A 32-year-old woman underwent fat grafting for facial rejuvenation with the MAFT procedure on her temple (R't/L't, 8.0/8.0 mL), tear trough/lid-cheek junction (R't/L't, 4.0/4.0 mL), lateral cheek (R't/L't, 6.0/6.0 mL), nasal dorsum (1.5 mL), nasolabial groove (R't/L't, 7.0/7.0 mL), and chin (4.0 mL) under intravenous general anesthesia. After 6 months of a single MAFT session, the volume restoration of all recipient areas was maintained effectively without any complication (left/right and upper/lower: pre-/post-MAFT) (**a**: AP view, **b**: oblique view, **c**: lateral view, **d**: upward- and downward-looking views)

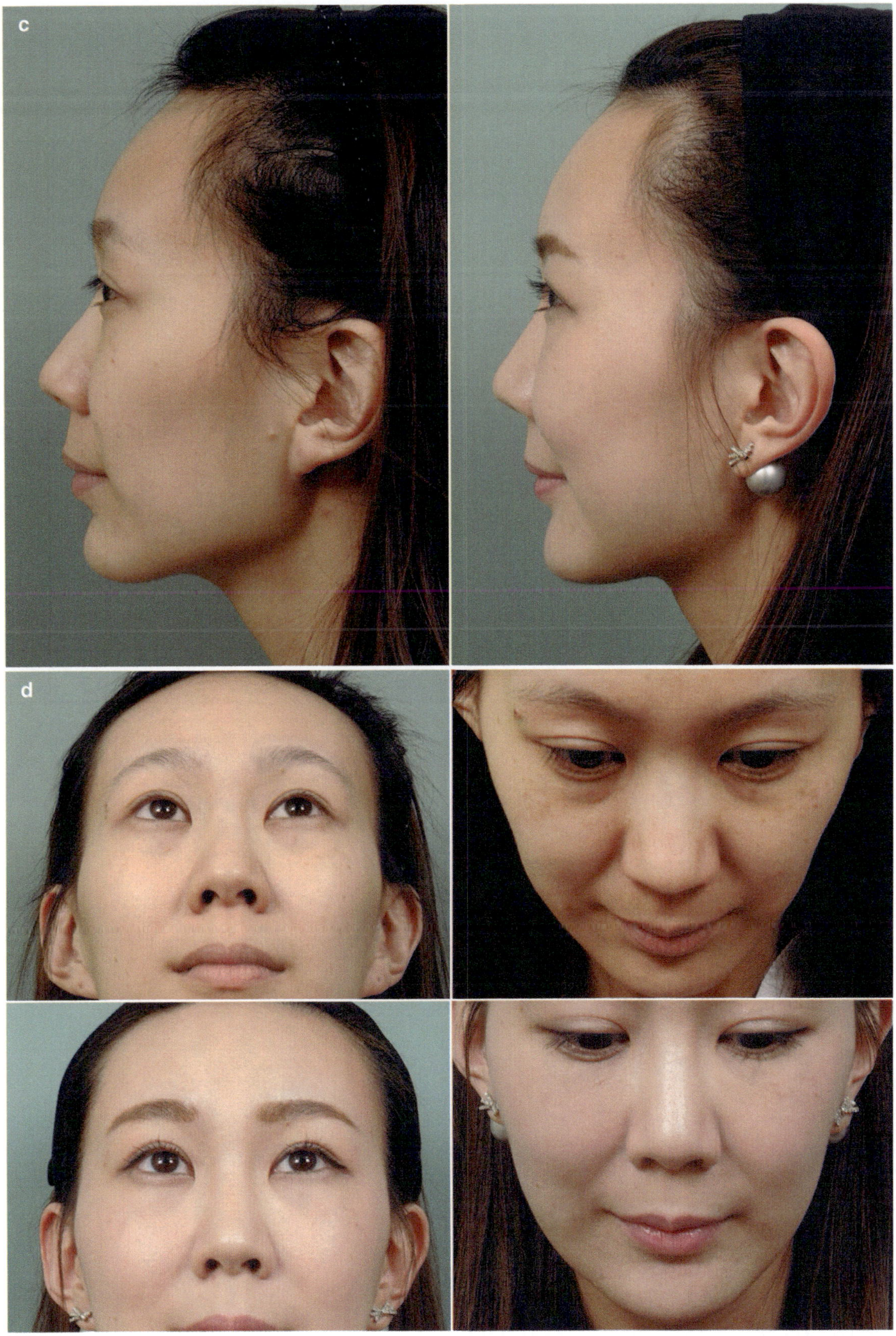

Fig. 12.6 (continued)

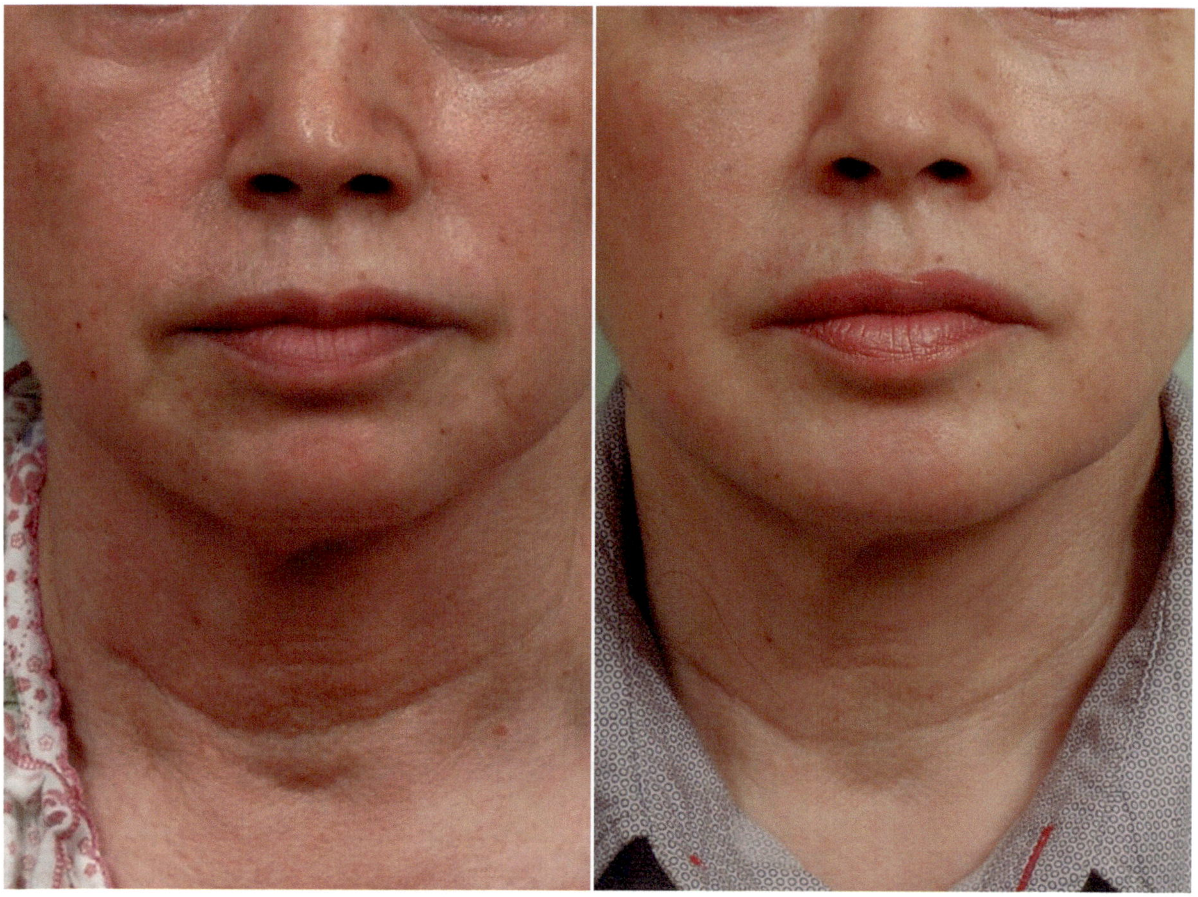

Fig. 12.7 A 65-year-old woman underwent fat grafting for her neck recontouring with the MAFT procedure. Under intravenous general anesthesia, a total of 8 mL of fat grafting was delivered at 1/60 mL per parcel over the mid-neck depressed area, and 4 mL was transplanted to obliterate the deepening of the neck lines. After 6 months of a single MAFT session, the volume restoration of the aging neck was maintained effectively with an acceptable improvement of the deepening neck lines (left/right: pre-/post-MAFT) (close-up AP view)

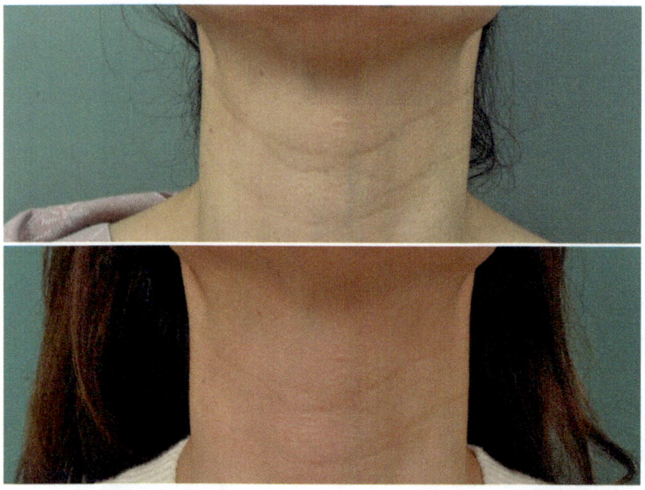

Fig. 12.8 A 51-year-old woman underwent fat grafting to improve her neck lines. The MAFT procedure was performed under intravenous general anesthesia with a total of 8.0 and 7.0 mL of fat for her upper- and lower-neck lines, respectively. After 7 months of a single MAFT session, her neck line demonstrated a shallow and rejuvenated appearance showing an acceptable improvement of deepening neck lines (upper/lower: pre-/post-MAFT) (close-up chin-up view)

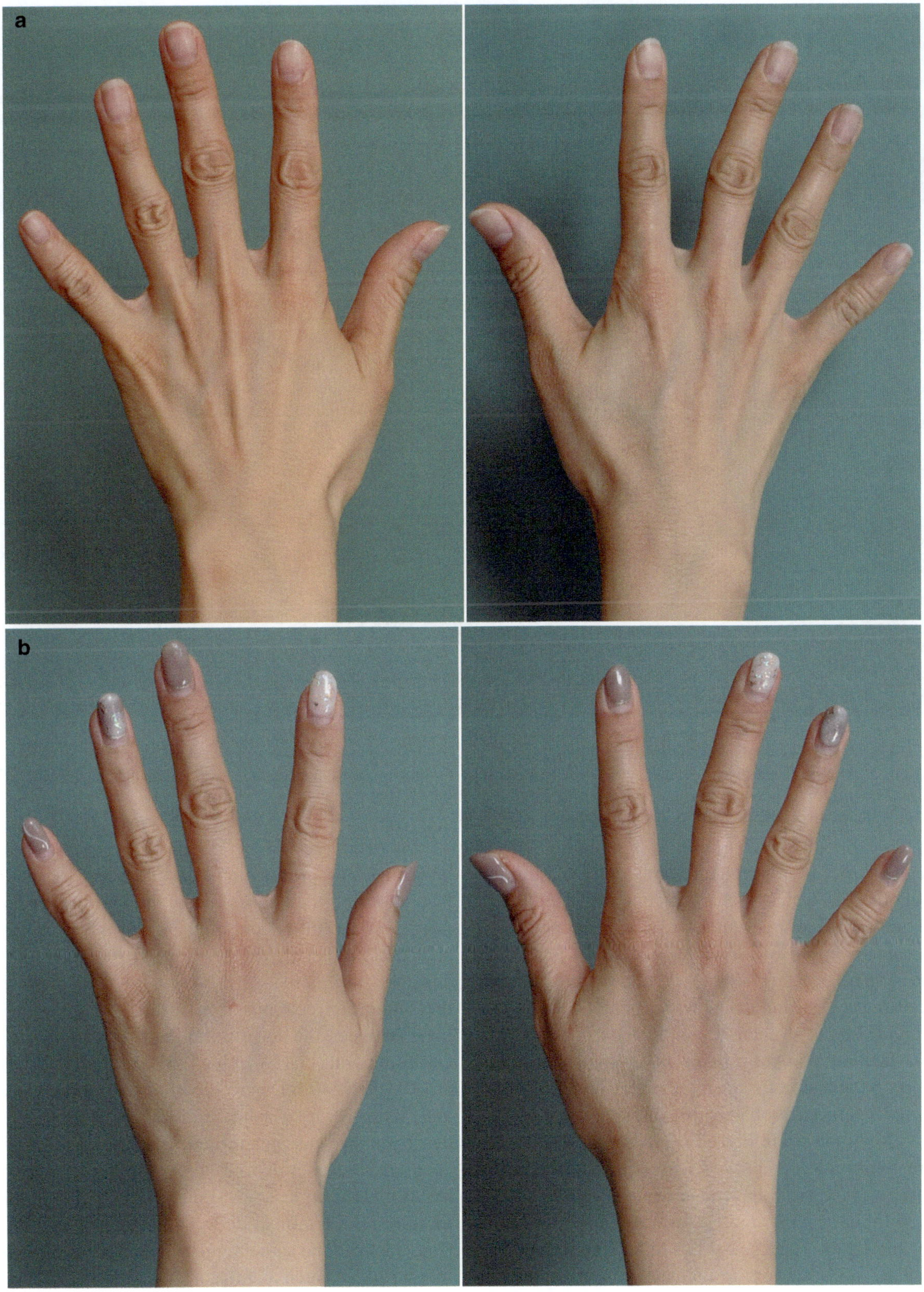

Fig. 12.9 A 52-year-old woman underwent fat grafting to rejuvenate her aging dorsal surface hands with skin laxity, visible extensor tendons, and varicosities. The MAFT procedure was performed for recontouring her dorsal hands, including the fingers, under intravenous general anesthesia with a total of 30 mL for each dorsal hand (15 mL for the dorsal hand area, 3 mL for the thumb, 4 mL for each of the index and middle fingers, and 2 mL for the ring and small fingers). Both dorsal surfaces of the hands showed a youthful appearance not only in terms of volume restoration but also in terms of the skin texture (**a**: pre-MAFT, **b**: post-MAFT)

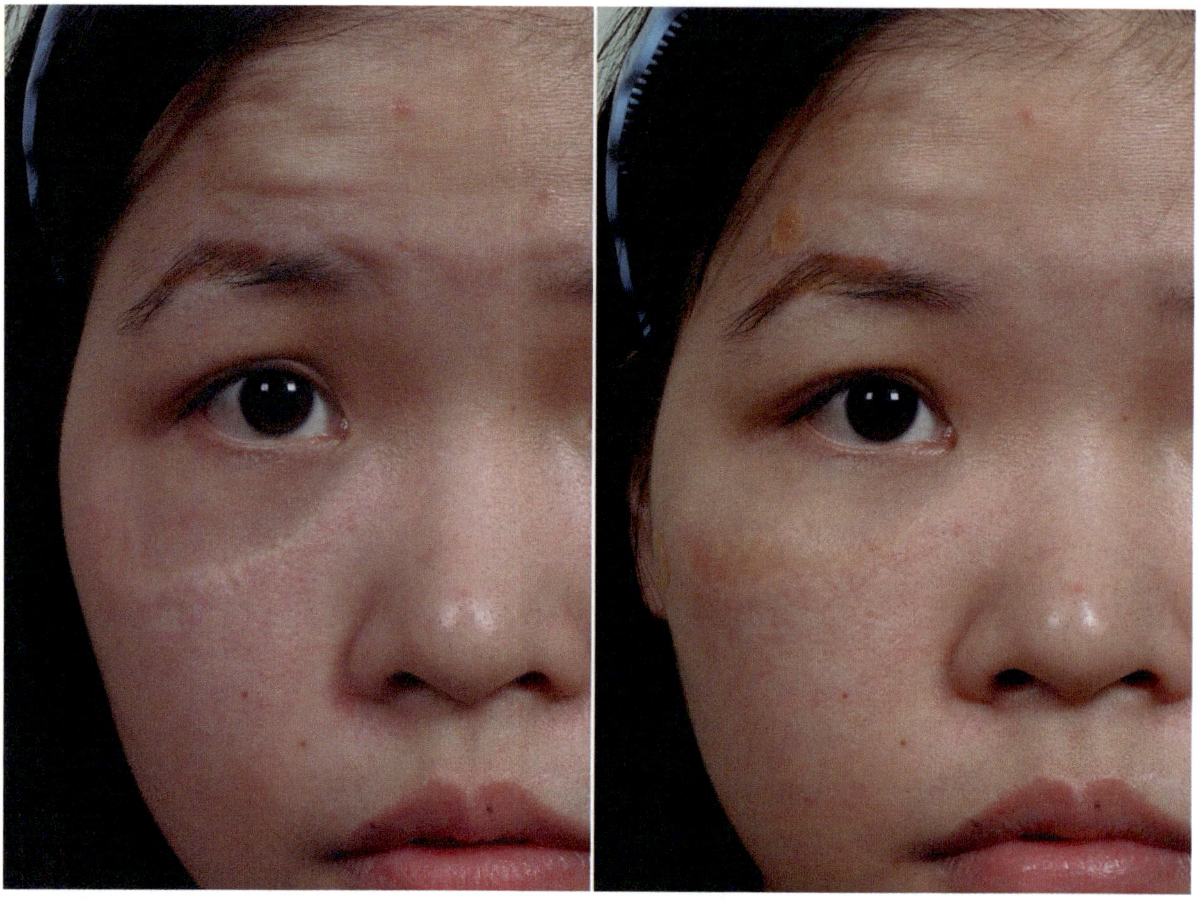

Fig. 12.10 This 22-year-old woman met an accident and had a visible scar over the right cheek with soft tissue deficiency cephalic to the scar. Implantation of 6.0 mL of fat was performed to recontour the soft tissue deficiency and rejuvenate the depressed scar. After 3 months of a single MAFT session, the scar was invisible (rejuvenating effect) with good volume restoration (recontouring) (left/right: pre-/post-MAFT) (close-up AP view)

tion over the skin. Laser treatment seems to have a limited effect, and other strategies, such as silicone gel/sheet, lasers (ablative, non-ablative, fractional, or radiofrequency), steroid or toxin injection, various filler injections, and other regenerative procedures (platelet-rich plasma injection or micro-needling) [34], cannot achieve sufficient volume restoration and tissue rejuvenation simultaneously. However, the MAFT technique might provide an alternative solution for these problems and offer a single effective strategy for dealing with soft tissue problems in a single procedure. The long-term effects shown in Fig. 12.10 illustrate sustainable volume restoration and improved scar appearance. Similar maneuvers for acne scars have also demonstrated promising results. Figure 12.11 shows a favorable long-term follow-up of intractable acne scars. For MAFT application in scar tissue, a 16-G (1.6 mm in diameter) blunt cannula is preferred to manipulate the touch fibrotic tissue. Each parcel size might be 1/60 or 1/90 mL through a nearby insertion.

Managing infraorbital dark circle is challenging. Recontouring the depression of the tear trough and deformity over the lid-cheek junction can improve the depressive darkness with the restoration of facial fullness. In addition, the increased thickness of the dermis [35] and the shading effect caused by a thin surviving fat grafting layer between the dermis and the orbicularis oculi muscle could alleviate dark circles [28]. An 18-G injection cannula with 1/240 mL per delivered fat parcel was adopted, and three-layer transplantation similar to our earlier descriptions was performed. Precise fat grafting was ensured because of the above-mentioned mechanism, resulting in a high survival rate. Satisfactory results were, therefore, demonstrated at long-term follow-up (Fig. 12.12). The same maneuver might be applied to treat fine wrinkles, such as infraorbital wrinkles or crow's feet. The rejuvenating effect noted after the MAFT technique reduced the depth of wrinkles and aging dyschromia and improved skin quality (Fig. 12.13).

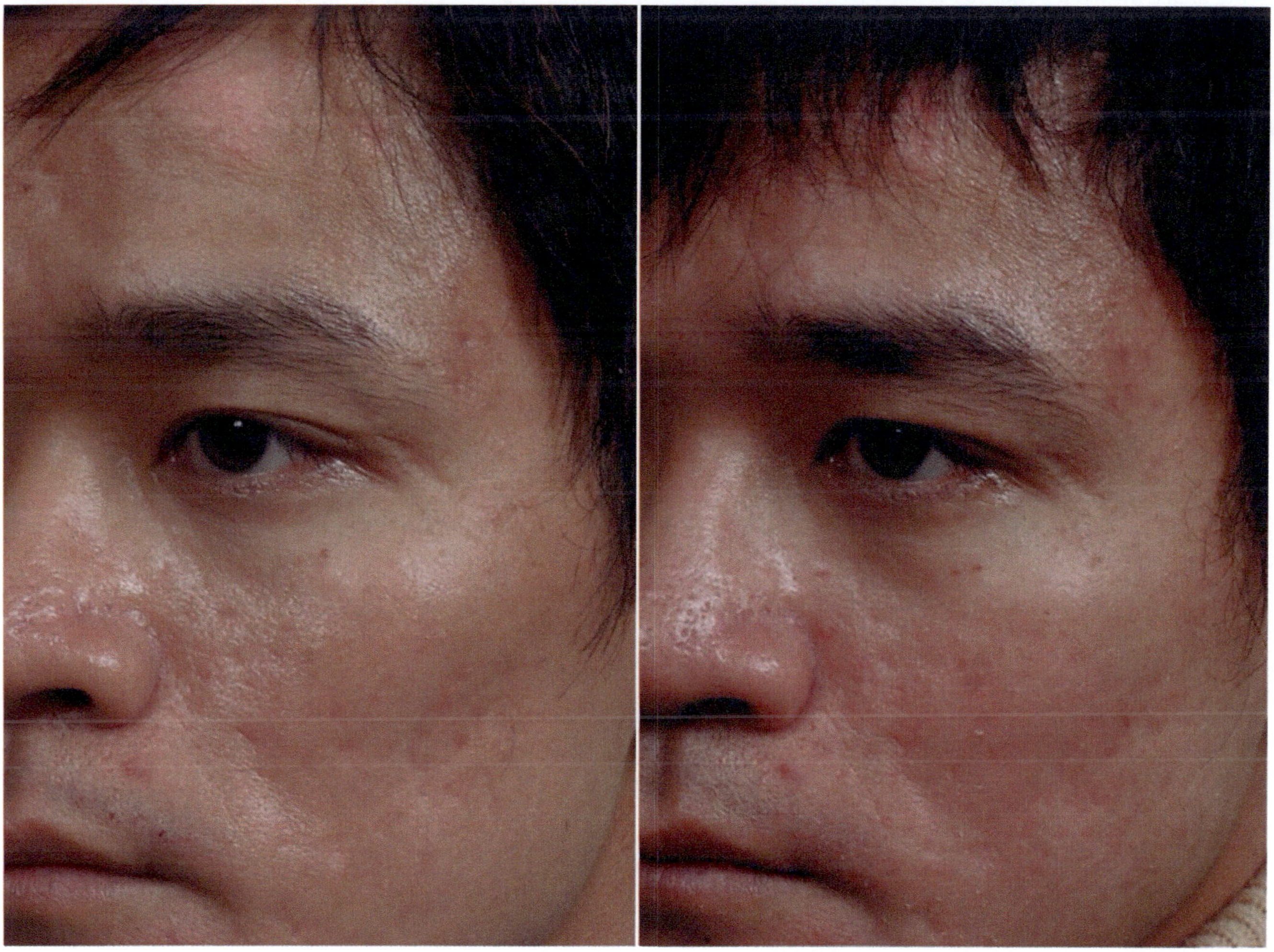

Fig. 12.11 This 29-year-old man had an acne scar over the temple and the cheek area following various treatments (laser and skin peeling). A single session of the MAFT procedure with implantation of 4.0 mL/6.0 mL of fat to the left temple/cheek area was performed. After 4 months of MAFT, the scar was invisible, and the skin texture improved (left/right: pre-/post-MAFT, close-up oblique view)

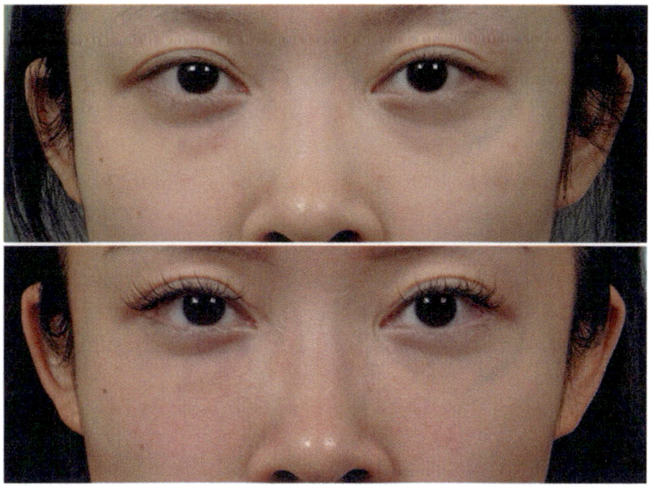

Fig. 12.12 This 24-year-old woman requested a comprehensive treatment of her infraorbital dark circle, mild eye bag, and a tear trough deformity. A single session of the MAFT procedure was performed by delivering 4.0 mL/3.5 mL of fat at 1/240 mL per parcel to the left and right tear trough/lid-cheek junction deformities, respectively. After 3 years and 10 months of MAFT, the grafted areas showed good results (upper/lower: pre-/post-MAFT, close-up AP view)

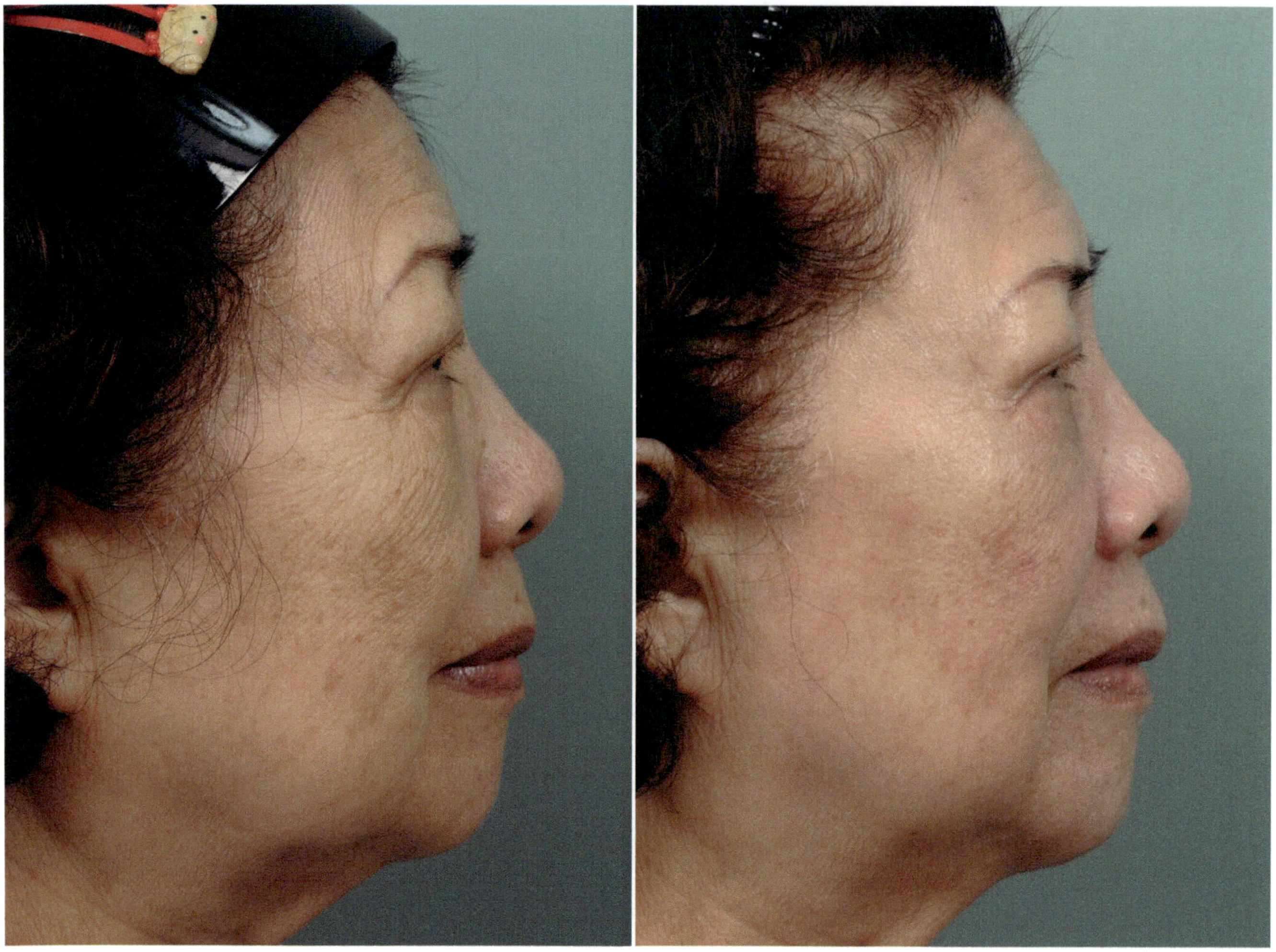

Fig. 12.13 This 80-year-old woman wanted to improve her lateral cheek wrinkles and crow's feet and restore the youthful skin texture. A single session of the MAFT procedure was performed by delivering 5 mL of fat to the crow's feet and 7 mL to the right lateral cheek area. After 8 months, an excellent outcome was noted. Restoration of the volume and rejuvenation of the skin are illustrated (left/right: pre-/post-MAFT, close-up lateral view)

12.5.5 Adjunctive Strategy for Lower Blepharoplasty

By avoiding the potential complications after fat grafting and ensuring a higher survival/retention rate, the MAFT technique might be an adjunctive procedure combined with transcutaneous or transconjunctival lower blepharoplasty, depending on the lower eyelid skin redundancy. In both situations, the appropriate orbital fat was removed, with the transcutaneous or transconjunctival fat removed first. Instead of variable manipulations of orbital fat by redraping or reposition/transposition, the authors preferred to immediately deliver fat grafting using the MAFT technique over the tear trough and the lid-cheek deformity. Lesser lower blepharoplasty-related complications, such as ectropion, and better recontouring of the tear trough and lid-cheek junction illustrated the indispensability of MAFT as an adjunctive procedure for lower blepharoplasty (Figs. 12.14 and 12.15)

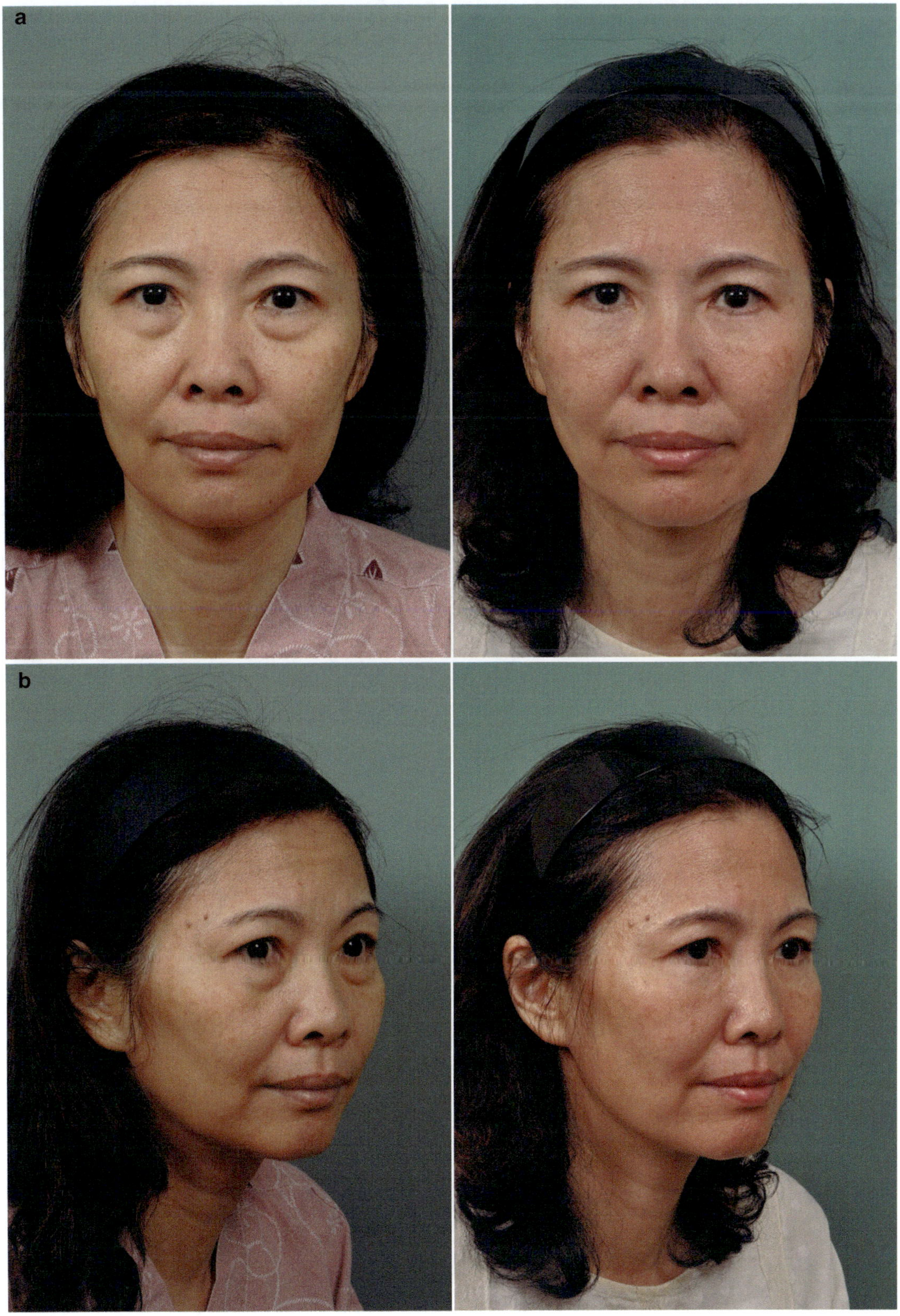

Fig. 12.14 This 52-year-old woman presented with a severe baggy lower eyelid, prominent tear trough, and lid-cheek junction. Skin-only transcutaneous lower blepharoplasty and MAFT for recontouring the tear trough deformity and lid-cheek junction were performed. The total resected orbital fat volumes were 0.23 g and 0.31 g (R't and L't, respectively), and the total amount of fat volumes grafted to the tear trough and lid-cheek junction were 2.0 and 2.0 mL (R't and L't, respectively). The improved appearance of her lower eyelid at 1 year and 8 months after a single MAFT and lower blepharoplasty is shown (**a**). The recontoured tear trough and lid-cheek junction with recontouring of the baggy lower eyelid are shown in the oblique view (**b**). Downward-looking view further illustrates that the contouring of the infraorbital area was well preserved (**c**) (left/right and upper/lower: pre-/post-MAFT) (**a**: AP view, **b**: oblique view, **c**: close-up downward-looking view)

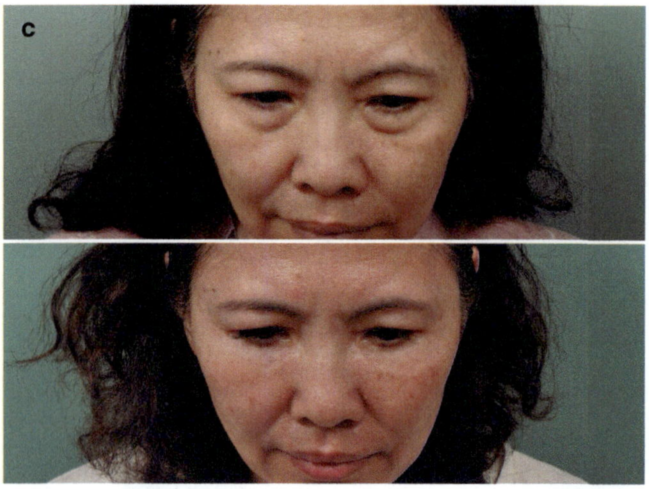

Fig. 12.14 (continued)

[28, 31]. A maneuver to deliver tiny parcels (1/240 mL per parcel) in multiple layers and recontour the tear trough deformity and lid-cheek junction depression with an 18-G injection cannula by the MAFT technique should be performed to improve long-term outcomes.

12.5.6 An Innovative Treatment of Gummy Smile

In the literature, the treatment of a gummy smile (defined as the exposure of >2–3 mm of the gum while smiling) includes the surgical intervention of the bony, gingival, or muscular abnormalities and noninvasive botulinum toxin injection. However, these strategies have not yet been established as definitive treatment modalities. In our previous work, we present for the first time, the development of a simple and consistent procedure based on the MAFT technique for treating gummy smile [36]. Parcel volume was meticulously delivered in the nasolabial groove (4.0–6.0 mL/side), ergotrid area (1.0–1.5 mL/side), and upper lip (1.0–2.0 mL) at a size of 1/120 mL per parcel. Favorable outcomes were achieved with sustainable long-term effectiveness, further confirming that this strategy of MAFT is

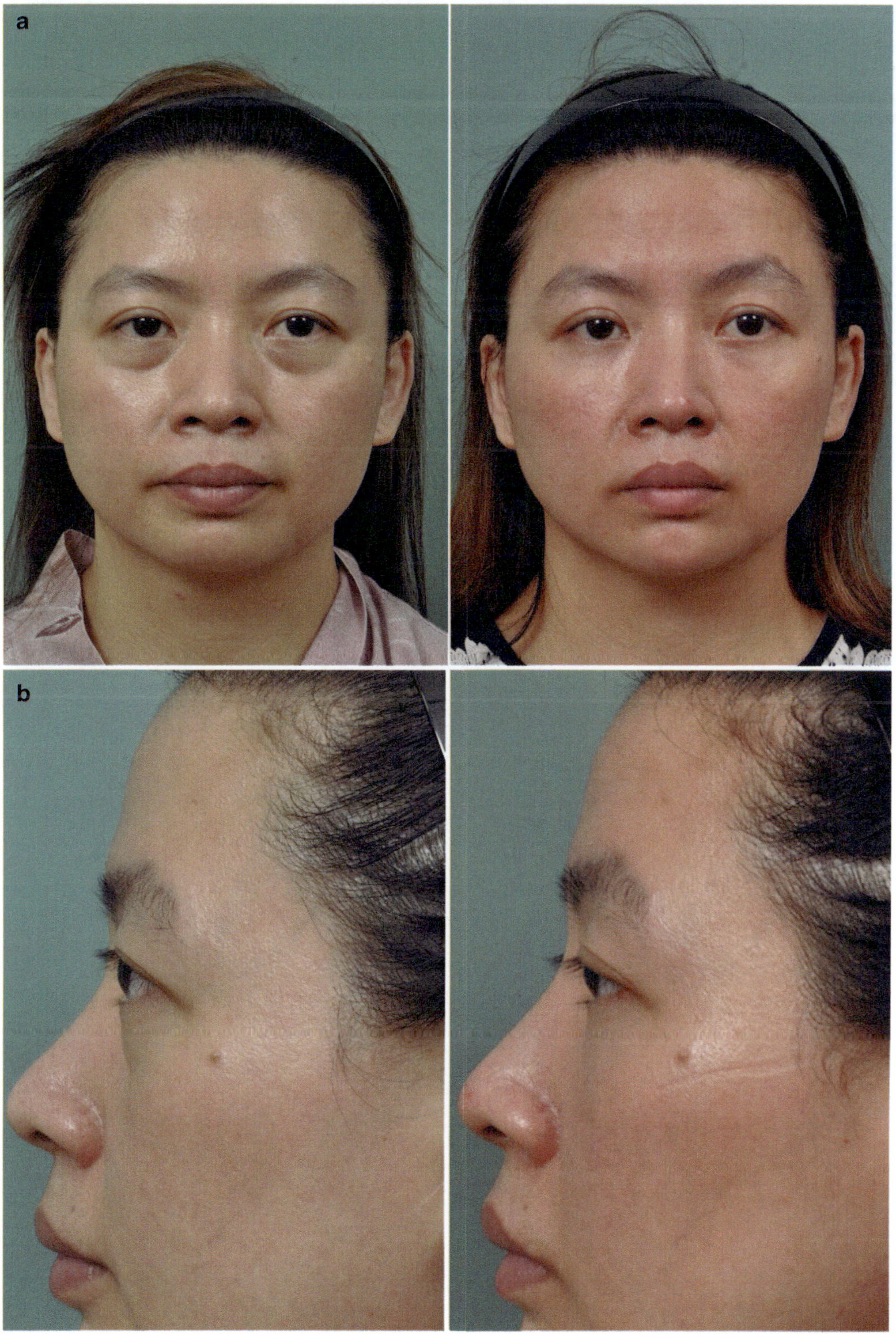

Fig. 12.15 This 40-year-old woman presented with a prominent baggy lower eyelid, deepening tear trough, and lid-cheek junction. Transconjunctival lower blepharoplasty and MAFT were performed to manage the baggy eyelid appearance and recontouring the tear trough deformity/deepening lid-cheek junction. The total resected orbital fat volumes were 0.22 g and 0.29 g (R't and L't, respectively), and the total amount of fat volumes grafted to the tear trough and lid-cheek junction were 3.0 mL and 3.0 mL (R't and L't, respectively). The improved appearance of her baggy lower eyelid at 10 months after a single MAFT and lower blepharoplasty is shown (**a**). The recontoured tear trough and lid-cheek junction with recontouring of the baggy lower eyelid are shown in the close-up lateral view (**b**). Downward-looking view further illustrates that the recontouring of the infraorbital area was well preserved (**c**) (left/right and upper/lower: pre-/post-MAFT) (**a**: AP view, **B**: close-up lateral view, **c**: downward-looking view)

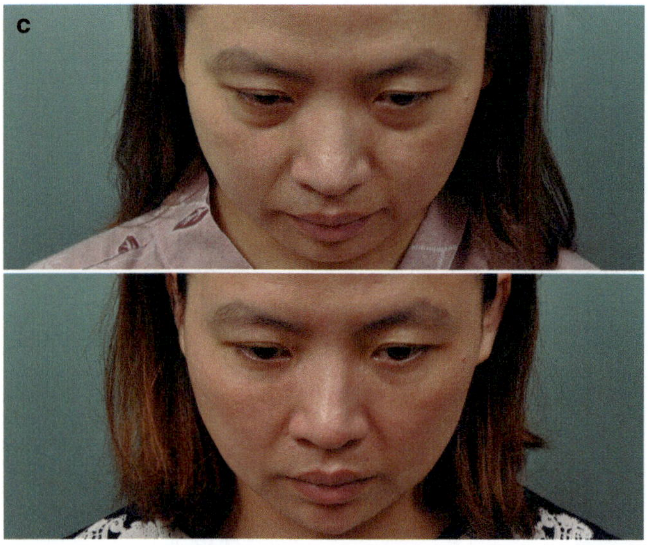

Fig. 12.15 (continued)

an innovative alternative for the treatment of a gummy smile (Fig. 12.16).

12.5.7 Breast Augmentation

Fat transfer for breast augmentation using liposuction from other body parts has become a popular and disruptive technique [8]. From the author's experience of handling more than 300 such cases, the long-term promising results suggest that this approach might be preferred.

The fundamental transplantation has emphasized using a 10-cc type MAFT-GUN [18] for delivering fat in two layers: the deep layer (retro-glandularly: supra-, intra-, and retro-pectoralis major muscle) and the superficial layer (subcutaneously). The lipoaspirated fat was harvested and processed in a way similar to MAFT for the

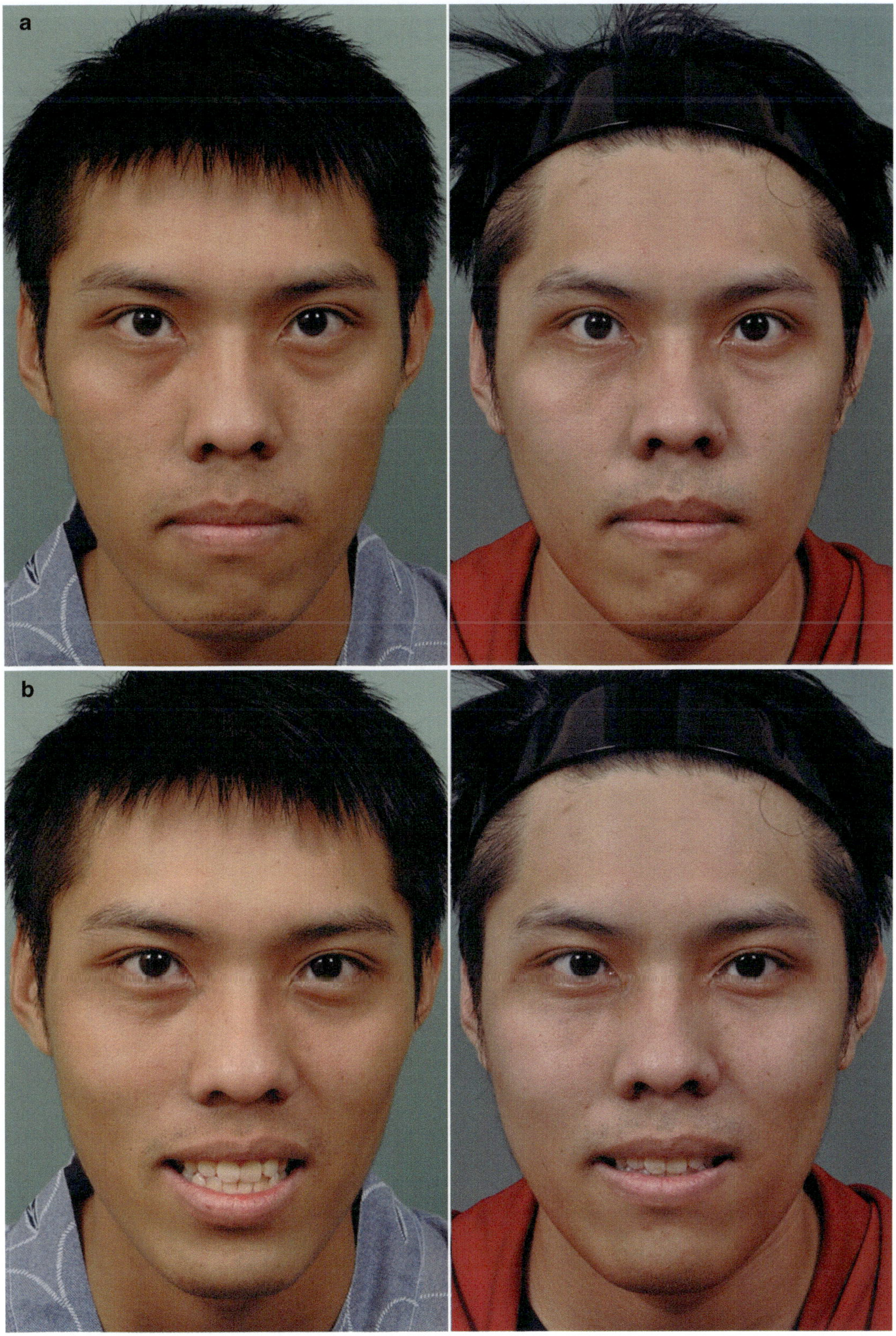

Fig. 12.16 This 24-year-old man presented with a severe gummy smile along with fat grafting. MAFT was performed on his nasolabial groove, ergotrid, and upper lip areas with the placement of a 14-mL fat graft (R't//L't side of the nasolabial groove, ergotrid, and upper lip: 5.0/5.0, 1.5/1.5, and 0.5/0.5 mL, respectively). After 3 years and 10 months of a single MAFT session, there was no significant change in his appearance, except a slightly increased ergotrid height (**a**). However, the gummy smile was improved (**b–d**). The overexposed gingiva was significantly improved by increasing the vertical length of the ergotrid and the thickness (width) of the upper lip

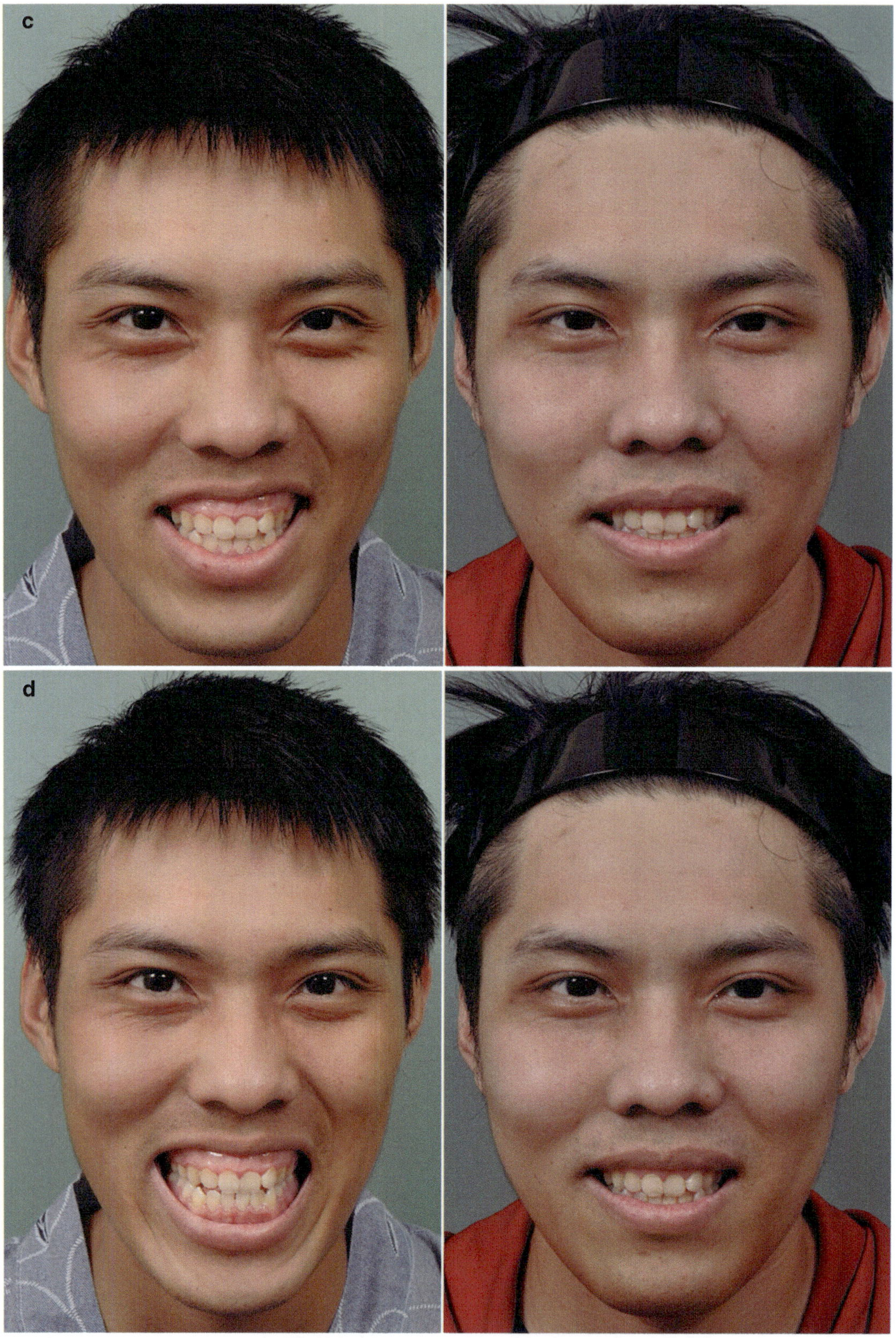

Fig. 12.16 (continued)

face. Another harvest strategy was adopted using water-jet-assisted liposuction (WAL) (Human Med AG, Schwerin, Germany) with 10 min decanting to condense the lipoaspirate. During transplantation, first, by using a blunt 14-G injection cannula through the lateral insertion site, an average of 80–120 mL (traditional liposuction with 1200 × *g* centrifugation for 3 min) of fat was transplanted into the deep layer evenly in each quadrate (Fig. 12.17). Second, an average of 60–100 mL of fat for the superficial layer was transplanted through peri-areolar incisions (3 and 9 o'clock) using a 16-G cannula. A 40–50% increase in the volume of transplanted fat was noted when WAL harvesting was used. Favorable results with a single session of MAFT were reported in approximately 60% of the total cases in our study [37]. The second session (4–6 months after the first MAFT) resulted in more satisfactory outcomes (>90%) for those cases who requested more enhancement of their breast size. No large oil cyst, macrocalcification, or other severe complications were reported, further demonstrating that the MAFT technique is feasible for aesthetic or reconstructive mammoplasty (Figs. 12.18, 12.19, 12.20, 12.21).

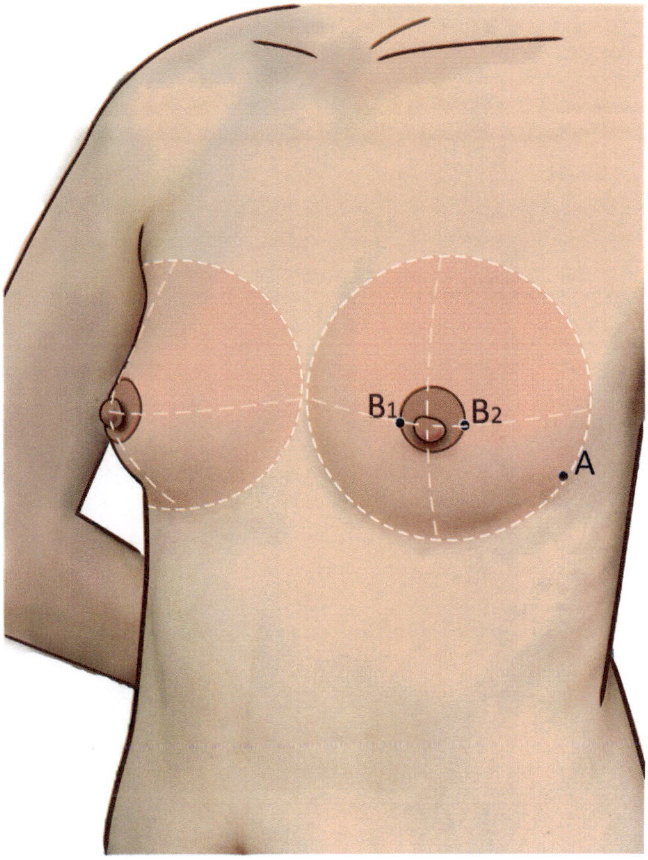

Fig. 12.17 The insertion sites of the MAFT procedure for augmentation mammoplasty were: point A (infra-mammary fold × anterior axillary line) for the retro-glandular deep layer, and points B1 and B2 (peri-areolar at 3 and 9 o'clock along the nipple-areolar complex border) for the superficial subcutaneous layer. The equally distributed volume of fat grafting in four quartans was transplanted both in deep and superficial layers, and the ratio of deep and superficial layer volume was around 60% to 40%

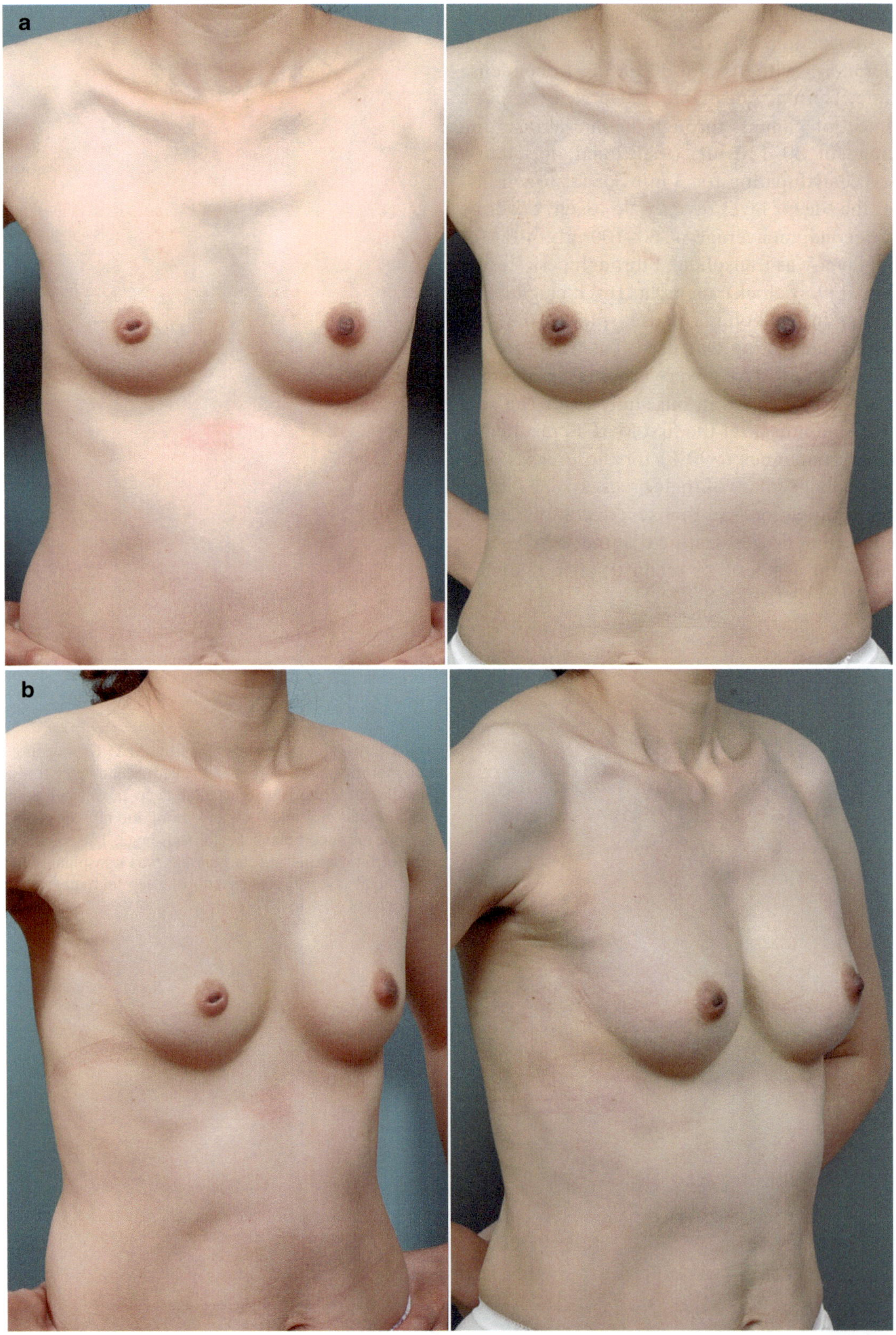

Fig. 12.18 This 40-year-old woman presented with breast asymmetry (the right breast smaller than the left, especially the lower pole) and right inverted nipple along with fat grafting. A single session of MAFT was performed for breast enlargement (R’t deep and superficial layer/L’t deep and superficial layer, 160 mL and 65 mL/150 mL and 30 mL, respectively, and 30 mL for the right inverted nipple). After 7 years of follow-up, nice breast contouring with sustainable volume was noted. The inverted nipple correction also showed satisfactory long-term results (left/right: pre-/post-MAFT) (**a**: AP view, **b**: oblique view, **c**: lateral view)

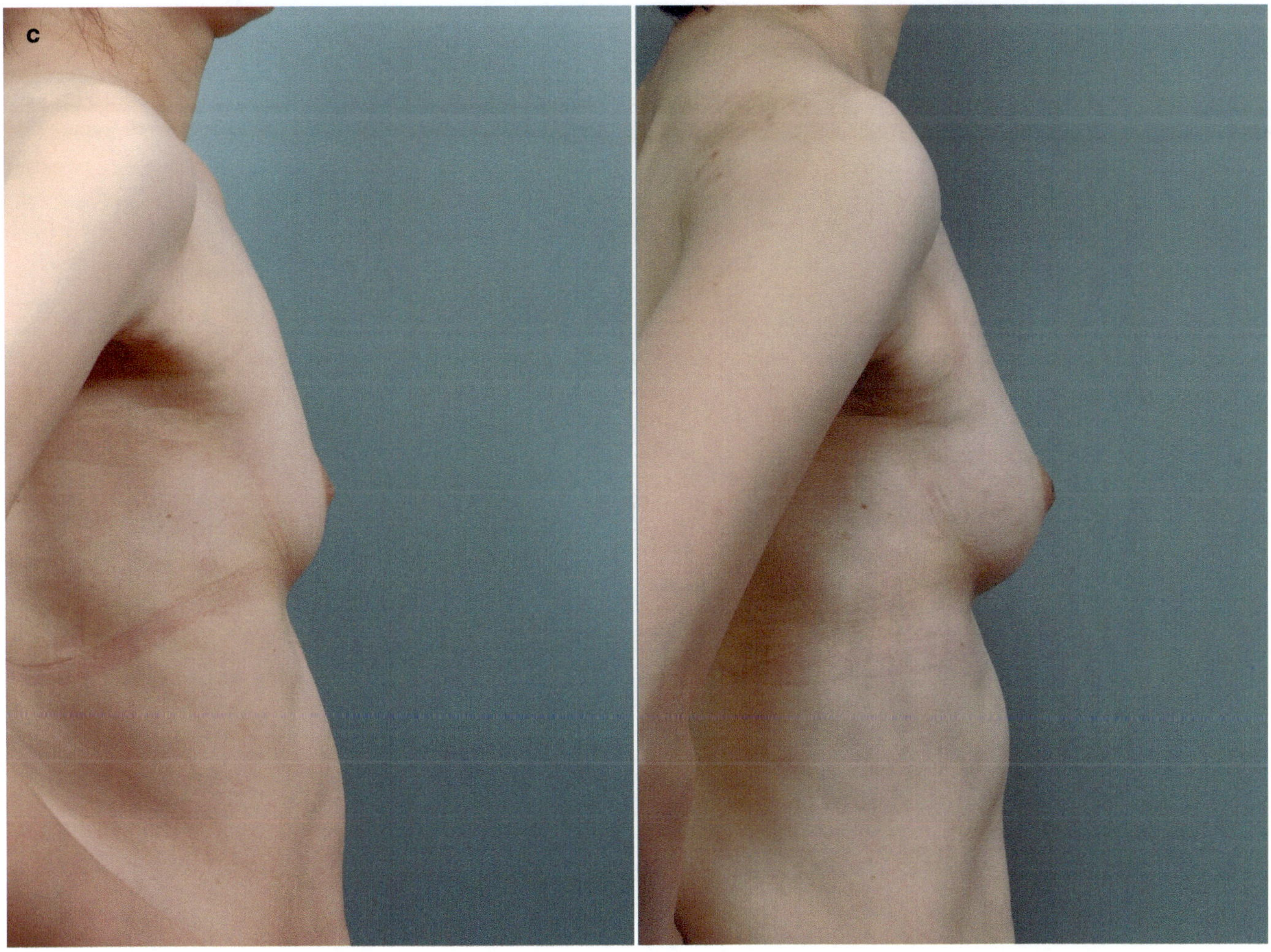

Fig. 12.18 (continued)

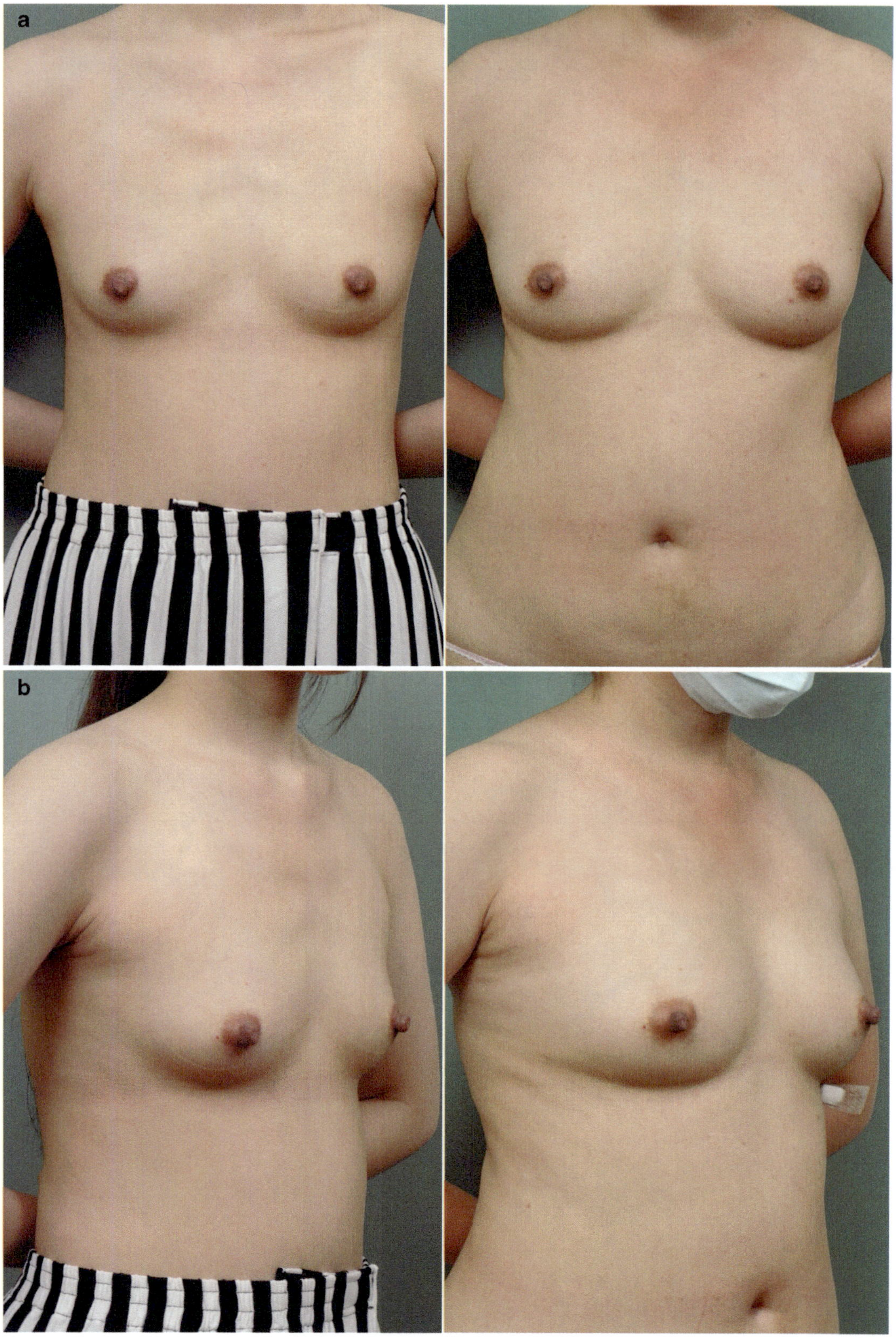

Fig. 12.19 This 30-year-old woman underwent the MAFT procedure, and water-jet assisted liposuction for the improvement of her breast size. In addition, a single session of MAFT was performed for breast enlargement (R't deep and R't superficial layer/L't deep and L't superficial layer, 180 mL and 140 mL/180 mL and 140 mL, respectively). At 4 years and 5 months of follow-up, the breast contour was well maintained with acceptable volume (left/right: pre-/post-MAFT) (**a**: AP view, **b**: oblique view, **c**: lateral view)

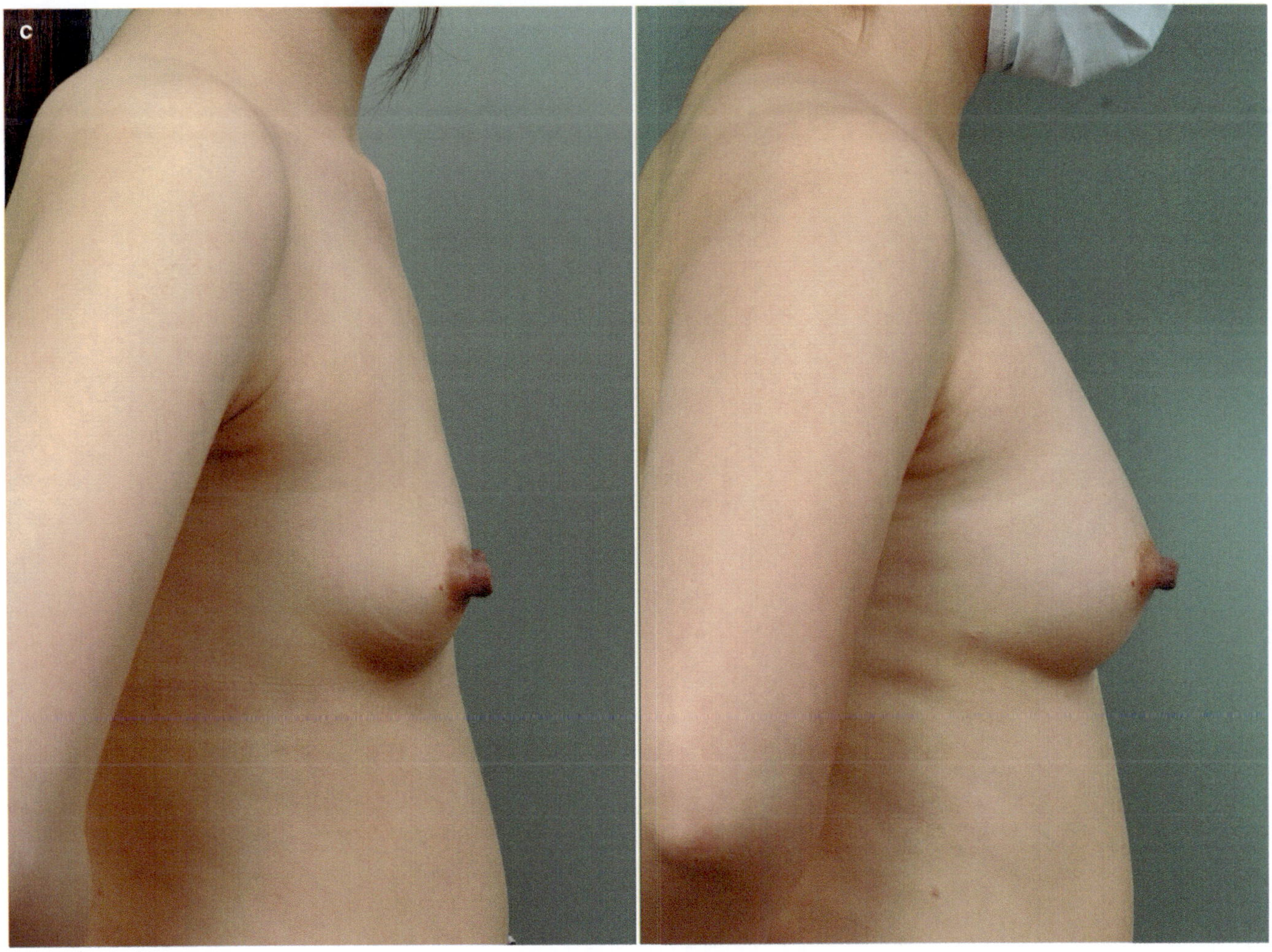

Fig. 12.19 (continued)

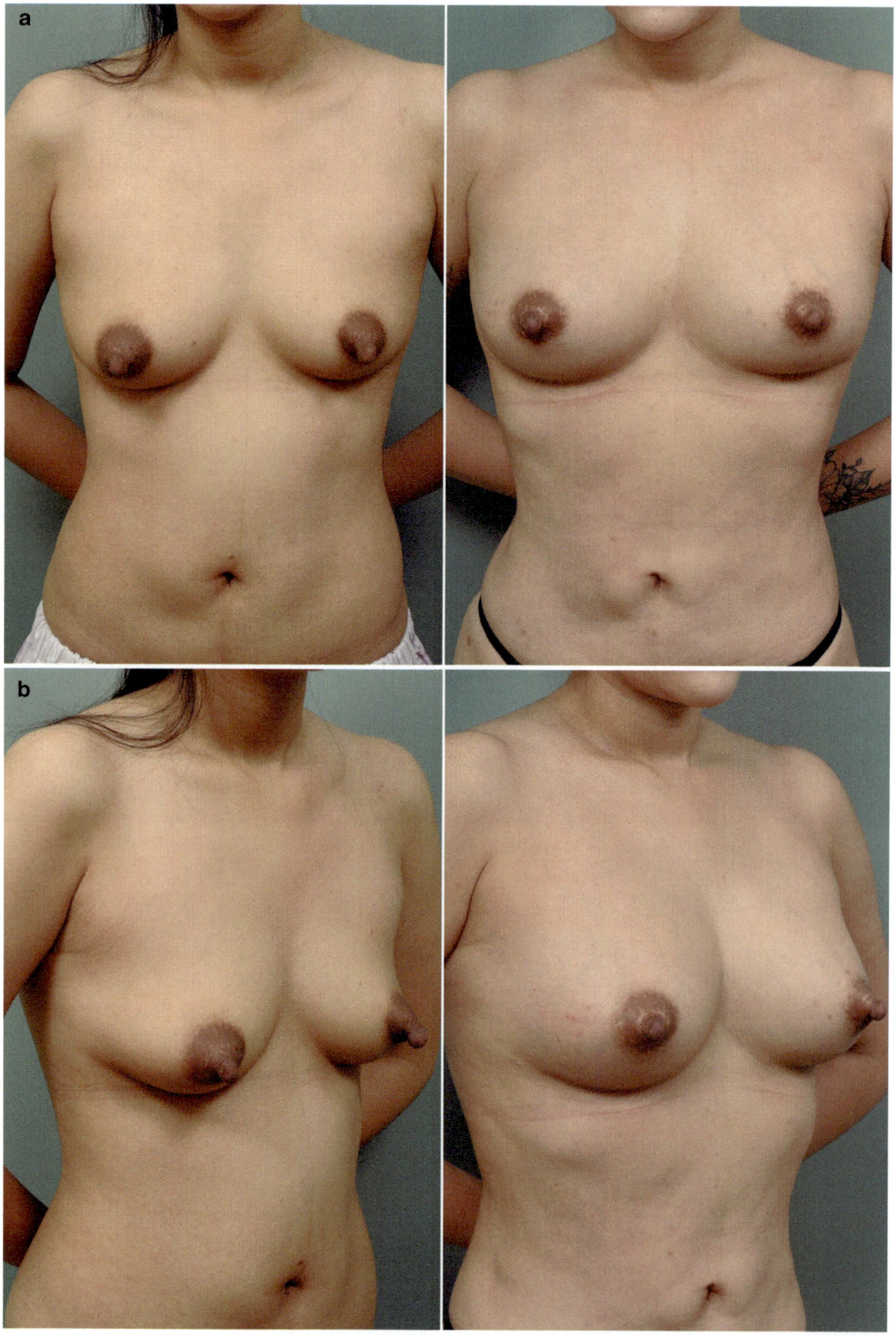

Fig. 12.20 This 35-year-old woman presented with hypoplastic breasts and grade II ptosis. In the first session, the MAFT procedure was performed with water-jet-assisted liposuction. A total of 360 mL/360 mL of fat was transplanted in the right and left breasts. After 8 months, another MAFT was performed (230 mL/360 mL fat transplantation for the right and left breasts, respectively) along with nippleplasty. After 2 years and 3 months of both sessions, the breast contour was well maintained with good projection and fullness (left/right: pre-/post-MAFT) (**a**: AP view, **b**: oblique view, **c**: lateral view)

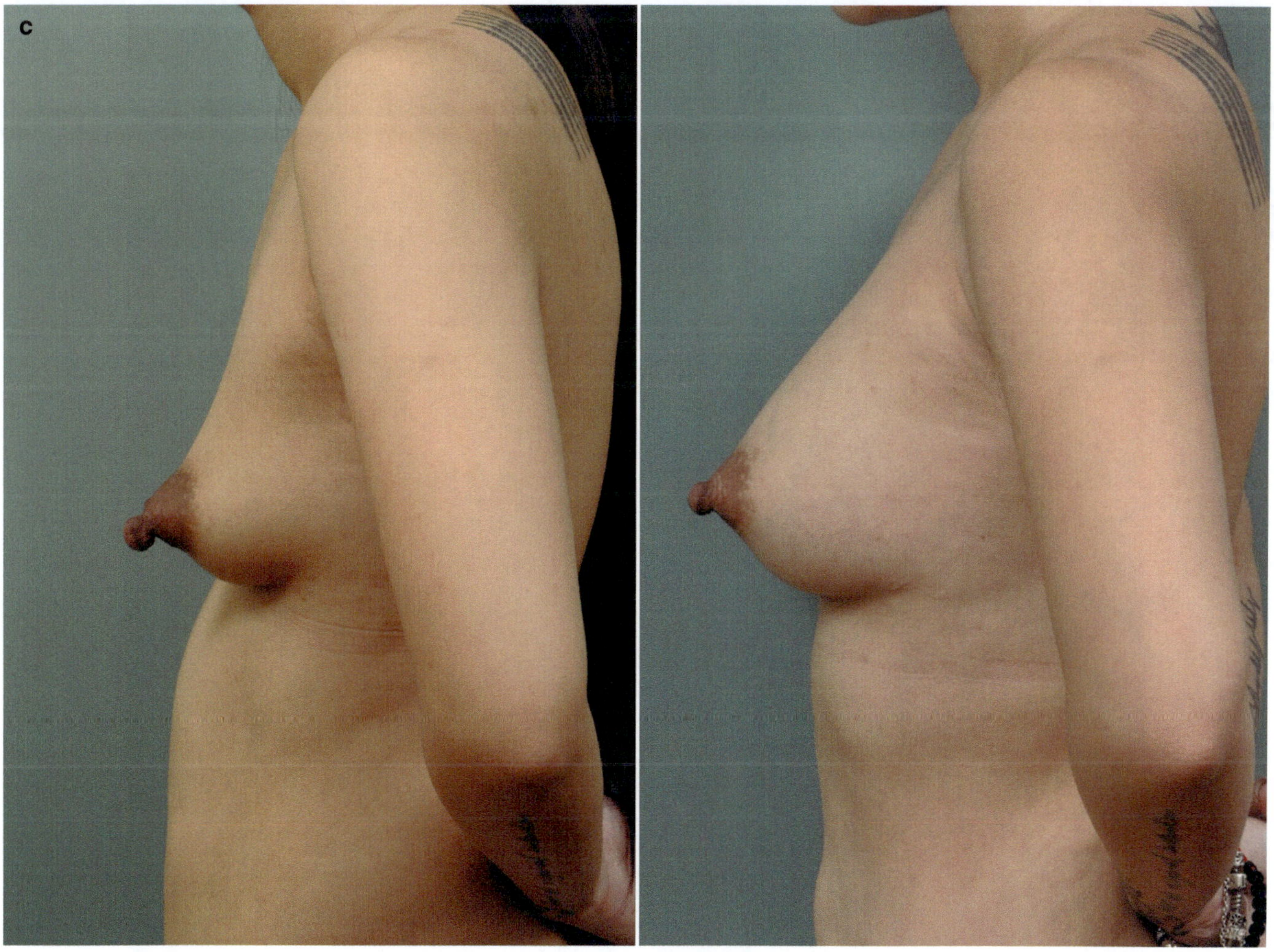

Fig. 12.20 (continued)

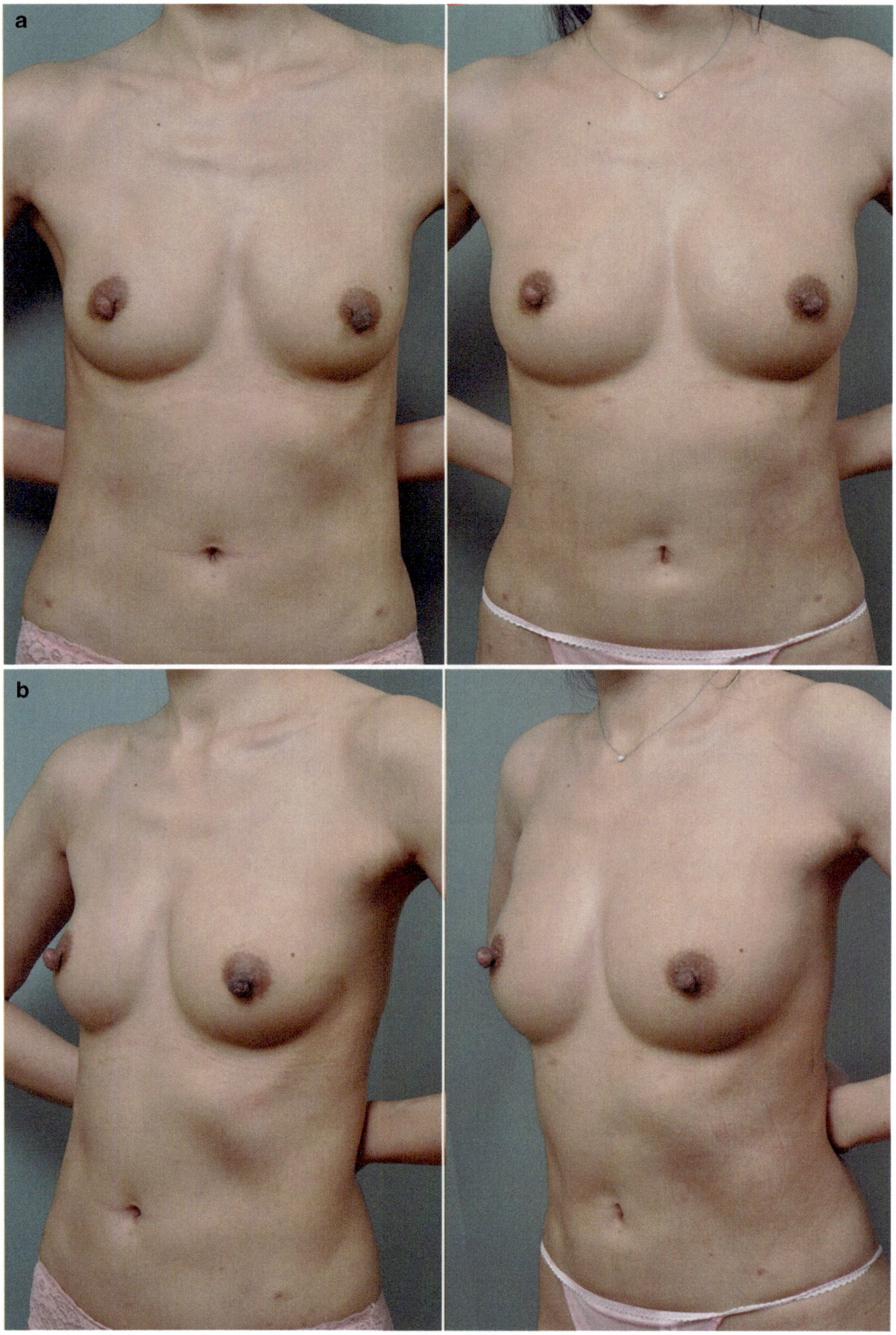

Fig. 12.21 This 37-year-old woman wanted enlargement of her breasts. In the first session, the MAFT procedure was performed with water-jet-assisted liposuction. The total fat volume transplanted was 220 mL/220 mL in the right and left breasts, respectively, at the first session. After 20 months, another touchup MAFT was performed (150 mL/140 mL fat transplantation for the right and left breasts, respectively) along with nippleplasty. After 6 months of the second procedure, the breast contour was well maintained with good projection and fullness (left/right: pre-/post-MAFT) (**a**: AP view, **b**: oblique view, **c**: lateral view)

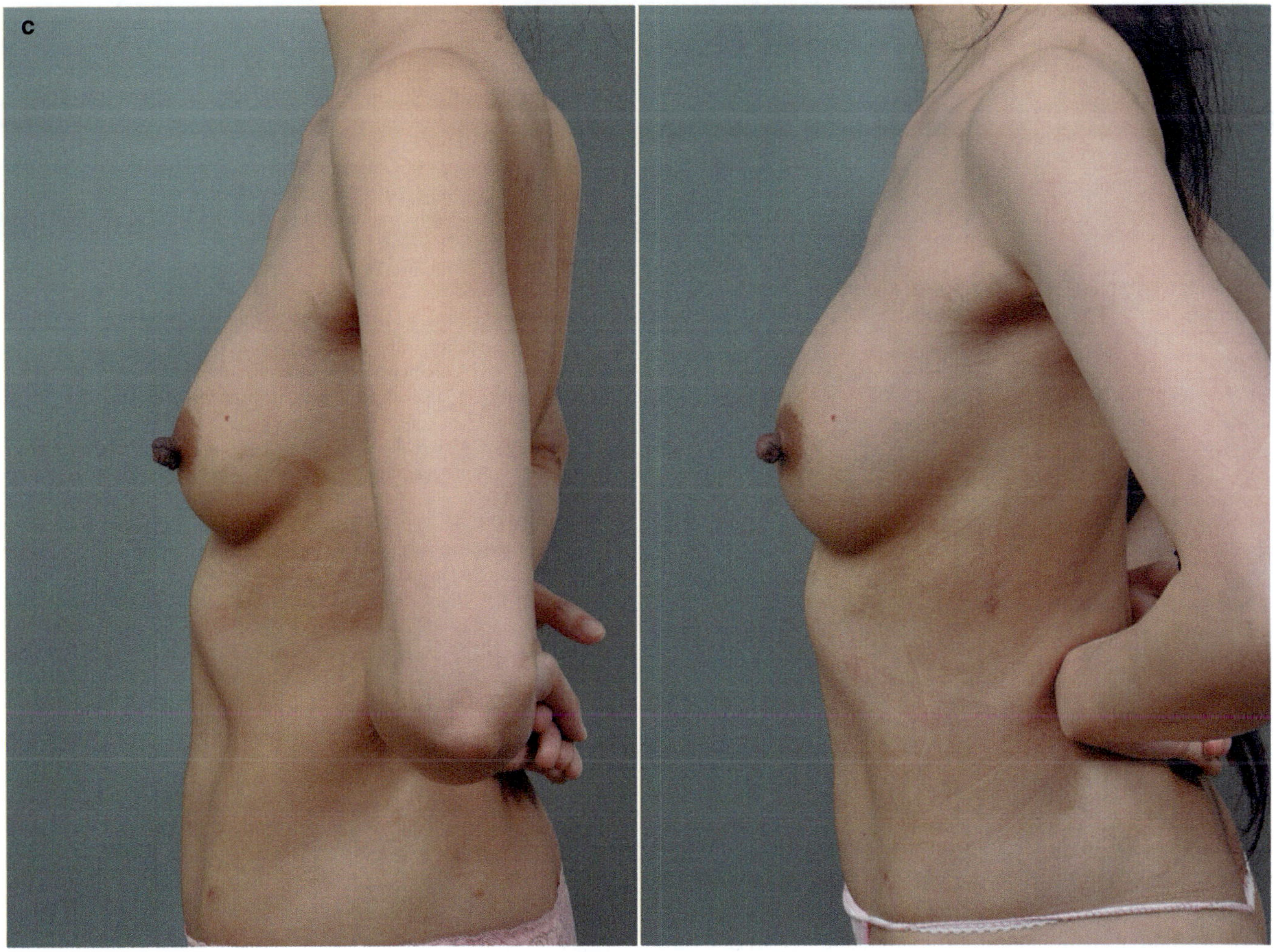

Fig. 12.21 (continued)

12.6 Discussion

In 1893 [1], Neuber concluded that *"Nach meiner Erfahrung heilen Fettstückchen, welche die Grösse etwa einer Bohne oder einer Mandel übersteigen, nicht mehr ein; aber bis zu dieser Grösse ist es mir fast immer geglückt. Je kleiner die Stückchen, desto sicherer der Erfolg."* In my experience, pieces of fat surpassing the size of a bean or an almond do not heal (are not taken), but up to this size, I was nearly always successful. The smaller the pieces, the more certain the success. The central dogma of the MAFT technique [30] originates from the concepts of Coleman [12] ("in some areas, such as the eyelid, the maximum placed should be closer to 1/30th or even 1/50th mL per withdrawal") and Carpaneda [13, 14]. ("The viable tissue was observed in the peripheral zone approximately 1.5 ± 0.5 mm from the edge of the graft.") The MAFT concept has been accepted since 2007 [30] and was further clarified by the works of Yoshimura, "Conclusive reports of 1500 μm (1.5 mm) from tissue surface (grafting edge) survived" [38, 39], and Khouri, "The Microribbon Model predicts that, in a typical human, fat injections larger than 0.16 cm (1.6 mm) in radius will have a region of central necrosis" [40]. Table 12.3 summarizes and highlights the concepts by conclusive mathematical calculations and further confirms the importance of total parcel numbers in 1 mL fat grafting. The injection frequency for 1 mL fat should be 30–240, which is the range of MAFT-GUN (60–240 triggers/mL), to obtain better clinical results.

Table 12.3 The summarized concepts by conclusive mathematical calculations of various theories in terms of the importance of total parcel numbers in 1 mL fat grafting

Author	Theory	No. of parcels per 1 mL	Size of parcel (mL)
Dr. Carpaneda [13, 14] 1993	1 mm≦radius≦2 mm	~30	1/30
Dr. Coleman [10–12] 2006	1/50 mL≦parcels≦1/30 mL	30 ~ 50	1/30 ~ 1/50
Dr. Khouri [40] 2014	Microribbon, Radius = 1.6 mm	58	1/58
Dr. Yoshimura [38, 39] 2014	Dynamic remodeling, radius = 1.5 mm	71	1/71
Lin et al. (authors) [23–33, 36, 37] 2007	MAFT (micro-autologous fat transplantation) technique	60 ~ 240	1/60 ~ 1/240

12.7 Conclusions

In summary, in the modern fat grafting era, there are many concepts to explore, techniques to refine, and research questions to investigate. As the final crucial step of fat grafting and transplantation, the concept of MAFT has shown its feasibility for clinical applications in aesthetic, reconstructive, and regenerative surgeries and demonstrated its indispensability in fulfilling the concepts proposed by various authors. The application of the innovative instrument, MAFT-GUN, fully demonstrates its implementation and proves its efficacy in facing challenges of fat grafting in the future.

References

1. Neuber F. Fettransplantation. Bericht über die Verhandlungen der Deutschen Gesellschaft für Chirurgie Zbl Chir. 1893;22:66.
2. Peer LA. Loss of weight and volume in human fat grafts: with postulation of a "cell survival theory". Plast Reconstr Surg. 1950;5:217–30.
3. Bames HO. Augmentation mammaplasty by lipotransplant. Plast Reconstr Surg. 1953;11:404–12.
4. Illouz YG. Une nouvelle technique pour les lipodystrophies localisées. Rev Chir Esthet. 1980;4:19.
5. Chajchir A, Benzaquen I. Liposuction fat grafts in face wrinkles and hemifacial atrophy. Aesthet Plast Surg. 1986;10:115–7.
6. Asken S. Facial liposuction and microlipoinjection. J Dermatol Surg Oncol. 1988;14:297–305.
7. Nguyen A, Pasyk KA, Bouvier TN, et al. Comparative study of survival of autologous adipose tissue taken and transplanted by different techniques. Plast Reconstr Surg. 1990;85:378–86; discussion 387–389.
8. Hultman CS, Friedstat JS. The ACAPS and SESPRS surveys to identify the most influential innovators and innovations in plastic surgery: no line on the horizon. Ann Plast Surg. 2014;2(6):S202–7.
9. Pu LL, Yoshimura K, Coleman SR. Future perspectives of fat grafting. Clin Plast Surg. 2015;42(3):389–94, ix-x.
10. Coleman SR. Facial recontouring with lipostructure. Clin Plast Surg. 1997;24(2):347–67.
11. Coleman SR. Structural fat grafting. Aesthet Surg J. 1998;18(5):386, 388
12. Coleman SR. Structural fat grafting: more than a permanent filler. Plast Reconstr Surg. 2006;118(3 Suppl):108S–20S.
13. Carpaneda CA, Ribeiro MT. Study of the histologic alterations and viability of the adipose graft in humans. Aesthet Plast Surg. 1993;17:43–7.
14. Carpaneda CA, Ribeiro MT. Percentage of graft viability versus injected volume in adipose autotransplants. Aesthet Plast Surg. 1994;18:17–9.
15. Khawaja HA, Hernandez-Perez E. Fat transfer review: controversies, complications, their prevention and treatment. Int J Cosmet Surg Aesthetic Dermatol. 2002;4:131–8.
16. Fournier PF. Microlipoextraction et microlipoinjection. Rev Chir Esthet Lang Fr. 1985;10:36–40.
17. Lin TM (2013) Dispenser device. US Patent 8,523,825 B2 Sept 2013.
18. Dermato Plastica Beauty Co., Ltd. Kaohsiung, Taiwan (2010) http:// https://www.maft-gun.com/maft-gun/?lang=en. Accessed 14 Feb 2021.
19. Chou CK, Lin TM, Chiou JH, et al. Influential factors in autologous fat transplantation-focusing on the lumen size of injection needle and the injecting volume. IPRAS J. 2012;8:25–7.
20. Fulton JE. Breast contouring with "gelled" autologous fat: a 10-year update. Int J Cosmet Surg Aesthetic Dermato. 2003;5:155–63.
21. Shu T, Lam SM. Liposuction and lipotransplants for facial rejuvenation in the Asian patient. Int J Cosmet Surg Aesthetic Dermato. 2003;5:165–73.
22. Coleman SR. The technique of periorbital lipoinfiltration. Oper Tech Plast Reconstr Surg. 1994;1:120–6.
23. Lin TM. Total facial rejuvenation with micro-autologous fat transplantation (MAFT). In: Pu LLQ, Chen YR, Li QF, et al., editors. Aesthetic plastic surgery in Asians: principles and techniques. 1st ed. St. Louis, MO: CRC Press; 2015. p. 127–46.
24. Chou CK, Lee SS, Lin TY, et al. Micro-autologous fat transplantation (MAFT) for forehead volumizing and contouring. Aesthet Plast Surg. 2017;41(4):845–55.
25. Lee SS, Huang YH, Lin TY, et al. Long-term outcome of micro-autologous fat transplantation to correct temporal depression. J Craniofac Surg. 2017;28(3):629–34.
26. Lin TM, Lin TY, Chou CK, et al. Application of microautologous fat transplantation in the correction of sunken upper eyelid. Plast Reconstr Surg Glob Open. 2014;2(11):e259.
27. Lin TM, Lin TY, Huang YH, et al. Fat grafting for recontouring sunken upper eyelids with multiple folds in Asians-novel mechanism for neoformation of double eyelid crease. Ann Plast Surg. 2016;76(4):371–5.
28. Huang SH, Lin YN, Lee SS, et al. Three simple steps for refining transcutaneous lower blepharoplasty for aging eyelids: the indispensability of microautologous fat transplantation. Aesthet Surg J. 2019;39(11):1163–77.
29. Kao WP, Lin YN, Lin TY, et al. Microautologous fat transplantation for primary augmentation rhinoplasty: long-term monitoring of 198 Asian patients. Aesthe Surg J. 2016;36(6):648–56.
30. Lin TM, Lin SD, Lai CS (2007) The treatment of nasolabial fold with free fat graft: preliminary concept of micro-autologous fat

transplantation (MAFT). Paper presented at the 2nd academic congress of Taiwan cosmetic association, Taipei, Taiwan, 7 May 2007.
31. Lin TM, Chou CK, Takahashi H (2019) Industry show case for transconjunctival lower blepharoplasty with micro-autologous fat transplantation (MAFT). Paper presented at the 17th International Federation for Adipose Therapeutics and Science (IFATS), Marseille, France, 4–7 December 2019.
32. Lin TM, Huang SH, Lin YN, et al. Fat grafting for facial contouring (nose and chin). Clin Plast Surg. 2020;47(1):91–8.
33. Lin YN, Huang SH, Lin TY, et al. Micro-autologous fat transplantation for rejuvenation of the dorsal surface of the aging hand. J Plast Reconstr Aesthet Surg. 2017;71(4):573–84.
34. González N, Goldberg DJ. Update on the treatment of scars. J Drugs Dermatol. 2019;18(6):550–5.
35. Covarrubias P, Cárdenas-Camarena L, Guerrerosantos J, et al. Evaluation of the histologic changes in the fat-grafted facial skin: clinical trial. Aesthet Plast Surg. 2013;37(4):778–83.
36. Huang SH, Huang YH, Lin YN, et al. Micro-autologous fat transplantation for treating a gummy smile. Aesthet Surg J. 2018;38(9):925–37.
37. Lin TM (2020) Refining technique of micro-autologous fat transplantation for aesthetic and reconstructive mammoplasty: a ten-year journey. Paper presented at the annual meeting of Asian Society of Breast Plastic and Reconstructive Surgery. Taoyuan, Taiwan. 25 Oct 2020.
38. Eto H, Kato H, Suga H, Aoi N, et al. The fate of adipocytes after nonvascularized fat grafting: evidence of early death and replacement of adipocytes. Plast Reconstr Surg. 2002;129(5):1081–92.
39. Kato H, Mineda K, Eto H, et al. Degeneration, regeneration, and cicatrization after fat grafting: dynamic total tissue remodeling during the first 3 months. Plast Reconstr Surg. 2014;133(3):303e–13e.
40. Khouri RK Jr, Khouri RE, Lujan-Hernandez JR, et al. Diffusion and perfusion: the keys to fat grafting. Plast Reconstr Surg Glob Open. 2014;2(9):e220.

Ultrasound-Assisted Liposuction (UAL) with VASER Technology in Body Contouring

13

Alberto Di Giuseppe, William W. Cimino, and Federico Giovagnoli

Contents

13.1 Introduction

Ultrasound-assisted liposculpture has developed as a modification of liposuction for body contouring and fat reduction. The effect of sound waves in the subcutaneous tissues is to emulsify the adiposities and, at the same time, preserve the vessels and nerves. The technique relies on the surgical use of ultrasonic energy, which allows the selective destruction or emulsification of adipose tissue producing a "cream" which is then aspirated. The procedure produces less edema and bruising than traditional liposuction and allows a great degree of skin retraction owing to "stimulation" of the dermis of the treated areas. Clinical application of ultrasound-assisted lipoculpturing is for primary lipodystrophy, secondary cases, and difficult areas such as inner thigh, circumferential thigh, calf, ankle, neck, chin, arm, flanks, and abdomen.

There is little blood loss, and a greater volume of fat tissue can be removed in one session. Ultrasound-assisted liposuction (UAL) can be used to treat obesity. This method

A. Di Giuseppe (✉)
University of Padova, Padova, Italy

W. W. Cimino
University of Purdue, Louisville, CO, USA
e-mail: cimino@gidbio.com

F. Giovagnoli
University Vita-Salute San Raffaele, Milan, Italy

© The Author(s), under exclusive license to Springer Nature Switzerland AG 2023
A. Di Giuseppe et al. (eds.), *Fat Transfer in Plastic Surgery*, https://doi.org/10.1007/978-3-031-10881-5_13

has been applied to surgery of the breast for volume reduction of the adipose components in the fatty breast (mostly in juvenile breast) and skin retraction after dermis stimulation for correction of minor degree ptosis. The main concept of this technique relies on the selective destruction of mainly the fluid fraction of the adipose tissue, which represents nearly 90% of all the adipose tissue volume. In contrast to traditional liposuction, which attacks and destroys mechanically all the structures of the dermis, which may have undesired side effects, UAL selectively destroys only the target adiposities and spares the supporting structures of the skin and dermis, such as vessels, nerves, collagenic matrix, and elastic fibers, thus conserving the elasticity of the treated areas [1].

Suction-assisted lipoplasty (SAL) is generally regarded as safe and effective with well-known and well-documented clinical results and potential complications. For this method of lipoplasty the fundamental mechanism of tissue removal is avulsion, that is, the fatty tissue is pulled into a hollow suction cannula by vacuum and then avulsed or torn by the movement of the cannula. Results and potential complications are a direct function of the avulsive/cutting process.

VASER-assisted lipoplasty (VAL) presents a fundamental change in the treatment and subsequent removal of targeted fatty tissues. The fatty tissue is first selectively emulsified using ultrasonic frequency vibrations delivered on a metal probe and then the emulsified tissues are removed using a gentle, minimally avulsive aspiration process. The VASER system's high selectivity for fatty tissue results in decreased overall damage to the tissue matrix comprising fat cells, vessels, nerves, structural tissues, and lymphatic tissue. VAL uses ultrasonic frequency vibrations to emulsify the fatty component of the tissue matrix but in a fundamentally different manner than earlier versions of ultrasonic instrumentation for lipoplasty, namely, ultrasound-assisted lipoplasty (UAL). The primary differences are that the USER system delivers significantly less power to the tissues while simultaneously increasing fragmentation/emulsion efficiency compared with UAL devices and eliminating the simultaneous aspiration feature of UAL devices [2, 3]. Ultrasonic energy was introduced to the plastic surgery community in the form of UAL in the late 1980s and early 1990s [4–9]. UAL technology was characterized by large-diameter (5-mm) ultrasonic cannulae, simultaneous aspiration and emulsion, continuous energy delivery, and sharp-edged elements at the cannulae ends, usually associated with the suction pert holes. The UAL technique was characterized by generally longer application times than those of SAL owing to low-efficiency ultrasonic cannulae designs to create the emulsion and low-efficiency aspiration owing to the small (2-mm) suction lumens in the ultrasonic cannulae. While both excellent results [10–15] and complications [16–21] were reported, there was no clear understanding of the causes for the excellent results or the complications other than the unfortunate general assignment to "ultrasonic energy." VASER technology uses ultrasonic frequency vibrations but delivers the energy to the tissues in a substantially different manner than UAL. In brief, VAL technology is characterized by small-diameter probes (2.2–3.7-mm diameter), pulsed delivery of the ultrasonic energy, and grooved probe designs that provide energy dissipation along the sides of the probes in addition to the front face of the probes. The VAL technique is characterized by procedure times similar to or shorter than typical SAL procedure times and incision sizes comparable to those of SAL. The amount of power delivered to the tissues and a method of measure during the efficiency of each probe design (including other ultrasonic devices) are summarized in the research published by Cimino [2].

13.2 Physical Principles

The standard acoustic spectrum comprises a wide range, going from 1 Hz (infra-sound) to 100 Hz (audible sound), to 10–100 mHz (ultrasound), which is too high to be perceived by the human ear. This frequency spectrum has been utilized in industrial application and medical application. The audible sound range varies from 100 to 10,000 Hz, and the medical application range varies from 105 to 107 Hz. Ultrasound waves are the result of the transformation of normal electric energy into high-frequency energy (over 16 kHz). These waves are produced by high-power ultrasonic generators, which produce and transform the electric energy into ultrasonic energy. The energy from the generator is transmitted to a piezoelectric quartz crystal or ceramic transducer and then transformed into mechanical vibrations that are amplified and transmitted. Other medical applications of ultrasound are dental cleaning, phacoemulsification for cataract surgery, diagnostic imaging, kidney calculi fragmentation, and in neurosurgery.

13.3 Cavitation (Bubble Mechanism)

Cavitation can be defined as a peculiar activity induced by the application of ultrasound waves in a liquid or liquid-like material that contains bubbles or pockets of gas or vapor. The consequences could be the production of stable or transient

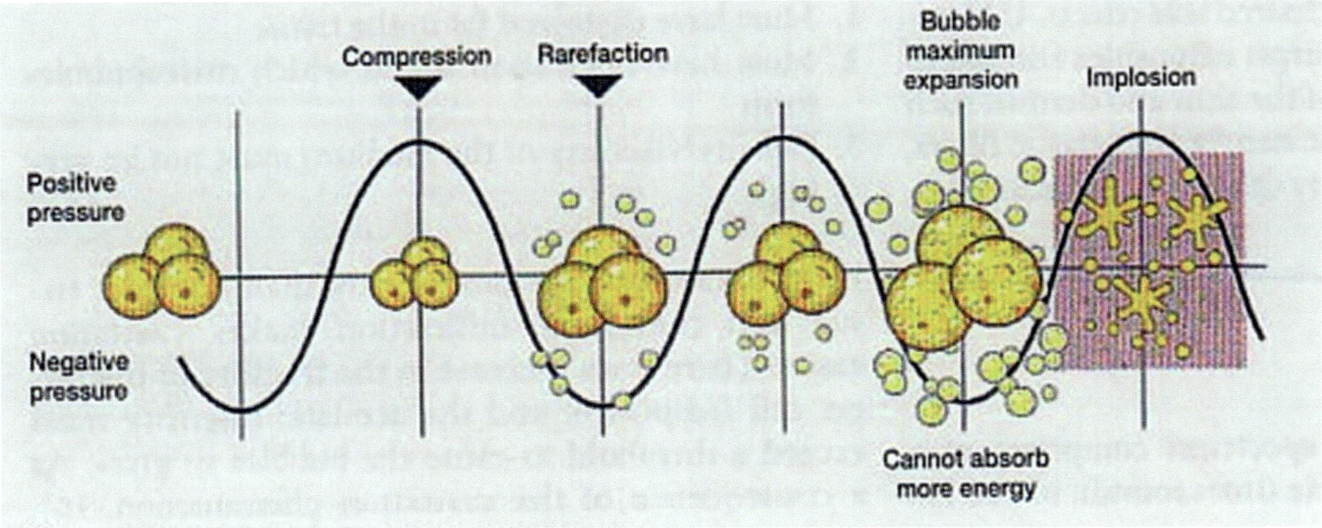

Fig. 13.1 Cavitation is the result of compression and rarefaction pressure. Photo taken from VASER Solta Medical User's Guide [24]. Courtesy of Bausch, ©2022 Bausch Health Companies Inc., All Rights Reserved

cavitation. In order to allow the cavitation to occur, the following conditions must be met:

1. Must have dissolved fat in the tissue.
2. Must have nucleation site at which microbubbles form.
3. Density/viscosity of the medium must not be very high.

Fatty tissues cavitate more easily than compact tissue, and tumescent infiltration makes cavitation easier. There is an increase in the fragility of the target cell (adiposity), and the acoustic intensity must exceed a threshold to cause the bubbles to grow. As a consequence of the cavitation phenomenon, implosion of bubbles results in cell disruption and fat emulsification.

In order to produce this mechanism, a special titanium probe attached to the piezoelectric transducer is able to convert the ultrasonic energy into the tip of the probe with 100% of the energy translated into ultrasonic energy at the tip of the probe. The cycle of ultrasound waves is passed through the titanium probe and causes an alternation of circular waves with nodal points of energy concentrated along the probe and into its tip. All of the energy will be concentrated in a perfectly functioning and efficient system.

The bubble mechanism (cavitation) causes the formation of bubbles of progressively increasing size in the fatty compartment. The alternation of positive and negative pressures, with an expansion and contraction at a frequency corresponding to the ultrasonic wave, leads to the progressive instability of the bubbles and microcavities already formed until their final explosion. Expansion and compression, implosion and explosion, related to alternate cycles, are the phenomena causing cavitation which leads to the final destruction of the target cells, the adiposity (Fig. 13.1).

13.4 Effects of UAL

The effects of UAL on adipose tissue are:

1. Micromechanical effect: The consequence of damage produced by the direct action of the ultrasonic waves and the organic intracellular molecules, which are violently moved and displaced in the extracellular space with subsequent breaking up of macromolecules.
2. Cavitation effect: The cellular fragmentation from cavitation determines the diffusion of the lipid matrix (fatty acids) of the adiposity into the inter cellular space where a stable emulsion with interstitial fluid and infiltration solution occurs. The formation of free radicals and the denaturation of the lipoprotein components of the cellular membranes may lead to a progressive lipolytic action, which will partially continue even after the ultrasonic treatment has ended.
3. Thermal effect: A hyperemic secondary mild effect is associated with the friction action of the titanic ultrasonic probe moving through the tissues. These effects are restrained and negligible if the technique is properly applied. Thermal effects do not cause any damage to cells, structures, or fluids involved and do not cause any undesired biological reaction, especially at the protein chain level (Figs. 13.2 and 13.3).

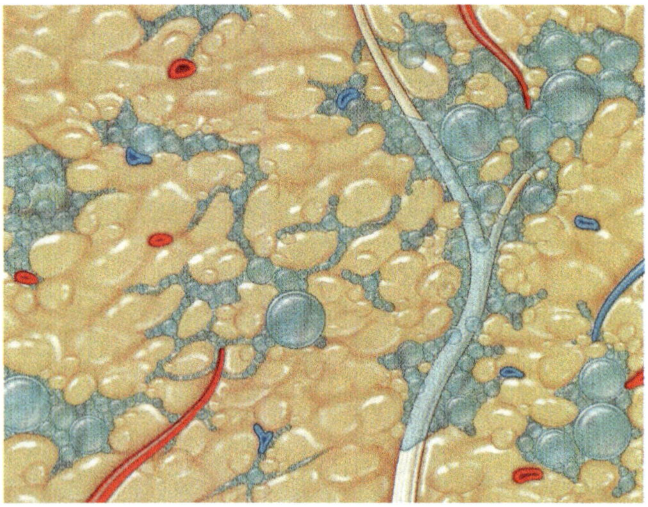

Fig. 13.2 The growth of the microbubbles to almost 200 μm forces the fat cell apart. As the microbubbles reach resonant size and collapse they dislodge fat cells from the tissue matrix. Photo taken from VASER Solta Medical User's Guide [24]. Courtesy of Bausch, ©2022 Bausch Health Companies Inc., All Rights Reserved

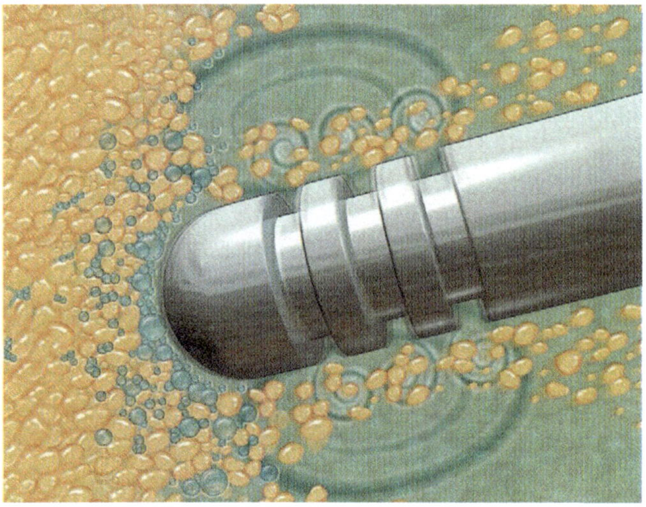

Fig. 13.3 Acoustic streaming is the result of powerful fluid forces that create swirling around the vibrating ultrasonic probe tip. This swirling causes the fat cells that have been dislodged from the matrix to mix the infusion fluid creating the emulsion. Photo taken from VASER Solta Medical User's Guide [24]. Courtesy of Bausch, ©2022 Bausch Health Companies Inc., All Rights Reserved

13.5 How VASER Ultrasound Works

There are millions of microscopic air bubbles in tumescent solution. When the bubbles are exposed to ultrasound energy, they expand and eventually collapse. These bubbles act as miniature crowbars to force the fat cells apart. Once the fat is loosened, it is mixed with the tumescent fluid to form an emulsion. Acoustic streaming causes intense localized swirling to further break up the fat into small clusters of cells. Small groups of cells are ideal for fat transfer (Fig. 13.4).

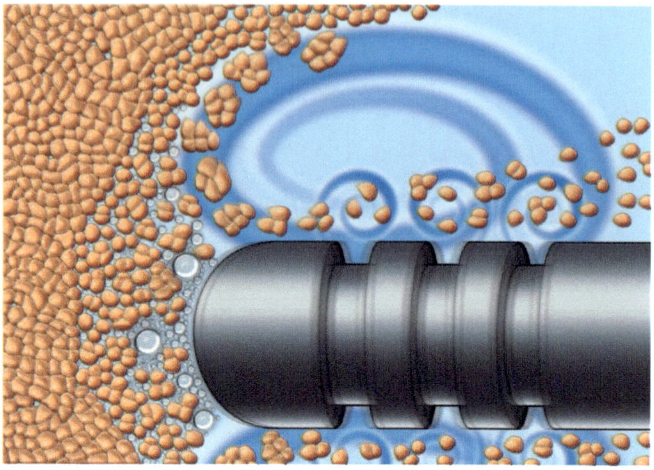

Fig. 13.4 VASER probe in action. Photo taken from VASER Solta Medical User's Guide [24]. Courtesy of Bausch, ©2022 Bausch Health Companies Inc., All Rights Reserved

13.6 VASER Technology: The VASER System

The technological advancements incorporated in VASER technology are designed to deliver the absolute minimum amount of vibratory energy to the tissues and still achieve the desired emulsification/fragmentation of the fatty component of the tissue matrix. These advancements include the use of small-diameter solid probes in the range of 2.2–3.7 mm, grooved probe designs to increase efficiency, pulsed delivery of the vibratory energy to reduce the average energy delivered, gentle aspiration cannulae designs to reduce tissue avulsion while maintaining aspiration performance, and refined instrumentation design with regard to size and weight to provide more artistic use and movement (Fig. 13.5).

The amount of energy delivered to the tissues by an ultrasonically vibrating probe is roughly proportional to the square of the diameter of the ultrasonic probe or cannulae, meaning that small reductions in diameter result in significant reductions in delivered energy. Therefore, the design objective is to utilize the smallest possible diameter probe that accomplishes the desired emulsification in a reasonable time. Generally speaking, probes smaller than about 2 mm in diameter are too flexible to control (unless they are very short), and probes larger than about 4 mm in diameter present excessive amounts of energy to the tissues. The performance tradeoff is that these smaller diameter probes have reduced contact area with the tissue and will therefore take longer to achieve the desired emulsification. To increase the efficiency of the emulsification process and to decrease the operative time, grooves have been added to the tips of the

Fig. 13.5 VASER system includes: (1) ultrasound amplifier (2) precision fluid management system (3) VentX infiltration and aspiration console (4) wireless foot switch system. Photo taken from VASER Solta Medical User's Guide [24]. Courtesy of Bausch, ©2022 Bausch Health Companies Inc., All Rights Reserved

probes to provide additional emulsification zones. With this grooved design, the tissue that contacts the sides of the probes will be emulsified in addition to the emulsification that occurs on the front surface of the probe, significantly increasing the emulsification efficiency and decreasing opera drive time.

Because the vibratory energy is dissipated from the sides of the probe as well as the front surface of the probe, the energy is less concentrated and results in a softer and more effective emulsion of the fatty tissue. The number of grooves determines the relative partitioning of the vibratory energy. Four identical 3.7-mm probes were evaluated, having zero, one, two, and three grooves. The results show that increasing the number of grooves moves more energy to the sides of the probe. Note also that the overall power and efficiency also rise, meaning that the probes with more grooves are not only more efficient but are also able to contact more tissue. The applicability of the different probe designs to different types of tissue (soft to fibrous) is discussed in detail in the next section on the VAL used in previous-generation UAL devices. The second form is pulsed mode, a form where the vibratory energy is delivered in short duration bursts, referred to as VASER mode.

Fig. 13.6 VASER Grooved probes. Photo taken from VASER Solta Medical User's Guide [24]. Courtesy of Bausch, ©2022 Bausch Health Companies Inc., All Rights Reserved

The advantage of VASER mode is that the peak energy is delivered in a short burst (the on condition) and is strong enough to emulsify fatty tissue, while the average energy delivered over time is reduced compared with that for the continuous mode. VASER mode delivers around ten bursts per second to the tissues, fast enough so that VASER mode is generally imperceptible to the surgeon and fast enough so that the resultant on/off effect is uniform, smooth, and repeatable with regard to emulsification of the tissue. If the burst rate is less than about ten times per second, then the on/off effect of the device begins to present a timing issue to the surgeon with regard to the forward and backward strokes of the device and also tends to produce less consistent results because of the physical spacing of the on periods in the tissue. Emulsified fatty tissue is more easily removed with aspiration cannulae than non-emulsified fatty tissue; therefore, only gentle aspiration is required. A series of emulsion cannulae, called the VentX cannulae, were designed for efficient emulsion removal and minimal trauma to the tissue matrix. The cannulae design features are discussed in detail in the next section (Fig. 13.6).

13.7 VASER Technique

The following guidelines cover safe and effective use and application of the VASER system for VAL including the infusion, emulsification, and aspiration phases. Probe selection for the various types of tissues (guideline 4) is discussed in detail following the 11 guidelines for infusion and emulsion. Aspiration is discussed after the section on probe selection.

13.8 Infusion and Emulsification

1. Infusion. Use sufficient and appropriate amounts of wetting solution. Make sure that the wetting solution is uniformly and evenly distributed in the intended fragmentation volume. A gentle firmness and fullness in the targeted area are desirable. Allow sufficient time for vasoconstrictive requirements (usually at least 10 min minimum for 1:1000000 concentrations). Apply wetting solution slightly beyond the marked boundaries and in all potential port locations. If you cannot achieve a sufficient distribution of fluid, then do not use the VASER system in that location. The targeted ratio of wetting solution into total aspirate out is approximately 1.5:1. If more than a 2:1 ratio is used, the VASER system begins to lose efficiency owing to the presence of excessive fluid. If less than a 1:1 ratio is used, then there will not be enough fluid to easily form an emulsion.
2. Incision protection. Skin ports are designed to protect the incision edges during the fragmentation phase with minimal incision size (approximately 4-mm incision) and also protect the incision edges during the suction phase. The skin ports should go in easily; otherwise, the incision is too small, or the port is hooked on the underside of the dermis.
3. Skin protection. Use a towel (wet or dry) to cover the skin near the skin port. This protects the skin in the accidental situation where the probe is levered into any exposed skin near the skin port. This is especially important across the lower abdomen when working in the upper abdomen and near creases (buttocks). Note: A single towel layer will not protect the skin from prolonged contact with the vibrating probe or if the probe is levered strongly into the skin through the towel. If prolonged contact or strong levering is unavoidable, use a triple-folded towel to ensure skin protection.
4. Probe selection. Use the correct probe. Adjust the setting so that the probe moves smoothly through the tissue. If the probe is dragging or struggling, then all of the energy will be used to penetrate the tissue, and very little emulsion will occur. In this case, move the amplitude up a notch or two. If this does not solve the problem, switch to a probe designed for more fibrous tissues.
5. Probe movement. Keep the probe moving at all times. Move it smoothly at a speed that the tissue and amplitude setting will allow without excessive pushing. Do not let the probe sit (vibrate) in one location. A speed just slightly slower than standard suction cannula movement is appropriate.
6. Torquing. Do not torque the probe! Move the probe in and out like spokes of a wheel, do not lever (torque) the probe sideways or up and down. The skin protector should not be used as a fulcrum. Torquing will lever the probe into the skin protector so that frictional rubbing can cause heat build-up in the skin protector.
7. Application time and surgical end points. Look for a loss of resistance to probe movement in all areas of the intended fragmentation volume as the primary indicator of the surgical end point. Initial application times can be based on 1 min of ultrasonic application (continuous or VASER mode) for every 200 mL of wetting solution infused at a site. This guideline usually results in only partial emulsification of the target volume. With experience, application times can rise to 1 min for every 100 mL of wetting solution. This guideline results in a more complete emulsification of the target volume. (These times are only suggestions; your requirements will be dictated by many factors that cannot be covered in this short space.) Special note: For the upper abdomen, use 1 min of ultrasonic application for every 200 mL of wetting solution infused until experience allows otherwise.
8. End-hits. Prevent end-hits or punching into the dermis from below. Place incisions so that probe movement is generally parallel to the skin. Do not try to go around tight corners – this may result in potential an end-hit or torquing of the probe.
9. Cross-tunneling. Cross-tunneling is highly desirable for more uniform fragmentation and to improve the subsequent aspiration performance.
10. Free air vibration. Keep the tip of the probe inside the patient at all times. Do not vibrate the probes in free air, or they may be subject to cracking.
11. Dry application. Do not reapply the VASER probe after a site has been aspirated (the site is dry).

13.9 Probe Selection

One of the most critical understandings necessary for successful VAL is the proper choice of the probe for the tissue type, tissue volume, and anatomical location. The key elements are (1) the diameter of the probe and (2) the number of grooves located at the tip of the probe. The smaller-diameter 2.9-mm probes all have three grooves for maximum efficiency and can be used on tissue from soft to fibrous because of their small diameter. The 3.7-mm probes may have one groove, two grooves, or three grooves and should be selected based on the tissue type and intended application. Tissue type ranges from very soft to very fibrous in five grades as shown in Table 13.1. The probe design and appropriate amplitude settings for each tissue type are shown. Table 13.1 shows additional guidelines for the selection and use of the various probes.

Table 13.1 Tissue type, VASER-probe selection, mode and amplitude setting. Photo taken from VASER Solta Medical User's Guide. Courtesy of Bausch, ©2022 Bausch Health Companies Inc., All Rights Reserved

Tissue type	Continuous mode?	VASER mode?	2.9 mm, 3-groove (any length)	3.7 mm, 3-groove (any length)	3.7 mm, 2-groove (any length)	3.7 mm, 1-groove (any length)
Very soft	Yes	Yes	60–80	70–80	70–80	–
Soft	Yes	Yes	60–80	70–90	70–90	70–80
Medium	Yes	Yes	70–90	80–90	80–90	70–90
Fibrous	Yes	2.9 mm only	80–90	–	80–90	80–90
Very fibrous	Yes	No	80–90	–	–	80–90

1. Do not vibrate the probes in free air. Always keep the tip of the probe, at least the distal 1–2 cm, in contact with tissue or fluids and inside the patient and skin port before initiating vibration. Vibrating the probes in free air can lead to cracking of the probe owing to unintended transverse vibrations.
2. The 3.7-mm probes are intended for rapid debulking and contouring of medium to large volumes of soft to fibrous tissue.
3. The 2.9-mm probes are intended for debulking and fine contouring smaller volumes of soft to extremely fibrous tissue, sensitive areas, and more superficial work.
4. General volume considerations. If a single anatomical site is expected to yield less than about 500 mL, then consider using the 2.9-mm probes. If the expected volume (single anatomical site) is estimated to be between 500 and 1000 mL, then consider using either the 2.9-mm probes or the 3.7 mm probes, depending on how fibrous the tissue is and the nature of the anatomical site. If the expected volume is over 1000 mL (single anatomical site), consider using the 3.7-mm probes.

For a given diameter, probes with more grooves (rings) fragment tissue more efficiently but do not penetrate fibrous tissues as easily because a significant amount of the vibratory energy is coupled from the sides of the probe as opposed to the front surface. Therefore, for a given diameter, probes with fewer grooves (rings) are more appropriate for fibrous tissues. If the tissue is too fibrous for a selected 3.7-mm probe design, select a probe with fewer rings. For comparison purposes, a three-ring 3.7-mm probe distributes approximately 58% of its energy from the sides and 42% from the front surface. A two-ring 3.7-mm probe distributes approximately 45% of its energy from the sides and 55% from the front surface. A one-ring 3.7-mm probe distributes approximately 35% of its energy from the sides and 65% from the front surface.

Smaller-diameter probes will penetrate fibrous tissues more easily than larger-diameter probes, irrespective of the number if rings. The 2.9-mm diameter probes all have three rings because their smaller diameter allows penetration of even the most fibrous tissues with three rings.

General recommendations: Use the 3.7-mm two-groove probes for most applications, 70–90% amplitude. Use the 3.7-mm three-groove probes for larger volumes of very soft tissue (70–90% amplitude) or the 3.7-mm one-groove probes for more fibrous tissues (70–90% amplitude). Use the 2.9-mm three-groove probes for smaller volumes, sensitive areas, fine contouring, or very fibrous tissues (60–80% amplitude). Ninety percent amplitude may be used with any 2.9- or 3.7-mm probe, but consider selecting a probe more appropriate for the "fibrousness" of the tissue if 90% amplitude is required for smooth, gliding motion. Use continuous mode for general use, if tissues are quite fibrous, or for higher-speed fragmentation. Use VASER mode for more delicate work, softer tissues, or for finer sculpting. VASER mode may be used for general use when probe selection is appropriate (Table 13.1).

13.10 Aspiration

Once a fatty area has been emulsified the next step is to remove the emulsified tissues and fluids using aspiration. In theory and in practice, almost any type of aspiration cannula will suffice. However, aggressive suction of an area treated with the VASER system is generally not required and will result in excessive trauma to the tissue matrix. Once appropriately treated with the VASER system the fatty tissue is largely in fluid form and does not require high-vacuum aggressive avulsion for removal. A gentle aspiration is sufficient and expedient and will result in decreased trauma to the tissue matrix. For this purpose, a gentle aspiration cannula, called Vent cannula, were developed to complement the emulsification of the VASER system. VentX cannula employ smaller port sizes for a selected cannula diameter than more traditional cannula and are less traumatic than traditional cannulae. The largely fluid emulsified tissues are re moved, while the tissue matrix is maintained. The port size (slot width and length) and the cannula diameter determine the application of the cannula, the speed of the cannula, and the aggressiveness of the cannula. The VentX cannulae use port sizes that maximize tissue and fluid removal speed while reducing suction trauma by balancing the total area of the ports with the cross-sectional area of the cannula. Port sizes with areas larger than the cross-sectional area of the cannula lumen, as in most traditional cannula designs, are unnecessarily traumatic.

The recommended approach to the selection of a Vent X cannula is to use a diameter one size larger than you would normally use because the port sizes on the VentX cannulae are smaller than those found on traditional suction cannula. The cannula descriptions that follow should help with cannula selection. The 4.6-mm SST-6 pattern is designed for rapid emulsion removal and debulking with small ports for reduced suction trauma. The port size on this cannula la is slightly smaller than the port size on a traditional 3.7-mm Mercedes cannula. The 3.7-mm SST-6 pattern is designed for general debulking and shaping with less suction trauma than a standard 3.7-mm Mercedes cannula. The port size on this cannula is slightly smaller than the port size on a traditional 3.0-mm Mercedes cannula. The 3.0-mm SST-6 pattern is designed for finishing and feathering, or slightly slower but less-traumatic debulking. The port size on this cannula is slightly smaller than the port size on a traditional 2.4-mm Mercedes cannula. In addition to smaller port sizes, these cannulae also include a precision continuously vented handle that creates a means by which the suction tube is continuously cleared during the aspiration process. This continuous venting has been named the VentX effect and decreases the available vacuum at the cannula by approximately 0.25–0.50 in. Hg. The VentX effect significantly increases the efficiency of aspiration when large amounts of fluid/tissue are in the suction tube. The suction tube will always appear to be empty because the VentX effect continually and rapidly empties the suction tube even when the cannula remains inserted in the patient. The VentX effect does not have a strong impact on suction performance during the final feathering and finishing steps because very little tissue/fluid is in the suction tube.

13.11 Tissue Selectivity

The VASER system spares the tissue matrix, removing primarily the fatty component, producing smooth, predictable results (Figs. 13.7 and 13.8).

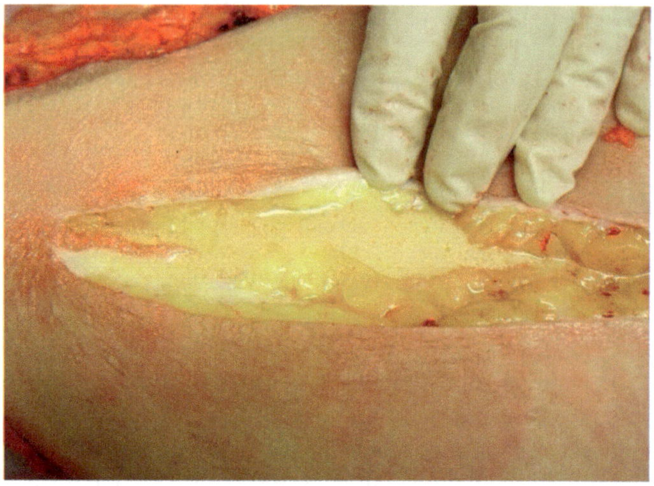

Fig. 13.7 Post ultrasound treatment

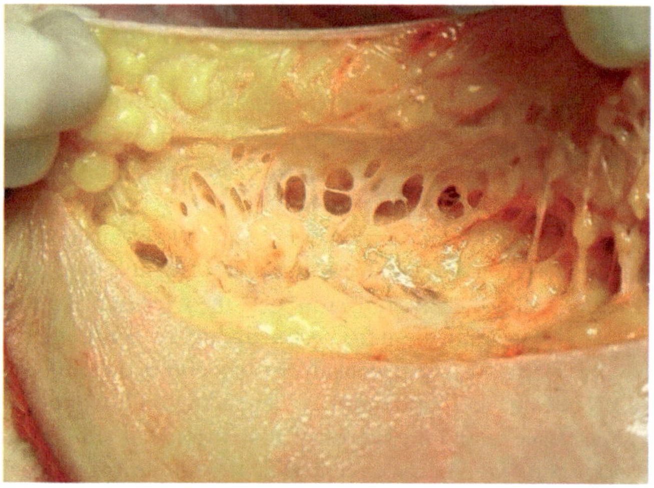

Fig. 13.8 Post aspiration

Fatty tissue is broken apart, and cells are suspended in infiltration fluid. Connective tissue is left intact after adipose tissue has been removed.

Fat cell structures are affected most by cavitation because the tumescent fluid containing the microbubbles can infiltrate between the cells.

Other tissues, such as blood vessels, muscles, fascia, and nerve bundles, have tighter interstitial cell junctions preventing the microbubbles from getting in between the cells.

Thus ultrasound action is uniquely tissue selective.

Ultrasound does not cavitate fat cells; it cavitates microbubbles in the tumescent fluid, infiltrated among the fat cells.

13.12 Skin Retraction: Endoscopic Evaluation

The application of the ultrasonic probe must be in the subdermal area in order to induce postoperative skin retraction. The duration of the procedure and the amount of energy required to liquefy the excess fat vary with the character of the tissue, the volume of the planned reduction, and the type of lipodystrophy. Hypertrophic lipodystrophy is easier to treat the mixed lipodystrophy or hyperplastic lipodystrophy. Less energy is required when the adiposity container has large volumes of fluid compared with tissues that are firm, and the fat content and the ratio of tissue to are less. For instance, in the treatment of inner thigh application of ultrasonic energy set at 65% power for 10–12 min is required to obtain a volumetric reduction of 250 mL and to give good skin stimulation. Clinical experience is helpful in establishing the time of application of the ultrasonic energy. The author applies up to 1 h of ultrasonic energy to defat the abdomen. The number and the location of access incisions also vary according to the treatment site. For the abdomen, usually, only a 2-cm suprapubic incision is required for good radial energy application. The addition of two 3-mm incisions in the upper lateral sides of the abdomen allows the removal of only the emulsion with a traditional small Teflon-covered cannula connected to a low-pressure (0.3 atm) vacuum. For the buttocks and lateral and posterior sides of the thighs, two 2 cm incisions located in the middle portion of the infragluteal fold on each side are utilized.

Tazi and Scheflan [22] introduced endoscopic evaluation of ultrasound-assisted liposculpture. The author began using this method with a Storz endoscopic system and camera. The tumescent technique was used in the inner thigh area and the abdomen. The instrument was placed in the superficial layer of the subcutaneous fat verified by needle depth. subcutaneous fat verified by needle depth. UAL was performed with criss-cross tunnels, recording the technique by video. An adjacent area was treated with standard liposuction. The results were:

1. Standard liposuction appears to be a more aggressive technique, with mechanical destruction of the subcutaneous tissue, including vessels, nerves, and supporting structures.
2. Ultrasonic liposculpting is a gentler, selective method, which is aggressive only in the fatty compartment of the body, sparing vessels, nerves, and elastic supporting fibers. Alterations in the tissue resulting from the use of SAL are a thickened der mal undersurface, markedly thickened vertical collagenous fibers, intact lymphatic vessels, and intact blood vessels.

Scheflan and Tazi [22] hypothesized that this horizontal and vertical thickening and shortening of the collagen in the dermis and ligamentous fibers is responsible for the remarkable skin tightening. The closer to the skin and the more complete the removal of fat from the immediate subdural space, the greater the skin-tightening effect. Although infrequent, significant complications such as thermal burns and skin necrosis are possible (Figs. 13.9 and 13.10).

Measuring skin retraction is very difficult, apart from clinical evidence. A new technique of developed-invisible, UV-activated micro-dot tatooc can help measure the skin retraction of each area of the patient skin (Figs. 13.11 and 13.12).

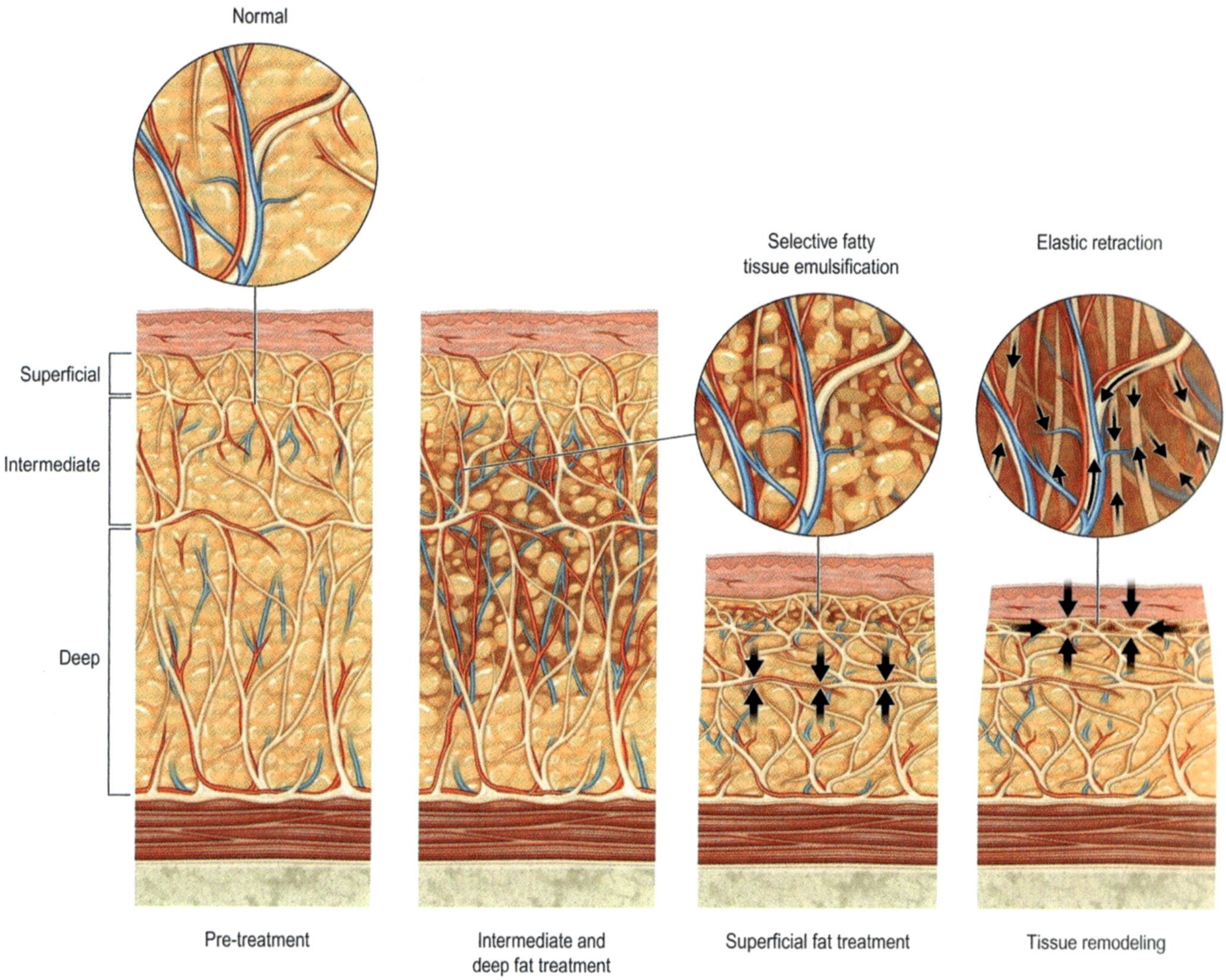

Fig. 13.9 Superficial fat treatment and tissue remodelling with elastic retraction: selective fatty tissue emulsification of the superficial layer, helps to expose and denude the elastic and connective structures of the subcutaneous tissues, by doing that in a safe way, as sparing vessels and nerves, tissue retraction capabilities are amplified

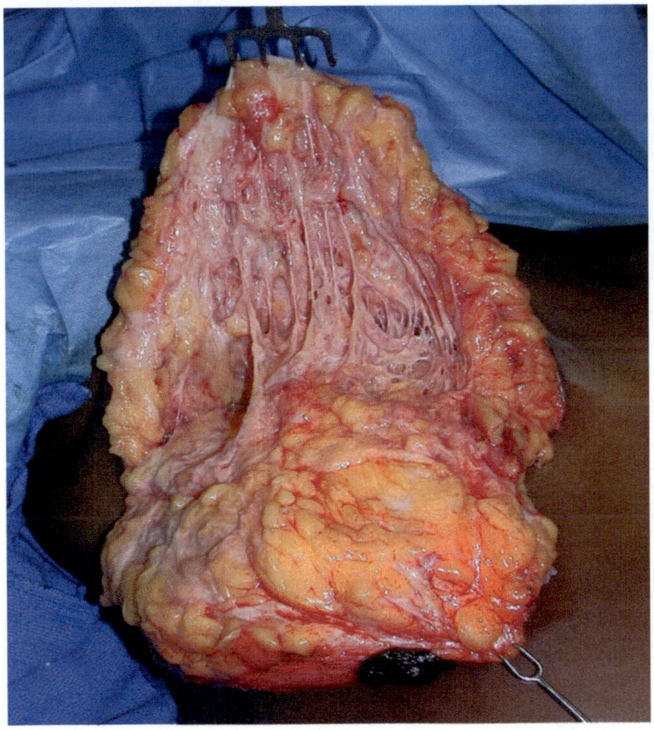

Fig. 13.10 Tissue selectivity: VASER of the upper pole of breast to reduce superficial fat component in breast reduction. Note the vertical connective and elastic fibres, nerves and vessels have been spared

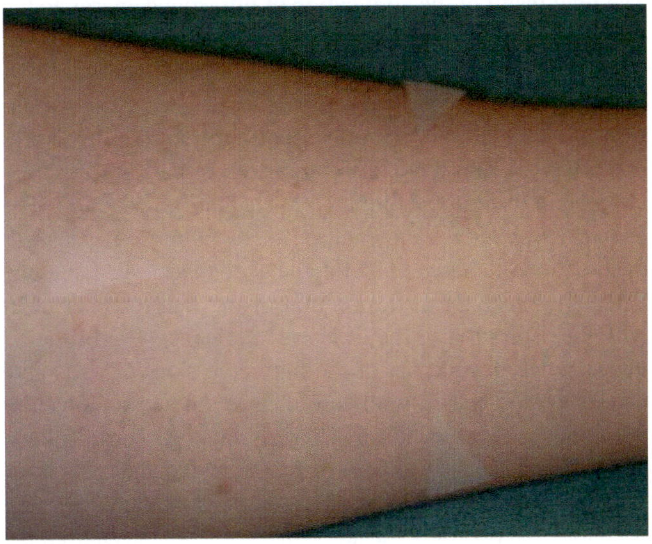

Fig. 13.11 Micro-dot tatooc

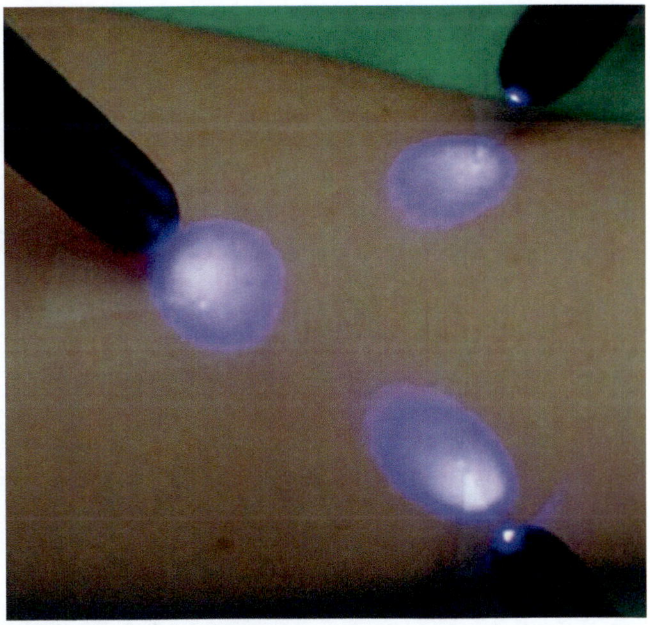

Fig. 13.12 Micro-dot tatooc measuring skin retraction

13.13 Autologous Fat Transfer with VASER Fat

Fat is actually utilized as a filler for breast, face, buttocks, and any other area to improve contouring and fill depressions and correct asymmetries. It gained popularity in the last decade, after proper techniques and technologies have been discovered and utilized to decrease trauma to fat cells and enhance survival rate and good long-term outcome. The principles of fat transfer are:

1. Remove fat from one area of the body and use it to enhance another area (buttocks, face, breast, etc.)
2. Alternative to fillers in small areas like the face
3. More cost-effective in large-volume transfer cases
4. More natural results
5. Correction of irregularities

VASER fat is good for fat transfer procedures: VASER ultrasonic energy dislodges fat cells from the tissue matrix, most of these fat cells remain viable, and emulsified fat is easy to re-inject.

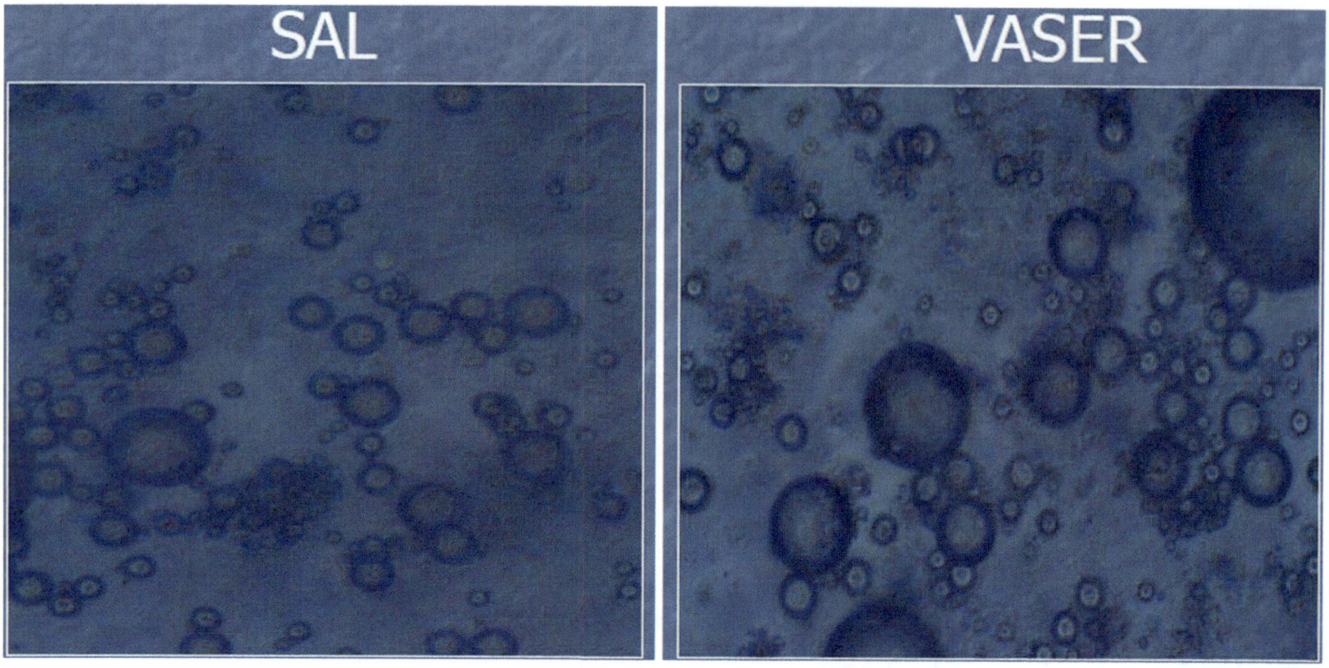

Fig. 13.13 Mature adipocytes

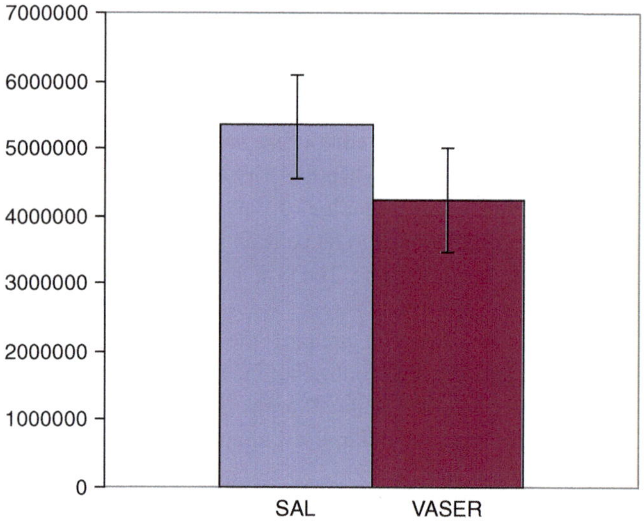

Fig. 13.14 Cell Yield 24 h after isolation

Fat viability study [23] has confirmed that VASER fat in any area of the body can be harvested as with standard suction-assisted liposuction, giving similar percentage of survival rate of the adipocyte and progenitor cells (stem cells) (Figs. 13.13 and 13.14).

13.14 Hematocrit

The final result of the application of ultrasound energy to the fat tissue compartment is the destruction of the adiposity and the creation of a fat emulsion represented by the liquid part of the fat fragmentation (as a result of the adiposity implosion) plus the fluid of the tumescent infiltration utilized preoperatively. This creamy emulsion characterizes the ultrasound technique compared to standard liposuction. The former is a gentle fragmentation of the adipose tissue with the creation of a fatty emulsion that is then aspirated, and the latter is a mechanical destruction of the fatty compartment with the aspiration of the destroyed elements. The former is the selective destruction of the fatty cell, since UAL is able to target only tissue with low density and low molecular cohesion, such as fat tissue. Vessels, nerves, elastic fibers, and connective tissue fibers of the subcutaneous structures are preserved. The latter is an unselective methodology and is unable to target only the adiposity, while all the anatomical components are attacked and destroyed, such as vessels and elastic fibers. This explains why UAL is accomplished with low blood loss, and the hematocrit drop after surgery is considerably reduced. This is clearly shown in clinical experience (Fig. 13.15).

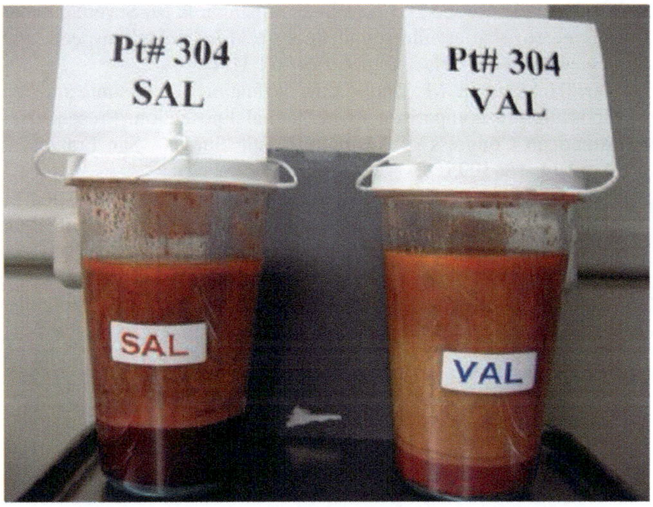

Fig. 13.15 Cannister of emulsified fat, derived from a SAL and VAL. Note much lower blood accumulation in the VAL solution

13.15 Optimizing Outcome

Main parameters to follow with VAL are:

Fluids

1. Infiltration. Must prefer Superwet technique with a ratio of 2:1 infiltration/aspiration
2. Make sure that fluid is even and uniform in the entire target volume
3. Wait 10 min minimum for vasoconstriction

Emulsification

1. Wait 1 min per 100 mL infused
2. Lower range of time for users and more delicate anatomic areas
3. Higher range of time for experienced users and less delicate areas
4. Higher values of time (2 min per 100 mL in) can be used with experience and specific application
5. Treat the volume, not a plane or zone
6. Use the correct probe

Aspiration

1. Minimize suction-induced trauma, use cannula designed for emulsion removal
2. Aspirate as much emulsified tissue/fluid as possible
3. Aspirate until final contour is achieved
4. Get the large early volume with larger emulsion cannula to shorten the procedure
5. Switch to smaller cannula when volume flow slows

13.16 VASER Results

VASER lipo started in 2001 when the first device was delivered for initial clinical research and to follow entered the US and worldwide market. In 2012 the new system was produced by Sound Surgical Technologies (SST Denver, Colorado). Since then, hundreds of thousands of procedures have been performed. Vaser physicians report the following VASER technology benefits:

- Fast patient recovery
- Less pain medication required
- Less blood loss
- Reduced need for re-treatments
- Smooth, predictable results
- Skin tightening
- Increased precision
- Reduced physician fatigue
- Smooth fat for re-injection

References

1. Di Giuseppe A. Ultrasonic assisted liposculpture. Presented at the World Congress on liposuction. San Francisco, May 3–5; 1996.
2. Cimino WW. Ultrasonic surgery: power quantification and efficiency optimization. Aesth Surg J. 2001;21(3):233–40.
3. Jewell ML, Fodor PB, de Souza Pinto EB, Al Shammari MA. Clinical application of VASER-assisted lipoplasty: a pilot clinical study. Aesth Surg J. 2002;22(2):131–46.
4. Scuderi N, Devita R, D'Andrea F, Vonella M. Nuove prospettive nella liposuzione la lipoemulsificazone. Giorn Chir Plast Ricostr ed Estetica. 1987;2(1):33–9.
5. Zocchi ML. Metodo di trattamento del tessuto adiposo con energia ultrasonica. Roma, Italy: Congresso dell Societa Italiana di Medicina Estetica; 1988.
6. Zocchi ML. New prospective in liposculpturing: the ultrasonic energy. Abs. 10th ISAPS Congress. Zurich, Switzerland; 1989.
7. Zocchi ML. Clinical aspects of ultrasonic liposculpture. Perspect Plast Surg. 1993;7:153–74.
8. Zocchi ML. Ultrasonic assisted lipoplasty. Clin Plast Surg. 1996;23(4):575–98.
9. Zocchi ML. Basic physics for ultrasound-assisted lipoplasty. Clin Plast Surg. 1999;26(2):209–20.
10. Maxwell GP. Use of hollow cannula technology in ultrasound-assisted lipoplasty. Clin Plast Surg. 1999;26(2):255–60.
11. Klohn RA. Liposuction with sonic sculpture: six years' experience with more than 600 patients. Aesthet Surg. 1996;16(2):123–8.
12. Rohrich RJ, Beran SJ, Kenkel JM, Adams WP Jr, Di Spaltro F. Extending the role of liposuction in body contouring with ultrasound-assisted liposuction. Plast Reconstr Surg. 1998;101(4):1090–102; discussion 1117–1119.
13. Gilliand MD, Commons GW, Halperin B. Safety issues in ultrasound-assisted large volume lipoplasty. Clin Plast Surg. 1999;26(2):317–35.
14. Di Giuseppe A. The harmonic lift: ultrasonically assisted skin remodeling. Int J Cosm Surg. 2000;2(2):125–31.
15. DiGiuseppe A, Santoli M. Ultrasound assisted breast reduction and mastopexy. Aesthet Surg. 2001;21:493–506.

16. Troilius C. Ultrasound-assisted lipoplasty: is it really safe? Aesthet Plast Surg. 1999;23:307–11.
17. Lack EB. Safety of ultrasonic-assisted liposuction (UAL) using a non-water-cooled ultrasonic cannula. A report of six cases of disproportionate fat deposits treated with UAL. Dermatol Surg. 1998;24:871–4.
18. Baxter RA. Histologic effects of ultrasound-assisted lipoplasty. Aesthet Surg. 1999;19:109–14.
19. Perez JA. Treatment of dysesthesias secondary to ultrasonic lipoplasty. Plast Reconstr Surg. 1999;103:1534.
20. Gerson RM. Avoiding end hits in ultrasound-assisted lipoplasty. Aesthet Surg J. 1997;17:331–2.
21. Grolleau JL, Rouge D, Chavoin JP, Costagliola M. Severe cutaneous necrosis after ultrasound lipolysis. Medicolegal aspects and review. Ann Chir Plast Esthet. 1997;42:31–6.
22. Tazi, H.,Scheflan, M.: Endoscopic evaluation of ultrasonic assisted liposculpture comparison to traditional liposuction. Presented at American Congress of Aesthetic Plastic Surgery, San Francisco, April 14–18, 1995.
23. Panetta NJ, et al. Tissue harvest by means of suction-assisted or third-generation ultrasound-assisted lipoaspiration has no effect on osteogenic potential of human adipose-derived stromal cells. Plast Reconstr Surg. 2009;124:65–73.
24. https://soltapracticeproposal.com/product-information/products/vaser-system/

Part III

Surgery

New Concepts for Safe Gluteal Fat Grafting

14

Maxim Geeroms, Lisa Ramaut, and Moustapha Hamdi

Contents

14.1 Introduction

Buttock augmentation through fat grafting has been performed for over three decades [1, 2] but has known a significant growth over the last years. Unfortunately, we were confronted with several reports of fatal pulmonary fat embolisms after gluteal fat grafting, which has made the safety of this procedure extremely relevant. Therefore, this chapter will not cover the aesthetic aspects of buttock enhancement but will concentrate on safety.

M. Geeroms (✉) · L. Ramaut · M. Hamdi
Department of Plastic and Reconstructive Surgery, Vrije Universiteit Brussel—Universitair Ziekenhuis Brussel, Brussels, Belgium

14.2 Gluteal Fat Grafting Today

From 2014 to 2018, we witnessed a 55.7% growth in procedures involving buttock augmentation [3]. The American Society of Aesthetic Plastic Surgeons (ASAPS) database taught us that approximately 94% of these augmentations are done with fat grafting [4]. Especially over the last 15 years, we have seen a strong increase in the amount of Pubmed-listed articles on this topic (Fig. 14.1). Due to growing emphasis on pronounced feminine curves [5–11] and due to popularized beauty trends on social media networks, the demand for buttock enhancement expanded from a mainly South American patient population to a global audience. Models and online influencers show off their buttocks, and buttocks have become the new breasts. For example, a "*belfie*" or a "bottom selfie" is a photographic self-portrait featuring the buttocks. The peach emoji, with its tender fleshy appearance, its soft subtle curvature and its seductive center seam, has become the international symbol of a plump derrière in our text messages. Intriguingly, only 7% of peach

© The Author(s), under exclusive license to Springer Nature Switzerland AG 2023
A. Di Giuseppe et al. (eds.), *Fat Transfer in Plastic Surgery*, https://doi.org/10.1007/978-3-031-10881-5_14

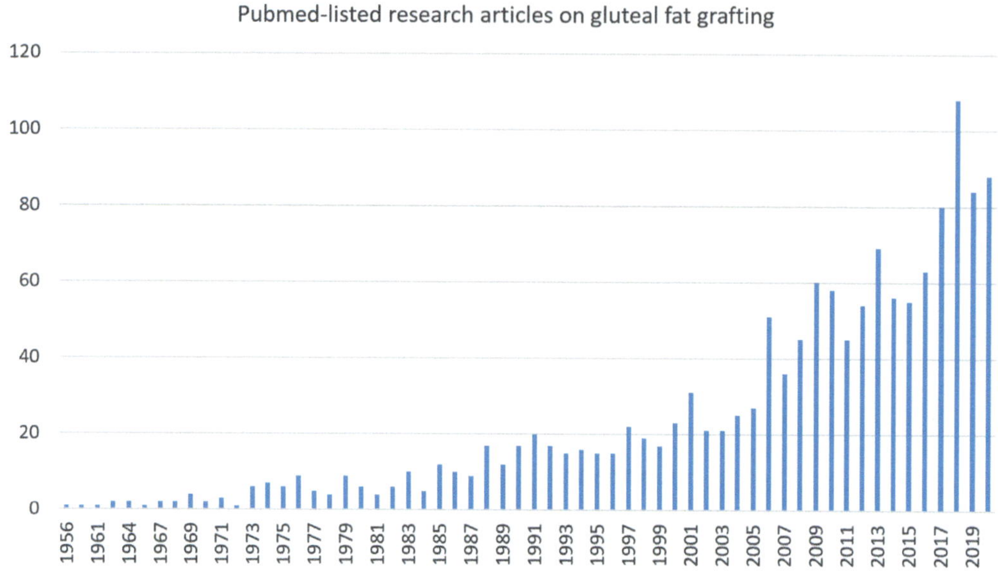

Fig. 14.1 Pubmed-listed research articles on gluteal fat grafting from 1956 to 2020

emojis refer to the actual fruit, while 60% of text messages containing a peach have a sexual or buttock-related connotation [12].

For the quality of care and patient safety, it is important that this trend is reflected in the plastic surgery training program. Trainees should have enough exposure to the gluteal fat grafting before performing the procedure autonomously. According to a survey among Brazilian board-certified plastic surgeons who are members of the *Sociedade Brasileira de Cirurgia Plastica*, 70% received training in this procedure during their residency. It is unsure what the exposure is of residents in other countries to gluteal fat grafting, let alone *hands-on* training, but it is likely a lower percentage since most gluteal augmentations are performed in Brazil (22% of all worldwide procedures, according to statistics provided by the International Society of Aesthetic Plastic Surgery, ISAPS) [3].

14.3 Risks of Gluteal Fat Grafting

Gluteal fat grafting is advantageous since no foreign material is introduced in the patient, recovery is fast, and it has the added benefit of improving the shape of the patients' fat depots through liposuction. However, fat grafting to the buttocks is not without risks [13–18].

Minor complications consist of:

- Seroma
- Edema
- Hematoma
- Erythema
- Pain
- Contour irregularities
- Undercorrection
- Asymmetry
- Hyperpigmentation
- Fat necrosis
- Oil cysts
- Sciatica
- Cellulitis
- Local infection
- Abscess formation

Major complications are rare but consist of:

- Fat embolism
- Deep venous thrombosis
- Anemia
- Symptomatic hypovolemia
- Sciatic nerve axonotmesis
- Necrotizing fasciitis
- Septic shock
- Death

A first case of probable fat embolism after gluteal fat grafting was reported in 1999 [19]. It was only 16 years later that the attention of the global plastic surgery community was drawn to the issue of fat embolism, after a concerning number of intraoperative and postoperative (within 24 h) deaths have been described [20, 21]. Cárdenas-Camarena's retrospective paper on autopsy reports after 22 fatal gluteal fat grafting procedures in Mexico and Columbia was an eye-opener, as it confronted the plastic surgery community with shocking findings after this aesthetic surgical procedure. It was demonstrated how macroscopic fat embolism was found in the gluteal veins, the inferior vena cava, the right cardiac cavities, the pulmonary arteries and the lung parenchyma

[21]. Interestingly, all dissections revealed intramuscular adipose tissue. There was no data on the total amount of procedures performed in these two countries during the studied period, which left us unsure about the incidence of deaths related to gluteal fat grafting. The lack of reported incidents and published case reports in the scientific literature between 1999 and 2015 is regrettable, as it could have warned our community earlier.

Following those reports, the Aesthetic Surgery Education and Research Foundation (ASERF) Task Force conducted a survey among members of the ASAPS and ISAPS [22]. Based on 2016 data, the risk for fatal fat embolism was estimated at 1:3448 (ranging between 1:12,351 and 1:6241 depending on the data source and methodology). Even though doubts have arisen about the data collection and analysis, the reported estimates are unacceptably high for aesthetic intervention. Regardless of the inherent limitations of the retrospective survey (i.e., recall bias, participation bias, and subject bias), this paper has raised awareness and put it into action several research projects [23, 24].

Cases have been described in which fat particles were present in the brain and kidney [25]. Patients with a right-to-left shunt, for example, across a patent foramen ovale, are at risk for these distant fat-induced infarctions [18, 26].

The reported catastrophic adverse events have contributed to the unfortunate reputation that is associated with gluteal fat grafting or the ill-named "Brazilian butt lift" [27] in the news. The British Association of Plastic, Reconstructive and Aesthetic Surgeons (BAPRAS) responded to the alarming reports with a statement in 2018 and drastically abandoned the gluteal fat grafting procedure [28]. It is highly questionable whether this draconian measure solves the problem, since a British moratorium on gluteal fat grafting might drive their patients into medical tourism, or they might be operated by "cosmetic surgeons" with little to no formal training, or undergo "little invasive" injections with foreign materials [29, 30].

Several cited authors in this chapter—who understood that a better understanding of the relevant anatomy, pathology of macroscopic fat embolism, and surgical technique would lead to a decreased mortality rate—should be congratulated and thanked for their scientific contribution in making gluteal fat grafting safer and not banned by other societies.

14.4 Subcutaneous-Only Fat Grafting

From Condé-Green's review in 2016, it became clear that 73.4% of gluteal fat grafting articles report intramuscular injection, solely in this plane or combined with subcutaneous fat grafting. In the same review, a higher, but non-significantly different, complication rate was observed with intramuscular injection [13]. The ASERF Task Force survey did demonstrate a significant risk increase for fatal pulmonary fat embolism when intramuscular injection was performed [22]. Cansancao's survey from 2019 showed that only 56% of the responders perform subcutaneous-only fat grafting (16% injected only in the intramuscular plane, and 28% targeted both the subcutaneous and intramuscular plane). The risk for fatal complications was 16 times higher after intramuscular grafting [14].

A Multi-Society Task Force for Safety in Gluteal Fat Grafting was established in 2017 by combined efforts from ASAPS, ISAPS, the American Society of Plastic Surgeons (ASPS), the International Society of Plastic and Regenerative Surgeons (ISPRES), and the International Federation for Adipose Therapeutics and Science (IFATS). The Task Force underlined that deceased patients after gluteal fat grafting had four findings in common: intramuscular fat, submuscular fat, damaged gluteal veins, and fat emboli in heart or lungs [31]. In not a single fatal case, fat had been found only in the subcutaneous plane [31–35]. The task force emphasized the ease of unintentionally penetrating the deep gluteal fascia and entering the muscle [32].

The intramuscular fat grafting was historically recommended based on the favorable vascularity of muscle tissue compared to subcutaneous adipose tissue, and on an animal experiment which showed the best results after intramuscular grafting following a 12-month observation [36]. Important to note is that this unfortunate scientific statement has not been reproduced in a clinical setting. Some surgeons still perform intramuscular fat grafting because it is believed that the muscular recipient site enhanced engraftment, or because the muscle represents additional recipient capacity and can accommodate extra volume. However, concerns or beliefs that grafting only in the subcutaneous layer would lead to suboptimal long-term outcomes have been countered by Cansancao's paper [37]. In a prospective study, he performed subcutaneous-only fat grafting to the buttocks. The adipose tissue layer was measured in 35 patients and showed an acceptable persistence of the enlargement after a 12-month follow-up. A quantitative and objective analysis of the subcutaneous tissue layer resulted in an 82% maintenance of thickness, from the immediate postoperative result to 1 year postoperatively. Note that the 18% resorption rate could possibly be an overestimation as the immediate postoperative result is followed by early resorption of the aqueous solution in the grafted material.

14.5 Scientific Evidence for Subcutaneous-Only Fat Grafting

Today, the evidence for isolated subcutaneous fat grafting has become so overwhelming, that papers recommending or mentioning intramuscular fat grafting as a potential option, even outside of theoretical danger areas or superficial under the deep gluteal fascia [21, 36, 38–47], should be corrected or contain a warning statement [48]. Other intraoperative guidelines described in this chapter are subordinate as they all serve the same purpose: staying safe, i.e., injecting in the subcutaneous plane (Fig. 14.2).

Different mechanisms and theories for fat embolism during gluteal fat grafting have been established. A direct and an indirect mechanism have been postulated based on cadaver studies [37, 49].

Fig. 14.3 Direct cannulation theory. Following a direct hit of the gluteal vessel, fat is injected intravascularly, with subsequent embolization. Drawing by Lisa Ramaut

14.5.1 Cannulation Theory

The cannulation theory is based on intravascular fat bolus injection directly into a gluteal vein with subsequent embolization (Fig. 14.3). The vascular pedicles are positioned under the gluteus maximus muscle, and both the superior and inferior gluteal veins are at risk for direct puncture. Ordenana studied the density, size, and topographic locations of the gluteal vessel network [50]. The superior and inferior gluteal veins are large vessels with a diameter of 7.6 and 13.7 mm on average, respectively [50]. Their submuscular branches are on average 3.9 and 3.5 mm wide. Gluteal vein tributaries in the muscle and subcutaneous tissues had smaller diameters and measured approximately 1.3 and 1.1 mm, respectively. A direct hit to a gluteal vein is more probable in the deeper layers as the veins are larger. The measurements were done on fresh cadavers from patients aged 63–76 years, which could have influenced the size of the vessels.

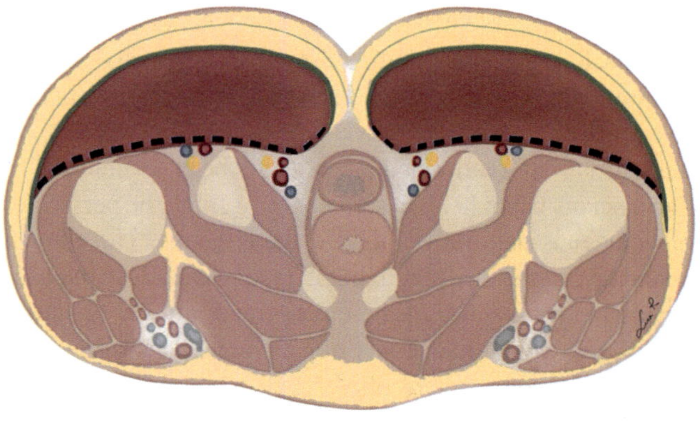

Fig. 14.2 A cross-section of the gluteal region contains, from the surface to deeper layers: the skin, the superficial subcutaneous tissue (yellow), the superficial gluteal fascia (thin green line), the deep subcutaneous tissue (yellow), the deep gluteal fascia (thick green line) and the gluteus maximus muscle (red). Fat grafting injections should aim for the subcutaneous tissues and stay superficial from the deep gluteal fascia. Note how the dashed black line indicates the non-existent fascia on the deep (anterior) side of the gluteal muscle. The large gluteal veins (blue) and arteries (red) lie deep in the gluteal muscles and give off branches through the muscle. Drawing by Lisa Ramaut

14.5.2 Laceration-Siphon Theory

The laceration-siphon theory implies that a large vein is ruptured, and fat gets siphoned into the vasculature (Fig. 14.4a, b). Because of the higher interstitial pressure inside the gluteus maximus muscle, fat particles follow a pressure gradient and can be squeezed into the veins [31, 50]. Adipose tissue will migrate towards the pulmonary and cardiac vessels and cause microscopic or macroscopic fat embolisms.

14.5.3 Deep Intramuscular Migration

Despite the growing awareness and the surgeons' conviction that fat is injected subcutaneously or "superficially in the gluteus muscle," which is distant from the large gluteal vessels, cases of fatal pulmonary fat embolisms continued to occur. This has prompted Del Vecchio to conduct a cadaver study which has led to the concept of "deep intramuscular migration": through direct visual inspection and endoscopic evaluation, it was clearly demonstrated how intramuscularly injected "proxy fat" (apple sauce as a surrogate for processed lipoaspirates) can and will migrate along the path of least resistance, which is deep, through the muscle, into the submuscular space (Fig. 14.5) [51]. Thus, intramuscularly grafted fat does not stay where it was initially injected. Upon dissection, fat squirts out from between muscle fiber, from beneath the muscle, and from around the gluteal vessels. Because of the deep gluteal fascia, which overlies the super-

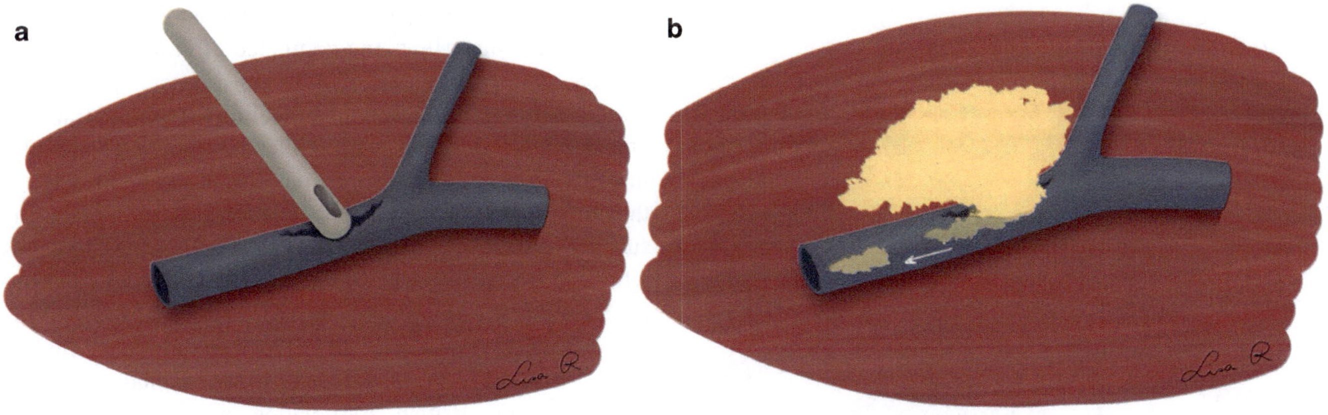

Fig. 14.4 Laceration-siphon theory. (**a**) Rupture of a gluteal vein, due to passage of the cannula, causes an entry for grafted fat particles into the circulation. (**b**) Because of the high intramuscular pressure, adipose tissue gets siphoned into the lacerated vein. Drawing by Lisa Ramaut

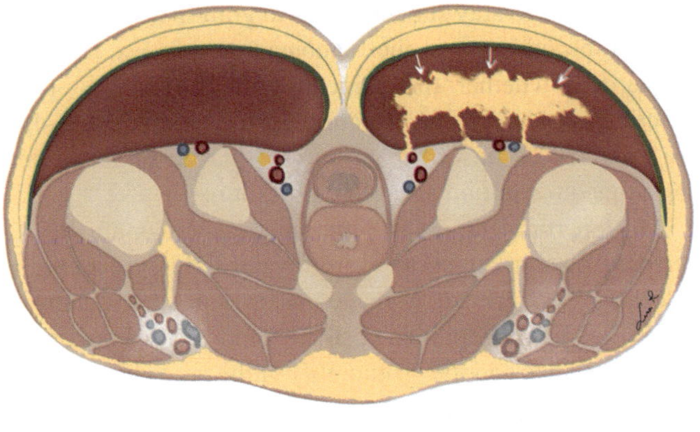

Fig. 14.5 Deep intramuscular migration. Intramuscular fat follows the path of least resistance and migrates towards areas with lower pressure, i.e., the submuscular region. The deep gluteal fascia (thick green line) is a backstop, preventing intramuscular fat from moving superficially towards the subcutaneous plane. As intramuscular pressure increases during intramuscular grafting, the robust deep gluteal fascia "pushes" the fat further down towards the critical structures. The deep migration through the muscle, which leads to submuscular fat pooling, is further enhanced by the nonexistence of a fascia against the undersurface of the gluteus maximus muscle, similar to the latissimus dorsi and pectoralis major muscles. This has clinical implications as it cannot function as a barricade against deep intramuscular migration. Drawing by Lisa Ramaut

ficial side of the gluteus maximus muscle, the intramuscular fat tissue cannot migrate back towards the subcutaneous tissue.

The deep gluteal fascia is a strong structure as it withstands the high pressure of subcutaneous-only fat grafting. The robust fascia protects us from intramuscular migration and sequential fat embolism and confirms the safety of subcutaneous fat grafting [51]. Wall Jr. confirmed in a similar experiment that subcutaneously injected fat does not exhibit deep intramuscular migration, even if the subcutaneous tissue pressure reaches extremely high pressures (up to 199 mmHg) and even in the case of a perforated deep gluteal fascia (to imitate accidental cannula perforations through the fascia) [52].

These two key studies demonstrate the strength of the deep gluteal fascia and its pivotal role in keeping subcutaneous fat subcutaneously, but also how it "pushes" intramuscular fat further deep towards the large gluteal vessels and sciatic nerve. The dissection photographs and videos of fat pouring out of the gluteus maximus muscle in the mentioned papers are useful for understanding the described concepts.

14.5.4 Venous Traction Theory

This laceration-siphon mechanism implicates a previous venous laceration during the expansion of the recipient site or during the actual fat grafting but without fat injection. Alternatively, the venous traction theory has been hypothesized by Del Vecchio. When intramuscular pressure increases, fat can migrate further deep, pool under the muscle, and expand the submuscular space. With an increasing volume of fat beneath the gluteal muscle, a longitudinal strain on the gluteal vein is exerted, leading to stretching and tearing and forming an entryway to the blood circulation (Fig. 14.6) [51–53].

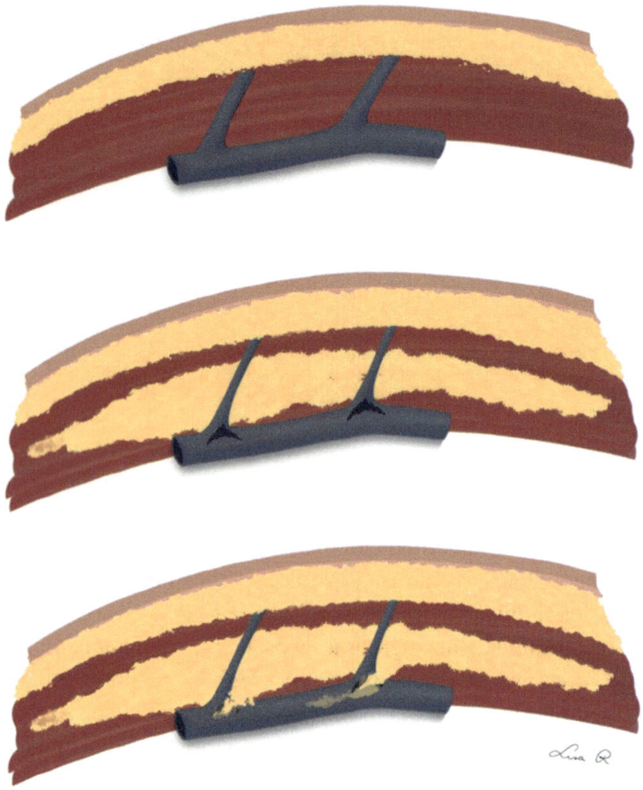

Fig. 14.6 Venous traction theory. Fat injected deep into the gluteal fascia can further migrate in the muscle towards the submuscular area. Submuscular pooling of the adipose tissue exerts a strain on the gluteal vein or its branches, leading to tearing and siphoning of the fat into the lacerated vessel. Drawing by Lisa Ramaut

14.6 New Guidelines Led to an Improved Safety Profile

Following the recommendations to graft subcutaneously, the ASERF Task Force survey was repeated in 2019 and sent again to ASAPS and ISAPS members. This second study corrected the ratio of fatal pulmonary fat embolisms to 1 out of 14,952 [54]. To date, the mortality numbers are unfortunately based on data gathered through anonymous surveys, and, thus, they remain mere estimates. A lower mortality risk after gluteal fat grafting procedures was also calculated by Cansancao, based on ISAPS statistics. He estimated the risk at 1:20,117 (0.005%) [14]. Non-fatal fat embolism occurred in 1 out of 9530 (0.01%) [14]. Interestingly, abdominoplasty, which was previously thought to be the highest-risk aesthetic procedure, carries a fatal complication risk of approximately 1 out of 13,000, which is mainly related to a different mechanism, i.e., deep venous thrombosis [22]. Thus, abdominoplasty and gluteal fat grafting have a similar mortality risk but a very different reputation.

The decreased risk for fatal fat embolisms in a period of only 2 years underlines the combined efforts of the plastic surgery community in raising awareness, conducting research, providing practical guidelines, offering training, and adopting surgeons who switched from intramuscular injection to subcutaneous injection. Rios showed that 94% of the participants were aware of the guidelines to inject in the subcutaneous layer only [54]. Eighty-six percent of surgeons who in 2016 performed intramuscular fat grafting switched to subcutaneous-only fat grafting [22, 54]. Only 0.8% of the surgeons in 2019 still injected in the deep muscle, compared to 13.1% in 2017 [54]. Sadibeen recently concluded from his survey among British Association of Aesthetic Plastic Surgeons (BAAPS) members that 96.9% of the surgeons inject only subcutaneously [55]. Overall, this proves the efficacy of the message that was sent out by several plastic surgery societies to stay subcutaneous to graft in a safe manner. It has to be said that, in these anonymous surveys, it cannot be verified whether the surgeon uses objective tools to describe the plane of fat injection. Consequently, these data are arbitrary because the majority of plastic surgeons perform this intervention as a blind procedure.

Most participants in the mentioned surveys are board-certified plastic surgeons. Unfortunately, this procedure is also being performed by inadequately trained providers in unlicensed or unaccredited cosmetic clinics. They are driven by the elegance and the profitability of the procedure, but can be deceived by the apparent ease of it. So we still have uncertainty about the real mortality and the current awareness among these "non-core physicians" [22, 56]. Even though 86% of female patients value board certification highly (the most important parameter for 54% of patients), there are still 20% of gluteal fat graftings performed by non-plastic surgeons, facial plastic surgeons, and dermatologists [22, 57].

14.7 Preoperative Guidelines

Apart from the standard history and physical examination, attention can be given to the presence of varicose veins as it increases the risk of venous lacerations with subsequent embolization. Symptoms from sciatic nerve compression or irritation should be noted since they can be associated with venous varicosities in proximity to the sciatic nerve [20].

Correctly informing the patient and including fat embolism as a potentially fatal complication in the informed consent, with an estimation of this risk, is correct preoperative management to allow well-informed and unbiased decision-making [32, 58]. A recent survey demonstrated that 78% of adult females underestimate the mortality risk related to gluteal fat grafting [57]. Seventy percent of women who were willing or neutral to undergo surgery were converted to unwilling after education about complications. No less than one out of ten patients would tolerate a mortality risk greater than 0.1%. Notably, patients with body dysmorphia and "social media dysmorphia" had more tendency to accept higher risk [57]. These results underline the general unawareness among our clientele as well as the patients' poor risk assessment capability. It also emphasizes our role as informers or educators during the preoperative consultation.

To avoid directly targeting the vessels, the preoperative marking of danger areas has been described in which fat grafting is discouraged or solely injected subcutaneously. Villanueva marks a danger triangle on each buttock, between the posterior superior iliac spine, greater trochanter, and ischial tuberosity, which delineates the regions of the main gluteal vessels and the sciatic nerve [38]. However, because of new insights from the discussed dissection studies, topographical thinking has become obsolete. The danger area should be redefined and encompasses the area beneath the deep gluteal fascia, which was corrected in a reply [48, 59].

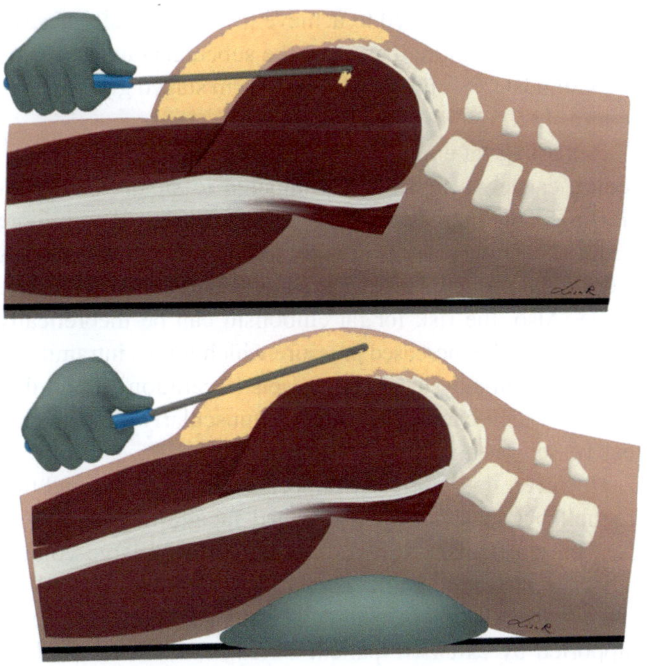

Fig. 14.7 Ergonomic advantage of jack-knife position, as depicted in the lower image. Fat grafting through an inferior incision while the patient is installed in a prone position with flexed hips, allows the practitioner to keep the dominant hand in a lower position, and points the tip of the cannula up into a more superficial plane. Drawing by Lisa Ramaut

14.8 Perioperative Guidelines

14.8.1 Positioning of the Patient

Villanueva recommended in his 2017 paper to position the patient in a prone jack-knife position with flexed hips while performing the gluteal fat grafting. Because this position allows lowering your dominant hand, it would direct the cannula into a superficial plane [38]. While this can help guide the cannula from the inferior gluteal incision (Fig. 14.7), it might create a false feeling of safety.

Authors have suggested certain angles which are safer than others from certain entry ports, and, as a general and logical observation, an obtuse angle between the patient's skin and the cannula is dissuaded [42, 60]. However, maintaining the cannula at a certain angulation is hard to control during the process of fat grafting, as a small vertical motion of the handpiece can change the angulation considerably [61].

14.8.2 Incisions

Risk-free access sites do not exist, as gluteal vein fat embolism has been seen regardless of the incision location. An often preferred [14] high intergluteal incision overlying the sacrum is considered safer because the sacrum can act as a ramp: its bony prominence steers the cannula into the subcutaneous plane and prevents it from pointing or bending down into the deeper tissues (Fig. 14.8).

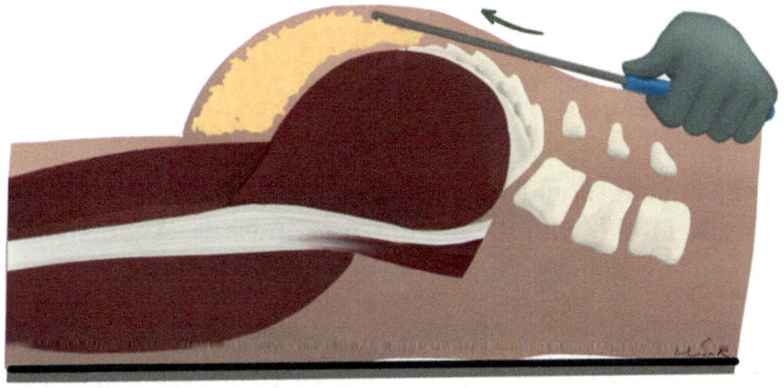

Fig. 14.8 The sacrum as a ramp: the bony prominence of the sacrum guides the cannula into the superficial tissues. Drawing by Lisa Ramaut

14.8.3 Volume

To overcome tissue loss and volume reduction due to fat graft resorption after gluteal fat grafting, overcorrection has been applied and advocated by clinicians, which means the injection of a volume that exceeds the requirement for correcting a volume depletion or contour defect. However, based on scientific evidence [62] and clinical experience [63], it has been concluded that overgrafting does not improve outcomes. On the contrary, increased interstitial pressure will restrict capillary perfusion to the adipocytes, leading to apoptosis, necrosis, and oil cyst formation. If a large volume is required to achieve the aesthetically desired result, which is more than what the subcutaneous tissue can accommodate, it is preferable to perform staged grafting procedures [32, 52].

Regarding fat embolism, a large-volume fat grafting procedure carries a higher risk compared to small-volume fat grafting. This increased risk is simply attributed to a more frequent cannula passage in order to inject the whole graft, as every axial movement can potentially penetrate the muscle fascia. Also, the risk for fat embolism can be theoretically attributed to the increased pressure which forces intramuscular fat into already lacerated vessels (laceration-siphon theory), or which can push the fat submuscularly where it can tear vessels (venous traction theory).

However, it is of greater importance to keep in mind that a small but inaccurately performed fat grafting procedure (i.e., intramuscular) is more hazardous than a large-volume subcutaneous-only graft. Therefore, the focus should be given to the position of cannula at every stroke. This belief is supported by extensive patient series in which no fatal fat embolisms were detected, even though volumes up to 4400 mL of subcutaneous fat were reported [49, 64]. In addition, autopsy reports of patients who died after gluteal fat grafting show relatively low volumes of fat grafts [25]. However, fat embolism and cardiac instability can arise suddenly during the operation, which urges the surgeon to stop the procedure.

14.8.4 Device

The cannula should be held parallel to the skin surface and muscle fascia, with the tip, directed upwards. Alternatively, an angled cannula can be used to aim for the superficial tissues.

Most surgeons prefer a cannula with a diameter of 3 or 4 mm [14]. Theoretically, it is easier to penetrate the deep gluteal fascia with a narrow cannula tip. ASERF found that using large-bore cannulas of 4.1 mm and more are correlated with less fatalities [22]. According to Rios' survey, 30% of ASAPS and ISAPS surgeons use cannulas of at least 4.1 mm, compared to 4% in 2016 [22, 54]. Importantly, changing surgical tools and starting to use 5-mm cannulas will not guarantee grafting in the subcutaneous plane.

Larger-bore cannulas have the additional advantage of being more rigid, allowing a more reliable orientation. Thin, long, flexible or bendable cannulas have a higher propensity to lead to flexibility misguidance: a slight blend of the instrument continues to curve in the patient toward the bend (Fig. 14.9) [23, 49]. By consequence, the cannula tip will be located in the deeper tissue planes, even when a superficial direction of the syringe is attempted by the misguided surgeon.

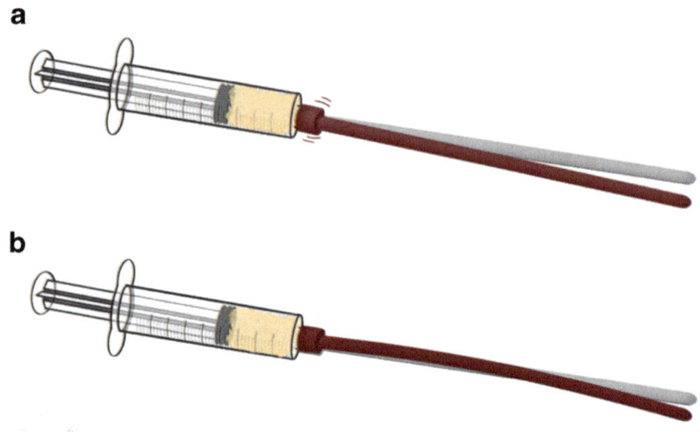

Fig. 14.9 (**a**) Manual fat graft injection by means of a syringe connected to a cannula is often performed, but a stable interface should be guaranteed. Loosening of the connection could angle the cannula towards deep tissues and lead to penetration of the deep gluteal facia. (**b**) The concept of flexibility misguidance. A slight bend in a cannula continues to curve further in the direction of the initial bend, leading to inadvertent deep injections. Drawing by Lisa Ramaut

If the injection is performed manually with syringes attached to a cannula, a stable connection should be ensured to avoid angulation of the cannula, and we should be aware of possible loosening or bending of connecting pieces (Fig. 14.9) [46]. Unstable interfaces than can pivot or angle at the junction with the syringe should be avoided.

Power-assisted devices for liposuction and fat grafting will be associated with less aggressive shoulder and elbow movements for axial motion, less fatigue-induced decrease in proprioception, and does not require the thumb generating fat flow by pushing the plunger of the syringe, which allows the surgeon to be completely focused on the orientation and direction of the cannula tip [38, 49]. The non-dominant hand is free to monitor the position of the cannula underneath the skin. These are valid arguments for the use of power-assisted devices in order to inject in a controlled manner. Although, deaths following grafting with automated devices have been reported as well.

14.8.5 Extensive Liposuction

Defining gluteal enhancement as making the buttocks bigger is a big understatement. As the surgeon's main objective is to accentuate the buttocks' shape instead of merely pumping volume to the buttocks, one has to take into account the patient's complete morphology. Through extensive liposuction of the waist and lower back, a visually enhanced contour of the upper gluteal region can be achieved, which can reduce the need for massive fat grafting. The first and most important donor site for the adipose tissue is by consequence the patient's flanks and lower back.

14.8.6 Ultrasound-Guided Gluteal Fat Grafting

We are taught to use our non-dominant hand to palpate and assess the location and the depth of the cannula tip at all times. However, autopsies and dissection courses have made us realize there is a discrepancy between what a surgeon claims and what is seen on dissection. Even though a complete subcutaneous injection has been attempted, fat can appear in the intramuscular or submuscular plane.

Ultrasound-assisted fat grafting has been described as a reliable technique to enable objective and real-time localization of the cannula tip and provide a means to verify subcutaneous injection [37, 65]. With little training, the ultrasound probe helps to distinguish the different layers under the skin. From superficial to deep, a first dense white line indicates the superficial gluteal fascia, and a second dense white line identifies the important deep gluteal fascia (Fig. 14.10) [65]. The acoustic shadow of the cannula on the portable screen can reliably indicate the cannula's position. It can warn us when the cannula tip is deeper than expected because of flexibility misguidance:

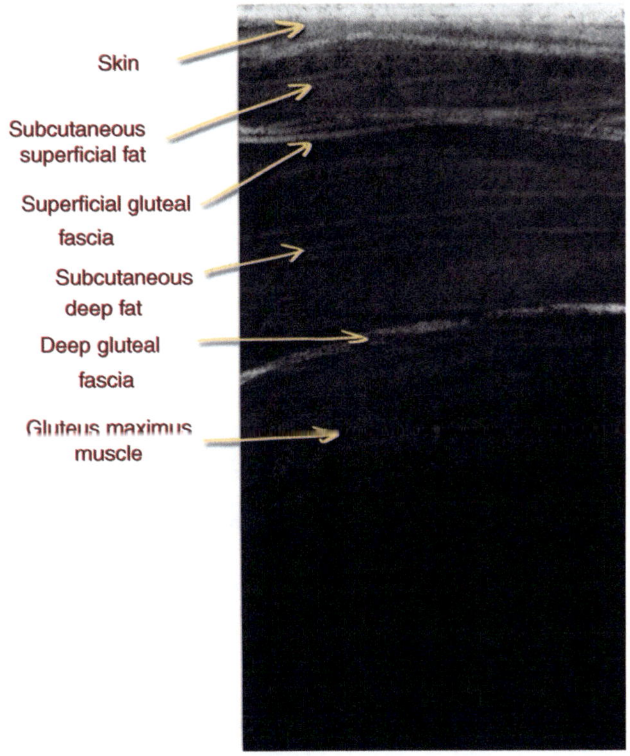

Fig. 14.10 With ultrasound the different layers from skin to muscle can be reliably identified. The important deep gluteal fascia is the second dense white line which is seen under the skin. If the acoustic shadow of the cannula starts below this level, the injection should be stopped in order to avoid intramuscular grafting

- the cannula is located in the superficial subcutaneous tissue if the acoustic shadow starts above the first dense white line (superficial gluteal fascia)
- the cannula is located in the deep subcutaneous tissue if the acoustic shadow starts between both white lines
- fat grafting should not be performed if the acoustic shadow starts below the second dense white line (deep gluteal fascia), as the cannula is positioned intramuscularly

Cansancao demonstrated the safety of ultrasound-guided fat grafting, as it did not lead to grafts unintentionally placed in the intramuscular plane in his series of 35 patients [37].

In addition, ultrasound allows us to evaluate the thickness and appearance of the subcutaneous fat layer [37, 66], which is useful for:

- Intraoperative measurements
  - To measure the depth of the deep gluteal fascia relative To the skin surface
  - To measure the expansion of the subcutaneous layer
- Postoperative follow-up
  - Resorption of the grafted fat
  - Appearance of cysts, necrosis, hematoma, or seroma

In general, the ultrasound device has been proven to be a helpful appliance for gluteal fat grafting, both for beginners as well as experts. The latter group can benefit from this device since it allows precise targeting of the superficial subcutaneous space to correct small deformities and concavities, or the deep subcutaneous space for diffuse reshaping and projection. The subcutaneous tissues and its fibrous septa can be evaluated and can be released with the cannula under ultrasound visualization.

However, an extra hand is required to hold the device, or a learning curve exists to simultaneously manipulate the cannula and the ultrasound probe yourself and interpret the ultrasound images. It inevitably slows down the procedure: the reported additional operative time was approximately 25 min [65]. Therefore, it is cumbersome to perform a complete procedure under ultrasound guidance.

Several brands exist that produce portable ultrasound devices with different resolutions and different prices. The device should have a high frequency to visualize superficial structures since great depth of penetration is not required. Some of them are wired and can be connected to smartphones or tablets. Other devices are wireless and can be introduced in a sterile cover, which increases intraoperative practicality and sterility.

14.8.7 "Smart" Sensing Cannula

Recent work by Turer and Rubin led to the innovation of a fat grafting cannula which measures electrical impedance in different tissue types [66]. Through this modality, the cannula tip can be reliably located in the subcutaneous fat or in the muscle throughout the procedure. The feedback from the device to the surgeon is both visual and audible (Fig. 14.11). The surgeon will be alarmed with red indicator lights and an audible signal as soon as the cannula touches the deep gluteal fascia and will stop us when we are misguided by the flexibility of the cannula or by our own proprioception. When connected to certain devices, the fat grafting flow can instantaneously be stopped as soon as the cannula tip enters the muscle.

Prototypes have been tested on ex vivo porcine models and on fresh cadavers [66]. With simultaneous ultrasound, it was confirmed that the smart sensing cannula is able to differentiate between fat and muscle with sub-centimeter accuracy. From the first demonstrations, the smart sensing cannula looks very easy to use and does not require additional skill such as simultaneously analyzing ultrasound images while performing ultrasound-guided fat grafting. It would also not add additional time to the intervention (Fig. 14.11).

While the smart sensing cannula is a very promising appliance for the future of gluteal fat grafting, its accessibility, usability, and safety should be studied in clinical trials. Time will tell whether the smart sensing cannula is cost-effective: what is the price of the cannula, and importantly, is it disposable or can it be sterilized and re-used?

The question arises whether this cannula is still reliable when a passage through the deep gluteal fascia has already occurred and fat grafts are located intramuscularly. The cannula tip might detect adipose tissue even though it is deep to the fascia, leading to no alarm signal, and leading to further injecting fat in the muscle.

14.8.8 Local Anesthesia

A series of 47 gluteal fat grafting procedures on awake patients has been described [67]. The patients could give immediate feedback to the surgeon when they experience pain. This would implicate that the cannula hits a non-anesthetized region, that is, the muscle. During deep gluteal fascia penetration, it is likely that sensitive nerves are being hit or stretched. The anesthetic protocol consisted of oral medication (hydrocodone 5 mg or acetaminophen 300 mg, and diazepam 10–20 mg), local infiltration of the access sites (lidocaine and epinephrine) and tumescent infiltration of the donor sites and gluteal subcutaneous tissues (for every liter of lactated Ringer solution: lidocaine 800–1000 mg with a maximum of 35 mg/kg body weight, epinephrine 1.5 mg, and sodium bicarbonate 12 mL) [67].

Additional advantages are the lack of inconveniences related to general anesthesia such as nausea, vomiting, possible overnight admission, deep venous thrombosis, aspiration, pneumonia, urinary retention, and corneal lesions [68].

Absorption of the tumescent solution into the gluteus muscles might theoretically numb the muscle and limit the usefulness of this technique [33]. However, in the authors'

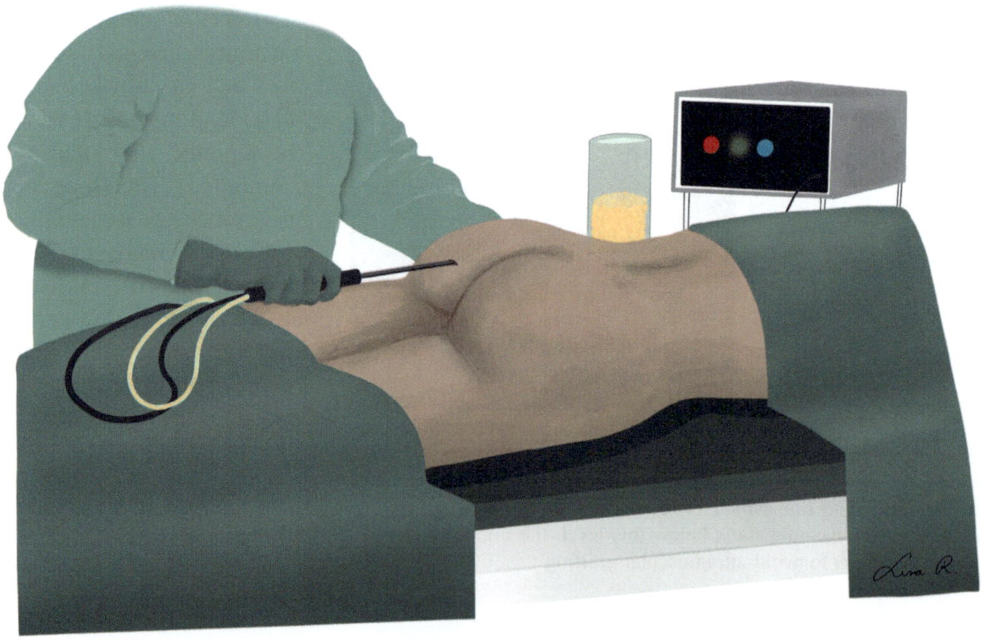

Fig. 14.11 The "smart" sensing cannula. When the cannula tip is located in the safe subcutaneous layer, a green light is shown. As soon as the cannula touches the deep gluteal fascia or enters muscle, the device gives a warning signal and shows a red light. The blue light indicates the cannula tip is not touching any tissue. Drawing by Lisa Ramaut

experience, the tumescent infiltration is inadequate to anesthetize the deep gluteal fascia and underlying muscle, and every hit with the fascia causes a pain response [67]. The thickness and firmness of the deep gluteal fascia also make it a relative barrier to diffusion [68].

Another limitation to this technique is the tumescent infiltration in the recipient site, which expands and distorts the buttock and might complicate a correct perioperative evaluation of the shape and projection. The authors recommend infiltrating the buttocks first so that the anesthetic product is absorbed and edema decreases by the time the fat grafting is done [67].

This approach under local anesthesia slows down the procedure because of the gentle injection and the waiting time for the tumescence to fully induce anesthesia. On the other hand, no time is spent on induction of general anesthesia, on patient position changes, or on awakening.

In our experience, not every patient is physically or mentally able to comfortably undergo liposuction and fat grafting awake, and often this is only clear during the procedure. Because of anxiety or pain, the liposuction and/or grafting volume can be limited, and the operation can be stopped before the desired result has been achieved.

14.9 Postoperative Guidelines and Administrative Guidelines

During the surgery and immediately postoperative, it is imperative to recognize signs of acute right-sided cardiac failure due to mechanical obstruction caused by fat embolism: hypoxemia, bradycardia, and (severe) hypotension [25]. This characteristic triad can arise during the process of gluteal fat grafting, which is especially true for macroscopic fat embolism, or can develop several hours later in the recovery area, which is more likely to happen in case of microscopic fat embolism [18, 25]. When fat embolism is suspected, inotropic and vasopressor support should be put into action as soon as possible. Dobutamine is the therapy of choice [69]. Aggressive resuscitation being our only available treatment at this moment, underscores the importance of prevention [69–71].

Safety in gluteal contouring should stay an item of ongoing study, given its frequency in today's plastic surgery practice and the severity of potential complications [72]. An online General Registry of Autologous Fat Transfer (GRAFT) has been established in which plastic surgeons from the USA, Brazil, and Canada can introduce patient demographics, procedural variables (graft volume, injection device, cannula length, cannula diameter, injection plane, …) and outcomes or complications (including fat embolism) [73]. The accurate analysis of these data can help us in understanding and limiting adverse events.

14.10 Alternatives for High-Volume Gluteal Fat Grafting

Gluteal augmentation by performing excisional and lifting techniques, by redistributing the patient's own tissues, or by using implants, reduces the risk of fat embolism related to intramuscular injection. Depending on the surgeon's preference, implants can be placed submuscularly, intramuscularly or subcutaneously, which have a different complication profile compared to fat grafting (among others: migration of the prosthesis, capsular contracture, seroma, hematoma, dehiscence or infection of the access site, unsightly scarring, thinning of native tissues, implant palpability or exposure) [16, 50, 74]. A hybrid technique combines the use of implants (for reliable projection) with superficial fat grafting (for hip widening and transitional fill) [44, 75]. Alternatively, suspension threads or loops have been described by Abboud to modulate the buttocks' footprint and to create fullness [11]. These options reduce the required fat volume and, consequently, the number of cannula passages. Theoretically, these strategies decrease the potential risk of inadvertently penetrating the deep gluteal fascia and can be discussed with the patient preoperatively [58].

However, gluteal enhancement through fat grafting remains the favored option among surgeons [14]. Liposuction of the waist and lower back, combined with gluteal grafting, allows to reshape the buttock and the silhouette of the patient, contributing to an overall improved shape of a body in harmony with a desirable waist-to-hip ratio [11]. In addition, fat grafting is an optimal tool to correct lateral hollowness or "hip dips," which cannot be improved with gluteal implants.

14.11 Conclusion

Gluteal fat grafting exhibited an increase in demand year after year, and expanded from Brazil to be a globally performed procedure. The main technical guideline in gluteal fat grafting to adhere to is the sole injection in the subcutaneous plane. The danger zone is the area beneath the deep gluteal fascia. Over the last few years, we have learned that one errant passage and inadvertent fat injection through the fascia can be sufficient for venous injury and fat embolization. Autopsy images showing macroscopic fat particles in the heart and lungs are carved into our minds [18, 21]. It is recommended to invest in cadaver dissection to truly understand the anatomy, technique and the required precision. For the same reasons, assisting the coroner in the autopsy after a fatal complication is a learning opportunity to become familiar with gluteal anatomy.

Plastic surgeons who want to provide optimal care in accordance with what is currently known regarding safety should never intentionally inject through the deep gluteal fascia. Studies have shown that an increasing percentage of plastic surgeons embrace the new guidelines and adapt their techniques, which has contributed to safer practice. However, gluteal fat grafting is most commonly performed as a blind procedure. An operation performed by a highly experienced surgeon with accurate knowledge of the gluteal and vascular anatomy, after proper patient selection and with the recommended tools and positioning, cannot guarantee a risk-free procedure. We should stay humble as we can inject in a deep (intramuscular) plane without realizing this.

Therefore, two helpful intraoperative risk-reduction tools are being discussed which can move us from cannula tip awareness to cannula tip confirmation. Technological advances will not bring the complication rate down to zero, but especially, the "smart" sensing cannula could be the game changer which can drastically improve outcomes, while not adding additional operation time.

References

1. Matsudo PK, Toledo LS. Experience of injected fat grafting. Aesthet Plast Surg. 1988;12(1):35–8.
2. Chajchir A, Benzaquen I. Fat-grafting injection for soft-tissue augmentation. Plast Reconstr Surg. 1989;84(6):921–34; discussion 35.
3. ISAPS international survey on aesthetic/cosmetic procedures; 2018. https://www.isaps.org/.
4. The American Society for Aesthetic Plastic Surgery's cosmetic surgery National Data Bank: statistics 2018. Aesthet Surg J. 2019;39(Suppl_4):1–27.
5. Wong WW, Motakef S, Lin Y, Gupta SC. Redefining the ideal buttocks: a population analysis. Plast Reconstr Surg. 2016;137(6):1739–47.
6. Wang J, Thornton JC, Bari S, Williamson B, Gallagher D, Heymsfield SB, et al. Comparisons of waist circumferences measured at 4 sites. Am J Clin Nutr. 2003;77(2):379–84.
7. Vartanian E, Gould DJ, Hammoudeh ZS, Azadgoli B, Stevens WG, Macias LH. The ideal thigh: a crowdsourcing-based assessment of ideal thigh aesthetic and implications for gluteal fat grafting. Aesthet Surg J. 2018;38(8):861–9.
8. Singh D. Universal allure of the hourglass figure: an evolutionary theory of female physical attractiveness. Clin Plast Surg. 2006;33(3):359–70.
9. Lewis DMG, Russell EM, Al-Shawaf L, Ta V, Senveli Z, Ickes W, et al. Why women wear high heels: evolution, lumbar curvature, and attractiveness. Front Psychol. 2017;8:1875.
10. Heidekrueger PI, Sinno S, Tanna N, Szpalski C, Juran S, Schmauss D, et al. The ideal buttock size: a sociodemographic morphometric evaluation. Plast Reconstr Surg. 2017;140(1):20e–32e.
11. Abboud M, Geeroms M, El Hajj H, Abboud N. Improving the female silhouette and gluteal projection: an anatomy-based, safe, and harmonious approach through liposuction, suspension loops, and moderate lipofilling. Aesthet Surg J. 2020;41:474.
12. Analysis of sampled English language tweets mentioning the peach emoji; 2016. https://blog.emojipedia.org/how-we-really-use-the-peach/.
13. Conde-Green A, Kotamarti V, Nini KT, Wey PD, Ahuja NK, Granick MS, et al. Fat grafting for gluteal augmentation: a systematic review of the literature and meta-analysis. Plast Reconstr Surg. 2016;138(3):437e–46e.
14. Cansancao AL, Conde-Green A, Gouvea Rosique R, Junqueira Rosique M, Cervantes A. "Brazilian butt lift" performed by board-certified Brazilian plastic surgeons: reports of an expert opinion survey. Plast Reconstr Surg. 2019;144(3):601–9.
15. Cardenas-Mejia A, Martinez JR, Leon D, Taylor JA, Gutierrez-Gomez C. Bilateral sciatic nerve axonotmesis after gluteal lipoaugmentation. Ann Plast Surg. 2009;63(4):366–8.
16. Sinno S, Chang JB, Brownstone ND, Saadeh PB, Wall S Jr. Determining the safety and efficacy of gluteal augmentation: a systematic review of outcomes and complications. Plast Reconstr Surg. 2016;137(4):1151–6.
17. Oranges CM, Tremp M, di Summa PG, Haug M, Kalbermatten DF, Harder Y, et al. Gluteal augmentation techniques: a comprehensive literature review. Aesthet Surg J. 2017;37(5):560–9.
18. Rapkiewicz AV, Kenerson K, Hutchins KD, Garavan F, Lew EO, Shuman MJ. Fatal complications of aesthetic techniques: the gluteal region. J Forensic Sci. 2018;63(5):1406–12.
19. Cardenas-Camarena L, Lacouture AM, Tobar-Losada A. Combined gluteoplasty: liposuction and lipoinjection. Plast Reconstr Surg. 1999;104(5):1524–31; discussion 32–3.
20. Astarita DC, Scheinin LA, Sathyavagiswaran L. Fat transfer and fatal macroembolization. J Forensic Sci. 2015;60(2):509–10.
21. Cardenas-Camarena L, Bayter JE, Aguirre-Serrano H, Cuenca-Pardo J. Deaths caused by gluteal lipoinjection: what are we doing wrong? Plast Reconstr Surg. 2015;136(1):58–66.
22. Mofid MM, Teitelbaum S, Suissa D, Ramirez-Montanana A, Astarita DC, Mendieta C, et al. Report on mortality from gluteal fat grafting: recommendations from the ASERF task force. Aesthet Surg J. 2017;37(7):796–806.
23. Wall S Jr, Del Vecchio D. Commentary on: report on mortality from gluteal fat grafting: recommendations from the ASERF task force. Aesthet Surg J. 2017;37(7):807–10.
24. Pronovost PJ, Ishii LE. Commentary on: report on mortality from gluteal fat grafting: recommendations from the ASERF task force. Aesthet Surg J. 2017;37(7):811–3.
25. Bayter-Marin JE, Cardenas-Camarena L, Aguirre-Serrano H, Duran H, Ramos-Gallardo G, Robles-Cervantes JA. Understanding fatal fat embolism in gluteal lipoinjection: a review of the medical records and autopsy reports of 16 patients. Plast Reconstr Surg. 2018;142(5):1198–208.
26. Teitelbaum S. Discussion: understanding fatal fat embolism in gluteal lipoinjection: a review of the medical records and autopsy reports of 16 patients. Plast Reconstr Surg. 2018;142(5):1209–11.
27. Del Vecchio DA, Rohrich RJ. A changing paradigm: the Brazilian butt lift is neither Brazilian nor a lift-why it needs to be called safe subcutaneous buttock augmentation. Plast Reconstr Surg. 2020;145(1):281–3.
28. BAPRAS Statement on 'Brazillian Butt Lift' Surgery. https://www.bapras.org.uk/media-government/news-and-views/view/bapras-statement-on-%27brazillian-butt-lift%27-surgery.
29. Namgoong S, Kim HK, Hwang Y, Shin SH, You HJ, Kim DW, et al. Clinical experience with treatment of Aquafilling filler-associated complications: a retrospective study of 146 cases. Aesthet Plast Surg. 2020;44(6):1997–2007.
30. Oam GS, Paddle A. Invited discussion on: clinical experience with treatment of aquafilling filler-associated complications: a retrospective study of 146 cases. Aesthet Plast Surg. 2020;44(6):2008–10.
31. Multi-Society Gluteal Fat Grafting Task Force issues safety advisory urging practitioners to reevaluate technique. https://www.surgery.org/sites/default/files/Gluteal-Fat-Grafting-02-06-18_0.pdf.

32. Multi-Society Task Force for Safety in Gluteal Fat Grafting. Urgent warning to surgeons performing fat grafting to the buttocks (Brazilian Butt Lift or "BBL"). https://www.surgery.org/sites/default/files/URGENT-WARNING-TO-SURGEONS.pdf.
33. Del Vecchio D. Common sense for the common good: staying subcutaneous during fat transplantation to the gluteal region. Plast Reconstr Surg. 2018;142(1):286–8.
34. Chopan M, White JA, Sayadi LR, Buchanan PJ, Katz AJ. Autogenous fat grafting to the breast and gluteal regions: safety profile including risks and complications. Plast Reconstr Surg. 2019;143(6):1625–32.
35. Teitelbaum S. Discussion: clinical implications of gluteal fat graft migration: a dynamic anatomical study. Plast Reconstr Surg. 2018;142(5):1196–7.
36. Guerrerosantos J, Gonzalez-Mendoza A, Masmela Y, Gonzalez MA, Deos M, Diaz P. Long-term survival of free fat grafts in muscle: an experimental study in rats. Aesthet Plast Surg. 1996;20(5):403–8.
37. Cansancao AL, Conde-Green A, David JA, Vidigal RA. Subcutaneous-only gluteal fat grafting: a prospective study of the long-term results with ultrasound analysis. Plast Reconstr Surg. 2019;143(2):447–51.
38. Villanueva NL, Del Vecchio DA, Afrooz PN, Carboy JA, Rohrich RJ. Staying safe during gluteal fat transplantation. Plast Reconstr Surg. 2018;141(1):79–86.
39. Toledo LS. Gluteal augmentation with fat grafting: the Brazilian buttock technique: 30 years' experience. Clin Plast Surg. 2015;42(2):253–61.
40. Ali A. Contouring of the gluteal region in women: enhancement and augmentation. Ann Plast Surg. 2011;67(3):209–14.
41. Murillo WL. Buttock augmentation: case studies of fat injection monitored by magnetic resonance imaging. Plast Reconstr Surg. 2004;114(6):1606–14; discussion 15–6.
42. Alvarez-Alvarez FA, Gonzalez-Gutierrez HO, Ploneda-Valencia CF. Safe gluteal fat graft avoiding a vascular or nervous injury: an anatomical study in cadavers. Aesthet Surg J. 2019;39(2):174–84.
43. Ghavami A, Villanueva NL. Gluteal augmentation and contouring with autologous fat transfer: part I. Clin Plast Surg. 2018;45(2):249–59.
44. Rosique RG, Rosique MJ. Deaths caused by gluteal lipoinjection: what are we doing wrong? Plast Reconstr Surg. 2016;137(3):641e–2e.
45. Wolf GA, Gallego S, Patron AS, Ramirez F, de Delgado JA, Echeverri A, et al. Magnetic resonance imaging assessment of gluteal fat grafts. Aesthet Plast Surg. 2006;30(4):460–8.
46. Ghavami A, Villanueva NL, Amirlak B. Gluteal ligamentous anatomy and its implication in safe buttock augmentation. Plast Reconstr Surg. 2018;142(2):363–71.
47. Cardenas-Camarena L, Duran H, Robles-Cervantes JA, Bayter-Marin JE. Critical differences between microscopic (MIFE) and macroscopic (MAFE) fat embolism during liposuction and gluteal lipoinjection. Plast Reconstr Surg. 2018;141(4):880–90.
48. Villanueva NL, Del Vecchio DA, Afrooz PN, Rohrich RJ. Reply: staying safe during gluteal fat transplantation. Plast Reconstr Surg. 2018;142(4):594e–5e.
49. Del Vecchio D, Wall S Jr. Expansion vibration lipofilling: a new technique in large-volume fat transplantation. Plast Reconstr Surg. 2018;141(5):639e–49e.
50. Ordenana C, Dallapozza E, Said S, Zins JE. Objectifying the risk of vascular complications in gluteal augmentation with fat grafting: a latex casted cadaveric study. Aesthet Surg J. 2020;40(4):402–9.
51. Del Vecchio DA, Villanueva NL, Mohan R, Johnson B, Wan D, Venkataram A, et al. Clinical implications of gluteal fat graft migration: a dynamic anatomical study. Plast Reconstr Surg. 2018;142(5):1180–92.
52. Wall S Jr, Delvecchio D, Teitelbaum S, Villanueva NL, Dayan E, Durand P, et al. Subcutaneous migration: a dynamic anatomical study of gluteal fat grafting. Plast Reconstr Surg. 2019;143(5):1343–51.
53. Donovan DL, Schmidt SP, Townshend SP, Njus GO, Sharp WV. Material and structural characterization of human saphenous vein. J Vasc Surg. 1990;12(5):531–7.
54. Rios L, Gupta V. Improvement in Brazilian Butt Lift (BBL) safety with the current recommendations from ASERF, ASAPS, and ISAPS. Aesthet Surg J. 2020;40(8):864–70.
55. Sadideen H, Akhavani MA, Mosahebi A, Harris PA. Current perceptions of 'Brazilian butt lift' (BBL) surgery in the UK: a BAAPS-led survey of BAAPS members. J Plast Reconstr Aesthet Surg. 2020;73(11):1966–75.
56. Singer R. Commentary on: improvement in Brazilian Butt Lift (BBL) safety with the current recommendations from ASERF, ASAPS, and ISAPS. Aesthet Surg J. 2020;40(8):871–3.
57. Fadavi D, He W, Kraenzlin F, Darrach H, Shetty P, Xun H, et al. Risk and reward: public perception of gluteal fat grafting safety. Aesthet Plast Surg. 2020;44(5):1628–38.
58. Nahai F. Acceptable risk: who decides? Aesthet Surg J. 2017;37(7):852–3.
59. Villanueva NL, Del Vecchio DA, Afrooz PN, Rohrich RJ. Reply: staying safe during gluteal fat transplantation. Plast Reconstr Surg. 2018;142(4):596e.
60. Ramos-Gallardo G, Orozco-Renteria D, Medina-Zamora P, Mota-Fonseca E, Garcia-Benavides L, Cuenca-Pardo J, et al. Prevention of fat embolism in fat injection for gluteal augmentation, anatomic study in fresh cadavers. J Investig Surg. 2018;31(4):292–7.
61. Turer DM, Rubin JP. Commentary on: safe gluteal fat graft avoiding a vascular or nervous injury: an anatomical study in cadavers. Aesthet Surg J. 2019;39(2):185–6.
62. Eto H, Kato H, Suga H, Aoi N, Doi K, Kuno S, et al. The fate of adipocytes after nonvascularized fat grafting: evidence of early death and replacement of adipocytes. Plast Reconstr Surg. 2012;129(5):1081–92.
63. Mineda K, Kuno S, Kato H, Kinoshita K, Doi K, Hashimoto I, et al. Chronic inflammation and progressive calcification as a result of fat necrosis: the worst outcome in fat grafting. Plast Reconstr Surg. 2014;133(5):1064–72.
64. Pane TA. Experience with high-volume buttock fat transfer: a report of 137 cases. Aesthet Surg J. 2019;39(5):526–32.
65. Cansancao AL, Conde-Green A, Vidigal RA, Rodriguez RL, D'Amico RA. Real-time ultrasound-assisted gluteal fat grafting. Plast Reconstr Surg. 2018;142(2):372–6.
66. Turer DM, Qaium EB, Lawrence AM, Clark WW, Rubin JP. A smart sensing cannula for fat grafting. Plast Reconstr Surg. 2019;144(2):385–8.
67. Chia CT, Theodorou SJ, Dayan E, Tabbal G, Del Vecchio D. "Brazilian butt lift" under local anesthesia: a novel technique addressing safety concerns. Plast Reconstr Surg. 2018;142(6):1468–75.
68. Lalonde DH. Discussion: "Brazilian butt lift" under local anesthesia: a novel technique addressing safety concerns. Plast Reconstr Surg. 2018;142(6):1476–7.
69. Peña W, Cardenas-Camarena L, Bayter-Marin JE, McCormick M, Duran H, Ramos-Gallardo G, et al. Macro fat embolism after gluteal augmentation with fat: first survival case report. Aesthet Surg J 2019;39(9):NP380-NP3.
70. Lari A, Abdulshakoor A, Zogheib E, Assaf N, Mojallal A, Lari AR, et al. How to save a life from macroscopic fat embolism: a narrative review of treatment options. Aesthet Surg J. 2020;40(10):1098–107.
71. Teitelbaum S, Benharash P. Commentary on: how to save a life from macroscopic fat embolism: a narrative review of treatment options. Aesthet Surg J. 2020;40(10):1108–10.
72. Nahai F. No "Quick Fix" for this: an update on the Brazilian Butt Lift. Aesthet Surg J. 2020;40(8):928–30.

73. Plastic Surgery Foundation. General Registry of Autologous Fat Transfer (GRAFT). http://www.thepsf.org/research/clinical-impact/general-registry-autologous-fat-transfer.htm.
74. Mendes J Jr, Mendes Maykeh VA, Frascino LF, Zacchi FFS. Gluteal implant-associated anaplastic large cell lymphoma. Plast Reconstr Surg. 2019;144(3):610–3.
75. Aslani A, Del Vecchio DA. Composite buttock augmentation: the next frontier in gluteal aesthetic surgery. Plast Reconstr Surg. 2019;144(6):1312–21.

Gluteal Fat Grafting: Technology, Techniques, and Safety

15

Onelio Garcia Jr and Pat Pazmiño

Contents

15.1 History of Gluteal Fat Grafting

Surgeons have been grafting fat for over 100 years. The first description of a fat grafting procedure was attributed to German surgeon Gustav Neuber in 1893. He described the transfer of adipose tissue from the arm for correction of an osteomyelitis-related, contour deformity in the infra-orbital region [1]. Two years later, another German surgeon, Vincent Czerny, published a description of a procedure in which he transferred a moderate-sized lipoma from the buttock to the breast, to correct asymmetry resulting from a partial mastectomy [2]. In 1920, Sir Harold Gillies described using fat grafts to correct various facial wounds and their sequelae [3].

Gluteal enhancement by means of fat grafting has been performed sporadically for over 40 years but gained popularity among plastic surgeons following the description of the modern version of the procedure by Toledo [4]. Initially, this was a strictly subcutaneous operation with a relatively low associated morbidity and mortality. Typical fat transfer volumes also appear to have been relatively modest in comparison to the average fat transfer volumes associated with the current version of the operation. The procedure was coined the "Brazilian Butt Lift" [5] around the turn of the century and soon after became one of the most sought-after aesthetic, body contouring procedures in the United States [6].

The aesthetic preferences of American women have evolved from the fit, athletic, "hard bodies" of the 1990s to fuller, more voluptuous figures after the turn of the century. Many women now also choose to embrace ethnic backgrounds that find wider hips and fuller buttocks aesthetically attractive [7]. Social media has also played a major role in promoting these concepts.

As fat transfer volumes increased to a level that could potentially overwhelm the subcutaneous recipient site, surgeons began grafting deeper into the well-vascularized glu-

O. Garcia Jr (✉) · P. Pazmiño
Division of Plastic Surgery, University of Miami, Miller School of Medicine, Miami, FL, USA
e-mail: oni@ogarciamd.com; author@miamia.com

© The Author(s), under exclusive license to Springer Nature Switzerland AG 2023
A. Di Giuseppe et al. (eds.), *Fat Transfer in Plastic Surgery*, https://doi.org/10.1007/978-3-031-10881-5_15

teal musculature. These techniques of "intramuscular grafting" allowed for large-volume fat transfers to be injected into a well-perfused recipient site, theoretically resulting in a higher percentage survival of the grafted fat. Surgeons with expertise in gluteal contouring surgery developed techniques for intramuscular fat grafting [8–10] and delineated what were thought to be "safe areas" for fat grafting within the gluteal musculature [11]. Since then, detailed MRI venography studies of the gluteal venous system suggest that there really is no "safe zone" for fat grafting in the intramuscular or submuscular plane [12]. Brazilian Butt Lift Surgery was the fastest-growing aesthetic surgical procedure between the decade 2010 and 2020 [6]. Unfortunately, with the increase in cases, came an unprecedented increase in the mortality associated with the procedure.

In 2015, Cárdenas-Camarena et al. reviewed the experience of plastic surgeons in Mexico and Colombia over the past 10 and 15 years, respectively, and identified 13 pulmonary fat emboli (PFE) deaths in Mexico and 9 PFE deaths in Colombia after gluteal fat grafting [13]. After analyzing the autopsy findings, they found that all the deaths were associated with intramuscular fat grafting. For the first time, the danger of grafting fat into the gluteal muscles was recognized, and these authors recommended avoiding fat graft injections into the deep gluteal muscle.

In July 2016, the ASERF Gluteal fat grafting survey was sent to all active members of the Aesthetic Society and the International Society for Aesthetic Plastic Surgery (4843 questionnaires). A total of 692 surgeons responded (14.3%) to 198,857 cases. The survey placed the mortality associated with BBL surgery somewhere between 1/2351 and 1/6241 [14]. The Aesthetic Surgery Education and Research Foundation (ASERF) created the Gluteal Fat Grafting Task Force to study the factors involved in the high mortality rate associated with BBL surgery. Under the direction of Peter Rubin MD, the Intersociety Gluteal Fat Grafting Task Force began a series of cadaver studies to identify the surgical techniques and anatomic factors that contributed to the unacceptably high BBL mortality rate.

The Task Force convened in Miami in 2017 to create an appropriate anatomic model to determine safe BBL techniques. Both authors served as consultants to the Task Force and participated in all the cadaver studies. The central hypothesis for the study was that "safe approaches for gluteal fat grafting could be delineated by modeling the operation in cadavers." Once the model was validated, two separate sessions were scheduled with invited surgeons (Fig. 15.1a, b) who performed their typical BBL techniques under video fluoroscopic monitoring and used color-coded fat to examine their graft dispersion. Following completion of the grafting sessions, the buttocks were anatomically dissected, and the colored fat dispersion was documented with high-resolution photography. (Fig. 15.2a, b).

The authors wish to extend their gratitude to all the invited surgeons who traveled nationally and internationally to participate in the gluteal fat grafting study. Dr. Rubin ran a very tight schedule, and the sessions were long and intense. Their participation resulted in significant contributions to our understanding of gluteal fat grafting and the events that could result in pulmonary fat emboli, associated with the procedure. All invited surgeons, who planned to fat graft only into the subcutaneous space, were able to do so, even when injecting large volumes (>1000 mL) into each buttock (Fig. 15.3a, b). The three most important safety recommendations derived from these cadaver studies were: (1) Fat grafting only into the subcutaneous space, (2) Not injecting at a downward angle, (3) Using a stiff injection cannula (>4 mm) [15].

ASERF repeated the gluteal fat grafting survey online in May 2019. This time, 5048 invitations were sent, and 572 responses (11.3%) were received from plastic surgeons reporting on 29,843 cases. This survey calculated the 2019 BBL mortality rate to be 1/14,921, which, if accurate, would have surpassed the safety profile of abdominoplasty (1/13,193 at the time of the survey) [16].

By this time, it had become increasingly clear that the mortality from BBL was intimately related to pulmonary fat emboli (PFE) [17, 18], and the physiologic and clinical differences between macro fat emboli and micro fat emboli syndrome were well described in the literature [19, 20]. Micro fat emboli syndrome carries reported mortality in the 10–30% range. A delay in treatment of the syndrome is associated with a 35% mortality [21], which may be a frequent occurrence in the plastic surgery population undergoing outpatient surgery. Macro pulmonary fat emboli following BBL surgery, however, is almost always a fatal event. A 2018 publication involving 16 BBL deaths found that 50% of the patients died on the operating table while the other 50% died within 3 h of the surgery [22]. Since that report, there have been three reported cases of survival following macro fat emboli associated with BBL surgery (two patients with pulmonary macro fat emboli and the other with a cerebral fat embolism, in a patient with a patent foramen ovale [23, 24]. The most important commonalities among these survival cases were extremely rapid access to intensive care at a tertiary care facility with in-house specialists and the availability of extracorporeal

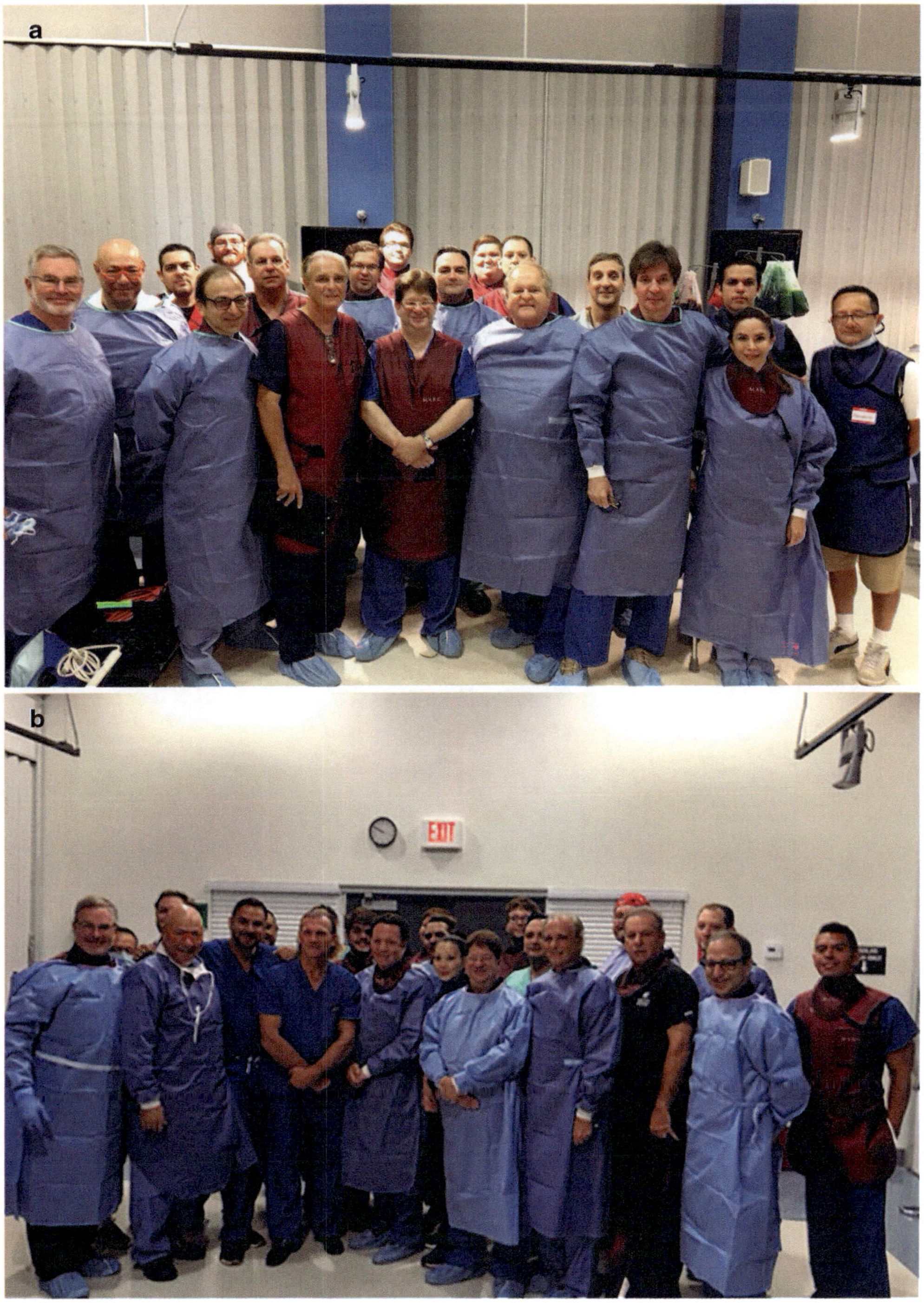

Fig. 15.1 (**a**) Session 1 Principal Investigator: J Peter Rubin, Co-Chair Gluteal Fat Grafting Task Force: Dan Mills Invited injectors: Ricardo Rodriguez (USA), Tino Mendieta (USA) and Arturo Ramirez Montanana (Mexico) Consultants to Gluteal Fat Grafting Task Force: Pat Pazmino, Onelio Garcia Jr Resident Assistants: David Turer, David Gerth Observers: James Fernau, Roger Khouri, Richard D'Amico (**b**) Session 2 Principal Investigator: J Peter Rubin Invited injectors: Dan Del Vecchio (USA), Ash Ghavani (USA, Raul Gonzalez (Brazil), and Tino Mendieta (USA) Consultants to the Gluteal Fat Grafting Task Force: Pat Pazmino, Onelio Garcia Jr Resident Assistants: David Turer, David Gerth Observers: James Fernau, Roger Khouri, Richard D'Amico, Ricardo Rodriguez

membrane oxygenation (ECMO) devices. Despite the heroic, life-saving measures, all three surviving patients suffered varying degrees of associated serious sequelae from their fat emboli episode. Since the majority of BBL surgeries take place at outpatient surgery centers and office-based surgical facilities, it appears highly unlikely that these conditions, necessary for the slimmest chance at survival, will present themselves even rarely. Currently, it is fair to say that the large amounts of fat seen in the heart, lungs, and major vessels during the autopsies of BBL-Associated Fat Emboli deaths are incompatible with life (Fig. 15.4a–c).

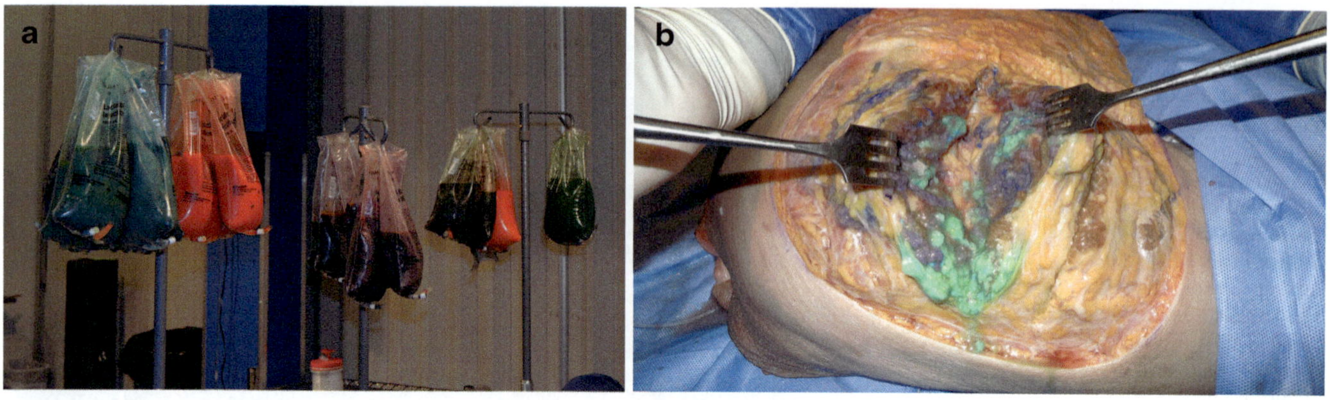

Fig. 15.2 (**a**) Color-coded fat prepared for injection. (**b**) Color-coded fat was injected into tissues. Note the lack of dye bleed into surrounding tissues which allowed for accurate assessment of the fat dispersion

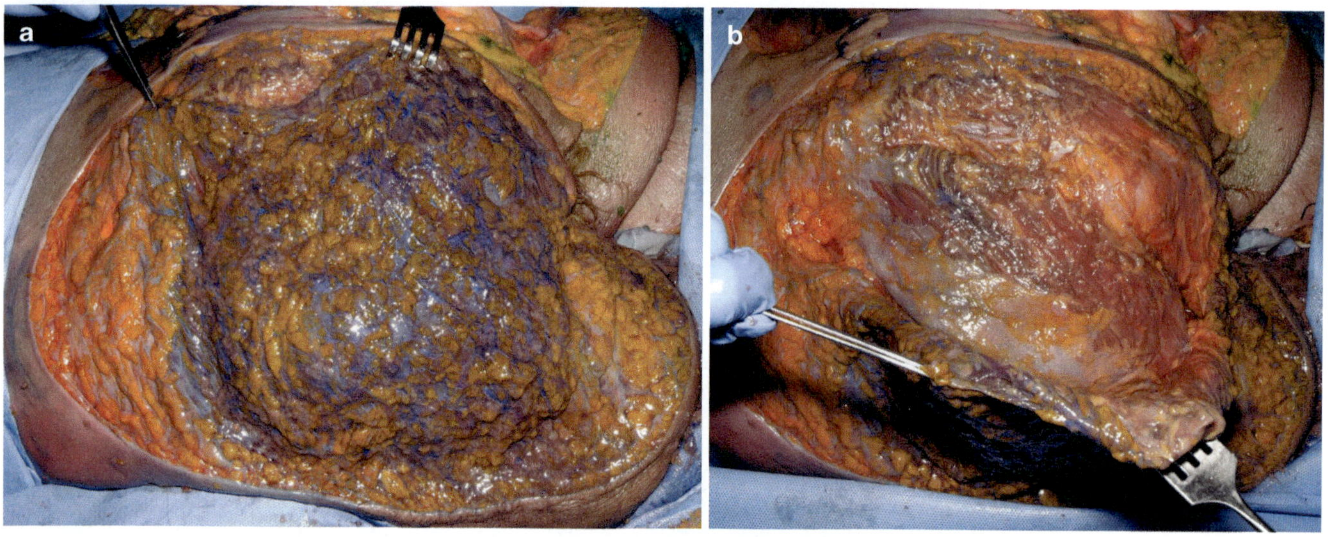

Fig. 15.3 (**a**) All invited injectors stayed in the subcutaneous plane, even when injecting large volumes (>1000 mL/buttock). (**b**) Note that all color-coded fat is superficial to the muscle fascia

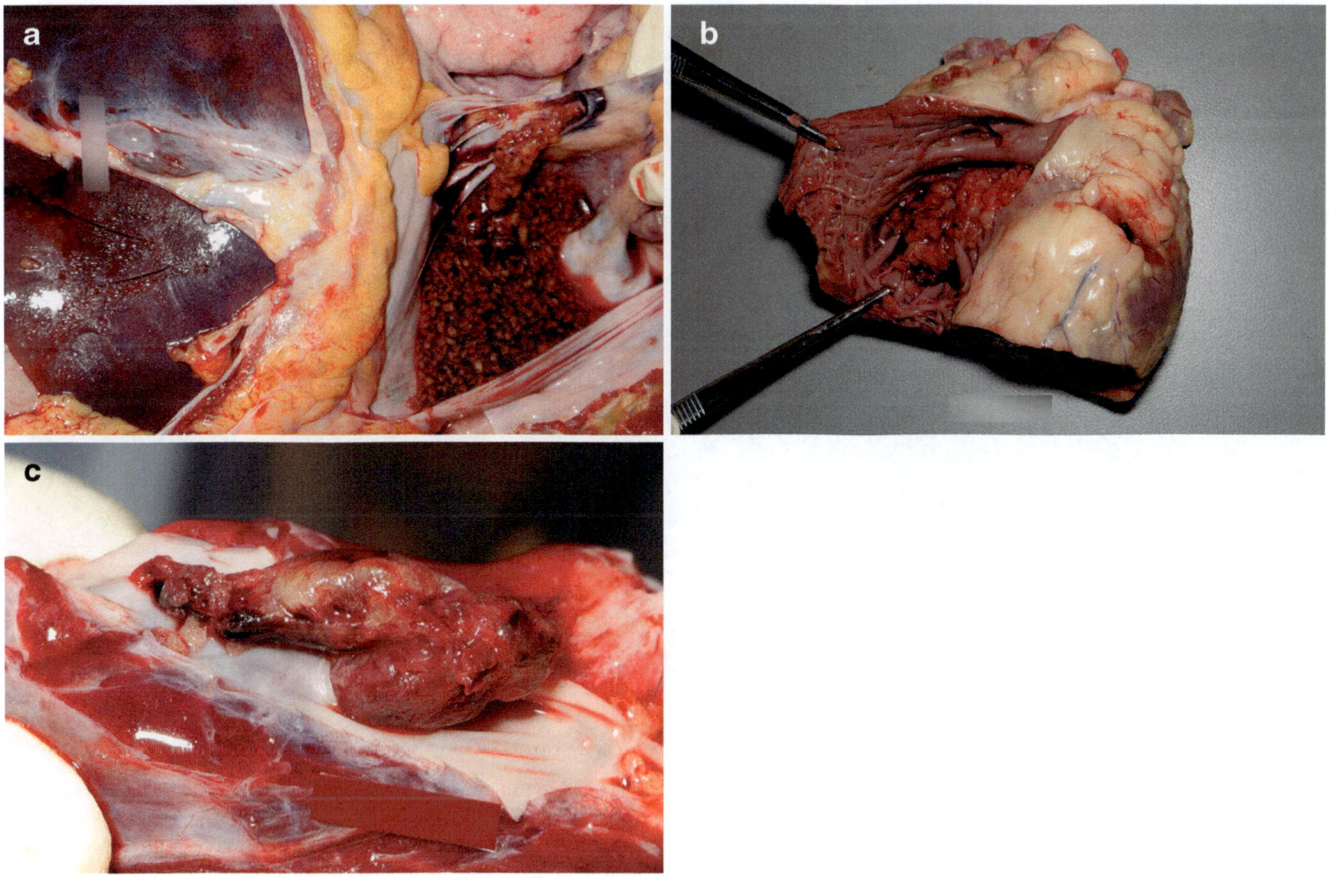

Fig. 15.4 (**a**) Shows fat exciting cut portion of the vena cava. (**b**) Displays gross globular fat within the right atrium. (**c**) Fat graft within pulmonary artery

15.2 The High BBL Mortality Rate in South Florida

A total of 29 BBL-associated deaths were recorded in Florida between January 2010 and April 2022. In the time frame between 2010 and 2016, an average of one BBL-associated death per year was recorded. Between 2017 and 2018, 8 deaths were recorded, which raised alarms and prompted the FL BOM to convene a special session to discuss BBL surgery and associated mortality.

In April 2019, the Florida Board of Medicine (FL BOM) held a special session to discuss the high mortality rate of BBL surgery in the state. One of the authors (OG) appeared before the FL BOM on behalf of the Florida Society of Plastic Surgeons (FSPS), and the results of the latest ASERF survey along with recommendations from the Task Force, were shared with the board members. The FL BOM drafted the Florida BBL surgery Rule, which prohibits surgeons from injecting fat grafts into the gluteal muscles or the submuscular space. In addition, the ruling notes that physicians who inject fat grafts into the gluteal muscle during BBL procedures are subject to disciplinary action, including the potential loss of their medical licenses. Fat graft injection, strictly restricted to the subcutaneous space, was considered the standard of care for BBL surgery in 2019 by the FL BOM [25].

It was both unfortunate and unexpected that neither the FL BOM rule nor the ASERF recommendations were able to improve BBL-associated mortality in Florida. The fact is that the years following the safety recommendations and the BOM rules have been the deadliest so far. There have been 14 deaths recorded in Florida since the ASERF guidelines and 12 deaths since the FL BOM "subcutaneous injection only" rule was enacted. The worst by year by far was 2021 when a total of six BBL PFE deaths and two nonfatal BBL PFE cases were recorded in South Florida alone (Fig. 15.5).

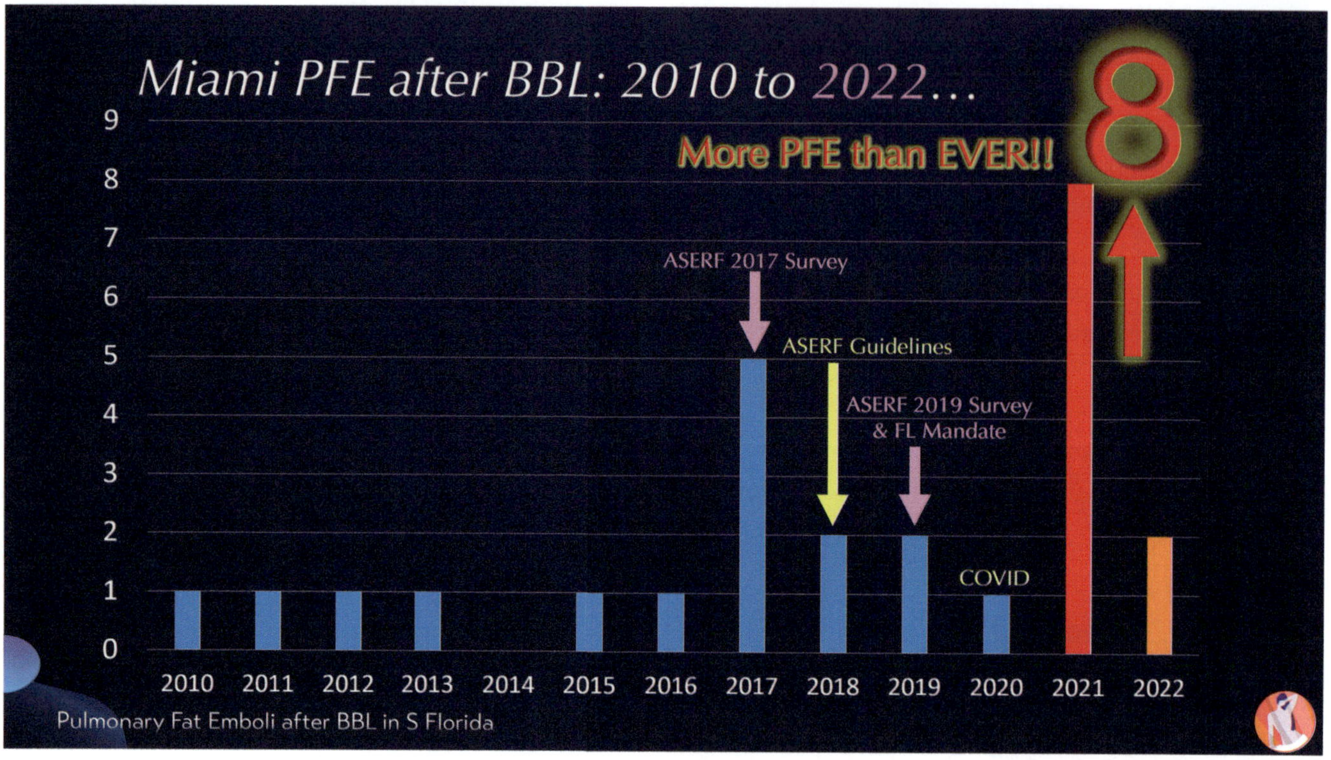

Fig. 15.5 There have been 29 FPE BBL-related deaths in South Florida from 2010 to 2022
Fourteen of the deaths have occurred since the ASERF guidelines
Twelve deaths have occurred since the ASERF 2019 survey that depicted a better mortality rate than abdominoplasty and the FL BOM BBL mandate
2021 was the deadliest year so far, with at least eight deaths confirmed
In 2021, two FPE after BBL patients survived, but these are not included in this data set
It is important to note that there have been more post-BBL deaths in South Florida (due to anemia and hypovolemic shock, DVT, sepsis, cardiac arrest, intra-abdominal organ damage, etc.), but that the focus of this paper is only FPE-associated BBL deaths

15.3 Lessons Learned in the Miami Morgue

In an effort to understand the factors associated with BBL mortality in South Florida, the authors developed a relationship with the Miami-Dade County Medical Examiner's Office. The Chief Medical Examiner was Dr. Emma Lew, and by 2017, she had become the foremost authority on BBL-associated fat emboli deaths, having performed more autopsies on these patients than any of her peers. She developed a special interest in these cases and was invited to the Gluteal Task Force cadaver injection sessions so that she could witness the mechanism of injury resulting in the fat embolus during BBL procedures. She, in turn, invited the authors to observe the gluteal dissections at the Miami-Dade Medical Examiner's Office.

Once the Medical Examiner team determined the cause of death, a detailed gluteal dissection was performed to determine if there was any evidence of vascular injury, particularly to the gluteal veins, and to note the dispersion and location of the fat grafts. The authors were present in 11 of the autopsies (22 hemi-buttocks were dissected). The anatomic commonalities noted in these cases are the result of their observations during the post-mortem gluteal dissections.

Fat grafting into the muscle during BBL surgery always resulted in fat grafts under the deep gluteal fascia and also below the muscle belly of the gluteus maximus. Del Vecchio et al. described in dynamic cadaveric studies that fat graft placed under the deep gluteal fascia could migrate through the muscle belly of the gluteus maximus and reach the intramuscular vessels or the submuscular vascular plexus [26, 27].

Another common finding in the post-mortem gluteal dissections was the presence of free fat graft around the gluteal vessels (Fig. 15.6). Typically, the grafts appeared as intramuscular strings of fat throughout the muscle, at multiple sites and depths, suggesting multiple passes of the cannula through the muscle and the injection of fat graft as the cannula is withdrawn (Fig. 15.7a, b). These findings are not consistent with intramuscular fat being the result of one or two inadvertent passes of the cannula, or subcutaneous fat graft spilling into the muscle. The findings suggest that either the surgeon had intentionally performed an intramuscular gluteal fat graft or that the surgeon was not aware of the precise location of the cannula tip at the time of fat graft injection.

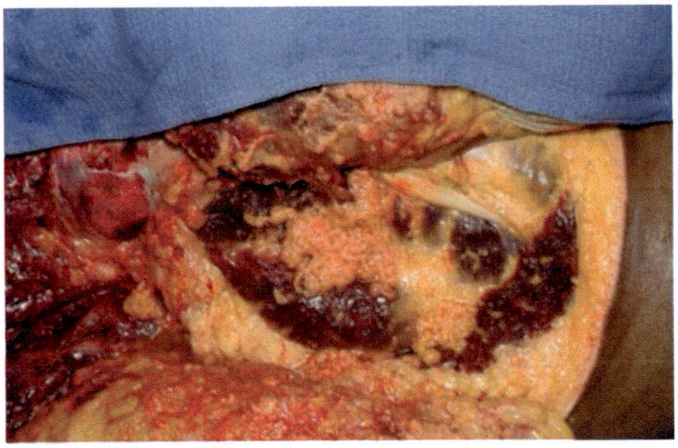

Fig. 15.6 Fat grafts in proximity to the Gluteal Veins

Another commonality relates to the location where the surgical procedure was performed. Out of the 29 BBL-associated deaths (2010–2022), 25 surgeries (86.2%) were performed at the low-budget, high-volume clinics that have become commonplace in South Florida. Two of the deaths were attributed to an outlier American Board of Plastic Surgery (ABPS) surgeon who has had three deaths from body contouring procedures (2 BBL-associated), plus numerous other serious complications also from body contouring procedures (Table 15.1).

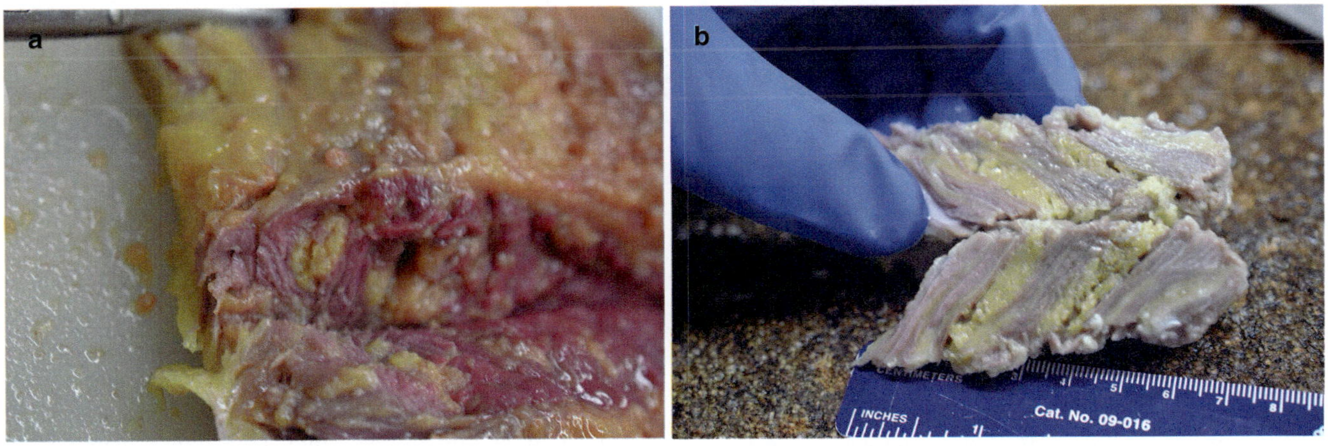

Fig. 15.7 (**a**) Numerous cannula tunnels within the muscle are laden with fat grafts. (**b**) Close-up image of gross pathology specimen of gluteus maximus muscle, displaying the fat grafts within the multiple injection cannula tracts

Table 15.1 Demographics of South Florida BBL FPE fatalities

Surg date	Age (decade)	BMI	ABPS surgeon	Budget clinic	Cause of death
2010.12	30–p40	24.4	No	Yes	FPE
2011.06	30–40	29.3	No	Yes	FPE
2012.03	30–40	23.4	Yes	Yes	FPE
2013.07	50–60	31.1	No	Yes	FPE
2015.07	50–60	32.2	Yes	Yes	FPE
2016.05	20–30	27.8	No	Yes	FPE
2017.03	30–40	29.8	Yes	Yes	FPEX
2017.03	20–30	25.5	BE	Yes	FPE
2017.04	30–40	22.8	BE	Yes	FPE
2017.06	30–40	30	No	Yes	FPE
2017.12	40–50	30.4	No	Yes	FPE
2018.06	30–40	29.1	Yes	Yes	FPE
2018.09	50–60	34.3	Yes	No	FPE
2019.05	20–30	30.2	Yes	Yes	FPE
2019.08	20–30	27.33	Yes	No	FPE
2020.09	40–50	28.2	Yes	Yes	FPE
2021.01	30–40	27.42	Yes	Yes	FPE
2021.01	20–30	30	Yes	Yes	FPE
2021.03	40–50	42	No	Yes	FPE
2021.05	30–40	36.9	Yes	Yes	FPE
2021.06	30–40	36	Yes	Yes	FPE
2021.06	40–50	37.8	Yes	Yes	FPE
2022.02	30–40	30	Yes	Yes	FPE
2022.04	40–50	42	Yes	Yes	FPE
AVG	37	31.0			

FPE Deaths: 25
ABPS: 17 17/25 = (68%)
Deaths in budget clinics/total 23/25 = (92%)

15.4 Optimizing Safety in BBL

Despite the potential dangers associated with this surgery, both authors agree that plastic surgeons must not abandon this operation. Gluteal fat grafting is a powerful tool that can augment tissue, correct deformities, and create impressive results that cannot be produced any other way. High patient demand for this procedure will continue. If board-certified plastic surgeons stop performing this procedure, interested patients will simply go to the non-board-certified, non-core practitioners who do not have training in this procedure [28]. As researchers and patient advocates, plastic surgeons must study this technique and determine how gluteal fat grafting can be performed safely and consistently.

Ideally, surgeons should have a consistent and reliable way to always confirm the position of their cannula tip during gluteal fat grafting. Furthermore, surgeons should have a reproducible method to prove that they only injected fat subcutaneously and to document that they never injected fat into the gluteal musculature. Keeping an ultrasound video recording of the surgery serves as important medicolegal protection for the surgeon by providing irrefutable proof that all the fat grafting was performed only into the subcutaneous space. In the last 3 years, ultrasound equipment has become portable, wireless, and affordable, opening the door for its use in the sterile field of the surgical suite. Ultrasound-visualization can be used with any cannula style or injection system [29]. Real-time intraoperative ultrasound visualization can help the surgeon perform fat harvesting and accurate fat grafting into the unique spaces of the subcutaneous region. This will not only make for a safer surgeon, but a better surgeon—a surgeon who can manipulate subcutaneous anatomy not appreciable without ultrasound.

15.5 Intraoperative Ultrasound Systems

Real-time intraoperative ultrasound can be used with any cannula or liposuction/fat grafting system. However, when a syringe fat grafting system is used, both of the surgeon's

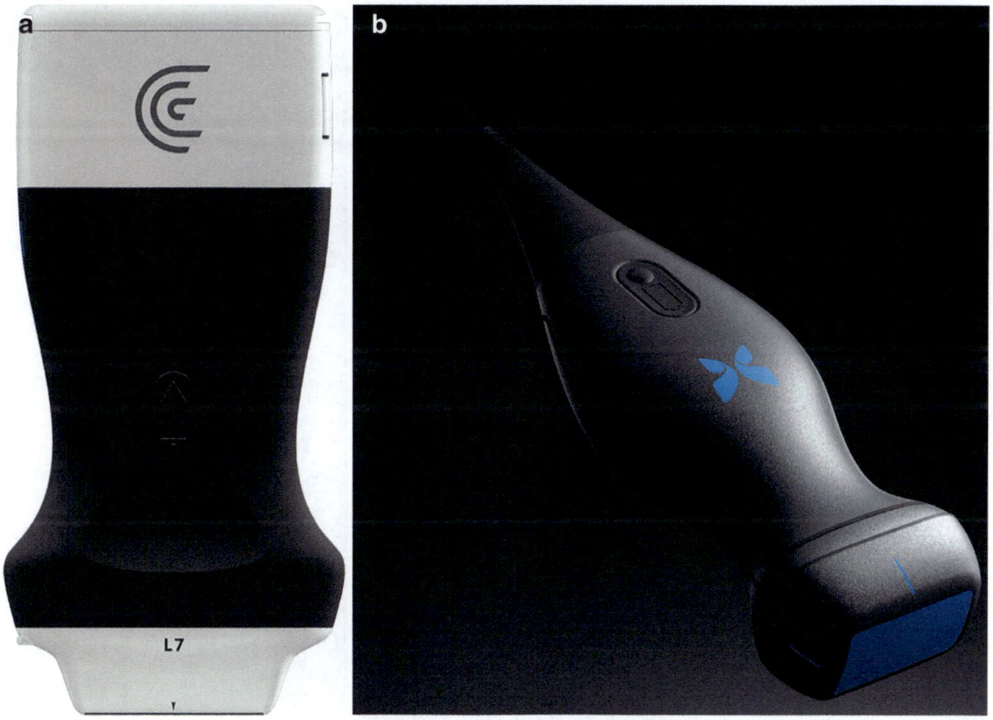

Fig. 15.8 (**a**) The Clarius L7 is a 4–13 MHz high-frequency linear, portable, waterproof, wireless ultrasound probe that can stream a high-resolution ultrasound video to Apple OS or Android phones or tablets. (**b**) The Butterfly iQ+ is a 1–10 MHz wired ultrasound probe that connects with a cable to Apple iPhones or iPad tablets

hands are occupied. One hand must hold the syringe while the other pushes the plunger to inject the fat. In this scenario, the surgical assistant or scrub tech often controls the sterile ultrasound probe, making coordination with the injecting surgeon difficult. To allow the surgeon to control the fat grafting system and the ultrasound probe simultaneously, a power-assisted liposuction system (PAL, MicroAire Charlottesville, VA) is used in conjunction with a peristaltic pump for controlled propulsion of the fat graft. In this manner, the surgeon can inject fat via expansion vibration lipofilling [30] with one hand and control the ultrasound probe with the other hand.

One of the authors (PP) has used nine different ultrasound systems during gluteal fat grafting procedures over the past 10 years. Currently, the two most common inexpensive and portable ultrasound systems for real-time intraoperative ultrasound are available: the Clarius ultrasound (Clarius, $3400) or the Butterfly iQ+ (Butterfly, $2399 with $199 annual subscription) (Fig. 15.8a, b). The Clarius L7 is a 4–13 MHz high-frequency linear, portable, waterproof, wireless ultrasound probe (maximum depth of 11 cm) that can be placed entirely in a sterile probe cover and can stream a high-resolution ultrasound video to Apple OS or Android phones or tablets. The Butterfly iQ+ is a 1–10 MHz wired ultrasound probe (maximum depth 30 cm) that connects with a cable to Apple iPhones or iPad tablets. Both the Clarius and Butterfly systems will upload their data to the cloud so that ultrasound still images and video can be accessed on a computer or added to a patient chart.

15.6 Gluteal Anatomy and Ultrasound

The pelvic bony framework, gluteal muscles, gluteal fat, and skin have been well described in our literature [31, 32]. Ultrasound can help us accurately delineate and manipulate the subcutaneous zone. Cadaver dissections have unequivocally identified two well-defined gluteal fasciae (Fig. 15.9). The external surface of the gluteus maximus muscle is covered with a fascial plane (the deep gluteal fascia). The Multi-Society Task Force has recommended surgeons never fat graft beneath this deep gluteal fascia [15]. However, there also exists a second fascial layer (the superficial gluteal fascia) within the subcutaneous zone above the deep gluteal fascia and below the dermis. The superficial gluteal fascia is thick, irregular, impregnated with fat, and can only be appreciated in an open dissection or with ultrasound visualization.

The superficial gluteal fascia is analogous to Scarpa's fascia in the anterior abdomen and divides the subcutaneous zone into two subcutaneous spaces: the superficial sub-

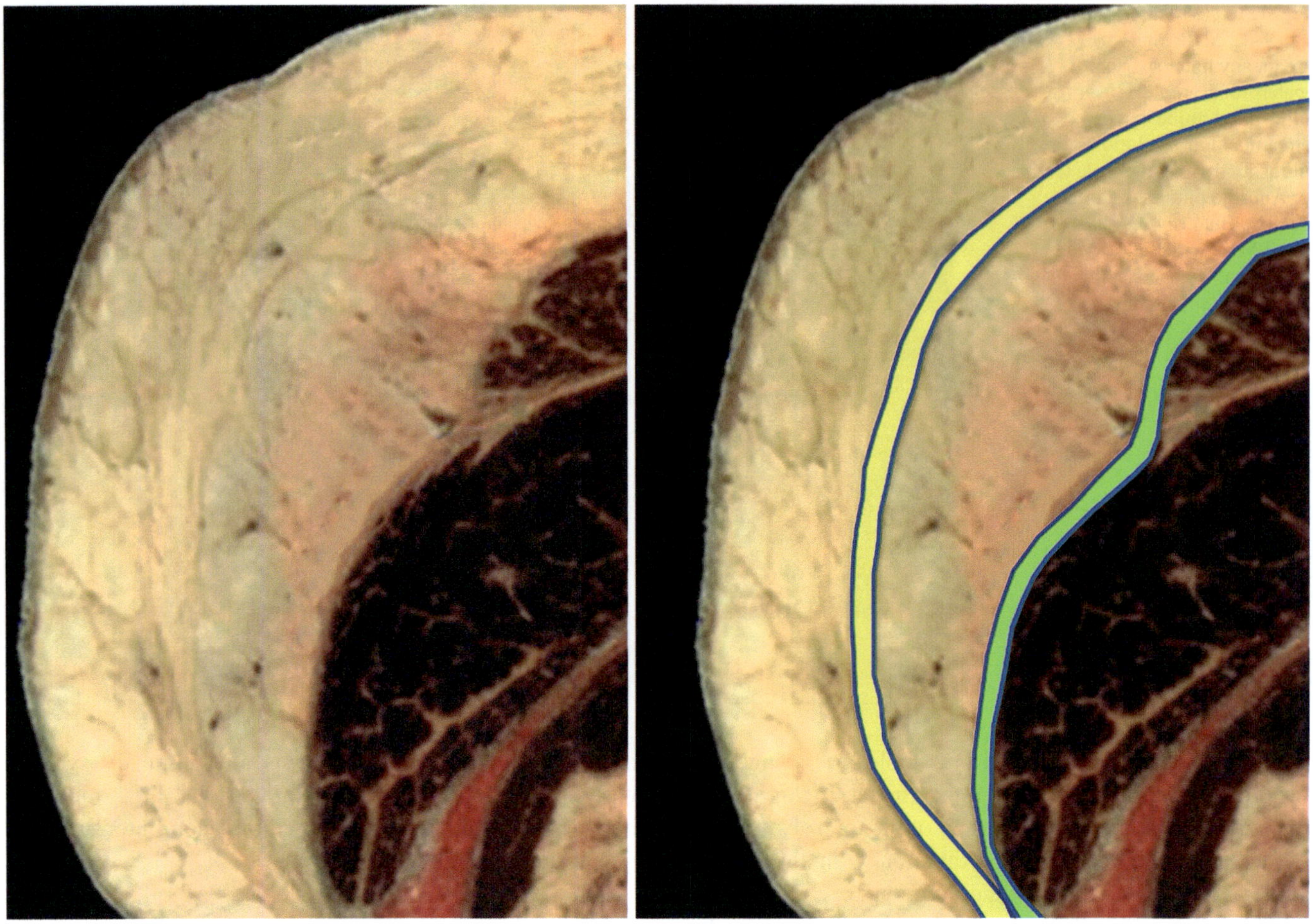

Fig. 15.9 Well-delineated superficial and deep gluteal fasciae

cutaneous space (between the dermis and the superficial gluteal fascia) and the deep subcutaneous space (between the superficial gluteal fascia and the deep gluteal fascia) [33] (Fig. 15.10). More important clinically, if the superficial gluteal fascia remains intact, it can retain the fat graft that is specifically injected above or below it, like the casing of a sausage. Fat graft injected into the deep subcutaneous space (above the deep gluteal fascia and below the superficial gluteal fascia) can create excellent volume and central dome projection, similar to a subfascial implant. Fat graft precisely injected into the superficial subcutaneous space (above the superficial gluteal fascia and below the skin) can correct superficial contour deformities and depressions. The consistent accurate injection of the fat graft into either the superficial or deep subcutaneous spaces can only be performed with real-time intraoperative ultrasound visualization.

It is also important to remember that the entire subcutaneous zone (including both the superficial and deep gluteal spaces) ranges in thickness from 1 cm (outer hips) to 3–4 cm (central gluteal dome). This thickness may also vary significantly between patients. This means that gluteal surgeons must graft in a thin space under a curving dome of varying thickness. This small variable target may account for the inadvertent deep intramuscular injections by well-intentioned surgeons performing gluteal fat grafting without ultrasound visualization.

A surgeon can use real-time intraoperative ultrasound to not only avoid an intramuscular fat graft injection but to accurately target fat graft into the superficial or deep subcutaneous spaces. Neither of these techniques are possible without ultrasound. Real-time intraoperative ultrasound during a Brazilian Butt Lift can not only make fat grafting safer but much more powerful and accurate as well.

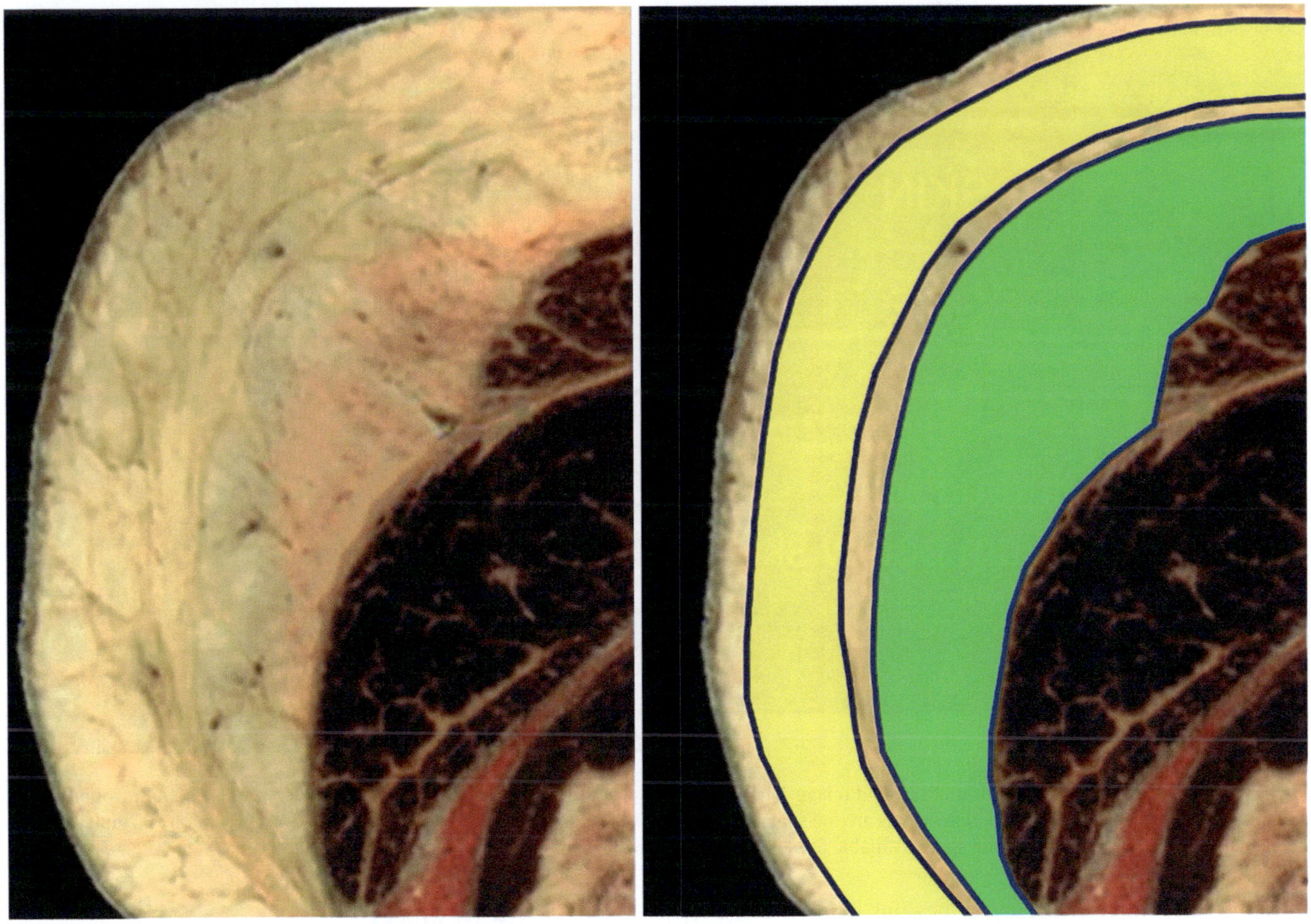

Fig. 15.10 The superficial (yellow) and the deep (green) gluteal subcutaneous spaces, divided by the superficial gluteal fascia

15.7 Real-Time Intraoperative Ultrasound Setup

Once the liposuction has been completed and the harvested fat collected, surgeons can add real-time intraoperative ultrasound visualization to their preferred fat graft injection technique. The following technique will be described using the EVL injection technique with the peristaltic pump for fat graft propulsion.

An Android or iOS tablet is mounted on an IV pole facing the gluteal area and the surgeon. The portable wireless ultrasound probe and a Bluetooth computer mouse are placed into a sterile probe cover (6″ × 48″ Soft Flex Probe Cover REF 20-PC648 Advance Medical Designs Marietta, GA) and brought onto the field. The ultrasound probe is placed on the skin over the first treatment area (central mound), and the sterile computer mouse is used to adjust the ultrasound probe's depth of field, gain, mode, and contrast. Then, recording of the ultrasound-guided fat grafting procedure begins.

The surgeon will hold the MicroAire handle and cannula with one hand and the ultrasound probe with the other to visualize the cannula tip and control the propulsion of the aqueous fat graft with a foot pedal. The order of gluteal fat grafting begins centrally to establish projection over the central gluteal dome, then laterally to the hips, and finally, into the supragluteal and medial gluteal regions. The cannula is inserted through the supragluteal or intergluteal incisions and advanced into the treatment zone until it is visualized by the ultrasound. To create significant central dome projection, the cannula tip is placed with ultrasound guidance in the deep subcutaneous space (under the superficial gluteal fascia and above the deep gluteal fascia), and fat graft is precisely injected. The cannula tip is always visualized, and care is taken never to place the cannula tip below the deep gluteal fascia, as this would result in an

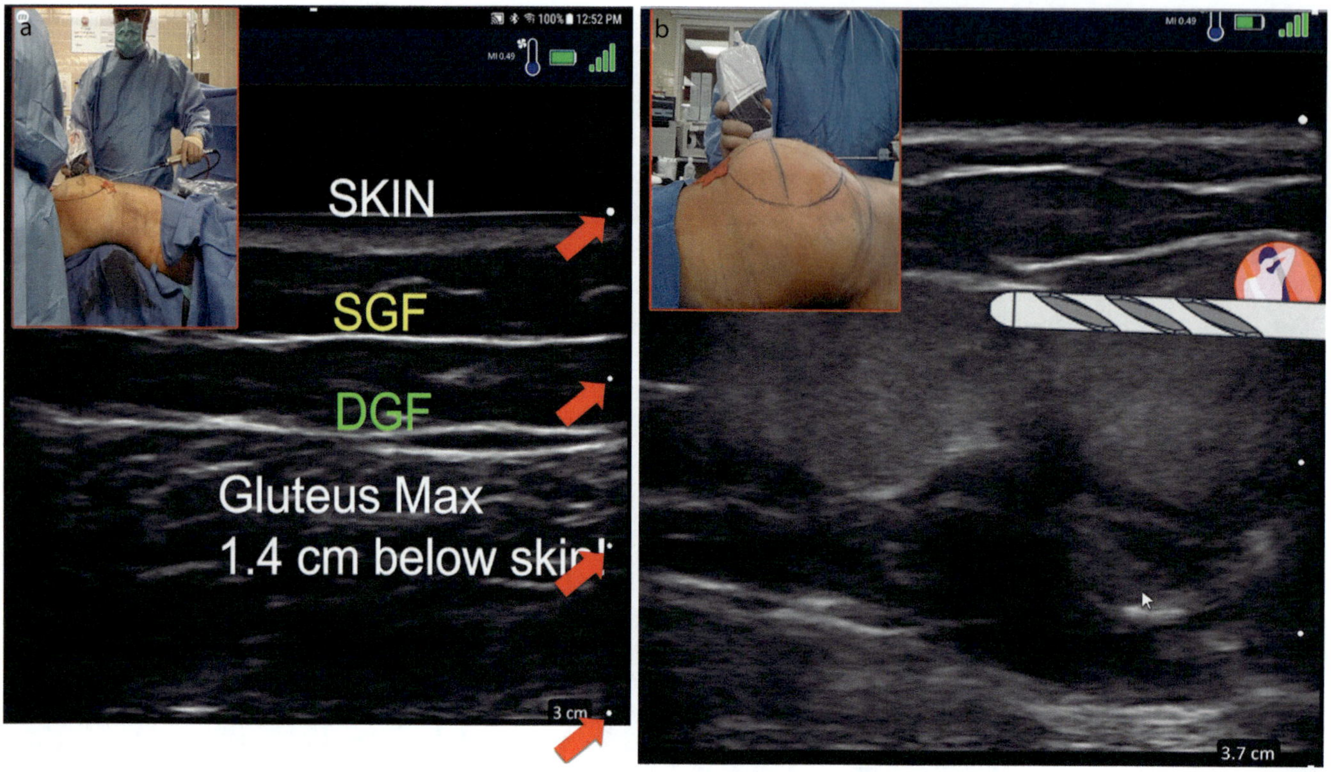

Fig. 15.11 (**a**) Typical, high-definition ultrasound image. The superficial and deep gluteal fascia is well delineated, as are the superficial and deep gluteal spaces. Note that the total space available for subcutaneous fat grafting measures only 1.4 cm in depth. (**b**) This ultrasound image displays the fat grating cannula in the deep subcutaneous gluteal space, just below the superficial gluteal fascia

intramuscular injection. Any adhesions within the deep subcutaneous space can be visualized and released, allowing for even distribution of the fat graft throughout this zone. To correct superficial skin depressions and asymmetries, the superficial subcutaneous space (above the superficial gluteal fascia and below the skin) is specifically addressed. Ultrasound-guided fat separation and release of adhesions are first performed in the superficial subcutaneous space, followed by controlled fat injection. Care is taken to keep the superficial gluteal fascia intact, as this will maintain separate deep and superficial subcutaneous spaces and prevent blowout irregularities. In this manner, the two anatomical subcutaneous spaces can be individually addressed in each anatomical area (Fig. 15.11a, b).

Real-time intraoperative ultrasound-guided fat grafting allows the surgeon to consistently avoid penetrating the deep gluteal fascia and prevent an inadvertent intramuscular fat graft injection. It also lets the surgeon accurately manipulate the structures of the subcutaneous region and precisely fat graft into the deep or superficial subcutaneous spaces. Ultrasound video of the entire fat grafting process can easily be created and saved to serve as definitive documentation that at no time was there an intramuscular injection. Real-time intraoperative ultrasound allows for precise fat grafting into the subcutaneous spaces and can keep the patient and the surgeon safe.

15.8 Patient Cases

15.8.1 Case #1

A 27-year-old, G0P0, African American female presents with lipodystrophy of the abdomen, waist, sides, flanks, and thighs and loss of volume, asymmetry and ptosis of the gluteal areas and hips, bilaterally. The patient was evaluated and on physical examination, and it was noted that she had significant lipodystrophy of the anterior abdomen, waist, flanks, lower back, and sacral area, and volume loss of the outer hips and gluteal ptosis.

The patient received an Ultra-BBL, a Brazilian Butt Lift with fat grafting under real-time intraoperative ultrasound visualization. The patient received fat separation without suction and fat extraction under suction of the abdomen, waist, flanks, lower back, and sacral areas. Care was taken to empty the waist and flanks and the suprasacral triangle concavity. She then received 1000 cc of fat graft per side. Next, 700 cc of fat graft was placed in the deep subcutaneous space (above the deep gluteal fascia and below the superficial gluteal fascia) to create gluteal volume, central dome projection, and supragluteal contour. Release of adhesions throughout the deep and superficial gluteal spaces at the outer hips was performed, taking care to leave the superficial gluteal fascia intact. She then received 300 cc of fat graft to the superficial subcutaneous space (above the superficial gluteal fascia and below the skin) at the outer hips, bilaterally. Further adhesion separation was performed after fat grafting to ensure even distribution of the fat graft in both spaces. The patient is shown 9 months postoperatively and is satisfied with the result (Fig. 15.12).

The patient's desired waist-to-hip ratio was achieved more by diminishing the circumference of the torso at the waist than by the addition of fat graft to the outer hips. Once the ratio was established, fat graft could be used to emphasize the point of maximum hip projection and create a smooth transition to the upper hip superiorly and to the thigh inferiorly. Using the ultrasound to measure the thickness of the subcutaneous zones in the outer thigh could aid in fat extraction and produce a more symmetric result.

15.8.2 Case #2

A 36-year-old, G1P1, African American female presents with lipodystrophy of the abdomen, waist, sides, flanks, and thighs, prominent flank creases, and loss of volume, asymmetry, and ptosis of the gluteal areas and hips, bilaterally.

The patient received an Ultra-BBL, a Brazilian Butt Lift with fat grafting under real-time intraoperative ultrasound visualization. The patient received fat separation without suction and fat extraction under suction of the abdomen, waist, flanks, lower back, and sacral areas. Care was taken to empty the waist and flanks and the suprasacral triangle concavity. VASER Liposelection was used above and below the flank creases and used to disrupt the flank creases, themselves. The patient received 800 cc of fat graft per side. First, 700 cc of fat graft was placed in the deep subcutaneous space (above the deep gluteal fascia and below the superficial gluteal fascia) to create gluteal volume, central dome projection, and supragluteal contour. She then received 100 cc of fat graft to the superficial subcutaneous space (above the superficial gluteal fascia and below the skin) at the outer hips, bilaterally. Further adhesion separation was performed after fat grafting to ensure even distribution of the fat graft in both spaces. The patient is shown 12 months postoperatively and is satisfied with the result (Fig. 15.13).

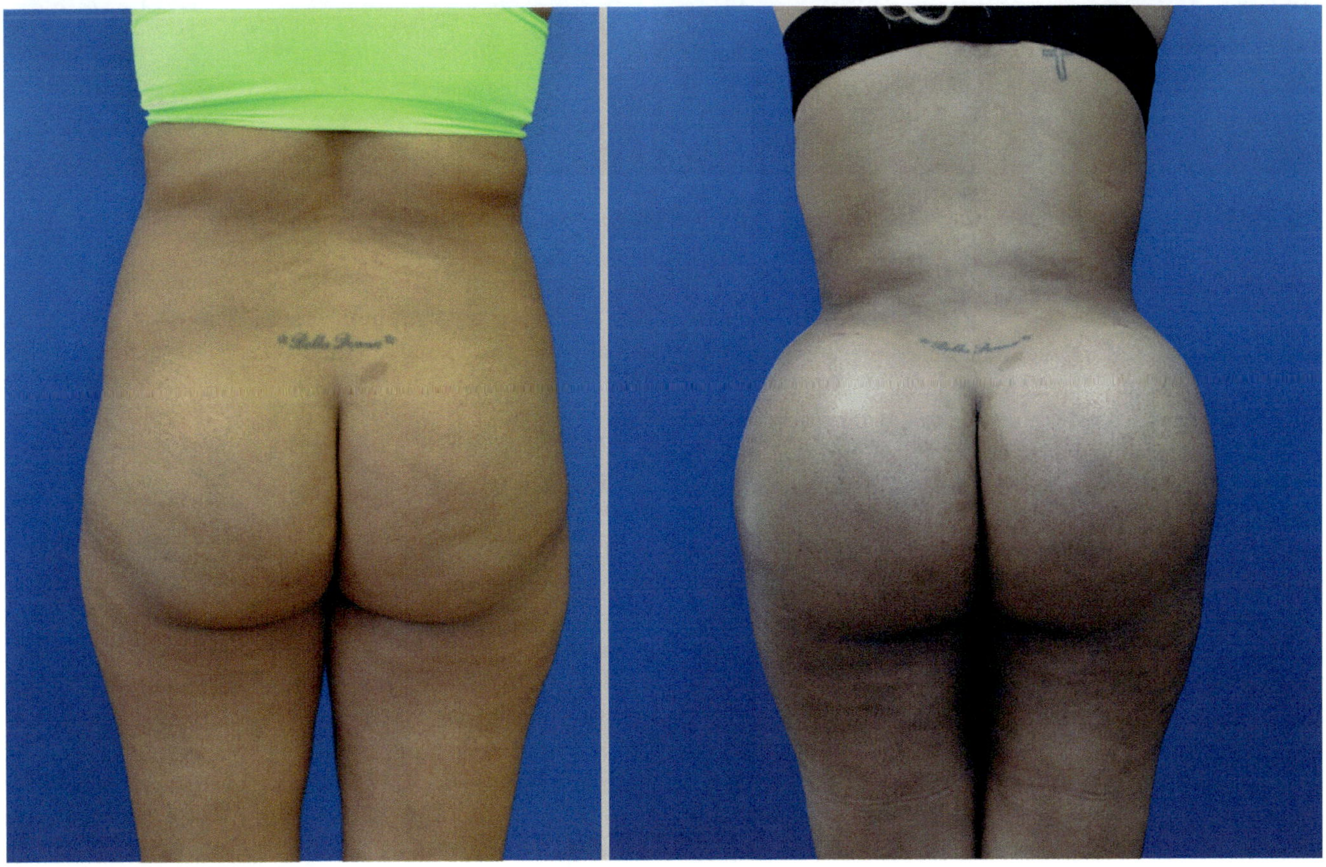

Fig. 15.12 Case 1. A 27-year-old female, 9 months post-Ultra-BBL procedure (1000 mL of fat grafted into each buttock)

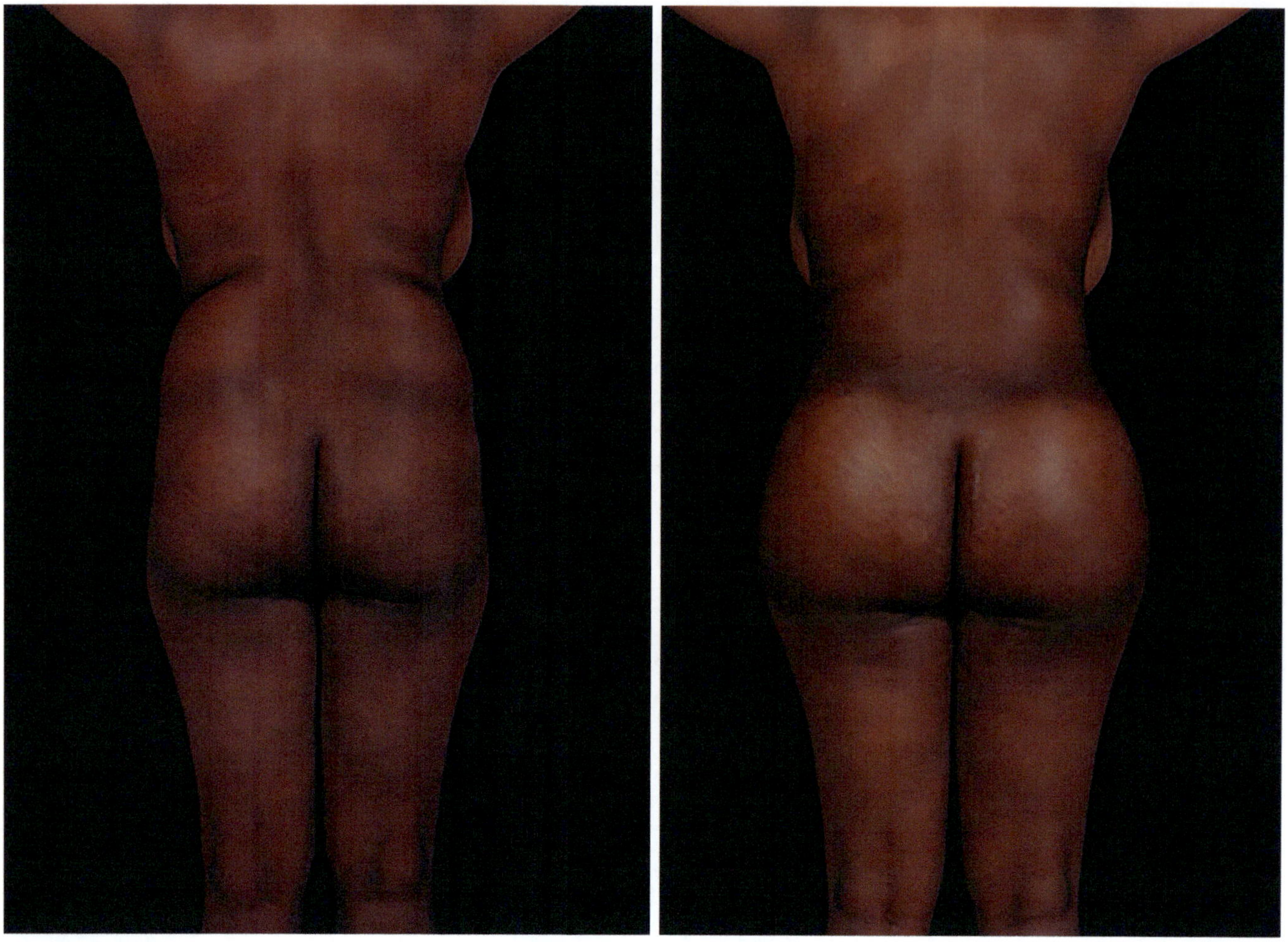

Fig. 15.13 Case 2. A 36-year-old female, 12 months post-Ultra-BBL procedure (800 mL of fat grafted into each buttock)

15.8.3 Case #3

A 28-year-old, G2P2, Caucasian female presents with lipodystrophy of the abdomen, waist, sides, flanks, and thighs, mild flank creases, and loss of volume, asymmetry and ptosis of the gluteal areas and hips, bilaterally.

The patient received an Ultra-BBL, a Brazilian Butt Lift with fat grafting under real-time intraoperative ultrasound visualization. The patient received fat separation without suction and fat extraction under suction of the abdomen, waist, flanks, lower back, and sacral areas. Care was taken to empty the waist and flanks and the suprasacral triangle concavity. The patient received 1000 cc of fat graft per side. First, 800 cc of fat graft was placed in the deep subcutaneous space (above the deep gluteal fascia and below the superficial gluteal fascia) for the creation of gluteal volume, central dome projection, and supragluteal contour. She then received 200 cc of fat graft to the superficial subcutaneous space (above the superficial gluteal fascia and below the skin) at the outer hips, bilaterally. Further adhesion separation was performed after fat grafting to ensure even distribution of the fat graft in both spaces. The patient is shown 16 months postoperatively and is satisfied with the result (Fig. 15.14).

15.8.4 Discussion

A 2020 article placed the mortality rate of abdominoplasty at 1/13,000 and that of BBL at 1/20,000 [34]. When the online fat grafting survey was repeated by ASERF in 2019 it was notable that most of the respondents were aware of the 2017 recommendations [16]. The great majority of the survey participants were aware of the dangers of intramuscular gluteal fat grafting and 84% of respondents were trying to inject fat with a subcutaneous only technique. The survey asked the respondent if they were aware of any BBL-associated deaths since the previous survey in 2017. Plastic surgeons participating in the survey reported on almost 30,000 patients and found two fatal Pulmonary Fat Emboli in the United States

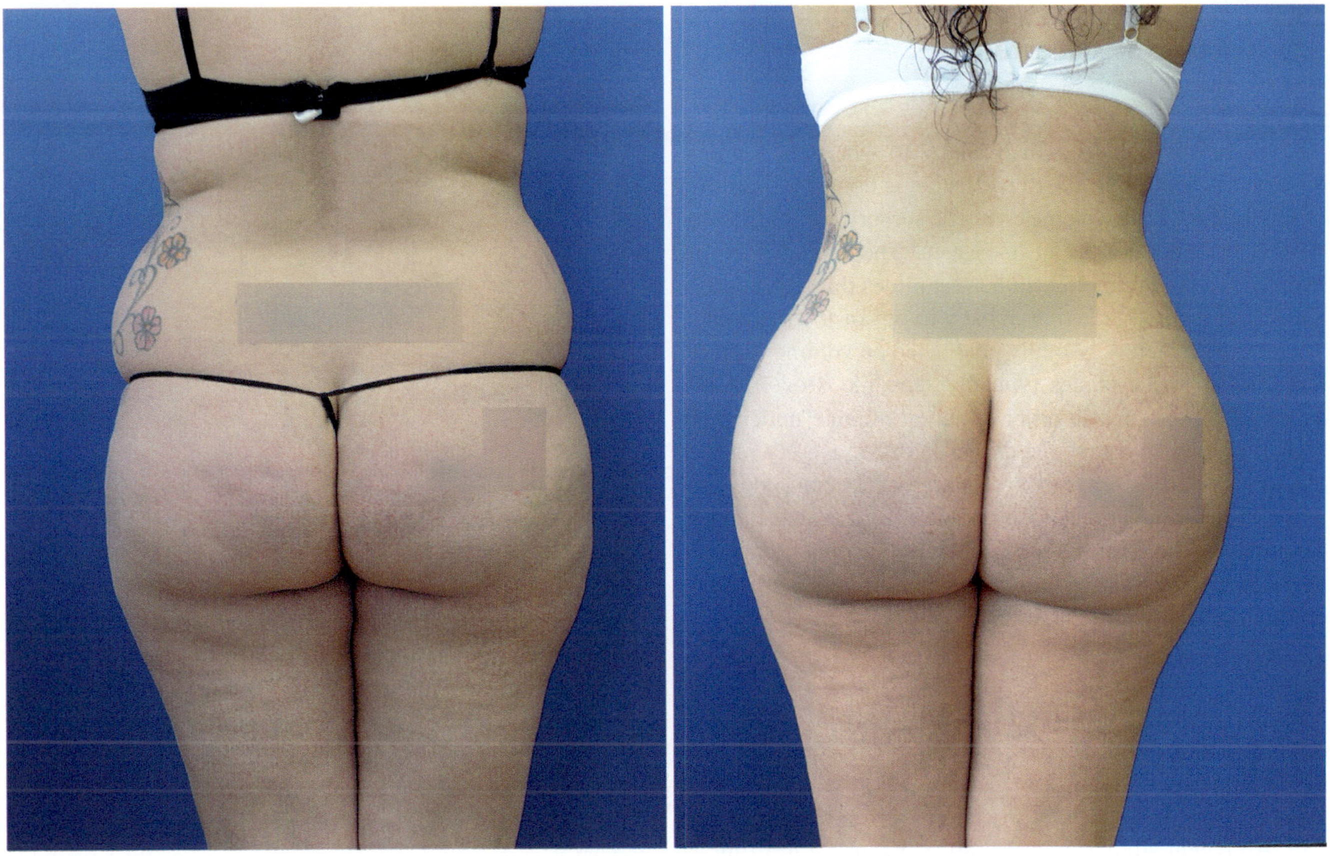

Fig. 15.14 Case 3. A 28-year-old female, 16 months post-Ultra-BBL procedure (1000 mL of fat grafted into each buttock)

and internationally during the 24 months between the surveys, (2017–2019) [16]. Since the survey data reported a BBL-related mortality of approximately 1/15,000, many felt that the safety profile of the procedure had surpassed that of abdominoplasty and that the problem maybe had been solved by surgeon education and the dissemination of safety guidelines. Unfortunately, this does not describe the South Florida experience where at least eight autopsies were documented as BBL-associated, pulmonary fat emboli, in that same time period. In 2020, Nahai cautioned the specialty on the danger of drawing definite conclusions on BBL-associated mortality based solely on anonymous surveys [35].

Although the autopsy-documented PFE deaths after BBL are startling numerator, it is difficult for us to describe the denominator in this ratio, as many of these cases are performed in low-budget, high-volume clinics that do not report their numbers, or the surgeon who performed the procedure. Nevertheless, the comparison to abdominoplasty is not favorable. According to the statistics from the Aesthetic Society, plastic surgeons perform abdominoplasty surgery at least four times more commonly than BBL surgery. Approximately 243,000 abdominoplasties were reported in the 2020–2021 statistics, while approximately 61,000 buttock augmentations were reported during the same period [36] (the buttock data included both fat grafting and implants). However, in South Florida, abdominoplasty-related deaths are rare in comparison to BBL-related deaths. For example, there have been 25 BBL-related deaths in Florida between 2010 and 2022. Applying the 4:1 ratio of abdominoplasties to BBLs reported in the Aesthetic Society statistics, one would expect approximately 100 abdominoplasty-related deaths in the area during that same period for the mortality rates to be comparable! There have been 12 BBL-related deaths in South Florida in the past 3 years (2019–2022). In a procedure with a higher reported mortality rate, which is performed at least four times more commonly than BBL surgery, one should expect approximately 48 deaths during that same 3-year period. The numbers are not even close. Abdominoplasty-related deaths in the area are most likely in the single-digit numbers over the same period, acknowledging the fact that some of the deaths were the result of diagnosed pulmonary emboli and were not pursued by the Coroner's Office. It is because of the current uncertainty regarding the actual number of BBL surgeries performed and the inability to collect valid data, that some authors have suggested creating a registry for BBL surgeries [37].

15.9 Why Is Florida an Outlier in BBL-Associated Mortality?

It is hard to dismiss a glaring commonality among these BBL-associated deaths, which relates to where the surgery took place. South Florida, in particular, is home to a number of low-budget, high-volume cosmetic surgery clinics, mostly owned and operated by businessmen. These clinics are in fierce competition to offer the lowest prices through internet and social media marketing, mostly to out-of-town patients. The pricing structure is totally predicated on volume surgery. Surgeons in these clinics typically schedule 8 or 10 BBL surgeries in one day. In order to stay profitable under their budget pricing structure, the clinics have to severely restrict the surgical times for their procedures. A recent expose on these clinics by National Geographic [38] revealed that the surgeon of record runs several operating rooms simultaneously, with the surgical assistants performing significant, critical portions of the surgeries.

15.10 Lessons Learned from the Florida BBL Experience

BBL surgery is a "blind" procedure performed by tactile feel. All recent studies point to subcutaneous fat grafting as the only safe technique to avoid Pulmonary fat emboli complications [39–43]. As previously mentioned, the autopsy findings in these BBL-associated deaths are not consistent with a "few inadvertent passes" of the grafting cannula into the muscle. Assuming the surgeons involved in these deaths were genuinely trying to inject only into the subcutaneous space, then the only logical explanation for the post-mortem findings is that they completely lost tactile perception as to the location of the tip of their injection cannula or were so inexperienced with the procedure that they could not recognize the proper anatomical space.

Common findings in the post-mortem gluteal dissections included a typical zone of injury adjacent to the gluteal vessels and numerous fat grafts injected at multiple levels within the muscle and in the submuscular space. The application of pressure to the infra-gluteal fold resulted in the flow of free fat grafts from the inferior muscle into the submuscular space during some of the dissections. (Fig. 15.15). This may explain why in several of the BBL deaths, the operating surgeons noted that the patient was arrested at the end of the surgery, shortly after turning the patient from the prone to supine position. This usually benign maneuver could place significant pressure on the lower buttocks allowing the free fat grafts to migrate beneath the muscle into the "danger zone."

In addition to the anatomical findings, the Florida experience has taught us that BBL-associated Macro Pulmonary Fat Emboli, is for all intents and purposes, incompatible with life (with the notable exception of the three reported outlier cases that required ECMO and significant intensive care). BBL-associated deaths at low-budget, high-volume clinics are related to relatively short surgical times, numerous cases attributed to a single surgeon in one day, and delegation of critical portions of the surgery to non-trained personnel.

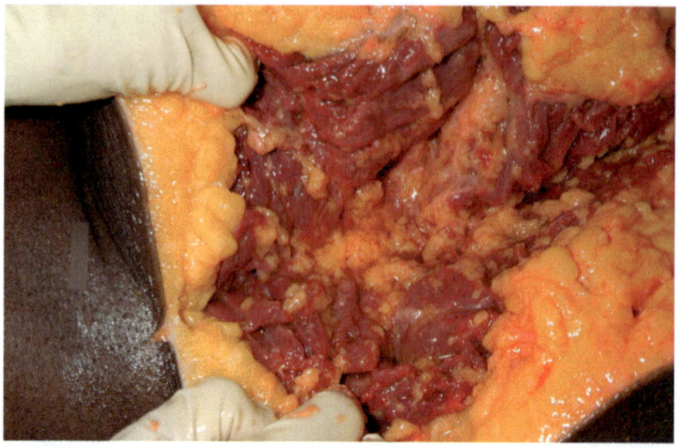

Fig. 15.15 Pressure on the inferior buttocks resulted in the rapid flow of free fat grafts into the submuscular space in BBL patients who had undergone extensive intramuscular fat grafting

The incidence of serious complications due to the lipo-harvest is significantly higher for BBL cases than for body contouring procedures performed only with liposuction. Emergency departments in South Florida have seen a sharp rise in serious complications from lipo-harvest. These include bowel perforations, bladder injuries, pneumothorax, liver lacerations, major vascular injuries, and acute anemic or hypovolemic episodes. The great majority of these serious BBL complications originate at low-budget, high-volume clinics.

15.11 A Way Forward

At the time of this writing, the Florida Board of Medicine is in the process of creating permanent rules to address BBL safety. Who should be performing these procedures and under what circumstances? Are plastic surgeons even capable of consistently staying in the subcutaneous tissue when fat grafting the buttocks? Several questions arise from the findings in our report. Will the FL BOM consider the procedure too risky and issue new mandates to curtail or ban BBL surgery in the state?

It does not set a good precedent for our specialty to have a state board of medicine dictate the procedures that we, as specialists, are allowed to perform. Both of the authors feel strongly that when this procedure is properly performed, it is associated with high patient satisfaction while maintaining an acceptable safety profile, equivalent to other common body contouring procedures. For that to happen, organized plastic surgery would need to take an active role in hands-on training for surgeons. The successful implementation of the Ultrasonic-Assisted Liposuction Task Force over 25 years ago serves as a good historical precedent in that respect [44]. Impedance cannulas have also been shown to be quite accurate and can be used as a valuable teaching tool [45].

Surgical technique in BBL surgery can also be easily monitored by hand-held intraoperative ultrasound to confirm the precise location of the injection cannula in "real-time" during the procedure. Ultrasound-Assisted BBL Surgery is gaining popularity among some experienced plastic surgeons. The procedure is extremely precise, safe, and easily taught to a trained plastic surgeon. The typical images have high resolution and are anatomically accurate (Fig. 15.16).

Recently, some published studies have suggested that the high volume of daily cases undertaken by surgeons at the low-budget, high-volume clinics, may be a significant factor in the high mortality associated with BBL surgery in South Florida [46]. The authors are in complete agreement that the extremely high volume of cases that the clinic surgeons are attempting on a daily basis is related to the high mortality. However, simply restricting the number of BBL surgeries allowed daily will not by itself solve the problem. Since significant portions of the BBL surgeries at the clinics are performed by the surgical assistants, the surgeon fatigue factor is not as much of an issue as for those individuals in private practice who personally perform their surgeries. Furthermore, it is the short anesthesia times for many of these cases that is truly alarming since they are indicative of surgical technique at a speed that is at best, imprecise and at worst, unsafe. Unfortunately, simply restricting the number of BBL surgeries per day will not translate to an increase in surgical times for these procedures. The low-budget, high-volume clinics are run under a business model that is predicated on volume surgery. Their current fee structure does not allow for a BBL surgery to take any longer, since the profit margin is so narrow that the clinic would lose money on those cases.

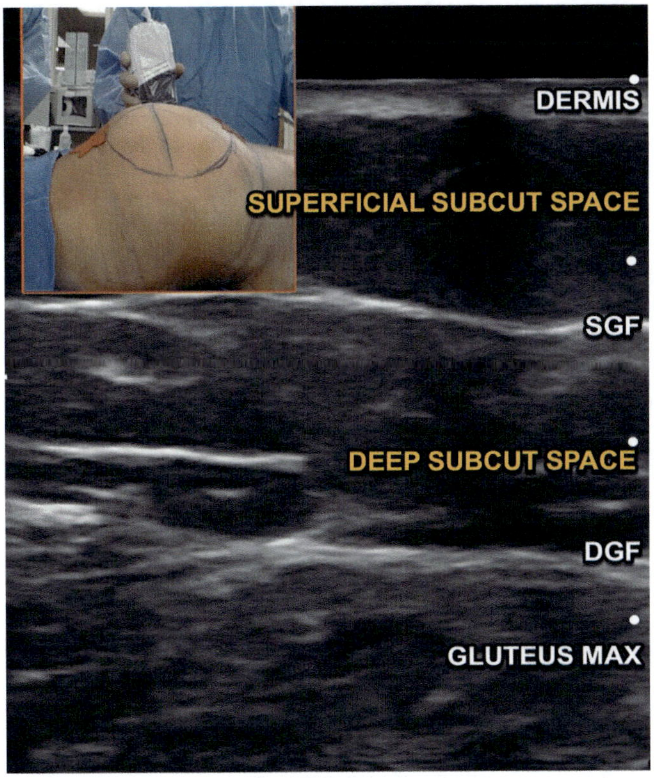

Fig. 15.16 High-definition ultrasound image displaying the dermis, superficial subcutaneous space, superficial gluteal fascia (SGF), deep subcutaneous space, deep gluteal fascia (DGF), and gluteus maximus muscle. Florida Board of Medicine regulations require that all surgeons performing BBL surgery utilize ultrasound guidance to ensure that all fat grafting is performed only above the deep gluteal fascia (DGF)

In June 2022, the FL BOM convened a special session to discuss the impact of their 2019 BBL "subcutaneous injection only" rule and evaluate the factors that have contributed to an increase in mortality since the rule went into effect. In addition to the previous BBL ruling, a new emergency rule was enacted that requires the use of ultrasound guidance for BBL surgery and will require the ultrasound video to be kept as a part of the patient's medical records. Also, as part of the new rule, the number of daily BBL surgery cases per surgeon will be restricted. This new FL BOM BBL rule is expected to go into effect June 14, 2022, almost 3 years to the date of the original "subcutaneous injection only" rule [47]. During this meeting, the FL BOM placed sanctions on the medical licenses of surgeons who performed BBL surgery, in patients who died from macro pulmonary fat emboli and whose autopsy reports revealed irrefutable evidence of intramuscular fat grafting.

15.12 Conclusions

The authors feel confident that we now know the events that lead to fatal complications in BBL surgery and the techniques necessary to avoid them.

As a result of the post-mortem gluteal dissections, the authors have been able to correlate commonalities in the anatomic findings with the mechanism of injury, resulting in macro fat emboli during BBL surgeries [48].

The BBL surgery experience is a prime example of how the working environment affects the surgical outcomes. As it pertains to this surgery, it is not only "who does it," but also "where it takes place," that directly affects the outcomes. The South Florida mortality data indicates that 92% of the BBL-associated deaths were related to low-budget, high-volume clinics, and 17 (68%) of those involved board-certified plastic surgeons working in that environment. It is apparent that board certification alone cannot overcome the unsafe working environment associated with the South Florida budget clinics.

It is incomprehensible that in the past 3 years since the ASERF recommendations were disseminated to all plastic surgeons and the FL BOM "subcutaneous injection only" mandate went into effect, the mortality for this procedure has significantly increased. However, except for the outlier surgeon, all the other BBL deaths originated at the low-budget, high-volume clinics. There has not been a single BBL-related death since the mandate attributed to a board-certified plastic surgeon, working in a traditional private practice or academic setting. The authors realize that South Florida is an outlier as it relates to BBL-associated mortality. We are confident that we have identified the root of the problem locally.

We are hopeful that the new FL BOM, BBL surgery rules, will make this surgery safer in our state. However, if the competition between the high-volume, low-budget clinics remains predicated solely on pricing, it is doubtful that the surgical times for BBL procedures at these places, will increase to more acceptable, safer levels.

All the major national and international plastic surgery societies (the Aesthetic Society, ASPS, ISAPS, IFATS) and our state society have supported the safety measures under consideration by the FL BOM. At the time of this writing, a small group of board-certified plastic surgeons in South Florida and the owners of the budget clinics have retained legal firms to challenge the BOM ruling on the grounds that it constitutes restraint of trade.

The vast majority of BBL patients are young, healthy women. Most have families and young children. It is unconscionable for us to allow the status quo to continue as it relates to BBL surgery in South Florida. This is a purely elective, aesthetic, surgical procedure, with an associated mortality rate that is on the rise, despite sound evidence that it can be performed safely, simply by employing the recommended techniques. The Florida Society of Plastic Surgeons fully supports the FL BOM emergency ruling and will continue to assist the BOM in creating a safer environment for BBL surgery patients in our state.

Acknowledgments Emma Lew MD, Chief Medical Examiner Dade County, for her assistance and support with the post-mortem gluteal dissections.

J Peter Rubin MD, for his exceptional leadership as principal investigator for the Gluteal Fat Grafting Task Force Cadaver Studies. Tough group to control!

References

1. Neuber F. Fettransplantation Bericht uber die Verhandlungen der Deutscht Gesellsch Chir. Zentralblatt fur Chirurgie, 22, 66. - References - Scientific Research Publishing; 1893. https://www.scirp.org/(S(oyulxb452alnt1aej1nfow45))/reference/ReferencesPapers.aspx?ReferenceID=1812808. Accessed 9 Oct 2022.
2. V Czerny Plastischer Ersatz der Brustdruse durch ein Lipom. Zentralbl Chir 1895;27:72. https://cir.nii.ac.jp/crid/1570572701408398848. Accessed 9 Oct 2022.
3. Plastic surgery of the face: based on selected cases of war injuries of the face including burns by Harold D. Gillies: Fair Hardcover (1920) 1st Edition | Springhead Books. https://www.abebooks.com/first-edition/Plastic-surgery-face-based-selected-cases/30996452102/bd. Accessed 9 Oct 2022.
4. Toledo LS. Gluteal augmentation with fat grafting: the Brazilian buttock technique: 30 years' experience. Clin Plast Surg. 2015;42(2):253–61. https://doi.org/10.1016/J.CPS.2014.12.004.
5. Society's singing a new tune: the derriere is where it's at—Chicago Tribune. https://www.chicagotribune.com/news/ct-xpm-2006-07-05-0607040287-story.html. Accessed 9 Oct 2022.
6. The Aesthetic Society. National databank aesthetic plastic surgery. Aesthet Surg J. Published online 2020.
7. Roberts TL, Weinfeld AB, Bruner TW, Nguyen K. "Universal" and ethnic ideals of beautiful buttocks are best obtained by autologous micro fat grafting and liposuction. Clin Plast Surg. 2006;33(3):371–94. https://doi.org/10.1016/J.CPS.2006.05.001.
8. Mendieta CG. Gluteal reshaping. Aesthet Surg J. 2007;27(6):641–55. https://doi.org/10.1016/J.ASJ.2007.09.001.
9. Ali A. Contouring of the gluteal region in women: enhancement and augmentation. Ann Plast Surg. 2011;67(3):209–14. https://doi.org/10.1097/SAP.0B013E318206595B.
10. Murillo WL. Buttock augmentation: case studies of fat injection monitored by magnetic resonance imaging. Plast Reconstr Surg. 2004;114(6):1606–14. https://doi.org/10.1097/01.PRS.0000138760.29273.5D.
11. Mendieta C, Stuzin JM. Gluteal augmentation and enhancement of the female Silhouette: analysis and technique. Plast Reconstr Surg. 2018;141(2):306–11. https://doi.org/10.1097/PRS.0000000000004094.
12. Turin SY, Fracol M, Keller E, et al. Gluteal vein anatomy: location, caliber, impact of patient positioning, and implications for fat grafting. Aesthet Surg J. 2020;40(6):642–9. https://doi.org/10.1093/ASJ/SJZ260.
13. Cárdenas-Camarena L, Bayter JE, Aguirre-Serrano H, Cuenca-Pardo J. Deaths caused by gluteal lipoinjection: what are we doing wrong? Plast Reconstr Surg. 2015;136(1):58–66. https://doi.org/10.1097/PRS.0000000000001364.
14. Mofid MM, Teitelbaum S, Suissa D, et al. Report on mortality from gluteal fat grafting: recommendations from the ASERF task force. Aesthet Surg J. 2017;37(7):796. https://doi.org/10.1093/ASJ/SJX004.
15. Mills D, Rubin JP, Saltz R. Urgent warning to surgeons performing fat grafting to the buttocks. Multi-society task force for safety in gluteal fat grafting. Published 2018. https://mail.google.com/mail/u/1/#search/urgent+warning+/FMfcgxvwzvPzKNXNrVFzJgDKNMmPBBrL. Accessed 9 July 2022.
16. Rios L, Gupta V. Improvement in Brazilian butt lift (BBL) safety with the current Recommendations from ASERF, ASAPS, and ISAPS. Aesthet Surg J. 2020;40(8):864–70. https://doi.org/10.1093/ASJ/SJAA098.
17. Cárdenas-Camarena L, Durán H, Robles-Cervantes JA, Bayter-Marin JE. Critical differences between microscopic (MIFE) and macroscopic (MAFE) fat embolism during liposuction and gluteal Lipoinjection. Plast Reconstr Surg. 2018;141(4):880–90. https://doi.org/10.1097/PRS.0000000000004219.
18. Garcia O. Commentary on: the potential role of corticosteroid prophylaxis for the prevention of microscopic fat embolism syndrome in gluteal augmentations. Aesthet Surg J. 2020;40(1):90–2. https://doi.org/10.1093/ASJ/SJZ196.

19. Wang HD, Zheng JH, Deng CL, Liu QY, Yang SL. Fat embolism syndromes following liposuction. Aesthet Plast Surg. 2008;32(5):731–6. https://doi.org/10.1007/S00266-008-9183-1.
20. Levy D. The fat embolism syndrome. A review. Clin Orthop Relat Res. 1990;261(261):281–6. https://doi.org/10.1097/00003086-199012000-00034.
21. Habashi NM, Andrews PL, Scalea TM. Therapeutic aspects of fat embolism syndrome. Injury. 2006;37(Suppl 4):S68. https://doi.org/10.1016/J.INJURY.2006.08.042.
22. Bayter-Marin JE, Cárdenas-Camarena L, Aguirre-Serrano H, Durán H, Ramos-Gallardo G, Robles-Cervantes JA. Understanding fatal fat embolism in gluteal Lipoinjection: a review of the medical records and autopsy reports of 16 patients. Plast Reconstr Surg. 2018;142(5):1198–208. https://doi.org/10.1097/PRS.0000000000004904.
23. Peña W, Cárdenas-Camarena L, Bayter-Marin JE, et al. Macro fat embolism after gluteal augmentation with fat: first survival case report. Aesthet Surg J. 2019;39(9):NP380–3. https://doi.org/10.1093/ASJ/SJZ151.
24. Wolfe EM, Weber LE, Wo LM, et al. Two cases surviving macro fat emboli complications following gluteal fat grafting. Aesthet Surg J. 2022;42(8):902–6. https://doi.org/10.1093/ASJ/SJAC063.
25. 22065771, 64B8ER19-1—Florida Administrative Rules, Law, Code, Register—FAC, FAR, eRulemaking. https://www.flrules.org/Gateway/View_notice.asp?id=22065771. Accessed 10 Oct 2022.
26. del Vecchio DA, Villanueva NL, Mohan R, et al. Clinical implications of gluteal fat graft migration: a dynamic anatomical study. Plast Reconstr Surg. 2018;142(5):1180–92. https://doi.org/10.1097/PRS.0000000000005020.
27. Wall S, Delvecchio D, Teitelbaum S, et al. Subcutaneous migration: a dynamic anatomical study of gluteal fat grafting. Plast Reconstr Surg. 2019;143(5):1343–51. https://doi.org/10.1097/PRS.0000000000005521.
28. Rapkiewicz AV, Kenerson K, Hutchins KD, Garavan F, Lew EO, Shuman MJ. Fatal complications of aesthetic techniques: the gluteal region. J Forensic Sci. 2018;63(5):1406–12. https://doi.org/10.1111/1556-4029.13761.
29. Cansancao AL, Condé-Green A, Vidigal RA, Rodriguez RL, D'Amico RA. Real-time ultrasound-assisted gluteal fat grafting. Plast Reconstr Surg. 2018;142(2):372–6. https://doi.org/10.1097/PRS.0000000000004602.
30. del Vecchio D, Wall S. Expansion vibration lipofilling: a new technique in large-volume fat transplantation. Plast Reconstr Surg. 2018;141(5):639e–49e. https://doi.org/10.1097/PRS.0000000000004338.
31. Shermak MA. Body contouring. Plast Reconstr Surg. 2012;129:6. https://doi.org/10.1097/PRS.0B013E31824ECD24.
32. Mendieta CG. The art of gluteal sculpting. Published online 2011:379.
33. Stecco C, Hammer W, Vleeming A, de Caro R. Subcutaneous tissue and superficial fascia. In: Functional atlas of the human fascial system. Published online January, Vol. 1, pp. 21–49; 2015. doi:https://doi.org/10.1016/B978-0-7020-4430-4.00002-6.
34. Rohrich RJ, Savetsky IL, Avashia YJ. Assessing cosmetic surgery safety: the evolving data. Plast Reconstr Surg Glob Open. 2020;8(5):2643. https://doi.org/10.1097/GOX.0000000000002643.
35. Nahai F. No "Quick Fix" for this: an update on the Brazilian Butt Lift. Aesthet Surg J. 2020;40(8):928–30. https://doi.org/10.1093/ASJ/SJAA117.
36. The Aesthetic Society. Aesthetic plastic surgery National Databank 2021, vol. 42. Aesthet Surg J; 2022. p. 1–18.
37. Nahai F. Acceptable risk: who decides? Aesthet Surg J. 2017;37(7):852–3. https://doi.org/10.1093/ASJ/SJX087.
38. Watch trafficked with Mariana van Zeller Season 2 Episode 1 black market surgery online. https://www.nationalgeographic.com/tv/shows/trafficked-with-mariana-van-zeller/episode-guide/season-02/episode-01-black-market-surgery/vdka25475912. Accessed 10 Oct 2022.
39. Turer DM, Rubin JP. Commentary on: safe gluteal fat graft avoiding a vascular or nervous injury: an anatomical study in cadavers. Aesthet Surg J. 2019;39(2):185–6. https://doi.org/10.1093/ASJ/SJY302.
40. O'Neill RC, Hanson SE, Reece E, Winocour S. Safety considerations of fat grafting in buttock augmentation. Aesthet Surg J. 2021;41(Suppl 1):S25–30. https://doi.org/10.1093/ASJ/SJAB092.
41. Garcia O, Pazmiño P, Turer D, Rubin J. Buttock fat grafting insights from the multi-society task force (cadaver demo). Presented at: Plastic Surgery, The Meeting 2018; October 1, 2018, Chicago, IL. In: Buttock Fat Grafting Insights from the Multi-Society Task Force (Cadaver Demo). Presented at: Plastic surgery, the meeting 2018; October 1, 2018, Chicago, IL. ASPS; 2018.
42. Kalaaji A, Dreyer S, Vadseth L, Maric I, Jönsson V, Haukebøe TH. Gluteal augmentation with fat: retrospective safety study and literature review. Aesthet Surg J. 2019;39(3):292–305. https://doi.org/10.1093/ASJ/SJY153.
43. Pane TA. Experience with high-volume buttock fat transfer: a report of 137 cases. Aesthet Surg J. 2019;39(5):526–32. https://doi.org/10.1093/ASJ/SJY191.
44. Fredricks S. Analysis and introduction of a technology: ultrasound-assisted Lipoplasty task force. Clin Plast Surg. 1999;26(2):187–204. https://doi.org/10.1016/S0094-1298(20)32604-3.
45. Turer DM, Qaium EB, Lawrence AM, Clark WW, Rubin JP. A smart sensing cannula for fat grafting. Plast Reconstr Surg. 2019;144(2):385–8. https://doi.org/10.1097/PRS.0000000000005866.
46. del Vecchio D, Kenkel JM. Practice advisory on gluteal fat grafting. Aesthet Surg J. 2022;42(9):1019–29. https://doi.org/10.1093/ASJ/SJAC082.
47. Florida Board of Medicine. 64B8-9.009: Standard of Care for Office Surgery—Florida Administrative Rules, Law, Code, Register—FAC, FAR, eRulemaking. Florida Administrative Code & Florida Administrative Registrar. Published July 12, 2022. Accessed 8 Oct 2022. https://www.flrules.org/gateway/ruleno.asp?id=64B8-9.009
48. Pazmiño P, Garcia O. Brazilian butt lift–associated mortality: the South Florida experience. Aesthet Surg J. Published online August 11, 2022. doi: https://doi.org/10.1093/ASJ/SJAC224.

Buttock Reshaping: Principles and Techniques Using Vaser Device and Fat Grafting

16

Alberto Di Giuseppe and Federico Giovagnoli

Contents

16.1 Introduction

Concepts about gluteal shaping have changed radically in the last 20 years. Up to the year 2000, the majority of girls and women were looking for a reduction of large flanks, large buttocks, love handles, etc. (Fig. 16.1)

Liposuction was invented as a technique for excess fat removal. At that time, no consideration about gluteal shaping or an integrated approach to sculpting the body figures was understood. So surgical reduction of the "bumpy" areas was the main goal of any liposuction or body contouring procedure.

The ideal shape was a small round buttock, with no major curves or projection. Since the beginning of the century, new concepts about buttock shape and volume have been imported from South America, especially from Brazil.

In those areas of the world, buttock shape, projection, and roundness have always been very important features of female appearance.

Also, culture from African countries played a role in the change of perspective in buttocks ideal contour (Fig. 16.2).

Fig. 16.1 Ideal shape was a small round buttock, with no major curves or projection along the 1980s and 1990s

From Fig. 16.2, the dimension of the buttocks and the overall volume is much bigger than the tiny figures in models and fashion in the 1980s and 1990s.

An important role has been played also by celebrities such as Kim Kardashian, whose profile introduced new ideals in body frame: tiny waist line, large curved lateral buttocks with significative projection.

A. Di Giuseppe (✉)
Department of Plastic surgery, University of Padova, Padova, Italy

F. Giovagnoli
University Vita-Salute San Raffaele, Milan, Italy

© The Author(s), under exclusive license to Springer Nature Switzerland AG 2023
A. Di Giuseppe et al. (eds.), *Fat Transfer in Plastic Surgery*, https://doi.org/10.1007/978-3-031-10881-5_16

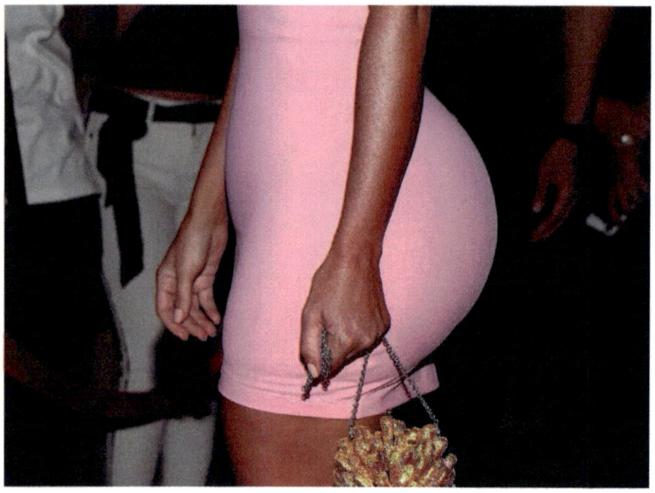

Fig. 16.2 African buttocks

Those new dimensions are better achievable by African bodies and South American bodies, due to the initial characteristics of the body frame. But the new concept and ideals have been exported all over the world, and become new parameters for many youngers, who have the tendency to copy celebrities and new fashion tendencies.

Buttocks enhancement was achieved with implants in the last 30 years, mainly starting in Brazil and Mexico, with Gonzalez-Ulloa [1], Robles [2], and Vergara [3]. The technique utilizes silicone-shaped implants of different dimensions, which are positioned submuscular, intramuscular, or subfascial in the buttock area.

Minding some variability of aesthetic results, in not experienced surgeons, the technique has never reached vast popularity among the occidental regions. Therefore, I reserve buttocks implants only for thin patients, where it is impossible to achieve good quantity of fat from donor areas.

The introduction of extended fat transfer from multiple body areas, combined with shaping and sculpturing techniques, has gained popularity with patients and plastic surgeons.

Growing demand for the so-called Brazilian buttock or Brazilian lift has hit the market worldwide.

16.2 Frame Evaluation

A better understanding of the buttocks anatomy and derived shape came with the book *The Art of Gluteal Sculpting* by Dr Mendieta from Miami [4].

Frame analysis includes:

1. Asymmetry (mild, moderate, and severe).
2. Shape (V, round, square or A): will determine the areas where liposuction is indicated.
3. Point C represents the degree of depression (none, mild, moderate, and severe) to decide if fat transfer is needed. Skin laxity and wrinkling must be evaluated as well if present at points A, B or C.
4. Sacral height must be in proportion with the intergluteal crease length and should be shorter. Eventually, this will require liposuction in order to define the V zone (Figs. 16.3, 16.4, 16.5, 16.6).

Those are the most important parameters to be considered for properly planning gluteal sculpting.

This is a 3D procedure similar to what is already presented with thigh contouring. It means adding volume, increasing buttocks size, and enhancing shape and projection. The final goal is to achieve more harmonic buttocks, reshaping the frame, improving the transition zones between the muscle and the body frame, equalizing the volume in all four quadrants, changing the muscle shape when indicated, and finally increasing or enlarging the buttock, but only as the last step of the procedure.

The volume of fat transfer for buttock enhancement is considerably higher than those used by most surgeons initially. In my experience, fat grafting in a small framed patient will require around 350–450 cc of tissue. Those numbers will increase to 700–900 cc in a medium-frame patient.

Larger patients could require 1000–1500 cc of fat for a harmonic result.

Fat transferred is needed for reshaping, recontouring (50%), and proper augmentation (50%). The first portion of fat to be transferred is utilized to correct the defects and depressions and to achieve a better aesthetic final frame.

Fat removal or liposuction technique is variable: SAL (Suction-Assisted Lipoplasty) was used initially by surgeons. Blood loss and surgical trauma to tissue are higher with this technique, and potential complications as well (loose skin, minor skin retraction and depression).

Ultrasound-assisted liposuction (VASER) allows similar volumes of fat removal and harvesting with a minor incidence of complication rate.

Body-jet liposuction seems to preserve 10–20% more fat than the other modalities.

Power-assisted liposuction, radiofrequency, or laser-assisted liposuction all seem to lose around 10% of the transferred fat. Minding the quality of the fat (more or less fibrotic), only between 60% and 70% of the total aspirated

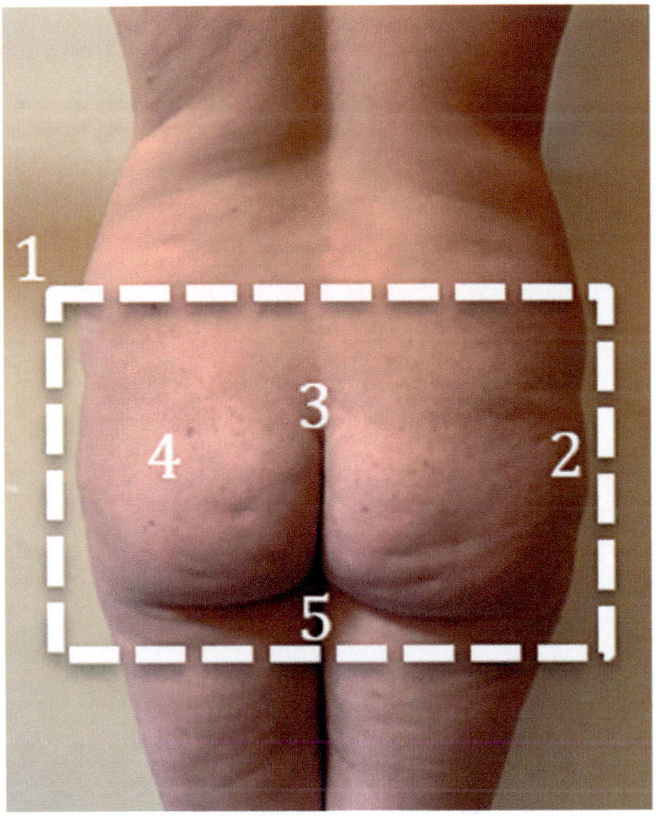

Fig. 16.3 Square buttocks. Shape (1), Point C (2), V zoze (3), gluteal muscle (4), lower inner gluteal leg junction (5)

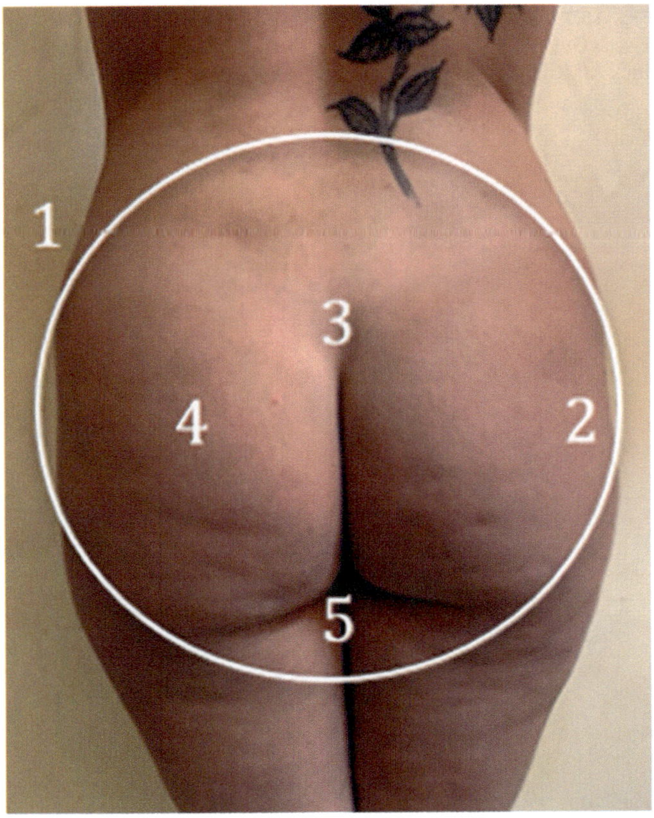

Fig. 16.4 Round buttocks with asymmetry. Shape (1), Point C (2), V zoze (3), gluteal muscle (4), lower inner gluteal leg junction (5)

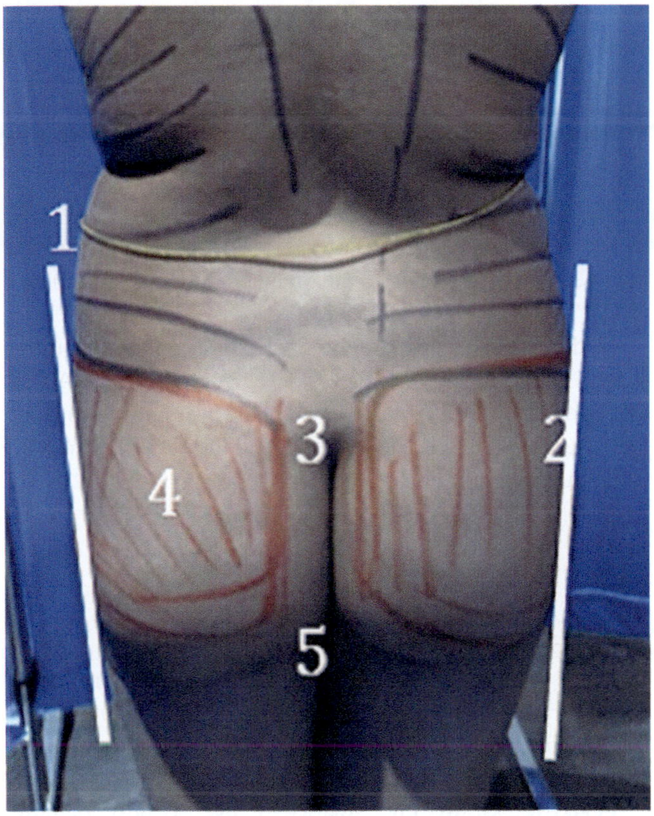

Fig. 16.5 V shaped buttocks. Shape (1), Point C (2), V zoze (3), gluteal muscle (4), lower inner gluteal leg junction (5)

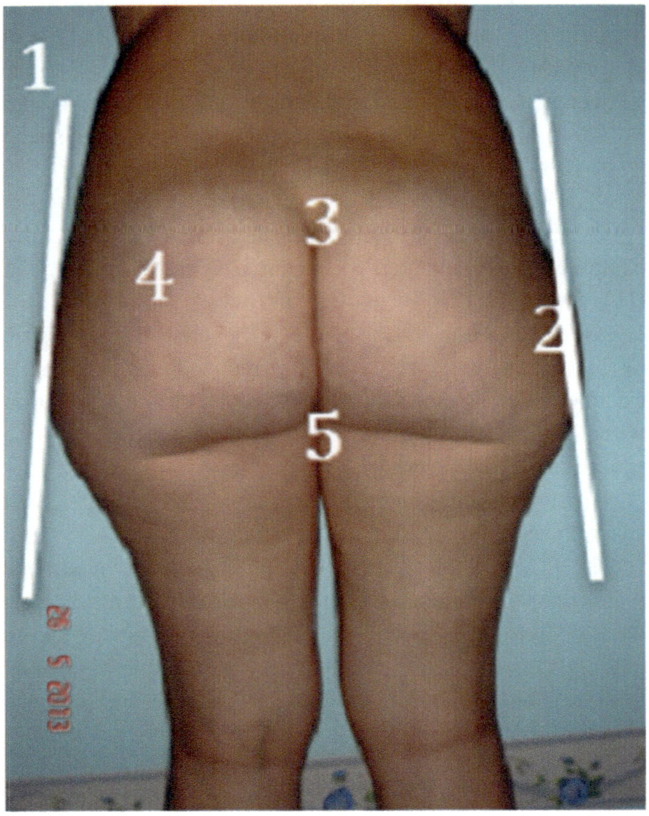

Fig. 16.6 A shaped buttocks. Shape (1), Point C (2), V zoze (3), gluteal muscle (4), lower inner gluteal leg junction (5)

will be turned into transferable fat. This means that 1000 and 4000 cc of supernatant fat will turn into 600 and 2500 cc of good fat.

16.3 Technology

In my experience, I have always used ultrasound-assisted liposuction, being one of the pioneers in this technology [5–7].

In the 1990s, I did use the SMEI first-generation ultrasound device. However, since 2001, I started using VASER third-generation ultrasound device, and I am still using the latest VASER system, the PowerX, and the VASER smooth™ for cellulite treatment.

16.3.1 Parameters

- VASER: In my practice, I mostly used VASER 3.7 mm 3 and 5 ring probes for emulsification and standard cases. Power setting is normally 80% of the total in the thighs and buttocks area. I use VentX cannula 3.7 mm with the full power of aspiration (Fig. 16.7).
- POWER X: In my practice I use Triport cannula 3 mm × 20 cm, VentX SST-6 cannula 3.7 mm × 26 cm, Mercedes cannula 4 mm × 30 cm. This is what I use for delicate tissue or in pre-treated fat or fat harvesting; for deep medium to soft tissue or for rapid debulking, I use the 4 or 5 mm Mercedes cannula. The rotation angle is normally 180 degrees for superficial treatment or 360 for deeper. Rotation speed in deeper layer is between 340 and 420 rpm; in superficial layer, I use a slower rotation speed of 180–260 rpm (Fig. 16.8).

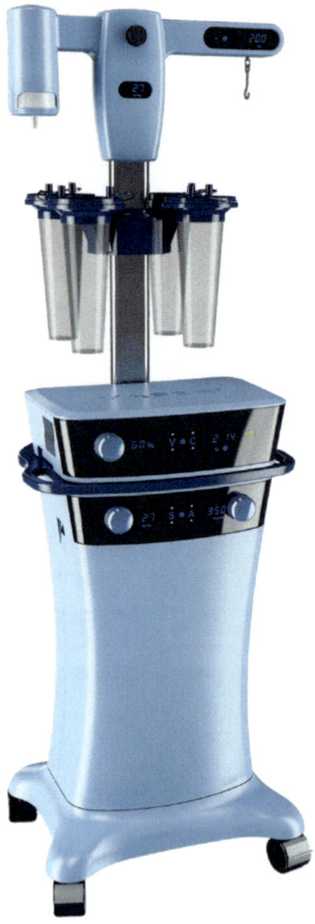

Fig. 16.7 VASER system. Photo taken from VASER Solta Medical User's Guide [8]. Courtesy of Bausch, ©2022 Bausch Health Companies Inc., All Rights Reserved

VASER system functioning has already been fully explained in the chapter "Ultrasound-Assisted Liposuction (UAL) with VASER Technology in Body Contouring."

PowerX is a power-assisted device where different degree of rotation and movement is used to reduce fat cells from the treatment area. PowerX can be used alone or in combination with VASER lipo during the evacuation part of the procedure.

The two devices can be used together to better define and shape the body and extract fat for micro fat graft.

The PowerX lipo system utilizes a powered handpiece that causes the attached cannula to rotate based on predetermined angle and speed settings, allowing physicians to quickly debunk large areas of fat. The unique rotational motion generated by the system and multiple user settings enable optimal control during the procedure and reduce aspiration time and physician fatigue.

Rotational motion of the cannula allows physicians to remove fatty tissue all the way around the cannula, rather than just above or below the orifice.

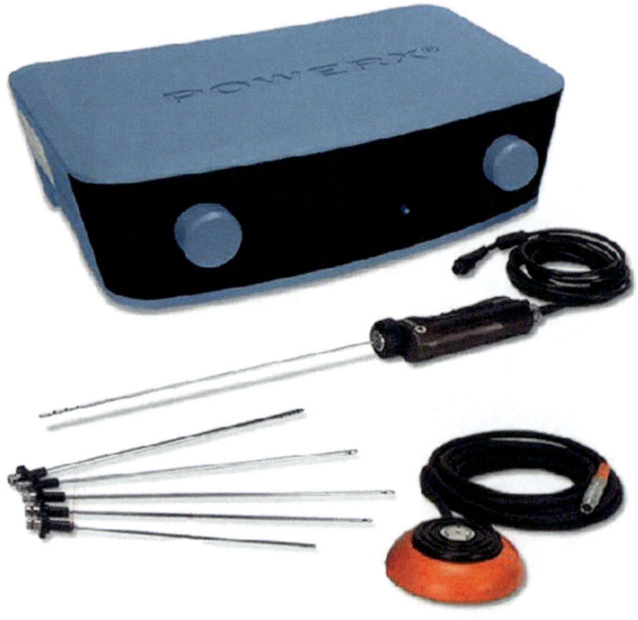

Fig. 16.8 PowerX system. Note rotation system from 90 to 360 degrees for better fat removal and sculpting. Photo taken from VASER Solta Medical User's Guide [8]. Courtesy of Bausch, ©2022 Bausch Health Companies Inc., All Rights Reserved

As in similar systems, a powered handpiece speeds up traditional liposuction cases and reduces physician fatigue.

I did test other systems on the market, and my impression is that this one has minimal handpiece vibration.

VentX handpiece provides continuous, uninterrupted suction and prevents tube clogging.

In addition to speeding up the suction process, this reduces fat cell trauma if the aspirate is harvested for fat grafting procedures.

A wide range of settings allows tailoring the system as desired. Superficial work is normally performed with a 180 degrees rotation angle. This is reserved for the high-definition technique.

More aggressive large-volume fat removal is achieved using a 360 degrees rotation angle in the deeper fat tissue.

The PowerX lipo system utilizes a rotational movement of the cannula to quickly and effectively treat fatty areas while minimizing trauma to the surrounding tissue. A surgeon has control over the rotation angle (how far the cannula rotates), and the rotation speed (how fast the cannula rotates).

In deeper tissue, a higher rotation angle and speed allow for efficient debulking of larger areas. For precision sculpting, a lower rotation angle and speed can be chosen to treat more superficial layers.

The PowerX lipo system also includes a specific Jog mode that allows the physician to orient the openings of any cannula tip to a desired starting point or "home position" prior to the start of the rotation. This ensures that the cannula orifice is positioned correctly inside the body to reduce contour irregularities and allow for more superficial sculpting applications.

16.4 Cellulite

Since 2015, I have introduced the VASER smooth™ device for the treatment of cellulite, mainly on the buttocks, in my practice.

As well known, two are the major causes of cellulite: the first is a hardening of the connective tissue support system of fat. These are referred to as connective tissue (fibrous) septae. These are made of collagen. Over time, these septae harden and contract. The contraction results in most of the dimpled appearance that creates cellulite. Additionally, the contraction leads to a blockage of the blood vessels and lymphatic system. The blood vessel blockage further hardens the septae. The lymphatic blockage leads to thick, swollen appearing skin. The second major cause of cellulite is protrusion of fat cells into the lower part of the skin known as the reticular dermis. This protrusion results in the worsening of the dimpling.

16.4.1 Hormonal Factors

Hormones play a dominant role in the formation of cellulite. Estrogen may be the important hormone to initiate and aggravate cellulite. However, there has been no reliable clinical evidence to support such a claim. Other hormones, including insulin, adrenaline and noradrenaline, thyroid hormones, and prolactin, are all believed to participate in the development of cellulite.

16.4.2 Genetic Factors

There is a genetic element in individual susceptibility to cellulite. Researchers have traced the genetic component of cellulite.

16.4.3 Predisposing Factors

Several factors have been shown to affect the development of cellulite. Sex, race, biotype, distribution of subcutaneous fat, and predisposition to lymphatic and circulatory insufficiency have all contributed to cellulite.

16.4.4 Lifestyle

A high stress lifestyle will cause an increase in the level of catecholamines, which have also been associated with the evolution of cellulite. Certain dieting practices can also diminish the level of these chemicals, and decreasing body fat typically results in the appearance of cellulite diminishing.

The VASER smooth Kit is an accessory to the VASER lipo system and is intended for the fragmentation and emulsification of subcutaneous fatty tissues for aesthetic body contouring. This includes the treatment of superficial fatty tissue and fibrous septae responsible for uneven skin appearance.

Objective is to enable the surgeon to emulsify fat, cut the bands ultrasonically, and allow the surgeon to lipo-shift the viable VASER fat into the treated area, smoothing out the affected area (Fig. 16.9).

The orientation of the handpiece is designed to help the surgeon to know where the fragmentation edge is located.

The probe expressively designed for cellulite treatment is the V-probe (showed in Fig. 16.10a), which I personally introduced in the past for gynecomastia treatment. This helps break the fibro-septa emerging from the underlying muscle or from the fascia layer.

The second probe (showed in Fig. 16.10b) is designed to emulsify with the tip and the lateral part of the shaft and to break the fibrotic superficial septa with the medium part of the shaft, which presents a pointed side, designed to cut instead of emulsify (showed in Fig. 16.10c).

In my personal technique, I normally use a 3.7 mm 2-ring Vaser probe, the Vaser power set at 70–80% of the total power, by working in a continuous mode.

For beginners, it should be advisable to drop the power to 60% and using the VASER mode initially to prevent burns or skin necrosis, when working close to the subcutaneous tissues.

In my experience, with more than 8000 body contouring procedures, never experienced such a kind of complication.

Regarding power X, I set up the angulation at 180 degrees and at 260 rpm (rotation per minute speed).

VASER smooth device must be used only by a real experienced surgeon, as it is designed as a special probe, which emulsify on the top of the shaft while it cut the fibrotic zone of adherence of the cellulitis with a central part of the probe.

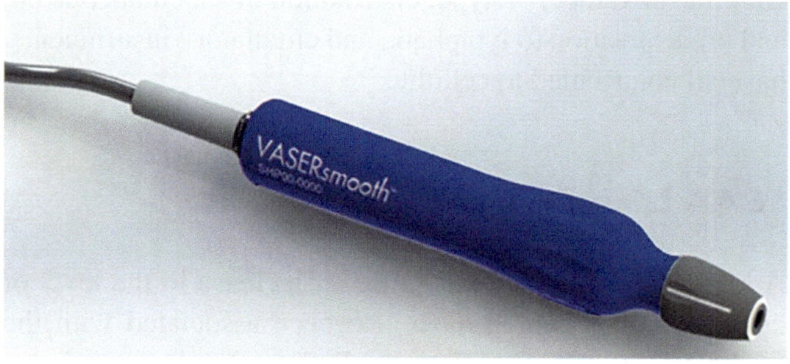

Fig. 16.9 VASER smoothTM handpiece. Photo taken from VASER Solta Medical User's Guide [8]. Courtesy of Bausch, ©2022 Bausch Health Companies Inc., All Rights Reserved

This cutting edge must be manipulated with great care: some small perforator vessels could emerge from the underlying fascia, which vascularize part of the thigh, which the cutting probe could sever.

Thus causing problems of vascularization of the thigh (Fig. 16.10).

I present two cases of accomplishing VASER lipo, cellulite treatment with VASER smooth, and fat transfer to C point (lateral trochanter depression).

Case 1

The first patient is a 43-year-old woman presenting a severe degree of cellulite, body asymmetry with different positions of gluteal fold, and lipoditrophy. Buttock frame A moderate with C point depression. Surgical plan: reshaping lateral flanks with VASER lipo, cellulite treatment with VASER smooth, and fat transfer to C point at the gluteal depression (Fig. 16.11).

Case 2

The second patient is a 43-year-old woman whose buttock frame is square, cellulite is diffused all over the buttocks, the lateral thighs and lateral flanks are too much represented, and the C point presents moderate depression (Fig. 16.12).

As a comment, VASER smooth has much-improved skin surface quality and appearance in initial and medium degree of cellulites, in buttocks and thighs. The double action of the VASER smooth probe allows a gentle breaking of the fibro-septa and, simultaneously, a lipo-shifting of the emulsified fat, which prevents the recurrence of the skin contracture and smooths the surface and the contour of the treated areas.

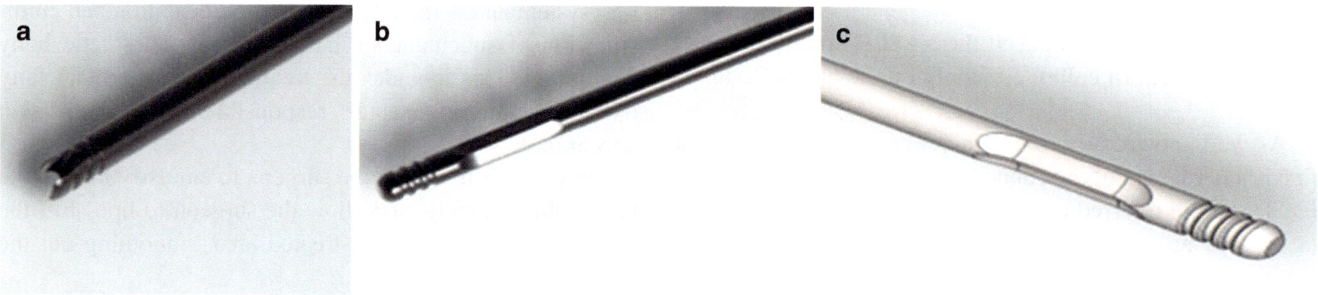

Fig. 16.10 VASER smoothTM probes. (**a**) 4.5 mm 2-groove (V-probe). (**b**) 3.7 mm 3-groove. (**c**) Side style rendering. Photo taken from VASER Solta Medical User's Guide [8]. Courtesy of Bausch, ©2022 Bausch Health Companies Inc., All Rights Reserved

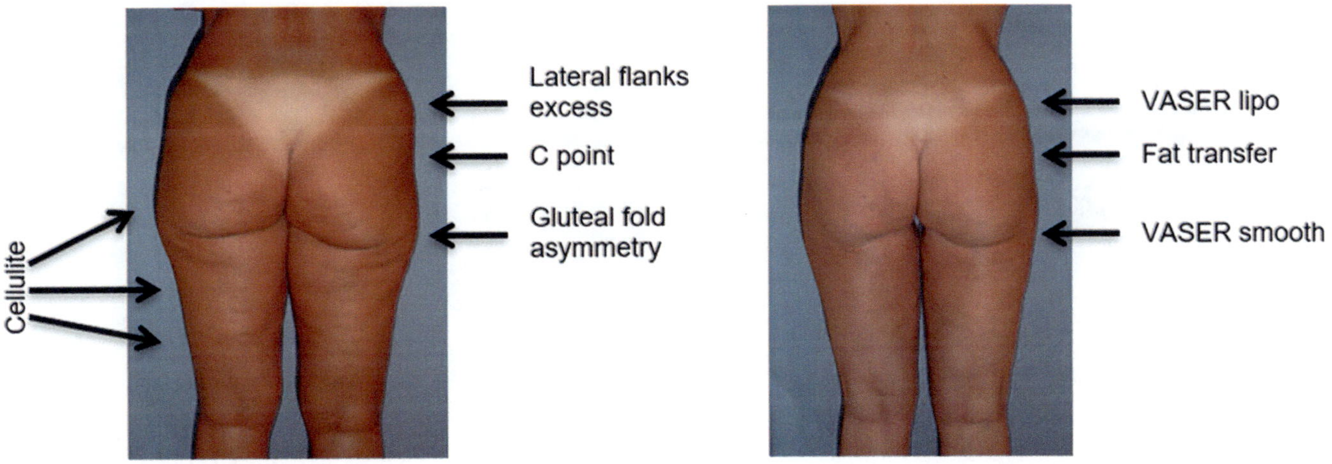

Fig. 16.11 Pre and 6 months post op

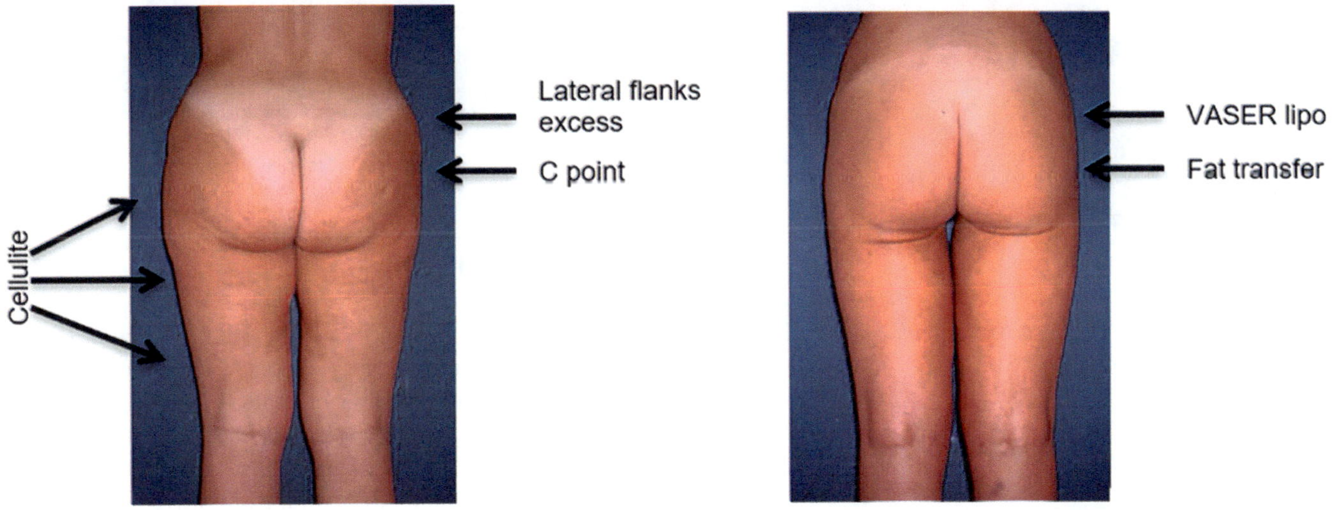

Fig. 16.12 Pre and 6 months post op

16.5 Fat Harvesting

With my technique, I emulsify fat with VASER liposuction; then I use the REVOLVE™ system to wash, centrifuge, and concentrate fat, removing oil, triglycerides, blood, and fluid excess. Extra fluid is collected in a standard disposable canister, or a washable canister can treat emulsified fat (Fig. 16.13).

I already reported about the functioning of the REVOLVE™ system in the chapter "Breast Fat Augmentation: Step-by-Step Technique and Ultrasound Contouring of Donor Area" (Fig. 16.14).

Fat ready to be transferred is collected in 60 mL syringes and left in a rack for further decantation. To follow, fat is transferred with a connector to 10 mL syringes, in order to decrease depression on the fat cells over infiltration (Fig. 16.15).

I use a 3 mm cannula with one hole for fat deposition in a multilayer fashion. The consistency of fat harvested with the REVOLVE™ system is thicker and does contain less fluid: as a consequence, it is more difficult to be transferred, as less aqueous, but it stays longer and gives less post-op swelling.

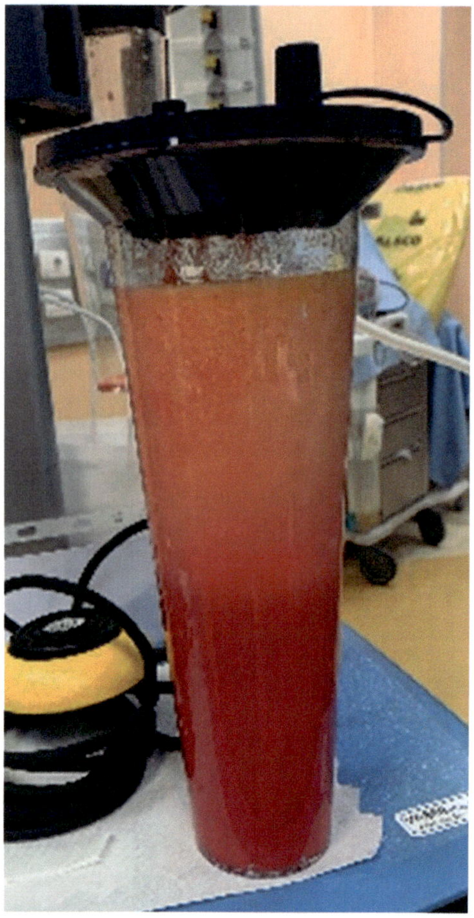

Fig. 16.13 Canister

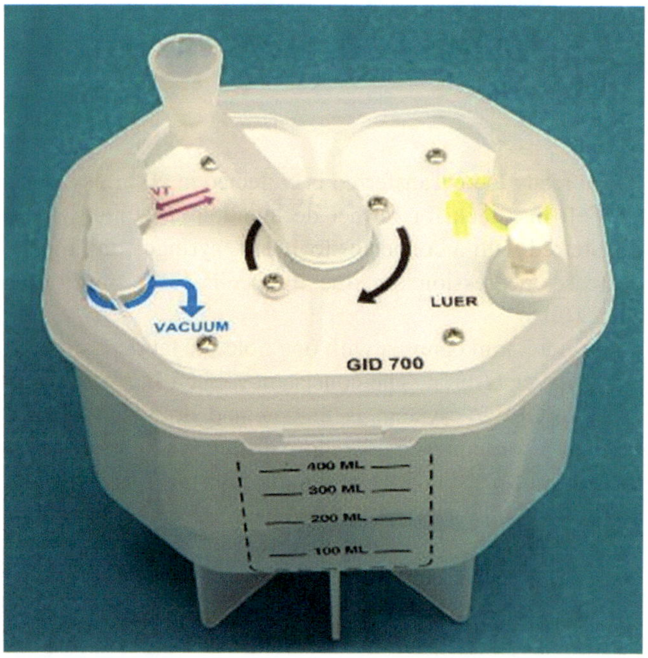

Fig. 16.14 REVOLVE™ system

Fig. 16.15 60 mL syringes ready for transferring

16.6 Gluteal Sculpting

Case 3: Medium-Framed Patient with a Sufficient Fat
A 45-year-old woman, A shape buttock, with good anatomy of the gluteal region, dimension, and volume. She presents carency on the C point, excess in the lateral flanks and mild lipodystrophy of the trochanter. The surgical plan will enhance lower lateral buttock volume, improve the transition zone, fill the C point, and reduce flanks and torso.

VASER plus REVOLVE™ were used. A total of 750 cc of fat were placed to better define the lateral contour of the gluteal region and mildly enhance the total projection of the muscle. Due to age, the patient required better contouring and shape defects without affecting buttock volume enhancement.

One-year post-op result shows a change into a more rounded A frame (Figs. 16.16 and 16.17).

Case 4: Medium-Framed Patient with a Sufficient Fat
A 36-year-old woman, round shape buttock, with asymmetry of flanks and trochanter height. The patient complained mainly about a flat buttock shape, looking for enhancement and projection.

VASER liposuction to abdomen, flanks, torso, banana fold, and anterior thighs and REVOLVE™ system to harvest 800 cc of total fat, 350 cc on each buttock for augmentation and 100 cc and more used for reshaping and contouring.

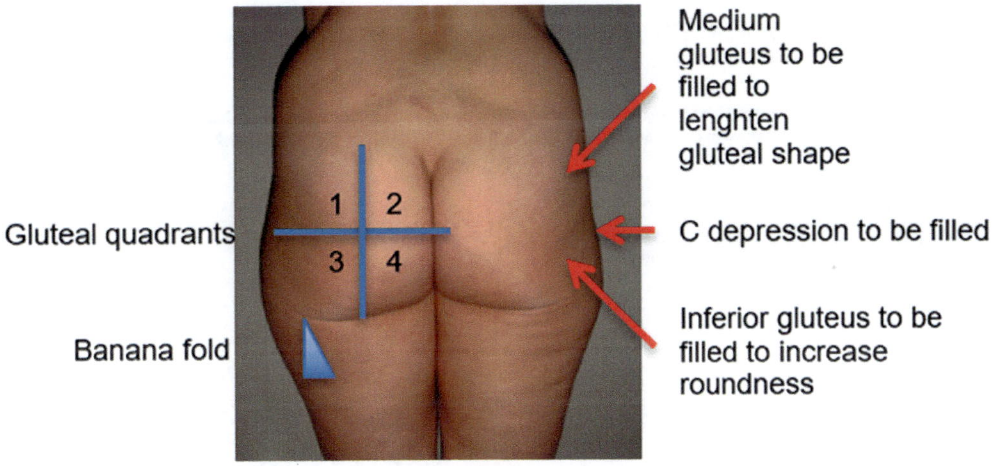

Fig. 16.16 Analysis and planning of buttock contouring

One-year post-op result shows a similar round frame with better lateral buttock projection (Fig 16.18).

Case 5: Medium-Framed Patient with Large Amount of Fat

A 25-year-old patient, square shape buttock, 74 kg weight, complaining flat buttocks, back rolls and abdominal and flanks excess. She does present short buttocks, no C point depression, but the total height of the gluteal muscle is deficient. Surgical plan: VASER liposuction to torso, flanks, upper and lower abdomen. 1400 cc of total fat harvested, 700 per side, to lengthen buttocks and achieve more roundness.

One-year post-op result shows an A round frame achieved (Fig. 16.19).

Case 6: Medium-Framed Patient with Sufficient Amount of Fat

A 39-year-old patient with a round frame, complaining of back rolls and looking for better definition and more roundness of buttocks. VASER liposuction to torso, flanks, and abdomen. A total of 1200 cc of fat was harvested. 600 cc for each side for reshaping, recontouring, and enhancing projection.

One-year post-op result (Fig. 16.20).

Case 7: Medium-Framed Patient with Large Amount of Fat

A 24-year-old woman, round frame, complaining of cellulites, large unshaped buttocks, 3D thigh excess, and large waist.

Surgical plan was: VASER liposuction of torso, flanks, and anterior thighs. VASER smooth for cellulites treatment buttocks and posterior legs. Fat transfer to C depression, to lower and medium buttocks for contouring, reshaping, and enhancing.

1000 cc of fat transferred (Figs. 16.21 and 16.22).

Case 8

A 38-year-old patient, with slim frame but unhappy about the contouring of flanks, poor definition and flat buttocks. Droopy buttocks, and banana fold deformity.

The surgical plan included flanks and abdomen definition, liposuction to harvest fat and fat transfer to the buttocks to increase shape and contouring—more definition of the abdominal area to enhance athletic shape (Fig. 16.23).

Case 9

A 44-year-old patient, two pregnancies, overweighted, with bulge and apron of the abdominal wall. On the back, rolls, heavy flanks, back rolls and torso laxity. Complaining also for flat and unshaped buttocks. Surgery consisted of VASER lipo of upper and lower abdomen, back rolls, and tors. A total of 4500 mL was aspirated. 600 mL of fat infiltrated each buttock. Pre-op photos, 6 months post-op photos, and photos taken at the end of surgery are presented. Note the great skin retraction of the abdominal skin which involved mainly the lower abdominal apron and bulge, avoiding the need for an abdominoplasty. In addition, superficial VASER was performed to enhance skin tightening capabilities, obtaining a unique effect of skin tightening (Fig. 16.24).

Case 10

A 33-year-old girl, 65 kg weight, 165 cm height for stubborn fat deposit in the upper and lower abdomen and hip deficiency. From the analysis of the body frame, there is a lack of convexity line of the oblique view of the patient abdomen-hips relationship. Heavy flanks and hip indentation are the oblique features of this patient. From the back view, heavy flanks are connected with back rolls, and the axillary area lacks definition. Regarding buttocks, they look flat and not rounded: the body frame clearly indicates the need for hip compensation with fat and enhancement of the fullness and projection of both buttocks (Figs. 16.25 and 16.26).

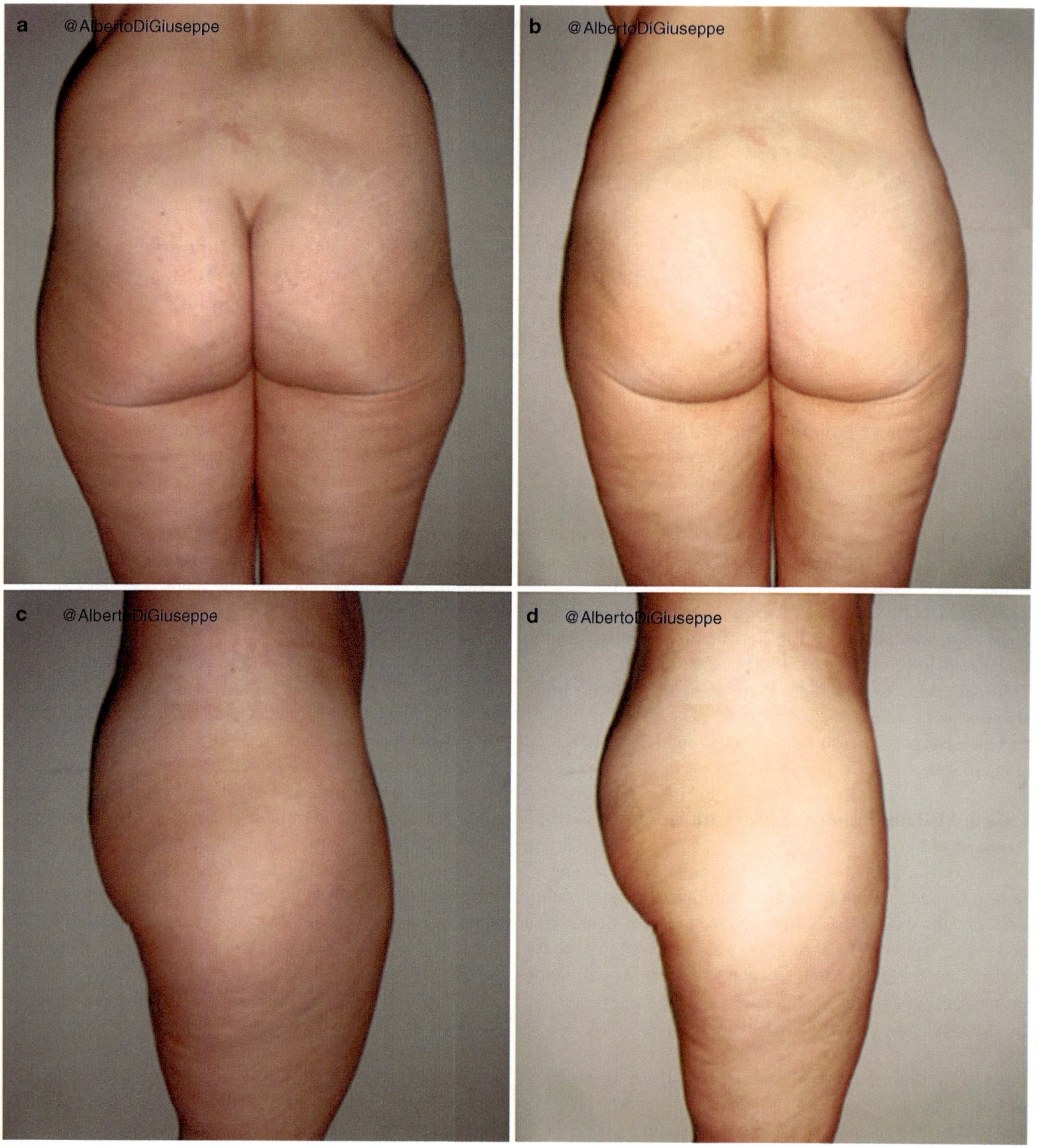

Fig. 16.17 45-year-old woman, A shape buttock. VASER to flanks, thighs, and buttocks. 750 cc of total fat transfer (C point, transition zone, buttock lateral and medium regions). Pre op (**a**, **c**). 1 year post op (**b**, **d**)

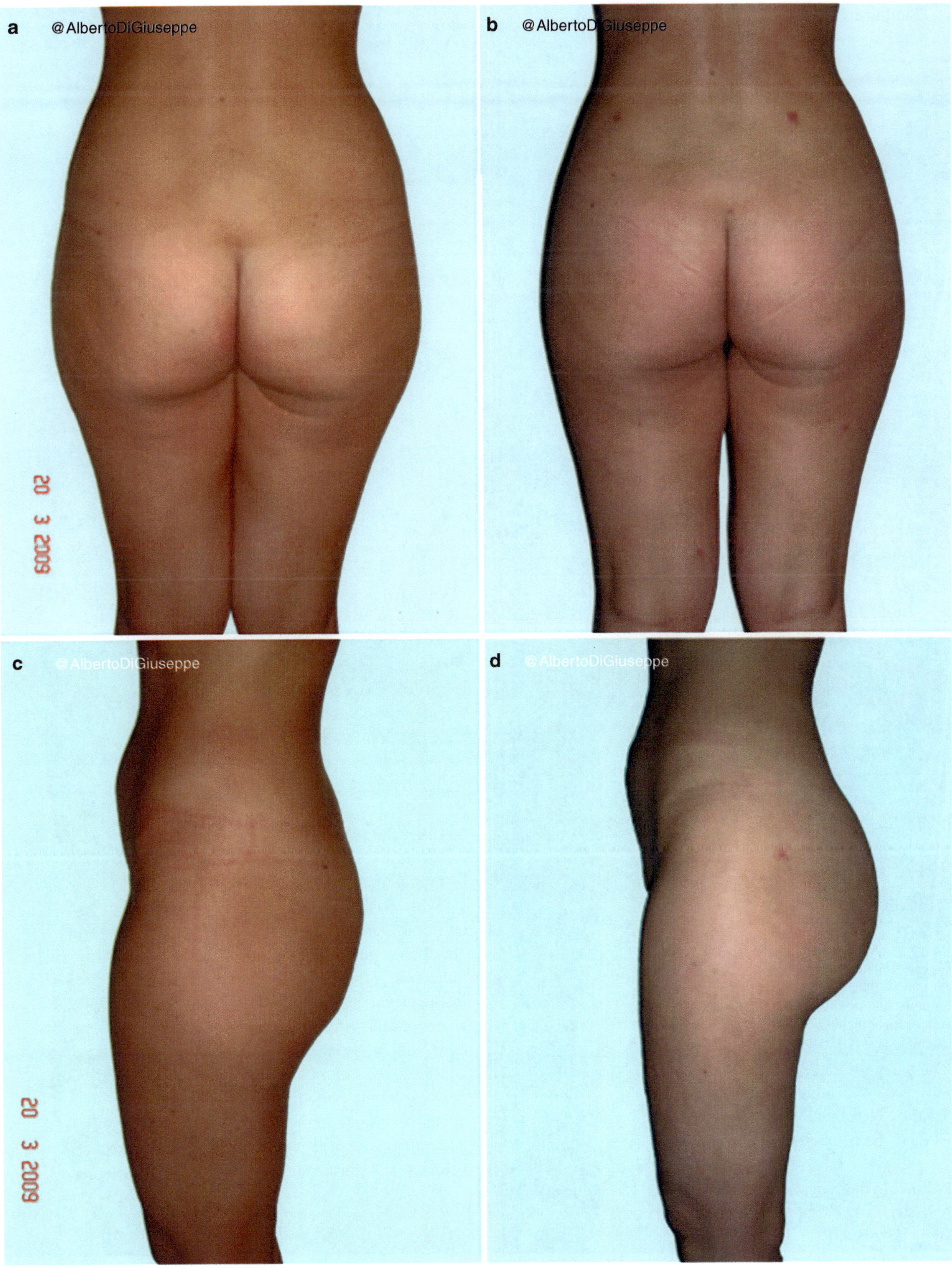

Fig. 16.18 36-year-old woman, round shape buttock. VASER to abdomen, flanks, torso, banana fold, and anterior thighs. 800 cc of total fat transfer. Pre op (**a**, **c**). 1 year post op (**b**, **d**)

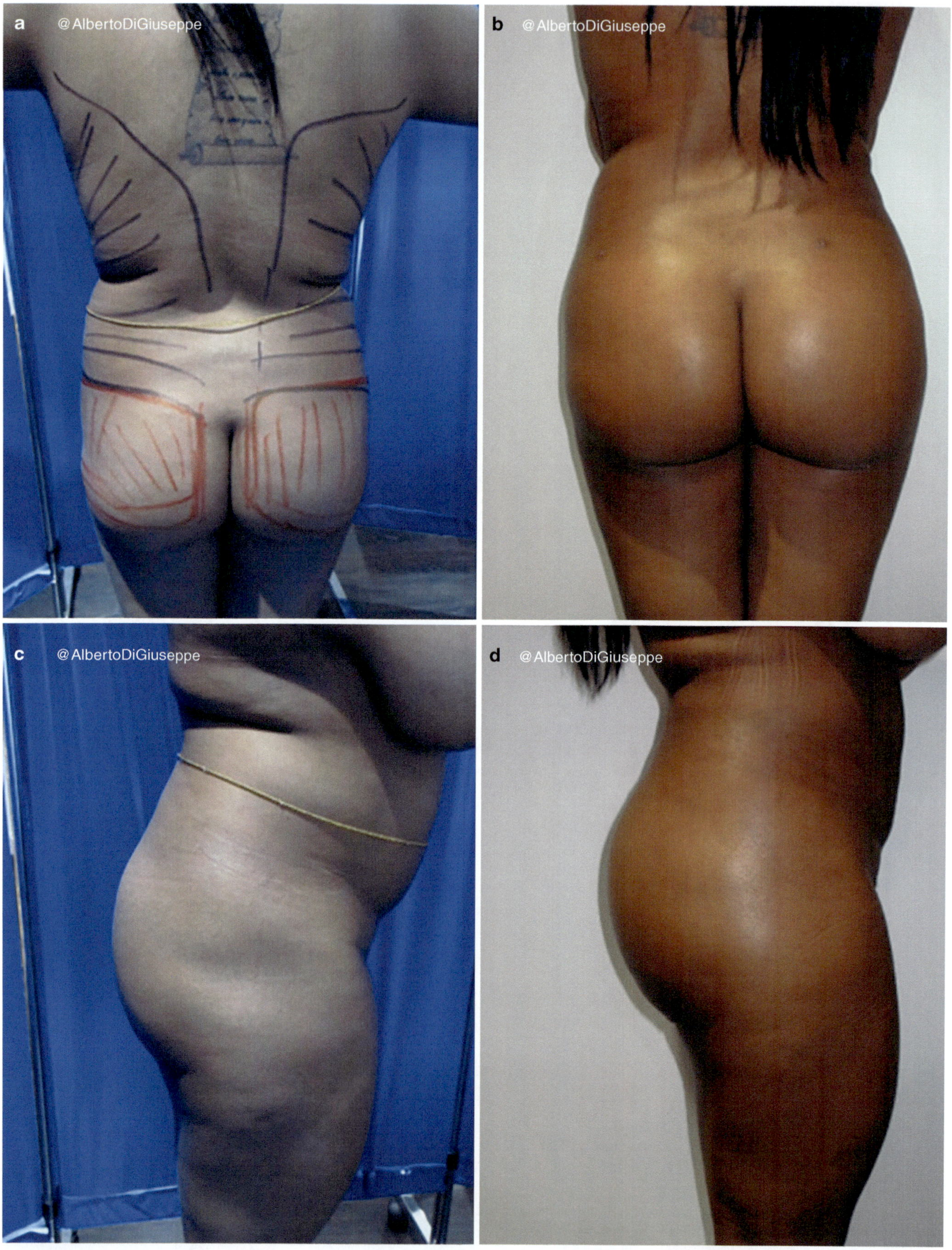

Fig. 16.19 36-year-old woman, round shape buttock. VASER to abdomen, flanks, torso, banana fold, and anterior thighs. 800 cc of total fat transfer. Pre op (**a**, **c**). 1 year post op (**b**, **d**)

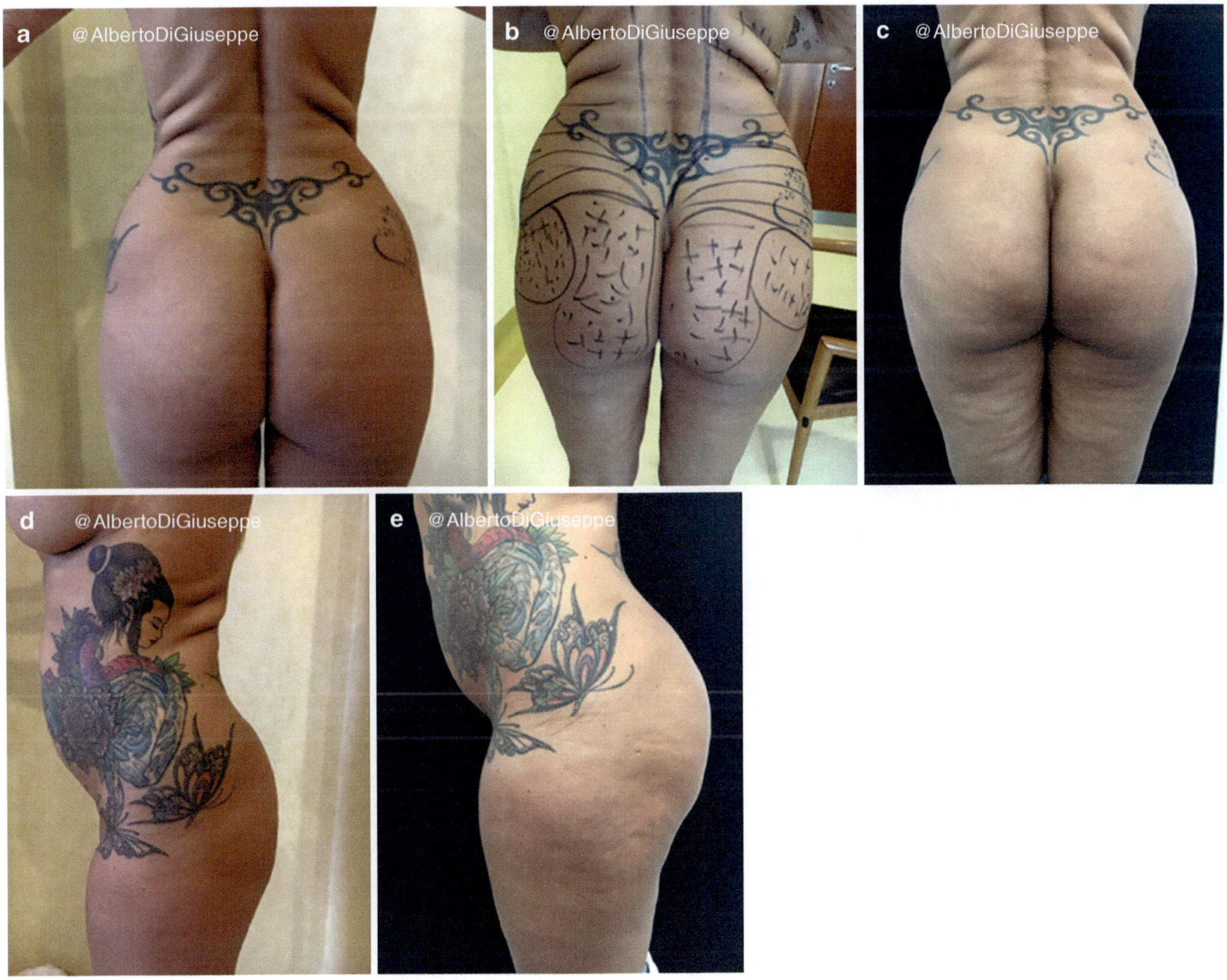

Fig. 16.20 39-year-old woman, round shape buttock. VASER to torso, flanks, and abdomen. 1200 cc of total fat transfer. Pre op (**a**, **b**, **d**). 1 year post op (**c**, **e**)

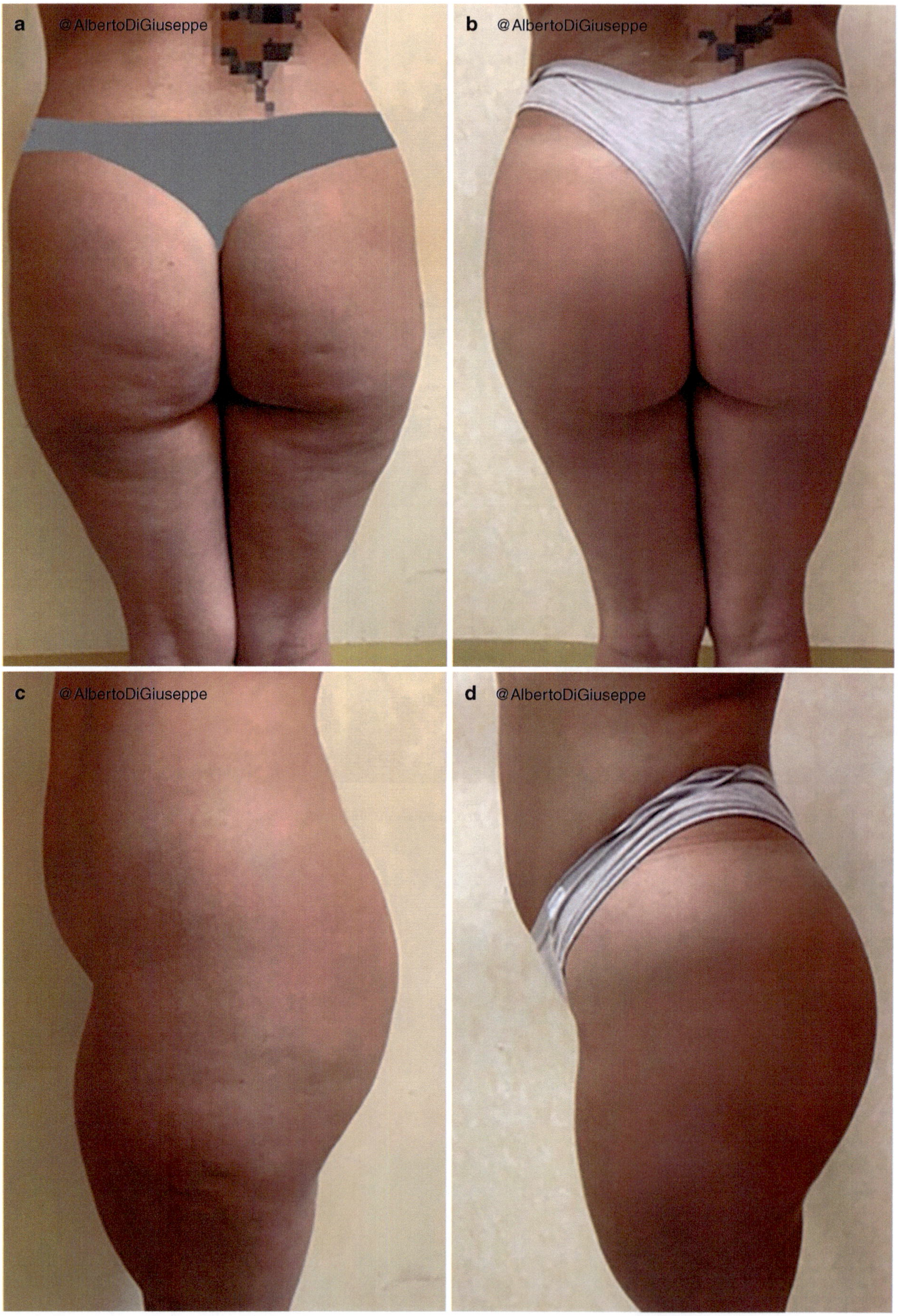

Fig. 16.21 24-year-old woman, round shape buttock. VASER to torso, flanks, and anterior thighs. VASER smooth for cellulite. 1000 cc of total fat transfer. Pre op (**a**, **c**). 1 year post op (**b**, **d**)

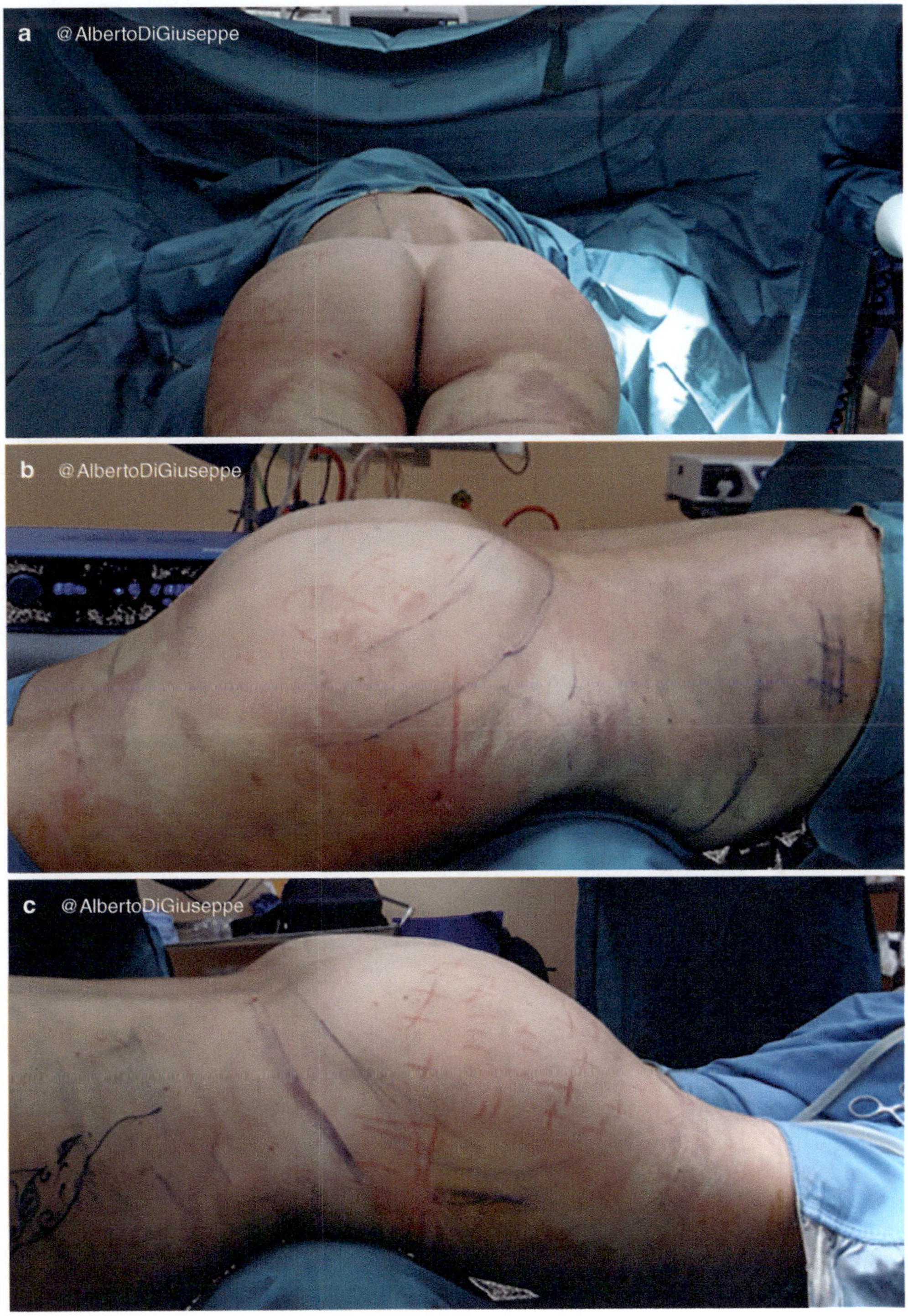

Fig. 16.22 24-year-old woman, round shape buttock. VASER to torso, flanks, and anterior thighs. VASER smooth for cellulite. 1000 cc of total fat transfer. Immediate post-op result in posterior (**a**), lateral (**b**) and oblique-upper (**c**) view

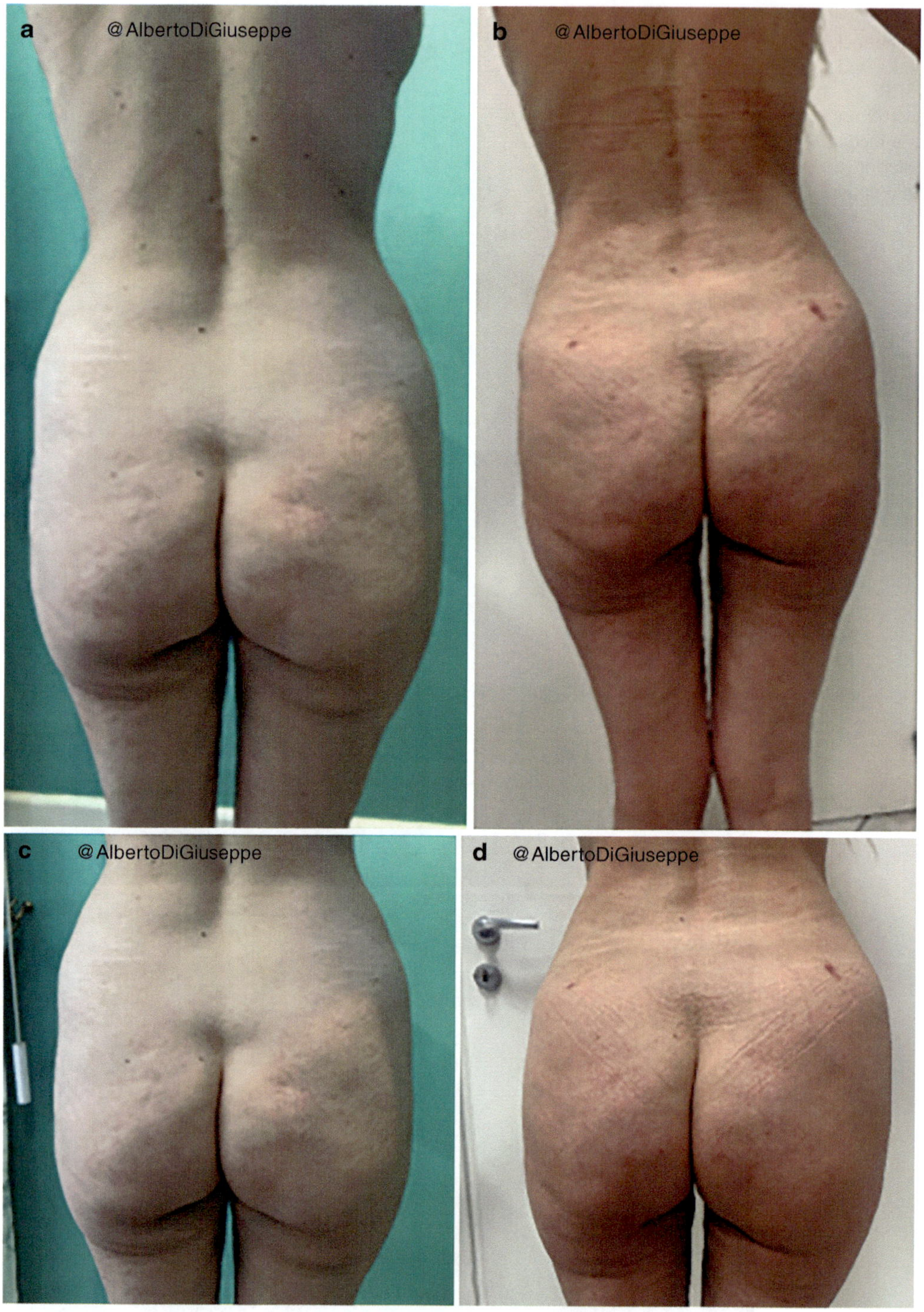

Fig. 16.23 38-year-old patient, slim frame, but unhappy about contouring of flanks, poor definition, and flat buttocks. Droopy buttocks and banana fold deformity. (**a**, **c**, **e**) pre op. (**b**, **d**, **f**) 6 month post op

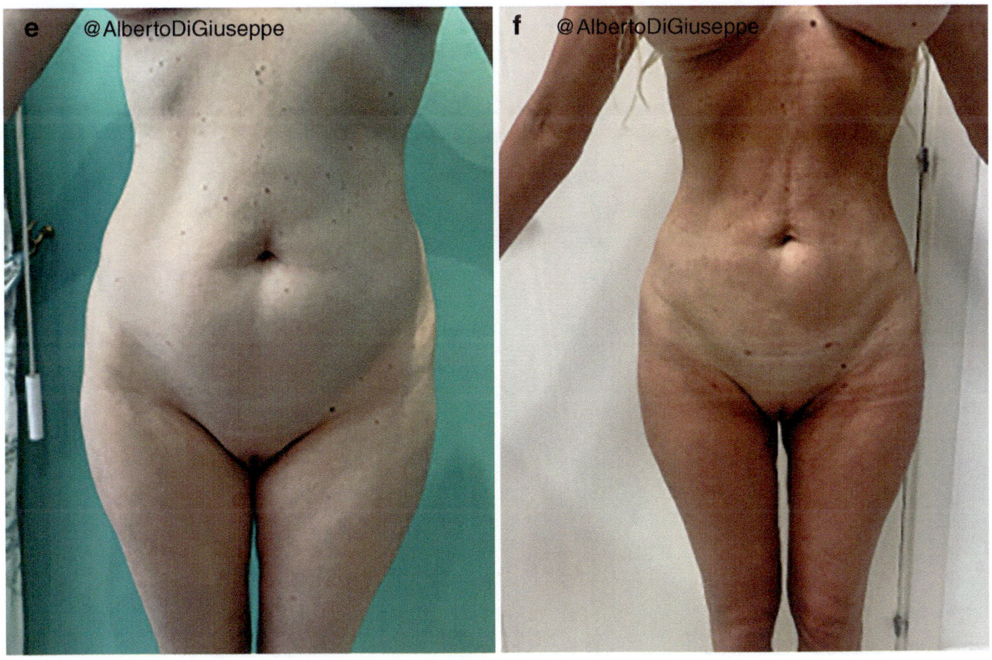

Fig. 16.23 (continued)

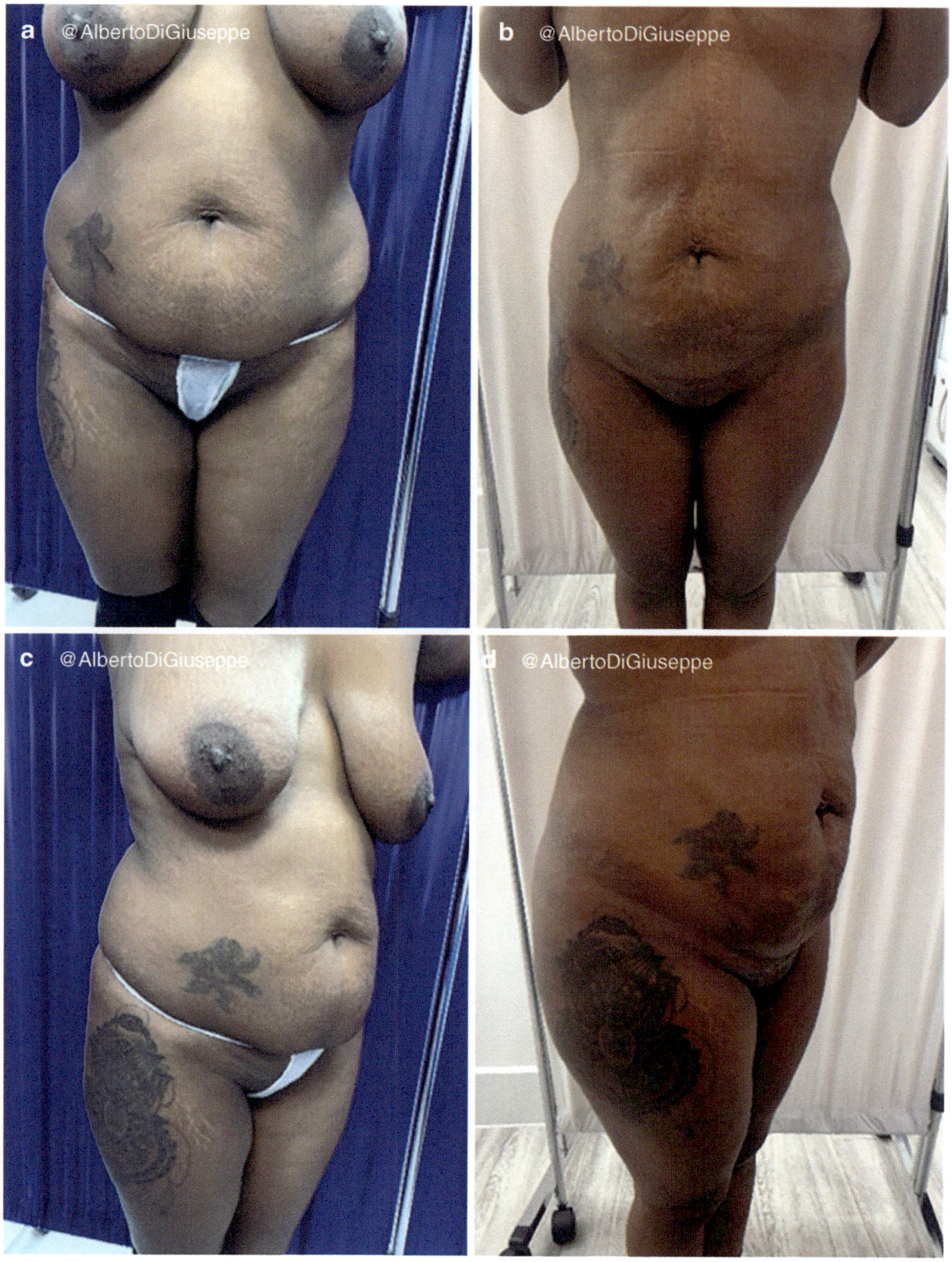

Fig. 16.24 44-year-old patient, two pregnancies, overweighted, with bulge and apron of the abdominal wall. On the back, rolls, heavy flanks, back rolls, and torso laxity. Complaining also for flat and unshaped buttocks. (**a**, **c**, **e**, **g**) pre op. (**b**, **d**, **f**, **h**) 6 month post op. (**i**, **l**) Immediate post op

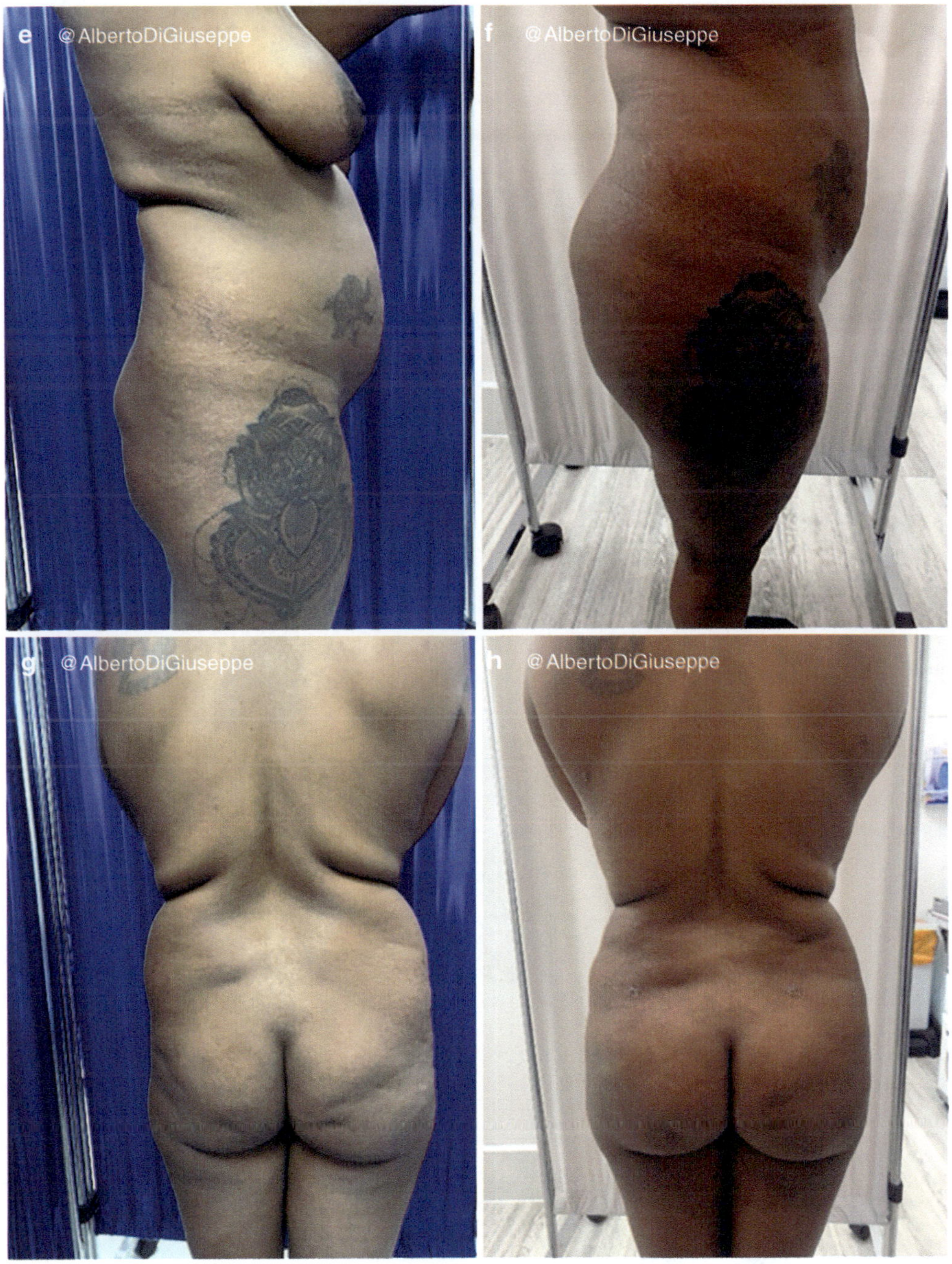

Fig. 16.24 (continued)

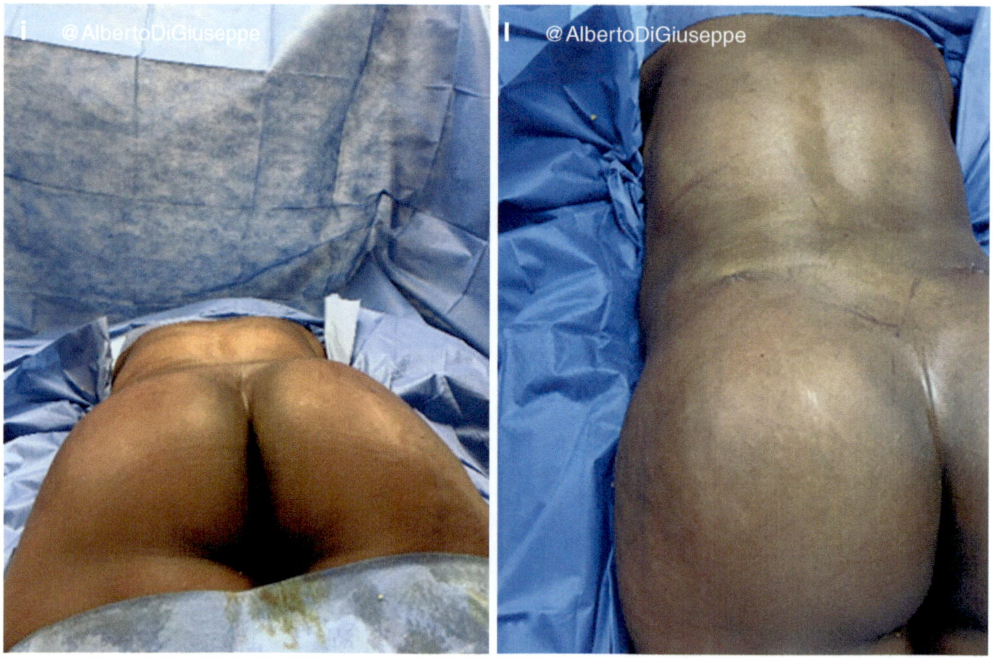

Fig. 16.24 (continued)

Fig. 16.25 Analysis of body frame

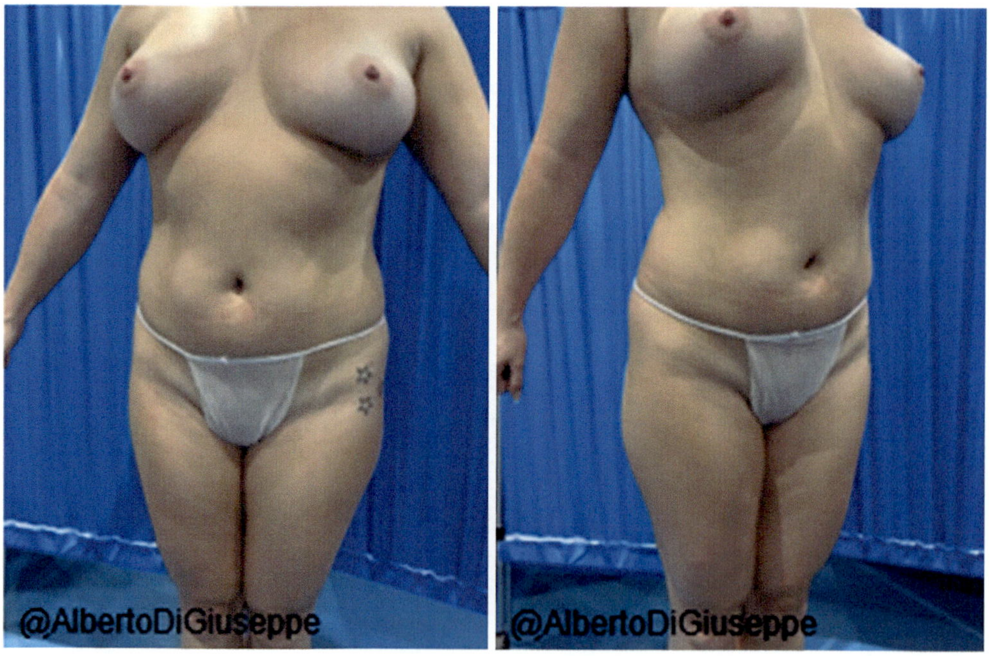

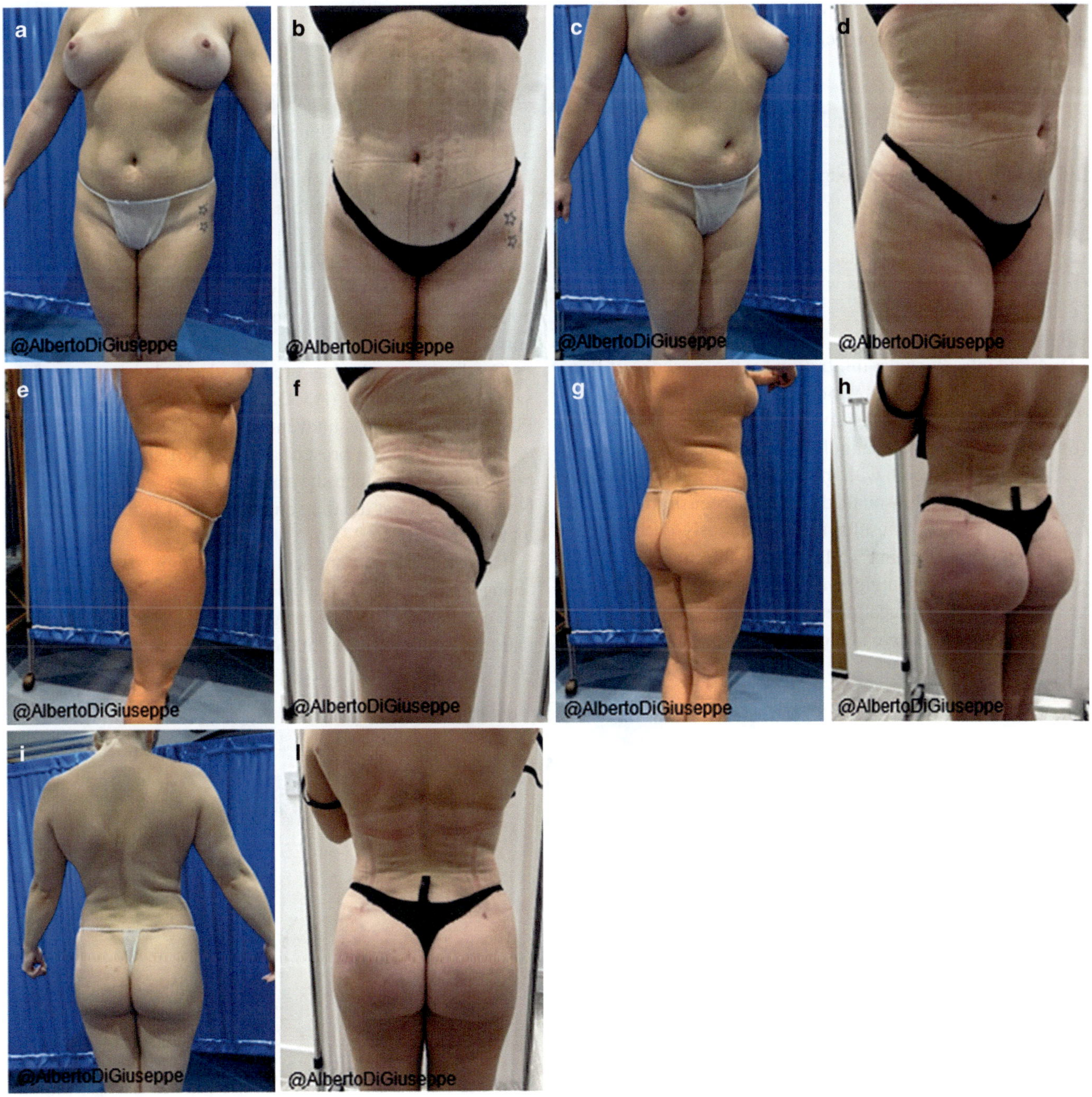

Fig. 16.26 Pre post VASER liposuction upper and lower abdomen, anterior and posterior flanks, back rolls, and trunk. (**a**, **c**, **e**, **g**, **i**) pre op. (**b**, **d**, **f**, **h**, **l**) 6 months post op

16.7 Safety Issues

16.7.1 Zone of Adherence

In buttock shaping and fat transfer, the zone of adherence and vascular danger zone must be considered. Adherence zones are mainly localized at the gluteal fold. They are vertical connective structures localized at the gluteal fold, which are in intimate connection with the dermal attachment of the buttocks.

Those vertical connections should not be destroyed during surgery. This 2 cm large area at the gluteal fold must be preserved from any surgical vertical work.

In case of accidental distraction, the buttocks will drop over the fold, or the gluteal fold itself will be lowered, creating severe damage, nearly impossible to be corrected, except with an open surgical approach.

16.7.2 Vascular Danger Zone

In buttocks enhancement, extreme care must be followed in directing fat transfer:

(a) Using blunt instruments.
(b) 2–3 mm large cannulas with an open place on the lower side of the cannula.
(c) Infiltration subdermal.
(d) Avoid intramuscular infiltration for the presence of large veins with no valves, which could eventually result in a fat collection and a massive fat embolus.
(e) Always direct the cannula in a horizontal fashion, by checking the depth with the contralateral hand. Recently a superficial ultrasound device has been introduced in the market and is useful for continuously checking the depth of the cannula while progressing into the buttock area. Thus preventing uncontrolled penetration of the fascia and the gluteal muscle (where the gluteal vessels and the sciatic nerve are located).

16.7.3 Complications

The first complication I did encounter in my series was an infection sustained by *Streptococcus*. The patient had BBL under general anesthesia, and donor areas were upper and lower abdomen, posterior flanks, and trunk.

The patient had the standard preparation, with betadine shower a few hours before surgery, then prepping with chlorhexidine.

Intra op, she had an IV full dose of AMOXICLAV 500/125 (amoxicillin 500 mg and clavulanic acid 125 mg). She had 500 per side fat transfer to buttock for enhancement and hip correction.

On discharge, she has prescribed a 5 days course of the same antibiotic orally (AMOXICLAVE per os, twice a day). Post-op recovery was uneventful until day 5 post-surgery.

Patient started to present a darker left side. This was the nurse's diagnosis when she saw the photos sent by the patient. No temperature at the time and no redness in the hip and buttock areas (Fig. 16.27).

On day 6, the patient sent photos again, showing quite a change in the color of the buttocks (Fig. 16.28).

The area has become progressively red and warmer; the patient signed with a permanent marking per the outer limits

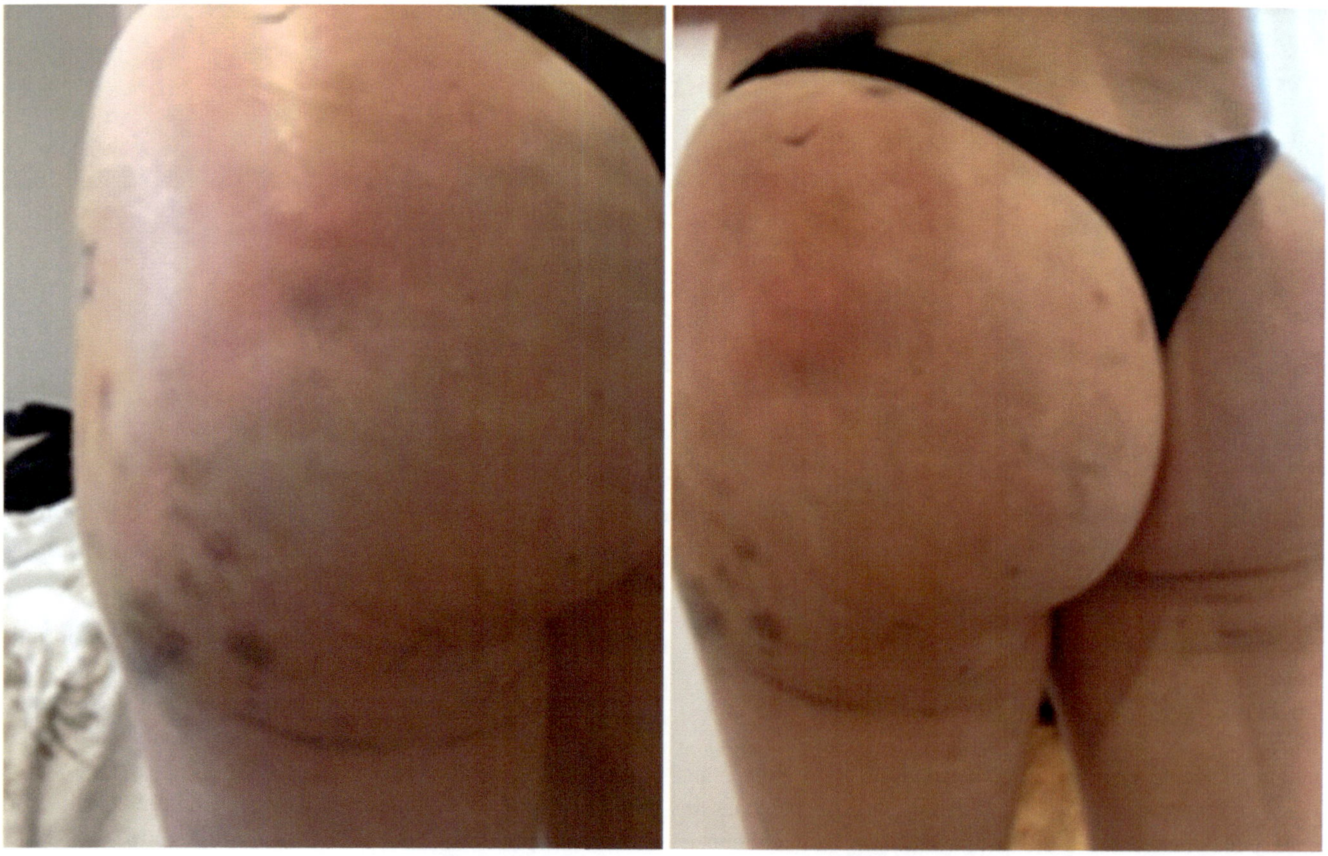

Fig. 16.27 Buttock presentation of day 5 after surgery

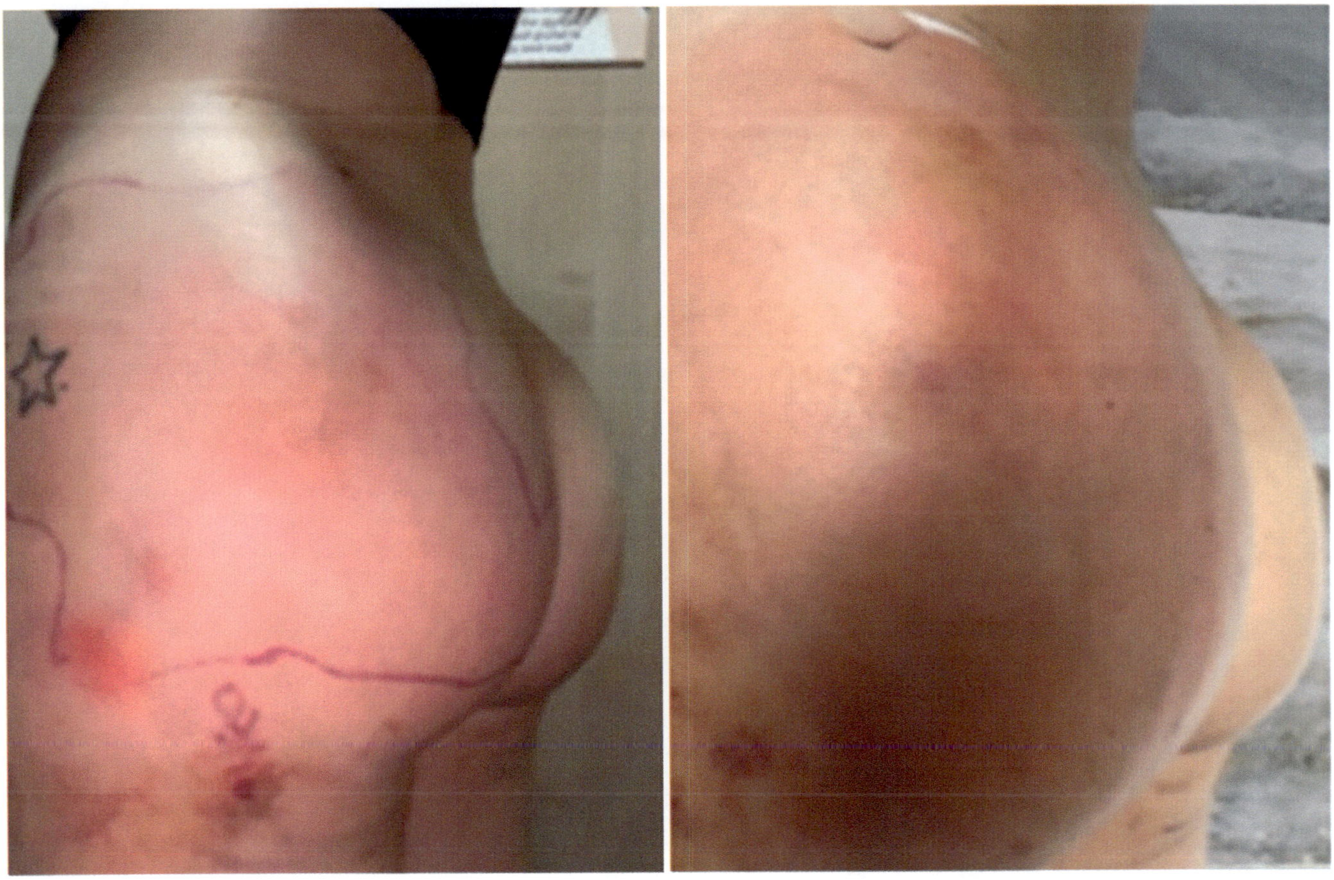

Fig. 16.28 Buttock presentation on day 6 after surgery. Note the progressive increase of redness with the marking in violet of the outer margin

of the redness of the day. At the same time, an area of bumpiness and swelling was noted on the center of the hip.

On day 7 post-surgery, the swelling becomes a palpable, warm and consistent bump. Yellow stuff (pus-like) started to ooze from the incision site. The patient was feeling unwell and was admitted to the hospital with the diagnosis of cellulitis. On admission, a swab was taken from the site where pus was oozing out.

At the same time, IV antibiotics were started, AMOXICLAVE 500/125 twice a day. The patient presented a temperature of 37.8 on admission.

The further report of the swab culture confirmed an infection sustained by *Streptococcus*. Sensitive to the same antibiotics utilized before (Fig. 16.29).

The patient had 3 days of treatment with IV antibiotics in the hospital with the resolution of pirexia, a progressive diminishing of the swelling and the redness, as shown in Fig. 16.30.

She was discharged with an indication to continue antibiotics (AMOXICLAVE) for further 5 days.

The patient was further reviewed 3 weeks after surgery by her original surgical team.

Figure 16.31 shows the perfect symmetry achieved despite the left-side cellulitis.

No contouring defect, tissue is still soft and no swelling (Fig. 16.31).

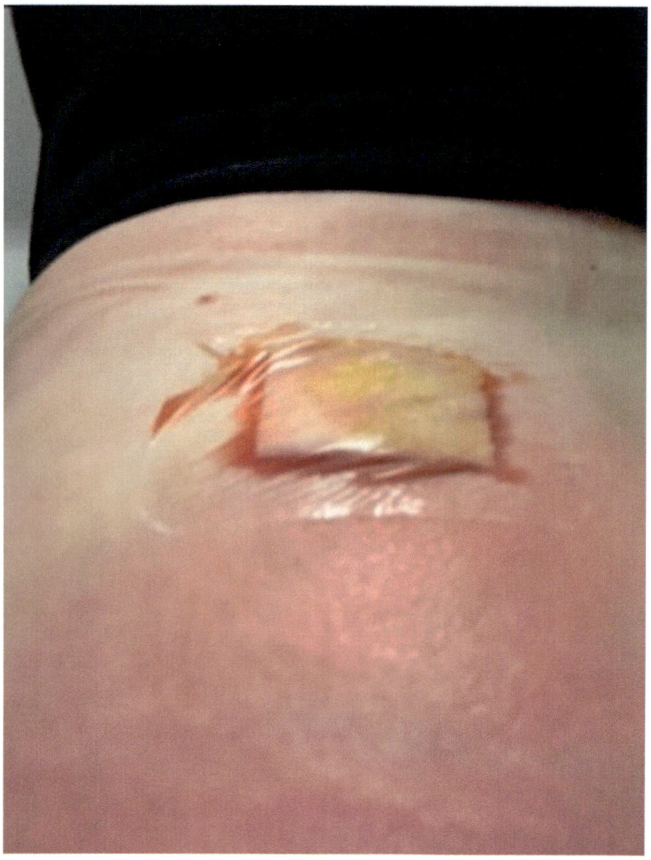

Fig. 16.29 Pus oozing out

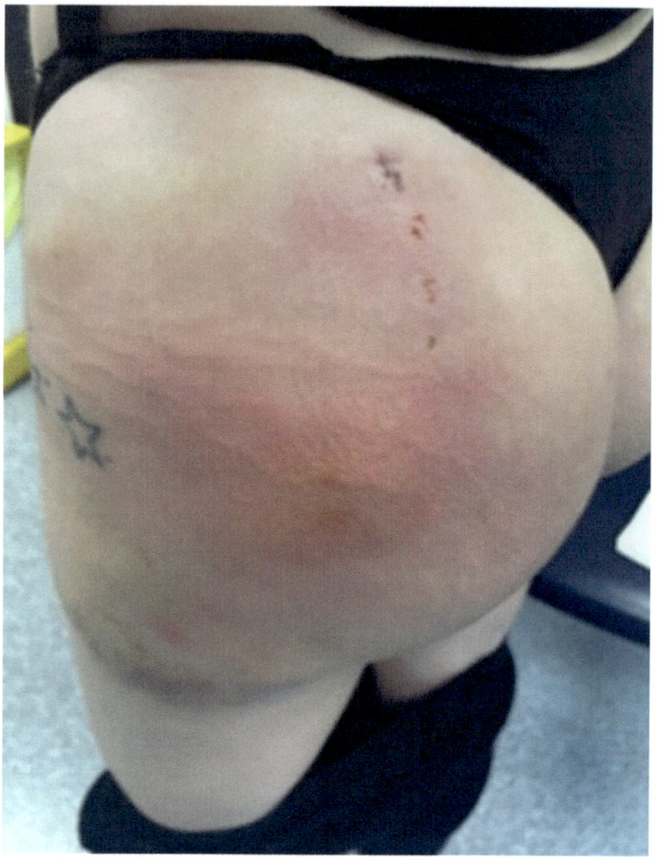

Fig. 16.30 Day 11th post surgery: patient discharged from hospital after IV antibiotics. Please note clearance of redness, termination of the oozing. Apireptic

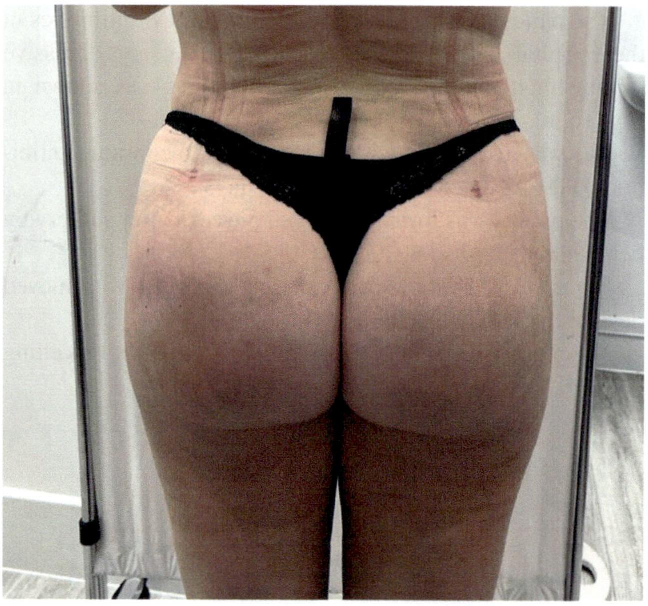

Fig. 16.31 Perfect symmetry achieved despite the left side cellulitis. No contouring defect, tissue still soft and no swelling

References

1. Gonzalez-Ulloa M. Gluteoplasty: a ten-year report. Aesthetic Plast Surg. 1991;15:85–91.
2. Robles JM, Tagliapetra JC, Grandi MA. Gluteoplastia de aumento: implante submuscular. Cir Plast Inerolatinoamer. 1984;10
3. Vergara R, Marcos M. Intramuscular gluteal implants. Aethetic Plast Surg. 1996;20:259–62.
4. Mendieta C. The art of gluteal sculpting. Quality Medical Publishing, Inc; 2011.
5. Di Giuseppe A. Liposuction: principles and practice. Berlin: Springer; 2006.
6. Di Giuseppe A. Body contouring. Art, science and clinical practice. Berlin: Springer; 2010.
7. Di Giuseppe A, Shiffman MA, Contorneado corporal. Arte, ciencia y practical clinica. Amolca Venezuela. 2013;
8. https://soltapracticeproposal.com/product-information/products/vaser-system/

Anatomy of the Male Torso in Relation to Body Contouring: Abdomen, Flanks, and Arms

17

Huseyin Kandulu

Contents

H. Kandulu (✉)
Huseyin Kandulu Private Clinic, Istanbul, Turkey
e-mail: huseyin@kandulu.com

© The Author(s), under exclusive license to Springer Nature Switzerland AG 2023

A. Di Giuseppe et al. (eds.), *Fat Transfer in Plastic Surgery*, https://doi.org/10.1007/978-3-031-10881-5_17

Fig. 17.1 The desired ideal athletic male body from the front view

17.1 Introduction

The shape and size of the human form are determined by the skeletal framework, the amount and distribution of subcutaneous fat, and muscle mass and tone. Other contributing factors include poor posture associated with lack of exercise, diet, bad habits, and weakness. Both the static and dynamic states of the human body should be defined to more clearly elucidate the human external body form. Physical features that are attractive in women include the convexities of the chest, hips, and lumbar lordosis. On the other hand, the features that are attractive in men include well-developed muscle mass, especially in the arms, chest, and abdomen (Fig. 17.1).

In order to achieve a more appealing body image, many people in modern society want to remove excess and/or unwanted fat. Body contouring is a technique used that removes a limited amount of adipose tissue to achieve a more aesthetically pleasing body shape. After determining the best surgical option for the patient wishing to have this technique, a detailed physical examination and careful evaluation of the patient's anatomy are required. It is especially important to note any anomalies in the patient's anatomy, such as deviation from the ideal contour, scars, asymmetry, skin tone, and elasticity, including any taut or excess skin. The surgeon must have a comprehensive knowledge of muscular anatomy in order to shape the human body, and he/she must be able to "look through the eyes of an artist" so that the superficial musculature can be ideally visualized. The majority of people with a normal body mass index can achieve an athletic and toned appearance by emphasizing the main muscle groups with high-definition body contouring. In this section of the review, the anatomy of the abdomen, flanks, chest, and arms will be discussed in detail. It is our hope that this section of the review will help surgeons to identify the muscular structures that create natural bumps in the body, termed "positive areas." Further, surgeons should be able to detect the shadows ("negative areas") created by the natural fat excess in the recesses between these natural bumps. In the following sections, we emphasize the examination techniques and light techniques that can be used to analyze this topographic anatomy.

17.2 Trunk

The trunk muscles include both the muscle groups arranged in the front and back of the abdominal and chest wall muscles (Fig. 17.2). This muscular body wall connects the rib cage to the pelvis. The trunk muscles have various effects on the trunk, including extension, flexion, rotation, and lateral bending. The prominence of the trunk muscles, especially in men, provides an athletic, strong, and healthy appearance. In men, the posterior trunk muscles form a V-shaped or triangular wedge, while the pectoralis major muscle over the chest and the rectus abdominis muscle between the abdomen's tendinous intersections form a convex muscle mass.

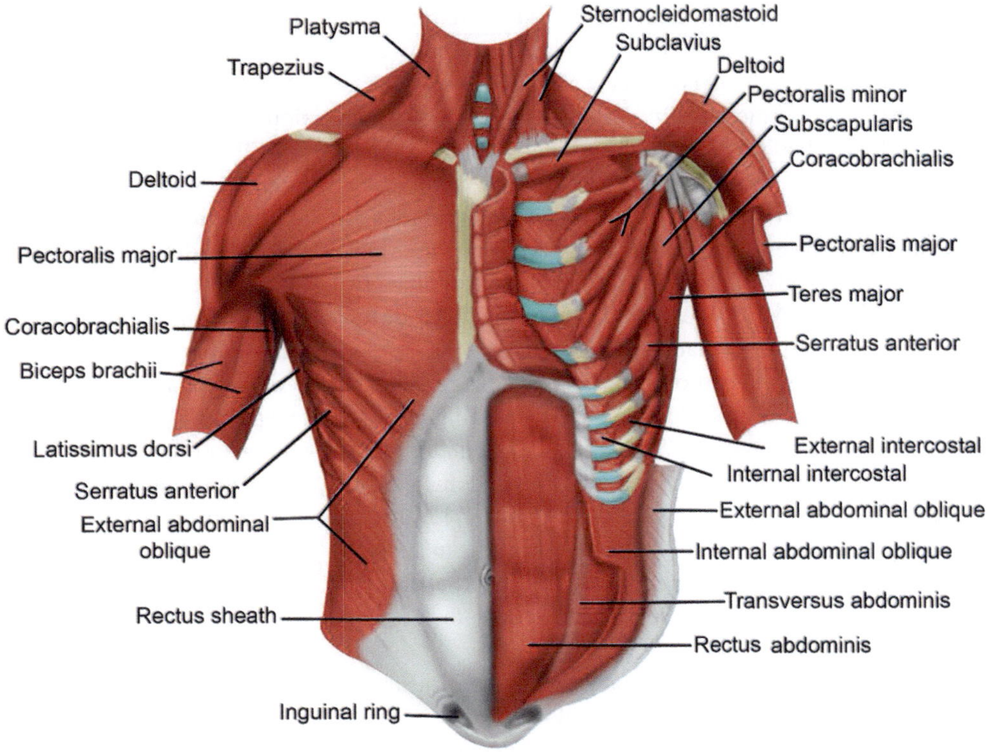

Fig. 17.2 Anterior trunk muscles

17.3 Abdomen

Since aesthetic body contouring surgery requires a general approach to body contour, it is first necessary to evaluate the shaped area and surrounding structures. For example, significant weight loss and obesity can cause abdominal deformation, which affects many parts of the body. In addition, when choosing an appropriate surgical strategy, predisposing factors must be examined, including skin laxity, localized or generalized lipodystrophy, musculoaponeurotic flaccidity, and stretch marks.

While the vast majority of women desire a flat abdomen, the vast majority of men want a well-defined "six-pack" (Fig. 17.3). However, the genes that develop the rectus abdominis produce a very small proportion of male "six packs," which also require a combination of healthy eating habits and exercise. Even athletes, who undergo intense physical exercise, may still have unwanted fat. Because of this unwanted fat, it can be very difficult to achieve a prominent muscle appearance. In the past, liposuction was only used to remove unwanted fat from the body, but now, it is also used for body contouring.

Abdominoplasty ("tummy tuck") is a procedure that is used to reduce excess skin and fat around the abdomen and strengthen the abdominal wall muscles. Due to the change in artistic body shape perception, current approaches to tummy tuck surgery have also changed, and now, it is often performed in combination with an advanced liposuction method.

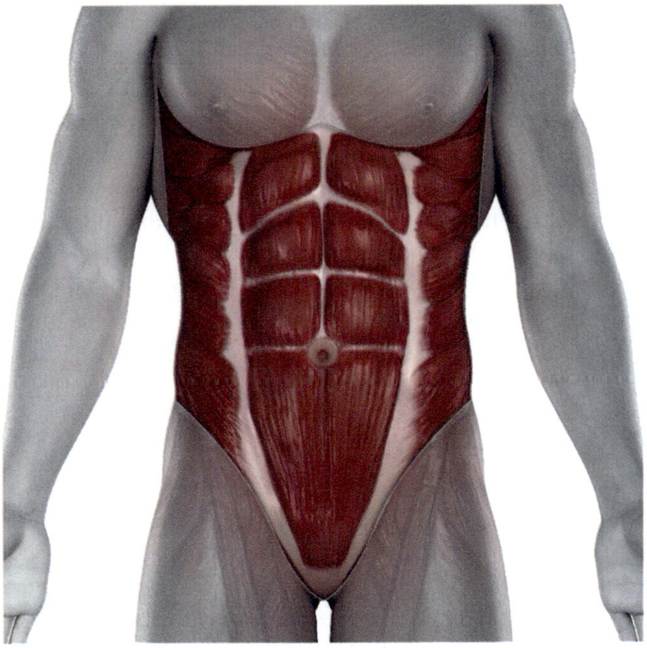

Fig. 17.3 The anterior abdominal wall. Rectus abdominus and six pack, which are the most important muscles of the anterior abdominal wall in body shaping

However, surgical techniques in this field will not be discussed here, as they are the subject of a separate chapter. Abdominal fat should be removed from the muscles to obtain oblique muscles, anterior serratus, a six-pack abdomen, and taut skin. Fat at the level of the tendinous intersections, anatomical lines, and muscle margins should be removed to emphasize muscle volume. The main purpose of body contouring in the male patient is to create an athletic and muscular appearance (termed a "V shape"), regardless of the patient's body shape. However, the patient's body type and body mass index are very important in body contouring. Therefore, the relevant sportive background, natural muscle structure, and current body morphology are very important. The abdominal body offers a wide range of opportunities for body contouring. All of these procedures should be developed with appropriate diet and exercise.

The abdomen is the part of the trunk that connects the rib cage and pelvis, and an abdominal wall consisting of skin, fascia, and muscle covers the abdominal cavity and internal organs. Unlike the thorax and pelvis, the abdomen lacks skeletal support, except for the spinal column and lower ribs in the back. The anterolateral abdominal wall (from superficial to deep) includes skin, superficial fascia, external oblique muscle, internal oblique muscle, transversalis fascia, and parietal peritoneum.

Men's skin is less prone to overstretching and looseness than women. This is most likely due to its thickness, and that it is not affected by pregnancy or the preponderance of female hormones. In men, excessive excess skin is usually only seen in patients who have experienced significant weight loss or fluctuations. With age, men's abdominal fat increases, and there is a corresponding decrease in subcutaneous fat. In addition, aging leads to the infiltration of fat between the muscles. Taken together, aging causes a decrease in lean body mass and an increase in body mass index, and this redistribution of fat has a major effect on the appearance of the abdomen.

The superficial fascia is located between the dermis and muscles. It is divided into a superficial fatty layer called the Camper fascia and a deep membranous layer called the Scarpa's fascia, which includes another adipose tissue layer deep within.

The camper fascia is the superficial fatty layer of the anterior abdominal wall that is divided by the fibrous septa, which connects the dermis to the deeper membranous layer. The camper fascia extends laterally from the xiphoid process to the seventh and tenth costal margins and lower to the inguinal ligaments. It continues down the groin as the subcutaneous fat of the thigh. In males, it combines with the Scarpa's fascia to form the dartos tunic of the scrotum, and it extends beyond the pubic symphysis.

Scarpa fascia makes up the membranous layer, and consists of connective tissue and elastic fibers. It is located below the Camper fascia and above the external oblique muscle. The oblique fibrous septa loosely connect the underlying external oblique aponeurosis and the rectus sheath. Scarpa's fascia is attached medially to the linea alba and the inguinal symphysis, and extends superficially to the scrotum, where it forms the dartos fascia. Scarpa's fascia plays an important role in the healing process of abdominal incisions. The preservation of Scarpa's fascia reduces seroma formation, which is the most common complication associated with abdominoplasty. In addition, the protection provided by Scarpa's fascia during abdominoplasty also provides good aesthetic results. Liposuction preferably removes the fats in the deep adipose layer (the third layer). The morphology of the adipose tissue and its location directly affects the surgical intervention. The goal of surgical intervention is to smoothly remove this fatty tissue so that a permanent and natural result is obtained, and so that the person's natural infrastructure is revealed. Although some believe that this may occur with the removal of superficial fat tissue, we suggest that deep level liposuction should always be applied, therefore avoiding unwanted complications. It is important to note that the desired athletic body shape in men, especially below the umbilicus, is achieved when the fat is removed under both fascia planes until the skin flap thickness is reduced to 0.5–1 cm.

17.4 Muscles

The muscles of the abdominal wall are divided into two groups as follows: (1) three smooth muscles located laterally and stacked on top of each other, and (2) two vertical muscles near the body's midline. The abdominal wall consists of three smooth muscles, the external oblique, the internal oblique, and the transverse abdominis, which flex and rotate the trunk (Fig. 17.4). Since the fibers of these muscles are cross-linked with each other, these three muscles strengthen the abdominal wall and reduce the risk of herniation.

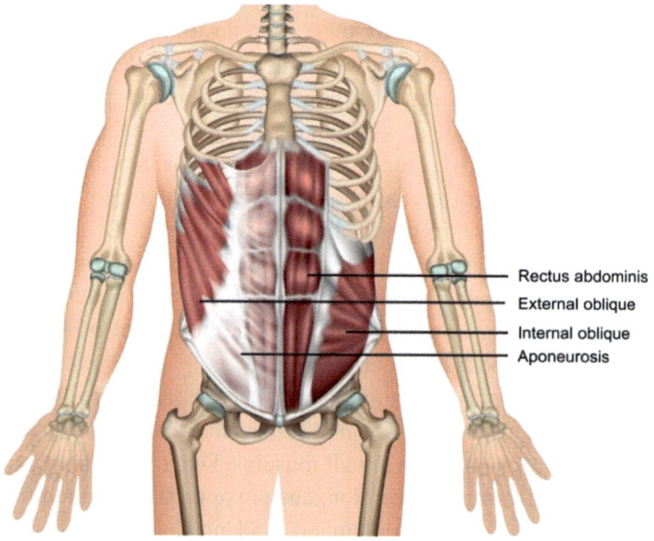

Fig. 17.4 Anterior abdominal wall muscles

17.5 External Oblique

The external oblique is the largest, thickest, and most superficially located anterolateral abdominal wall muscle. The external oblique muscle originates from the outer and lower surfaces of the lower 7 ribs and consists of the upper thoracic part and the lower flank part. The upper attachments of the muscle may be continuous with the serratus anterior, pectoralis major, or latissimus dorsi. The thoracic part of the external oblique consists of parallel and long muscle bundles extending from the outer surfaces of the ribs to their aponeuroses in the semilunar line. In the area where it enters its aponeurosis, the anterior end of the thoracic part of the external oblique is irregular, and its lowest fibers form a transitional zone in the waist region above the flank part. The flank part extends posteriorly from the outer surfaces of the lower ribs to the anterior half of the iliac crest. The flank pad creates a fleshy trapezoid around the waist between the ribs and pelvis. It includes the flank part of the external oblique on the anterior portion, and the flank fat pads on the posterior portion. Even in individuals with very good muscle development, the bilateral oblique muscle fibers are not visible on the convex flank pad. The most posterior fibers that pass vertically through the underlying two ribs to enter the iliac crest form a posterior free border, and do not enter the thoracolumbar fascia. These posterior fibers form the anterior border of Petit's lower lumbar triangle, which is typically covered with a fat pad. The external oblique becomes aponeurotic in the medial midclavicular line and in the inferior spinoumbilical line. The aponeurosis passes in front of the rectus abdominis and decussates with the aponeurotic fibers of the internal oblique, external oblique, and transversus abdominis, which comes from the side opposite the midline. The tip of the tenth rib forms the base of the rib cage and the upper border of the external oblique abdomen. Athletic people often have a small triangular depression along the semilunar line in this area. In subdermal liposplasty, depression in this area is used as a checkpoint to define the anterolateral abdominal wall. The inguinal ligament is formed from the lower edge of the external oblique aponeurosis, which extends between the anterior superior iliac spine and the pubic tubercle. The external oblique muscle is the most important muscle group to highlight the V-shape in the waist region in males. Therefore, surgery should both emphasize and clarify this region. During the examination, the patient is told to inflate his abdomen outward, and the lower border of the muscle should be able to be felt by hand. Although this is often sufficient, beginners will see a shadowing in the lower border of the muscle due to the bright light source projected onto the muscle from the opposite direction. Due to the natural protrusion of the muscle, this shadowing will be on the lumbar triangle. The lumbar triangle should be marked by the surgeon, and all fat within this triangle should be removed during the surgery in order to define the lower edge of this muscle. Although this negative area is evident in women with an athletic body, a woman's external oblique muscle is not as strong as that in a man. For this reason, the operation should not be performed in this area in female patients if the triangle cannot be detected (Figs. 17.5 and 17.6).

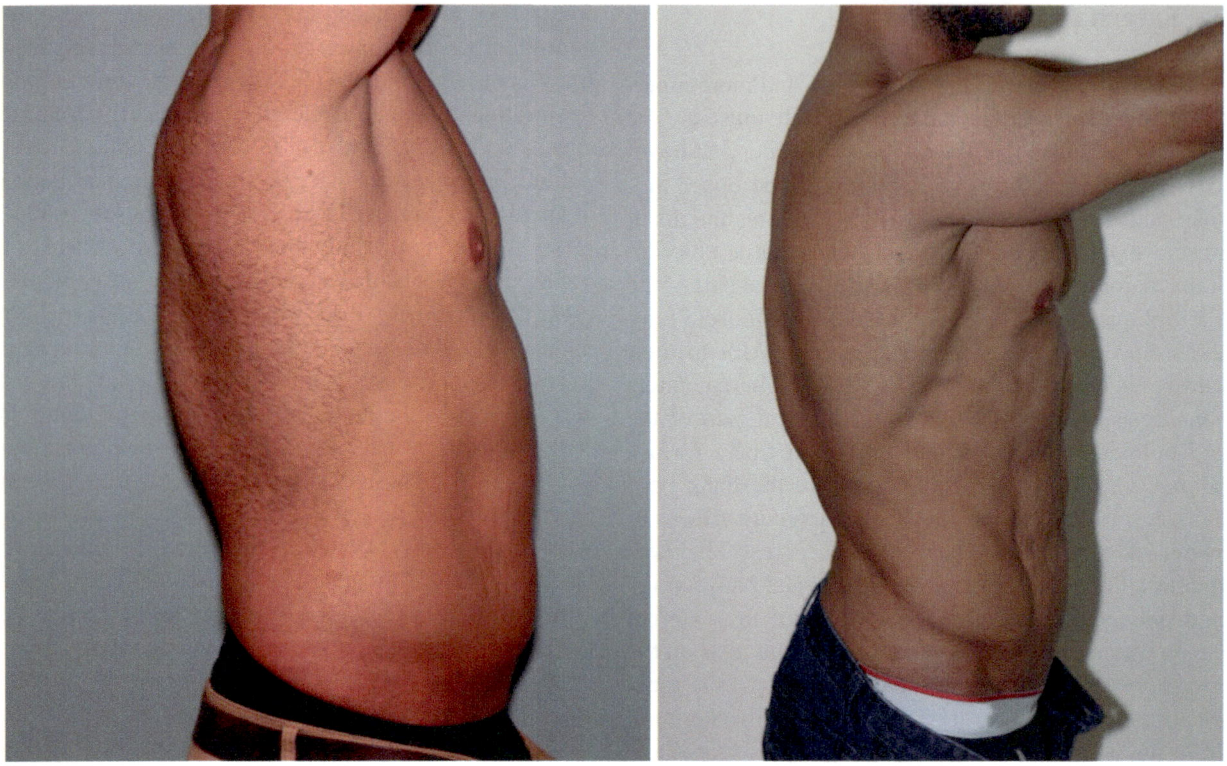

Fig. 17.5 Pronounced external oblique muscle after removing excess adipose tissue from the natural spaces around the muscle

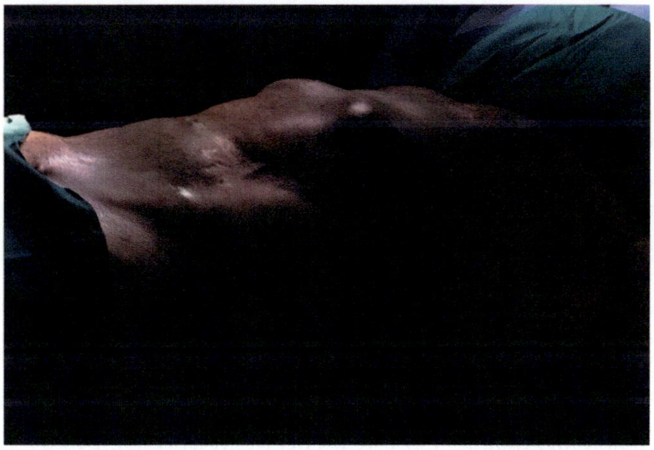

Fig. 17.6 Light techniques are important in advanced body shaping. Bright areas show natural protrusions, while shadows reveal natural dimples between muscles. Here, the external oblique muscle is bright

17.6 Internal Oblique

The internal oblique muscle is the middle layer of the anterolateral abdominal wall muscles. It arises from the lateral iliac crest, iliopsoas fascia, and lumbodorsal fascia. As opposed to the external oblique muscle, the internal oblique muscle's fibers extend from the lower/lateral to the upper/medial to accommodate the lower 5 rib cartilage. Posteriorly, some fibers from the internal oblique are attached to the thoracolumbar fascia. Below, it takes a more diagonal course, and enters the inguinal tubercle. It joins the aponeurosis of the transversus abdominis muscle to form the adherent tendon in the lower part. Like the external oblique muscle, the internal oblique muscle also contributes to the rectus sheath. The fibers adjacent to the inguinal ligament extend downward and medially along the spermatic cord in men. Since the internal oblique muscle is in the deep plane, it does not naturally protrude on the surface, so it does not require special work.

17.7 Transversus Abdominis

The transversus abdominis is the deepest of the anterolateral abdominal muscles. It stems from the lateral one-third of the inguinal ligament deep into the iliopectineal arch, the anterior two-thirds of the inner lip of the iliac crest, the thoracolumbar fascia between the iliac crest and the twelfth rib, and the lower six costal cartilage. Transversus abdominis fibers extend across the lateral abdominal wall anteriorly to the midline and inserts into the linea alba, xiphoid process, and symphysis pubis by the rectus sheet. The lower tendinous fibers originating from the inguinal ligament arise inferomedially over the inguinal canal, and join with the internal abdominal oblique aponeurotic fibers to form the conjoint tendon.

17.8 Rectus Abdominis

The rectus abdominus is a long, double vertical muscle located on either side of the midline. The rectus abdominis is narrow and thick at the bottom and wider and flat in the upper abdomen. It extends from the symphysis of the pubis to the costal cartilage of the fifth to seventh ribs. It has three horizontal lines known as "tendinous intersections" that contribute to well-developed abdominal muscles. These fibrous bands are usually located at the level of the xiphoid process, around the umbilicus, and in between the xiphoid process and umbilicus. The arcuate line is located on a fourth tendinous intersection, which is visible below the umbilicus marks. Tendinous intersections may be in line with the opposite side, or they may be at different levels, giving an asymmetrical appearance to the anterior abdominal muscles. The upper borders of the rectus abdominus may continue along the lower edge of the pectoralis major. The lateral border of the rectus abdominis appears in the abdomen as a vertical groove between the ninth costal cartilage and the inguinal tubercle. This linea semilunaris (Spigelian line) corresponds to a line drawn from the midpoint of the clavicle to the middle of the thigh. The ninth ribs are farther apart than the inguinal tubercles, giving the linea semilunaris its curved shape. In males, the linea semilunaris starts as a depression just below the medial of the nipple, moves downward, and expands into a triangular area above the inguinal ligament. In the linea semilunaris, the tendons of the lateral abdominal muscles meet with the sheath surrounding the rectus abdominis muscle, which is also known as the rectus sheath. The lateral abdominal muscles include the external oblique, internal oblique, and transverse abdominis. The linea semilunaris forms the lateral borders of the rectus abdominis muscles bilaterally, while the linea alba separates the right and left rectus abdominis muscles in the midline. The linea semilunaris, tendinous intersections, and linea alba are structures that make possible the visibly well-developed muscular male abdomen that athletes desire, and the fleshy protruding abdominal segments known as "six-packs." The muscle bulk is emphasized by the removal of fat at the tendinous intersections, the anatomic lines, and the muscular borders. The rectus sheath is an aponeurosis formed by the five muscles of the abdomen. The rectus sheath is divided into anterior and posterior layers for the majority of its length, and covers the rectus abdominis muscle. The anterior layer consists of the outer oblique aponeuroses and half of the inner oblique muscles, while the posterior layer consists of the aponeuroses of the transversus abdominis and the other half of the inner oblique muscles. The arcuate line is the lower part of the posterior rectus sheath, which lies approximately in the middle of the umbilicus and pubic symphysis. Therefore, only the anterior layer of the rectus sheath is located below the arcuate line. Oblique abdominal muscles can be found on both sides of the rectus abdominis muscle. After a thorough explanation of anatomy, it is important to know how to determine this desired muscle group, which is known and desired by all men as a "six pack." The first and most important point to keep in mind is that if the palpable muscle group cannot be determined, the surgeon should not attempt to clarify these muscles with surgery. Another important point is that these operations require a surgeon with advanced liposuction experience. The main purpose of the surgery is to remove the excess fat and reveal the natural infrastructure. For this reason, it is not recommended to shape the fat while clarifying the abdominal muscles. In such cases, although the result on the operating table seems satisfactory, an unnatural and non-permanent result will occur within a short time. To avoid this problem, in the drawing phase prior to the surgery, the frame of the muscle groups should be determined and drawn. During the surgery, it is recommended that the excess fat be removed *after* applying liposuction, therefore revealing the whole structure. The drawing should be done while the patient is standing up and pulling his abdomen in. Then, if the surgeon or an assistant shines a light source from the opposite direction, the semilunar lines and line alba in the muscle shadows can easily be seen. The patient should then be instructed to perform a breathing maneuver and tighten the abdominal muscles. The lowest horizontal tendinous intersection is often located 0.5–1 cm above the umbilicus level. It should be kept in mind that the mid-level abdominal muscle group follows the serratus anterior, and therefore, determining the lateral border of the semilunar line and drawing it towards the line alba will be useful for the surgeon. Also, the uppermost rectus abdominis may be located under the ribs in some patients, and in such cases, these muscles should not be emphasized. Further, it is known that the uppermost rectus abdominus can also sit on 5–7 ribs in a much smaller structure. In cases such as these, the drawings should be done with light shining on the patient, and/or by asking the patient to tighten his muscles. Surgeons should realize that this type of surgery is dynamic, and there will be a minimum of ±1 cm vertical movement between the contraction and relaxation movements of the muscles. This distance is defined as a "transition zone." The most important transition zones are on the gluteus medius muscle in women and on the pectoral muscle in men. However, while performing the six-pack definition, especially in patients with loss of skin elasticity, the patient can be put in a crunch position, and the surgeon can determine whether the lines were correctly drawn in the standing and lying positions. Although surgical techniques are not the subject of this chapter, it is important to note that there is no need to make permanent incisions on the trunk while clarifying these lines, especially in the abdominal region, and it is necessary to work with surgical equipment designed for hiddden areas (Figs. 17.7, 17.8, 17.9).

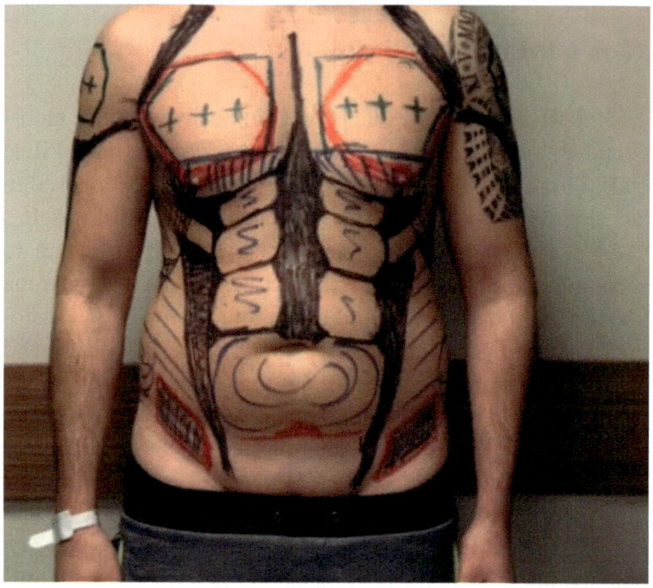

Fig. 17.7 Drawing frames around the muscle groups using different colors

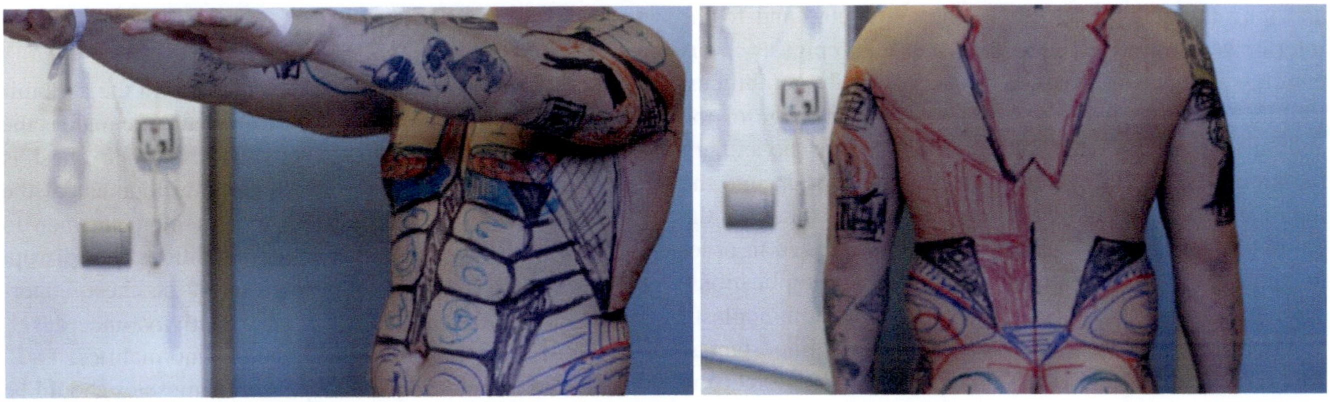

Fig. 17.8 Areas requiring deep liposuction are the black color frames, the red color borders, the green color fat transfer areas, and the dark blue color areas. The light blue color indicates the areas that require superficial liposuction

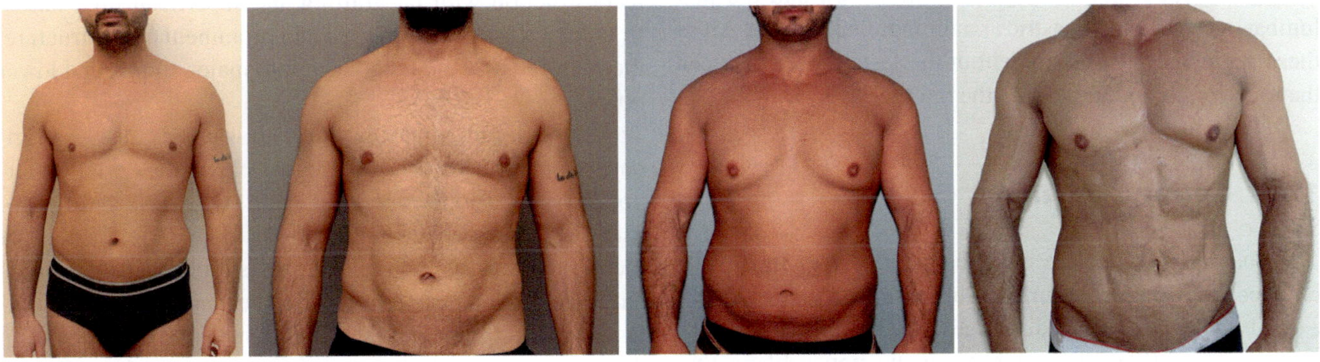

Fig. 17.9 Pronounced abdominal muscle groups after the fat has been removed from the natural space

17.9 Pyramidalis

The pyramidalis is a triangular muscle located on either side of the linea alba. The pyramidalis stems from the pubic symphysis and the pubic crest, and contracts as it moves upward and settles in the linea alba, located between the umbilicus and groin. The rectus abdominis and pyramidalis muscle, together with the lateral abdominal muscles (external oblique, internal oblique, and transverse abdominis), form the anterolateral abdominal wall. The pyramidalis muscle acts in conjunction with the rest of the abdominal muscles, contributing to various abdominal wall functions, including motions that increase intra-abdominal pressure when necessary (e.g., labor, forced expiration, defecation).

17.10 Blood Supply, Lymphatics, and Nerves of the Abdominal Wall

The arterial supply of the anterior abdominal wall can be divided into three areas. Zone I is the central anterior abdominal wall in the region of the epigastrium and rectus abdominis, and is supplied by the superior and inferior epigastric vessels. Zone II is the lower anterior abdominal wall below zone I, which is mainly supplied by the superficial epigastric, superficial external pudendal, and superficial circumflex iliac arteries. Zone III is the lateral of zone I, and is supplied by the lumbar, musculophrenic, lower intercostal, and subcostal arteries. Venous drainage from the abdominal wall enters the superior vena cava (SVC) via the inner breast, intercostal, and long thoracic vessels above the umbilicus, and to the inferior vena cava (IVC) via the superficial epigastric, circumflex iliac, and pudendal veins below the umbilicus. Similar to venous drainage, the lymphatic drainage of the abdominal wall above the umbilicus drains into the axillary lymph nodes, and those below the umbilicus drain into the inguinal lymph nodes. The vascularization status of the entire abdominal wall should be well characterized in order to achieve optimal results and to prevent complications. Vascular regions should be well defined for abdominoplasty and liposuction. Lymphatic drainage is equally important as vascular drainage, and care should be taken to protect the inguinal lymph node basins, especially during the dissection and removal of skin flaps.

The muscles and skin of the anterior abdominal wall innervate the ventral rami of the sixth to eleventh intercostal nerves, the subcostal nerve (twelfth thoracic), and the first lumbar nerve (iliohypogastric and ilioinguinal nerves). All of these segmental nerves run within the fascia layer between the transversus abdominis and the internal oblique.

17.11 Serratus Anterior

The quadrangular muscle emerges anteriorly as finger-like bundles from the outer surfaces of the upper 8–9 ribs, and extends to the vertical, medial edge of the scapula (Fig. 17.2). It consists of three parts, including the upper fibers that emerge from the 1–2 ribs and intercostal fascia, the middle fibers from the 3–6 ribs, and the lower fibers that are interdigitated with the upper five fibers of the external oblique. In muscular individuals, the lowest 3–4 finger-like bundles of the serratus anterior can be seen on the lateral chest wall at the lower edge of the pectoralis major. Since the finger-like bundles of the serratus anterior are thicker and more prominent than the external oblique fibers, they are more easily distinguishable. The remaining part of the serratus anterior is hidden at the top of the chest wall by the pectoralis major and posteriorly by the latissimus dorsi. The serratus anterior may be distinguished under the latissimus dorsi muscle, where it bulges out. The prominence of the serratus anterior plays an important role in body shaping in male patients, as it is important in accentuating the middle V in the male body. The most important part of the serratus anterior is the bundle, which extends superficially on the 8–9 ribs, just above the semilunar line. When the topographic drawing of this area is made, the patient should be told to put his hands on his waist and squeeze his waist, while the frame of the muscles can be marked with both manual and light examination.

17.12 Pectoralis Major

In addition to the rectus abdominus muscles in the abdominal region, another attractive feature of an athletic male is the well-defined pectoral region. An attractive male chest is strongly associated with the shape and development of the pectoralis muscles. Common causes that motivate men to have chest shaping surgery include conditions that disrupt normal anatomy, such as gynecomastia in the chest, abnormal fat accumulation, lack of physical activity, underdevelopment of chest muscles, or congenital abnormalities.

The surface anatomy of the pectoralis muscles should be revealed in the ideal male chest. Although the pectoral muscle has its own anatomical reflection, the chest should be considered as a whole, and the surrounding areas should be treated as well. Good results can be obtained for male chest reconstruction with fat grafting and superficial liposuction. In order to obtain a developed and prominent chest structure, the pectoralis should be divided into major, upper, and lower poles.

The pectoralis major, a thick fan-shaped muscle, originates from three parts, including the clavicular part, the sternocostal part, and the abdominal part (Figs. 17.2 and 17.10). The clavicular part originates from the anterior surface of the medial half of the clavicle, the sternocostal part originates from the anterior surface of the sternum (costal cartilage of the 1–6 ribs), and the abdominal part originates from the anterior layer of the rectus sheath, and joins laterally and attaches to the great tubercle of the humerus by a bilaminar tendon. The fibers originating from the clavicular part pass laterally and downward until they are adjacent and parallel to the deltoid fibers. Here, the muscle bundle of the clavicular part passes over other parts coming from the sternocostal and abdominal parts. The sternocostal muscle bundles pass almost horizontally, while the abdominal muscle bundles pass upward and laterally, integrating deeply into other parts. The pectoralis major muscle functions to perform adduction and internal rotation in the shoulder joint. The upper part of the muscle is often straight, while the lower part has more mass and provides a smooth convexity. The lower edge of the muscle is either straight and horizontal, or it can slope slightly downward medially to the side. The pectoral fat pad located near the nipple contributes to the volume and anterior convexity of the chest. The deltopectoral fossa is located between the clavicular part and the deltoid. There is a depression on the sternum between the right and left pectoralis major muscles in the midline, and this depression is more prominent in muscular individuals. In an aesthetically pleasing male chest, the upper pole has a

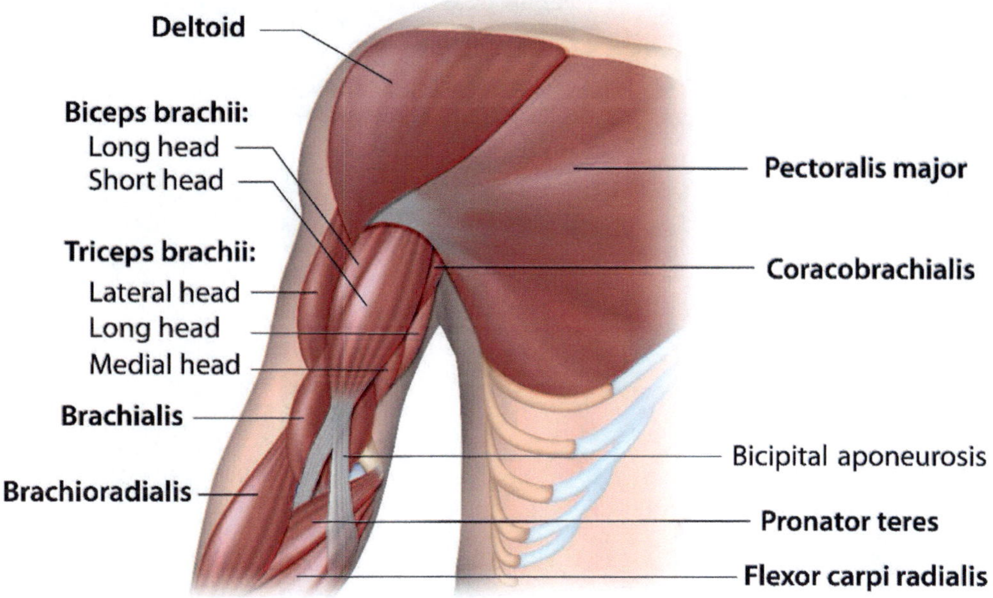

Fig. 17.10 Upper arm and superolateral abdominal wall. The deltoid, triceps, and biceps muscles, which are the most important muscles of the upper arm in body shaping

greater volume than the lower pole, due to its greater muscle mass. In the techniques used for male chest contouring, fats towards the upper abdomen and axilla are removed, volume is increased by fat grafting to the clavicular part, and the borders of the sternocostal part are made more prominent. Around 1 cm below the nipple, there is a horizontal line (pectoral line) along the lower border of the pectoralis muscle. Subdermal and intermediate liposuction should be aggressively used within the triangular area defined by the pectoral line and the lateral border of the rectus abdominis muscle (subpectoral triangle). Further, the surgeon must shape the triangular area between the lateral border of the pectoralis muscle and the lateral border of the latissimus dorsi muscle (pectoral-latissimus triangle), although beyond the pectoral region, in order to extend the concavity to this limit. The two most important muscles in the masculine male body are the pectoralis major and the deltoid muscle. For this reason, it is not recommended to operate on the abdominal area without first intervening in these two regions while creating the masculine "V." Regarding the pectoral muscle, the muscle margins in the relaxed and contracted positions of the muscle should be marked. The area between these two marks is known as the "transition zone." Since the transition zone is a dynamic area, it should undergo careful liposuction, taking into account the skin structure and age of the patient. There is a possibility of sagging breast appearance in cases with excessive fat resection from this area. Of note, the pectoral-latissimus dorsi triangle is located between the outer border of the latissimus muscle, the outer border of the pectoral muscle, and the line joining the lower border of the latissimus from the nipple. The pectoral-latissimus triangle, which is of great importance in the clarification of the upper body triangle, should have so much liposuction to the extent that almost no fat remains. Since the fullness in the male breast should be in the upper 2/3 of the pectoral muscle, fat transfer should be applied to this area under and into the muscle. This intervention ensures that the desired nipple position is achieved, which is often lateral and slightly downward in athletic males (Figs. 17.11 and 17.12).

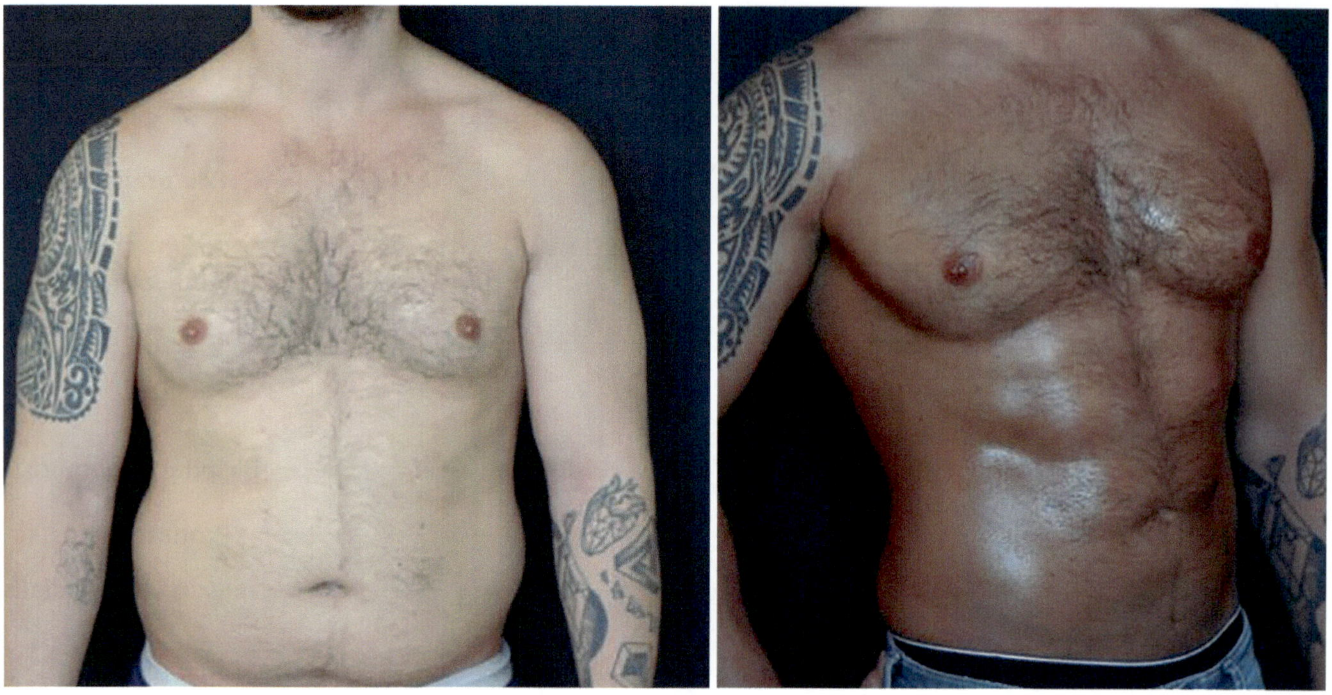

Fig. 17.11 Change in chest structure after pectoral muscle fat transfer. The deltoid and pectoral muscles should be worked on first in order to achieve the masculine "V"

Fig. 17.12 The insufficient masculine appearance of the patient, who had previously undergone gynecomastia operation, was remedied via pectoral intramuscular fat transfer

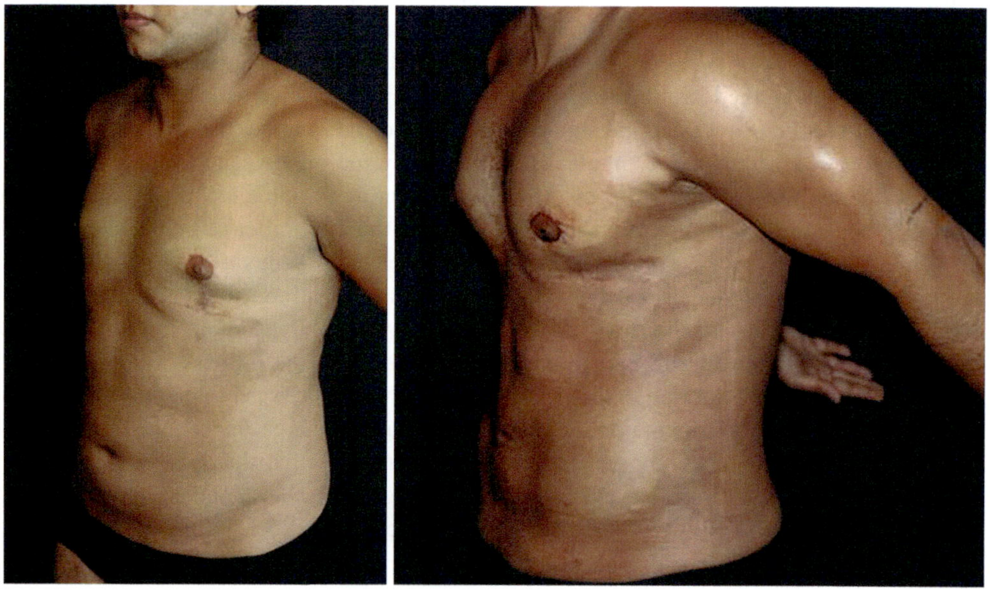

17.13 Latissimus Dorsi

The latissimus dorsi is a large, flat, triangular muscle that merges into a narrow tendon that overlies the lumbar region and lower thorax. It originates from the spinous processes of the 7th–12th thoracic vertebrae, all of the lumbar and sacral vertebrae, the posterior third of the iliac, and the spinous processes of the outer surfaces of the lower three ribs (Fig. 17.13). The latissimus dorsi passes from the midline to the arm like a cape. A tendon attaches the muscle bundles converging towards the axilla to the humerus. The muscle bundles come together and wrap around the teres major and form a hanger

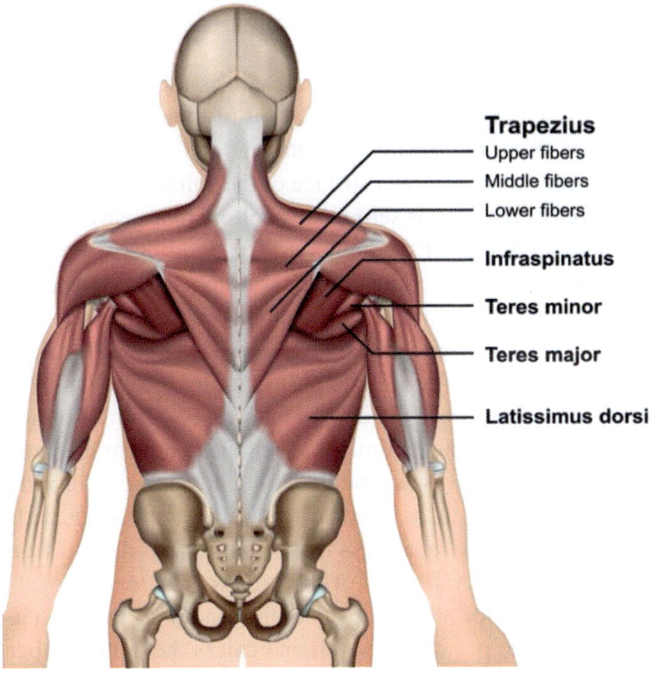

Fig. 17.13 Posterior trunk muscles. The most important muscle of the posterior trunk in body shaping is the latissimus dorsi

that forms the posterior wall of the axilla. Both of these latissimus dorsi muscles form an inverted triangle or "V" shape. The tendon originating from the latissimus dorsi forms a line or curve from the midline to the top of the ilium through the thoracic vertebrae. While the shape of the muscle is normally prominent in the middle and upper part, the part of the muscle below is hidden by the dorsal fat pad, which also hides the lower part of the outer oblique. At the drawing stage, the patient should press his hand down on the surgeon's shoulder to determine the lateral border of the latissimus. In the male patient, a certain amount of fat should remain on the latissimus muscle in order to make the dorsal triangle prominent. Therefore, the patient should lean his body to the right and left while standing in anatomical position in order to evaluate the presence of excess adipose tissue (Fig. 17.14).

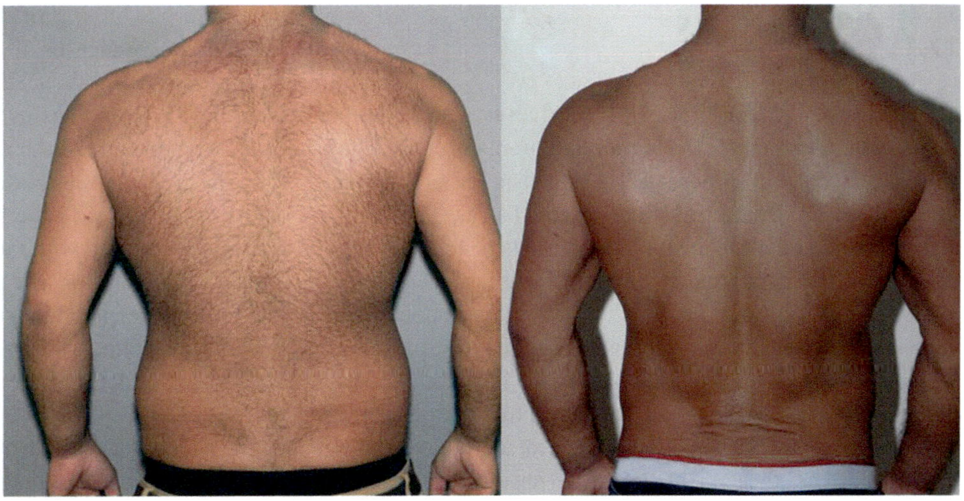

Fig. 17.14 The expected masculine V in the posterior view with the latissimus dorsi and lumbar muscle pronounced

17.14 Arm

The arm is one of the most difficult areas in body contouring due to the high risk of contour irregularity of the underlying fat structures (mostly superficial fat), the difficulty of achieving symmetry in bilateral treatment, and the thickness of the arm skin. While arm shape is determined by each patient's muscle structure and mass, there are three main muscles responsible for arm shape, including the deltoid, triceps, and biceps (Fig. 17.10). In both males and females, a well-defined arm is not straight but rather has curves. The deltoid and biceps muscles clearly define the curves in the forearm and lateral arm. However, it is more difficult to define and form the contour of the back of the arm due to its unique fat distribution. More prominent musculature in a man's arm is considered to be athletic and healthy. Fat should be removed in some areas to improve the anatomical structure, but should be grafted in others. As in all other regions, understanding the upper limb anatomy is of utmost importance in a surgical setting. It is necessary for the surgeon to know which plane he/she is in and the neurovascular structures nearby in order to minimize complications and ultimately increase patient satisfaction and survival.

17.15 Skin

The skin of the medial arm is thinner than it is around other body parts, and therefore, it is prone to premature wrinkling. The subcutaneous adipose tissue consists of areolar and lamellar layers separated by superficial fascia. The areolar layer is located under the skin and consists of fat cells separated by connective tissue arches. The lamellar layer consists of horizontally located fat cells and a large amount of connective tissue. The back of the arm consists of a prominent lamellar layer. Patients with increased fat have an increased thickness in the lamellar layer compared to the areolar layer. This increase is predominantly concentrated in the upper and middle third of the arm in the posterior-outer region. The back of the arm is also more prone to skin excess and looseness due to the large fat storage capacity in the area.

17.16 Deltoid

The deltoid is a triangular muscle consisting of an anterior part originating from the lateral third of the clavicle, a middle part coming from the acromion, and a posterior part stemming from the lower surface of the spine of the scapula. All three of these parts settle in the deltoid tuberosity, located in the middle part of the humerus. The anterior part is separated from the pectoralis major by a deltopectoral triangle and groove. A superficial liposuction procedure in this triangle will make both the deltoid and the pectoral muscles more prominent. The middle part is multipennate and contains three septa from the deltoid tubercle and four intramuscular septa from the acromion. When viewed from the front, the middle part gives the shoulder a round appearance, and it inserts from below, in contrast to the front and back parts. The posterior part is a mass that separates the deltoid from the long and lateral heads of the triceps. The tendon of the deltoid's anterior part passes over the tendon of the clavicular part of the pectoralis major, and enters the deltoid tubercle on the anterior surface of the humerus. The other two parts of the deltoid insert into the lateral side of the humerus. The shape of the deltoid slightly changes when the arm is rotated medially or laterally. In slim and muscular individuals, the cross-linking of the muscle bundles creates a segmented appearance on the side of the shoulder. The round deltoid shape is enhanced during high-definition lipoplasty from the incision sites in the anterior and posterior axillary folds. The midline of the deltoid should be marked first with the arm in 90 degrees abduction. At the point where the deltoid muscle ends on the humerus, there is a small natural negative area filled with fat at the midline between the triceps and biceps. Anterior and posterior rotation movements and biceps and triceps boundaries can be easily determined, and the muscle frame should be marked. When the arm is in 90-degree abduction in a posterior view, the angle made by the posterior axillary line and the arm is expected to be 90 degrees in women and 60 degrees in men. All of the fat should be removed from the marked frame in order to clarify the deltoid muscle. In addition, 100–150 cc of fat should be injected into the muscle from the posterior (Fig. 17.15).

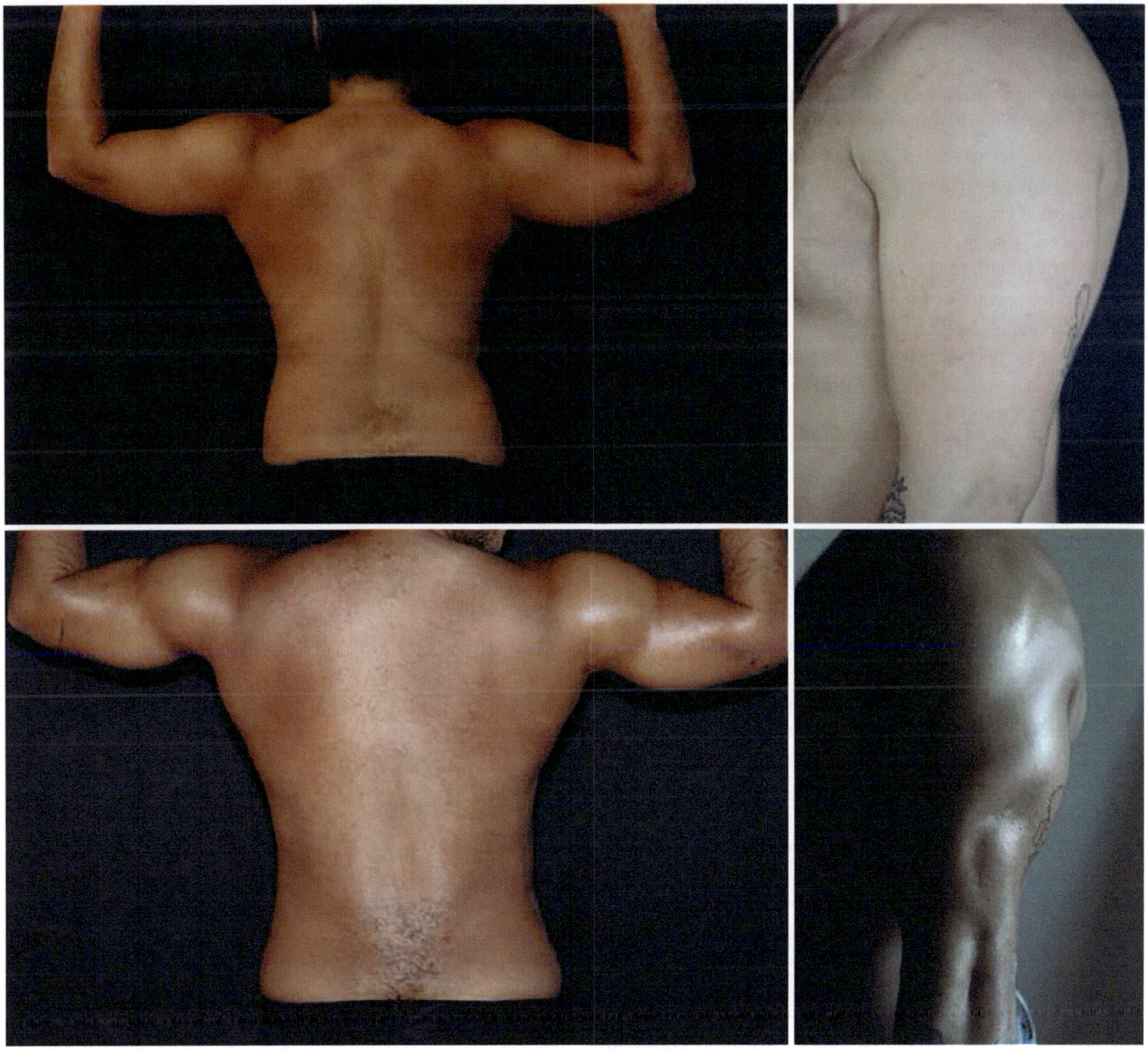

Fig. 17.15 The definition of the deltoid and triceps muscles after fat transfer and liposuction

17.17 Triceps

The large triceps muscle, which fills the majority of the extensor compartment of the upper arm, consists of three heads, the long, the lateral, and the medial. The long and lateral heads of the triceps provide the superficial shape of the rear upper arm. The medial head is deeper and contributes to the arm's volume and thickness. The long head of the triceps stems from the infraglenoid tubercle of the scapula, crosses the shoulder joint, and forms the main mass of the posterior upper arm. The lateral head originates from the proximal posterior surface of the humerus, and can easily be seen as a mass near the deltoid on the lateral side of the arm. The medial head has a broad origin and originates from the medial intermuscular septum, the lateral intermuscular septum, and proximally from the humerus, just below the radial nerve's groove. The medial head is covered with lateral and long heads, and can only be seen distally in the humerus. The medial head inserts the tendons of all three triceps' heads into the posterior aspect of the proximal olecranon. A cylindrical form containing part of the triceps' medial head can be seen in the lower half of the medial side of the arm between the biceps brachii and the long head of the triceps. In this region, the surgeon can make prominent the groove between the biceps and the triceps' medial head anteriorly, and posteriorly, the groove between the medial and long heads of the triceps. When viewed from the side in a 90° position, there are folds in the proximal and distal regions of the athletic posterior arm. A well-developed triceps creates a convex area in the posterior middle, while the triceps tendon straightens the distal posterior arm, and the triceps proximal insertion creates a curvature to the axilla. This curvature is more pronounced in men. At 90 degrees of abduction, a vertical parallel angle, called the "youth angle," is defined between the shoulder and the lower border of the arm. The narrower the angle, the more pronounced the triceps muscle mass. The arm should be brought into hyperextension for marking the triceps muscle and separating all of its structures. Thus, the contraction of the triceps muscle groups is ensured, and the natural shadows can be easily detected. Liposuction in these areas, called "negative areas," is often sufficient for definition.

17.18 Biceps Brachii

The biceps brachii gets its name from its two proximally attached parts. The long head of the biceps originates from the supraglenoid tubercle of the scapula, while the short head originates from the coracoid apex of the scapula. The long and short heads of the muscle moving on the humerus insert into the radius tuberosity through the biceps' tendon. The two heads of the biceps brachii, which usually appear as one bulging form on the upper arm, can only be seen as two separate structures medial to the cephalic vein. Proximally, the part of the biceps overlapping the pectoralis major and deltoid is distally covered by only fascia and skin. At the medial border, the biceps brachii is associated with the coracobrachialis, brachial vessels, and the median nerve, while at the lateral border, it is associated with the deltoid and brachioradialis. A groove forms between the biceps brachii and pectoralis major with abduction and external rotation. This groove should be emphasized in body contouring in order to clarify the separation between the muscles. Since the convex form on the anterior of the upper arm is mostly formed by the mass of the biceps brachii, a well-developed biceps brachii symbolizes strength, athleticism, and an aesthetically ideal upper arm form. If one desires to clarify the biceps muscle with fat injection, the most practical and useful approach is for the surgeon to reach into the muscle (while protecting the veins and nerves) with a 2 mm incision in the medial, and to inject fat into the medial part (Fig. 17.16).

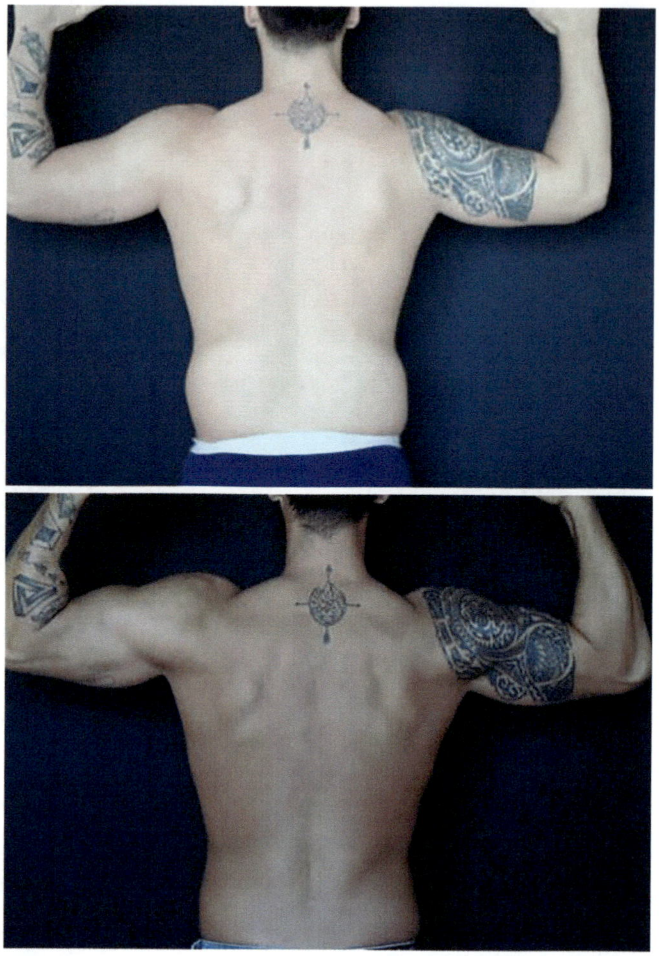

Fig. 17.16 Fat transfer is the most important part of the male body shaping operation. These images show the change after injections were made into the deltoid, biceps, and trapezius muscles

17.19 Brachialis

The brachialis is a large muscle that is located behind the biceps brachii in the humerus. The brachialis emerges from the lower half of the anterior surface of the humeral shaft, begins on both sides of the deltoid entrance, and passes through the elbow joint into the ulna's coronoid process. The junction of the brachialis and the deltoid can be seen as a depression or groove in the superolateral origin of the brachialis. The medial fibers of the brachialis form the base of a concave cavity bounded near the elbow by the anterior bicipital aponeurosis and posterior by the medial head of the triceps. This is where the brachial vessels are located. When viewed from the front, the brachialis creates a lateral protrusion from the back of the biceps' brachia.

17.20 Blood Supply, Lymphatics and Nerves of the Arm

The arterial supply of the upper limb begins with the subclavian artery. The subclavian artery is called the axillary artery once it passes through the lateral edge of the first rib. The axillary artery passes from the axilla towards the humerus and branches off first to the anterior and posterior circumflex humeral arteries, followed by the subscapular artery, its largest branch. As it passes through the teres minor, it becomes the brachial artery. Venous drainage of the upper limb occurs via two large veins. The first is the basilic vessel, which consists of the radial and ulnar vessels, and the second is the cephalic vessel, which stems from around the hand and progresses into the axillary vein.

All of the nerves in the upper limb are supplied by the brachial plexus, which consists of the anterior rami of the spinal nerve levels from C5 to T1. The nerve roots originating from C5 to C7 form the musculocutaneous nerve, and these move down distally in the upper arm, and deeply pierce the coracobrachialis in an upwards manner. The musculocutaneous nerve that travels between the biceps brachii and the brachialis muscle eventually becomes a lateral cutaneous nerve when it passes through the lateral side of the biceps' tendon. The musculocutaneous nerve provides motor innervation to the biceps brachii, brachialis, and coracobrachialis and gives sensory innervation to the radial side of the forearm. The musculocutaneous nerve consists of the nerve roots of the C5 and C6 axillary nerves. The axillary nerve provides motor innervation to the deltoid muscle and teres minor as well as sensory innervation via the upper lateral cutaneous nerve of the arm.

It is not uncommon for the arm's sensory nerves to be damaged during brachioplasty surgery. The most commonly damaged cutaneous sensory nerves during brachioplasty surgery are the medial brachial cutaneous nerves and the medial antebrachial cutaneous nerves. Reliable estimation of the location of the sensory nerves affects incision options and gives limited dissection depth in arm lift surgery.

Body Fat Grafting Contouring in the Male Patient

18

Alberto Di Giuseppe and Federico Giovagnoli

Contents

18.1 Art and Anatomy

The merging of art and anatomy is perhaps most evident and glorious in the works of Leonardo da Vinci. His anatomical manuscripts are filled with detailed and intricate drawings that elegantly reveal muscular anatomy, symmetry, and human proportions [1]. His knowledge of anatomy was derived from his own independent dissections and research and is elegantly and beautifully portrayed in many of his renowned paintings. Michelangelo similarly celebrated human physicality through his work as an artist and sculptor. David represents one of the most recognized sculptures of the Renaissance and showcases the human body in all its strength, athleticism, and youthful beauty. The muscular definition of deltoid, biceps, pectoralis major, rectus abdominis, and external obliques is clearly visible in Michelangelo's David in Florence, Italy.

The sense of artistry and the understanding of the body anatomy are represented by the drawings of Michelangelo, who painted the Sistina Chapel at the Vatican Museum in Rome.

So it is confirmed that the sense of an athletic body figure, and the shaping of the muscle edges, in order to obtain a defined frame, was well understood already in sixteenth century.

The ideal female abdomen presents anatomical differences with the male body: length and consistency of the muscle is different, and interdigitations among rectus muscles are different (Figs. 18.1, 18.2, 18.3 and 18.4).

Abdomen in women is not just flat: it is a combination of convexities and concavities that follows the anatomy of the bone and muscles underneath. The female torso is characterized by a perfect combination of curves and projections that influence the well-known waist-to-hip ratio. Achieving an ideal waist-to-hip ratio and seamless smooth contours in this zone are important objectives in order to achieve female attractiveness. Achieving the correct degree of definition is a challenge. In women, the definition must be appropriate to obtain athletic, slim, and feminine results, while in men, a harder, more muscular appearance is preferred.

A. Di Giuseppe (✉)
University of Padova, Padova, Italy

F. Giovagnoli
University Vita-Salute San Raffaele, Milan, Italy

© The Author(s), under exclusive license to Springer Nature Switzerland AG 2023
A. Di Giuseppe et al. (eds.), *Fat Transfer in Plastic Surgery*, https://doi.org/10.1007/978-3-031-10881-5_18

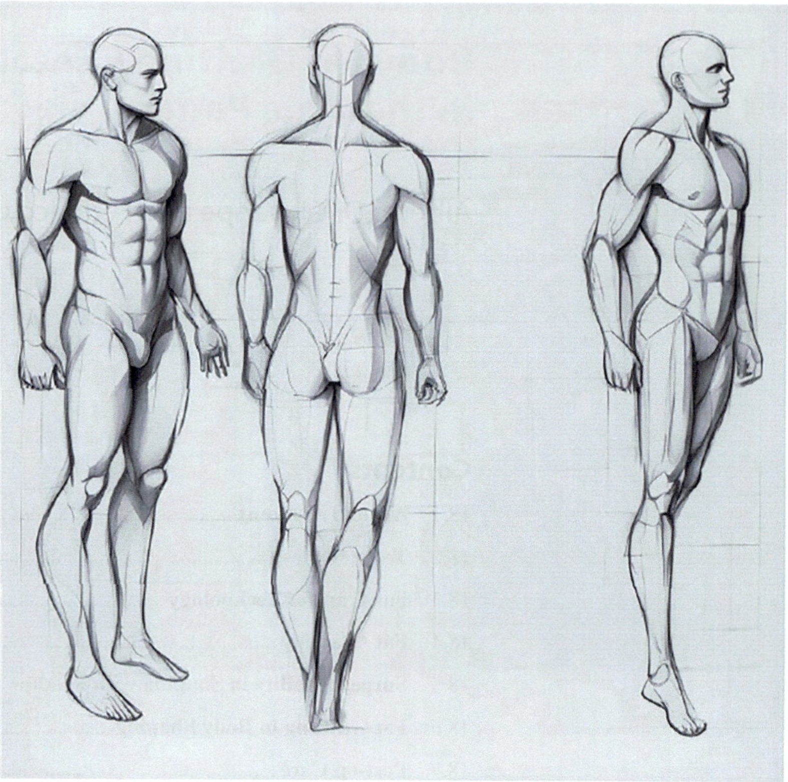

Fig. 18.1 Drawings of the ideal proportions of the human body

Fig. 18.2 Man abdominal and thorax muscle

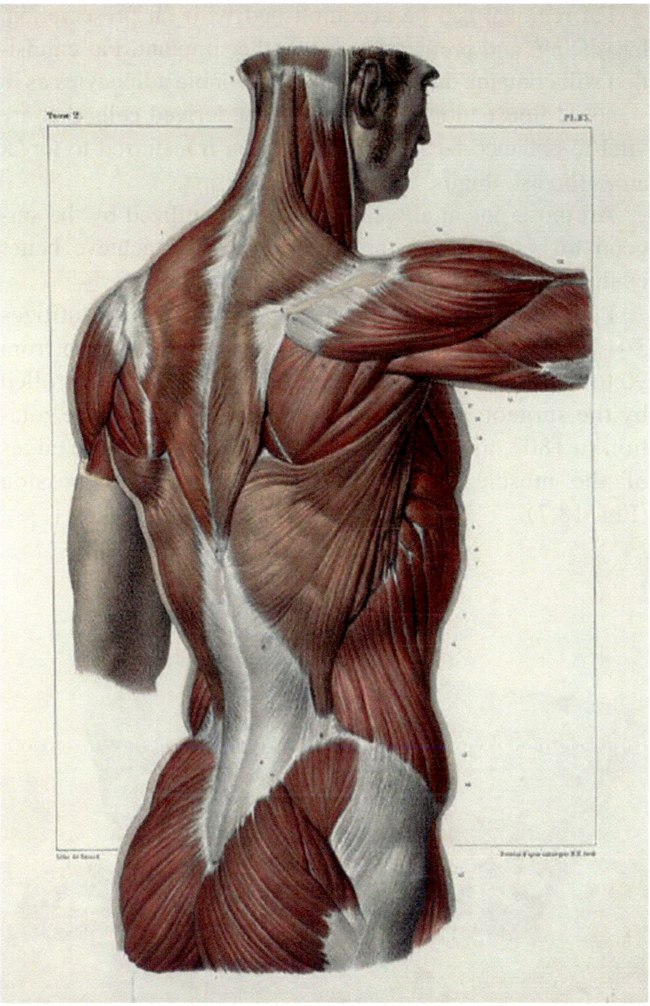

Fig. 18.3 Dorsal and back muscles

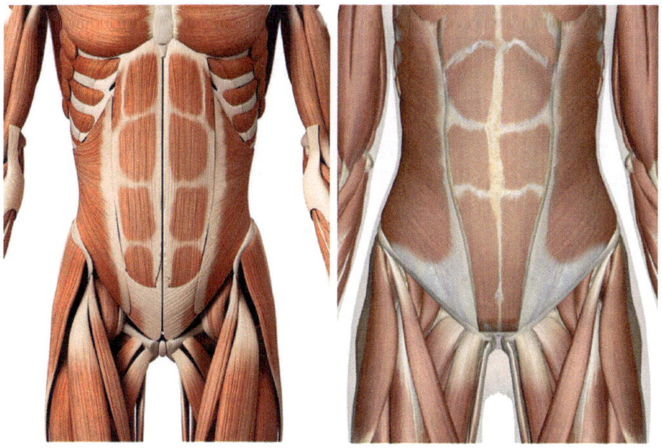

Fig. 18.4 Male and female abdominal muscles

18.2 Technologies

The technologies I did use for sculpturing men and female body are:

- Vaser™ ultrasound-assisted lipoplasty
- PowerX, Power-assisted lipoplasty
- Revolve™ cannister for fat harvesting

New technologies have entered the market in the last decade, as power-assisted technology, ultrasound technology, and laser technology, in order to facilitate fat removal, reduce trauma, and improve skin retraction. I have personally been one of the pioneers of ultrasound-assisted liposuction since they appeared in the early 2000s and contributed to the realization of the new vaser device, more advanced and safer than previous ultrasound-assisted devices in commerce until 20 years ago. Vaser™ is an acronym of vibration amplification sound at resonance and is a third generation of ultrasound-assisted liposuction that uses ultrasound energy to melt fat tissue (Fig. 18.5).

Ultrasound technology emulsifies the fat for removal in such a delicate way that preserves as much of the tissue of the matrix as possible while emulsifying the desired amount of fatty tissue. Special titanium probes deliver the ultrasound energy to fat tissue, after proper infiltration with tumescent solution, allowing the selective destruction of only the fat tissue through three mechanisms: (1) cavitation, (2) mechanical, and (3) thermal (Fig. 18.6).

In cavitation effect, the Vaser™ probes vibrate at ultrasonic frequencies creating compressive and rarefactive forces around the grooved tip; cavitation microbubbles expand and then implode, releasing energy that disrupts the adipocyte architecture until an emulsion of fat and fluid is formed.

Mechanical disruption of fat occurs at the tip of the probe, where the vibrating metal surface comes in contact with the adipocytes. Cavitation and mechanical disruption of adipose tissue occur due to relative fragility of this tissue compared to other tissues such as vessels, muscle, and nerves. This is why this process is safe and effective, spears the connective and vascular network of the skin and subcutaneous tissue, allows a safe subcutaneous superficial undermining of the skin, and is thus a major skin retraction of tissue. The thinnest is the dermal tissue, the highest is the skin retraction, and liposuction is the term to define vaser smoothening of the fat. Careful emulsification, not an aggressive fat removal, is selective where needed in two different layers: subcutaneous for superficial definition and deep dermal for volume removal [3].

Minding the advantages of this technology, the plastic surgeon must be able to achieve superior results in terms of definition and shaping.

With superficial fattening and careful undermining, he can achieve a thin layer of skin flap, ideal for superior skin retraction.

Deeper fat removal can be done in a less traumatic mode, leaving less scar tissue, with less chances of seroma formation and induration and asymmetry.

Fat removal can be accomplished with fat preservation, harvesting, and preparation for further implant. Fat emulsified still contains the same amount of viable adipocytes as in standard liposuction, with same stem-derived cells that are vital to enhance fat survival rate when transferred to target areas (breast, thighs, buttocks, face, etc.).

All those potential weapons must be utilized by the surgeon to better define the body shape and achieve better contouring.

PowerX is a power-assisted device, which utilizes 3–4 mm Mercedes cannula with a range of rotation from 90 to 360°. The speed of suction can be set and controlled by the surgeon as well. In my experience, I set the rotation to 180°, thus maximizing the aspiration at the edges of the muscle where I wanted to create depression (Fig. 18.7).

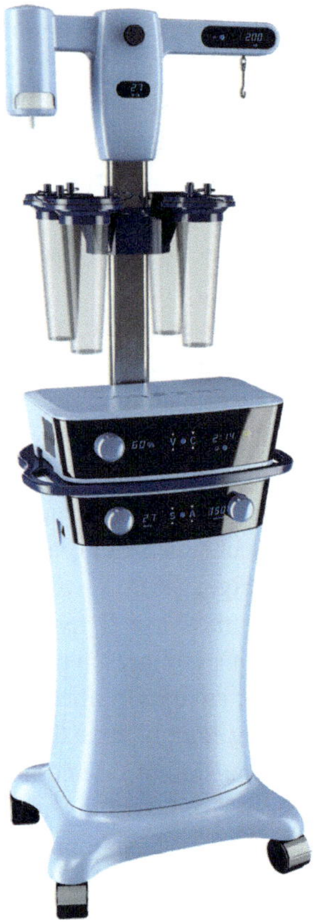

Fig. 18.5 Vaser system (Photo taken from VASER Solta Medical User's Guide [2]. Courtesy of Bausch, ©2022 Bausch Health Companies Inc., All Rights Reserved)

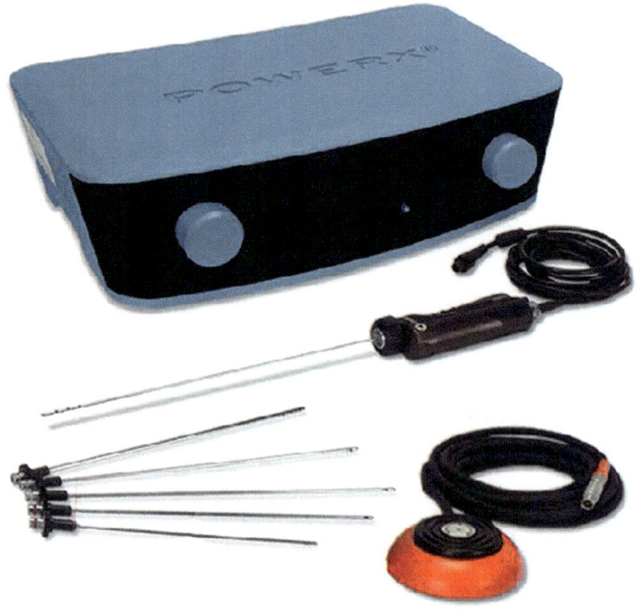

Fig. 18.7 PowerX system (Photo taken from VASER Solta Medical User's Guide [2]. Courtesy of Bausch, ©2022 Bausch Health Companies Inc., All Rights Reserved)

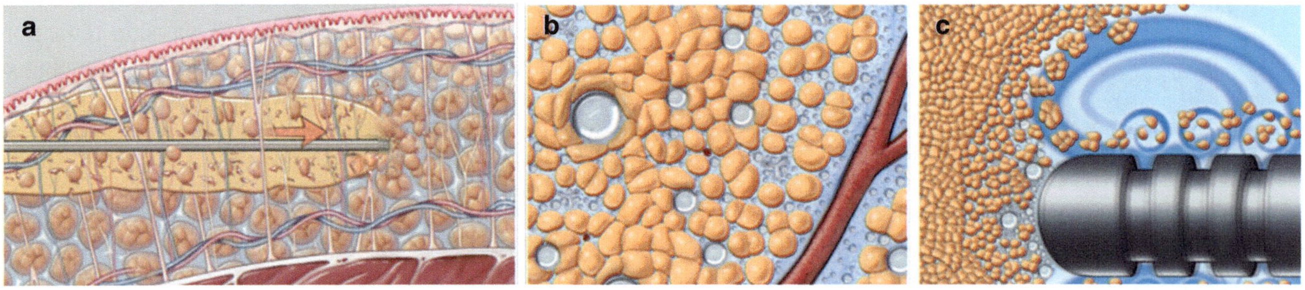

Fig. 18.6 Fragmentation (**a**), cavitation (**b**), acoustic streaming (**c**) (Photo taken from VASER Solta Medical User's Guide [2]. Courtesy of Bausch, ©2022 Bausch Health Companies Inc., All Rights Reserved)

18.3 Fat Transfer Technology

In the last 4 years, I used the REVOLVE™ system by LifeCell.

An integrated, high-volume fat processing system designed by adipose transplantation pioneers and founders of the International Federation for Adipose Therapeutics and Science.

It is an autologous fat tissue [AFT] collection device and tissue delivery system. It collects autologous adipose tissue and rapidly washes and delivers the tissue for body contouring.

REVOLVE™ System is a sterile single-use disposable tissue canister used for harvesting, filtering, separating, concentrating, and transferring autologous tissue components for reintroduction to the same patient during a single surgical procedure for repair, reconstruction, or replacement of integumentary or musculoskeletal tissues.

For additional volume of tissue, canister may be filled and emptied a second time during the same surgery on the same patient.

It is an all-in-one closed system, for harvest, filter, active wash, and remove strands.

Mesh filter designed to filter and strain lipoaspirate during the harvesting phase and active mechanical washing process.

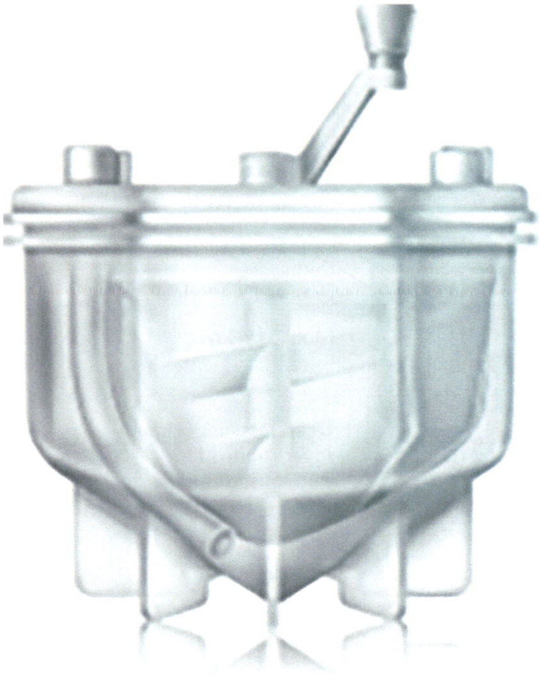

Fig. 18.8 REVOLVETM system (Photo taken from VASER Solta Medical User's Guide [2]. Courtesy of Lifecell Inc.© 2022, All Rights Reserved)

It rapidly removes tumescent, reduces cell debris, eliminates free oil, and finally concentrates adipose tissue (Fig. 18.8).

The system contains a prepeller for active washing, it does enhances washing process for high-quality fat and removes collagen strands to minimize injection syringe clogging.

Fat is extracted by syringes. It is possible to direct load syringes for injection from canister to minimize steps.

Extraction with catheter tip or luer lock syringe is performed (Fig. 18.9).

The system is a high-volume fat processing system: it processes up to 700 mL of lipoaspirate in less than 10 min. It is easier to use, requiring fewer steps to process large quantities of adipose tissue.

It can control critical fat grafting variables and is sterile, single-patient, disposable device.

It is a closed system that minimizes tissue handling and exposure to outside air. It can control critical fat grafting variables and is sterile, single-patient, disposable device.

Clinical result shows that REVOLVE™ system processed a higher percentage of adipose tissue compared with the two other methods: decantation and centrifugation. It yielded significantly less blood cell debris and a lower percentage of free oil. It has significantly higher fat graft retention (73.2%) than decantation (37.5%) and similar to centrifugation (67.7%) and maintained fat in physiological conditions (pH and osmolality) (Figs. 18.10 and 18.11).

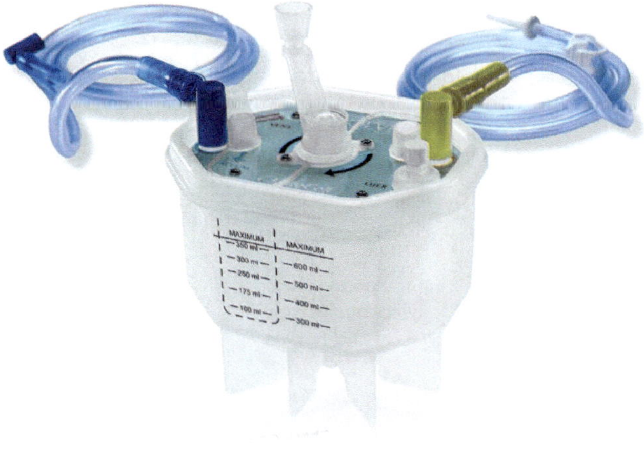

Fig. 18.9 REVOLVETM system syringes for extracting fat (Photo taken from VASER Solta Medical User's Guide [2]. Courtesy of Lifecell Inc.© 2022, All Rights Reserved)

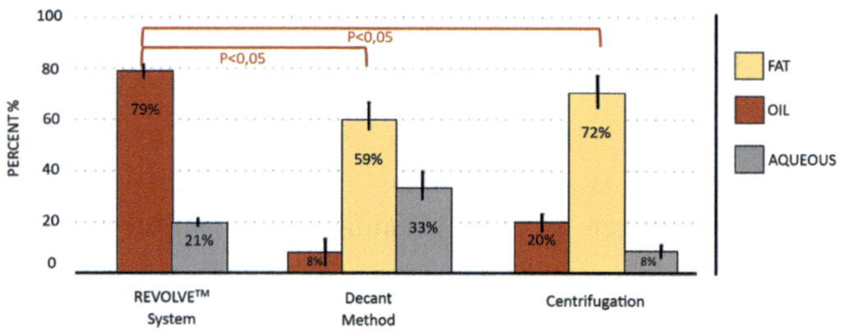

Fig. 18.10 REVOLVE System minimized potential inflammatory response for transplanted tissue by significantly removing red blood cells, fatty acids and debris (Photo taken from VASER Solta Medical User's Guide [2]. Courtesy of Lifecell Inc.© 2022, All Rights Reserved)

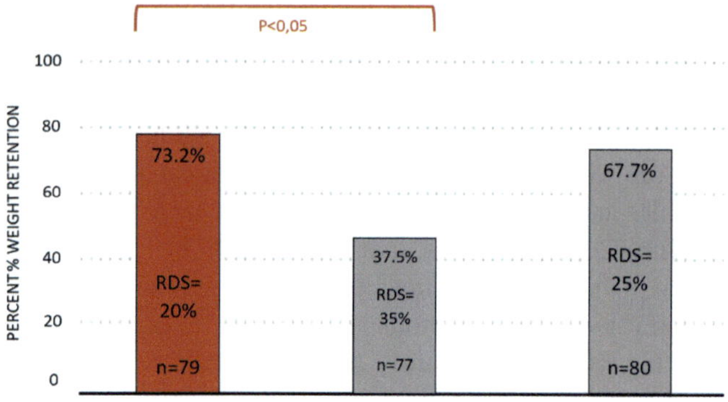

Fig. 18.11 Higher fat graft retention than decantation and similar to centrifugation (Photo taken from VASER Solta Medical User's Guide [2]. Courtesy of Lifecell Inc.© 2022, All Rights Reserved)

18.4 Fat Anatomy

The subcutaneous tissue is divided into three layers: a superficial adipose tissue layer, an intermediate membranous layer, and a deep adipose tissue.

The thickness and proportion of those layers vary throughout the body depending on the anatomic region: the abdominal has a prominent fascial plane, making it easier to distinguish between the two fatty layers.

In the leg, there is an attenuated fibrous membrane that separates from the muscle fascia. In the trunk, the adipose tissue is similar, as in gross appearance and density and structure (Fig. 18.12).

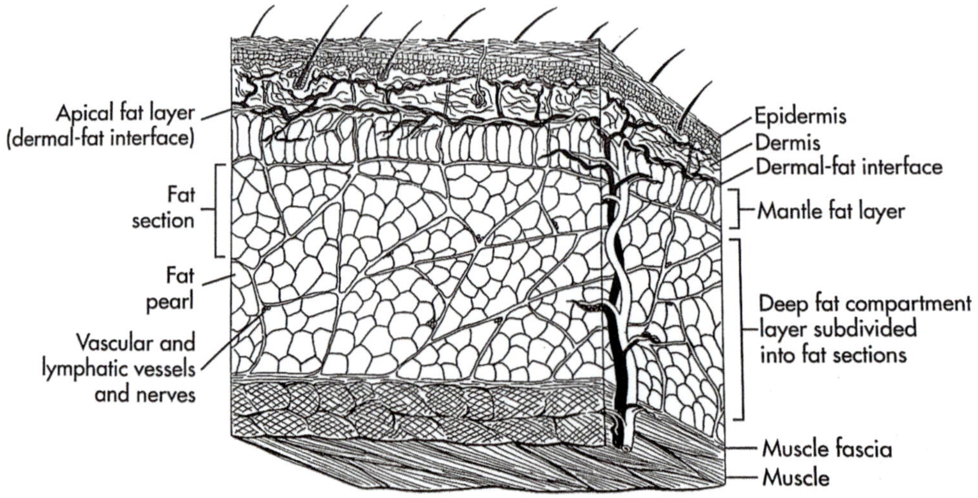

Fig. 18.12 Fat anatomy [4]

18.5 Surgeon Ability in Shaping with Shadow and Depression Lines

The introduction of vaser allows efficient and safe emulsification of fat in superficial and deep layers while preserving vascular and neural structures.

This is a new chisel in surgeon hands to sculpture the body as a work of the artist, working in all subcutaneous planes, by adding and subtracting fat with delicate instruments and refined techniques. The subdermal plane is no longer a taboo area; controlled deformities are desirable; and muscular definition is attainable through lipoplasty by revealing the underlying anatomy.

The full understanding of muscle anatomy is essential to try to recreate a muscle-shaped body and enhance definition and contouring [5, 6] (Fig. 18.13).

The lines of body are described as natural curves that are defined by muscle, bone frame, and fat deposit. Posture is another component to assess the natural body appearance, and man and woman are different of course. Lipoplasty can address only subcutaneous tissue and is not a solution for abdominal protrusion secondary to muscle weakness or intraabdominal fat. This will require muscle exercise and diet to compensate. The achievement of symmetry and proportion is the goal of any body contouring procedure. The man of Vitruvio, designed by Leonardo da Vinci, remains the golden reference of human proportions. In the recent age, new trends in fashion have defined new more athletic figure, more toned, even in woman, answering to the so popular attended gym all over the world.

The surgeon view and sense of artistry are essential in planning a good shaped body. He must create in his mind a tridimensional view of the new body to shape: ideally must thin the subcutaneous tissues as much as to display and reveal the superficial musculature, removing fat and highlighting major muscle groups. The salient features of the muscular anatomy relevant to body contouring are outlined such as origin, insertion, orientation, form created, and relationship to adjacent muscle groups.

Main muscles involved are the rectus muscles, the oblique muscle, the pectoralis in the front appearance, the latissimus dorsi, the lumbaris, and the gluteus in the back.

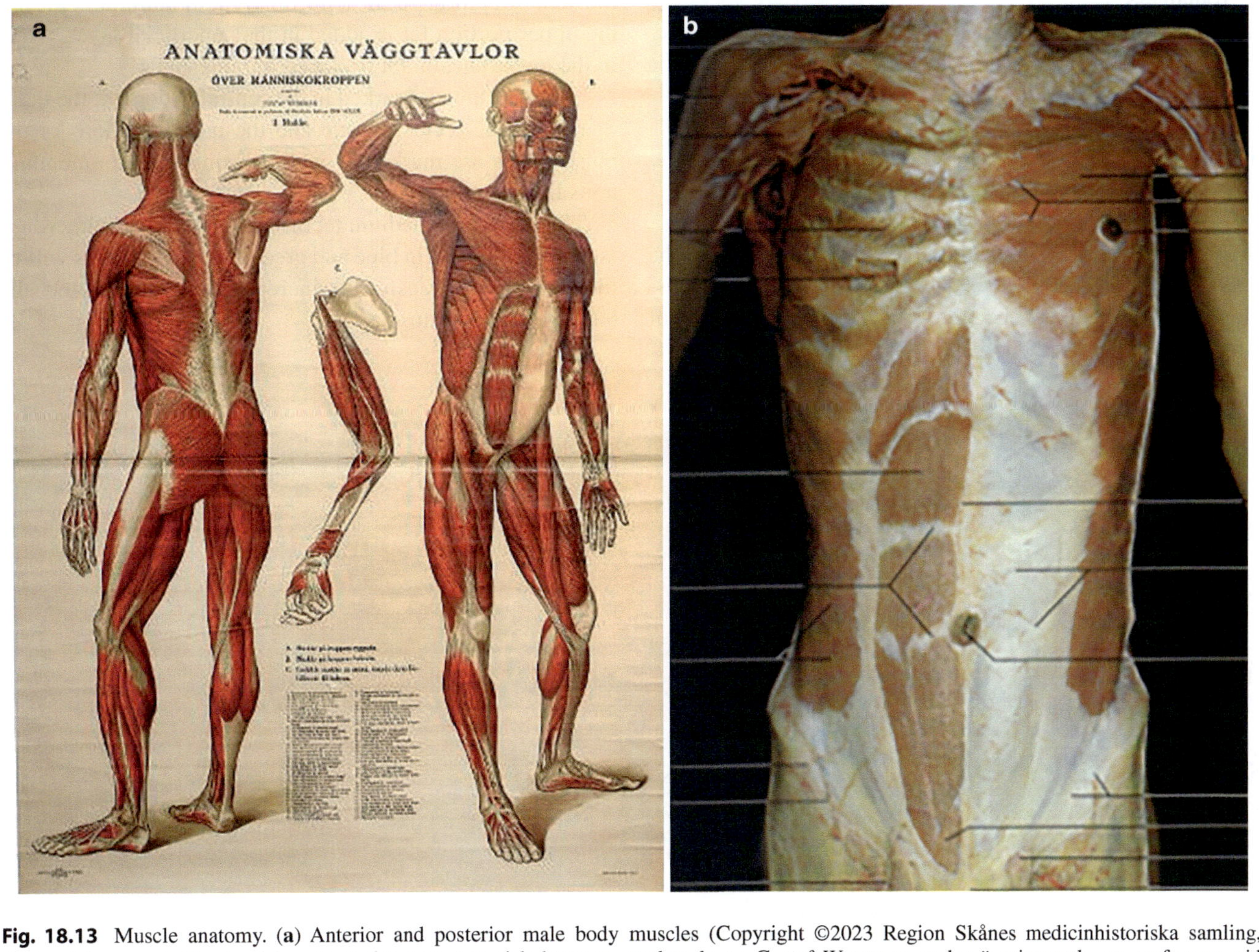

Fig. 18.13 Muscle anatomy. (**a**) Anterior and posterior male body muscles (Copyright ©2023 Region Skånes medicinhistoriska samling, Anatomisk plansch;"Anatomiska väggtavlor över människokroppen utarbetade av Gustaf Wennman under överinseende av professorn vid Karolinska Institutet Erik Mülle. II. Muskler" MS13640) (**b**) Rectus abdomini muscles with interdigitations defining the six

The muscle edges and the intermuscular digitations must be outlined to better define an athletic abdomen, together with the linea alba and linea semilunaris recreation.

Lines of transition between muscles must be obtained with more suction in a superficial manner. The main concept is removing and revealing what is underneath in normal-weighted patients, while traditional liposuction focuses in removing fat in overweight patients. The sculpturing surgeon tries to reveal the underlying musculature and body anatomy; shaping is a combination of removal to reveal what is underneath and adding in areas of deficiency to give more curvature.

It is a mixture of light and shadow and depression and concavities to give the impression of a more defined and toned body [7–11].

18.6 Fat Grafting in Body Shaping

Indications for fat transfer in men body shaping are:

1. enhancing buttock frame and increasing volume and roundness
2. enhancing muscle volume where lacking (pectoralis, rectus abdomini, biceps, and triceps muscles)

18.6.1 Enhancing Buttock Frame and Increasing Volume and Roundness

In man, fullness is desirable in the upper pectoral region, the deltoid, and the biceps, and even in the medium gluteus muscle: fat is grafted to enhance the volume, giving an image of a more sculptured and defined muscle region.

An example of this high-definition surgery in man is given [12, 13].

Case 1

This 32-year-old athletic body was looking for enhancment of shaping, definition and more bulky, six pacs, and abdominal muscle shape.

A mix of fat graft and extra anatomical resection, followed by multilayer fat grafting, and the creation of negative spaces (shadows) is the key for obtaining an attractive shaped athletic body frame (Fig. 18.14).

In Fig. 18.14a, the pre op appearance of this 32-year-old man is shown; patient was looking for more athletic definition all over his upper trunck, removal of his mild gynecomastia, definetly enhancing volume of his pectoralis and rectus abdominis muscles.

Expecially he wanted his six pacs really evident as well as the inguinal ligament and the oblique muscles.

In Fig. 18.14b, planning of a high-definition surgery from the frontal view is shown. The areas of biceps muscle are marked, the area of male gland reduction is circled, the area of superficial and deep fat transfer for pectoralis is marked: the area of fat transfer is defined in red, minding that mostly the upper portion fat is added in the superficial layer, while in the lower portion fat is placed intramuscularly.

The linea alba is marked in the central abdomen; the outer margin of the rectus muscle and the interdigitations of the muscle area are marked as well defining the semimembranous line.

The upper and medium rectus abdomini muscle interdigitations are marked in blue and green. Those two areas will be filled with fat transfer to increase the volume and the bulkyness of the muscle.

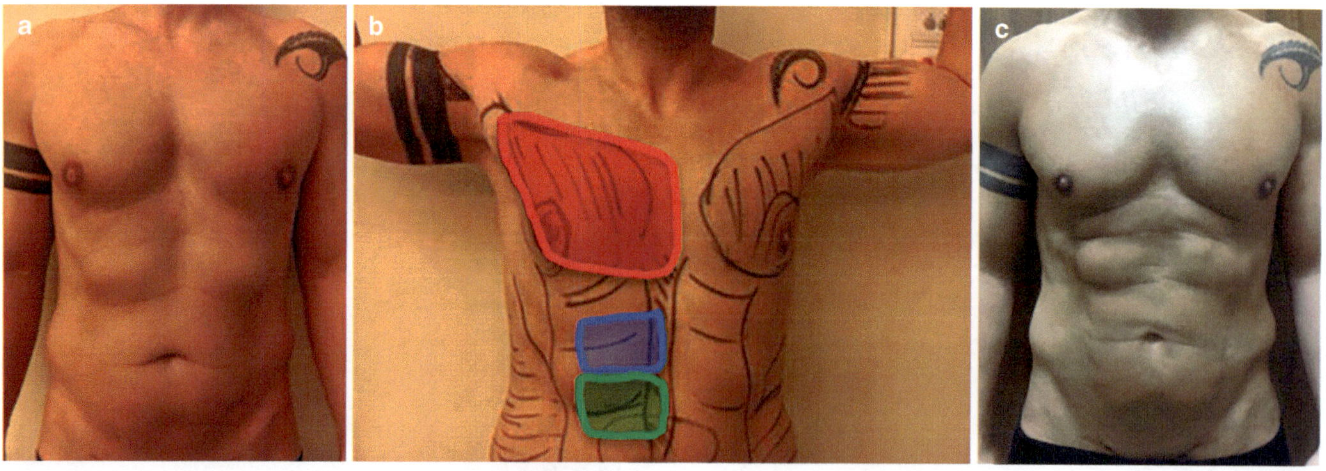

Fig. 18.14 (**a**) Pre op. (**b**) Planning; red zone pectoralis enhancement, blue upper rectus abdomini muscle, green medium rectus abdomini muscle. (**c**) Post op 6 months. Pre op and post op oblique view. (**f**) Pre op lateral view. (**g**) Marking of areas of fat removal and fat adding (in red). (**h**) Post op appearance

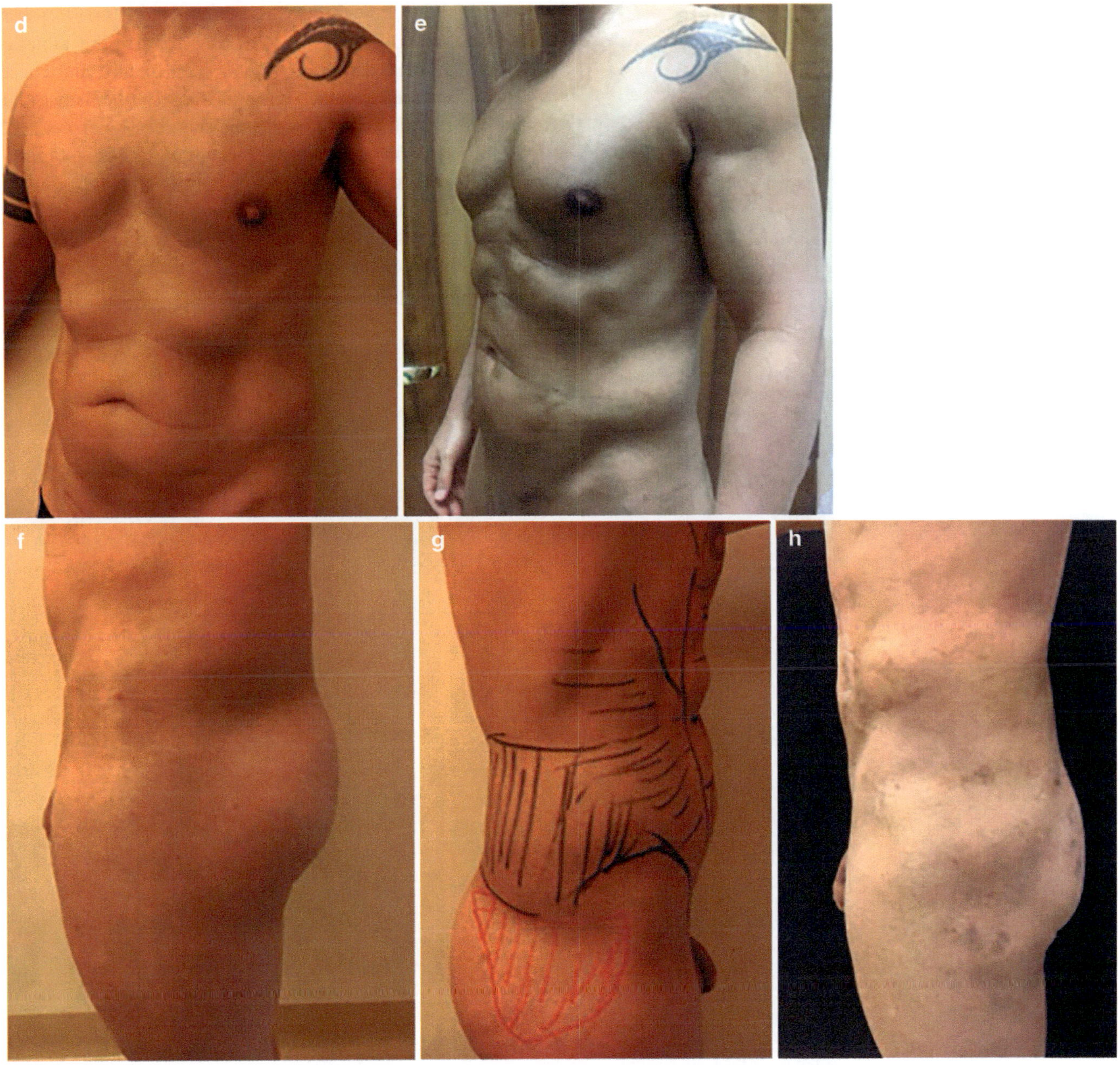

Fig. 18.14 (continued)

The margin of the oblique muscles, the inguinal ligament to create a marked depression, and the extra fat surrounding the navel are circled, and finally, the lateral flanks are lined where fat needs to be removed.

"Planning" of lateral side: margin of the oblique muscle outlined, edge of latissimus muscle, the inguinal ligament, and in red the outer medium gluteus zone to be grafted with fat to increase roundness.

Result of definition of an athletic young body is shown, after linea alba and linea semilunaris deepening and oblique muscle and inguinal ligament being enhanced.

Note the concavities and shadows created, give an image of an athletic well-shaped body with enhanced muscle projection and definition (Fig. 18.14).

The gluteal region is approached in a distinct way: lateral flanks are reduced in an aggressive way; the inguinal ligament is enhanced and glutes' prominence reduced, while the outer medium gluteus muscle is filled with fat in order to improve the roundness of the area and compensate a muscle deficiency (Fig. 18.14f–h).

18.6.2 Enhancing Muscle Volume Where Lacking

Case 2

Other case of men body definition: 25-year-old man with a standard body frame, mildly overweight. Was looking for a modeling job and asked a really shaped athletic frame.

VaserTM and PowerX devices were used for achieving a better defined contour. Fat was transferred to the pectoralis area to enhance and better define the contouring.

Post op 6 months (Fig. 18.15).

Case 3

This 32-year-old patient was looking for a more athletic body shape. High definition of abdomen, flanks, and chest was accomplished together with fat transfer to enhance the upper chest volume and improve the shape.

Figure 18.16b shows the planning of the surgery. Fig. 18.16c shows the early post-op appearance. Figure 18.16d and e shows the immediate post-op result. Figure 18.16f shows 6 months post-op appearance of the new frame with enhanced abdomen and thorax.

1 year post-op (Fig. 18.16).

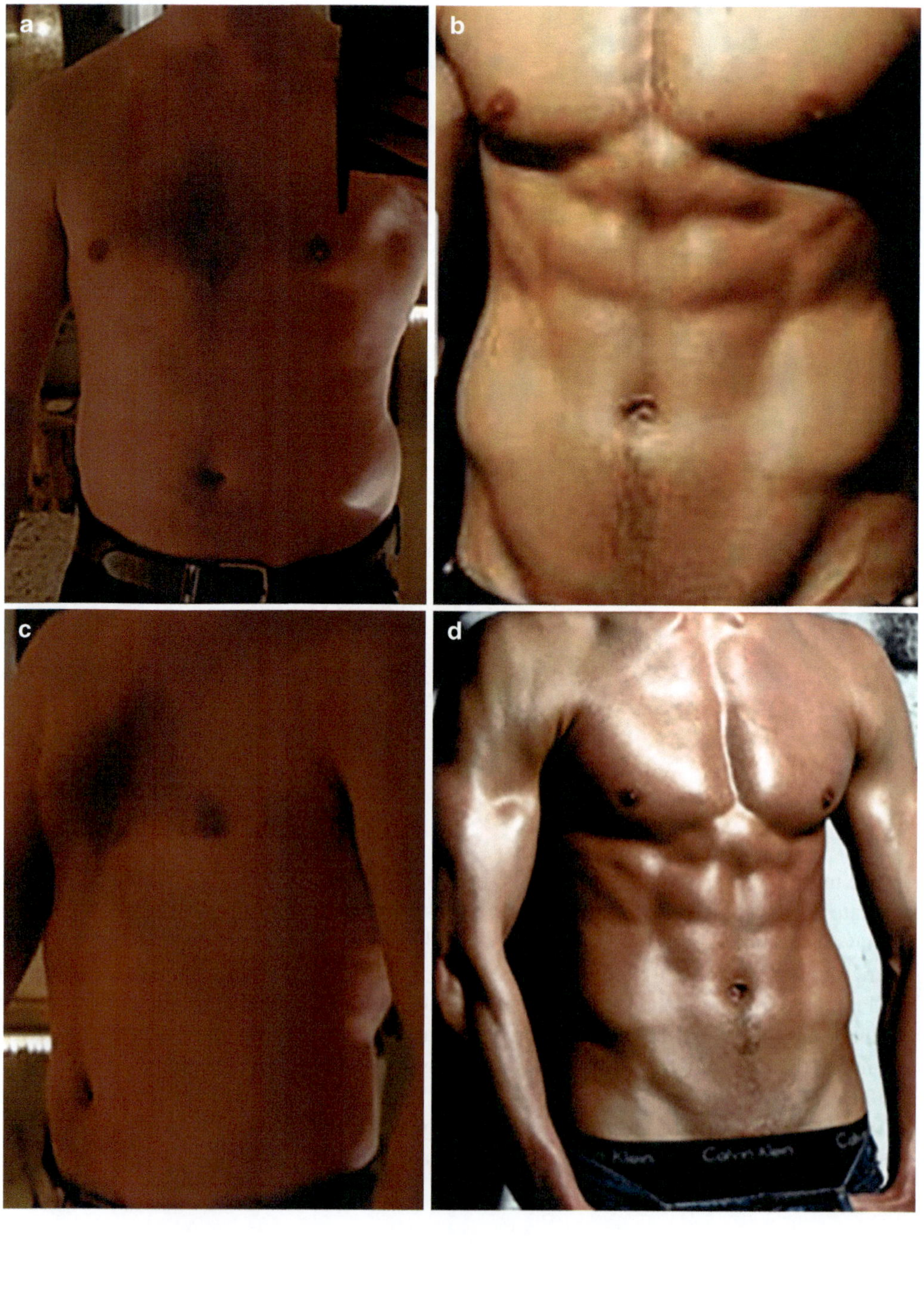

Fig. 18.15 Body high definition surgery. (**a**, **c**) Pre op appearance. (**b**, **d**) Post op 6 months with fat transfer to the pectoralis area

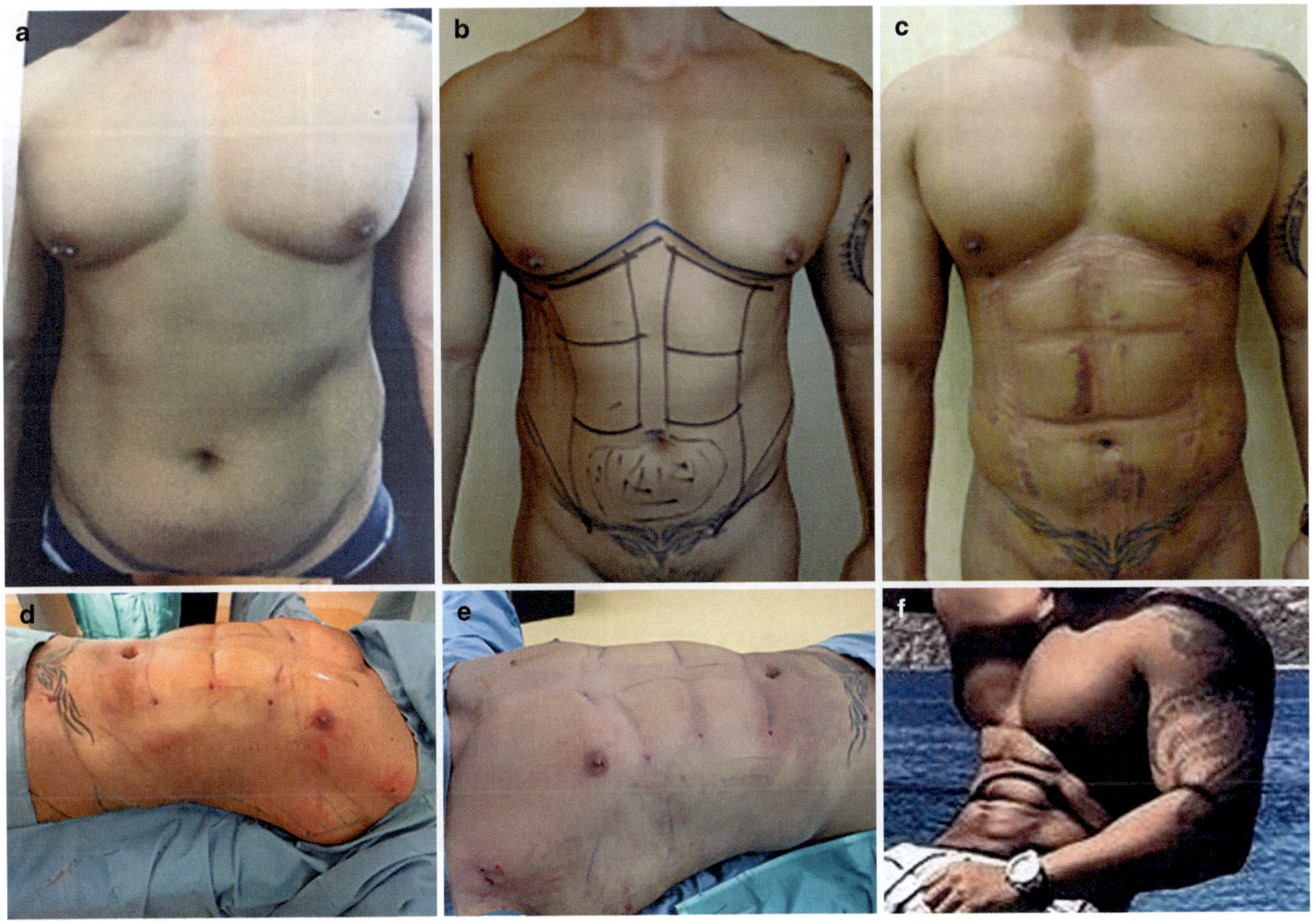

Fig. 18.16 (**a**) Pre op. (**b**) Planning. (**c**) Pos op 1 year. (**d, e**) Immediate post op. (**f**) Six months post op

Case 4
This 42-year-old oriental patient was looking for an athletic appearance. Unfortunately his abdominal muscles were not particularly strong and he did not spend much time in gym for fitness.

So fat transfer was used to enhance the volume and the bulky appearance of the rectus muscle.

1-year post op (Fig. 18.17).

18.6.2.1 Pectoralis Enhancement

Pectoralis enhancement is a compelling technique in body shaping. Chest shape sculpturing must follow the lines of the pectoralis major muscle. A clear depression must be created on the lateral chest, starting from the axilla, and projecting down laterally to the nipple. Pectoralis muscle is a fan-shaped large triangle muscle. It terminates at the level of the nipple.

In order to achieve consistent nice results, the area must be treated with fat transfer in both the superficial and the deep layers. In Fig. 18.18a, the area of fat transfer, superficial and deep are drawn. In Fig. 18.18b, incision lines for placing the fat are indicated. Usually 2 or 3 mm large cannula are indicated for fat transfer to male trunk.

1-year post op (Fig. 18.18).

Case 5
A 28-year-old patient requiring pectoralis enhancement. A total of 180 mL of fat were transferred to the chest area. 40 mL on each side were placed on the deep muscle, which does not allow more than this limit. The remaining 140 mL of fat were diffused on the superficial layer. Pre- and post-op results are presented in Fig. 18.19.

Post op 6 months (Fig. 18.19).

Case 6
A 35-year-old patient, mildly overweighted, with loose skin at chest and abdomen, and poor muscle shape and strength. A program of fitness muscle reinforcement was adviced, complimentary after surgery. Chest wall and abdominal muscle were further reintegrated with fat transfer as well to improve the final shape and outcoming.

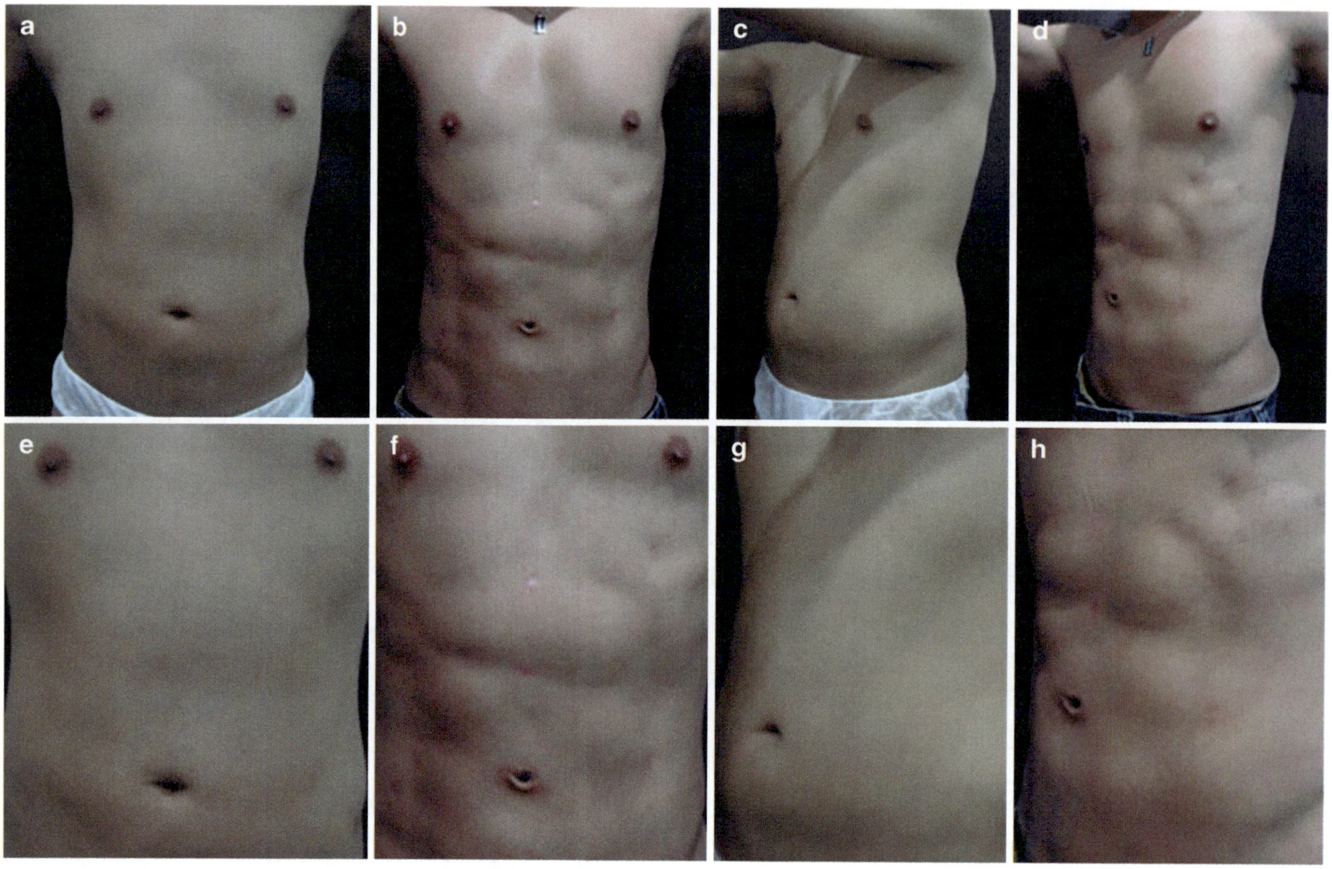

Fig. 18.17 (**a**, **c**, **e**, **g**) Pre op. (**b**, **d**, **f**, **h**) 1 year post op

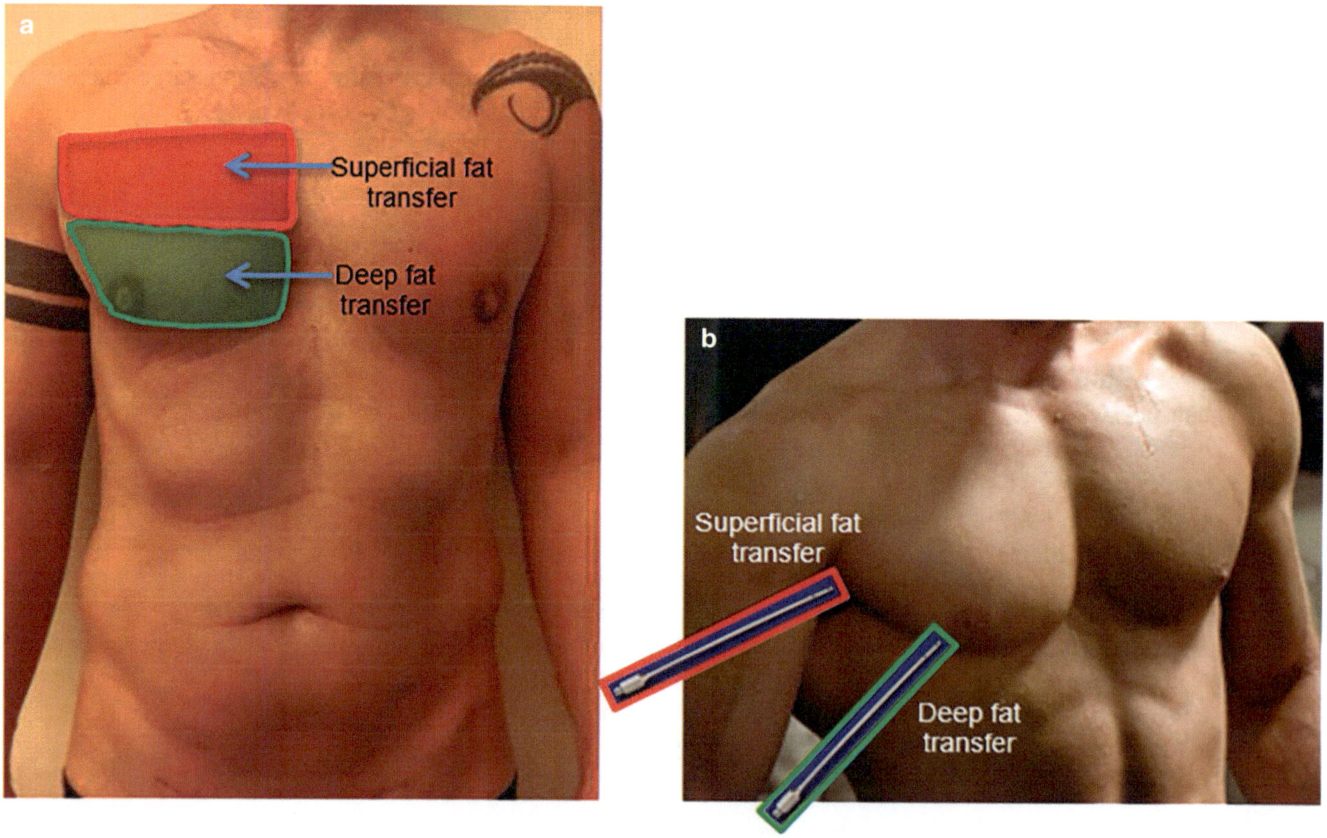

Fig. 18.18 (**a**) Areas of fat transfer. (**b**) Incision lines and direction of cannulas (2–3 mm large)

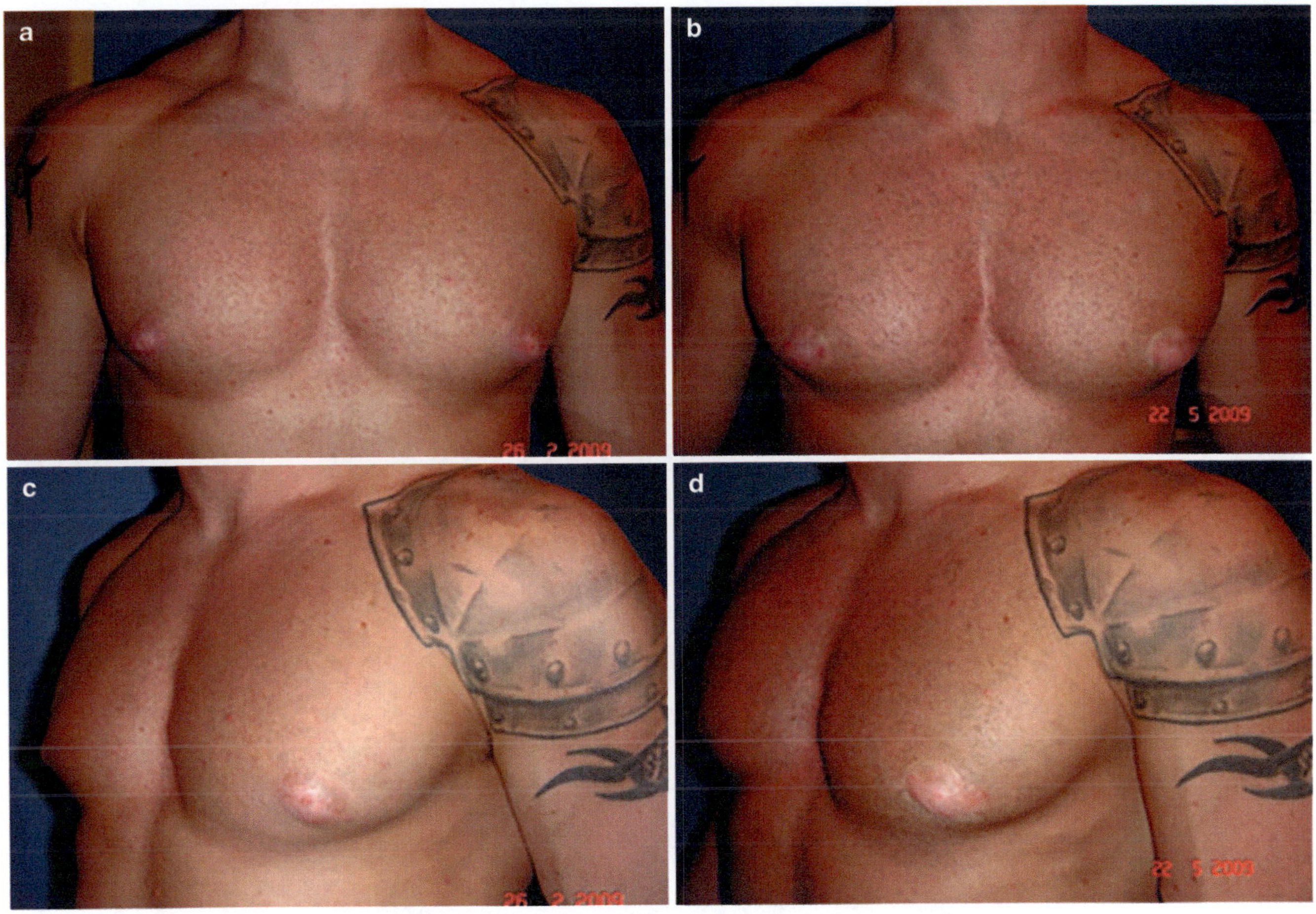

Fig. 18.19 Pectoralis enhancement with 180 ml of fat, superficial and deep layers. (**a**, **c**) Pre op. (**b**, **d**) Post op 6 months

Keeping in shape by constant training is essential to maintain the proper nice outcome. It is easy to put on weight if the patient stops training as well as losing the muscle tone and strength.

Post op 6 months (Fig. 18.20).

Case 7

High definition can be accomplished even in the older patient, with senile elastosis skin appearance. This 58-year-old patient presented with a mild form of gynecomastia, fat excess on the upper abdomen, in an otherwise well-trained person. The lower part of the abdomen presents one of the most tricky points in any loose older skin: a double ply (see red arrows). Using a good technology (Vaser, superficial ultrasound emulsification, and thus skin retraction) in expertised hands will give this nice redraping in the body frame, with a nicer look, a more athletic appearance, and a better aesthetic result.

6-month post-op (Fig. 18.21).

18.7 Post-op Care

Post-op care includes:

1. Compressiver garment for 6–8 weeks
2. Special foam to further compression of the abdominal region, for 2–3 weeks
3. MLD (manual lymphatic drainage) twice a week, for 4 weeks at least, until swollen is settled
4. TED stockings and 14 days microheparine cycle
5. Antibiotics, 7 days, pain medication is needed (Figs. 18.22 and 18.23).

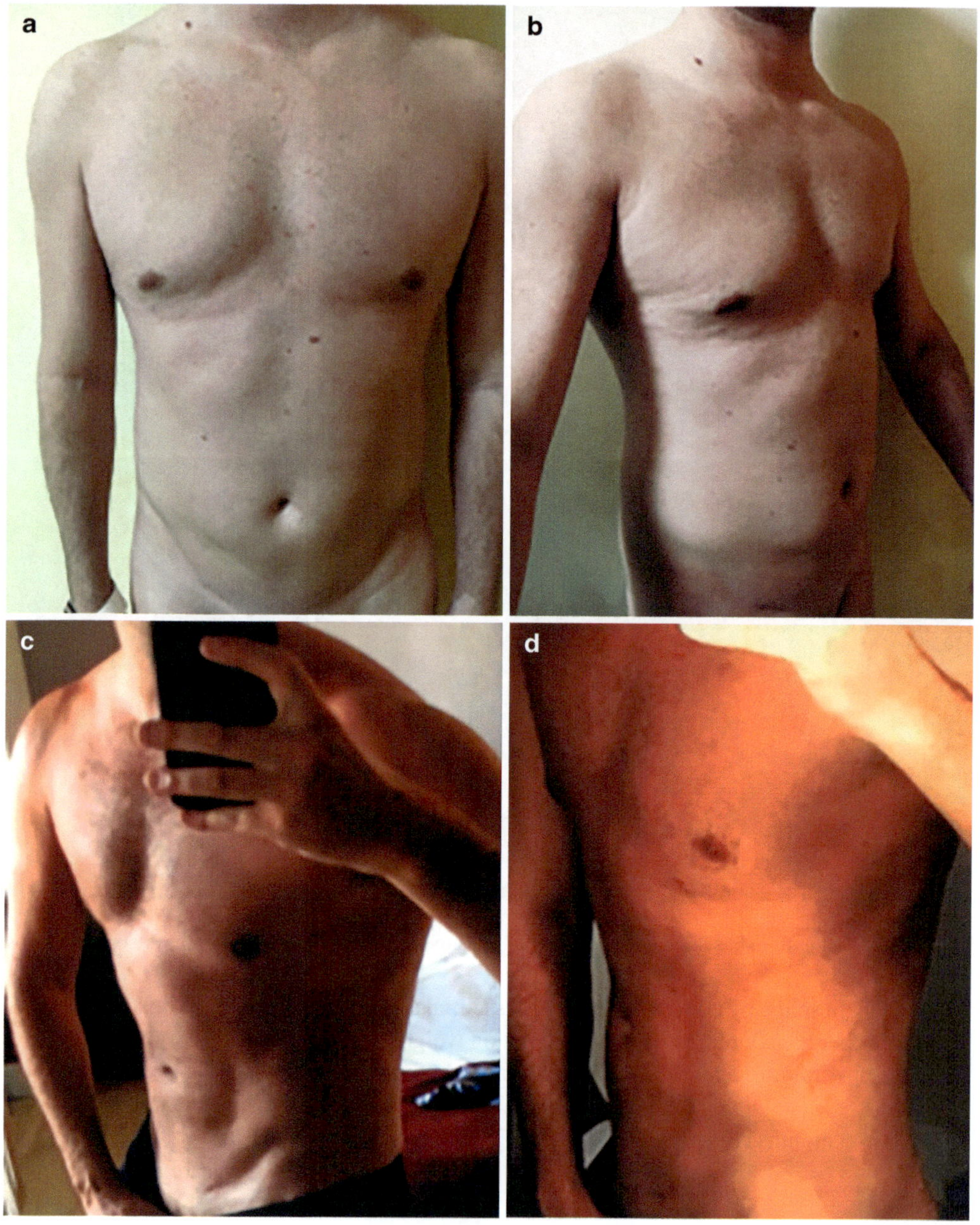

Fig. 18.20 (**a**, **b**) Pre op. (**c**, **d**) Post op 6 months after high definition of trunk and fat transfer to the chest wall. Complimentary training program and diet for enhancing quality of result

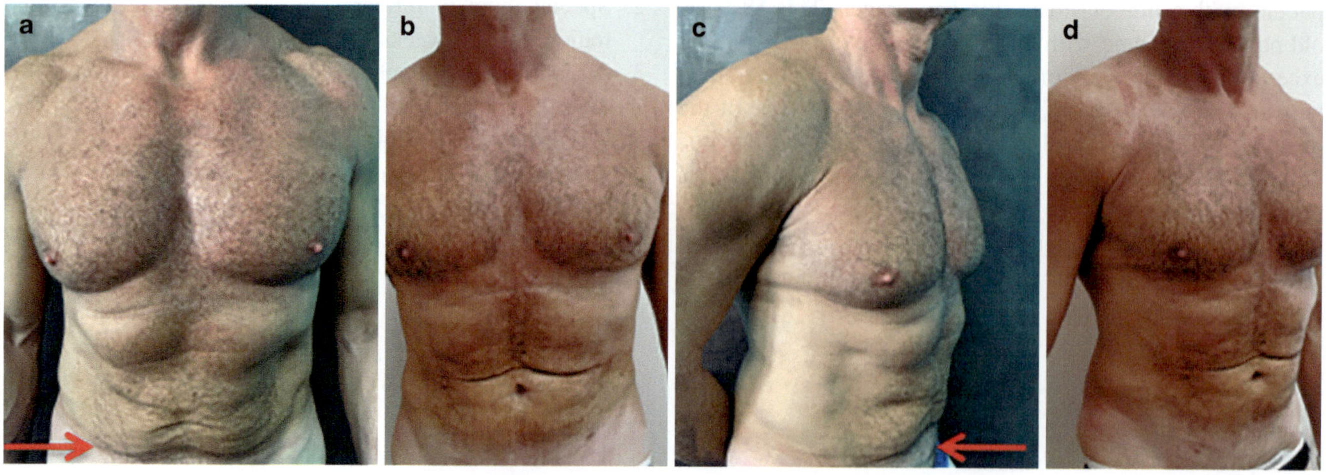

Fig. 18.21 (**a**, **c**) Pre op; see red arrows which indicate skin excess with double ply in loose skin in older patient. (**b**, **d**) Post op 6 months appearance after Vaser body shaping to upper chest, gynecomastia, lower abdomen and flanks. Note skin reraction and redraping of the lower abdomen

Fig. 18.22 Special foam for compression

Fig. 18.23 Special compression garment for high definition

References

1. Simblet S. Anatomy of art. In: Anatomy for the artist. New York: DK Publishing; 2001.
2. https://soltapracticeproposal.com/product-information/products/vaser-system/
3. Cimino WW. Ultrasound-assisted lipoplasty hasic physics, tissue interactions, and related results/complications. In: Prendergast PM, editor. Aesthetic medicine. Art and techniques. Berlin/Heidelberg: Springer; 2011.
4. Hoyos AE, Prendergast PM. High definition body sculpting. Berlin Heidelberg: Springer; 2014.
5. Hoyos AE, Millard JA. VASER-assisted high definition liposculpture. Aesthet Surg J. 2007;27(6):594–604.
6. Hoyos AE, Prendergast PM. High definition body sculpturing, art and advanced lipoplasty techniques. Berlin/Heidelberg: Springer; 2014.
7. Williams PL. Gray's anatomy: the anatomical basis of medicine and surgery. 38th ed. Edinburgh: Churchill Livingstone; 1995. p. 811.
8. Goldfinger E. Human anatomy for artists. The elements of form. Oxford: Oxford University Press; 1991. p. 289.
9. Singh D, Young RK. Body weight, waist to hip ratio, breast, and hips: roles in judgements of female attractiveness and desirability for relationships. Ethol Sociobiol. 1995;16:483–507.
10. Cornelissen PL, Toveé MJ, Bateson M. Patterns of subcutaneous fat deposition and the relationship between body mass index and waist-to-hip ratio: implications for models of physical attractiveness. J Theor Biol. 2009;256(3):343–50.
11. Nahas FX. An aesthetic classification of the abdomen based on the myoaponeurotic layer. Plast Reconstr Surg. 2001;108(6):1787–95.
12. Shiffman MA, Di Giuseppe A. Liposuction. Principles and practice. Berlin: Springer; 2016.
13. Di Giuseppe A, Giovagnoli F, Di Giuseppe S, Ronconi D. Vaser body contouring achieving a more defined shape. The art of body contouring. London: Intechopen; 2019. p. 105–122.

Body Fat Grafting Contouring in the Female Patient

19

Alberto Di Giuseppe and Federico Giovagnoli

Contents

19.1 Introduction

Fat transfer in body contouring has many indications:

- Banana fold deformity treatment
- Correction of asymmetries secondary to surgery
- Muscle edges definition
- Correction of primary asymmetries
- Treatment of cellulitis
- Post-traumatic defect correction
- Enhance body frame by increasing areas of deficiency

Fat transfer is the best solution for all these indications. It is a unique element to be added to the body to solve, of help solving, complicate cases, and secondary defects post–previous surgery.

Liposuction, intended as a technique to remove fat, has evolved in liposculpture, intended as a technique to shape and add fat if, when and where is necessary.

A. Di Giuseppe (✉)
University of Padova, Padova, Italy

F. Giovagnoli
University Vita-Salute San Raffaele, Milan, Italy

Evolution of technique and advancement in materials have deeply increased the rate of survival of fat, which initially (in the 90's) was around 20–30% of the initial infiltration.

Nowadays (2010–2020), depending on the technique utilized for fat extraction, fat harvesting, fat decanting, fat cleaning and finally fat infiltration, fat survival rate ranges from 80% to 95%, depending on the author studies reported.

On top, the discovery of the stem cell, defined as "turbo" cells which boost the survival rate of the implant and fat, has made a substantial enhancement in the quantity of fat survival, and in the quality of fat itself.

In my surgical experience, I have mainly utilized ultrasound-assisted liposuction (VASER™) for fat emulsification and extraction and REVOLVE™ cannister, for fat harvesting and preparation to infiltration. I have acquired a 15 years of clinical experience with this technique, after having used different tools and devices in the past, with less satisfactory results [1–3].

Recently (2020), I started utilizing BeautiFill™ by LipoLife, produced by Alma Laser (Israel) (Fig. 19.1).

LipoLife streamlines the surgical process by offering a complete, all-in-one solution covering all stages of liposuction, skin tightening, and fat grafting procedures. It is the

© The Author(s), under exclusive license to Springer Nature Switzerland AG 2023

A. Di Giuseppe et al. (eds.), *Fat Transfer in Plastic Surgery*, https://doi.org/10.1007/978-3-031-10881-5_19

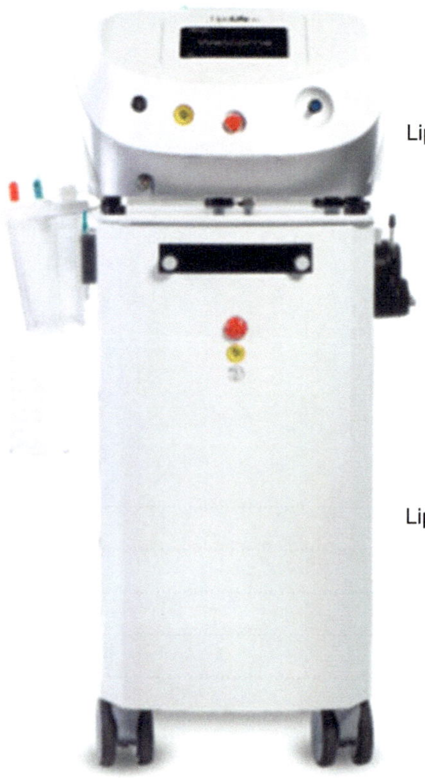

Fig. 19.1 BeautiFill™ by LipoLife (Courtesy of Alma Lasers, ©2022 BeautiFill™ by LipoLife, All Rights Reserved)

first integrated solution of its kind combining multiple modalities in a compact, portable, and neatly organized system. Simultaneous lasing and suctioning simplify the procedure and significantly reduce the treatment time, thermal energy and a unique radial emission fiber allow for effective, homogenous skin tightening, and the collection of high vitality fat permits successful, immediate fat grafting.

This is the third generation of Alma Laser's liposuction technology offering better safety and clinical success.

19.2 Simultaneous Lasing and Suction

It features an integrated cannula and fiber interface that allows for simultaneous lasing and suction. This unique interface allows physicians to pass through the tissue more easily and remove fat cells more quickly and efficiently. As laser energy is delivered to the tissue, the laser cauterizes blood vessels, minimizing bleeding, swelling and body trauma such as hematomas, edema, and bruising as well as minimizing postoperative pain.

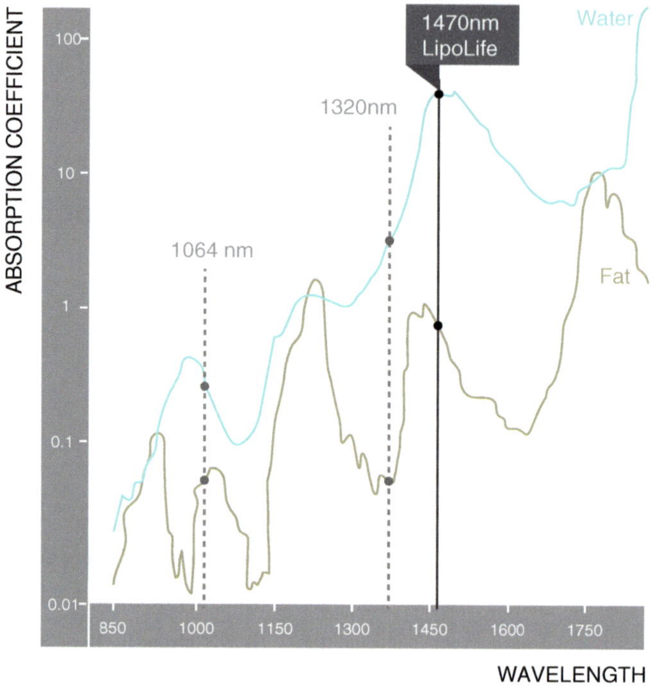

Fig. 19.2 Selective absorption of the 1470 wavelength in water compared to fat (Courtesy of Alma Lasers, ©2022 BeautiFill™ by LipoLife, All Rights Reserved)

Simultaneous lasing and suction create a one-step process that significantly reduces the amount of physical effort required to perform liposuction procedures, especially in areas with fibrous tissue.

LipoLife uses the 1470 nm wavelength which specifically targets water in soft tissue and features a higher absorption rate in water and fat than all other competing devices. This allows for optimal fat removal with minimal tissue damage. Laser energy is delivered at a low power density for maximum safety, minimizing the risk of burns (Fig. 19.2).

Briefly, the LipoLife device is operated with a thin radial fiber laser (400 or 600 μm) at 1470 nm wavelength that targets water and is integrated into a small suction cannula using an Angel tip (360° annular probe) for combined gentle laser treatment and careful suction in a single-in-serial step. The LipoLife device involves simultaneous work of laser (CW mode) and suction. Power settings are a function of the liposuction area: 10–13 W (chin, arms, and knees), 13 W (abdomen, back), and 15 W (thighs, hips, and buttocks). The laser energy is conveyed into the fat layer using a 3 or 4 mm cannula which incorporates a 400- or 600-μm optical single-use fiber (Angel). Given its water absorption, radial emission, and relatively low irradiance on the fat tissue channeled via Mercedes perforated-tip cannula, the 1470 nm wave-

Fig. 19.3 Angel radial emission fiber (Courtesy of Alma Lasers, ©2022 BeautiFill™ by LipoLife, All Rights Reserved)

length laser fiber enables gentle fat tissue as shown in Fig. 19.2. LipoLife system: integrated LipoDiode (left), Angel radial emission fiber (middle), and selective absorption of the 1470 wavelength in water compared to fat collection/yield and provide increased relative ratios of viable adipocytes in the harvested tissue [9] (Fig. 19.3).

Autologous fat transfer for soft tissue contouring has become increasingly popular in recent years. The immediate availability of the patient's own fat, its natural integration into host tissues, and its permanent correction potential make it a highly instrumental application. LipoLife achieves a fat vitality rate of greater than 95%.

The procedure is quick and convenient with a fully equipped kit, allowing you to easily restore patient tissue volume.

With LipoLife's unique lasing technology, fat cells gently undergo lysis to a semi-liquid state, generating two distinct phases: a significant adipose phase with highly viable and small uniform fat cells and a negligible blood phase, with no oil phase present as with mechanical liposuction. The preservation of fat cells with high vitality combined with a filtering mesh allows for quality fat grafting and successful autologous fat injection, with longer term results.

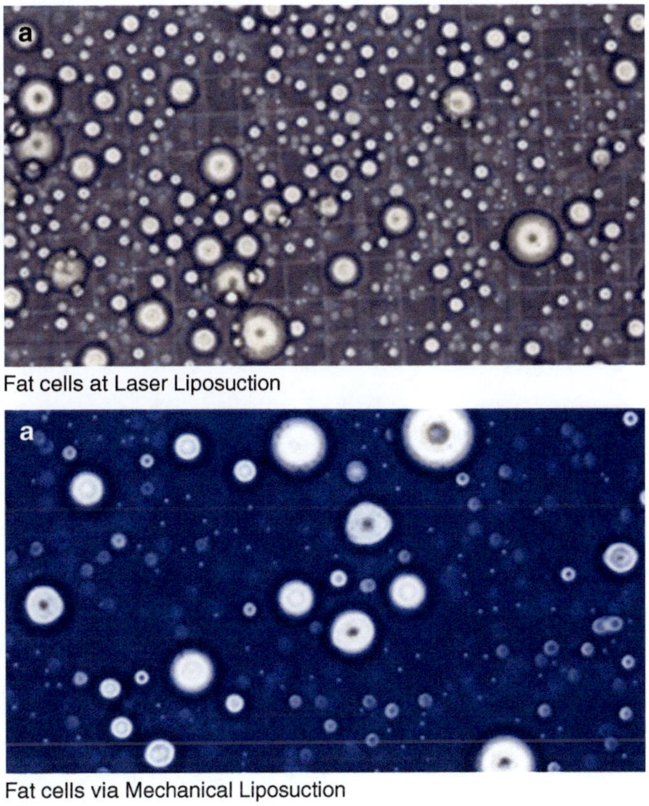

Fig. 19.4 Fat viability. (a) Mechanical liposuction, (b) laser liposuction (Courtesy of Alma Lasers, ©2022 BeautiFill™ by LipoLife, All Rights Reserved)

A study conducted in a single center in Israel in the year 2018, at the Assaf Harofeh Medical Center, did analyze laser-assisted derived fat cell. The main target was fat viability after emulsification with Lipolife laser (Fig. 19.4).

19.3 Study Evaluation (Table 19.1)

Lipoaspirate sample was analyzed with mechanical method, enzymatic method, and migration method.

The summary of the cell count and viability result is shown in the diagram in Table 19.2.

Table 19.1 Study evaluation

Method	Purpose
Cell viability count analysis (Tryphan blue staining)	Counting the viable SVF cells from the lipoaspirate
FACS	Phenotype analysis of ASC cells will by the following markers: CD34, CD73, CD90, CD105—positive cell surface markers, support adipose tissue growth CD45—negative marker
Immunohistochemistry	Histological analysis of harvested lipoaspirate fat
Immunofluorescence (optional)	Identification of ASC cells with specific markers; the transcription factors Nanog and Oct4
RT-PCR analysis (optional)	Gene expression analysis of vascular endothelial growth factor (VEGF-A), CD31, and von-Willebrand factor (vWF) in the ASCs to confirm adipose tissue origin
Proliferation assay (XTT)	Cell proliferation assessment in ASC cultured cells

Table 19.2 Summary of cell count

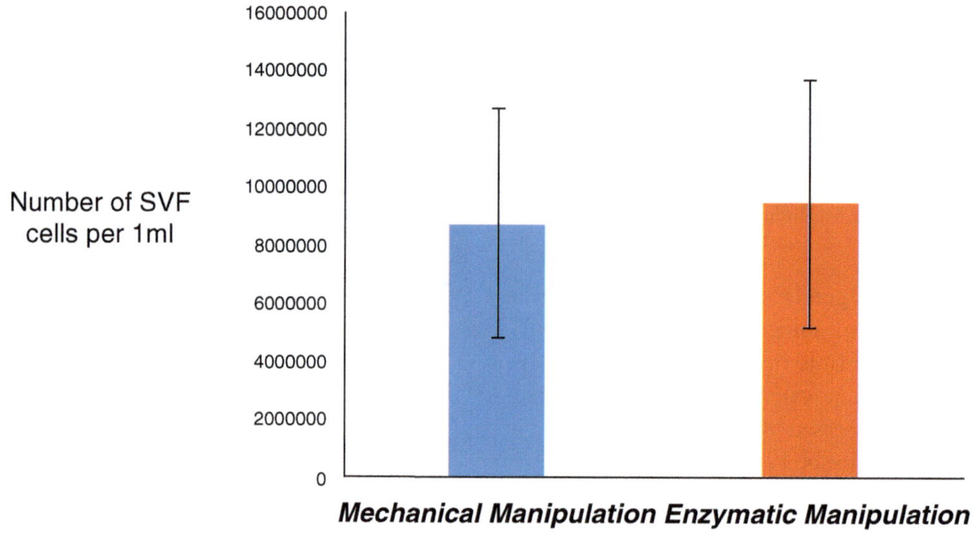

	Mechanical	Enzymatic
No. of SVF cells	87×10^6	9.4×10^6
SVF cell viability (%)	97%	98%
No. of ASC cells	1×10^6	1.2×10^6

19.4 Skin Tightening Benefits

In addition to achieving effective lipolysis, the thermal energy generated by the 1470 nm diode laser contracts existing collagen and elastin fibers and stimulates the formation of new collagen for firmer, tighter looking skin. This is particularly important for delicate areas such as the face and neck as well as for areas of cellulite. In effect, LipoLife targets cellulite precisely at its source within the underlying structural tissue. Laser energy is directed beneath the skin, separating the skin from adipose tissue, effectively breaking up pockets of cellulite and restructuring the skin for a smoother appearance.

19.5 Face Tightening Lipotight Cannula

In addition to the skin tightening benefits, it offers a dedicated cannula with a flat multiuse laser fiber, designed exclusively for skin tightening of the face and neck. Laser energy is delivered directly underneath the skin above the hypodermis to treat areas of skin laxity such as nasolabial folds, jowls, and neck. The skin tightening cannula may also be used for post-facelift touchups including addressing any asymmetry or loose skin (Fig. 19.5).

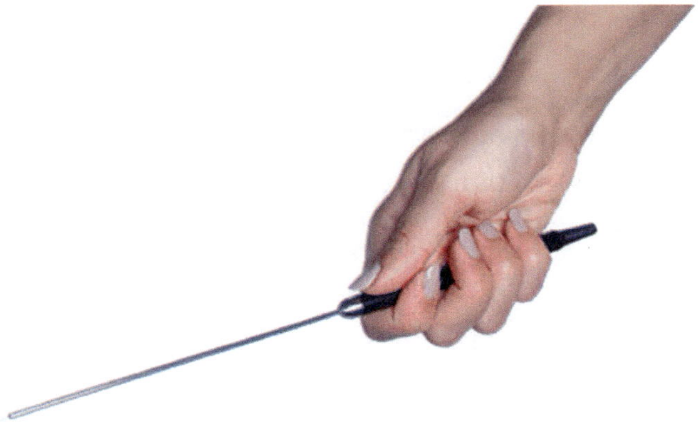

Fig. 19.5 Lipotight™ cannula (Courtesy of Alma Lasers, ©2022 BeautiFill™ by LipoLife, All Rights Reserved)

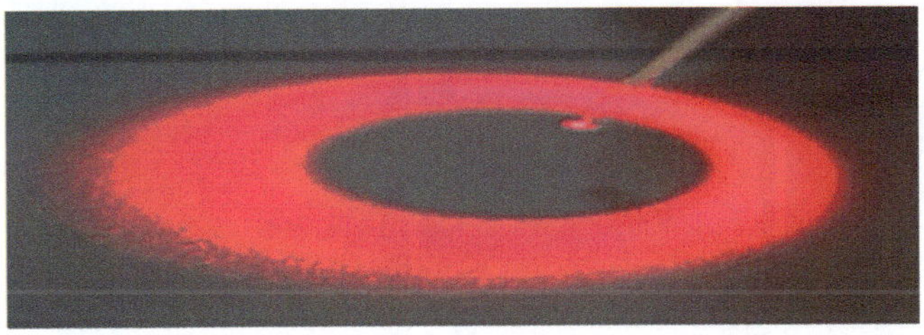

Fig. 19.6 Radial fiber integrated into a Mercedes perforated-tip cannula which emits energy circumferentially in a 360° pattern (Courtesy of Alma Lasers, ©2022 BeautiFill™ by LipoLife, All Rights Reserved)

19.6 Skin-Safe Unique Angel™ Emitting Fiber

It is a radial fiber integrated into a Mercedes perforated-tip cannula that emits energy circumferentially in a 360° pattern. Radial lasing is a low-power density approach with multiple benefits including minimizing the risk of burns, distributing energy evenly throughout the treatment area, removing fat cells more quickly, and covering greater tissue area, allowing for faster treatment speed. The fiber's radial emission pattern allows for safe, homogenous, and more efficient skin tightening during the procedure (Fig. 19.6).

19.7 Liposense™ Temperature Control

LipoLife features a temperature-sensing cannula that continuous lymonitors subdermal temperatures during laser-assisted lipolysis, body contouring, and small area procedures. Laser energy is delivered to the tissue until the target temperature is reached, achieving precise, uniform treatment. The sensor provides instant temperature feedback allowing the system to deliver the appropriate level of laser energy to the tissue while maximizing procedure safety Intuitive software allows you to define a maximum temperature limit for the procedure. If subdermal temperature exceeds the limit, the system will automatically stop, preventing thermal tissue damage. LipoSenseTM temperature control enables you to perform precise body contouring with greater efficiency and safer results.

I have utilized this new system in 45 cases in 2020: face and neck tightening, gynecomastia, abdomen, flanks and buttocks shaping, and fat transfer to breast and buttocks and secondary correction of post-liposuction defects.

My comments to my initial experience with this new device are positive as clinical results in terms of skin tightening (which is superior to any other system tried before, especially in the neck area).

The second excellent goal achieved by this device is the high-quality fat ready for transfer, with a fast integrated system, and an excellent long-term survival rate.

19.8 Clinical Cases

19.8.1 Banana Fold Deformity: Treatment

The so-called "banana fold deformity" is a double plication of the subgluteal fold following an improperly treated outer thigh and subgluteal areas. It is caused by multiple factors: breaking the vertical fibers of the infragluteal fold, residual fat of the subgluteal fold, heavy residual trocanter, and heavy inner buttocks.

Treatment of the area is quite a complicated issue which first needs a full understanding of the specific problem of the area.

In the case presented (Fig. 19.7) in Fig. 19.7a, it is noticeable the heaviness of the outer trochanter, due to insufficient removal in the first stage. Also the subgluteal fold appears heavy and not smooth enough according to the remaining thigh-buttock area.

Treatment was performed with VASER 2.9 mm probe, 2 rings, after tumescent infiltration, with power reduced to 60% of total. Gentle and careful emulsification of fat in the superficial fat layer was performed with reduced power, in order to thin the subgluteal flap. Also the outer thigh (trochanter) was treated in a similar way (Fig. 19.7b surgical plan) [4].

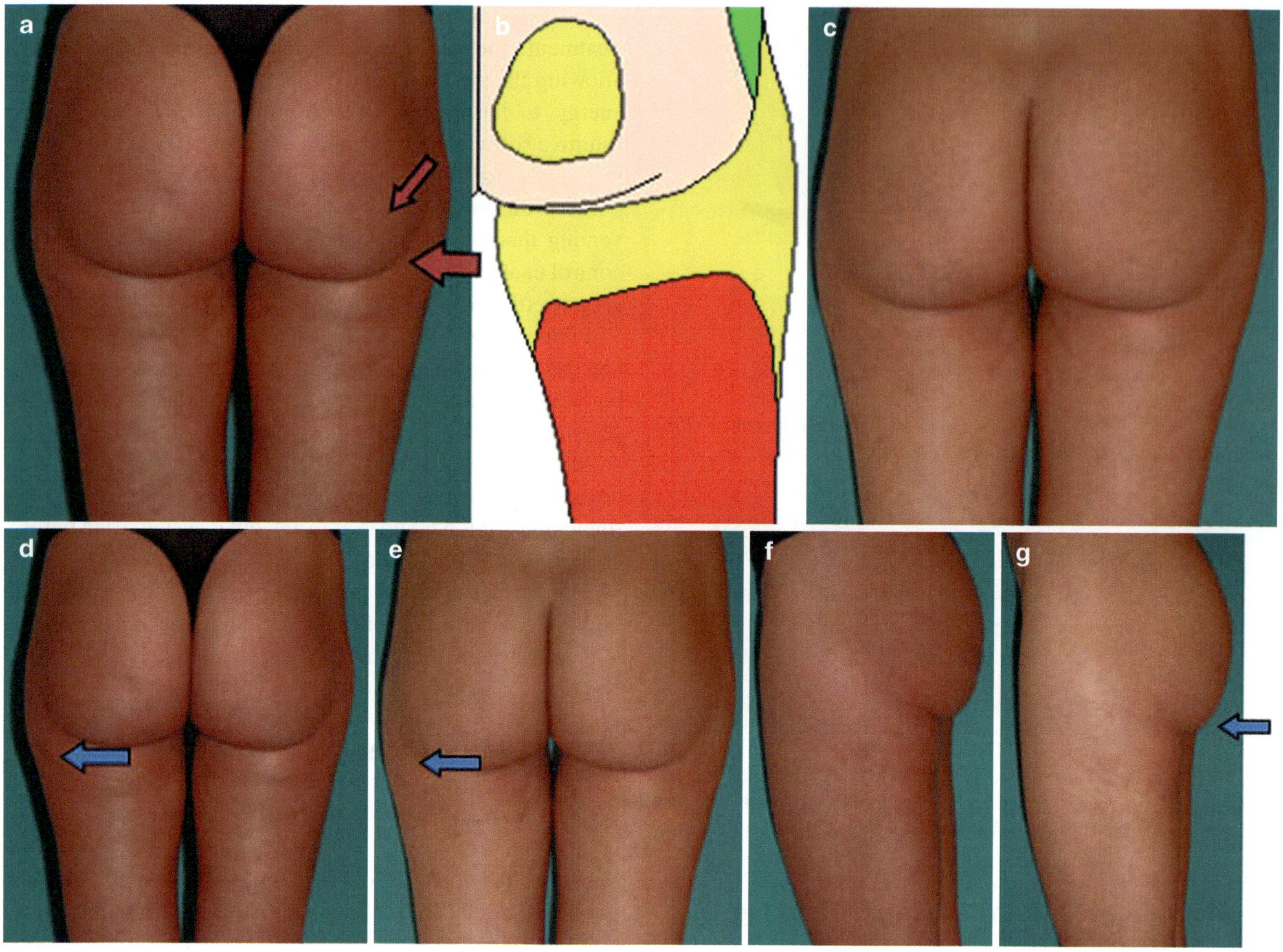

Fig. 19.7 Banana fold deformity post liposuction. (**a**, **d**) Pre op, (**b**) planning, (**c**, **e**) immediate post op, (**f**) pre op lateral view, (**g**) post op lateral view. Pre and 1 year post op photos

Pre- and 1-year post of photos.

19.8.2 Correction of Asymmetries Secondary to Surgery

Case 1
This 43-year-old patient had two previous liposuctions with other surgeons, combined with breast surgery.

The result was unpleasant: in Fig. 19.8a, lots of irregularities in shape of abdomen and flanks are visible, with multiple dimples, irregularities, depressions, together there were areas of fat excess.

Surgical plan (Fig. 19.8b) consisted in VASER secondary liposculpture, with superficial sculpturing of the abdominal muscle edges and fat transfer in selected areas to correct the dimples and defects from the previous surgery.

Pre- and post correction results are shown in Fig. 19.8c and d.

Pre- and 1-year post-op photos.

Case 2
This 32-year-old patient sustained two previous superficial liposuction to contour minor shape defect of the outer buttocks, outer thigh, gluteal fold, inner lower thigh, and knees. She was left with several areas of dimples, depression, and irregularities (see blue arrows in Fig. 19.9a). She underwent fat transfer in several areas to improve the contour and to smooth the surface as well. No further liposuction was performed as the patient was scared [5].

Pre- and 1-year post-op photos.

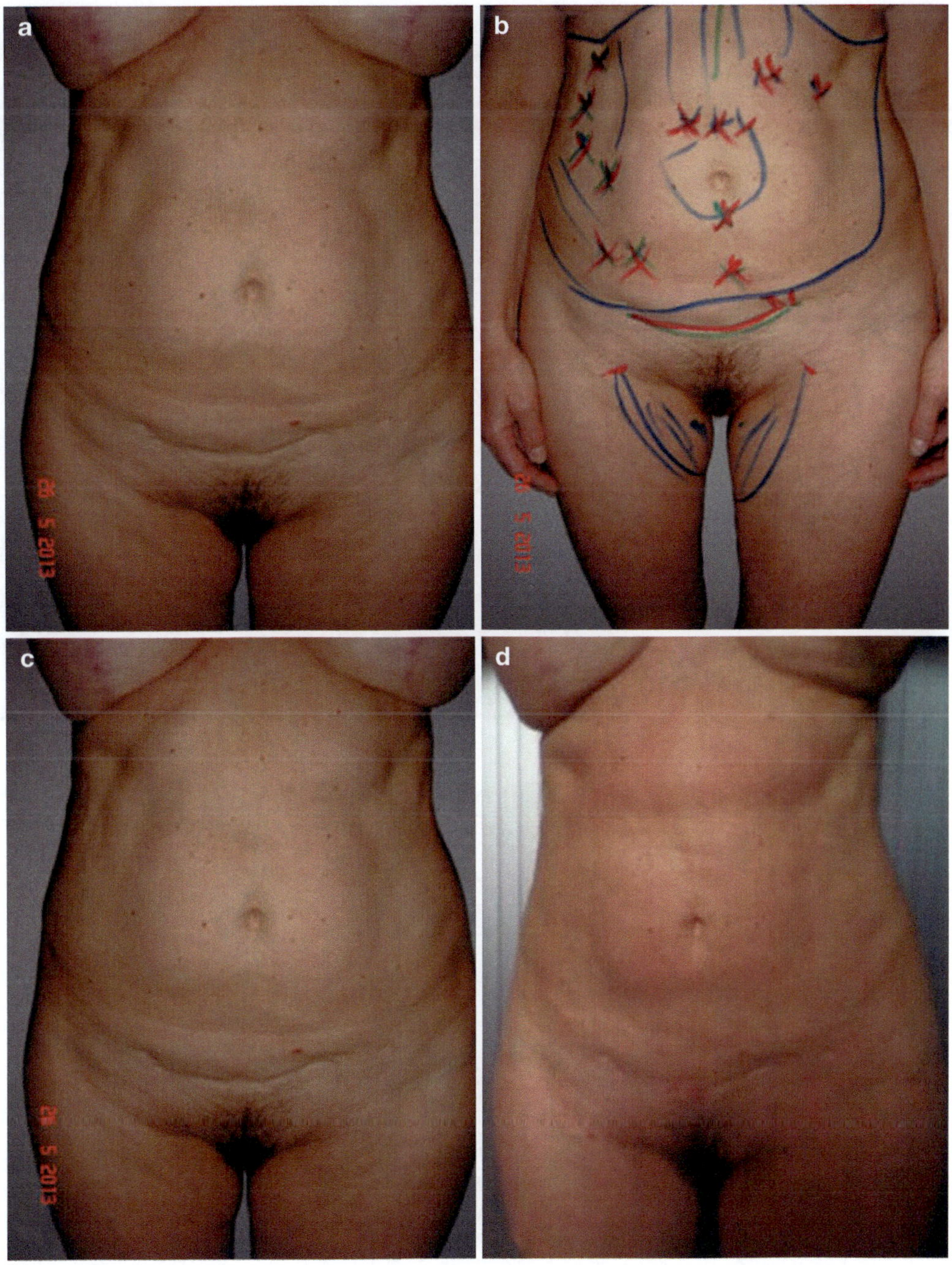

Fig. 19.8 Correction of asymmetries secondary to surgery. (**a**, **c**) Pre op, (**b**) planning, (**d**) post op. Pre and 1 year post op photos

19.8.3 Muscle Edges Definition

Combining fat extraction and fat adding has become the routine in the majority of body contouring cases management. In the case of Fig. 19.8, there is an overweighted figure in the abdomen and flank areas, while a deficiency in the outer thighs and lack of definition in the abdominal muscle edges. VASER body contouring liposculpture was performed, combined with the definition of the vertical linea alba and outer edge of the rectus abdomini muscles. Fat was integrated into the outer thighs and buttocks to better shape and round those areas [6].

Pre- and 1-year post-op photos (Fig. 19.10).

Fig. 19.9 Irregularities of contour after liposuction of thighs and legs. See blue arrows (**a**). patient had fat transfer in multiple areas to compensate. Pre and 1 year post op photos, back (**a**) and front (**b**)

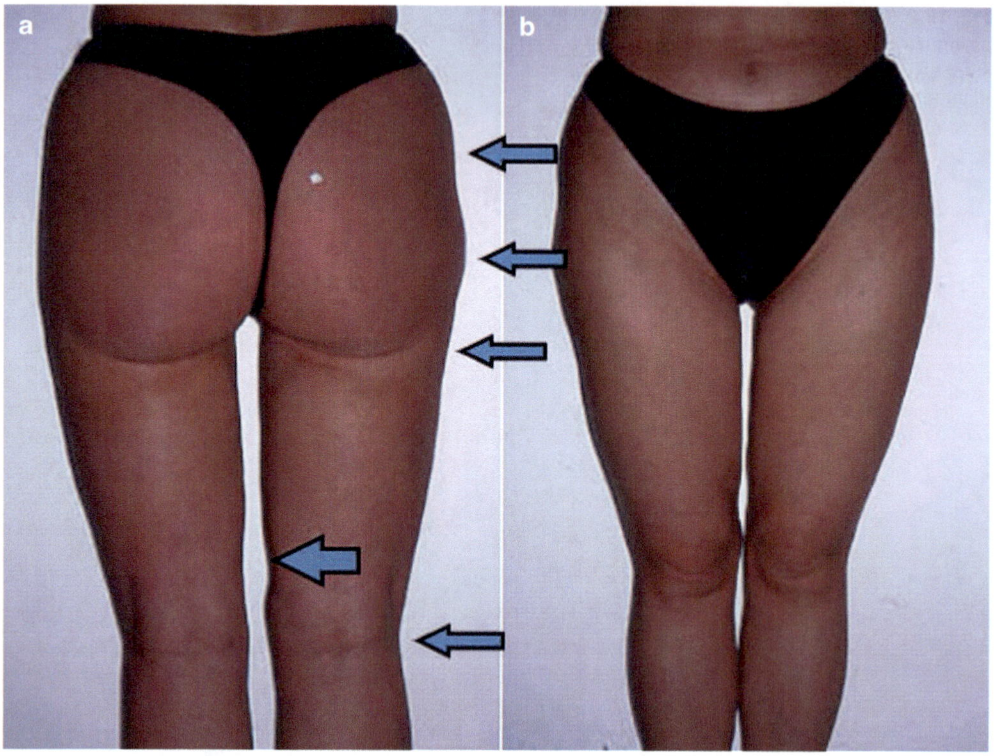

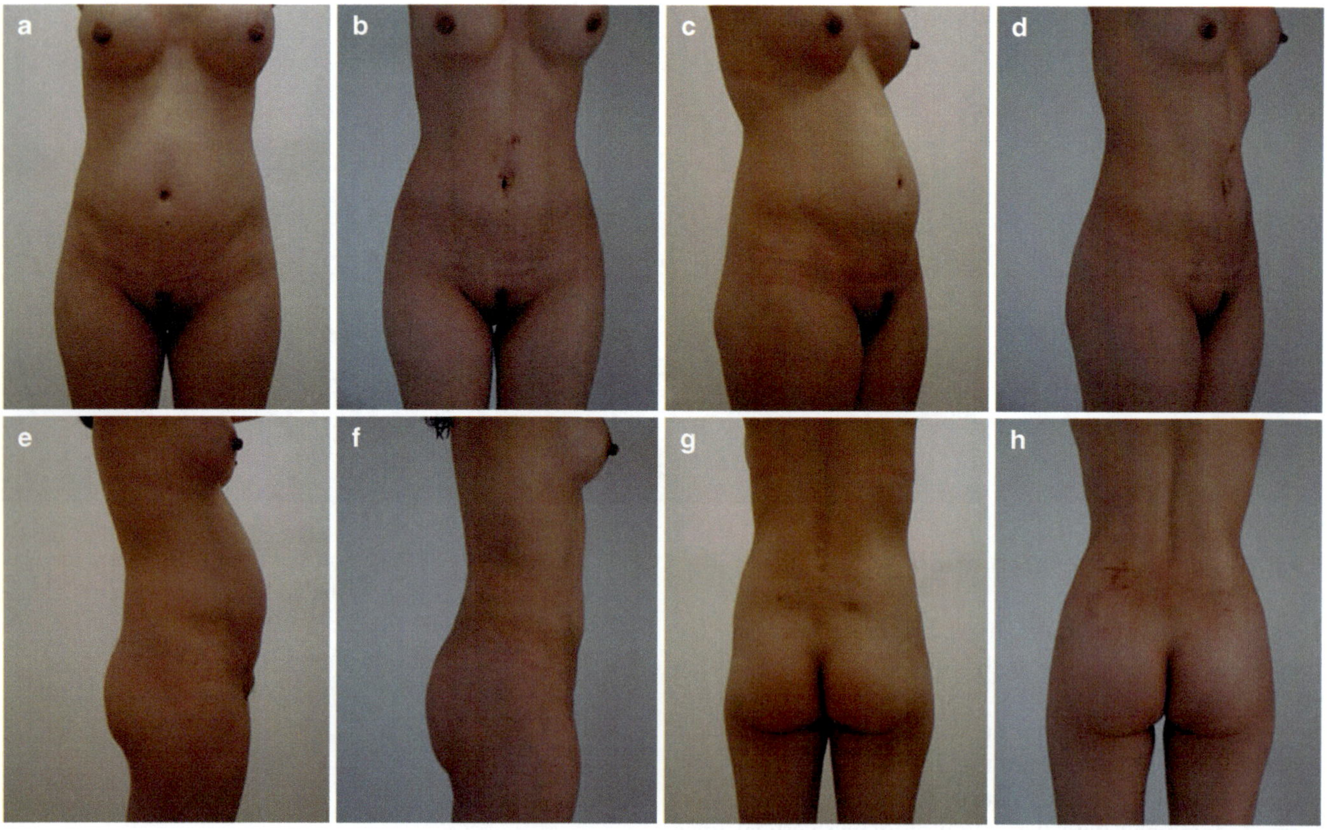

Fig. 19.10 Improve definition of muscle edges and contouring. (**a**, **c**, **e**, **g**) Pre op; (**b**, **d**, **f**, **h**) post op. Pre and 1 year post op photos

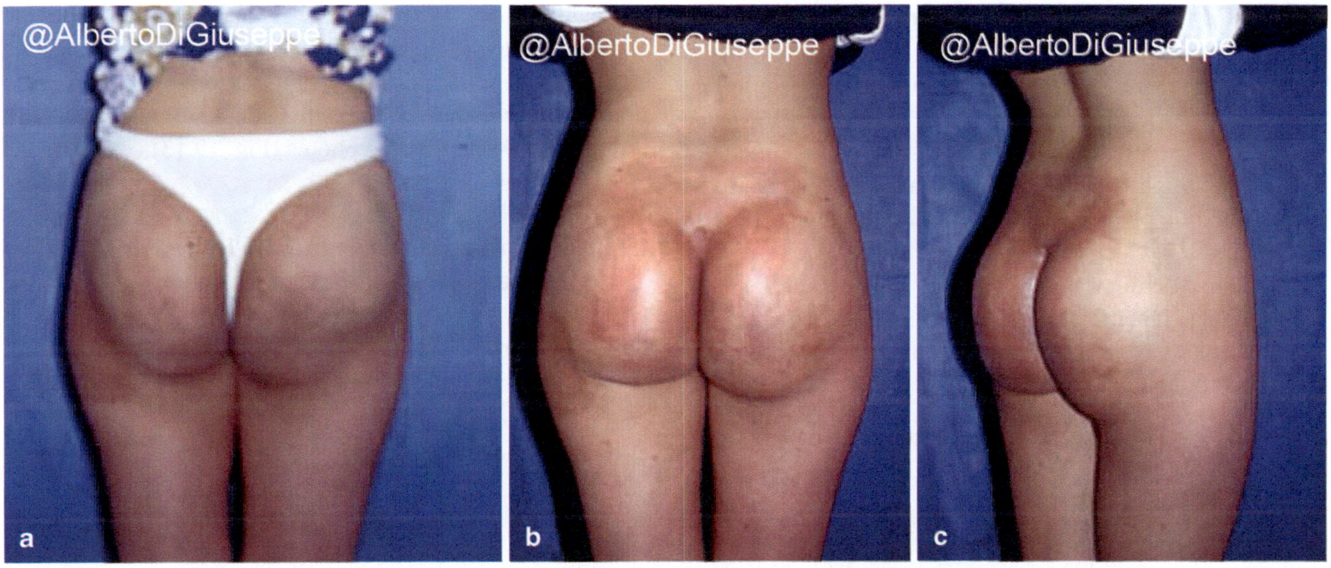

Fig. 19.11 Correction of primary asymmetry of gluteal, buttocks and external thigh. (**a**) Pre op deformity, (**b**) post op correction with fat transfer, (**c**) post op oblique view. Pre and 1 year post op photos

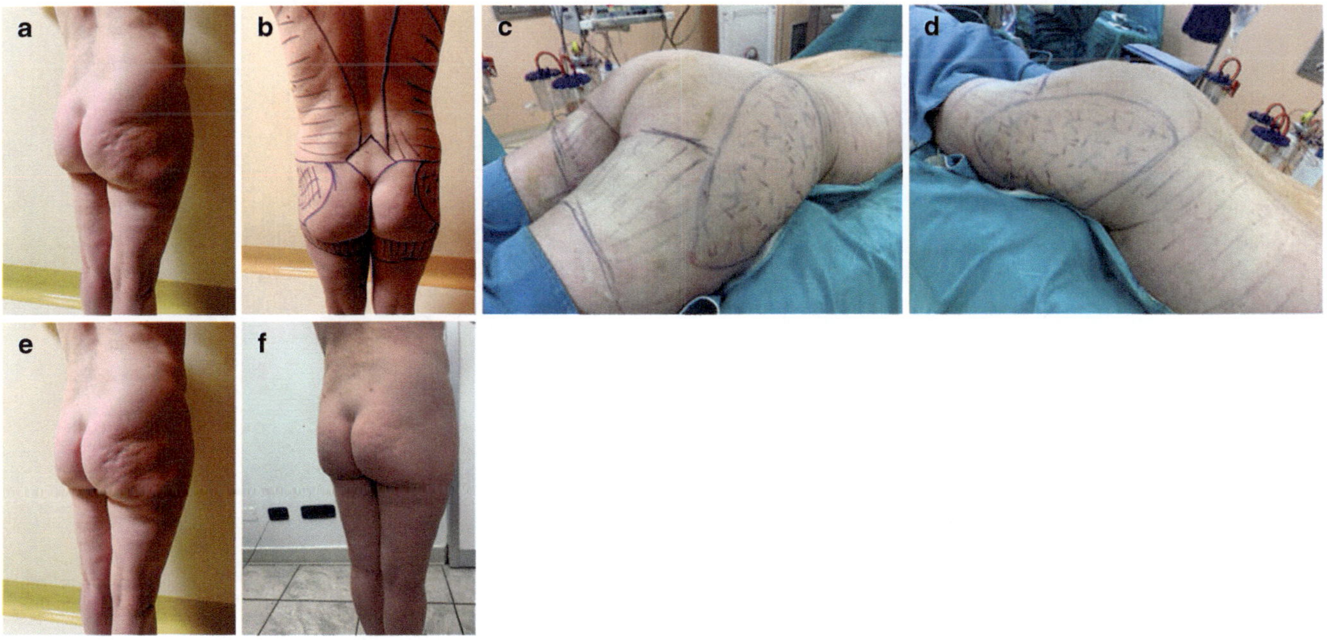

Fig. 19.12 Treatment of cellulitis. (**a**, **e**) pre op, (**b**) planning, (**c**, **d**) immediate post op, (**f**) post op. Pre and 1 year post op photos

19.8.4 Correction of Primary Asymmetries

This patient presented a primary deformity of the external gluteal fold bilaterally, interesting partially the medial gluteal muscle, as well as the fat layer.

This case was treated with BeautiFill device, and 400 ml of fat transfer for each side. Pre- and post-op photos are 8 months apart.

Pre- and 1-year post-op photos (Fig. 19.11).

19.8.5 Treatment of Cellulitis

Cellulitis treatment is one of the most controvert issues in esthetic medicine.

This case of gluteal cellulites plus lipodistrophy of flanks and buttock deficiency was treated with BeautiFill with laser liposuction of posterior trunk and flanks and abdomen, and fat transfer to lateral, central and medial buttocks, to correct cellulitis and contour deficiency.

Pre- and 1-year post-op photos (Fig. 19.12).

19.8.6 Posttraumatic Defect Correction

This case concerns a 45-year-old patient with important asymmetry of the thighs and buttocks, celluliis of the same areas, and mostly an important depression at the left side of the external thigh, due to a previous trauma, which caused a mild herniation of the tensor fascia lata muscle and thus an asymmetry.

Vaser liposuction was used to shape and contour the outer thighs, flanks, inner thighs and buttocks.

Superficial Vaser undermining and fat emulsification further allowed the treatment of cellulitis.

Finally, around 150 ml of fat were harvested and infiltrated on the left outer thigh, in order to correct and compensate the area of deficiency. Liposhifting of the fat on top of the depression helped in shaping properly the dimples of this area [7, 8].

Pre- and 1-year post-op photos (Fig. 19.13).

19.8.7 Enhance Body Frame by Increasing Areas of Deficiency

This 38-year-old patient presents the following body problems:

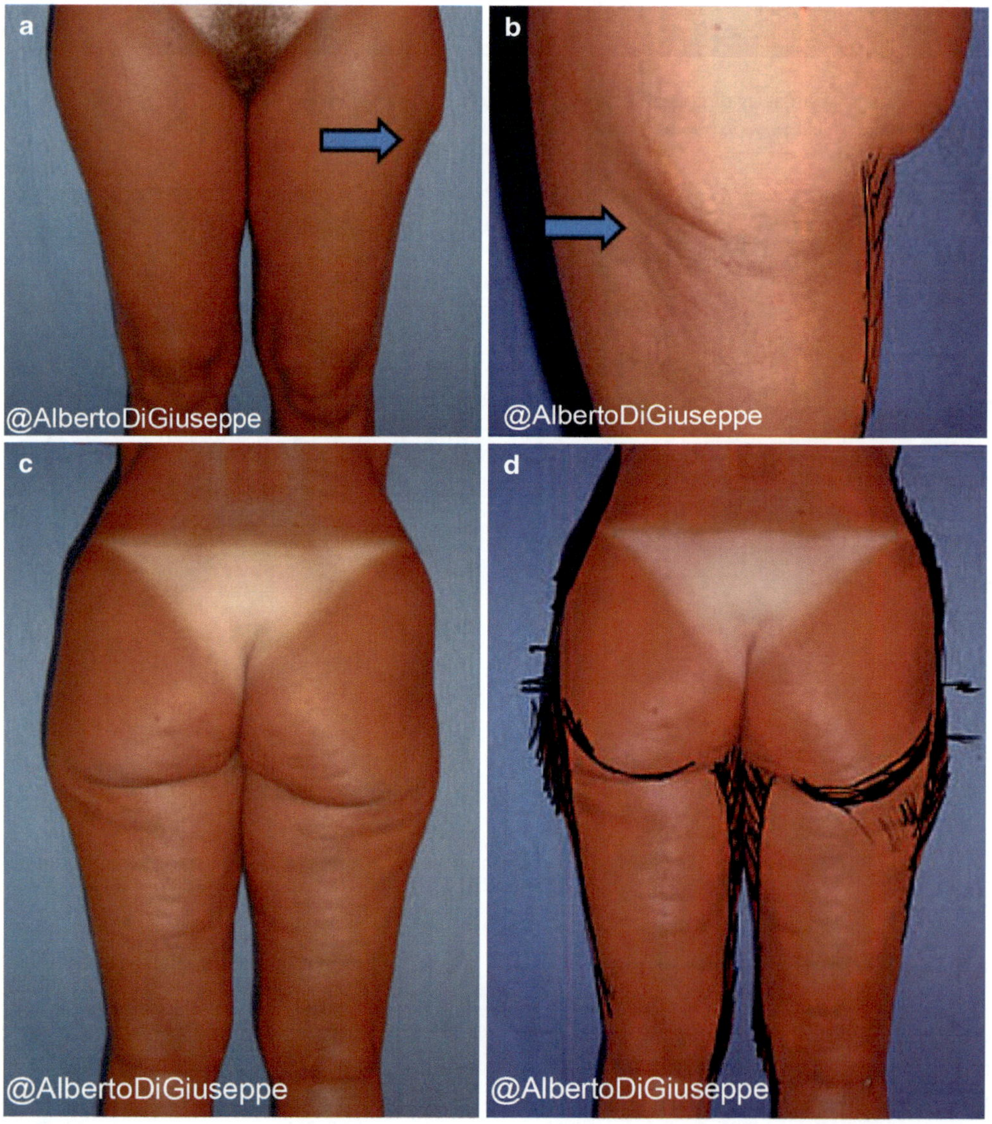

Fig. 19.13 45 year patient with lipodystrophy, cellulitis and post traumatic depression of the left lateral outer thigh. See blue arrows in front and lateral views (**a**, **b**). (**c**) Note important asymmetry of the gluteal fold. (**d**) Pre op planning, with correction of asymmetries, and fat transfer to integrate the area of depression. (**e**, **f**) Pre and post op after Vaser liposuction and fat transfer to left outer thigh. Front view. (**g**, **h**) Pre and post op after Vaser liposuction and fat transfer to left outer thigh. Posterior view. (**i**, **l**) Pre and post op after Vaser liposuction and fat transfer to left outer thigh. Lateral view. Pre and 1 year post op photos

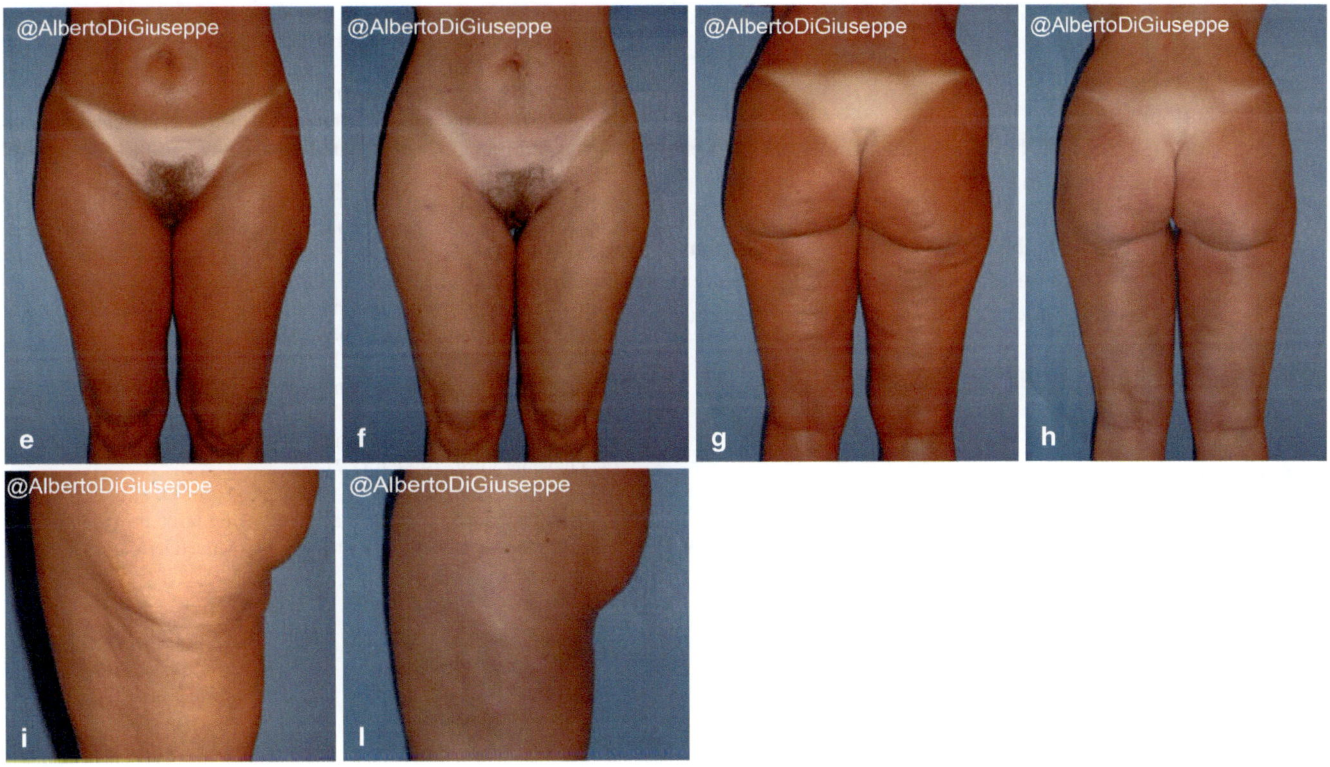

Fig. 19.13 (continued)

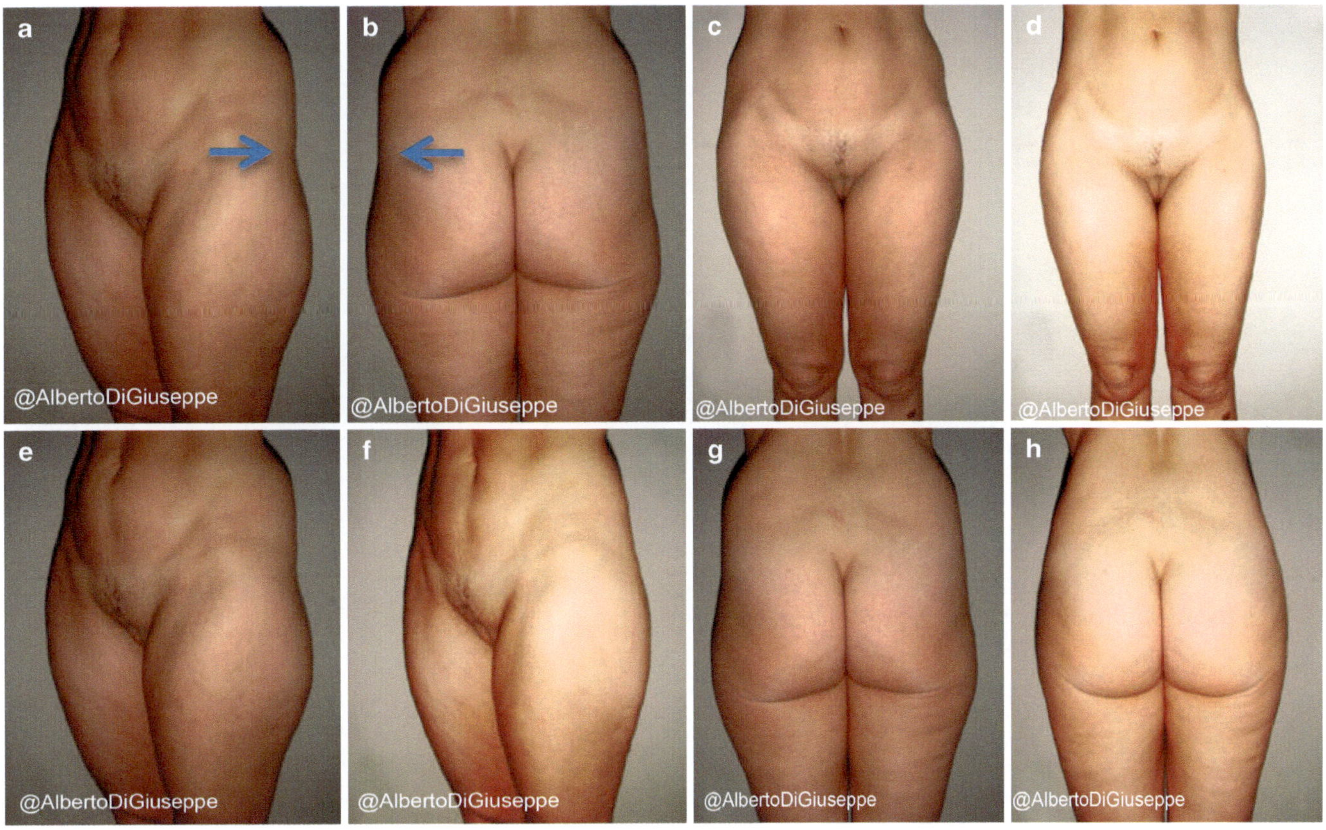

Fig. 19.14 Body asymmetry and depression of the outer buttock. (**a**, **b**) Body frame analysis. Blue arrows indicate areas of depression. (**c**, **d**) Pre and post op after Vaser liposuction and fat transfer. Front view. (**e**, **f**) Pre and post op after Vaser liposuction and fat transfer. Oblique view. (**g**, **h**) Pre and post op after Vaser liposuction and fat transfer. Posterior view. Pre- and 1-year post-op photos

1. flanks lipodystrophy with asymmetry in shape and volume
2. outer thighs excess
3. square buttocks shape
4. depression in the areas between outer thigh and lower flanks (see blue arrows in Fig. 19.13a, b).

After this analysis of the body frame, surgical plan included Vaser liposuction of all those areas and fat transfer to the depressed area, in order to achieve a nicer profile and contour.

Pre- and 1-year post-op photos (Fig. 19.14).

References

1. Shiffman MA, Di Giuseppe A. Body contouring. Art, science, and clinical practice. Springer-Verlag Berlin Heidelberg; 2010.
2. Shiffman MA, Di Giuseppe A. Liposuction. Principles and practice. Springer-Verlag Berlin Heidelberg; 2016.
3. Shiffman MA, Di Giuseppe A. Contorneado Corporal. Arte, Ciencia y Practica Clinica. Amolca; 2013.
4. Pereira LH, Sterodimas A. Correction for the iatrogenic form of banana fold and sensous triangle deformity. Aesthet Plast Surg. 2008;32(6):923–7.
5. Pereira LH, Nicaretta B, Sterdimas A. Correction of liposuction sequelae by autologous fat transplantation. Aesthet Plast Surg. 2011;35(6):1000–8.
6. Di Giuseppe A, Giovagnoli F, Di Giuseppe S, Ronconi D. Vaser body contouring achieving a more defined shape. The art of body contouring. London: IntechOpen; 2019. p. 105–122.
7. Shiffman MA, Blugerman G. Fat shifting for the treatment of skin indentations. Liposuction. Principles and practice. Springer-Verlag Berlin Heidelberg; 2016.
8. Saylan Z. Liposhifting: treatment of postliposuction irregularities. Int J Cosmet Surg. 1999;7(1):71–3.
9. Levenberg A, Scheinowitz M, Sharabani-Yosef O. Higher cell viability and enhanced sample quality following laser assisted liposuction versus mechanical liposuction. J Cosmet Dermatol Sci App. 2015;5(3):238–45.

Three-Dimensional Thigh Contouring the Role of Fat Grafting

20

Alberto Di Giuseppe and Federico Giovagnoli

Contents

20.1 Introduction

In 1487, Leonardo da Vinci published his celeber drawing of the Vitruvian Man (conserved in Galleria dell'Accademia, Venice, Italy) [1].

Vitruvius, the human body was the source of all geometry. As a man of his time, Vitruvius followed the Pythagorean tradition and considered nature the source of all knowledge, transcribed in the form of the perfect number 10. To Vitruvius, the human body is nature's prime work, incorporating the only creature with ten fingers, and the ability to think, create, and manufacture objects. The Vitruvian man does not move to produce geometry, "flat on his back and passive, he is made e to produce it." The active agents, as Vitruvius tells it, are the compass and the set square. The results produced are meaningful through the action of the architect and the culture to which he belongs. Leonardo da Vinci rediscovered the Vitruvian man in the Renaissance and created its most famous descriptive drawing in which man is not static at all, but standing and in motion. Leonardo called this system a Canon of Proportions in which 16 possible positions for the human body are systematized in the same drawing, generated by variations on two main positions overlapped and inscribed within the cosmological symbols of the circle and the set square. It must be no coincidence either that Leonardo's Canon of Proportions depicts 16 actions. Leonardo's interpretation of the Vitruvian man is standing, following the Roman theology of the feet.

To follow, Leonardo realized 50 sheets of drawings of the human body, made between about 1485 and 1510–1515, which are based on the artist's own anatomical dissections and show his evolving understanding of physiology. The drawings demonstrate, as well, Leonardo's progress from technical mastery of his subject to consummate draftsmanship.

In the primitive conditions of the late fifteenth century, and with no medical training, this astonishing man acquired a knowledge of human anatomy far in advance of the medical profession of his day.

From those drawings (Figs. 20.1 and 20.2), it is clear the concept, that Leonardo emphatized, of drawing and measuring the proportion of thighs, knees, and legs in a circumferential way. He did use some measurement and proportion numbers to transform into mathematics his view of the esthetics.

A. Di Giuseppe (✉)
University of Padova, Padova, Italy

F. Giovagnoli
University Vita-Salute San Raffaele, Milan, Italy

© The Author(s), under exclusive license to Springer Nature Switzerland AG 2023

A. Di Giuseppe et al. (eds.), *Fat Transfer in Plastic Surgery*, https://doi.org/10.1007/978-3-031-10881-5_20

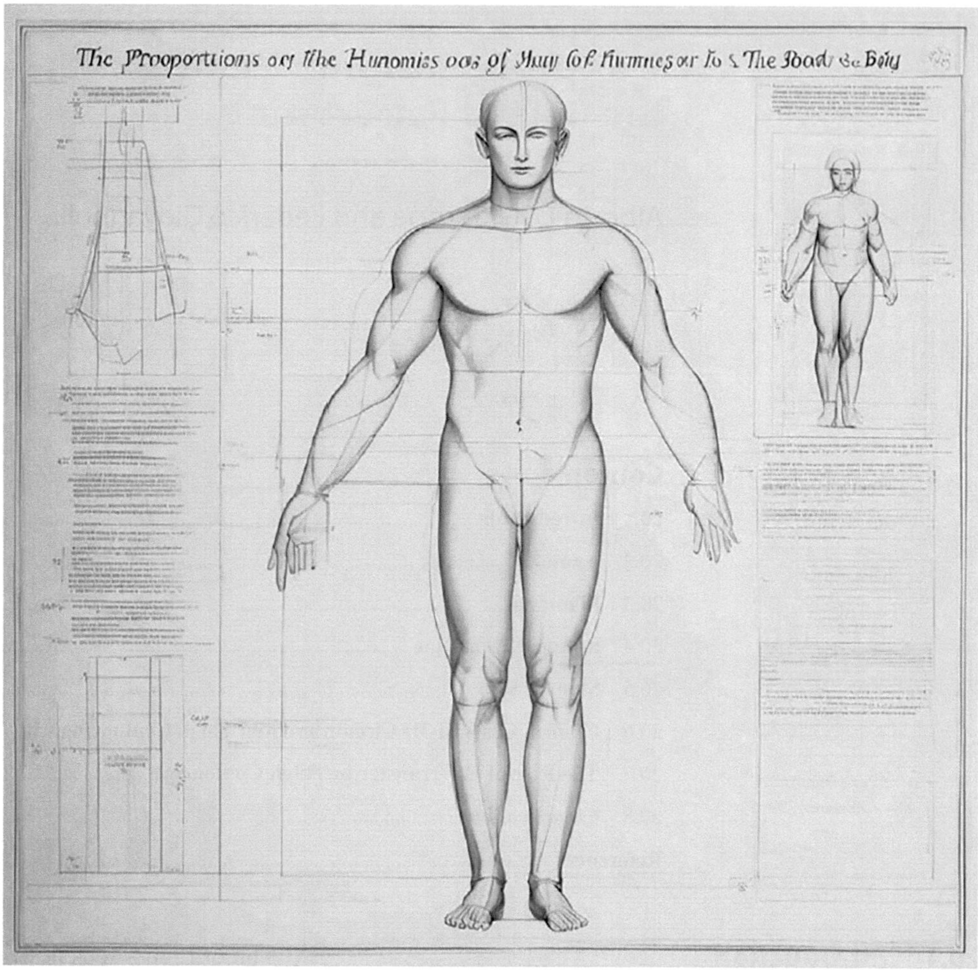

Fig. 20.1 The proportions of the human body

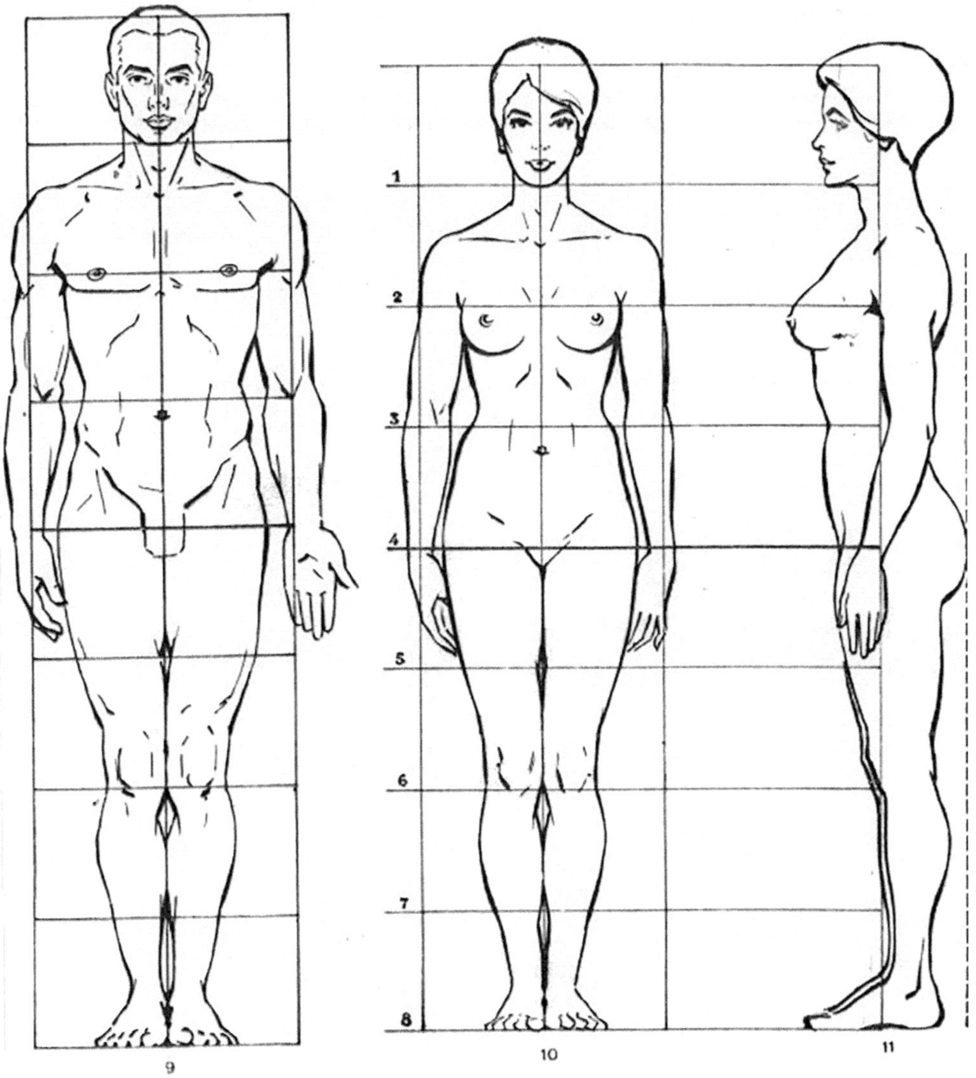

Fig. 20.2 Drawings of ideal proportions [2]

In body contouring of thighs, for ages the proportion considered was only the anteroposterior view (Fig. 20.3).

At that time, in planning correction of the body figure, I was considering only the anterior and the posterior parts of the thighs. No matter how the full dimension of the entire thigh was looking. This direction was given by my teachers and masters in plastic surgery at the time (1980–1995). This perspective was deeply influenced by the drawings of Gerald Illouz who published the "Zone of adherence" in body [3].

He described a total of five zones of adherence from trunk to lower extremity: lateral gluteal depression, gluteal crease, distal posterior thigh, mid-medial thigh, and inferolateral ileo-tibial truck (Fig. 20.4).

Those zones were considered to be the "forbidden areas" in thighs and buttocks contouring: the reason was that the fat layer in those areas is really minimal, and an excessive aspiration in these areas could cause severe deformities, difficulty to be corrected. Minding those anatomical issues, and in consideration of the diameter of the cannulas utilized at that time by Illouz (6 mm cannula), this was a wise advise to surgeons to prevent deformities, irregularities, dimples, and depression. For ages, plastic surgeons world wide have treated thighs without considering in a 3D way, as scared of potential secondary damages.

The lateral view, and the oblique view of thighs, introduces the need for a three-dimensional artistic understanding of the

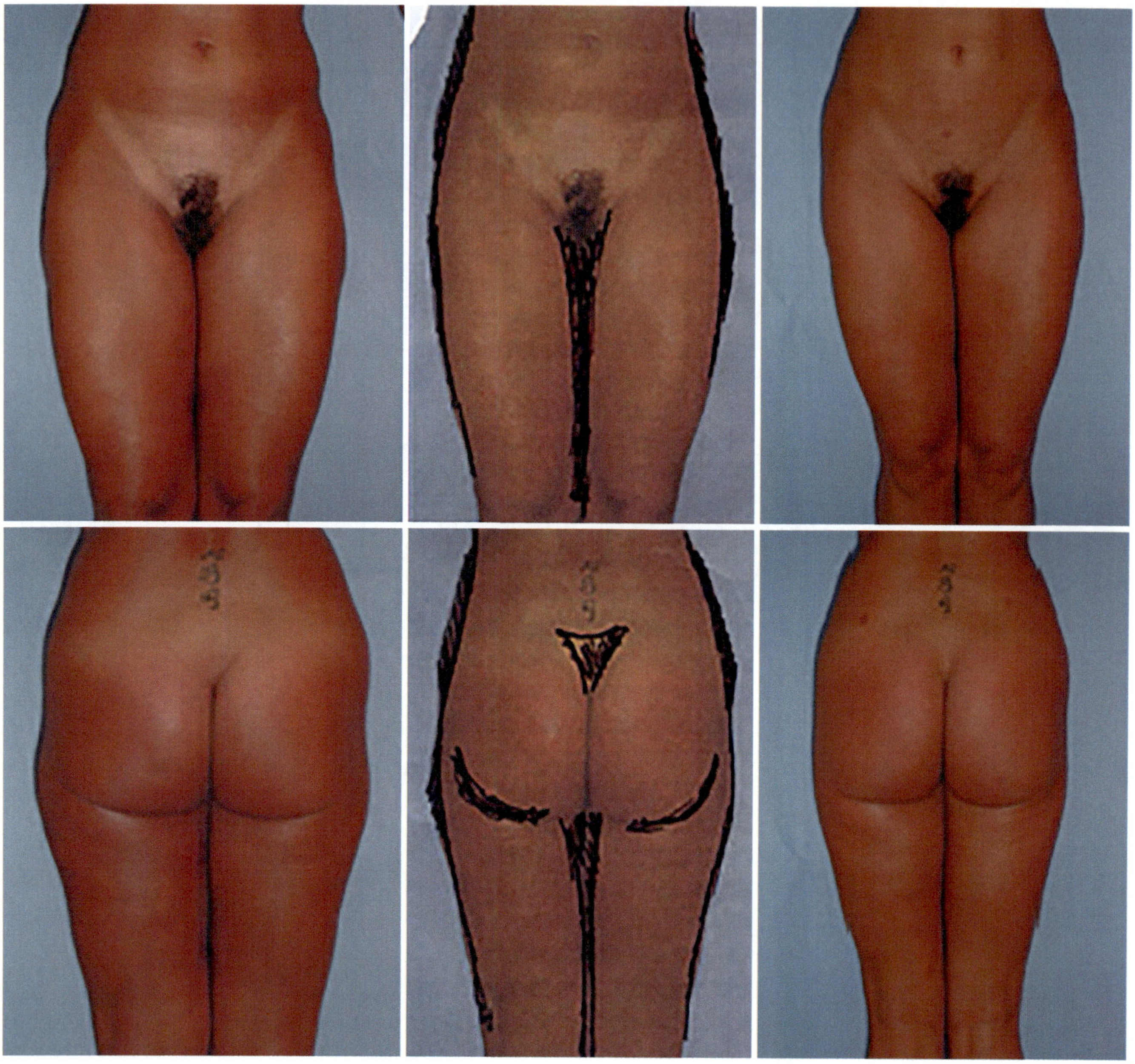

Fig. 20.3 Pre-op photos, planning and 6-month post-op result. Only anterior and posterior view shown

anatomical and surgical adipose layers when performing body contouring procedures. A combination of superficial subcutaneous undermining and deeper fat removal must be accomplished to obtain a nice esthetic and proportioned result, depending on the level of thighs where the surgeon is working (Fig. 20.5).

Since year 2000, by using an ultrasound-assisted device (VASER) for liposculpture, I have started treating the anterolateral portion of the thighs, considering the earlier difficulties in addressing that area. Since then, also my pre-op photos and planning include the lateral and the oblique views in order to have a full understanding of the body dimension and proportion. And I'm not scared any more to fully treat, three-dimensionally, the whole area [4].

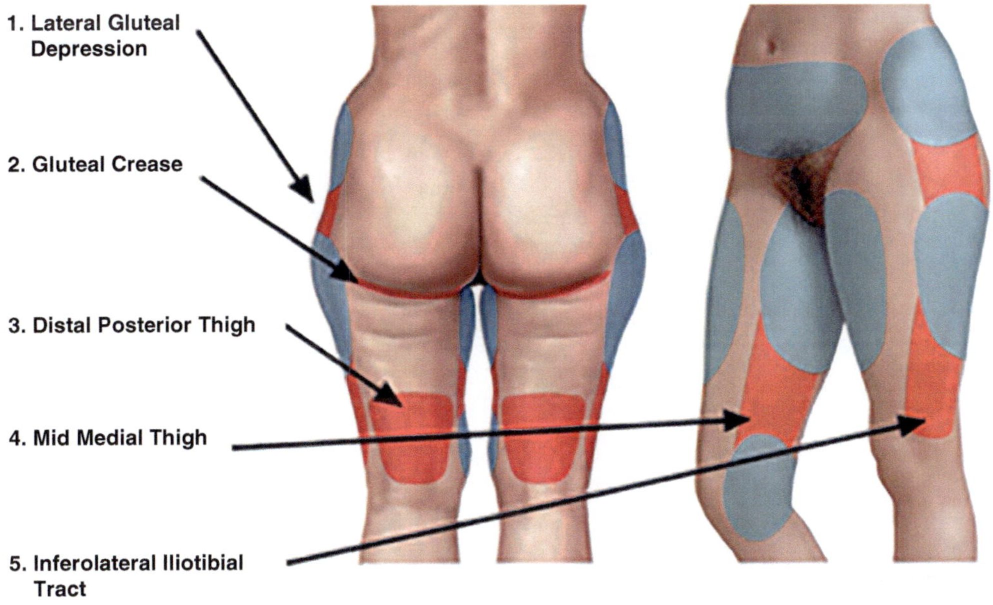

Fig. 20.4 Zone of adherence introduced by G. Illouz

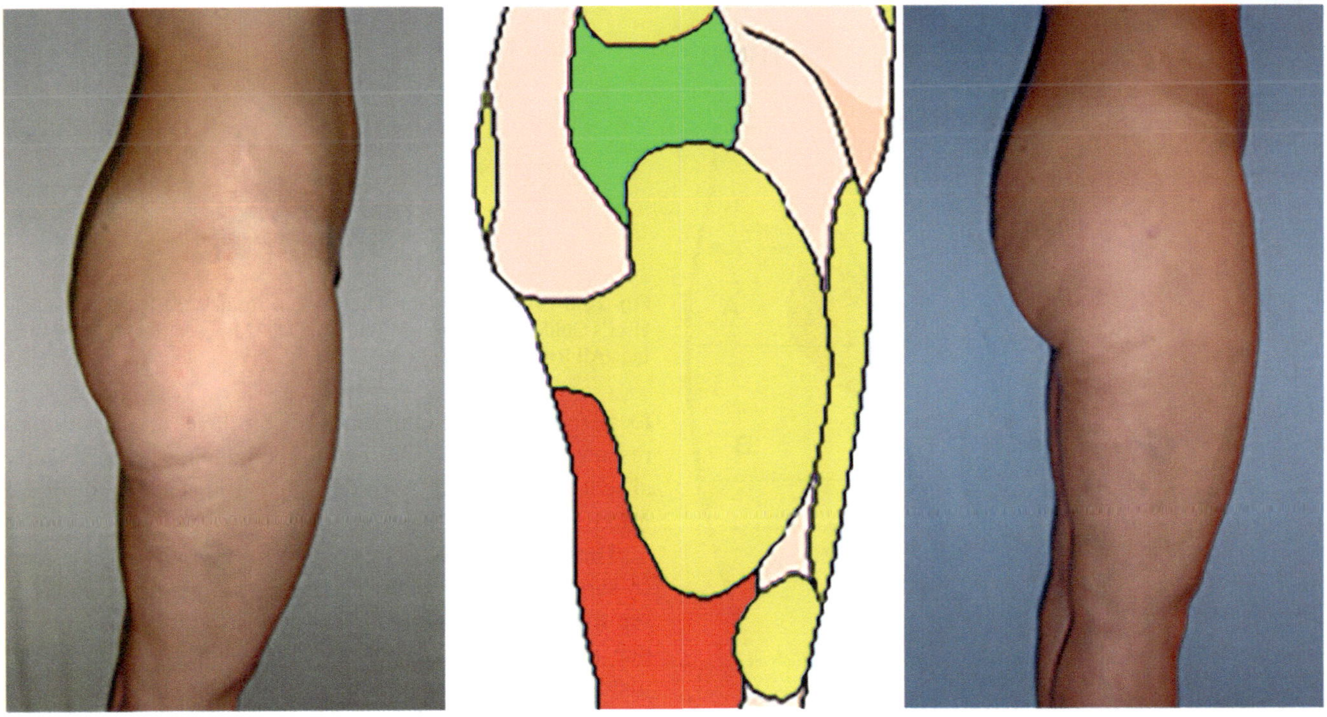

Fig. 20.5 Side-lateral view of thighs. This further projection introduces the concept of 3D contouring, in addition to anteroposterior view previously considered

20.2 Anatomy

Fat distribution in lower extremities and thigh is quite variable, depending on the area:

1. anterior surface: under the skin envelope, there is an areolar fat layer of variable thickness, then underlying fascia, muscle layer, and then femur (bone);
2. external surface: under the skin envelope, the fat layer is minimally represented. This is one of the forbidden areas in correspondence with the deeper fascia lata (distal posterior thigh);
3. posterior surface: under the skin envelope, there is a distinct fat layer, with several myocutaneous fibrotic attachment to the overlying tissue, as in the infragluteal crease (Fig. 20.6).

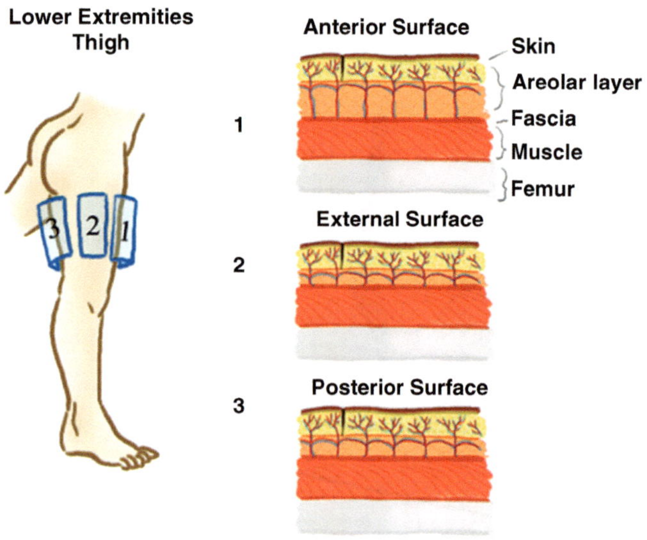

Fig. 20.6 Anatomy of fat layer distribution anterior, external and posterior surface of thigh

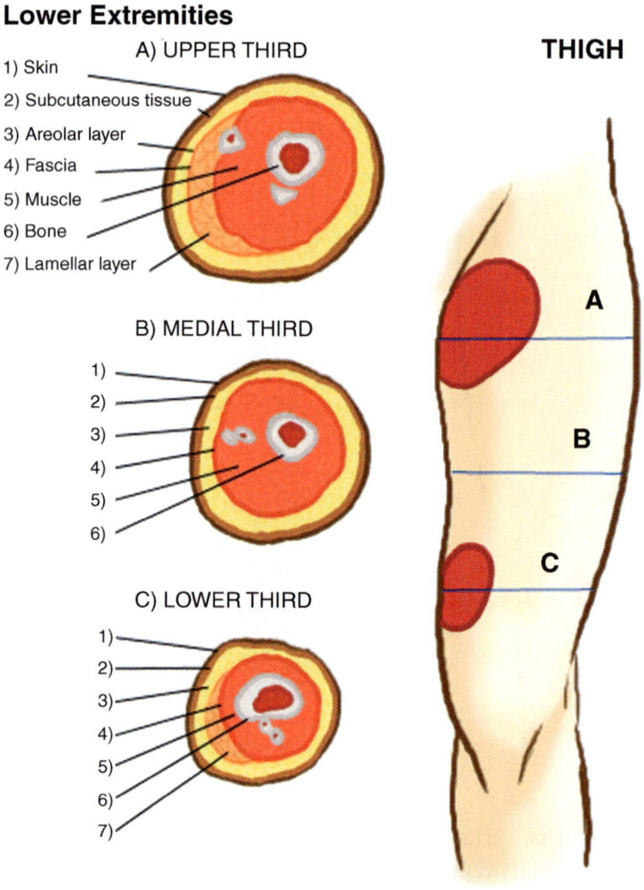

Fig. 20.7 Fat distribution in upper, medial, and lower third of thigh

A further distinction of the fat layers affects upper (A), medial (B) and lower third (C) of thigh (Fig. 20.7).

Fat distribution in antero-medial thigh presents few differences on the thickness of the lamellar layer. This is localized underneath the subcutaneous layer and is well represented in the upper third (A). The lamellar layer is nearly absent in the medial third (B). And it appears again in the lower third (C), but less represented.

Fig. 20.8 VASER System (Photo taken from VASER Solta Medical User's Guide [5]. Courtesy of Bausch, ©2022 Bausch Health Companies Inc., All Rights Reserved)

Those anatomical variations must carefully be analyzed by surgeon when planning and performing his thigh contouring liposuction. The medial third must be treated carefully, due to the absence of the lamellar layer. Only superficial undermining must be addressed in this layer to avoid unpleasent complications as depressions and irregularities.

The upper third (A) presents a thicker lamellar layer that must be addressed properly to reduce thickness. The subcutaneous layer could be quite thin by aging: here too, a cautious superficial undermining is the key to avoid secondary irregularities and dimples.

In order to accomplish those needs, I believe surgeon must use a delicate instruments which could exalt skin retraction capability.

And I did utilize VASER ultrasound device in order to safely treat those areas and achieve good skin retraction. The key to success is to find a proper level of emulsification of the fat layers in the different areas of thigh (Fig. 20.8) [6].

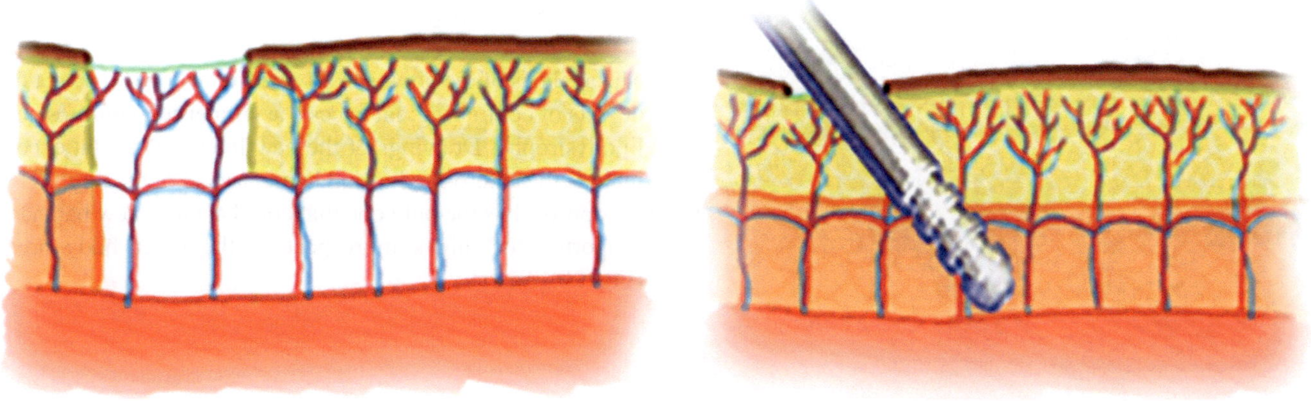

Fig. 20.9 Level of emulsification (Photo taken from VASER Solta Medical User's Guide [5]. Courtesy of Bausch, ©2022 Bausch Health Companies Inc., All Rights Reserved)

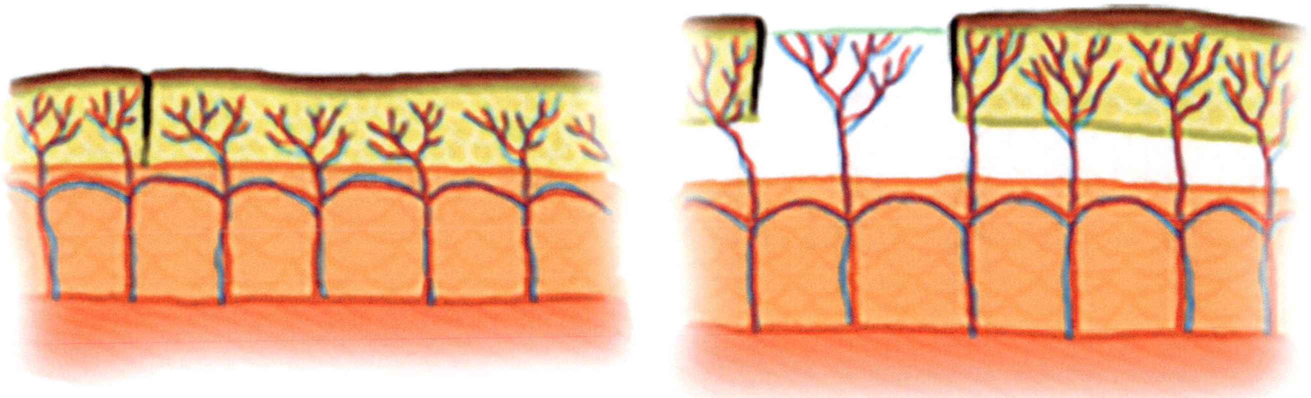

Fig. 20.10 Level of undermining (Photo taken from VASER Solta Medical User's Guide [5]. Courtesy of Bausch, ©2022 Bausch Health Companies Inc., All Rights Reserved)

I do use 3.7 mm 2 or 3 rings VASER titanium probe with continous administration of energy, with power around 70–80% of total. In alternative, VASER mode could be utilized with energy setting still at 70–80% of total power. VASER mode allows delivery of 50% of energy at the same amount of time, thus reducing the total energy administrated and thus potential complications. It is ideal for beginners, to prevent mistakes and overemulsification of fat tissue (Fig. 20.9).

In order to achieve a better skin retraction, tissues must be carefully undermined from the subcutaneous areolar layer, by gently preserving the vascular network and the connective structure which supports the overling tissues. Ultrasound energy is ideal for this target, as it is selective only for fat tissue. And it does spare all others structures (vessels, nerves, elastic fibers, and connective tissue), which need to stay as essential for the stability of the tissue structure (Fig. 20.10) [7].

The sequence of surgical plan must be as follow:

(a) Tumescent infiltration of the deeper layer.
(b) Tumescent infiltration of the superficial layer (until peau d'orange (Fig. 20.11) aspect is achieved).
(c) VASER ultrasound undermining of the superficial level with minimal emulsification (VASER mode or continous mode depending on surgeon ability and experience).
(d) VASER ultrasound deeper tissue emulsification (lamellar layer essentially).
(e) VENTX aspiration or other system liposuction with reduced power (15 mhg).
(f) Tissue compression to favor tissue adhesion after cleaning of emulsified fat (Fig. 20.12).

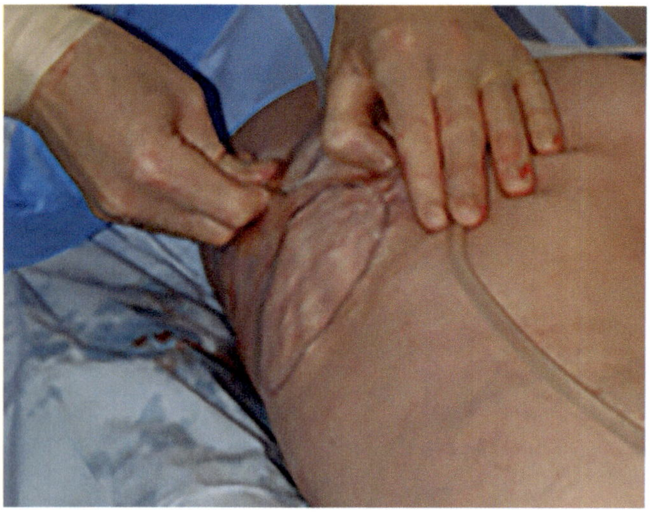

Fig. 20.11 Peau d'orange appearance

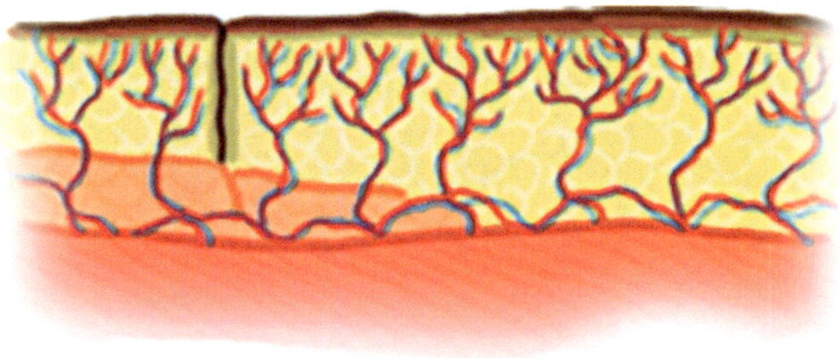

Fig. 20.12 Tissue adhesion after cleaning of emulsified fat (Photo taken from VASER Solta Medical User's Guide [5]. Courtesy of Bausch, ©2022 Bausch Health Companies Inc., All Rights Reserved)

20.3 Planning

The analysis of the photos in all direction allows to detect the defects of contouring: fat excess, asymmetries, celulites, and lipodystrophy. Drawings of the new lines and the fat excess include the antero-medial and lateral aspect of the thigh. This 3D approach is the only one that could achieve new balanced proportion of thighs in respect to the other body areas (Fig. 20.13).

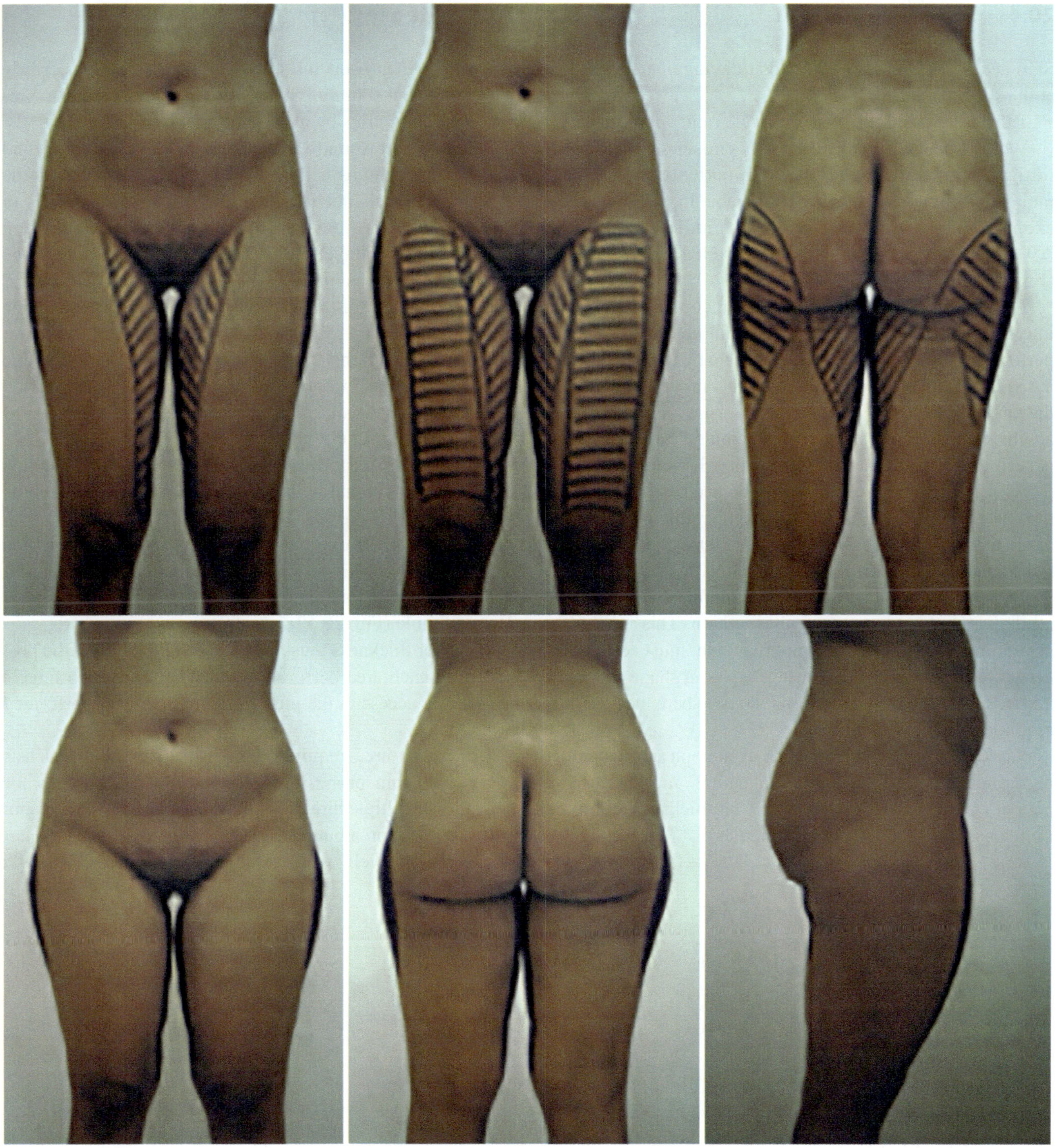

Fig. 20.13 Planning of 3D thigh contouring

20.4 Surgical Technique

Skin access incisions are planned at the inguinal crease and at the suprapatella.

Tumescent infiltration will be in 2:1 ratio appoximately.

VASER probe must be carefully chosen depending on initial thickness of the fat tissue of patients. I do recommend 3.7 mm 2 rings VASER probe in all the areas, exept the knees area and the inner-middle thigh, where 2.9 mm 2 rings probe should be electively selected (Fig. 20.14a, b).

After waiting 10 min for proper functioning of adrenaline, an extensive subdermis undermining of all the areas is performed. This process must be delicate by using continous VASER mode, and it normally takes 3–4 min to complete all the target areas.

Once undermining is completed, emulsification of the fat is performed in a uniform pattern with a 3.7 mm probe.

Areas where more caution must be used are: the patella, the inner-medial thigh, and the inner upper thigh. In all those areas, a careful aspiration of emulsified fat with low pressure must be performed using a 3.0 mm VENTX cannula.

Over correction, depression, irregularities, and dimples, all derive from a too aggressive aspiration. This is why the aspiration phase is the most important one: must address the final result and contouring, must give proper shape and lines, and reduce thickness of the fat layer where needed and required.

Sense of artistry and proportion of surgeon are mandatory requirments as well as a perfect planning, after analysis of patient characteristics (skin texture, fat thickness, celulites, dimples, irregularities, and asymmetries).

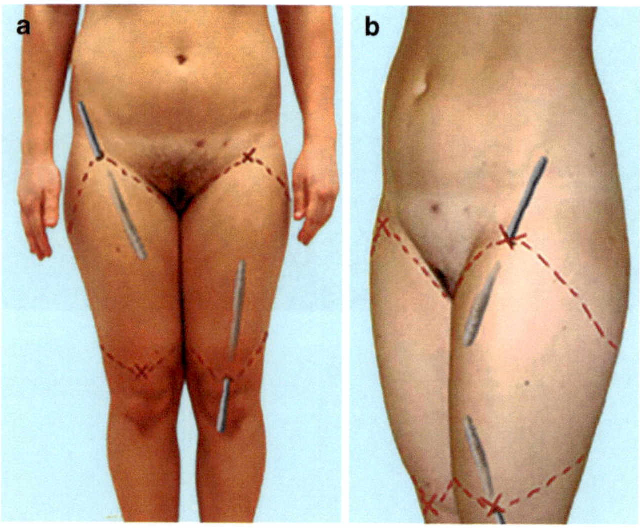

Fig. 20.14 (**a**) Surgical access through inguinal and patella incision, (**b**) Full undermining of antero-medial lateral thighs and knees, using 2.9 or 3.7 mm probe 2 rings

20.5 Step by Step

The patient has been infiltrated anteriorly and posteriorly with tumescent solution. Skin ports have been placed to prevent friction injuries by ultrasound probe (Fig. 20.15).

VASER ultrasound administrated on the inner and antero-medial thighs with a 3.7 mm probe 2 grooves for the superficial emulsification, using 70–80% of power in VASER mode or continous mode depending on surgeon experience. Deep layer emulsification by using same probe always in continous mode. 60–120 s ultrasound delivery for 100 mL of TLA infiltrate in each area is a proper guideline (Fig. 20.16).

In posterior thighs, extreme caution must be taken at the subgluteal crease, where strong connective ligaments do support the gluteal and must be respected to avoid drop of the area (Fig. 20.17).

Aspiration is performed with a 3.7 mm VENTX cannula on the deep layer and with a 3 mm VENTX cannula on the superficial layer (Fig. 20.18).

The final thickness and contouring of the reduced flap must be verified with the same cannula without aspiration. It will allow a nice redraping of the skin evenly, by using a 3 mm cannula (Fig. 20.19).

Fat layer thickness must be checked at the end of the procedure in each area performed. Pinch test with two fingers will allows access to the precision of the surgery performed (Fig. 20.20).

The knees area requires finer instrumentation and less energy: 2.9 mm probe, VASER mode, 3 mm cannula for aspiration. Multiple directions must be followed to reach not only the knees area but also the antero-medial and the anterior thigh as well with a single access at the patella (Fig. 20.21).

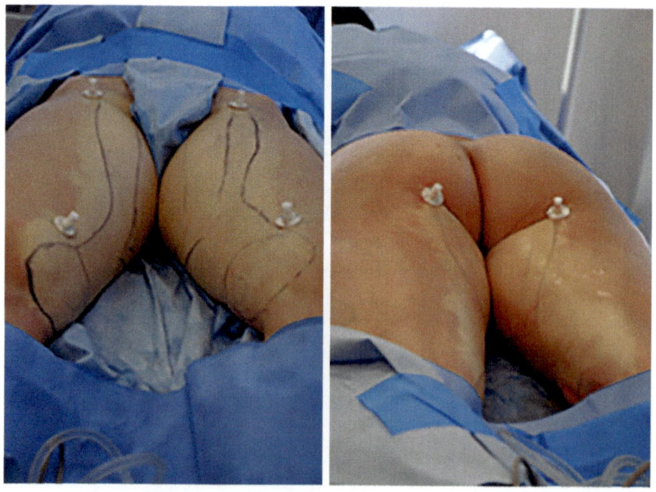

Fig. 20.15 Tumescent infiltration completed and skin ports on site

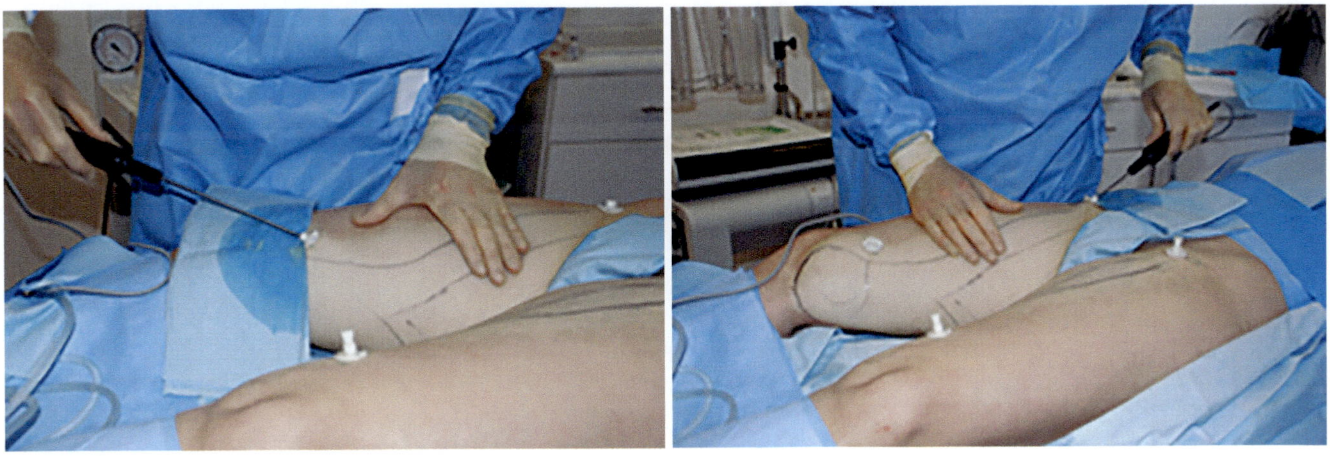

Fig. 20.16 VASER antero-medial thighs

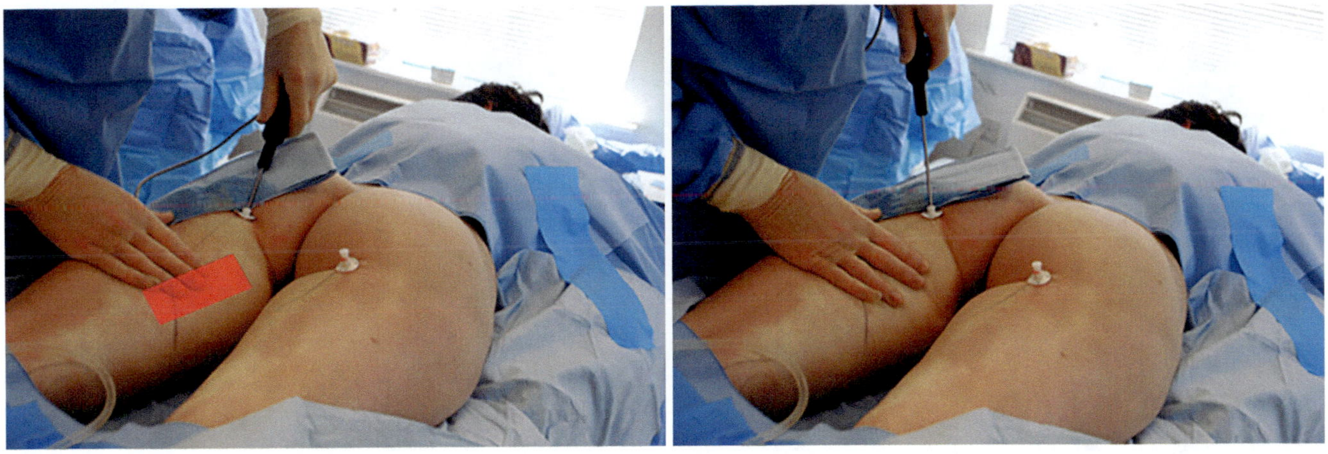

Fig. 20.17 VASER in postero-medial thigh. In red caution area

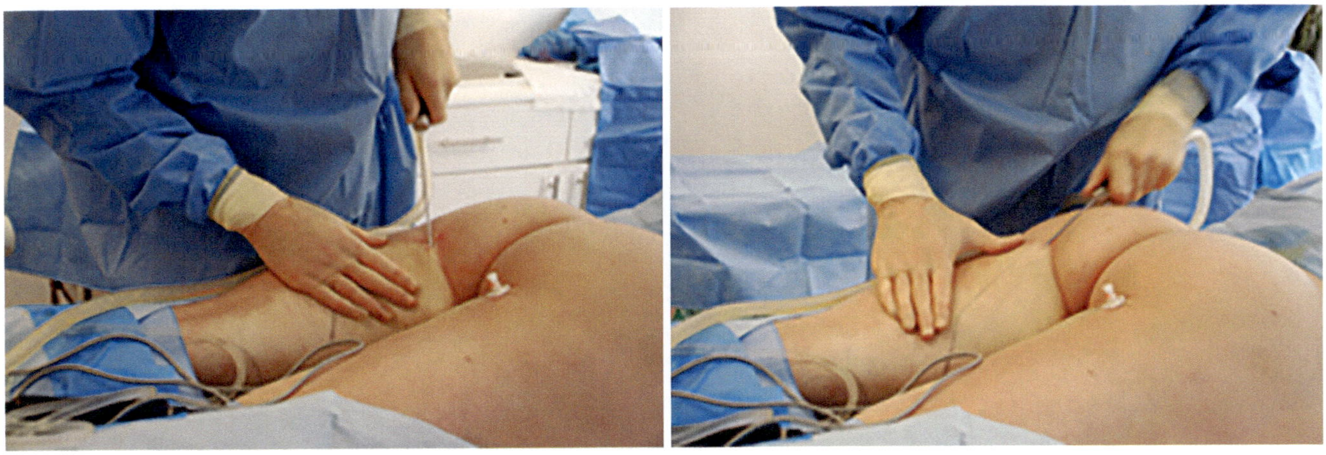

Fig. 20.18 Aspiration 3.7 mm cannula for deeper layer, 3.0 mm VENTX for superficial

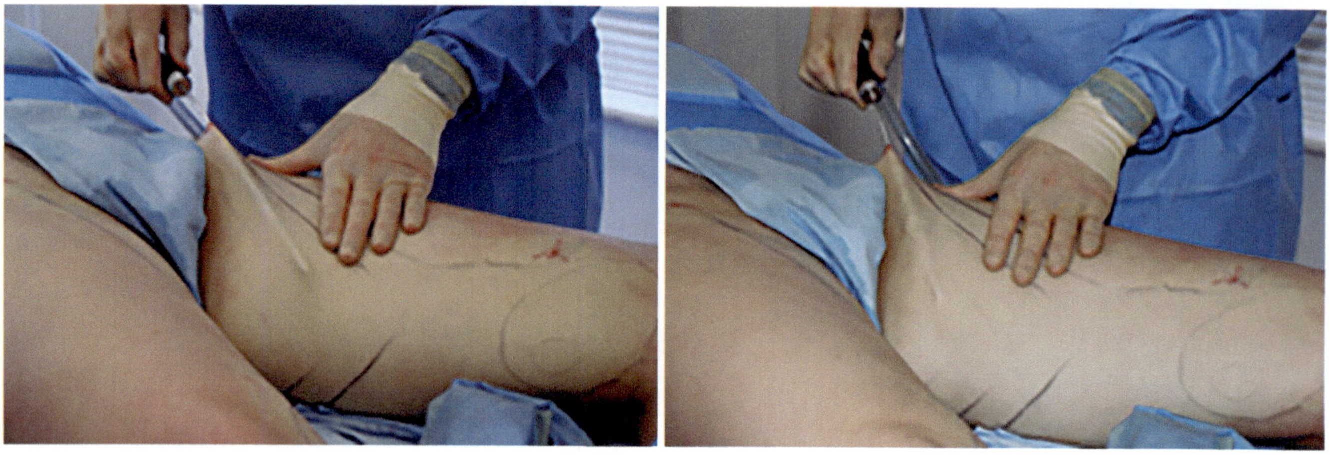

Fig. 20.19 Redraping the skin evenly (3 mm cannula)

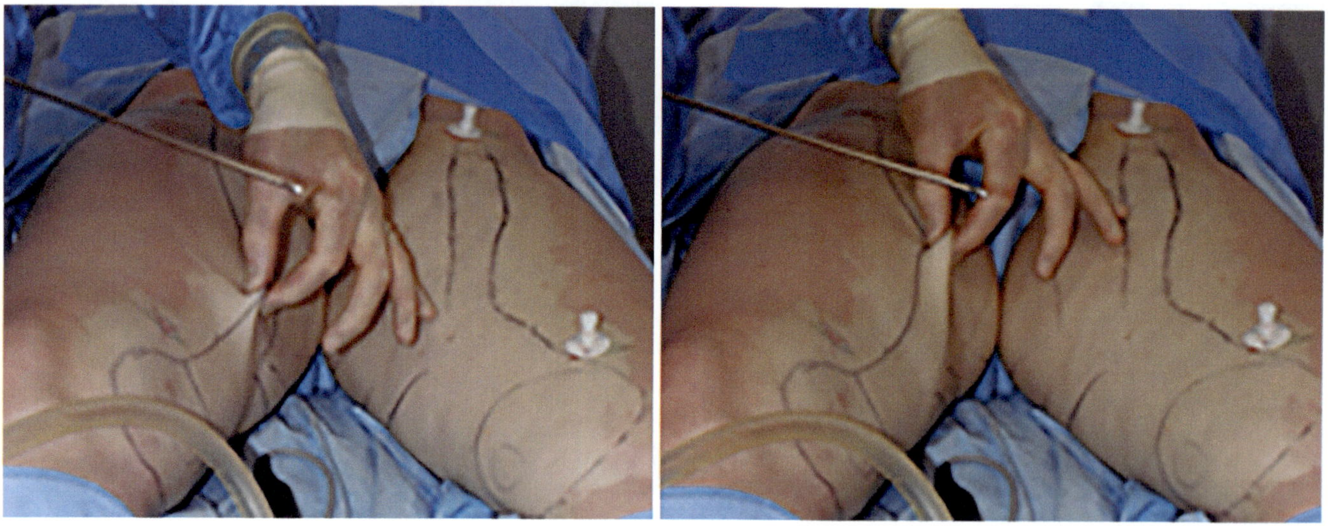

Fig. 20.20 Pinch test

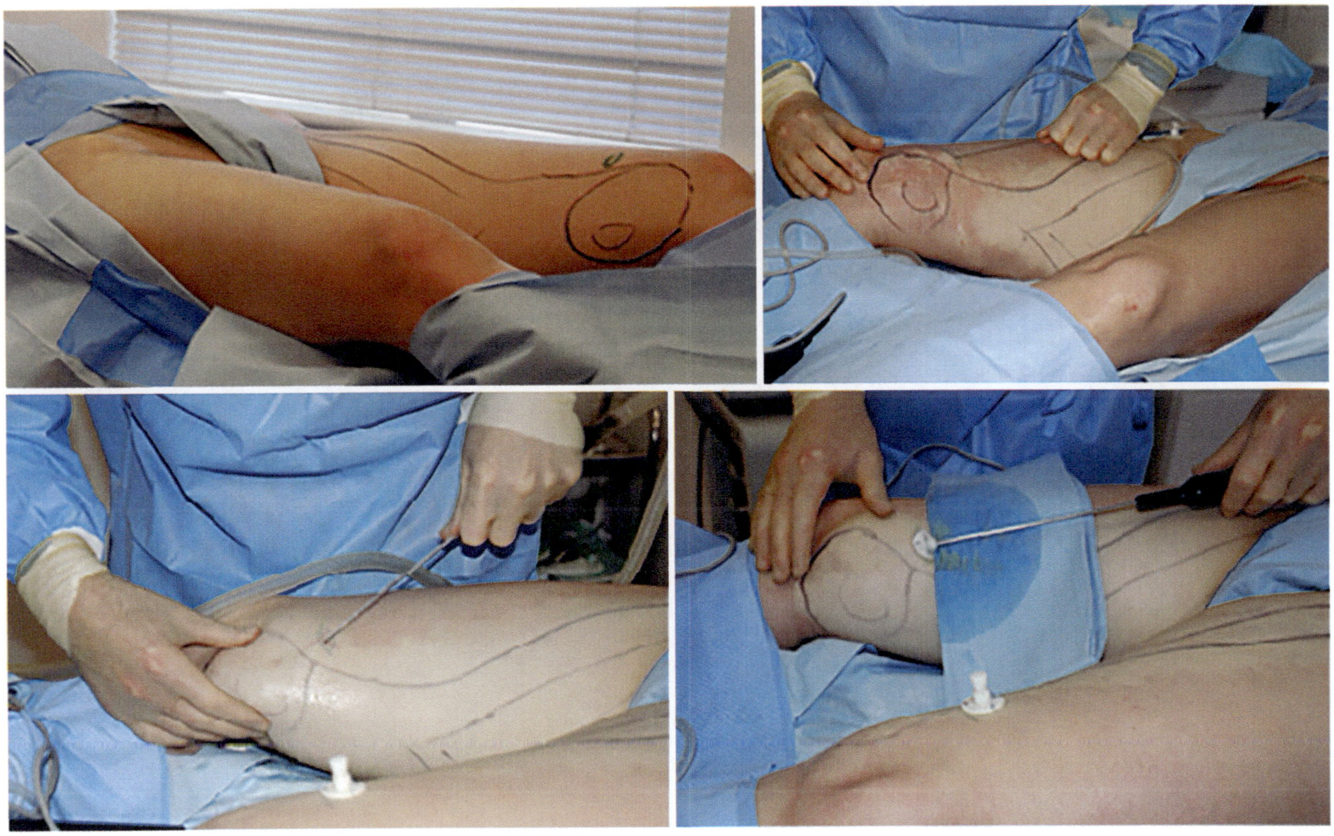

Fig. 20.21 Knees area: multiple area approach with single incision

20.6 Clinical Cases of 3D Circumferential Thigh Contouring with VASER

Those are clinical examples of pre op, planning and 6 months post-op appearance in body contouring reshaping, at abdomen flanks, outer thigh, inner thigh, and anterior thigh (3D circumferential thigh) (Figs. 20.22 and 20.23).

Case 1

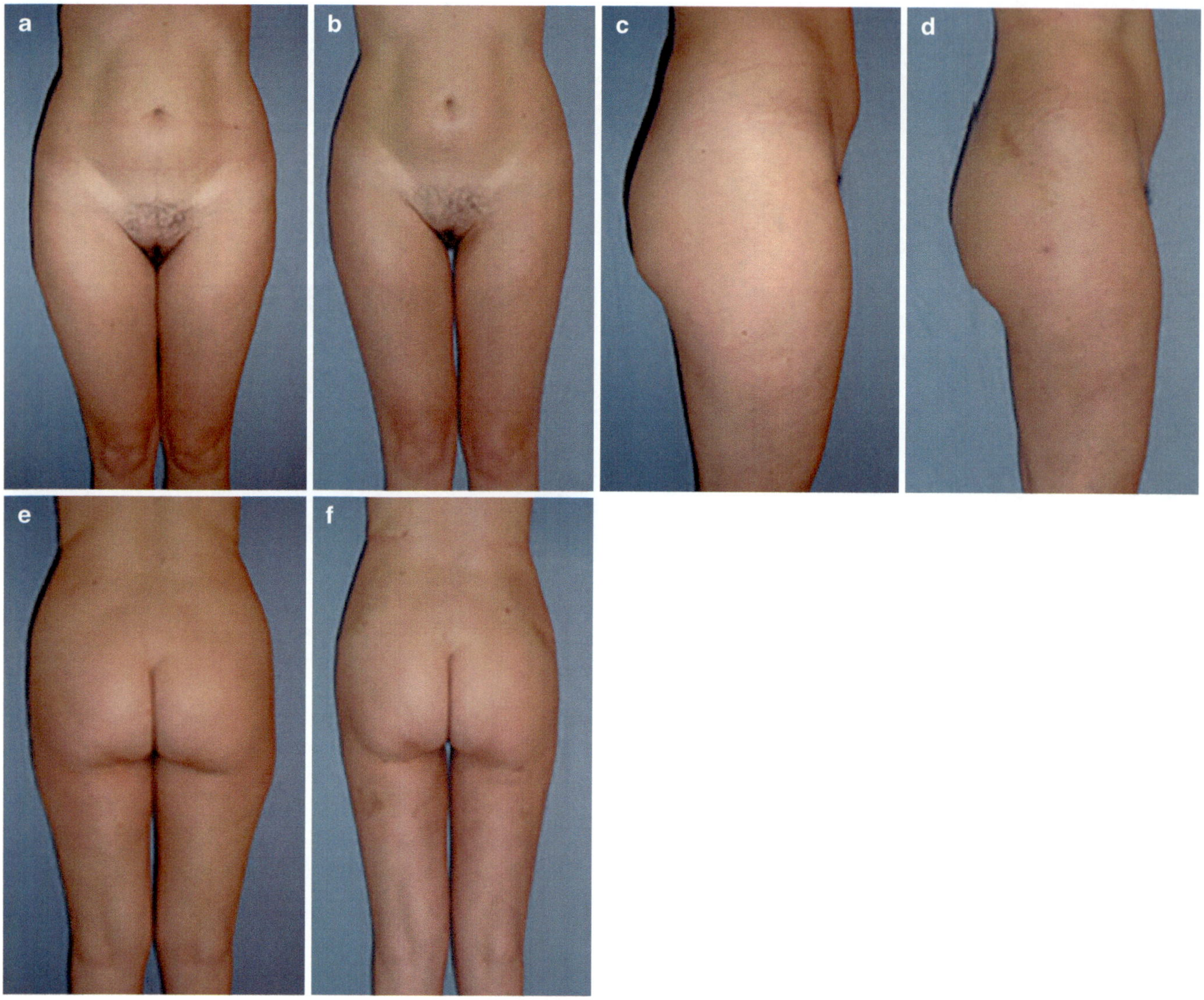

Fig. 20.22 Anterior, posterior, and lateral views. (**a**, **c**, **e**) Pre op, (**b**, **d**, **f**) 6 months post op after 3D circumferential thigh VASER

Case 2

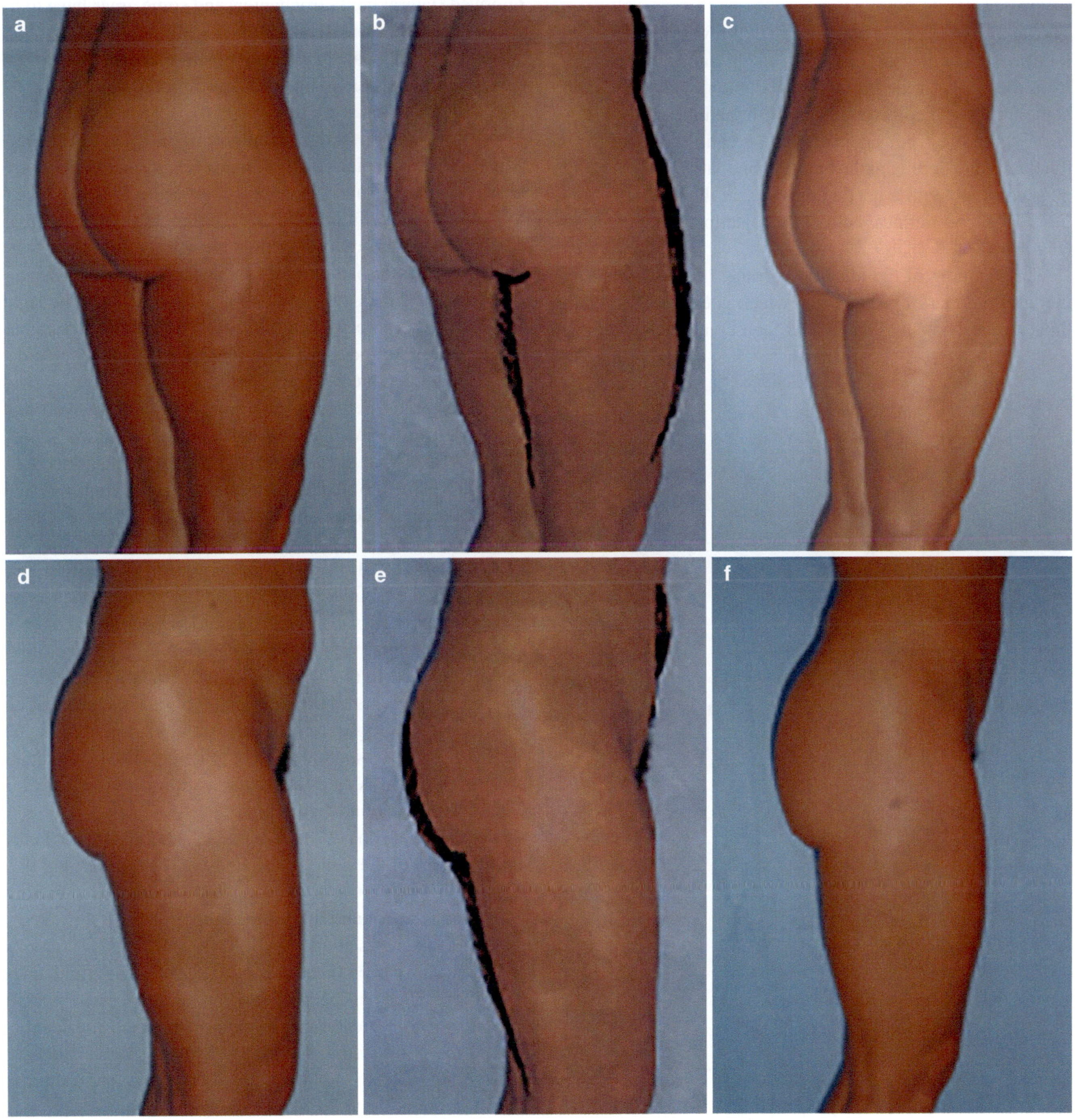

Fig. 20.23 Pre op (**a**, **d**), planning (**b**, **e**), 6 months post op (**c**, **f**) oblique view

20.7 The Role of Fat Transfer in Thigh Contouring

The analysis of female body figure will reveal the ideal contour lines that enhance the beauty and harmony. C lines goes from the axilla to the breast and abdomen, with a concavity at the midline in correspondance with the umbelicus. S lines are a curvature from lateral thorax, creating a concavity at the flanks and ending at the buttocks and lateral thigh. S lines does continue with a gentle convexity at the knee-calf junction. R curve defines buttocks projection and shape [8] (Fig. 20.24).

The planning of the ideal body frame of the lateral thigh includes the marking of the line which goes from the lateral flank to the outer trochanter and then continous and terminates along the lateral thigh at the knee. This line should be ideally a gentle S curve, where some areas have to be removed (lateral flank, trochanter) and some areas have to be added (lateral gluteal depression) (Fig. 20.25a).

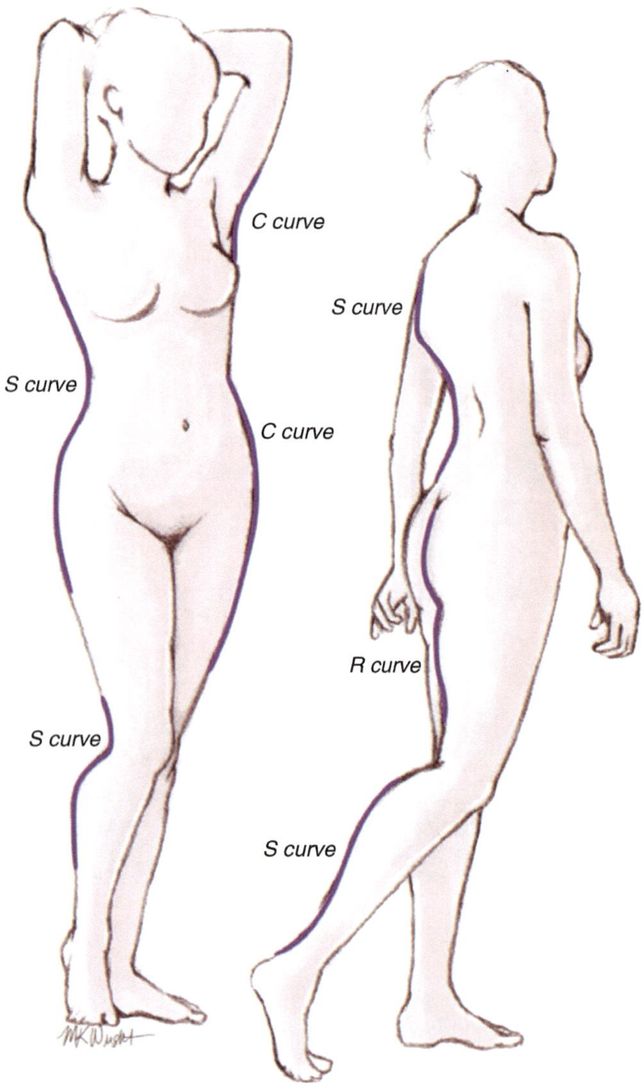

Fig. 20.24 C curve, S curve, R curve (Photo taken from VASER Solta Medical User's Guide [5]. Courtesy of Bausch, ©2022 Bausch Health Companies Inc., All Rights Reserved)

So, the final ideal shape results from a combination of fat substraction (VASER liposuction) and fat addition (fat transfer).

All information about technique of fat transfer are forwarded to the chapter named "Breast fat augmentation: step-by-step technique and ultrasound contouring of donor area."

Surgeon does often encounter asymmetries in body areas when planning is contouring procedures.

The following patient presents a rather consistent asymmetry in the flanks area as well in the trochanter. The right flank appears more rounded and full, the lateral trochanter and the lower buttocks are definitely larger than the left side. At the lateral gluteal depression, there is a more marked deformity on the left side. The inner thighs do convert and touch themselves without leaving an ideal "window space" which should be in the ideal frame of inner thighs. There is a lack of roundness of the lower buttocks, which are heavy and saggy. This attitude is mostly due to heavyness of middle buttocks.

In order to fix all those esthetic issues, a liposculpture of flanks, abdomen, buttocks, and thighs has been planned, with body reshaping by substraction and addiction of fat.

The abdominal region was treated with high definition VASER in order to reshape and enhance the edges of abdominal muscle (rectus abdominis and oblique muscles). The anterior lateral thighs were reduced and reshaped with 3D thigh VASER liposuction. Skin tigthening after superficial undermining and deeper fat emulsification is a major advantage of the ultrasound technology.

Body proportions of thigh, buttocks, abdomen, and flanks are often part of the same frame study of the female fig. I like to treat all those areas at the same time, in order to give the best esthetic result and final ideal frame.

In the case presented (Figs. 20.26 and 20.27), buttocks and thighs have been reshaped together with flanks and anterior thighs. Fat has been transfered to the lateral gluteal depression, to recreate a natural curvature and the ideal S line of the body frame. The final result is a lighter body, a curved natural shaped buttock-thigh complex, and a thinner circumferential thigh. Mastering body frame liposculpture is nowadays a combination of fat removal and fat adding.

Surgeon must be prepared to use both techniques in order to achieve the best esthetic result as the patients are nowadays more and more demanding: the internet generation has figured that with the latest technologies, surgeon could achieve any possible change in the body appearance and pretend perfection without understanding the human limits of a surgical result, which depends in large part on the initial condition.

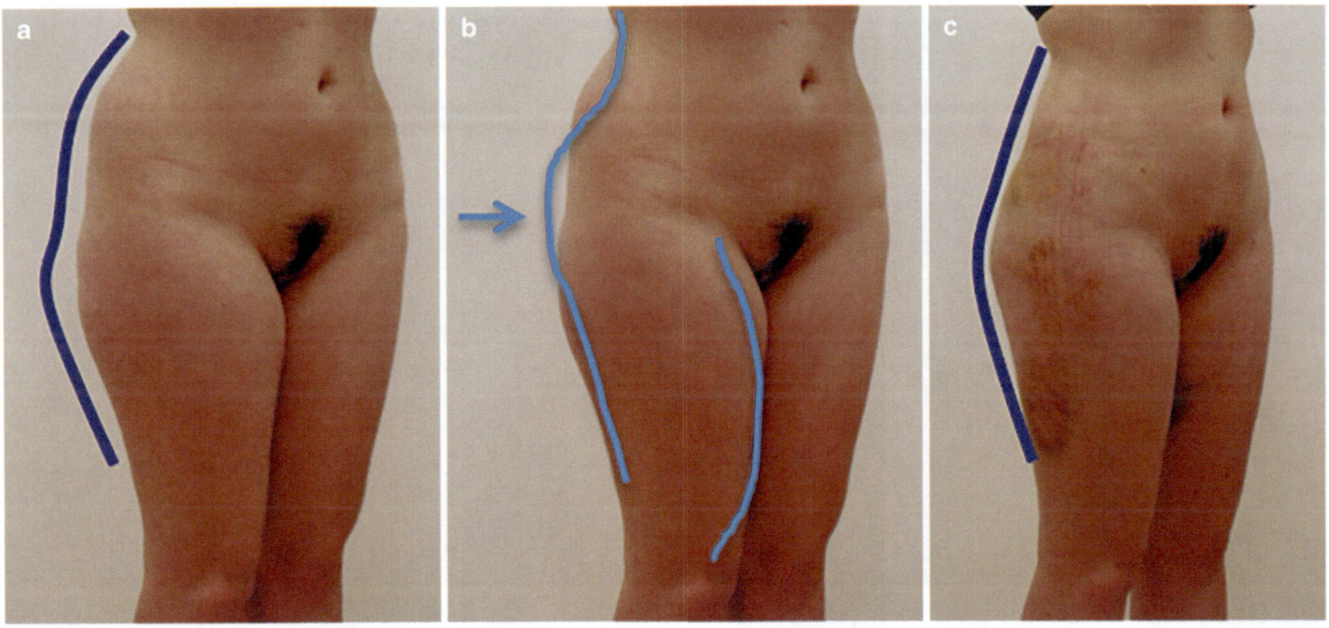

Fig. 20.25 (**a**) Pre-op line, (**b**) planning correction with fat removal and fat addition, arrow indicates fat transfer area, (**c**) new line after liposculpture and fat transfer (10 days after surgery)

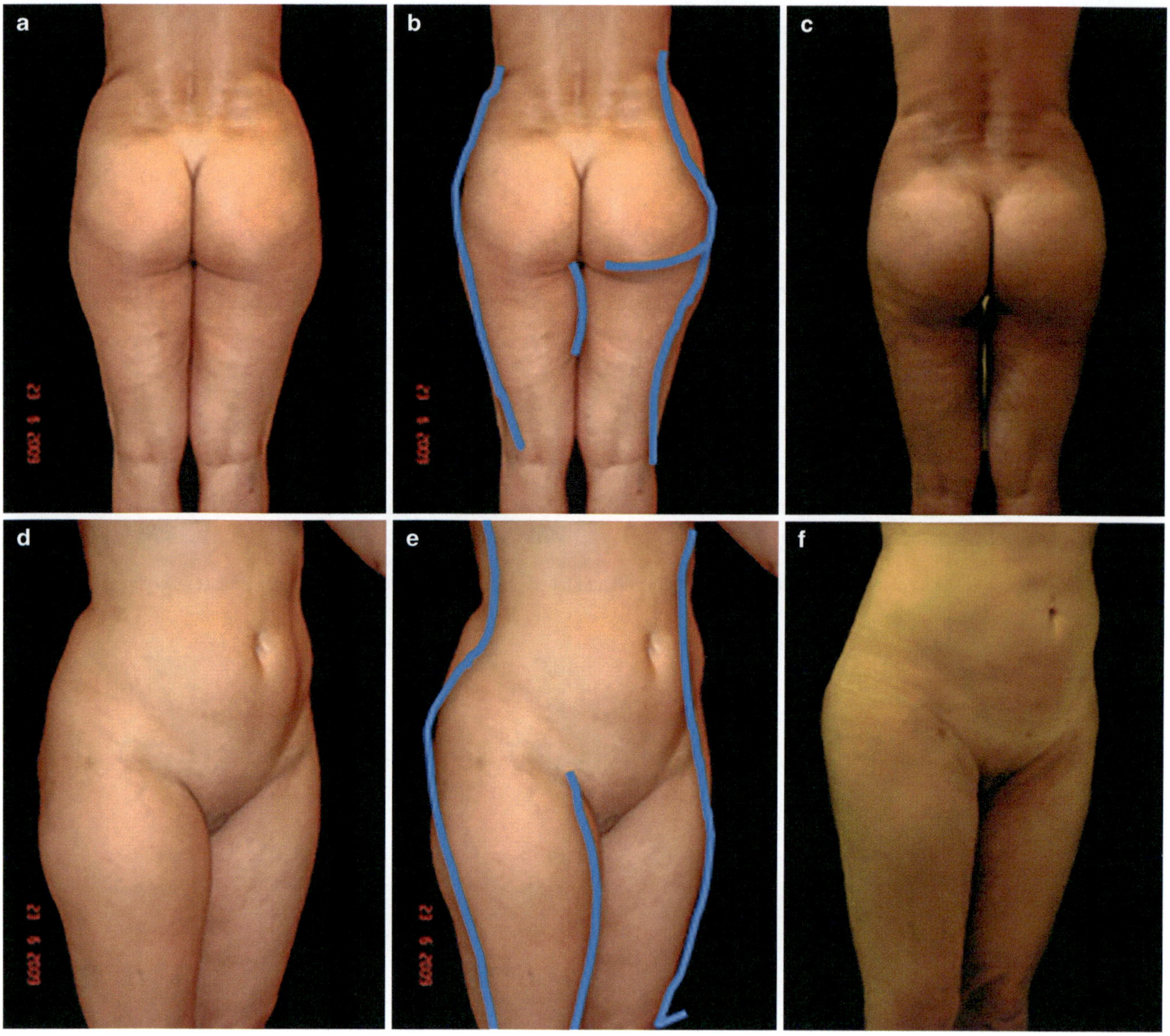

Fig. 20.26 Pre op (**a**, **d**, **g**, **i**), planning (**b**, **e**), post op 6 months (**c**, **f**, **h**, **j**)

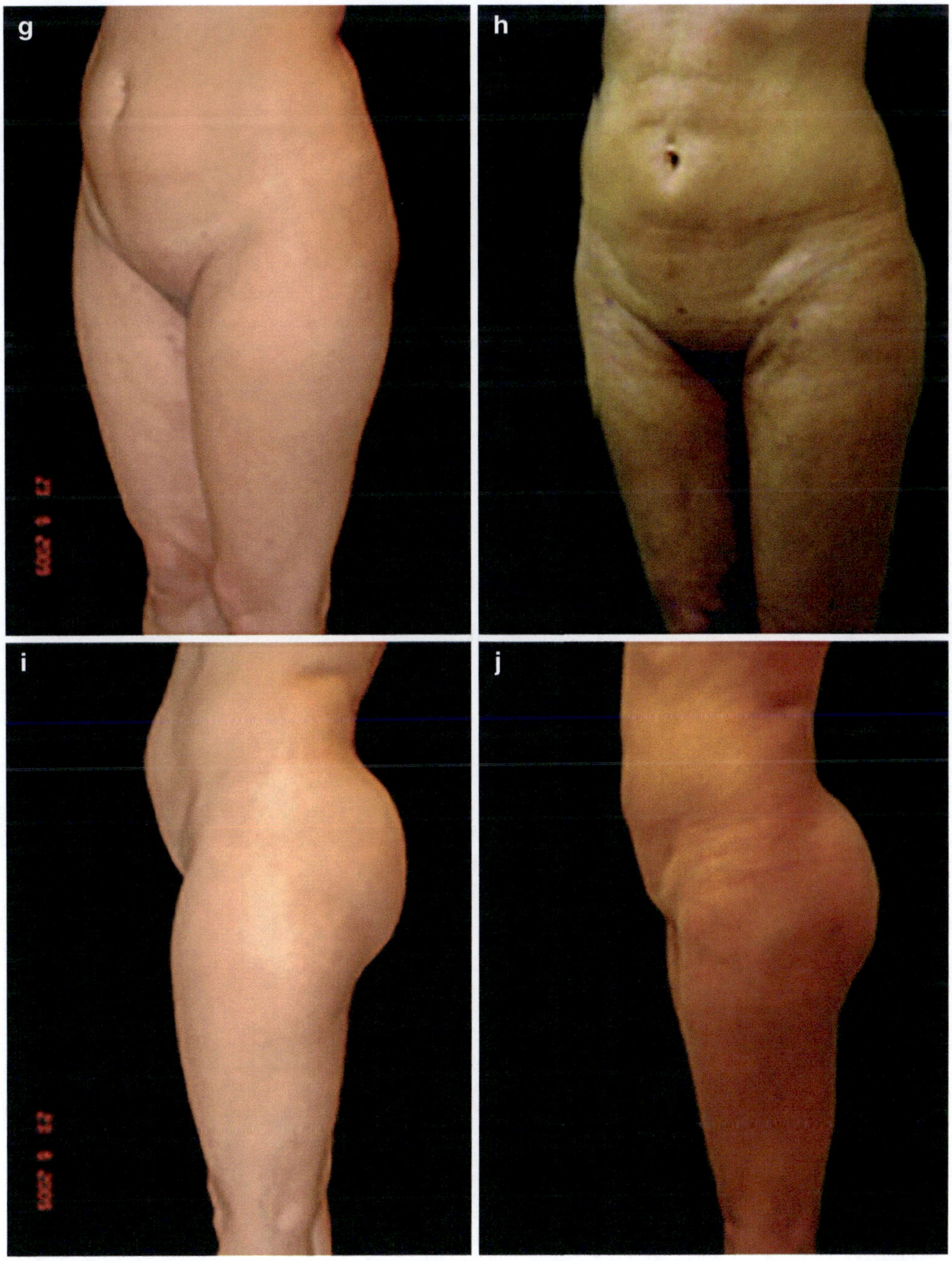

Fig. 20.26 (continued)

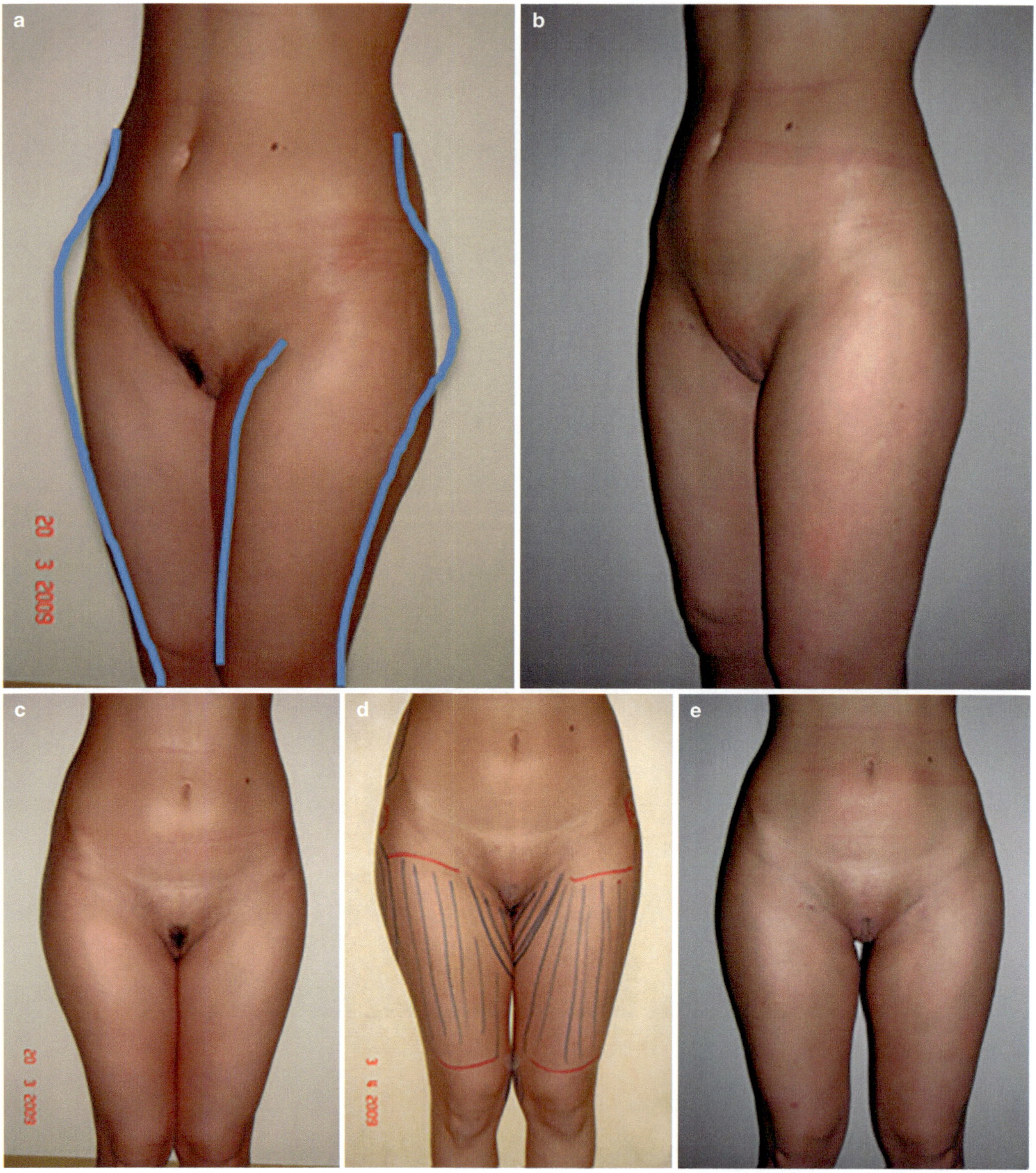

Fig. 20.27 Planning ideal curves drawn with concavity (**a**), convexity, zone of fat transfer at the gluteal depression, final result (**b**), pre op (**c**, **f**, **i**), planning (**d**, **g**, **j**), post op 6 months (**e**, **h**, **k**)

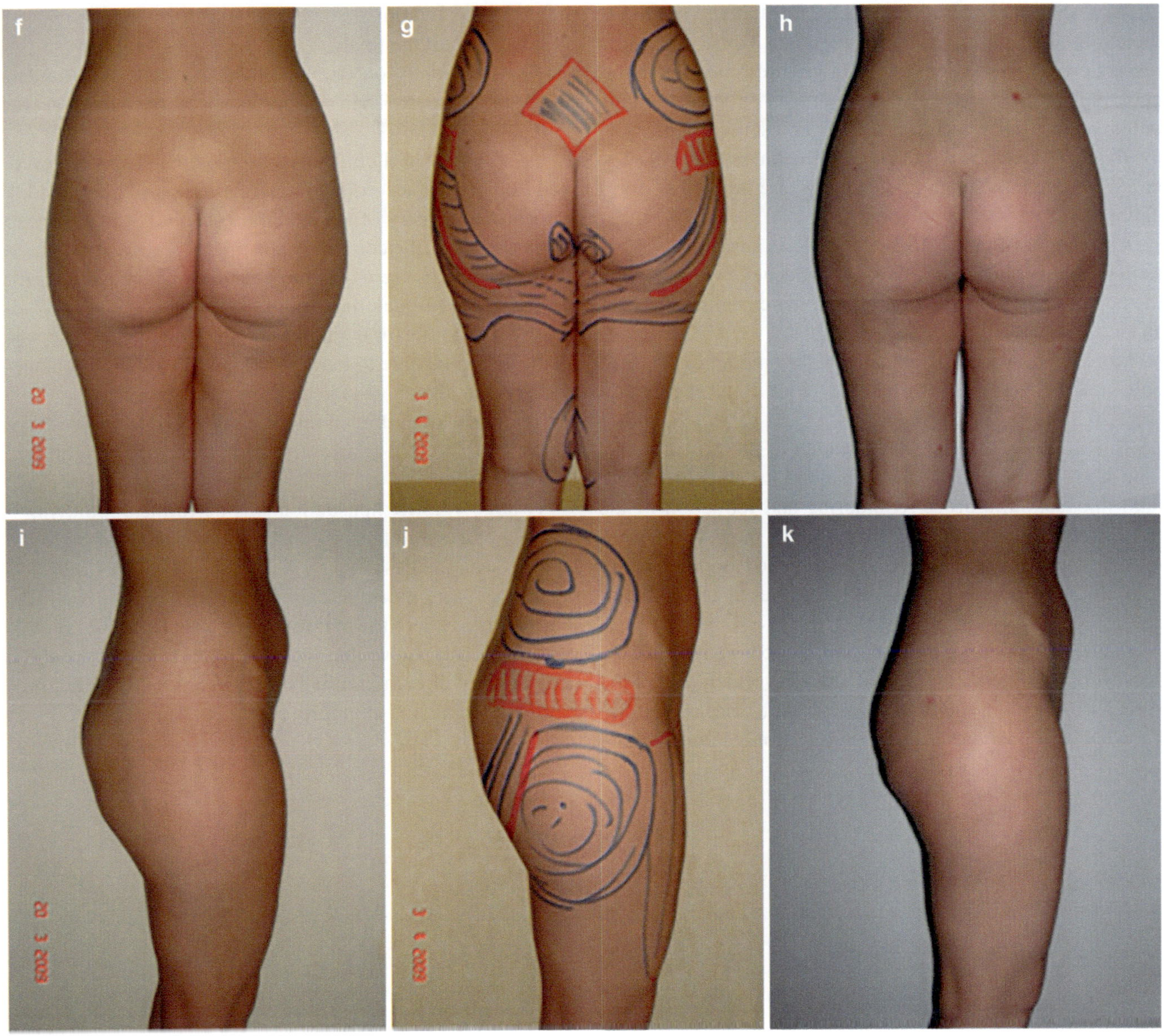

Fig. 20.27 (continued)

3D VASER liposuction is somehow different from three-dimentional: as described by Leonardo da Vinci, in order to respect proper proportions, in the thighs area, the plastic surgeon must not only reduce the dimension circumferentially but he must also respect the anatomy and sense of beauty of the thigh. Which means reducing the thickness of the tissues, in a different way, minding the area where he works. And I define it 3D VASER liposuction.

As a matter of fact, VASER lipo allows surgeons to work as in the deeper layer, to debulk and reduce the volume, as in the superficial layer to sculpture and better define the final shape.

More it is essential in enhancing skin retraction in difficult area as the inner thigh, or in older skin, saggy skin (Fig. 20.28).

This is a 54-year-old patient: quality of the skin and skin tone are related to the age of the patient. Likely with a standard liposuction, a proper contouring and retraction of the tissue would have been impossible. Photos show the 3D contouring of buttocks and thighs. Lateral and oblique views (Fig. 20.28c–h) show the final result achieved in a difficult case.

I personally believe those type of patient can be treated succesfully only with the help of a device, which allows the surgeon to work in different layer of the tissue: and to my knowledge, ultrasound liposuction is the only device which helps. I used 3.7 mm diameter probe with 2 or 3 rings: power is set at 80% of the total.

I work initially on the superficial layers to fully undermine the tissues. Then I approach the deeper layer for debulking.

As a final step, I clinically observed the shape achieved, and I do correct according to the final proportion required.

Sense of artistry of the surgeon is capital at this point to properly assess the contour of the thigh.

At the end, I use a 3.7 mm cannula with no suction, for verifying the full untermining of the tissues.

In order to obtain the highest skin retraction, plastic surgeon should be able to harvest a flap with a thin subcutaneous layer in all the areas treated.

Tridimensional thigh contouring is possible also in large cases. This is a 32-year-old patient, 78 kg of weight, and 158 cm height.

A large VASER liposuction was performed, removing 5 litres of emulsified fat. Liposuction was extended also to the abdomen and flanks.

In this case, greast attention was devoted to debulk the deeper layer where the majority of fat was located.

Final result shows that the dimention of the entire thigh was reduced circumferentially achieving a much better proportion with a good contouring (Fig. 20.29).

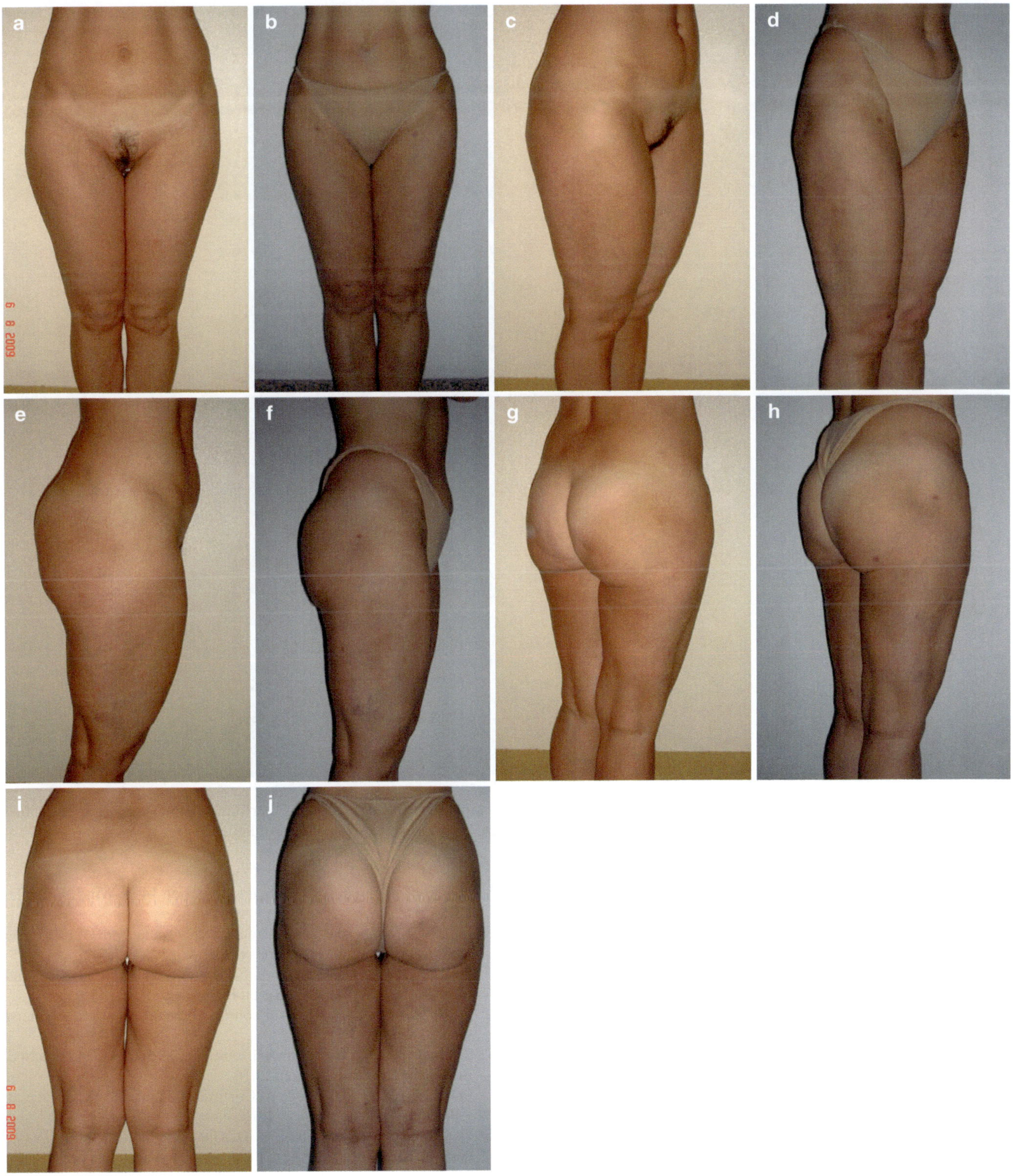

Fig. 20.28 This is a 54-year-old patient: quality of the skin and skin tone is related to the age of the patient. Pre op (**a**, **c**, **e**, **g**, **i**), 6-month post op (**b**, **d**, **f**, **h**, **j**) oblique view

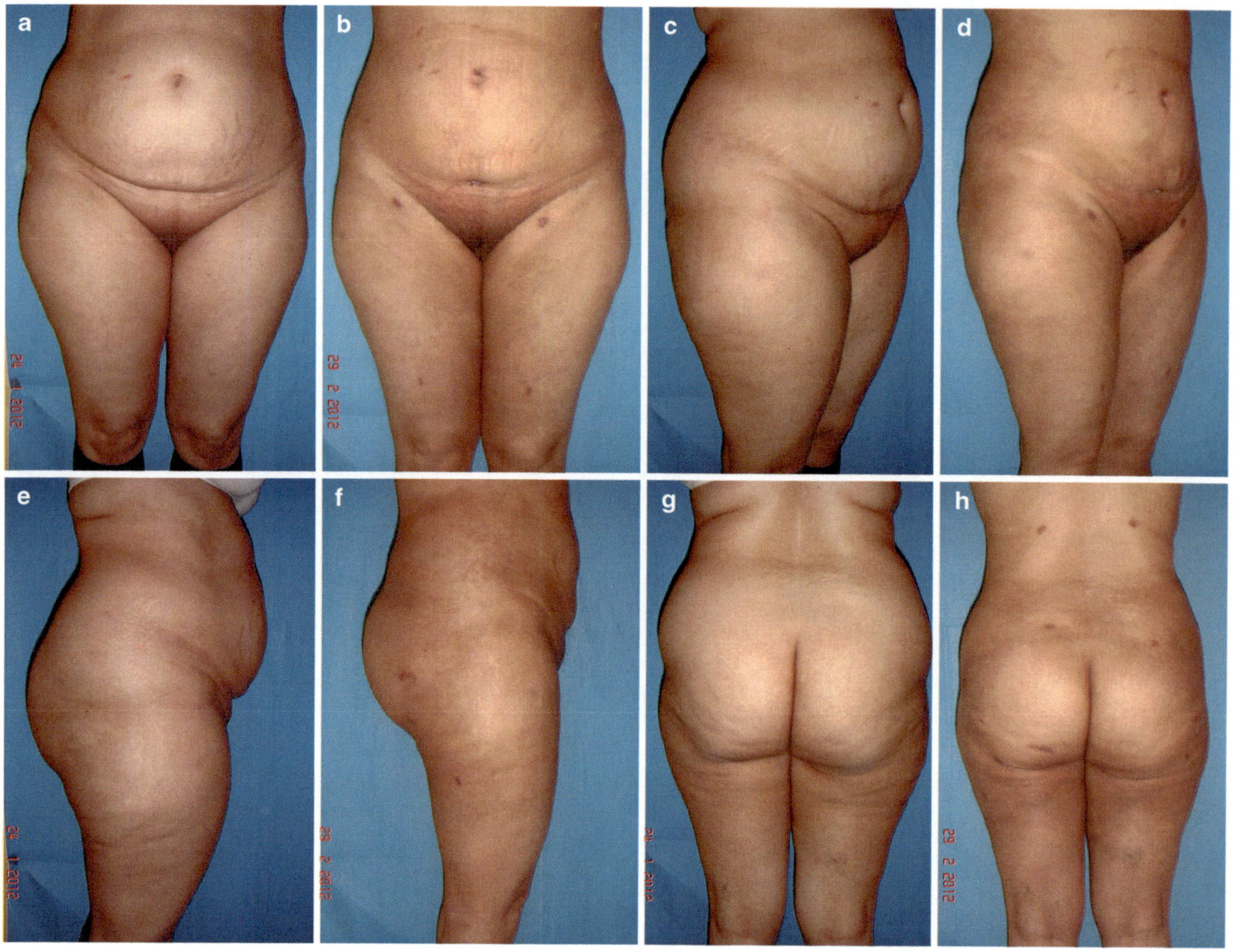

Fig. 20.29 This is a 32-year-old patient, 78 kg of weight, 158 cm height. A large VASER liposuction was performed, removing 5 L of emulsified fat. Liposuction was extended also to abdomen and flanks. Pre op (**a**, **c**, **e**, **g**), 6 months post op (**b**, **d**, **f**, **h**) oblique view

20.8 Complications

Complications of thighs liposuction have been already discussed in chapter "Body fat grafting contouring in female patient."

Essentially they relate to irregularities, depression: management of complication has been discussed in the mentioned chapter.

References

1. da Vinci L. Vitruvian Man 1487. Gallerie dell'Accademia, Venice, Italy.
2. da Vinci L. Drawing human body proportion 1490. Windsor Castle, UK.
3. Illouz GI. Anatomic zone of adherence. Annual meeting of The American Society. Plast Reconstr Surg. 1982;
4. Shiffman MA, Di Giuseppe A. Body contouring. Art, science, and clinical practice. Springer, Berlin Heidelberg; 2010.
5. https://soltapracticeproposal.com/product-information/products/vaser-system/.
6. Shiffman MA, Di Giuseppe A. Liposuction. Principles and practice. Berlin Heidelberg: Springer; 2016.
7. Shiffman MA, Di Giuseppe A. Contorneado corporal. Arte, Ciencia y Practica Clinica. Amolca; 2013.
8. Di Giuseppe A, Giovagnoli F, Di Giuseppe S, Ronconi D. Vaser body contouring achieving a more defined shape. The art of body contouring. London: IntechOpen; 2019. p. 105–122.

Breast Augmentation with Fat. Patient Selection and Guidelines

21

Alberto Di Giuseppe and Federico Giovagnoli

Contents

21.1 Introduction

Quality of aesthetic result in plastic surgery mostly derives from proper selection of patient indication.

Limitations are often more important, even to avoid to select patient with poor indication to the type of surgery selected. First limitation, in breast augmentation with fat transfer, is given by the amount of the donor sites.

Structural fat grafting started in 1895 with Czerny [1] who transplanted a lipoma to reconstruct a breast defect.

Lexer [2] used resected fat to treat bilateral breast defects. Because of the limited success achieved with these techniques, they were abandoned.

After liposuction was introduced in 1983 by Illouz [3] and Fournier [4], the idea of removing a good amount of fat from an area offered to surgeon a new opportunity to use fat as a potential source of an autologous filler.

Initial poor results were due to poor understanding of technique needed to harvest properly fat and to transfer to recipient area.

A. Di Giuseppe (✉)
University of Padova, Padova, Italy

F. Giovagnoli
University Vita-Salute San Raffaele, Milan, Italy

In 1987, Bircoll [5, 6] presented a paper regarding the utilization of autologous fat for breast augmentation.

The American Society of Plastic Surgeons (ASPS) wrote a position paper as an ad hoc committee on new procedures issued an opinion that fat grafting to the breast could interfere with breast cancer detection and therefore not be performed [7].

As a result, the fat grafting procedure was condemned for the potential risk with any scientific evidence. As a consequence, to follow there was a world-wide moratorium to fat grafting to breast.

In 2007, Coleman [8] published a review of 17 patients who underwent structural fat grafting to breast and were followed with serial photography. Results were successful, and the author could show that the overall quantity of the fat transferred was maintained over a 7–12 years follow-up time. Coleman utilized syringe technique to transfer significant volume of fat in a single session.

To follow, Baker and Khouri [9] suggested to inject smaller volumes of fat in serial grafting sessions, in a pre-expanded recipient breast. This latest technique seemed to work nicely on a very small breasted woman with cup A or B initial volume. As the final result is proportioned to the initial condition, a 100 mL fat volume transferred could make a real difference in a cup A or B patient, and on the contrary be undetectable in a C, D cup, or even bigger breast size.

© The Author(s), under exclusive license to Springer Nature Switzerland AG 2023
A. Di Giuseppe et al. (eds.), *Fat Transfer in Plastic Surgery*, https://doi.org/10.1007/978-3-031-10881-5_21

A remarkable amount of different studies emerged between 1987 and 2007 from mostly Italy (Zocchi 2005 [10], Grisotti 2006 [11]), Rigotti 2007 [12]), and France (Delay 2005 [13]). Those scientific paper reported experience on fat grafting for breast augmentation (Zocchi), breast reconstruction (Grisotti), for the treatment of radiation damage to the chest wall and reduction of breast capsule contracture (Rigotti), and finally on soft tissue coverage of breast implants.

Position in North America has been always more cautious than in Europe, Japan, and China: in 2008–2009, the Plastic Surgery Education Foundation (PSEF) and the American Society of Aesthetic Plastic Surgery (ASAPS) finally published in the official journal *Plastic and Reconstructive Surgery* a new statement which sounds as follow: fat grafting may be considered for breast augmentation and correction of defects related with medical conditions and previous breast surgeries; however, results are dependent on technique and surgeon expertise. Fat grafting to the breast could potentially interfere with breast cancer detection: however, no evidence was found that strongly suggests this interference could lead to confusion and wrong diagnosis. Radiological studies suggest that imaging technologies (ultrasound, mammography, and MRI) can identify the grafted fat tissue, microcalcifications, and suspicious lesions, and biopsies may be performed if needed for additional clarification.

After this statement, the use of fat for breast enhancement and reconstruction has been accepted worldwide and is now constantly practiced.

21.2 Donor Area

Ideally, the desire of any patient is to remove fat from undesired, stubborn area of fat deposits, and transfer into poorly shaped breast, lacking volume, and projection. However, a potential complication of any technique of liposuction could be overaspiration and creation of deformities of the donor sites.

Large amount of fat should be generally harvested in order to obtain proper quantity of fat, suitable for transferring and increasing breast aesthetics. Patient with previous liposuction could be missing adequate quantity of donor fat. Thin patient, on the other hand, could be either lacking enough fat volume to give a satisfactory augmentation.

In my experience, by using VASER ultrasound system and the REVOLVE™ harvesting system, we can obtain 300–350 mL of pure fat, ready to transfer, by emulsifying around 1000 mL of fat, after tumescent infiltration. It means that around 3500 mL of emulsification is needed to acquire between 800 and 900 mL of harvested fat, in order to obtain a 2 cup size increase in each breast.

In clinical practice, this will require multiple area liposuction, as abdomen, flanks, posterior trunk, external thighs, of circumferential thighs. Knees and arms contain minimal amount of fat, and they do not increase significantly the final result (Figs. 21.1 and 21.2). Two typical cases of patients undergoing breast fat transfer are shown in the underlying photos.

From the first patient, a total of 400 mL of pure fat was obtained, mainly from abdomen and flanks. This patient has no further good area of fat excess for transferring. A 400 mL of total fat harvested, will result in 200 mL for side breast increase, which means from an A cup to a B/C cup size increase.

In the second case, in a heavier patient, 950 mL were harvested from abdomen, flanks, and trunk. 475 mL for side will mean around 2/3 cup size increase.

Those parameters are my guidelines to forecast the final volume of breast increase in clinical situation, which is in proportion with initial breast volume size. In clinical practice, I do use pre- and post-op photos of other treated patients, to show the final volume increase that the patient is going to achieve: this is important to verify patient expectations and eventually reject those unsuitable for this technique, as in the case of lack of proper amount of donor fat, to achieve the result they do want.

Case 1

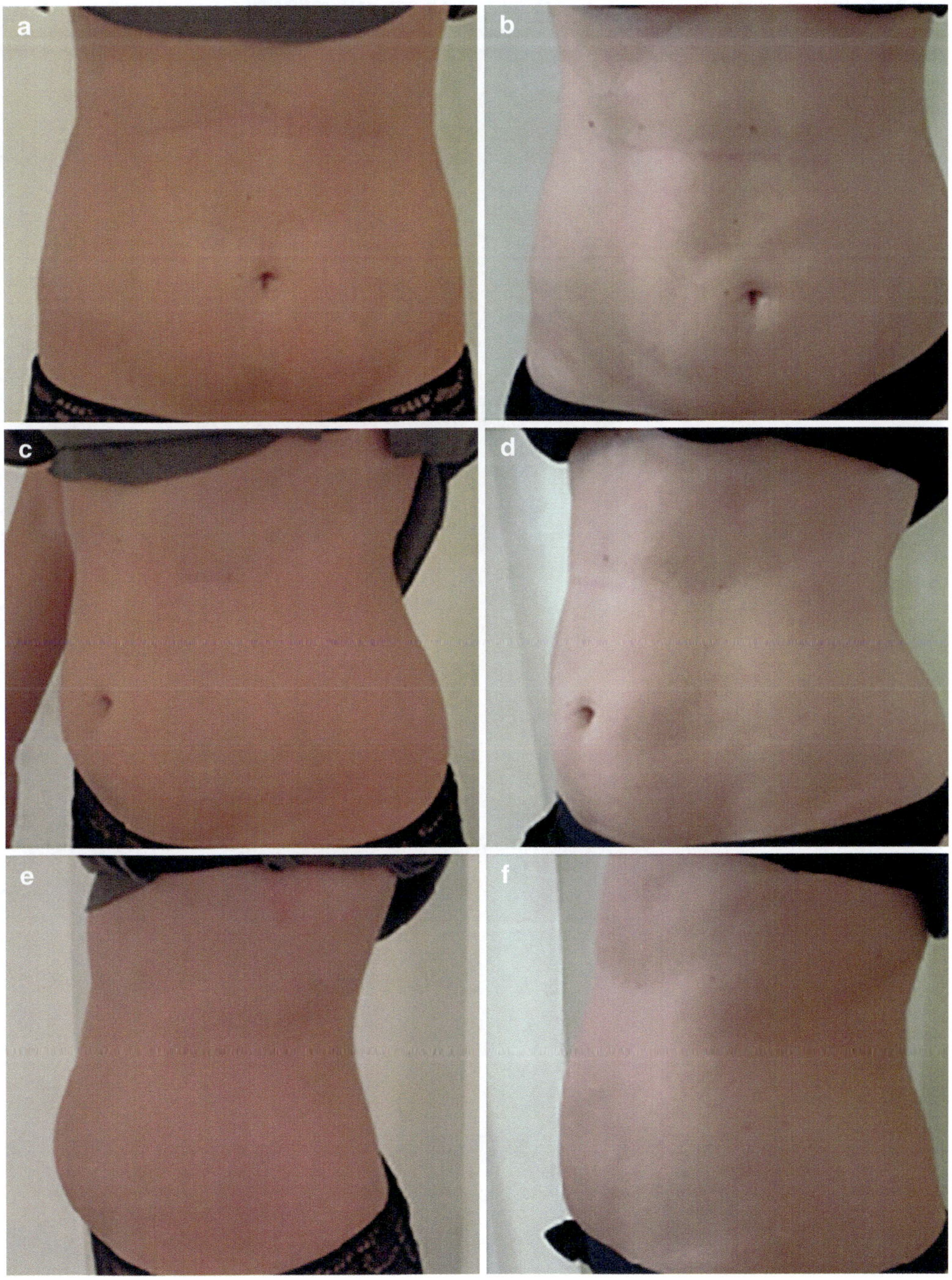

Fig. 21.1 400 mL of pure fat harvested from upper lower abdomen and lateral flanks. Pre op (**a**, **c**, **e**). Post op 6 months (**b**, **d**, **f**)

Case 2

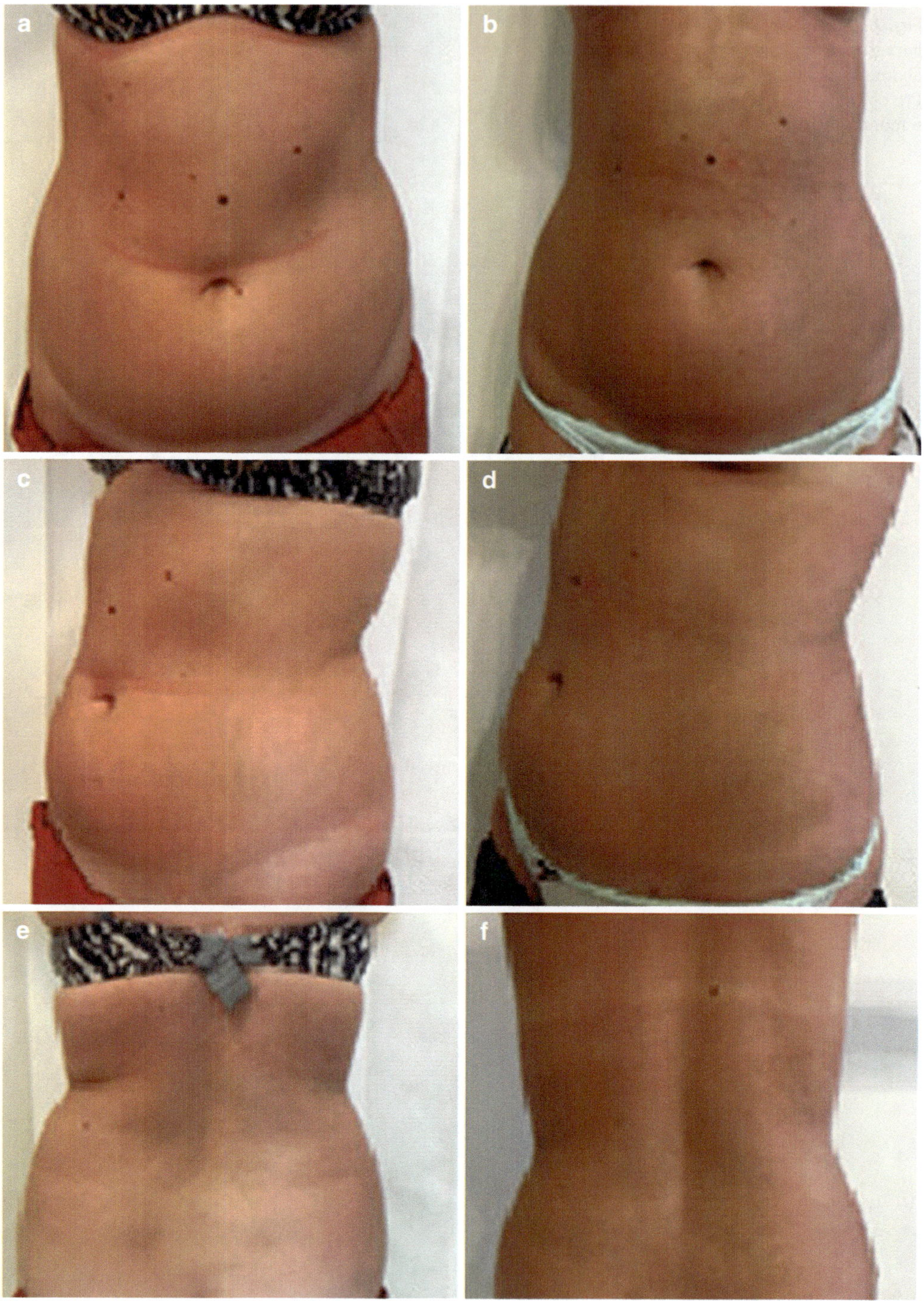

Fig. 21.2 950 mL of pure fat harvested from upper lower abdomen, flanks, and trunk. Pre op (**a**, **c**, **e**). Post op 6 months (**b**, **d**, **f**)

21.3 Oncology

Similarly to breast augmentation with implants, any patient undergoing breast fat transfer must have a proper initial consultation regarding her breast health condition.

Palpable lump, masses, fibrocystic condition, postpartum important variation of the breast consistency, past medical family history of strong hereditary breast cancer, and irradiated breast: all those suspicious conditions must be investigated before undergoing any breast surgery, which will involve an implant of fat transfer.

GP (General Practitioner) and breast clinic are normally monitoring breast clinical situation of all patients in many occidental countries where a national health system (NHS) does follow on a regular base breast conditions of patients (the UK, Germany, France, Italy, Spain, etc.).

Radiological studies are requested in those suspicious patients: ultrasound scan is the simplest, not invasive test required. Further investigations could require mammograms or MRI.

In case of doubt, biopsies could be performed to check breast pathologies.

I will exclude from a breast enhancement with fat, any patient with suspicious breast lump, cyst, major fibrocystic disease, calcifications with diagnostic doubt. I do regularly ask the opinion of a senologist or an oncologist, or the GP himself, when there is a case with an unclear pattern of breast morphology.

For years, plastic surgeons were worried to use fat grafting for breast because of the speculation that transplanted fat might die and cause lumps or calcifications that would interfere with the detection of breast cancer.

Bircoll's paper reported more than 600 breast fat transfer cases with a calcification rate of 1.4%. I did monitor the first 200 cases I performed with breast fat transfer with pre- and post op ultrasound scan, at 1–3 years interval, with the supervision of a senologist of the university department of radiology of my University. Suspicious cases were further sent to an oncologist for a second opinion.

Most of the patients even with a small aliquot of fat transferred in each pass can necrose and develop small fat cyst and calcification. Those small calcifications are benign, can be easily detected by a radiologist as they appear small and numerous. The calcification connected with breast cancer, are typically more dense, larger, and less numerous.

The simple appearance of calcification does not imply an increase of breast cancer risk.

Cancer detection is the safety issue and not the incidence of calcifications.

Safety guidelines, according to National Breast Cancer Organisation, do include now fat grafting to breast as a standard approved procedure which must be monitored overtime. By patient herself, with self-examination, by regular check by a physician, GP or senologist, and by planned mammograms on a 1-year or 5-year interval.

Mammograms are the favorite test by radiologist for differentiation of cancer from benign lesions of the breast. Ultrasound and MRI are less invasive and traumatic (MRI significantly more expensive).

Fat transfer has been utilized in the past decade in breast reconstruction after cancer removal. Indications were: asymmetry after lumpectomy, hybrid fat transfer in combination with implant reconstruction, and unique breast reconstruction with fat in multiple stages. Those arguments are due to be treated by other authors in this book.

21.4 Breast Initial Volume (Figs. 21.3, 21.4, and 21.5)

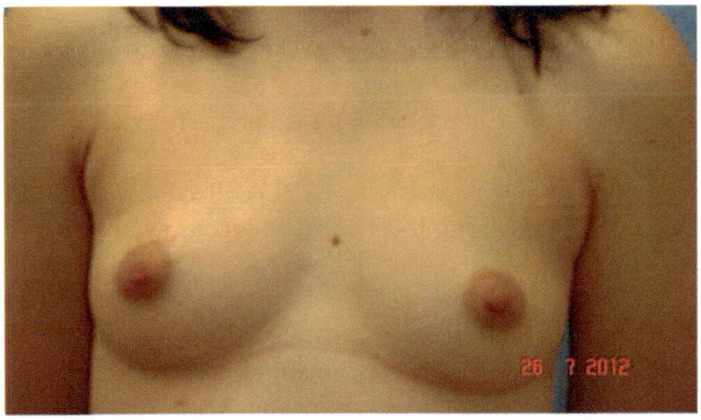

Fig. 21.3 Cup A

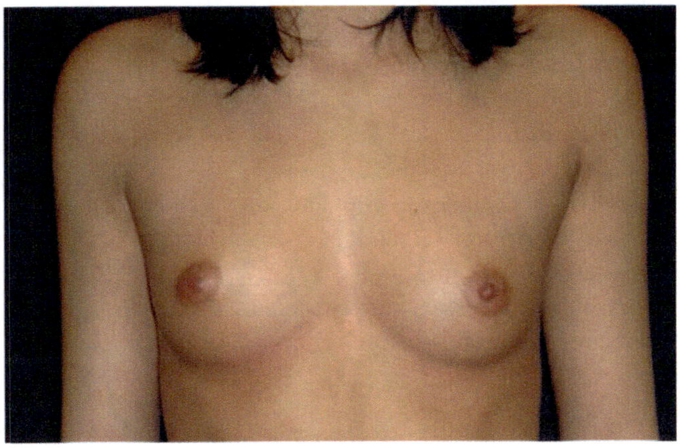

Fig. 21.4 Cup B

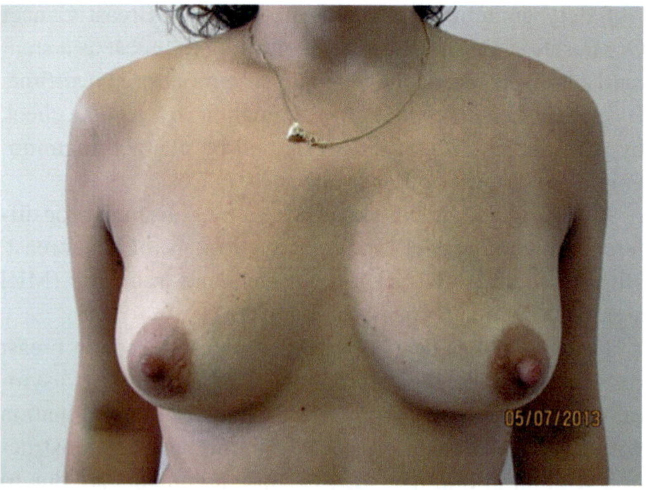

Fig. 21.5 Cup C

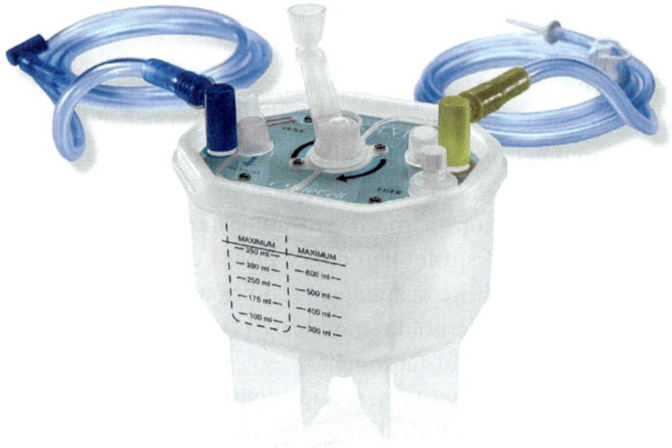

Fig. 21.6 REVOLVE™ system (Courtesy LifeCell Corporation, an AbbVie company. © 2021 AbbVie. All rights reserved)

21.5 Preferred Fat Transfer System

In the past 4 years, I used the REVOLVE™ system by LifeCell.

An integrated, high-volume fat processing system designed by adipose transplantation pioneers and founders of the International Federation for Adipose Therapeutics and Science.

It is an autologous fat tissue (AFT) collection device and tissue delivery system. It collects autologous adipose tissue and rapidly washes and delivers the tissue for body contouring.

REVOLVE™ System is a sterile single-use disposable tissue canister used for harvesting, filtering, separating, concentrating, and transferring autologous tissue components for reintroduction to the same patient during a single surgical procedure for repair, reconstruction, or replacement of integumentary or musculoskeletal tissues.

For additional volume of tissue, canister may be filled and emptied a second time during the same surgery on the same patient.

It is an all-in-one closed system, for harvest, filter, active wash, and remove strands.

Mesh filter designed to filter and strain lipoaspirate during the harvesting phase and active mechanical washing process.

It rapidly removes tumescent, reduces cell debris, eliminates free oil, and finally concentrates adipose tissue (Fig. 21.6).

The system contains a propeller for active washing, it does enhances washing process for high-quality fat and removes collagen strands to minimize injection syringe clogging.

Fat is extracted by syringes. It is possible to direct load syringes for injection from canister to minimize steps.

Extraction with catheter tip or luer lock syringe is performed (Fig. 21.7).

The system is a high-volume fat processing system: it processes up to 700 mL of lipoaspirate in less than 10 min. It is easier to use, requiring fewer steps to process large quantities of adipose tissue.

It can control critical fat grafting variables and is sterile, single-patient, disposable device.

It is a closed system that minimizes tissue handling and exposure to outside air. It can control critical fat grafting variables and is sterile, single-patient, disposable device.

Clinical result shows that REVOLVE™ system processed a higher percentage of adipose tissue compared with the two other methods: decantation and centrifugation. It yielded significantly less blood cell debris and a lower percentage of free oil. It has significantly higher fat graft retention (73.2%) than decantation (37.5%) and similar to centrifugation (67.7%) and maintained fat in physiological conditions (pH and osmolality) (Figs. 21.8 and 21.9).

Breast initial volume is the parameter of reference to plan fat enhancement.

In Fig. 21.3, an A cup patient is shown. Potentially this patient could arrive at a maximum of a B/C cup final result, minding an average of 200 ml of fat transferred for side, which is the average of donor area achievable.

In Fig. 21.4, a B cup patient is shown. Potentially this patient could arrive at a maximum of a C/D cup final result, minding an average of 300–350 mL of fat transferred for side.

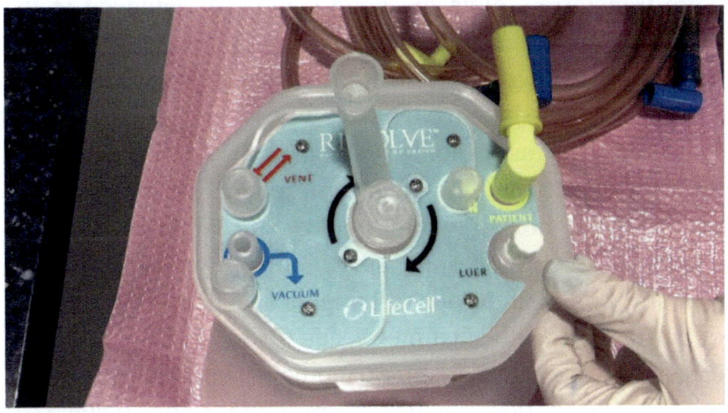

Fig. 21.7 REVOLVE™ system syringes for extracting fat (Courtesy LifeCell Corporation, an AbbVie company. © 2021 AbbVie. All rights reserved)

In Fig. 21.5, a C cup patient is shown. Potentially this patient could arrive at a maximum of a D/DD cup final result, minding an average of 450–500 mL of fat transferred for side.

In the past, I used the Crisalix Virtual 3D and 4D imaging system for planning and showing patient a potential result of a breast enhancement with fat. This system is well known and utilized worldwide for breast augmentation and other body contouring and facial procedures. Of course there are few other systems on the market with similar characteristics. I did abandon the system, as 3D or 4D image result is less natural than the real one I will achieve eventually after surgery.

When I discuss final volume achievable, I normally state as follow:

1. Fat absorption rate is around 20% of the volume transferred, minding the technique utilizes VASER system and REVOLVE™ harvesting device (350€ cost for a single device).
2. Patient does not compress enhanced breast with strong compression bra. Fat does not like compression that will eventually lead to necrosis and absorption. A light bra is necessary in the post-op period.

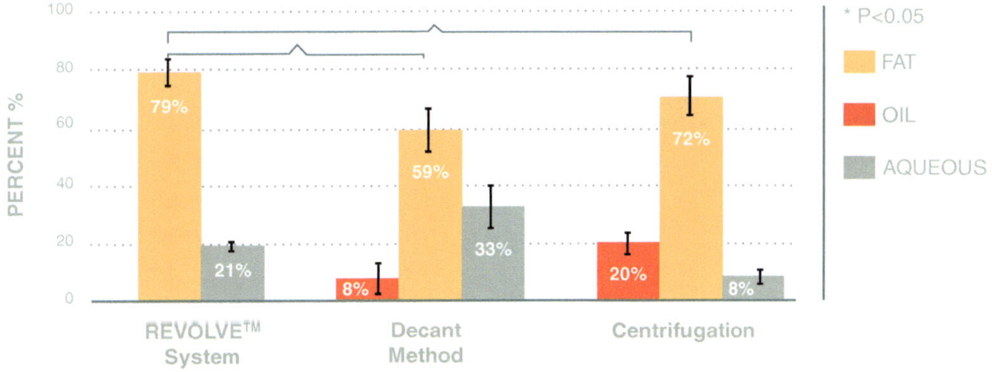

Fig. 21.8 REVOLVE System minimized potential inflammatory response for transplanted tissue by significantly removing red blood cells, fatty acids and debris (Courtesy LifeCell Corporation, an AbbVie company. © 2021 AbbVie. All rights reserved)

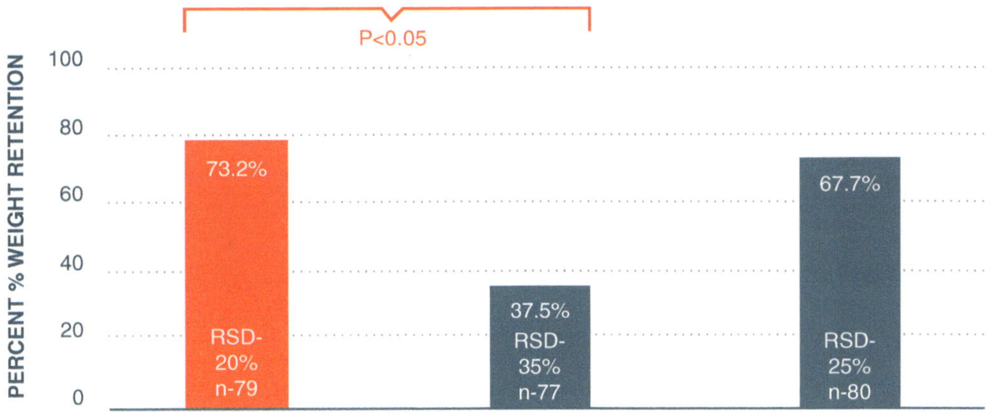

Fig. 21.9 Higher fat graft retention than decantation and similar to centrifugation (Courtesy LifeCell Corporation, an AbbVie company. © 2021 AbbVie. All rights reserved)

3. Fat transferred must be preserved at list for the first 3 months after surgery, until the process of full revascularization is completed. This will mean no diet, no burn-out, and gym exercise as well.

Maintaining volume in the future will depend on this main factor: weight maintenance over time.

It is clear that any significant variation of weight, in terms of losses, will determine a partial or total loss of the transplanted fat, as well as a fat reduction all over the body, affecting the external fat and or the visceral fat.

Hormonal changes, as due to menopause or pregnancy, could affect as well the final volume maintenance.

I make sure that the patient is fully aware of those potential changes connected with weight changes and hormonal variations, which are totally independent from surgical technique and surgeon skill.

First case (Fig. 21.10) is the A cup breast size that turned into a full B/small C cup after 235 mL per side of fat transferred.

Second case (Fig. 21.11) is the B cup breast size that turned into a full C/small D cup after 360 mL per side of fat transferred.

Third case (Fig. 21.12) is the C cup breast size that turned into a full D/DD cup after 400 mL per side of fat transferred.

First case

Second case

Third case

Another excellent indication for fat transfer to breast is in volume asymmetry, or in correcting secondary cases.

Fourth case is a patient of 30 years age, who underwent, when 18 years, a mastopexy. She ended up with a volume asymmetry and a mild asymmetry in shape. Patient underwent breast fat augmentation, and 500 mL and 550 mL of fat were transferred. Note the good symmetry achieved in the 8 months post-op follow-up. Final volume was a DD/E cup starting from an asymmetrical B cup size (Fig. 21.13).

Fourth case

In the fifth case, another example of volume asymmetry after mastopexy and implants is presented. Thirty-two-year-old woman, who had two mastopexy and implants, with two different surgeons. Despite a decent result was achieved, she was still complaining for missing cleavage at the inner side of both breast, with asymmetry, and diverting nipples. In this case, plan was to fill the inner and upper quadrants, to fulfill patient expectations, which were quite high. Instead of changing again breast implants, which could cause further gravity and lateralization of breasts, solution was to transfer 250 mL of fat mainly to the inner quadrants to improve cleavage and fullness of the top (Fig. 21.14).

Fifth case

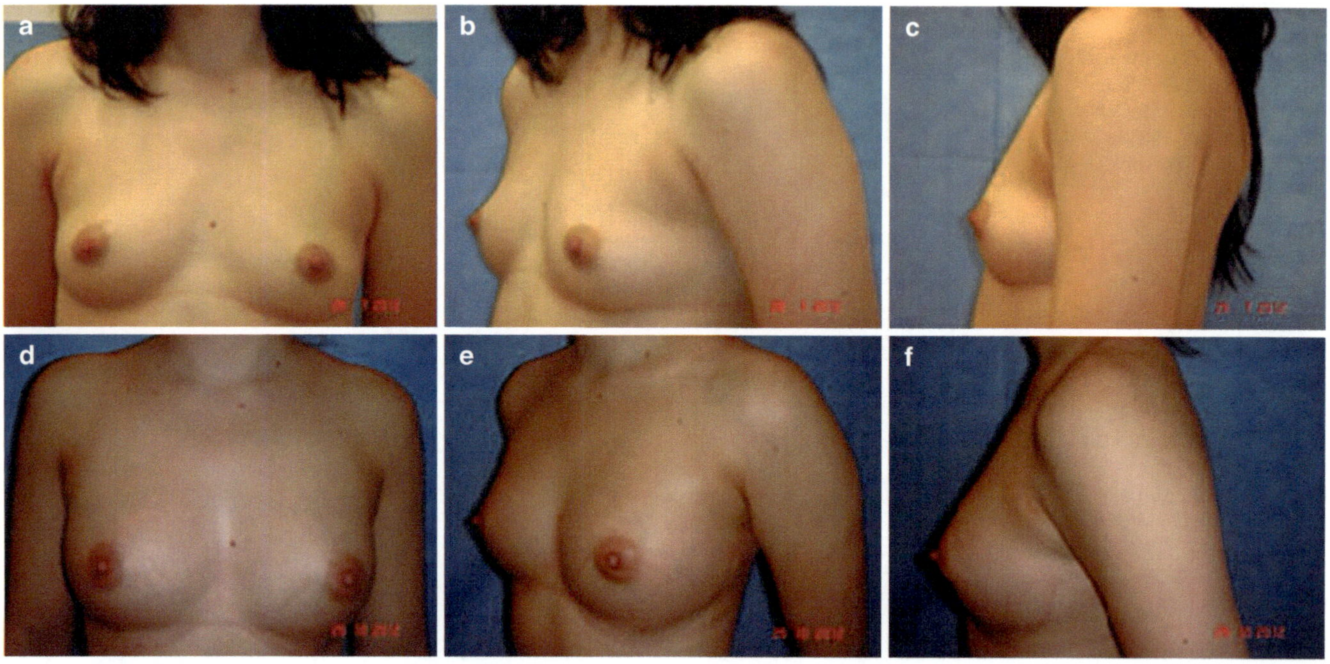

Fig. 21.10 A cup breast. Pre op (**a**, **c**, **e**), post op 6 months after 235 mL per side fat transferred (**b**, **d**, **f**). Full B/small C cup achieved

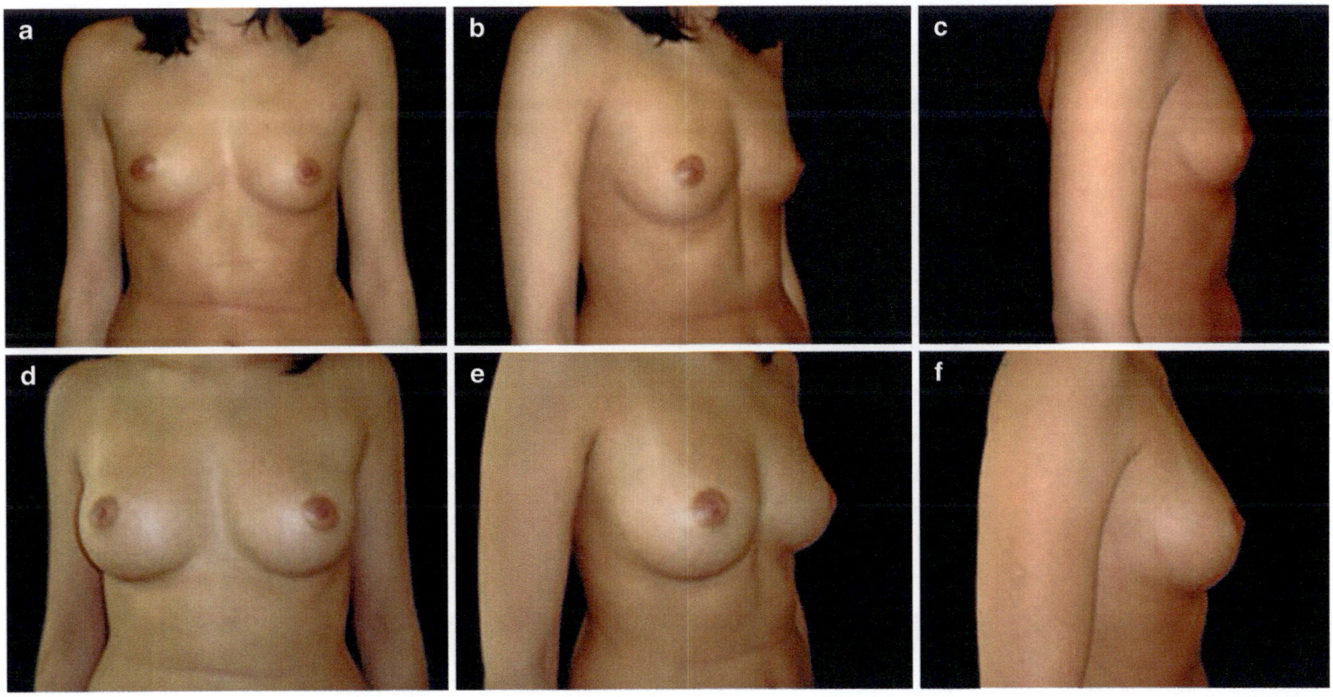

Fig. 21.11 B cup breast. Pre op (**a**, **c**, **e**), post op 6 months after 360 mL per side fat transferred (**b**, **d**, **f**). Full C/small D cup achieved

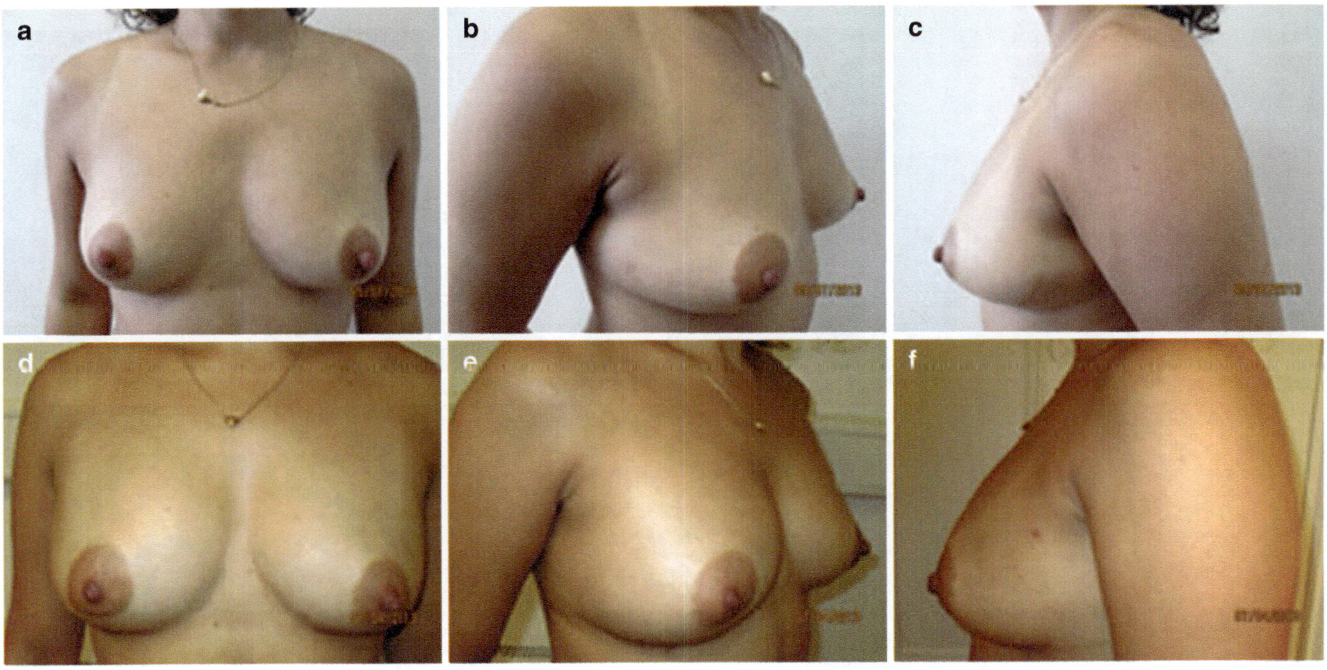

Fig. 21.12 C cup breast. Pre op (**a**, **c**, **e**), post op 6 months after 400 mL per side fat transferred (**b**, **d**, **f**). Full D/DD cup achieved

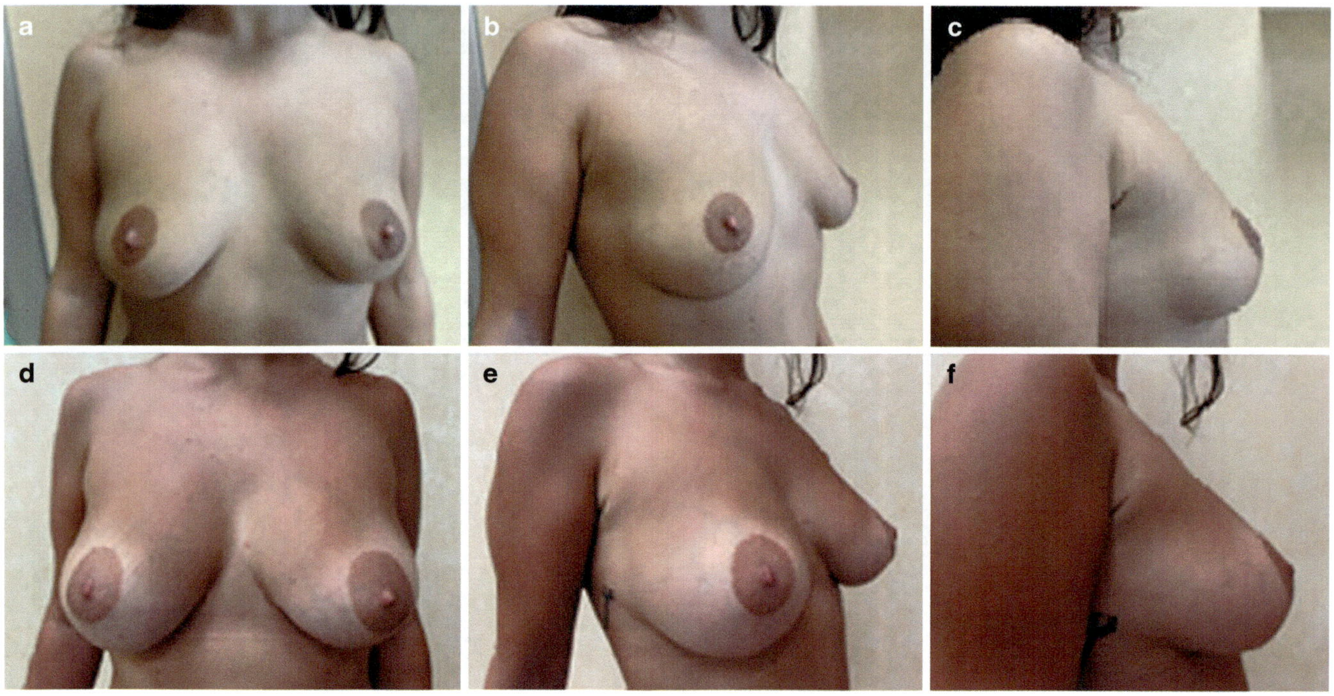

Fig. 21.13 Breast volume asymmetry after mastopexy, B cup size, pre op (**a**, **c**, **e**). Post op 6 months after 500 and 550 mL of fat transferred (**b**, **d**, **f**) to improve symmetry. DD/E cup size achieved

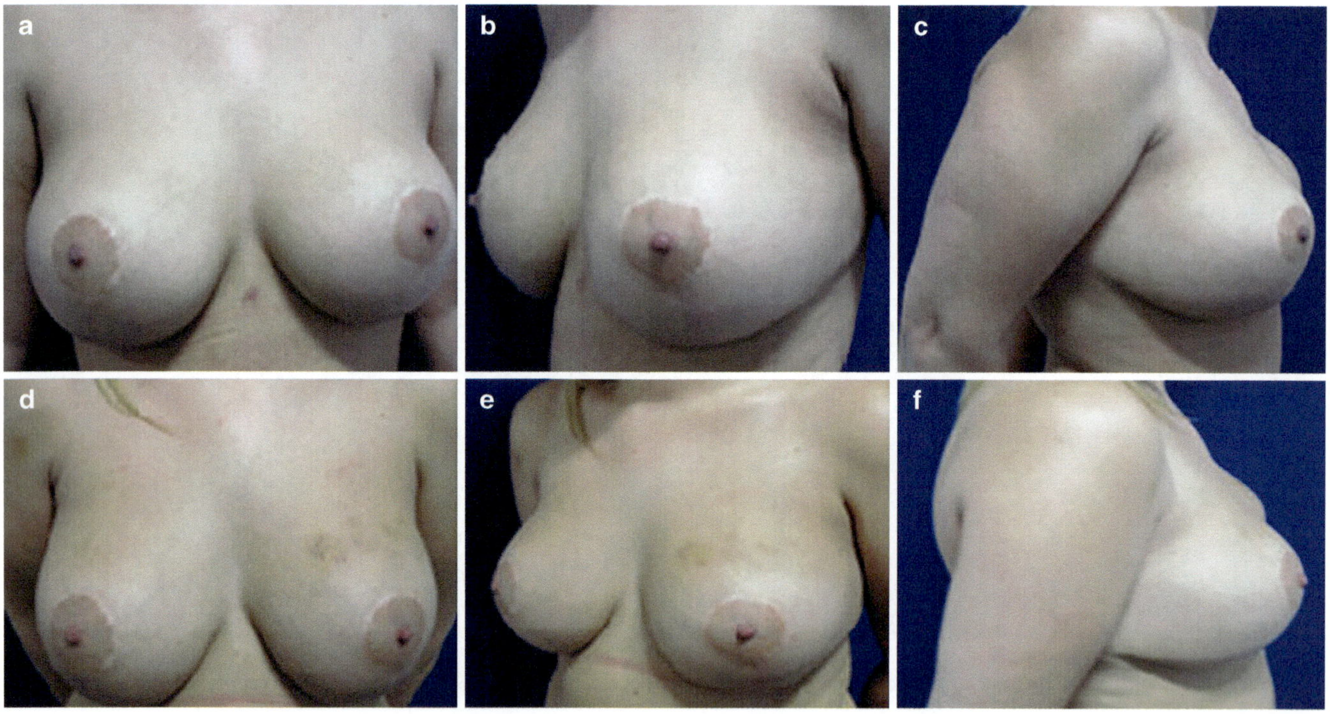

Fig. 21.14 Breast volume asymmetry after mastopexy, missing cleavage, and top fullness, C/D cup size, pre op (**a**, **c**, **e**). Post op 6 months after 250 mL of fat transferred (**b**, **d**, **f**). D/DD cup size achieved. Cleavage improved

21.6 Breast Shape

Breast shape, breast position, and aging are all factors that will influence the indication for breast fat transfer.

With fat, it is impossible to achieve the same volume of an implant as the final result depends on the total volume of fat that can be harvested from patient.

Fat has a different consistency in comparison with silicon: it is softer, cannot achieve same projection of an implant. Careful selection of patient is mandatory in order to avoid misunderstanding and complain from patient, concerning the goals achieved or the targets not achievable.

Breast initial position and shape is another factor that influences patient selection. Initial nipple position should not be exceeding 22 cm from middle clavicle.

The sixth case is a 32-year-old patient with emptiness of the top quadrants of the breast, post pregnancy, and lactation. She had 250 ml of fat transferred on each side, nearly positioned on the upper quadrants to fill the loss of volume (Fig. 21.15).

Sixth case

The seventh case is a 40-year-old black woman, with breast ptosis and loss of top fullness after breast feeding. Nipple initial position was located at 22 cm from midclavicle. This is still a good indication for fat transfer, minding that patient is looking for a natural result, with improved upper quadrant volume, but no change in projection and position of the nipple. 320 mL of fat was transferred bilaterally (Fig. 21.16).

Seventh case

Aging is the last factor to consider in the patient selection. Skin texture, nipple position, and skin redraping capability are the factors that could limit indications.

Middle-aged women would rather prefer a natural looking result and a mild volume increase, in order to restore general breast volume lost by progressive atrophy, due to menopause and hormonal changes.

In my experience, those patients are finally happier than the younger population, as their expectations are lower.

Eighth case shows a 62-year-old woman, with still good skin tone, good positioning of the NAC nipple areolar complex, and a generalized atrophy of all breast (A/B cup size). 250 mL of fat were transferred bilaterally. One-year post-op photos show that overall nice result has been achieved, with natural outcome, and even improved skin texture, probably due to stem cell regenerative capability of neovascularization (Fig. 21.17).

Eighth case

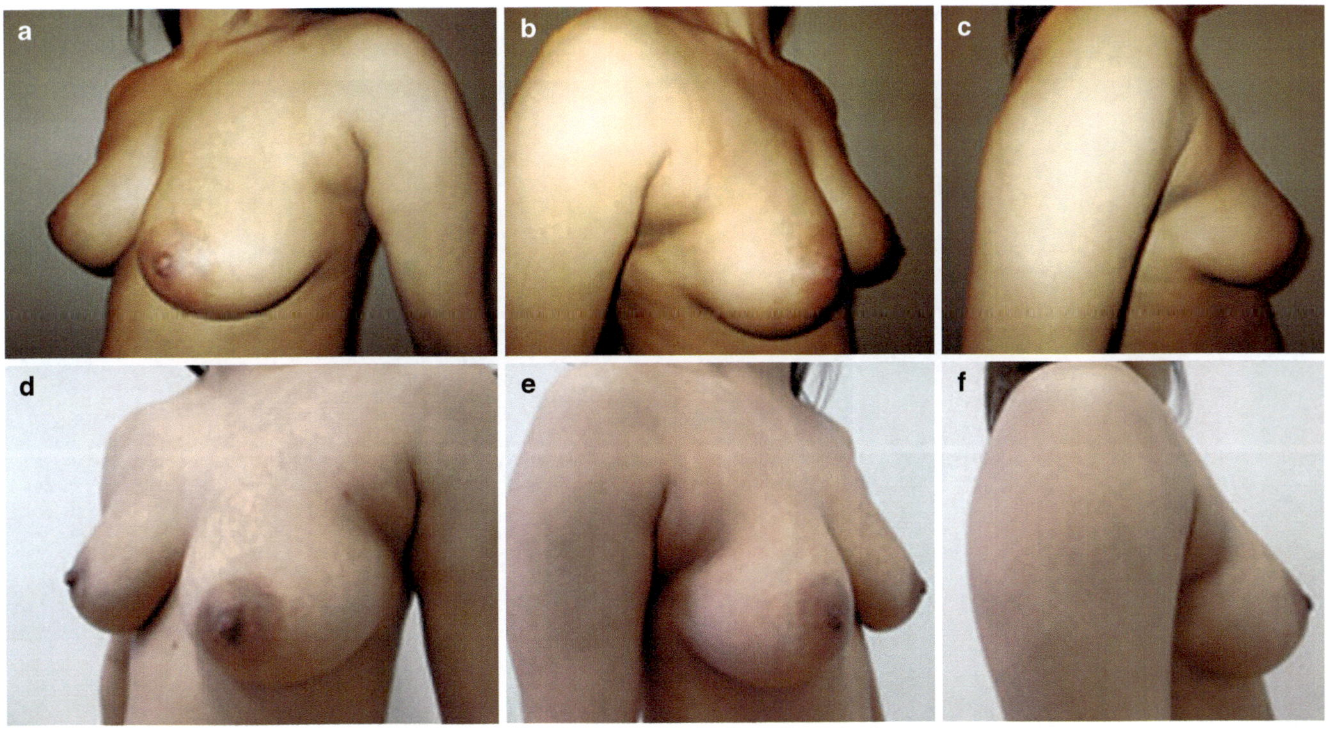

Fig. 21.15 Breast volume C cup size, mild ptosis, and emptiness of the upper quadrants. Pre op (**a**, **c**, **e**). Post op 6 months after 250 mL of fat transferred (**b**, **d**, **f**) transferred mainly on the upper quadrants

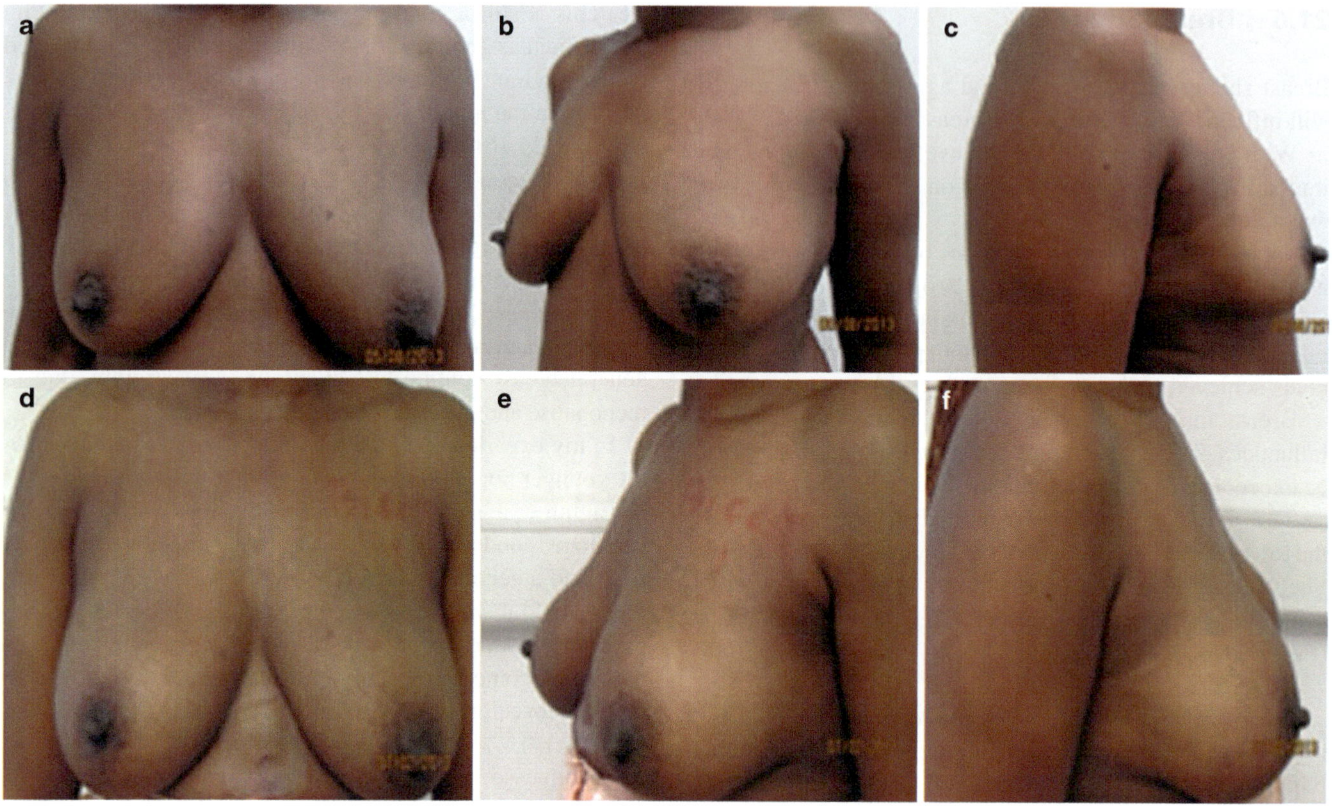

Fig. 21.16 Breast ptosis 22 cm from midclavicle and post pregnancy lack of volume on top quadrants, C cup size, pre op (**a**, **c**, **e**). Post op 6 months after 320 mL of fat transferred (**b**, **d**, **f**) mainly to upper quadrants to improve fullness

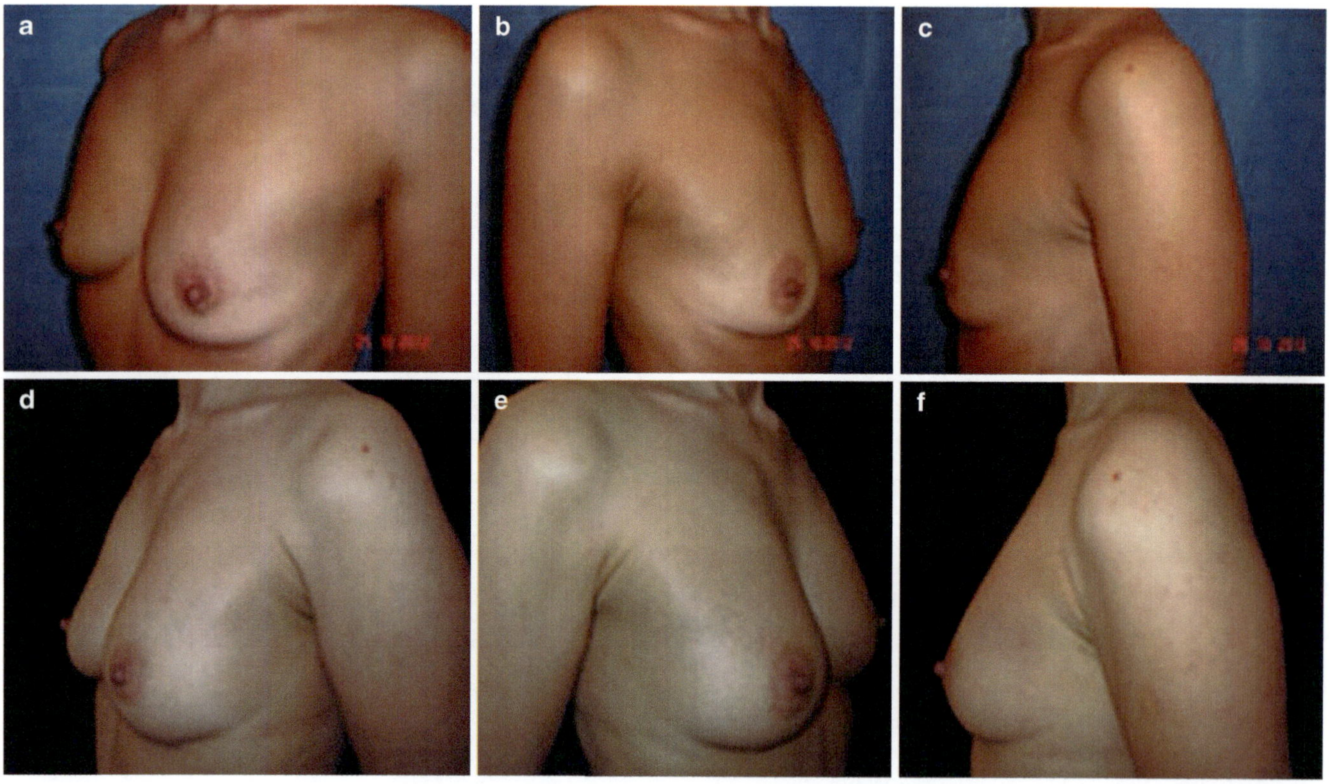

Fig. 21.17 62-year-old patient with breast atrophy after menopause. Pre op (**a**, **c**, **e**). Post op 6 months after 250 mL of fat transferred (**b**, **d**, **f**)

21.7 Contraindication

Absolute contraindication to breast fat transfer is extremely skinny patients where donor fat is unavailable. Also patients addicted to gym work out have a poor indication, as they might burn out the transferred fat immediately after surgery. I normally warn patients to refrain from any intense physical activity for the first 3 months after surgery.

Another obvious contraindication is represented by patients who are looking for massive volume increase, which is clearly unreachable with any volume of fat, even if available. Projection of top of breast is another impossible goal achievable with fat transfer. Patient willing to have top fullness and very rounded breasts should be indicated only for breast surgery with ultrahigh projection implants.

Otherwise, there will be surely unsatisfied results achieved.

One of the mistakes that surgeon could make initially with breast fat transfer is not to select properly patients. This case presents a 43-year-old woman with ptotic breast, with nipple at 23–24 cm from middle clavicle and hanging breasts, with atrophy after weight loss. This lady presents a clear contraindication for fat transfer, as fat has not the consistency and the ability to lift breast even partially, at might happen with an implant. In this case, a hybrid breast augmentation and mastopexy could be performed (periareolar of vertical scar) combining a submuscular implant with fat transferred to achieve better contour and natural looking result (Fig. 21.18).

Hybrid breast augmentation is a new trend that is rapidly expanding in the day practice of a plastic surgeon, as it combines the benefits of an implant (consistency, shape, and projection) with the advantages of softness, naturalness, and avoiding the typical stigmata of implants in skinny patients (visibility on top breast, on lateral breast, where muscle is absent of minimally represented, unnatural appearance in few instances, tendency to capsule contracture). In those patients, hybrid breast augmentation does play an important role (Fig. 21.19).

Fat is useful for preventing and treating breast capsule contracture, as demonstrated by Rigotti [12]. The role of stem cell in promoting a neo-vascularization of the capsule around the implant helps softening the interaction between tissue and foreign body, thus decreasing the strength of the foreign body reaction.

The action of stem cells works as regenerative agent in stimulating the growth of a softer capsule, with minor retraction and pain and deformity associated. Fat transfer has been found to be extremely useful also for releasing severe degree of capsule contracture (Baker III–IV), thus ameliorating the overall appearance of the constricted breast, and decreasing the often associated pain.

However, there is a growing tendency in those days to accomplish breast fat augmentation with fat transfer, especially in secondary cases, where over the revision surgery, the surgeon often encounter problems with capsule contracture of the previous implant.

Risk of recurrence of capsule contracture cannot be underestimated, and the fat transferred could potentially help preventing the formation of a new tight capsule and a firm contracture.

Smoothening the interface between implant and soft tissues is another benefit which fat transfer can help with: especially in thin patients, or in patient with subcutaneous thinning, this represents the first option nowadays.

In the past, plastic surgeon did use alloplastic material, as porcine-derived mesh, or acellular human tissue matrix, derived from cadaveric tissue, which exhibits regenerative

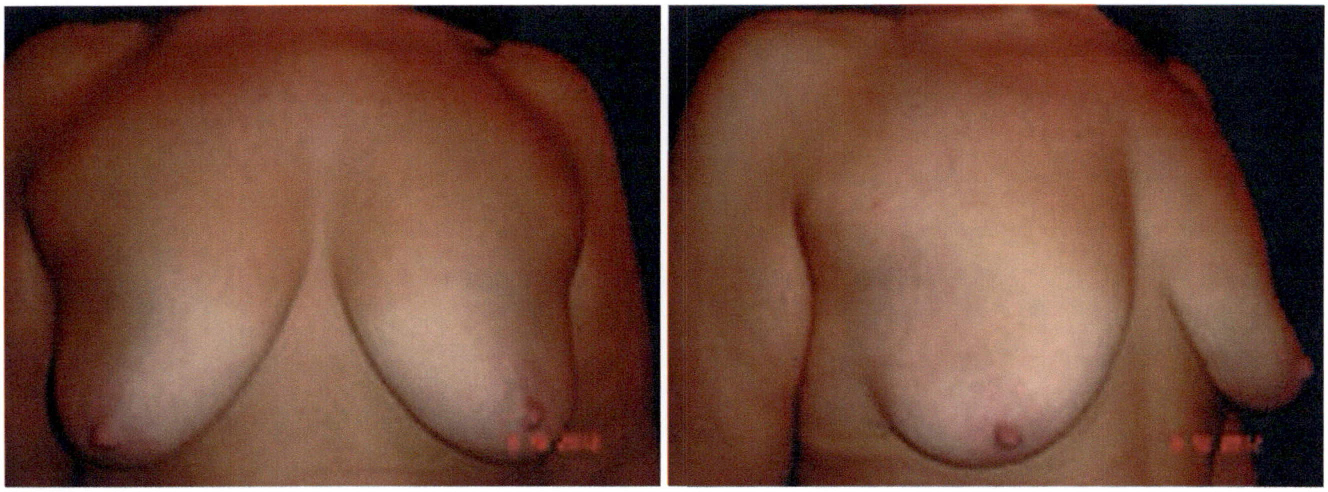

Fig. 21.18 Patient with mild ptosis (nipple 23–24 cm from midclavicle). Contraindicated for fat transfer

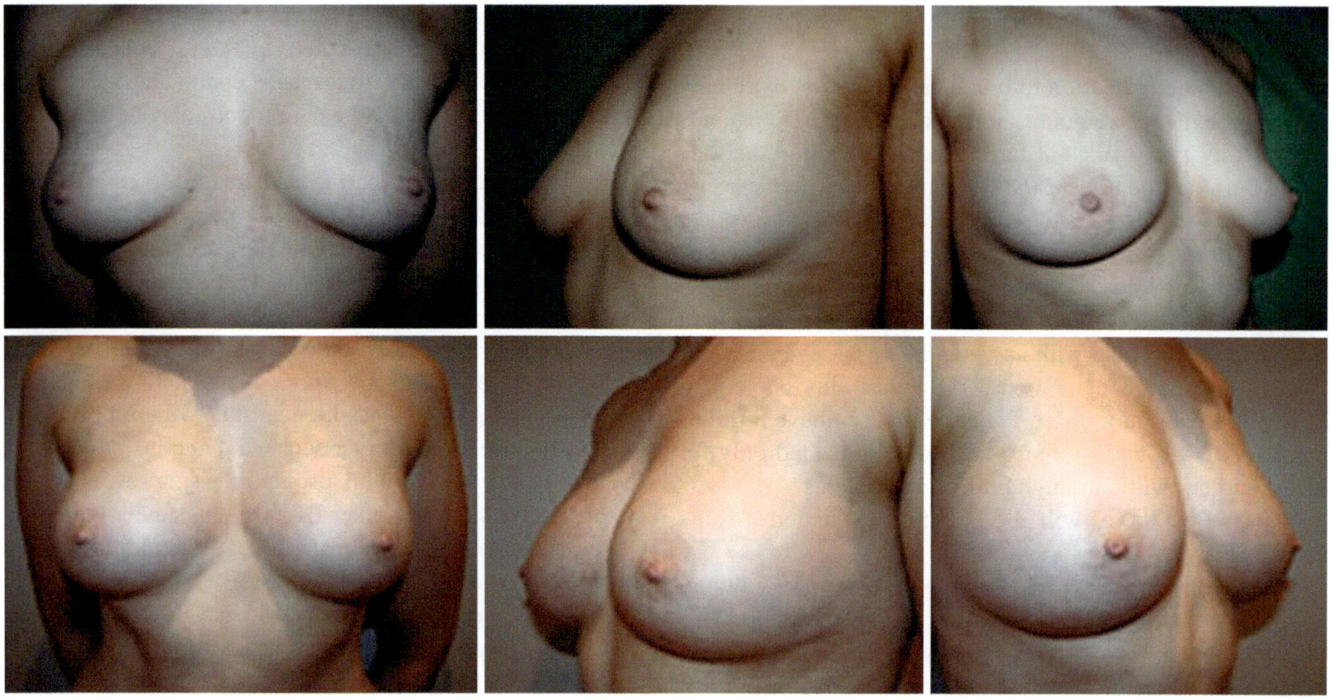

Fig. 21.19 Hybrid breast augmentation. 250 cc moderate plus subpectoral implants and 200 mL of fat transferred per side. Post op 6 months

properties. All donor cell and allergic epitopes are removed, leaving a collagen scaffold, growth factor receptors, and vascular channels that aid in tissue regeneration with minimum scar and fibrosis.

The acellular matrix generates physiological cues that mimic the native tissue microenvironment. Acellular dermal matrix is a soft connective tissue graft generated by a decellularization process that preserves the intact extracellular skin matrix.

Fat transfer has the capability to vicariate and largely substitute any need for cellular matrix or other porcine-derived mesh, in cases of thinning of the subcutaneous tissue in the breast region.

References

1. Czerny V. Plastischer Erzats de Brustdrüse durch ein Lipom. Zentralbl Chir. 1895;27:72.
2. Lexer E. Die freien Transplantationen, vol. 1. Stuttgart: Ferdinand Enke; 1919.
3. Illouz YG. Body contouring by lipolysis: a 5-year experience with over 3000 cases. Plast Reconstr Surg. 1983;72:591–7.
4. Fournier PF, Otteni FM. Lipodissection in body sculpturing: the dry procedure. Plast Reconstr Surg. 1983;72:610–9.
5. Bircoll M. Cosmetic breast augmentation utilizing autologous fat and liposuction techniques. Plast Reconstr Surg. 1987;79:267–71.
6. Bircoll M, Novack BH. Autologous fat transplantation employing liposuction techniques. Ann Plast Surg. 1987;18:327–9.
7. ASPS ad hoc Committee on New York procedures. Report on autologous fat transplantation. Plast Surg Nurs. 1987;7:140–1.
8. Coleman SR, Saboeiro AP. Fat grafting to the breast revisited: safety and efficacy. Plast Reconstr Surg. 2007;119:775–85.
9. Baker DC, Khouri RK. Breast augmentation with Brava external expansion and fat grafting: a prospective study. Presented at the annual meeting of the American Society of Aesthetic Plastic Surgery, Orlando, FL, April 2006.
10. Zocchi ML, Zuliani F, Nava M, et al. Bicompartmental breast lipostructuring. Presented at the seventh international congress of aesthetic medicine, Milan, Oct 2005.
11. Grisotti A. Lipostructure, of course in the body and breast. Presented at the Seventeenth Annual American Alpin Wirkshop in Plastic Surgery, Sun Valley, Idaho, Feb 2006.
12. Rigotti G, Marchi A, Galiè M, et al. Clinical treatment of radiotherapy tissue damages by lipoaspirate transplant: a healing process mediated by adipose stem cells (ASCs). Plast Reconstr Surg. 2007;119:1409–22; discussion 1423–1424.
13. Delay E, Delaporte T, Sinna R, et al. Breast implant alternatives. Ann Chir Plast Esthet. 2005;50:652–72.

Total Breast Reconstruction with "Deflating Technique"

22

Franco Bassetto and Martina Grigatti

Contents

22.1 Introduction

In the treatment of breast cancer, breast reconstruction represents an integral part of the treatment. The alternatives currently available are autologous tissue transfer, such as flaps or fat grafting, implants, or a combination of both.

The breast is a subcutaneous tissue influenced by gravity, aging, and posture; for this reason microsurgical tissue transfer represents the first choice in reconstruction to achieve long-lasting natural results based on the replace-like-with-like principle. However, this technique requires an experienced surgeon with microsurgical skills, longer hospitalization, extra scarring, and morbidity in the donor site. Sometimes for some patients, autologous tissue for a microsurgical approach is not available or simply not wanted by all patients.

Alloplastic breast reconstruction with implants generates fewer scars, no donor site morbidity, and less operating time. However, it increases the risk of infection, capsular contraction, animation deformity, implant migration, poor aesthetic results, for example, with a deficient lower pole and unnatural contours. Moreover, the challenge with implant-based reconstruction is the soft tissue coverage.

Lipofilling, in the past, is considered dangerous in breast cancer for its stem properties; however, it is now a universally accepted technique to correct soft tissue deficiencies or to improve contour in breast reconstruction.

In this chapter, we will describe the total breast reconstruction with the "deflating technique" after mastectomy. This method provides a protocol of tissue expansion with serial deflation-lipofilling.

Indication The challenge is to compensate the volume loss after mastectomy, where the skin flaps are completely without support previously provided by Cooper's ligament and the breast glandular tissue.

This method could be a choice especially for small to moderate sized breast patients. It is indicated for motivated patients who do not want or who cannot have a microsurgical reconstruction, but who want a natural effect without an implant (or with one as small as possible).

Technique Different techniques are described in the literature, the concept in common is the breast reconstruction after mastectomy with fat grafting combined with internal tissue expansion.

It is a multistep procedure. It starts with the preoperative marking in the upright position of the dissection limit and the existing and the new inframammary fold.

Surgery is normally performed under general anesthesia. The expander is placed subcutaneously in a prepectoral position, which is the most rational place as the breast is a subcutaneous structure. It is inserted through the inframammary fold, while the existing mastectomy scar is left intact in secondary reconstruction, in order to preserve the skin envelope and to avoid the scar reopening that would compromise the lipofilling procedure as the wound needs to heal again.

F. Bassetto · M. Grigatti (✉)
Department of Neurosciences, Clinic of Plastic Surgery, Padua, Italy
e-mail: franco.bassetto@unipd.it

© The Author(s), under exclusive license to Springer Nature Switzerland AG 2023

A. Di Giuseppe et al. (eds.), *Fat Transfer in Plastic Surgery*, https://doi.org/10.1007/978-3-031-10881-5_22

The expander in a deflated condition should be placed lower than the existing inframammary fold to support the lower pole expansion. A closed suction drain must be placed in the pocket and removed when the output is minor of 30 cc for 24 h. Antibiotic prophylaxis is also necessary postoperatively.

Expansion is normally started at 2 weeks postoperatively and performed on weekly basis with saline or air in outpatient setting.

The expansion process generates a well-vascularized capsule, and fat injection is performed in the space between skin and this capsule.

Eight weeks after the onset of expansion, lipofilling with Coleman's technique is gently performed, paying attention not to create excessive tissue turgor, which could compromise fat graft survival. Donor sites are thigh, buttock area, and abdomen.

The fat grafting could be delayed until 6 month after the end of adjuvant therapy.

Although the traditional concerns regarding fat viability, especially after radiotherapy, it is demonstrated that injecting fat after reducing the expander pressure is conductive to its integration in the recipient site (Fig. 22.1).

Lipofilling uses lipoaspirate material (liquefied), and it injects numerous minuscule particles of fat at different levels to maximize the surface area of contact between the fat graft and the recipient site (Fig. 22.2).

Diffusion of the fat in the prepectoral pocket is prevented by the expander and his capsule, which acts as an anchor. The capsule has a newly generated vascular plexus in the outer layer, and it represent the support for the fat graft.

Fat grafting is combined with gradual emptying of the expander during each course (i.e., 50 mL of fluid removed and 70 mL of fat injected).

In this way, the breast volume remains constant, but the consistency of the breast changes gradually.

Fat grafting sessions are performed with an interval of 2–3 months in a day surgery setting until the result is reached. At that time, the expander is removed.

Even if this technique allows total breast reconstruction using only fat graft, it is possible, in case of bigger breast or when more volume is needed, to add a core volume and projection with an implant (smaller than in a traditional implant-based reconstruction) inserted after expansor removal to provide extra projection, volume, and shape.

In this case, implant's impact in surrounding tissues is minimized with the restoration of a thick subcutaneous autologous fat-grafted tissue.

Complications are rare or unsevere. The most frequent are seroma and hematoma and expander infection.

Advantages The advantages are many. The method is safe with minimal scar and it requires short minimal invasive procedures in day surgery or outpatient setting, with a very simple equipment and low costs. The donor site morbidity is minimal, and the procedure does not affect the patient function with a minor disruption of patient professional and personal life. The consistency of the breast is natural, and the fat graft has a positive effect on the mastectomy scar. Finally, this method allows to convert to breast reconstruction with an implant at every stage.

Disadvantages The disadvantages are few, but they are worthy of attention in choosing this technique of breast reconstruction. This method in fact requires many procedures, a long-time treatment, so that the patient must be really motivated. It requires also the right patient, with an appropriate body structure with a sufficient volume of the donor site for fat grafting. Moreover, another important disadvantage is the unpredictable fat resorption rate.

Conclusion The rapid restoration of breast volume with the expander and the gradual change in breast consistency and shape motivate the patient to continue the treatment and encourage the return to professional and intimate life.

This fact, together with all the advantages listed above, make this technique a valid choice of breast reconstruction in selected patients.

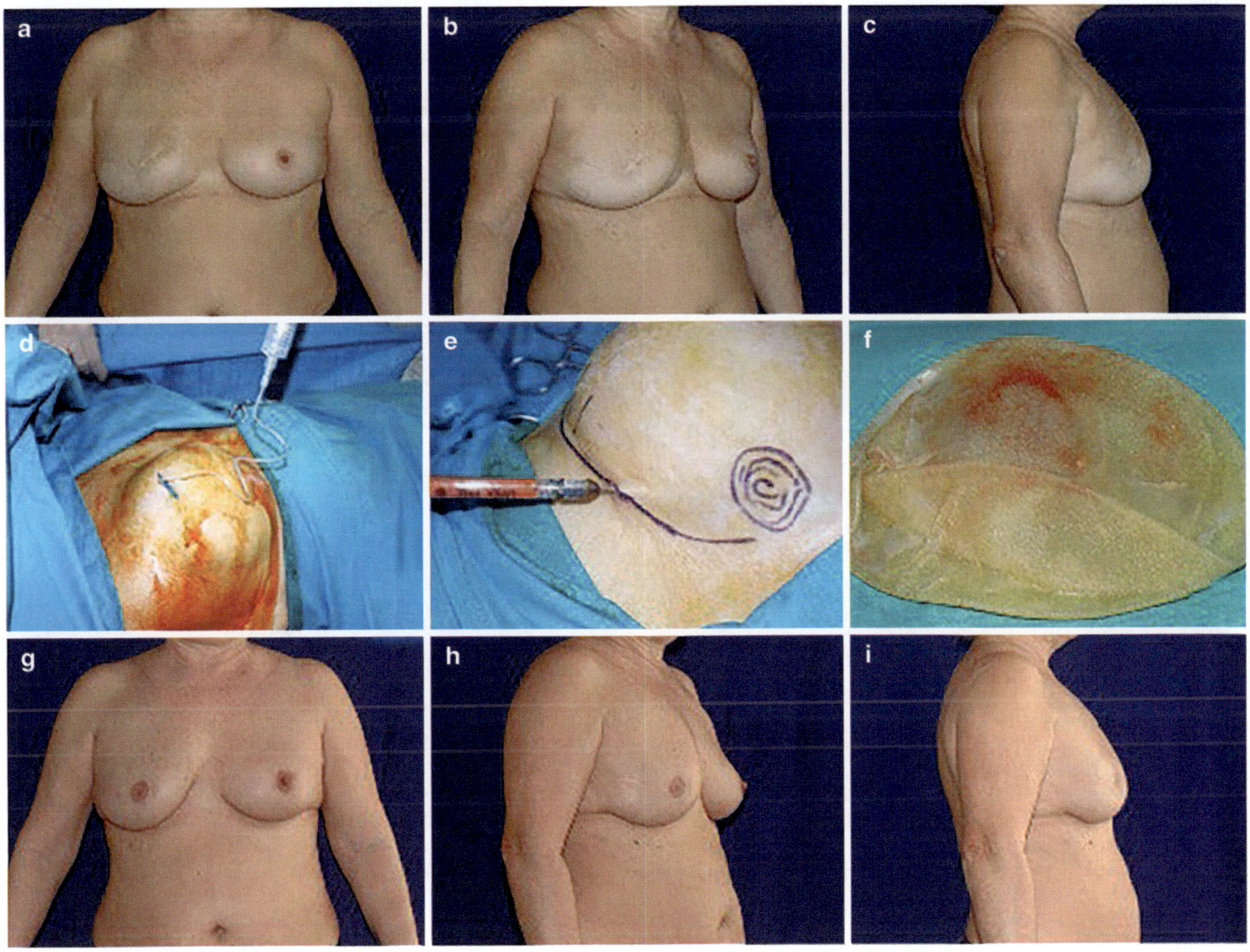

Fig. 22.1 Sequential reconstruction using fat grafts. Preoperatory images (**a**–**c**). Progressive deflating of breast expander (**d**–**f**). Follow-up after 2 years and 3 sessions of adipose tissue grafts (**g**–**i**)

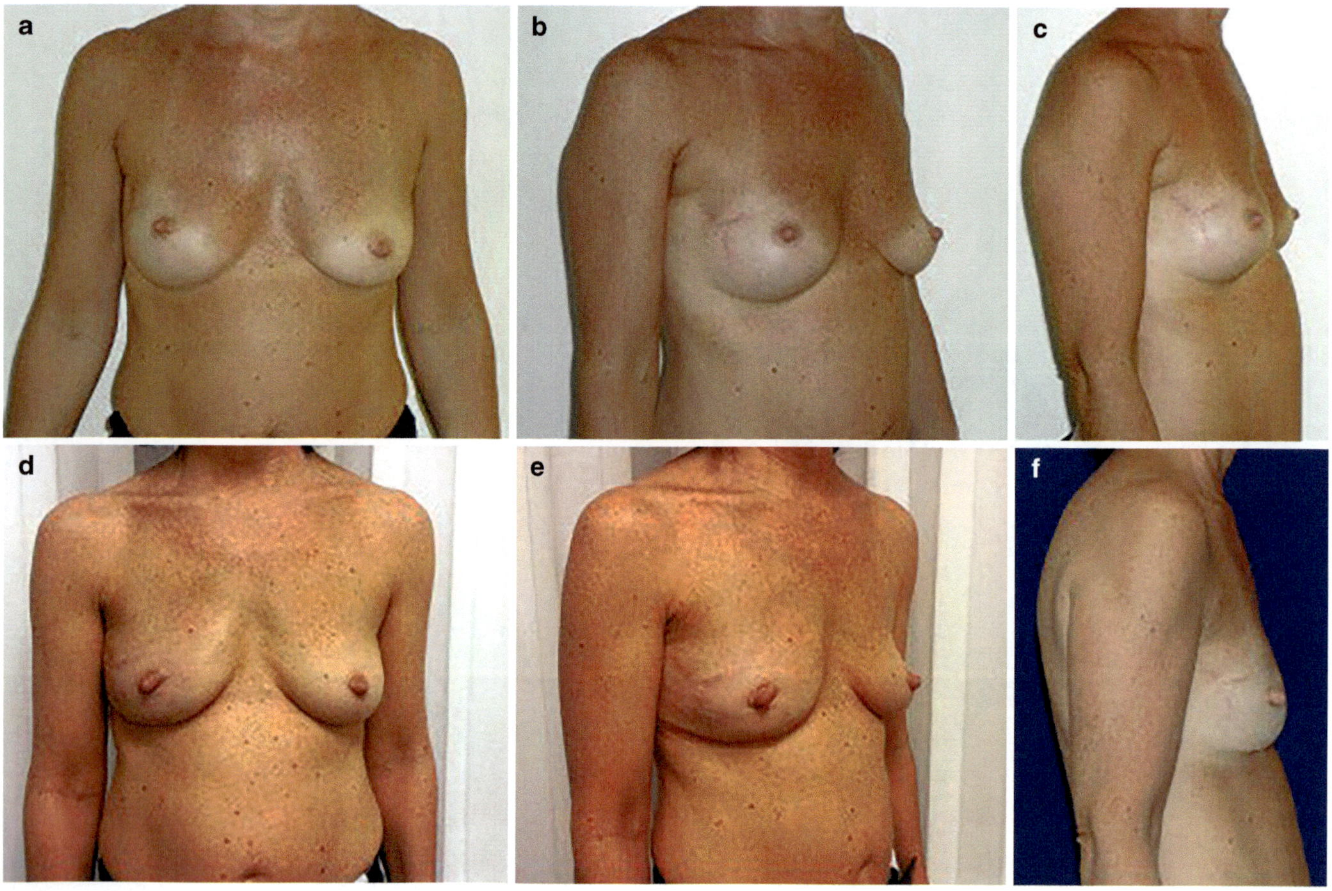

Fig. 22.2 Sequential reconstruction using fat grafts. Preoperatory images (**a–c**). Follow-up after 3 years and 4 sessions of adipose tissue grafts (**d–f**)

Bibliography

1. Sommeling CE, Van Landuyt K, Depypere H, Van den Broecke R, Monstrey S, Blondeel PN, Morrison WA, Stillaert FB. Composite breast reconstruction: implant-based breast reconstruction with adjunctive lipofilling. J Plast Reconstr Aesthet Surg. 2017;70(8):1051–8.
2. Stillaert FBJL, et al. The prepectoral, hybrid breast reconstruction: the synergy of lipofilling and breast implants. Plast Reconstr Surg Glob Open. 2020;8:e2966.
3. Kolasinski J. Total breast reconstruction with fat grafting combined with internal tissue expansion. Plast Reconstr Surg Glob Open. 2019;7(4):e2009.

Total Fat Grafting Breast Augmentation for a Harmonious Reshaping

23

Caterina Gardener and Vincenzo Vindigni

Contents

23.1 Introduction

Fat grafting in cosmetic surgery is a widespread technique used for many purposes. Due to its manageability and safety, fat transfer offers many possibilities and clinical applications, which are more and more expanding.

In the panorama of breast surgery, it represents a valid and alternative method to conventional silicone implants. Although breast implants have become indispensable tools for plastic surgeons, they remain foreign bodies for human organisms, and they are also associated with several complications and discomfort. Implants can develop capsular contraction, rupture, displacement, infection, and they need to be changed sooner or later during a lifelong period. Because of this and because plastic surgeons are always aiming of the best outcome for each patient, considering his/her expectations and history, fat grafting has become a significant option. In particular, breast can definitely improve in terms of size and shape with autologous fat transfer.

A sort of autologous fat transfer on the breast was practiced for the first time in 1895 by Czerny [1], who implanted a lumbar lipoma into a breast after mastectomy to avoid asymmetry. After almost a century, Bircoll was the first author in 1987 who described the technique of autologous fat injection for breast augmentation using liposuction [2]. Several years after, Coleman and Delay introduced the term "lipomodelling" [3]. Since then fat transfer has been more and more used both in cosmetic and in reconstructive surgery to obtain the highest quality results, in order of safety and effectiveness.

Breast augmentation obtained by autologous tissues has peculiar features, it is distinguished by a pleasing and anatomical shape, breast results soft to the touch, and it feels natural and comfortable. Scars are almost invisible and, in addition, removal of unwanted fat tissue from one part of the body for enhancement of another represents a captivating option for patients. For all of these reasons, breast augmentation with lipofilling is continuously more requested.

C. Gardener (✉) · V. Vindigni
University of Padua, Department of Neuroscience, Clinic of Plastic and Reconstructive Surgery, Padua, Italy
e-mail: vincenzo.vindigni@unipd.it

© The Author(s), under exclusive license to Springer Nature Switzerland AG 2023

A. Di Giuseppe et al. (eds.), *Fat Transfer in Plastic Surgery*, https://doi.org/10.1007/978-3-031-10881-5_23

23.2 Indications

Autologous fat grafting is a safe procedure, without risk of developing an immune response, its technical steps are easy, and it offers the possibility of being repeated and perfected if needed.

Because of this fat transfer is characterized by a wide spread of surgical indications. It is suitable to obtain primary breast cosmetic augmentation, in patients desiring a modest increase. Augmentation, for example, also in transfeminine patients. It can be used to correct congenital or acquired deformities such as micromastia, tuberous breast (Fig. 23.1), asymmetry, or defects caused by Poland syndrome, to add volume to the underdeveloped breast. It is important to notice that this type of conditions represent the ideal recipient tissue, a virgin tissue, with no scars or fibrosis from previous surgery.

And it also offers a valid solution to restore volume loss after implants removal (Fig. 23.2).

Fat grafting for filling defects after reconstruction with implants or flaps post breast cancer are arguments of other chapters of the book.

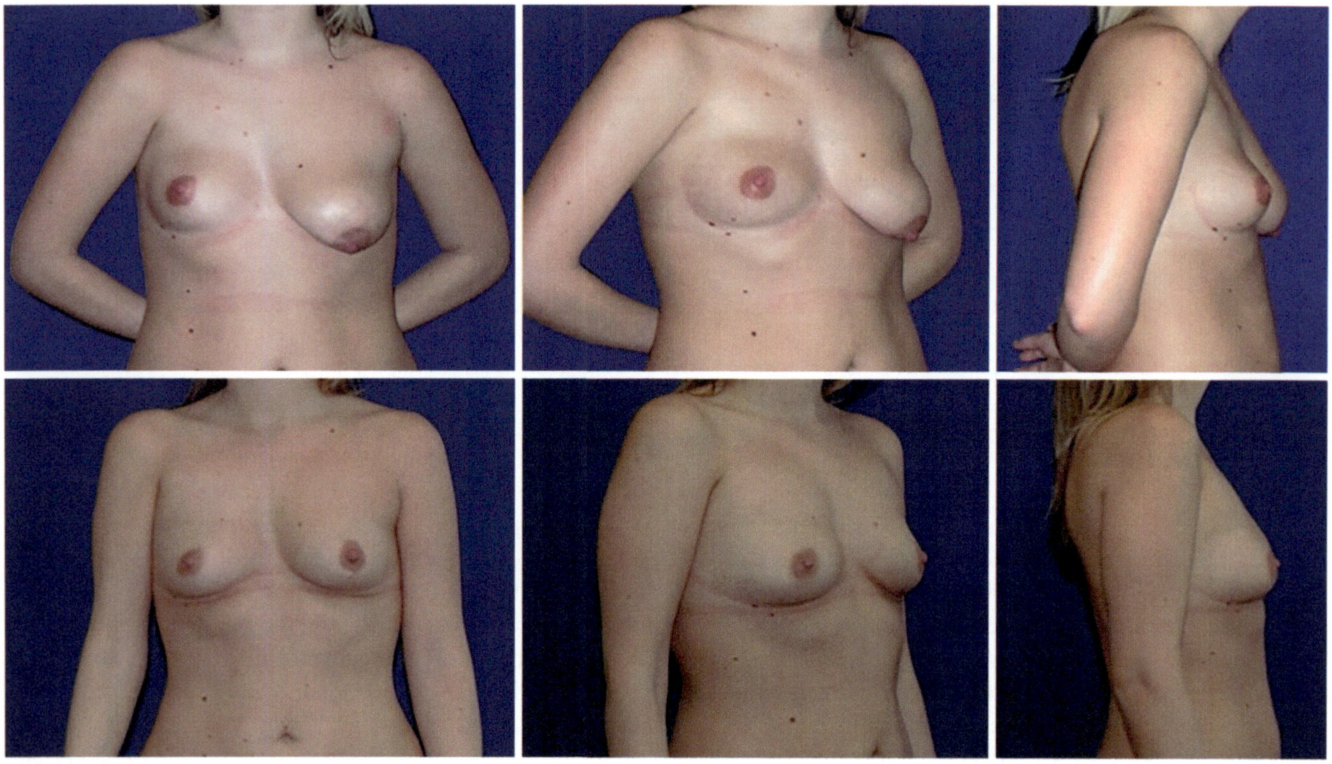

Fig. 23.1 Pre (up) and postop (down) pictures of a patient affected by breast asymmetry. On the right breast, augmentation with lipofilling was performed, mastopexy on the left side

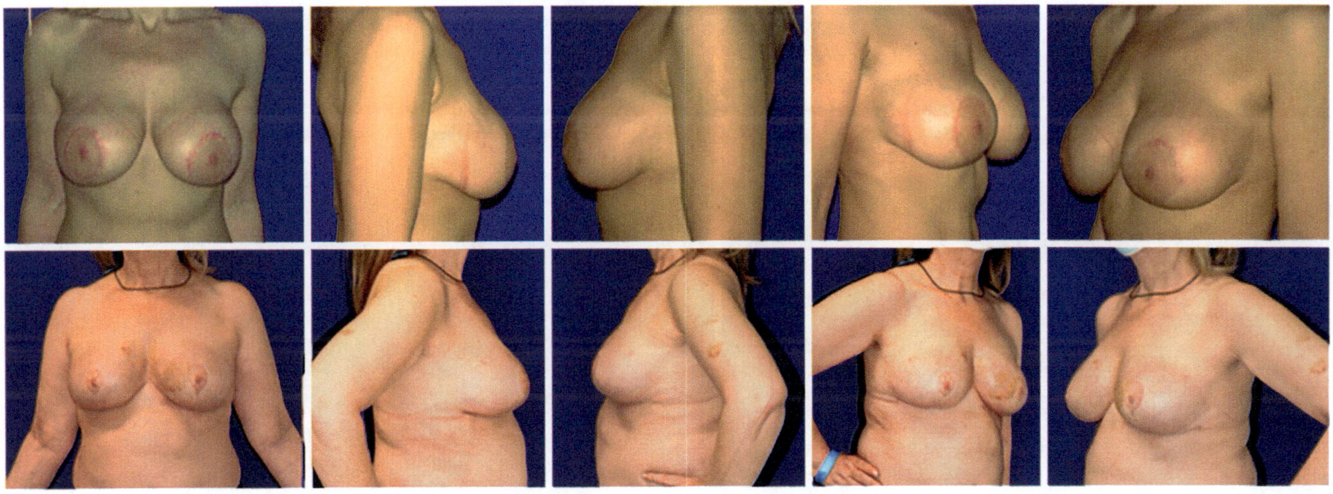

Fig. 23.2 Removal of breast implants and replacement of lost volume with fat grafting. Before the procedure in 2015 (up) and after the procedure in 2021 (down)

23.3 Before Surgery

Before planning fat transfer for breast augmentation, it is important to evaluate patient's past medical history, social, and health status. Surgeons should consider risks and benefits of surgery, keeping in mind that lipofilling is a procedure not recommended in smokers or in people affected by bleeding disorders.

It is mandatory also to discuss in depth patient's goals for the operation, in order to understand and achieve patient's expectations. Breast augmentation reached by this technique is moderate and not too projected. Promised results must be obtainable, and they depend on the physical conformation of the patient. Evaluation of donor areas should be made to predict the degree of correction, being conscious that for one cup augmentation an average of 250–400 cc of fat per side is required. In addition, a total of about 200–250 cc represent the traditional upper limit of fat transfer for each mammal; this limit can be higher if assisted by external expanders [4].

Before surgery patients should be well informed about the possible loss of volume due to reabsorption and that their body contour will change. Because unlike silicone implants, lipofilling results are not so predictable, and patients must also be warned that multiple procedures should be needed to achieve a satisfying result. Usually 1–5 sessions are required, with an average of 3 [5].

Before fat transfer, some surgeons mandate a mammogram or MRI, to both exclude underlying pathologies, to study anatomy in depth, and to establish a baseline for later comparison [6, 7].

Learned societies agree on the following points:

- To not perform fat transfer on patients at high risk for breast cancer
- To conduct preoperative radiological checkup
- To conduct regular postoperative follow-up [8]

23.4 Procedure

For reaching the desired result, a careful evaluation of breast symmetry and skin quality must be conducted. Unlike silicone implants, this technique allows not only augmentation of size but also allows to remodel the intermammary distance cleft, to correct shape asymmetries, and to volumize the superior medial pole for a better décolleté. But, unlike silicone implants, it is a time-consuming procedure.

The body offers several areas to harvest fat, most useful and handy areas are lower abdomen, buttocks, and thighs. Often abdomen can represent a single donor area.

On the other hand, the thigh is less fibrous, contains less septae, and has a relatively low blood supply and is therefore less susceptible to develop postoperative collections [9]. Anyway, it has been noticed that the choice of donor sites does not affect the survival of the adipocytes [10].

External volume expansion (EVE), as the BRAVA system, has shown to improve results [11, 12] (Fig. 23.2). Despite the use of BRAVA was not very widespread, neither its indications were standardized, in past years, it has been an effective tool for enhancing fat transfer in breast surgery [13] (Fig. 23.3).

Preoperative marking in standing position is fundamental to ensure surgery understanding by the patient and the operating team. It is also important to mark sites of incision with the patient upright in order to place them in natural body creases.

Most surgeons practice the wet technique and hand suction with blunt tipped cannulae after waiting about 15/20 min to harvest fat. But power-assisted liposuction and lipofilling [14, 15] represent a very useful and practiced option because they minimize the surgery time.

Once the fat has been collected either centrifugation [6, 16], mechanical decanting, or graft "washing" is needed to purify it. Particles more than 3 mm should not be transferred

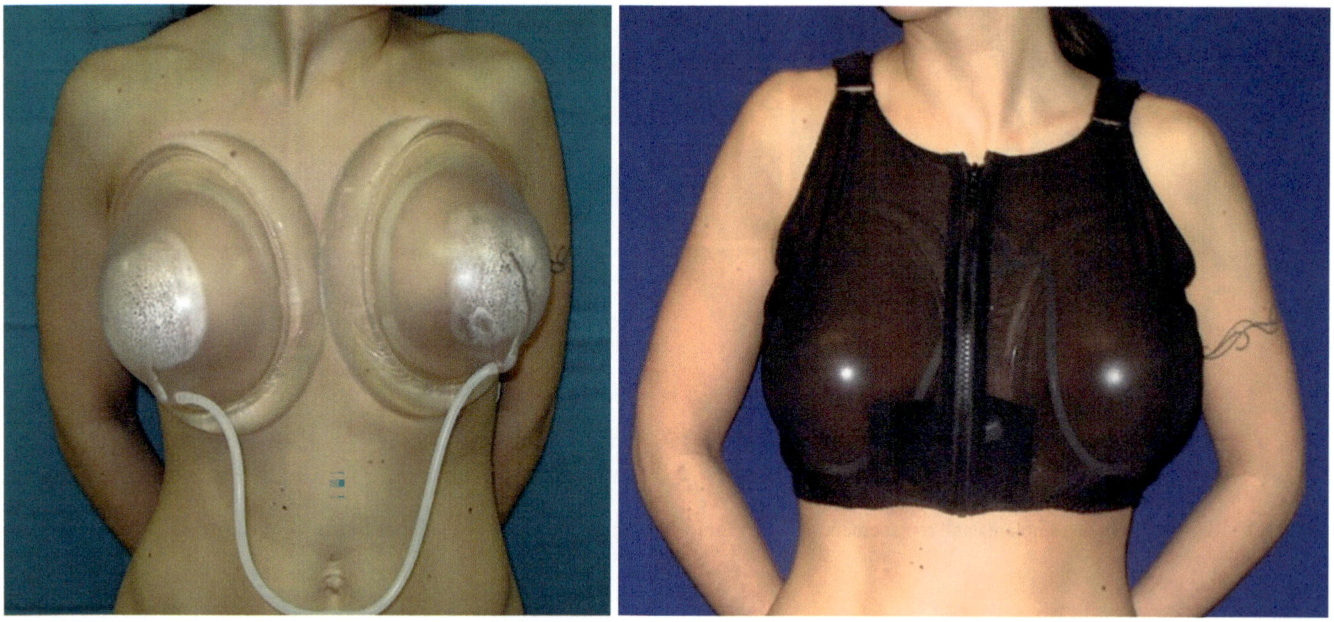

Fig. 23.3 External volume expansion treatment (EVE) before surgery

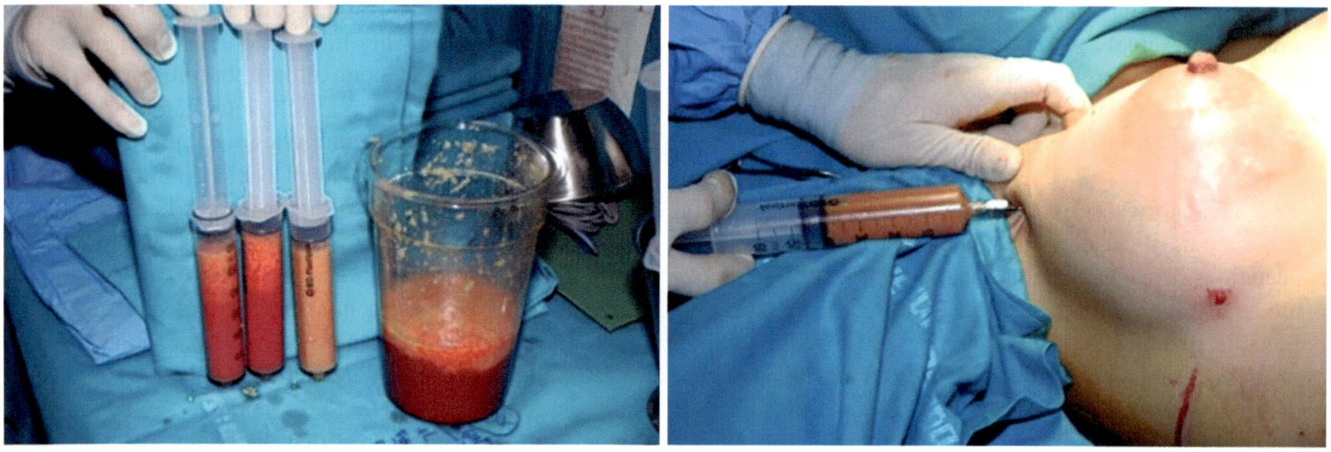

Fig. 23.4 Fat decanting after harvesting (right), fat injection technique (left)

as this increases the chance of developing necrosis or cysts (Fig. 23.4).

Although the methods of fat harvesting, processing, and injection all have an impact on successful clinical outcome, it is described that harvesting and processing have less impact than the method of injection [17].

In fact, it is fundamental to inject fat meticulously, without haste and taking the necessary time to do it. Fat has to be placed in very small amounts as the needle is gradually withdrawn, to maximize the surface area to revascularization. Coleman suggests approximately 0.25 cc of fat with each ejection of the cannula [6].

Injecting large volumes of fat could jeopardize the vascularity of the graft and cause contour irregularities [3].

To inject fat, small incisions are practiced in the inframammary fold, in natural creases or periareolar for the inner half [18]. Cannulae should be moved with fan-shaped wide movements. The fat must be distributed in different layers: most of it has to be grafted submammarly into the pectoralis major muscle, which provides volume and projection. Injection in breast parenchyma enhances overall breast volume, but not too much fat can be grafted at this level to not compromise graft survival. Placement in more superficial layers instead allows to control shaping of the breast and to correct specific deformities [6].

Zocchi described the "bicompartmental grafting technique": one part of the graft is placed subglandular/prefascial and the other subcutaneous, avoiding breast parenchyma

[19]. During surgery, it is useful to do a check with the patient seated at an angle of 45° to enable surgeons to evaluate the symmetry and shape under gravity effect.

Once the graft is distributed and incisions are closed with sutures, the breast has to be molded with hands using lubricant such as Vaseline [19].

23.5 Postoperative Time

Once surgery is concluded, it is crucial to minimize sheer forces on the breast to allow engraftment and revascularization of the grafted adipocytes. For this purpose and to provide protection to healing tissues, patients should wear surgical brasseries day and night.

On donor sites, compressive dressing is usually applied to prevent seroma and hematoma formation.

The amount of fat injected depends on the indication of the procedure, but a part of this volume will be lost within the first 4–6 months. The precise quantity that will be reabsorbed is not known and every case is different. For this reason, it is better to wait at least 4 months before a second procedure of fat grafting if needed (Fig. 23.5).

If the procedure is not performed correctly, both early and late complications can occur. Most common complications are edema, seroma, swelling, bruising, and infection. Necrosis of variable portion of the graft is possible, and it can present as a palpable lump, or more often as an oil cyst or calcification on mammogram. Late complications can be contour irregularities, deformation of the donor site, as well as hypersensitive nipples and breasts [20]. Anyway, most of complications can be avoided by increasing fat survival rate, goal obtainable by practicing this procedure meticulously.

Calcifications in some cases can interfere with breast cancer surveillance, but no study has shown an increase in the incidence of breast cancer due to this surgical practice [8].

Most of unsatisfactory results can be avoided by previous and accurate interviews with patients, who must have been well informed of the possibility of having to undergo more procedure to obtain the final result.

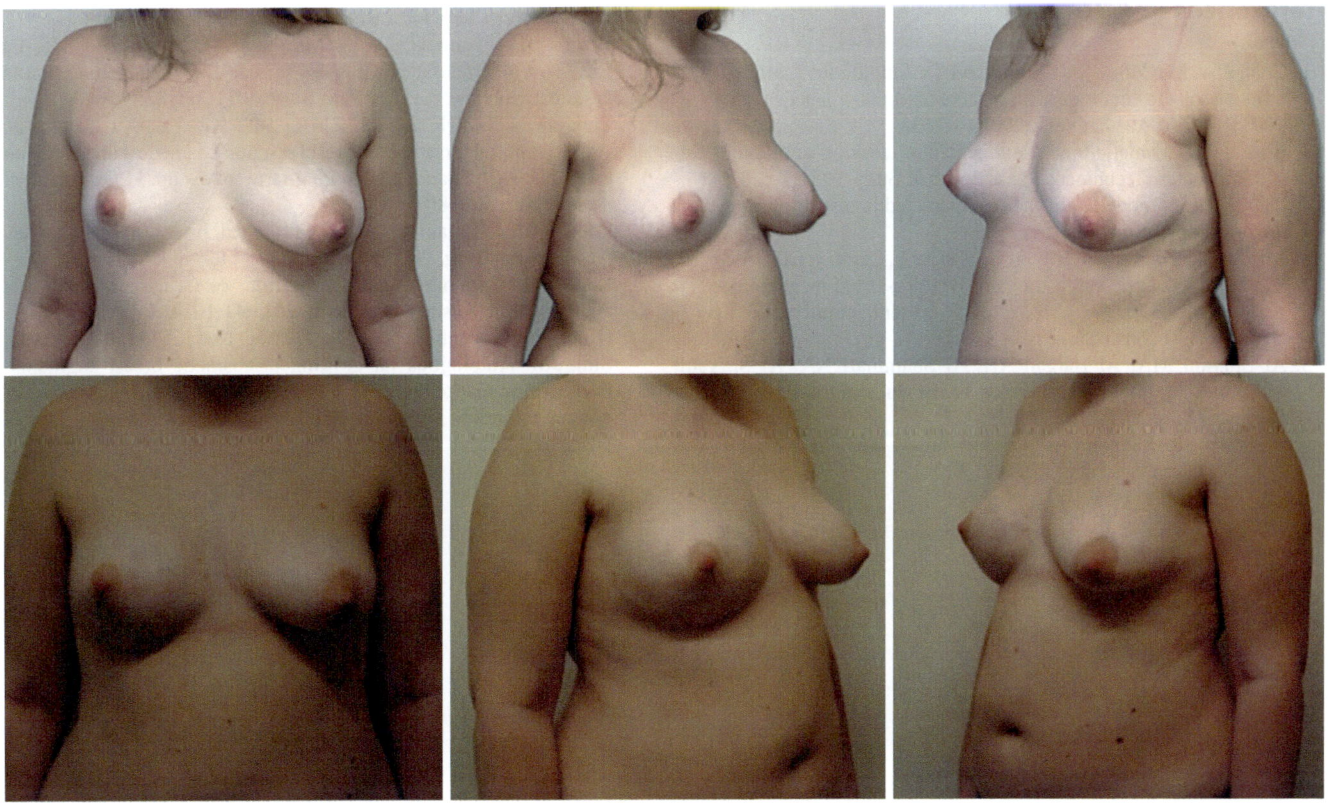

Fig. 23.5 Patient treated with EVE before surgery and transfer of 350 cc of fat on the right breast. Before surgery (up) after 1 year (down)

23.6 Conclusion

Fat grafting for breast augmentation is a procedure characterized by high level of patients' satisfaction [21] and minimal complications rate. It offers good cosmetic results with low costs, and breast will be perceived as its own part of the body with natural consistency. In addition, it provides secondary benefits such as liposuction of desired areas, remodelling patient's silhouette.

If practiced by expert surgeons, it represents an easy technique with high level of reproducibility.

The length of surgery time constitutes a major limiting factor such as a major disadvantage is that in some cases it could be necessary to repeat the procedure several times in order to achieve the desired result. But if on the one hand, it is true that sometimes more interventions may be necessary, on the other hand, once obtained, results are stable over time. Patients will not have the problem of implants' rupture neither the need of implants substitution during lifetime, saving money and unnecessary hospitalization.

References

1. Chan CW, McCulley SJ, Macmillan RD. Autologous fat transfer—a review of the literature with a focus on breast cancer surgery. J Plast Reconstr Aesthet Surg. 2008;61:1438.
2. Bircoll M. Cosmetic breast augmentation utilizing autologous fat and liposuction techniques. Plast Reconstr Surg. 1987;79:267.
3. Delay E. Lipomodeling of the reconstructed breast. In: Surgery of the breast: principles and art; 2012.
4. Khouri RK, Khoari RK, Rigotti G, Marchi A, Cardoso E, Rotemberg SC, et al. Aesthetic applications of Brava-assisted megavolume fat grafting to the breasts: a 9-year, 476-patient, multicenter experience. Plast Reconstr Surg. 2014;133:796.
5. Al Sufyani MA, Al Hargan AH, Al Shammari NA, Al Sufyani MA. Autologous fat transfer for breast augmentation: a review. Dermatol Surg. 2016;42:1235.
6. Coleman SR, Saboeiro AP. Primary breast augmentation with fat grafting. Clin Plast Surg. 2015;42:301.
7. Simonacci F, Bertozzi N, Grieco MP, Grignaffini E, Raposio E. Procedure, applications, and outcomes of autologous fat grafting. Ann Med Surg. 2017;20:49.
8. Voglimacci M, Garrido I, Mojallal A, Vaysse C, Bertheuil N, Michot A, et al. Autologous fat grafting for cosmetic breast augmentation: a systematic review. Aesthet Surg J. 2015;35:378.
9. Niechajev I, Ševćuk O. Long-term results of fat transplantation. Plast Reconstr Surg. 1994;94:496.
10. Rohrich RJ, Sorokin ES, Brown SA. In search of improved fat transfer viability: a quantitative analysis of the role of centrifugation and harvest site. Plast Reconstr Surg. 2004;114:1945.
11. Del Vecchio DA, Bucky LP. Breast augmentation using preexpansion and autologous fat transplantation: a clinical radiographic study. Plast Reconstr Surg. 2011;127:2441.
12. Oranges CM, Striebel J, Tremp M, Madduri S, Kalbermatten DF, Schaefer DJ. The impact of recipient site external expansion in fat grafting surgical outcomes. Plast Reconstr Surg Glob Open. 2018;6:e1649.
13. Liu W, Luan J. Recent advances on the application of negative pressure external volume expansion in breast plastic surgery. Aesthet Plast Surg. 2018;42:112.
14. Abboud MH, Dibo SA, Abboud NM. Power-assisted liposuction and lipofilling: techniques and experience in large-volume fat grafting. Aesthet Surg J. 2020;40:180.
15. Mangialardi ML, Ozil C, Lepage C. Cosmetic breast augmentation using power-assisted liposuction and lipofilling: a prospective study. Aesthet Plast Surg. 2021;45:1975.
16. Coleman SR, Saboeiro AP. Fat grafting to the breast revisited: safety and efficacy. Plast Reconstr Surg. 2007;119:775.
17. Rosing JH, Wong G, Wong MS, Sahar D, Stevenson TR, Pu LLQ. Autologous fat grafting for primary breast augmentation: a systematic review. Aesthet Plast Surg. 2011;35:882.
18. Zheng DN, Li QF, Lei H, Zheng SW, Xie YZ, Xu QH, et al. Autologous fat grafting to the breast for cosmetic enhancement: experience in 66 patients with long-term follow up. J Plast Reconstr Aesthet Surg. 2008;61:792.
19. Zocchi ML, Zuliani F. Bicompartmental breast lipostructuring. Aesthet Plast Surg. 2008;32:313.
20. Kontoes P, Gounnaris G. Complications of fat transfer for breast augmentation. Aesthet Plast Surg. 2017;41:1078.
21. Groen JW, Negenborn VL, Twisk JWR, Ket JCF, Mullender MG, Smit JM. Autologous fat grafting in cosmetic breast augmentation: a systematic review on radiological safety, complications, volume retention, and patient/surgeon satisfaction. Aesthet Surg J. 2016;36:993.

Breast Fat Augmentation. Step-by-Step Technique and Ultrasound Assisted Liposuction for Contouring of Donor Area

24

Alberto Di Giuseppe and Federico Giovagnoli

Contents

24.1 Introduction

In 2009, Panetta et al. [1] published on *Plastic Reconstructive Surgery Journal* a paper on the effects on the third generation ultrasound liposuction (VASER) on adipose tissue. The paper analyzed the fact of ultrasound energy delivered by VASER and then suctioned through the VentX system on the lipoaspirate for autologous fat transfer (AFT) as a source material for adipose-derived regenerative cells (ADRCs), already known as stem cells. Panetta analyzed the viability of adipose tissue (fat cell) obtaining data which shows that the fat lipoaspirated from VASER ultrasound device contains viable fat cells as well as standard liposuction. In this study, he found also a mixture of cell types within the ADRCs obtained from the adipose tissue. New guidelines were suggested to make the utilization of the VASER ultrasound system and the VentX aspiration system for body contouring procedures more effective (Fig. 24.1).

A. Di Giuseppe (✉)
University of Padova, Padova, Italy

F. Giovagnoli
University Vita-Salute San Raffaele, Milan, Italy

Adipocyte viability of unprocessed samples ranges between 85.1 ± 11%.

VASER aspirate adds aqueous content of 20.7 ± 2% in free lipids content of 5.29 ± 1.8%.

After processing in the pure graft system, aqueous, and free lipids are reduced to 16.5 ± 3.5% and 0.63 ± 0.5%.

After processing in the Celution System by Cytori, the mean viability of ADRCs cells was found to be 87.4 ± 4.58%.

Post freezing viability of ADRCs dropped to 11%, after cryo preservation 50%.

A standard VASER treatment followed by low-vacuum aspiration using the VentX system, achieved as raw fat viability of 85%.

Also, by maintaining the overall cell population of the lipoaspirate (mature adipocytes and the "loosely adherent" cell population: peri-adipocytes, stem cells, stromal cells, etc.), the success rate of the implanted tissue matrix is enhanced.

In addition to the use of harvested fat for immediate reinjection, adipose tissue is a significant source of stem cells for both enhancement of autologous fat grafts and for use in other developing clinical treatments.

As with direct autologous fat injections, the viability of the regenerative cells derived from the harvested adipose tis-

© The Author(s), under exclusive license to Springer Nature Switzerland AG 2023

A. Di Giuseppe et al. (eds.), *Fat Transfer in Plastic Surgery*, https://doi.org/10.1007/978-3-031-10881-5_24

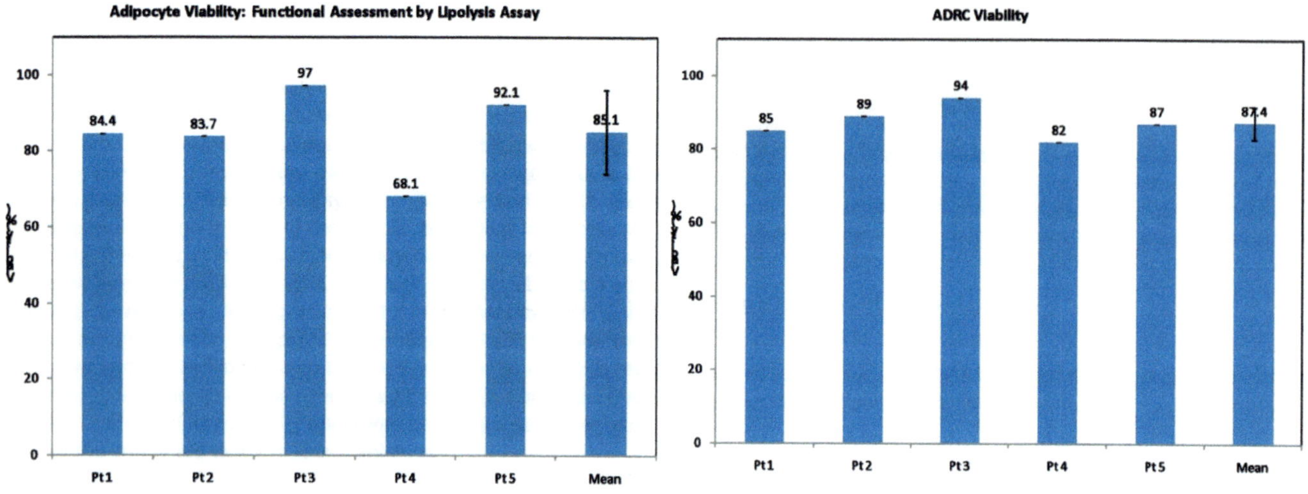

Fig. 24.1 The two diagrams show the adipocyte viability and the ADRC viability using VASER ultrasound liposuction. Results are similar to fat removed from standard liposuction

sue is of primary importance, both as fresh tissue and after freezing.

The field of clinical treatments using adult stem cells continues to expand and patients undergoing liposuction are increasingly interested in storing tissue samples containing stem cells.

These cells could then be tawed at some future date and utilized in the development of genetic- and tissue-based treatments.

The viability of the adipose-derived stem cells was higher than that for the overall tissue samples and over 87%.

More importantly, the viability of the adipose-derived stem cells was only slightly reduced to 75% after freezing using liquid nitrogen vapor.

In this application, the separation of the cells using the VASER System minimized tissue damage and improves the viability of the harvested cells.

The VASER Ultrasound system and the VentX Aspiration System provide an efficient means of harvesting adipose tissue without sacrificing tissue viability.

Guidelines includes:

1. Donor area: abdomen, flanks, and thighs
2. Tumescent infiltration of fluid: 2 mL infiltrated for every 1 mL removed
3. VASER performed with 3.7 mL probe at an amplitude of 60% in pulsed VASER mode
4. Aspiration occurred using VentX cannula diameter of 3.0 mm at a vacuum level of 15 in hg

In my clinical experience, I have followed those guidelines in my first 180 cases of breast fat augmentation using VASER liposuction from 2010 to 2013.

In August 2013, a letter was published on *Plastic Reconstructive Surgery Journal* written by Severiano Vilaboa and Ramon Llull (Stem center of Palma de Mayorca, Spain) [2]. They summarized the variables that make unpredictable the post-surgical volume loss, in AFT. Among those, lack of adequate vascularization, lack of regenerative cell within the fat graft, the traumatic handling of the fat during preparation and implantation, immunosystem-mediated cellular phagocytosis in the proinflammatory environment after grafting, or likely a combination of all of these effects.

They advocated returning fat graft to physiologic condition using washing: in other words, the long-term survival rate of the harvested fat was deeply connected also to the restoring of the physiologic condition of fat, which was too aqueous with previous technique utilized.

When I started performing fat extraction, I utilized as the Puregraft™ System, as the Acquavage system [3, 4]. Both the systems had the advantage to obtain a liquid fat, easy to be injected (Fig. 24.2).

The aqueous liquid content of grafts prepared using the different preparation methods can be seen for one representative sample. The numerical data of all samples were presented as average percentage of the total graft. Compared to unmanipulated control samples (33.21 ± 1.8%) and the samples prepared by gravity separation (24.5 ± 1.4%), the graft tissues prepared by centrifugation and Puregraft contained significantly less liquid ($p < 0.001$ for all comparisons). The mean liquid content of tissue prepared using Puregraft was 8.1 ± 0.9%, which was comparable to that of graft tissue prepared by centrifugation (5.2 ± 0.5%). In addition, the fluid content in grafts prepared by the Puregraft system may be adjusted to a wider range (from 2.67% to 20%), depending on the drain time and drain techniques (active or passive) used. Thus, physicians can control the graft viscosity to their desired consistency.

But as contra, swelling post surgery was impressive, and absorption or retention of real fat was about 60% of the total

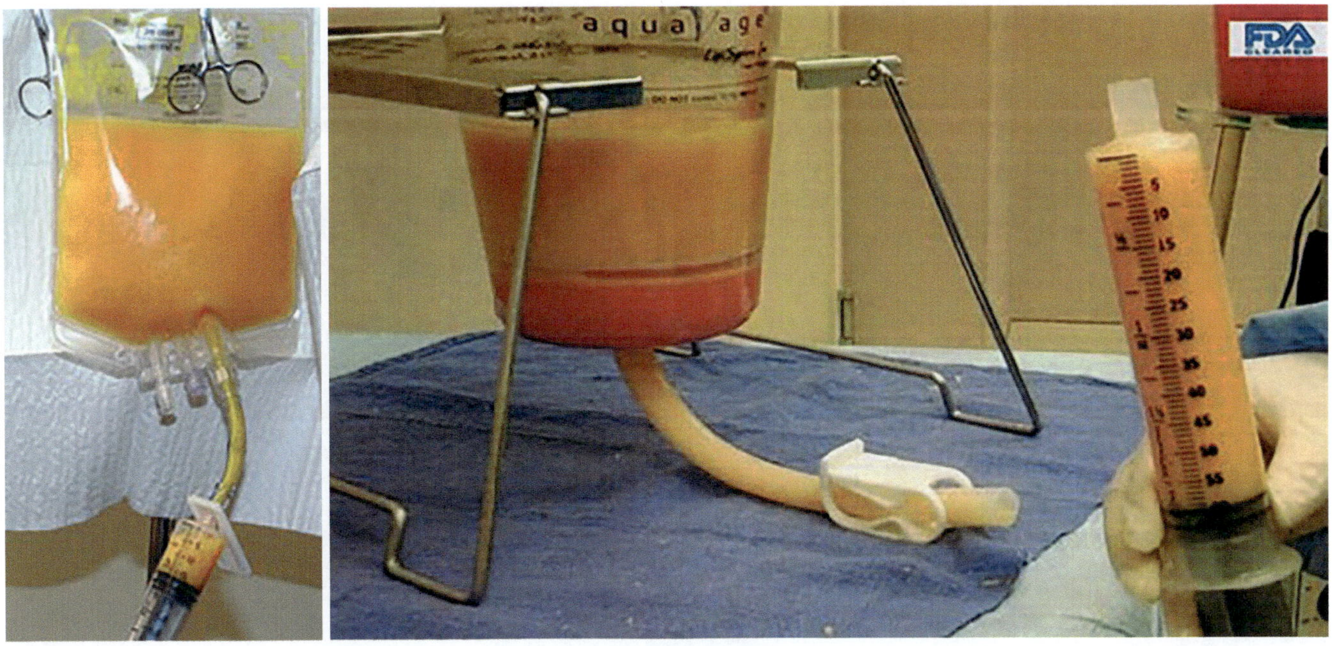

Fig. 24.2 Puregraft™ System—Acquavage system for fat harvesting and collection. Note the aqueous prevalence of the harvested fat

infused, in my clinical experience. As you can see in this clinical example of breast augmentation using Puregraft system, the swelling of the first post-op day is really impressive (Fig. 24.3b): but in 10–20 days, the edema is absorbed and final volume reached is much less that desired and expected. This could lead to dissatisfaction of both patient and surgeon in terms of expectations of achievable volume. Sometimes patient commented that part of the fat had already been absorbed, or the technique was wrong [5].

In order to reduce the aqueous component of harvested fat, a new system called REVOLVE™ (Allergan) was introduced on the market, and I have started to utilized in all major procedures of a fat harvesting and transfer performed in the past decade.

REVOLVE™ System (Fig. 24.4) is a sterile single-use disposable tissue canister used for harvesting, filtering, separating, concentrating, and transferring autologous tissue components for reintroduction to the same patient during a single surgical procedure for repair, reconstruction, or replacement of integumentary or musculoskeletal tissues.

For additional volume of tissue, canister may be filled and emptied a second time during the same surgery on the same patient.

The adipose tissue after sampling has little physiological conditions that enhance its fragility.

The increase in LDH and triglycerides (TGC) retard angiogenesis and engraftment fat.

The rapid return to plasma osmolarity (270–300 mOsm/L) through the wash shows the adipose tissue to the initial conditions with a physiological pH, osmolarity in the standard, removing hemoglobin, erythrocytes, and triglycerides.

This diagram shows the comparison among the contents of triglycerides (above, left), the release of lactate dehydrogenase (above, right), the hematocrit (below, left), and the osmolarity (below, right) in adipose tissue decanted compared to washed adipose tissue by the device REVOLVE™ (Fig. 24.5).

The grains contain fat for replanting adipose tissue and a variable amount of water.

1. A high percentage of water makes the graft semi-liquid and easy to inject, but entails a loss of important secondary volume.
2. A low percentage of water determines a graft too often to infiltrate and requires an excess pressure in the cannula and into the tissues (Fig. 24.6).

The percentage of water adipose graft should be between 10% and 20% to allow for easy installation and a minimal loss of volume.

1. A syringe of lipoaspirate, after 15/20 min still contains 20–30% of water content (longer than settling down = water).
2. A system of fluid filter or a centrifuge can be used to improve this result (Fig. 24.7).

REVOLVE™ System is a sterile single-use disposable tissue canister used for harvesting, filtering, separating, concentrating, and transferring autologous tissue components for reintroduction to the same patient during a single surgical

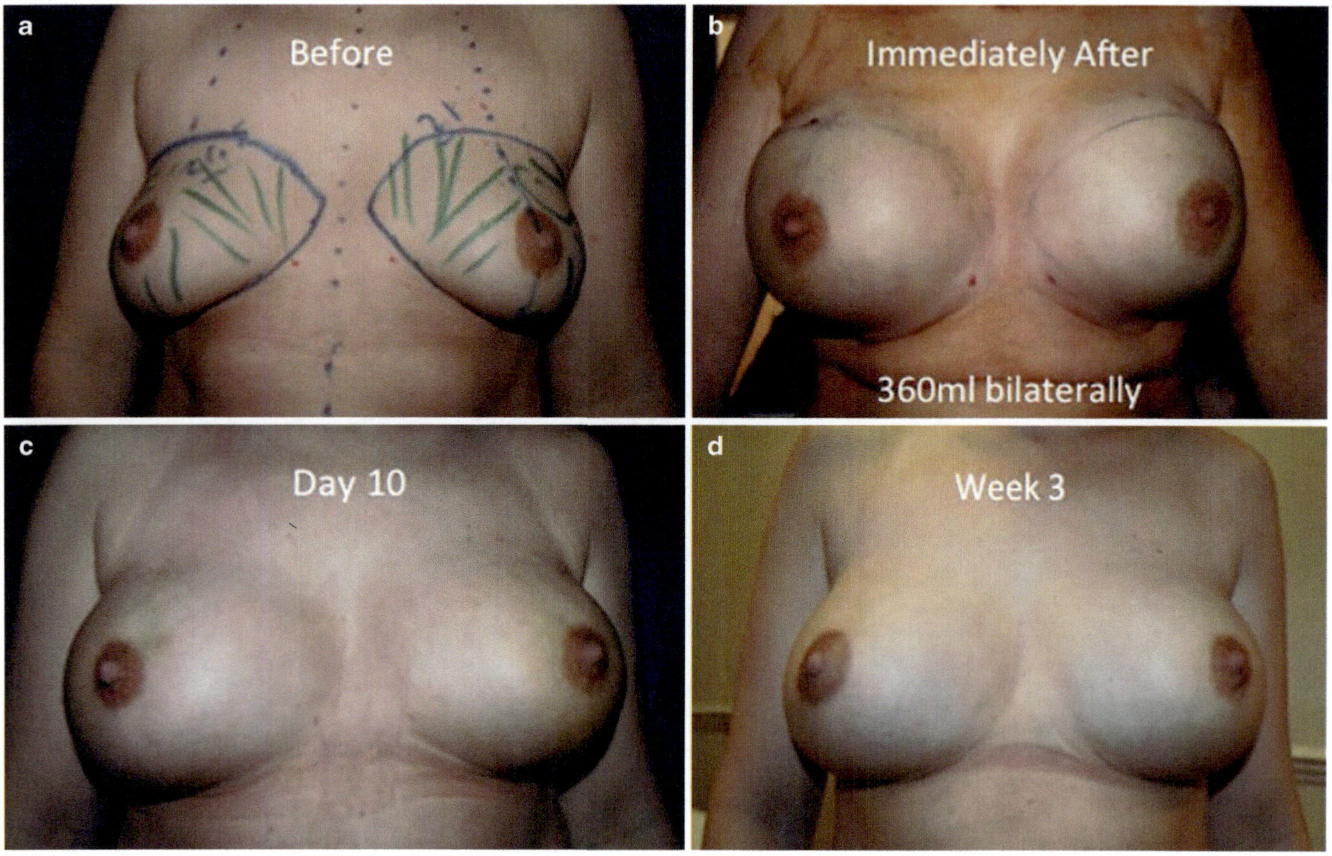

Fig. 24.3 (**a–d**) Breast augmentation with Puregraft™ System. Note massive swelling of breast 1 day after surgery (360 mL infused for side). Note at day 10, swelling is reduced and definitely absorbed at day 21

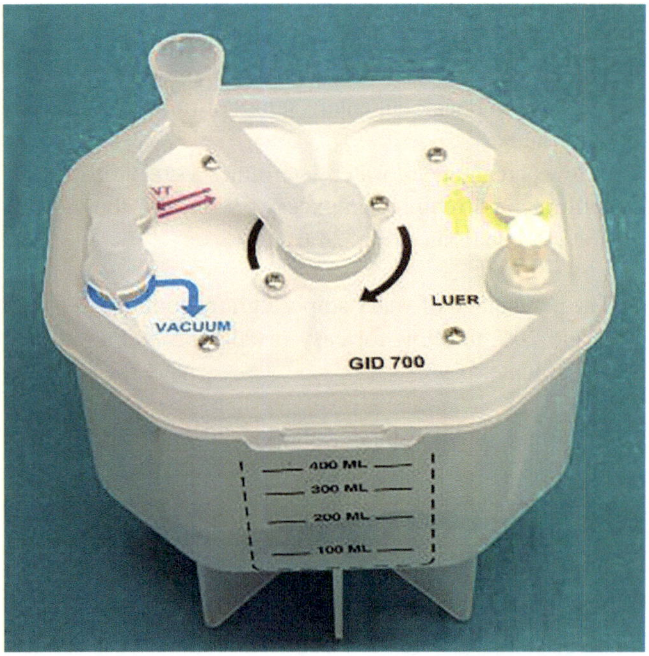

Fig. 24.4 The REVOLVE™ System. (Courtesy LifeCell Corporation, an AbbVie company. ©2021 AbbVie. All rights reserved [6])

procedure for repair, reconstruction, or replacement of integumentary or musculoskeletal tissues.

For additional volume of tissue, canister may be filled and emptied a second time during the same surgery on the same patient.

It is an integrated, high-volume fat processing system designed by adipose transplantation pioneers and founders of the International Federation for Adipose Therapeutics and Science.

REVOLVE™ System is an autologous fat tissue [AFT] collection device and tissue delivery system and collects autologous adipose tissue and rapidly washes and delivers the tissue for body contouring.

It is a sterile single-use disposable tissue canister used for harvesting, filtering, separating, concentrating, and transferring autologous tissue components for reintroduction to the same patient during a single surgical procedure for repair, reconstruction, or replacement of integumentary or musculoskeletal tissues.

For additional volume of tissue, canister may be filled and emptied a second time during the same surgery on the same patient.

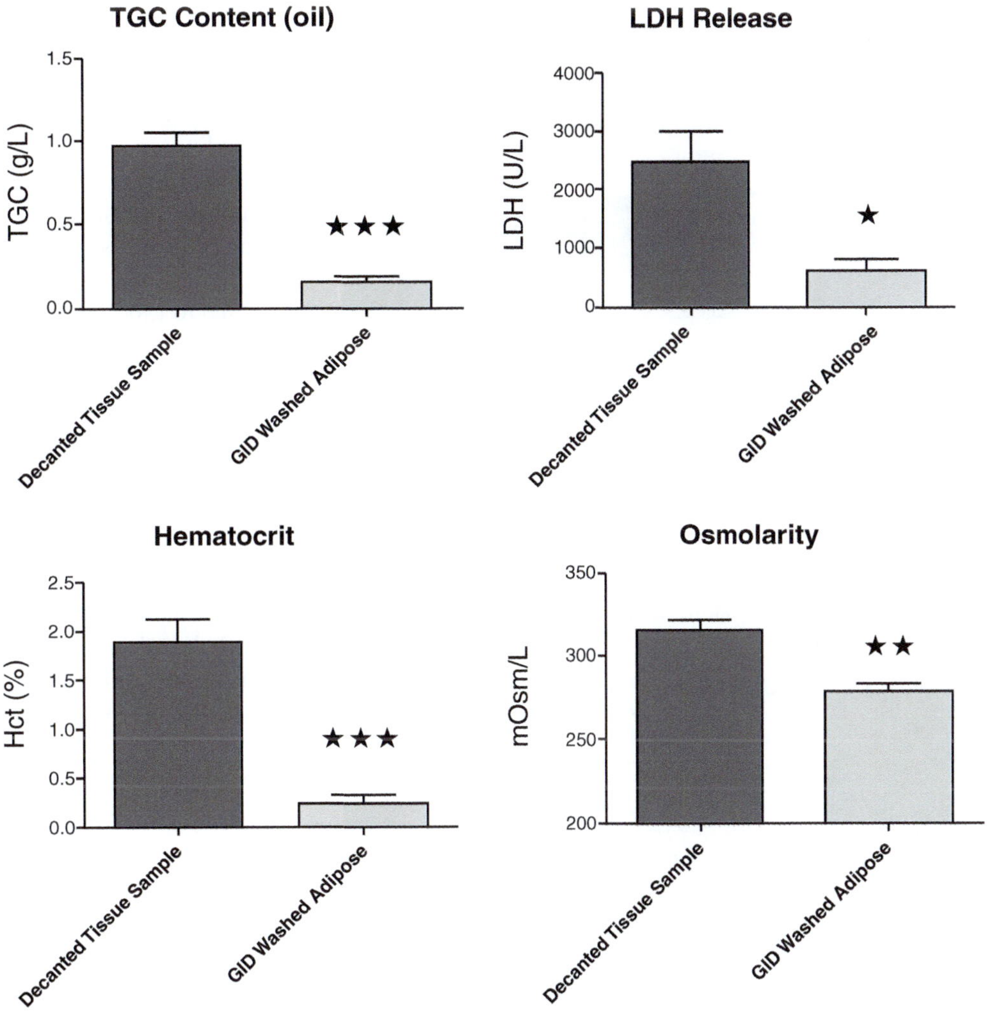

Fig. 24.5 Comparison of the contents [7]. (Courtesy LifeCell Corporation, an AbbVie company. ©2021 AbbVie. All rights reserved [6])

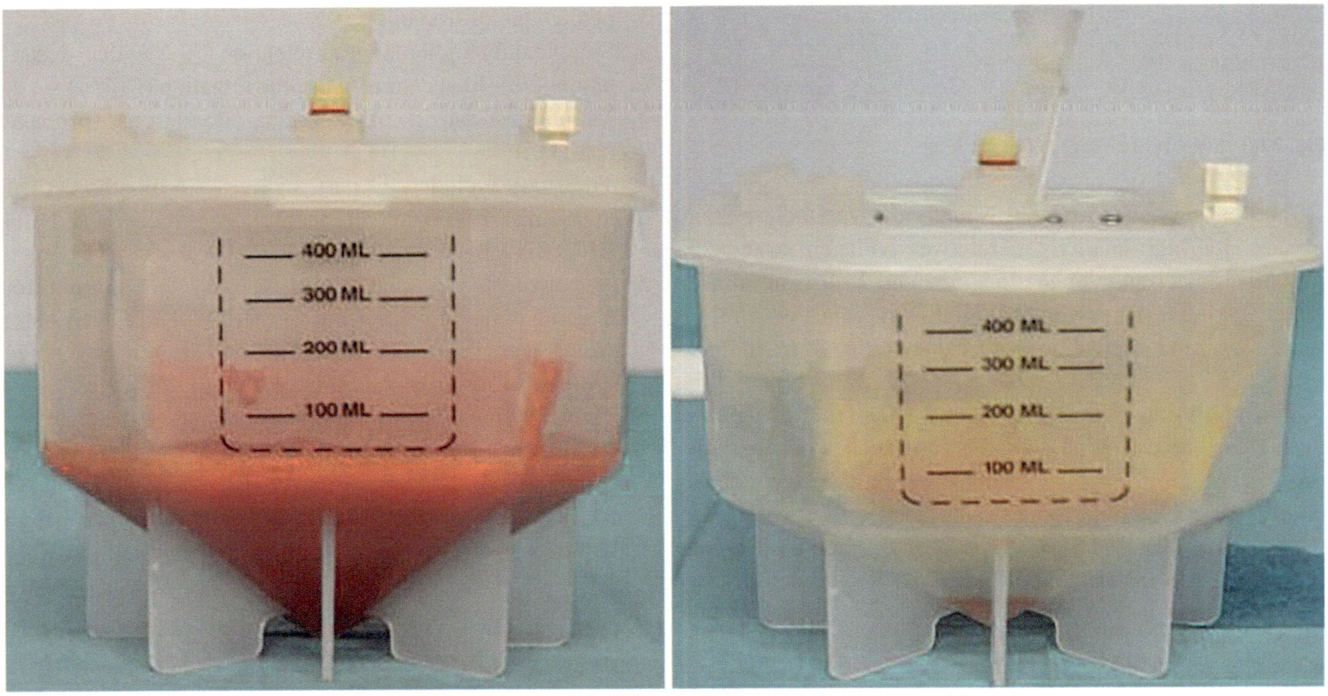

Fig. 24.6 Fat in the REVOLVE™ dispositive before and after washing

Fig. 24.7 Design of the fan. (Courtesy LifeCell Corporation, an AbbVie company. ©2021 AbbVie. All rights reserved [6])

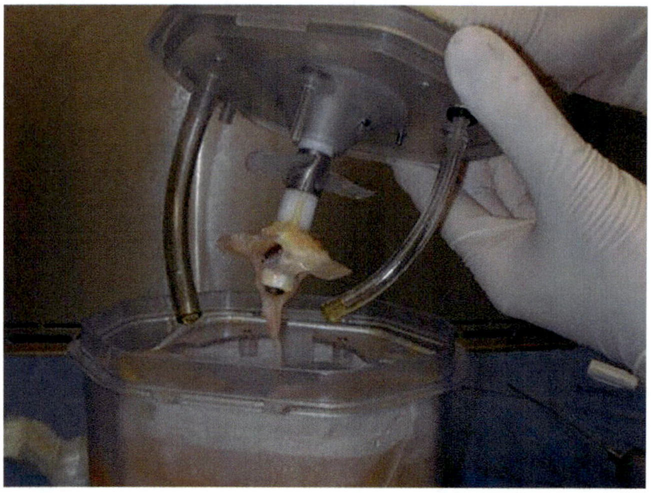

Fig. 24.8 Strands of collagen after washing

The propeller for active washing enhances the washing process for high-quality fat, removes collagen strands to minimize injection syringe clogging (Fig. 24.8), and allows direct load of syringes for injection from canister to minimize steps and extraction with catheter tip or Luer-Lock syringe.

The mesh filter is designed to filter and strain lipoaspirate during the harvesting phase and active mechanical washing process and rapidly removes tumescent, reduces cell debris, eliminates free oil, and concentrates adipose tissue.

The REVOLVE™ System can processes up to 700 mL of lipoaspirate in less than 10 min and it is easier to use, requiring fewer steps to process large quantities of adipose tissue and it is less time consuming than other competitive devices.

All-in-one integrated device: harvests, processes, and extracts lipoaspirate within one canister. Quickly processes fat, it filters during harvesting and actively washes and separates collagen strands. It is an integrated system that allows loading of syringes for injection directly from canister.

Processes a large volume of lipoaspirate with fewer steps compared to LipoVage® and Puregraft™ device. It is a time-efficient system that minimizes training, preparation, and clean up time, requiring only one operator, can be placed on the sterile field, and can be used multiple times for the same patient within the same procedure.

In a preclinical study, human fat injected into animals was processed using three different methods:

It is a decant method and centrifuge method (processed at 1200 g for 3 min).

Yielded a higher concentration of adipose tissue with significantly less debris and oil.

The system minimized potential inflammatory response for transplanted tissue by significantly removing red blood cells, fatty acids, and debris and obtaining yielded more predictable results.

Based on the preclinical study, it yielded significantly higher fat graft retention than decantation and similar to centrifugation.

REVOLVE™ System processed a higher percentage of adipose tissue compared with the two other methods: decantation and centrifugation [7, 8].

1. Yielded significantly less blood cell debris and a lower percentage of free oil.
2. Significantly higher fat graft retention (73.2%) than decantation (37.5%) and similar to centrifugation (67.7%).
3. Maintained fat in physiological conditions (pH and osmolarity).
4. The study included 37 patients over a 6 months period, including 13 Coleman technique patients, and 24 REVOLVE™ System patients.
5. The average rate of fat transfer was 1.77 cc/min for Coleman and 4.69 cc/min for REVOLVE™ System, which was a statistically significant difference between groups.
6. The study included a total of 118 patients in the centrifugation and 103 patients in the REVOLVE™ System group.
7. REVOLVE™ System processed more fat per minute than centrifugation, decreasing operating time and possibly OR costs.

24.2 Step by Step

Patient candidate for breast fat augmentation:

1. Adequate fat donor areas
2. Realistic expectations for volume enhancement (1–2 sizes)
3. Natural results expected
4. Highly motivated patients with full understanding of pro and contra of the procedures
5. Patients with pathologies of breast to be excluded (ultrasound before surgery required)
6. Patient with high oncologic risk to be excluded
7. Patient needs to maintain weight in order to retain the transferred fat
8. Gym and diet could consistently decrease the amount of fat transferred (as in all diet)

24.3 Breast Planning and Body Planning

Planning of breast includes traditional measurements as distance from mid-clavicle to nipple, from nipple to inframammary fold, and from nipple to midline. Breast is commonly divided into four quadrants (Fig. 24.9).

For body contouring, I mark the outline of the areas to be considered for fat removal. Markings are in front, posterior, and both lateral projection (Fig. 24.10). I mark the outer contour of the fat excess, the lines of transition between areas. I mark in the red line the areas not to be touch. I try to draw lines of concavity and convexity in order to recreate a good shape and proportion with the body. With experience, a plastic surgeon can predict the amount of fat he will be able to collect in a patient, with a good approximation. I normally guess that I will succeed increasing 1 cup size breast by using 200 mL of harvested fat (400 mL for each side to

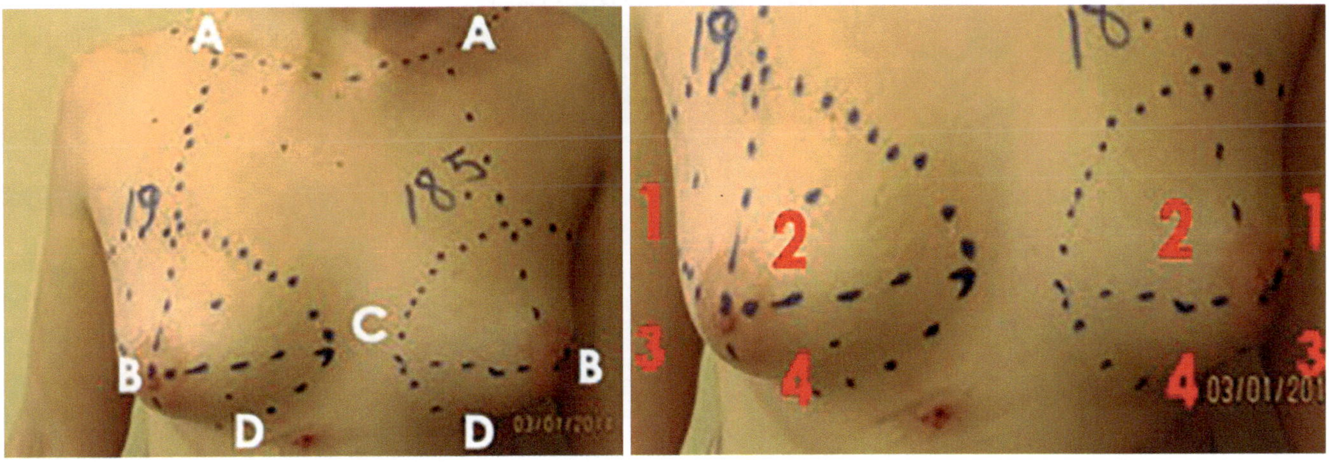

Fig. 24.9 Breast planning and measurements divided in four quadrants

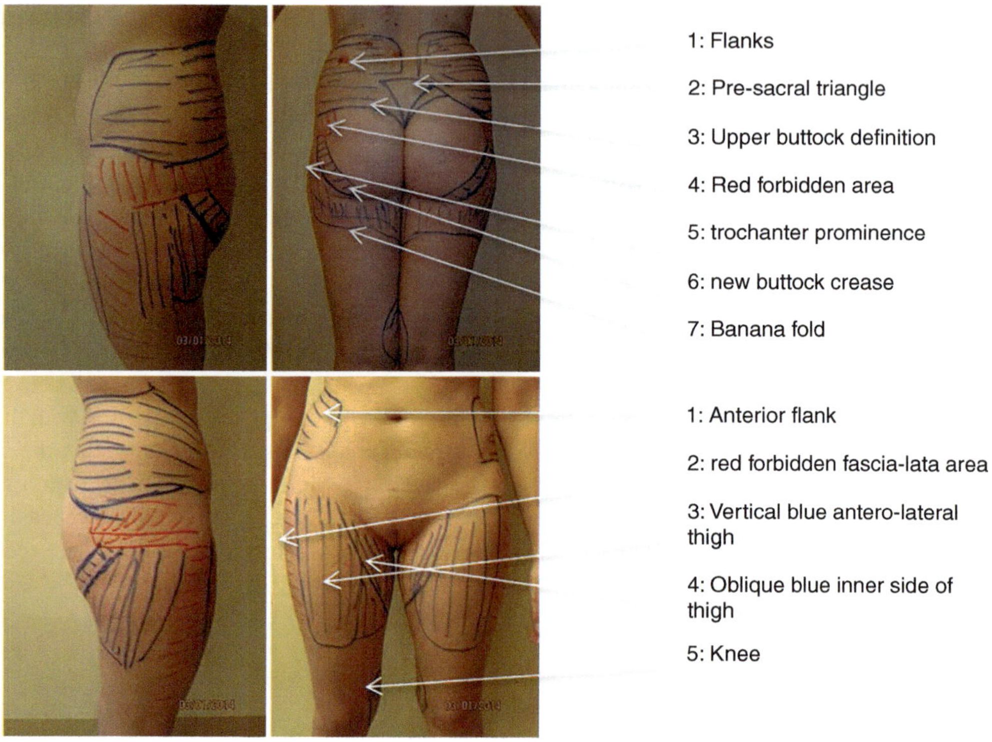

Fig. 24.10 Planning and marking body areas: in blue areas of fat removal, in red forbidden areas. Vertical blue lines: top line of prominence

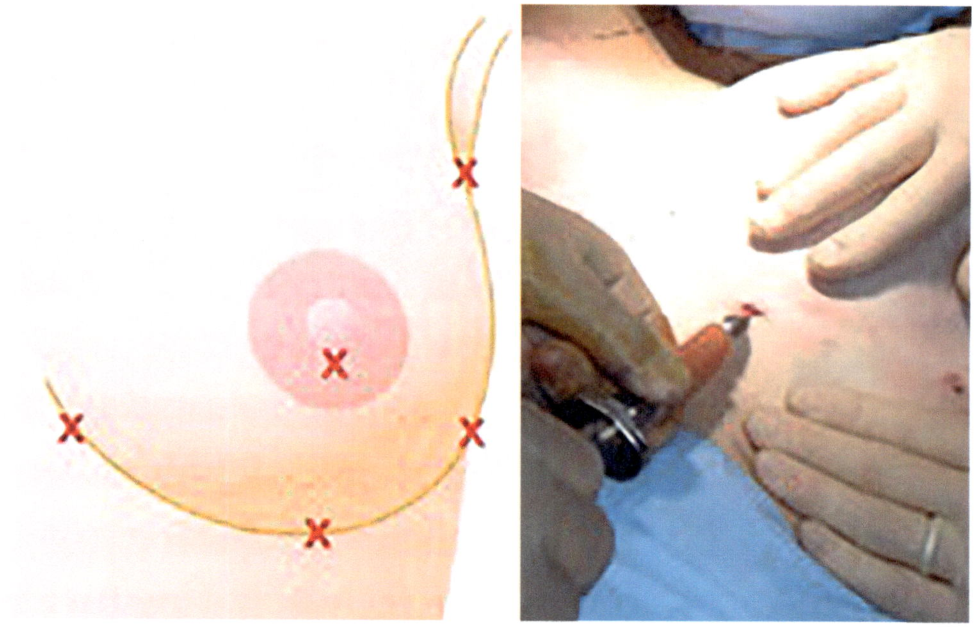

Fig. 24.11 Favorite incision lines of breast (submammary mostly preferred)

obtain a 2 cup size increase). Minding that the infiltration rate is normally twice than aspiration, I plan 2500 mL of aspirate (with 5000 mL infiltration to obtain around 1600–1800 of pure fat to be infiltrated), counting loss of debris of solution, blood, triglyceride, and oil (Fig. 24.11).

24.4 Advantages of Fat Grafting

No implant leakage, or deflation
No capsule contracture
No palpability

No visibility
Natural appearance
Natural feeling

24.5 Breast Cancer Detection

Calcifications and fat necrosis occur with every type of breast surgery: breast biopsy, implant procedure, radiation therapy, breast reconstruction, breast reduction, and liposuction of the breast.

Incidence rate of calcification can be as high as 50% of patients after 2 years.

Radiologists knows how to distinguish calcification of malignant causes from the benign calcification from fat necrosis.

Bircol [9] reported 1.4% of incidence rate of microcalcification.

Fat placed even in small aliquots can necrose and develop small fat cysts and calcifications.

Post-op follow-up with ultrasound scan or mammograms are recommended at 1 year and regular interval.

3D imaging helps long-term evaluation of fat survival.

Breast cancer detection remains the safety issue, not the incidence of calcifications.

I routinely ask to perform an ultrasound scan to any patient before undergoing fat transfer to breast. This simple investigation will allow to detect minor lumps or any suspicious pathology of breast. Mammograms of RMI are requested only in more complicated patients, as initial investigation. I frequently ask a second oncological opinion in challenging cases, and I refuse patient with strong hereditary history of family breast cancer.

24.6 Technique

1. Selection of patient candidate for fat transfer to breast
   Fair amount of fat to be transferred to breast.
   Realistic expectation of patient (1–2 breast size increase, no spectacular projection achievable).
   Good initial position of nipple (18–22 cm from midclavicle).
   Patient must give privilege to natural results achievable more than to the volume and projection.
   For this purpose, I do show patients a prospective achievable result using 3D imaging system (CRISALIX) (Fig. 24.12).
2. Selection of sites of fat harvesting
   Abdomen, flanks, thighs, knees, upper arm, back, trunk.
3. Selection of technique of fat removal
   Liposuction, VASER ultrasound assisted, laser lipolysis.
4. Selection of anesthesia
   Modified local tumescent plus IV sedation.
   Intercostal nerve blocks (high epidural).
   General anesthesia (Fig. 24.13).

In my experience, in close breast fat enhancement, I infiltrated fat into different levels by layering in order to better shape and define breast cone. The principle is to graft selectively from the chest wall to the skin, placing fat under the glandular tissue and subcutaneously. Skin incision is 1.5 cm large, and I normally start with midline inframammary incision.

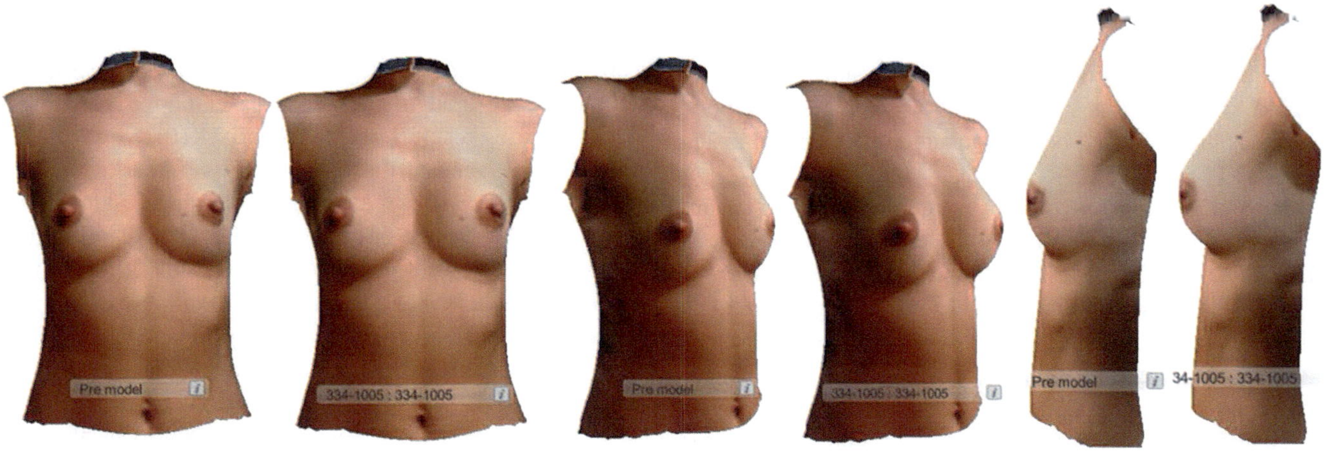

Fig. 24.12 3D imaging simulation

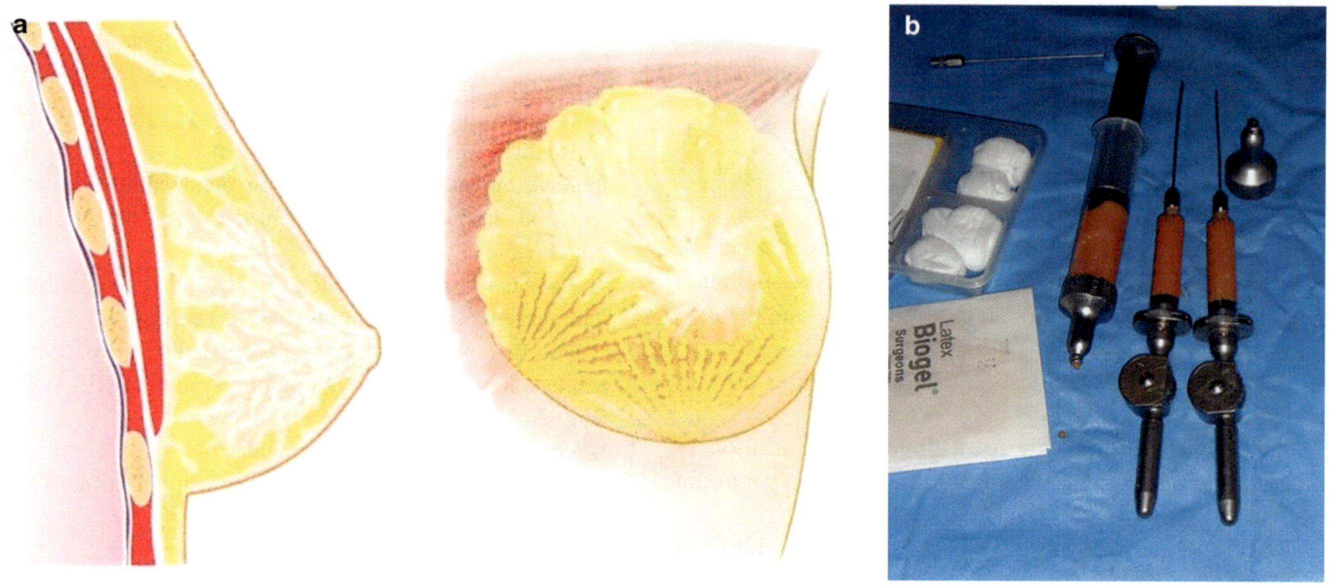

Fig. 24.13 (**a**) Level of fat infiltration. Subglandular and subcutaneous. (**b**) Special cytori 10 mL syringes

24.7 Operative Technique

Once the fat is emulsified with VASER liposuction, it is aspirated and collected in the REVOLVE canister. It could collect up to 350 mL of emulsified fat, which is then added with 250 mL of ringer lactate solution, prewarmed at a temperature of about 38 °C to optimize washing (Fig. 24.14).

The washing procedure is repeated three times alternating with the spinning to help separation and decanting of the emulsified washing fat. It is possible to collect up to 1000 mL of emulsified fat throughout three times aspirating and washing the emulsified fat. After processing the fat (washing, spinning, and decanting), it is possible to obtain a good quality of "grain fat" (Fig. 24.15).

Once fat is washed and spinned, the nurse or assistant will start the aspiration from REVOLVE canister by using 60 mL Luer-Lock syringes until the canister is empty. Syringes are collected on the rack and left them to decant (Fig. 24.16a, b).

Fat is then transferred from 60 mL syringes to 10 mL syringes to continue cleaning of fat, with minimal pressure, through a connector (Figs. 24.17 and 24.18).

Once the nurse has completed the processing of fat in the different areas (flanks, trunk, abdomen, thighs, etc.), the surgeon must concentrate on the sculpturing of the body areas which have gone under treatment.

Achieving a good symmetry and a nice contour is a must of the surgeon: the procedure is an association of removing fat from stubborn areas and sculpturing and recreating concavity and convexity lines of the natural ideal body shape. Superficial VASER liposuction helps contouring body and recreating nice lines, but talent of surgeon and ability in sculpturing the "body material" are essential for achieving a nice contouring result (Figs. 24.19 and 24.20).

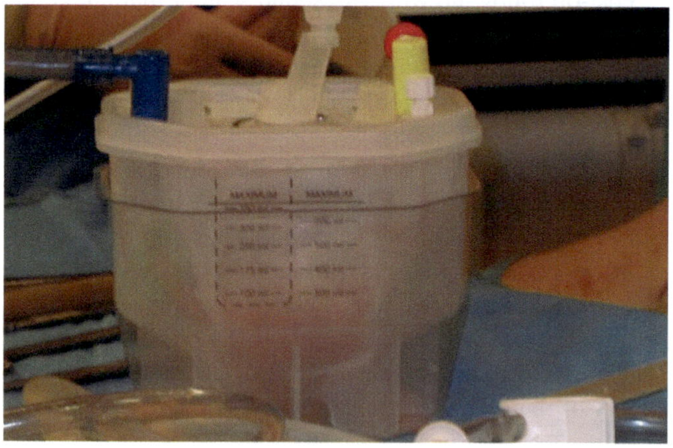

Fig. 24.14 REVOLVE canister filled with fat and then washed

Breast fat transfer utilizes 10 mL syringes with 2.7–3.0 mm cannula: this is the optimal compromise to lower the pressure of the syringe through the fat and thus diminishing damages to harvested fat. Also the utilization of a one hole 3.0 mm cannula is ideal, minding the "grain" structure and dimension of the harvested fat without the aqueous component. Less aqueous is the grain fat to transfer, larger has to be the cannula to be used. With the previous Acquavage system, fat transferred was more aqueous and easy to be injected with smaller cannulas, but the swelling was major and fat absorption more substantial (Fig. 24.21).

The filling process starts with a submammary 2–3 mm skin incision and progresses from the upper quadrants in

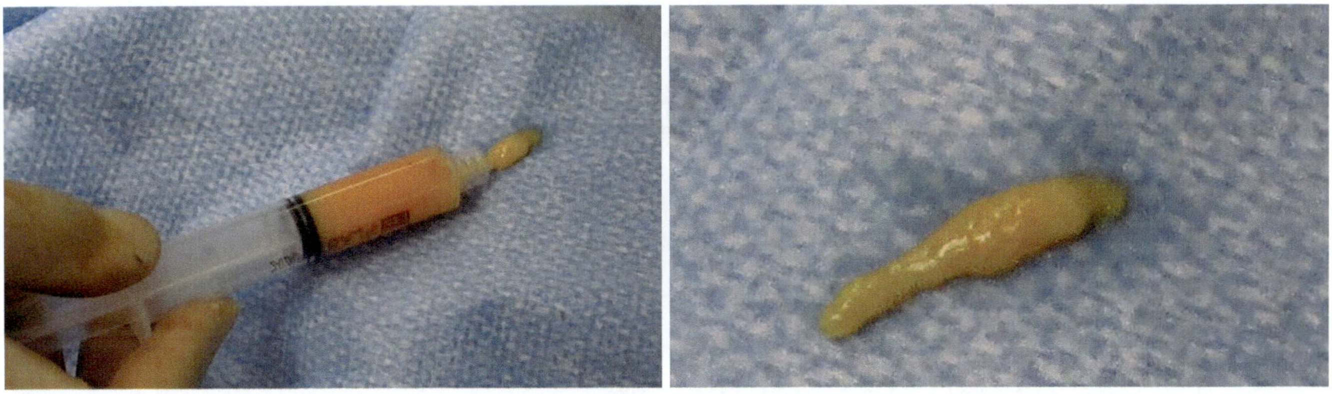

Fig. 24.15 Grain fat after processing with the REVOLVE canister. Note the consistency of fat and the little aqueous component

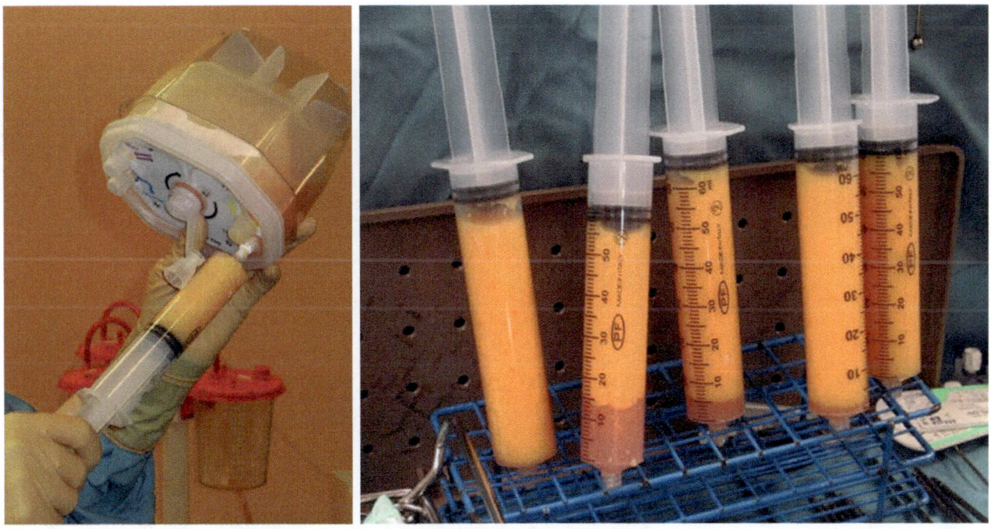

Fig. 24.16 (**a**) 60 mL Luer-Lock syringe to aspirate from REVOLVE canister. (**b**) Syringes collected on the rack and left for decanting

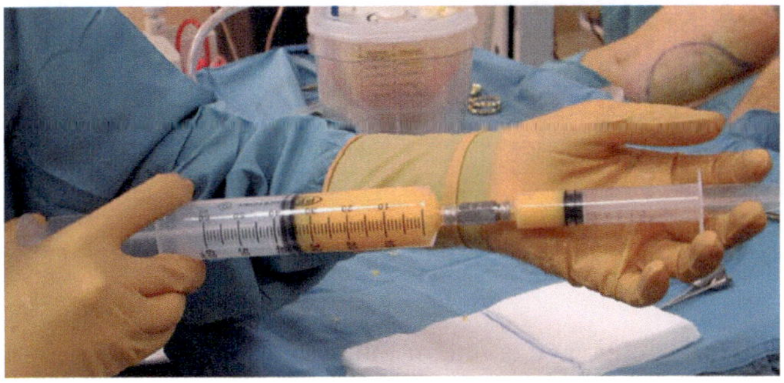

Fig. 24.17 Fat transfer from a 60 mL syringe to a 10 mL syringe

order to fill emptiness of the top breasts. With two experienced surgeons, the filling must be bilateral at the same time in order to speed up the process. A good coordination and a teamwork is essential to progress symmetrically and achieve a good aesthetic contouring. Nurse must record the total of fat available and give similar amount of fat to be filled side to side and quadrant to quadrant (upper, lower, internal, and external) (Fig. 24.22).

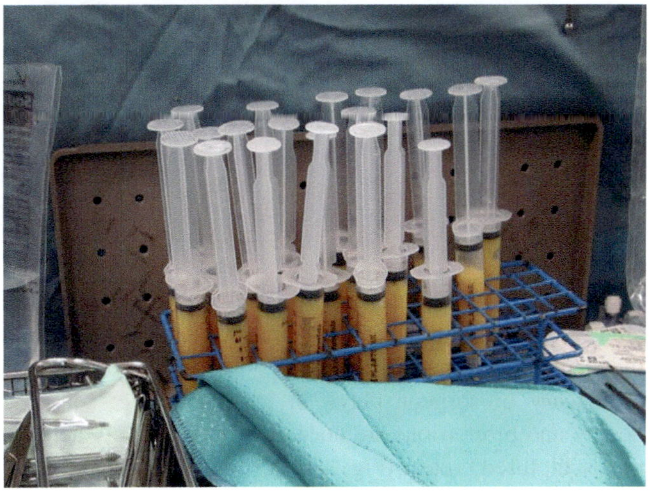

Fig. 24.18 60 mL syringes of fat ready for harvesting

In order to achieve an increase of nipple and breast projection, fat filling must be concentrated on the lower and medial quadrants.

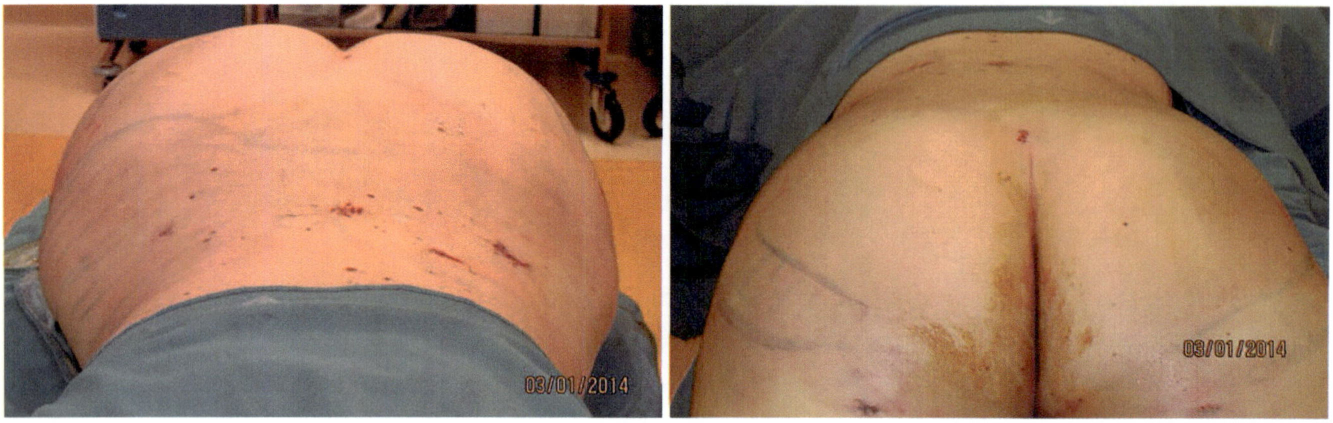

Fig. 24.19 Immediate post-op result after VASER contouring of flanks, trunk, outer thighs, adding also fat to better contouring and project flat buttocks

Fig. 24.20 Pre- and post-op comparison

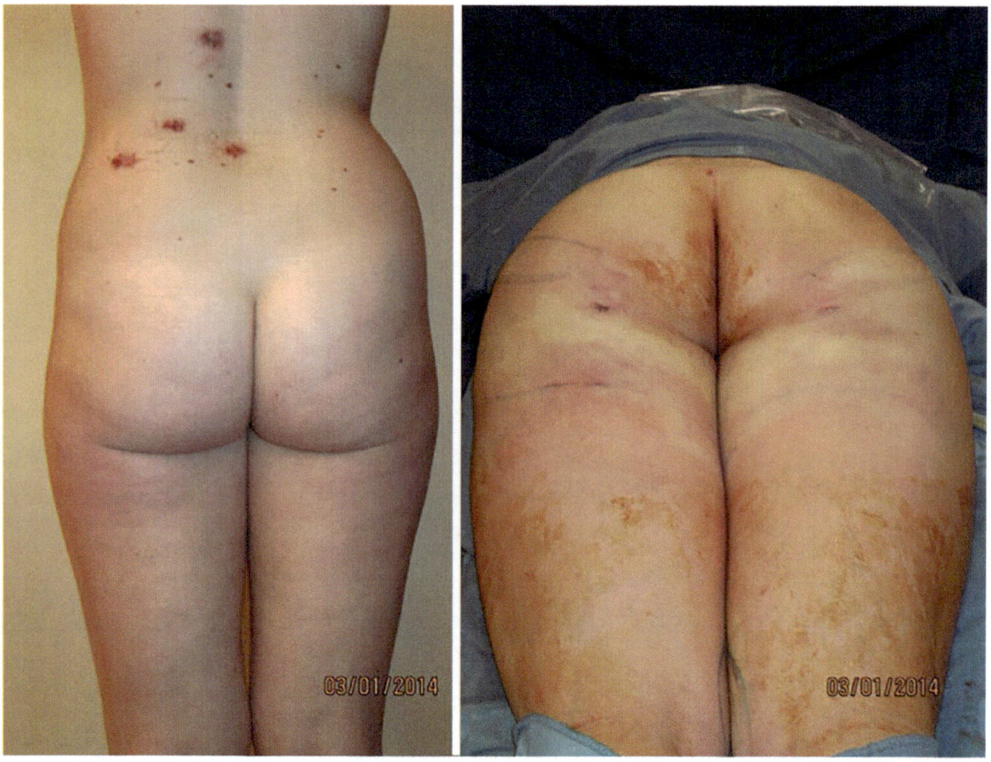

Upper pole fat transfer is performed after lower poles are already enhanced; this depends on the total fat available and the ideal volume distribution.

Artistry of surgeon plays an important role: sense of beauty and proportions are needed to progressively achieve nice sculptured breasts (Fig. 24.23).

Breast cleavage is one of the best targets and desires of women worldwide in breast aesthetic surgery. In order to improve cleavage, fat filling of the medial quadrant is essential. Surgeon must continuously check with his contralateral hand the proper positioning of the fat (Fig. 24.24).

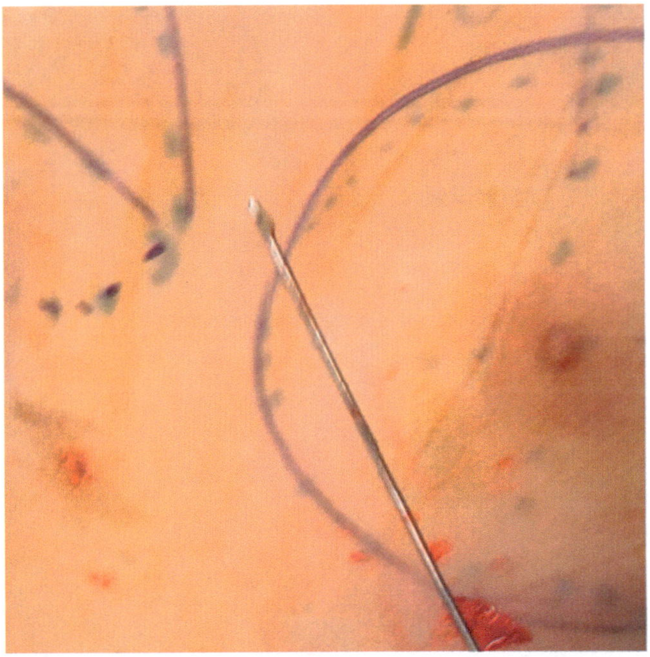

Fig. 24.21 3 mm one hole cannula

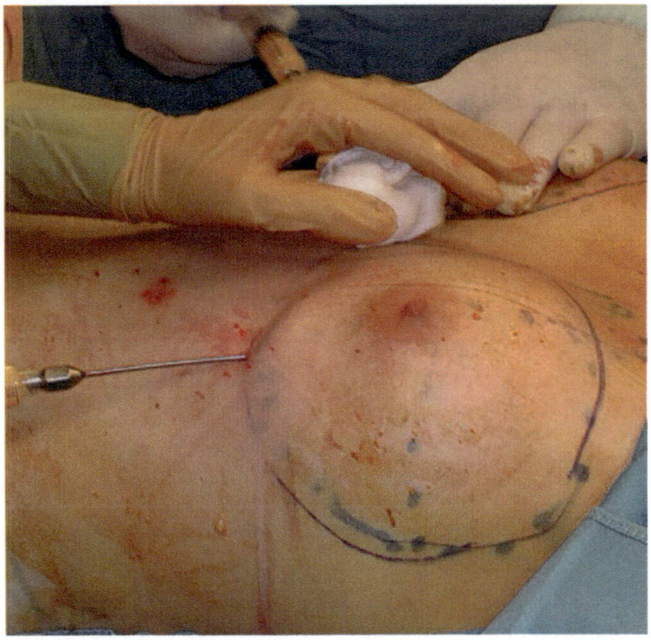

Fig. 24.23 Filling lower medial quadrants to increase breast projection

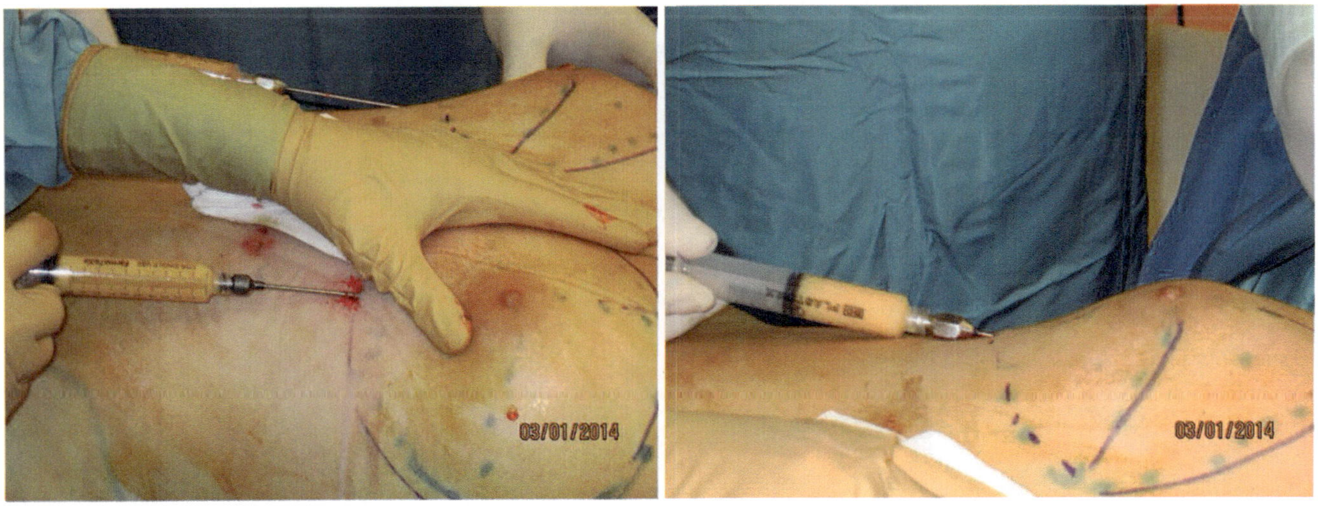

Fig. 24.22 Simultaneous bilateral breast fat augmentation, filling upper and lower quadrants as by original drawings

Filling the retro areolar space, always in a subglandular layer, is the next step to increase breast shaping in my own sequence. I normally use last syringes of fat to add those areas, after a good shape has been already achieved (Fig. 24.25).

Breast fat augmentation has a similar target than augmentation with implant: recreating a nice breast cone, with improved volume and projection and natural shape. In order to achieve those goals, surgeon must have precise ideas of body proportions and the desired shape. His talent and sense of proportion and beauty is essential in order to achieve a good aesthetic result (Figs. 24.26, 24.27, and 24.28).

A second incision at the sternum is sometime necessary to better fill the medial quadrant and improve breast cleavage.

A final assessment must include: symmetry of breast shape and volume, contouring of body flanks in harmony with increase breast shape. Post-op assessment is done through proper visualization from lower, lateral, and top breast position.

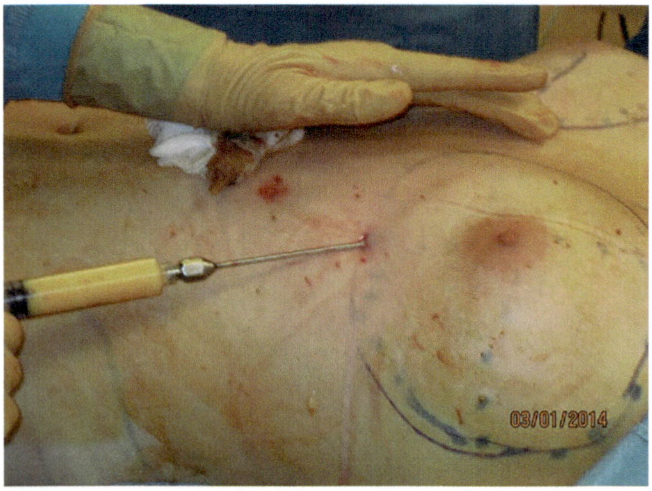

Fig. 24.24 Breast cleavage: filling medial breast quadrant

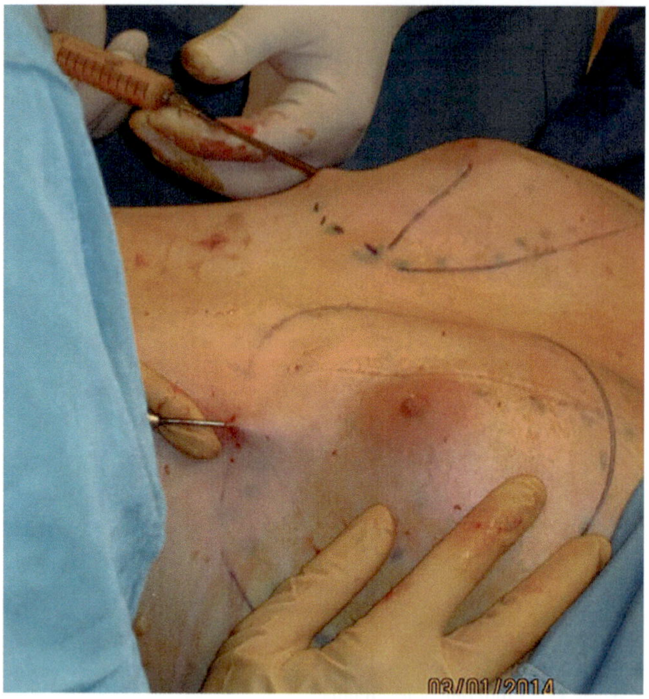

Fig. 24.25 Last step: retro areolar space filling

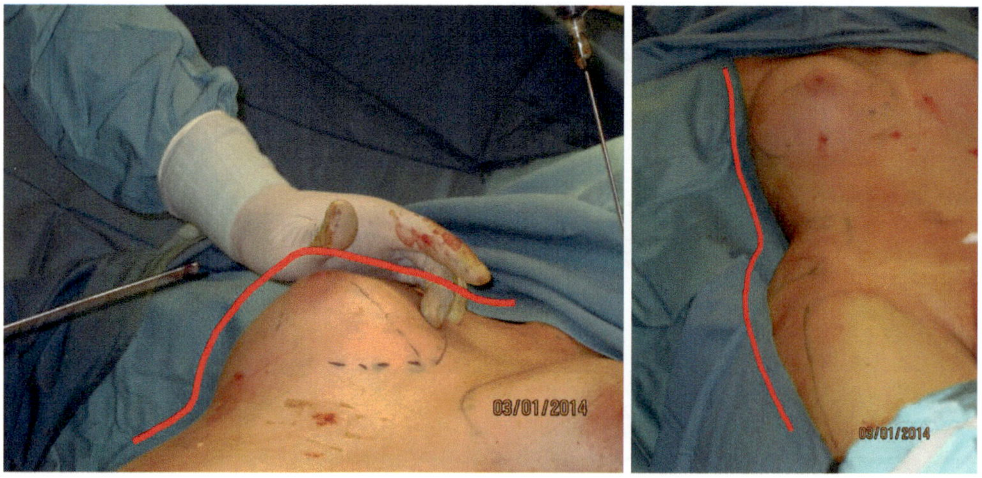

Fig. 24.26 Achieving a nice breast cone and body concavity and convexity lines

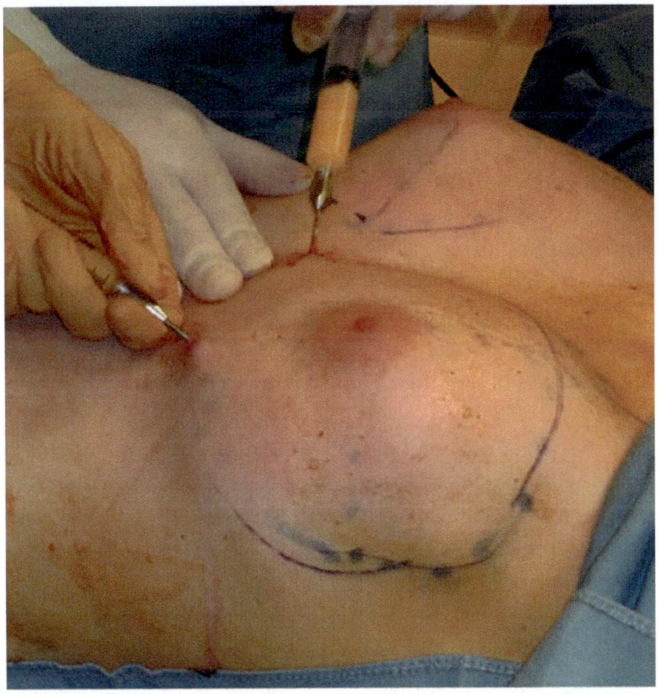

Fig. 24.27 Second incision in the sternum to complete medial quadrant filling

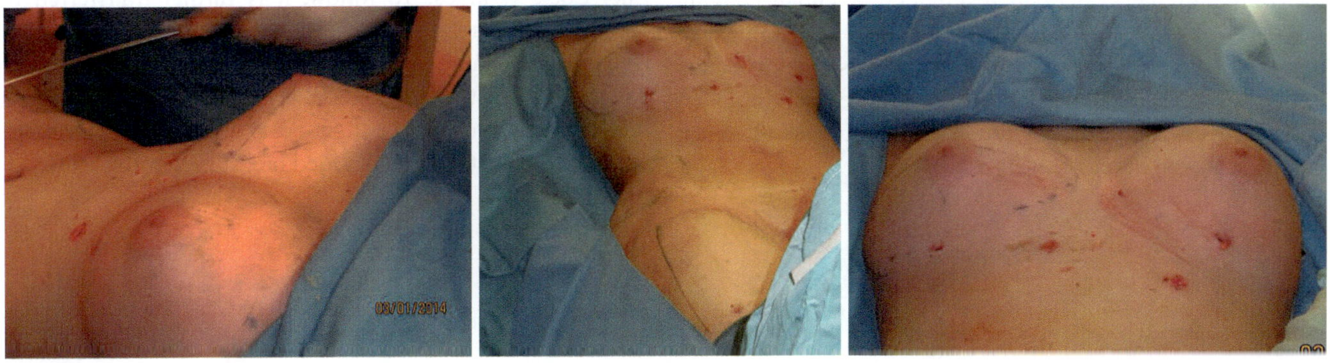

Fig. 24.28 Final assessment

24.8 Post-op Care

Dressing of breast includes a light cotton bra, which should not be tight as fat doesn't like compression.

The area of liposuction must be compressed with a tight garment for 6–8 weeks post op. I strongly recommend manual lymphatic massage twice a week for 6–8 weeks in order to reduce post-op swelling and improve the healing process (Fig. 24.29).

Post-op recommendations include: avoid sport intense activity for the first 3 months post op, in order to avoid to burn out transferred fat at the early stage. Moreover, avoid diet that will cause partial or total absorption of the transferred fat (Fig. 24.30). Images show pre- and post-op comparison of the patient presented in this step-by-step fat harvesting using REVOLVE™ System and breast enhancement.

Patient had at the same time fat transfer to reshape and enhance her buttocks. Photos 3 months post op are shown.

In the following photos, comparison of the 3D Crisalix planning and the final result is presented (Fig. 24.31).

Reality is always better than a computer planning simulation. It is difficult for a computer image system to give an idea of the soft natural looking result of breast enhanced with fat transfer. In my experience, I now use the 3D Crisalix system only for breast augmentation, where is shown to be an effective predictable device, but I drop it use in fat transfer patients, as planning is not reliable according to the final results achieved in reality.

Final body contouring shaping is the second must and goal to be achieved: circumferential thigs, abdomen, flanks, and posterior buttocks have been reduced and recontoured. Even superficial buttock cellulitis has been corrected by nanograft (Fig. 24.32).

Recently, I have introduced in my practice BeautiFill Alma Laser lipolysis. It is a laser-assisted liposuction that allows simultaneous lasing and suction, increases skin tightening, and reduces blood loss.

It utilizes a 1470 nm wavelength, with radial initial fibers, provide reduced treatment time, reduced thermal energy, and allows collection of high vitality fat with immediate fat grafting (Fig. 24.33).

Lipolife device includes (1) an infiltration of peristaltic pump with speed selection, (2) aspiration system with a rate selection, (3) a power setting in Watts (0.15 W), (4) total energy delivered countered, and (5) liposense temperature control.

I've started utilizing BeautiFill device for liposuction and fat transfer only in the last year.

In my experience, main advantage of the tool is the capability of a fast emulsification and collection of ready for transfer fat. Emulsified fat is aspirated in a canister with a microfilter: after decanting, fat is collected with 50–60 cc Luer-Lock syringes. To follow, it is further transferred to 10 mL syringes ready for fat transfer into breast (Fig. 24.34).

This is an example of application of BeautiFill fat transfer device in breast augmentation. The following photos are related to a 33 years old patient post partum breast asymmetry, empty breast with right-sided breast bigger than left.

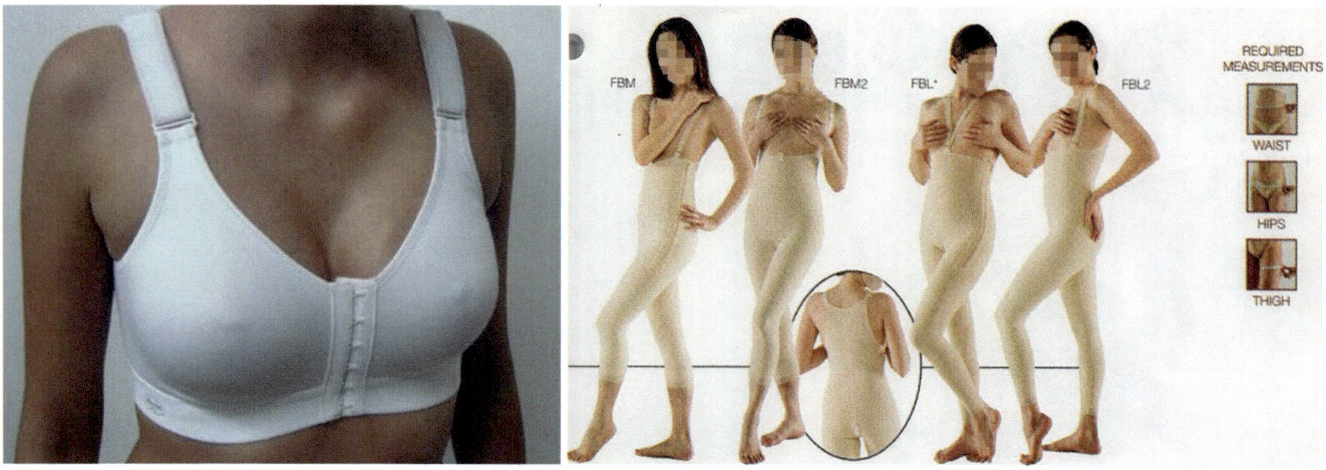

Fig. 24.29 Post-op light cotton bra and post-op compression garment. (Courtesy of Marena ComfortWear, Marena Group LLC, The Riverside Company)

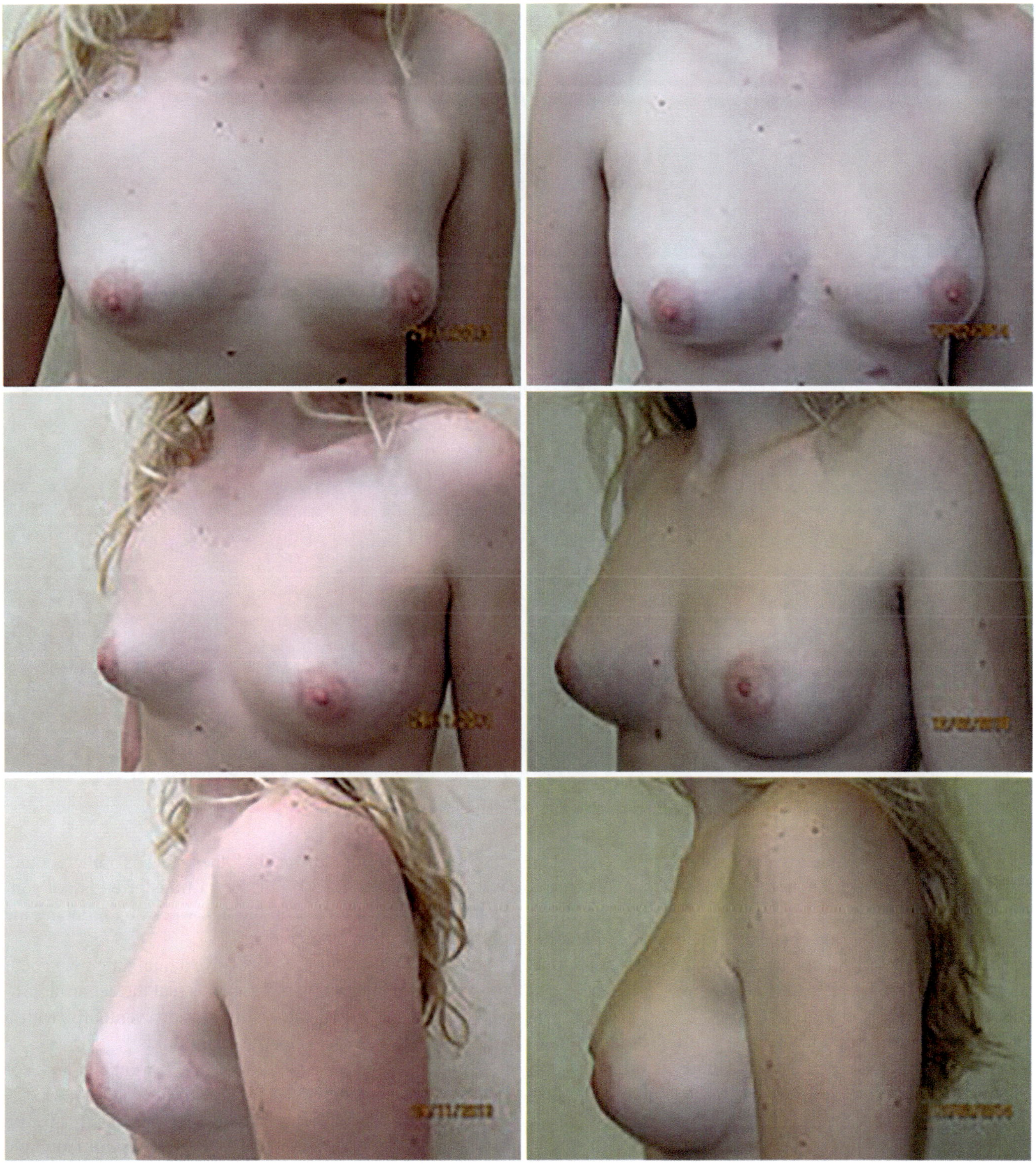

Fig. 24.30 24-year-old girl presenting with breast asymmetry of shape and sides, poor cleavage, cup A-B size. Breast fat augmentation, 360 mL for side added. Full C cup achieved, natural looking breast, undetectable surgical result, 3 months post surgery

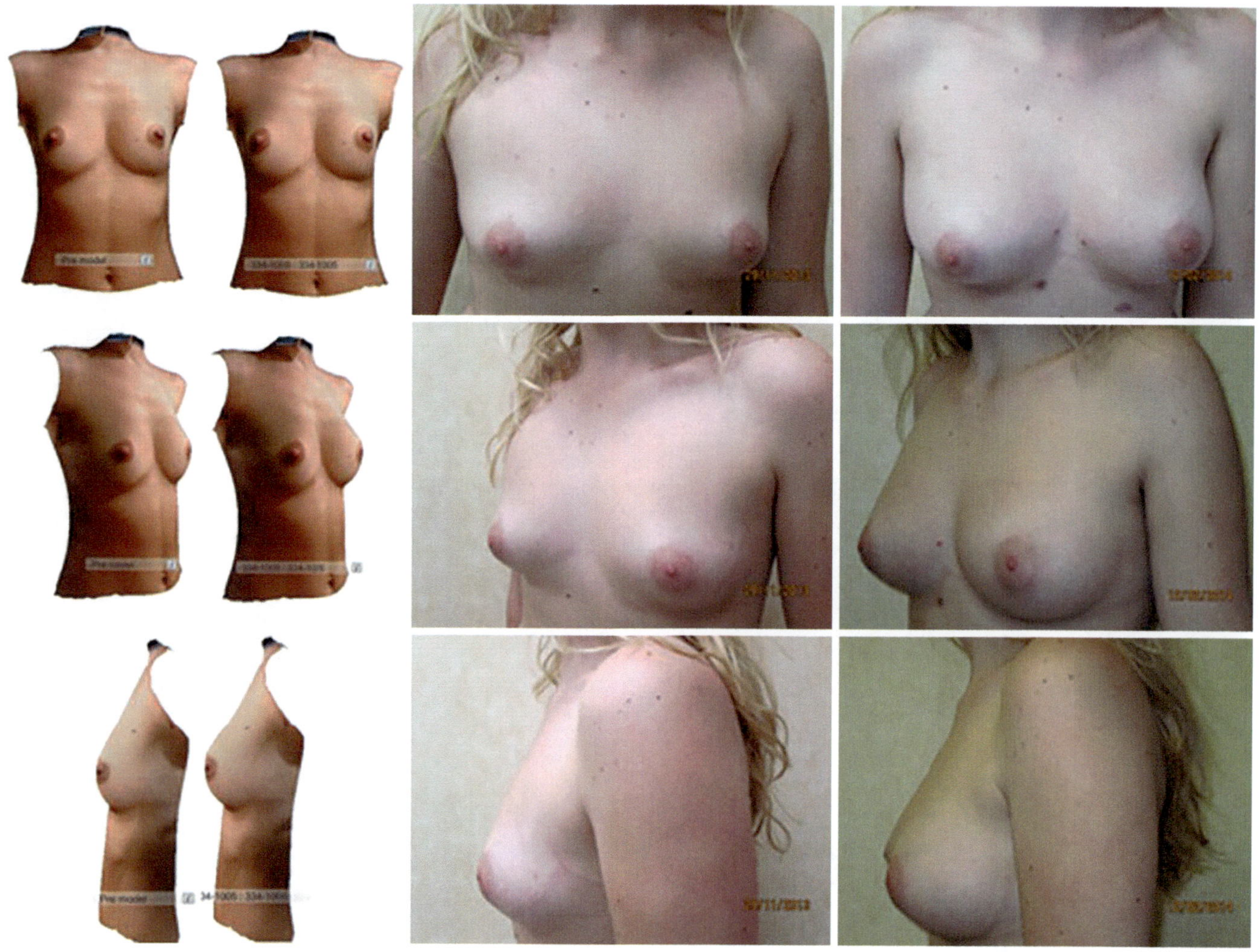

Fig. 24.31 Crisalix 3D imaging pre op planning and pre- and post-op 3 months reality pictures

Planning to remove fat from abdomen, flanks, and inner thighs, to be transferred into breast for enhancement (Fig. 24.35).

After infiltration of the donor areas with tumescent solution, fat is emulsified through laser fiber and at the same time suctioned into the special canister, where is collected, concentrated, and decanted (Fig. 24.36).

Canister contains up to 1000 mL of emulsified fat: after proper decanting, fat is transferred into 50 mL syringes. Ready for fat enhancement (Fig. 24.37). A total of 650 mL is harvested in this case.

This procedure allows a fast fat emulsification and simultaneous fat harvesting, which rather reduces the total time of surgery. Quality of fat, on the other side, is rather liquid. Which means around 30% of the fat transferred is not concentrated and will be lost in the first weeks by reducing the liquid and swelling component.

I need long follow-up terms to give a final judgment about this new technology. My impression is that the cannister provided by Alma Laser can be improved, eventually with a spinning system to help clearing extra fluids.

But the survival rate reported by Levenberg and coauthors [10] is around 95% (see Fig. 24.38).

This is the comparison of pre op and immediately post-op appearance after liposuction and fat harvesting from multiple areas and 350–300 mL of fat to breasts (Figs. 24.39 and 24.40).

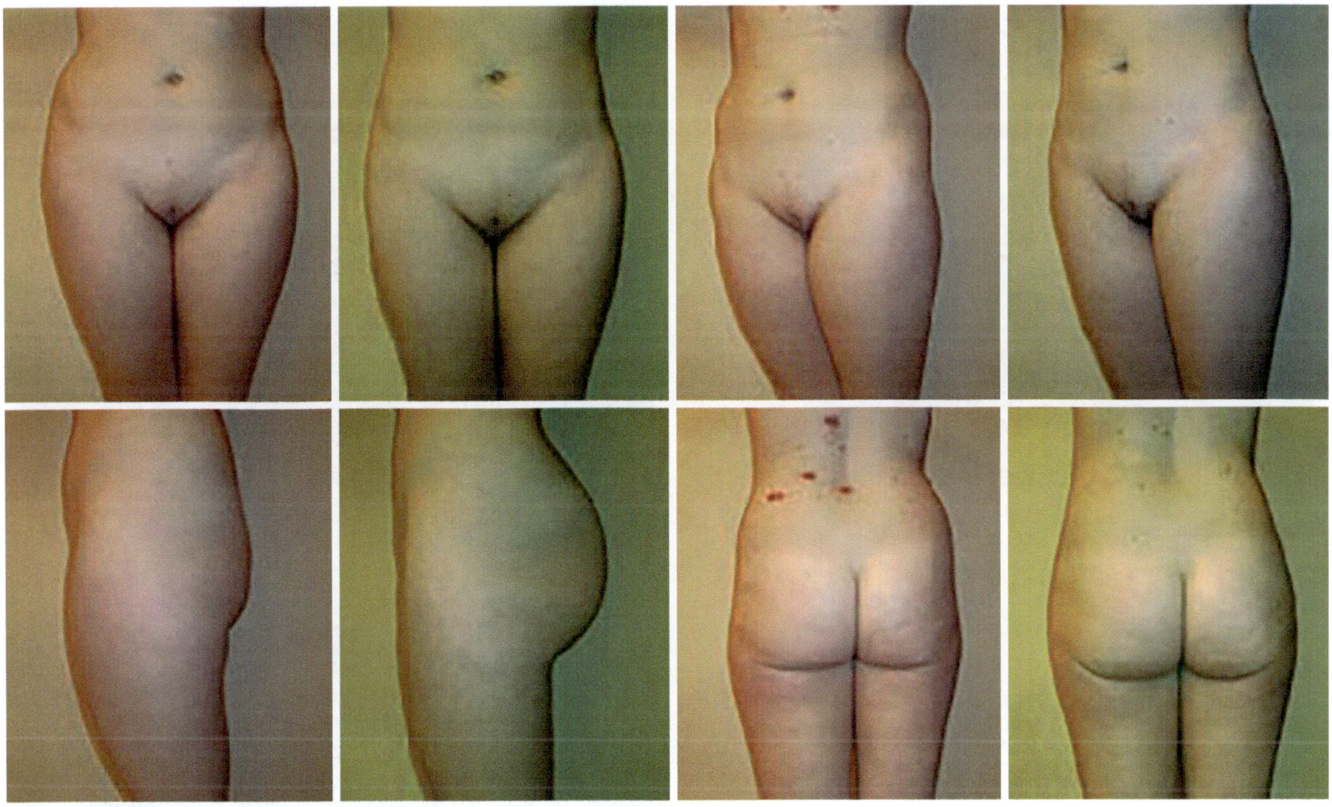

Fig. 24.32 24-year-old girl pre- and post-op body contouring and fat transfer to breast and buttocks. 3850 mL of fat aspirated, 1400 mL of fat harvested with REVOLVE. 720 mL of fat for breast enhancement and 700 mL for buttock enhancement

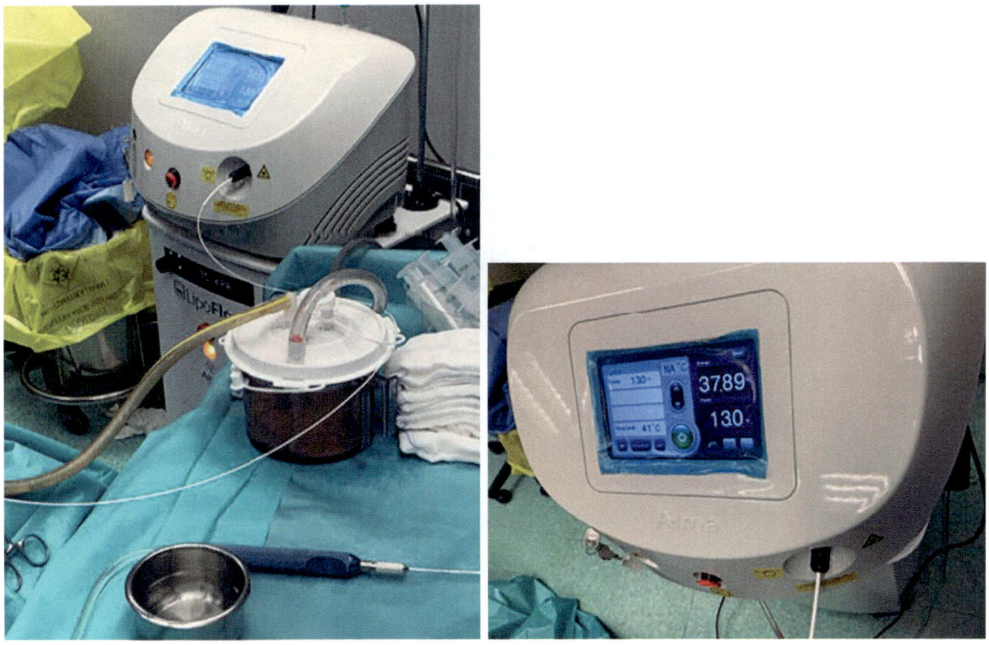

Fig. 24.33 Lipolife device

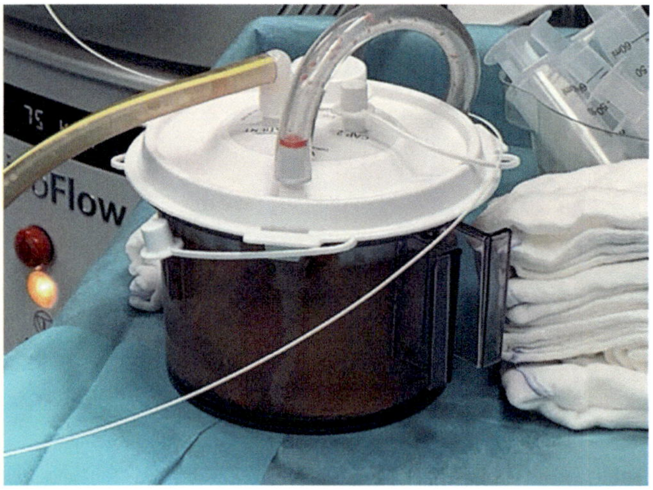

Fig. 24.34 Canister

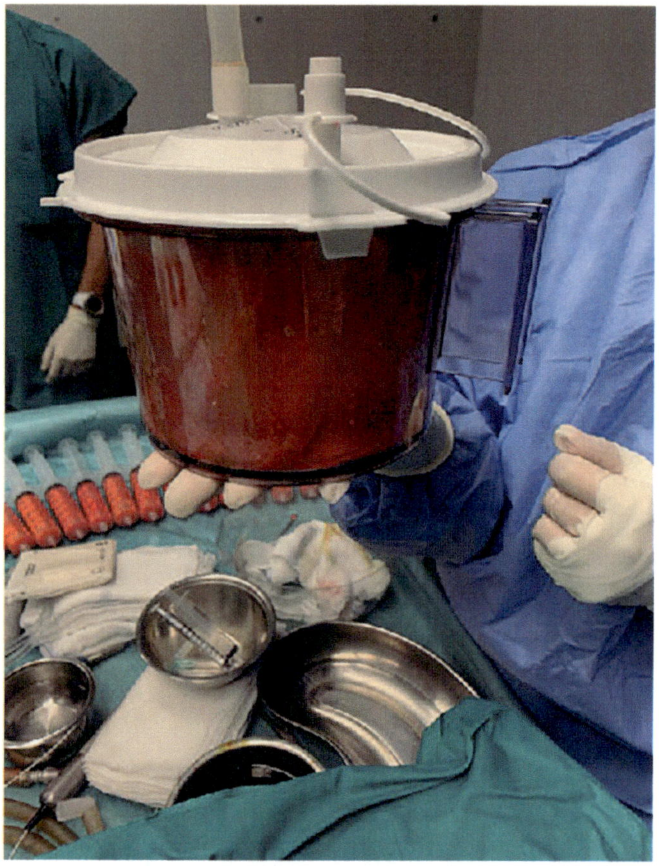

Fig. 24.36 Canister containing up to 1000 mL of emulsified fat

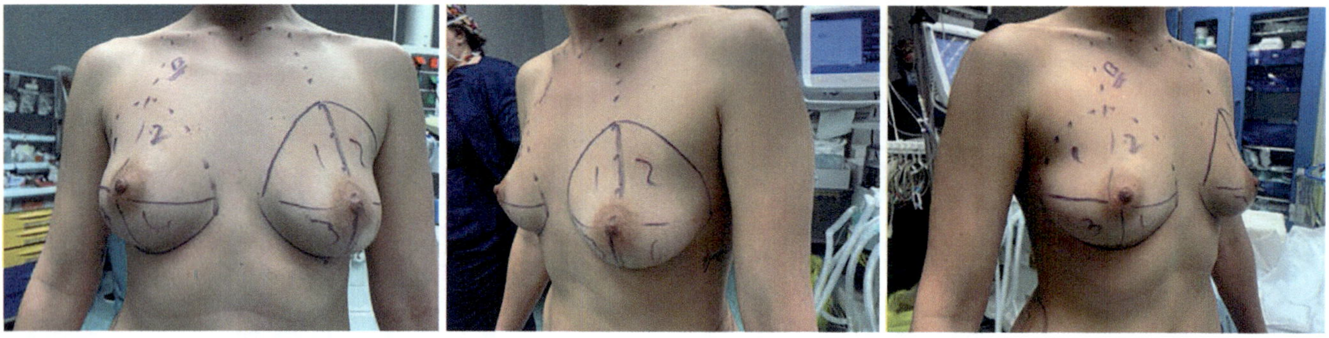

Fig. 24.35 33-year-old patient with breast asymmetry. Pre op and markings

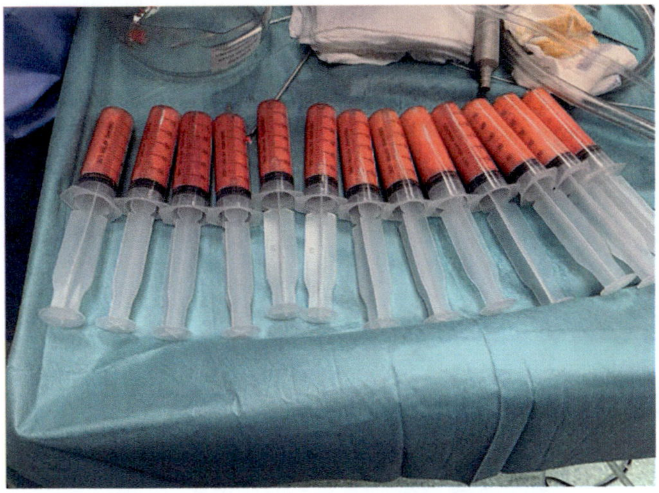

Fig. 24.37 13 syringes with 50 mL of fat ready for transferring (total 650 mL)

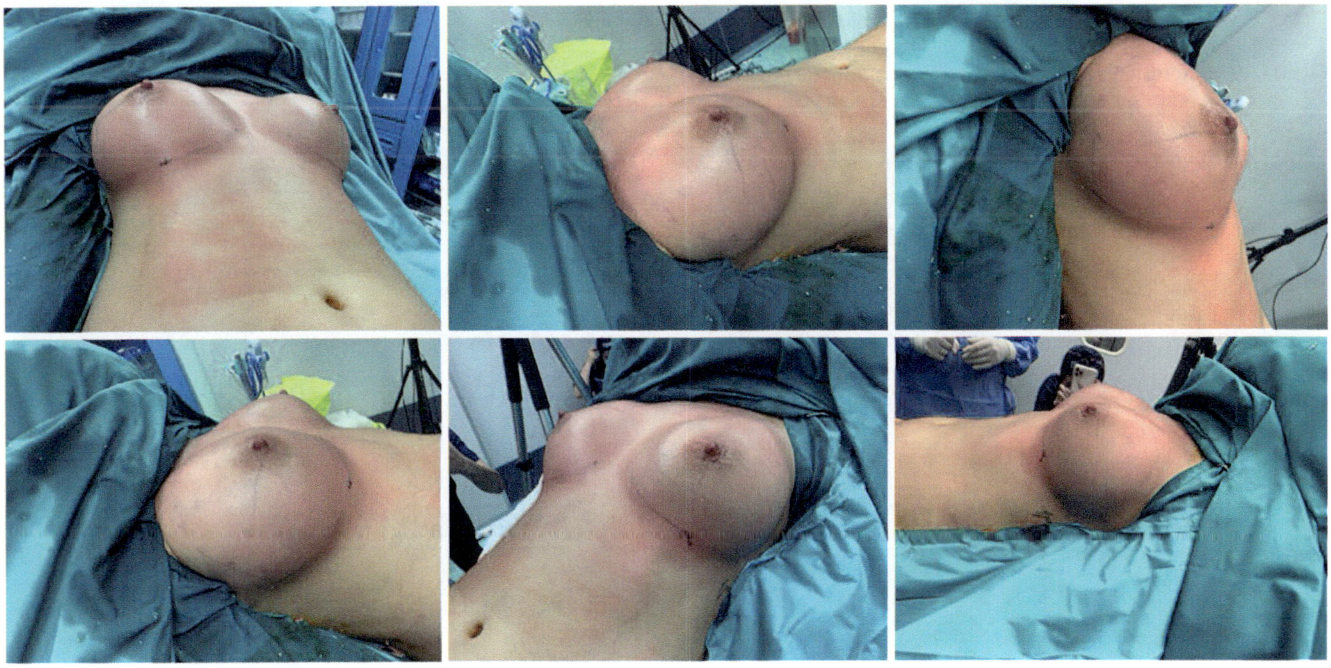

Fig. 24.38 Immediate post-op result after fat transfer (350 mL in right side and 300 mL in left side)

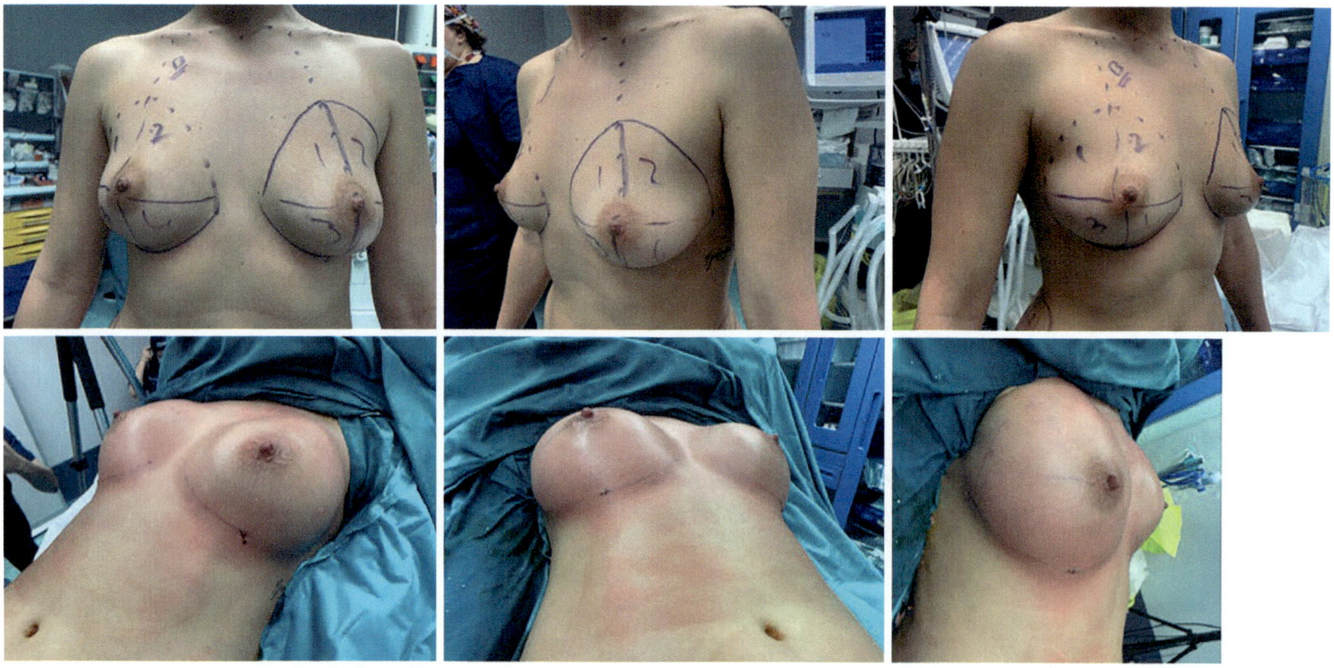

Fig. 24.39 Before and immediately post op fat transfer to breasts

Fig. 24.40 Before and after buttocks and outer thigh liposuction

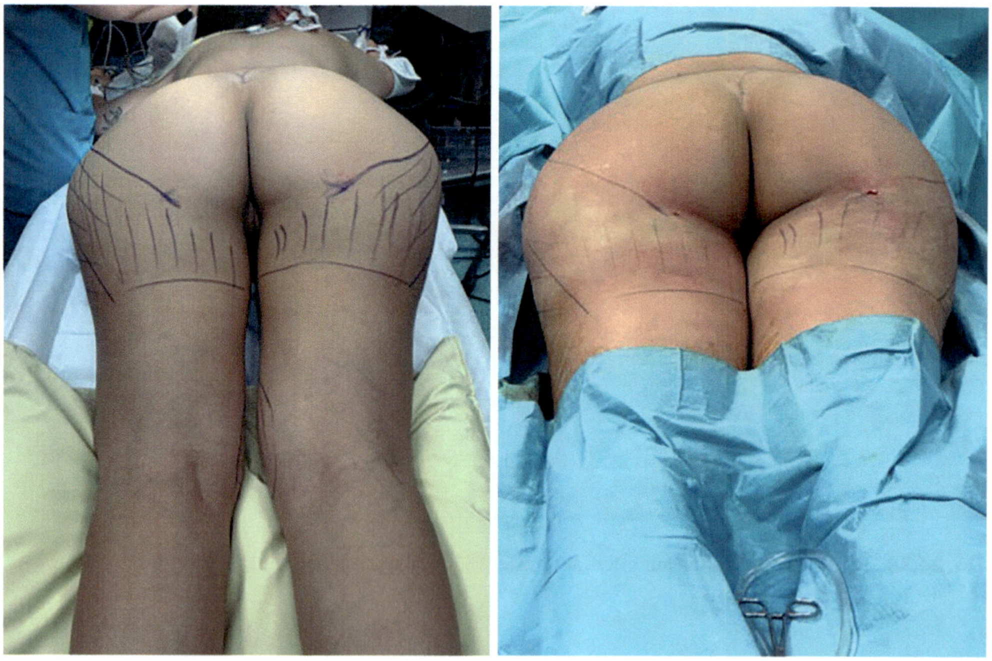

24.9 Clinical Cases

In the period 2012–2020, 440 patients underwent breast fat augmentation.

Average age of patients varies from 18 to 62 years.

Range of fat injected varies from 70 to 750 mL of fat for each side.

Main donor areas were: upper and lower abdomen, flanks, torso, lower arms, outer thigh, inner thigh, anterior thigh, and knees.

Maximum volume aspirated was 7.5 L.

Long-term follow-up varies from 6 months to 5 years.

All patients had pre-op breast ultrasound performed: patients with fibroadenoma, suspicious lump, past medical history of strong hereditary breast cancer, unsuitable patients as for lacking good donor area were not considered.

Patients with unreasonable expectation, in terms of final size or final shape, or projection, were not considered.

Patients with ptosis grade 2–3 or more were not considered as well.

In the past 3 years, hybrid breast augmentation with implants and hybrid mastopexy were performed, but those issues will be treated in different chapters.

24.9.1 Case Number 1

300 mL and 360 mL of fat transferred to right and left side in a mild asymmetry. 1-year post-op result that shows quite total persistence of fat transferred.

VASER and REVOLVE system utilized in this patient (Fig. 24.41).

24.9.2 Case Number 2

24-year-old patient, with mild asymmetry. 355 mL and 330 mL of fat transferred to match the two breasts.

Fat harvested from flanks and outer and inner thigh. 1-year post-op result that shows quite total persistence of fat transferred.

VASER and REVOLVE system utilized in this patient (Fig. 24.42).

24.9.3 Case Number 3

48-year-old patient, with mild ptosis and hypotrophy of both breasts post partum and breast feeding. Was looking for a natural result, refused implants, not willing high projected breast.

400 mL and 400 mL of fat transferred in total.

Fat harvested from flanks, outer and inner thigh and posterior flanks and torso.

1-year post-op result that shows quite total persistence of fat transferred.

VASER and REVOLVE system utilized in this patient (Fig. 24.43).

24.9.4 Case Number 4

35-year-old patient, with mild ptosis and emptiness on the top of breasts. Was looking for a natural result, refused implants, mainly looking for a top breast fullness.

360 mL and 360 mL of fat transferred in total.

Fat harvested from flanks, outer and inner thigh and posterior flanks and torso.

1-year post-op result that shows quite total persistence of fat transferred.

VASER and REVOLVE system utilized in this patient (Fig. 24.44).

24.9.5 Case Number 5

52-year-old patient, with empty breasts but with nipples still in good position. Was looking for a natural result, refused implants, mainly looking for a full breast fullness enhancement.

250 mL and 250 mL of fat transferred in total. Fat harvested from flanks and posterior flanks and torso.

1-year post-op result that shows quite total persistence of fat transferred.

VASER and REVOLVE system utilized in this patient (Fig. 24.45).

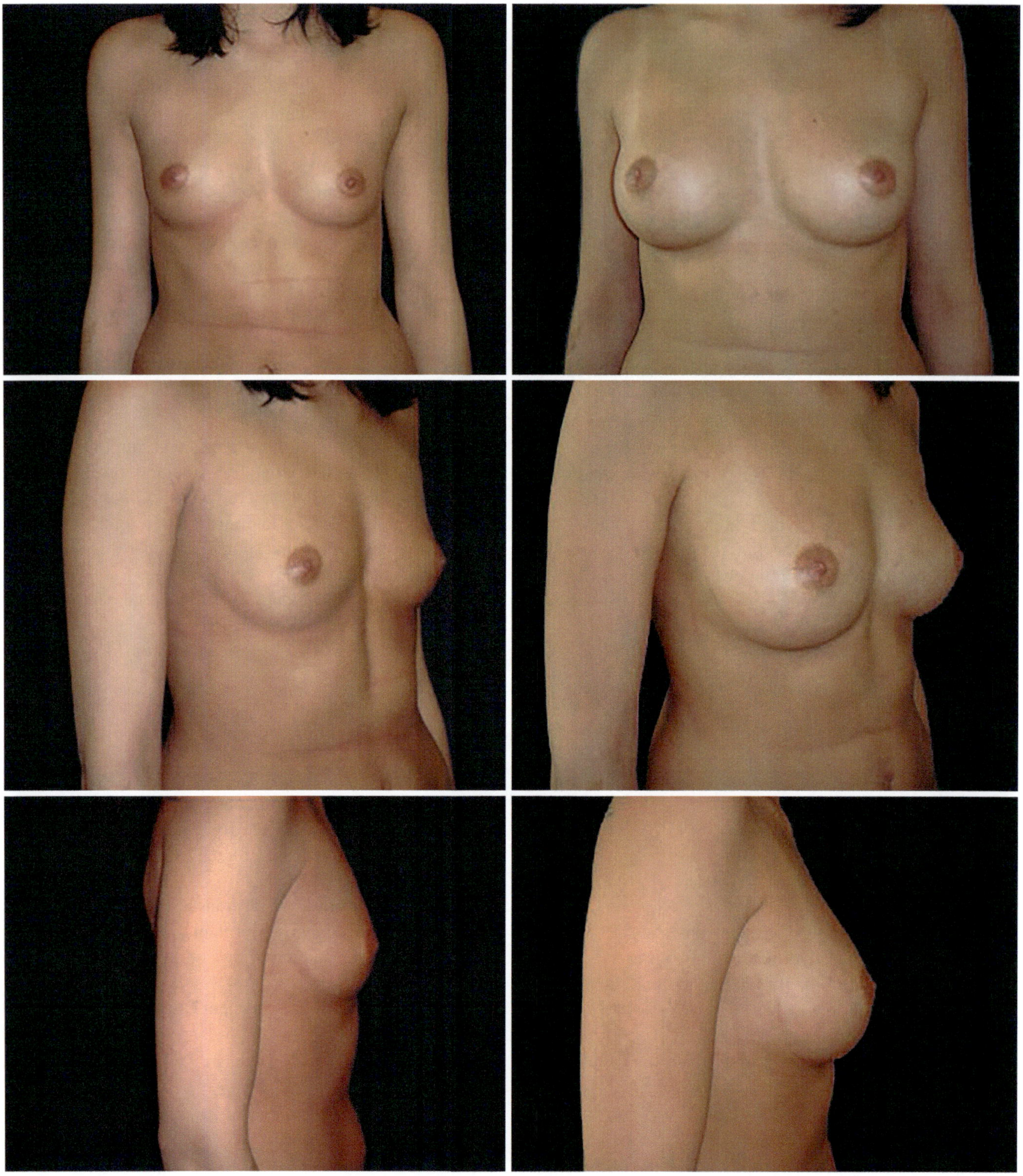

Fig. 24.41 28-year-old girl, mild breast asymmetry, fat transfer to breast, fat harvested from abdomen and flanks

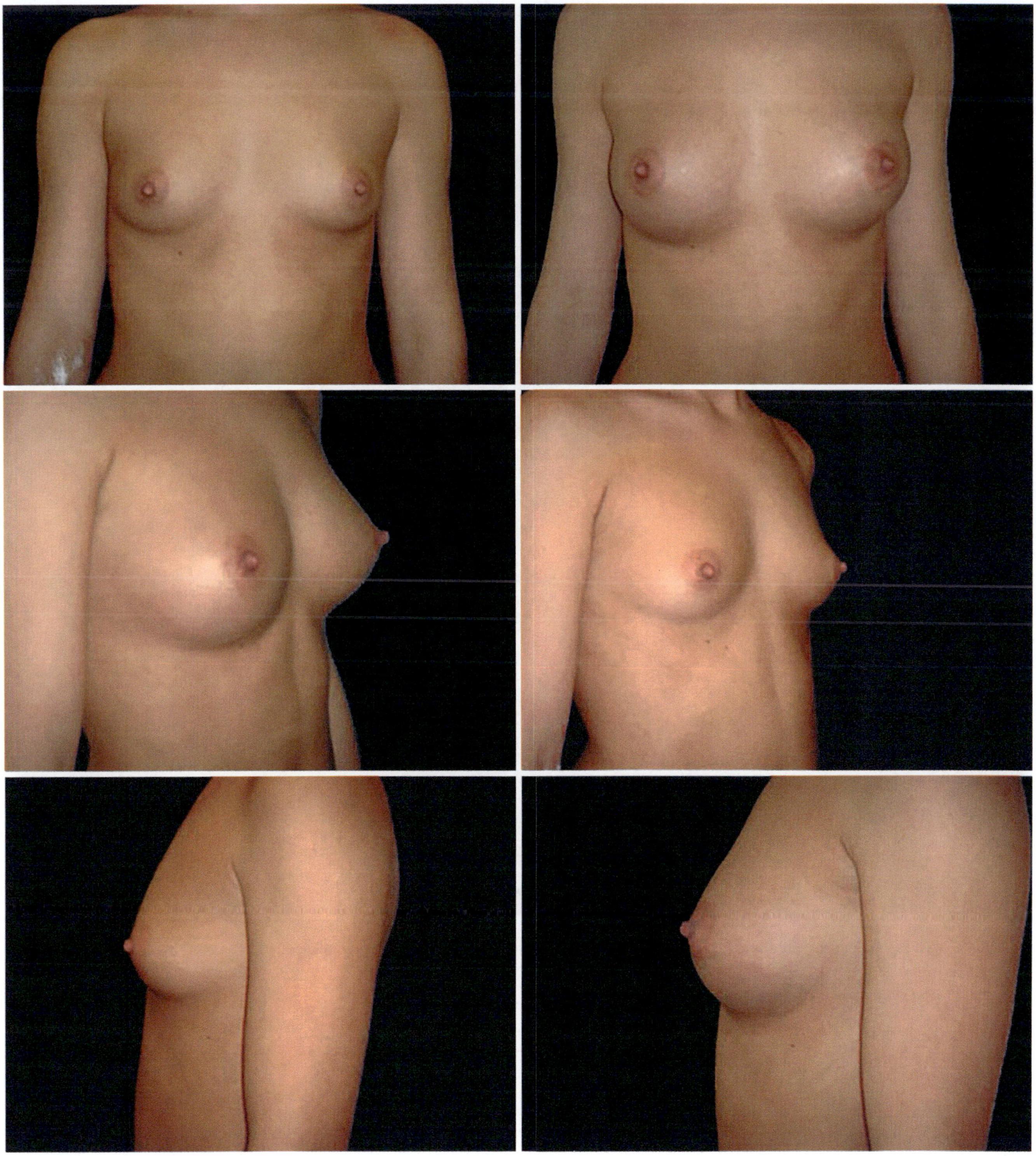

Fig. 24.42 24-year-old girl, mild breast asymmetry, fat transfer to breast, fat harvested from flanks and outer and inner thighs

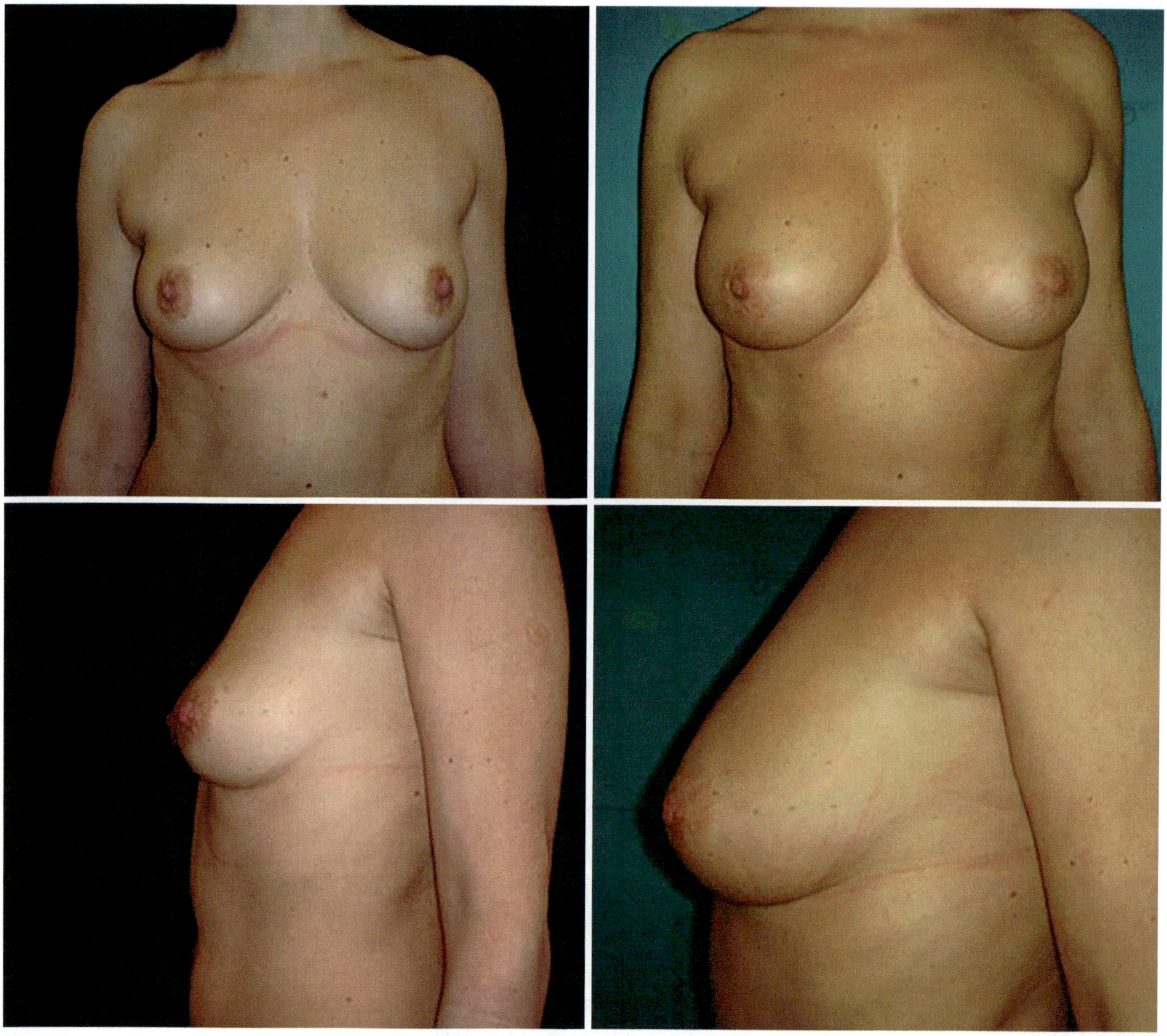

Fig. 24.43 48-year-old patient, mild breast asymmetry, fat transfer to breast, fat harvested from flanks, outer and inner thighs and torso

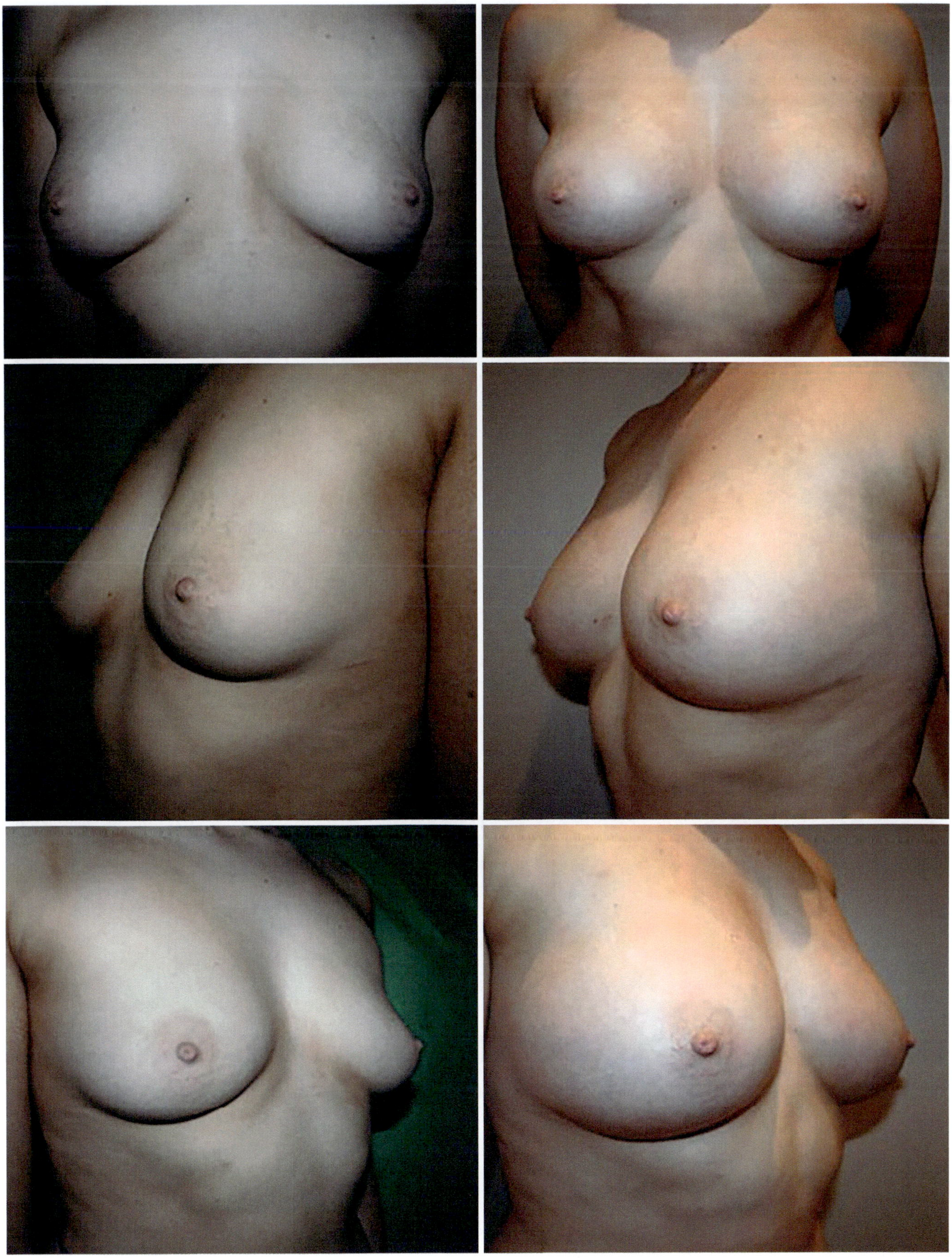

Fig. 24.44 35-year-old patient, mild breast asymmetry, fat transfer to breast, fat harvested from flanks, outer and inner thighs and torso

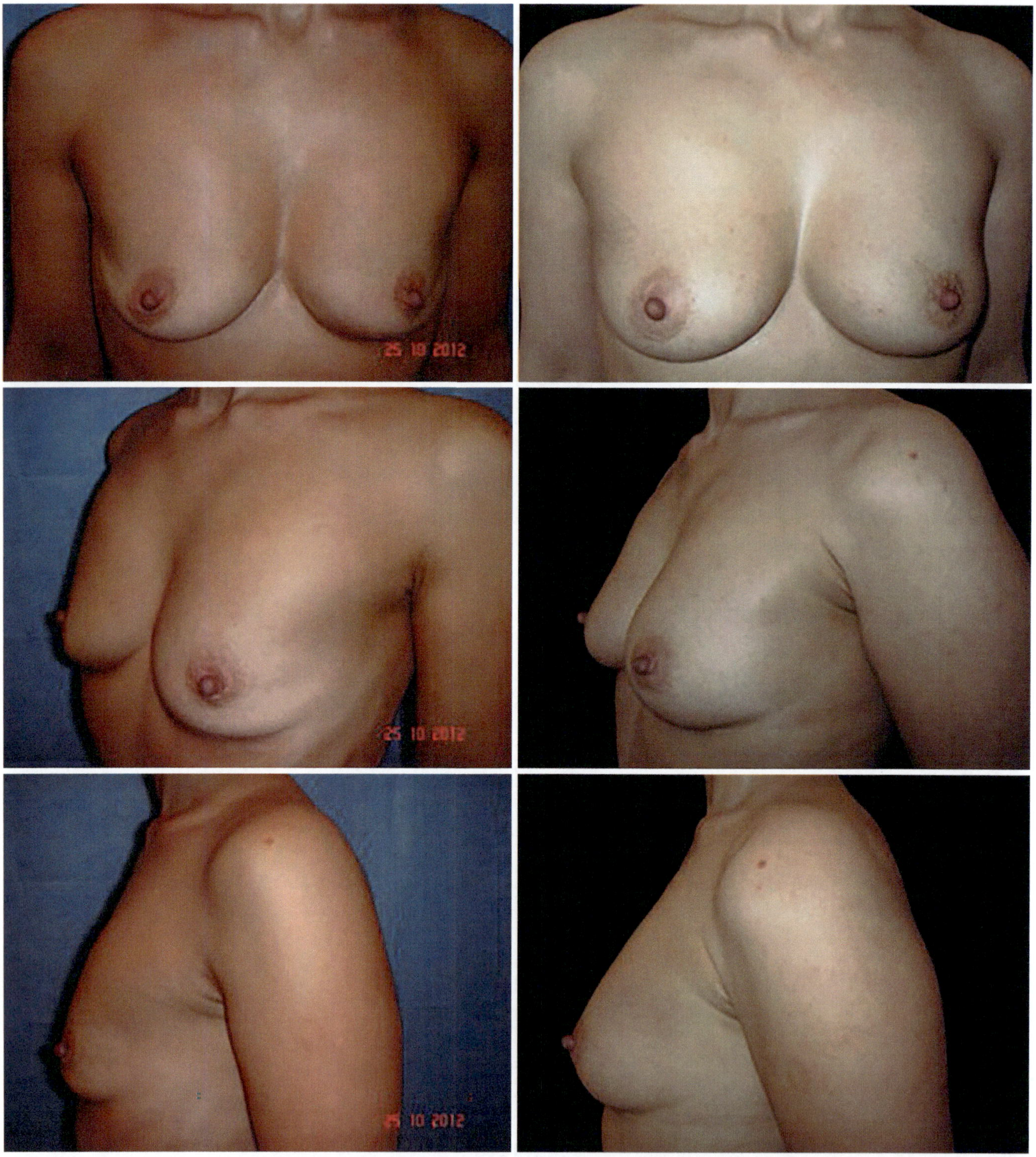

Fig. 24.45 35-year-old patient, mild breast asymmetry, fat transfer to breast, fat harvested from flanks, outer and inner thighs and torso

References

1. Panetta NJ, et al. Tissue harvest by means of suction-assisted or third-generation ultrasound-assisted lipoaspiration has no effect on osteogenic potential of human adipose-derived stromal cells. Plast Reconstr Surg. 2009;124:65.
2. Vilaboa S, Llull R, Mendel TA. Returning fat grafts to physiologic conditions using washing. Plast Reconstr Surg. 2013;132:323e.
3. Mestak O, et al. Centrifugation versus Puregraft for fat grafting to the breast after breast-conserving therapy. World J Surg Oncol. 2014;12:178.
4. Hanson SE, Garvey PB, Chang EI, Reece G, Liu J, Butler CE. A prospective pilot study comparing rate of processing techniques in autologous fat grafting. Aesthet Surg J. 2019;39(3):331–7.
5. Shiffman MA, Di Giuseppe A, Bassetto F. Stem cell in aesthetic procedures, art, science and clinical techniques. Springer; 2014.
6. REVOLVE Advanced Adipose System by LifeCell AbbVie. Instructions for use 2017.
7. Dos-Anjos Vilaboa S, Llull R, Mendel TA. Returning fat grafts to physiologic conditions using washing. Plast Reconstr Surg. 2013;132:323e–6e.
8. Dos-Anjos Vilaboa S, Navarro-Palou M, Llull R. Age influence on stromal vascular fraction cell yield obtained from human lipoaspirates. Cytotherapy. 2014;16:1092–7.
9. Bircoll M. Autologous fat transplantation: an evaluation of microcalcification and fat cell survivability following (AFT) cosmetic breast augmentation. Am J Cosmet Surg. 1988;5:283.
10. Levenberg A, Scheinowitz M, Sharabani-Yosef O. Higher cell viability and enhanced sample quality following laser-assisted liposuction versus mechanical liposuction. J Cosmet Dermatol Sci Appl. 2015;5:238–45.

Postmastectomy Total Breast Reconstruction by Serial Lipografting

25

Pietro Berrino and Valeria Berrino

Contents

25.1 Introduction

Fat transfer is currently widely used in breast reconstructive surgery for the treatment of tissue damage after radiation therapy [1], for site preparation before alloplastic reconstruction [2], for correction of contour irregularities after reconstruction using implants [3], for correction of defects conservative surgery [4, 5], and for refinements after autologous tissue reconstruction. Its use in total breast reconstruction after mastectomy is still limited because lipografting can only provide volume, but it cannot stretch the skin at the mastectomy site.

Candidates to total breast reconstruction by fat injection can be divided into two different groups:

Group 1: Women with preserved submammary fold and relaxed mastectomy site skin can obtain projection and ptosis by simple multiple fat injections. The presence of an untouched submammary fold is crucial because the skin above the fold is mammary skin, which is thinner, more pliable, and expandable than thoracic skin. The mammary skin filled by fat grafting stretches forward and bends over the preserved submammary crease, thus creating a fold and a natural looking breast.

Group 2: In these patients, the submammary fold has been violated and moved upward during the mastectomy procedure. The thoracic skin is stretched to the mastectomy scar. Lipofilling under the thick inelastic thoracic skin produces a fat chest wall, but cannot provide the projecting shape of a new breast. In this group of patients, therefore, expansion is needed in order to:

- Stretch the thoracic skin forward
- Change the thick thoracic skin into a thinner, more pliable, and complying envelope, suitable for lipofilling
- Create a new submammary fold in the proper position

Mastectomy site skin in Group 2 women can be expanded by external or internal expansion.

External expansion was described by Khouri and Rigotti [6] and Khouri and Del Vecchio [7, 8] who used serial lipografting to create a breast mound in patients who were preliminarily treated by external expansion using the BRAVA System (BRAVA Inc., Miami, FLO). The use of external expansion devices according to the manufacturer's protocol [9] led to a 28% withdrawal rate because of social embarrassment, discomfort, difficulties in wearing the device at night, and insomnia [10]. The external expander device appears to show poor compliance in postmastecomy patients due to the long and difficult pre- and postoperative period of external expansion.

In 2008 at the 19th Euraps Meeting, we first reported the use of internal expansion and serial lipografting, naming it "the REAL (Reverse Expansion And Lipofilling) technique" [11]. This presentation was awarded as "best paper" of the Meeting. Since 2007 we reported the procedure in several

P. Berrino (✉) · V. Berrino (✉)
Chirurgia Plastica srl, Genoa, Italy

© The Author(s), under exclusive license to Springer Nature Switzerland AG 2023
A. Di Giuseppe et al. (eds.), *Fat Transfer in Plastic Surgery*, https://doi.org/10.1007/978-3-031-10881-5_25

Meetings, and it has become a popular procedure, although reserved to a small subgroup of candidates [12, 13].

Selection of the candidates on the basis of the individual psychological attitude is as much important as correct assessment of the physical characteristics. The classification we have proposed for psychological assessment of the postmastectomy patients [14] proved to be crucial for the selection of the right candidate to total breast reconstruction by fat grafting.

Type 1 patients are highly motivated women who are willing to accept major operations in order to recover the body image they had before the mastectomy: typically they are candidates to autogenous tissue reconstruction by the abdominal flap. Type 2 women are more sophisticated highly demanding patients who have considered aesthetic surgery before mastectomy and badly accept additional scarring and multistage procedures. Typically, the choice for this group of woman is implant reconstruction with contralateral mammoplasty. Type 3 women are those who do not wish to undergo major procedures—and related complications—and prefer to undergo multistage minor procedures not affecting their social and family life. These women are typical candidates to total breast reconstruction by fat grafting.

25.2 Total Breast Reconstruction by Fat Grafting in Group 1 Women

Group 1 women show an untouched submammary fold and abundant skin above it. This condition can be seen after skin-sparing or nipple-sparing mastectomy in patients who did not receive an implant or who had the implant removed due to secondary complications. This group of patients does not generally include women who received radiation therapy, but RT is not by itself a contraindication to the procedure unless severe skin damage is present. Fat injections, in facts, often produce an amazing improvement in the quality of the radiated skin and can change a stiff skin area into a pliable, elastic skin envelope. Some of these patients may require scar revision and release due to scarred retracted areas that can jeopardize skin compliance.

Contraindications to the procedure are:

- Severely scarred or radiation damaged skin at the mastectomy site
- Insufficient fat deposits at the donor sites
- Patients unwilling to undergo multiple procedures or to accept a long period for completion of the reconstruction
- Patients with large contralateral breast unwilling to undergo reduction

Ambulatory lipofilling sessions are scheduled at about 3-month intervals, and they are carried out under sedation and local infiltration. The donor site is infiltrated with modified Klein solution (for each 500 cc of saline solution, 1 cc adrenaline, 12 cc xylocaine 2%, 2 cc sodium bicarbonate). Adipose tissue is suctioned using a 2.5 mm (inner diameter) cannula and a 10 cc Luer-lock syringe. The preferred donor sites are the flanks, lateral upper thighs, and inner knees. We prefer to avoid the abdomen as a donor site due to the high risk of skin irregularities and long-term skin emptiness.

Separation of fat and fluids is obtained by simple decantation. A 10-min period is sufficient for separation of fat from fluids and oil. The aspirated fat is transferred into 2.5 mL Luer-lock syringes. After infiltration of the recipient site with the described modified Klein solution, the fat is slowly injected on withdrawal of the plunger using 2 or 1.5 mm (inner diameter) cannulas, so as to release a fraction of milliliter at each passage. Fat is transplanted over the pectoralis major muscle and in the subdermal space, according to the principle of bicompartmental lipostructure described by Zocchi for breast augmentation [15].

Fat dispersion into healthy subcutaneous layers is crucial in order to obtain a satisfactory graft take. In spite of accurate fat dispersion, 12–15 mL of aspirate per minute can be injected, so that 200 mL are injected in about 15 min and each lipostructure session is completed in approximately 40 min.

During fat grafting sessions, contralateral mammoplasty can be performed, and local release of scarred tissue is also accomplished when needed. Nipple-areola reconstructions and final touch-ups are carried out during the last surgical stage.

Each patient has a clinical and photographical evaluation at 1 and 3 months after each surgical procedure and at 1, 3, 6, and 12 months after the last operation.

In our series, all Group 1 patients who completed the scheduled sessions showed good to excellent results except one woman who had a cranially displaced submammary fold which could not be repositioned by lipostructure alone. This woman was erroneously included in Group 1; she was actually a Group 2 patient with a dislocated submammary fold needing internal expansion. Almost all patients show oil cysts on sonography evaluation at the 1-year follow-up and no suspicious breast lump was observed [12]. Volume reduction of the fat graft occurring after each session was estimated to be approximately 30% at the 4-month follow-up. Volume appeared to be stable after 4 months, and long-term follow-ups after the final grafting session showed minimal fat resorption. Only a minority (approximately 15%) of patients requested further fat injection after more than 1 year from the final operative session.

25.2.1 Clinical Examples

25.2.1.1 Case 1

A 51-year-old woman presented for breast reconstruction after mastectomy. The submammary fold was preserved, skin was not tight, and the contralateral breast was small and showed minor glandular ptosis (Fig. 25.1a, b). Three lipografting sessions were carried out and 200 mL of aspirated fat were injected at each session (Fig. 25.1c, d). During the third session, contralateral periareolar mastopexy was carries out, and NAC reconstruction by contralateral grafts was performed (Fig. 25.1e, f).

25.2.1.2 Case 2

A 51-year-old woman presented for left side implant removal after nipple sparing mastectomy. She had chronic infection with drainage that had been treated conservatively for more than 12 months. She had developed severe scar retraction around the NAC. Contralaterally, she had a ptotic augmented breast (Fig. 25.2a). She was affected by multiple sclerosis, and she preferred to avoid major surgery under general anesthesia. The left infected implant was removed with relief of symptoms. The right implant was also removed (Fig. 25.2b, c). Four lipografting sessions were planned. Twenty days after the first lipostructure session, she showed healing of the

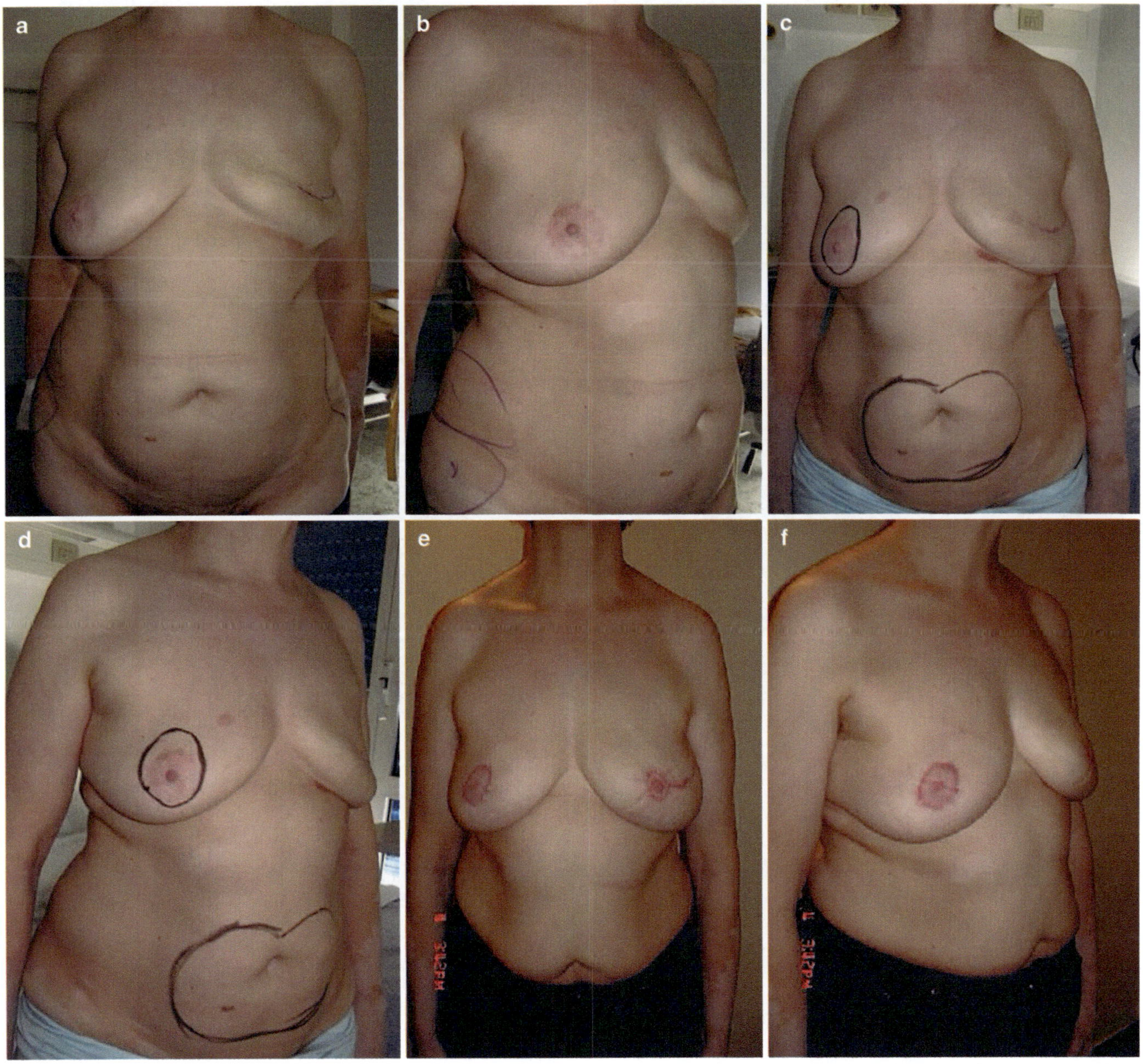

Fig. 25.1 Front (**a**) and lateral (**b**) view of a 51-year-old. Group 1 woman after total mastectomy, showing preserved submammary fold; front (**c**) and lateral (**d**) view after 2 lipofilling sessions, before right periareolar pexy, nipple/areola reconstruction by contralateral grafts, and final fat grafting; front (**e**) and lateral (**f**) view after the last operative stage

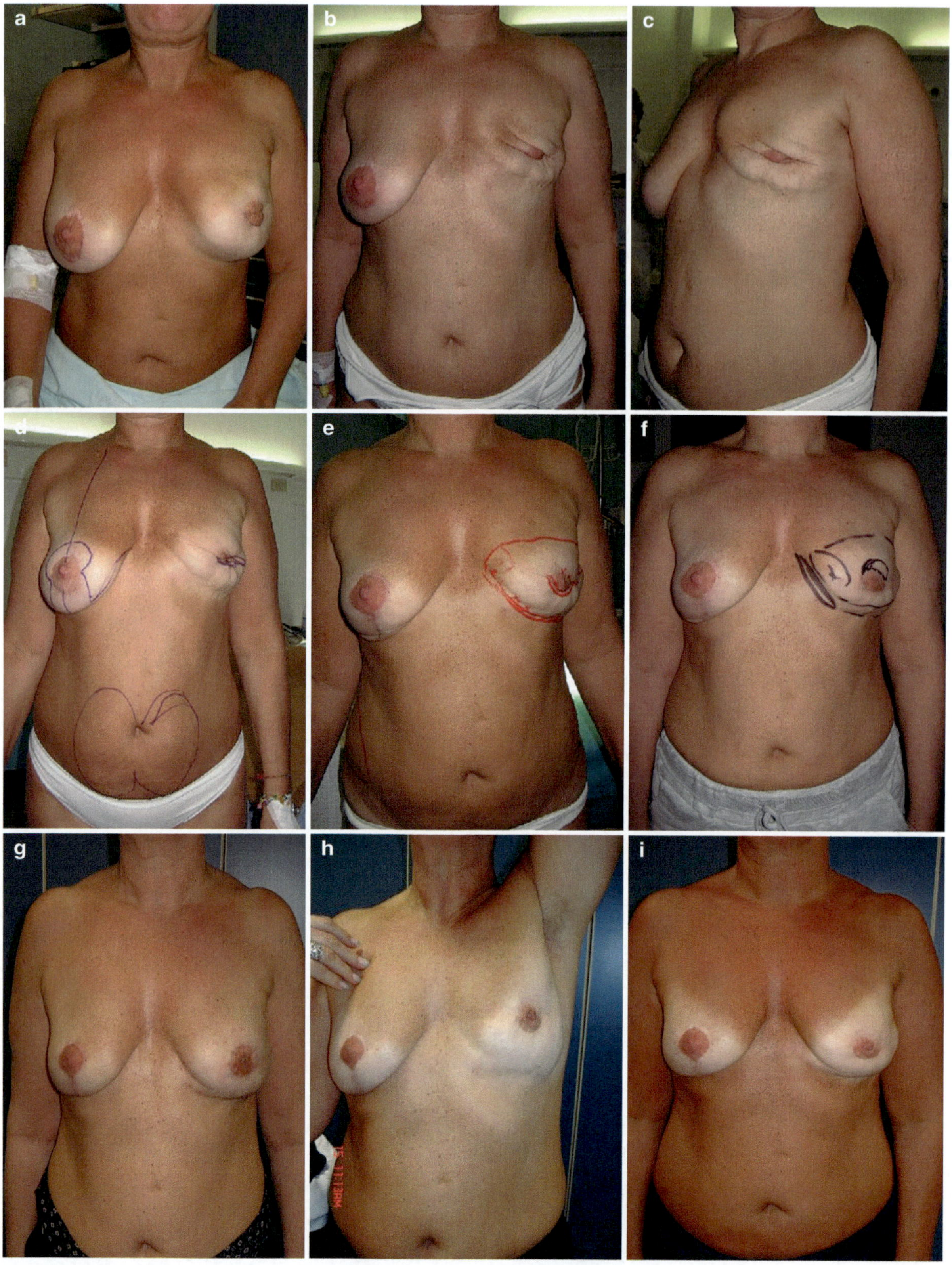

Fig. 25.2 Front view (**a**) of a 52-year-old woman, 2 years after left nipple sparing mastectomy and right periareolar pexy with subglandular implant positioning. A draining fistula from the left implant pocket to the periareolar scar is present since 1 year. Front (**b**) and lateral (**c**) view after bilateral implant removal and healing of the fistula. Front view (**d**) after the first session of lipografting and scar release on the left side. Right vertical mastopexy is planned and further scar release and fat grafting on the left side is planned. Front view before the third lipografting session (**e**). Front view before the fourth surgical session (**f**). Front view 40 days after the last session (**g**). Eleven months after the last session: complete release of the scarred tissue is observed (**h**). Nineteen months after the last fat grafting sessions the result is maintained (**i**)

periareolar fistulas. Revisions of the retracted scar and contralateral pexy were also carried out during lipografting sessions (Fig. 25.2d–f). All session were performed under local anesthesia and sedation. Full-volume restoration was obtained, and total release of the severely scarred and retracted tissue was achieved (Fig. 25.2g–i).

25.3 Total Breast Reconstruction in Group 2 Women

Group 2 women show violated inframammary fold. The abdominal skin is stretched cranially and reaches the mammary skin at the mastectomy scar. These patients obviously show various degrees of skin tightness and thickness; the pectoralis major muscle can be damaged or absent. Group 2 women can be selected to undergo breast reconstruction by internal expansion and serial lipografting according to the following inclusion criteria:

- Adequate adipose tissue reserve (based on regional fat deposits and on the volume to restore)
- Skin thickness of more than 1.5 cm at the recipient site, evaluated with the pinch test
- Absence of skin tension, severe scarring or RT damage at the mastectomy site
- Willingness to accept multistage procedure and to avoid the use of permanent implants

From a psychological point of view, the candidate for "REAL technique" is the patient that we have categorized as "type 3" in terms of expectations and psychological attitude [14]: a woman who is not willing to accept major procedures and associated possible complications, who is not willing to undergo long hospitalization and who expects a reasonable result through minor and well tolerated operations.

In patients who have an expander positioned during the mastectomy operation serial lipoinjections with expander deflation can be started. In patients who do not already have an expander, an expandable implant is positioned during the first operation. The lower border of the expander is positioned at or slightly below the future submammary fold. The expander is positioned submuscularly if the pectoralis major muscle is present. Lipografting is carried out into the subcutaneous layer and into the prepectoral fat layer (or above the chest wall if the muscle is not present). According to the inclusion criteria, thickness of the subcutaneous layer above the expander is at least 1.5 cm allowing appropriate fat graft dispersion. The technique for fat harvesting and injection is identical to that described for Group 1 patients. During the first session, after positioning the expander, 200–250 mL of aspirated fat is usually injected in the subdermal plane, so as the graft is not dispersed into the implant pocket. In the next 8–10 weeks after the first session, the expander is slowly inflated to the expected final volume. Care is taken not to create skin tension above the expander, in order to avoid fat compression and atrophy. After reaching the final expected volume, serial fat grafting is scheduled.

Three to five sessions of fat injection are planned, based on the volume to restore.

At each session, the expander is partially deflated by 100–250 mL, and the subcutaneous layer is grafted with aspirated fat. The volume of the fat graft injected in each session exceeds the milliliters of saline taken out from the expander by approx. one-third (e.g., if the expander is deflated by 200 mL, 260 mL of aspirate are injected). The amount of fat grafted at each session exceeds the desired volume by 25–30% because an approximate 30% volume loss after each procedure is to be expected.

Sessions are scheduled at about 3-month intervals. Contralateral adjustments are performed, if needed, during one of the fat grafting procedures.

During the last session, the expander is removed, nipple-areola complex reconstruction is usually performed, and final touch-ups are carried out.

Each patient has a clinical and photographical evaluation at 1 and 3 months after each surgical procedure and at 1, 3, 6, and 12 months after the last intervention.

All sessions are carried out under sedation and local infiltration of the already described modified Klein solution. Sessions requiring expander implantation and contralateral mammoplasty are performed under general anesthesia.

Fifteen percent of Group 2 patients selected for the REAL technique wished to interrupt the lipofilling schedule and to have the expander replaced by an anatomical permanent implant. These patients were finally given a "hybrid" reconstruction as described by Nava [16]. All the patients who completed the scheduled sessions showed good to excellent results.

Patients frequently showed minute oil cysts at sonographic evaluation at the 1-year follow-up.

As reported for Group 1 patients, approximately 30% of the grafted volume was lost in a 4-month period, while volume remained stable thereafter. Approximately 20% of the patient who underwent the REAL technique requested additional fat injections after 3–10 years from the last scheduled session.

25.3.1 Clinical Examples

25.3.1.1 Case 3

A 54-year-old woman presented after right modified radical mastectomy (Fig. 25.3a, b). She also showed a remarkable presternal contour defect due to pectus excavatum (PE) deformity. Two lipostructure sessions were preliminarily carried out in order to correct the PE deformity (Fig. 25.3c).

Three months after PE correction, a 450 cc breast expander (Siltex Contour Profile-Mentor Medical) was posi-

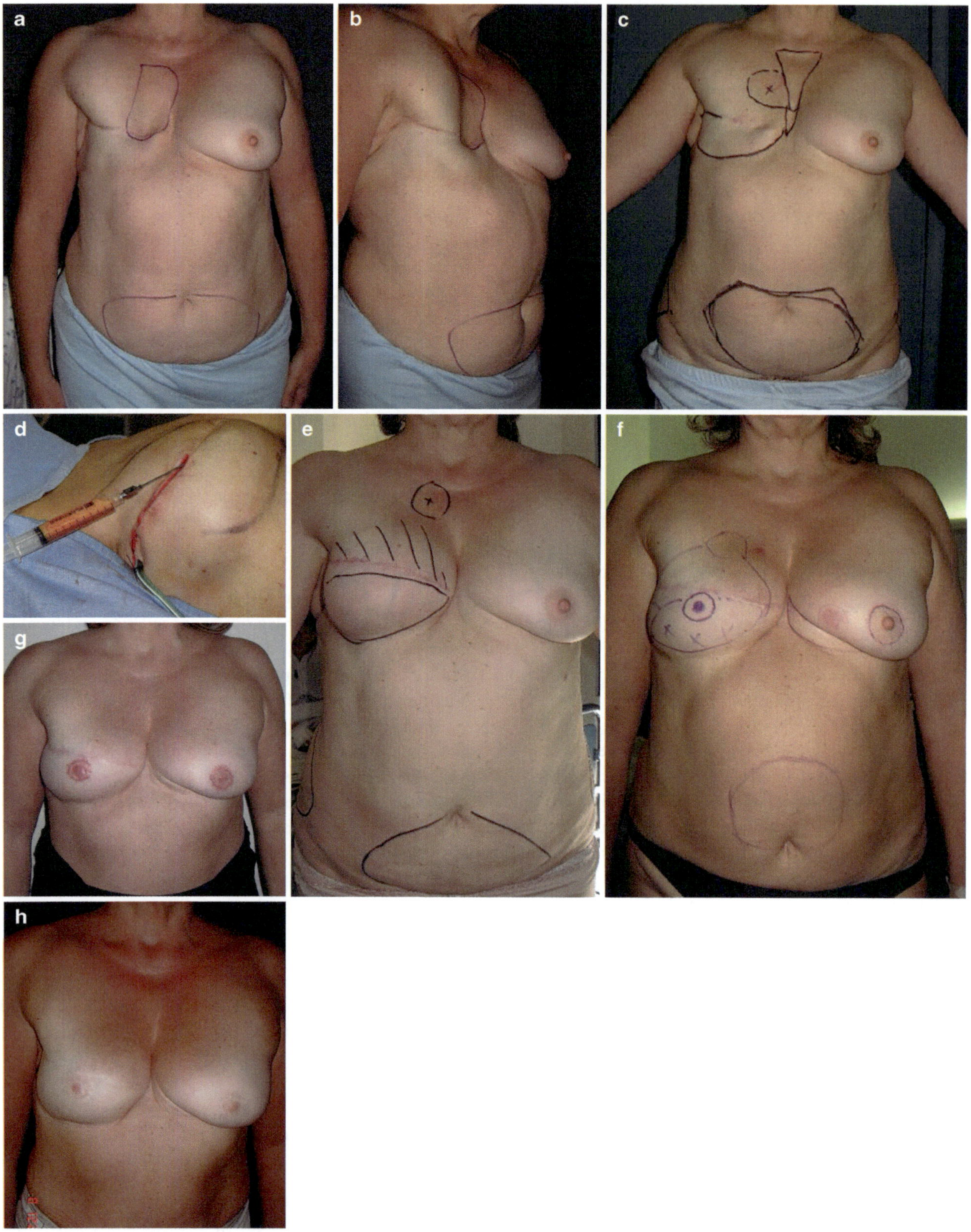

Fig. 25.3 Front (**a**) and lateral (**b**) view of a 54-year-old woman showing Pectus Excavatum deformity and right radical mastectomy. Front view (**c**) after correction of the Pectus Excavatum by two lipografting sessions; further presternal fat injection and expander positioning is planned. Intraoperative view (**d**) showing immediate subdermal fat injection after placement of the expander: the lower quadrants are mainly grafted. Front view (**e**) after the first lipografting session: implant deflation and lipografting is planned. Front view (**f**) after three REAL sessions, before the final stage: nipple/areola reconstruction, further lipografting and implant removal are planned. The outcome is shown 3 months (**g**) and 3 years (**h**) after the final surgical stage

tioned, which was filled with 200 cc of saline solution. In the same session, additional fat grafting into the presternal region and into the prepectoral space was carried out (Fig. 25.3d).

The expander was inflated to 450 mL in 8 weeks. In the next operative stage, the expander was partially deflated (−200 cc), and the prepectoral space was filled with 250 cc of fat (Fig. 25.3e); 2 months later, an identical session was scheduled. Fat grafts were harvested from the abdomen, the flanks and the peri-trochanteric regions bilaterally. Three months later, the patient underwent nipple-areola reconstruction, additional lipofilling and implant removal (Fig. 25.3f, g). At the 3-year follow-up, a stable result was observed (Fig. 25.3h).

25.3.1.2 Case 4

A 57-year-old patient presented at our observation after a right modified mastectomy. On clinical examination, she showed total mastectomy defect on the right side and a high degree of breast ptosis on the contralateral side (Fig. 25.4a).

Four surgical sessions were planned. During the first session, a 550 cc Siltex Contour Profile (Mentor Medical) breast expander was placed submuscularly, and it was filled with 200 cc of saline (Fig. 25.4b). Lipoinjections in the subdermal plane were also carried out during the first operation. The expander was filled to nominal volume in 7 weeks. Then, a total of 500 cc of fat were harvested and injected in the premuscular and in the subdermal plane at the recipient site in two sessions, after partial deflation of the implant (−200 cc at each session) (Fig. 25.4c, d).

In the last session removal of the expander, nipple-areola complex reconstruction and contralateral breast reduction were carried out and 150 mL of fat were injected (Fig. 25.4e, f). At the 4-year follow-up, the result was stable but depigmentation of the nipple-areola graft occurred (Fig. 25.4g).

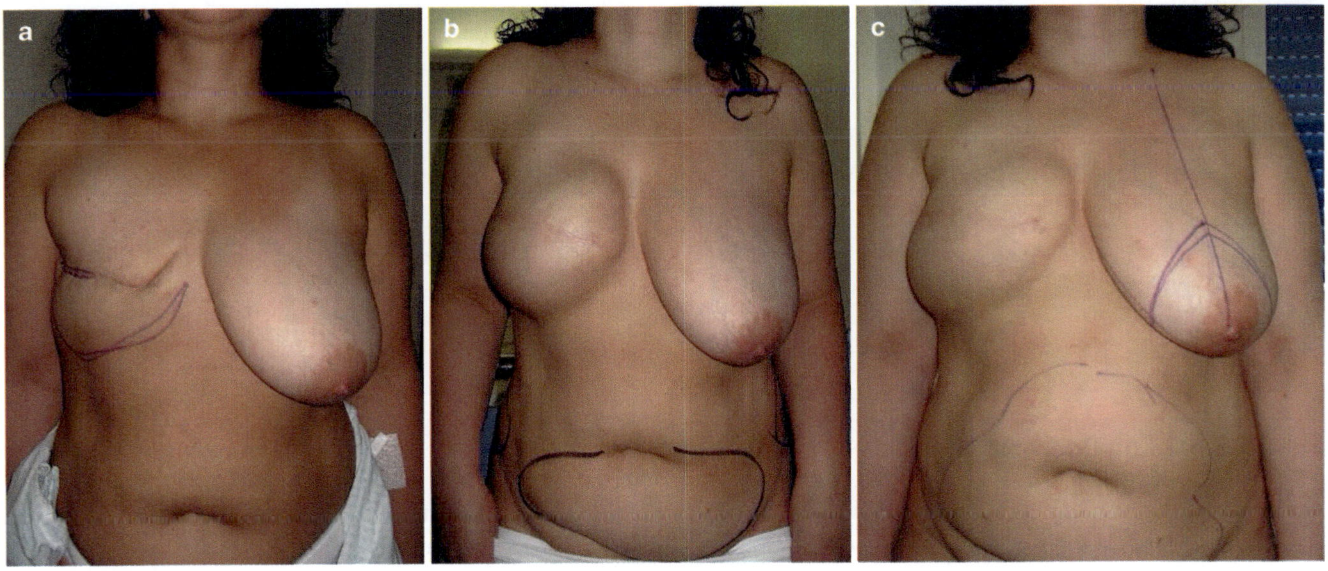

Fig. 25.4 Front view (**a**) of a 57-year-old woman after radical mastectomy: violated inframammary fold and scar retraction can be observed at the mastectomy site. Frontal view (**b**) after expander positioning and subdermal fat grafting. Frontal view (**c**) before the second stage: implant deflation, fat grafting and contralateral reduction mammoplasty is planned. Frontal view (**d**) before the third operative stage: total (−125 mL) expander deflation and 250 mL fat grafting is planned. Frontal view (**e**) before the final operative stage: expander removal, nipple/areola reconstruction by contralateral grafts, revision of the right periareolar scar and final lipografting is planned. Front view 1 year (**f**) and 4 years (**g**) after the final operative stage

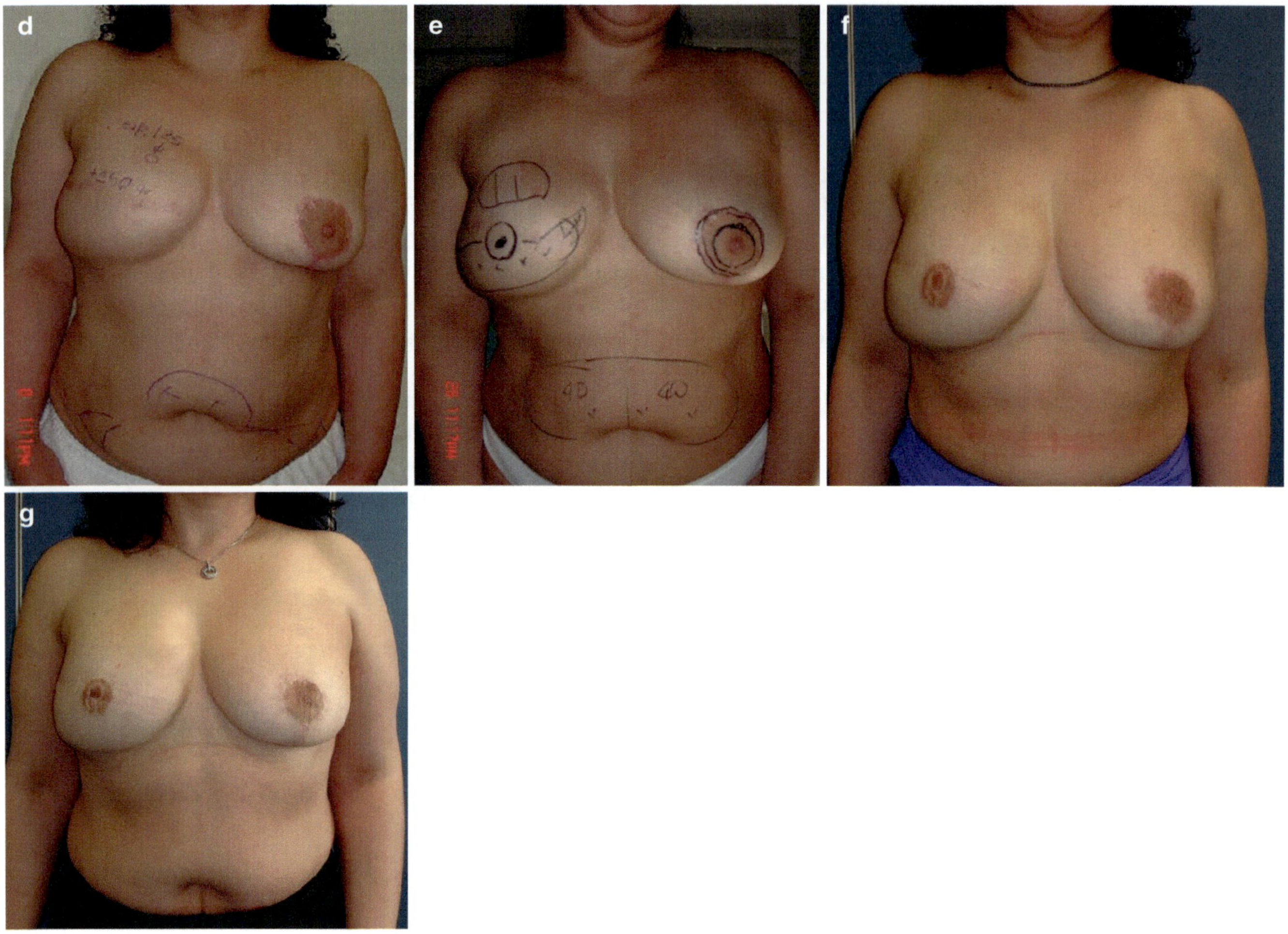

Fig. 25.4 (continued)

25.4 Final Considerations

Over the past 20 years, we experienced an impressive change in the choice of the technique for breast reconstruction, shifting from a majority of autogenous distant flaps in the 1980s and 1990s to a prevalence of implant-based breast reconstructions during the past 20 years. This change was due to the evolution of marketed devices and implantation techniques, but also to social and financial issues, balancing the clinical outcome, the risk of complications, and resource costs. However, nowadays and following the ALCL issue, there is a trend toward avoiding textured anatomical implants. In this new scenario, autogenous fat grafting meets all the requirements to be a valid alternative to the other techniques for total breast reconstruction in selected patients.

Total breast reconstruction after mastectomy can be reliably achieved in group 1 patient by lipostructure alone and in group 2 women by the REAL technique. Patients' selection is crucial in order to select the right procedure and to achieve satisfactory results. Women with preserved submammary fold and relaxed mastectomy site skin (Group 1 patients) can obtain outstanding results by simple multiple fat injections since the breast skin filled by fat grafting stretches forwards and bends over the preserved submammary. In women who had the submammary detached and moved upward during the mastectomy procedure (Group 2 patients), expansion is mandatory. Knowledge of this preoperative evaluation is necessary because misjudging the selection of patients can lead to unsatisfactory outcome.

Psychological attitude evaluation is as much important as evaluation of the mastectomy site conditions in order to properly select candidates. In fact, total breast reconstruction by fat grafting can be offered to women who wish to avoid major surgery and/or the use of permanent implants. However, patients must be aware that they have to accept a multistage ambulatory route often requiring more than 1 year to be completed.

References

1. Rigotti G, et al. Clinical treatment of radiotherapy tissue damage by lipoaspirate transplant: a healing process mediated by adipose-derived adult stem cells. Plast Reconstr Surg. 2007;119(5):1409–22.
2. Salgarello M, Visconti M, Tarallo E. Autologous fat graft in radiated tissue prior to alloplastic reconstruction of the breast: report of two cases. Aesthetic Plast Surg. 2010;34(1):5–10.
3. Berrino P, Berrino V. Chapter 19.4. Fat grafting with implants. In: Kronowitz SJ, Benson JR, Nava M, editors. Oncoplastic and reconstructive management of the breast. 3rd ed. Boca Raton: CRC Press; 2020.
4. Spear SL, Wilson HB, Lockwood MD. Fat injection to correct contour deformities in the reconstructed breast. Plast Reconstr Surg. 2005;5:1300–5.
5. Kanchwala SK, Glatt BS, Conant EF, Bucky LP. Autologous fat grafting to the reconstructed breast: the management of acquired contour deformities. Plast Reconstr Surg. 2009;124(2):409–18.
6. Khouri R, Rigotti G, Cardoso E, Baker T. Minimally invasive autologous total breast reconstruction by external expansion and serial lipografting: a preliminary experience. EURAPS. 2008.
7. Khouri R, Del Vecchio D. Breast reconstruction and augmentation using pre-expansion and autologous fat transplantation. Clin Plast Surg. 2009;36:269–80.
8. Del Vecchio D. Breast reconstruction for breast asymmetry using recipient site pre-expansion and autologous fat grafting a case report. Ann Plast Surg. 2009;62(5):523–7.
9. Khouri RK, et al. Brava and autologous tissue fat transfer is a safe and effective breast augmentation alternative: result of a 6 year, 81 patient prospective multicenter study. Plast Reconstr Surg. 2012;129(5):1173–87.
10. Howe BHL, et al. Efficacy of an external volume expansion device and autologous fat grafting for breast reconstruction following breast conserving surgery and total mastectomy: small improvements in quality of life found in a prospective cohort study. J Plast Reconstr Aesthet Surg. 2020;73:27–35.
11. Berrino P. Presented at 19th EURAPS meeting. Funchal (Portugal), May 29–31 2008.
12. Manconi A, De Lorenzi F, Chahuan B, Berrino V, Berrino P, Zucca-Matthes G, et al. Total breast reconstruction with fat grafting after internal expansion and expander removal. Ann Plast Surg. 2017;78(4):392–6.
13. Mineda K, Kuno S, Kato H, Kinoshita K, Doi K, Hashimoto I, et al. Chronic inflammation and progressive calcification as a result of fat necrosis: the worst outcome in fat grafting. Plast Reconstr Surg. 2014;133(5):1064–72.
14. Berrino P. Operative strategies in breast plastic surgery. SEE (Firenze). 2007.
15. Zocchi M, Zuliani F. Bicompartmental breast lipostructuring. Aesthetic Plast Surg. 2008;32:313–28.
16. Nava MB, Catanuto G, Rocco N. Hybrid breast reconstruction. Minerva Chir. 2018;73(3):329–33.

Hybrid Fat Transfer, Breast Implants, and Fat

26

Nicola Zingaretti, Giovanni Miotti, and Pier Camillo Parodi

Contents

26.1 Hybrid Breast Reconstruction

The goal of modern breast reconstruction is to restore a breast shape that is as close to "normal" as possible. Reconstructive procedures should be based on the replace-like-with-like principle. Breast reconstruction modalities are based on allogenic materials, autologous tissue transfer, or a combination of both.

The breast is a subcutaneous structure: gravity, aging, and posture influence breast volume distribution and its contour. Microsurgical autologous tissue transfer guarantees long-lasting and natural results, but despite its advantages, to date, implant-based breast reconstructions are out-numbering autologous reconstructions [1]. In autologous reconstructions the transplanted tissue adapts to its new environment because of its unique plasticity. In implant-based reconstructions, prosthesis do not adapt to its new surroundings, affect the remaining tissues, and they are the major manipulator of the aesthetic outcome. Lipofilling is now a universally accepted technique to correct soft-tissue deficiencies or to improve contour irregularities. Total prosthesis replacement with fat in breast reconstruction is desirable but is achievable at the moment only for small breast sizes [2].

Capsular contracture, with an incidence of 0.6–30%, is the most common complication following reconstructive breast surgery [3]. Nava et al. showed how radiotherapy increases the risk of complications by more than 40% in implant-based reconstructions, increasing the rate of capsular contractures to between 25% and 30% of patients [4]. Moreover, even the best implant-reconstructed breast will lack the soft-tissue envelope of natural breasts, leading to possible implant-related complications, such as implant visibility, rippling/wrinkling or depression of the skin [5]. To reduce these aspects soft tissue coverage through acellular dermal matrices (ADMs) and autologous fat grafting have been widely used with improvement of postoperative patients' satisfaction [6, 7].

In recent years, the concept of hybrid reconstruction has been developed. For hybrid breast reconstruction (HBR), we mean a technique that combines the use of autologous tissues and heterologous devices (prostheses or expanders).

Historically, the most common approach to hybrid reconstruction was the association of pedicled flap (i.e., latissimus dorsi, TRAM, etc.) or free flap (i.e., DIEP flap) and prosthetic implant [8–10]. In this approach, the transfer of well-vascularized soft tissue allows the reconstruction of natural breast ptosis and the addition of an implant provides the desired projection, but without being associated with complications related to flap transfer [11].

Nowadays, hybrid breast reconstruction combines breast implants with fat transfer to the breasts. This procedure is becoming increasingly popular with women and surgeons alike.

Fat is a filler with ideal properties: it naturally integrates into tissues, is autologous and in 100% biocompatible.

N. Zingaretti (✉) · G. Miotti · P. C. Parodi
Department of Medical Area (DAME), Clinic of Plastic and Reconstructive Surgery, Academic Hospital of Udine, University of Udine, Udine, Italy
e-mail: Piercamillo.parodi@uniud.it

© The Author(s), under exclusive license to Springer Nature Switzerland AG 2023
A. Di Giuseppe et al. (eds.), *Fat Transfer in Plastic Surgery*, https://doi.org/10.1007/978-3-031-10881-5_26

However, this is not the only function of lipofilling; fat is an active and dynamic tissue composed of several different cell types, including adipocytes and pre-adipocytes [12]. It displays regenerative potential and therapeutic effects that go beyond simple filling capability. Adipose-derived stem cells (ASCs) have a differentiation potential similar to that of other mesenchymal stem cells [13].

The advantages of combining fat grafting with prosthetic implants are:

- The ability of fat grafting to focally augment overlying soft tissue coverage and shape the breast beyond the limitations of an implant.
- Mask implant edges in thin patients and to improve the cleavage.
- Permit surgeons to select more superficial planes, rather than submuscular planes, for implant placement in patients with more limited soft tissue coverage [14, 15].
- Improve the texture of the irradiated skin by enhancing its vascular supply [13]. Literature has highlighted an important role of fat grafting in patients who underwent radiotherapy [16–18]. The chronic ischemic status of the irradiated tissue represented the rationale for the applicability of adipose-derived adult stem cell therapy, and the mechanisms of action of fat grafting were widely described in 2007 by Rigotti et al. [19].
- Improve the outcome for capsular contracture after breast implant reconstruction: capsular contraction is one of the most common causes of reoperation after breast augmentation or implant-based reconstruction, with a prevalence of 0.6–30% and is more frequent in patients who underwent radiotherapy [3, 4, 6, 16, 20]. The HBR protocol is associated with lower rate of capsular contracture, less breast pain at long follow-up times, and lower overall rates of revision surgery compared to standard expander-implant reconstruction [6, 16].

Calabrese et al. proposed a HBR protocol wherein fat grafting is performed during the course of the expander/implant reconstruction [16]: after unilateral mastectomy and immediate reconstruction with an expander, the patients were submitted to one to three serial deflation-lipofilling procedures. Positioning of a definitive prosthesis was performed during the last lipofilling session (Fig. 26.1). Although it is true that performing an intermediate lipofilling, or more, before the TE substitution determines a greater cost, as well as greater stress for the patient. However, as confirmed by several authors, one more intermediate procedure generates a better overall result of the reconstruction [1, 6]. This adjunct treatment permits us to obtain a result more natural and decrease the rate of future complications [13]. In fact, an additional procedure determines a greater engraftment of the fat with an increased subcutaneous thickness and less visibility of the prosthesis.

The surgical technique of hybrid breast reconstruction with fat is well described by Stillaert et al. [1], and it is characterized by four steps (Fig. 26.2):

- First: at the time of mastectomy a two-stage reconstruction is planned, and a tissue expander is inserted, preferably in prepectoral position.
- Second: about 2–4 weeks after surgery expansions are started.
- Third: a series of fat grafting to restore the subcutaneous tissue are performed, partially deflating the expander. Fat grafts are placed in the space between the skin and the capsule. This is a well-defined space that supports the survival of grafted fat. Fat grafting sessions are performed with a 3-month interval until an acceptable volume is obtained.
- Fourth: obtained the desired subcutaneous tissue thickness, an implant insertion to provide additional volume and core projection is performed. Implants used in this case are smaller than implants normally used without subcutaneous thickening derived by fat graft (Fig. 26.3).

Hybrid breast reconstruction can be performed also as immediate breast reconstruction and not only as two-stage technique [15]. In this case, a smaller implant has been chosen, and the desired volume is obtained through one or more planned lipotransfers.

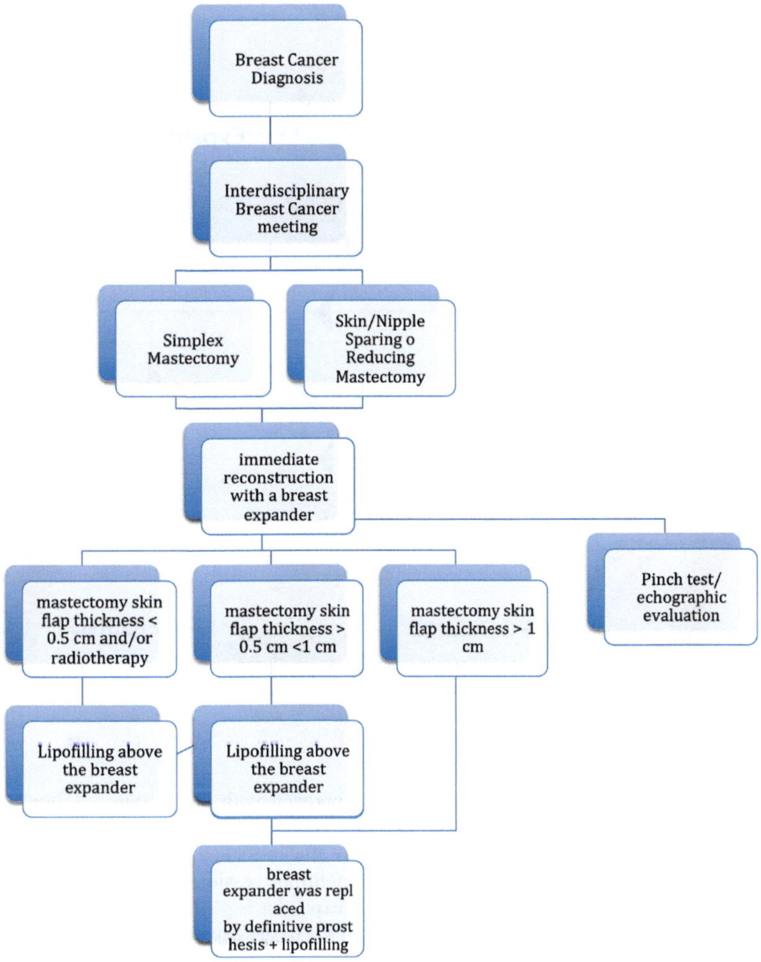

Fig. 26.1 The HBR protocol: the association of lipofilling and implant during the two-stage expander/implant procedure following mastectomy

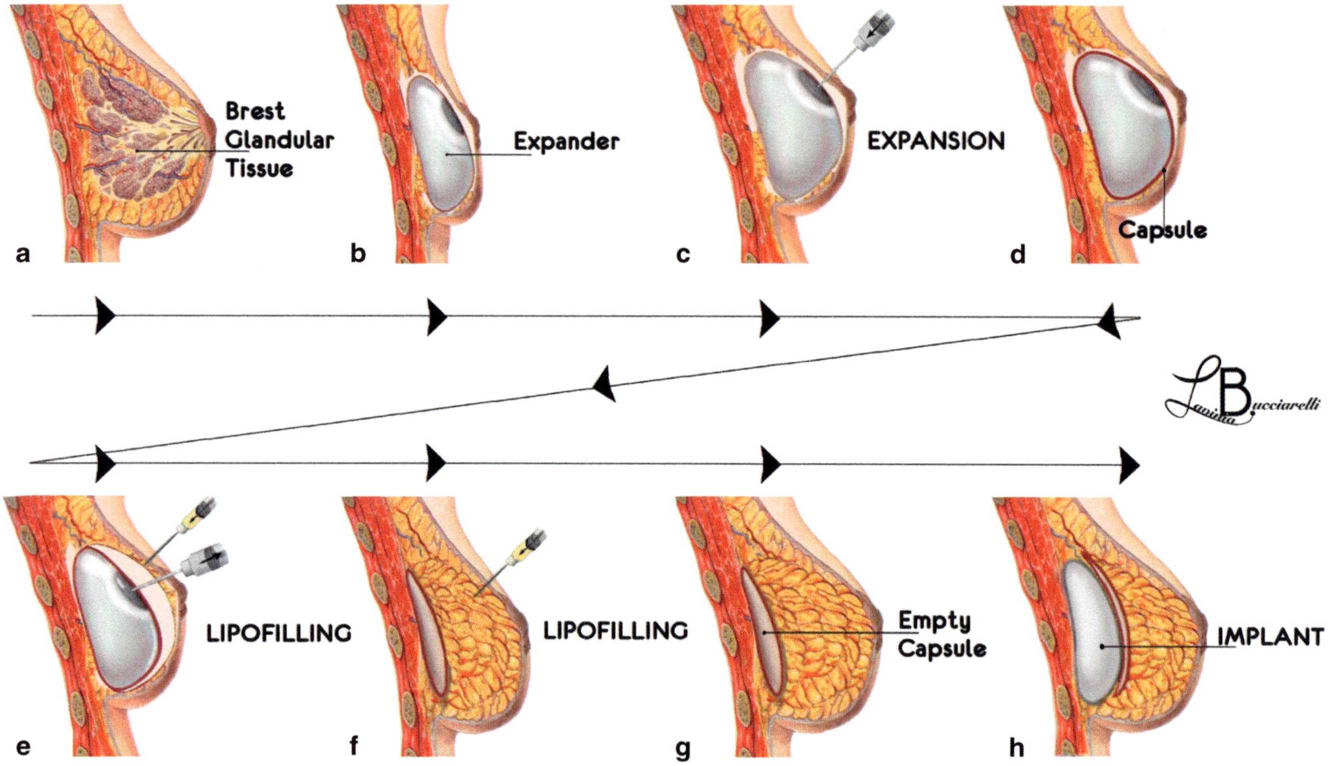

Fig. 26.2 Schematic drawing of the HBR approach with initial skin expansion followed by serial fat grafting sessions to augment the subcutaneous tissue layers. Depending on the skin thickness and previous radiotherapy, an average of 1–3 deflation-lipofilling sessions can be programmed. Normal breast anatomy (**a**). An expander is positioned in the prepectoral plane (**b**). By serial inflation of the tissue expander with saline (**c**), sufficient tissue cover was achieved for a second operative placement of a definitive breast implant. Capsule formation as a normal physiological response (red line) (**d**). Serial deflation and fat grafting in the subcutaneous plane (**e–g**). Removal of the expander with retained capsule and placement of definitive prosthesis (**h**). (Drawing ©Lavinia Bucciarelli, All Rights Reserved)

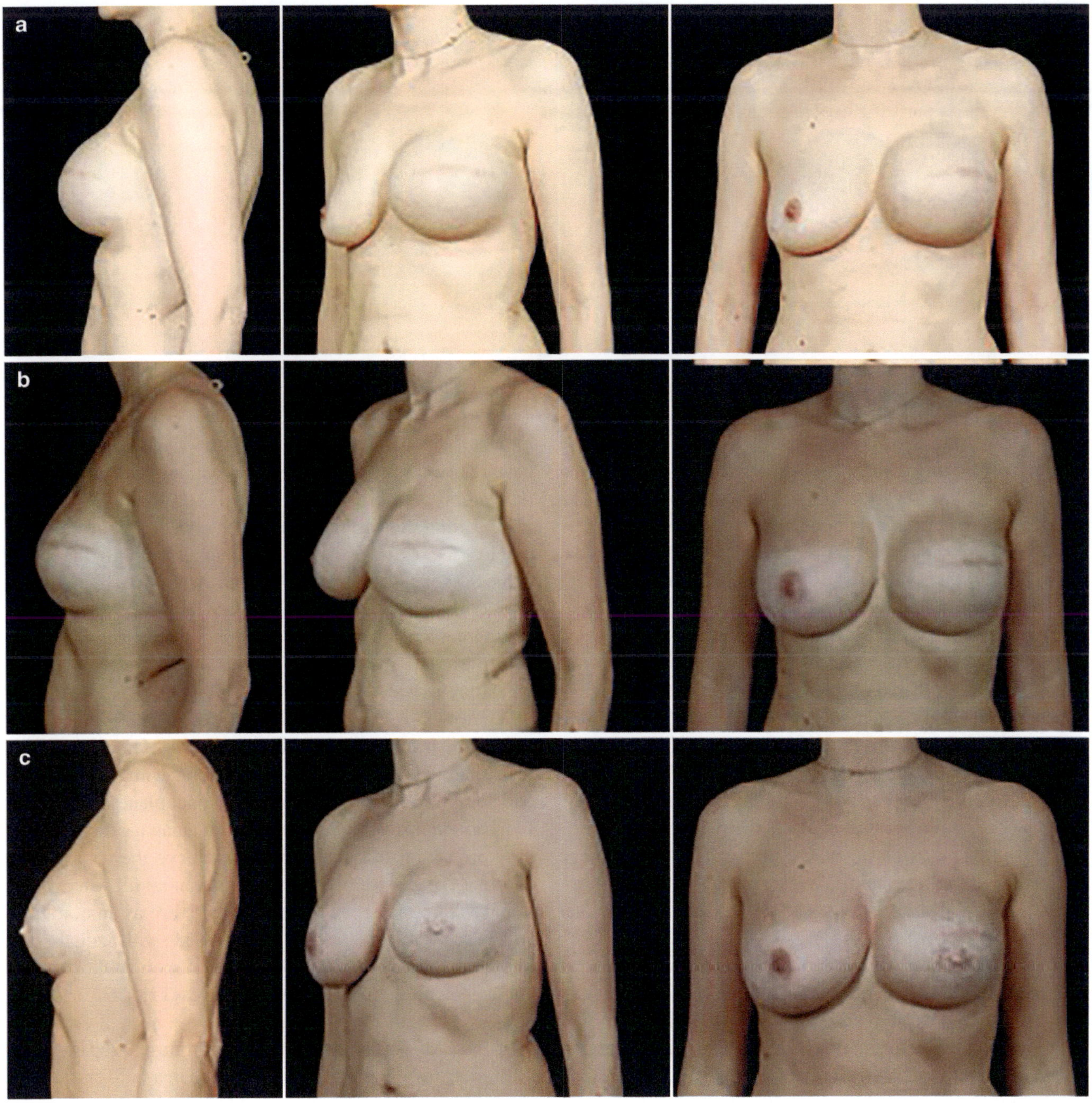

Fig. 26.3 (**a**) Preoperative pictures of a 42-year-old patient who underwent nipple-sparing mastectomy and expander immediate reconstruction. (**b**) Postoperative result after the first lipofilling session above the expander. (**c**) Postoperative result after 24 months from the second surgical step of expander/implant exchange procedure. The patient had one lipofilling session during the latter surgery

26.2 Hybrid Breast Augmentation

Even in breast implant augmentation, lipofilling has a well-recognized role [6, 16–18].

Fat grafting has emerged as a modality capable of addressing many of the limitations in techniques used in breast surgery. Kling et al. evidenced in their work how about one-fourth of American plastic surgeons use lipofilling in aesthetic breast surgery [21]. Cosmetic indications for fat grafting to the breast include primary breast augmentation without the need of a prosthetic device, volume restoration following implant removal, and concomitant injection alongside breast implant placement to disguise rippling and address visible step-offs that can sometimes be seen in thin

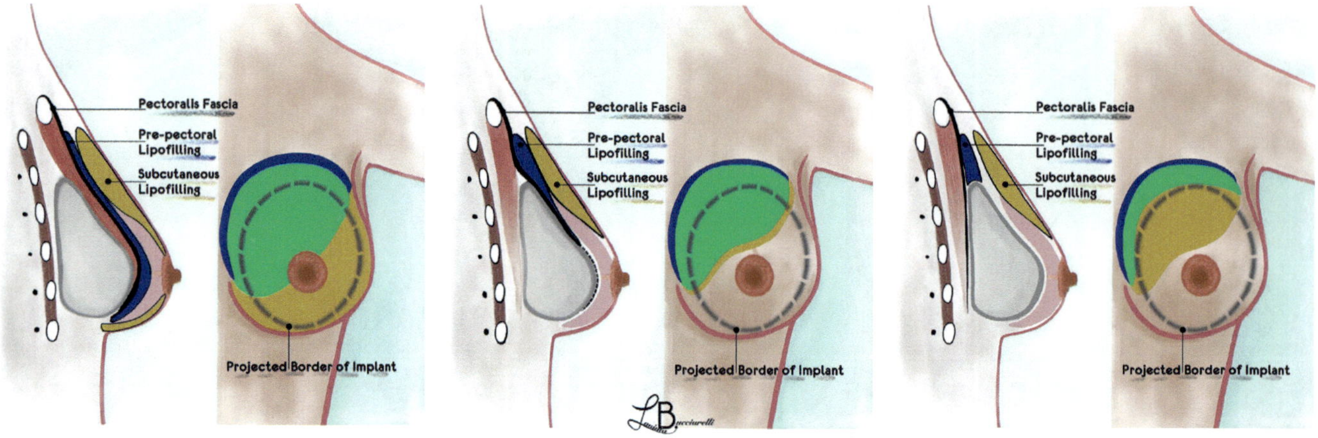

Fig. 26.4 Frontal and lateral views of the three potential routes of implant insertion in composite breast augmentation: the (*left*) submuscular, (*center*) subfascial, and (*right*) subglandular routes of insertion. (Drawing ©Lavinia Bucciarelli, All Rights Reserved)

patients with subglandular implant [22, 23]. Overall, fat grafting to the breasts for cosmetic purposes appears to be a generally safe, increasingly popular procedure with high levels of patients' satisfaction. It represents nowadays a fundamental, easy, fast, and effective complementary procedure integrating the traditional surgical approach to breast augmentation [24].

What prompted the increase in the use of hybrid reconstruction is the research of an adequate soft tissue coverage of the implant, particularly in thin patients with severe hypomastia. A paucity of breast tissue and subcutaneous fat can portend a poor cosmetic result despite excellent enhancement of volume and projection, due to the inability to hide the presence of the prosthesis under the patient's native tissue. This mismatch between size and soft tissue coverage results in implant palpability, edge visibility, and often rippling. This unnatural result often disconcerts patients and surgeons.

To date, prospective comparison of techniques (implant only vs composite/hybrid augmentation) is challenging. Additional studies retrospectively comparing composite breast augmentation to patients with similar BMI, ptosis, and breast size that underwent implant-only augmentation may further reinforce benefits of fat grafting with regards to minimizing sequelae of a deficient soft tissue envelope [25].

Historically, different techniques were proposed to give the best natural aspect to the augmented breast. Tebbets, in 2001, introduced the dual-plane technique: implants are placed in the subpectoral space, utilizing a dual plane pocket [26]. This technique increases overlying soft tissue coverage under varying amounts of the pectoralis major. Moreover, the introduction of highly cohesive, form-stable anatomic implants demonstrated decreased rates of wrinkling and rippling, as seen by Hedén [27]. And also, acellular dermal matrices have been proposed in aesthetic breast augmentation to improve soft tissue coverage and address soft tissue envelope complications [28]. Breast augmentation using only adipose tissue is not completely satisfactory. There is a natural limitation to the projection achieved with fat alone due to its soft nature: breast projection using only fat comes at the cost of a wide breast footprint, often wider than desired [29]. This technique remains a good option in limited and selected cases, when patients desire breast enhancement but do not want a prosthesis, also if this procedure gives limited breast projection.

The potential advantage of composite breast augmentation stems from the ability of the surgeon to impart customized shape and size change by working with two versatile and complementary media in two distinct planes (Fig. 26.4). In hybrid breast augmentation, one can manage the core volume projection of an implant and achieve the natural look and feel of fat, with overlay and preferential fill where additional volume is required. And if there are asymmetries of soft-tissue, volume can be addressed with equal-sized implants and differential fat transplantation [30]. Implants can be potentially placed in three spaces (Fig. 26.5):

1. *Subpectoral space*: pectoralis major muscle separates the implant from fat. Fat can be injected in a third space from within the pectoralis muscle and the superficial subcutaneous tissue. The deeper the implant insertion plane, the larger the volume and greater the footprint coverage of the fat overlay.
2. *Subfascial space*: the pectoralis fascia separates the implant from the fat, which can be placed in a third space from immediately above the fascia and the superficial subcutaneous tissue.
3. *Subglandular space*: implants lie above the pectoralis fascia and under the gland. The third space, where fat can be placed, is limited to the superficial subcutaneous tissue. Deeper injection risks migration of fat into the implant pocket.

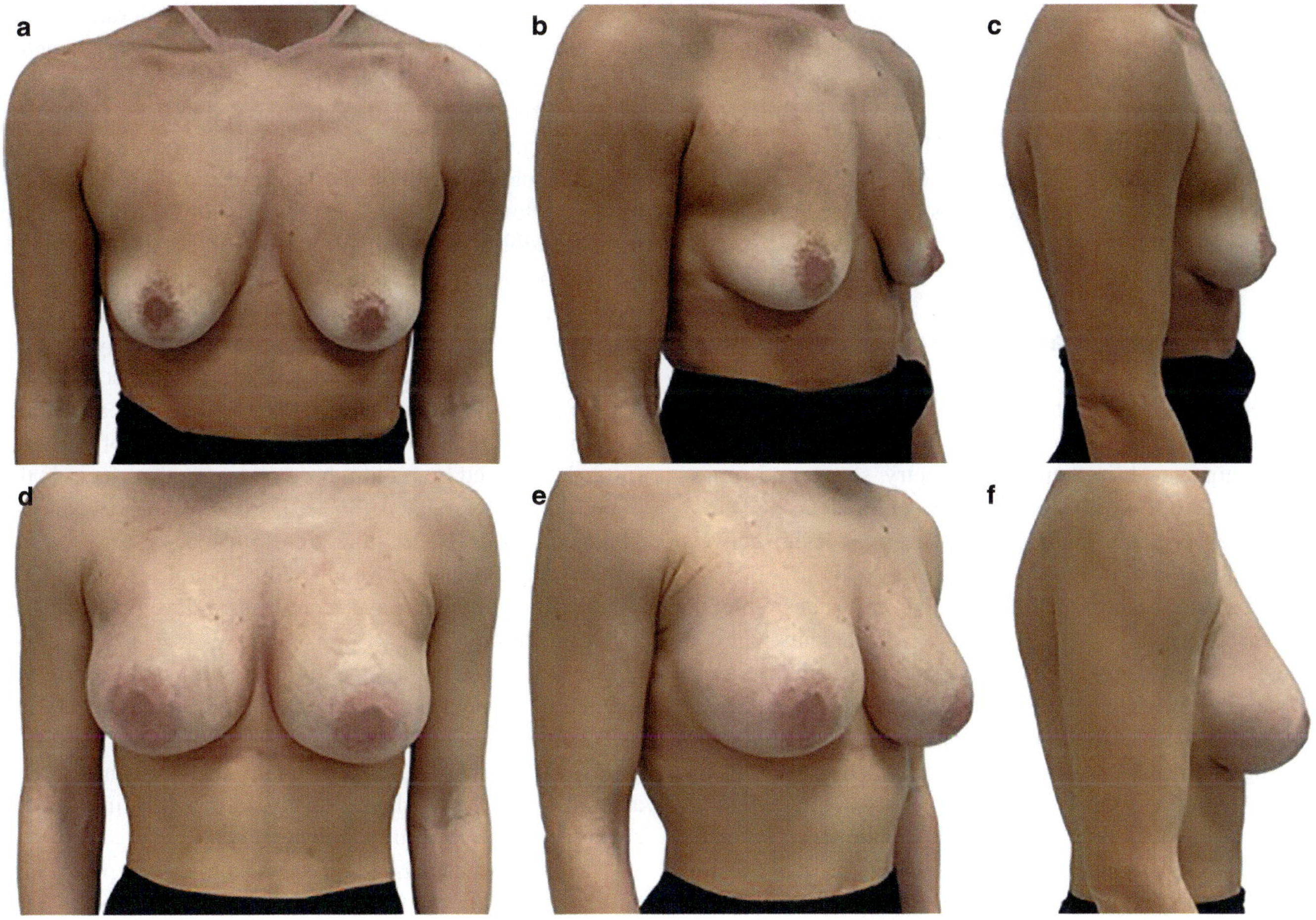

Fig. 26.5 (**a–c**) Preoperative view of a 30-year-old woman with bilateral mammary hypoplasia. (**d–f**) Postoperative view 12 months after subfascial breast augmentation and simultaneous lipofilling of the upper pole

Hybrid breast augmentation allows the use of subfascial and subglandular spaces for breast augmentation also in patients with inadequacy of soft tissue coverage, sparing the pectoralis muscle from overlay duties and avoiding potential animation and lateral drift deformities. Furthermore, by avoiding any retro-muscular dissection, patients rarely complain of pain being a significant issue in their postoperative course [20]. In patients with adequate breast soft tissue and favorable esthetics, there is no need for additional fat overlay. In this case, for Auclair et al., the best result can be obtained with implants alone [30].

There are some advantages in Hybrid Breast Augmentation:

- Combining breast implants with fat transfer allows a smaller size implant to be used since a portion of the desired volume will come from the patient's own fat.
- It enables the surgeon to increase volume while also correcting asymmetry between the breasts.
- It allows patients to further customize their result by making small adjustments with targeted fat transfers (i.e., patients wish to add a little extra volume to the upper part of the breast to accentuate cleavage).
- It allows a layer of natural fat to be transferred on top of the implants to provide a cushion, helping to disguise the implant and create a more natural look. It offers a special advantage to women who are older and/or have thinner overlying skin and tissues (small wrinkles or rippling of breast implants might show through their tissues because they are so thin).

Breast augmentation with fat has been condemned, during years, as causing calcifications and oil cysts that cannot be distinguished from malignancy, resulting in confusion, patient fear, and unnecessary biopsies. Current findings suggest that the incidence of cysts and calcifications in fat grafting is low and the radiologic ability to effectively discern them is technique-dependent [29, 31]. This problem is also reduced in hybrid breast augmentation than in fat augmentation due to low fat quantities transplanted.

Overall, implants and fat, working together, provide the best attributes each has to offer: the core volume projection of implants and the natural look and feel of fat.

As seen by Maximiliano et al. [32], hybrid breast augmentation technique may be limited by the volume of autologous fat that can be grafted, since there is an inverse relationship between the autologous fat graft volume and fat integration.

26.3 Breast Implants and Fat

As described by Bach et al., implants have been proposed as solution when autologous reconstructions failed to give a satisfactory result in terms of symmetry, shape, and projection [29]. But if the relationship between implants and autologous flaps has been largely studied, the relationship between them and fat graft is already under investigation.

Total breast reconstruction with fat only is desirable, but not reachable at the moment. The use of implants will continue to be necessary unless new technologies are discovered to stabilize the shape of full-fat reconstructed breasts, achieving better fat survival through fat enhancement techniques or through dedicated scaffolds to drive cell regeneration and shape definition [15].

The development of fat-injection techniques to refine the results of implant-based breast reconstructions or breast augmentation is changing the outcomes of breast surgery [15].

The advantages of fat grafting in implant-based reconstruction or breast augmentation are well described. Papadopoulos et al. observed better breast sensation, reduced foreign body sensation, reduced tension after radiotherapy, and reduced feeling of cold breast in patients treated with fat grafting. They also reported fat grafting can be used for pain relief due to capsular formation, regardless of the implant position to the pectoral muscle [33]. Roca et al. first evaluated the efficacy of autologous fat grafting in a porcine model as a treatment for capsular contracture [34]. Moreover, they confirmed that fat grafting, due to its regenerative properties, can ameliorate fibrotic damage in particular in implant-based reconstruction and successive radiation therapy. These results confirmed the trend of the past decade, when radiotherapy damage has been treated with autologous fat cell transplantation [35].

What implant has to be used with fat grafting is a debated argument. Maximiliano et al. obtained the ideal shape for hybrid breast augmentation using round implants and adding fat graft in a pattern similar to a cone shape on the upper portion of the implant [20]. In this way, they could calculate the accurate fat volume required for each of the three different implant projections. For example, for moderate profile implants, which have a proportionally larger base diameter, larger fat volumes were expected.

The evolution of silicone implants led to the introduction of new surfaces and viscoelastic gel properties. Smooth and micro/nanotextured implants were created with the aim to have very low roughness to avoid tissue ingrowth, minimize bacteria adhesion, and finally have the best natural aspect. Another reason strongly moved this evolution. During the past years, a new defined form of neoplasm, called BIA-ALCL, has been introduced. This rare form of lymphoma has been seen associated with macrotextured breast implants, but the physiopathology is still not completely clear to date. Chronic inflammation generated by these types of implants (probably caused by biofilm born on textured surface) is the main cause investigated [36, 37]. For this reason, smooth implants have been now re-evaluated, macrotextured implants have been removed in a lot of countries, and micro/nanotextured implants have been strongly developed as alternatives to smooth ones.

Breast augmentation recently had significant progress in aesthetic outcomes after the introduction of new-generation silicone gel implants and advances in fat grafting. Most surgeons associate fat grafting in the upper, medial and lateral areas of the breast, where thin tissue provides insufficient coverage and leads to implant visibility.

Fat grafting technique is a substantially safe procedure, but the result is unpredictable, with long-term retention rates varying between 10% and 80% [38].

If major complications occur in the first postoperative weeks, minor complications occur up to 6 months after surgery. The most common complication of fat grafting is fat necrosis, which can hesitate in oil cyst formation and calcifications. Palpable masses can develop due to fat necrosis.

The formation of cystic lesions, with the consequent functional and aesthetic manifestations to the patient, is mostly probably related to the massive transfer of fat clumps in the breast tissue, which led to fat necrosis and encapsulation [39]. Fat transplantation in the breast parenchyma might be another reason for this complication. Maintaining the viability of the harvested fat cells is crucial. The chances of survival are higher the less the fat graft is manipulated and the more quickly it is reinjected.

Ørholt et al. described an overall 2.1% of complications after hybrid breast augmentation [40]. In particular, they observed 1.6% of major complications (as seroma, hematoma, and infection) and 0.5% of minor complications (palpable cysts), but none of the patients underwent surgical revision.

References

1. Stillaert FBJL, Lannau B, Van Landuyt K, Blondeel PN. The prepectoral, hybrid breast reconstruction: the synergy of lipofilling and breast implants. Plast Reconstr Surg Glob Open. 2020;8:e2966.
2. Fabiocchi L, Semprini G, Cattin F, Dellachiesa L, Fogacci T, Frisoni G, Samorani D. "Reverse expansion": a new technique of breast reconstruction with autologous tissue. J Plast Reconstr Aesthet Surg. 2017;70(11):1537–42.
3. Pio Grieco M, Simonacci F, Bertozzi N, Grignaffini E, Raposio E. Breast reconstruction with breast implants. Acta Biomed. 2019;89(4):457–62.
4. Nava MB, Pennati AE, Lozza L, Spano A, Zambetti M, Catanuto G. Outcome of different timings of radiotherapy in implant-based breast reconstructions. Plast Reconstr Surg. 2011;128(2):353–9.
5. Nava MB, Catanuto G, Rocco N. Hybrid breast reconstruction. Minerva Chir. 2018;73:329–33.
6. Calabrese S, Zingaretti N, De Francesco F, Riccio M, De Biasio F, Massarut S, Almesberger D, Parodi PC. Long term impact of lipofilling in hybrid breast reconstruction: retrospective analysis of two cohorts. Eur J Plast Surg. 2020;43:257–68.
7. Vaia N, Lo Torto F, Marcasciano M, Casella D, Cacace C, De Masi C, Ricci F, Ribuffo D. From the "fat capsule" to the "fat belt": limiting protective lipofilling on irradiated expanders for breast reconstruction to selective key areas. Aesthet Plast Surg. 2018;42(4):986–94.
8. Momeni A, Kanchwala S. Delayed-immediate hybrid breast reconstruction-increasing patient input and precision in breast reconstruction. Breast J. 2019;25:898–902.
9. De la Torre JI, Fix RJ, Gardner PM, Vasconez LO. Reconstruction with the latissimus dorsi flap after skin-sparing mastectomy. Ann Plast Surg. 2001;46:229–33.
10. Lee HC, Lee J, Park S-H, Yoon E-S. The hybrid latissimus Dorsi flap in immediate breast reconstruction: a comparative study with the abdominal-based flap. Ann Plast Surg. 2020;86:394.
11. Momeni A, Kanchwala S. Hybrid prepectoral breast reconstruction: a surgical approach that combines the benefits of autologous and implant-based reconstruction. Plast Reconstr Surg. 2018;142(5):1109–15.
12. Domenis R, Lazzaro L, Calabrese S, Mangoni D, Gallelli A, Bourkoula E, Manini I, Bergamin N, Toffoletto B, Beltrami CA, Beltrami AP, Cesselli D, Parodi PC. Adipose tissue derived stem cells: in vitro and in vivo analysis of a standard and three commercially available cell-assisted lipotransfer techniques. Stem Cell Res Ther. 2015;6:2.
13. Serra-Renom JM, Munoz-Olmo JL, Serra-Mestre JM. Fat grafting in postmastectomy breast reconstruction with expanders and prostheses in patients who have received radiotherapy: formation of a new subcutaneous tissue. Plast Reconstr Surg. 2010;125:12–8.
14. Serratore F, Guerra M, Bucher S, Lay G, Ribuffo D. Outcomes of protective lipofilling on post-mastectomy radiotreated patients in immediate expander-implant breast reconstruction. ESJO. 2015;41:S5.
15. Nava MB, Catanuto G, Rocco N. How to optimize aesthetic outcomes in implant-based breast reconstruction. Arch Plast Surg. 2018;45(1):4–13.
16. Calabrese S, Zingaretti N, Zanin C, Fin A, Mura S, Parodi PC. Hybrid breast reconstruction: preliminary report. Plast Recontr Surg Glob Open. 2018;6:e1660.
17. Wan D, Rohrich RJ. Revisiting the management of capsular contracture in beast augmentation: a systematic review. Plast Reconstr Surg. 2016;137:826–41.
18. Del Vecchio DA. "SIEF"-simultaneous implant exchange with fat: a new option in revision breast implant surgery. Plast Reconstr Surg. 2012;130:1187–96.
19. Rigotti G, Marchi A, Galiè M, Baroni G, Benati D, Krampera M, Pasini A, Sbarbati A. Clinical treatment of radiotherapy tissue damage by lipoaspirate transplant: a healing process mediated by adipose-derived adult stem cells. Plast Reconstr Surg. 2007;119(5):1409–22.
20. Auclair E, Anavekar N. Combined use of implant and fat grafting for breast augmentation. Clin Plast Surg. 2015;42(3):307–14.
21. Kling RE, Mehrara BJ, Pusic AL, Young VL, Hume KM, Crotty CA, Rubin JP. Trends in autologous fat grafting to the breast: a national survey of the American Society of Plastic Surgeons. Plast Reconstr Surg. 2013;132(1):35–46.
22. Coleman SR, Saboeiro AP. Primary breast augmentation with fat grafting. Clin Plast Surg. 2015;42(3):301–6.
23. Davis MJ, Perdanasari AT, Abu-Ghname A, Gonzalez SR, Chamata E, Rammos CK, Winocour SJ. Application of fat grafting in cosmetic breast surgery. Semin Plast Surg. 2020;34:24–9.
24. Mione L, Caviggioli F, Vinci V, Lisa A, Barbera F, Siliprandi M, Battistini A, Klinger F. Fat graft in composite breast augmentation with Roun implants: a new concept for breast reshaping. Aesthetic Plast Surg. 2018;42:1465–71.
25. Salibian AA, Frey JD, Bekisz JM, Choi M, Karp NS. Fat grafting and breast augmentation: a systematic review of primary composite augmentation. Plast Reconstr Surg Glob Open. 2019;7(7):e2340.
26. Tebbets JB. Dual plane breast augmentation: optimizing implant-soft-tissue relationships in a wide range of breast types. Plast Reconstr Surg. 2001;107(5):1255–72.
27. Hedén P, Bronz G, Elberg JJ, Deraemaecker R, Murphy DK, Slicton A, Brenner RJ, Svarvar C, van Tetering J, van der Weij LP. Long-term safety and effectiveness of style 410 highly cohesive silicone breast implants. Aesthet Plast Surg. 2009;33(3):430–6.
28. Maxwell GP, Gabriel A. Acellular dermal matrix for reoperative breast augmentation. Plast Reconstr Surg. 2014;134(5):932–8.
29. Bach AD, Morgenstern IH, Horch RE. Secondary "hybrid reconstruction" concept with silicone implants after autologous breast reconstruction—is it safe and reasonable? Med Sci Mont. 2020;26:e921329.
30. Auclair E, Blondeel P, Del Vecchio DA. Composite breast augmentation: soft-tissue planning using implants and fat. Plast Reconstr Surg. 2013;132(3):558–68.
31. Bravo FG. Parasternal infiltration composite breast augmentation. Plast Reconstr Surg. 2015;135(4):1010–8.
32. Maximiliano J, Munhoz AM, Pedron M, de Oliveira ACP, Duarte DW, Neto R, Portinho CP, Collares MVM. Hybrid breast augmentation: a reliable formula for preoperative assessment of fat graft volume based on implant volume and projection. Aesthet Surg J. 2020;40(8):NP438–52.
33. Papadopoulos S, Vidovic G, Neid M, Abdallah A. Using fat grafting to treat breast implant capsular contracture. Plast Reconstr Surg Glob Open. 2018;6:e1969.
34. Roca GB, Graf R, da Silva Freitas R, Salles G Jr, Francisco JC, Noronha L, Maluf I Jr. Autologous fat grafting for treatment of breast implant capsular contracture: a study in pigs. Aesthet Surg J. 2014;34(5):769–75.
35. Nava MB, Catanuto G, Rocco N. How to optimize aesthetic outcomes in implant-based breast reconstruction. Arch Plast Surg. 2018;45:4–13.
36. Cordeiro PG, Ghione P, Ni A, Hu Q, Ganesan N, Galasso N, Dogan A, Horwitz SM. Risk of breast implant associated anaplastic large cell lymphoma (BIA-ALCL) in a cohort of 3546 women prospectively followed long term after reconstruction with textured breast implants. J Plast Reconstr Aesthet Surg. 2020;73(5):841–6.
37. Zingaretti N, Vittorini P, Savino V, et al. Surgical treatment of capsular contracture (CC): literature review and outcomes utilizing in revisionary surgery. Aesthetic Plast Surg. 2021;45:2036–47. https://doi.org/10.1007/s00266-021-02148-z.

38. Simonacci F, Bertozzi N, Pio Grieco M, Grignaffini E, Raposio E. Procedure, applications, and outcomes of autologous fat grafting. Ann Med Surg (Lond). 2017;20:49–60.
39. Kontoes P, Gounnaris G. Complications of fat transfer for breast augmentation. Aesthetic Plast Surg. 2017;41(5):1078–82.
40. Ørholt M, Larsen A, Hemmingsen MH, Mirian C, Zocchi ML, Vester-Glowinski PV, Herly M. Complications after breast augmentation with fat grafting: a systematic review. Plast Reconstr Surg. 2020;145(3):530e–7e.

27 Percutaneous Fasciotomy and Fat Grafting for the Correction of the Tuberous Breast Deformity

Patricia Gutierrez-Ontalvilla and Nina S. Naidu

Contents

27.1 Introduction

Tuberous breast deformity was first described by Rees and Aston in 1976 as a congenital malformation due to abnormal mammary development that becomes evident at puberty. This deformity results in a tubular shape of the breast resulting in hypoplasia of the lower quadrants, an elevated inframammary fold, herniation of the nipple–areola complex, and an increased areola diameter [1] (Fig. 27.1). This malformation frequently causes anxiety, lack of self-confidence, and isolation in teenagers [2]. The traditional solutions for correction have included breast remodeling techniques and the use of implants. However, correction of the tuberous breast with implants carries specific complications including asymmetry, implant displacement, rippling, capsular contracture, late seromas, and potential financial costs due to implant replacement [3]. Since Coleman's original description of lipostructure [4], we have implemented fat grafting as an option in the correction of tuberous breasts as it provides optimal results without the need for implants or flaps.

P. Gutierrez-Ontalvilla
Hospital La Fe, Unit of Plastic Surgery, Valencia, Spain
e-mail: patricia@dragutierrez.com

N. S. Naidu (✉)
Weill Cornell Medical Center, New York, NY, USA
e-mail: drnaidu@naiduplasticsurgery.com

© The Author(s), under exclusive license to Springer Nature Switzerland AG 2023
A. Di Giuseppe et al. (eds.), *Fat Transfer in Plastic Surgery*, https://doi.org/10.1007/978-3-031-10881-5_27

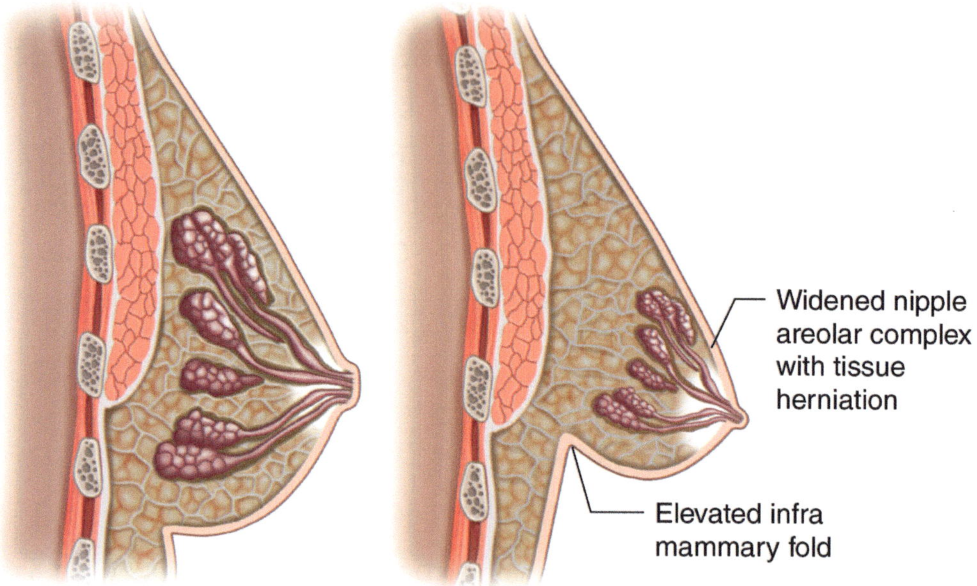

Fig. 27.1 Anatomy of the normal and tuberous breasts

27.2 Etiology

The pathology of the tuberous breast is generally congenital and has an unknown etiopathogenesis. There is a possible genetic role in the development of the tuberous breast deformity, with a possibility of parental consanguinity for such breast deformities [5]. Grolleau et al. have proposed a cause secondary to anomalies of the fascia superficialis, which involve strong adherence between the dermis and the muscular plane, restricting peripheral expansion of the breast during development at puberty [6].

27.3 Epidemiology

The exact incidence of the tuberous breast is unknown and would likely be difficult to quantitate due to the large number of women with mild deformity who may be unaware of their diagnosis. One retrospective analysis revealed a high prevalence of tuberous breast deformity in the general population (approximately 50%), and in particular in women requesting breast surgery [7].

27.4 Histopathology

Tuberous breast deformity is marked not only by thickening of the fascia superficialis but also fibrosis of the breast parenchyma fibrosis. The underlying pathogenesis includes significant differences in quantity and disposition of collagen fibers in patients with tuberous breasts as compared to normal breasts. Indeed, collagen fibers appear to be altered in both placement and quantity. These fibers sometimes appear to be assembled in bundles, contributing to the typical glandular firmness, fascial thickening, and general fibrosis [8].

27.5 Classification

The tuberous breast has different degrees of severity, and it is mandatory that the surgeon understand these varieties. By identifying the degree of deformity, the surgeon can design and perform the optimal surgical treatment for each patient.

There are two predominant classification systems within the literature.

27.5.1 Von Heimburg Classification [9]

Type I: Hypoplasia of the lower medial quadrant

Type II: Hypoplasia of the lower medial and lateral quadrants

Type III: Hypoplasia of the lower medial and lateral quadrants with deficiency of the skin in the subareolar region

Type IV: Severe breast constriction with minimal breast base

27.5.2 Grolleau Classification [6]

Type I: Hypoplasia of the lower medial quadrant

Type II: Hypoplasia of both lower quadrants

Type III: Hypoplasia of all four quadrants

27.6 History of Correction

Many surgical techniques have been proposed for correction of the tuberous breast deformity. Rees and Aston initially described radial incisions in the breast parenchyma for release of the constrictive bands with reduction of the nipple–areolar complex and placement of subglandular implants [10].

Ribeiro et al. have described a technique that uses an exclusively periareolar approach to divide the breast into two portions to disrupt the constricting ring and create an inferiorly based pedicle [11]. Mandrekas et al. proposed a similar approach, with a periareolar incision and transection of the constricting ring. Two vertical breast pillars are formed and folded together to provide volume to the lower pole of the breast [12].

The use of combined periareolar and inframammary incisions associated with breast implants was described by Teimourian and Adham [13]. Most patients with tubular breasts desire an increase in breast volume and improvement in shape, both of which are provided by this technique.

The current chapter presents a procedure for the correction of tuberous breasts with the use of percutaneous fasciotomies and fat grafting with areolar reduction [14]. This approach permits correction without the use of flaps and implants that might mandate future revision surgery.

27.7 Percutaneous Fasciotomy and Fat Grafting for the Correction of the Tuberous Breast Deformity

Prior to surgery, the breasts are marked with the patient in the standing position. Each breast is divided into four quadrants according to the method of Silva-Vergara to better assess the areas of deficiency and thus to estimate the quantity of fat to be injected (Fig. 27.2) [15].

The new inframammary fold is marked at a lower and more anatomic position. A circumareolar skin reduction is marked in all cases to allow for the reduction of the herniated tissue and to reduce the areolar diameter. The fat donor sites used are typically the abdomen, trochanteric zones, and inner thighs, although this varies depending upon both the individual patient's fat distribution and where the surgeon can most easily access the donor sites.

All surgical procedures are performed under general anesthesia. The nipple–areolar complex and areolar herniation are first reduced with the use of 38-mm marker to delineate the new areolar margin. The circumareolar incision is then closed in layers with 3-0 PDS sutures for the subcutaneous tissue, CV-3 Gore-Tex suture (Gore-Tex, W.L. Gore and Associates, Flagstaff, AZ, USA) placed in an interlocking round-block pattern to reduce the areolar diameter and a running 4-0 PDS for the cuticular margin (Figs. 27.3, 27.4, and 27.5). The fat donor sites are then infiltrated with saline solution containing epinephrine but without lidocaine (1000 cc saline solution with 1 cc epinephrine) in a 1:1 proportion to the amount of fat to be harvested.

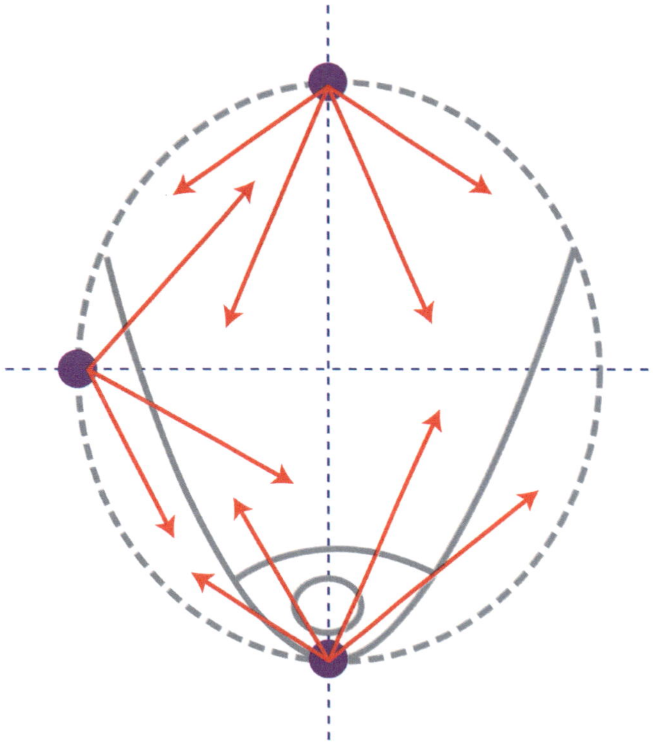

Fig. 27.2 Drawing representing planning for autologous fat grafting. Each breast is divided into quadrants and the cannula entry ports (located at the new inframammary fold, anterior axillary line, and the intersection of the upper quadrants) are designated. Introduction of the cannula through the areola is avoided so as to avoid harm to the glandular tissue

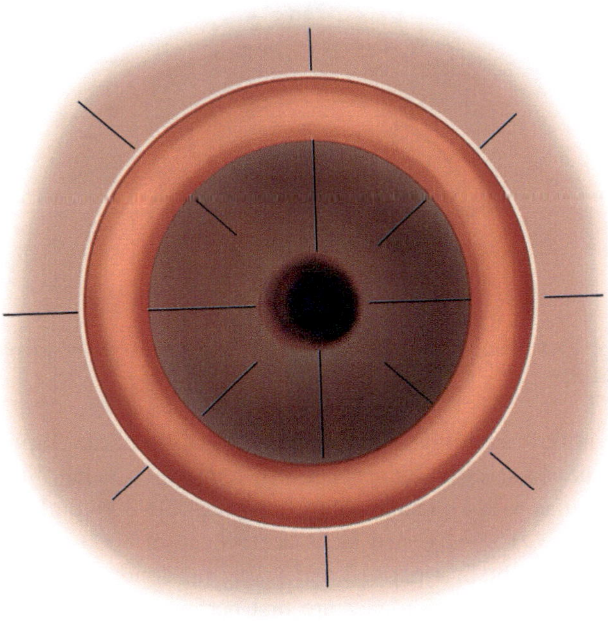

Fig. 27.3 Marked reduction of the nipple–areolar complex

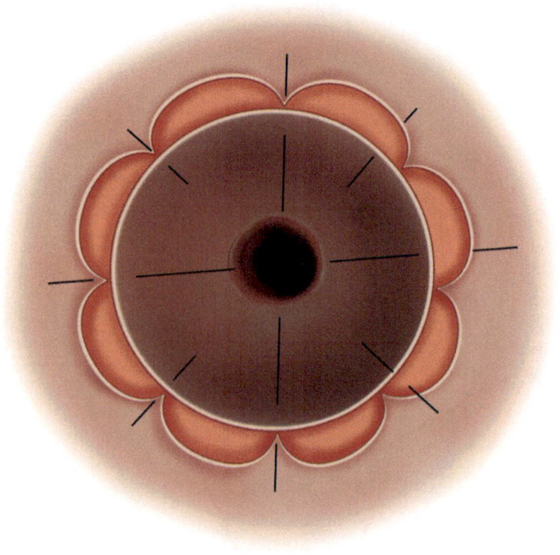

Fig. 27.4 Dermal closure of the reduced areola

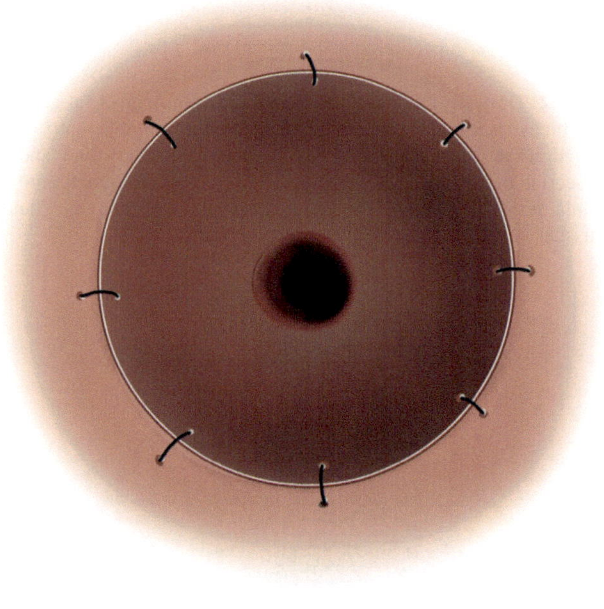

Fig. 27.5 Cuticular closure demonstrating reduced nipple–areolar complex

Following a tumescent period of 20–25 min, the adipose tissue is harvested by conventional liposuction using 3-mm cannulas with a vacuum pump, applying 250–300 mmHg of pressure. The harvested fat is centrifuged at 3000 rpm for 3 min to obtain a purified and viable fat donor graft pool. Percutaneous fasciotomies are first performed in the lower pole of the breast by making an incision with a 14-gauge

Fig. 27.6 A 2-mm Toledo cannula with a v-shaped tip for release of constricting bands (Black & Black Surgical, Tucker, GA, USA) (Courtesy of Black & Black Surgical, ©2022 Tucker, GA, USA, All Rights Reserved)

catheter followed by the use of a 2-mm-diameter Toledo cannula with a V-shaped tip to release the adherent tissues in the constricted lower pole (Black & Black Surgical, Tucker, GA USA) (Fig. 27.6).

Fat is then reinjected into the breasts through three designated sites located at the inframammary fold, anterior axillary line, and upper pole (Fig. 27.7) using 3-cc syringes with 9-cm-long, 1.6-mm-diameter Coleman style 2 concave, or straight blunt cannulas (Mentor, Santa Barbara, CA, USA). Fat grafting is performed in layers from deep to superficial, beginning above the pectoralis major muscle to the subcutaneous tissues, just until skin turgor is achieved to prevent ischemia of the grafted cells. The injection sites are covered

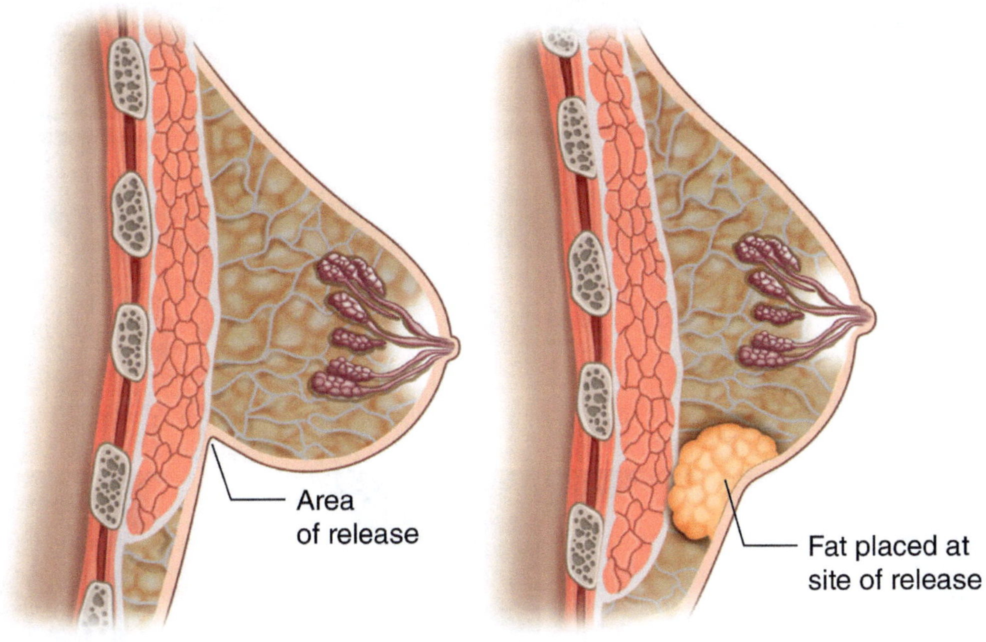

Fig. 27.7 Release of the lower breast pole and fat grafting technique

with Steri-Strips (3 M, Maplewood, MN, USA). Dressings consisting of paper tape are applied over and around the breast mound as a soft splint to minimize edema without constricting blood flow to the grafted tissue.

The postoperative protocol consists of an office visit at 7 days postoperative for removal of the dressings and evaluation. Suture removal from the liposuction sites is performed 2 weeks following surgery. Patients are evaluated at 1 month, 3 months, 6 months, and 12 months.

27.8 Prognosis/Results

Approximately 70% retention of volume is typically obtained by 3 months following each fat grafting procedure, as determined by physical examination and review of postoperative photographs. There is generally a high satisfaction rate secondary to reshaping of the breast mound, reduction of areolar herniation, and increased breast volume [14] (Figs. 27.8, 27.9, and 27.10).

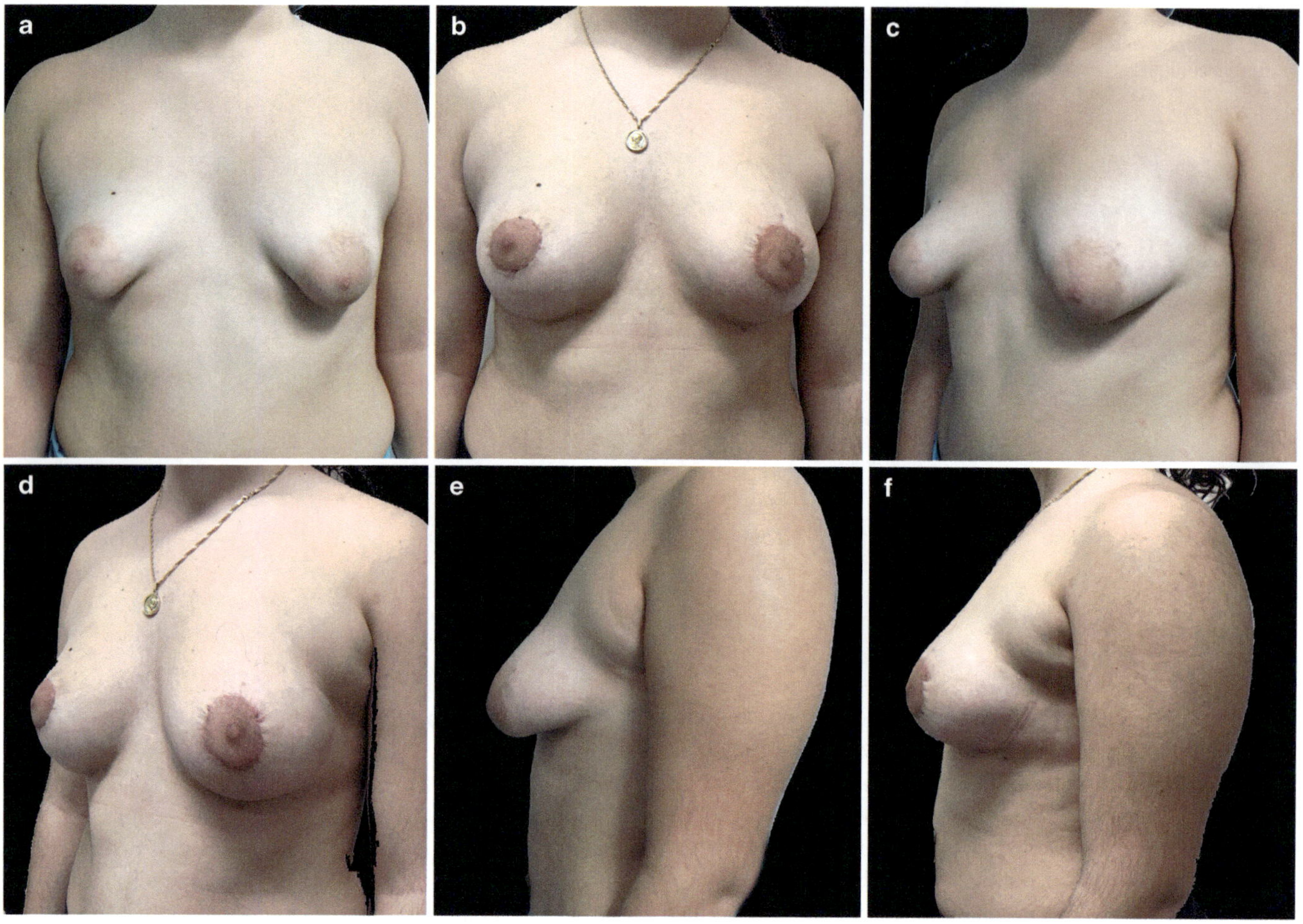

Fig. 27.8 (**a**–**f**): Preoperative and postoperative views of a 15-year-old girl who underwent two sessions of fat grafting with correction of the size and herniation of the nipple–areola complex. The patient is shown prior to surgery and 13 months following her second operative session

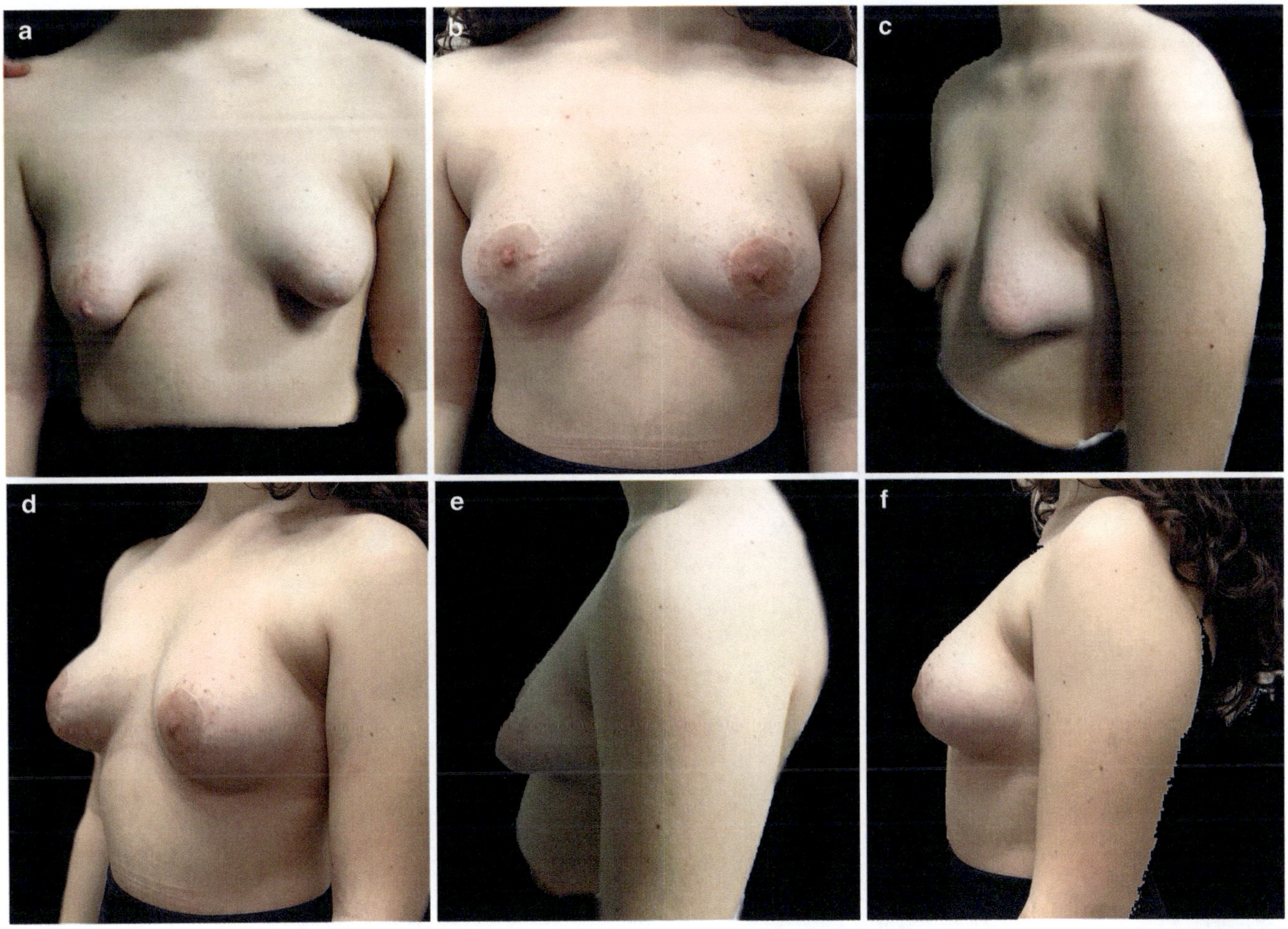

Fig. 27.9 (**a**–**f**): Preoperative and postoperative views of a 15-year-old girl who underwent two sessions of fat grafting with correction of the size and herniation of the nipple–areola complex. The patient is shown prior to surgery and 20 months following her second operative session

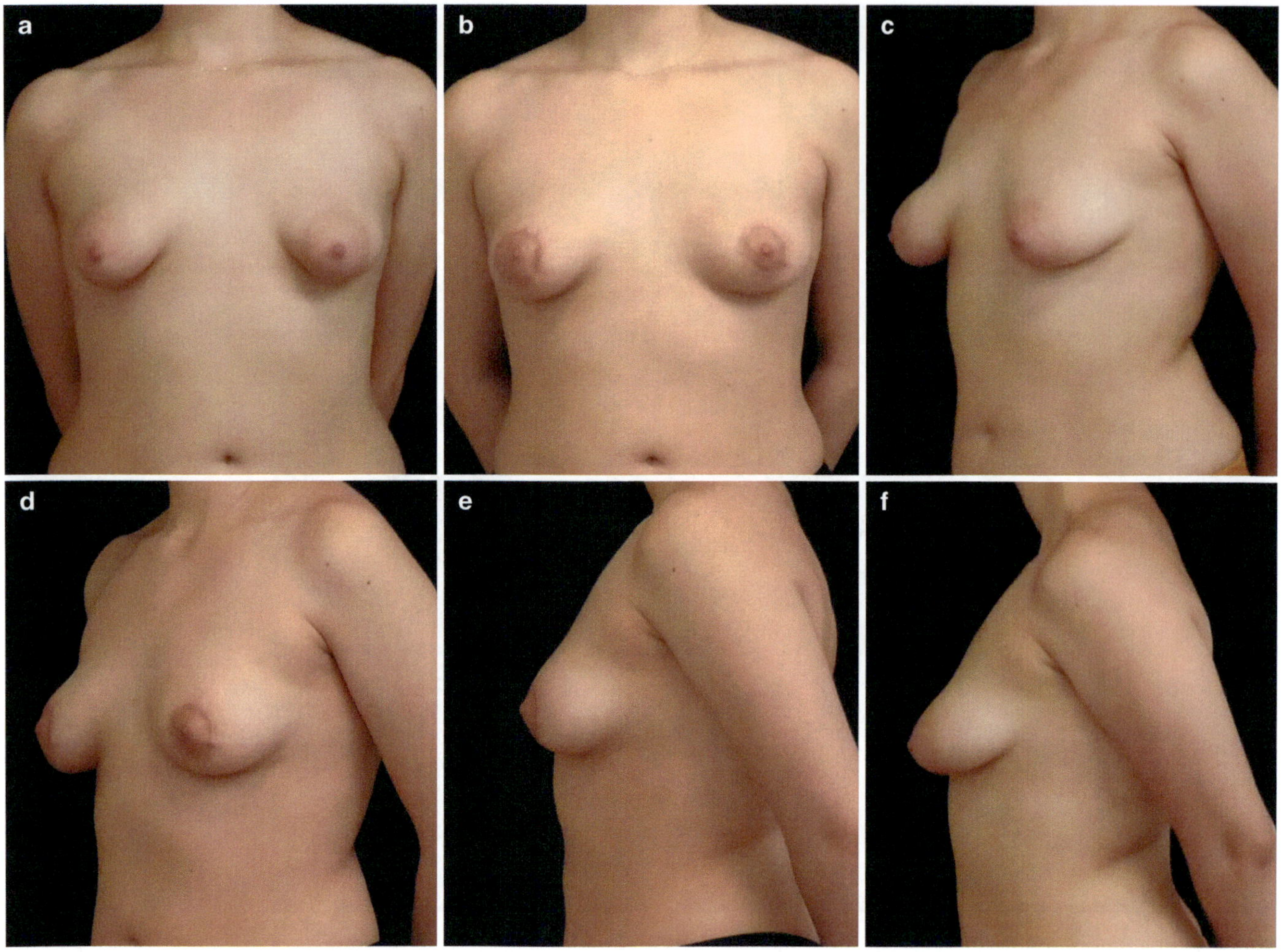

Fig. 27.10 (**a**–**f**): Preoperative and postoperative views of a 27-year-old woman who underwent one session of fat grafting with correction of the size and herniation of the nipple–areola complex. The patient is shown prior to surgery and 12 months following surgery

27.9 Complications

Severe complications including infection, dehiscence, or hematoma have not been observed [14]. Asymptomatic oil cysts (less than 10 mm in diameter) may appear and can be detected with routine ultrasound. Ultrasound examinations by a radiologist specializing in breast imaging are recommended as part of a standard protocol beginning at 6 months postoperative. Other complications include palpable, superficial, and painful nodules secondary to fat necrosis which can be excised under local anesthesia. Scar widening requiring revision and the need for additional volume are additional possibilities.

27.10 Conclusions

Autologous fat grafting in conjunction with percutaneous fasciotomy is a validated and versatile tool in aesthetic breast surgery. When used in conjunction with the reduction of the nipple–areolar complex and breast tissue herniation, it becomes especially useful for the treatment of tuberous breast deformity. This technique is especially applicable in teenage girls who suffer from the social stigmata of breast deformity, but for whom the placement of implants or use of flaps would be premature given the potential for further breast growth. It is a reliable method which produces excellent results with minimal complications.

References

1. Nahabedian MY. Breast deformities and mastopexy. Plast Reconstr Surg. 2011;127(4):91e–102e.
2. Latham K, Fernandez S, Iteld L, et al. Pediatric breast deformity. J Craniofac Surg. 2006;17:454–67.
3. Forster NA, Kunzi W, Giovanoli P. The reoperation cascade after breast augmentation with implants: what the patient needs to know. J Plast Reconstr Aesthet Surg. 2013;66:313–22.
4. Coleman SR, Saboeiro AP. Fat grafting to the breast revisited: safety and efficacy. Plast Reconstr Surg. 2007;119:775–85.
5. Dessy LA, De Santo L, Onesti MG, Fallico N, Mazzocchi M. Tuberous breast and predisposition to breast deformity in consanguineous. Breast J. 2018;24(1):51–4.
6. Grolleau JL, Lanfrey E, Lavigne B, Chavoin JP, Costagliola M. Breast base anomalies: treatment strategy for tuberous breasts, minor deformities, and asymmetry. Plast Reconstr Surg. 1999;104:2040–8.
7. Klinger M, Caviggioli F, Giannasi S, Bandi V, Banzatti B, Veronesi A, et al. The prevalence of tuberous/constricted breast deformity in population and in breast augmentation and reduction mammaplasty patients. Aesthetic Plast Surg. 2016;40(4):492–6.
8. Klinger M, Caviggioli F, Klinger F, Villani F, Arra E, Di Tommaso L. Tuberous breast: morphological study and overview of a borderline entity. Can J Plast Surg. 2011;19(2):42–4.
9. von Heimburg HD, Exner K, Kruft S, Lemperle G. The tuberous breast deformity: classification and treatment. Br J Plast Surg. 1996;49:339–45.
10. Rees TD, Aston SJ. The tuberous breast. Clin Plast Surg. 1976;3(2):339–47.
11. Ribeiro L, Canzi W, Buss A Jr, Accorsi A Jr. Tuberous breast: a new approach. Plast Reconstr Surg. 1998;101(1):42–50.
12. Mandrekas AD, Zambacos GJ, Anastasopoulos A, Hapsas D, Lambrinaki N, Ioannidou-Mouzaka L. Aesthetic reconstruction of the tuberous breast deformity. Plast Reconstr Surg. 2003;112(4):1099–108.
13. Teimourian B, Adham MN. Surgical correction of the tuberous breast. Ann Plast Surg. 1983;10(3):190–3.
14. Gutierrez-Ontalvilla P, Naidu NS, Blanco EL, et al. Autologous fat grafting with percutaneous fasciotomy and reduction of the nipple–areolar complex for the correction of tuberous breast deformity in teenagers. Aesthetic Plast Surg. 2020;44(2):264–9.
15. Silva-Vergara C, Fontdevila J, Weshahy O. Fat grafting technique, a paradigm shift in the treatment of tuberous breast. World J Plast Surg. 2018;7(1):72–7.

Postbariatric Breast Reshaping and Fat Grafting

28

Vincenzo Vindigni, Paolo Marchica, and Franco Bassetto

Contents

28.1 Introduction to Body Contouring and Breast Reshaping in Massive Weight Loss Patients

Massive weight loss patients (MWL) number is increasing, especially due to the development and the request of bariatric surgery, hence the problems related to redundant skin and soft tissues following massive weight loss. An important weight loss involves all the body districts, from the abdomen to the thighs, the arms, the face, and the breasts as well. A MWL patient, thus, asks for body contouring surgery in order to reshape his body morphology, to resolve functional and aesthetic problems. The redundant soft tissue, indeed, may cause functional impairment in wearing clothes, doing physical exercise, promoting the development of skin infection, and determining social concern with regard to the image of the own body. Therefore, an MWL patient deserves the chance to restore his body natural appearance at cost, however, of facing with scars.

The mammary gland is a very important anatomical site with regard to woman due to its role in the canons of beauty and, nowadays, is gaining importance for man as well. The MWL population usually presents a ptotic, deflated breast, medialized (sometimes lateralized) nipple-areola complex and chest or axillary rolls. Hence, MWL patients ask to the plastic surgeons to reshape the body, and the breast surgery is one of most requested procedures for both male and female patients.

V. Vindigni · P. Marchica (✉) · F. Bassetto
Clinic of Plastic and Reconstructive Surgery, University of Padua, Padua, Italy
e-mail: vincenzo.vindigni@unipd.it; franco.bassetto@unipd.it

28.2 Physiopathology of Breast Deformity, Surgical Indication, and Contraindication

Post-bariatric patients present a reduction in collagen, while the elastic fibers remain unchanged. This leads to a mismatch in these components ratio that finally determines a soft tissue ptosis which involves all body district, including the breast.

Traditional techniques are often inadequate in massive weight loss patients because of the extent of the deformity, the lack of structural volume or support, and the inelastic skin envelope. The traditional breast reshaping focuses on the skin envelope, which is resected, tightened, and redraped over the breast mound and none or minimal surgery is provided to the parenchyma. In MWL patient, instead, more attention has to be paid with regard to the glandular reshaping, to minimize the incidence of recurrent ptosis, which is so frequently relapsing, and to maintain the shape (Fig. 28.1).

© The Author(s), under exclusive license to Springer Nature Switzerland AG 2023

A. Di Giuseppe et al. (eds.), *Fat Transfer in Plastic Surgery*, https://doi.org/10.1007/978-3-031-10881-5_28

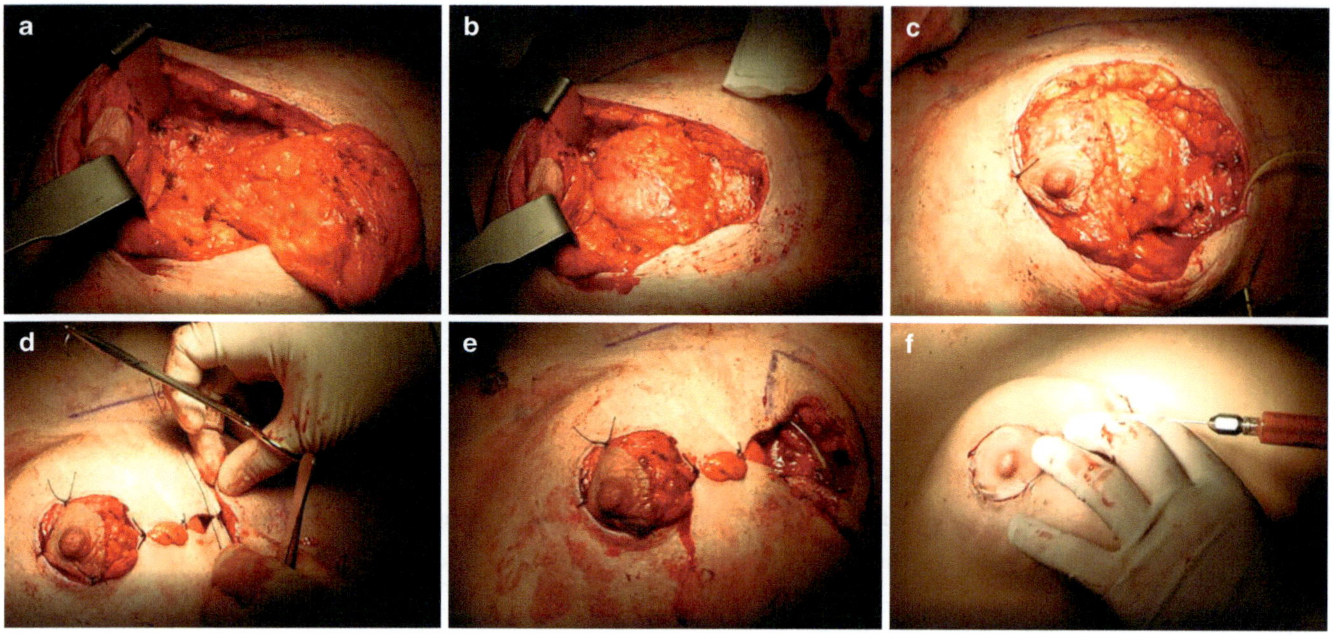

Fig. 28.1 Auto-augmentation using dermoglandular flap and fat grafts. The dermoglandular flap is prepared (**a**) and fixed in the desired position (**b**, **c**). The periareolar and vertical incisions are approximated (**d**). Excess of skin is removed (**e**). Fat grafts are used to improve the volume and the profile of the breast (**f**)

To understand the deformity of an MWL patient's breast, the surgeon should consider: the volume of the gland, that is usually reduced, thus the deflation, except for the cases of macromastia; the degree of ptosis, which should be mild, moderate, or severe, according to Regnault; the NAC position, which is frequently medialized; the condition of the skin envelope that is usually stretched and anelastic and less susceptible to skin retraction, thus it needs a skin resection accompanied to a parenchymal reshape to maintain the breast final shape; finally, the associated chest, axillary, and abdominal deformities (chest and axillary rolls, redundant upper abdomen skin) that may jeopardize the final result, if not considered and eventually treated.

The choice of the right technique is demanded to the surgeon who can take advantages from the Pittsburgh Rating Scale (PRS). This classification described by Song in 2005 recommends the options to consider for surgical improvement and it is applied to all body districts. With regard to the breast, it classifies a range of breast deformities occurring after MWL and provides an objective scoring system which is based on the severity of the contour deformities.

The PRS classifies as Grade 0, the patients which present normal breast findings and they do not need any surgery. The Grade 1 consists in Ptosis grade I and II or severe macromastia. These patients need a traditional mastopexy, reductive mammoplasty, or augmentation techniques depending on the single case. The Grade 2 presents a severe ptosis (grade III) or a moderate volume loss or a constricted breast. Such patients need to be addressed with a traditional mastopexy with or without an augmentation mammoplasty. Finally, the Grade 3 consists in severe lateral rolls with or without a severe volume loss and loose skin. These patients are the most difficult to deal with, and the breast mound needs a complex parenchymal reshaping, even with dermal suspension and sometimes an auto-augmentation techniques may be considered.

The Pittsburgh Rating Scale

Area		Scale	Preferred procedure
Breasts	0	Normal	None
	1	Ptosis grade 1–2 or severe macromastia	Traditional mastopexy, reduction, or augmentation techniques
	2	Ptosis grade 3 or moderate volume loss or constricted breast	Traditional mastopexy ± augmentation
	3	Severe lateral roll and/or severe volume loss with loose skin	Parenchymal reshaping techniques with dermal suspension, consider autoaugmentation

Song AY, Jean RD, Hurwitz DJ, Fernstrom MH, Scott JA, Rubin JP. A classification of contour deformities after bariatric weight loss: the Pittsburgh Rating Scale. Plast Reconstr Surg. 2005 Oct;116(5):1535–44; discussion 1545–6. https://doi.org/10.1097/01.prs.0000182606.92069.13.

Not all patients present with these deformities after MWL and each of them must be evaluated individually. Indeed, the surgical treatment of the breast must be tailored to the single case.

In addition to the breast physical findings described by the PRS, the surgical indications to breast reshape in MWL patients include a stable weight control within the previous 6 months and the achievement of a balance in the nutritional status. As a matter of fact, most of MWL patients had weight loss after bariatric surgery, thus they need of a postoperative recovery time to adjust their gastrointestinal physiology and resolve the eventually occurred nutritional deficits.

A globally considered guidelines to body contouring surgery after massive weight loss were released by BAPRAS (*British Association of Plastic, Reconstructive and Aesthetic Surgeons*) in 2017 and, to date, they are the most complete recommendations about this surgery. An adequate preoperative BMI seems to be one of the most important factors pivotal to obtain a successful outcome. Absolute contraindications to breast reshaping are very few and consist in active breast cancer, active skin infection, and systemic disorders, which preclude a general anesthesia. Most of the contraindications are classified as relative and consist in a BMI higher than 35 kg/m^2, while a BMI $\leq$ 30.0 kg/m^2 is recommended to obtain a stable and good result. Smoking is considered a negative prognostic factor in conditioning the outcomes, but it is not an absolute contraindication to surgery. Patients are advised to quit smoking at least 4 weeks before surgery and avoid tobacco use up to 4 weeks after surgery, to reduce the complication rate. Finally, before surgery, it is appropriate to avoid the assumption of medications that increase the risk of bleeding or cause an increase in thrombotic risk. A wound healing disorder and systemic disease may represent a relative contraindication; thus, they should be carefully investigated, and the patient stabilized before surgery.

Such a real query is whether the fat grafting is to be performed. Little asymmetries among the breast which are not solvable with parenchymal and skin reshaping techniques or cases with patient refusing breast implants or the ones who had little benefits obtained from auto-augmentation flaps may be the ideal candidates to fat grafting. Finally, its use in oncological patients is still debated by literature and last discussions and experts' positions are listed below.

28.3 Surgical Techniques and the Role of Fat Grafting

After the patient is individually staged and investigated for what it may concern the breast deformity, the technique is chosen and tailored to the single breast.

As discussed by Song, several techniques may be used to reshape the breasts, depending on breast physics and these may be divided into:

Augmentation techniques: a breast implant (silicone or polyurethane-coated silicone prosthesis) is placed in a subglandular, submuscular, subfascial, or dual plane position and sometimes associated with a mastopexy technique; this category allows to act on the breast mound and on the skin envelope (Fig. 28.2).

Mastopexy and reductive mammoplasty: the breast is lifted and reduced in volume, if needed and both the parenchyma and the skin envelope are reshaped to improve the aesthetic and functional outcome. A mild deformity with a minimal ptosis may be treated with a periareolar reduction mastopexy as described by Benelli. This technique allows a parenchymal reshape through a periareolar incision and performing a criss-crossing of the medial and lateral lower quadrants of the breast, below a superiorly based NAC flap. A breast implant may be positioned through the same access if needed. Benelli's technique has the advantages of hiding the scar along the areolar boundaries. However, the nipple-areola complex may be slightly repositioned due to a limited skin excision (crescent or donut excision) and an excessive resection may cause a widening of the scar and of the new areola. Hence, a round block suture may limit the tension on the wound edges, though determining a flattening of the new breast. Moderate deformities need to be addressed with more aggressive techniques, which consequently need longer and more visible scars such as vertical incision techniques (Lassus, Lejour, etc.). A vertical technique incision involves a periareolar and a vertical inferior pole incisions, which allow to treat a major ptosis and a better parenchymal reshaping. It may be combined with an auto-augmentation technique or a breast implant positioning or, if needed, it allows to perform a reductive mammoplasty (Fig. 28.3). However, it has the problem of a late aesthetically pleasant shape because an excessive projection of the upper pole is almost always present in the first 2–3 months postoperatively and the NAC may initially be facing down or present a bottoming out deformity. Finally, a severe deformity may only be addressed with an inverted-T incision pattern to excise horizontal and vertical skin excess and allow to reshape a severe deformed breast. Different techniques may be used, and among them the superior pedicle technique by Pitanguy is the best known. These techniques are characterized by a wide freedom of resection in each quadrant of the breast with a better chance to adequately reshape the gland, at cost of more extended scar (periareolar, vertical, and horizontal along the inframammary fold).

Autoaugmentation techniques, with eventual dermal suspension: these techniques include the procedures that take advantages of the auto-prostheses concept (part of breast parenchyma is used in order to increase the volume and the projection of the mammary gland, as the inferior pedicle technique described by Ribeiro) and may be combined with dermal or dermoglandular flaps that are anchored to the chest wall (Fig. 28.1).

Fat grafting represents an additional auto-augmentation technique (Fig. 28.1). The most recent International Expert

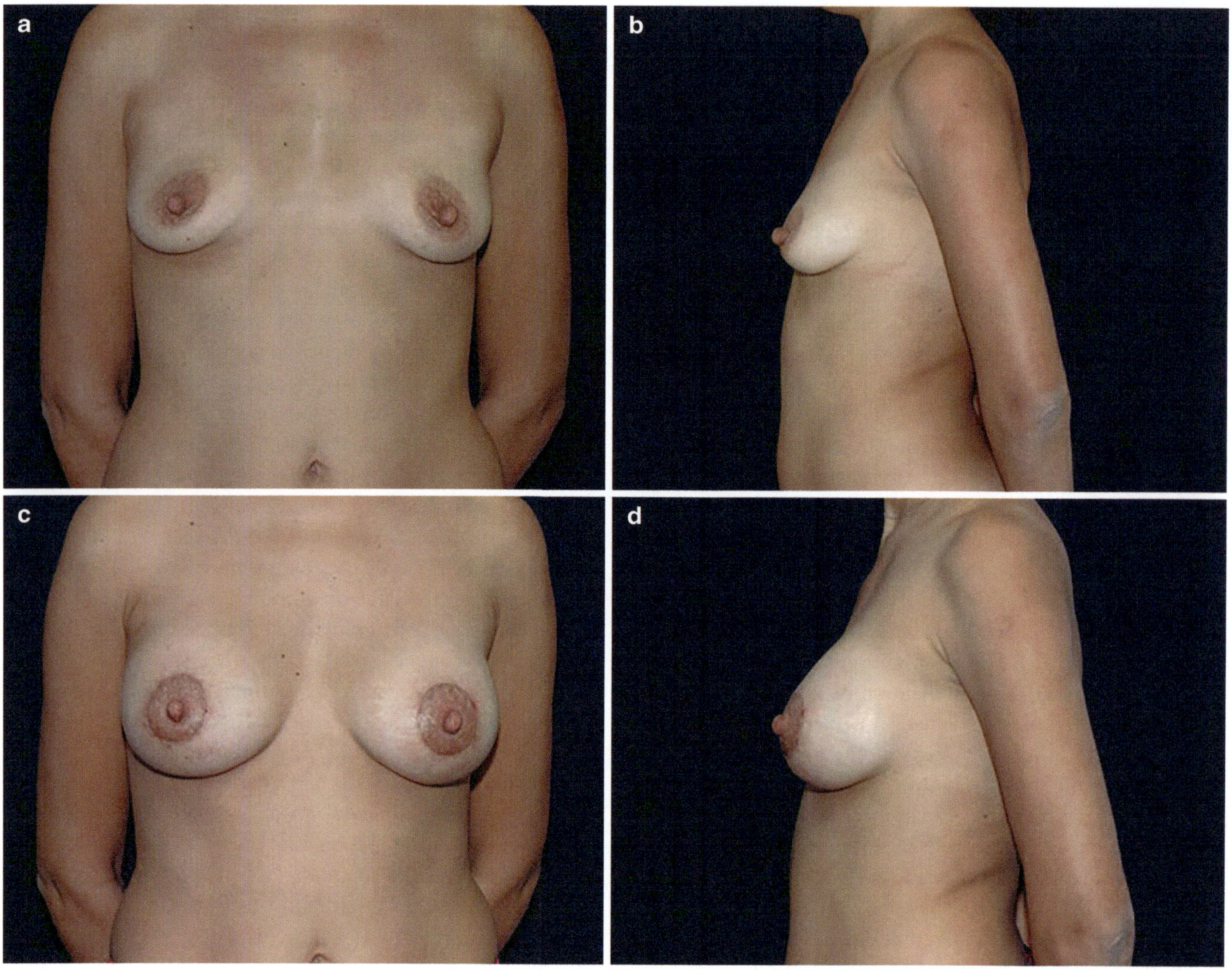

Fig. 28.2 Breast augmentation with subglandular round prostheses. Preoperatory images (**a**, **b**) and postoperatory results after 1 year (**c**, **d**)

Panel Consensus on Fat Grafting of the Breast in 2019 produced a document in which the authors prepared a set of ten "key statements" that have been discussed in a dedicated face-to-face session during the meeting. Despite this consensus conference presented some limitations (each key statement was not assessed with a standardized methodology and only the Level of Evidence (LoE) according to Oxford Criteria was evaluated) and remained expression of a group of expert and not guidelines for clinical practice, it provided some advice to plastic and reconstructive surgeons approaching fat grafting. We report three of them, which in our opinion represent the most important statements emerged from this consensus, with regard to the use of fat grafting in aesthetic breast surgery and fat grafting oncological safety:

- "Fat grafting is a useful tool in aesthetic breast surgery in association to implants (composite/hybrid breast augmentation) or for minor refinements. Fat grafting could be also considered as an alternative to implants for breast augmentation with adequate patient information about the achievable results and the number of sessions needed" (LoE IV).
- "Oncological concerns have risen with the use of fat grafting for breast reconstruction. There is an evident contradiction between experimental and clinical findings about the oncological safety of fat grafting. The very complex interplay "in vivo" is not simulated in "in vitro" studies. Fat grafting could be considered a safe procedure after breast-conserving treatment and mastectomy for breast cancer treatment with an adequate postoperative surveillance. Prospective clinical trials with adequate follow-up are awaited to definitely confirm the safety of fat grafting in BRCA-mutated patients (both following risk-reducing surgery and surgery for the treatment of breast cancer)" (LoE III).

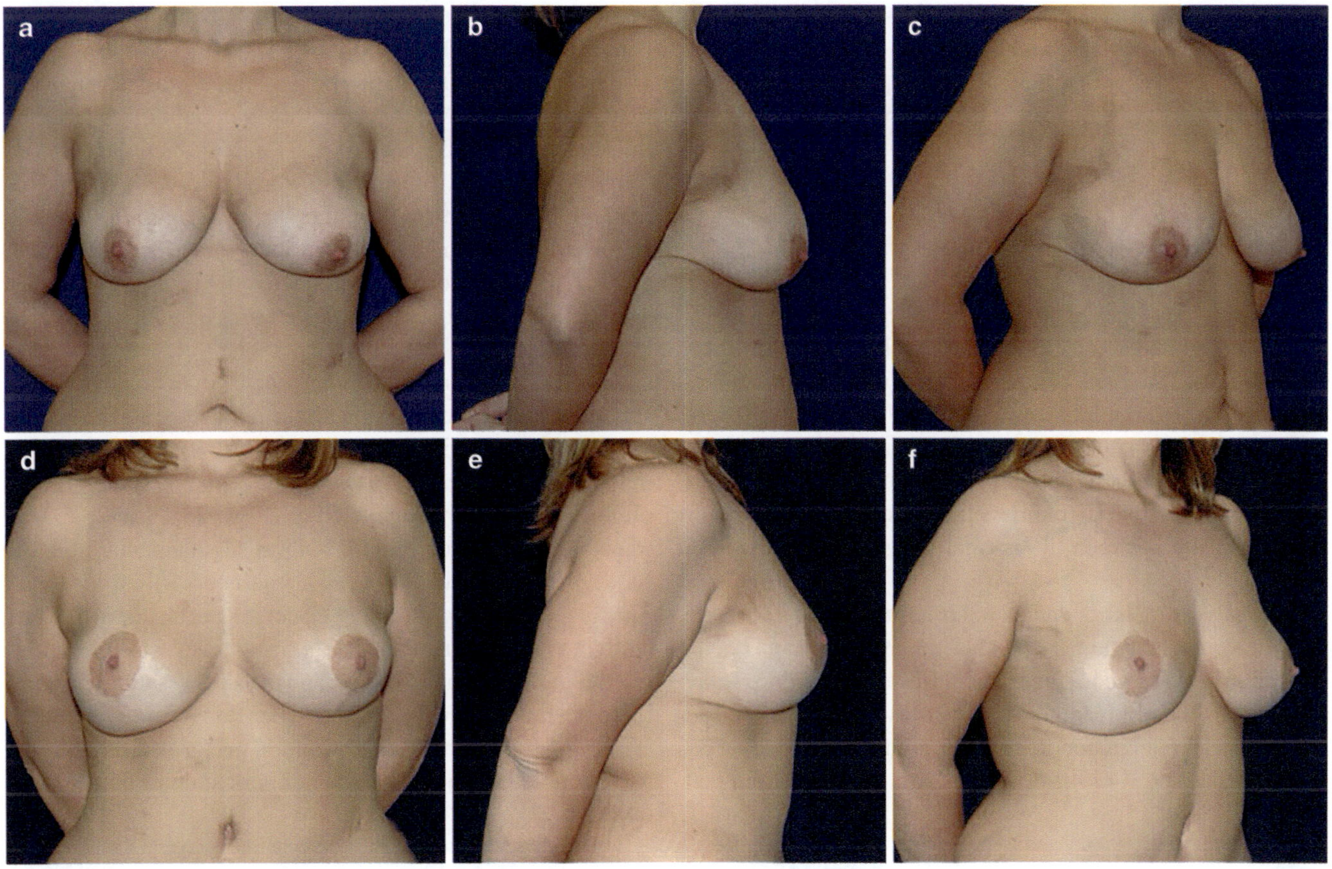

Fig. 28.3 Mastopexy with vertical scar. Preoperatory images (**a**–**c**). Postoperatory results after 2 years (**d**–**f**)

- "Fat grafting is a safe procedure when considering the impact on breast cancer detection and surveillance if a proper technique is used and the surveillance is granted by dedicated breast imaging specialists" (LoE III).

Why has the fat grafting reached so much importance in breast surgery? Molecular and cellular benefit of fat transplantation was first described by Coleman in the face, followed by its application in other body districts. Subsequently, Yoshimura discussed lipofilling for breast augmentation, based on its cellular composition. Indeed, grafted fat tissue contains adipogenic progenitor cells (adipose stem cells or ASCs) and also multipotent stem cells able to differentiate into different tissues. Some of these cells may be able to build the environment to allow the fat engrafting, such as endothelial cells and vascular mural cells, promoting angiogenesis and fat survival. Finally, the ASCs produce angiogenic growth factor in response to hypoxia, thus they are able to stimulate angiogenesis in a damaged site, such as happens for irradiated tissue (e.g., lipofilling after radiation therapy). In patients such as MWL, adipose tissue transplantation, therefore, plays a trophic role in the restoration of a heavily damaged tissue first by weight gain and then by massive weight loss, perhaps associated with nutrients depletion.

Macroscopic benefits of fat grafting may be described as the capability of lipofilling in defining the details of a reconstructed breast, where the implants or an autologous flap do not completely achieve the desired shape. Indeed, it is very difficult to deal with breast asymmetries solely through the use of implants. Fat grafting results effective where the implants fail. Finally, as for oncological patients, even MWL patients often present a reduced skin flap thickness which corresponds to a visible implant borders when placed in a subglandular plane. Therefore, the solution may be represented by a submuscular pocket. However, despite submuscular implant placement provides additional tissue coverage and support to the prosthesis and reduces capsular contraction rate, it often requires pectoralis major muscle release from lower sterno-costal insertions, thus violating an anatomic structure, and it could be associated with an annoying, albeit initial, discomfort, with a more painful postoperative course, with a higher risk of bleeding and the risk of animation of the pectoralis major muscle with an upward thrust of the prosthesis, which in patients practicing intense physical activity to maintain weight loss may be more noticeable and therefore less acceptable. Hence, the attempt to continue using the subglandular implant required the aid of fat transplantation. Indeed, fat grafting plays a role in camouflaging and masking the implant borders when these are placed in the more anatomic subglandular plane, especially but not limited to, the upper and medial breast quadrants, reducing the unnatural appearance and the palpability of the prosthesis.

28.4 Fat Grafting Technique

Autologous fat grafting may be performed under general anesthesia or under local anesthesia with or without sedation. Primary donor site is often referred to the abdomen, but other sites are available as arms, flanks, or gluteal area. These sites are quite available in MWL patients and easier to exploit compared with non-MWL patients. No significant differences are reported in literature with regard to fat viability for any donor site between the normal weight and overweight patients.

The harvesting is conducted, after infiltration of a variable mixture of normal saline, epinephrine, and local anesthetic depending on surgeon preference (local anesthetics may modulate isolated preadipocytes viability rates, thus not all surgeons use them in donor site infiltration), by manual aspiration with 2- or 3- or 4-millimeters cannula (increased cell viability is demonstrated when using a 4-mm cannula compared with a 2- or 3-mm devices). The attached syringe may vary from 10 to 60 cc, and the generated pressure inside it may influence the ASCs availability and viability as reported in literature (lower pressures <250 mmHg increase cell viability and survival). Centrifugation (usually 3000 rpm for 3 min as described by Coleman) or decantation eventually followed by washing and filtration may be performed. Several studies report that adipocyte counts are significantly greater in decanted lipoaspirates compared with centrifuged lipoaspirates.

Delivering is performed in the site of injection, usually in subcutaneous space and sometimes additional injection into subpectoral or retroglandular spaces. 1–2 mm Coleman cannulas and 10 cm^3 syringes are used, though some authors prefer to use 12/14 gauge needles (Figs. 28.4 and 28.5). Fat transfer is performed blindly both in subcutaneous, subpectoral, or retroglandular spaces at the end of the procedure. Subcutaneous fat grafting is easier to perform with either straight or blunt cannulas, and it presents less risk of impairing the implant. On the other hand, it is necessary to be careful to do not damage the implant when delivering subpectoral or retroglandular fat. This grafting is performed through the use of curved, blunt, and large-caliber cannulas that are first

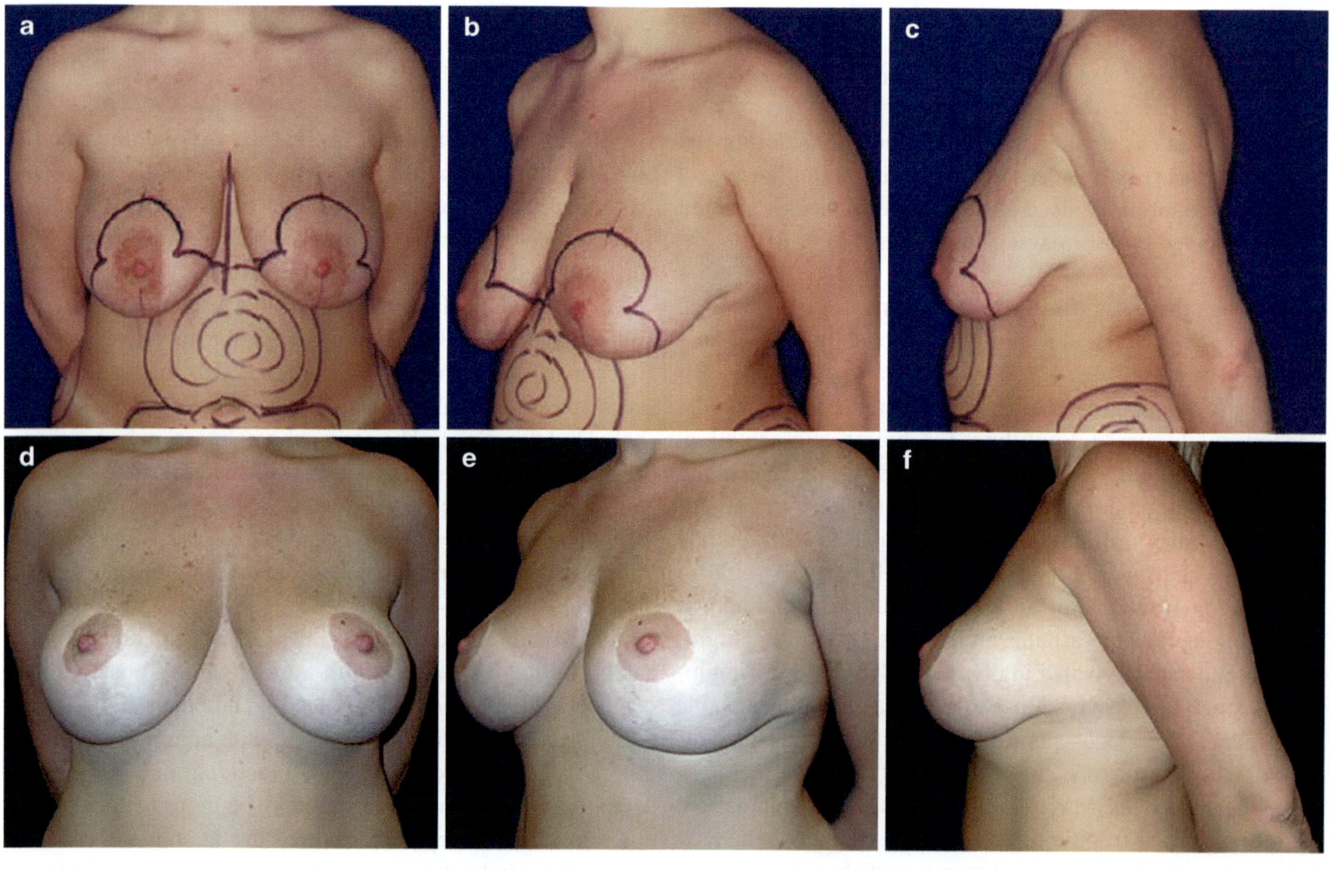

Fig. 28.4 Auto-augmentation using dermoglandular flap and fat grafts. Preoperatory images and surgical project (**a–c**). Postoperatory results after 2 years (**d–f**)

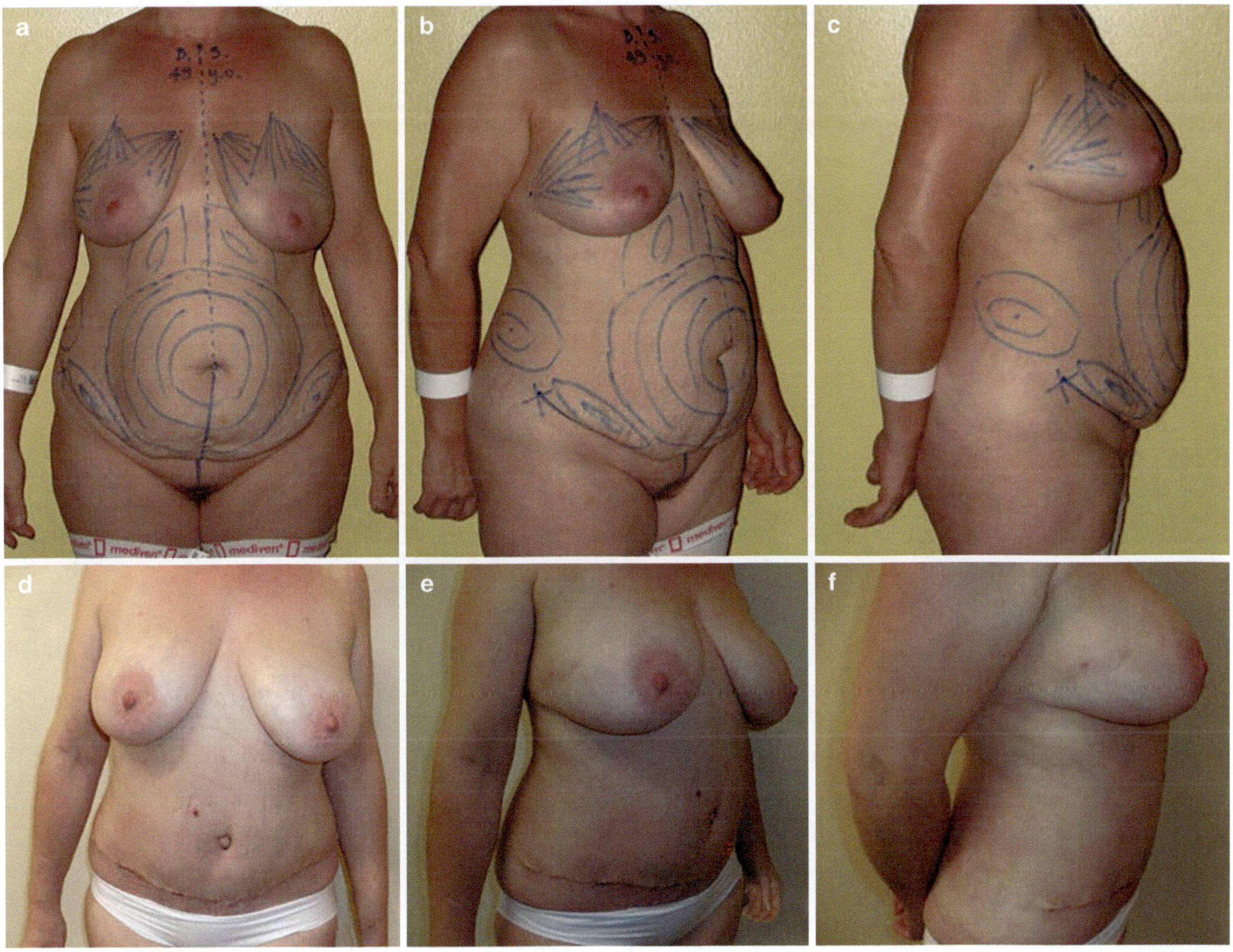

Fig. 28.5 Auto-augmentation using dermoglandular flap and fat grafts and abdominoplasty. Preoperatory images and surgical project (**a–c**). Postoperatory results after 2 years (**d–f**)

advanced on the ribs, which represent a safe point of reference to avoid deepening into the thorax, and then made to proceed retroglandular or subpectoral, around the implant, also through the help of the sensory feedback that the prosthesis provides to the advancement of the blunt cannula. This procedure is not without risks for the prosthesis and requires a certain degree of expertise, but it provides good results. The use of fat grafting in subpectoral space does not allow to place a suction drain in the implant pocket.

Adipocyte viability seems to be influenced by the cannula caliber (increased with 2.5-mm cannula), but no differences have been found with regard to needle gauge, as reported by literature.

After fat injection, dressing should be maintained with a mild compression to favor the fat survival with a medical bra for 4 weeks, night, and day. Antibiotic therapy should be preoperatively administered to prevent infection and continued postoperatively if prostheses are used.

28.5 Fat Grafting and Complications

Fat grafting complications are common to all districts but if considered for the breast they are burdened by the oncological risk and cancer surveillance, as lipofilling may potentially interfere with breast cancer detection, as previously discussed. Most common complications are palpable indurations and cysts, ecchymosis after treatment, and pain, while less common are represented by hematoma, infection, and seroma (Fig. 28.6).

Radiological changes frequently occur. As reported by Expert Panel Consensus on Fat Grafting of the Breast in 2019, autologous fat grafting may be considered a safe procedure in aesthetic breast surgery and after breast cancer surgery with an adequate postoperative surveillance.

A recent systematic review discussed the complications occurring after fat grafting for cosmetic breast augmentation, thus without considering other procedures or medical

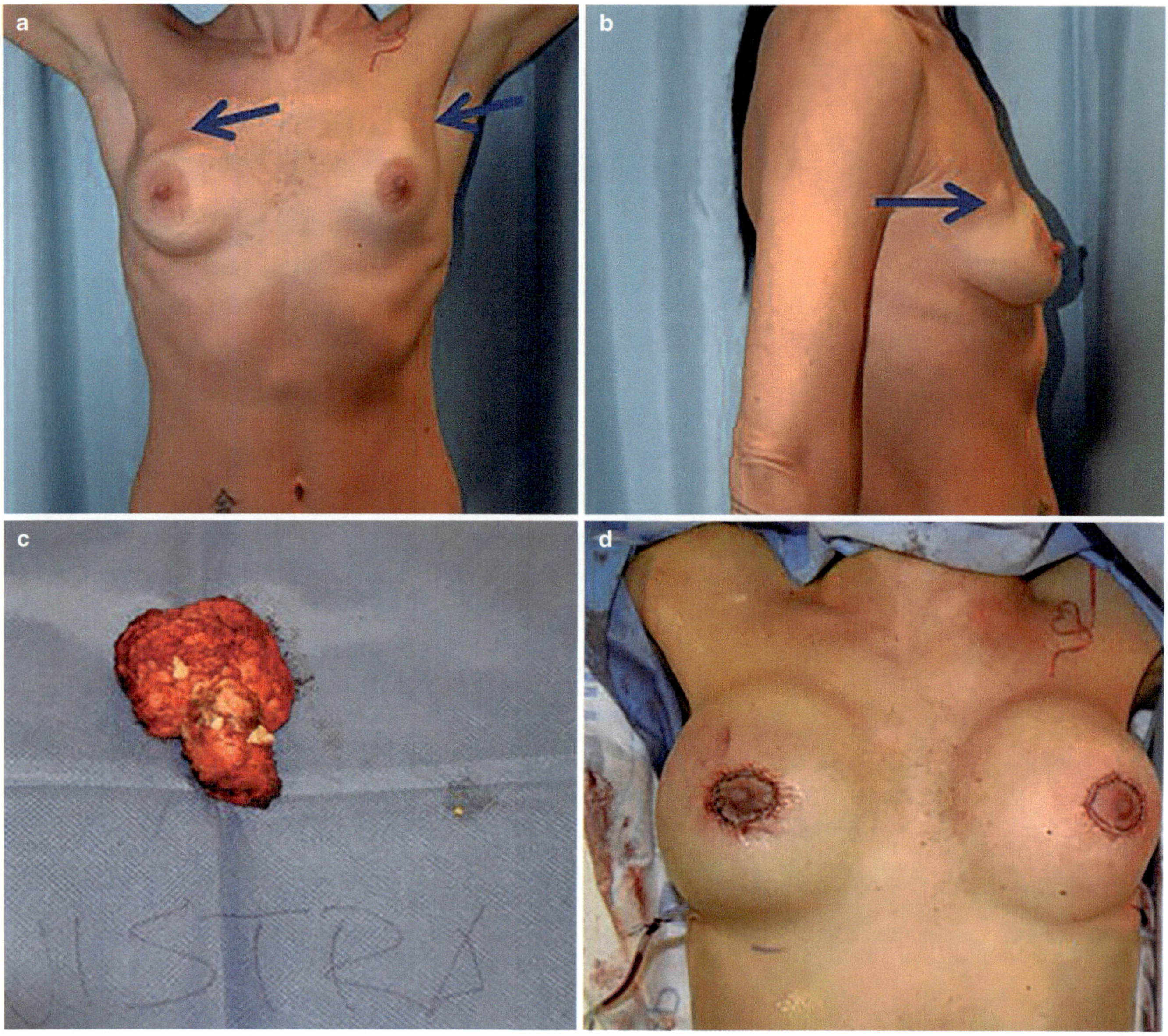

Fig. 28.6 Complication of fat grafts. Blue arrows point area of liponecrosis (**a**, **b**). The total removal of necrotic fat grafts (**c**) required total reconstruction whit prostheses (**d**)

condition (breast reconstruction with radiotherapy, flap-based reconstruction, and implant-based breast augmentations) associated with fat grafting that may jeopardize the outcomes and overestimate the complications rate. The sample included 2073 patients, evaluating the rate of major and minor complications. Among major complications, 0.5% patient developed hematoma, 0.6% developed infection, and 0.1% developed seroma. None of them required revision surgery. Minor complications rate was 2.0% for palpable cysts, most of them treated with aspiration. Radiologic changes were also investigated: oil cysts 6.5%, calcification 4.5%, and fat necrosis 1.2%. Only 3.2% need biopsy to exclude malignant changes, and none of them had therapeutic consequences, thus configuring fat grafting safe and effecting as an alternative to breast augmentation. Donor site infection, fat embolism, and pneumothorax are considered rare complications, while donor site deformation may be ascribed to a poor surgical technique. No studies are available with regard to fat grafting complications rate in MWL patients compared to non-MWL patients.

28.6 Conclusion

Nowadays, the autologous fat grafting is considered a natural filler able to provide volume enhancement, contour correction, and regenerative capability, and it results in a promising method even in MWL patient in breast contouring, able to achieve a cosmetic augmentation and shape restoration, with satisfactory satisfaction rates.

Further Reading

1. Agha-Mohammadi S, Hurwitz DJ. Nutritional deficiency of post-bariatric surgery body contouring patients: what every plastic surgeon should know. Plast Reconstr Surg. 2008;122:604–13.
2. American Society of Plastic Surgeons. Fat transfer/fat graft and fat injection: ASPS guiding principles [Internet] Arlington Heights: American Society of Plastic Surgeons; 2009. http://www.plasticsurgery.org/Documents/medical-professionals/health-policy/guiding-principles/ASPS-Fat-Transfer-Graft-Guiding-Principles.pdf. Accessed 23 Jan 2021.
3. Auclair E, Anavekar N. Combined use of implant and fat grafting for breast augmentation. Clin Plast Surg. 2015;42(3):307–14. vii
4. BAPRAS. UK commissioning guide: massive weight loss body contouring. London: British Association of Plastic, Reconstructive and Aesthetic Surgeons, The Royal College of Surgeons of England; 2017. http://www.bapras.org.uk/docs/defaultsource/commissioning-and-policy/2017–draft-for-consultation–body-contouring-surgery-commissioning.pdf?sfvrsn=0. Accessed 23 Jan 2021.
5. Beidas OE, Rubin JP. Breast reshaping after massive weight loss. Clin Plast Surg. 2019;46(1):71–6.
6. Benelli L. A new periareolar mammaplasty: the "round block" technique. Aesthet Plast Surg. 1990;14(2):93–100.
7. Bircoll M. Cosmetic breast augmentation utilizing autologous fat and liposuction techniques. Plast Reconstr Surg. 1987;79(2):267–71.
8. Cheriyan T, Kao HK, Qiao X, Guo L. Low harvest pressure enhances autologous fat graft viability. Plast Reconstr Surg. 2014;133(6):1365–8.
9. Coleman SR. Facial recontouring with lipostructure. Clin Plast Surg. 1997;24(2):347–67.
10. Coleman SR, Saboeiro AP. Fat grafting to the breast revisited: safety and efficacy. Plast Reconstr Surg. 2007;119(3):775–85. discussion: 786–7.
11. Czerny V. Plastischer ersatz der brustdruse durch ein lipom. Verhandl Deutsche Ges Chir. 1895;24:216–7.
12. Fulton JE. Breast contouring with "gelled" autologous fat: a 10-year update. Int J Cosmet Surg Aesthetic Dermatol. 2003;5(2):156–63.
13. Gentile P, Di Pasquali C, Bocchini I, et al. Breast reconstruction with autologous fat graft mixed with platelet-rich plasma. Surg Innov. 2013;20(4):370–6.
14. Groen JW, Negenborn VL, Twisk JW, et al. Autologous fat grafting in cosmetic breast augmentation: a systematic review on radiological safety, complications, volume retention, and patient/surgeon satisfaction. Aesthet Surg J. 2016;36:993–1007.
15. Hönig JF, Frey HP, Hasse FM, Hasselberg J. Autoaugmentation mastopexy with an inferior-based pedicle. Aesthet Plast Surg. 2009;33(3):302–7. https://doi.org/10.1007/s00266-009-9310-7.
16. Lancerotto L, Stecco C, Macchi V, Porzionato A, Stecco A, De Caro R. Layers of the abdominal wall: anatomical investigation of subcutaneous tissue and superficial fascia. Surg Radiol Anat. 2011;33(10):835–42.
17. Lassus C. A technique for breast reduction. Int Surg. 1970;53(1):69–72.
18. Lejour M. Vertical mammaplasty and liposuction of the breast. Plast Reconstr Surg. 1994;94(1):100–14.
19. Lohsiriwat V, Curigliano G, Rietjens M, Goldhirsch A, Petit JY. Autologous fat transplantation in patients with breast cancer: "silencing" or "fueling" cancer recurrence? Breast. 2011;20(4):351–7.
20. Macchi V, Tiengo C, Porzionato A, Busetto L, Morra A, Martini R, Bassetto F, De Caro R. Anatomical remodelling of the anterior abdominal wall arteries in obesity. Clin Hemorheol Microcirc. 2014;57(3):255–65.
21. Masuda T, Furue M, Matsuda T. Novel strategy for soft tissue augmentation based on transplantation of fragmented omentum and preadipocytes. Tissue Eng. 2004;10:1672–83.
22. Matarasso A, Pitanguy I. The keel resection/pitanguy reduction mammaplasty. Oper Tech Plast Reconstr Surg. 1996;3(3):156–69.
23. Matsumoto D, Sato K, Gonda K, Takaki Y, Shigeura T, Sato T, Aiba-Kojima E, Iizuka F, Inoue K, Suga H, Yoshimura K. Cell assisted lipotransfer: supportive use of human adipose-derived cells for soft tissue augmentation with lipoinjection. Tissue Eng. 2006;12:3375–82.
24. Miranville A, Heeschen C, Sengenes C, Curat CA, Busse R, Bouloumie A. Improvement of postnatal neovascularization by human adipose tissue-derived stem cells. Circulation. 2004;110:349–55.
25. Nava MB, Blondeel P, Botti G, Casabona F, Catanuto G, Clemens MW, De Fazio D, De Vita R, Grotting J, Hammond DC, Harris P, Montemurro P, Mendonça Munhoz A, Nahabedian M, Pompei S, Rancati A, Rigotti G, Salgarello M, Scaperrotta G, Spano A, Stan C, Rocco N. International expert panel consensus on fat grafting of the breast. Plast Reconstr Surg Glob Open. 2019;7(10):e2426.
26. Ørholt M, Larsen A, Hemmingsen MN, Mirian C, Zocchi ML, Vester-Glowinski PV, Herly M. Complications after breast augmentation with fat grafting: a systematic review. Plast Reconstr Surg. 2020;145(3):530e–7e.
27. Pang JH, Coombs DM, James I, Fishman J, Rubin JP, Gusenoff JA. Characterizing breast deformities after massive weight loss: utilizing the Pittsburgh rating scale to examine factors affecting severity score and surgical decision making in a retrospective series. Ann Plast Surg. 2018;80(3):207–11.
28. Pavan C, Marini M, De Antoni E, Scarpa C, Brambullo T, Bassetto F, Mazzotta A, Vindigni V. Psychological and psychiatric traits in post-bariatric patients asking for body-contouring surgery. Aesthet Plast Surg. 2017;41(1):90–7.
29. Pitanguy I, Gontijo de Amorim NF, Radwanski HN. Contour surgery in the patient with great weight loss. Aesthet Plast Surg. 2000;24:406–11.
30. Pitanguy I. Surgical treatment of breast hypertrophy. Br J Plast Surg. 1967;20:78.
31. Planat-Benard V, Silvestre JS, Cousin B, Andre M, Nibbelink M, Tamarat R, Clergue M, Manneville C, Saillan-Barreau C, Duriez M, Tedgui A, Levy B, Penicaud L, Casteilla L. Plasticity of human adipose lineage cells toward endothelial cells: physiological and therapeutic perspectives. Circulation. 2004;109:656–63.
32. Regnault P. Breast ptosis: definition and treatment. Clin Plast Surg. 1976;3:193–203.
33. Rehman J, Traktuev D, Li J, Merfeld-Clauss S, Temm-Grove CJ, Bovenkerk JE, Pell CL, Johnstone BH, Considine RV, March KL. Secretion of angiogenic and antiapoptotic factors by human adipose stromal cells. Circulation. 2004;109:1292–8.
34. Report on autologous fat transplantation. ASPRS Ad-Hoc Committee on New Procedures, September 30, 1987. Plast Surg Nurs. 1987;7(4):140–141.
35. Ribeiro L, Accorsi A Jr, Buss A, Marcal-Pessoa M. Creation and evolution of 30 years of the inferior pedicle in reduction mammaplasties. Plast Reconstr Surg. 2002;110(3):960–70.
36. Ribeiro L. A new technique for reduction mammaplasty. Plast Reconstr Surg. 1975;55:330.
37. Sjöström L, Lindroos AK, Peltonen M, Torgerson J, Bouchard C, Carlsson B, Dahlgren S, Larsson B, Narbro K, Sjöström CD, Sullivan M, Wedel H, Swedish Obese Subjects Study Scientific

Group. Lifestyle, diabetes, and cardiovascular risk factors 10 years after bariatric surgery. N Engl J Med. 2004;351(26):2683–93.
38. Song AY, Jean RD, Hurwitz DJ, et al. A classification of contour deformities after bariatric weight loss: the Pittsburgh rating scale. Plast Reconstr Surg. 2005;116:1535–44.
39. Toy JW, Rubin JP. Post-bariatric reconstruction. In: Neligan PC, Rubin JP, editors. Plastic surgery, vol. 2. 4th ed. London: Elsevier; 2017. p. 709–30.
40. Waked K, Colle J, Doornaert M, et al. Systematic review: the oncological safety of adipose fat transfer after breast cancer surgery. Breast. 2017;31:128–36.
41. Yoshimura K, Sato K, Aoi N, et al. Cell-assisted lipotransfer for cosmetic breast augmentation: supportive use of adipose-derived stem/stromal cells. Aesthet Plast Surg. 2007;32(1):48–55.
42. Zuk PA, Zhu M, Ashjian P, De Ugarte DA, Huang JI, Mizuno H, Alfonso ZC, Fraser JK, Benhaim P, Hedrick MH. Human adipose tissue is a source of multipotent stem cells. Mol Biol Cell. 2002;13:4279–95.
43. Zuk PA, Zhu M, Mizuno H, Huang J, Futrell JW, Katz AJ, Benhaim P, Lorenz HP, Hedrick MH. Multilineage cells from human adipose tissue: implications for cell-based therapies. Tissue Eng. 2001;7:211–28.

29 Brachioplasty in Overweight Patients: The Fat Grafting Role

Franco Bassetto and Paolo Marchica

Contents

29.1 Introduction to Brachioplasty and Fat Grafting

The arm shape is determined by the muscular mass and the fat distribution. An aesthetic pleasant contour in men is characterized by a more pronounced musculature. Contrarywise, as regard to women, a "muscular" arm in considered a disfiguration, while slender curves with smaller muscles are considered aesthetically acceptable. Moreover, the fat structures in the area are characterized by the high level of complexity: the anterior, lateral, and medial aspects of the arm present a single fat distribution type as areolar fat, while the posterior aspect of the arm, as described by Avelar, seems to be composed by two fat distribution types, respectively, deep lamellar fat and superficial areolar fat, divided by the superficial fascia.

Thus, the arms are considered one of the most challenging areas, and the risk of contour irregularities is very high and to achieve symmetry when a bilateral treatment is required is often tricky.

Arm-related deformity is an emergent complaint by plastic surgery patients, due to the increasing of people requesting surgery after massive weight loss or senile aging. The brachioplasty is the body contouring procedure used to perform the reshaping of the upper arm, from the shoulder to the elbow.

It is addressed to patients presenting upper arm soft tissue relaxation, which causes functional and aesthetic problems. The procedure allows to improve the cosmetic appearance of the arms and the symptoms related to the excess skin (rashes, blistering, irritation, and limited mobility, which may aid in exercise and further weight loss) by removing the heavy hanging tissue. Surgery is considered the main solution to improve the arm shape, but it is inseparable from creating a scar. A proper restoration of the shape and a good quality scarring are pivotal to obtain a satisfactory outcome. Indeed, scarring process is the most important issue that the patient has to understand and accept.

The story of brachioplasty started in the first half of twentieth century, and the procedure was later developed by Pascal, Le Louarn, Gusenoff, and Rubin. As reported by recent literature, the main difference among the techniques stands, to date, in the position of the scar, which may be located along biceps brachii sulcus on the medial aspect of the arm or even on the

Supplementary Information The online version contains supplementary material available at https://doi.org/10.1007/978-3-031-10881-5_29.

F. Bassetto · P. Marchica (✉)
Clinic of Plastic and Reconstructive Surgery, University of Padua, Padua, Italy
e-mail: franco.bassetto@unipd.it

© The Author(s), under exclusive license to Springer Nature Switzerland AG 2023
A. Di Giuseppe et al. (eds.), *Fat Transfer in Plastic Surgery*, https://doi.org/10.1007/978-3-031-10881-5_29

posteromedial surface of the arm. These two different sites imply that the scar is noticeable from a front view with abducted arms (biceps brachii sulcus) or from a back view with adducted arms (posteromedial aspect of the arm).

Fat grafting history is instead older (1890 ca.); however, its effective use dates back to the 1980s when several positive reports of fat grafting were described in literature by Coleman and colleagues. These authors, indeed, started talking about "lipostructure" and "a technique to fill depression." Nowadays, fat grafting may support major surgery procedures as brachioplasty, due to his role in correcting small imperfections and improving scarring combined with primary brachioplasty surgery or in revision surgery.

29.2 Pathophysiology of Upper Arm Deformity, Surgical Indication, and Contraindication

The soft tissue ptosis and the loss of skin elasticity may be caused by either a massive weight loss after obesity condition or by senile skin aging. These conditions lead to the so-called "bat wing" deformity and to skin rolls and wrinkles formation.

Female patients seem to be affected more than males as the aesthetic discomfort related to the arms is more perceived by the women. However, men requiring arm reshape are increasing as well.

It is important to correctly address the patient to a tailor-made procedure to obtain a satisfactory result. To achieve this outcome, it is necessary to identify the ideal patient and provide the proper surgical indication. The *British Association of Plastic, Reconstructive and Aesthetic Surgeons (BAPRAS)* guidelines may be helpful to identify the ideal candidates among massive weight loss patient, as, to date, they are the more reliable recommendations as regards body contouring surgery. Indeed, *BAPRAS* recommends a BMI $\leq$ 30.0 kg/m^2, weight stability of 12 months, and the absence or the resolution of physical and psychological disorders that may jeopardize the outcomes. The authors found similar results in their studies, as they found that performing brachioplasty in a patient with BMI $\leq$ 30.0 kg/m^2 is pivotal to reduce the negative effects of unmodifiable risk factors (diabetes mellitus, surgical weight loss modality, older age, smoking history, and chronic anemia).

Conversely, no specific recommendations exist as regards senile skin aging, and the surgical indication to brachioplasty procedures comes from patient's request related to aging tissues causing personal discomfort.

Literature suggests some classification methods to fit the surgery to each patient, but these are limited to massive weight loss patients. The classification by Song et al. [1] seems to be the most widespread and globally accepted, and it is known as the Pittsburgh Rating Scale (PRS), which identifies three classes of patients affected by soft tissue deformity of the upper arms. The difference among the three classes resides in the amount of fat tissue deposits and the skin excess. Therefore, in grade 1 PRS patients with adiposity and good skin tone, liposuction will represent the only treatment. In grade 2 PRS patients, in which the skin is loose and hanging and no severe adiposity is encountered, surgical excision with traditional brachioplasty will be indicated. Finally, in grade 3 PRS patients, both brachioplasty and liposuction will be needed as loose and hanging skin is present as severe adiposity. However, this classification is subjected to modification, due to the interindividual variability of the patients.

A more detailed classification was described in 2007 by El Khatib [2] and, as the PRS, it addresses each type of patient to a tailor-made procedure, based on the adipose tissue deposits and the skin ptosis, which is measured from the biceps brachii sulcus to the lowermost border of the pendulous skin on the medial aspect of the arm. Five different stages are considered: stages 1 and 2 include patients with minimal adipose tissue deposit (<300 mL lipoaspirate) with no ptosis (1) or patients with moderate fat deposits and grade 1 ptosis (<5 cm) (2) and treatment consists in circumferential liposuction; stages 2b and 3 include patients with severe fat deposits and grade 2 ptosis (5–10 cm) (2b) or patients with severe fat deposits (>500 mL lipoaspirate) and grade 3 ptosis (>10 cm) and the suggested treatment consists in traditional brachioplasty with or without liposuction of the lower posterior and medial arm; finally, stage 4 patients have minimal or no fat deposits and a grade 3 ptosis, thus the treatment consists in brachioplasty procedure only.

Both the classifications do not consider fat grafting. Several years later, they were Abboud and colleagues who described combining liposuction and fat grafting to reshape the upper arm. However, their technique can be successfully used to correct mild ptosis because patients who underwent revision surgery after liposuction were the ones with the worse arm deformities. Thus, to extend the use of liposuction to moderate ptosis (thus beyond the grade 2a of El Khatib classification) may lead to unsatisfying results.

The other indications to surgery include the achievement of weight stability and the adjustment of caloric and nutritional intake, especially if the patient had bariatric surgery to obtain the weight loss. Indeed, the aforementioned surgical weight loss modality showed worse outcomes when brachioplasty and other body contouring surgeries are performed. Physical exercise is strongly recommended to achieve and maintain the target weight and reduce the weight regain phenomenon.

Contraindications to surgery are distinguished as relative and absolute. If a BMI $\leq$ 30.0 kg/m^2 is recommended, a BMI higher than 35 kg/m^2 does not represent an absolute contraindication, but it is related to worse outcomes due to the increase of complications and a worse scarring quality. Thus, it is better to address the patients to surgery after reducing their BMI, unless their general conditions do not allow to obtain values lower than 35. An active smoker status is considered a relative contraindication, and it is strongly recommended to quit smoking at least 4 weeks before surgery to avoid complications. Furthermore, it is recommended to avoid the assumption of drugs that could increase the bleeding risk (e.g., ASA and NSAID) or to replace anticoagulant (e.g., coumadin) with low-molecular-weight heparin. Care must be taken when a hormonal therapy is taken by the patient (estroprogestinal steroids). Patients presenting wound healing disorder must be carefully investigated before surgery, whether they are on an individual basis or due to the administration of corticosteroid therapy. However, it is considered a relative contraindication. The same goes for any systemic diseases representing high-risk situations when patients undergo to major surgical procedures and for the conditions leading to a higher risk of lymphedema (axillary lymphadenectomy or radiotherapy).

The real absolute contraindications to upper arm reshaping surgery consist in noticeable upper limb lymphedema, arterial, or venous insufficiency of the upper limbs, active skin infections, and, finally, systemic disorders that may preclude a general anesthesia.

Indications to fat grafting, associated to arm liposuction, are asymmetry of fat and muscles distribution among the arms and the need of enhancement of the muscular groups through "liposculpting" liposuction. The target muscle groups to be enhanced are the deltoid area and the biceps and triceps brachii areas. Contraindications are the same of brachioplasty.

29.3 Preoperative Skin Marking

Preoperative photographs and skin markings are performed with the patient in an upright standing position and with abducted arms (Figs. 29.1 and 29.2). Liposuction marking is different with regard to muscular masses in men and women. Male patients need to have enhanced the muscular groups, thus the biceps brachii, the triceps brachii, and the deltoid are statically and dynamically marked. Women receive markings to biceps brachii and deltoid area only. Triceps in not marked because the ideal female arm does not have an enhanced muscular mass in that area. The boundary areas will receive fat suction to obtain a negative effect (depression) and to enhance the muscle masses, which will be better defined through fat grafting to obtain a positive effect (addition). Furthermore, visible extra fat is also marked when present. This will be extracted through liposuction and eventually grafted where needed.

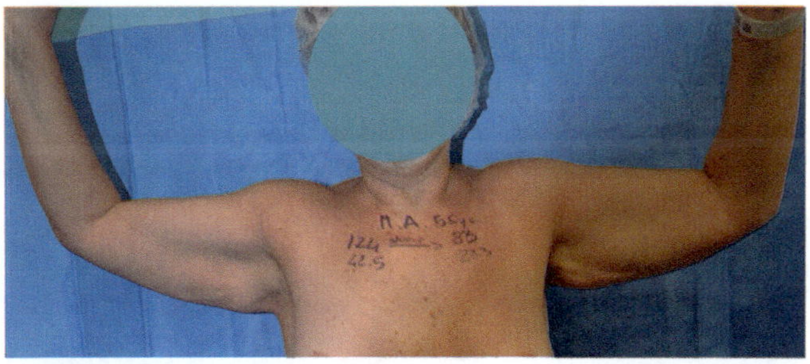

Fig. 29.1 Preoperative photograph: front view. The case: a 50-year-old female patient, from 124 kg (BMI 42.5 kg/m^2) reached 83 kg (BMI 28.3 kg/m^2) due to bariatric surgery (sleeve gastrectomy)

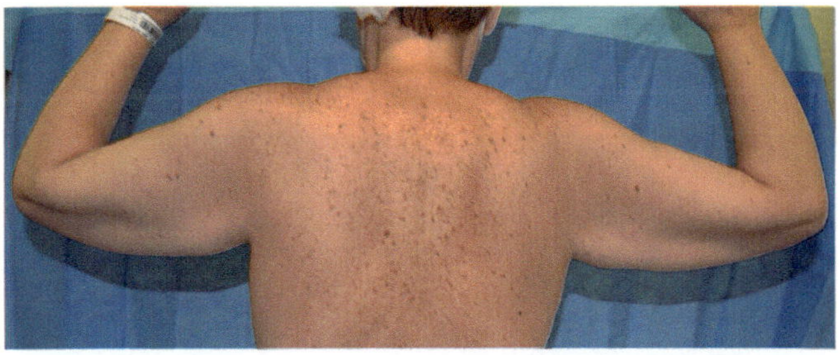

Fig. 29.2 Preoperative photograph: back view

Subsequently, excess skin is estimated through repeated pinch tests, and the site of the future scar is identified on the posteromedial aspect of the arm (along a line connecting the medial epicondyle of the humerus and the insertion point of the long head of triceps brachii into the armpit) or along biceps brachii sulcus, depending on the chosen technique and on the patient's preferences. Finally, the incision lines are marked based on the previous pinch tests. The incision may also be extended to the armpit and the lateral chest, if the soft tissue excess is also found in these areas and further contouring is needed. Thus, a Z plasty may be planned if there is a risk of scar retraction in the axilla.

29.4 The Technique

The procedure is performed under general anesthesia, with the patient in the supine position and abducted arms.

Antibiotic therapy is administered intraoperatively, unless particular scenario in which it is necessary to prosecute with a postoperative prophylaxis.

The first step consists of liposuction. Through multiple stab incisions (typically made in the posterior axillary crease, in the anterior axillary fold, and in the elbow), tumescent solution is bilaterally infiltrated (Fig. 29.3). The composition of the solution is 500 mL of normal saline and 0.5 mg epinephrine (1:1000 epinephrine) for each arm. Subsequent liposuction is performed with a ratio of infiltration to suction of 1:1. Surgeon performs bilateral suction-assisted lipectomy

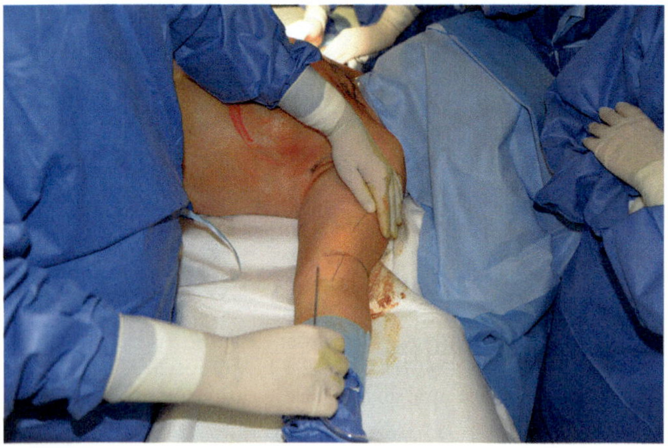

Fig. 29.3 Infiltration of tumescent solution

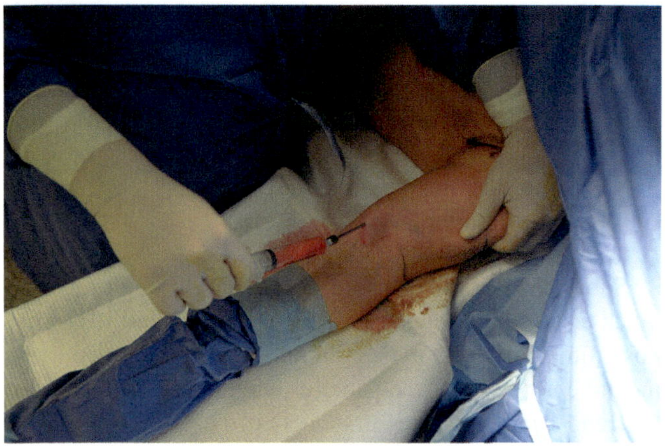

Fig. 29.4 Fat harvesting

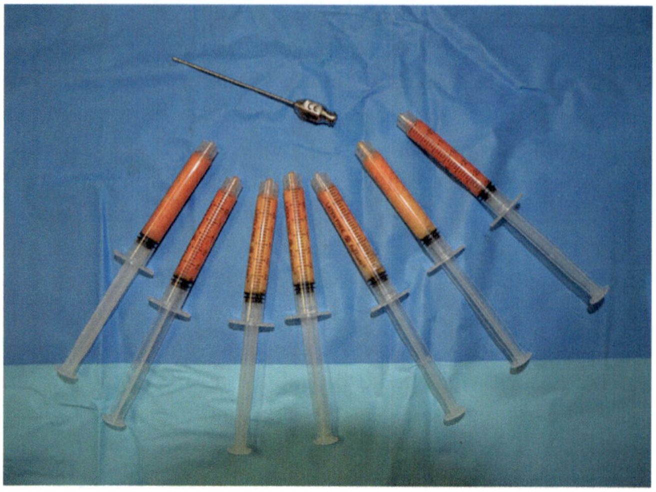

Fig. 29.5 After decanting, the fat is ready to be grafted

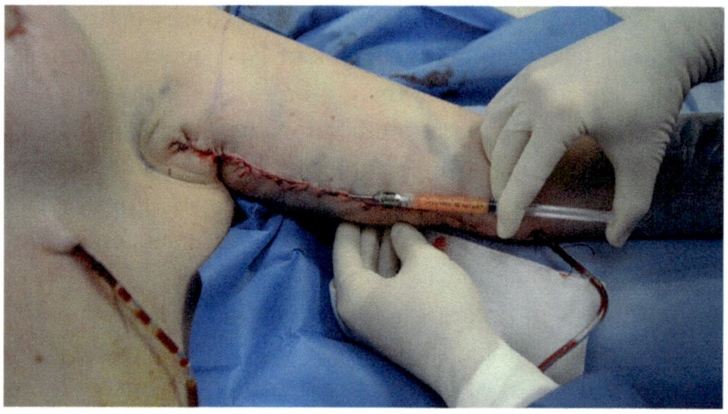

Fig. 29.6 The supernatant fat is extracted and injected subcutaneously

(SAL) to remove the adipose tissue from the target areas (Fig. 29.4), starting in the deep layer and ending in the superficial layer, to achieve skin retraction and contouring. SAL is performed on the posterior and posteromedial aspects of the arm, where the fat deposits are more relevant and on the boundaries of the muscular groups to achieve contour definition.

Straight 3- and 4-mm cannulas are used to perform SAL. Afterward, while decanting is achieved in order to then graft the fat, surgical resection is started. The anteromedial incision of the arm is carried on until the superficial fascia. This trick helps avoid damaging deep structures because it spares the deep laminar fat which covers the deep fascia and protects superficial sensitive nerves (medial antebrachial and medial brachial cutaneous nerves) and the lymphatic network, thus reducing the risk of seroma formation or postoperative lymphedema onset. Dissection is then performed from the anteromedial to the posterolateral direction. Repeated pinch test helps to estimate intraoperatively the excess skin and confirms the preoperative markings and the posterolateral incision is finally carried on with, resecting the redundant soft tissues. A suction drain is usually placed. A final layered suture is accomplished, once faced the skin margins, with resorbable stitches #3/0 (*polyglactin 910* or *glyconate*) and subcuticular running suture #4/0 (*glyconate*) allows to close the skin. The contralateral arm is addressed with the same technique with repeated assessing of the symmetry.

After decanting, the fat is ready to be grafted (Fig. 29.5). The supernatant fat is extracted and injected subcutaneously (some authors perform intramuscular injection to provide more bulk to the muscular masses, others perform finishing touches to the shape of the resected arm to obtain a more natural profile (Fig. 29.6) with straight or curved 3-mm cannulas where contouring is needed, especially in deltoid region and other areas of asymmetry, comparing both the arms. A volume between 50 and 100 mL is injected to obtain liposculpture. As reported in the literature, a more availability of stem cells is provided from fat tissue of the inner thighs and abdomen, thus if a combined liposuction is performed (e.g., abdomen or thighs), the harvested fat may be injected into the arm sites.

29.5 Postoperative Therapy

Surgical wounds may be treated with different devices. Traditional dressings are represented by sterile taping that properly distribute tension on the faced margins, reducing the single-point tension developed by the thread in the suture bites. Tension reduction is pivotal to obtain a faster healing, to reduce ischemia of the skin margins, and to achieve an optimal scarring, with a physiological collagen production. Alternatively, incisional (or single use) negative pressure wound therapy (iNPWT or sNPWT) is rapidly emerging providing mechanobiological aid to the wound healing process. Otherwise, a tape-based device that acts as a zip (*Zip® Surgical Skin Closure System*, ZipLine Medical, Campbell, CA) may be employed with satisfying results in reducing tension on the surgical margins.

Patient has to wear a compressive sleeve garment from the time of surgery up to 3 months postoperative. This girdle supports lymphatic drainage and reduces postoperative edema.

Dressings are maintained until complete wound healing (about 2 weeks after surgery), and the suction drain may be removed if collected fluids are less than 50 cc/24 h, usually on the first postoperative day.

The areas treated with liposuction and liposculpture may present ecchymosis due to the trauma on the soft tissues, thus they may benefit from treatment with local or systemic drugs with anti-edema action (escin, bromelin, etc.). No antibiotic therapy is needed postoperative, unless complications. Common painkillers may help to control postoperative pain. Tightness sensation is the most complained issue by the patients during upper limbs' movements. Manual lymphatic drainage may be started 2–3 weeks after surgery to resorb extra fluids and edema. The patient is advised to reduce at minimum the upper limb physical activities for the first weeks. Mild physical activities are encouraged after 1–2 weeks. Driving and working activities are allowed after 3 weeks, and light physical exercise may be gradually started after 4 weeks. It is important that the patient do not smoke within the first 4 weeks after surgery because vasoconstriction induced by tobacco's components is critical to increase surgical complications, delayed wound healing, infections, and furthermore it limits fat engraftment after liposculpture.

A proper scar treatment is advised because a good-quality scarring is strongly related to a satisfied patient and may benefit from a physical treatment. Indeed, the mechanical effect provided by massages or taping or self-drying silicone gel products allow the correct orientation of the newly deposited collagen fibers by the scars' fibroblasts. Finally, it is important to protect the scar tissue from the UV rays, which are responsible for trophic changes of the scar. Thus, it is mandatory to avoid direct sunlight exposition if not utilizing a highly protective sunscreen.

29.6 Complications

Brachioplasty's complications reported in the literature range from 22% to 56%, depending on the considered case series. This wide range is attributable to different techniques and evaluation criteria. However, the most common complications are found in all cases series, such as postoperative edema, lymphedema, poor scarring, seromas and hematomas, infections, nerve injuries, under or overresection with subsequent persisting heavy hanging tissue or a "pinched appearance" of the arm, and asymmetry. Patients often complain with regard to scars. Indeed, poor scarring represented by scar hypertrophy, cheloids, or scar atrophy or diastasis is quite common and is also burdened by an observer-dependent variability, thus a scar may be differently accepted by each patient, with great differences between men and women. Scar treatment may benefit from corticosteroid injection in case of hypertrophy, or lipofilling through revision surgery in case of scar atrophy or diastasis. Hence, fat grafting also finds an indication in revision surgery. Postoperative arm asymmetry may benefit from liposculpture as well. Finally, major complications are less represented (deep vein thrombosis, severe anemization, and systemic infection).

Fat grafting may also take part in causing complications, even if more rarely, such as soft tissue abscesses, bruises, hematomas, seromas, asymmetry, and infections. Macrofat embolism is a rare but severe major complication and needs to be considered a surgical emergency.

29.7 Conclusion

Brachioplasty is an evolving technique and its demand is constantly increasing, due to the increase of massive weight loss patients and to a closer look at skin aging by men and women. A careful selection of the candidates to surgery makes it possible to achieve better results. Liposuction role remains debated as regards conditioning the outcomes. Literature is divided between authors bearing an increased risk and those claiming a better result with an acceptable risk. However, in our opinion, liposuction is a safe and effective method of body contouring. Furthermore, it may be helpful in primary and revision surgery, thanks to the possibility of "liposculpting." Indeed, the subtractive effect and the eventual additive purpose after fat grafting to define body contour allow surgery to have greater attention to the arm shape and muscular masses details.

References

1. Song AY, Jean RD, Hurwitz DJ, Fernstrom MH, Scott JA, Rubin JP. A classification of contour deformities after bariatric weight loss: the Pittsburgh rating scale. Plast Reconstr Surg. 2005;116(5):1535–44.
2. El Khatib HA. Classification of brachial ptosis: strategy for treatment. Plast Reconstr Surg. 2007;119(4):1337–42.

Further Reading

3. Abboud MH, Abboud NM, Dibo SA. Brachioplasty by power-assisted liposuction and fat transfer: a novel approach that obviates skin excision. Aesthet Surg J. 2016;36(8):908–17.
4. Abramson DL. Minibrachioplasty: minimizing scars while maximizing results. Plast Reconstr Surg. 2004;114:1631–4; discussion 1635–7.
5. Avelar J. Regional distribution and behavior of the subcutaneous tissue concerning selection and indication for liposuction. Aesthetic Plast Surg. 1989;13:155–65.
6. BAPRAS. UK commissioning guide: massive weight loss body contouring. London: British Association of Plastic, Reconstructive and Aesthetic Surgeons, The Royal College of Surgeons of England; 2017. https://www.bapras.org.uk/docs/default-source/commissioning-and-policy/2017%2D%2Ddraft-for-consultation%2D%2Dbody-contouring-surgery-commissioning.pdf?sfvrsn=0. Accessed 1 Nov 2020.
7. Bourin P, Bunnell BA, Casteilla L, et al. Stromal cells from the adipose tissue-derived stromal vascular fraction and culture expanded adipose tissue-derived stromal/stem cells: a joint statement of the International Federation for Adipose Therapeutics and Science (IFATS) and the International Society for Cellular Therapy (ISCT). Cytotherapy. 2013;15:641–8.
8. Breiting LB, Lock-Andersen J, Matzen SH. Increased morbidity in patients undergoing abdominoplasty after laparoscopic gastric bypass. Dan Med Bull. 2011;58:A4251.
9. Chamosa M, Murillo J, Vázquez T. Lipectomy of arms and lipograft of shoulders balance the upper body contour. Aesthetic Plast Surg. 2005;29(6):567–70.
10. Chia CT, Theodorou SJ, Hoyos AE, Pitman GH. Radiofrequency-assisted liposuction compared with aggressive superficial, subdermal liposuction of the arms: a bilateral quantitative comparison. Plast Reconstr Surg Glob Open. 2015;3(7):e459.
11. Coleman SR. Facial recontouring with lipostructure. Clin Plast Surg. 1997;24:347–67.
12. Cuomo R, Grimaldi L, Nisi G, Zerini I, Giardino FR, Brandi C. Ultraportable devices for negative pressure wound therapy: first comparative analysis. J Invest Surg. 2021;34(3):335–43.
13. Czerny V. Plastischer Erzats de Brustdruse durch ein Lipom. Zentralbl Chir. 1895;27:72.
14. Duscher D, Pollhammer MS, Wenny R, Shamiyeh A, Schmidt M, Huemer GM. Barbed sutures in body-contouring: outcome analysis of 695 procedures in 623 patients and technical advances. Aesthetic Plast Surg. 2016;40:815–21.
15. Gusenoff JA, Coon D, Rubin JP. Brachioplasty and concomitant procedures after massive weight loss: a statistical analysis from a prospective registry. Plast Reconstr Surg. 2008;122:595–603.
16. Holländer E. Die kosmetische Chirurgie (S.669–712, 45 Abb.). In: Joseph M, editor. Leipzig: Verlag van Veit; 1912. p. 690–691.
17. Hoyos A, Perez M. Arm dynamic definition by liposculpture and fat grafting. Aesthet Surg J. 2012;32(8):974–87.
18. Khansa I, Harrison B, Janis JE. Evidence-based scar management: how to improve results with technique and technology. Plast Reconstr Surg. 2016;138(3 Suppl):165S–78S.
19. Illouz YG. The fat cell "graft": a new technique to fill depressions. Plast Reconstr Surg. 1986;78:122–3.
20. Levi K, Ichiryu K, Kefel P, Keller J, Grice J, Belson O, Storne E, Safa. Mechanics of wound closure: emerging tape-based wound closure technology vs. traditional methods. Cureus. 2012;8(10):e827.
21. Lockwood T. Brachioplasty with superficial fascial system suspension. Plast Reconstr Surg. 1995;96:912–20.
22. Marchica P, Bassetto F, Pavan C, Marini M, Raimondi AM, Gardener C, Grigatti M, Pagani A, Brambullo T, Zocchi M, Vindigni V. Retrospective analysis of the predictive factors associated with good surgical outcome in brachioplasty in massive weight loss patients. J Plast Surg Hand Surg. 2022;56(6):326–34. Epub ahead of print.
23. Masson IF, de Oliveira BD, Machado AF, Farcic TS, Júnior IE, Baldan CS. Manual lymphatic drainage and therapeutic ultrasound in liposuction and lipoabdominoplasty post-operative period. Indian J Plast Surg. 2014;47(1):70–6. https://doi.org/10.4103/0970-0358.129627.
24. Monstrey S, Middelkoop E, Vranckx JJ, Bassetto F, Ziegler UE, Meaume S, Téot L. Updated scar management practical guidelines: non-invasive and invasive measures. J Plast Reconstr Aesthet Surg. 2014;67(8):1017–25.
25. Michaels JV, Coon D, Rubin JP. Complications in postbariatric body contouring: postoperative management and treatment. Plast Reconstr Surg. 2011;127:1693–700.
26. Miotto G, Ortiz-Pomales Y. Arm contouring: review and current concepts. Aesthet Surg J. 2018;38(8):850–60.
27. Neuber GA. Fettransplantation. Chir Kongr Verhandl Deutsche Gesellschaft für Chir. 1893;22:66.
28. Padoin AV, Braga-Silva J, Martins P, et al. Sources of processed lipoaspirate cells: influence of donor site on cell concentration. Plast Reconstr Surg. 2008;122:614–8.
29. Pascal JF, Le Louarn C. Brachioplasty. Aesthetic Plast Surg. 2005;29:423–9.
30. Regnault P. Brachioplasty, axilloplasty, and pre-axilloplasty. Aesthetic Plast Surg. 1983;7:31–6.
31. SICPRE—Società Italiana di Chirurgia Plastica Ricostruttiva ed Estetica. Prospetto informativo sull'intervento di lifting delle braccia. https://www.sicpre.it/wp-content/uploads/2015/12/22.pdf. Accessed 1 Nov 2020. From: "*Il consenso informato consapevole in chirurgia plastica estetica*". Azzena B, Casadei A, Fasano D, Fraccalanza E, Parodi PC, Schiavon M, Siliprandi L. New Magazine Editore; 2012.
32. Simonacci F, Bertozzi N, Grieco MP, Grignaffini E, Raposio E. Procedure, applications, and outcomes of autologous fat grafting. Ann Med Surg (Lond). 2017;20:49–60. Published 2017 Jun 27. https://doi.org/10.1016/j.amsu.2017.06.059.
33. Sisti A, Cuomo R, Milonia L, Tassinari J, Castagna A, Brandi C, Grimaldi L, D'Aniello C, Nisi G. Complications associated with brachioplasty: a literature review. Acta Biomed. 2018;88:393–402.
34. Strong AL, Cederna PS, Rubin JP, Coleman SR, Levi B. The current state of fat grafting: a review of harvesting, processing, and injection techniques. Plast Reconstr Surg. 2015;136(4):897–912.
35. Temourian B, Malekzadeh S. Rejuvenation of the upper arm. Plast Reconstr Surg. 1998;102:545–51; discussion 552–3.
36. Toy JW, Rubin JP. Post-bariatric reconstruction. In: Neligan PC, Rubin JP, editors. Plastic surgery, vol. 2. 4th ed. London: Elsevier; 2017. p. 709–30.
37. Zomerlei TA, Neaman KC, Armstrong SD, Aitken ME, Cullen WT, Ford RD, Renucci JD, VanderWoude DL. Brachioplasty outcomes: a review of a multipractice cohort. Plast Reconstr Surg. 2013;131(4):883–9.

Lower Eyelid Blepharoplasty and Midface: Liposculpture and Biorevitalization

30

Domenico De Fazio

Contents

Face is composed of multiple structures that with aging tend to undergo an alteration of mass, skeletal proportion, and atrophy, with the redistribution of subcutaneous adipose tissue. Furthermore, we can appreciate a general loss of suspension by the supporting structures, an excess and a folding of the skin. Schaverien and colleagues' work describes the adipose tissue compartments in the face, explaining their location in well-defined compartments, separated from each other by strips of fibrous tissue that prevent their movement into other compartments.

This new description of face ligaments and adipose tissue anatomy of the face allows us to better understand what happens during the aging process, in which we attend contrasting phenomena of lipo-atrophy and lipo-hypertrophy of neighboring areas. The bone and cartilage system acts as a support while the superficial muscle-aponeurotic system, thanks to the fibrous interlobulary septa of the adipose tissue, connects directly to the skin to ensure its stability.

The autologous fat transplantation restores volume to altered features, using patient own fat and thus excluding the possibility of any allergic reactions, which may otherwise occur with the use of other fillers.

Face lipofilling consists in the infiltration, in precise and well-defined areas, of adipose tissue usually taken from places such as abdomen, thighs, buttocks, or knees; by doing so, it restores an appearance of apparent youth to faces which were signed by age, fatigued, and stress. Furthermore, it can also ameliorate congenital defects.

The lower eyelid is composed externally by very thin skin, sometimes even the thinnest of our body, and it can assume a brownish to yellow discoloration or appear as a transparent veil. Skin places on the orbital and eyelid part of the orbicularis muscle, which is bounded laterally by the lateral canthal tendon and medially by the medial canthal tendon. Below to the orbicularis muscle we have the orbitary fat, divided into septa belonging to the medial or lateral lodge. The aging process of the eyelids leads to the formation of the so-called tear trough, which can also be worsened by the presence of wrinkles of the eyelid skin. This phenomenon is often associated with the loss of fat volume which becomes atrophic. Also, due to the weakening of the skin caused by the thinning of the dermal layer, the malar bags may become even more noticeable. Furthermore, we may also attend a tarsal laxity of the lower eyelid that will compromise even more the adhesion of the eyelid to the eyeball (Fig. 30.1).

Therefore, in all cases of aging process involving a laxity of the lower eyelid and subsequent formation of tear trough and malar bags, we may correct it with a shortening of the horizontal eyelid and a blepharoplasty technique. This latter can be carried out either with a cutaneous or myocutaneous flap to reduce the excess of the eyelid skin.

The prominent malar fatty bags should then be removed by carrying out the pressure maneuver on the eyeball with closed eye, to highlight the prolapse of the adipose tissue. Once cutting the septum, the portion of the fat to be removed

D. De Fazio (✉)
Private Practice, Milan, Italy

© The Author(s), under exclusive license to Springer Nature Switzerland AG 2023

A. Di Giuseppe et al. (eds.), *Fat Transfer in Plastic Surgery*, https://doi.org/10.1007/978-3-031-10881-5_30

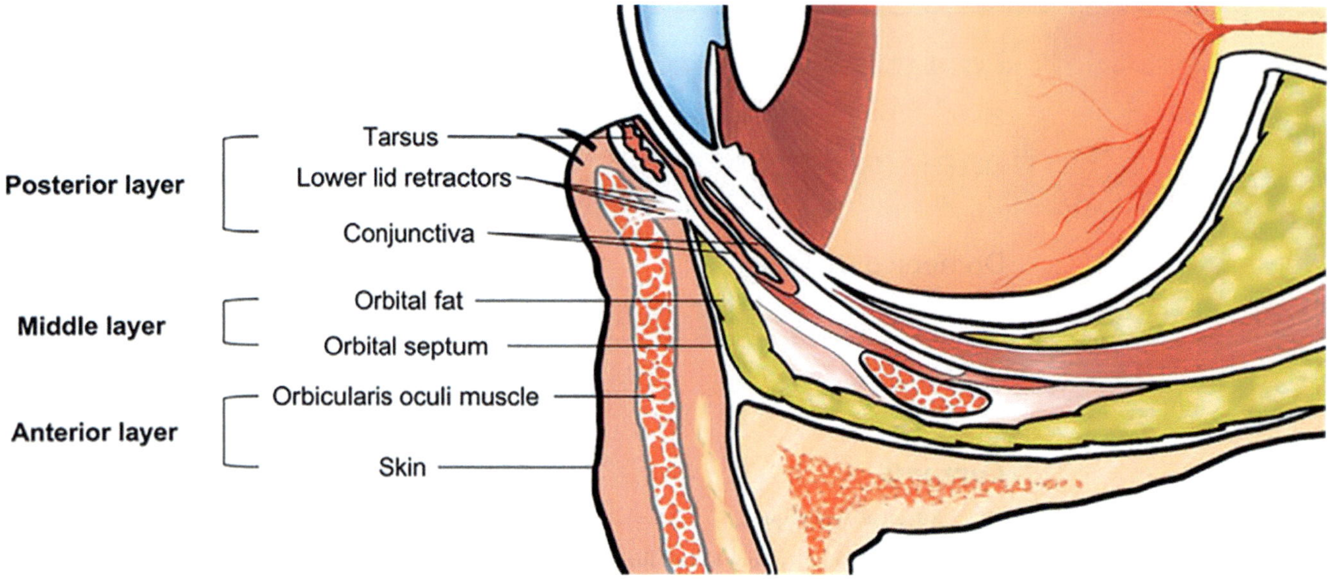

Fig. 30.1 Lower eyelid anatomy

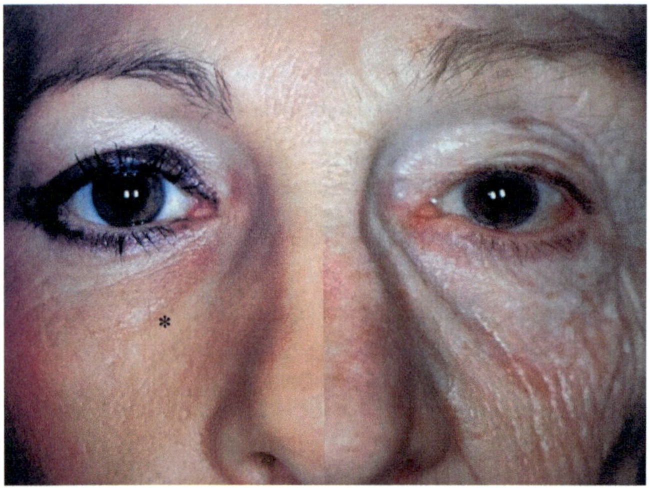

Fig. 30.2 Aging

is clamped off with a small Klemmer and an accurate hemostasis is performed before releasing the reduced bag (Fig. 30.2).

In the past two decades, a careful study of the aging process of the periorbital region has led surgeons to understand that in most cases patients need a refill of lost volumes, in order to restore the youthful appearance of this area. Indeed, after the removal of the malar bags during lower blepharoplasty, there was a sunken appearance of the lower eyelid, which in anyway would have given the desired rejuvenation appearance. Therefore, the volumization effect that can be done with lipofilling technique allows in many cases to solve the problem of the periorbital aging. The paper published in 2013 by P. Tonnard et al. introduced a new concept of periorbital lipofilling technique with microfat and nanofat.

Surgery indications for the correction of tear through can be either genetic reasons or aging process, as very often demonstrated by photos of the same patient at an early and advanced age. The loss of periorbital adipose tissue very often highlights areas with excess fatty bags. The surgeon's assessment should distinguish where there is a need to remove adipose tissue by trans-conjunctival blepharoplasty and where to graft adipose tissue. The surgeon's assessment will have to take into account the fact that lipofilling of depressed areas in many cases is sufficient to solve the imperfection without having to remove the fat bags (Fig. 30.3).

Fig. 30.3 Aging

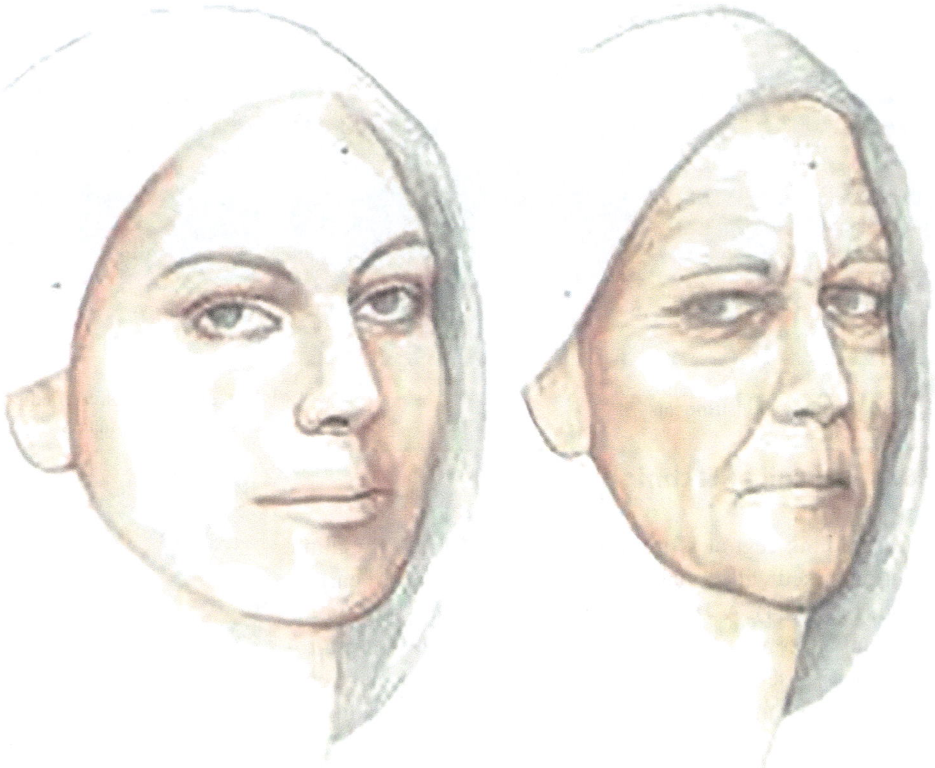

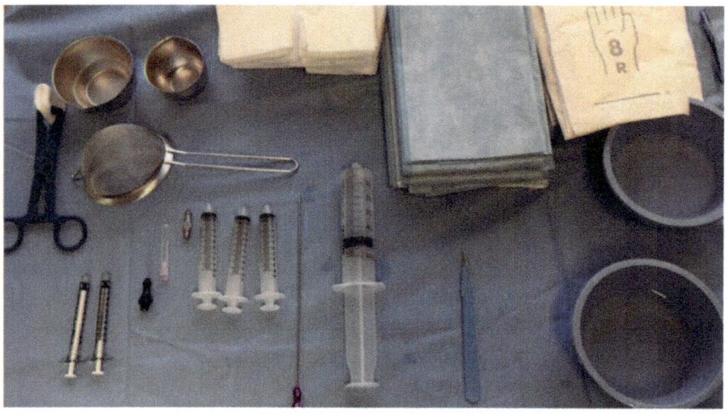

Fig. 30.4 Surgical instruments

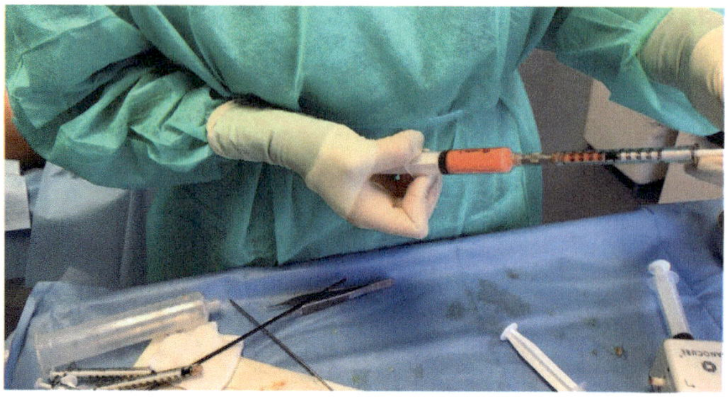

Fig. 30.5 Transfer of adipose tissue between syringes with connectors of different diameters

30.1 Surgical Technique

The sampling region is infiltrated with 1000 mL of saline solution mixed with 1 vial of adrenaline and 4 vials of carbocaine (Fig. 30.4). A 19G needle is used to access the tissue and then a 1.5 mm diameter infiltration cannula with multiple holes is used which is connected to a 10 cc Luer-Lock syringe. The same cannula is then used to remove adipose tissue connected to the 10 cc Luer-Lock syringe. By aspirating adipocytes through 1 mm holes, it is possible to harvest very small droplets of adipose tissue called micro-fat. The characteristics of this adipose tissue are ideal for carrying out a volumetric correction of the eyelids or malar region. In particular, the characteristics of adipose tissue filtered through small holes eliminates the risks of creating visible and evident fat accumulations in the eyelids compartments. Indeed, the risk to end up with a visible and palpable fat graft, especially at the level of the lower eyelid, is one of the most concrete risk of this method. There are different types of devices on the market to obtain the micro and nanofat once taken with standard method connected to a syringe. Already Tonnard and authors had described a method of connectors for syringe with decreasing diameters that allows to treat the adipose tissue with multiple passages from one syringe to another progressively reducing the diameter of the connector in order to obtain a more fluid graft and reducing cellularity (Figs. 30.5 and 30.6).

By transforming the adipose tissue into a nanofat, we can obtain a regenerative graft that has lost its volumizing function, proper of the normal adipose tissue, but maintains the regenerative ones.

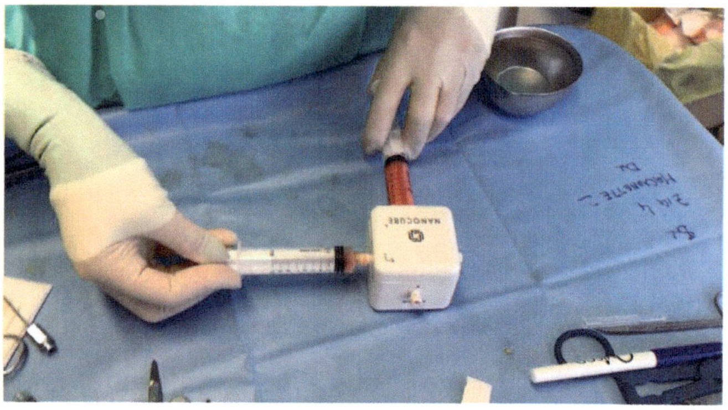

Fig. 30.6 Particular device enabling partial filtration of fibrous material in a closed sterile system, in order to obtain micro, nano, or macro fat

30.2 Types of Fat Grafting Either for Volumizing or Regenerative Results

Type of fat grafts	Parcel size	Injection site
Macrofat	<2.4 mm	Deep fat compartments
Microfat	<1.2 mm	Superficial fat compartments
Nanofat	<400–600 ù	• Subdermal • Intradermal • Microneedling

Preoperative drawing is carried out with the patient standing in upright position, in order to mark exactly the sinking above the lower orbital margin (Fig. 30.7). Access is performed with an 18G needle about 2 cm from the lateral margin of the orbit and 1 cm below to the lower orbital mar-

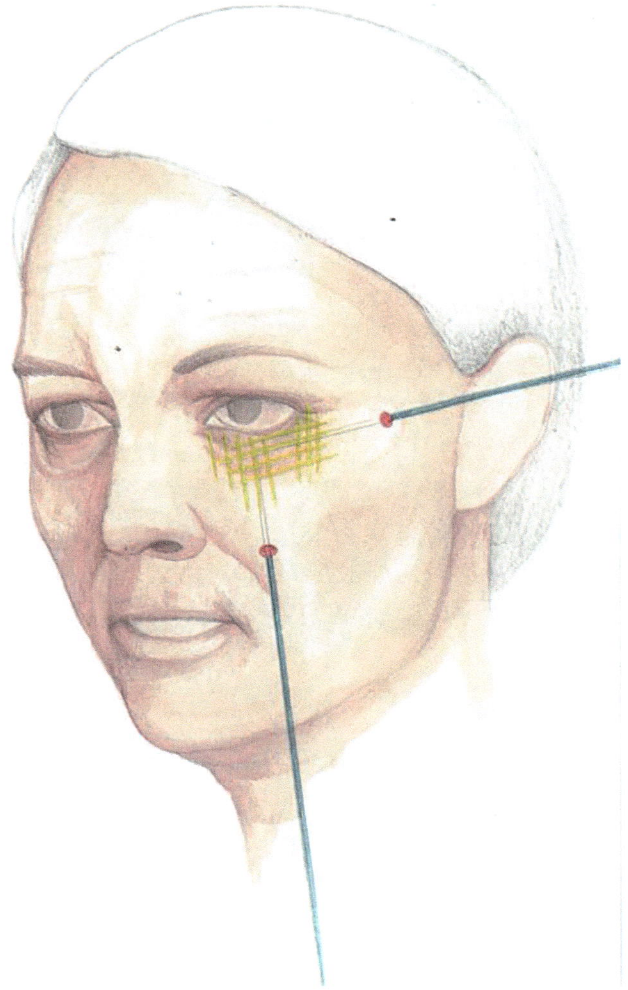

Fig. 30.7 Cross-fat injection

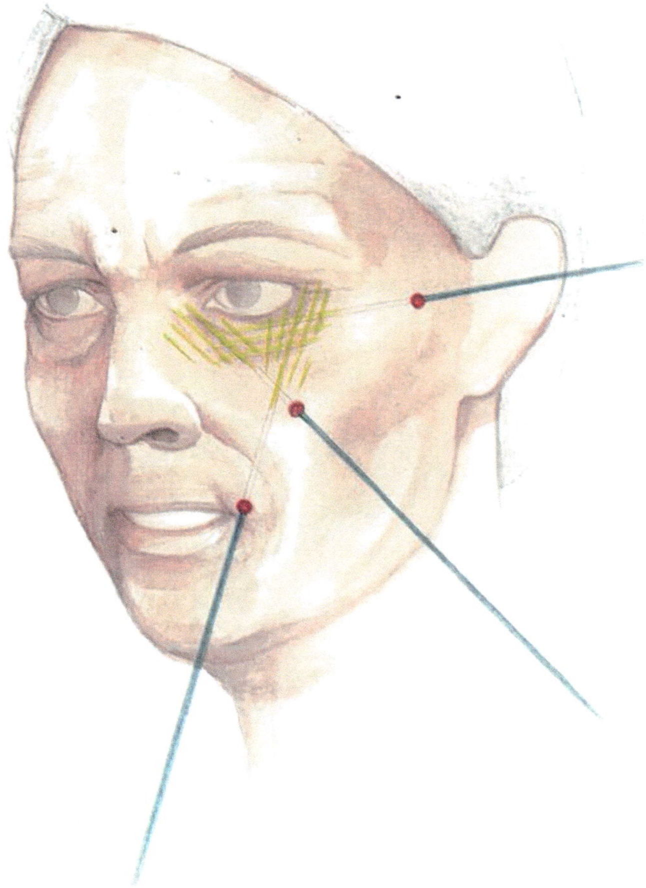

Fig. 30.8 Access points

gin. Then, multiple crossings are made in the deep subcutaneous compartment by using a 19 Gauge lipofilling cannula connected to a 1 cm Luer-Lock syringe. The correction of the lower eyelid also includes tear trough deformity and it extends inferiorly, medially and laterally by grafting microfat.

Volumes of microfat regularly grafted at the level of the lower eyelids range from 1 to 4 cc on average.

The intervention is performed in sedation and local anesthesia by infiltrating the region of interest with carbocaine plus adrenaline.

The graft is inserted deeply being able to lay the cannula in contact with the inferior orbital bone margin and gradually it superficializes without, however, risking to make evident any cutaneous irregularities of the adipose graft.

The lipofilling of the lower eyelids should not be overcorrected to avoid a result with excessive volume of fat. The graft is then carried out in the lower eyelid under the orbicolaris oculi muscle in the SOOF compartment (Figs. 30.8, 30.9, 30.10, and 30.11).

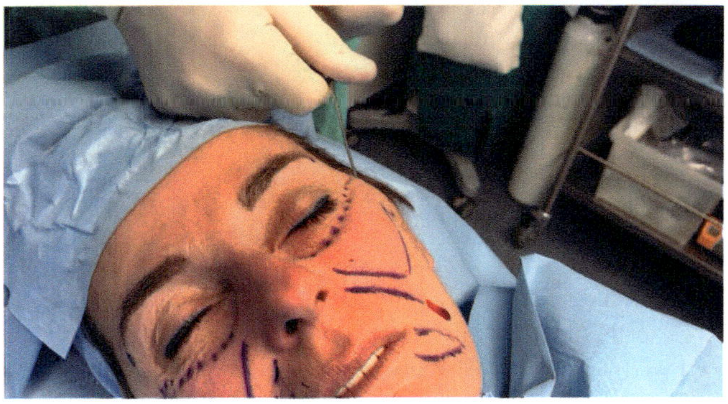

Fig. 30.9 Lateral margin access of lower eyelid

Tonnard's technique involves the transformation of adipose tissue into nanofat which can be injected with needles of 25 gauge using either and intradermal, subdermal, or needling technique.

This adipose tissue that has lost its volumizing characteristics maintains a regenerative effect for the eyelid skin. On average, from 0.5 to 2 cc of nanofat are used for the lower eyelid (Fig. 30.12).

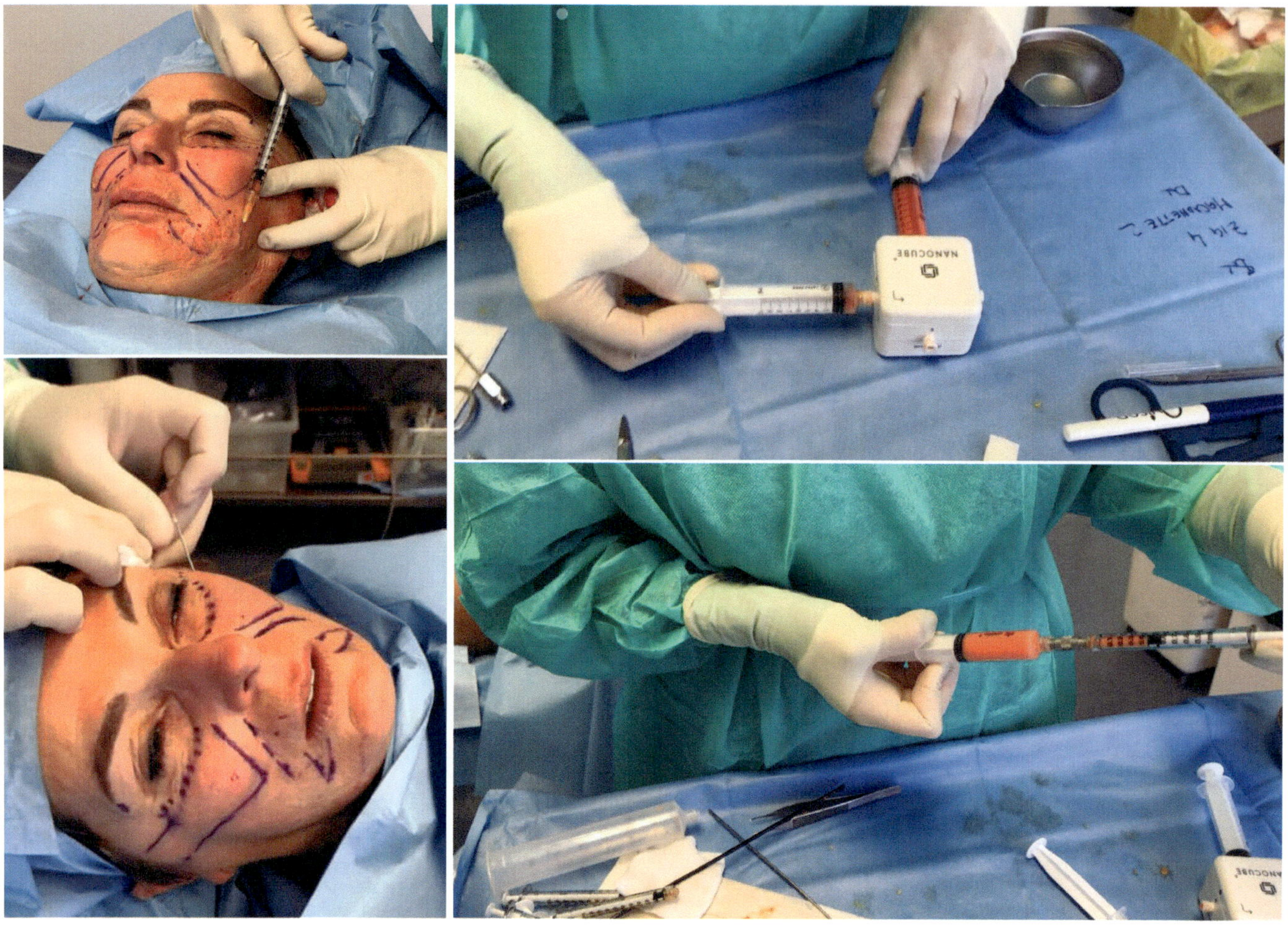

Fig. 30.10 Microfat and nanofat technique

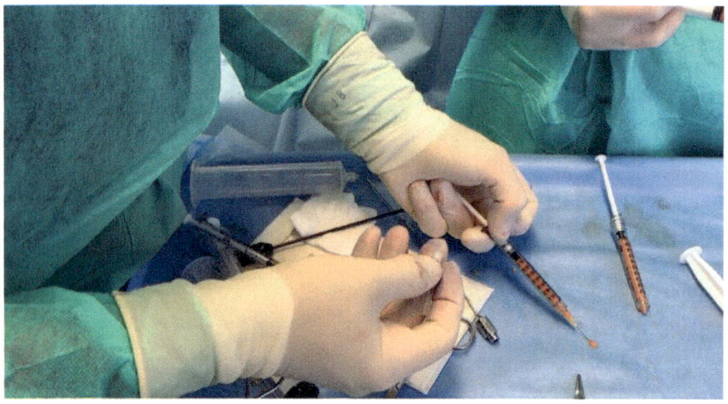

Fig. 30.11 Nanofat

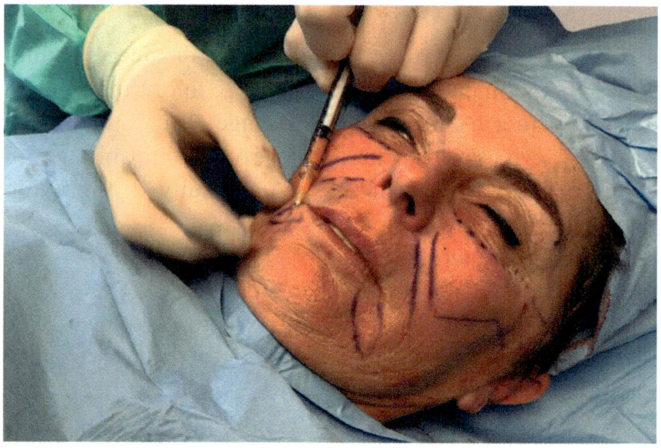

Fig. 30.12 Nanofat intradermal implantation technique

Nanofat has a regenerative effect and it is used with needling technique with accurate device for the skin of the malar region (Figs. 30.13, 30.14, 30.15, 30.16, and 30.17).

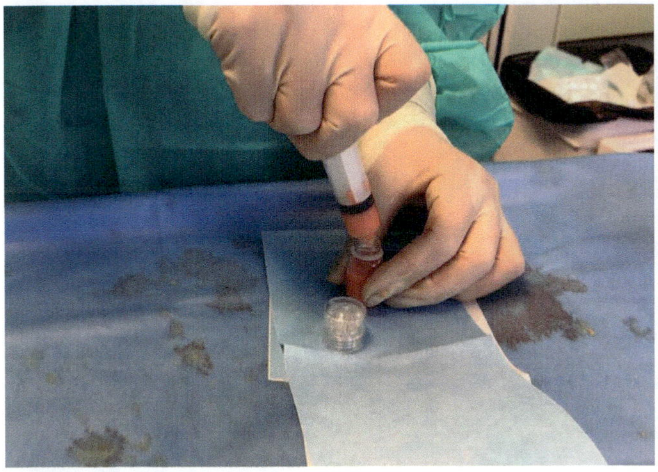

Fig. 30.13 Nanofat

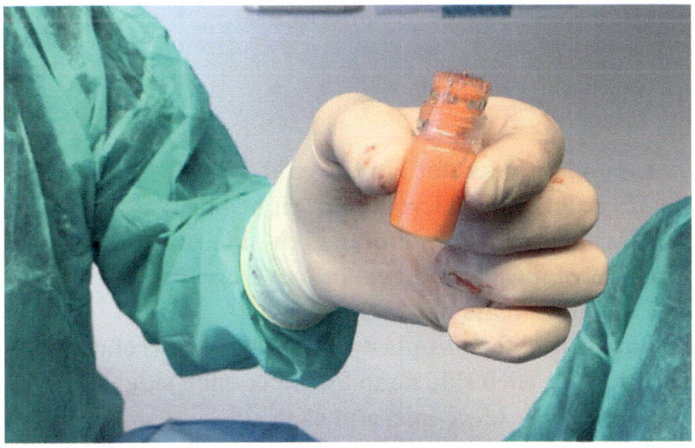

Fig. 30.14 Device's filling with nanofat

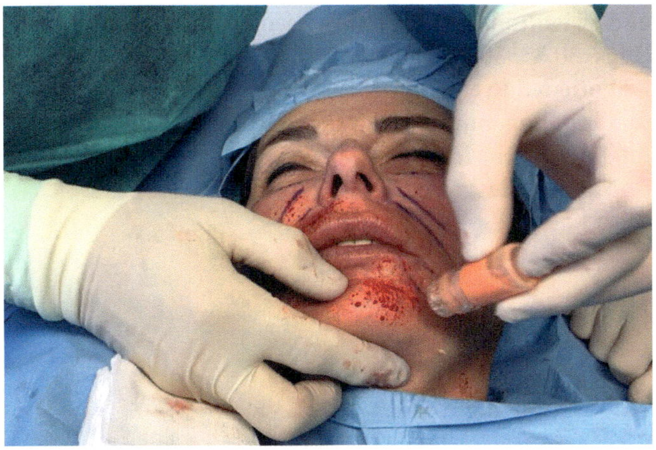

Fig. 30.15 Microneedling with nanofat step 1

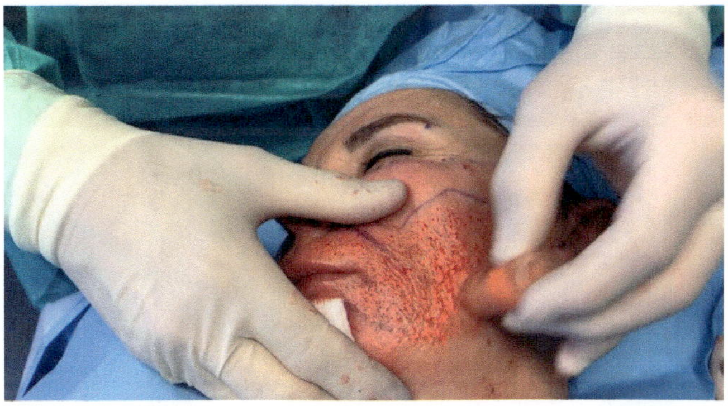

Fig. 30.16 Step 2

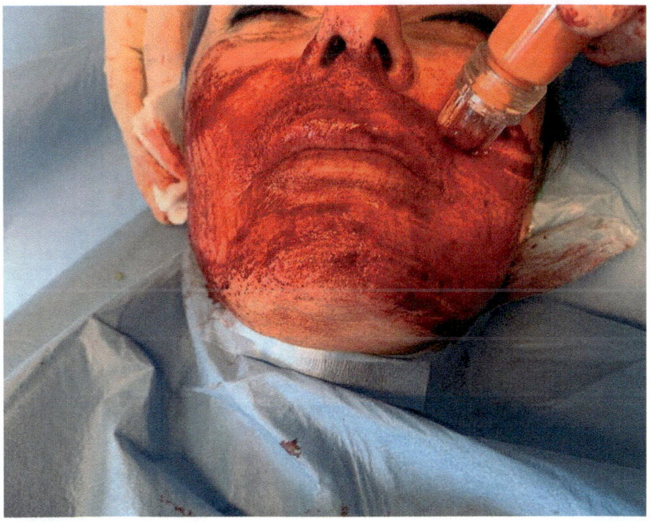

Fig. 30.17 Step 3

30.3 Clinical Cases

Patient 45 years old 21.34 BMI

Lipofilling microfat lower eyelid 2.5 cc each side nanofat 1.5 cc each side

Marionette 2.5 cc microfat each side

Nasiolabial fold 2.5 cc microfat each side (Figs. 30.18, 30.19, 30.20, and 30.21)

Patient 55 years old 21.23 BMI

Lipofilling microfat cheekbone 4 cc each side, cheek 30 cc nanofat each side, marionette 2 cc microfat each side Eyebrow 1.5 cc microfat each side and 2 cc nanofat (Figs. 30.22, 30.23, 30.24, and 30.25)

Patient 59 years old 23.21 BMI

Lipofilling microfat lower eyelid 2 cc each side nanofat 1.5 cc each side

Marionette 1.5 cc microfat each side

Nasiolabial fold 4 cc microfat right side and 5 cc microfat left (Figs. 30.26, 30.27, and 30.28)

Patient 66 years old 26.18 BMI

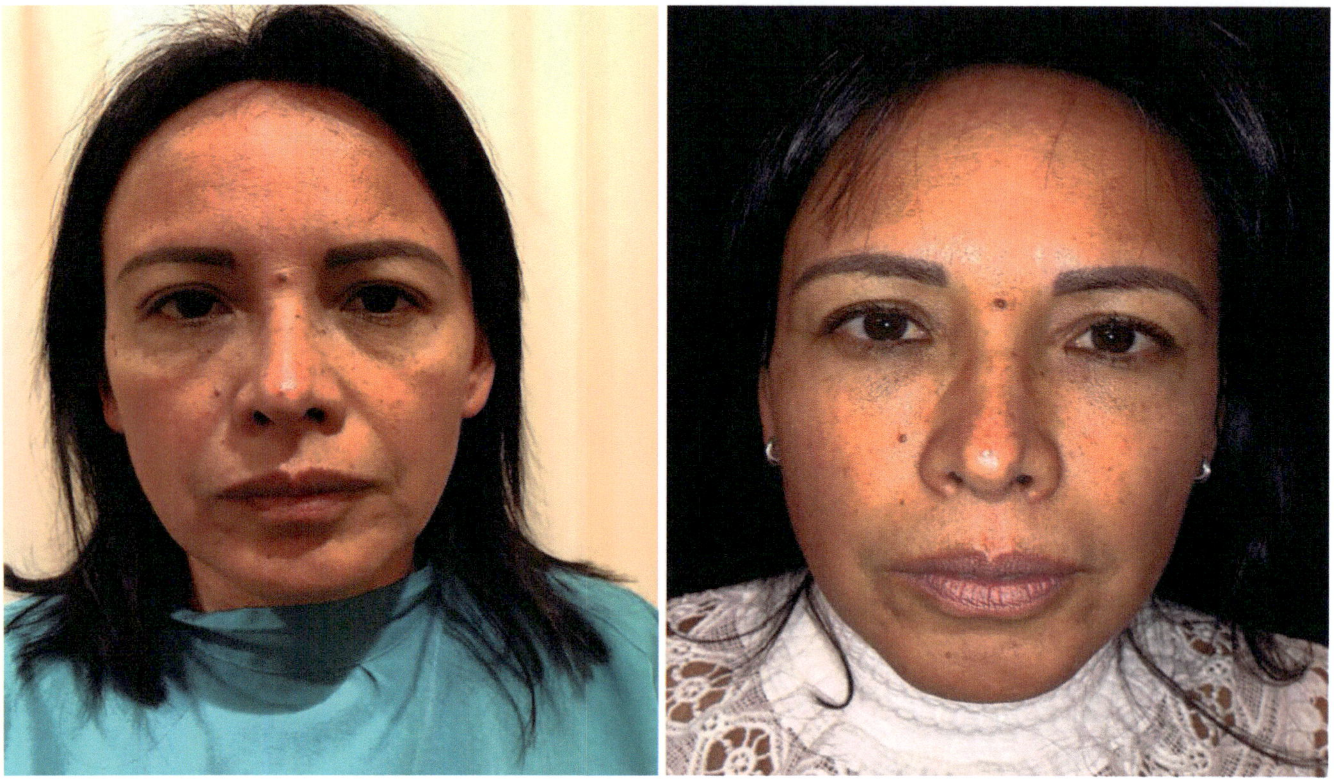

Fig. 30.18 Pre and post 2 years

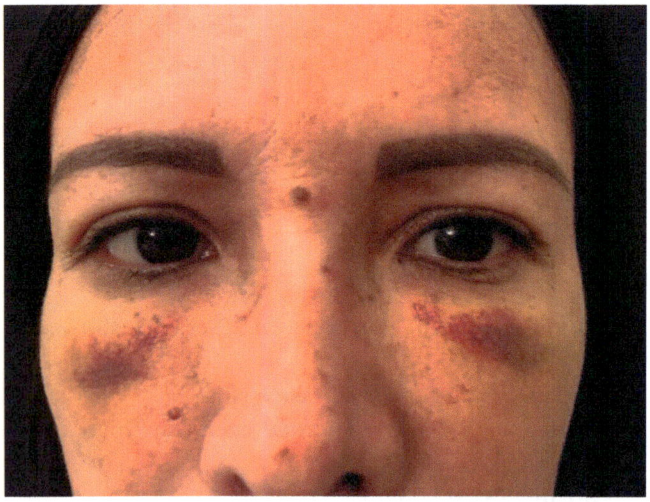

Fig. 30.19 15 days later

Lipofilling microfat lower eyelid 2.5 cc left side nanofat 1.5 cc left side, microfat 1 cc right side

Marionette 2 cc microfat each side, nasiolabial fold 3 cc microfat each side, malar area 5 cc (Fig. 30.29)

Patient 45 years old 21.30 BMI

Superior blepharoplasty lipofilling microfat lower eyelid 1.5 cc each side nano fat 1.5 cc each side

Marionette 1.5 cc microfat each side

Nasiolabial fold 3 cc left and 2 cc right microfat, superior lip 6 cc nanofat (Figs. 30.30, 30.31, and 30.32)

Patient 43 years old 21.48 BMI

Lipofilling malar scar microfat 3 cc and nanofat 2.5 cc and "rigottomies" with 18G needle to relax the fibrosis (Figs. 30.33, 30.34, and 30.35)

Patient 54 years old 18.36 BMI

Lipofilling microfat lower eyelid 1.5 cc each side nanofat 1.5 cc each side

Marionette 2.5 cc (Figs. 30.36 and 30.37)

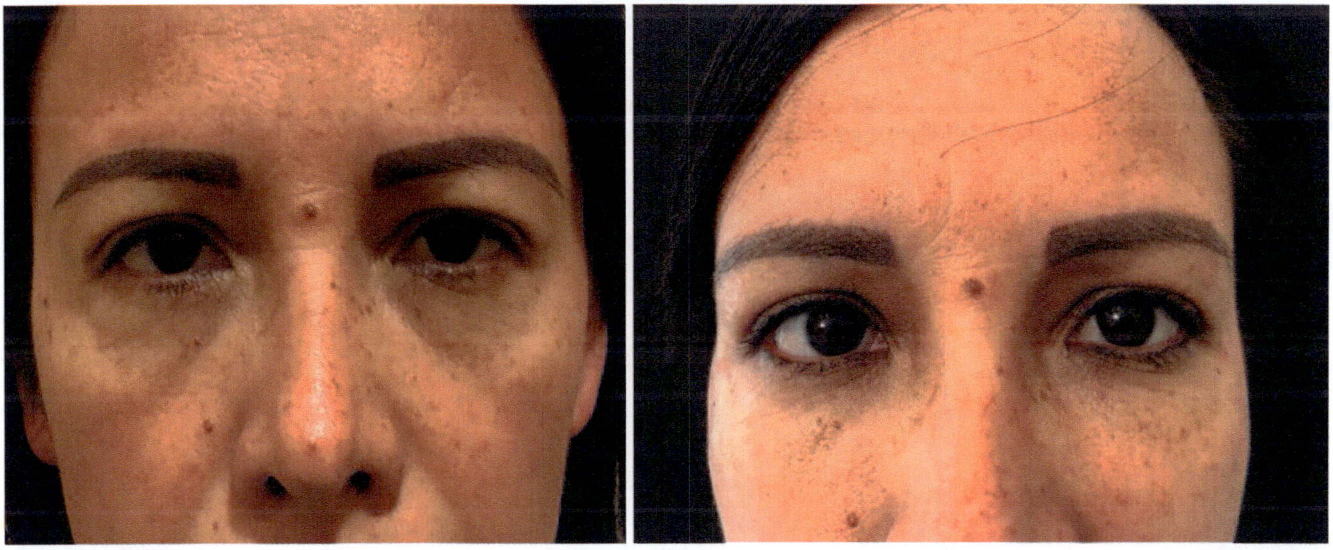

Fig. 30.20 Pre and post 2 years

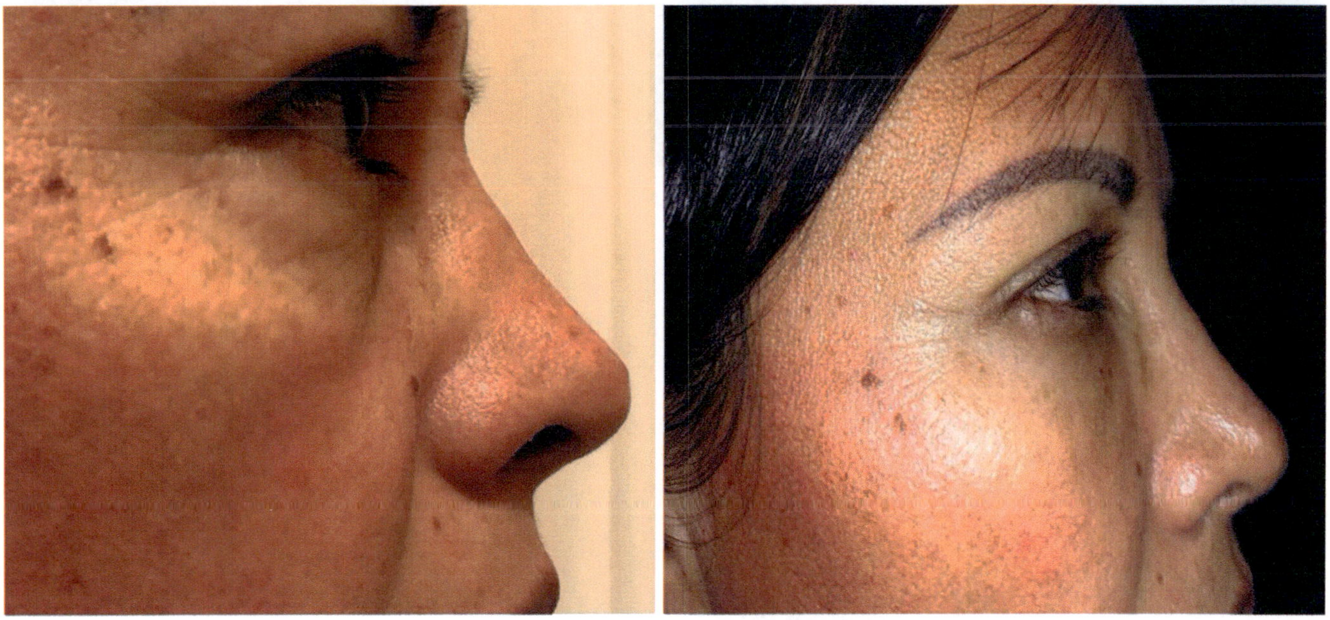

Fig. 30.21 Pre and post 2 years

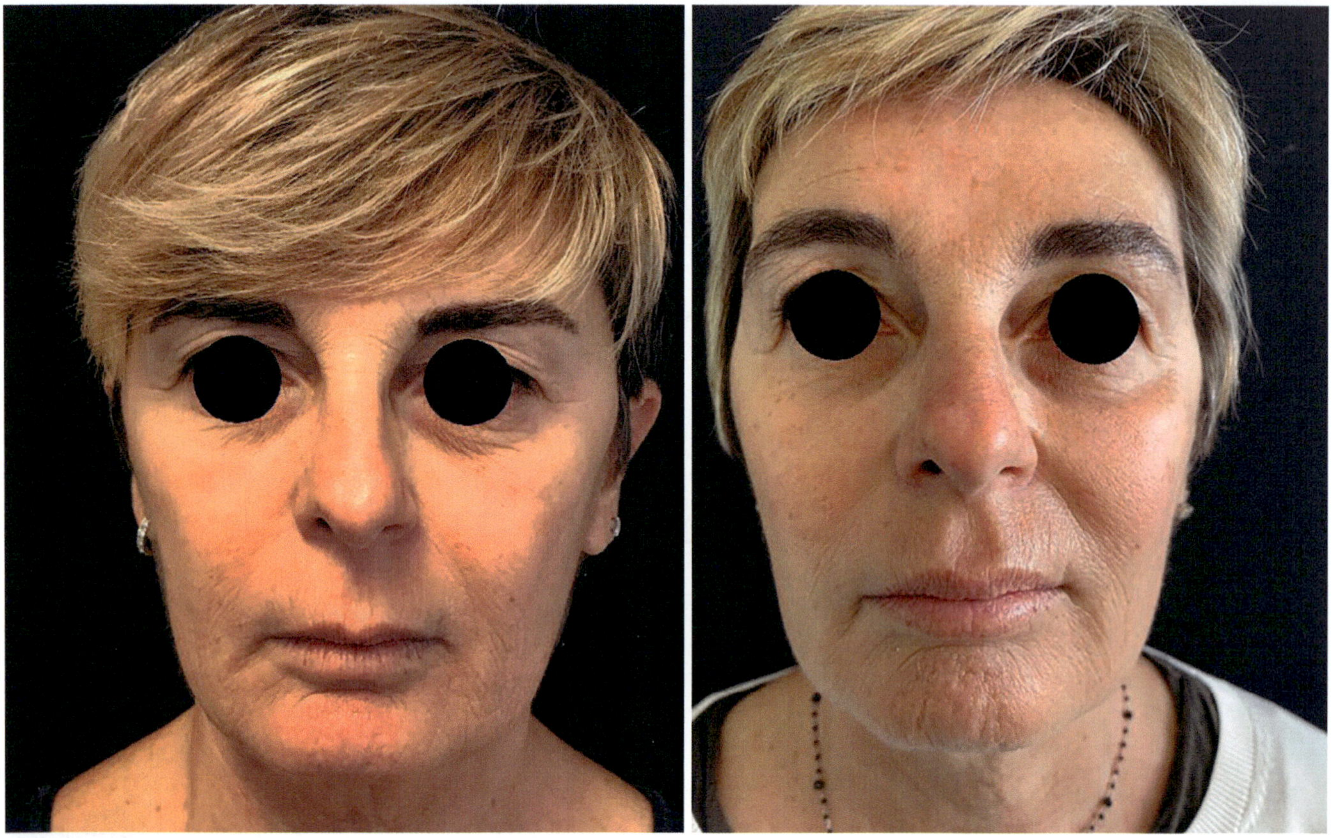

Fig. 30.22 Pre and post 1 year

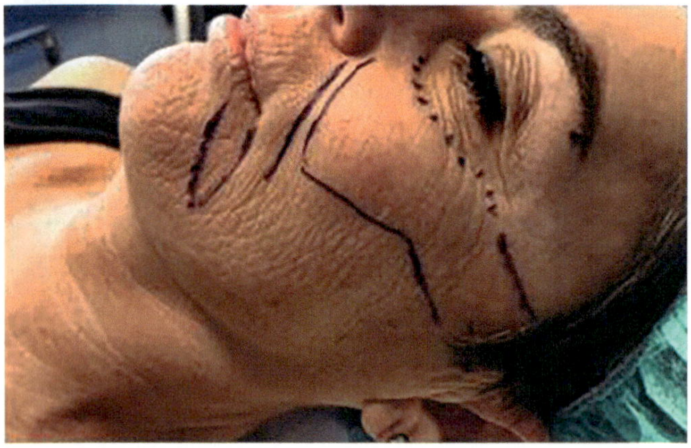

Fig. 30.23 Pre

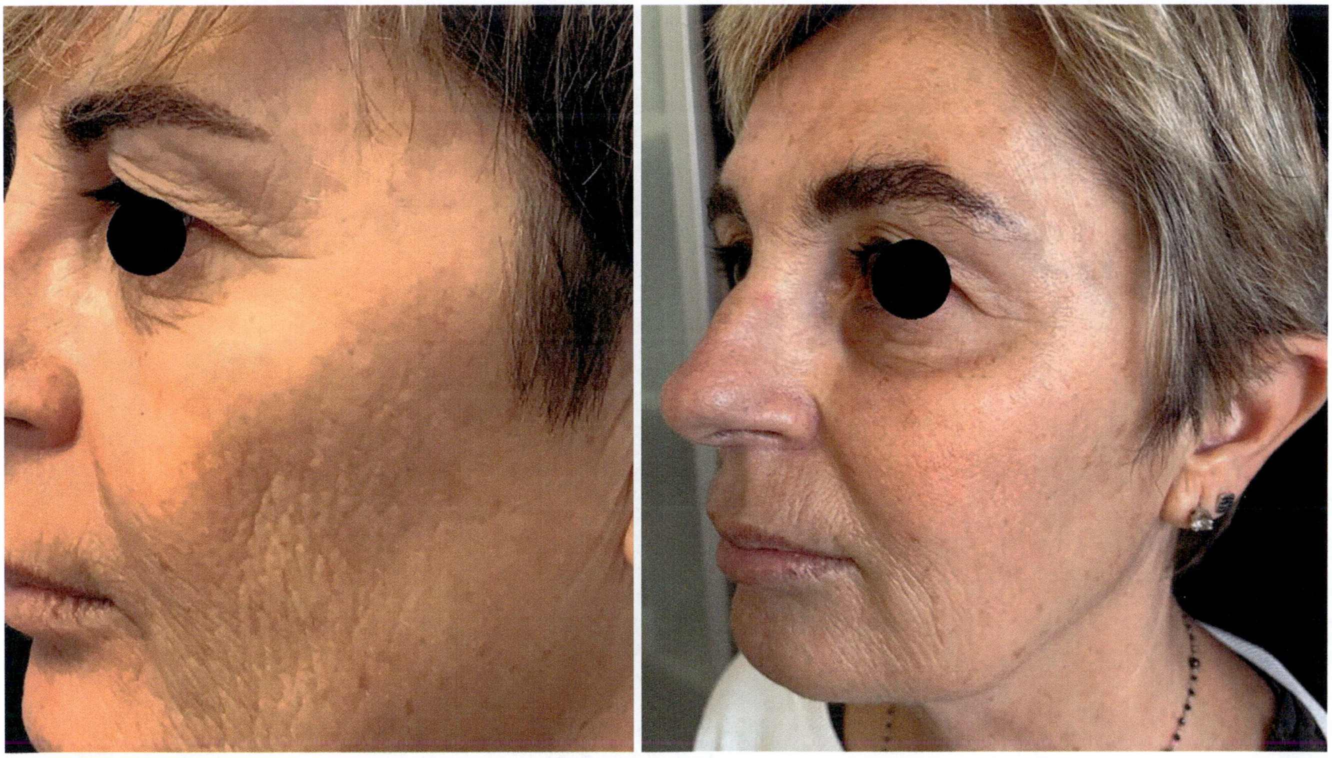

Fig. 30.24 Pre and post 1 year

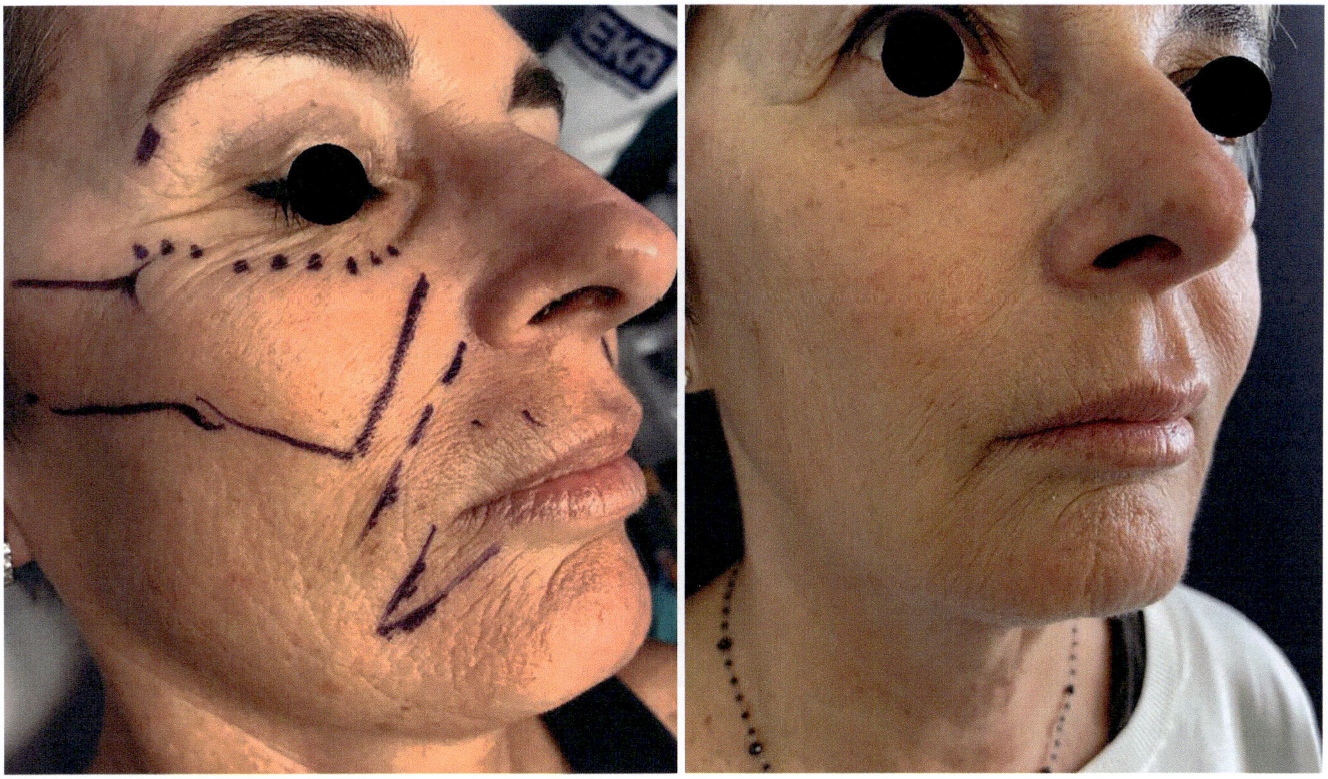

Fig. 30.25 Pre and post 1 year

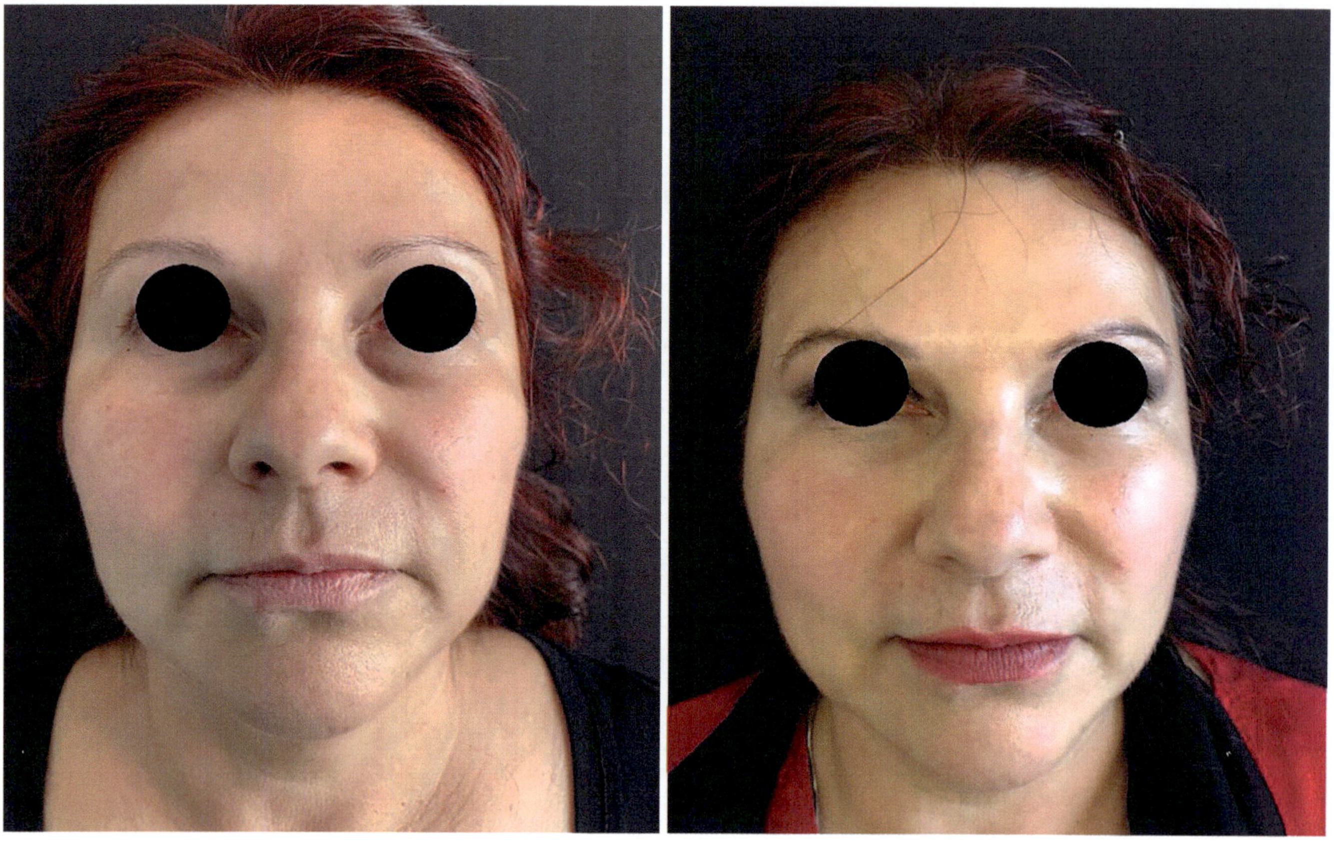

Fig. 30.26 Pre and post 12 months

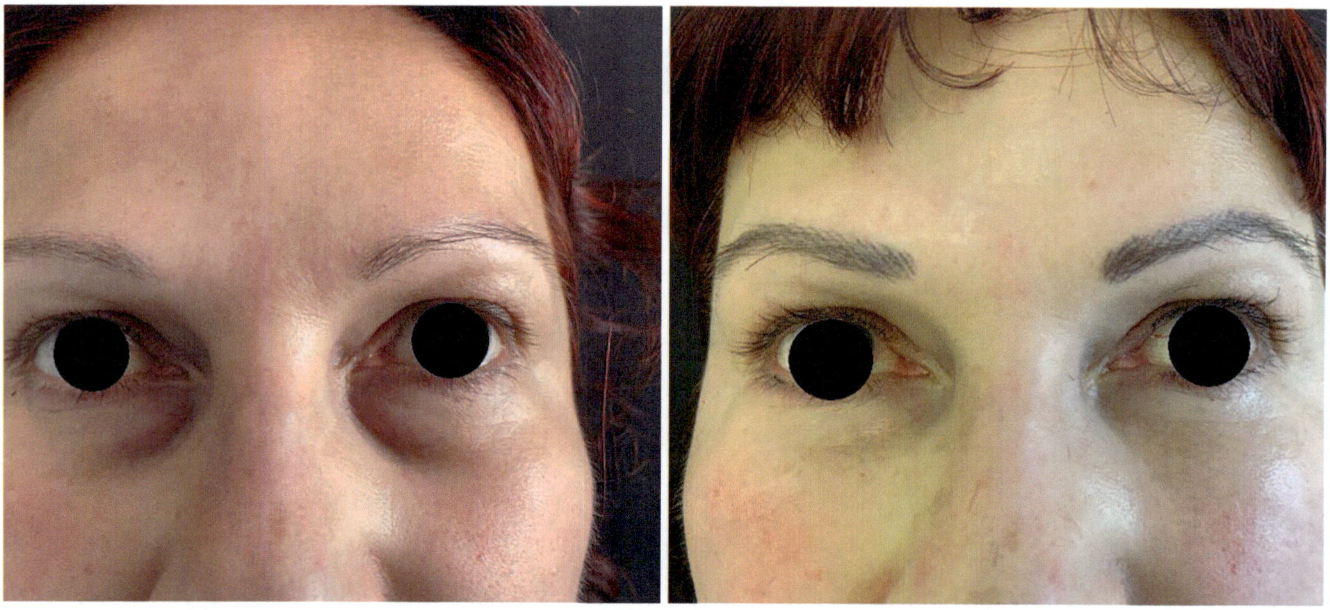

Fig. 30.27 Pre and post 12 months

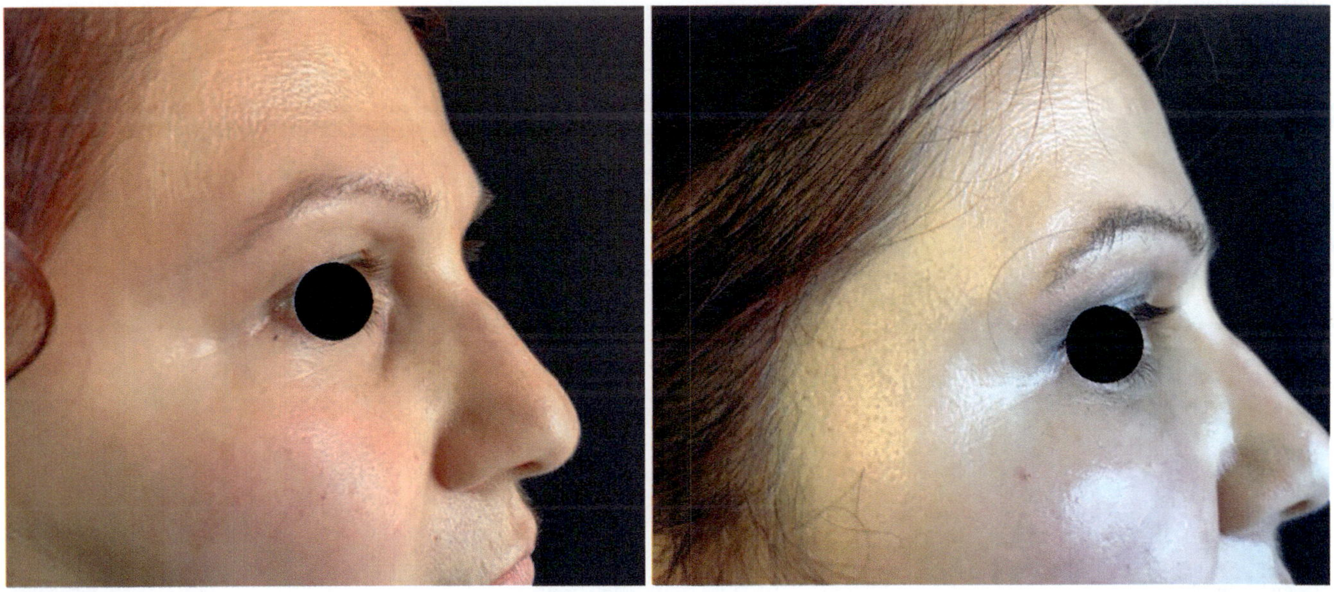

Fig. 30.28 Pre and post 12 months

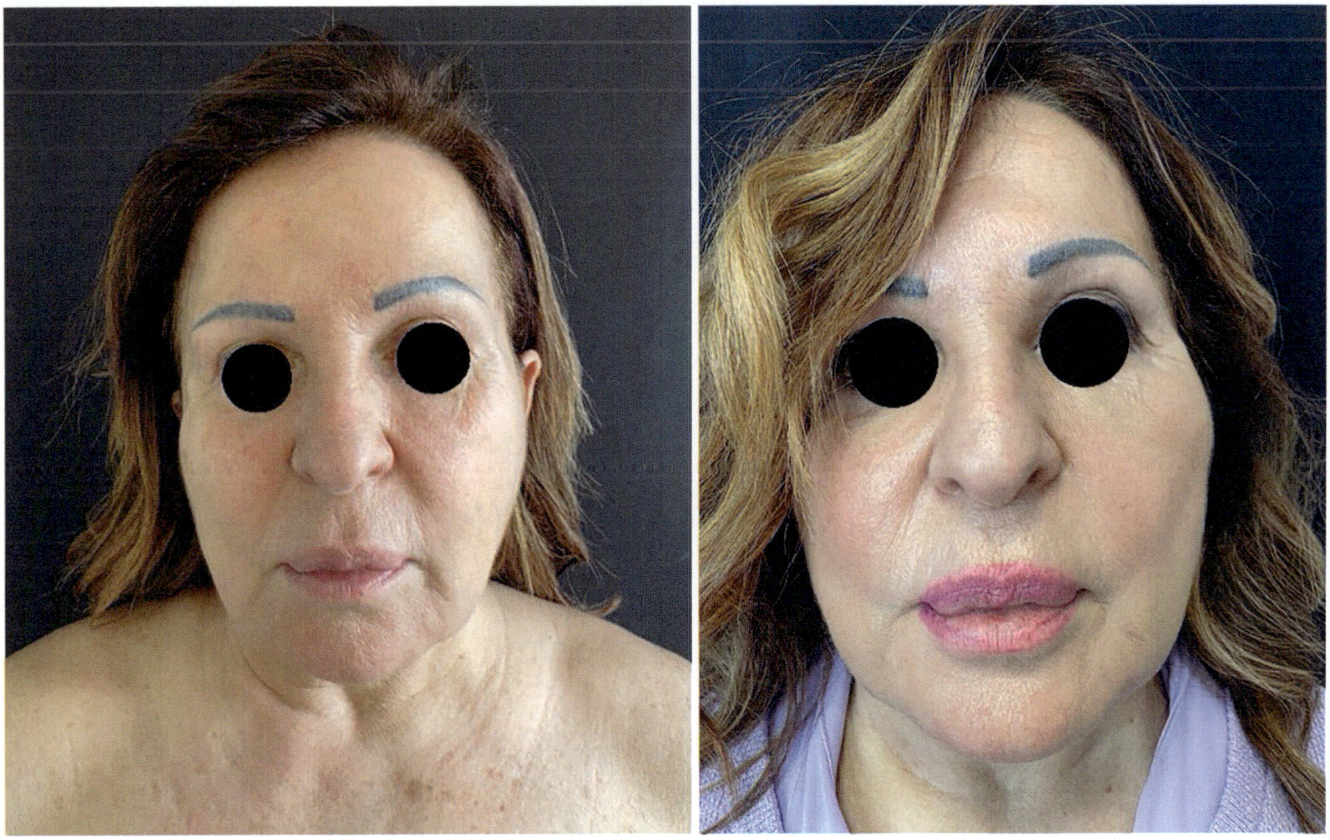

Fig. 30.29 Pre and post face lipofilling

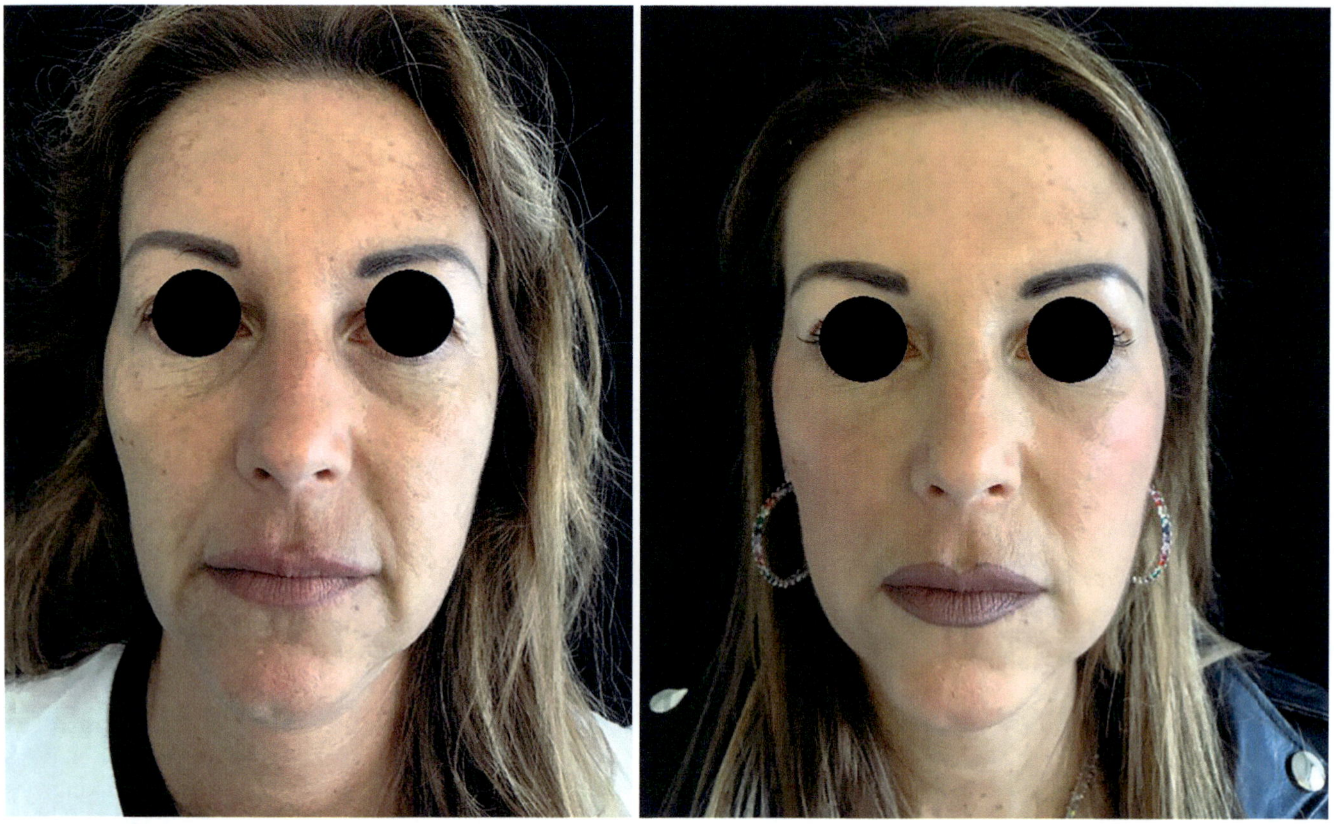

Fig. 30.30 Pre and post 2 years

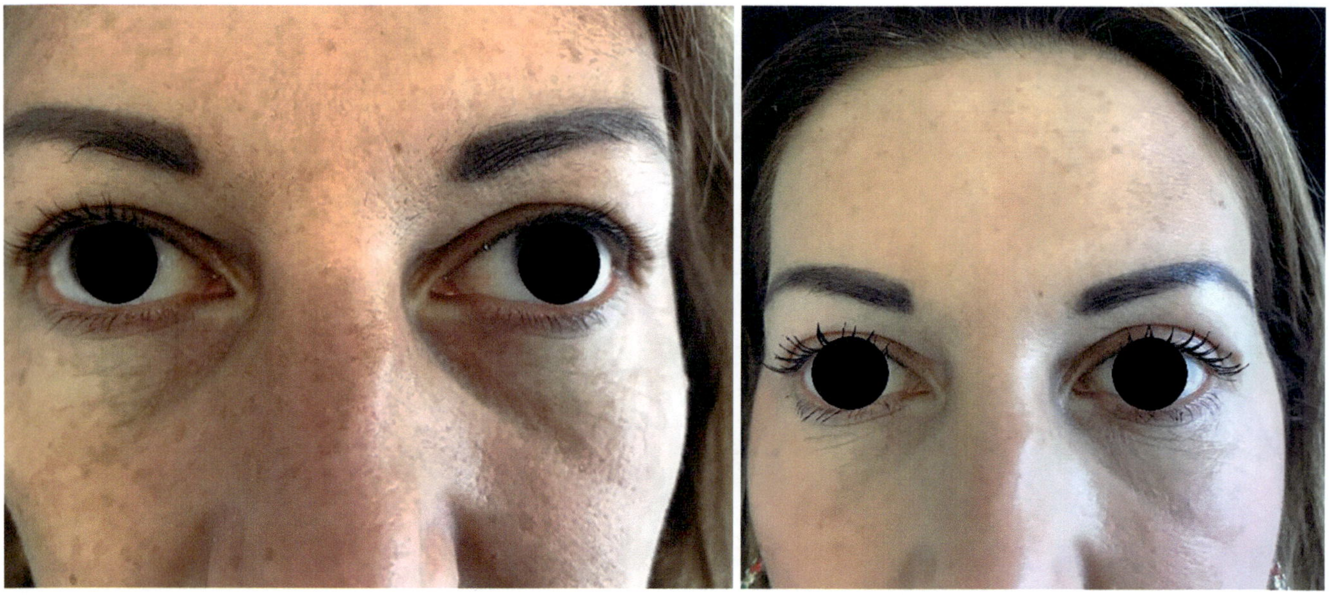

Fig. 30.31 Pre and post

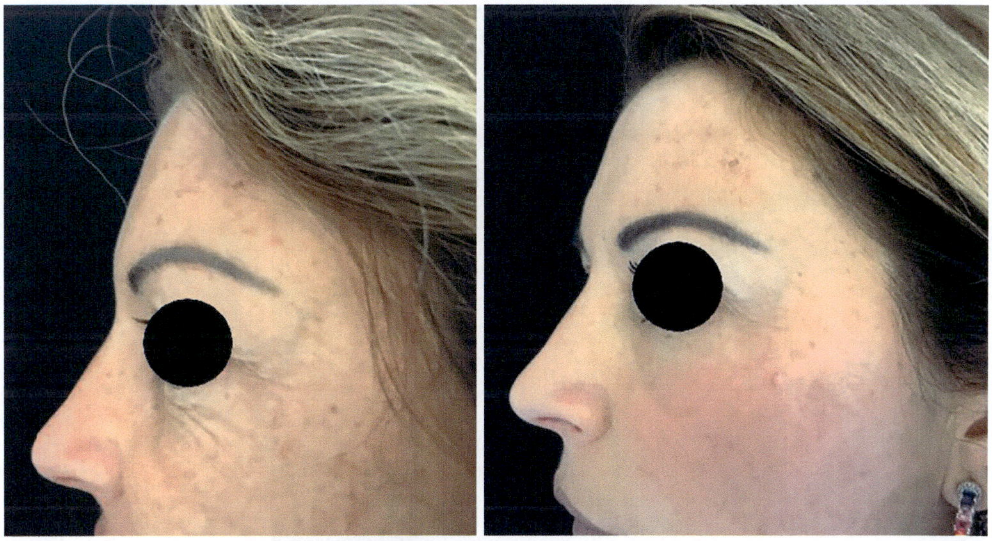

Fig. 30.32 Pre and post

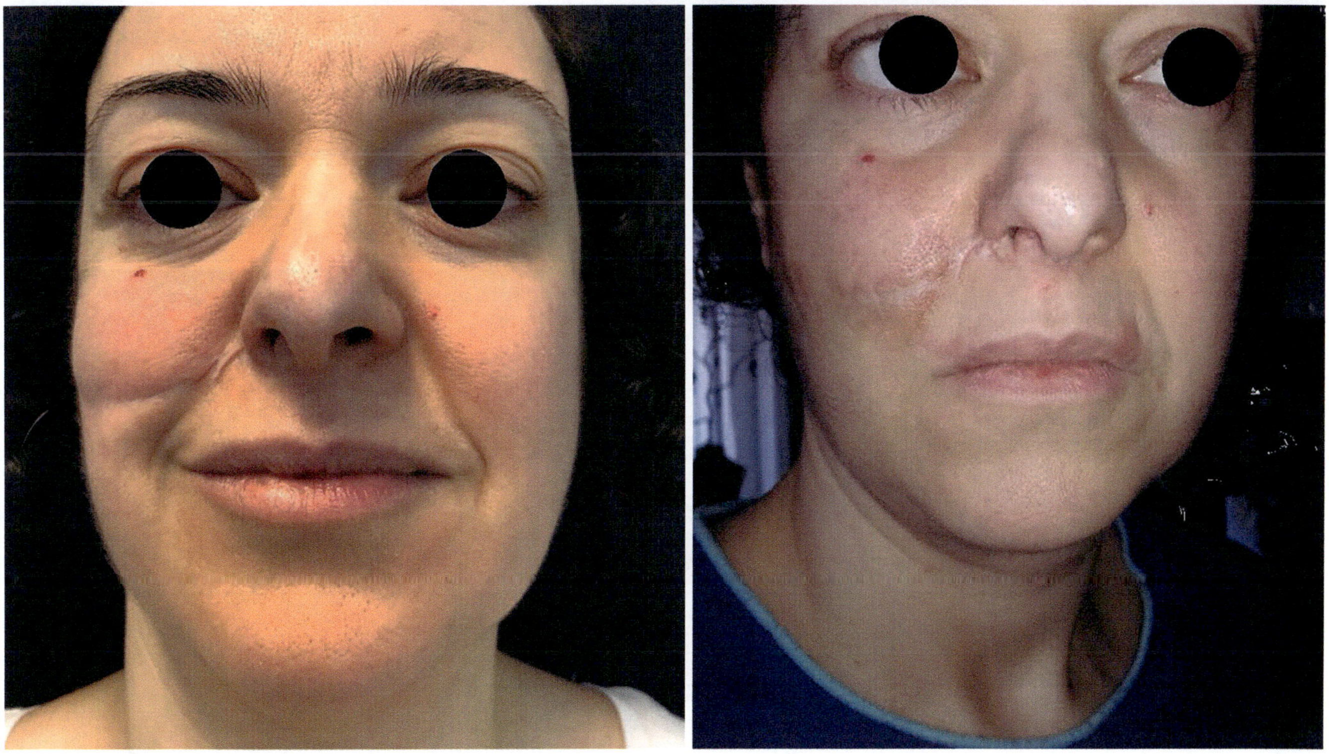

Fig. 30.33 Pre and post at 6 months

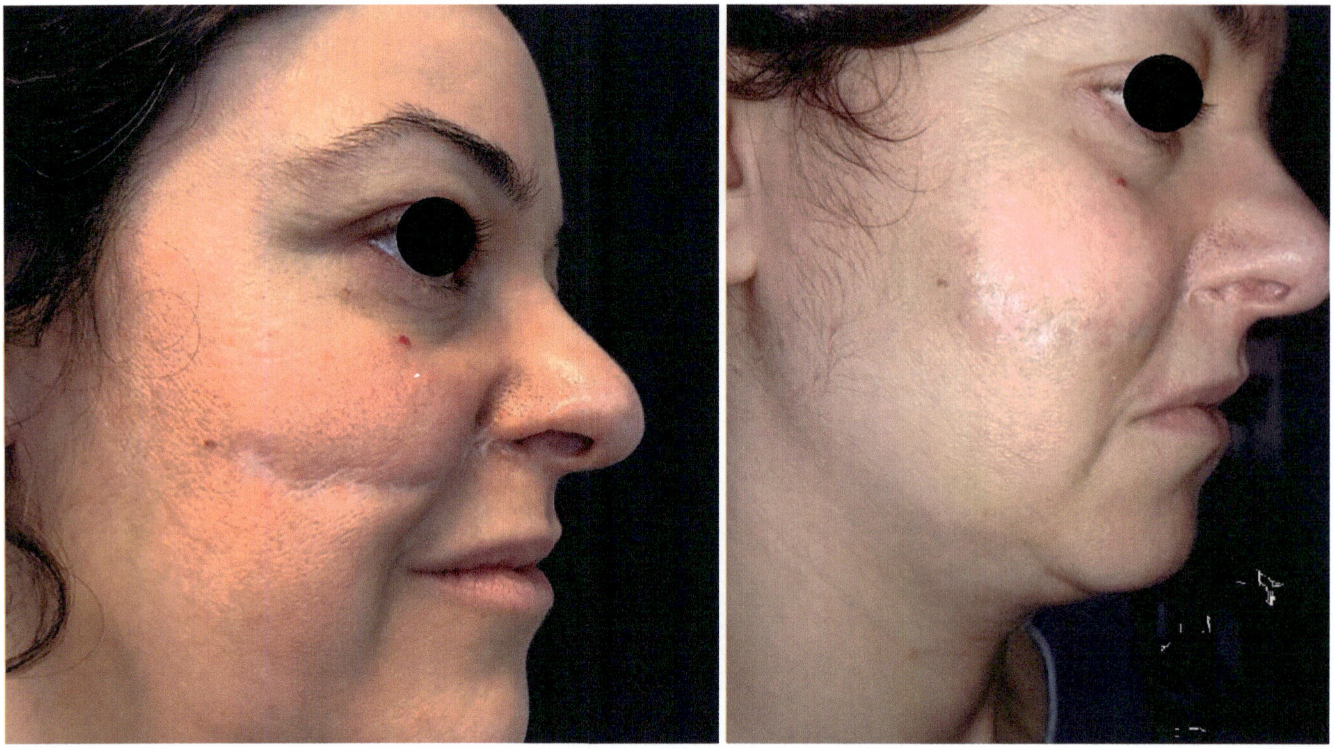

Fig. 30.34 Correction of the depressed area

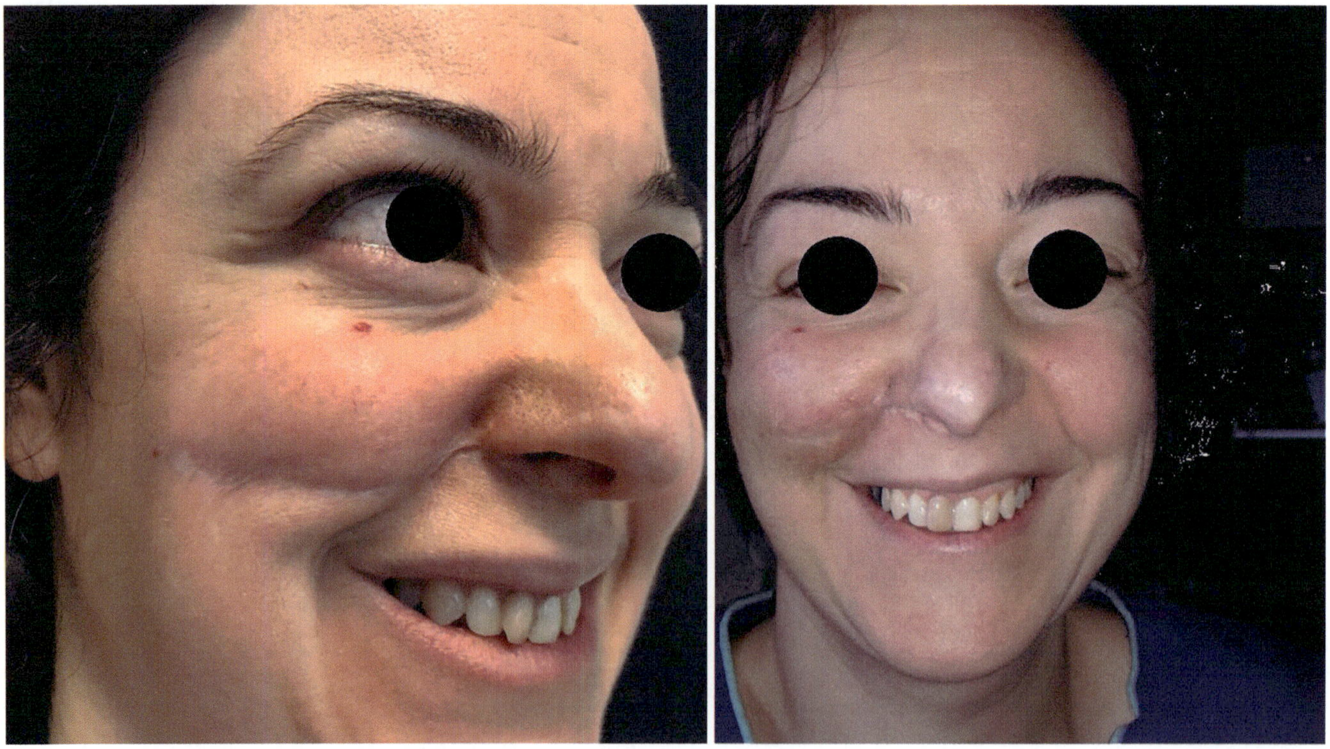

Fig. 30.35 Correction of the skin retraction

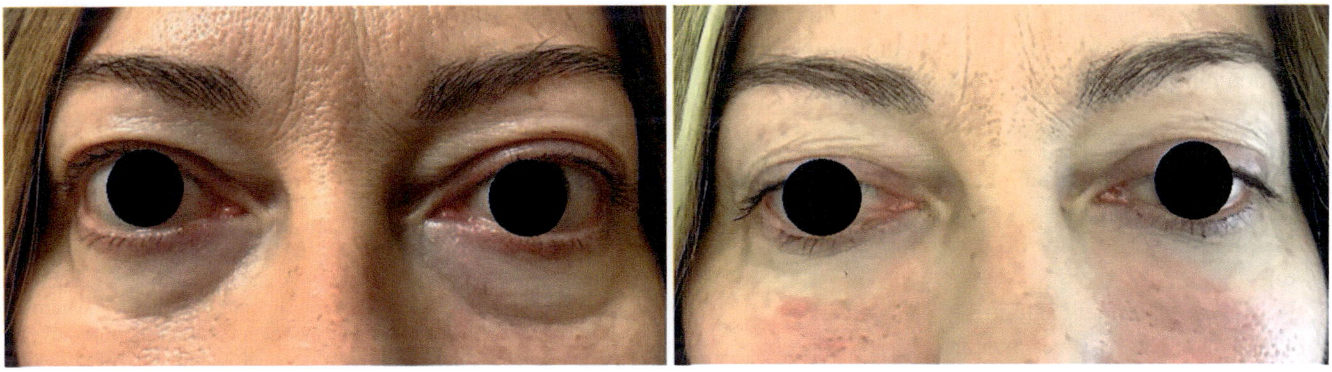

Fig. 30.36 Pre and post 2 year

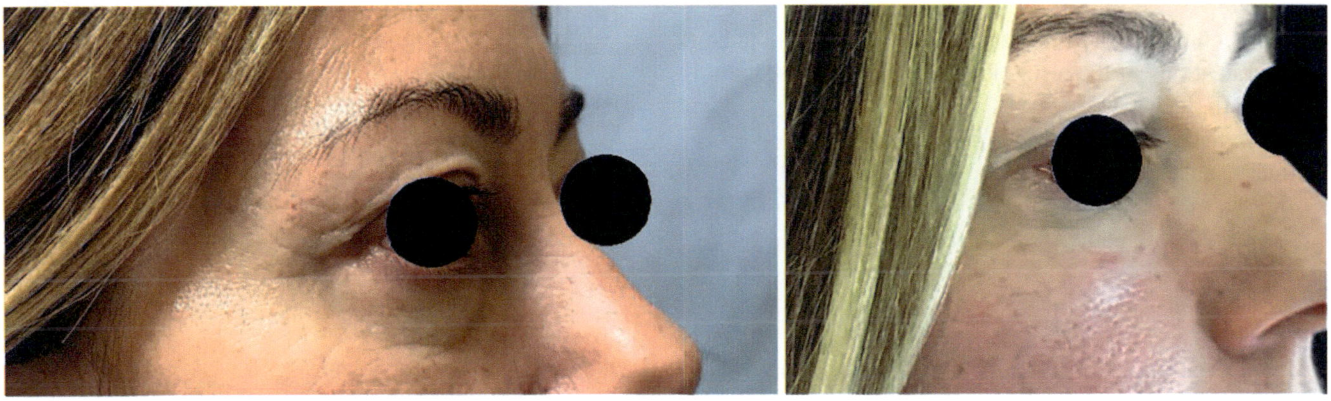

Fig. 30.37 Pre and post 2 year

30.4 Complications

Postoperative hematoma and infections are usually not a recurrent neither relevant complications. Even asymmetries are not typical of lipofilling technique; in most cases, they are already present in the preoperative, and it must be clarified very well with the patient if and how they can be corrected.

On the other hand, edema and ecchymosis are always present in the first 2–3 weeks postoperative. Furthermore, at the level of the eyelids, the edema can also persist even longer; in this case, it is often prescribed a lymphatic drainage to accelerate the healing process.

Very rare are nerve injuries because of the use of blunt cannulas that will protect from many of these complications and even if paralysis may sometimes occur, it will resolve within 60 days.

It is always necessary to avoid the presence of palpable and evident spaghetti-like appearance of injected adipose tissue, especially in the eyelids and lips areas. However, this is nowadays avoided with the use of much smaller and blunt cannulas.

Rare cases of fat embolisms are described, therefore it is always recommended to inoculate the adipose tissue with a retraction movement of the cannula, especially at risk areas.

Pearls and Pitfalls

The advantages of this technique are numerous:

- There is no possibility of developing allergies neither rejection.
- Results are extremely natural.
- The transferred fat is permanent.
- The intervention invasiveness is reduced thanks to the use of Coleman micro-cannulas.
- Surgery can be performed under local anesthesia in day hospital.
- This technique can be repeated several times without contraindications, if the result achieved in the first session is not sufficient.
- Immediately after treatment patients are able to perform their usual activities.
- In case of eyelid lipofilling, it is recommended to use an extremely fine method to extract the fat tissue.

Disadvantages are:

- Possible swelling and redness in the treated area.
- The need sometimes to reoperate two or three times in order to obtain and stabilize the desired result. Indeed, we can appreciate an unavoidable resorption of about 20–30% of the adipose tissue grafted in the first 3 months. However, it is extremely recommended not to hypercorrect the area, especially the in face region. Indeed, it is always better and easier to add some volume rather than remove the excess of adipose tissue.

Further Reading

Avelar RL, Goelzer JG, Azambuja FG, De Oliveira RB, Pase PF. Use of autologous fat graft for correction of facial asymmetry stemming from Parry-Romberg syndrome. Oral Surg Oral Med Oral Pathol Oral Radiol Endod. 2010;109(2):e20–5.

CAPITOLO 72—LIPOFILLING DEL VOLTO 13.

Carraway JH, Mellow CG. Syringe aspiration and fat concentration: a simple technique for autologous fat injection. Ann Plast Surg. 1990;24(3):293–6.

Coleman SR. Long-term survival of fat transplants: controlled demonstrations. Aesthetic Plast Surg. 1995;19(5):421–5.

Coleman SR. Facial recontouring with lipostructure. Clin Plast Surg. 1997;24:347–67.

Coleman SR. Structural fat grafts: the ideal filler? Clin Plast Surg. 2001;28(1):111–9.

Coleman SR. Structural fat grafting. St Louis: Quality Medical Publishing; 2004.

Coleman SR. Structural fat grafting: more than a permanent filler. Plast Reconstr Surg. 2006;118(3 Suppl):108S–20S.

De Fazio D, et al. Autologous fat transfer for the treatment of HIV-related face lipoatrophy: a long follow-up experience [Abstract 87]. 6th international workshop on adverse drug reactions and lipodystrophy in HIV, Washington, DC, 2004a.

De Fazio D, et al. Long-term follow-up of graft hypertrophy after autologous fat transfer for HIV-related face lipoatrophy (hamster syndrome 1 year later) [Abstract 90). 6th international workshop on adverse drug reactions and lipodystrophy in HIV, Washington, DC, 2004b.

De Fazio D, et al. Long-term follow-up of graft hypertrophy after autologous fat transfer for HIV-related face lipoatrophy (hamster syndrome 1 year later) [Abstract 90). 6th international workshop on adverse drug reactions and lipodystrophy in HIV, Washington, DC, 2004c.

De Fazio D, et al. Chapter 55. Autologous fat transfer. In: Shiffman MA, editor. Facial fat hypertrophy in patients who receive autologous fat tissue transfer. Springer; 2010.

Eder H. Importance of fat conservation in lower blepharoplasty. Aesthetic Plast Surg. 1997;21(3):168–74.

Fitzgerald R, Graivier MH, Kane M, Lorenc ZP, Vleggaar D, Werschler WP, Kenkel JM. Facial aesthetic analysis. Aesthet Surg J. 2010;30(Suppl):25S–7S.

Foyatier JL, Mojallal A, Voulliaume D, Comparin JP. Clinical evaluation of structural fat tissue graft (Lipostructure) in volumetric facial restoration with face-lift. About 100 cases. Ann Chir Plast Esthet. 2004;49(5):437–55.

Loeb R. Naso-jugal groove leveling with fat tissue. Clin Plast Surg. 1993;20(2):393–400.

Moore JH, Kolaczynski JW, Morales LM, et al. Viability of fat obtained by syringe suction lipectomy: effects of local anesthesia with lidocaine. Aesthetic Plast Surg. 1995;19(4):335–9.

Moseley TA, Zhu M, Hedrick MH. Adipose-derived stem and progenitor cells as fillers in plastic and reconstructive surgery. Plast Reconstr Surg. 2006;118(3 suppl):121S–8S.

Piasecki JH, Gutowski KA, Lahvis GP, et al. An experimental model for improving fat graft viability and purity. Plast Reconstr Surg. 2007;119(5):1571–83.

Rohrich RJ, Joel E. Chapter 29 - the subcutaneous fat compar tments and their role in facial rejuvenation - Pessa. In: Coleman SR, Mazzola RF, editors. Fat injection from filling to regeneration. St Louis: Quality Medical Publishing; 2009.

Rohrich RJ, Sorokin ES, Brown SA. In search of improved fat transfer viability: a qualitative analysis of the role of centrifugation and harvest site. Plast Reconstr Surg. 2004;113(1):391–5.

Schaverien MV, Pessa JE, Rohrich RJ. Vascularized membranes determine the anatomical boundaries of the subcutaneous fat compartments. Plast Reconstr Surg. 2009;123(2):695–700.

Shiffman MA, Mirrafati S. Fat transfer techniques: the effect of harvest and transfer methods on adipocyte viability and review of the literature. Dermatol Surg. 2001;27(9):819–26.

Sperber SM, Dawid IB. Barx1 is necessary for ectomesenchyme proliferation and osteochondroprogenitor condensation in the zebra- fish pharyngeal arches. Dev Biol. 2008;321(1):101–10.

Sterodimas A, Huanquipaco JC, De Souza FS, Bornia FA, Pitanguy I. Autologous fat transplantation for the treatment of Parry-Romberg syndrome. Plast Reconstr Aesthet Surg. 2009;62(11):e424–6.

Szabo-Rogers HL, Geetha-Loganathan P, Nimmagadda S, Fu KK, Richman JM. FGF signals from the nasal pit are necessary for normal facial morphogenesis. Dev Biol. 2008;318(2):289–302.

Ten Cate AR. Oral histology. Development and structure. St Louis: Mosby; 1998.

Von Heimburg D, Hemmrich K, Haydarlioglu S, et al. Comparison of viable cell yield from excised versus aspirated adipose tissue. Cells Tissues Organs. 2004;178(2):87–92.

Yoshimura K, Matsumoto D, Gonda K. A clinical trial of soft tissue augmentation by lipoinjection with adipose-derived stromal cells. Presented at the international fat applied technology society (IFTAS) third annual meeting, Charlottesville, VA, 2005.

Zoumalan RA, Larrabee WF Jr. Anatomic considerations in the aging face. Facial Plast Surg. 2011;27(1):16–22.

Fat Grafting to Treat Genital Lichen Sclerosus

31

Massimiliano Brambilla

Contents

Key Messages

- Fat graft may improve tissue quality in LS.
- Microfatgraft percutaneous release reduces scar retraction.
- Multilayer fat graft (micro- and nanofatgraft with or without PRP) is fundamental for the treatment of different levels.
- Quality and quantity of fat graft is key for a successful treatment.

M. Brambilla (✉)
Plastic Surgery Unit, Department for the Health of Women, Children and Newborn, IRCCS Policlinico Mangiagalli Hospital Foundation, Milan, Italy

University of Milan, Milan, Italy
e-mail: dr@massimilianobrambilla.it

31.1 Introduction

Lichen sclerosus (LS) is a scripting inflammatory dermatosis affecting anogenital region caused by autoimmune response whose mechanism is still uncertain. Family history may be seen. The association between LS and autoimmune disease is common in women (thyroid, bowel, skin diseases, rheumatoid arthritis) and less in men [1–4].

There is evidence that LS is an immunogenic disease, given increased expression in genes responsible for immune response [5], with increased proinflammatory cytokines (IL-1, IL-7, IL-15, IFN-γ, and TNF-α) and decreased anti-inflammatory cytokines (TNF-β), suggesting a mediation by a T-helper type 1 (T_h1) course of action [6].

Autoimmune and genetic tissue targets are as follows:

– ECM1 is increased; as an autoantigen for humoral autoimmunity serving as a "biological glue" especially at the dermal–epithelial junction; it disrupts the scaffold structure, interfering with collagenase activity [7–10].

© The Author(s), under exclusive license to Springer Nature Switzerland AG 2023
A. Di Giuseppe et al. (eds.), *Fat Transfer in Plastic Surgery*, https://doi.org/10.1007/978-3-031-10881-5_31

- TNF-a and IL-6 are increased, determining upstreaming Treg cell function [7].
- miR-155 promotes fibroblast proliferation and reduces Treg cells activity [5, 11–13].
- Elastic fibers are decreased.
- Endothelial ECM1 is decreased and induces hyalinized dermal vessels.
- P53 is increased and induces ischemic stress and increased the risk of malignancy.

Histologically, the lesions are characterized by lymphatic infiltrate in a sclerotic dermis with thickened hyalinized collagen and homogenized collagen fibrils, hyperkeratosis and orthokeratosis may be seen frequently. In early stages, a confined superficial involvement may be found (Fig. 31.1), while in later stages a full-thickness one is seen with sclerosis from the dermis [14–16] (Fig. 31.2).

The diagnosis is based upon clinical evidence, and there is a lack of unanimity regarding the reliability of histological biopsy, especially in the early stages [17]. The loss of structural anatomical architecture determines developing lesions leading to itching, bruising, burning sensation, stenosis, and sexual dysfunction. The sclerosis process determines progressive vulvar stenosis. With progressive narrowing of the vaginal introitus, fusion, and resorption of the labia minora and clitoral phimosis. The inflammatory process may be the cause of malignant transformation so that periodical vulvoscopy is crucial to exclude degeneration into vulvar cancer [18].

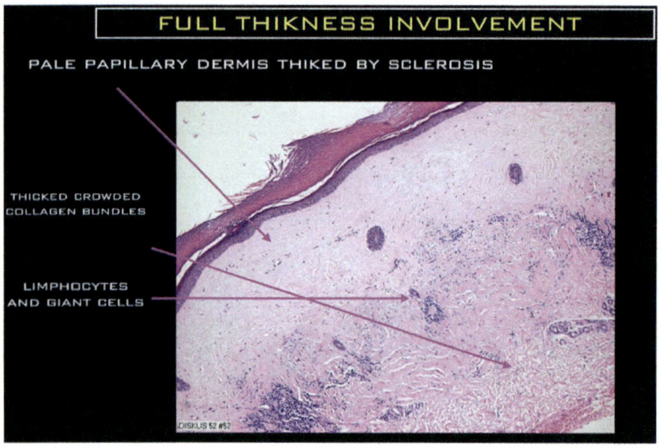

Fig. 31.2 Lichen sclerosus: full thickness involvement

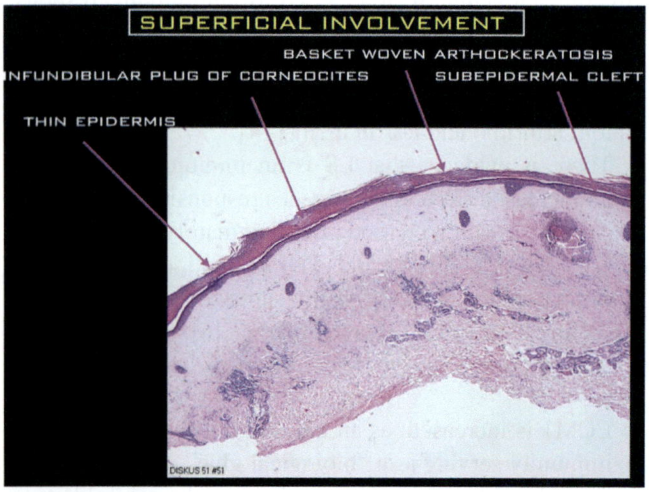

Fig. 31.1 Lichen sclerosus: superficial involvement

Lichen sclerosus may determine severe impact on psychosexuology [19]. There is actually no consensus regarding a classification of lichen sclerosus, thus recent publications attempt to define grading according to anatomical modifications [20–23]. Current local treatments of LS are topical steroids, emollients, topical calcineurin inhibitors, topical androgens, progesterone, and estrogens [24]. Recently, lasers, TCA, LED, and carboxitherapy have been suggested as possible therapies [25–27]. In case of vaginal dryness, vaginal treatments are advisable with local hormones, bioidentical hormones, or lasers.

31.2 Fat Graft and Lichen Sclerosus

The benefits of fat grafts are both regenerative [28–32] and mechanical.

The specific use of fat graft for the treatment of LS has been reported in the literature [33, 34].

As well as a preliminary report to the use of nanofatgraft as a unique procedure [35].

The combination of fat graft and prp has been reported as well with an emphasis on the anti-inflammatory role of PRP [36].

The combination of microfatgraft and nanofatgraft has been an attempt for the treatment of GSM [37].

Tissue regeneration takes place when the fat graft induces neovascularity and fibroblast production of "soft" collagen type 1 and 2 [38]. Reducing fibrosis [39] interferes positively with local autoimmune response [40]. Moreover, fat graft reduces inflammatory response [41] and pain through pain fiber modulation [42].

Mechanical expansion through tissue expansion is achieved with the interruption of the scar and the interposition of fat clusters. This procedure is performed through percutaneous scar release ("rigottomy") [43, 44] or during a surgical amplification of the stenosis (open wound scar release) and before flap or skin graft reconstruction. Regenerative treatments do not exclude all the other "standard" treatments such as the use of topical corticosteroids in case of severe inflammations, topical and general hormones administration, and modified dietary supply.

31.3 Before Surgery and Preoperative Procedures

Accurate general and genital anamnesis, vulvoscopy, images recording, and subjective reports have to be given. Appropriate general and local hormones are given if possible as well as diet modifications.

If presence of an inflammation is found, appropriate local treatment is applied according to inflammatory cause (corticosteroid, antimycotic, and antibacterial treatment).

Preoperative exams are suitable for sedation or spinal, or general anesthesia is given.

31.3.1 Procedure

Procedures are carried out under local anesthesia plus sedation or under spinal or general anesthesia on a one-day surgery basis according to the skill and experience of the surgeon and anesthetist.

An antibiotic prophylaxis is given 1 h preoperatively. The patient is set in a lithotomy Trendelenburg position. There is no need of catheter, but if a surgical anterior vulvar amplification is planned, a catheter is set in place; otherwise, an extemporary catheterization will be performed at the end of the procedure.

31.4 General Considerations on Fat Graft for Lichen Sclerosus

Fat graft varies according to lichen grade, local inflammatory condition, and subjective report.

Quality, quantity, tissue level of grafting, grafting modality, and eventually combination with prp are crucial to achieve best results.

31.4.1 Fat Cluster Dimension and Fat Quality

In the treatment of delicate tissues such as the ones in LS, cluster dimension plays an important role [45, 46].

Due to the fragility of the tissues it is advisable to use microfatgraft for subdermal to fascial fat graft, while nanofatgraft and/or prp has to be used for intradermal injection [47, 48].

Centrifuged and more concentrated microfatgraft is used in early stages while decanted and diluted microfatgraft is indicated in later stages with sclerotic tissues due to better tissue expansion and cell spreading into tissues.

Nanofatgraft that is a more regenerative than anti-inflammatory treatment may be injected into the dermis if low inflammation is found while PRP, which has more anti-inflammatory than regenerative, is advisable in inflammatory conditions.

31.4.2 Fat Harvesting

Fat harvesting may be performed where more is convenient, according to the patient's wishes.

Infiltration is performed, and an average of 100 cc of fat is collected via a 14-gauge multihole cannula and 10 mL syringes with low negative pressure [49].

The fat is transferred into 60 mL syringes and washed with saline solution (30% of saline and 70% of harvested fat) in order to decrease lidocaine potential negative effect on cells viability [50], and then decanted for 5/10 min. Then fat is transferred into 5 mL syringes.

If a more dense preparation is needed, the fat is centrifuged for 3 min at 3000 rpm.

Infiltration of the vulvar area of 10–20 cc of saline + 0.5 mL of lidocaine + adrenaline 1:500,000 is carried out with a 5 cc syringe.

31.5 Vulvar Infiltration

Infiltration of the vulvar area of 10–20 cc of saline + 0.5 mL of lidocaine + adrenaline 1:500,000 is carried out with a 5 cc syringe.

After 5–10 min, the procedure can be performed having achieved vascular constriction.

31.5.1 Injection Site and Level of Injection

Multilayer injection is advisable. Microfatgraft is done from the dermis to the fascia and nanofatgraft or prp is injected into the dermis.

Vaginal microfatgraft and nanofatgraft may be combined in case of vaginal atrophy.

It is advisable to treat the entire vulvar area even if the extension of lichen sclerosus seems to be limited to a specific area. This is to promote better vulvar vascularization.

31.6 Volume of Fat Graft

Injected volume varies according to fibrosis and distention tissue capability.

1–2 cc of microfatgraft + 0.5 cc of microfatgraft or 0.5 prp is the average quantity for the treatment of 10 cm^2.

31.7 Injection Technique

In early lichen sclerosus stages, with moderate tissue sclerosis, a blunt 21 gauge cannula is used for microfatgraft spreading of the fat into the tissues with fan, multilevel technique.

A 27 gauge needle is used to inject nanofatgraft or prp into the dermis (Fig. 31.3).

In later stages when vulvar stenosis appears, microfatgraft is performed with an 18 gauge sharp needle. The needle is inserted at least 2 cm away from the area to be treated and is

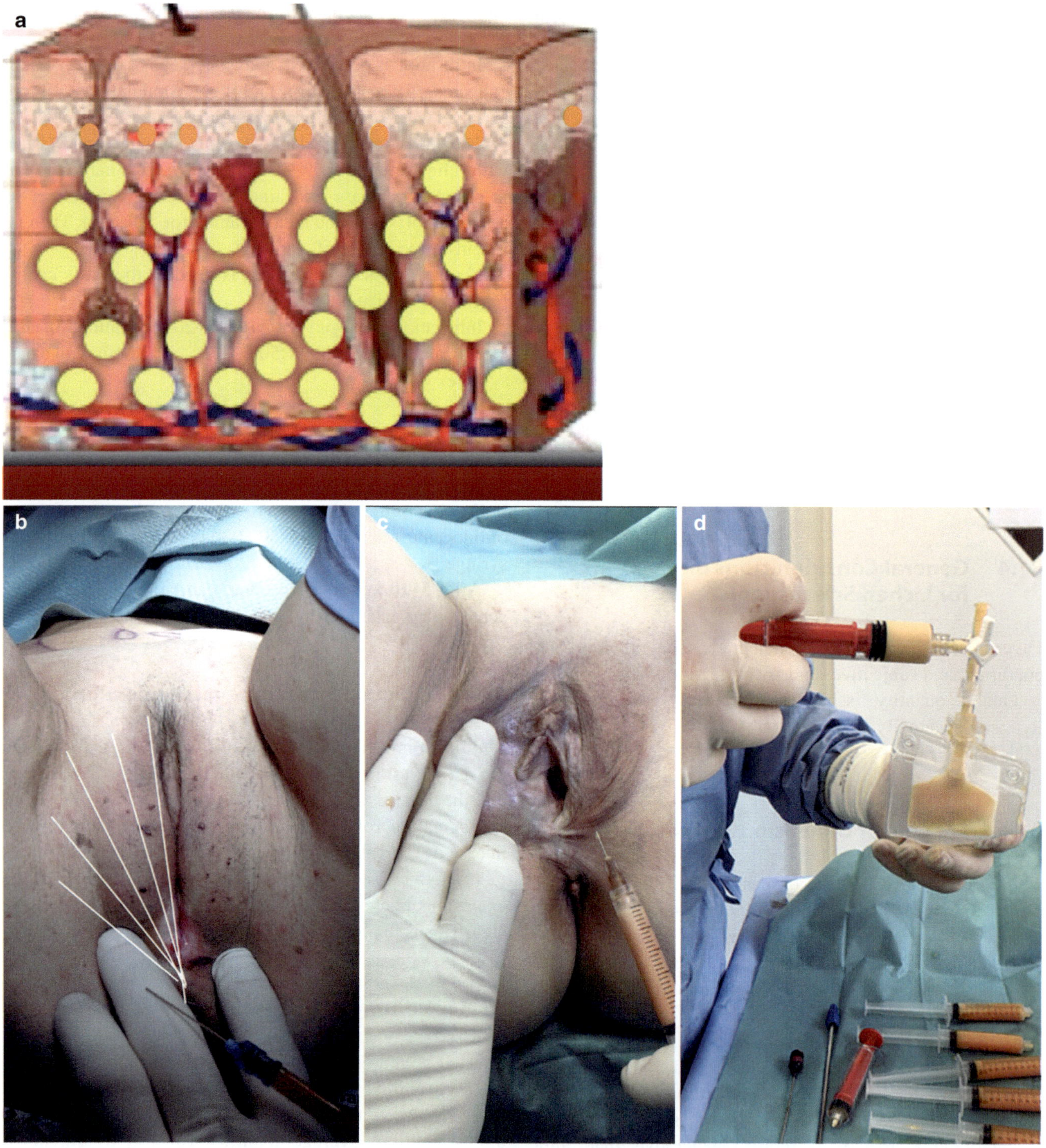

Fig. 31.3 Multilevel micro and nanofatgraft: a diagram of injection. Microfatgraft (yellow), nanofatgraft (orange) (**b**) microfatgraft with 21 gauge blunt cannula c intradermal nanofatgraft with 27 gauge needle (**d**) nanofat preparation

used as a cutter, with its tip performing small and multiplanar cuts. The needle moves with small fan multilevel movements in order to avoid wide dead spaces and simultaneous fat injection is performed. The needle may delicately subcise the dermis creating a subcisional net (Fig. 31.4).

In case of severe stenosis, the procedure can be performed at the same time of the surgical posterior wall amplification and tissue reconstruction with perineal island tunneled flaps [51, 52].

The procedure can be performed "open wound" before suturing the medial margins of the flaps (Fig. 31.5) together or after the suture. Severe amplifications may be covered with full-thickness skin grafts. The skin graft will be treated after 3 months with a microfatgraft and a nanofatgraft, reducing retraction and gaining elasticity. This procedure can be performed during the additional fat graft sessions planned for the treatment of stenosis.

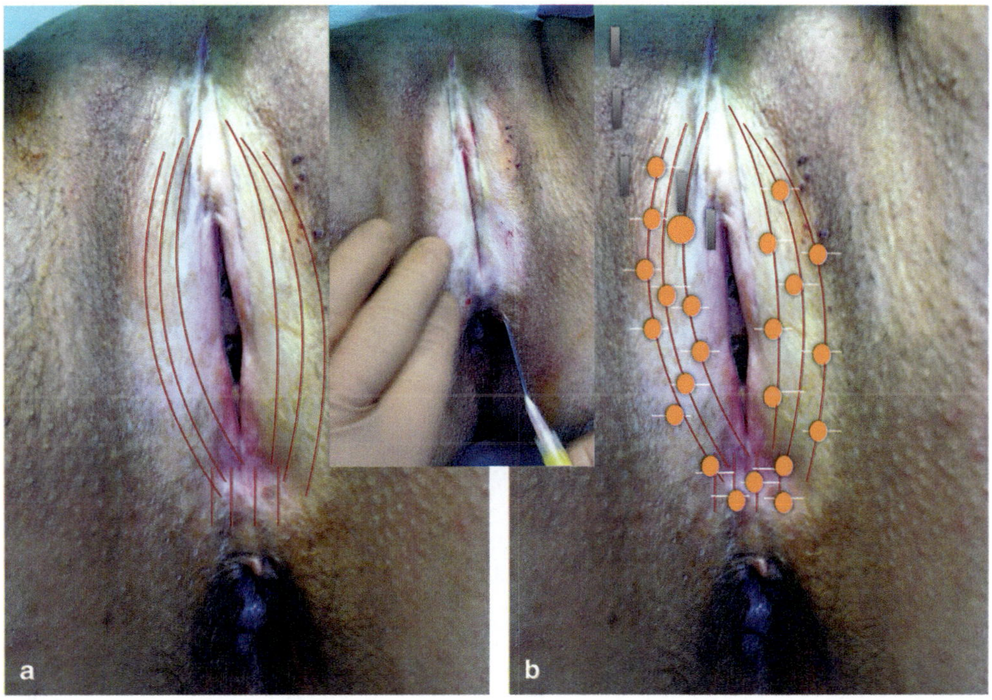

Fig. 31.4 Microfatgraft + percutaneous scar release (rigottomy): a tension lines b scheme of tension line fragmentation with interposition of microfatgraft clusters (orange)

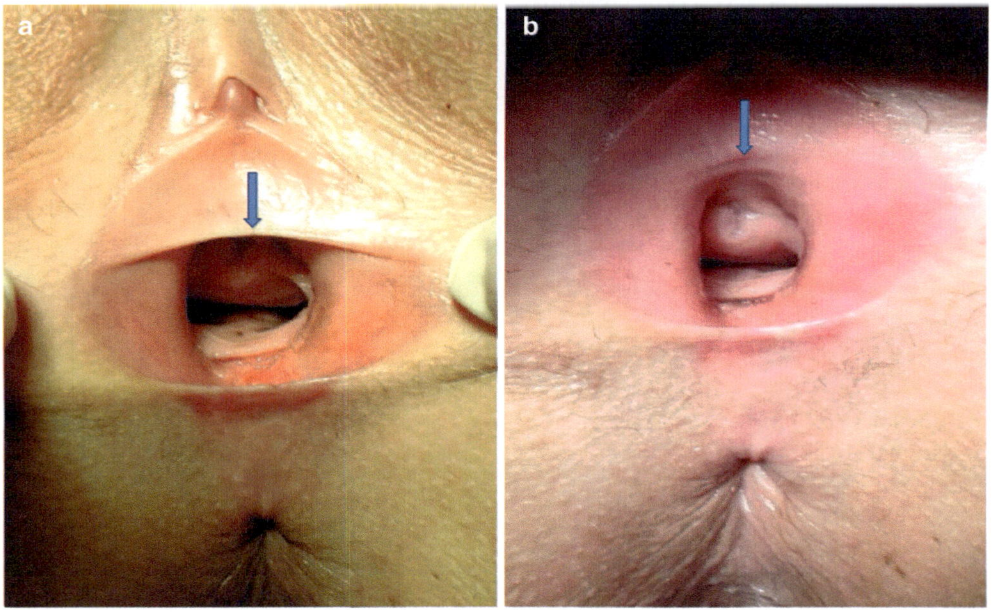

Fig. 31.5 Lichen sclerosus: micro + nanofatgraft. (**a**) Preoperative image, (**b**) postoperative image at 1 year, superior web release

31.7.1 Immediate Postoperative Care

Polydine ointment is applied to the vulva, and a Kemicetine (chloramphenicol) egg is inserted into the vagina. A cold pad is applied to the vulva, the catheter is removed (if inserted), and the patient may be discharged on the same day of the procedure or after one night of observation.

Postoperative pain can be controlled easily with nonmorphinic analgesics (Paracetamol [acetaminophen]), and oral antiedema (Bromelain) and hydrating genital creams are prescribed for 1 month. Blood fluidifying medication is prescribed for 15 days. Patients must avoid sexual activity, cycling, and riding horses for 30 days. Pelvic floor rehabilitation is an essential part of the treatment and starts 1 week after the procedure with finger massage only, followed by 2 weeks after the operation with tissue expansion with inflatable dilators.

31.8 Number of Procedures and Final Results

In early stages, a single treatment of 5–15 cc of microfatgraft with blunt 21 gauge cannula injection +2/4 cc of nanofatgraft or PRP can achieve a good result, in elasticity and trophism and a second treatment may be necessary 4–6 months after the first treatment. Rarely, a third treatment is needed. Results are stable and long lasting in the majority of the cases (Figs. 31.5, 31.6, 31.7, and 31.8).

Lichen sclerosus stenosis requires percutaneous microfatgraft + nanofatgraft or PRP treatment, followed if necessary at 4 months by posterior commissure amplification with transposition of the perineal flaps plus fat grafting (Fig. 31.9).

Severe vulvar stenosis that are considered urgent procedures will need first a surgical amplification with the interposition of tunneled labia majora flaps (Fig. 31.10),

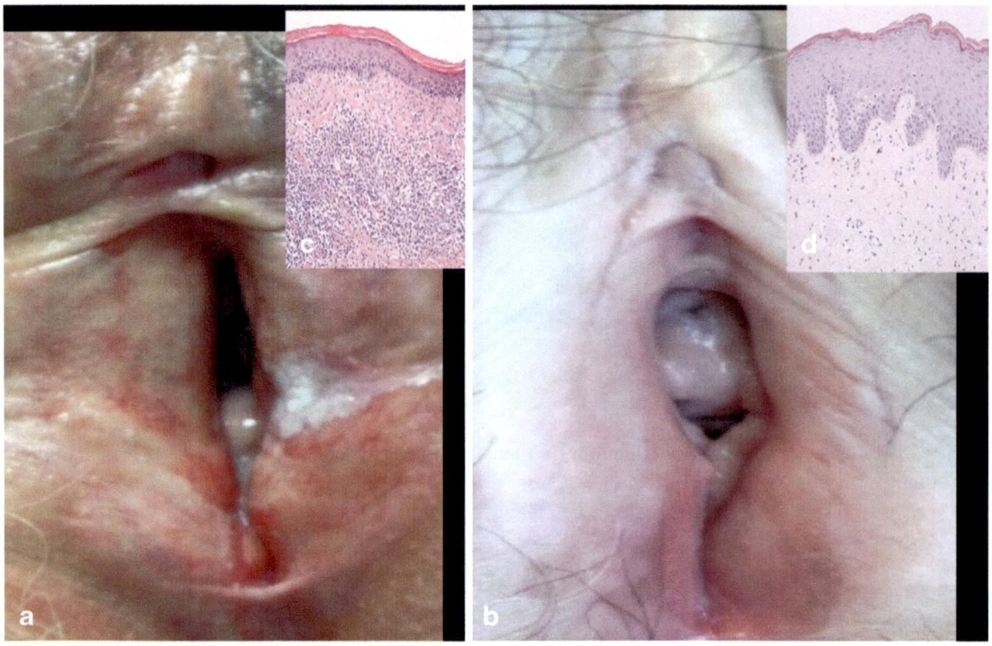

Fig. 31.6 Lichen sclerosus: microfatgraft + rigottomy + nanofatgraft. (**a**) Preoperative image, (**b**) postoperative image at 1 year, superior web release, (**c**) preoperative histology, (**d**) postoperative histology

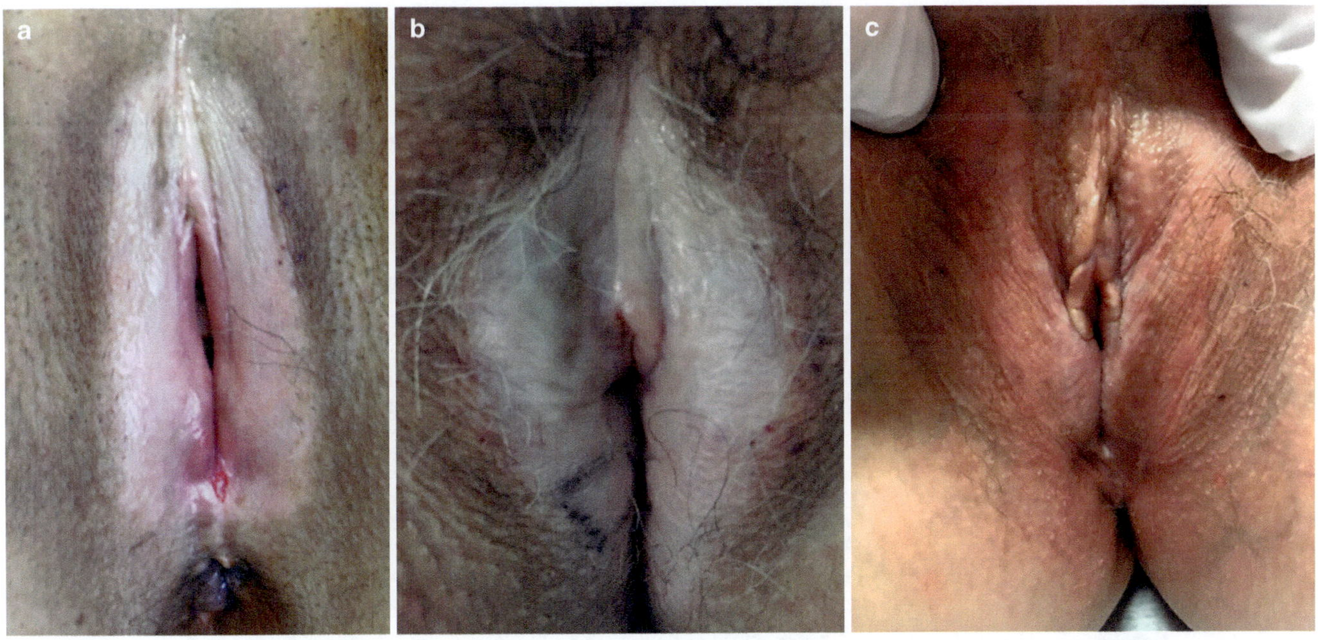

Fig. 31.7 Lichen sclerosus: microfatgraft + rigottomy + nanofatgraft. (**a**) Preoperative image, (**b**) postoperative image at 1 year, (**c**) postoperative image at 4 years after second treatment

Fig. 31.8 Lichen sclerosus: micro + nanofatgraft. (**a**) Preoperative image, (**b**) postoperative image at 4 year, superior web release

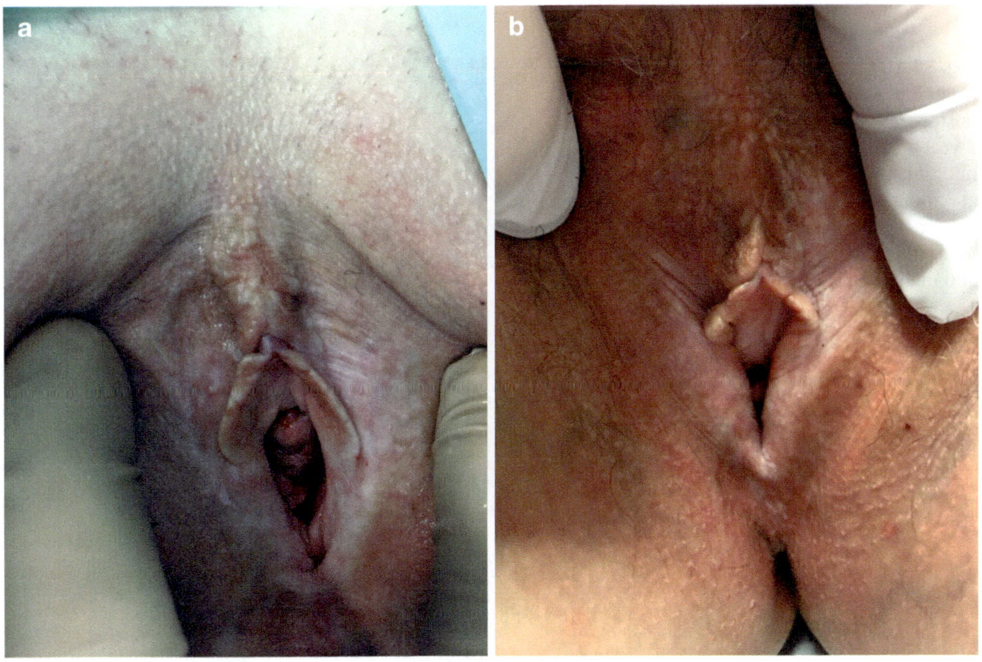

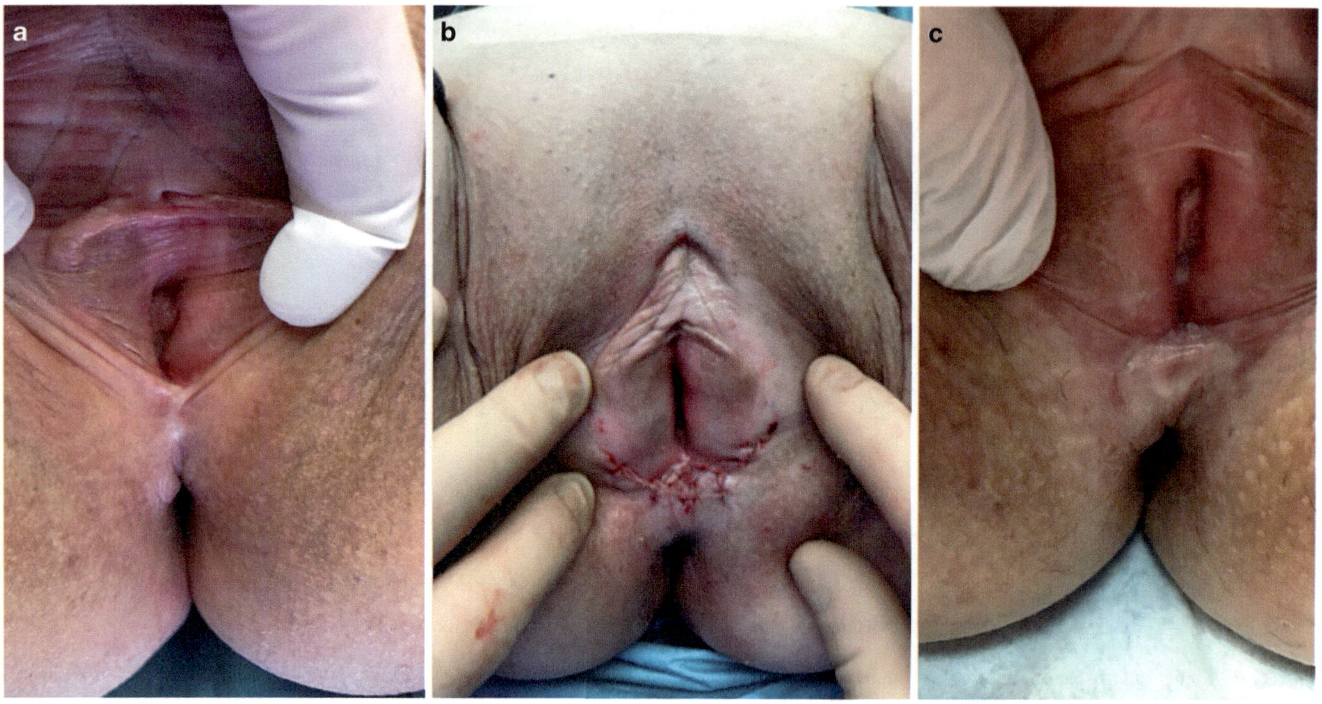

Fig. 31.9 Lichen sclerosus: (**a**) preoperative image, (**b**) immediate post operative (**c**) postoperative image at 2 years, after rigottomy micro/nanofatgraft followed at 3 months by double perineal island flaps transposition + rigottomy micro/nanofatgraft followed at 6 months by nanofatgraft

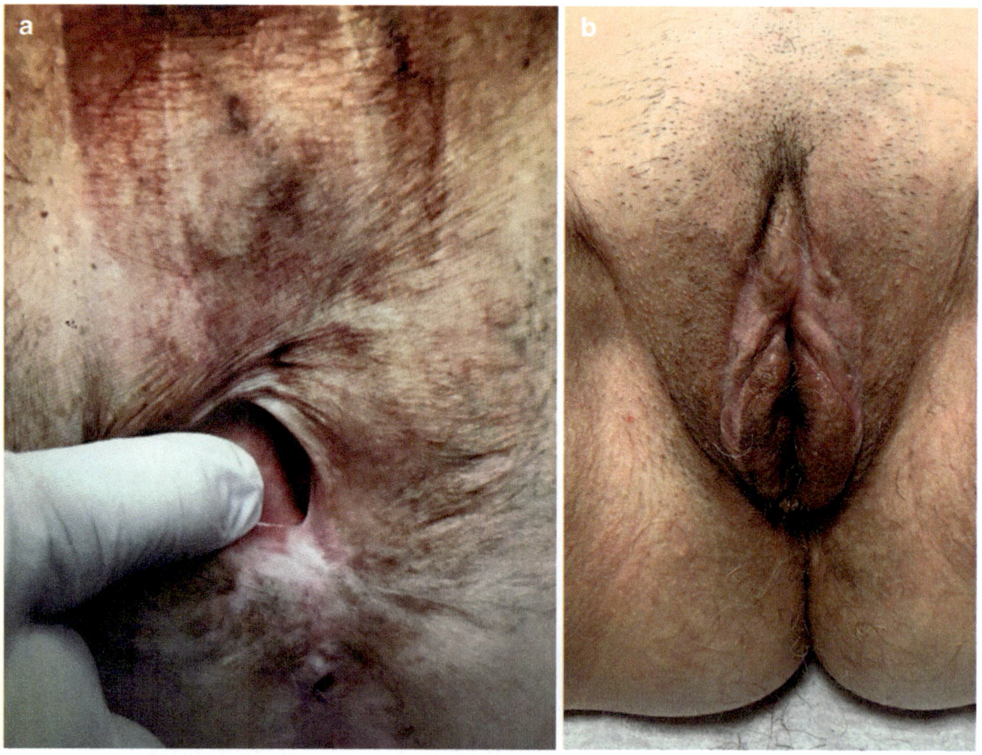

Fig. 31.10 Lichen sclerosus: (**a**) preoperative image, (**b**) postoperative image at 2 years, after rigottomy micro/nanofatgraft followed at 3 months by double labia majora island flap + rigottomy micro/nanofatgraft followed at 6 months by nanofatgraft

or full thickness skin graft (Fig. 31.11), or a combination of both procedure if needed (Fig. 31.12). Simultaneous open-wound microfatgrafts are advisable in order to amplify and promote regeneration (Fig. 31.13). In order to enhance vulvar skin grafts elasticity after 3–6 months from surgery, it is advisable to perform two procedures of superficial microfatgraft and nanofatgraft at 4-month intervals.

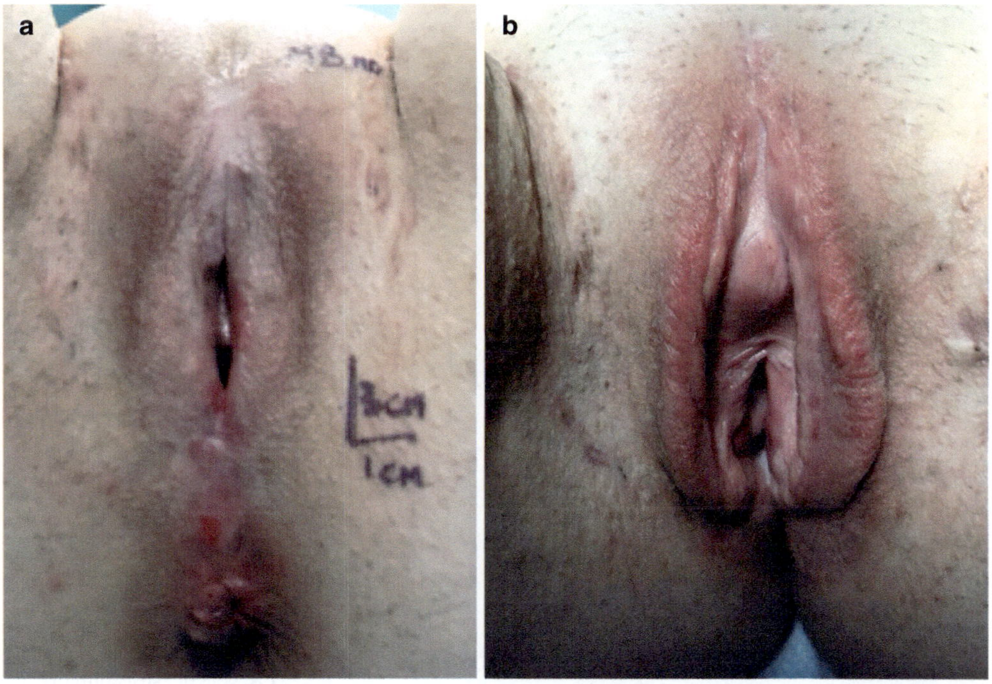

Fig. 31.11 Lichen sclerosus: (**a**) preoperative image, (**b**) postoperative image at 2 years, after rigottomy micro/nanofatgraft followed at 3 months by labia minora skin graft reconstruction + rigottomy micro/nanofatgraft followed at 6 months by nanofatgraft

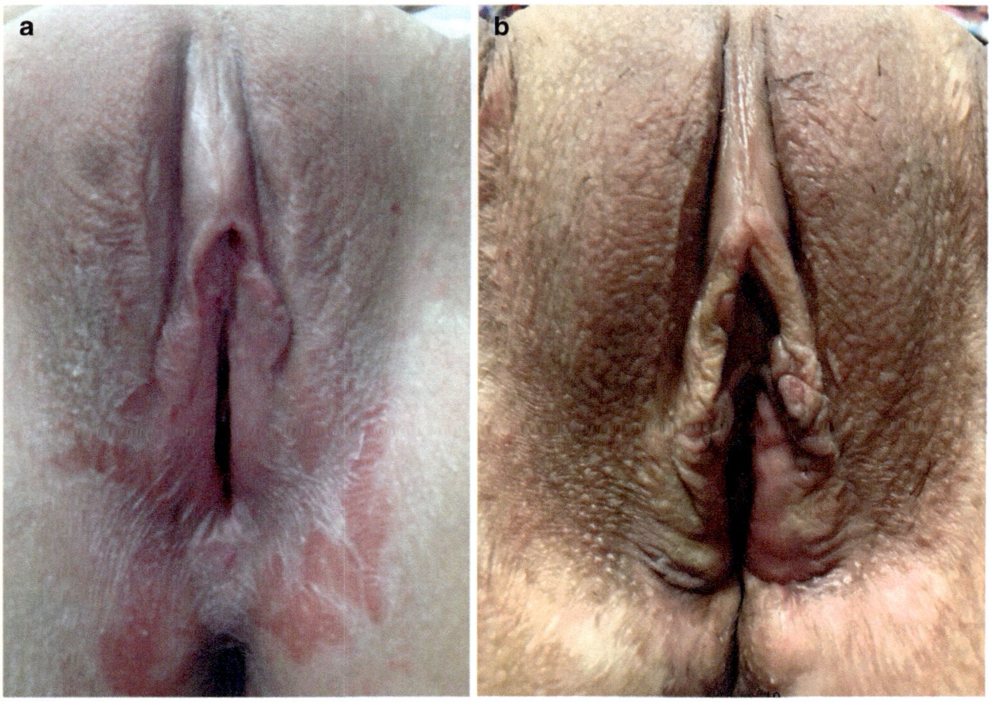

Fig. 31.12 Lichen sclerosus: (**a**) preoperative image, (**b**) postoperative image at 2 years, after rigottomy micro/nanofatgraft followed at 3 months by labia minora skin graft reconstruction and double perineal island posterior commissure reconstruction + rigottomy micro/nanofatgraft followed at 6 months by nanofatgraft

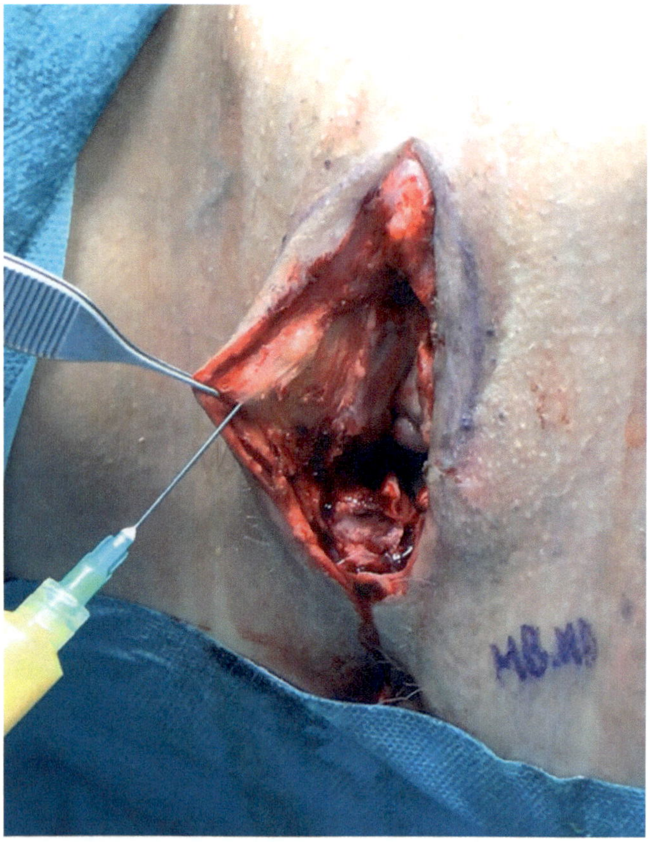

Fig. 31.13 Open wound rigottomy microfatgraft

31.9 Summary of Instruments and Equipment

- 14-gauge 140 mm multihole cannula to harvest the microfatgraft
- 10 cc Luer Lock syringes
- 27-gauge needle
- 18-gauge 5 cm needle (for vulvar stenosis)
- 18-gauge 9 cm needle (for vaginal stenosis)
- 5 cc Luer Lock syringe
- 500 cc of saline plus 40 mL lidocaine plus 1 mL 1:500.000 adrenaline for donor-site and recipient-site hydrodissection
- 100 cc saline for the fat washing procedure
- Centrifuge
- Nanofatgraft filtration system or prp kit

31.10 Complications

The immediate complications of fat grafting include infection, hematoma (will decrease efficacy of the treatment), skin/mucosal necrosis due to overinjection or excessive undermining, and edema. Painful calcifications following fat graft necrosis may occur due to excessive fat graft. Pulmonary embolism and death have never been reported, and thus to reduce this fatal complication, it is advisable to avoid the treatment of patients affected by vulvar and vaginal varix.

31.11 Conclusions

If performed with a good knowledge of anatomy, instrumentation, and biological fat properties, the treatment of vulvar LS is a safe and very successful procedure.

References

1. Kreuter A, Kryvosheyeva Y, Terras S, Moritz R, Mollenhoff K, Altmeyer P, et al. Association of autoimmune diseases with Lichen sclerosus in 532 male and female patients. Acta Derm Venereol. 2013;93:238–41.
2. LS genetic susceptibility is strengthened by a significant positive association with genes regulating HLA class II antigens, which are critical regulators of humoral immunity. Int J Biol Sci. 2019;15(7):1429–1439. https://doi.org/10.7150/ijbs.34613.
3. Tran DA, Tan X, Macri CJ, Goldstein AT, Fu SW. Lichen sclerosus: an autoimmunopathogenic and genomic enigma with emerging genetic and immune targets Lichen sclerosus. In: Genetic and rare diseases; 2019.
4. Oyama N. Current understanding of molecular-based immunopathology in Lichen sclerosus. Int Trends Immunity. 2015:6.
5. Edmonds E, Barton G, Buisson S, Francis N, Gotch F, Game L, et al. Gene expression profiling in male genital lichen sclerosus. Int J Exp Pathol. 2011;92:320–5.
6. Terlou A, Santegoets LAM, van der Meijden WI, Heijmans-Antonissen C, Swagemakers SMA, van der Spek PJ, et al. An autoimmune phenotype in vulvar Lichen sclerosus and lichen planus: a Th1 response and high levels of MicroRNA-155. J Invest Dermatol. 2012;132:658–66.
7. Sercu S, Zhang M, Oyama N, Hansen U, Ghalbzouri AEL, Jun G, et al. Interaction of extracellular matrix protein 1 with extracellular matrix components: ECM1 is a basement membrane protein of the skin. J Invest Dermatol. 2008;128:1397–408.
8. Smits P, Ni J, Feng P, Wauters J, Van Hul W, Boutaibi ME, et al. The human extracellular matrix gene 1 (ECM1): genomic structure, cDNA cloning, expression pattern, and chromosomal localization. Genomics. 1997;45:487–95.
9. Edmonds EVJ, Oyama N, Chan I, Francis N, McGrath JA, Bunker CB. Extracellular matrix protein 1 autoantibodies in male genital Lichen sclerosus. Br J Dermatol. 2011;165:218–9.
10. Fujimoto N, Terlizzi J, Aho S, Brittingham R, Fertala A, Oyama N, et al. Extracellular matrix protein 1 inhibits the activity of matrix metalloproteinase 9 through high-affinity protein/protein interactions. Exp Dermatol. 2006;15:300–7.
11. Collagen 1,3,5 are increased and promote deposition and development of sclerotic and hyalinized tissues.
12. Godoy CAP, Teodoro WR, Velosa APP, Garippo AL, Eher EM, Parra ER, et al. Unusual remodeling of the hyalinization band in vulval lichen sclerosus by type V collagen and ECM 1 protein. Clinics. 2015;70:356–62.

13. Martin P, Teodoro WR, Velosa APP, de Morais J, Carrasco S, Christmann RB, et al. Abnormal collagen V deposition in dermis correlates with skin thickening and disease activity in systemic sclerosis. Autoimmun Rev. 2012;11:827–35.
14. McPherson T, Cooper S. Vulval lichen sclerosus and lichen planus. Dermatol Ther. 2010;23:523–32.
15. Regauer S, Liegl B, Reich O. Early vulvar lichen sclerosus: a histopathological challenge. Histopathology. 2005;47:340–7.
16. Hasegawa M, Ishikawa O, Asano Y, Sato S, Jinnin M, Takehara K, et al. Diagnostic criteria, severity classification and guidelines of lichen sclerosus et atrophicus. J Dermatol. 2018;45:891–7.
17. Niamh L, Naveen S, Hazel B. Diagnosis of vulval inflammatory dermatoses: a pathological study with clinical correlation. Int J Gynecol Pathol. 2009;28:554–8.
18. Bleeker MCG, Visser PJ, Overbeek LIH, van Beurden M, Berkhof J. Lichen sclerosus: incidence and risk of vulvar squamous cell carcinoma. Cancer Epidemiol Biomark Prev. 2016;25:1224–30.
19. Haefner HK, Aldrich NZ, Dalton VK, Gagné HM, Marcus SB, Patel DA, et al. The impact of vulvar Lichen sclerosus on sexual dysfunction. J Womens Health (Larchmt). 2014;23:765–70.
20. Lynch PJ, Moyal-Barracco M, Scurry J, Stockdale C. 2011 ISSVD terminology and classification of vulvar dermatological disorders: an approach to clinical diagnosis. J Low Genit Tract Dis. 2012;16:339–44.
21. Almadori A, Zenner N, Boyle D, Swale V, Reid W, Maclean A, Butler PEM. Development and validation of a clinical grading scale to assess the vulvar region: the vulvar architecture severity scale. Aesthet Surg J. 2020;40(12):1319–26. https://doi.org/10.1093/asj/sjz342.
22. Veronica Boero, Carlo A. Liverani, Fabio Parazzini, Massimiliano Brambilla, Ermelinda Monti, Roberto Tozzi, Filippo Murina, Eugenia Di Loreto, Giada Libutti, Paolo Vercellini. A new proposal for a clinical classification of vulvar lichen sclerosus. J Genit Tract Diseases. in press.
23. Nilanchal S, et al. Treatment options in vulvar lichen slcerosus: a scoping review. Cureus. 2021;13(2):e13527.
24. Kirtschig G, Becker K, Günthert A, et al. Evidence-based (S3) guideline on (anogenital) lichen sclerosus. J Eur Acad Dermatol Venereol. 2015;29:1–43.
25. Balchander D, Nyirjesy P. Fractionated CO2 laser as therapy in recalcitrant lichen sclerosus. J Low Genit Tract Dis. 2020;24:225–8.
26. Hobson JG, Ibrahim SF, Mercurio MG. Recalcitrant vulvar lichen sclerosus treated with erbium YAG laser. JAMA Dermatol. 2019;155:254–6. https://doi.org/10.1001/jamadermatol.2018.4461.
27. Terras S, Gambichler T, Moritz RKC, Stücker M, Kreuter A. UV-A1 phototherapy vs clobetasol propionate, 0.05%, in the treatment of vulvar lichen sclerosus: a randomized clinical trial. JAMA Dermatol. 2014;150:621–7.
28. Gimble JM, Katz AJ, Bunnell BA. Adipose-derived stem cells for regenerative medicine. Circ Res. 2007;100:1249–60.
29. Zuk PA, Zhu M, Ashjian P, et al. Human adipose tissue is a source of multipotent stem cells. Mol Biol Cell. 2002;13:4279–429.
30. Leto Barone AA, Lee KS, et al. Immunomodulatory effects of adipose-derived stem cells: fact or fiction? Biomed Res Int. 2013;2013:383685.
31. Traktuev D, Li J, et al. Secretion of angiogenic and antiapoptotic factors by human adipose stromal cells. Circulation. 2004;109:1292–8.
32. Magalon J, Velier M, Simoncini S, François P, Bertrand B, Daumas A, Benyamine A, Boissier R, Arnaud L, Lyonnet L, Fernandez S, Dignat-George F, Casanova D, Guillet B, Granel B, Paul P, Sabatier F. Molecular profile and proangiogenic activity of the adipose-derived stromal vascular fraction used as an autologous innovative medicinal product in patients with systemic sclerosis. Ann Rheum Dis. 2019;78(3):391–8. https://doi.org/10.1136/annrheumdis-2018-214218. Epub 2019 Jan 5. PMID: 30612118.
33. Boero V, Brambilla M, Sipio E, et al. Vulvar lichen sclerosus: a new regenerative approach through fat grafting. Gynecol Oncol. 2015;139:471–5.
34. Almadori A, Hansen E, Boyle D, et al. Fat grafting improves fibrosis and scarring in vulvar lichen sclerosus: results from a prospective cohort study. J Low Genit Tract Dis. 2020;24:305–10.
35. Tamburino S, Lombardo GA, Tarico MS, Perrotta RE. Arch the role of nanofat grafting in vulvar lichen sclerosus: a preliminary report. Plast Surg. 2016;43(1):93–5. https://doi.org/10.5999/aps.2016.43.1.93. Epub 2016 Jan 15.
36. Casabona F, Priano V, Vallerino V, Cogliandro A, Lavagnino G. New surgical approach to lichen sclerosus of the vulva: the role of adipose-derived mesenchymal cells and platelet-rich plasma in tissue regeneration. Plast Reconstr Surg. 2010;126(4):210e–1e. https://doi.org/10.1097/PRS.0b013e3181ea9386.
37. Menkes S, SidAhmed-Mezi M, Meningaud JP, Benadiba L, Magalon G, Hersant B. Microfat and nanofat grafting in genital rejuvenation. Aesthet Surg J. 2021;41(9):1060–7. https://doi.org/10.1093/asj/sjaa118. PMID: 32386063.
38. Garza RM, Paik KJ, Chung MT, Duscher D, Gurtner GC, Longaker MT, Wan DC. Studies in fat grafting: part III. Fat grafting irradiated tissue—improved skin quality and decreased fat graft retention. Plast Reconstr Surg. 2014;134(2):249–57.
39. Maria AT, Toupet K, Maumus M, Fonteneau G, Le Quellec A, Jorgensen C, Guilpain P, Noël D. Human adipose mesenchymal stem cells as potent anti-fibrosis therapy for systemic sclerosis. J Autoimmun. 2016;70:31–9. https://doi.org/10.1016/j.jaut.2016.03.013. Epub 2016 Apr 1.PMID: 27052182.
40. Maria AT, Toupet K, Bony C, Pirot N, Vozenin MC, Petit B, Roger P, Batteux F, Le Quellec A, Jorgensen C, Noël D, Guilpain P. Antifibrotic, antioxidant, and immunomodulatory effects of mesenchymal stem cells in HOCl-induced systemic sclerosis. Arthritis Rheumatol. 2016;68(4):1013–25. https://doi.org/10.1002/art.39477.
41. Evans BGA, Gronet EM, Saint-Cyr MH. How fat grafting works. Plast Reconstr Surg Global Open. 2020;8(7):e2705.
42. Alessandri-Bonetti M, Egro FM, Persichetti P, More. The role of fat grafting in alleviating neuropathic pain: a critical review of the literature. Plast Reconstr Surg Global Open. 2019;7(5):e2216.
43. Khouri RK, Smit JM, Cardoso E, Pallua N, Lantieri L, Mathijssen IMJ, Khouri RK Jr, Rigotti G. Percutaneous aponeurotomy and lipofilling: a regenerative alternative to flap reconstruction? Plast Reconstr Surg. 2013;132(5):1280–90. https://doi.org/10.1097/PRS.0b013e3182a4c3a9. PMID: 23924652.
44. Pignatti M, Bruti M, Rigotti G. Saving one step in lipostructure: M-AGO (magic needle). Plast Reconstr Surg. 2001;108(6):1833–4. https://doi.org/10.1097/00006534-200111000-00087.
45. Magalon J, Daumas A, Veran J, Magalon G, Rossi P, Granel B, et al. Autologous adipose tissue-derived cells: are we talking about adipose derived stem cells, stromal vascular fraction, or Coleman fat grafting? Cell Transplant. 2015;24:2667–8.
46. Alharbi Z, Opländer C, Almakadi S, et al. Conventional vs. micro-fat harvesting: how fat harvesting technique affects tissue-engineering approaches using adipose tissue-derived stem/stromal cells. J Plast Reconstr Aesthet Surg. 2013;66:1271–8.
47. Tonnard P, Verpaele A, Peeters G, et al. Nanofat grafting: basic research and clinical applications. Plast Reconstr Surg. 2013;132:1017–26.
48. Uyulmaz S, Sanchez Macedo N, Rezaeian F, Giovanoli P, Lindenblatt. Nanofat grafting for scar treatment and skin qual-

ity improvement. Aesthet Surg J. 2018;38(4):421–8. https://doi.org/10.1093/asj/sjx183. PMID: 29365061.
49. Charles-de-Sá L, Gontijo de Amorim NF, Dantas D, Han JV, Amable P, Teixeira MV, de Araújo PL, Link W, Borojevich R, Rigotti G. Influence of negative pressure on the viability of adipocytes and mesenchymal stem cell, considering the device method used to harvest fat tissue. Aesthet Surg J. 2015;35(3):334–44. https://doi.org/10.1093/asj/sju047.
50. Wang WZ, Fang X-H, Williams SJ, Stephenson LL, Baynosa RC, Khiabani KT, et al. Lidocaine-induced ASC apoptosis (tumescent vs. local anesthesia). Aesthet Plast Surg. 2014;38:1017–23.
51. Brambilla M. Chapter 47. Fat graft of the genital area in fat injection: from filing to regeneration. In: Coleman S, Mazzola R, Pee L, editors. Thieme.
52. Brambilla M. A regenerative approach to treat vulvar and vaginal scarring trends and techniques in aesthetic plastic surgery, 2021.

Female Genital Mutilation: A Surgical Approach to Reshaping

32

Aurora Almadori and Stefania de Fazio

Contents

32.1 Introduction

Female genital mutilation (FGM) consists in multiple procedures involving the partial or total removal of the external female genitalia or other injury to the female genital organs, whether for cultural or any other nontherapeutic reasons. The strong negative connotation of the word "mutilation" has been suggested and adopted by WHO to emphasize the gravity of the practice and to underline that is a violation of girls' and women's human rights [1].

32.1.1 Prevalence

It is estimated that 100 to 140 million girls and women in the world today have undergone some form of FGM, and two million girls are at risk from the practice each year [2]. Although these figures are only estimate and the true incidence is difficult to assess, they do indicate the massive scale of this human rights violation [3]. The majority of the affected women live in Africa (Fig. 32.1), followed by Middle East and Asia. It has also been reported to be practiced in India by the Daudi Bohra Muslims. Nowadays women with FGM are also found in Europe, Australia, New Zealand, Canada, and the USA because of increased migration flows.

A. Almadori
Department of Plastic Surgery, Royal Free London NHS Foundation Trust Hospital, London, UK

Centre for Nanotechnology and Regenerative Medicine, Division of Surgery and Interventional Science, University College London, London, UK

S. de Fazio (✉)
Rome, Italy

32.1.2 The Origins of FGM

It is not known when or where the tradition of FGM originated. Multiple hypotheses have been formulated. It has been proposed that FGM was common in ancient Egypt and among the ancient peoples of the Arab world; that FGM started with the arrival of Islam in some parts of sub-Saharan Africa; that it commenced during the slave trade when black slave women entered ancient Arab societies; and that the practice developed independently among certain ethnic groups in sub-Saharan Africa as part of puberty rites [2, 4].

32.1.3 Classification

Four types of FGMs have been described, namely: type I, which consists of excision of the clitoris; type II, which involves the excision of the clitoris with scraping or removal of the labia minora; type III, which consists of excision of part or all of the external genitalia and the stitching together of the vaginal opening; and type IV, which encompasses all other practices on female genitalia not covered in the

© The Author(s), under exclusive license to Springer Nature Switzerland AG 2023
A. Di Giuseppe et al. (eds.), *Fat Transfer in Plastic Surgery*, https://doi.org/10.1007/978-3-031-10881-5_32

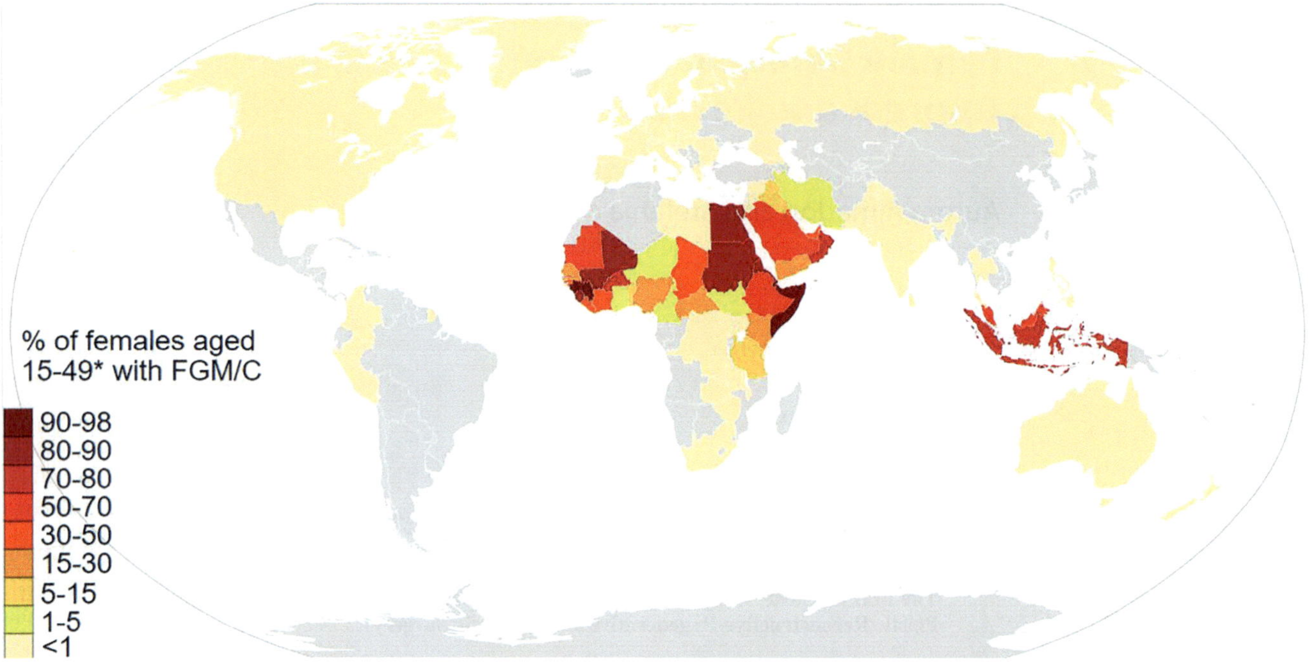

Fig. 32.1 Map showing the percentage of women and girls aged 15–49 years (unless otherwise stated) who have undergone FGM/C according to the March 2020 Global Response report (2020 Global Response report FGM world map, based on the EndFGM EU network report, Open Access CC BY-SA 4.0)

previously described types, including pricking, piercing, cauterization, or incision of the clitoris or labia; scraping of the tissue surrounding the vaginal orifice or cutting of the vaginal wall; and the introduction of corrosive substances or herbs into the vagina to cause bleeding, tightening, or narrowing (Fig. 32.2) [6].

32.1.4 Reasons for Performing FGM

Multiple reasons are responsible for FGM being practiced. The social pressure is an important aspect for perpetuating the practice, with uncut girls experiencing rejection by their group or family.

The majority of reasons are sociocultural and related to traditions. These include ensuring virginity, hence marriage; rite of passage to adulthood; hygienic reasons as external genitalia are considered dirty; to avoid damages to penis or fetus as clitoris is considered dangerous. FGM is also believed to make a girl beautiful. There are also communities performing FGM for religious reasons believing that removing the external genitalia is necessary to make a girl spiritually clean. In Muslim societies that practice FGM, people believe that it is required by the Koran. However, FGM is not mentioned in the Koran, although it is frequently carried out by communities in the genuine belief that it is part of their religion [2]. In addition to that, FGM can be performed for the belief that uncut women have an overactive and uncontrollable sex drive so that she is likely to lose her virginity prematurely, to disgrace her family and damage her chances of marriage, and to become a menace to all men and to her community as a whole. It is also believed that the tight vaginal orifice of an infibulated woman enhances male sexual pleasure, in turn preventing divorce or unfaithfulness [2].

32.1.5 Complications

The consequences of FGM can be both physiological and psychological and also affect the sexual life. The extent of the complications is mostly affected by the type of FGM performed.

32.1.5.1 Health Implications

The range of health complications associated with FGM is wide and some are severely disabling. The immediate physical complications include: pain due to the operation being performed with crude instruments and without anesthetic; injury to the adjacent tissues; bleeding, which is the most common and life-threatening complication; shock (hemorrhagic or neurogenic shock); urine retention that often leads to urinary tract infection; infection, which is very common and can prevent the wound healing potentially resulting in an abscess, fever, ascending urinary tract infec-

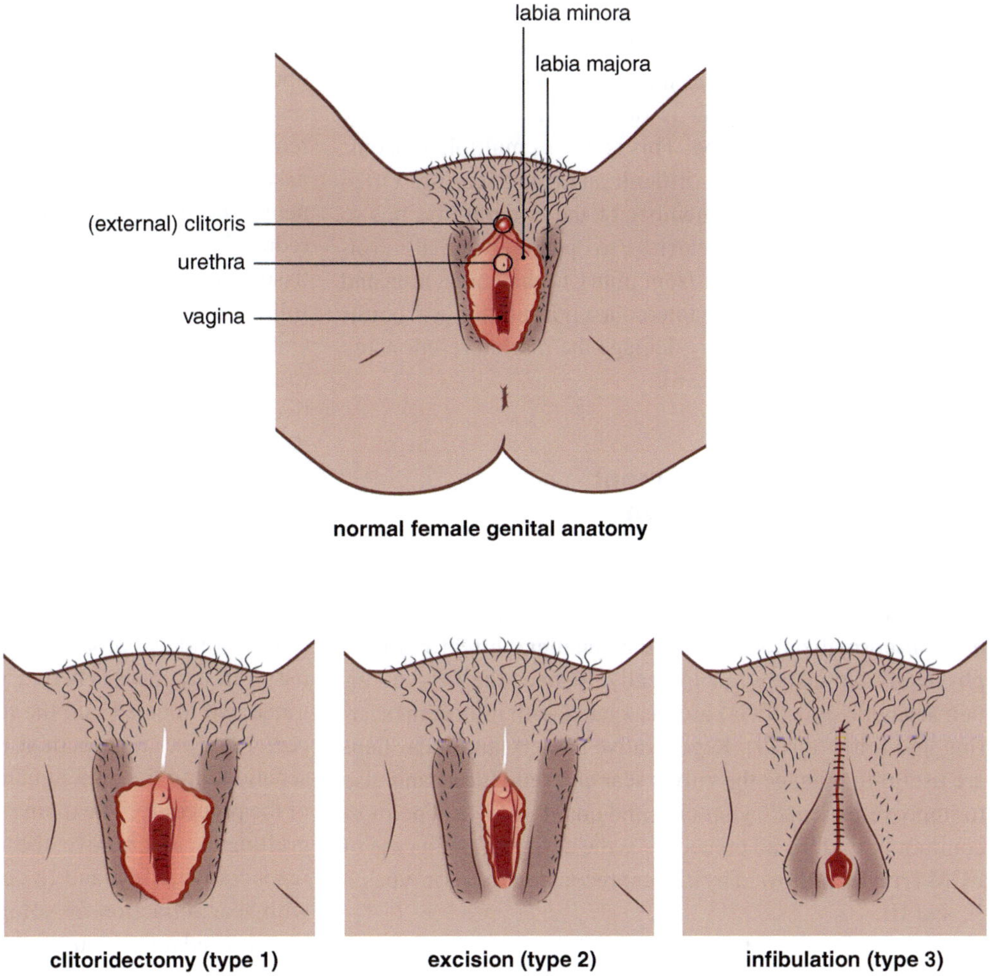

Fig. 32.2 Classification of FGM. Type I refers to excision of the clitoris. Type II refers to excision of the clitoris with partial or total removal of the labia minora. Type III refers to excision of part or all of the external genitalia and stitching or narrowing of the vaginal opening (infibulation). Type IV includes all other harmful procedures to the female genitalia for nonmedical purposes, e.g., pricking, piercing, incising, scraping, and cauterizing the genital area. (Reproduced from [5])

tion, pelvic infection, tetanus, gangrene, or septicemia; recurrent urinary tract infection, particularly when the normal flow of urine is obstructed; infertility as result of pelvic infections; keloid scar formation; cysts and abscesses on the vulva; painful clitoral neuromas; dysmenorrhea and hematocolpos; calculus formation in the vagina; recto-vaginal or vesico-vaginal fistulae; dyspareunia; sexual dysfunction; problems during childbirth; difficulties in performing an examination during labor; prolonged and obstructed labor, leading to tearing of the perineum and higher risk of hemorrhage, fistula formation, uterine rupture or prolapse, and even harm to the new born (including stillbirth). In particular, childbirth represents a big challenge for infibulated women, with high maternal mortality rates due to labor complications [2, 7].

32.1.5.2 Psychosocial Implications

The experience of genital mutilation has been associated with a range of mental and psychosomatic disorders. A recent systematic review of the literature and meta-analysis found that of the 16 studies included, 14 reported an association between FGM and at least 1 adverse mental health outcome [8]. Reports include the short-term disturbances such as sleeplessness, nightmares, loss of appetite, weight loss, or excessive weight gain, as well as panic attacks, difficulties in concentration and learning, and other symptoms of posttraumatic stress disorder. As they grow older, women may develop feelings of incompleteness, loss of self-esteem, depression, chronic anxiety, phobias, panic, or even psychotic disorders. Many women suffer in silence, unable to express their pain and fear.

FGM is commonly performed when girls are young and uninformed and is often preceded by acts of coercion and violence by parents and relatives. It is also positively reinforced by friends and the other members of the community, causing contrast feelings. Girls are generally conscious when the painful operation is undertaken as no anesthetic or other medication is used. This can be traumatic and lead to psychosocial problems particularly post-traumatic stress disorder.

32.1.5.3 Sexual Implications

Sexual problems as a result of FGM can affect both partners, from fear of the first sexual intercourse onward, and create great anxiety. Dyspareunia is one of the most commonly reported aspects after FGM. This is due to multiple reasons. Sexual intercourse can be difficult and cause tearing. Often the husband, or another member of the community, has to perform an incision after marriage to open the vaginal introitus. Vaginismus may result from injury to the vulval area and repeated vigorous sexual intercourse. Inhibition of coitus because of fear of pain may damage the marital relationship and even lead to divorce [2, 8].

32.2 Surgical Management and Reconstruction

Women with FGM can be offered different reconstructive procedures, depending on their clinical problems. The anatomical normalization will have a beneficial effect not only physically but also psychologically. Besides the loss of certain aesthetic units, FGM are associated with important scarring and retraction [9]. Regenerative surgery and local flaps are useful to manage the vulvar scar and retractions and also to improve patients' symptoms and quality of life. The most commonly performed procedure is the defibulation in case of FGM type III, followed by clitoral reconstruction for types I, II, and III.

32.2.1 Regenerative Surgery

Regenerative therapies based on the use of adipose stem cells for tissue repair and regeneration are promising to ameliorate the scar tissue and improve the volumetric contouring of the vulvar anatomical structures, with an important effect on patients' quality of life. Its antiscarring effect has been extensively evaluated in multiple clinical studies involving different conditions, such as hypertrophic scar, scleroderma, lichen sclerosus, and radiation-induced fibrosis [10–12].

Almadori et al. were the first to propose the use adipose stem cell-based techniques in women with FGM to improve the scarred tissues and to reshape the vulvar volumes [13].

Recently, the adipose tissue, deriving from the embryonic mesenchyme like the bone marrow, has been reevaluated as an important organ after the discovery of its content of a subpopulation of multipotent mesenchymal stem cells called adipose-derived stem cells (ASCs). Based on in vitro experiments and clinical studies, ASCs have been applied in various clinical fields as regenerative therapy of different tissues. By secreting growth factors, they manage to control and stimulate the regeneration of damaged skin cells. It has been seen that ASCs promote and speed up the healing of wounds, ulcers, and skin defects and improve skin fibrosis through the activation of neighboring cells. ASCs, thanks to their differentiating ability but also paracrine, have been considered optimal for tissue regeneration in the case of hypertrophic scars [14]. While the use of adipose-derived stem cells requires cell expansion via cell culture and use of collagenase for separation of the ASCs for the rest of the components, the use of fat grafting is more easily available with reduced costs. The first isolated experiences with the adipose tissue graft date back to the last century. Since the late 1990s, following the standardization of the technique by Sydney Coleman, fat grafting has become one of the best techniques to correct volume defects secondary in congenital, degenerative, traumatic, or aging diseases [15]. In 2001, Patricia Zuk described the presence of mesenchymal stem cells in the adipose tissue, the adipose-derived stem cells (ASCs), and since then numerous studies have evaluated these mesenchymal progenitor cells, extending their areas of application well beyond plastic surgery [16].

The lipofilling technique, or fat grafting or lipostructure, is a consolidated procedure that consists in removing adipose tissue in the areas where a good presence of fat is concentrated and injected into the recipient site. The surgery can be performed in an outpatient basis, under local anesthesia, or sedation, and the patient can be discharged on the same day of the procedure. The main rationale for the application of fat grafting is to improve the quality of the skin through its regenerative effect and to correct the deformities associated with scar retraction or volume loss. Being minimally invasive, it is advantageous compared to standard corrective surgery. Adipocytes are suggested to have wider biological effects than other fillers and may offer more durable results. The operation takes place in three phases, the first is the extraction of fat from a donor area. Belly, thighs, and hips are usually excellent donor site. The second phase is the processing of the extracted fat using a centrifuge. The aim of centrifugation is not only to remove undesired materials, such as anesthetic liquids, and oil with cellular debris, but also to concentrate the fraction rich in ASCs. With this process, the fat is purified and then made available for the third and final phase, the grafting on the vulvar tissues.

A variety of surgical techniques have been described to optimize the final product. The different harvesting, processing, and infiltration techniques can influence the number, viability, and differentiation capacity of the ASCs and the viscosity of the lipoaspirate. The choice of one method over another usually depends on the area to be treated and the indication for treatment (large or small volume fat grafting, volumetric or trophic effect).

32.2.2 Local Flaps

Local flaps are an invaluable tool in vulvar reconstruction, particularly after FGM where the main aim is to reduce the

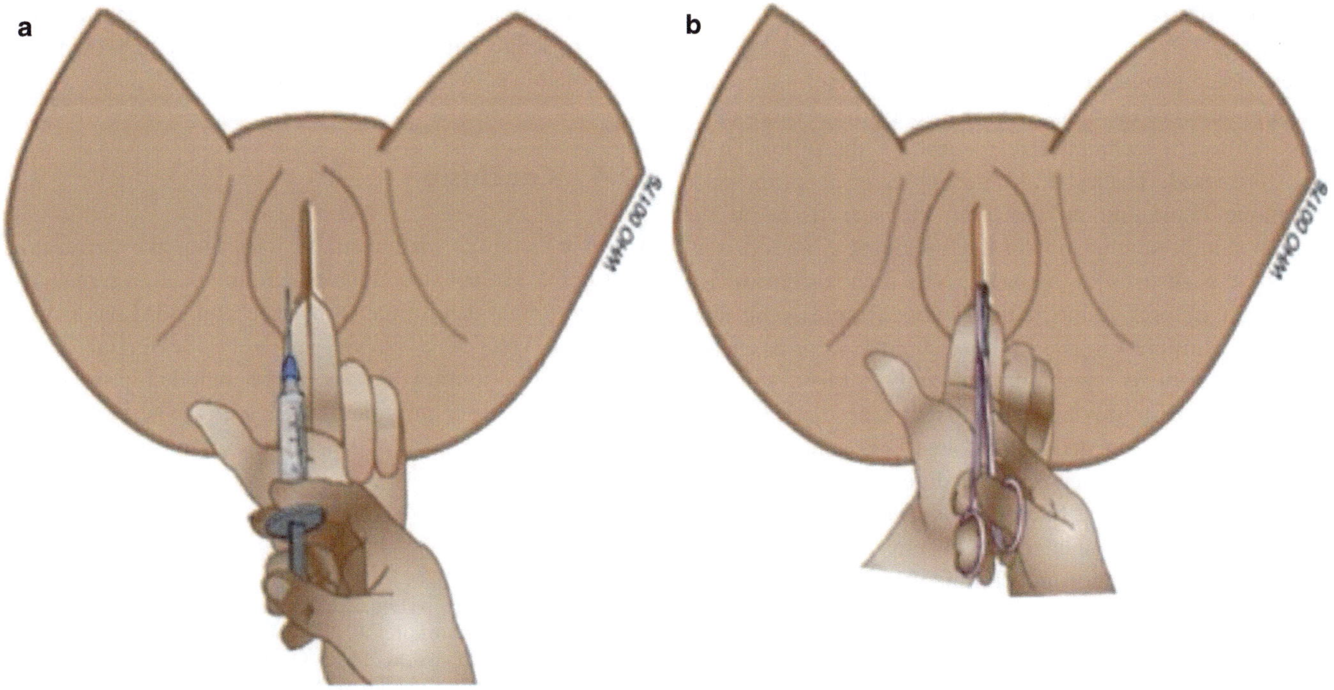

Fig. 32.3 Defibulation. The defibulation procedure consists in (**a**) injecting local anesthetics in the area to be incised; (**b**) incision interiorly from the clitoral area to the vaginal opening. Particular attention should be paid in order not to cut the underlying tissues. The edges are then sutured. (Reproduced from [2])

tension of the scar or reconstruct structures such as clitoris and labia minora. Accurate knowledge of the vulvar anatomy is required to safely reconstruct the vulvar area. Flaps are categorized by blood supply into axial flaps, which are supplied by a named artery, and random pattern flaps, which rely on the vascular plexus of the dermis and subcutaneous tissue. The type of primary movement of the flap can further subdivide random pattern flaps into three categories: advancement, rotation, and transposition. Advancement and rotation flaps recruit adjacent lax tissue and move in either a linear or arced motion to fill in the primary defect. Transposition flaps recruit noncontiguous donor tissue that is incised and lifted over intact skin and placed into the primary defect or scar. This allows the use of skin from areas of less tension and directs incision tension vectors more favorably. Transposition flaps most commonly used include Z-plasty [17]. The Z-plasty can improve cosmesis of scars crossing relaxed skin tension lines and be helpful to release scar contractures by redistributing the tension. This transposition flap is one of the mostly used in scars revision. Potential indications include improvement of scar cosmesis by changing the direction of the scar, particularly those crossing relaxed skin tension lines and releasing scar contractures by redistributing the tension, which is important in those near free margins with visible deformity. This technique creates two flaps from three equal limbs and two same-degree angles and ultimately results in a lengthening of the scar [17].

The use of V-Y advancement flaps is also valuable in vulvar reconstruction, being a simple and easy procedure with few complications in most cases [18]. It is more indicated for moderate- to large-sized defects as less tension is placed on the wound compared to the transposition flap [19]. However, in the vulvar area, the transposition flaps are associated with better aesthetic results with less scarring compared to the V-Y; however, expertise in surgical dissection of the vessels is necessary for transposition flaps [20]. Both are associated with overall low complication rates.

32.2.3 Defibulation

Women with type III who are pregnant or who suffer long-term complications should undergo defibulation. Defibulation is a surgical procedure wherein a vertical incision is made on the scar to expose the introitus and create new labia majora [21]. It is performed with local anesthesia (Fig. 32.3a), but often patients are very worried and they might even re-experience the traumatic event of infibulation, therefore in certain cases a sedation can be indicated to have the patient more relaxed. The procedure consists in a vertical incision anteriorly, exposing both the introitus and urethra (Fig. 32.3b). The raw edges are sutured separately.

32.2.4 Clitoral Reconstruction

After female genital cutting type I, II, or III, there is usually an amputation of the clitoris at the level of the corpora clitoridis. Different techniques to reconstruct the clitoris have been proposed. Thabet et al. and Foldes et al. proposed a technique consisting in dissecting the scar tissue at the clitoral stump, mobilization of the remaining clitoris dissection of the suspensory ligament, and then anchoring the mobilized clitoral stump to the bulbocavernosus muscles [22, 23]. Multiple variants to this technique were described to improve patients' symptoms and quality of live, with particular focus on improving sexual function, aesthetical aspect of the overall vulvar and clitoral area, and enhanced body image [24–27]. However, the procedure is associated with a high complication rate. Main complications include wound dehiscence, hematoma, infection, hyperesthesia of the clitoris, and keloid formation and chronic pain [28]. To this date, official guidelines and algorithms are lacking. Mayor international bodies such as WHO do not recommend clitoral reconstruction because of a lack of supporting evidence and high complication rate, and the Royal College of Obstetricians and Gynaecologists currently recommends against because there is not enough evidence to support its risk-to-benefit ratio [29].

32.3 The Role of Plastic Surgeons in FGM and the Example of SICPRE (Italian Society of Plastic Reconstructive-Regenerative and Aesthetic Surgery)

The role of plastic surgery is to restore form and function with an approach from the less invasive to the most invasive. Plastic surgeons, having specific reconstructive and regenerative expertise, represent key healthcare practitioners when providing care to women with FGM. So far, FGM management has been performed mainly by gynecologists and urologists. A multidisciplinary approach including psychiatrists/psychologists, gynecologists, urologists, and plastic surgeons would be ideal to provide optimized FGM care and management. Particularly, involving a plastic surgeon should be encouraged in order to provide the patients with the most up-to-date reconstructive options.

The Italian Society of Plastic Reconstructive-Regenerative and Aesthetic Surgery (SICPRE) recently instituted, within the Female and Male Plastic Surgery Chapter, a subchapter entirely dedicated to FGM. The aim is to support training and research in vulvar surgery specifically applied to FGM. The authors of the present chapter are the referents for the FGM topic created in 2018 within the SICPRE. Plastic surgery societies worldwide are encouraged to follow this example creating dedicated FGM sections in order to spread the knowledge among plastic surgeons and to provide the best possible care to FGM patients.

32.4 Conclusion

Multiple techniques are available to reshape the vulvar area after FGM. Patient selection and communication remain the most important determinants in successful FGM reconstruction. The role of a plastic surgeon is essential in the reconstruction journey and should be included in the multidisciplinary team to provide patients with the best options, in safety.

References

1. World Health Organization. Division of Family Health. Female genital mutilation: report of a WHO technical working group, Geneva, 17–19 July 1995. World Health Organization; 1996. https://apps.who.int/iris/handle/10665/63602.
2. https://www.who.int/gender/other_health/Studentsmanual.pdf.
3. Changing a harmful social convention: female genital mutilation/cutting. Innocenti Digest. 2005;12.
4. Jones SD, Ehiri J, Anyanwu E. Female genital mutilation in developing countries: an agenda for public health response. Eur J Obstet Gynecol Reprod Biol. 2004;116(2):144–51. https://doi.org/10.1016/j.ejogrb.2004.06.013. PMID: 15358454.
5. Kawous R, Allwood E, Norbart E, van den Muijsenbergh METC. Female genital mutilation and women's healthcare experiences with general practitioners in the Netherlands: a qualitative study. PLoS One. 2020;15(7):e0235867. https://doi.org/10.1371/journal.pone.0235867.
6. World Health Organization. Female genital mutilation report of a WHO Technical Working Group Geneva, 17–19 July 1995, WHO document: WHO/FRH/WHD/96.10.
7. Klein E, Helzner E, Shayowitz M, Kohlhoff S, Smith-Norowitz TA. Female genital mutilation: health consequences and complications—a short literature review. Obstet Gynecol Int. 2018;2018:7365715. https://doi.org/10.1155/2018/7365715. PMID: 30116269; PMCID: PMC6079349.
8. Abdalla SM, Galea S. Is female genital mutilation/cutting associated with adverse mental health consequences? A systematic review of the evidence. BMJ Glob Health. 2019;4(4):e001553. https://doi.org/10.1136/bmjgh-2019-001553.
9. Caroppo E, Almadori A, Giannuzzi V, Brogna P, Diodati A, Bria P. Health care for immigrant women in Italy: are we really ready? A survey on knowledge about female genital mutilation. Ann Ist Super Sanita. 2014;50(1):49–53.
10. Almadori A, Griffin M, Ryan CM, et al. Stem cell enriched lipotransfer reverses the effects of fibrosis in systemic sclerosis. PLoS One. 2019;14(7):e0218068.
11. Almadori A, Hansen E, Boyle D, Zenner N, Swale V, Reid W, Maclane A, Butler PEM. Fat grafting improves fibrosis and scarring in vulvar Lichen sclerosus: results from a prospective cohort study. J Low Genit Tract Dis. 2020;24(3):305–10. https://doi.org/10.1097/LGT.0000000000000520. PMID: 32205767.
12. Griffin MF, Drago J, Almadori A, Kalavrezos N, Butler PE. Evaluation of the efficacy of lipotransfer to manage radiation-induced fibrosis and volume defects in head and neck oncology. Head Neck. 2019;41(10):3647–55.

13. Almadori A. Fat grafting and adipose stem cells to treat vulvar scarring and fibrosis post female genital mutilation (FGM). In: Kaalaji A, editor. Plastic and aesthetic regenerative surgery and fat grafting: clinical application and operative techniques. Springer Nature; 2021. (in press).
14. James I, Bourne D, Silva M, et al. Adipose stem cells enhance excisional wound healing in a porcine model. J Surg Res. 2018;229:243–53.
15. Coleman SR. Facial augmentation with structural fat grafting. Clin Plastic Surg. 2006;33:567–77.
16. Zuk PA. The adipose-derived stem cell: looking back and looking ahead. Mol Biol Cell. 2010;21(11):1783–7. https://doi.org/10.1091/mbc.e09-07-0589.
17. Bednarek RS, Sequeira Campos M, Ramsey ML. Transposition flaps. [Updated 2021 Feb 15]. In: StatPearls [Internet]. Treasure Island: StatPearls Publishing; 2021. https://www.ncbi.nlm.nih.gov/books/NBK500028/.
18. Hand LC, Maas TM, Baka N, Mercier RJ, Greaney PJ, Rosenblum NG, Kim CH. Utilizing V-Y fasciocutaneous advancement flaps for vulvar reconstruction. Gynecol Oncol Rep. 2018;26:24–8. https://doi.org/10.1016/j.gore.2018.08.007. PMID: 30186930; PMCID: PMC6122364.
19. Di Donato V, Bracchi C, Cigna E, et al. Vulvo-vaginal reconstruction after radical excision for treatment of vulvar cancer: evaluation of feasibility and morbidity of different surgical techniques. Surg Oncol. 2017;26(4):511–21.
20. Giannini A, Di Donato V, D'Oria O, et al. The V-Y gluteal fold advancement flap: outcomes following radical surgery for vulvar malignancies. Int J Gynecol Obstet. 2021;152:421–4.
21. Nour NM, Michels KB, Bryant AE. Defibulation to treat female genital cutting: effect on symptoms and sexual function. Obstet Gynecol. 2006;108(1):55–60. https://doi.org/10.1097/01.AOG.0000224613.72892.77.
22. Thabet SM, Thabet AS. Defective sexuality and female circumcision: the cause and the possible management. J Obstet Gynaecol Res. 2003;29:12–9.
23. Foldès P, Cuzin B, Andro A. Reconstructive surgery after female genital mutilation: a prospective cohort study. Lancet. 2012;380:134–41.
24. Ouedraogo CM, Madzou S, Simporé A, et al. Clitoral reconstruction after female genital mutilation at CHU Yalgado of Ouagadougou, Burkina Faso. About 68 patients operated. J Gynecol Obstet Biol Reprod (Paris). 2016;45:1099–106.
25. O'Dey DM. Die komplexe Rekonstruktion der Vulva nach weiblicher Genitalverstümmelung/Genitalbeschneidung. Urologe A. 2017;56(10):1298–301.
26. Chang CS, Low DW, Percec I. Female genital mutilation reconstruction: a preliminary report. Aesthet Surg J. 2017;37:942–6.
27. Mañero I, Labanca T. Clitoral reconstruction using a vaginal graft after female genital mutilation. Obstet Gynecol. 2018;131:701–6.
28. Sharif Mohamed F, Wild V, Earp BD, Johnson-Agbakwu C, Abdulcadir J. Clitoral reconstruction after female genital mutilation/cutting: a review of surgical techniques and ethical debate. J Sex Med. 2020;17(3):531–42. https://doi.org/10.1016/j.jsxm.2019.12.004.
29. RCOG release: updated guidelines provide clarity for healthcare professionals on the care of women with FGM. London: Royal College of Obstetricians and Gynecologists; 2015.28.Berg RC.

Periocular Fat Grafting

33

Mario Pelle-Ceravolo

Contents

The authors have no commercial associations or financial disclosures that might pose or create a conflict of interest with information presented in this article. No funding was received for the work presented in this article.

Key Messages

Diluting the fat to improve results in periocular fat grafting.

A study on 200 patients showed very low complication rate and high patients' satisfaction rate.

The upper and lower eyelids contain a certain amount of fat that for many years was removed through the classical blepharoplasty technique. In many patients, however, the eyelids can take benefit from the addition of new volumes.

The work by Lambros [1] powerfully remarks how much volume depletion can do in terms of improving facial aging consequences. Many consequences of aging, such as tear-trough deformities, sunken orbits, and naso-jugal depressions, can be treated effectively with fat grafting to the deflated areas, thereby obviating complex fat flaps or procedures that require transposition of skin, muscle, and fat (e.g., midface lift). In the past decade, fat grafting to the face has become a routine procedure in the operating room and may be performed independently or in conjunction with techniques such as facelift or blepharoplasty.

In many patients, the empty-appearing upper eyelid emphasizes the roundness of the globe. Lipofilling of the upper orbit can apparently lower the supratarsal fold and transform round, hollow eyes into pleasant, almond-shaped eyes. Volume deflation in the lower eyelid area yields a skeletal appearance with more visible fat pads, an accentuated tear-trough deformity, an elongated lower lid, a deeper lid-cheek junction, loss of malar prominence, and more noticeable projection of the lower globe (Fig. 33.1).

Primary indications for periocular fat grafting are round and hollow eyes, age-related deflation, sunken eyelids and iatrogenic deformities, such as in blepharoplasty with overzealous fat removal. Nonsymptomatic eye prominence and a negative vector (i.e., a condition in which the anterior surface of the cornea lies anterior to the malar prominence) also are indications for periocular fat grafting, and we frequently encounter these features in our practice in southern European countries. Fat transfer to the periorbital region in these patients can increase the projection of the upper and lower

Supplementary Information The online version contains supplementary material available at https://doi.org/10.1007/978-3-031-10881-5_33.

M. Pelle-Ceravolo (✉)
Department of Plastic Surgery, Anglo-American Hospital, Rome, Italy

Department of Plastic Surgery.University of Padua, Padua, Italy

© The Author(s), under exclusive license to Springer Nature Switzerland AG 2023

A. Di Giuseppe et al. (eds.), *Fat Transfer in Plastic Surgery*, https://doi.org/10.1007/978-3-031-10881-5_33

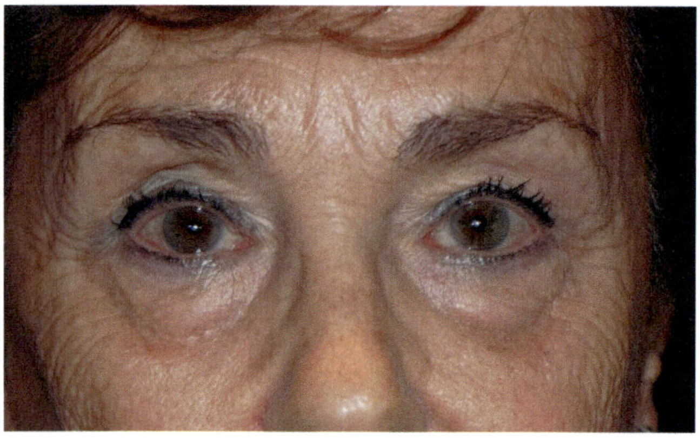

Fig. 33.1 This 62-year-old woman shows the typical age-related changes due to volume deflation in her upper and lower eyelid area, i.e., visible fat pads, accentuated tear-trough deformity, elongated lower lid, depression on her lid-cheek junction, and loss of malar prominence with negative vector

orbit, thereby creating the appearance of globe retrusion to an aesthetically appealing plane.

Consensus is lacking on the details of periorbital fat grafting, including choice of donor sites, fat harvesting, fat preparation (e.g., washing, filtering, and centrifuging), and fat injection tools [2–11]. Investigators generally agree on the following: (1) some amount of volume loss after fat grafting is the rule and (2) multiple lipofilling sessions often are necessary to achieve a stable result [12, 13].

Fat grafting is performed less frequently in the periocular region than in other facial zones. The risk of blindness with periocular fat transfer represents a fearful occurrence, but many surgeons also have expressed concerns that this procedure can yield contour irregularities, including nodules and masses, that are a source of patient dissatisfaction [3, 9]. Unfavorable contour results can occur relatively frequently, especially in the lower eyelid, which has thin teguments. Uneven accumulation of grafted periocular fat is difficult to conceal—especially in patients who are older and have excess skin—because of an insufficient native fat pad and the subtle orbicularis. After periocular fat transplantation, noticeable adipose masses or nodules in the eyelids can be a cause for great concern in patients.

When contour irregularities occur after this procedure, the only effective treatment is surgical removal of the grafted fat. In our hands, steroid injections yield little or no improvement. The aim of this chapter is to describe our technique of periocular fat grafting, with which we have obtained consistently satisfactory results and few contour irregularities. We use preparations of "properly diluted fat (PDF)" [14] in which centrifuged fat is recombined with fluid rich in stromal vascular factors and diluted with saline such that the following was approximately true: (mL of centrifuged fat) ÷ (mL of PDF) = 0.7.

33.1 Preparing the Fat

The donor site is determined for each patient based on presence of fat, patient preference, and operating position. The inner knee is the most common donor site, followed by the upper abdomen, external thighs, the lower abdomen, and the flanks.

Donor areas are infiltrated with a solution containing 10 mL of 10% bupivacaine, 90 mL of normal saline, and 0.5 mg of epinephrine. The infiltration volume differed among patients and was based on the amount of fat harvested.

In the recipient area, a small volume of a solution containing 10 mL of 10% bupivacaine, 30 mL of normal saline, and 0.4 mg of epinephrine is infiltrated into the supraorbital and infraorbital foramen. The remaining recipient areas are infiltrated with a solution of saline and epinephrine (1:100,000) to induce vasoconstriction. There is some debate among researchers regarding the effects of anesthetic drugs on harvested fat [15], but our preference was to avoid unnecessary risks of damaging the fat.

Fat is harvested by means of a 0.5-mm, multiple-hole cannula attached to a 10-mL syringe under light aspiration pressure. Lipoaspirates are centrifuged at 1000 rpm for 2 min. The bottommost 1 mL of reddish fluid in the syringe is extracted and retained. This infranatant fluid comprises blood and infiltrate and is thought to contain a high density of stromal vascular factors as advised by Dr. Rigotti, through personal communication. In a recently published paper by Gontijo-de-Amorim et al. [16] in which Dr. Rigotti was a coauthor, washed fat that had been enriched with the pellet of centrifuged lipoaspirate (i.e., the solids immediately beneath the reddish fluid layer) was better retained in the face at 1 year, compared with unenriched fat. Any remaining reddish liquid and the uppermost oily layer (i.e., the layers flanking the centrifuged fat fraction) then are discarded.

A series of 10-mL syringes then are prepared for fat transfer by adding 7 mL of centrifuged fat, 1 mL of the reddish fluid (presumably enriched in stromal vascular factors), and 2 mL of normal saline, so that centrifuged fat constituted 70% of the PDF and 30% of the PDF comprised the diluent (i.e., saline plus the reddish infranatant) (Video 33.1).

33.2 Injecting the Fat

The recipient site is prepared for fat transfer by making a network of crisscrossing tunnels with an empty 1.2-mm cannula (Figs. 33.2 and 33.3). We confirm that there is no bleeding, which indicates that no large vessel has been damaged. Immediately before fat injection, we blend the contents of the 10-mL syringe with gentle shaking. The contents are then

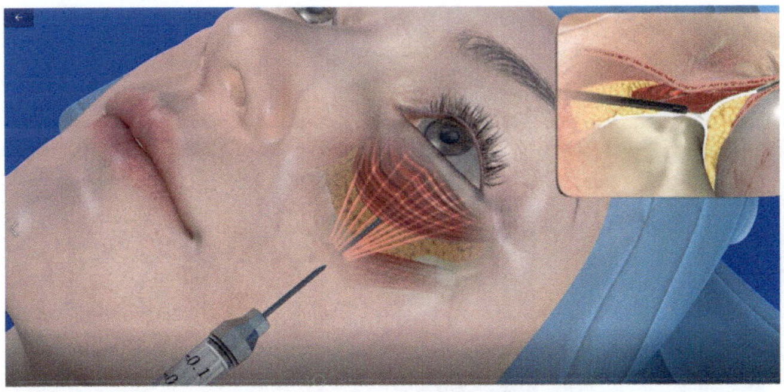

Fig. 33.2 A 1.2 mm cannula creates multiple tunnels following a vertical vector into a plane beneath the orbicularis muscle

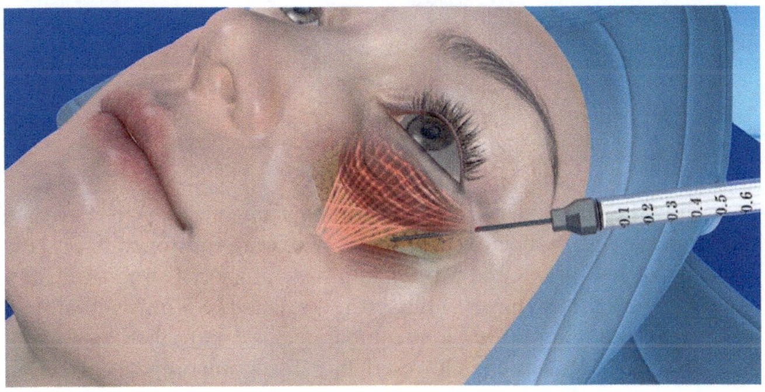

Fig. 33.3 The same cannula creates other tunnels into the same plane following a horizontal vector

transferred to 1-mL Luer-lock syringes, and PDF is injected into the recipient site by means of a 1.2-mm cannula with a single 1-mm hole positioned laterally at its end. PDF transfer is carried out immediately after transfer to each 1-mL syringe to avoid separation of the centrifuged fat and diluent.

For treatment of the upper eyelid, injections are made at the midpoint of the eyebrow and at the upper-lateral canthal area. Treatment of the lower eyelid involves an injection in the cheek area—1 cm medial to the lateral canthus and 2 cm lower than the upper border of the malar bone—and an injection 1 cm lateral and caudal to the lateral canthus. These two entry points enable passage of the cannula in a crisscrossing, fan-like pattern, which yields even distribution of fat.

Injection of the PDF is carried out slowly and in small quantities as the cannula is withdrawn. Because the lipoaspirate is collected at the donor site through 0.5-mm holes and then is diluted, it has a fluid-like consistency on injection. Thus, it readily flows through the 1.2-mm cannula without clogging. Due to the fluidity of the fat and the presence of the tunnels, the fat distributes smoothly into the areas avoiding the need for numerous passes. The syringe is passed 5–7 times to infiltrate 1 mL of fat (Video 33.2). The procedure is completed within 15–20 min for the four eyelids.

All patients receive PDF injection in the submuscular plane and in the supraperiosteal or supraseptal area. No patient undergoes injection in the subdermal plane. When treating deep tear-trough deformities, some fat is injected into the medial portion of the orbicularis muscle, which, in this area, is firmly attached to the bone. In older patients, extreme care is exercised to avoid perforating the thin orbicularis. The injection entry points are not sutured but covered with Steri-Strips (3M, St. Paul, MN). In our experience, it is difficult to calculate the precise amount of fat needed to correct a specific area of depression or deflation. We recommend gaining experience in this procedure and exercising good clinical judgment. Given the expectation of postoperative volume loss owing to resorption, we generally overcorrect by 20–30%; however, it should be emphasized that this is an approximation.

33.3 Postoperative Care

To avoid placing pressure on the lipofilled sites, no ice or dressings are applied postoperatively. Antibiotic therapy (amoxicillin [1000 mg] and clavulanic acid [125 mg]) is given for 4 days. Patients are advised to sleep in the supine position for 3 days, to avoid sun exposure until the ecchymoses had resolved and to refrain from physical exercise for 2 weeks.

All patients receive follow-up on the first postoperative day. Because much of the injected liquid volume and edema usually dissipates by 24 h, we evaluate patients on the first postoperative day and carry out revisional aspiration of any localized fat accumulation with the same cannula used for infiltration. The latter maneuver has been carried out only twice in the last 10 years.

In the authors' clinical practice, 35% of patients present with prominent eyes (defined as >20 mm projection, as measured with a Hertel exophthalmometer). Ascertainment of eye prominence or protrusion is complicated and can be imprecise. Because the eye globe is surrounded by the upper and lower orbit (i.e., the malar area), its prominence is assessed relative to either or both of these structures. In our assessment of eye prominence, we distinguish orbital protrusion from malar protrusion, i.e., negative vector by utilizing the superolateral orbit or the most anterior point of the malar area, respectively, as reference points [17].

A large orbital prominence (i.e., exophthalmos) may coexist with a large or small malar prominence. Both orbital protrusion and malar retrusion are common in our patient practice because of the dolichocephalic anatomy typical of patients living in southern European countries. We perform fat grafting routinely in these individuals to improve insufficient protrusion at the orbital and malar levels (Fig. 33.4a–f).

By increasing the projection of the eyebrow and the upper eyelid, the orbital prominence is made less obvious (Fig. 33.5a–c). Lipotransfer to the lower lid and the lid-cheek junction fills out the palpebro-jugal depression transforming

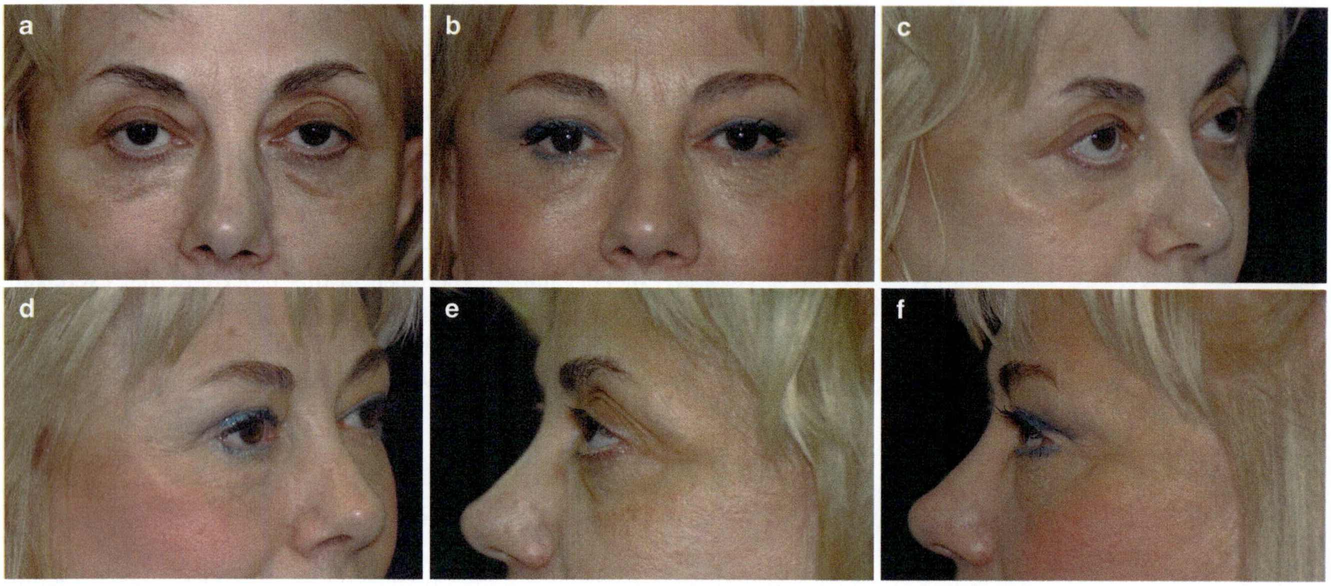

Fig. 33.4 (**a**, **c**, **e**) This 57-year-old woman presented with excessively round eyes and asymmetrically hollow upper orbits with excess skin. On profile view, exophthalmos (23 mm by Hertel exophthalmometry) and negative malar prominence (4 mm negative vector) were evident. She had scleral show bilaterally (3 mm on the right and 2 mm on the left). The patient received 5.2 mL of PDF (3.64 mL of centrifuged fat) in the left upper lid and 4 mL of PDF (2.8 mL of fat) in the right upper lid. In addition, 4.5 mL of PDF (3.15 mL of fat) was injected in each lower lid. She also underwent upper blepharoplasty with skin resection. Canthopexy was associated. (**b**, **d**, **f**) Two years postoperatively, the eyes appear less round and protuberant, scleral show is improved substantially, and the asymmetry has been corrected. No ptosis treatment was carried out. Fat grafting to the upper eyelid creates the illusion that the supratarsal fold has been lowered, thereby simulating the correction of the upper eyelid ptosis. Fat grafting into her lower lid contributed with canthopexy to treat her scleral show

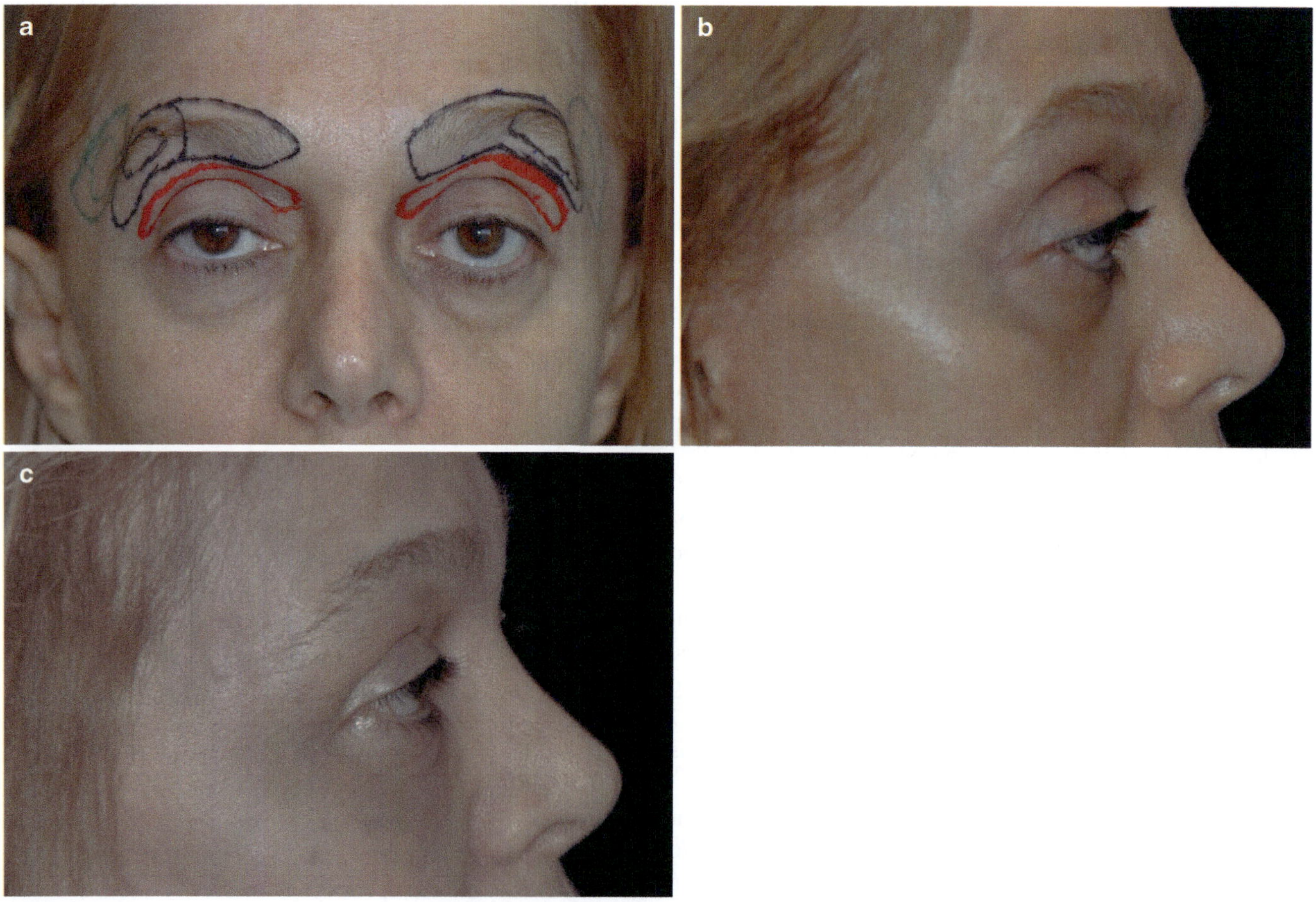

Fig. 33.5 (**a**) This 65-year-old woman presented with round, hollow-appearing and protruding eyes bilaterally. The markings on the areas to be grafted are carried out. (**b**) Before the injection of 4.5 mL of PDF (i.e., 3.15 mL of fat) on each upper eyelid and 7 mL (5.9 mL of fat) of PDF into each brow area. (**c**) Six months postoperatively: Brow projection has been enhanced and supratarsal fold has been apparently lowered by the grafted fat. The result is a visible decrease of her eye prominence

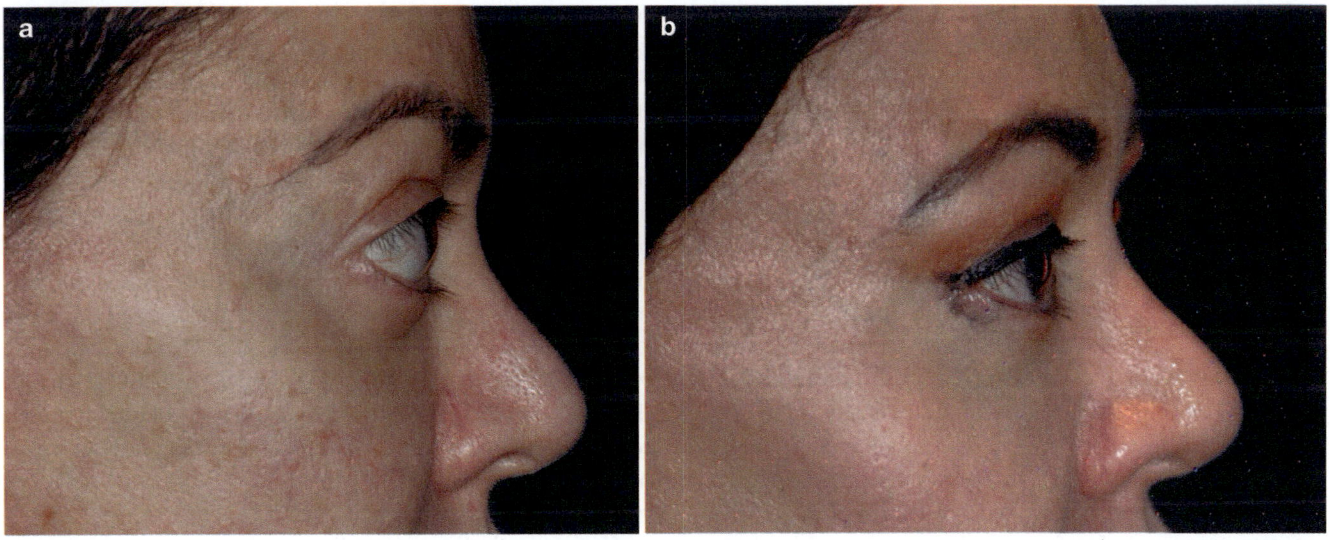

Fig. 33.6 (**a**) This 54-year-old woman presented with bilateral eye protrusion (24 mm on left and 23 on right by Hertel exophthalmometry), negative malar prominence (3 mm negative vector), and 3 mm of scleral show on right and 2 mm on left. The globe protrusion was exacerbated by deflation of the lower eyelids. The patient received 6 mL of PDF (i.e., 4.2 mL of centrifuged fat) in each brow, 4.5 mL of PDF (i.e., 3.15 mL of fat) in the upper part of each eyelid, and 5 mL of PDF (i.e., 3.5 mL) in the lower part of each eyelid. She also received a facelift without sub-SMAS undermining, but she did not undergo canthopexy or placement of a malar prosthesis. (**b**) Two years postoperatively, the globes are no longer prominent, scleral show has definitely improved and the eyes have a pleasing appearance. The rims of the lids are raised postoperatively because the lower eyelid has been straightened and is no longer concave. The SOOF was improved only by fat grafting

it into a flat surface. Less skin is then needed to cover the lower globe. The newly created excess skin together with the improved support provided by the injected volume raises the lid rim and reduces the amount of scleral show (Figs. 33.4a–d and 33.6a, b) Furthermore, increasing the projection of the malar area and of the lower eyelid corrects the negative malar prominence to a positive one, making the lower globe appear less protrusive (Figs. 33.4e, f and 33.6a, b).

33.4 A Clinical Study

A retrospective study was conducted in 200 consecutive patients who underwent injection of PDF in the periocular area (i.e., the upper and lower eyelids, and in some patients, the brows) from January 2011 to January 2017. This study was conducted in accordance with guiding principles set forth in the Declaration of Helsinki. All patients provided written consent for the procedure.

Patients were included who had (1) hollow-appearing upper and/or lower eyelids owing to age or to genetic or iatrogenic factors (170 patients; 85%) and/or (2) prominent eyes (70 patients; 35%). Patients with unilateral eye deformities or posttraumatic cases were not purposely included in this study.

All procedures were performed in the operating room, either under local anesthesia with sedation (140 patients; 70%) or under general anesthesia (60 patients; 30%). A total of 152 patients (76%) received additional facial treatments in the same surgical session. We included patients who also received upper blepharoplasty only when tissue excision was limited to the skin and there was no undermining of the orbicularis; similarly, we included patients who also underwent lower blepharoplasty only when the preseptal portion of the orbicularis was avoided during the operation.

Follow-up visits were planned for days 1, 4, and 10 and for months 1, 3, 6, and 12. Photographs typically were obtained on postoperative day 10 and at subsequent follow-up visits. Results of periocular grafting with PDF were evaluated by meticulously comparing preoperative photographs with those obtained postoperatively at 6 months and 1 year. Patients who presented for follow-up at 1 year but not at 6 months were excluded from the study. Visual inspection of the photographs was carried out by three examiners: the operating surgeon, an independent plastic surgeon with no working relationship with the primary surgeon, and a secretary from the administrative department of our hospital. Each examiner completed a chart to assess change in the facial aesthetic and contour irregularities 6 months and 1 year postoperatively. Because many patients in the series underwent additional aesthetic procedures, the examiners were asked to consider only results of the fat grafting procedure in their assessment.

Eighteen of the 200 patients were men (9%) and 182 were women (91%). The mean age of the patients was 52.3 years (median, 50.1 years; range, 26–73 years). The upper eyelids were treated in 171 patients, the lower eyelids in 138, and the eyebrows in 62.

Injection volumes varied from 1.5 mL of PDF (1.05 mL centrifuged fat) to treat a minimal localized depression to 36 mL of PDF (27.5 mL of centrifuged fat) to address the four eyelids and the eyebrows. Injection volumes of PDF were 3–6.5 mL for each eyebrow, mean being 4.5 mL, 1.5–5.2 mL for each upper eyelid mean being 3.7 mL and 1.5–4.2 for each lower lid mean being 2.7 mL. Aspiration of accumulated fat to revise the aesthetic result 1 day postoperatively was carried out in two patients.

A total of 164 patients (82%) presented for follow-up 6 months postsurgically and were photographed; 83 patients (41.5%) presented and were photographed 1 year after the operation. Of the 164 patients who returned for the 6-month visit, photographs of four patients were excluded because the images were of insufficient quality for analysis. Similarly, of the 83 patients who returned for the 1-year appointment, four patients were excluded from analysis owing to inadequate photographic resolution.

No complications were noted intraoperatively. All patients experienced swelling immediately postoperatively as a result of the dilution of injected fat and the intentional overcorrection. By 24 h after surgery, swelling generally was observed to decrease abruptly, with limited edema and some ecchymoses in most patients. The short duration of the edema was attributed, at least in part, to the small number of passes of the syringe (5–7 times for each mL of fat), which elicited only limited, local trauma. In three patients, yellowish skin discoloration was noted 7–10 days after surgery; this resolved spontaneously by 1 month postoperatively.

Postoperative complications are summarized in Table 33.1. No patient experienced obvious hematoma or seroma. Temporary upper lid ptosis of limited degree occurred in four patients; this effect subsided spontaneously in 4–8 weeks. Transient ptosis is known to occur occasionally after fat transfer to the upper eyelid [1] and may result from trauma to the elevator muscle caused by the cannula as well as the extra weight of the fat injected into the upper eyelid.

Two patients (both of whom had undergone upper and lower eyelid skin excision through a "pinch maneuver") experienced limited unilateral chemosis. This condition was treated with steroid ointment and drops and resolved within 1 week. In two patients, visible masses became visible in one of the lower eyelids by 2–3 months postoperatively. Both patients requested treatment by surgical removal of the mass. For one patient, this was conducted by a direct approach through lower lid skin for the other patient, removal of the mass was achieved by means of a lower-lid blepharoplasty incision. In two patients, the aesthetic results were deemed very good at 6 months and 1 year postoperatively, but a volume excess (upper eyelid for one patient; lower eyelid for the other) occurred subsequently. For both cases of excess fullness, the cause was attributed to gain in body weight (5–7 kg) [19]. Of these patients, one returned to her former body weight and the eyelid fullness subsided without treatment; the other similarly lost the excess weight, but still required removal of some grafted fat from the left upper eyelid. We noted that the rim of the left upper lid was lifted as a result of removal of excess fat.

Table 33.1 Complications in the study population

Complication	No. (%) of patients
Hematoma or seroma	0 (0)
Temporary upper-eyelid ptosis	4 (2)
Chemosis	2 (1)
Visible masses	2 (1)
Volume excess in treated area after body weight increase	2 (1)

Examiner findings regarding aesthetic improvement and contour irregularities are presented in Tables 33.2 and 33.3, respectively. Results of our evaluation of 6-month-postoperative photographs indicated that 154 of 160 patients (96.2%) had an excellent to moderate improvement in the periocular aesthetic, based on the averaged findings of three evaluators; six patients (3.7%) were regarded as having little or no improvement.

In assessments of photographs taken 1 year postoperatively, the averaged findings of the three examiners were that 74.3 of 79 patients (94%) had excellent to moderate periocular improvement and 4.6 patients (5.8%) had little or no improvement. All 79 patients evaluated 1 year postsurgically also had been evaluated 6 months after surgery.

On average, the three examiners noted contour irregularities in the 6-month-postoperative photos of 5.3 of 160 patients (3.3%) and in the 1-year-postoperative photographs of 2.6 of 79 patients (3.2%). The two patients in whom excess transferred fat was removed in a secondary procedure were omitted from the photographic evaluations.

The findings of the three evaluators were similar; this might be explained by the good clarity of the photographs, which enabled easy scoring of aesthetic results. The 79 patients who were photographed at follow-up visits 6 months and 1 year postoperatively were closely examined by the operating surgeon. Sixty of these patients had no detectable change in volume from the 6-month to the 1-year visits, whereas 12 had some degree of volume depletion during this time. A similar likelihood of volume retention was found in patients evaluated at 1-year and 2-year postoperative visits. Hence, we agree with other researchers who maintain that stable results of fat grafting cannot be assessed until 2 years postoperatively.

We did not assess patient satisfaction directly in this study. The results of our retrospective review of medical records suggested that most patients were satisfied because

Table 33.2 Assessments of aesthetic improvement from patient photographs

Evaluator	Total no. of patients evaluated	No. (%) of patients with excellent improvement	No. (%) of patients with moderate improvement	No. (%) of patients with little or no improvement
6 months postoperatively				
Surgeon[a]	160	104 (65)	51 (31.8)	5 (3.1)
Doctor[b]	160	110 (68.7)	45 (28.1)	5 (3.1)
Administrative assistant	160	106 (66.2)	46 (28.7)	8 (5)
1 year postoperatively				
Surgeon[a]	79	50 (63.2)	25 (31.6)	4 (5)
Doctor[b]	79	53 (67)	22 (27.8)	4 (5)
Administrative assistant	79	51 (64.5)	22 (27.8)	6 (7.5)

[a] Performed the operation
[b] No professional relationship with the operating surgeon

Table 33.3 Evaluations of contour irregularities from patient photographs

Evaluator	Total no. of patients evaluated	No. (%) of patients without contour irregularity	No. (%) of patients with visible contour irregularity
6 months postoperatively			
Surgeon[a]	160	155 (96.8)	5 (3.1)
Doctor[b]	160	156 (97.5)	4 (2.5)
Administrative assistant	160	153 (95.6)	7 (4.3)
1 year postoperatively			
Surgeon[a]	79	76 (96.2)	3 (3.7)
Doctor[b]	79	76 (96.2)	3 (3.7)
Administrative assistant	79	77 (97.4)	2 (2.5)

[a] Performed the operation
[b] No professional relationship with the operating surgeon

out of the 200 patient's records only three contained a notation about a patient's complaint for partial or unsatisfactory result.

33.5 The Tetrapod of PDF

For the past two decades, autologous fat has been utilized widely as a soft-tissue filler [1, 2, 5, 6, 10, 13, 20]. Fat grafting in the periorbital area is particularly challenging because of thin skin and a scarcity of subcutaneous tissue in this region. Even in expert hands, complications of periorbital fat grafting are common [14, 19], especially those involving subcutaneous nodules that may require challenging surgical maneuvers for removal.

When we began performing periocular fat grafting in our clinical practice, we adhered carefully to techniques described in the literature, including utilization of larger-hole cannulas for harvesting and smaller-hole cannulas for injecting, extensive centrifugation of the lipoaspirate, and delivery of concentrated fat. We found that contour irregularities were common (about 15%) with these conventional methods and were a primary cause of patient concern. We sought to achieve better aesthetic results of periocular fat grafting. Toward this aim, we incorporated dilution of centrifuged fat into our technique. In our hands, transfer of PDF is crucial to avoid the accumulation of fat clusters that result in unfavorable contour postoperatively. Moreover, PDF transfer appears to elicit less edema and ecchymosis because fewer passes of the syringe are made. Currently, our periocular fat grafting technique requires attention given to (1) the proper plane, (2) the fat fluidity, (3) tunnelization, and (4) dilution of fat.

33.5.1 The Proper Plane

Some surgeons suggest that, in addition to submuscular injection, fat should be injected into the muscle or under the skin. We avoid these maneuvers because these planes are likely to yield contour irregularities, in our experience. We consider the submuscular plane to be the safest for injection of fat.

To treat the upper eyelid, we inject PDF under the orbicularis and above the periosteum in the upper area; we also inject PDF above the septum in the upper lid. In patients who receive treatment of the brow and upper lid in the same session, we also wrap fat around the supraorbital rim. This transfer pattern positions fat on the anterior surface of the supraorbital rim and yields an even cleavage plane. We never intentionally inject fat under the septum. To address the lower eyelid, we transfer fat under the orbicularis and over the septum; in the lower orbital area, fat is delivered above the periosteum.

When additional procedures, such as canthopexy or facelift, are planned for the same surgical session as the periocular fat grafting, we avoid undermining the supraperiosteal plane to preserve its integrity to receive the fat graft. Removal of transconjunctival retroseptal fat can be performed in conjunction with fat grafting to the palpebro-malar groove if the integrity of the septum is preserved (Fig. 33.7a–e).

The upper and lower eyelids should be regarded as parts of an aesthetic unit that also comprises the brow and malar areas. Therefore, when volume is delivered to a circumscribed area, such as the lower eyelid, the fat must be "feathered out" toward the malar region, thereby creating a smooth transition (Fig. 33.8a, b).

33.5.2 Fat Fluidity

If the holes of the harvesting cannula are larger than those of the injecting cannula (e.g., 2 mm vs 1 mm), the fat cells may cluster, forming small, solid masses. During expulsion of fat, these masses can clog the holes of the injecting cannula, requiring the surgeon to increase pressure on the plunger and force the cluster into the recipient site. When fat is transferred to the eyelid in this manner, a mass can form that is not immediately evident, owing to edema in the intraoperative and early postoperative periods. By a few weeks postoperatively, a contour irregularity can become visible at this site, causing concern in the patient and precluding a favorable aesthetic result.

To avoid this complication, several years ago we designed a harvesting cannula with multiple 0.5-mm holes. Nowadays, many manufacturers produce similar cannulas. On aspiration, this cannula reduces fat into very fine particles, resembling a liquid. Subsequent delivery of fat is carried out through a single 1-mm hole of the injecting cannula. Moreover, the fat harvested through 0.5-mm holes is diluted prior to delivery, which enhances its fluidity. In our hands, this technique involves fast, smooth distribution of PDF in the periocular region with almost no risk of bolus formation.

33.5.3 Tunnelization

Before fat injection, we prepared the recipient site with a network of crisscrossing tunnels by inserting an empty cannula. We perform tunneling for several reasons. First, if a vessel is perforated by the tunneling cannula, some bleeding is likely to occur, informing the surgeon of the increased risk of fat embolism. In addition, we have found empirically that injecting PDF [20] into a network of preformed channels facilitates its smooth and rapid distribution into the recipient site. We inject 1 mL of PDF into a site that has been prepared

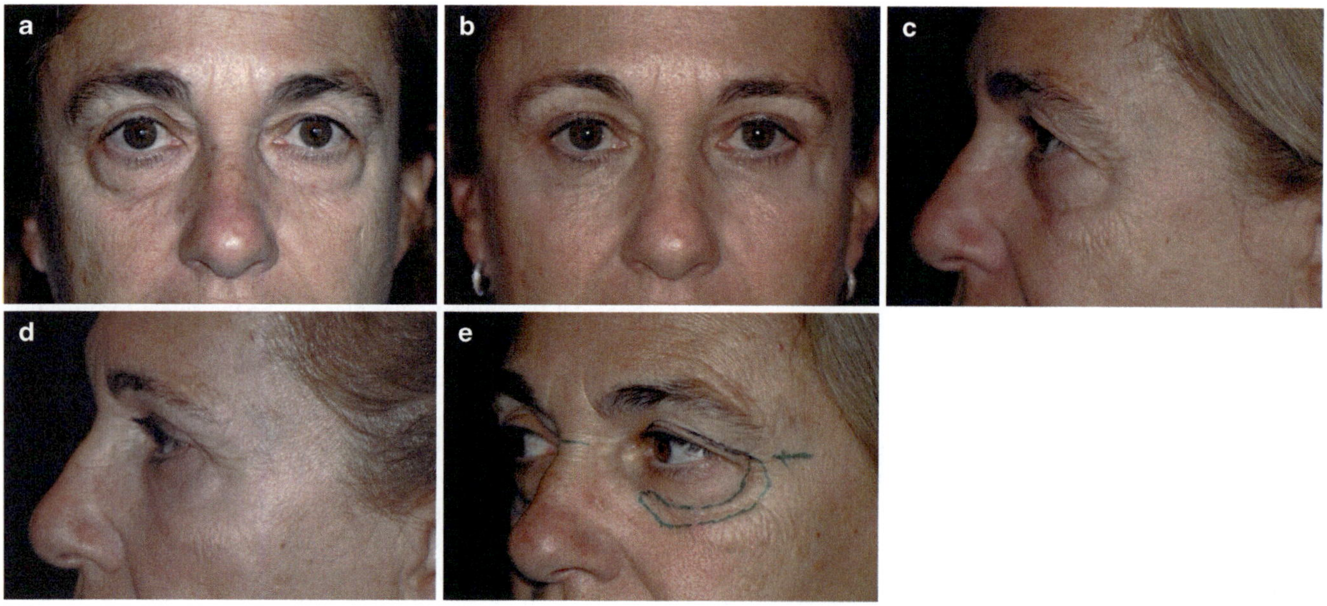

Fig. 33.7 (**a, c**) This 54-year-old woman presented with excess skin of the upper lid; the right upper lid appears more hollow than the left. Fat hernias were evident on the lower lids, with deep depressions below this area. Negative prominence of the malar region also was noted (6 mm negative vector). It was predicted that fat removal from the lower lids alone would create an obvious concavity, accentuating the negative vector. The patient underwent removal of excess skin from the upper lids and delivery of 1.5 mL of PDF (i.e., 1.05 mL of centrifuged fat) into the upper lid on the right side. A moderate amount of fat was excised transconjunctivally through a retroseptal approach, and 3 mL of PDF (i.e., 2.1 mL of centrifuged fat) was injected into the preseptal plane at the level of each lid-cheek junction. Some fat also was injected into the malar areas. Canthopexy was associated. (**b, d**) One year postoperatively, the result is satisfactory, and the negative malar prominence has been corrected to neutral. (**e**) Preoperative markings of the lid-cheek junction, where fat delivery was planned

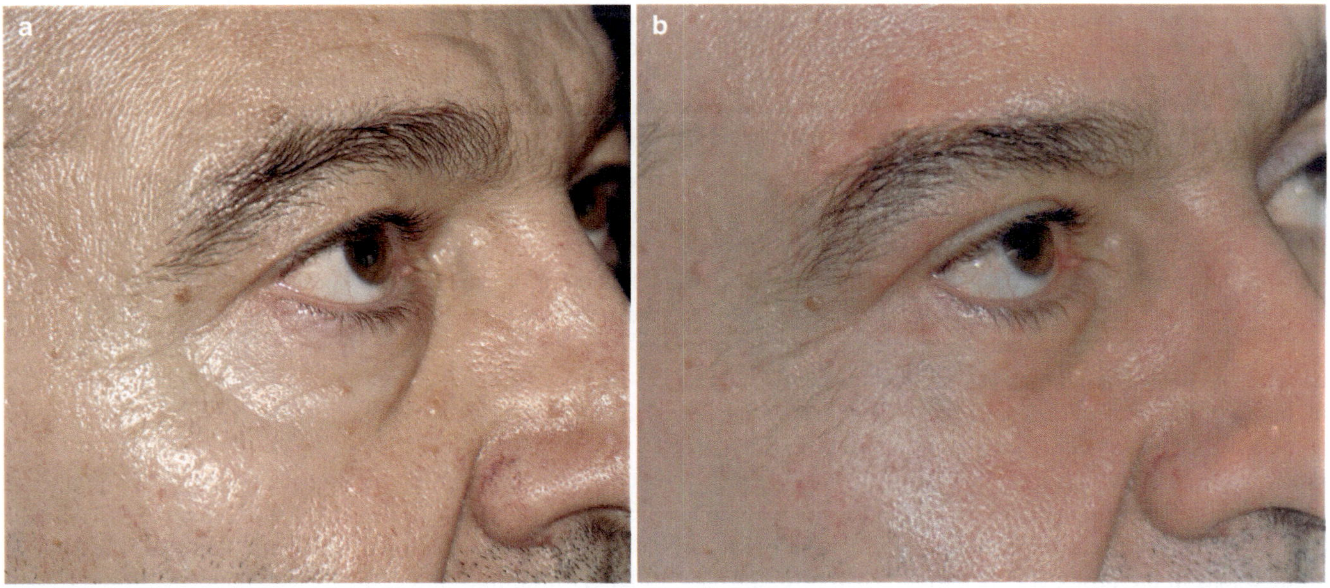

Fig. 33.8 (**a**) A 56-year-old man with brow ptosis, obvious depression over the cheek-lid junction and not enough fat to be transposed. (**b**) One year after surgery in which direct transcutaneous brow lift was carried out together with conservative fat removal through transconjunctival approach and 5 mL (i.e., 3.5 mL of fat) of PDF injected into to each the lid-cheek junction

through tunneling, with five or seven passes of the injecting cannula (Video 33.2). This technique is less traumatic than injecting with far more numerous passes per mL of fat, as suggested by other authors. Fewer passes also decrease the risk of penetrating a vessel with the cannula and is associated with less edema.

33.5.4 Dilution of Fat

Fibrin and other hematic components in lipoaspirate induce aggregation of fat particles, similar to the agglutination of red blood cells. It is a visible phenomenon that the likelihood and extent of aggregation increase proportionate with the concentration of fat cells. Removing all fluids from the aspirate—by filtration or centrifugation—concentrates the fat cells and yields a denser substance that may get trapped in the holes of the injecting cannula. We employ light centrifugation (2000 rpm for 2 min), which we have found to be adequate to separate fat from the oily supernatant and the infranatant blood and anesthetic fluid. We resuspend each 7-mL aliquot of centrifuged fat in a diluent of 1 mL of reddish infranatant liquid and 2 mL of saline [14]. This is an essential part of our technique because it improves the fluidity of the fat such that light pressure on the syringe plunger is sufficient to distribute the fat uniformly (and without nodules) in the tunneled recipient matrix.

In the lower eyelid and malar area, fat grafting with this technique also yields the favorable result of decreased scleral show (Figs. 33.4a–d and 33.6a, b). By flattening the surface of a concave lower eyelid, scleral show can be mitigated without the need for additional skin. Further support can be achieved by lipofilling the malar area. In our hands, periocular fat grafting alone usually is sufficient to correct a bowed lower lid and scleral show in many primary patients, especially when fibro-ligamentous structures and orbicularis muscle tone are adequate. Patients with a history of facial surgery generally have less favorable results of fat grafting alone, owing to the presence of fibrous and retracted tissues in the facial area and less fat take. In these patients, more complex procedures must be added to fat grafting.

We have observed that patients in our practice who undergo periocular grafting with PDF generally are satisfied, even if the volume improvement is only moderate. This finding was unexpected, and we attribute it, at least in part, to the patient being unable to predict the outcome and being willing to accept a less profound aesthetic outcome in exchange for a painless and scarless procedure.

At 1 year postoperatively, 79 of the 200 patients returned for follow-up. It is not unreasonable to interpret this finding as a lack of major complications in the 121 patients who did not present for follow-up. The presence of contour irregularities likely would have motivated these patients to express their dissatisfaction. Adipose tissue masses, especially in the lower lid, are noticeable and certainly unfavorable. However, we acknowledge that we cannot rule out complications in this patient subset. The unknown fate of this cohort constitutes a weakness of this study.

To avoid contour irregularities caused by fat nodules, we advocate only moderately overcorrecting facial concavities.

We assert that the risk of a moderately suboptimal result is preferable to the risk of a visible facial mass. The lower eyelid is most at risk of contour irregularities, followed by the upper lid. The eyebrow area is a more forgiving site for fat transfer because thicker tissues in this area help conceal irregularities.

The only disadvantage we have noted with our technique is that it is challenging to evaluate the aesthetic results of fat transfer intraoperatively when the fat being delivered is diluted. Nevertheless, intraoperative assessments of fat placement are unreliable with concentrated fat as well because of edema caused by local infiltration and the trauma of numerous passes of the cannula. We predetermine the amount of fat to be delivered; intraoperative findings are rarely cause for deviation from this plan.

Gradual loss of volume after fat grafting is not a complication but the rule. It is impossible to predict the extent of volume loss in the postoperative period. Therefore, the surgeon should exercise caution to avoid excessive overcorrection, especially in the lower eyelid where fat transfer can create the appearance of malar bags Overcorrection can be especially problematic in patients who are younger; these individuals typically have better volume retention and are likely to experience body-weight gain over time—which can exacerbate facial fullness after fat transfer. We strongly recommend that surgeons discuss the following with patients: (1) progressive fat resorption is common, and additional grafting may be needed to replace lost volume, (2) increase in body weight can dramatically alter the aesthetic result because the transplanted fat behaves as if it still is part of the donor site [19].

This study had several limitations. The study design was retrospective and nonrandomized; therefore, patient selection bias may have influenced the results. In addition, the examiners were asked to address only the results of periocular fat grafting, even though many patients in this series underwent additional aesthetic facial treatments, including blepharoplasty and facelift. These additional procedures may have influenced the perceived results.

33.6 Conclusions

Periocular fat grafting is considered to be difficult and risky, especially when the lower lid is treated. We demonstrate herein that recipient-site tunneling and injection of PDF is a simple and reliable technique that yields good contour in the periocular region (Figs. 33.9, 33.10, 33.11, and 33.12). We do not suggest that PDF is inferior or superior to non-diluted fat with regard to volume retention, but we do assert that delivery of PDF is unlikely to yield contour irregularities. Of 200 patients who underwent periocular fat grafting with this technique, only two presented to our office requiring surgical correction of poor contour. Injection of PDF is safe and effective for diverse patients, including those with prominent eyes, a negative vector, primary scleral show, small fat hernias, and palpebro-malar grooves.

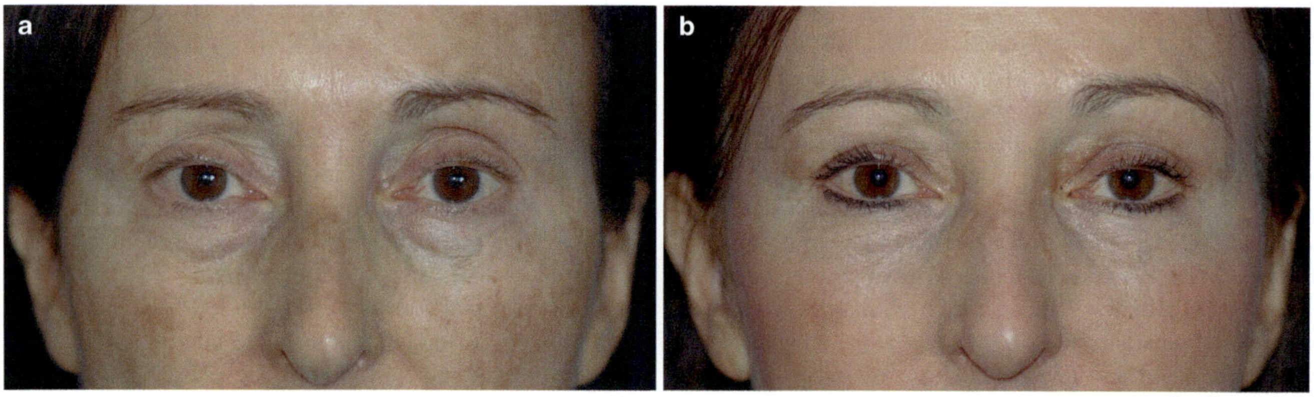

Fig. 33.9 (**a**) A 62-year-old woman with asymmetric hollow eyes. Candidate for face lifting, upper blepharoplasty, and upper eyes fat grafting. (**b**) One year after face lifting, upper blepharoplasty and 5 mL of PDF (i.e., 3.5 mL of fat) on her right upper lid and 6.5 cc of PDF (i.e., 4.55 mL of fat) on left upper lid

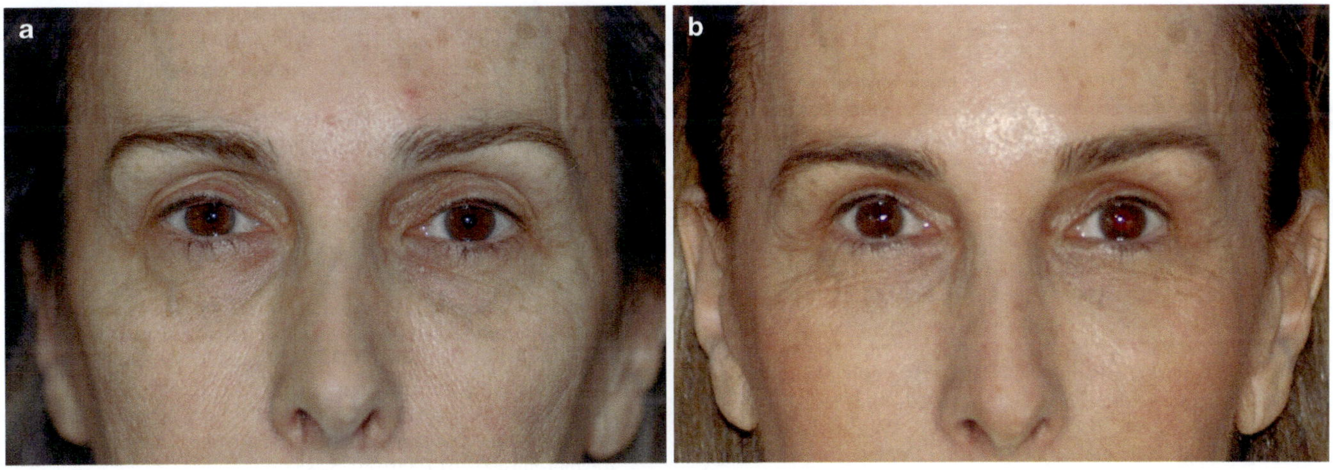

Fig. 33.10 (**a**) A 65-year-old woman with hollow eyes. Candidate for face lifting, malarplasty and eye surgery. (**b**) One year postoperatively after face lifting, malarplasty with silicone implants, upper blepharoplasty, and periocular fat grafting. 3 mL of PDF (i.e., 2.1 mL of fat) were injected into her right upper lid, 4 mL (i.e., 2.8 mL of fat) into her left upper lid, and 1.5 mL (i.e., 1.05 mL of fat) into each tear trough area

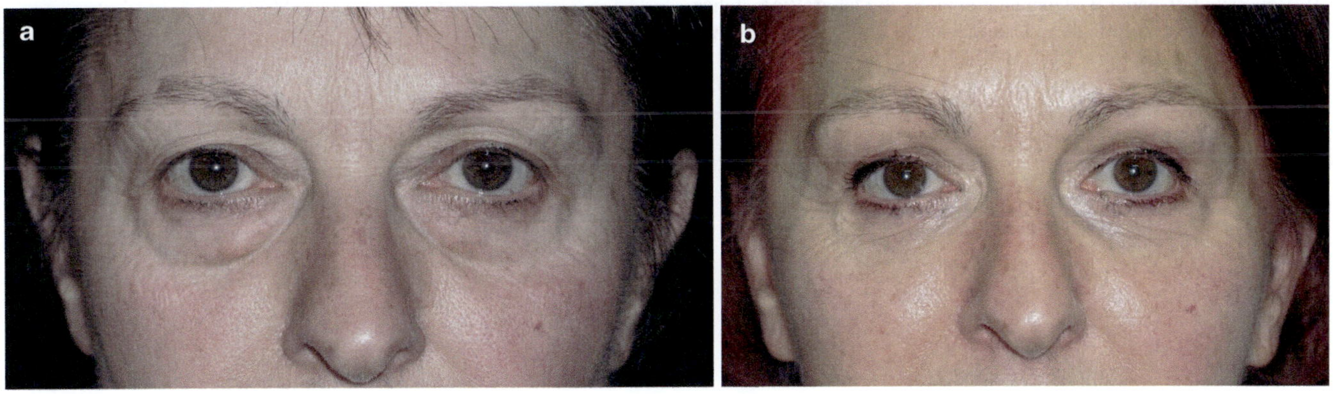

Fig. 33.11 (**a**) A 66-year-old woman with considerable excess skin on her upper lids, moderately prominent eyes (20 Hertel), visible fat pad on lower lids, and marked tear-trough deformity. (**b**) One year postoperatively after upper and lower blepharoplasty with 2.5 mL of PDF (i.e., 1.75 mL of fat) on each upper lid and 2 mL of PDF (i.e., 1.4 mL of fat) into each tear-trough area

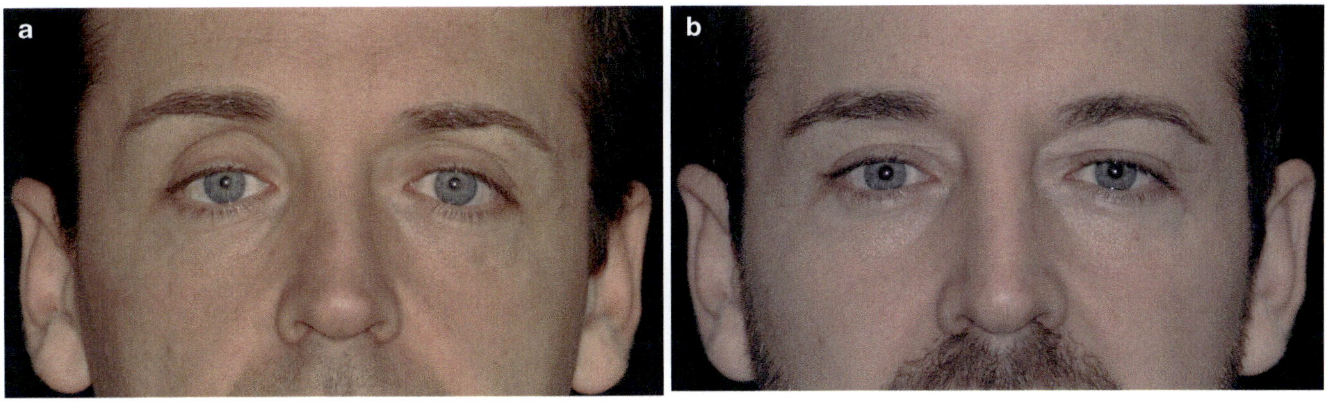

Fig. 33.12 (**a**) This 34-year-old man presented with concerns of round, hollow-appearing eyes and asymmetry. He received 3.5 mL of properly diluted fat (i.e., 2.45 mL of fat) on the right side and 2.5 mL of PDF (i.e., 1.75 mL of fat) on the left. The patient did not undergo treatment to correct the ptosis of the right upper lid. (**b**) One year postoperatively, the eyes protrude less and have a pleasing appearance. The apparently lower supratarsal fold creates the illusion of ptosis correction

References

1. Lambros V, Amos G. Three-dimensional facial averaging: a tool for understanding facial aging. Plast Reconstr Surg. 2016;138(6):990e–82e.
2. Coleman SR. Facial recontouring with liposculpture. Clin Plast Surg. 1997;24(2):347–67.
3. Holck DE, Lopez MA. Periocular autologous fat transfer. Facial Plast Surg Clin North Am. 2008;16(4):417–27.
4. Pu LL, Coleman SR, Cui X, Ferguson RE Jr, Vasconez HC. Autologous fat-grafts harvested and refined by the Coleman technique: a comparative study. Plast Reconstr Surg. 2008;122(3):932–7.
5. Sykes JM, Tapias V, Pu LL. Autologous fat grafting viability: lower third of the face. Facial Plast Surg. 2010;26(5):376–84.
6. Zielins ER, Brett EA, Longaker MT, Wan DC. Autologous fat grafting: the science behind the surgery. Aesthet Surg J. 2016;36(4):488–96.
7. Kakagia D, Pallua N. Autologous fat grafting: in search of the optimal technique. Surg Innov. 2014;21(3):327–36.
8. Sinno S, Wilson S, Brownstone N, Levine SM. Current thoughts on fat grafting: using the evidence to determine fact or fiction. Plast Reconstr Surg. 2016;137(3):818–24.
9. Boureaux E, Chaput B, Bannani S, et al. Eyelid fat grafting: indications, operative technique and complications; a systematic review. J Craniomaxillofac Surg. 2016;44(4):374–80.
10. Khouri RK Jr, Khouri RK. Current clinical applications of fat grafting. Plast Reconstr Surg. 2017;140(3):466e–86e.
11. Guyuron B, Majzoub RK. Facial augmentation with core fat graft: a preliminary report. Plast Reconstr Surg. 2007;120(1):295–302.
12. Coleman SR. Long-term survival of fat transplants: controlled demonstrations. Aesthetic Plast Surg. 1995;19(5):421–5.
13. Lam SM, Glasgold RA, Glasgold MJ. Fat harvesting techniques for facial fat transfer. Facial Plast Surg. 2010;26(5):356–61.
14. Pelle-Ceravolo M, Angelini M. Properly diluted fat (PDF): an easy and safe approach to periocular fat grafting. Aesthet Surg J. 2019;40:sjz039.
15. Shoshani O, Berger J, Fodor L, et al. The effect of lidocaine and adrenaline on the viability of injected adipose tissue—an experimental study in nude mice. J Drugs Dermatol. 2005;4(3):311–6.
16. Gontijo-de-Amorim NF, Charles-de-Sá L, Rigotti G. Mechanical supplementation with the stromal vascular fraction yields improved volume retention in facial lipotransfer: a 1-year comparative study. Aesthet Surg J. 2017;37(9):975–85.
17. Pelle-Ceravolo M, Botti G. Midface and neck aesthetic plastic surgery, vol. 1–2. Parma: Acta Medica Edizioni; 2013.
18. Lambros V. Fat grafting: a growing problem? Plast Reconstr Surg. 2018;141(2):527–8.
19. Coleman SR. Structural fat grafting: more than a permanent filler. Plast Reconstr Surg. 2006;118(3 Suppl):108S–20S.

Head and Neck Reconstruction with Autologous Fat: The Versatility of Autologous Fat Graft in Correction of Facial Deformities and Functional Head and Neck Disorders

34

Riccardo Cipriani and Valentina Pinto

Contents

34.1 Overview

In the past 30 years, autologous fat grafting has increased its application in both aesthetic and reconstructive procedures.

Autologous fat grafting has been performed as whole grafts since the 1890s and as injectable transplantation (lipofilling) since the 1920s; fat grafting is a simple, effective, and reproducible technique, with a high satisfaction rate and few disadvantages or complications. However, it is only within the past 20 years that the popularity of autologous fat grafting has increased its application also for facial contouring, both in aesthetic and reconstructive or malformative surgery. Autologous adipose tissue represents the ideal filler for facial soft tissue augmentation and reshaping: it is readily available and inexpensive to harvest, it is autologous and therefore lacks a host immune response, it is safe and noncarcinogenic and it is acquired with a minimally invasive procedure.

Various clinical indications in head and neck surgery are ideal to be treated with autologous fat grafting: congenital craniofacial malformations, posttraumatic/iatrogenic or post oncological facial asymmetries and burn disfigurements. More recently, autologous fat tissue grafting has been successfully also used to regenerate atrophic or fibrotic skin for a large number of clinical conditions such as radio dermatitis, burning scars, systemic autoimmune connective tissue diseases (scleroderma and different types of morphea). In the past decade, the regenerative potential of the fat graft is also applied for functional disorders as velopharyngeal incompetence or vocal folds paralysis. In craniofacial malformations, the combined procedures using orthognathic surgery and autologous fat graft provide a more customizable treatment, optimizing the results and facial contouring. For functional and autoimmune disorders, the regenerative potential related to adipose-derived stem cells and growth factors plays a key role in increase in skin elasticity and vascularization in sclerotic tissues, with functional and aesthetic improvement (Table 34.1).

Several reports were published in the early twentieth century and the technique gathered popularity among the scientific community; however, over the following years, autologous fat grafting fell out of favor for both the unpredictable results due to the variable reabsorption rate and the flourishing of pedicled and microvascular-free flaps, particularly for post-oncologic reconstruction. The real problem in fat grafting was the variability of results, particularly in terms of fat absorption rate (25–90% after 6 months) and the subjectivity of the quantification methods. These efforts were characterized by the use of large amounts of injected fat to correct volume defects. However, this approach may result in the accumulation of large volumes of fat, inducing necrosis in cells that are remote from sources of nutrition, compromising the final result.

In recent years, there has been a resurgence in the use of autologous fat grafting in head and neck region for aesthetic enhancement of the face, for the treatment of congenital deformities associated with soft-tissue atrophy and/or asymmetry and for reconstructing the morphology in head and neck oncologic patients.

R. Cipriani · V. Pinto (✉)
Plastic Surgery, IRCCS Azienda Ospedaliero-Universitaria di Bologna, Bologna, Italy

Plastic Surgery, Azienda Ospedaliero Universitaria di Modena, Modena, Italy

© The Author(s), under exclusive license to Springer Nature Switzerland AG 2023

A. Di Giuseppe et al. (eds.), *Fat Transfer in Plastic Surgery*, https://doi.org/10.1007/978-3-031-10881-5_34

Table 34.1 Different applications of autologous fat grafting in the aesthetic, reconstructive and other surgical procedures

Reconstructive surgery		Aesthetic surgery		Other indications	
Facial contouring in head and neck post traumatic conditions	**LIPOSTRUCTURE**	*Facial contouring (zygoma, lips, nasolabial folds, chin, mandibular profile, lids, nose)*	**LIPOSTRUCTURE**	*Cranio-facial malformations*	**LIPOSTRUCTURE + NANOFAT**
Facial contouring in head and neck free flap reconstruction	**LIPOSTRUCTURE**	*Facial biorevitalization and rejuvenation (zygoma, lips, nasolabial folds, chin, mandibular profile, lids, cheek)*	**LIPOSTRUCTURE + NANOFAT**	*Post-traumatic or post-surgical scars*	**LIPOSTRUCTURE + NANOFAT**
Facial contouring in head and neck local flap reconstruction	**LIPOSTRUCTURE**	*Volumetric enlargement (zygoma, lips, chin)*	**LIPOSTRUCTURE + NANOFAT**	*Velopharyngeal incompetence VPI*	**LIPOSTRUCTURE + NANOFAT**
Facial burns	**LIPOSTRUCTURE + NANOFAT**	*Sequelae of rhinoplasty or other aesthetic surgery*	**LIPOSTRUCTURE + NANOFAT**	*Vocal folds paralysis*	**LIPOSTRUCTURE + NANOFAT**
Sequelae of radiotherapy	**LIPOSTRUCTURE + NANOFAT**			*Autoimmune connective tissue diseases Systemic sclerosis disease and different types of Morphea*	**LIPOSTRUCTURE + NANOFAT**
Sequelae of tracheostomy	**LIPOSTRUCTURE + NANOFAT**			*Pharyngo-cutaneous fistulae*	**LIPOSTRUCTURE + NANOFAT**

In this latter application, autologous fat grafting can be used alone or as an adjunct to pedicled and free flap reconstructions to enhance the final aesthetic result by correcting residual contour deformities. In addition, autologous fat grafting can play a role in healing irradiated head and neck tissue damage.

The treatment of head and neck cancer, primarily squamous cell carcinoma, has benefitted from significant advances in our understanding of the natural course of the disease, as well as of prognostic factors, diagnostic techniques, therapy combinations (including chemotherapy, radiotherapy, and surgery), and advances in reconstructive microsurgery techniques.

Despite these advances, however, it is still a challenge to preserve and/or restore defects of form and function resulting from the treatment of advanced tumors. The need for a safety margin in tumor resection and radical radiotherapy can result in altered form and function that severely compromises patients' quality of life.

Now that overall survival has improved after cancer treatment, and the quality of life is of increasing interest.

Defects differ depending on the type of tumor and its location, and there are generally three types: contour defects and irregularities, asymmetries, and atrophic skin. These defects are generally caused by surgery and radiotherapy and appear to be unavoidable, even in patients in whom microsurgical reconstruction has been performed. It has been described therapeutic irradiation doses (>50 Gy) could cause irreversible (subcutaneous) tissue damage.

The main objective of fat injection in post-oncologic patients is to improve skin quality (elasticity, vascularization) and secondly to restore volume loss. Usually, it is possible to achieve an improvement in the quality of irradiated skin following autologous fat grafting, as well as other important functional improvements in phonation, and swallowing.

There has been limited experience in treating the sequelae of head and neck cancer treatment using fat tissue grafting. Nonetheless, these publications [1, 2] have indicated that this procedure is associated with a low rate of procedure-related complications and a high rate of mid-term postoperative benefit on aesthetic and functional parameters.

The theoretical risk associated with the injection of adipose-derived stem cells into a post-oncologic tissue bed has not been confirmed by the research conducted so far; hence, this reconstructive approach, under the proper indication, can be offered to patients previously treated for malignancy. However, continued averseness and studies are needed to rule out potential long-term oncologic risks. Before treatment, it could be indicated to obtain clearance from the patients' oncologists. Autologous fat grafting should be performed at least 6 months from the end of surgical and radiation treatment, in order to roll out residual tumor or early recurrence and also to allow enough time for the tissues to heal following oncologic treatment and allow a clear appreciation of the extent of the defect following resolution of acute posttreatment edema.

The fat grafting technique is based on surgeon preference. Different techniques are described in the literature, and it is hard to comment on the complication rates and effectiveness based on technique. There is also conflicting evidence regarding cell populations present within the graft based on the processing technique.

Autologous fat grafting seems to be less effective when the underlying bony structure has been removed, while it works as useful adjunct to correct minor contour defects after bone-free flap or graft reconstruction. When fat is injected into facial soft tissues not supported by bone, the volume will take the path of least resistance. For example, in cheek deformities following maxillectomy, some of the increase in volume will be apparent intraorally or within the maxillectomy cavity because the oral mucosal tissues are softer and more pliable, whereas irradiated skin is stiff and resistant to changing shape. In our clinical practice, we use adipo-fascial-free flaps for substantial defects and we reserve autologous fat grafting for smaller defects or as an adjunct to bone or soft-tissue-free flap reconstruction. We feel that adipo-fascial-free flaps provide a more predictable restoration of facial contouring, often in a single stage. Serial autologous fat grafting may have a role in patients who have sizable defects but are not good candidates for microvascular-free flap reconstruction.

In addition to its utility in treating aesthetic defects, autologous fat grafting has recently attracted attention because of important biological properties that underlie its potential for regenerative medicine [3, 4]. These studies support the idea that adipose tissue enhances angiogenesis by secreting various multipotential stem cell growth factors. This could be critical for previously irradiated areas, which suffer irreversible tissue damage, decreased vascularization, and altered blood vessel morphology, leading to tissue ischemia. Fat grafting could promote increased vascularization and revitalization of the interstitial tissue, resulting in improved elasticity and tissue quality.

34.2 Surgical Procedure

The procedure of head and neck lipostructure consists of three distinct phases performed in a single surgical time:

1. Fat Tissue Harvesting
2. Fat Tissue Processing
3. Lipotreated Tissue Injection

34.2.1 Fat Tissue Harvesting

Each patient was examined preoperatively in orthostatic position with the pinch test, to precisely evaluate and draw the area to be lipoaspirated.

The ideal donor site, also in children, is represented by the abdomen because of its ease of access and availability; it is unique and median (for which we do not need to take bilateral sample as from the hips, thighs, and buttocks to avoid morpho-volumetric asymmetries). Further, the abdominal area as donor site allows to place the patient in supine position and easily allows the injection time.

In very thin children, the supragluteal region and inner knees are the best donors of adipose tissue, but requires bilateral sampling to preserve the contouring of the buttock; furthermore, they require to move the position with the child's pronation-supination. In case of severe asymmetries and great volumetric soft tissue deficit, multiple donor sites are evaluated.

The first step is the infiltration in the donor site of a solution composed of Epinephrine 2 mcg/mL in saline solution (1:500), using a 17-Gauge cannula connected to the Luerlock® syringe.

After 10 min, lipoaspiration of an adequate amount of fat tissue (depending on the clinical defect of the patient) is performed in a standardized fashion of the liposuction through a millimetric skin incision. Fat is harvested from the selected area using manual suction with a 20 cc syringe Luerlock® and a 3 mm 13-Gauge blunt cannula.

34.2.2 Fat Tissue Processing

Processing of the harvested adipose tissue is carried out in a closed Luerlock® syringe system.

The processing mechanism consists of a washing procedure, using the Ringer's lactate solution and a gradual sedimentation by gravity and shakering, as a method to remove nonviable components of the lipoaspirate and impurities (blood, oil, and cell debris), producing an injectable fluid containing purified fat graft and many pericytes and mesenchymal stem cells. Thanks to this procedure, fat tissue is microfragmented gently and washed from proinflammatory oil and blood residues, without enzymes or other additives.

Furthermore, to obtain a gradual reduction in the adipose tissue clusters, we usually use Tulip's TRUE NanoFat System (TNF™); nanofat obtained is ideal to promote tissue pliability restoration and remodeling scars thanks to the high concentration of viable regenerative cells.

Processed microfat and nanofat micro-aliquots, further reduced by a sizing transfer 2.4 mm, 1.4 mm, or 1.2 mm, are to be placed in selected anatomical areas, as vocal folds, paraglottic space, palatoplasty scars, intradermal plane, in multiple tunnels to enhance graft neovascularization, and tissue regeneration.

34.2.3 Lipotreated Tissue Injection

The processed purified fat graft is then injected in head and neck area, following the principles of structural fat grafting. Multiple access sites and a fan-like pattern technique using 1.4 mm and 1.8 mm blunt cannulas were used to transfer small aliquots of fat into various depths of the soft tissue (from the dermis to the muscular fascia or muscle). To obtain a volumetric improvement, a three-dimensional network of tunnels is created to improve the contact between the graft and the native adipose tissue, in order to maximize its vitality and its clinical effects. In case of volumetric restoration, we usually perform a 30% overcorrection to obtain symmetry with contralateral side because of the unpredictable loss of volume over time.

In selected case (functional disorders and fibrotic tissues), traditional fat grafting can be combined to nanofat [5] to promote tissue pliability restoration and remodeling scars thanks to the high concentration of viable regenerative cells.

Nanofat micro-aliquots are to be placed in the superficial plane (intradermic layer or submucosal plane), in multiple tunnels to enhance graft neovascularization and tissue regeneration, injecting with 27G and 30G needles, connected to a Luerlock® system syringe.

34.3 Clinical Applications

34.3.1 Craniofacial Malformations

Patients with congenital cranio-facial malformations are generally treated with traditional surgery, as orthognathic surgery and/or additional procedures (genioplasty, alloplastic implants, injectable absorbable fillers) with satisfactory results. However, despite achieving skeletal symmetry with orthognathic procedures, noticeable facial asymmetry may persist in soft tissues, and these conditions can be satisfactorily treated with lipofilling refinement procedures.

Lipostructure procedure depends on the proper features of the malformation, including facial asymmetry in patients with hemifacial microsomia and the post-surgical sequelae of cleft lip and palate.

Treacher Collins syndrome is a rare autosomal dominant congenital disorder, featured by severe hypoplasia of the facial skeletal particularly the maxilla, mandible, and zygoma, resulting in hypoplasia of the overlying soft tissues. Abnormal position and shape of the ear are common, often accompanied by conductive hearing loss as well as ophthalmic abnormalities such as downward slanting of palpebral fissures.

Multistep lipostructure in this kind of patient can contribute to achieve an acceptable facial contouring (Fig. 34.1). Malformative patients, often had undergone multiple reconstructive surgeries, presenting adherence to the underlying tissue at the scar after skeletal surgery or cleft repair, so need of scar releasing and rigottomies before fat transplantation. Nanofat injection could optimize the soft tissue quality.

In the case of cleft lip and palate sequelae, autologous fat grafting can improve scarring and increase upper lip volume. In these cases, fat graft can restore paranasal hypoplasia too. Further, autologous fat graft is an optimal option for velopharyngeal incompetence, often related to cleft palate.

34.3.2 Velopharyngeal Incompetence

Velopharyngeal incompetence (VPI) is a multifactorial deficit of resonant speech control, nasalence, and sometimes loss of optimal intraoral pressure to obtain orally directed speech sounds, leading to marked relationship disorders, especially in scholar age. Etiological causes can be structural and genetic, functional and neurogenic, mechanical interference, and acquired factors.

Nowadays, the most common cause is the cleft palate, isolated or associated to cleft lip.

Traditional surgical pharyngoplasty is mandatory in severe conditions, when the gap is too large and velopharyngeal closure is inferior to 50%, and when the contraction of the velum is absent.

In mild and moderate cases, the augmentation of the lateral and posterior pharyngeal walls (named Passavant's crest) with autologous fat graft can be sufficient to compensate mild cases of velopharyngeal incompetence and resolve hypernasal speech and air leak.

Different materials have been described during the past decades, but nowadays the gold standard remains autologous fat graft.

We usually perform autologous fat grafting in cleft patients, about 5 years old, underwent to speech therapy without satisfactory results in the resonance impairments. The surgical procedure requires general anesthesia. The most frequent donor site in children is the abdomen, followed by the thigh and the hips; autologous fat is processed by washing and decantation closed technique, without centrifugation. The selected autologous fat was injected into posterior pharyngeal wall (midline, Passavant's crest, soft palate, palatopharyngeal arch, uvula) using blunt 18G cannula with microtunnel technique. Usually, cleft patients are treated with a single injection; the amount of fat injected in the posterior pharyngeal wall ranged from 6 to 12 mL (Fig. 34.2). Microfat can be combined to nanofat injec-

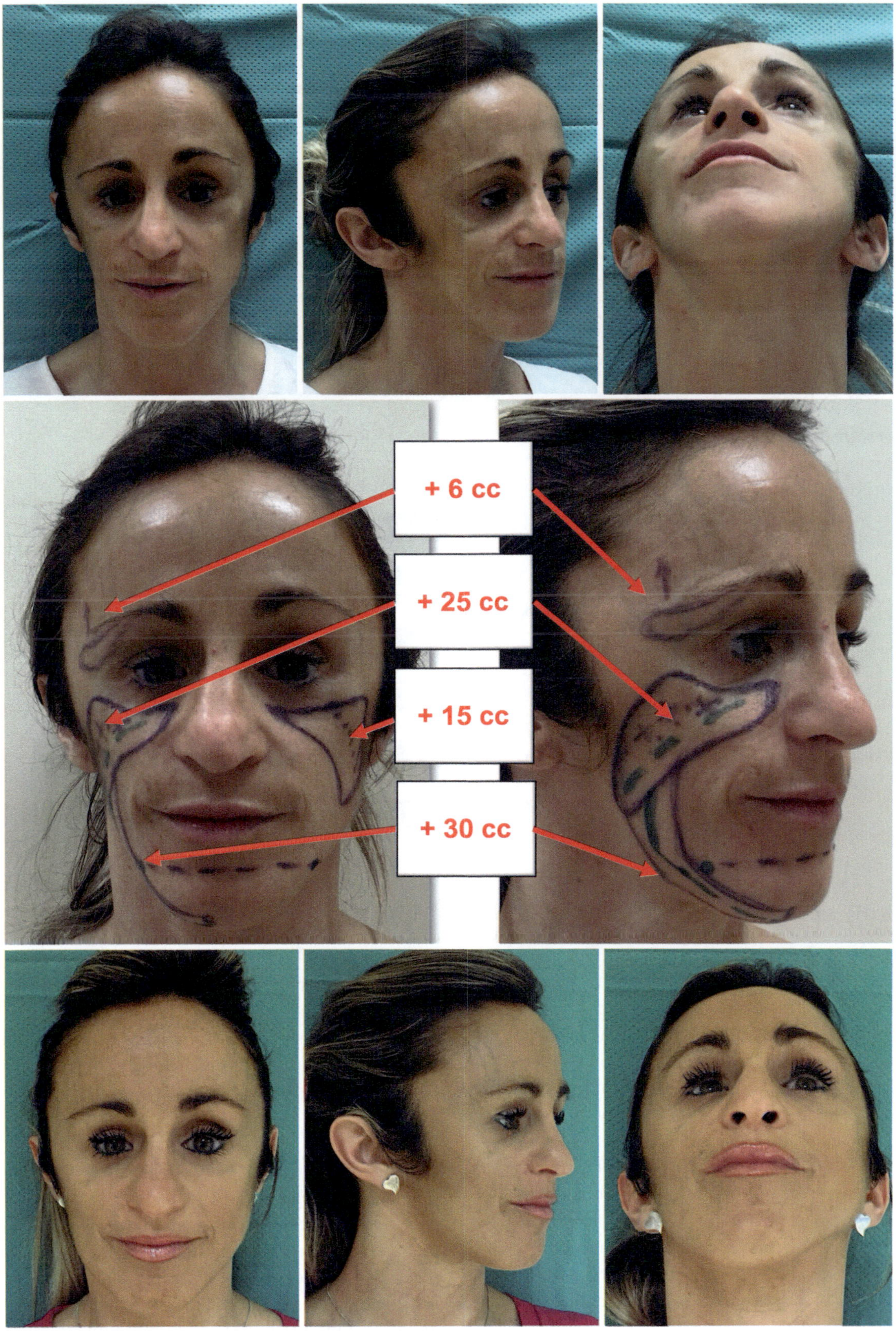

Fig. 34.1 Full face fat grafting in Treacher Collins syndrome (frontal, lateral, and submental-vertix view). A 33-year-old woman affected by Treacher-Collins syndrome, with a severe facial asymmetry even after multiple orthognatic surgeries, characterized by right hemifacial hypoplasia. Postoperative outcome after one session of structural grafting procedures (follow-up time: 14 months). Note the improvement of facial symmetry, malar area, and mandibular profile

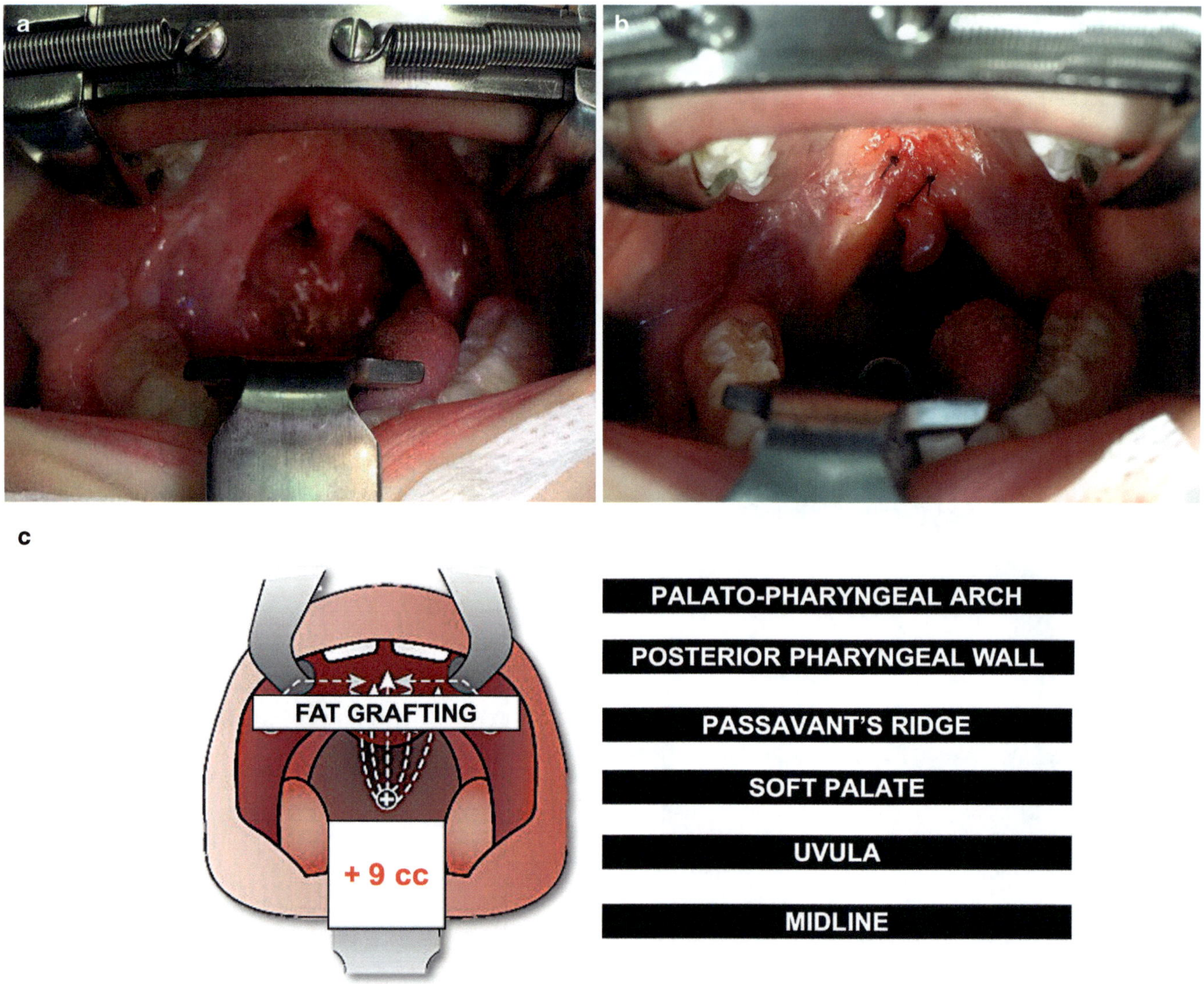

Fig. 34.2 Fat grafting in velopharyngeal incompetence in cleft patient. (**a**) Preoperative view. (**b**) Immediate postoperative view after autologous fat grafting. (**c**) Main injection sites for velopharyngeal incompetence

tion to promote palatal mobility and vibration. We inject with 27G and 30G needles, connected to a Luerlock® system syringe for palatal surgery. The treated patients presented good postoperative intelligibility and phonation, reduction of hypernasality, and improvement of postoperative nasalence. Postoperative nasofibroscopy is mandatory to assess the degree of velopharyngeal closure during phonation and feeding and to guide speech therapist for postoperative rehabilitation. Regenerative potential of fat grafting in comparison to other injectable material presents a higher amount of stem cells, favors tissue regeneration, and does not compromise velopharyngoplasty and other traditional surgical treatments when autologous fat grafting does not determine expected outcomes.

34.3.3 Systemic Sclerosis

Systemic sclerosis (SSc) is a rare systemic autoimmune connective tissue disease of unknown etiology, characterized by cutaneous and visceral fibrosis; it is a heterogeneous pathology, due to an immunological overactivation of the fibroblasts that led to a progressive cutaneous and internal organ fibrosis. Skin involvement is constant and can be variable. Patients with systemic sclerosis often complain of aesthetic and functional concerns; facial scleroderma consists of aesthetic disfigurement and limited facial expression with a mask-like stiffness of the face. Further, the loss of elasticity and the thickening of the skin in the perioral area, furrowing and narrowing of the oral aperture, leading to mouth opening reduction that interferes considerably with life's basic functions such as eating, speaking, oral hygiene, and professional dental care. Furthermore, the clinical evolution of the disease, induces salivary gland fibrosis and reduced saliva production, leading to the pathognomonic xerostomia (dry mouth). Then, facial manifestations of systemic sclerosis are extremely disabling and severely impair the patients' self-image and compromise their quality of life.

There are limited treatment options for scleroderma xerostomia and microstomia, as stretching exercises, Intense Pulsed Dye light and CO_2 laser, with ineffective results.

Thanks to a filling effect and regenerative potentials due to their adipose-derived stem cells (ASCs) content, autologous fat grafting represents nowadays an effective therapeutic alternative of the symptoms and of the clinical presentation of the disease.

Mouth opening handicap is a pathognomonic feature in the systemic sclerosis due to the alterated skin and mucosal biophysical properties (Rodnan skin score). Lipofilling treatment in plays a key role in SSc patients' satisfaction, both for the aesthetic effects (improvement of the patients' appearance) and for the functional aspects related to possible changes in the collagen pattern and content (Fig. 34.3).

34.3.4 Vocal Folds

Vocal fold scarring is the cause of severe dysphonia and phono-asthenia and represents a therapeutic challenge. The ideal surgical solution should concurrently provide vocal fold augmentation and re-establishment of tissue elasticity. Patients affected by severe dysphonia related to extensive vocal fold scarring due to previous oncological surgery, with glottal insufficiency, phono-asthenia and a gap no greater than 3 mm caused by unilateral vocal fold paralysis or vocal fold atrophy, represent ideal candidates to fat grafting. The procedure requires general anesthesia and injection have to be performed via direct microlaryngoscopy using a transoral approach. The adipose tissue is harvested as conventional procedure and then processed as decantation technique for obtaining microfat for vocal fold augmentation. Traditional fat grafting is then combined to nanofat (processed by using Tulip's TRUE NanoFat System (TNF™) to promote tissue pliability restoration and remodeling scars thanks to the high concentration of viable regenerative cells. Processed microfat and nanofat micro-aliquots, further reduced by a sizing transfer 1.2 mm, are to be placed in the vocal fold and paraglottic space in multiple tunnels to enhance graft neovascularization. We inject with 27G and 30G needles connected to a Luer-lock system syringe for phonosurgery.

Regenerative surgery by using adipose fat grafting improves the voice quality and the perceived swallowing capability. Postoperative objective videolaryngostroboscopic evaluation analyzed voice handicap index (VHI) and maximal phonation time (MPT), showing improved glottic closure despite a limited amount of fat resorption. The regenerative treatment reporting also a subjective satisfaction of the treated patients. Alternative treatment for vocal fold augmentation is the injection of calcium hydroxyapatite (CaHA); the key point of fat transplant in comparison to other injectable substances is its regenerative potential related to adipose-derived stem cells and growth factors. Further, the combination of nanofat can be a promising adjunct to traditional fat augmentation to improve elasticity of the delicate multilayered structure of the vocal fold and to enhance its vibratory capabilities during eloquence.

Therefore, fat grafting is an effective and biological filler also for permanent vocal fold injuries.

The goal is to obtain a three-dimensional network of viable adipocytes that will receive more nourishment, reducing the necrosis areas and, then, the risk of reabsorption and oily cists.

Thanks to its low invasiveness, low rate of complications and its regenerative potential, autologous fat graft can be considered a current, safe and effective technique for a more customizable facial contouring, restoring volume loss and regenerate local tissues in selected head and neck functional disorders.

34.3.5 Immediate Postintervention Situation

Facial lipostructure presents a low complication rate and a short hospitalization time (one-day surgery procedure).

There were no significant surgical complications, either from the fat donor sites or the reconstructed areas. Short lasting mild edema, facial ecchymosis, and bruising are frequent and these conditions are spontaneously resolved within 2 weeks. Compression dressing is applied on the donor site for 1 week and then compressive garments for at least one month. Heparan sulfate ointment is indicated to reduce ecchymosis in the donor sites. Major or persistent complications such as a vascular occlusion phenomenon, nervous injuries infections, fat cysts, and liponecrosis are uncommon. Fat embolism syndrome and acute pulmonary embolism from deep vein thrombosis present a low risk considering the small caliber of the blunt cannula used for lipoaspiration, the proper prophylactic therapy, and fast mobilization. No local or systemic infectious complications related to the procedure were recorded. Patients can require a second procedure. Antibiotic therapy (Amoxicillin) is administered for 5 days.

The easy procedure, the low invasiveness, and the short learning curve are the key points of the facial lipostructure. Volume improvement and facial reshaping, with a customized and natural look outcome of the fat grafting, is the goal of the treatment of the craniofacial malformations.

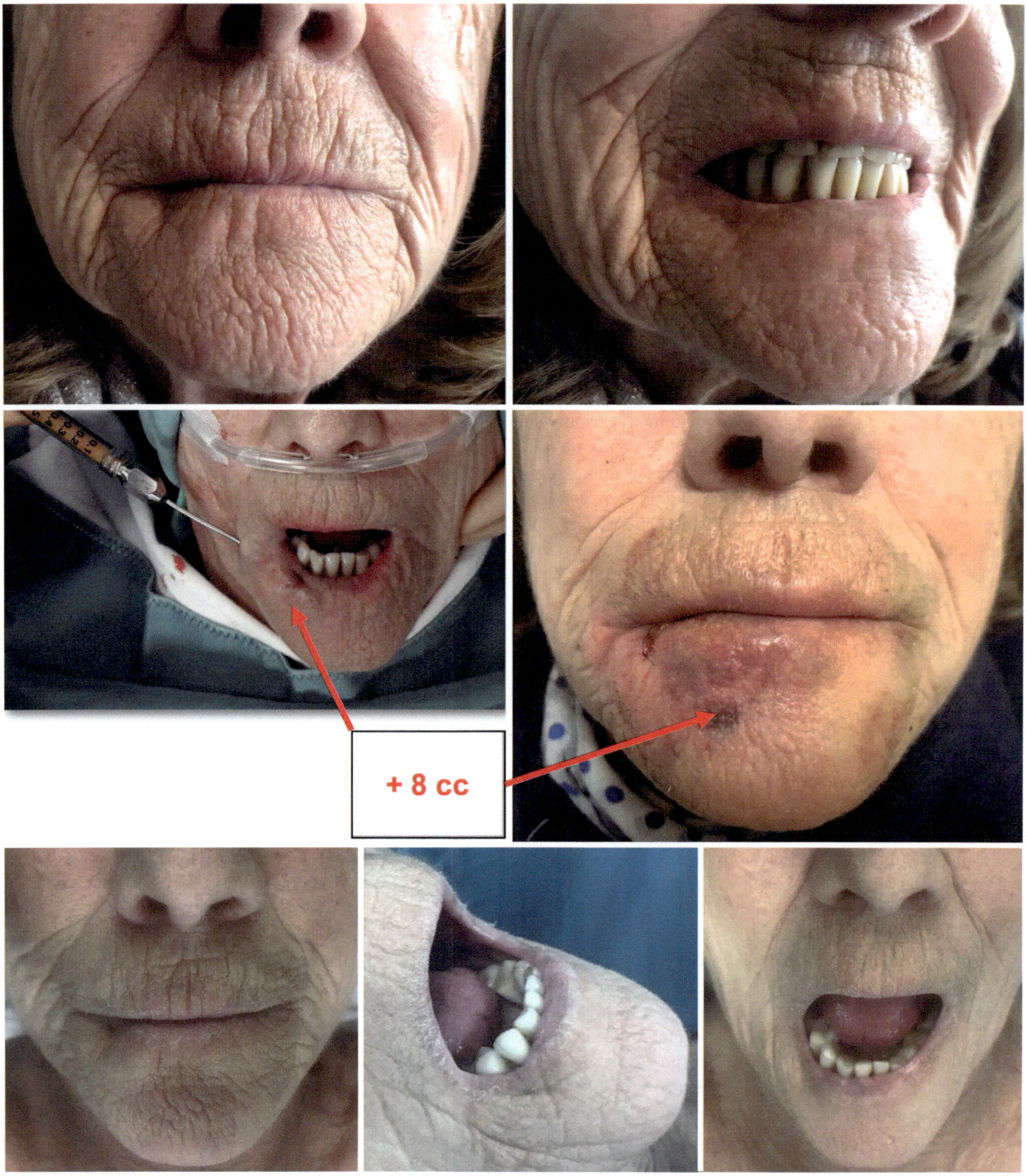

Fig. 34.3 Fat grafting in systemic sclerosis (xerostomia), with evident perilabial fibrosis and difficult mouth opening. Preoperative, intraoperative, and immediate/long-term postoperative view. Significant improvement in mouth opening capacity 6 months after autologous fat transfer (8 cc)

Irregularities can be observed when fat is injected too superficially or in an area with thin skin.

Regenerative capacity is the winning card in systemic autoimmune connective tissue disease. In these cases, skin elasticity and local vascularization can improve and thanks to the beneficial effects of the combined use of autologous fat graft and platelet-rich plasma (PRP).

A second step of fat graft usually is required to optimize the profile defect. The reason for a further surgical fat injection could be underestimation of the required soft tissue volume, high unpredictable resorption rate of the grafted fat, or inadequate recipient site, especially in cases of scars, burns, or autoimmune fibrosis. In clinical practice, surgical procedures designed to augment soft tissues are the most commonly performed after orthognathic surgery and reconstructive surgery, and fat graft seems to be a reliable option to produce a more natural result for correction of maxillomandibular asymmetry and other cranio-facial sequelae (Fig. 34.4). It allows the surgeon the opportunity to sequentially fine-tune and refine the final contours and volume of the facial soft tissue, respecting the proper anatomy of facial subunity resulting in improved facial symmetry. Improvement of skin quality of the treated region is generally observed a few months after fat grafting. Regarding facial contouring, we consider that injection respecting aesthetic facial subunits anatomy is the key point for a successful outcome. Further, every anatomic facial area presents a different rate of fat integration with native tissues, as the malar and lateral cheek regions seem to be highly favorable for fat grafting, in comparison to the upper and lower lips subunits.

34.3.6 Late Postintervention Situation

Data and parameters must be collected for every patient: demographic data, clinical data, the number of surgical sessions, the total and average amount of harvested/injected fat, and preoperative and postoperative imaging and photographs. These features are significant to evaluate postoperative outcomes, complications, reabsorption, eventual need to optimize result with other type of surgery. The final assessment of the fat grafting will be performed at least 6 months after surgery, when adipose tissue transplant can be considered as stable.

For clinical follow-up, 3D photographic reports and MRI are suggested. In functional disorders, specific evaluation is needed to assess clinical outcomes.

Autologous adipose tissue graft is a well-established method with several clinical applications in plastic surgery and finds increasing application in head and neck surgery. The autologous adipose tissue has the main advantage of being completely biocompatible and easily available, also in pediatric patients or in severe autoimmune disease, without morbidity for the donor site and without significant additional costs for the operating room.

Adipose tissue can be safety used to restore volumetric soft tissue deficit in congenital cranio-facial malformations, post traumatic/iatrogenic or post oncological facial asymmetries, and burn disfigurements, as well as to promote skin regeneration and quality improvement in radio dermatitis, burning scars, systemic autoimmune connective tissue dis-

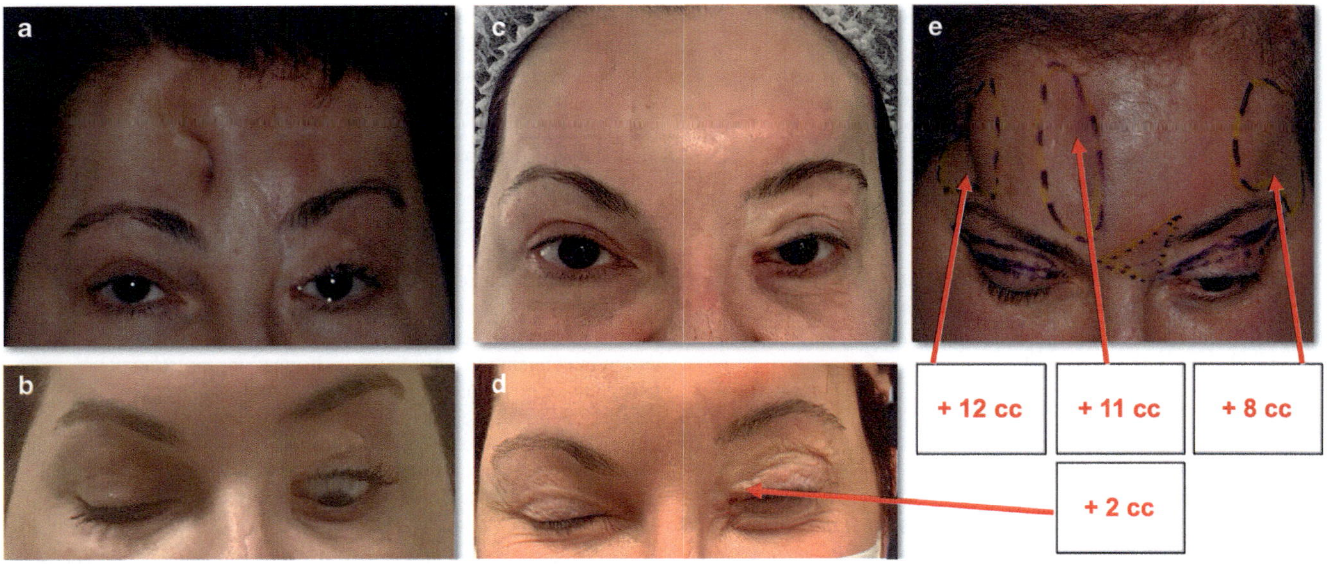

Fig. 34.4 Fat grafting in posttraumatic case. (**a**) Preoperative view of right forehead posttraumatic depression and contour irregularities. (**b**) Preoperative view of posttraumatic retracting left upper lid and lagophtalmos. (**c**) 2 years post op after one session of fat grafting (right forehead depression). (**d**) 1 year post op after full thickness skin graft and consecutive 2 cc of fat grafting for left upper lid. (**e**) Intraoperative planning of fat graft

eases or functional disorders as velopharyngeal incompetence or vocal folds paralysis.

Despite the well-known clinical advantages of the autologous fat graft, the main limit remains the unpredictable viability of transferred fat. Several studies showed greater tissue viability and a lower percentage of contaminants in fat tissue washed and filtrated within a closed system. Fat grafting is suitable and can improve the quality of life in patient treated for head and neck diseases, due to the possible benefits in mobility, aesthetic outcomes, and skin quality.

Questions

First Block		
[Question 1]	***What are the main advantages of autologous fat grating?***	The autologous fat is readily available and inexpensive to harvest, it is "self" and therefore lacks a host immune response, it is safe and noncarcinogenic and it is acquired with a minimally invasive procedure
[Question 2]	***Which is the main downside of autologous fat grating?***	It is the variability of results, particularly in terms of fat absorption rate (25–90% after 6 months), hardly to estimate due to the subjectivity of the quantification methods
[Question 3]	***What are the main goals of lipofilling after oncologic demolition in head & neck?***	The main objective of fat injection in post-oncologic patients is to improve skin quality (elasticity, vascularization) and secondly to restore volume loss. Usually is possible to achieved an improvement in the quality of irradiated skin
[Question 4]	***Is fat grafting associated to and increased risk for oncologic recurrence?***	The theoretical risk associated with the injection of adipose-derived stem cells into a post-oncologic tissue bed has not been confirmed by the research conducted so far; hence, this reconstructive approach, under the proper indication, can be offered to patients previously treated for malignancy

Second Block		
[Question 1]	***What is the role of lipofilling in cleft lip and palate sequelae?***	Cleft lip sequelae: autologous fat grafting can improve scarring and increase upper lip volume. Cleft palate sequelae: autologous fat graft is an optimal option for velopharyngeal incompetence
[Question 2]	***When is dedicated to treat velopharyngeal incompetence with lipofilling?***	Mild and moderate cases. The augmentation of the lateral and posterior pharyngeal walls (named Passavant's crest) with autologous fat graft can be sufficient to compensate mild cases of velopharyngeal incompetence
[Question 3]	***Why fat grafting is so useful for treating systemic sclerosis?***	Because lipofilling carry both filling effect and regenerative potentials due to their adipose-derived stem cells (ASCs) content
[Question 4]	***Among all of the patients affected by dysphonia, which represent the ideal candidates to fat grafting therapy?***	Patients affected by severe dysphonia, related to extensive vocal fold scarring due to previous oncological surgery, with glottal insufficiency, phono-asthenia and a gap no greater than 3 mm caused by unilateral vocal fold paralysis or vocal fold atrophy

Third Block		
[Question 1]	***Regarding facial contouring, what should be keep in mind to achieve a satisfactory result?***	We believe that the key point for a successful outcome is that injections respect aesthetic facial subunits anatomy
[Question 2]	***What are the main applications of lipofilling in head and neck?***	Adipose tissue can be safety used to restore volumetric soft tissue deficit in congenital cranio-facial malformations, posttraumatic/iatrogenic or post oncological facial asymmetries, and burn disfigurements, as well as to promote skin regeneration and quality improvement in radio dermatitis, burning scars, systemic autoimmune connective tissue diseases or functional disorders as velopharyngeal incompetence or vocal folds paralysis
[Question 3]	***What are the methods to quantify the results?***	3D photographic reports and MRI
[Question 4]	***Why seems to be wise processing the harvested tissue before the reinjection?***	Several studies showed greater tissue viability and a lower percentage of contaminants in fat tissue washed and filtrated within a closed system

Acknowledgments The authors would like to thank dr. Federico Giorgini for his collaboration in the drafting of the chapter.

Conflicts of Interest The authors declare that there are no conflicts of interest regarding the publication of this paper.

Data AvailabilityThe data used to support the findings of this study are available from the corresponding author upon request.

References

1. Clauser LC, Tieghi R, Galiè M, Carinci F. Structural fat grafting: facial volumetric restoration in complex reconstructive surgery. J Craniofac Surg. 2011;22(5):1695–701.
2. Mazzola RF, Cantarella G, Torretta S, Sbarbati A, Lazzari L, Pignataro L. Autologous fat injection to face and neck: from soft tissue augmentation to regenerative medicine. Acta Otorhinolaryngol Ital. 2011;31(2):59–69.
3. Coleman SR. Structural fat grafting: more than a permanent filler. Plast Reconstr Surg. 2006;118(3, supplement):108S–20S.
4. Khouri RK, Rigotti G, Cardoso E, Biggs TM. Megavolume autologous fat transfer: part I. Theory and principles. Plast Reconstr Surg. 2014;133:550–7.
5. Tonnard P, Verpaele A, Peeters G, Hamdi M, Cornelissen M, Declercq H. Nanofat grafting: basic research and clinical applications. Plast Reconstr Surg. 2013;132(4):1017–26.

Reverse Expansion Technique for Breast Reconstruction After Skin-Sparing and Nipple-Sparing Mastectomy

35

Cipriani Riccardo, Pinto Valentina, and Fabiocchi Luca

Contents

In the past 30 years, autologous fat grafting has increased its application both in the aesthetic and reconstructive procedure.

Although autologous fat grafting in the context of breast reconstructive surgery is mainly related to surgical revisions of breast reconstructions with prostheses or autologous flaps, or for refinements, in this chapter we present a peculiar breast reconstruction technique based exclusively on adipose tissue transplantation, without prostheses or autologous flaps.

Fat grafting as ancillary technique for prosthetic or flap-based reconstructions allows improvement in esthetic results and optimizes natural look result both in terms of breast shape and touch. Further advantage is the reduction of adipose storage in nonesthetic places reporting greater patients' satisfaction. The proposed autologous technique for breast reconstruction, so-called "*reverse expansion,*" consists of a lipostructure multisteps procedure after a breast expander insertion and inflation, using an exclusive fat tissue transplantation (lipostructure) and a progressive breast expander deflation [1].

C. Riccardo · P. Valentina (✉)
Plastic Surgery, IRCCS Azienda Ospedaliero-Universitaria di Bologna, Bologna, Italy

F. Luca
Plastic Surgery, Azienda Ospedaliero Universitaria, Modena, Italy
e-mail: luca@fabiocchi.it

35.1 Surgical Technique

"*Reverse expansion*" represents a current autologous breast reconstruction requiring the combined use of a skin expander and multiple lipofilling sessions [1].

The first step is the traditional placement of a breast tissue expander under the major pectoral muscle at the time of the skin sparing (SS) mactectomy or nipple skin sparing (NSS) mactectomy because of breast cancer. After a minimum period of 6 months from mastectomy, autologous breast reconstruction can start with a progressive deflation of the expander and a gradual fat grafting in the obtained breast pocket (Fig. 35.1).

The procedure is usually performed under general anesthesia; adipose graft can be harvested by using a traditional blunt liposuction cannula (2.5 mm); and a previous tissue infiltration is by using adrenaline 1 mg in 1000 cc of saline solution. Fat tissue is harvested from common fat donor sites such as the abdomen, gluteus, hips, and lumbar region. Harvested fat is collected in 50-mL syringes through the Lipokit machine® *(Medikan Ltd, Kangnam, South Korea),* which allows suction and then centrifugation. The centrifugation lasts for 3 min at 4000 rpm. The proper filter of this kind of syringes allows to completely separate the oily component form the adipose tissue in a closed circuit, avoiding air contact.

The purified adipose tissue is then injected in the recipient site by using 3 mL syringes and a Coleman cannula (Table 35.1).

© The Author(s), under exclusive license to Springer Nature Switzerland AG 2023

A. Di Giuseppe et al. (eds.), *Fat Transfer in Plastic Surgery*, https://doi.org/10.1007/978-3-031-10881-5_35

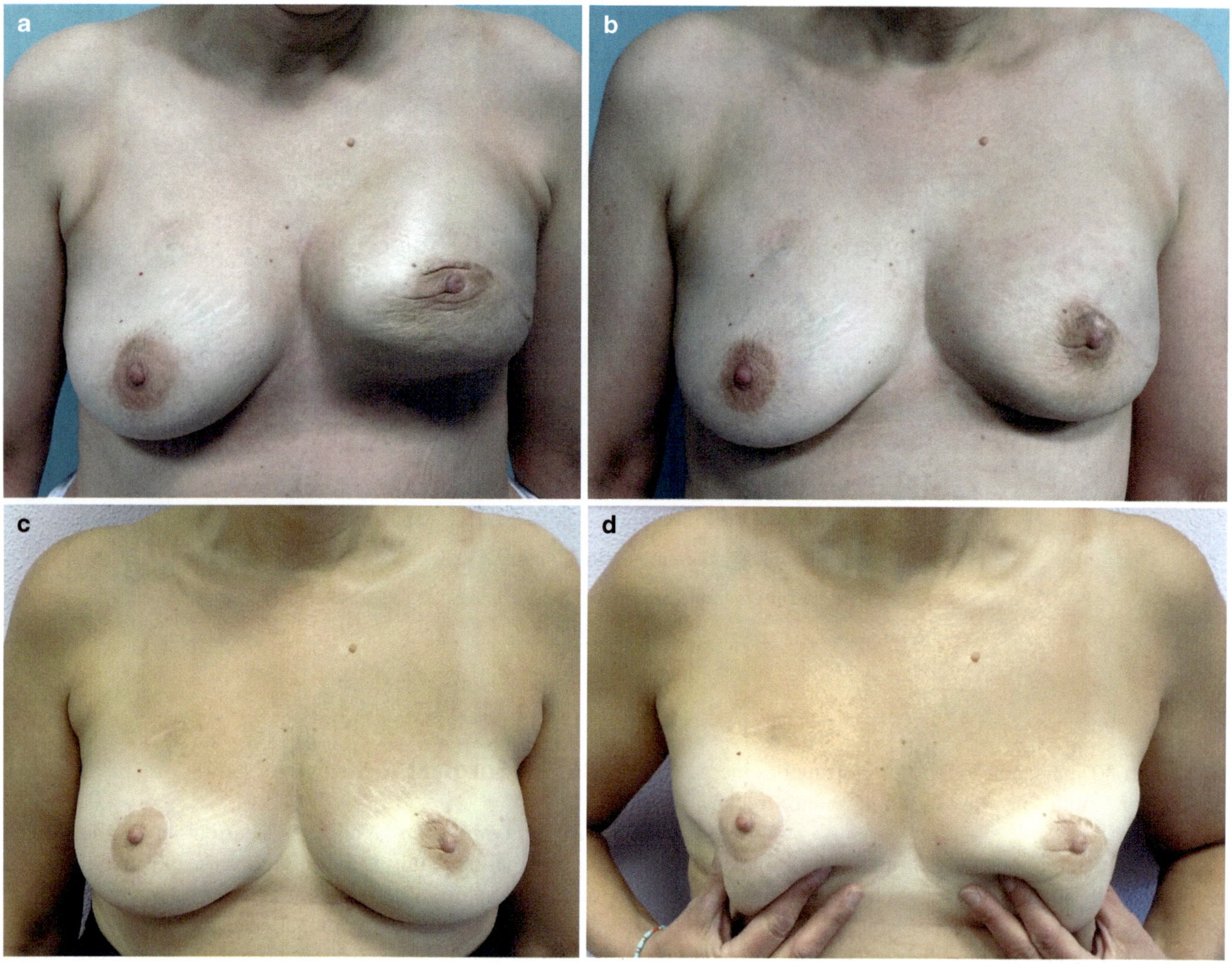

Fig. 35.1 "Reverse expansion" total breast reconstruction in monolateral left NSS mastectomy. Two sessions of fat grafting (mean grafted volume for each session: 307.5 cc). (**a**) Preoperative view. (**b**) Postoperative first session of fat grafting; skin expander still inside gradually deflated. (**c**) 2 years post-op second session after skin expander removal and further fat graft in the left breast and contralateral mastopexy. (**d**) Stable symmetry and natural texture and elasticity after 2 years from the final session

At the beginning of every surgical session, once the amount of fat graft to be harvested and injected is determined, the skin expander is deflated, by removing a similar volume of saline solution. Thanks to this approach, the fat graft has the chance to root without pressure, while simultaneously, the partially deflated skin expander helps in keeping the skin straight and the breast reconstruction stable over the time.

For each surgical step, an average of 300 cc of fat graft is usually injected, being able to reach a maximum of 480 cc. Complete breast with reverse expansion cannot be performed in one surgical session; it needs about 3 or 4 surgical steps (4 for breast underwent to previous radiotherapy).

The external appearance of "orange peel skin" will be indicative of excessive interstitial pressure and will require interruption of the procedure. According to these advices, it is possible to obtain a complete breast with about 3 lipostructure sessions, removing the tissue expander during the last step. Needling and Rigottomies are useful to release scar tissues; drains or removal of the capsular envelope are not needed: the capsular layers collapsing are functional to the definition of a natural breast ptosis.

Of course, the number of sessions will depend on the volume of the harvested fat, the volume of the injected fat tissue, the volume of the breast to be reconstructed, and the elastic potentials of pre-radiotreated skin. Therefore, the previously irradiated patients require a higher number of sessions as their breast tissues are less expandable than the normal ones. The average volume of transplanted fat, higher in the radiotreated patients, is better explained if we consider that the percentage of living fat graft is lower when it is compressed by any sort of pressure. A rigid, nonelastic,

Table 35.1 "Reverse expansion" for breast reconstruction: sequential steps in fat graft harvesting, processing, and injection

"Reverse expansion" surgical procedure	
1	Skin incision with a 11 blade (peripheral area of the donor site)
2	Injection of adrenaline/saline solution (1 mg/1000 cc SS) in the selected area
3	Waiting 10 min
4	Harvesting fat tissue in the selected area by using a 2.5 mm blunt cannula with multiple elliptical holes, connected with a 50 cc Luer Lock FPU (fat processing unit) single-use syringe (Lipokit machine® kit—Medikan Ltd., Kangman, South Korea)
5	3 min—4000 rpm fat graft centrifugation
6	Injection of the processed fat tissue to reconstruct breast using 3 mL single-use syringes (about 300 cc/surgical session)

radiotreated skin with poor blood supply may cause higher fat loss than a normal one, and therefore, to achieve the same amount of living fat, a higher volume needs to be injected.

During the last lipostructure session, the skin expander is removed, thus accomplishing a complete breast reconstruction using autologous tissue alone without requiring the use of implants or autologous flaps. In the same surgery, contralateral breast modeling (mammary reduction/mastopexy) and, when possible, nipple reconstruction can be performed. The minimum time between two consecutive surgical steps has to be about 4 months.

35.2 Key Points of the Procedure

Key points of the *"Reverse expansion"* procedure are the real factors limiting the engraftment of adipose tissue and its viability. It is important at this point to clarify that when adipose tissue is transplanted into subcutaneous recipient, the interstitial pressure is automatically increased. When, following the progressive adipose graft, this pressure exceeds the diffusion pressure of the capillaries of 16 mmHg, there will be no more engraftment because neoangiogenesis has not yet occurred and the transplanted fat will no longer receive nourishment. Therefore, one of the main factors limiting engraftment is precisely the *capillary diffusion pressure*. The rational of this concept is that fat centrifugation on defined parameters (3 min at 4000 rpm), modified than Coleman technique, allows to obtain a drier adipose tissue, with a lower aqueous component, avoiding to stress excessively the adipocytes and promoting their viability. This strategy leads to graft a greater amount of adipose tissue before reaching the limit of the interstitial pressure of the 16 mmHg, value that acts as a biological factor limiting engraftment. For this purpose, the breast tissue expander plays a key role; according to basic notions of the plastic surgery, an expanded tissue develops a richer vascular network, showing itself more elastic and compliant. These features, then, allow a greater amount of fat grafting in the expanded recipient site before interstitial pressure increases over the engraftment limit of 16 mmHg.

A further factor determining effective engraftment is the fat injection technique; the favorite procedure provides fat transplantation using a retrograde injection (0.1 mL/lipoaspirate for step) by using 3 mL syringes, connected to blunt cannula with 1.5–2 mm caliber. Grafted fat must be homogeneously distributed in all the breast area to be reconstructed. This allows not only a better esthetic result but also a great reduction of complications (Fig. 35.2).

The goal is to obtain a three-dimensional network of viable adipocytes that will receive more nourishment, reducing the necrosis areas and the risk of reabsorption and oily cists.

The periprosthetic capsule can work as an excellent guide for the transplantation of the fat graft.

1. [Question 1]	***What is the "Reverse expansion" technique for breast reconstruction?***	"Reverse expansion" represents a current autologous breast reconstruction technique requiring the combined use of a skin expander and multiple lipofilling sessions
2. [Question 2]	***How many surgical sessions need to complete breast reconstruction?***	Complete breast with reverse expansion cannot be performed in one surgical session; it needs about 3 or 4 surgical steps (4 for breast underwent to previous radiotherapy)
3. [Question 3]	***What are the biological key points of the procedure?***	One of the main factors limiting engraftment is precisely the capillary diffusion pressure. The rational of this concept is that fat centrifugation on defined parameters (3 min at 4000 rpm), modified than Coleman technique, allows to obtain a drier adipose tissue, with a lower aqueous component, avoiding to stress excessively the adipocytes and promoting their viability. This strategy leads to graft a greater amount of adipose tissue before reaching the limit of the interstitial pressure of the 16 mmHg, value that acts as a biological factor limiting engraftment
4. [Question 4]	***What is the favorite fat injection technique of fat grafting?***	The favorite fat injection procedure provides fat transplantation using a retrograde injection (0.1 mL/lipoaspirate for step) by using 3 mL syringes, connected to blunt cannula with 1.5–2 mm caliber

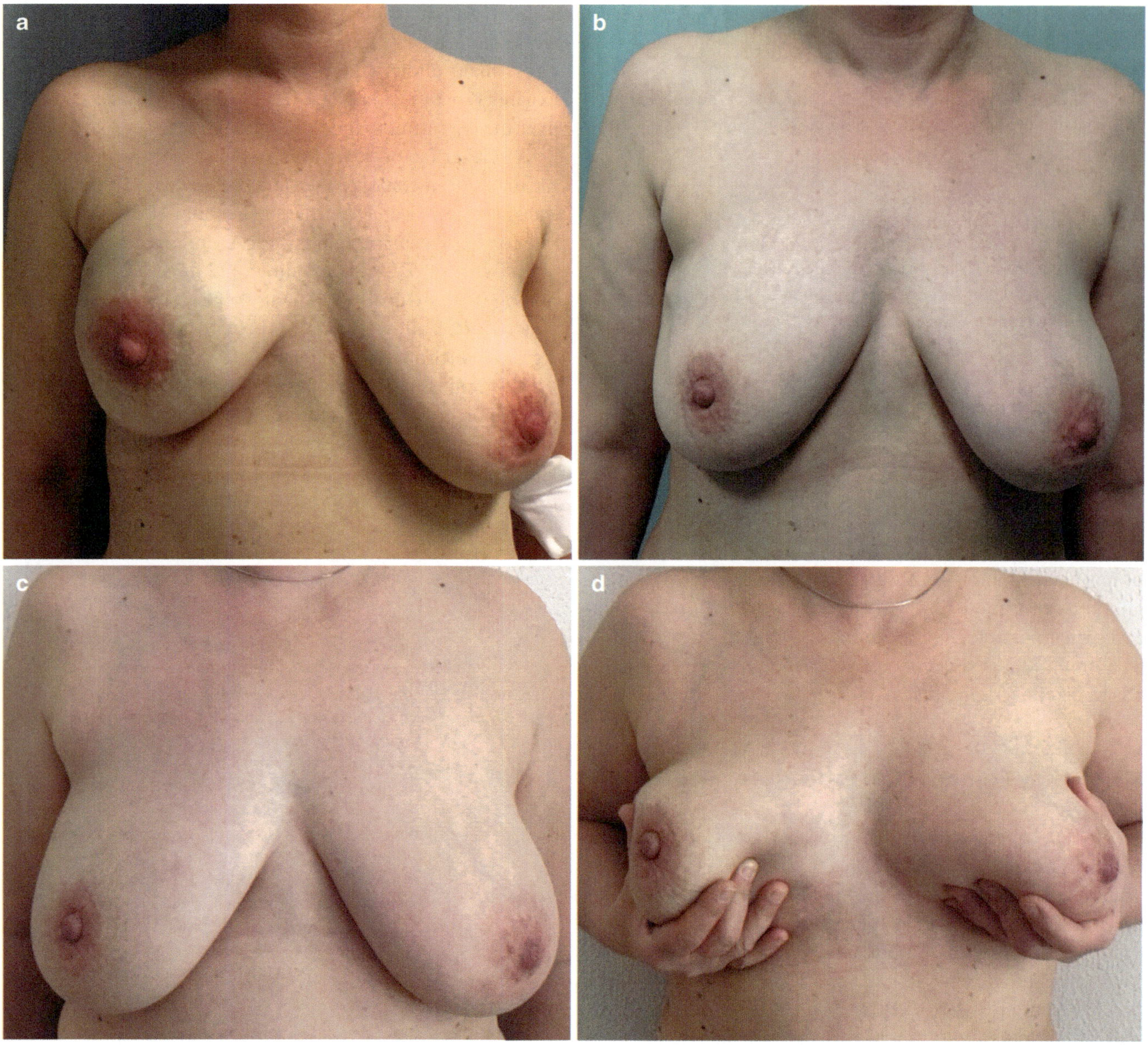

Fig. 35.2 "Reverse expansion" total breast reconstruction in monolateral right NSS mastectomy. Two sessions of fat grafting (mean grafted volume for each session: 370 cc). (**a**) Preoperative view. (**b**) Postoperative first session of fat grafting; skin expander still inside gradually deflated. (**c**) 2 years post-op second session after skin expander removal and further fat graft in the left breast without contralateral surgery. (**d**) Stable symmetry and natural texture and elasticity after 2 years from the final session

35.3 Immediate Postintervention Situation

35.3.1 Advantages

The *"Reverse expansion,"* as all the lipostructures techniques, presents a low complication rate and a short hospitalization time (1-day surgery procedure). No hematomas, infections, vascular, or nervous injuries were reported in the treated patient. No significant further surgical complications either from the donor site or the injected site were reported. Mild edema and bruising were frequent during the first post-operative week. Major complications such as fat embolism syndrome and acute pulmonary embolism from deep vein thrombosis present a low risk considering the small caliber of the blunt cannula used for lipoaspiration, the proper prophylactic therapy, and fast mobilization. Other advantages are an easy learning curve compared to other breast reconstruction methods with autologous tissue, a natural and soft texture breast look, a remodeling of the silhouette following liposuction and the scars lackness, and otherwise present in the donor areas in the classic breast reconstruction with autologous tissue.

Volume improvement and the mechanical effect of the fat graft can be immediately seen, while improvement of skin quality of the reconstructed breast is generally observed a few months after fat grafting (Fig. 35.3, 35.4). Compression dressing is applied on the donor site for 1 week and then compressive garments for at least 1 month. Antibiotic therapy (Amoxicillin) is administered for 5 days. Heparan sulfate ointment is indicated to reduce ecchymosis in the donor sites.

The long time and the multisteps procedure to complete breast reconstruction are the main disadvantages of this autologous technique. Further, the absence of fat tissue to be harvested is a major limitation to complete breast reconstruction in thin patients. In these cases, and after a surgical evaluation, patients can be treated with a combined approach using implants and fat grafting or autologous flaps.

Autologous fat tissue is the ideal filler, which is readily available and biocompatible. It has no major rejection risks in the recipient site, and furthermore, it is affordable. It can be easily harvested from many storage sites by simply using a liposuction cannula. Further, lipostructure allows body reshaping through the elimination of any possible nonesthetic fatty donor sites [2].

The first techniques using fat tissue in plastic surgery for breast reconstruction and body reshaping were introduced about one century ago. The first surgeon to use these procedures was Gustav Neuber in 1893, when he described the use of a forearm lipoma to correct face defects after tuberculosis. The autologous fat tissue transplantation progressively improved as a technique by the introduction of liposuction, described by Fisher in 1974.

In the literature, the risk of cancer relapse stimulation after lipofilling has been much debated in the past few years because of the presence of growth factors in the recipient site [3]. In 2012, Petit et al. published a study based on a group of patients who underwent lipofilling after mastectomy because of an intraepithelial neoplasm [4]. Petit hypothesized an augmented risk of cancer relapse in these patients. Few months later, because of the small sample of patients, to its extreme dishomogeneity, and to a sometimes ineffective preoperative stadiation, Petit himself admitted that it was not possible to reach any definite conclusion on the basis of the sole available data. Recently, Waked et al. published a meta-analysis considering a large number of papers concerning lipofilling. Despite a broad number of possible pro-oncogenic adipokines in a lipofilling recipient site, no clinical study has pointed out a significant increase in cancer relapse risk. Detractors of the lipofilling technique criticise this method because of the extreme variability in fat tissue survival. Despite many proposed techniques, it is currently unclear which is the most effective.

Many papers that describe many possible collections, processing, and injection techniques exist in the literature. The

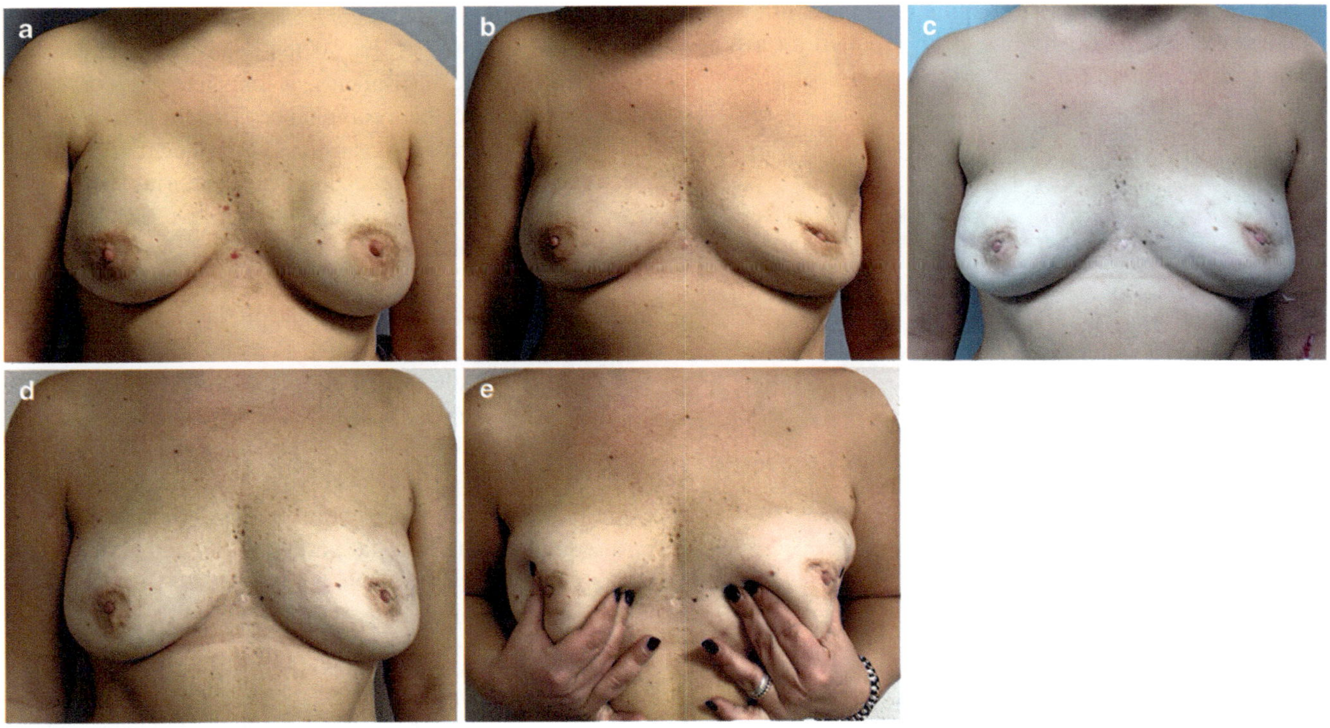

Fig. 35.3 "Reverse expansion" total breast reconstruction in bilateral NSS mastectomy. Three sessions of fat grafting (mean grafted volume for each session: 340 cc/right–324 cc/left). (**a**) Pre-operative view. (**b**) Post-operative first session of bilateral fat grafting; skin expander still inside gradually deflated. (**c**) Post-op second session after skin expander removal and further bilateral fat grafting. (**d**) 1 year post-op third session of bilateral fat grafting. (**e**) Stable symmetry and natural texture and elasticity after 1 year from third fat grafting

reverse expansion technique has been described in 2016 by Stillaert et al. in seven cases [5], and later, Fabiocchi et al. [1] presented their experience with the procedure in their cohort of patients. At present, it has been demonstrated how fat tissue survival percentage does not depend on the collection technique. Despite this, it is surprising how the management of the recipient site has never been considered as a study point to improve the percentages of surviving fat tissue. Compression removal over the implanted fat in the recipient site, which means pre-expansion, allows significant growth of the living fat tissue volume, improving the so-called graft~~o~~-capacity ratio. Some authors performed the mammary pre-expansion of the recipient tissue using the BRAVA system, which is applied after mastectomy 10 h a day, for 4 weeks. Pre-expansion allows for an ischemic reduction and then a percentage increase of living fat tissue. Then, through pre-expansion, a neo-angiogenesis stimulation is possible, followed by a better and faster revascularization of the grafted fat.

Reverse expansion allows to achieve the same neoangiogenic effects and increase of living fat using an implanted breast expander as well. The recipient site preparation, in fact, may be performed through a standard breast expander implant after mastectomy, which is much more comfortable and easier to manage for the patient in comparison with any external instrument such as BRAVA. Moreover, a breast expander stretches the skin faster than BRAVA, allowing ample reconstructions in a shorter time. Our mean grafted fat volume was 304.9 mL. Khouri et al. reported an average grafted fat volume of 225 mL.

1. [Question 1]	***Which are the most common complications after a surgery for "Reverse expansion"?***	The "Reverse expansion" presents a low complication rate. Mild edema and bruising are the most common minor complications during the first postoperative week
2. [Question 2]	***How long is the hospitalization after a surgery for "reverse expansion"?***	The hospitalization time is very short (usually day surgery/1 day surgery)
3. [Question 3]	***What does the postoperative care include?***	Fast mobilization and proper prophylactic. Antibiotic therapy (Amoxicillin) for 5 days. Heparan sulfate ointment is indicated to reduce ecchymosis in the donor sites
4. [Question 4]	***Is fat grafting dangerous in post-oncologic tissues?***	This specific aspect is very important and still under evaluation. So far, no evidence of an increased recurrence of malignancy has been reported in large data studies. It is accepted by the scientific community that fat grafting is safe after oncologic demolition

35.4 Late Postintervention Situation

Data and parameters must be collected for every patient: demographic data, clinical data, the number of surgical sessions, the total and average amount of harvested/injected fat, and preoperative and postoperative imaging. These features are important to evaluate postoperative outcomes, complications, reabsorption, and eventual recurrence.

The final assessment of the fat grafting will be performed at least 6 months after the last surgical step, when adipose tissue transplant can be considered as stable.

In selected cases, the clinical results of this autologous breast reconstruction are comparable to those obtained with other autologous procedures, which are performed more often, such as the DIEP flaps. Breasts reconstructed by *"Reverse expansion"* procedure are completely made up of adipose tissue, and this condition led to a volumetric modification following patient's weight changes just like a natural breast.

For clinical follow-up, we suggest an annual mammographic and ultrasound assessment; some patients underwent to reverse expansion can develop dystrophic microcalcifications in the grafted fat tissue, which required a stricter ultrasound follow-up, leaving MRI and biopsies for unclear cases, as for any breast cancer patient's follow-up.

Several procedures have been described for complete autologous breast reconstruction over the years. DIEP flap is nowadays considered the gold standard, but is not always possible to perform it, due to the need for microsurgical skills and specific instruments, a long learning-curve, the ideal patients, and their compliance. The fat grafting combined to reverse breast expansion is considered a safe, simple, and not invasive procedure, not requiring a peculiar learning curve, ensuring a fast postoperative recovery with excellent and stable results.

Thanks to these favorable features, our approach consisting of the "reverse expansion" is the preferred surgical option for an autologous complete breast reconstruction and it represents a viable alternative to microsurgical DIEP flap to obtain a natural breast shaping, avoiding major complications.

Although our clinical results are encouraging, "reverse expansion" still needs an optimization of the procedure to ensure greater long-term fat graft viability and a higher follow-up, necessary to evaluate both esthetic results and long-term oncologic safety (Figs. 35.5 and 35.6).

1. [Question 1]	***When the result obtained with the reverse expansion can be considered stable?***	There is an individual variability, but generally after 6 months the result is definitive

2. [Question 2]	***What is the evolution of breast reconstruction over the years?***	Breasts reconstructed by "Reverse expansion" are composed by adipose tissue and undergo volumetric modification following patient's weight changes
3. [Question 3]	***What is the proper follow-up after reverse expansion?***	Mammographic and ultrasound assessment every year
4. [Question 4]	***What are the main upside of reverse expansion?***	It is a safe, simple, and not invasive procedure. It does not require a peculiar learning curve. Fast postoperative recovery. Stable and pleasant results

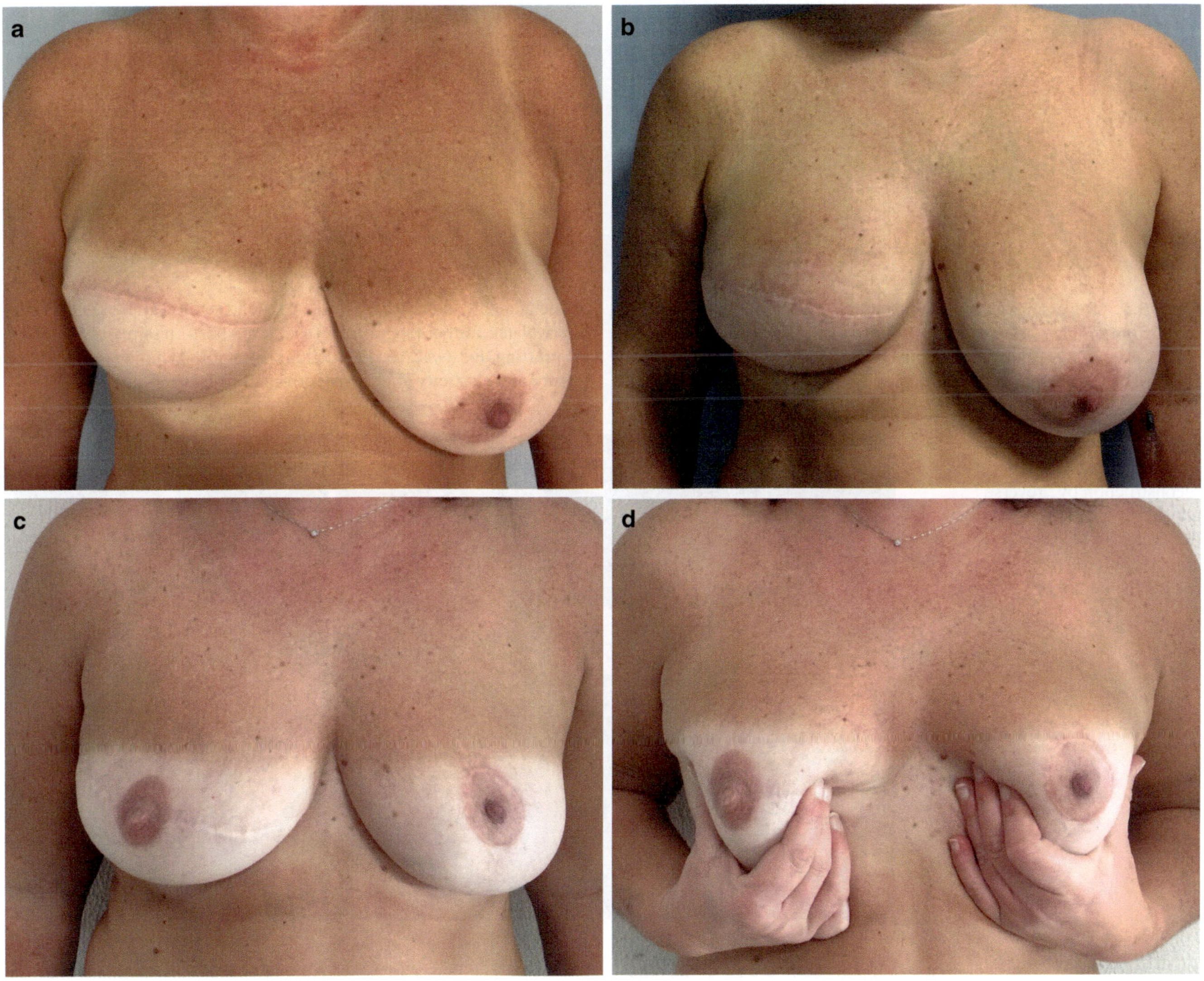

Fig. 35.4 "Reverse expansion" total breast reconstruction in right SS mastectomy. Two sessions of fat grafting (mean grafted volume for each session: 320 cc). (**a**) Preoperative view. (**b**) Postoperative first session of fat grafting; skin expander still inside gradually deflated. (**c**) 2 years post-op second session after skin expander removal and further fat graft, NAC reconstruction, and contralateral mammary reduction. (**d**) Stable symmetry and natural texture and elasticity after 2 years from completing reconstruction

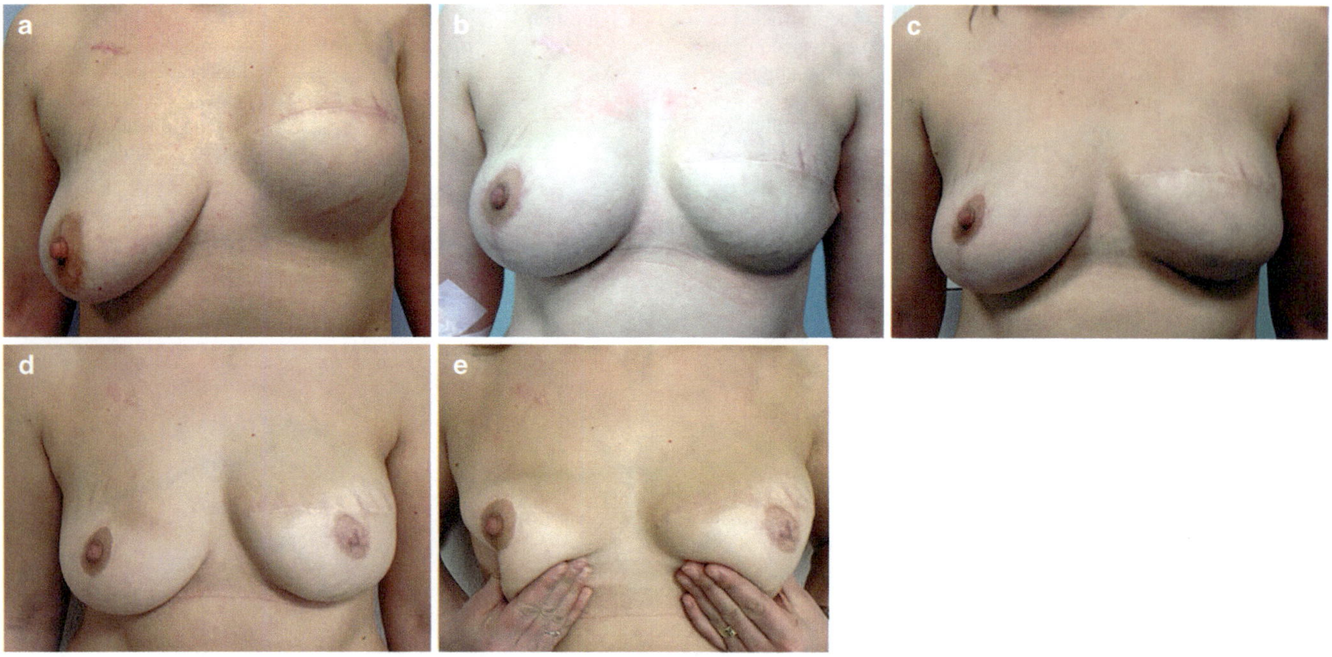

Fig. 35.5 "Reverse expansion" total breast reconstruction in left SS mastectomy. Three sessions of fat grafting (mean grafted volume for each session: 329 cc). (**a**) Preoperative view. (**b**) Postoperative first session of left fat grafting and contralateral mastopexy; skin expander still inside gradually deflated. (**c**) Post-op second session after skin expander removal and further fat grafting. (**d**) 2 years post-op third session of left fat grafting and NAC reconstruction. (**e**) Stable symmetry and natural texture and elasticity after 2 years from third fat grafting

Fig. 35.6 Breast MRI after total right breast reconstruction by "reverse expansion". Imaging shows us the exclusive adipose component of the reconstructed breast

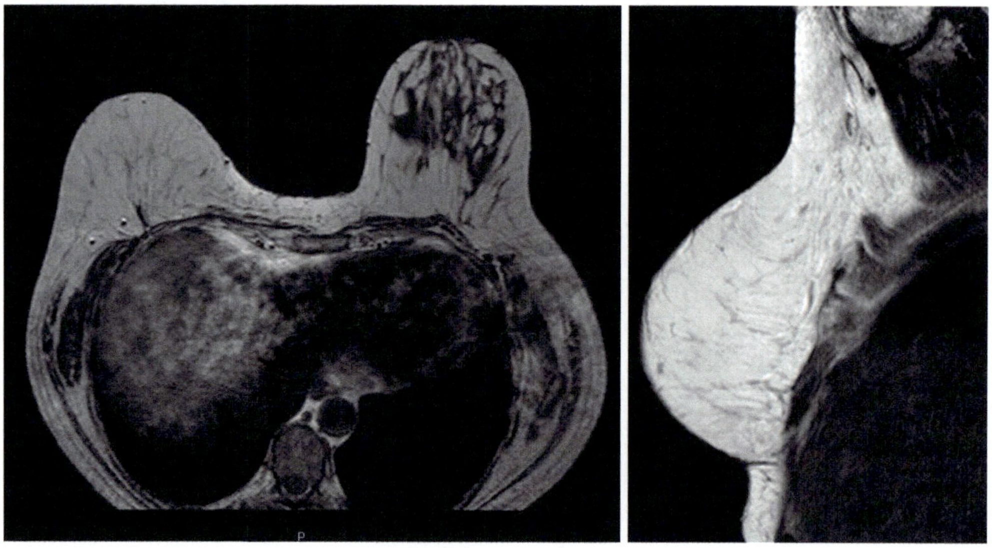

Data Availability The data used to support the findings of this study are available from the corresponding author upon request.

Conflicts of Interest The authors declare that there are no conflicts of interest regarding the publication of this paper.

References

1. Fabiocchi L, Semprini G, Cattin F, Dellachiesa L, Fogacci T, Frisoni G, Samorani D. Reverse expansion: a new technique of breast reconstruction with autologous tissue. J Plast Reconstr Aesthet Surg. 2017;70(11):1537–42.
2. Illouz YG, Sterodimas A. Autologous fat transplantation to the breast: a personal technique with 25 years of experience. Aesthet Plast Surg. 2009;33:706–15.
3. Waked K, Colle J, Doornaert M, Cocquyt V, Blondeel P. Systematic review: the oncological safety of adipose fat transfer after breast cancer surgery. Breast. 2017;31:128–36.
4. Petit JY, Botteri E, Lohsiriwat V, et al. Locoregional recurrence risk after lipofilling in breast cancer patients. Ann Oncol. 2012;23(3):582–8.
5. Stillaert FB, Sommeling C, D'Arpa S, et al. Intratissular expansion-mediated, serial fat grafting: a step-by-step working algorithm to achieve 3D biological harmony in autologous breast reconstruction. J Plast Reconstr Aesthet Surg. 2016;69(12):1579–87.

Simultaneous Body Liposuction and Breast Remodelling by Fat

36

D. Fasano, G. Gasparini, and G. Fasano

Contents

36.1 Introduction

Since the introduction in the early 1980s by Illouz [1, 2] and Fournier [3], liposuction has become one of the most performed procedures in plastic surgery, permitting in many cases impressive remodeling of the trunk and limbs.

Soon after, the idea of reintroducing part of the lipoaspirate was developed to correct unfavourable results of liposuction [3, 4] and defects of the face [5]. The idea of fat grafting was not new at that time but had been previously nearly abandoned because of the inconsistent long-term results [6].

The possibility to obtain easily and readily an abundant quantity of fat was very attractive and stimulated a raised interest in fat grafting with an increased use to accomplish various reconstructive and aesthetic objectives. The regenerative potential of autologous fat transfer was then demonstrated by the successful treatment of radiation tissue damages, scars, and ulcers [7–10].

Concerning the breast, the use of fat grafting was initially forbidden by the ASPS in 1987 because of issues related to safety and efficacy.

The introduction of a rigorous systematic technique of fat harvesting, preparation, and reintroduction by Coleman, defined as "lipostructure", has led to positive results for the face and body [11, 12]. Based on this favourable experience, Coleman decided to perform fat grafting to the breast reporting his results in 2007 [13]. In the same year, the ASPS Fat Graft Task Force was created to assess the efficacy and safety of lipofilling in breast surgery and establish evidence-based recommendations to promote its use. The Task Force conclusion was favourable and the moratorium condemning the procedure was reversed. In 2009, the ASPS Fat Graft Task Force stated that "Fat grafting may be considered for breast augmentation and correction of defects associated with medical conditions and previous breast surgeries; however, results are dependent on technique and surgeon expertise" [14].

Concern about radiologic sequelae and surveillance radiologic difficulties after lipofilling has been solved by the distinction between suspicious fine, linear and pleomorphic microcalcifications and macrocalcifications related to fat necrosis as observed after fat transfer [15, 16].

Fat transfer to the breast has been more and more frequently performed with many now well-established indications in both aesthetic and reconstructive surgery, leading to an important changing in the way of performing breast surgery.

Actual indications include breast augmentation, improvement of results of breast augmentation with a prosthesis, replacement of implants, treatment of tuberous breast, cor-

D. Fasano (✉)
Plastic Surgeon, Private Practice, Bologna, Italy
e-mail: daniele.fasano@fastwebnet.it

G. Gasparini
Plastic Surgeon, Private Practice, Parma, Italy

G. Fasano
Resident Plastic Surgeon, University of Siena, Siena, Italy

© The Author(s), under exclusive license to Springer Nature Switzerland AG 2023

A. Di Giuseppe et al. (eds.), *Fat Transfer in Plastic Surgery*, https://doi.org/10.1007/978-3-031-10881-5_36

rection of lumpectomy and radiotherapy, breast reconstruction after mastectomy [13, 17–19].

We have personally verified in our experience that lipomodeling improves the volume, shape, projection, feel, and silhouette of the breast.

Because of this, we have been attracted by the possibility of improving other different and peculiar breast defects, in particular when an important remodeling of the lower limbs and the trunk was required. The simultaneous treatment of defects in different areas of the body taking advantage of what is removed to improve one area without discharging it by injecting into another deficient area is one of the fundamental principles of plastic surgery (the Robin Hood principle of Millard). The contemporary presence of some adiposities requiring liposuction and of a breast defect is a perfect indication.

Breast asymmetry, augmented intermammary distance, Poland syndrome, pectus excavatum are some peculiar conditions that can be successfully treated by fat grafting above all when liposuction is indicated.

36.1.1 Asymmetric Breasts

Asymmetric breasts are a real challenge for the plastic surgeon, in particular when one breast has a good fullness and a nice shape, while the other is hypoplastic, stenotic, or morphologically unsatisfactory. Traditionally, an implant is inserted in the underdeveloped breast. Often, but not always, the initial result is satisfactory, but, with time, asymmetry of shape and volume becomes evident.

In these cases, lipomodeling can yield a breast very similar to the contralateral, with stable results and a maintained symmetry, with particular evidence of ptosis and volumetric changes due to weight variations. Depending on the degree of asymmetry and hypoplasia, one or more fat transfer sessions will be needed for an optimal result.

36.1.2 Reduction of the Intermammary Distance

In cases of augmented intermammary distance or medial breast asymmetry, fat transfer can be very useful both as a complement of an augmentation mammaplasty with implants and as a unique procedure to reduce the intramammary distance and improve the "dècolleté". This specific condition can be corrected uniquely by this approach [20, 21].

36.1.3 Poland Syndrome

The treatment of Poland syndrome remains a challenge to the plastic surgeon. The traditional treatment consists of combining a latissimus dorsi flap with an implant. Lipomodeling appears to be very useful in these cases, if an adequate quantity of fat is available, obtaining a satisfactory reconstruction of the anterior axillary pillar simulating the lateral border of the pectoralis major and an appropriate and stable breast volume. Often repeated procedures, every 3 or 4 months, are necessary to achieve a satisfactory result. The advantages of the technique are that the procedures are simple, leading to an improvement of the donor sites in the correct indications, avoiding the unfavourable side effects of flaps and implants. Lipomodeling can revolutionize the management of Poland syndrome [22, 23].

36.1.4 Pectus Excavatum

Pectus excavatum is a complex deformity involving hollowing of the anterior sternocostal wall. It usually has little or no functional impact and, in most cases, the problem is essentially morphologic and aesthetic, with considerable deformity of the bust if the condition is very marked or lateral. Custom-made silicone implants were traditionally used to correct the deformity with a high risk of extrusion.

Fat transfer is an additional tool with many advantages. It can provide a satisfactory one-stage correction in cases of mild deformity; in cases of more severe defects, multiple procedures are necessary if adequate donor sites are available [19].

36.1.5 Mastopexy

The satisfactory experience of fat transfer in many difficult cases of breast plastic surgery has yielded frequent use in mastopexy. It can be an excellent option to provide upper pole fullness if enough fat is available and the patient does not require a significant augmentation, maintaining a natural breast. The breast is shaped by the mastopexy and fat grafting helps create a natural transition from the chest wall to the upper pole of the breast [24].

36.2 Personal Approach

Fat transfer has changed the modern approach to breast surgery, permitting remarkable results with minimal scar.

However, there are still some unclear aspects in the behavior of fat transfer concerning the retention rate. In a systematic review of volume retention after autologous fat grafting in cosmetic breast augmentation referred to eight articles, average volume retention of 62.4% after a mean follow-up of 16.6 months [25].

Volume retention rates after fat transfer to the breast vary according to different techniques and methods used for evaluation, ranging from 37 [26] to 65% [27].

This uncertain behavior has been confirmed by our experience. Believing that actually it is difficult to augment the volume of the breast with fat in one shot in a way comparable to the implants, and several procedures can be eventually required. Then, the variable decrease in volume is to be considered in the first 3 or 4 months, leading to a quite unpredictable final result.

Of consequence, the woman can be disappointed by the augmentation achieved even if an evident improvement of other areas of the body is obtained by liposuction. For this reason, we do not actually suggest fat transfer to achieve just a breast augmentation.

On the contrary, in our experience, the most effective and predictable results are obtained in the presence of volumetric asymmetries, malformations, and morphologic anomalies, when the primary goal is not the volumetric increase. An adequate improvement of shape is much more consistently reached than an important augmentation, above all in one step.

The satisfaction rate in these cases is high, above all if adiposities are present in other areas and are likely to be well improved by liposuction. For these reasons, attention has been focused particularly on these problems.

In any case, we try to preserve an additional donor site for a subsequent procedure, such an adiposity considered minor from the patient or the posterior lumbar area that can be very useful as secondary donor site.

36.3 Personal Technique (Table 36.1)

The preferred indication for fat transfer to the breast is when a patient, with an indication of body recontouring and liposuction, requires breast remodeling. It was believed that in these cases the patients are satisfied combining the correction of two or more morphologic defects, and the surgeon takes advantage of one of the best principles of plastic surgery by using expandable undesired tissue in excess to improve deficient areas (Robin Hood's principle of Millard).

Regular preoperative exams are ordered, and in all cases, preoperative mammography or ultrasound was recommended according to the age to exclude pathology and to record the breast preoperative morphology. A postoperative control will be made after 6 months to record the new breast morphology for future follow-up (Table 36.1).

The procedure is performed on an outpatient basis or with a one-night stay according to the cases. General anesthesia or local anesthesia plus sedation is generally chosen. Perioperative antibiotic prophylaxis is administered with 2 g of cephalosporin i.v., and it is repeated if the procedure is extended more than 3 h.

A meticulous preoperative marking of the areas to be treated was made.

Table 36.1 Personal technique

• All patients underwent preoperative mammography or ultrasound according to the age to exclude pathology and to record the breast preoperative morphology. A postoperative control is made after 6 months
• Anesthesia: general; rarely local + sedation
• Antibiotics: cephalosporin iv
• Preoperative markings of the areas to be treated
• Donor site infiltration: wet infiltration (1–1). In every 1000 cc ringer lactate, add 1 cc adrenalin. If local anesthesia is needed, lidocaine 2% is added at 0.05%
• Donor site: depending on the patient request for remodelling
• Quantity of harvested fat: about 5 cc out of 10 cc was useful for reinjection. Add generally 30/40% due to resorption
• Modality of harvesting: syringes 10 cc Luer lock; cannula 3 mm Coleman
• Preparation: centrifugation for 1 min
• Position of the patient: supine and then sitting to control
• Injection: several incisions are used in creating multiple microtunnels; inject multiple sites; multiple directions and planes. Initially, 10 cc syringes are utilized, while for refinements 1 cc syringes are used. Small quantities of fat are transferred in each tunnel to be surrounded by well-vascularized tissue. The fat is injected slowly in a retrograde way by the withdrawal of the cannula
• Sites of injection: initially in the retroglandular space to add volume and then in the subcutaneous plane to reshape the breast morphology. The fat transfer begins from deep to superficial. Rigottomies are preliminarily done when needed
• A moderate hypercorrection is obtained previewing fat resorption in the following 3/4 months. 30/40% more is to be injected
• The injection is stopped before any blanching appearance. It is important to underline that hypercorrection does not mean in any case hypergrafting
• Dressing of the breast: just to protect. Dressing of the donor site: garment as in liposuction

36.3.1 Fat Harvesting

In our experience and according to several clinical evidence in the literature, no preferred adipose tissue donor site was evident; depending on patient morphology and desires in terms of body remodelling, fat can be harvested from different donor sites such as the abdomen, hips, thighs, and buttocks [28–31].

The fat harvesting technique was essentially the same in all procedures and followed the non-traumatic method recommended by Coleman. Donor site infiltration was made according to the superwet procedure (1:1 principle) adding 1 cc of adrenaline to every 1000 cc of ringer lactate. In the case of general anesthesia, local anesthetic was not used based on its negative effects on ASC founded in some studies [32]; if the procedure was done in sedation, lidocaine 0.05% or 0.1% was added.

High-powered machines were not used, and fat was harvested using Coleman connected to a closed low-pressure aspiration system.

The quantity of fat to be harvested depends on the objective of the treatment: reshaping, localized augmentation, or global augmentation. In our experience, corresponding with the prevailing literature, the volume retention rate is at most 70% [25]. Because of this, 30/40% excess fat has to be injected to obtain a satisfactory result. It is well known that after centrifugation almost 50% of the fat is useless. Considering all these aspects, if 100 cc of fat has to be injected, between 250/300 cc of fat has to be harvested. It is evident, therefore, that the procedure is time-consuming, and a standardized and well-organized team is important to reduce the operative time.

36.3.2 Fat Processing

Fat tissue was always processed by centrifugation in 10 cc syringes (3300 rpm for 1 min), and after removing the superior oily portion and the inferior serum bloody portion, the intermediate pure fat portion was ready to be injected. The procedure can be time demanding considering that in some cases 4–10 centrifugation cycles can be necessary to obtain an adequate quantity of graft. Despite this, in our experience, this is the first choice fat processing procedure in terms of predictability of fat take and long-lasting volume maintenance.

36.3.3 Fat Injection

Fat injection was made following the preoperative plan, without anesthetic solution administration to the recipient site. Several incisions were made to obtain a multidirectional fat deposit. At the beginning, 10 cc. Luer lock syringes and Coleman small injection cannulas were used following the "micro and multi tunnel" technique with the fat released under light pressure in a multilayer fashion, as previously described [13, 19].

If expansion is needed, "rigottomies" are extremely important: a 16 gauge needle is inserted percutaneously in multiple sites with radial multidirectional oscillations to cut the fibrotic tissues and expand the donor site permitting a better insertion of fat tissues [7, 10].

Fat was injected initially into the retroglandular plane to augment the volume and in the subcutaneous plane to shape the new breast as desired; the intra and retromuscular planes were injected if necessary [33, 34]. The mammary gland was avoided, but, in our opinion, the absolute avoidance of glandular injection seems to be illusory.

When most of the procedure was done, the patient was seated to evaluate the result, and the appropriate corrections were marked. The final touch-up was performed with a 1 cc syringe.

In all the patients treated, a moderate overcorrection (30–40% more) was obtained, avoiding excessive tension of the skin envelope and the consequent ischemia of the underlying fat.

Clinical experience and an adequate learning curve of the technique are fundamental requirements for the procedure's success and predictability.

36.3.4 Postoperative Care

At the end of the surgery, a compressive garment was worn in the donor site; compression of the breast was strictly avoided.

The donor sites were treated as usual after a liposuction procedure. A light bra was recommended for 1 month; after this period edema and bruising were generally resolved.

Massages and breast manipulations were contraindicated.

36.4 Results

The results were considered stable after 3–4 months. In the referred indications, the satisfaction rate has been high. Secondary procedures have been uncommon in our experience because of the high satisfaction rate and the undesired request for another procedure.

36.4.1 Complications

Donor site complications appear to be minimal and related to the liposuction technique.

Local complications of lipofilling of the breast include the palpable area of induration, fat necrosis, oil cysts, calcifications, and infections.

In a systematic review of 42 articles published up to 2014, the reported complication rates were low (<15%); lipofilling can lead to initial complications, including infections, hematoma, inflammatory reactions, and wound dehiscence. Very few media to long-term complications have been reported, but the average follow-up was limited to 22 months [35].

In a review of 24 studies, 207 complications were reported and fat necrosis accounted for 62% of all complications [36].

In another review of 22 articles, 3565 patients with follow-up periods ranging from 12 to 136 months were included. The complication rate was 17.2% and the most frequent was indurations (33.3% of all complications), followed by persistent pain (25%) and hematoma (16.4%), fat necrosis (6.6%). Mammograms revealed microcalcifications (9%) and macrocalcifications (7%) [25].

Case 1 (Fig. 36.1) 21 years old, female, 167 cm height, 70 kg, with a weight loss of 20 kg by diet, stable for 1 year. She presented with a breast asymmetry with right breast ptosis and left breast relative hypoplasia, with a presumed difference of 150 cc. She had localized fat deposits in the inner thighs and skin laxity. The planned treatment was harvesting fat from the inner thighs, inner thigh lift according to Pascal—LeLouarn, right round block mastopexy, and left breast 250 cc fat injection according to the presented technique. Postoperative result at 1 year.

Case 2 (Fig. 36.2) 31 years old, female, 160 cm height, 72 kg weight, with a weight loss of 25 kg by diet, stable since 1 year. She presented with breast ptosis with an empty skin envelope and important fat deposits in the lateral thighs and the inner thighs with skin laxity. She did not want implants, but just remodeling of the breast. The planned treatment was harvesting of fat and liposuction of the lateral and inner thighs with an inner thigh lift according to Pascal—LeLouarn, superior pedicle vertical mastopexy and fat transfer in the medial quadrants to reduce the intermammary distance with an injection of 130 cc per side and in the upper quadrant to achieve more fullness of the upper pole with an injection of 100 cc per side. Postoperative result at 1 year (Fig. 36.2).

Case 3 (Fig. 36.3) 26 years old female, height cm 166, weight 63 kg.

Diagnosis: Left Poland syndrome, right ptotic breast with severe asymmetry.

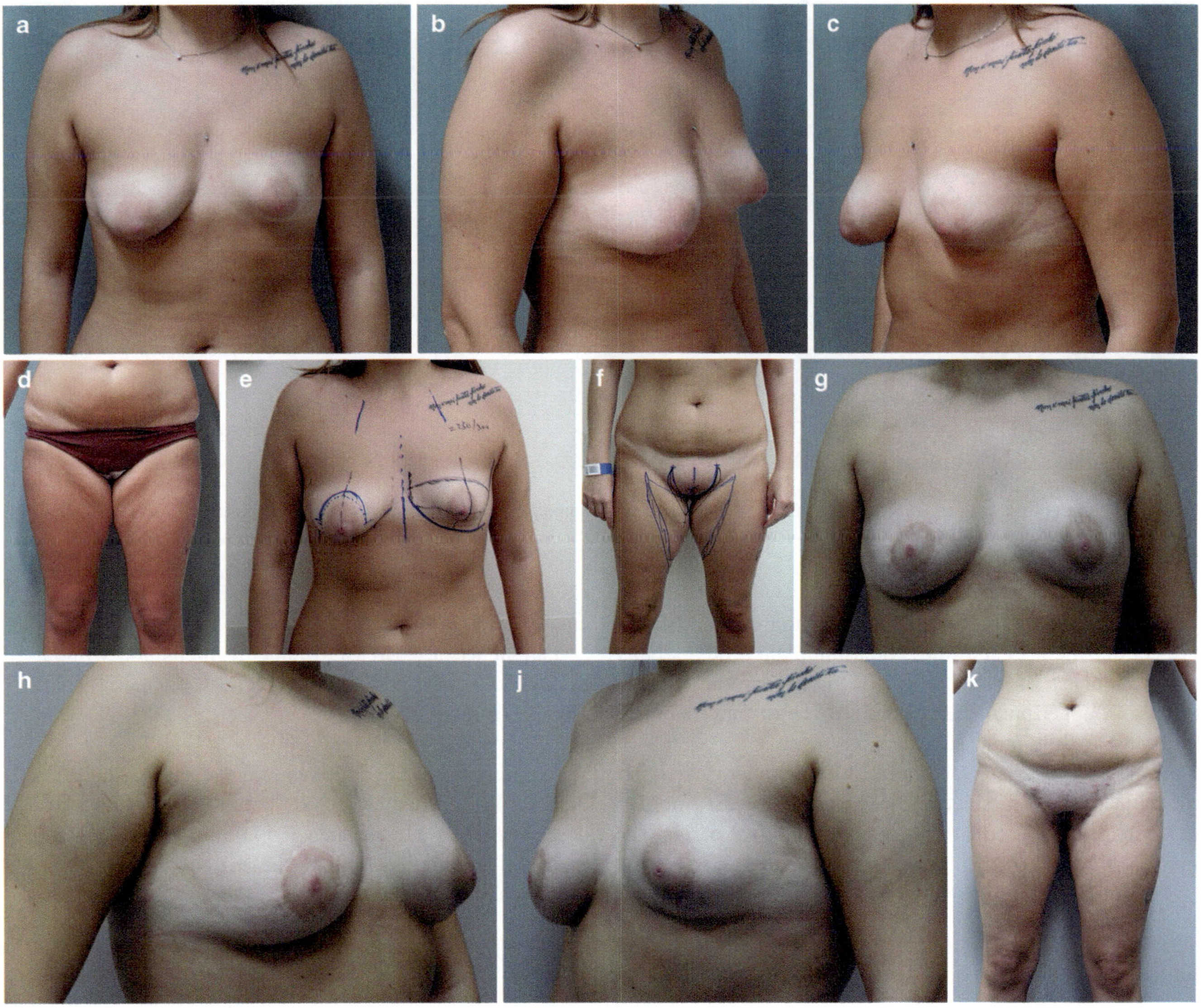

Fig. 36.1 Breast asymmetry with right breast ptosis and left breast hypoplasia, with fat deposits in the inner thighs and skin laxity. (**a–d**) Preoperative. (**e, f**) Operative plan. (**g–k**) Result a 1 year

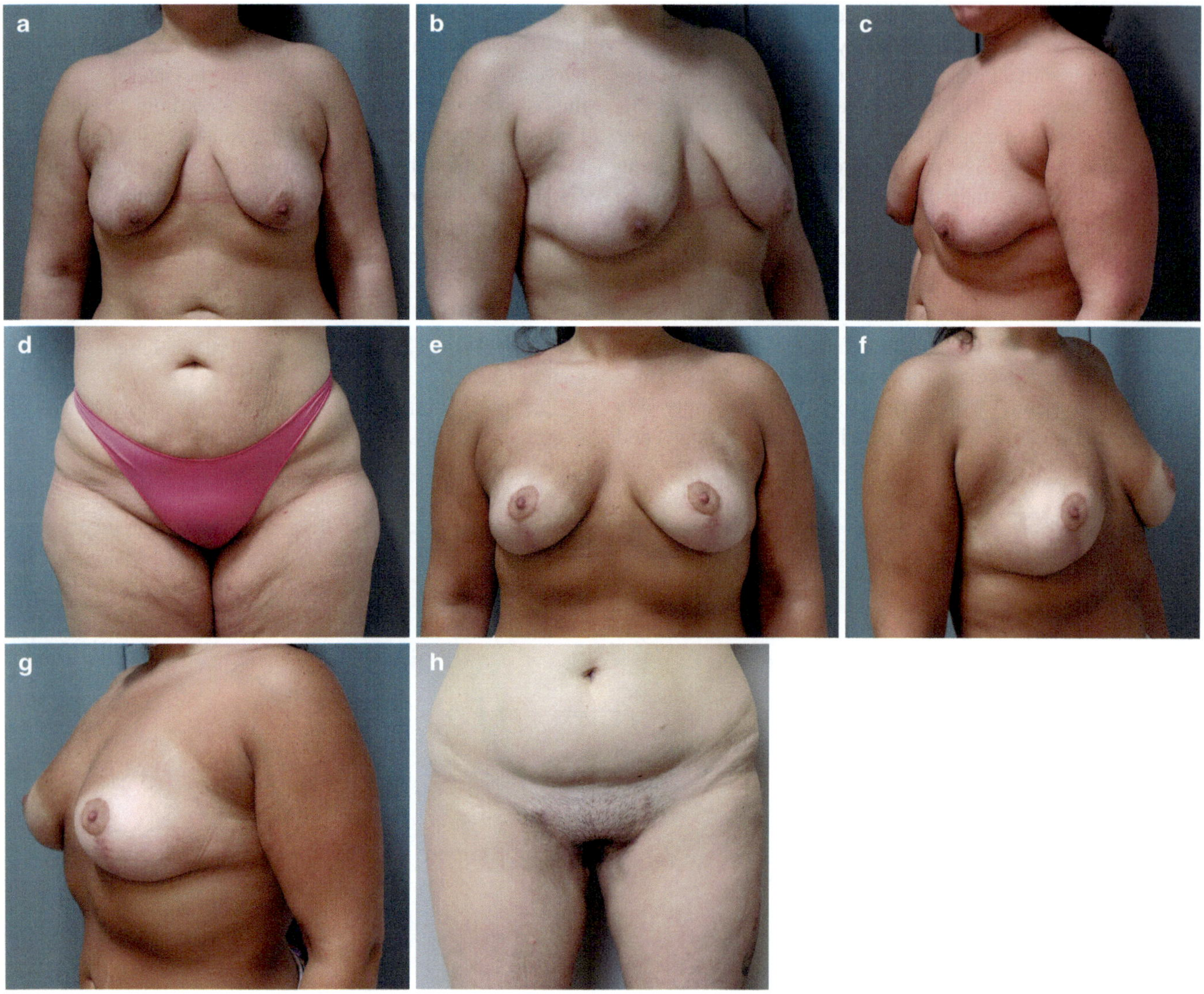

Fig. 36.2 Breast ptosis with an empty skin envelope after 25 kg loss and important fat deposits in the lateral thighs and in the inner thighs with skin laxity. (**a–d**) Preoperative. (**e–h**) Result at 10 months

The planned treatment was:

Left side: Periareolar approach, insertion of a round full projection silicone gel textured breast implant (360 mL) in a sub-glandular pocket, and fat grafting to the anterior axillary pillar and upper/inner breast pole. 290 cc of pure fat were injected basically in the subcutaneous plane.

Right side: Periareolar approach, insertion of a round medium projection silicone gel textured breast implant (220 mL) in a type I dual plane pocket, and final periareolar mastopexy.

Fat donor site: Abdomen, flanks, and inner thigh (900 cc).

Case 4 (Fig. 36.4) 34 years old man (transgender), height cm 178, weight 82 kg.

Patient wishes: Breast volume augmentation without an implant to improve shape and volume despite hormonal therapy and trunk remodelling.

Planned treatment was:

Liposuction of the abdomen, flanks, and back. 1900 mL of fat tissue were collected, prepared according to the described technique and injected into the breast (320 mL in the right side and 340 mL in the left one), basically in the subcutaneous plane.

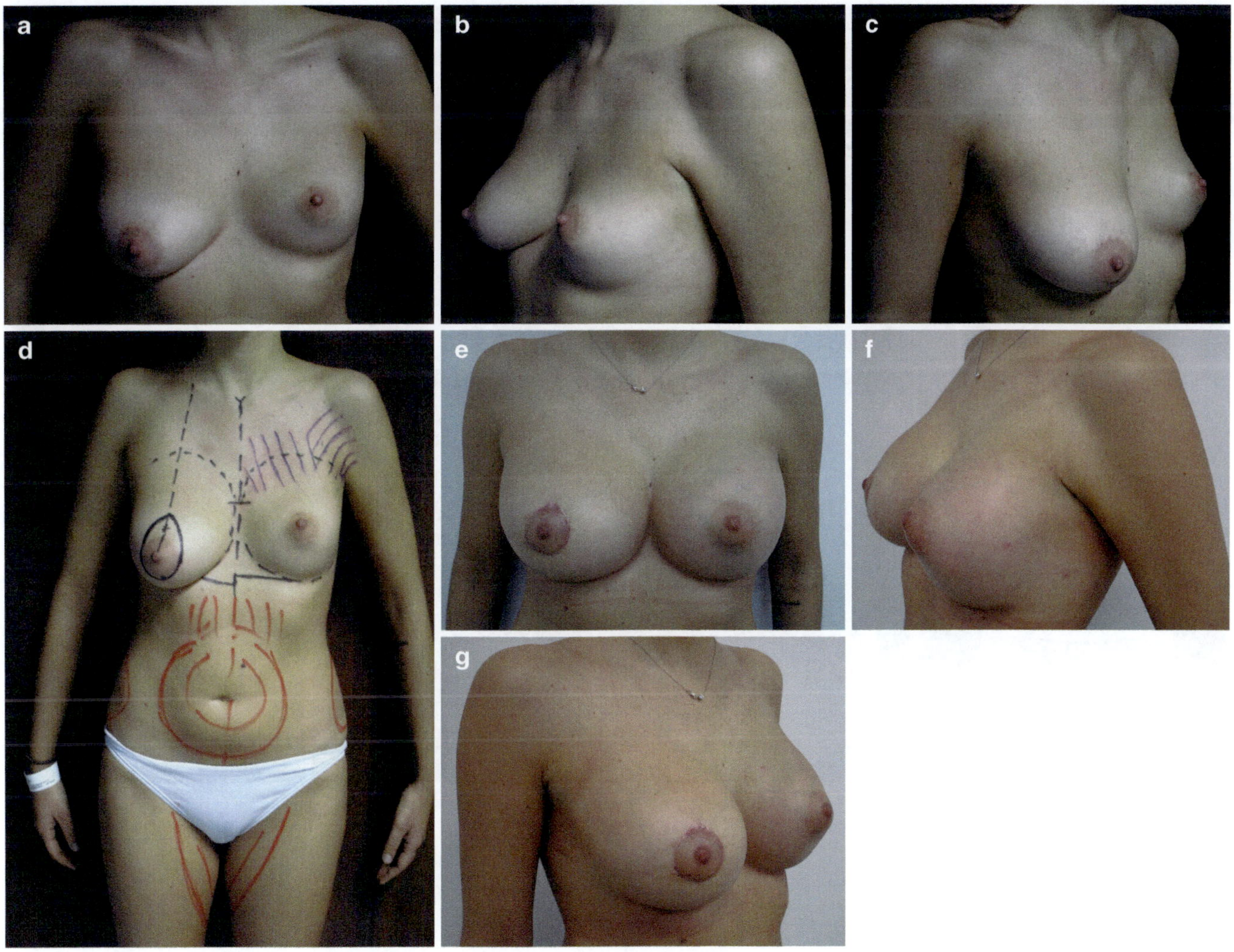

Fig. 36.3 Left Poland syndrome, right ptotic breast wity severe asymmetry. (**a**–**c**) Preoperative. (**d**) Operative plan. (**e**–**g**) Result at 4 months

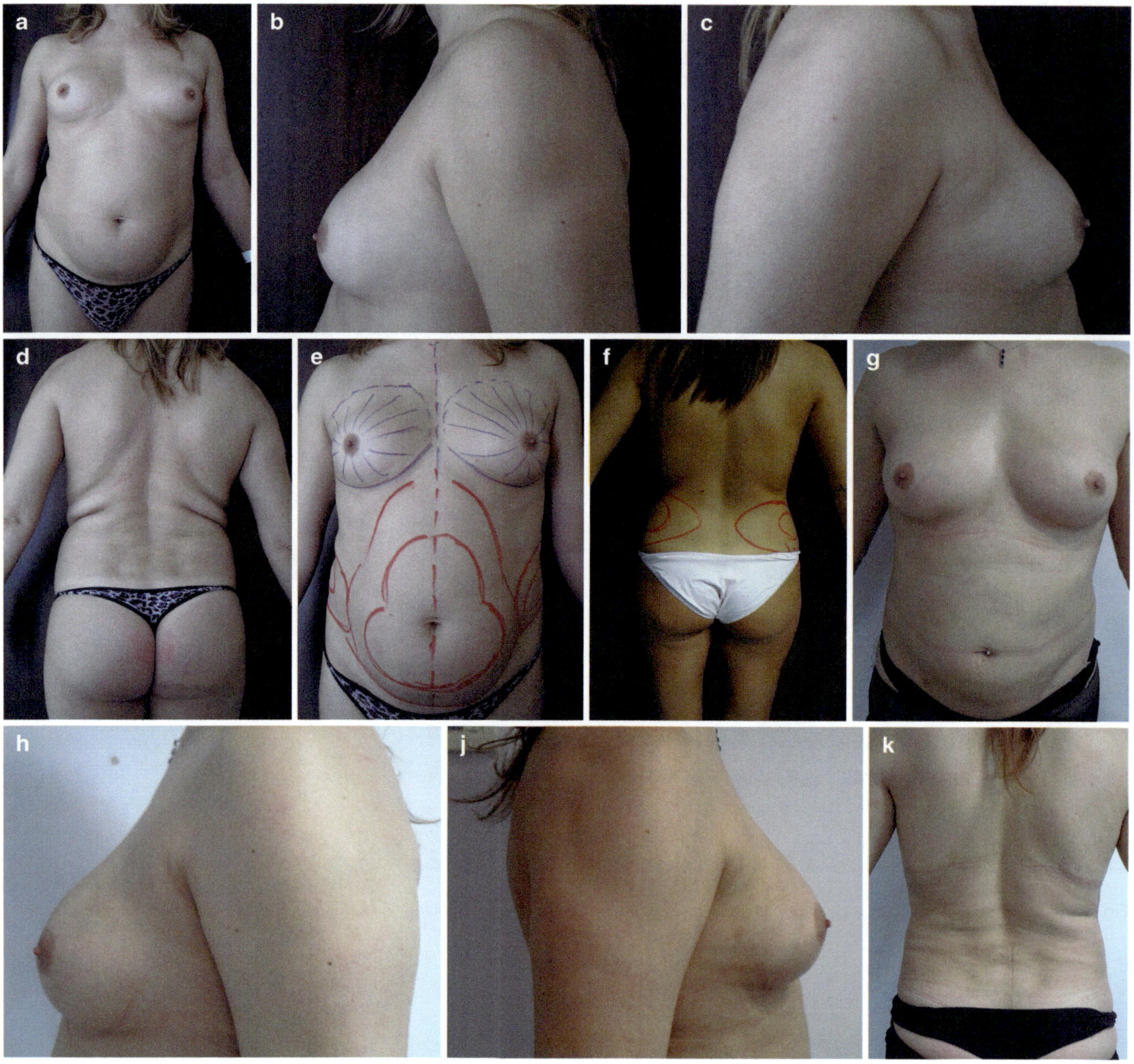

Fig. 36.4 Transgender looking for breast volume augmentation without implants and trunk remodelling. (**a**–**d**) Preoperative. (**e**, **f**) Operative plan. (**g**–**j**) Result at 4 months

References

1. Illouz YG. Une nouvelle technique pour les lipodystrophies localisées. La revue de chir esthet de langue francaise. 1980;4:19.
2. Illouz YG. Body contouring by lipolysis: a 5-year experience with over 3000 cases. Plast Reconstr Surg. 1983;72:591–7.
3. Fournier PF. Microlipoextraction et microlipoinjection. Rev Cir Esthe't Langue. 1985;10:36–40.
4. Illouz YG. The fat cell "graft": a new technique to fill depressions. Plast Reconstr Surg. 1986;78:122–3.
5. Chajchir A, Benzaquen I. Fat-grafting injection for soft-tissue augmentation. Plast Reconstr Surg. 1989;84:921–3.
6. Mazzola RF, Mazzola IC. History of fat grafting: from ram fat to stem cells. Clin Plast Surg. 2015;42:147–53.
7. Rigotti G, Marchi A, Galiè M. Clinical treatment of radiotherapy tissue damage by lipoaspirate transplant: a healing process mediated by adipose-derived adult stem cells. Plast Reconstr Surg. 2007;119:1409–22.
8. Klinger M, Marazzi M, Vigo D. Fat injection for cases of severe burn outcomes: a new perspective of scar remodeling and reduction. Aesthet Plast Surg. 2008;32:465.
9. Khouri RK Jr, Khouri RK. Current clinical applications of ft grafting. Plast Reconstr Surg. 2017;140:466–86.

10. Rigotti G, Chirumbolo S. Biological morphogenetic surgery: a minimally invasive procedure to address different biological mechanisms. Aesthet Surg J. 2019;39:745–55.
11. Coleman SR. Long term survival of fat transfers: controlled demonstrations. Aesthet Surg J. 1995;19:421–5.
12. Coleman SR. Facial recontouring by lipostructure. Clin Plast Surg. 1997;24:347–67.
13. Coleman SR, Saboeiro AP. Fat grafting to the breast revisited: safety and efficacy. Plast Reconstr Surg. 2007;119:775–85.
14. Gutowski KA, Baker SB, Coleman SR, et al. Current applications and safety of autologous fat grafts: a report of the ASPS fat graft task force. Plast Reconstr Surg. 2009;124:272–80.
15. Veber M, Tourasse C, Toussoun G. Radiologic findings after breast augmentation by autologous fat transfer. Plast Reconstr Surg. 2001;127:1289–99.
16. Poulagam SR, Poulton T, Mamounas EP. Long-term clinical and radiologic results with autologous fat transplantation for breast augmentation: case reports and review of the literature. Breast J. 2006;12:63–5.
17. Zocchi M, Zulani F. Bicompartmental breast lipostructuring. Aesthet Plast Surg. 2008;32:313–28.
18. Illouz YG, Sterodimas A. Autologous fat transplantation to the breast: a personal technique with 25 years of experience. Aesthet Plast Surg. 2009;33:706–15.
19. Delay E, Garson S, Tousson G, Sinna R. Fat injection to the breast: technique, results, and indications based on 880 procedures over 10 years. Aesthet Surg J. 2009;29:360–76.
20. Serra Mestre JM, Peñuela RF, Foti V, D'Andrea F, Serra-Renom JM. Breast cleavage remodeling with fat grafting: a safe way to optimize symmetry and to reduce intermammary distance. Plast Reconstr Surg. 2017;140:665e.
21. Bravo FG. Parasternal infiltration composite breast augmentation. Plast Reconstr Surg. 2015;135:1010–8.
22. Pinsolle V, Chichery A, Grolleau JL, Chavoin JP. Autologous fat injection in Poland's syndrome. J Plast Reconstr Aesthet Surg. 2008;61:784–91.
23. Delay E, Sinna R, Chekaroua K, Delaporte T, Garson S, Toussoun G. Lipomodeling of Poland's syndrome: a new treatment of the thoracic deformity. Aesthet Plast Surg. 2010;34:218–25.
24. Walters J, Bourn L, Tessler O, Patterson C, Khoobehi K. Single staged mastopexy with autologous fat grafting: an alternative to augmentation mastopexy with implants. Aesthet Surg J. 2020;40:NP152–8.
25. Groen JW, Negenborn VL, Twisk JW, Ket JC, Mullender MG, Smit JM. Autologous fat grafting in cosmetic breast augmentation: a systematic review on radiological safety, complications, volume retention and patient/surgeon satisfaction. Aesthet Surg J. 2016;36:993–1007.
26. Spear SL, Pittman T. A prospective study on lipoaugmentation of the breast. Aesthet Surg J. 2014;34:400–8.
27. Lin JY, Song P, Pu LLQ. Management of fat necrosis after autologous fat transplantation for breast augmentation. Plast Reconstr Surg. 2018;142:665E–73E.
28. Rohrich RJ, Sorokin ES, Brown SA. In search of improved fat transfer viability: a quantitative analysis of the role of centrifugation and harvest site. Plast Reconstr Surg. 2004;113:391–5.
29. Gir P, Brown SA, Oni G, et al. Fat grafting: evidence-based review on autologous fat harvesting, processing, reinjection, and storage. Plast Reconstr Surg. 2012;130:249–58.
30. Kakagia D, Pallua N. Autologous fat grafting: in search of the optimal technique. Surg Innov. 2014;21:327–36.
31. Lin JY, Wang C, Pu LLQ. Can we standardize the techniques for fat grafting? Clin Plast Surg. 2015;42:199–208.
32. Keck M, Zeyda M, Gollinger K, Burjak S, Kamolz LP, Frey M, Stulnig TM. Local anesthetics have a major impact on the viability of preadipocytes and their differentiation into adipocytes. Plast Reconstr Surg. 2010;126:1500–5.
33. Del Vecchio D, Rohrich RJ. Classification of clinical fat grafting: different problems, different solutions. Plast Reconstr Surg. 2012;130:511–22.
34. Rosing JH, Wong G, Wong MS, Sahar D. Autologous fat grafting for primary breast augmentation: a systematic review. Aesthet Plast Surg. 2011;35:882–90.
35. Voglimacci M, Garrido I, Mojallal A, Vaysse C, Bertheuil N, Michot A, Chavoin JP, Grolleau JL, Chaput B. Autologous fat grafting for cosmetic breast augmentation: a systematic review. Aesthet Surg J. 2015;35:378–93.
36. Agha RA, Fowler AJ, Herlin C, Goodacre TE, Orgill DP. Use of autologous fat grafting for breast reconstruction: a systematic review with meta-analysis of oncological outcomes. J Plast Reconstr Aesthet Surg. 2015;68:143–61.

Penis Enhancement and Reshaping with Autologous Fat

37

Littara Alessandro Giuseppe and Melone Roberto

Contents

37.1 Introduction

Male genital image is correlated, albeit not in a necessarily linear manner [1], to overall body image, psychosocial variables and sexual health [2]; in turn, sexual health is correlated to genital image [3]. Concern over genital endowment has archaic roots [4, 5]. It typically emerges during adolescence [6, 7] and is triggered more by comparison among men than by the fear of not satisfying the partner [8]. Discomfort over genital size can influence satisfaction and men's sexual function and push them to look for surgical and non-surgical solutions for penis alteration.

37.2 Penile Size

The remarkable differences in the penile measurements reported by various authors can be explained by the methodological differences and the variety of the characteristics, even ethnic, of the populations studied (Tables 37.1 and 37.2). Furthermore, these measurements were rarely conducted on statistically adequate samples. The availability of regulatory data per defined population would be essential not only for diagnostic and therapeutic purposes but also to reassure patients who display feelings of inadequacy [1, 7, 15, 46] and to manufacture correctly sized prophylactics [20]. Penis size is an anthropometric measurement [19] and is correlated to anthropometric measurements such as height, weight, and body mass index (BMI) [19, 29]. These measurements are intercorrelated [29] and they are polygenic traits subject to multifactorial influences [31] (Figs. 37.1 and 37.2).

Men's satisfaction and sexual function is influenced by discomfort over genital size which leads to seek surgical and non-surgical solutions for penis alteration. We report the results of a retrospective study of 355 cases of enlargement phalloplasty. We found a significant improvement in circumference at rest at 2, 6, and 12 months post-surgical procedure (all $p < 0.0001$). This study is clinically relevant due to the large cohort of patients.

This study was registered on 04/04/2017 (ISRCTN number: ISRCTN60774878), and 355 men participated in this retrospective clinical study. They came to our center in Milan (Italy) for a cosmetic phalloplasty. The patients' medical history was gathered, and they underwent a medical examination that included an objective examination of the external genitals and the prostate, routine blood tests, basal penile ultrasound scan to verify the presence of nodules, plaques, or

L. A. Giuseppe (✉) · M. Roberto
Sexual Medicine Center, Blumar Medica, Milan, Italy
e-mail: info@robertomelone.it

© The Author(s), under exclusive license to Springer Nature Switzerland AG 2023

A. Di Giuseppe et al. (eds.), *Fat Transfer in Plastic Surgery*, https://doi.org/10.1007/978-3-031-10881-5_37

Table 37.1 Global published data of mean penile size (excluding self-reported measurements)

First author, year	Country	*N*	Age (years; range)	FPL (cm)	SPL (cm)	EPL (cm)	FPG (cm)	EPG (cm)
Loeb (1899) [9]	Germany	50	17–35	9.41				
Schonfeld (1942) [10]	USA	71	18–19		13.11		8.50	
		54	20–25		13.02			
Kinsey (1948) [11]	USA	2770	20–59	9.07	15.05			
Aimani (1985) [12]	Nigeria	320	17–23	8.16			8.83	
Bondil (1992) [13]	France	905	17–91	10.07	16.74			
Da Ros (1994) [14]	Brazil	150				14.05		11.92 Proximal 11.05 Distal
Wessells (1996) [15]	USA	80	21–82	8.85	12.45	12.89	9.71	12.3
Smith (1998) [16]	Australia	184				15.71		
Chen (2000) [17]	Israel	55	21–78	8.03	12.05	13.06		
Adaikan (2000) [18]	Singapore	208		8.05		12.03	8.0	
Ponchietti (2001) [19]	Italy	3300	17–19	9.0	12.5		10.0	
Schneider (2001) [20]	Germany	111	18–19	8.60		14.48	9.68	
		32	40–68	9.22		14.48	9.02	
Sengezer (2002) [21]	Turkey	200	20–22		8.98			
Shah (2002) [22]	UK	104	17–84		13.0			
Spyropoulos (2002) [23]	Greece	52	19–38		12.18			
Son (2003) [24]	Korea	123	19–27	6.9	9.6		8.5	
Savoie (2003) [25]	USA	124	59.1 (avg.)	9.0	13.0			
Pereira (2004) [26]	Portugal	498	20:26	9.85	15.14		9.39	
Awwad (2005) [27]	Jordan	271	17–83	9.3	13.5		8.98	
		109	22–68	7.7	11.6	11.8		
Mehraban (2007) [28]	Iran	92	20–40		11.58		8.66	
Promodu (2007) [29]	India	500	18–60	8.21	10.88	13.01	9.14	
Kamel (2009) [30]	Egypt	949			12.9		8.9	
		78		11.2			8.8	
Nasar (2011) [31]	Egypt	1000		8.37	13.77		10.48	
Khan (2012) [32]	Scotland	609	16–90	10.2	14.3			
Söylemez (2012) [33]	Turkey	2276	18–39	8.95	13.98		8.89	
Chen (2014) [34]	China	5196		6.5	12.9		8.0	
		311≈				12.9		10.5
Shalabi (2015) [35]	Egypt	2000	22–40		13.84			
Veale (2015) [36]	UK	15,521	17–91	9.16		13.12	9.31	11.66
Habous (2015) [37]	Saudi Arabia	778	20–82			12.53/14.34	11.50	
Salama (2016) [38]	Egypt	239		7.4		11.8	8.7	11.3
Hussein (2017) [39]	Afghanistan	223		9.8	12.6			

FPL flaccid penile length, *SPL* stretched penile length, *EPL* erect penile length, *FPG* flaccid penile girth, *EPG* erect penile girth, *(avg.)* average

Table 37.2 Global published data of mean penile size (self-reported measurements only)

First author, year	Country	*N*	Age (range)	FPL (cm)	SPL (cm)	EPL (cm)	FPG (cm)	EPG (cm)
Richters (1995) [40]	Australia	156				15.99		
Bogaert (1999) [41]	USA	935	30 (avg.)	10.41	16.4	16.4	9.65	12.57
		4187	30 (avg.)	9.83		15.6	9.40	12.19
Harding (2002) [42]	UK	312				15.25	12.55	
Schaeer (2012) [43]	Middle East	804				15.6		
Herbenick (2013) [44]	USA	1661	17–91			14.15		12.23
Shaeer (2013) [45]	USA	1133	52.38 (avg.)	13.1	15.6	16.3	10.6	

FPL flaccid penile length, *SPL* stretched penile length, *EPL* erect penile length, *FPG* flaccid penile girth, *EPG* erect penile girth, *(avg.)* average

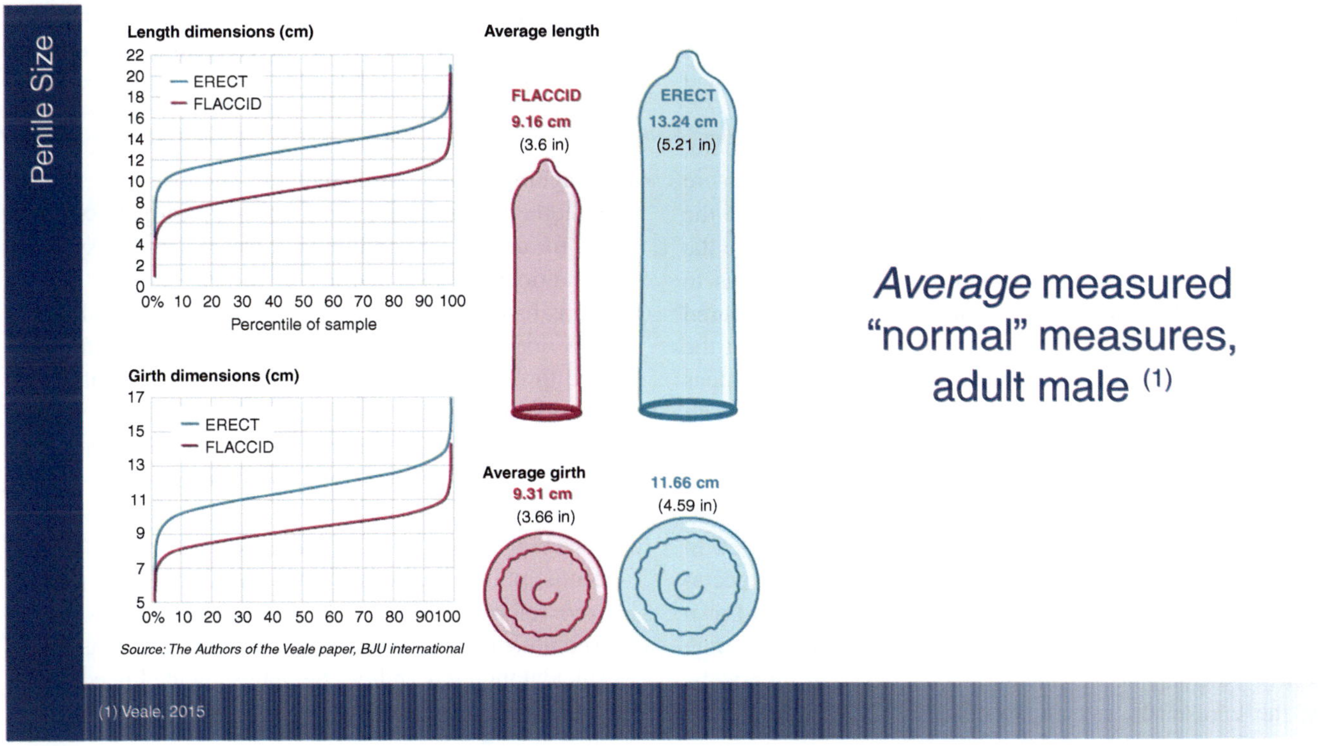

Fig. 37.1 Average measured normal measures, in adult male

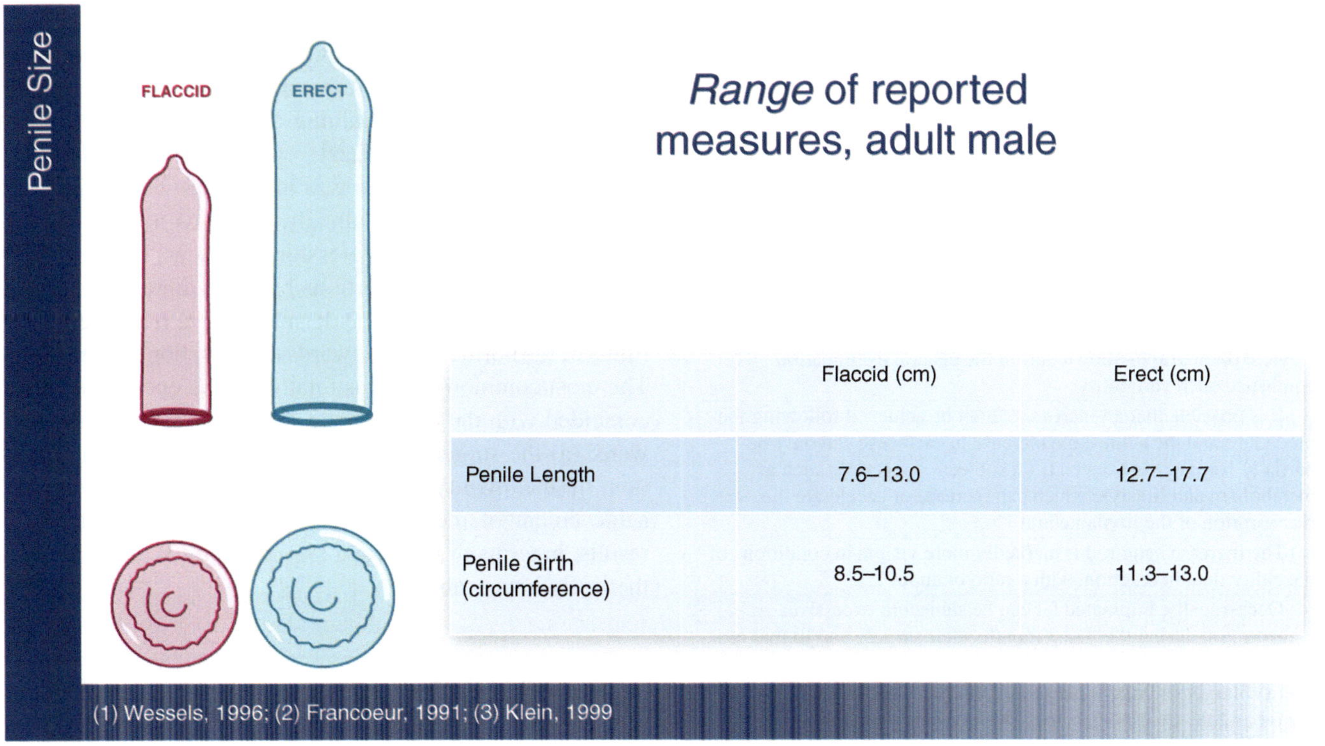

Fig. 37.2 Range of reported measure, adult male

lesions in the internal tissues of the penis and measurement of the circumference of the penis at rest (flaccid). All patients signed the informed consent to undergo the procedure and for the video to be published.

Measurement was always performed in the same room, by the same operator, and using the same flexible measure after a brief introductory interview, performed to put the patient at ease. The measurement was performed before the ultrasound scan to avoid variations caused by changes in temperature. The circumference of the penis was measured at rest at mid-shaft. In all cases (100%), we found that the measurements were coherent with the morphometric values of reference of adult men according to Wessels and Ponchietti [15, 19] and this information was shared with the patients. After measuring height and weight using routine methods employed in the clinical setting, the general medical examination continued with an in-depth interview conducted in order to investigate the patients' motivations and his expectations, discuss the foreseen method and the results and provide in-depth answers to the patients' questions. A meeting between the patients and the anesthetist occurred separately. At the end of the general examination, patients received instructions to be followed the night before and the morning prior to the surgical operation. In addition, we gave our availability to answer the patients' questions at any time until the procedure. The information summarized in Table 37.3 was also discussed with all the patients during the general medical examination.

Table 37.3 Information regarding the enlargement phalloplasty discussed with the patients during their general examination

Elements discussed by the physician with the patients during the general medical examination
(a) The estimated results given by our center (+20–35% circumference) refer to an increase between a minimum and maximum obtained from a historic average of all the patients operated. The availability of a vast collection of pre- and post-surgical photographs shown during the general examination confirmed such variability
(b) It is possible that an increase cannot be achieved following the procedure and the achievable increase in each case can only be partially foreseen and depends on subjective variables such as metabolism and lifestyle which can increase or accelerate the reabsorption of the implanted fat
(c) The increase acquired is markedly more visible in conditions of flaccidity than in erection, with a ratio of about 2:1
(d) Occasionally, implanted fat can be subject to excessive reabsorption during the first 3 months after surgery and in that case, if the patient wishes, a new definite transplant can be performed
(e) In enlargement phalloplasty the different consistency between the fat and the cavernous bodies causes a change in the tactile consistency of the penis; along the shaft, such change is progressive so that no "steps" are felt, and there is no variation in the quality of the erection or local sensitivity
(f) Exceptionally, nodularity can occur in the implanted fat; such nodularity is, however, transitory and almost always resolves spontaneously

The enlargement cosmetic phalloplasty candidate is a healthy and potent man with no congenital or acquired abnormalities or urogenital diseases. In this study, exclusion criteria were:

(a) Coagulopathies, cardiopathies, neoplasies, chemo-radiotherapy, infections in progress, prior pelvic surgeries for urogenital conditions or trauma, severe systemic conditions and psychiatric conditions
(b) Unrealistic expectations; patients who requested results superior to those declared by the center or who felt entitled to obtain the maximum penile increase within our historic series were excluded
(c) Revision surgery; patients requesting a re-operation because of the failure of a previous cosmetic phalloplasty were excluded
(d) True hypoplasia (micropenis) defined as length <2.5 percentile points according to Mondaini [6] (these patients were referred to an andrology center)
(e) Significant anxiety, distorted body image, a history of suicidal thoughts and/or attempted suicide linked to presumed genital inadequacy with psychogenic sexual dysfunction.

The time that elapsed between the first examination and the surgical procedure was 2–6 months. Among the motivations for seeking this surgical procedure, the most frequently cited by patients were psychological discomfort in homosocial situations, discomfort towards women—almost always linked to one or more devaluing observations made during intimacy, the desire to "dazzle" women, the well-founded perception that genital size was incoherent with their body, the desire to improve an already generous natural endowment for narcissistic or professional reasons, the desire for better correlation or proportions between dimensions at rest and during erection, and the desire to move from the lowest limits of the normal range towards the morphometric median. The most common concerns relative to the operation, which coincided with the patient expectations from the procedure, were: (a) the surgery "beyond suspicion," (b) the preservation of the quality of erection and local sensitivity, (c) achievement of the mathematical average of the declared results, in terms of the circumference, and (d) the result aesthetically impeccable.

37.3 Anesthesia

The choice of anesthesia for cosmetic phalloplasty must be in line with the criteria of clinical adequacy, minimum invasiveness, and rapid discharge. Among the different choices of anesthesia, a vast array of scientific documentation [47] exists to support the decided clinical advantages of sedation methods associated with local and loco-regional anesthesia techniques. On the basis of such scientific support, we have opted for the following anesthesia protocol.

37.3.1 Sedation

Premedication: Midazolam 0.04–0.05 mg/kg.

Induction: Fentanyl 0.7–0.8 g/kg + Propofol 0.8–1.6 mg/kg.

Maintenance: Propofol 0.3–0.5 mg/kg/h.

Only in rare cases ($n = 6$) it was necessary to use additional amounts of Propofol (0.5–0.8 mg/kg) and/or Fentanyl (0.4–0.8 g/kg) to guarantee adequate sedation.

37.3.2 Local Anesthesia

Anesthesia in the pubic and penile region was executed by the surgeon using deep infiltration in the zone of the suspensory ligament of the penis and the cutaneous/sub-cutaneous zone affected by the surgical aggression:

Lidocaine 2%, 20 mL.

Mepivacaine/carbocaine 2%, 10 mL (total solution 30 mL).

10 mL of the above-mentioned solution was used in its pure form for cutaneous and deep peri-nervous infiltration, while the same was diluted in 230 mL of 0.9% sodium chloride with 1 mg epinephrine (1/250,000) for infiltration in the subcutaneous region where adipocytes will be harvested. In our experience, such procedure resulted in being fully ideal for allowing surgical treatment, devoid of complications and major side effects, widely liked by patients and guaranteed brief protected discharge times (180 ± 30 min).

37.4 Surgical Procedure

37.4.1 Fat Harvesting and Purification

Prior to the operation, the patients were photographed while standing. The operation began after disinfection of the skin, with the harvesting of the adipose tissue. This was performed by explanting fat bilaterally from the thighs if the patient was tendentially thin and from the periumbilical region if the patient was normo-weight or overweight and from the suprapubic region if there was any localized adiposity. This latter area of harvesting permitted, in certain cases, the reduction of the suprapubic adipose panniculus (suprapubic lipectomy) rendering the point of insertion of the penis deeper and visually increasing the length of the external portion of the penis.

Thereafter infiltration of the donor site was performed with a tumescent solution. After a few minutes of waiting, necessary to consolidate the vasoconstrictor effect of the epinephrine, adipose explant was performed using a thin cannula (2 mm) and a 10 cc Luer-lock syringe (Fig. 37.3). The quantity of fat explanted varied from subject to subject on the basis of the volume to be filled, but it was never less than 80 mL. That volume comprised infiltration material which was then removed by decantation first and then centrifugation later. Such a process of purification is of primary importance since it determines the percentage integration of fat in the penis. In our surgical center, we first performed the decantation through sedimentation of each 10 cc syringe in such a way as to put the harvested material through an initial process of purification. Each syringe was filled with fat again and each time the infiltration material was removed, repeating the decantation by sedimentation process many times. Once a seemingly stable mixture was obtained, the syringes of crudely purified fat underwent centrifugation for 2 min at 1000 rpm (Fig. 37.4). Reducing the time and the number of rpm, with respect to the original Coleman's technique, which involves centrifugation for 3 min at 3000 rpm, the integrity of the adipose globules whose integrity is in turn responsible for the good integration of the fat was safeguarded.

37.4.2 Fat Transfer

Once the penile elongation operation had been performed, the test tubes containing the purified adipose material were extracted. They typically contained three layers: the most superficial was oily, the middle contained the purified fat and the lower was made up of blood and infiltration material [48]. The inferior and superior layers were eliminated and the purified material was implanted. Two mini-incisions of about 4 mm were made close to the pubo-penile junction at 10 o'clock and 2 o'clock, respectively. The purified adipose material contained in the 10 cc syringes was decanted using a specific connector into 2.5 cc syringes better suited for the job. A blunt-tip cannula 2 mm in diameter was used for the implantation. The purified fat was then implanted into the subdartoic space (Fig. 37.5) taking care of the tunnel using the cannula and arranging the implant symmetrically. The space addressed was relatively avascular, and, as a result, the formation of localized hematomas was rarely observed. In the few cases where hematomas were observed, bandaging was applied.

Adverse events are summarized in Table 37.4. The quantity to be implanted varied considerably depending on the space to be filled, also considering that 30% of the implant would be resorbed within the first/second month. Therefore, we decided to inject a modestly superior quantity to take into account its predicted partial resorption. At the end of the implantation, the surgical wound was closed and sutured

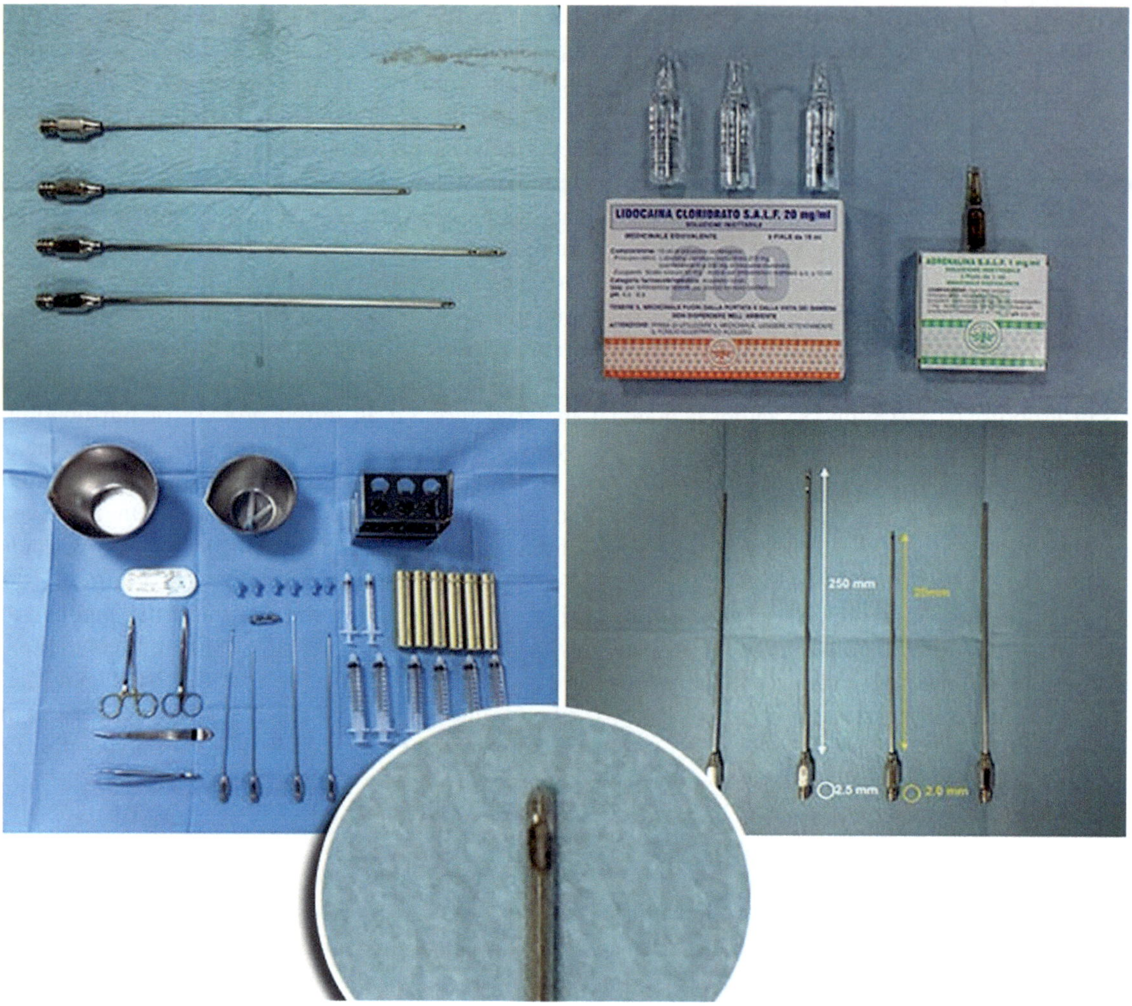

Fig. 37.3 Surgical tray. Adipose explant was performed using a thin cannula (2 mm) and a 10 cc Luer-lock syringe

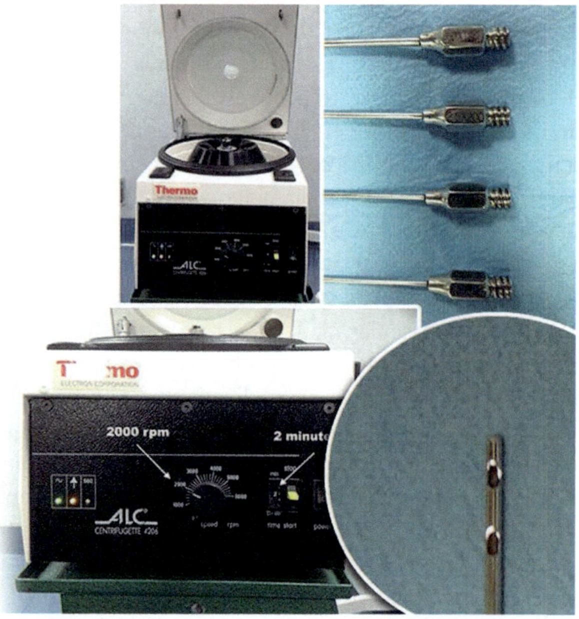

Fig. 37.4 Once a seemingly stable mixture was obtained, the syringes of crudely purified fat underwent centrifugation for 2 min at 1000 rpm

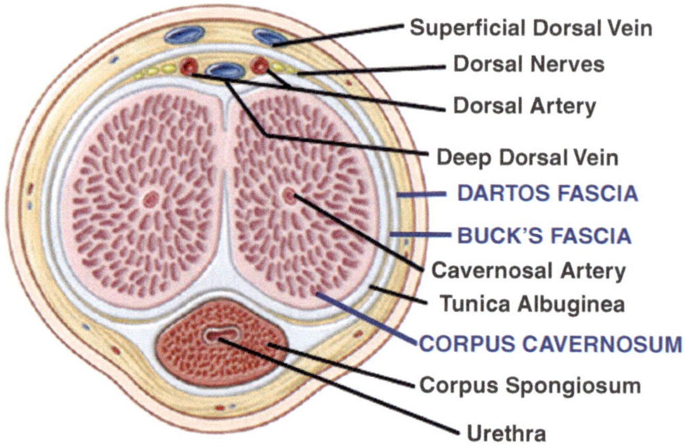

Fig. 37.5 The purified fat is implanted into the subdartoic space, between Dartos fascia and Buck's fascia

Table 37.4 Summary of adverse events

Adverse events	(%)
Loss of erectile function	0
Decrease of erectile function (temporary)	0
Penile edema	8
Long-standing hematoma	4
Seroma	3
No increase in girth	7
Fat loss (>30%)	32
Fat nodules, fat lumps	3
Fat migration	1
Sclerosing lipogranuloma	1
Loss of sensation (mild)	2
Fibrosis	0
Deep infection	2
Paradoxical penile shortening	0
Penile deformity	3
Penile asymmetry	4

using resorbable thread, a maneuver of manual "kneading" of the penis was then performed [49] to aid in the uniform distribution of the implanted fat, and finally, a cohesive elastic bandage of adequate thickness was applied. The bandage has the important function of preventing the formation of crude asymmetries caused by posture and/or frequent erections during the first month after the operation. In fact, statistically at least 30 days are needed for the implant to be consolidated and the fat integrated, and it is useful to limit the movement of the fat during this period using the elastic bandage. At the end of the operation, a modestly compressive dressing was applied, and ice locally. The patient was discharged that evening with directions for medical therapy at home and adequately informed of the recovery period. In particular, the patient was urged to abstain from sexual and masturbatory activity for 30 days. The duration of the operation was recorded from the moment of sedation until the final suture and it was about 40 min.

Table 37.5 Baseline descriptive statistics of patients' demographics

	Age (years)	Weight (kg)	Height (cm)	Baseline circumference at rest (cm)
Number of values	355	355	355	354
Minimum	19	56	167	5
25% Percentile	29	68	173	7.5
Median	36	74	178	8.3
75% Percentile	46	81	181	9.1
Maximum	63	99	192	13.3
Mean	38.08	75.13	177.4	8.377
Std. deviation	10.81	8.969	5.126	1.213
Std. error of mean	0.5737	0.476	0.2721	0.06445
Lower 95% CI of mean	36.95	74.2	176.8	8.25
Upper 95% CI of mean	39.21	76.07	177.9	8.504

37.4.3 Statistical Analysis

Penis circumference data was analyzed using a one-way analysis of variance (ANOVA) with Dunnett's post-hoc test for comparison of each time point with baseline. Data are presented as mean ± standard error of the mean (SEM). A $p < 0.05$ was considered significant (Table 37.5).

37.5 Results

Following surgical procedure, circumference at rest significantly increased at 2 (11.5 ± 0.09), 6 (11.36 ± 0.09), and 12 (11.06 ± 0.1) months, compared to baseline (8.3 ± 0.06) (all $p < 0.0001$) (Figs. 37.6, 37.7, 37.8, and 37.9).

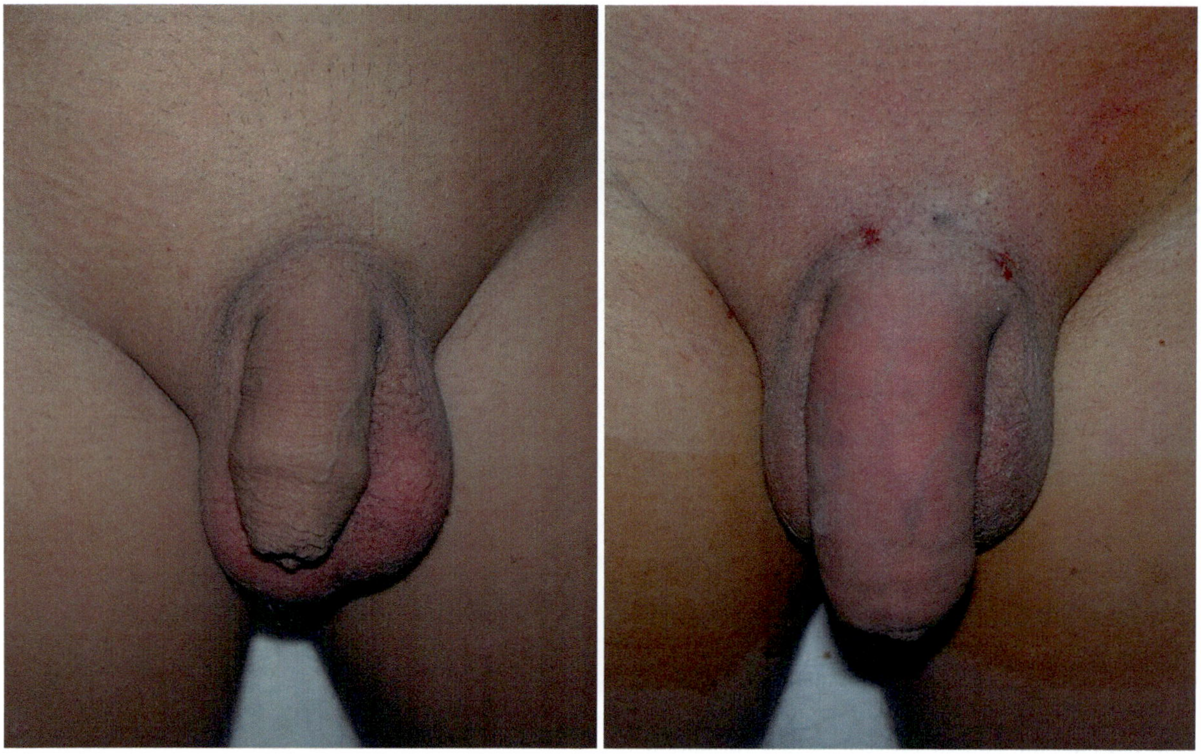

Fig. 37.6 Results: immediately post-op result

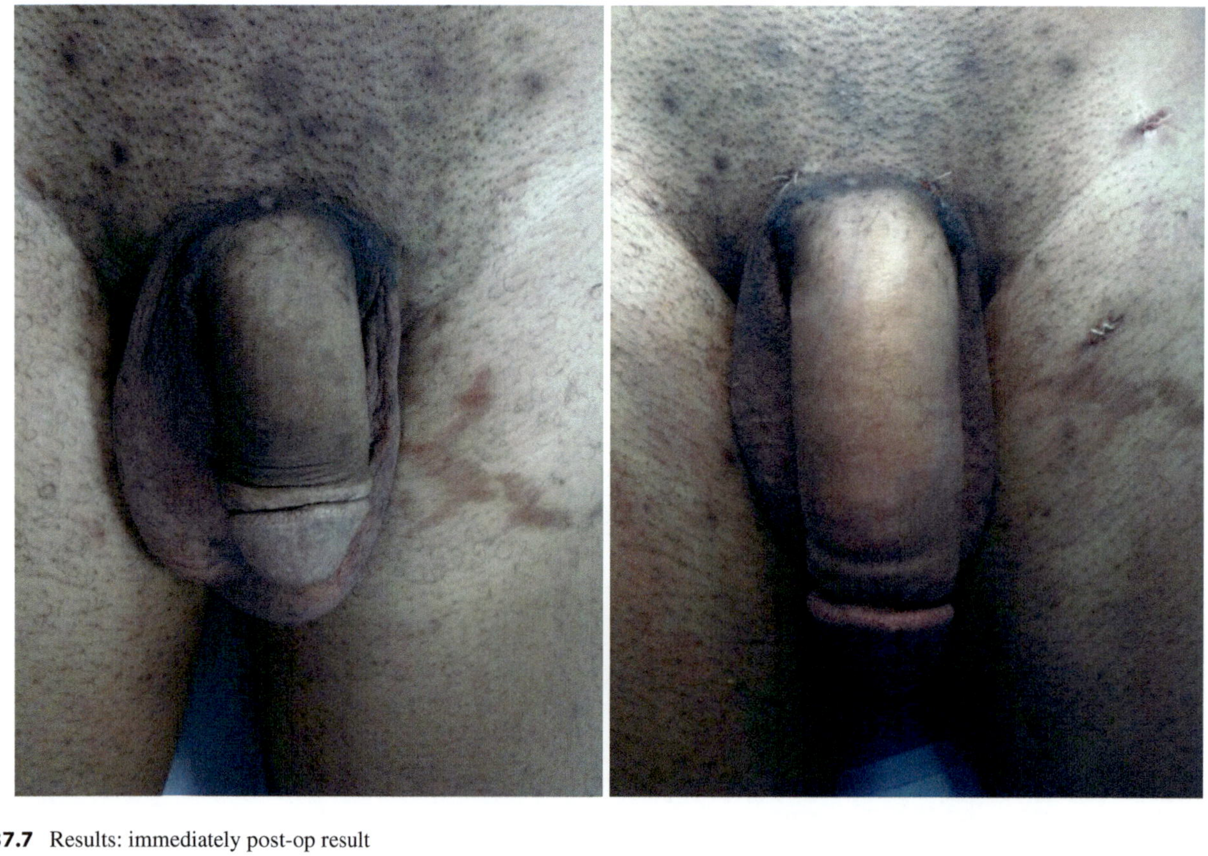

Fig. 37.7 Results: immediately post-op result

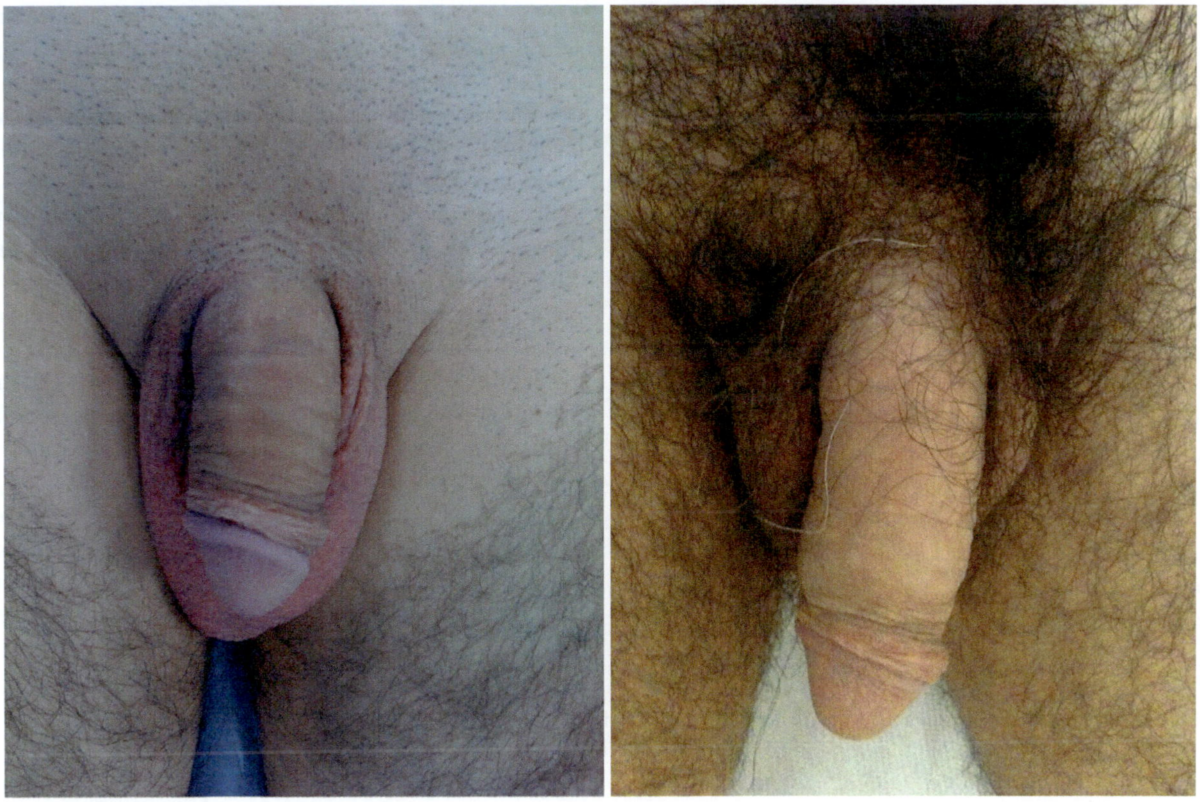

Fig. 37.8 Result: post-op after 12 months

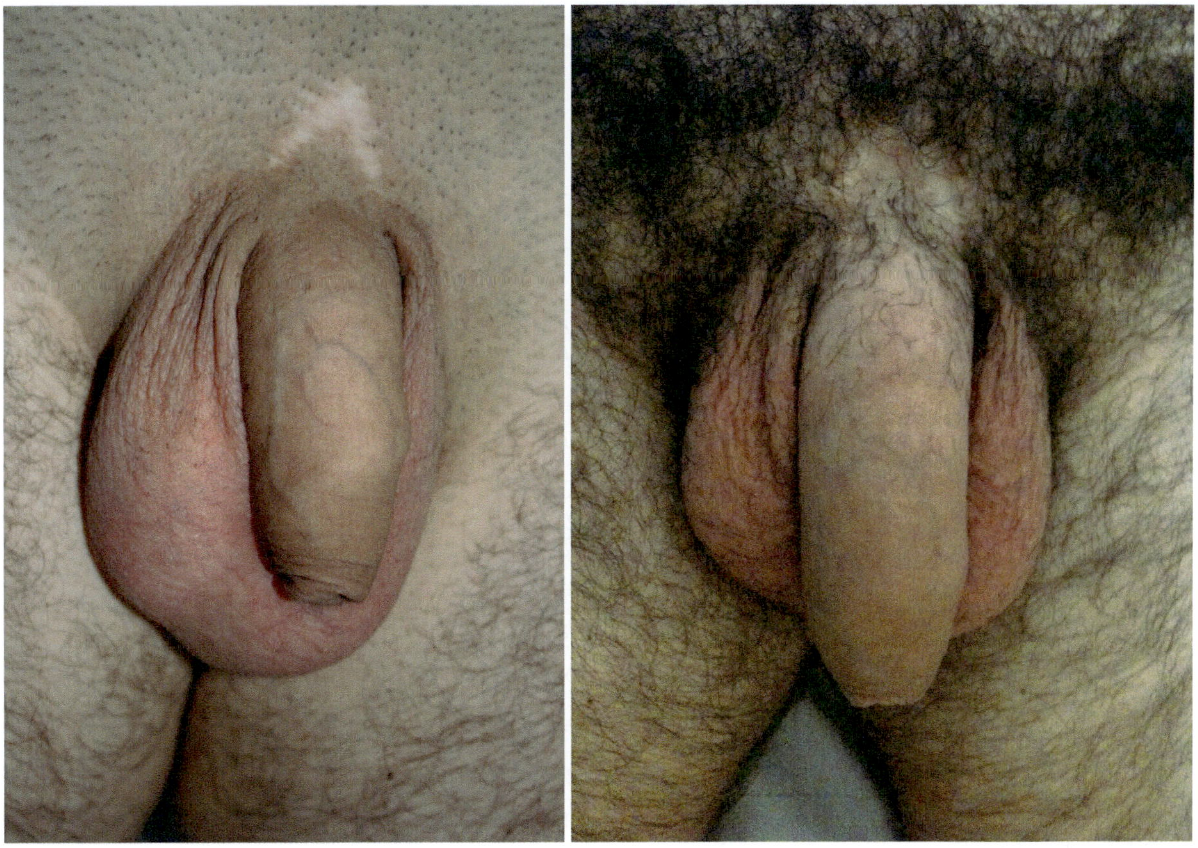

Fig. 37.9 Result: post-op after 12 months

37.6 Discussion

We found that cosmetic phalloplasty significantly improves circumference at rest at 2, 6, and 12 months post-surgery.

The growing demand for autologous fat transplant (AFT) beginning at the end of the 1980s is linked to the advent of liposuction. The current methods of fat transfer were popularised and extensively described by Sydney Coleman [48, 50, 51] who in 1986 began to transplant fat in iatrogenic deformities from liposuction and subsequently in the face. AFT is today a widely tested procedure, appreciated by patients and very widespread among plastic surgeons even for reconstructive surgery [52–58] despite no consensus has been reached regarding the best technique or its success rate.

The fat injection is the most common technique of penile girth enhancement. The fat harvested from the patient is implanted into the subdartoic space with the objective to symmetrically and uniformly increase the circumference of the penis [59].

The inhomogeneities of the surgical techniques and the selection criteria of the patients render it difficult to compare the results obtained by our center with those found in literature and reported from other clinics (an overview of surgical techniques employed for phalloplasty and result obtained is summarized in Table 37.6).

In our experience, cosmetic phalloplasty has evolved in time moving in the direction of increased safety. Patients who undergo combined elongation and girth enhancement phalloplasty are particularly satisfied compared to those who undergo a single operation which is probably linked to the availability of an overall greater penile volume [72]. In Italy, there is no validated test for measuring patients' satisfaction in cosmetic penoplasty, and the absence of a measurement of patients' satisfaction is a limitation of our study.

In line with other authors, we believe that, even in its relative simplicity, cosmetic phalloplasty requires a profound knowledge of anatomy and surgical technique and that the selection of candidates is a fundamental and essential element together with

Table 37.6 Overview of result obtained by phalloplasty surgical techniques

	First author, year	*N*	Method	Follow-up (months)	Gain in length (cm)		Gain in girth (cm)	
					Flaccid	Erect	*Flaccid*	Erect
	Austoni (2002) [60]	39	Autologous saphenous graft	9				1.1–2.1 (∅)
	Perovic (2006) [61]	204	Biodegr. scaffolds coated w/ autologous fibroblasts	24 (*N* = 84)			*3.15*	
	Shaeer (2006) [62]	1	Superficial circumflex iliac artery island flap	6			*9.5*	8.5
	Bin (2009) [63]	20	Saphenous grafts, PTFE artificial vessel patches	1–5 years			*1.0–2.3*	1.5–3.0
	Jin (2011) [64]	69	Biodegr. scaffolds coated w/ autologous fibroblasts	1, 3, 6			*4.01*	2.92
	Alei (2012) [65]	69	Porcine dermis graft	6.12			*3.1*	2.4
Referral action	Wessels (1996) [66]	*12*	*SLD (N = 12) + VYP, FI (N = 10)*					
	Alter (1997) [67]	*30*	*SLD + VYP, FI*					
	Klein (1999) [6]	*58*[a]	*SLD (N = 10), FI (N = 1), SLD + DFG (N = 6), SDL, FI (N = 41)*	*12.2*	*3.0*	*0.76*	2.52	*2.03*
	Roos (1994) [8]	260	SLD, Y (Z/M flap)	4 (*N* = 100)	*4.0*			
	Shirong (2000) [68]	52	SLD + scrotal flap/skin graft, VYP	6 (*N* = 20)	*3.5–6.5*			
	Spyropoulos (2005) [69]	11	SLD, VYP (*N* = 5) SLD, DFG (*N* = 3) SBL, SLD (*N* = 2)		*1.6*		*2.3/2.6*	
	Li (2006) [70]	42	SLD (*N* = 42) + silicone spacer (*N* = 27), VYP (*N* = 17)	16	*1.3 (stretched)*			
	Panfilov (2006) [49]	88	SDL, FI (*N* = 31) FI (*N* = 57)	12	*2.42*		*1.0–4.0*	
	Mertziotis (2013) [71]	82	SLD + VYP, DFF (*N* = 35)	12	*1.92*		*2.21*	
			Circumcision ligamentolysis, DFF (*N* = 47)		*2.11*		*2.0*	
	Monreal (2015) [72]	259°	SLD, FI (*N* = 148) FI (*N* = 127)	6 (*N* = 160) 12 (*N* = 87)	*3.1 3.2*		*1.7 1.6*	
	Xu (2016) [73]	23	SLD + DFT		*2.27*		1.67	

[a]Data obtained from questionnaires administered in several surgical centers by at least 10 different surgeons
°Total number of procedures: 275

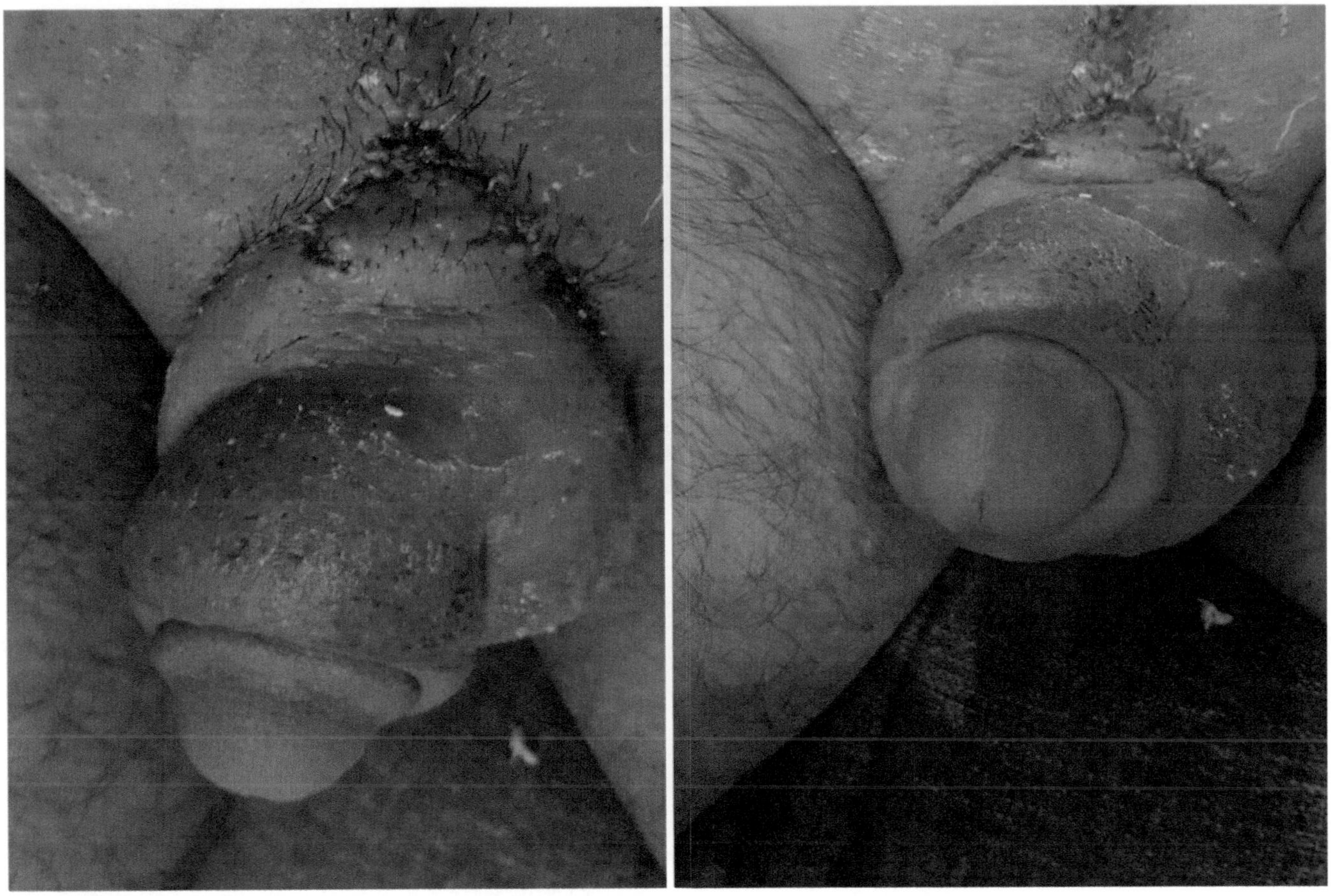

Fig. 37.10 Complication: prepuce edema

scrupulous information regarding not only the operation and the obtainable results but also post-surgical conduct since resuming of sexual activity prior to 30 days after the operation can compromise the results (Figs. 37.10, 37.11, 37.12, 37.13, 37.14, 37.15, 37.16, and 37.17).

While confirming that cosmetic phalloplasty very rarely produces spectacular results and that there is an objective necessity to improve the stability of the fat in time, we retain that the data from our center show that the surgical technique we utilize is safe, repeatable, and produces concrete and measurable results. Finally, the operation, last resort to improve the patient's discomfort, can considerably improve the patient's self-esteem and improve the quality of his sex life and, in turn, his relationships.

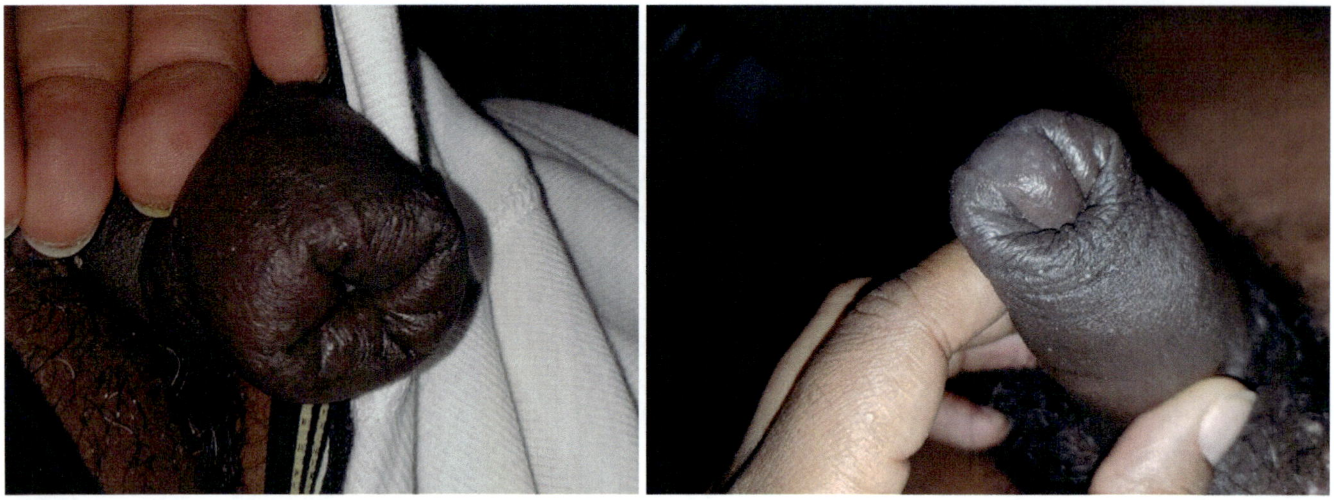

Fig. 37.11 Complication: prepuce edema

Fig. 37.12 Complication: botched fat grafting

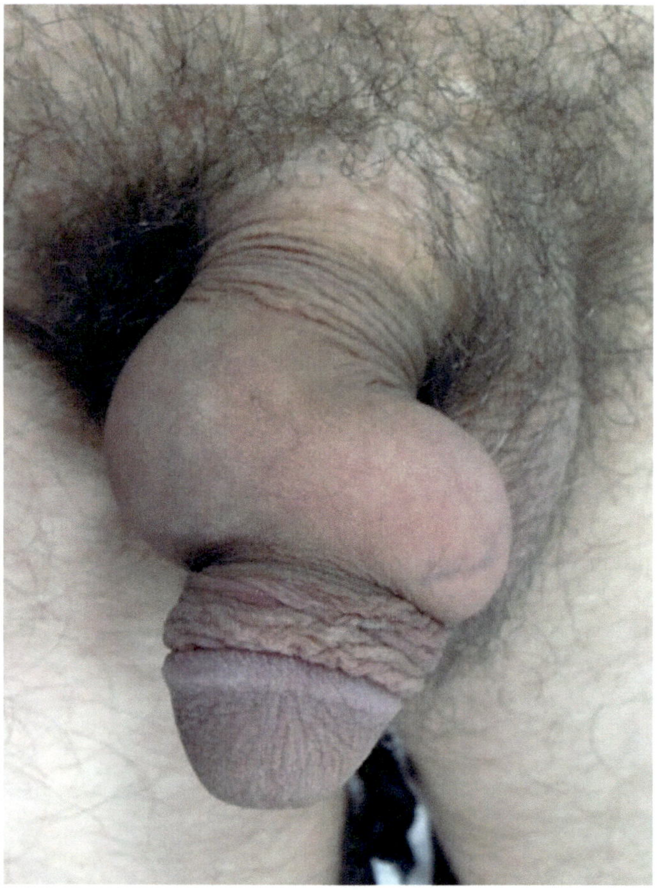

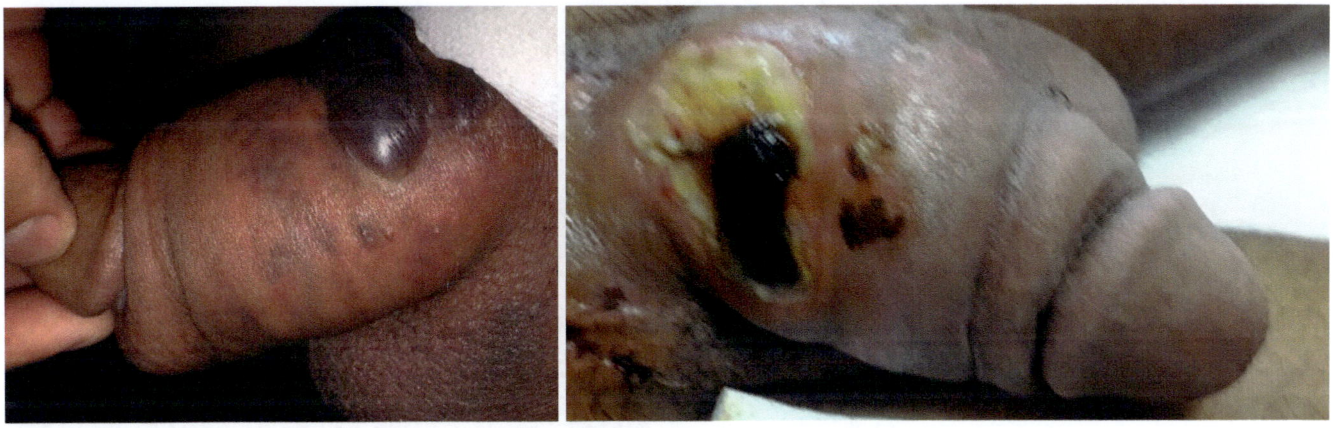

Fig. 37.13 Complication: infection

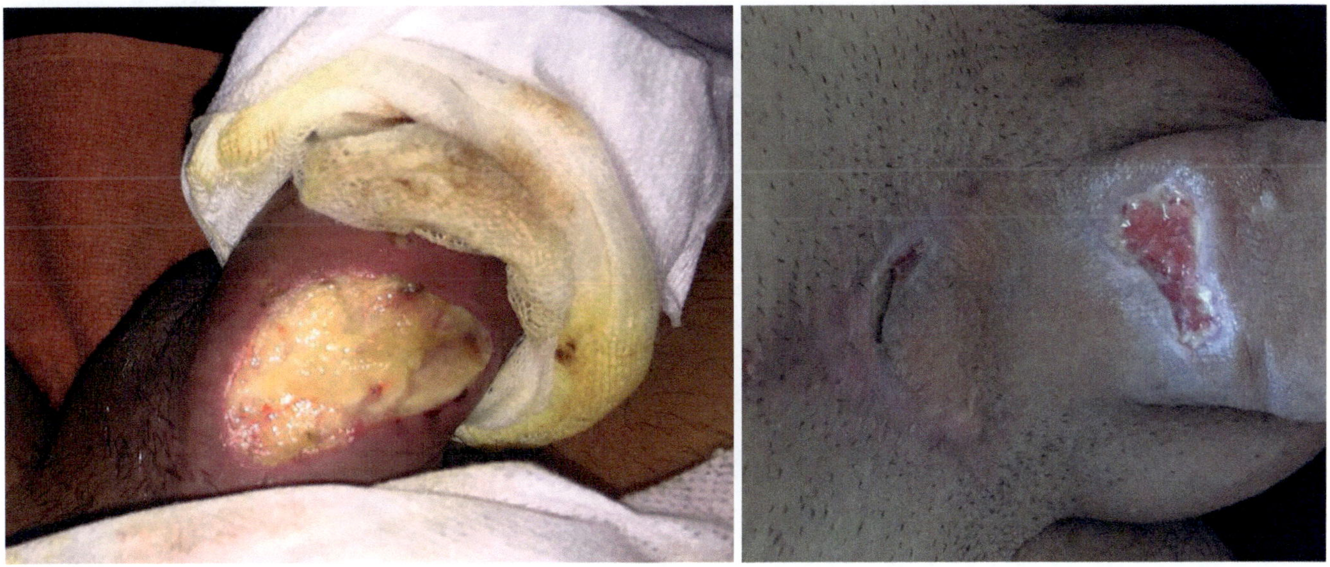

Fig. 37.14 Complication: infection after 15 and 45 days

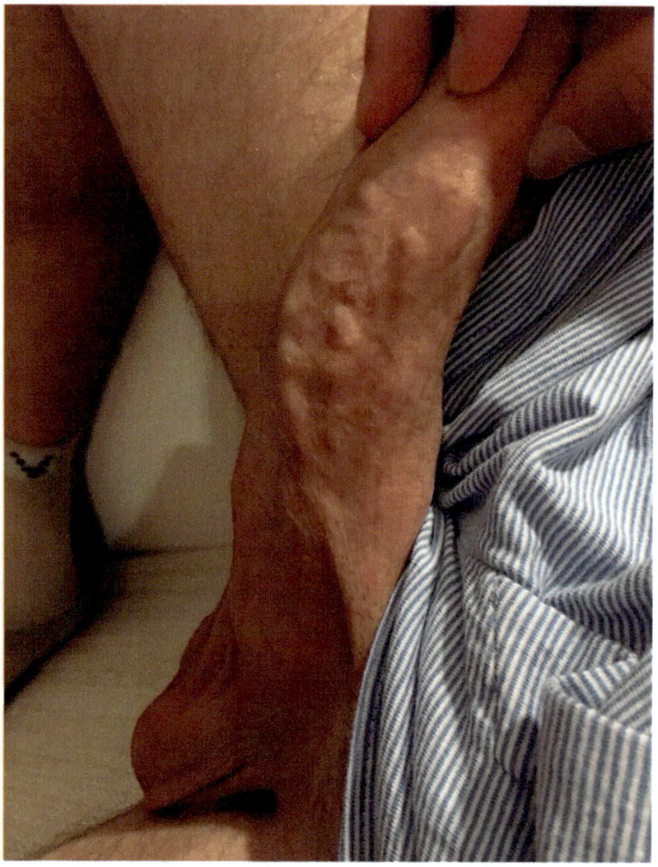

Fig. 37.15 Complication: oil cysts

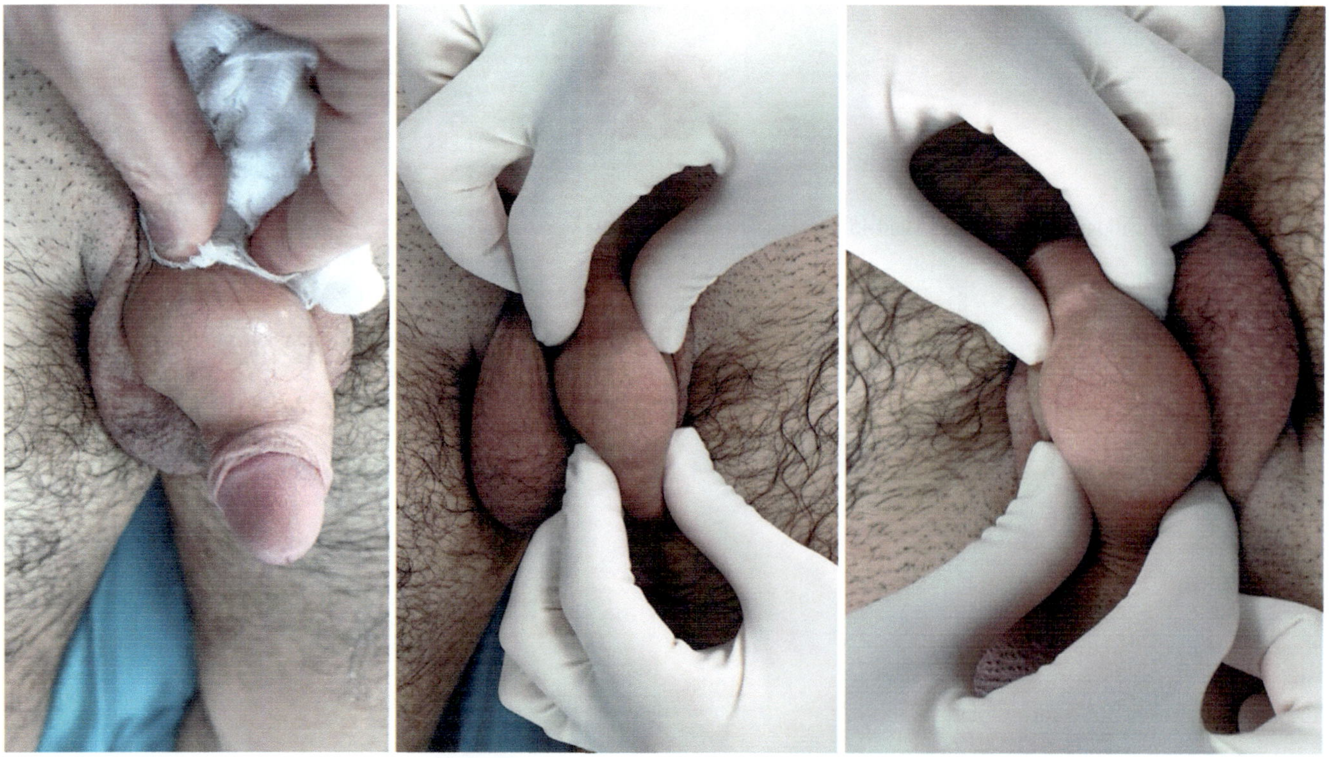

Fig. 37.16 Complication: hyaluronic acid nodules

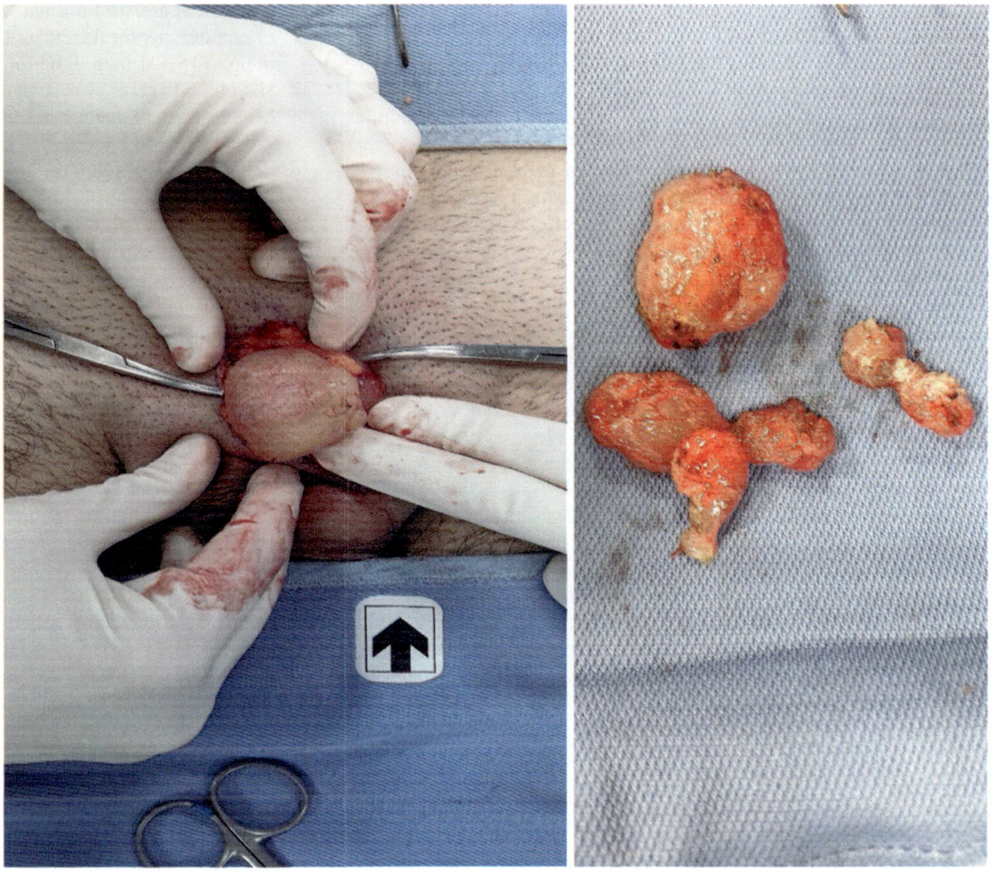

Fig. 37.17 Complication: hyaluronic acid incapsulated nodules, and after removal

37.7 Conclusions

The limited literature regarding cosmetic phalloplasty consists of studies performed using diverse surgical techniques and candidate selection criteria that include patients who should in fact be excluded (e.g., men with psychiatric conditions, namely body dysmorphic disorder) or whose existing conditions (e.g., failure of previous phalloplasty and trauma) make it impossible to compare results. If we consider the lack of universally shared morphometric values, we see how this niche of cosmetic surgery suffers from an inevitable lack of methodological rigor. In the present study, we show the efficacy of cosmetic phalloplasty in a large cohort of patients up to 1-year follow-up. In addition, we describe in detail inclusion and exclusion criteria for patient selection and technical aspects of our surgical procedure which ensure reproducibility of our findings and should be adopted in future clinical studies of cosmetic phalloplasty. We are confident that this study will encourage other authors to publish their experiences with cosmetic phalloplasty and that the method we have described in this article will contribute to the consolidation of a standard for this type of surgery.

Conflict of Interest Statement The authors declare no competing interests.

Permission I confirm that the permission has been obtained from the original copyright holder.

References

1. Lever J, et al. Does size matter? Men's and women's views on penis size across the lifespan. Psychol Men Masculinity. 2006;7(3):129.
2. Davis SN, et al. The Index of Male Genital Image: a new scale to assess male genital satisfaction. J Urol. 2013;190(4):1335–9.
3. Shaeer O, Shaeer K. Impact of penile size on male sexual function and role of penile augmentation surgery. Curr Urol Rep. 2012;13(4):285–9.
4. Francoeur RT, Perper T. A descriptive dictionary and atlas of sexology. Santa Barbara: Greenwood Press; 1991.
5. Cleaveland P, et al. The history of penile enhancement—to cut a short story long. BJU Int. 2015;115:33.
6. Klein R. Penile augmentation surgery: San Francisco, CA. Electron J Hum Sex. 1999;2:1999.
7. Mondaini N, et al. Penile length is normal in most men seeking penile lengthening procedures. Int J Impot Res. 2002;14(4):283.
8. Roos H, Lissoos I. Penis lengthening. Int J Aesthet Restor Surg. 1994;2:89–96.

9. Loeb H. Harnrohren capacitat und tripperspritzen. Munch Med Wochenschr. 1899;46:17.
10. Schonfeld WA, Beebe GW. Normal growth and variation in the male genitalia from birth to maturity. J Urol. 1942;48(6):759–77.
11. Kinsey AC, et al. Sexual behavior in the human male. Philadelphia: Saunders; 1948.
12. Ajmani M, et al. Anthropometric study of male external genitalia of 320 healthy Nigerian adults. Anthropol Anz. 1985;43:179–86.
13. Bondli P, et al. Clinical study of the longitudinal deformation of the flaccid penis and of its variations with aging. Eur Urol. 1992;21:284–6.
14. da Ros C, et al. Caucasian penis: what is the normal size. J Urol. 1994;151(Pt 2):323A–5A.
15. Wessells H, et al. Penile length in the flaccid and erect states: guidelines for penile augmentation. J Urol. 1996;156(3):995–7.
16. Smith AM, et al. Does penis size influence condom slippage and breakage? Int J STD AIDS. 1998;9(8):444–7.
17. Chen J, et al. Predicting penile size during erection. Int J Impot Res. 2000;12(6):328.
18. Munding MD, et al. Pilot study of changes in stretched penile length 3 months after radical retropubic prostatectomy. Urology. 2001;58(4):567–9.
19. Ponchietti R, et al. Penile length and circumference: a study on 3300 young Italian males. Eur Urol. 2001;39(2):183–6.
20. Schneider T, et al. Does penile size in younger men cause problems in condom use? A prospective measurement of penile dimensions in 111 young and 32 older men. Urology. 2001;57(2):314–8.
21. Sengezer M, et al. Accurate method for determining functional penile length in Turkish young men. Ann Plast Surg. 2002;48(4):381–5.
22. Shah J, Christopher N. Can shoe size predict penile length? BJU Int. 2002;90(6):586–7.
23. Spyropoulos E, et al. Size of external genital organs and somatometric parameters among physically normal men younger than 40 years old. Urology. 2002;60(3):485–9.
24. Son H, et al. Studies on self-esteem of penile size in young Korean military men. Asian J Androl. 2003;5:185–9.
25. Savoie M, et al. A prospective study measuring penile length in men treated with radical prostatectomy for prostate cancer. J Urol. 2003;169(4):1462–4.
26. Pereira NM. Estudo do tamanho do pénis na população portuguesa. Rev Int Androl. 2004;2:15–21.
27. Awwad Z, et al. Penile measurements in normal adult Jordanians and in patients with erectile dysfunction. Int J Impot Res. 2005;17(2):191.
28. Mehraban D, et al. Penile size and somatometric parameters among Iranian normal adult men. Int J Impot Res. 2007;19(3):303.
29. Promodu K, et al. Penile length and circumference: an Indian study. Int J Impot Res. 2007;19(6):558.
30. Kamel I, et al. Comparing penile measurements in normal and erectile dysfunction subjects. J Sex Med. 2009;6(8):2305–10.
31. Nasar TA, et al. Relationship between penile size and body measurements. Hum Androl. 2011;1(2):48–50.
32. Khan S, et al. Establishing a reference range for penile length in Caucasian British men: a prospective study of 609 men. BJU Int. 2012;109(5):740–4.
33. Söylemez H, et al. Relationship between penile size and somatometric parameters in 2276 healthy young men. Int J Impot Res. 2012;24(3):126.
34. Chen X, et al. A comprehensive, prospective study of penile dimensions in Chinese men of multiple ethnicities. Int J Impot Res. 2014;26(5):172.
35. Shalaby M, et al. Penile length-somatometric parameters relationship in healthy Egyptian men. Andrologia. 2015;47(4):402–6.
36. Veale D, et al. Am I normal? A systematic review and construction of nomograms for flaccid and erect penis length and circumference in up to 15,521 men. BJU Int. 2015;115(6):978–86.
37. Habous M, et al. Erect penile dimensions in a cohort of 778 Middle Eastern men: establishment of a nomogram. J Sex Med. 2015;12(6):1402–6.
38. Salama N. Consultation for small-sized penis in the Egyptian males: a case control study. Am J Mens Health. 2016;10(3):220–7.
39. Hussein NS, et al. Reference range of flaccid and stretched penile lengths of adult males in Baghdad: a cross-sectional study. Arab J Urol. 2017;15(1):68–73.
40. Richters J, et al. Are condoms the right size(s)? A method for self-measurement of the erect penis. Venereology. 1995;8(2):77–81.
41. Bogaert AF, Hershberger S. The relation between sexual orientation and penile size. Arch Sex Behav. 1999;28(3):213–21.
42. Harding R, Golombok SE. Test-retest reliability of the measurement of penile dimensions in a sample of gay men. Arch Sex Behav. 2002;31(4):351–7.
43. Shaeer O, Shaeer K. The Global Online Sexuality Survey (GOSS): Ejaculatory function, penile anatomy, and contraceptive usage among Arabic-speaking Internet users in the Middle East. J Sex Med. 2012;9(2):425–33.
44. Herbenick D, et al. Erect penile length and circumference dimensions of 1661 sexually active men in the United States. J Sex Med. 2014;11(1):93–101.
45. Shaeer O, Shaeer KZ. The Global Online Sexuality Survey (GOSS): The United States of America in 2011 penile size and form among English speakers. Hum Androl. 2013;3(2):46–53.
46. Alter GJ. Penile enhancement. Aesthet Surg J. 1996;16(4):226–30.
47. Siaarti C. Raccomandazioni clinico organizzative per l'anestesia in day surgery. Minerva Anestesiol. 2000;66:915–26.
48. Pu LL, et al. Autologous fat grafts harvested and refined by the Coleman technique: a comparative study. Plast Reconstr Surg. 2008;122(3):932–7.
49. Panfilov DE. Augmentative phalloplasty. Aesthet Plast Surg. 2006;30(2):183–97.
50. Coleman SR. Long-term survival of fat transplants: controlled demonstrations. Aesthet Plast Surg. 1995;19(5):421–5.
51. Zeltzer AA, et al. Sharp-needle intradermal fat grafting (SNIF). Aesthet Surg J. 2012;32(5):554–61.
52. Illouz YG, Sterodimas A. Autologous fat transplantation to the breast: a personal technique with 25 years of experience. Aesthet Plast Surg. 2009;33(5):706–15.
53. Dickinson BP. Autologous fat transfer national consensus survey: trends in techniques for harvest, preparation and application; and perception of short and long term results. Plast Reconstr Surg. 2006;119(1):323–31.
54. Kling RE, et al. Trends in autologous fat grafting to the breast: a national survey of the American Society of Plastic Surgeons. Plast Reconstr Surg. 2013;132(1):35–46.
55. Gutowski KA, Force AFGT. Current applications and safety of autologous fat grafts: a report of the ASPS fat graft task force. Plast Reconstr Surg. 2009;124(1):272–80.
56. Leopardi D, et al. Autologous fat transfer for breast augmentation: a systematic review. ANZ J Surg. 2014;84(4):225–30.
57. Claro F, et al. Applicability and safety of autologous fat for reconstruction of the breast. Br J Surg. 2012;99(6):768–80.
58. Villani F, et al. Current applications and safety of autologous fat grafts: a report of the ASPS Fat Graft Task Force. Plast Reconstr Surg. 2010;125(2):758–9.
59. Dillon B, et al. Penile size and penile enlargement surgery: a review. Int J Impot Res. 2008;20(6):519.
60. Austoni E, et al. A new technique for augmentation phalloplasty: albugineal surgery with bilateral saphenous grafts—3 years of experience. Eur Urol. 2002;42(3):245–53.

61. Perovic SV, et al. New perspectives of penile enhancement surgery: tissue engineering with biodegradable scaffolds. Eur Urol. 2006;49(1):139–47.
62. Shaeer O, Shaeer K. Penile girth augmentation using flaps "Shaeer's augmentation phalloplasty": a case report. J Sex Med. 2006;3(1):164–9.
63. Bin Y, et al. A comparative study on two kinds of surgical procedures of penile corpora cavernosa augmentation. J Plast Reconstr Aesthet Surg. 2009;62(3):357–64.
64. Jin Z, et al. Tissue engineering penoplasty with biodegradable scaffold Maxpol-T cografted autologous fibroblasts for small penis syndrome. J Androl. 2011;32(5):491–5.
65. Alei G, et al. Original technique for penile girth augmentation through porcine dermal acellular grafts: results in a 69-patient series. J Sex Med. 2012;9(7):1945–53.
66. Wessells H, et al. Complications of penile lengthening and augmentation seen at 1 referral center. J Urol. 1996;155(5):1617–20.
67. Alter GJ. Reconstruction of deformities resulting from penile enlargement surgery. J Urol. 1997;158(6):2153–7.
68. Shirong L, et al. Modified penis lengthening surgery: review of 52 cases. Plast Reconstr Surg. 2000;105(2):596–9.
69. Spyropoulos E, et al. Augmentation phalloplasty surgery for penile dysmorphophobia in young adults: considerations regarding patient selection, outcome evaluation and techniques applied. Eur Urol. 2005;48(1):121–8.
70. Li C-Y, et al. Penile suspensory ligament division for penile augmentation: indications and results. Eur Urol. 2006;49(4):729–33.
71. Mertziotis N, et al. Is V–Y plasty necessary for penile lengthening? Girth enhancement and increased length solely through circumcision: description of a novel technique. Asian J Androl. 2013;15(6):819.
72. Monreal J. Composite augmentation phalloplasty: personal experience after 275 patients. Plast Aesthet Res. 2015;2(1):27.
73. Xu L, et al. Augmentation phalloplasty with autologous dermal fat graft in the treatment of "small penis". Ann Plast Surg. 2016;77:S60–5.

Total Facial Rejuvenation Through Lipofilling: Anatomic and Regenerative Fat Grafting

38

Steven R. Cohen, Sarah Patton, and Tunc Tiryaki

Contents

38.1 Introduction

Advancements in understanding facial aging have paved the way for new insights into the gradual loss of facial soft tissue and bone volume. As we age, photometric changes have shown the deflation in the soft tissue of the face by Lambros [1]. Certain areas of the facial skeleton lose broad surface areas of bone, while the other components of the soft tissue envelope do not experience shrinkage [2, 3]. Rohrich and Pessa [4] clarified the anatomy of the superficial and deep fat compartments. Facial fat is more tightly clustered in the superficial compartments above the muscles, mainly serving to support skin function, providing stem cells and their associated factors and paracrine functions. In contrast, below the musculature, the fat is more loosely organized in the deep facial compartments, permitting the muscles of facial expression to glide.

Aging occurs continually as an evolution of the interdependent processes our cells undertake throughout our lifetime. The first two decades of one's life are predominated by growth and development, followed by a gradual decay of all our tissues until death. Various anatomic and histologic changes occur individually and interdependently in all areas, including the skin, fat compartments, and the underlying bone. Early intervention has the potential to slow the aging

Supplementary Information The online version contains supplementary material available at https://doi.org/10.1007/978-3-031-10881-5_38.

S. R. Cohen (✉) · S. Patton
FACES+ Plastic Surgery, Skin and Laser Center, San Diego, CA, USA

Division of Plastic Surgery, University of California, San Diego, c, CA, USA
e-mail: scohen@facesplus.com

T. Tiryaki
Cellest Clinic, Istanbul, Turkey

Cadogan Clinic, London, UK

© The Author(s), under exclusive license to Springer Nature Switzerland AG 2023
A. Di Giuseppe et al. (eds.), *Fat Transfer in Plastic Surgery*, https://doi.org/10.1007/978-3-031-10881-5_38

of cells, replace losses of tissue and prolong the "disease" free interval. Stem and regenerative cells have been found within fat and have been used to regenerate aging tissues by Rigotti [5], Cohen [6], and others [7–10]. Aesthetic procedures utilizing fat grafting, stromal vascular fraction (SVF)-enriched fat grafting, platelet-rich plasma (PRP), exosomes and growth factors are some of the few therapies that have demonstrated neoangiogenic and tropic effects (to some extent) in virtually all patients [5, 11].

38.2 Injectable Tissue Replacement and Regeneration

Injectable Tissue Replacement and Regeneration (ITR^2) is a concept that attempts to answer the following questions; can a dynamic model be used to determine losses in specific fat compartments and bone? If so, can we then replace and/or regenerate the absent tissues to reverse or reduce the effect of facial aging?

The technique of replacing losses in fat with like material placed into the anatomic compartments of the losses, along with regenerative therapies to improve blood supply and physiologic function of the skin, is a new concept in facial fat grafting. The procedure has been standardized yet remains individualized to the patient's specific clinical findings.

38.2.1 Procedure

1. Diagnose volume loss by evaluating the surface topography of the face.
2. Address specific anatomic losses of the skin, deep and superficial facial fat compartments, and bone.
3. Replace the deficits of fat and bone with autogenous fat grafts. The fat grafts are sized specifically for the compartment in which they are inserted for replacement and regeneration.

The regenerative cells are obtained through mechanical fragmentation of the fat graft. A Nanocube (Lipocube, Inc., London, UK) is used to create a cell-optimized nanofat, which is a cell aggregate product derived from adipose tissue. Nanofat is a type of mechanically reduced adipose that preserves stromal vascular fraction cells while eliminating fat cells [12–15]. It can be small enough to inject through a 27-gauge needle. Different devices make different types of nanofat. Nanocube has previously been shown to increase the regenerative cell count [16]. Nanofat has no ability to augment tissue. Its sole purpose is regeneration, whether injected, microneedled or delivered as a biocream [17].

38.3 Evaluation of the Patient

While seated in an upright position, the patient is marked with a white makeup pencil (Video 38.1). The assessment begins from the top by noting the scalp hair quality and/or loss to determine if a restorative and regenerative approach might be beneficial. The thickness of the epidermis, dermis, and subcutaneous tissues are evaluated. In the upper third of the face, the degree of bone recession in the glabella and along the supraorbital rims is analyzed. The degree of photodamage is noted. If deeper rhytids are present, they are noted for the possible use of the sharp-needle intradermal fat grafting (SNIF) technique [18]. Temporal depression is associated with deep compartment fat loss, while protrusion of the temporal vein indicates superficial fat loss. Often both are present. There is an inspection of the upper and lower eyelids along with the periorbital region. Loss of fullness of lateral brow, loss of convexity of the skin caudal to the eyebrow, and the depth of the supra-tarsal fold are all noted.

In the inferior orbital, rim recession and prominence of the intra-orbital fat are assessed. The tear trough and lid cheek junction are evaluated. Lateral and medial portions of the sub-orbicularis oculi fat (SOOF) are noted. The degree of proptosis of the eyeball is viewed from the vortex point. The extent of senile enophthalmos and the lid-to-pupil position is inspected.

Within the middle third of the face, the zygomatic arch is outlined, indicating the superior arch corresponding to the inferior temporal region. Deep medial fat compartments of the cheek are noted, along with the degree of buccal hollowing. The nose is assessed for any aesthetic deformity and/or pyriform recession due to aging. Lips and peri-oral tissues are inspected for thinning and rhytids.

The lower third of the face is evaluated at the marionette basin, the chin, and the labiomental fold. Chin texture can be improved with nanofat microneedling and/or topical delivery of a nanofat biocream in conjunction with a fractional laser or alone. Lateral to the mandibular retaining ligaments, the curvature of the prejowl area proceeds along the inferior border of the mandible and gonial angle. Varying degrees of masseteric hypertrophy may be present. The projection of the chin is evaluated. Inspection of the neck and décolletage assesses the degree of subcutaneous loss, deep and fine rhytids, and the severity of sun damage.

38.4 Surgical Procedures

38.4.1 Preoperative Preparation and Anesthesia

One day before the procedure, the patient is given oral prophylactic antibiotics if planned to receive local anesthesia. The antibiotics can also be administered intravenously before the procedure. Patients are given the choice of anesthesia. Typically for ITR[2], local or intravenous anesthesia with tumescent lipoharvest is utilized unless the patient is also having other facial procedures.

38.4.2 Adipose Tissue Harvest

If only fat grafting is being performed, the surgery itself takes approximately 45 min to an hour (see Video 38.1). If sufficient fat is available, the fat is harvested from areas of patient preference or any area of excess subcutaneous fat. The patient is prepped and draped under sterile conditions. The tissue harvest begins with a 14 G needle puncture followed by infiltration of tumescent fluid (500 mL of Ringer Lactate with 25 mg lidocaine and one vial of epinephrine [1:1000]). A 12-holed cannula, with openings measuring 2.5 mm in diameter (Marina Medical, Davie, Florida), is inserted into a slightly dilated 14 G needle hole. Fat is aspirated using a 60-mL syringe with a lock. Generally, 120 mL of fat is removed. The punctures are often closed by secondary intention or by Dermabond (Ethicon, Bridgewater, New Jersey, US) wound adhesive.

38.4.3 Fat Processing

The tumescent fluid is decanted from the removed fat, and the fat is rinsed with Ringer lactate to eliminate blood contamination. According to new research on fat preparation and degree of engraftment, most filtration systems and centrifugation are unnecessary and add to the time and cost of the procedure [19, 20]. Simple washing, gravity separation, and decantation to remove the tumescent solution are necessary to process the fat. After cleaning, a portion of the fat, Millifat (2 mm parcel size), is set aside to restore the deep fat compartment loss and facial bone recession. The remaining fat is transferred into 20-mL syringes, then processed into Microfat and Nanofat using the Nanocube (Lipocube, Inc., London, UK). The Nanocube has four ports used to resize the fat with different size cutting screens and emulsification (Video 38.1). Alternative systems that process fat grafts into various sizes can also be used.

38.4.4 Delivery Techniques

According to the topographic assessment, fat grafts are assigned to anatomic locations in the face according to their parcel size of Millifat (2–2.5 mm), Microfat (1 mm), and Nanofat (500 µm and less) (see Video 38.1). Millifat is first placed in the deepest compartments of the face, and as one progresses superficially Microfat, then Nanofat are utilized.

Figure 38.1 displays the options for the 18 G puncture sites, which can be reused for each of the three sized fat grafts whenever possible. The safe volume recommendations and sites for fat grafting are shown in Fig. 38.2.

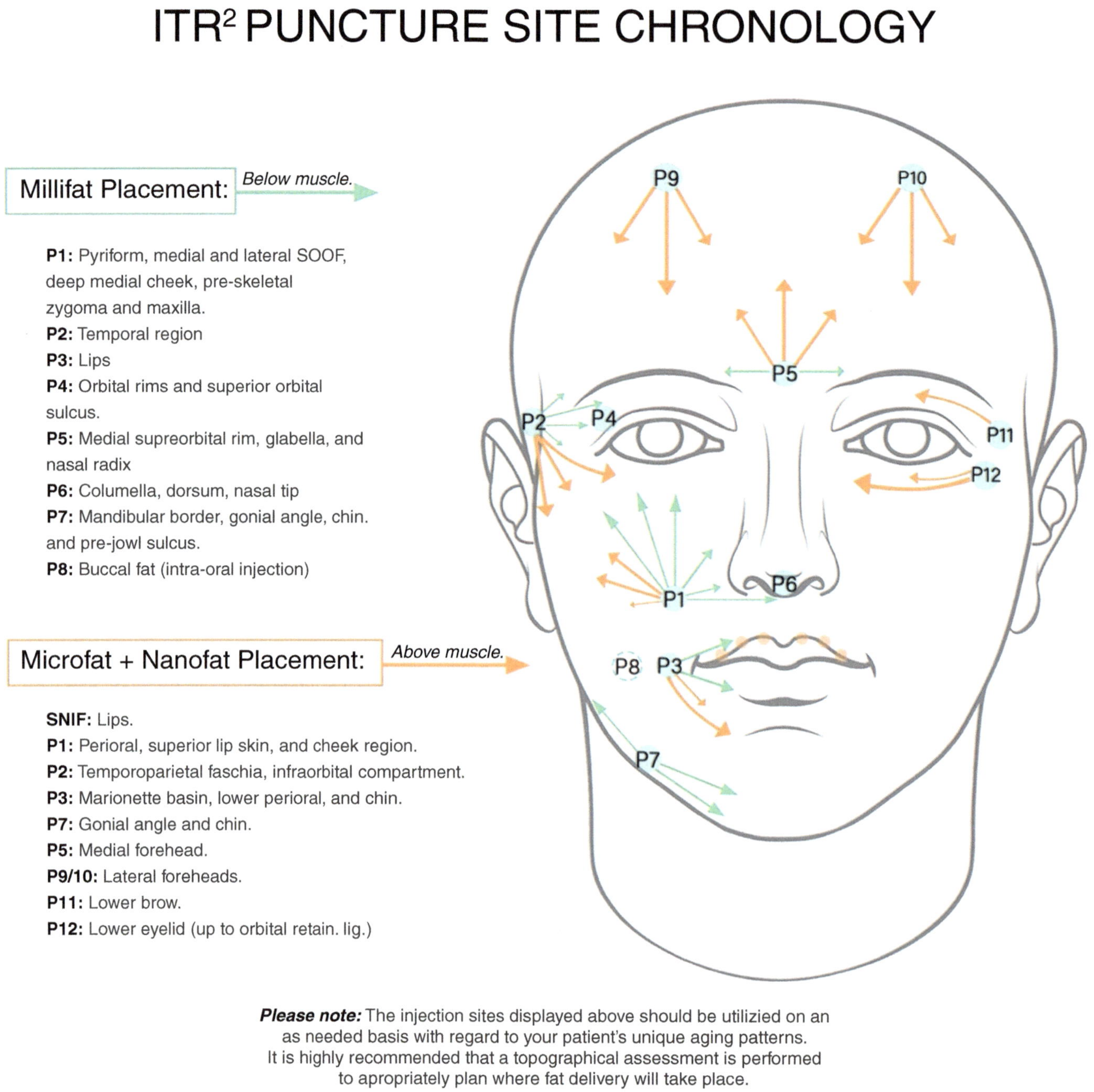

Fig. 38.1 Puncture site chronology and injection vectors used in ITR² fat grafting. (*From* Cohen SR, Womack H, Ghanem A. Fat Grafting for Facial Rejuvenation through Injectable Tissue Replacement and Regeneration. Clinics in Plastic Surgery. 2020;47(1):31–41; with permission)

38.5 Millifat (2–2.5 mm Parcels)

Millifat is placed through an 18 G needle puncture at the nasolabial fold lateral and superior to the oral commissure. Millifat fills the areas of bone recession in the pyriform region. The cannula is positioned upward to graft the deep medial fat compartment and the medial and lateral SOOF. Next, the deep temporal region is grafted along with the pre-periosteal lateral supraorbital brow. An 18 gauge needle puncture may be made at the lateral tail of the eyebrow, and an 18 gauge blunt cannula with a side port is inserted and guided behind the orbicularis into the pre-periosteum along and inside the inferior orbital rim. Injection of the medial orbital rim with 0.5% lidocaine with 1:100,000 epinephrine is used to constrict the supraor-

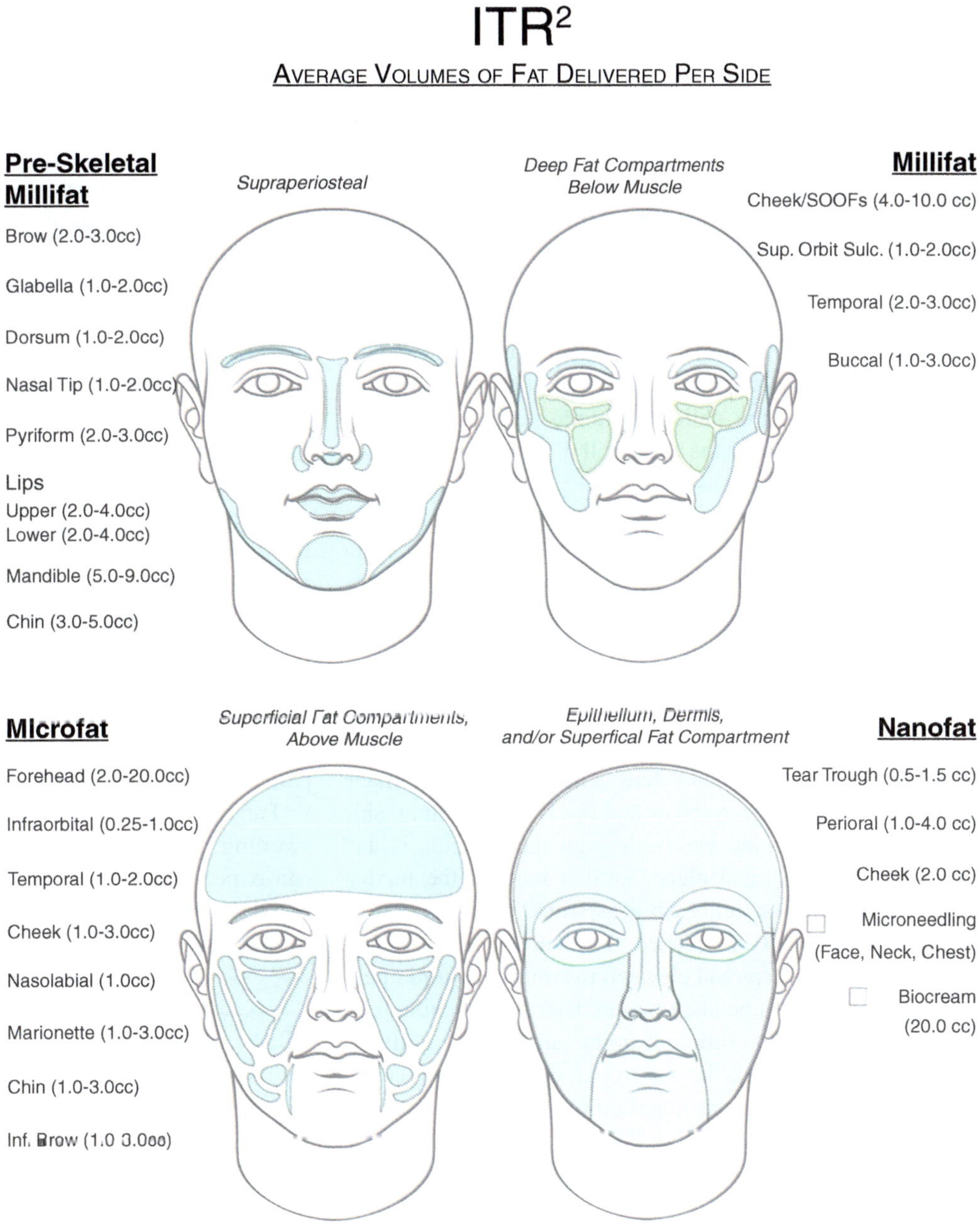

Fig. 38.2 The recommended safe volume ranges with corresponding anatomic fat grafting locations. (*From* Cohen SR, Womack H, Ghanem A. Fat Grafting for Facial Rejuvenation through Injectable Tissue Replacement and Regeneration. Clinics in Plastic Surgery. 2020;47(1):31–41; with permission)

bital and supratrochlear vessels as a precaution to avoid retinal artery embolism. The upper and lower hemi-lip are injected with Millifat at the commissure.

The glabella and nasal radix are injected from a puncture in the central glabella, approximately 1.5–2 cm above the nasofrontal junction. Another puncture at the nasal tip is the entry point for grafting the nasal dorsum, tip, and columella. The nose is first injected with 0.5% lidocaine with 1:100,000 epinephrine to constrict nasal vessels as a precaution for embolism.

Modest retrogenia can be improved with Millifat grafting. Grafting is carried out at the pre-skeletal level, crossing to the opposite side of the chin and injecting retrograde with a Cellbrush (Cytori Lorem, San Diego, California), which is a thumb guided delivery device. Fat is also grafted in the area just lateral to the mandibular ligament to camouflage the pre-jowl sulcus. Fat is grafted along the mandibular border through the same puncture. Millifat is placed along the border of the inferior mandibular and into the gonial angle. This positioning can be used to define the jawline, disguise jowls, or lower an obtuse mandibular angle (see Fig. 38.1, P7). Fat can be grafted in small amounts along the inferior chin to provide additional vertical height if needed. Lastly, the labiomental fold can be softened with fat.

When indicated the buccal fat compartment the same needle puncture used to graft the lips. Instead of tunneling subcutaneously, the cannula is directed submucosally and from there into the buccal compartment are given intravenous clindamycin. It is important not to overfill this area because it is a lowlight area. Place small amounts of Millifat in the buccal compartment, while frequently re-inspecting the area.

38.6 Microfat (1-mm Parcels)

Mircofat when needed is grafted superficially above the muscle for thinning of the perioral skin. This area reuses the needle incision in the nasolabial fold for the upper lip and the oral commissure incision for the marionette basin and chin tissue. To address the philtral columns (cupids bow) and rhytids perpendicular to the white roll of the upper and lower lips, the SNIF technique is used [18].

The puncture site from the deep temporal compartment (P2) is also used to graft the superficial temporal fat compartment with Microfat.

Bilaterally, at the midpupillary line, two more needle incisions are made at the hairline or just inside of it to graft the upper 2/3's of the forehead. It is important to be pre-periosteal and not in the superficial tissue above the galea where the blood vessels run. I avoid injection of the medial brow.

To graft the upper eyelid sulcus, a needle incision is made at the lateral tail of the eyebrow and the 18 Gauge blunt, side port cannula is guided *gently* through the orbicularis and along the pre-periosteal plane, until it reaches the medial orbit. The syringe or Cellbrush is aspirated before injection. Generally, 1–1.5 mL are placed and then the lid margin is grasped with a forceps and elevated to simulate eyelid opening so the sulcus can be inspected to determine if enough fat has been placed. Two different points are used to graft the lower eyelid. The first is the tear trough point, and the second being just lateral to the nasojugal groove (P12). If necessary, microfat is placed in droplets only below the orbital retaining ligament. One must be very conservative, and rarely is microfat needed in this area.

In order to restore a uniform silhouette of the lower face, microfat may be utilized in the superficial fat compartments of the chin, jawline, lateral superior gonial angle, and submental crease.

38.7 Nanofat (500 μm Parcels)

Nanofat is typically utilized in three ways, with the SNIF technique, topically with microneedling, or with a topical biocream. Previously, the authors prepared nanofat with a gradual emulsification technique described by Tonnard [12]. As of 2018, ITR^2 nanofat has been prepared with the Nanocube. The mechanical digestion within the Nanocube produces a matrix-rich group of regenerative cells [16]. When delivered through the SNIF technique for dermal rhytids, this cellularly optimized nanofat is injected using a 25 G cannula attached to a finger-activated grafting device, 3-mL Celbrush (Cytori, San Diego, California), or an automatic grafting device, Lipopen (Juvaplus, Neuchâtel, Switzerland). Finally, nanofat is delivered mechanically with a microneedling device into the face, neck, and décolletage. A portion of the nanofat, 5–20-mL, is saved and combined with a transdermal liposomal carrier to form a topical nanofat biocream (neo-U [Aries Biomedical, San Diego]).

38.8 Postoperative Care

Postoperative care consists of analgesia, nonsteroidal anti-inflammatory medications, and arnica for bruising. No direct application of ice is permitted. If there is excessive swelling, the patient is treated with a tapering oral steroid regimen as needed. For patients undergoing facelift surgery or laser resurfacing, a specific skincare regimen is followed. The products contain matrikine (tripeptides and hexapeptides) ingredients that help to modulate extracellular matrix repair and neocollagenases (Alastin Skincare, Carlsbad, California) [21]. Preoperative prophylactic antivirals are prescribed to patients with a history of herpes simplex.

Patients are expected to experience some bruising and swelling of the lips for approximately 5–10 days. Patients can expect facial swelling and mild ecchymosis for about 3–5 days. 15% of patients may take longer than a few weeks for the bruising and swelling to subside.

38.9 Outcomes and Complications

Patients treated with nanofat microneedling and/or nanofat biocream (neo-U) in conjunction with fractional lasers of different wavelengths have shown significant improvement in aesthetic outcomes with faster healing than those without [17].

Patients who receive facelifts with ITR^2 experience an increase in facial volume by 45% in a month; at 7–12 months, the facial volume decreases to approximately 25–30% but improves again to 74% at 18–24 months [19]. These findings suggest that the usage of ITR^2 with facelift surgery may reverse tissue decay that is experienced with aging.

Using ITR^2 as a standardized technique in patients not having facelifts or other procedures has recently been studied. Twenty-nine female patients with a mean age of 52.9 years were analyzed for mid-facial volume changes after autologous fat transfer with ITR^2. Preliminary evidence showed a dynamic improvement in facial volume in patients under 55 years of age regardless of the amount of fat grafted and with no effect of weight change on the outcome. Patients over 55 having isolated ITR^2 showed a gradual decrease in mid-facial volume over the 19 months to an improvement of 30%. Those under 55 years of age rebounded to a nearly 80% improvement in mid-facial volume at 19 months after treatment [22].

Complications from ITR^2 have been rare and related to excessive fat grafts in the lower eyelids in 2 cases. Transconjunctival or transcutaneous lower blepharoplasty with the removal of fat has taken care of these infrequent complications. The authors are currently only using matrix-rich nanofat above the orbital retaining ligament, not micro-fat. Since adopting ITR^2, the authors have not experienced any overgrowth with patient weight gain.

The senior author (SRC) has had one case of partial, unilateral blindness, most likely from a fat embolism into the supra-orbital artery from an unnoticed tear in the vessel in the lower third of the forehead. Most likely, the fat was sucked into the torn vessel during diastole. The large 18-gauge cannula would not have been able to directly inject fat into a vessel of this size. The patient was treated immediately with direct ocular pressure and high-dose steroids and brought to the emergency room for ophthalmology consultation. Partial salvage of vision was possible, although for all practical purposes the patient is blind in one eye. Our present regimen is to inject the area around the supra-orbital and supra-trochlear vessels with 0.5% Lidocaine with epinephrine to vasocontrict these vessels and to avoid the lower forehead medially. Above the lower third of the forehead, the vessels become superficial and it is quite safe to inject pre-periosteally in this area.

38.10 Case Examples

38.10.1 Patient 1

Our first patient is a 39-years-old female who demonstrates signs of natural aging over the course of 6 years that can be seen in Fig. 38.3a, b. Undergoing ITR^2, this patient received a total of 34 mL of fat to the face. Beginning with millifat, approximately 3 cc were placed deep fat compartments into the temporal regions, 3 cc into the right cheek, 5 cc into the left cheek, 1 cc into the nasal bridge, 1 and 2 cc into the right and left nasolabial folds, respectively. Microfat was then placed more superficially into the tear troughs, nasolabial folds, marionette lines and into the upper and lower lips. Nanofat was used throughout. Note the replacement of fat loss in the deep and superficial fat compartments in Fig. 38.3c, in our patient 6 months postoperatively. Figure 38.3d, e show the appearance of our patient 2-years and 4-years, respectively post-ITR^2. Skin improvement and fat retention can be appreciated.

38.10.2 Patient 2

Our next patient is a 67-year-old women who demonstrates fat loss in deep and superficial fat compartments, modest sun damage, moderate skin laxity, with dermal and epithelial thinning. The plan for this patient included a high SMAS facelift along with pinch upper and lower blepharoplasties and ITR^2. Figure 38.4a shows our patient preoperatively. The red overlays in Fig. 38.4b represent the areas of deep fat loss that will be treated with millifat, while the orange overlays represent the areas of bone loss that will also be replaced with millifat. The blue overlays in Fig. 38.4c indicate the areas of superficial fat loss that will be treated with micro and/or nanofat. Finally, nanofat was microneedled into the skin of the entire face. The patient shows significant improvements within the cheeks and jowls regions that can be noted 6 months postoperatively in Fig. 38.4d.

38.10.3 Patient 3

A 52-year-old woman, seen in Fig. 38.5a, presents with concerns of periorbital aging, loss of facial volume, modest sun damage and mild laxity in the jowls and neck. Orbital expansion created a subtle appearance of pseudoptosis and "senile" enophthalmos. During ITR^2, this patient received 58.5 total mL of fat that was placed into the forehead, temporal regions, periorbital, perioral, midface, pyriform, gonial angles. Note the improvement in the global position from the intraorbital fat grafting, post-ITR^2 and skin-only upper/lower pinch blepharoplasties in Fig. 38.5b.

38.10.4 Patient 4

Our final patient is a 60-year-old woman who shows facial aging in Fig. 38.6a, c, e. This patient underwent a facelift with high SMAS, sub-mental platysmal plication, periorbital, temporal, deep compartment, and superficial compartment fat grafting via ITR^2. A total of 22 mL of fat was used throughout the face. Note the improvements in the lower eye pigmentation, perioral texture, nasojugal fold and the lid cheek blending in our patient 18 months post-op in Fig. 38.6b, d, f.

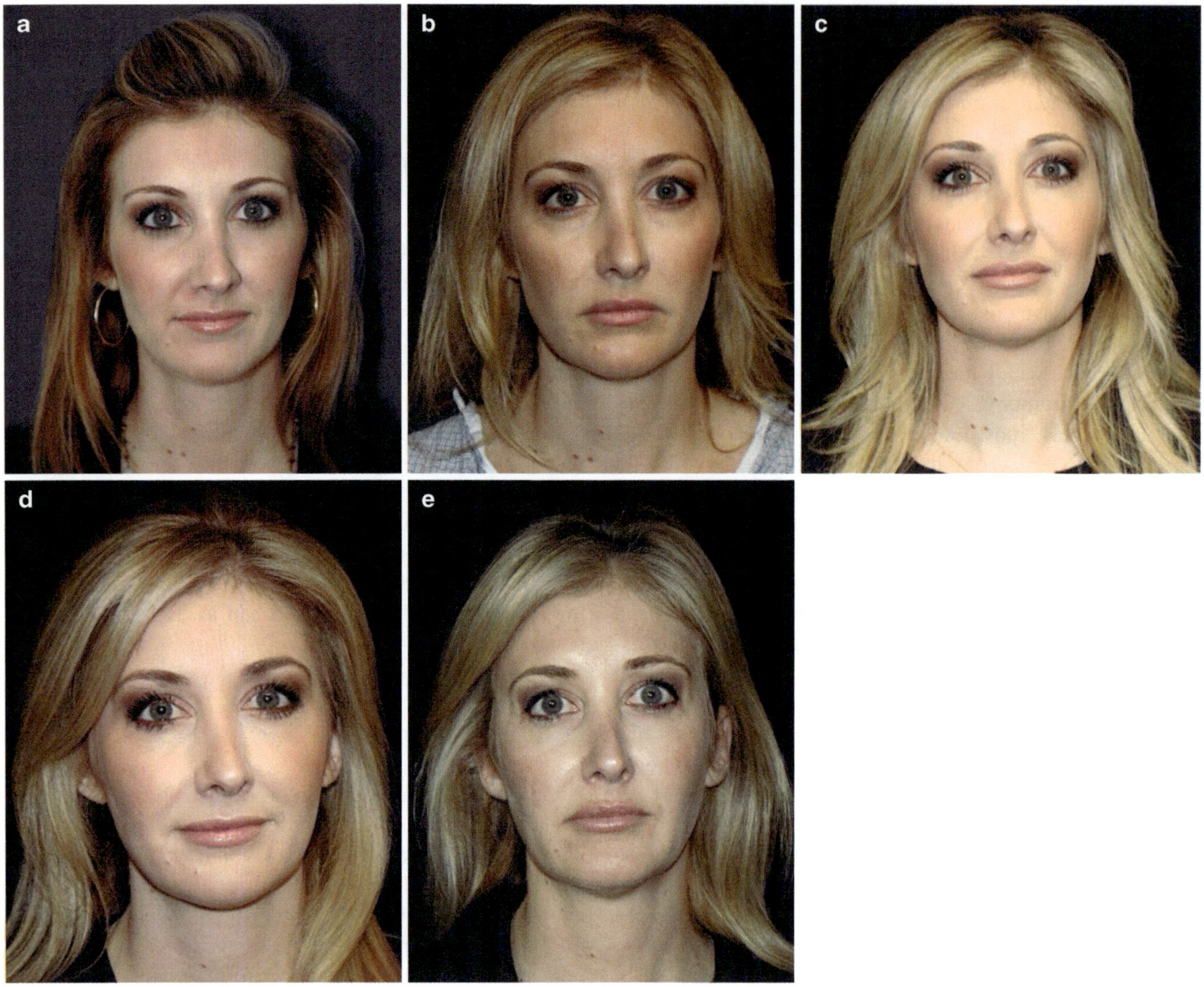

Fig. 38.3 A 39-year-old patient shown preoperatively in (**a**), and again after 6 years of aging in (**b**). (**c**) Shows the patient 6 months post ITR^2. (**d**, **e**) Show the patient 2- and 4-years post ITR^2. (*From* Cohen SR, Womack H, Ghanem A. Fat Grafting for Facial Rejuvenation through Injectable Tissue Replacement and Regeneration. Clinics in Plastic Surgery. 2020;47(1):31–41; with permission)

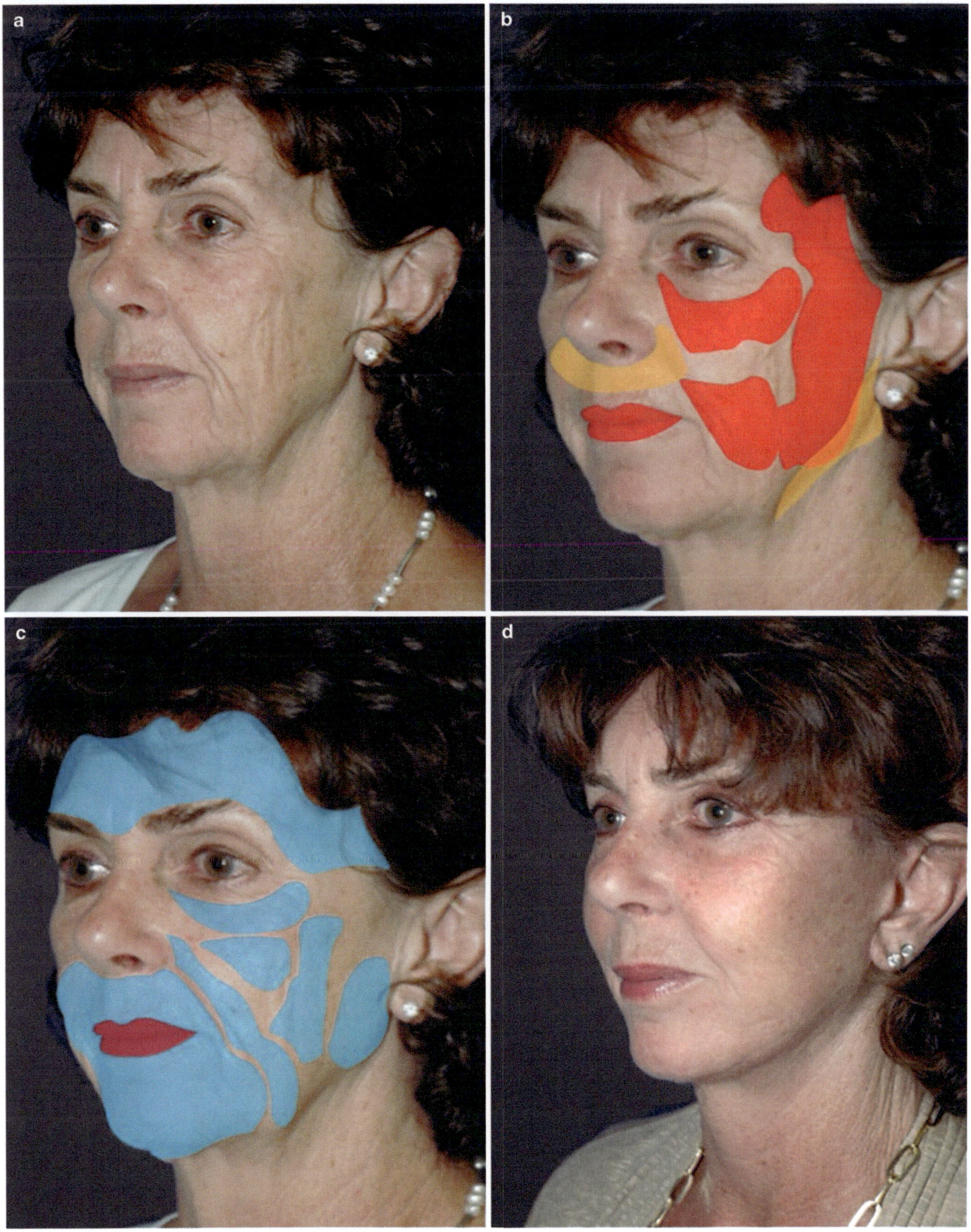

Fig. 38.4 A 67-year-old patient presents preoperatively in (**a**). The red and orange overlays in (**b**) represent the areas that plan for millifat placement, while the overlays in (**c**) represent the areas that will be treated with microfat and nanofat. (**d**) Shows the patient 6 months postoperatively after ITR[2], a deep plane face and neck lift, with upper and lower blepharoplasties. (*From* Cohen SR, Womack H. Injectable Tissue Replacement and Regeneration: Anatomic Fat Grafting to Restore Decayed Facial Tissues. Plas Reconstr Surg Glob Open. 2019;7(8):e2293; with permission)

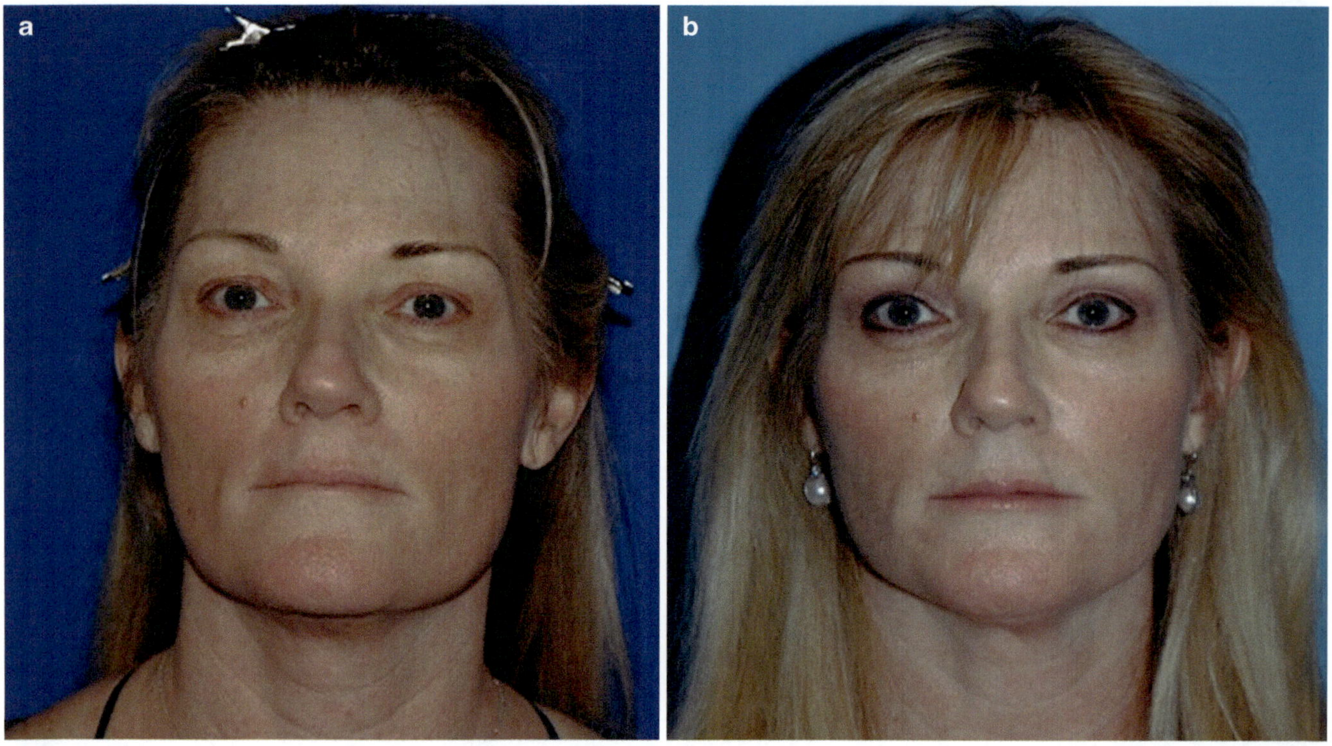

Fig. 38.5 A 52-year-old patient is shown preoperatively in (**a**), then 1-year post-ITR[2]. (*From* Cohen SR, Womack H. Injectable Tissue Replacement and Regeneration: Anatomic Fat Grafting to Restore Decayed Facial Tissues. Plas Reconstr Surg Glob Open. 2019;7(8):e2293; with permission)

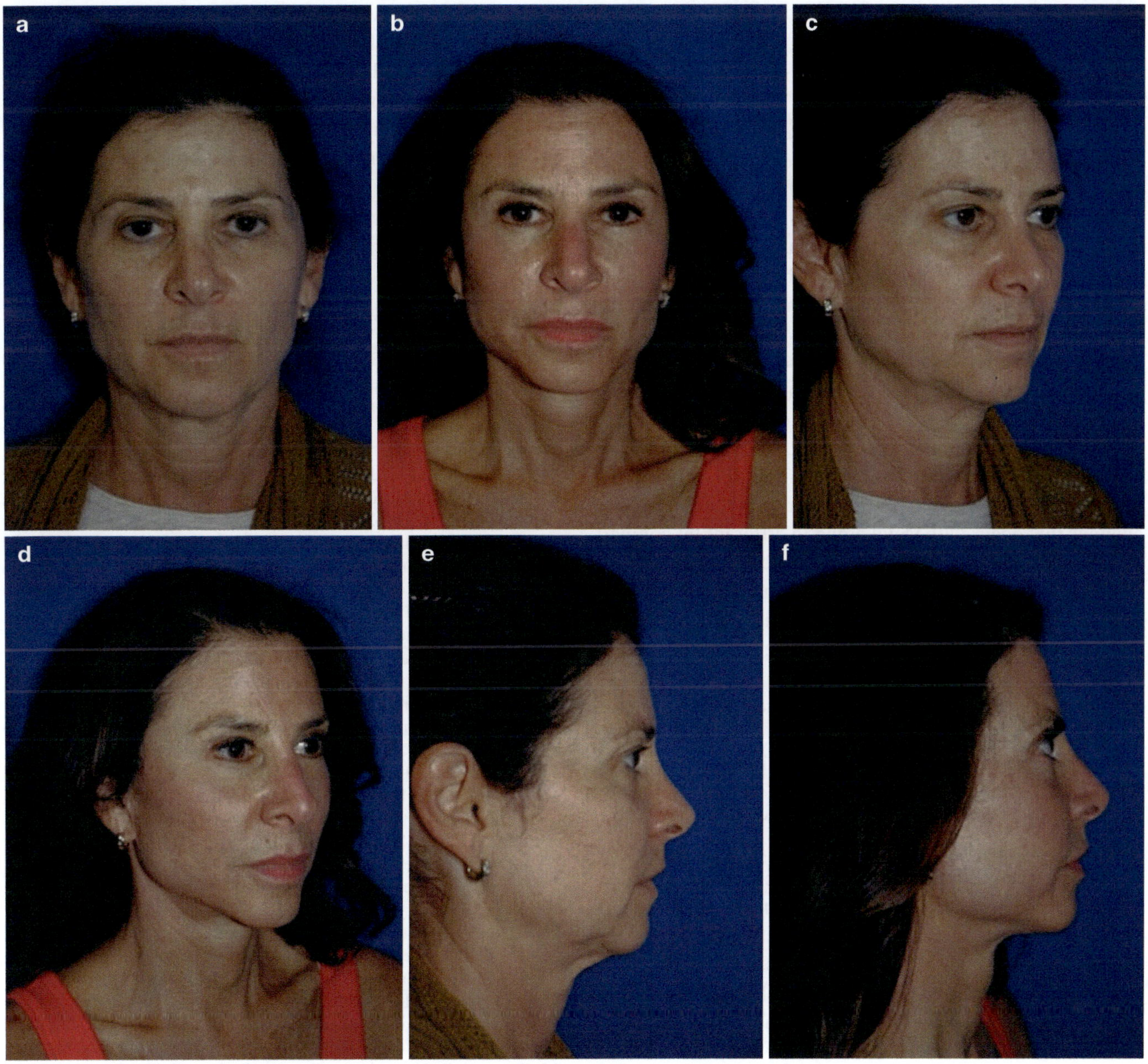

Fig. 38.6 A 60-year-old patient is shown preoperatively in (**a**, **c**, **e**). The patient is shown again 18 months post ITR^2 and face lift in (**b**, **d**, **f**). (*From* Cohen SR, Hewett S, Ross L, et al. Regenerative Cells for Facial Surgery: Biofilling and Biocontouring. Aesthetic Surgery Journal. 2017;37(S3): S16-S32; with permission.)

38.11 Discussion

Currently, most surgeons and dermatologists inject fat aesthetically as if it were a natural filler. In the senior author's opinion, fat can be replaced in an anatomically correct fashion to avoid poor aesthetic results that can occur with weight gain as well as replace lost tissues with like tissues. Nanofat is used in the superficial fat, intradermally and in the lower eyelid to regenerate tissue because there is no structural requirement. There are two important benefits to utilizing an anatomical and regenerative approach: (1) it addresses the actual anatomical changes that occur with aging and (2) neoangiogenesis improves tissue health, possibly delaying atrophy while reconstructing the functional matrix.

Although this approach may seem complex, it is simple, standardized and routine to perform. Like the artistic understanding of anatomy required for high-definition liposuction techniques, ITR^2 is based on observing the anatomical changes that occur with facial aging in different fat compartments, skin, and bones. Treatment is directed for all tissues that have decayed from epithelium to bone. Fat grafts of two sizes address structural changes, superficial fat loss, and skin thinning. ITR^2 can be combined with other procedures on the eyelids and face.

It is possible to follow the facial tissues as they progress from growth and development to decay. The authors' concept involves intervention at the earliest sign of decay to prevent the rapid decline of these changes by stimulating the tissues with PRP and/or cellularly optimized nanofat. Skincare, lasers, and other energy-based devices are used to treat sun damage. Skincare products containing matrikine ingredients are used to clear the extracellular matrix of debris [21]. Fillers and other aesthetic products used for beauty enhancement have little to no effect on tissue health. Patients that require more than one filler are excellent candidates for their first ITR^2 treatment for facial volume loss.

38.12 Summary

ITR^2 presents a dynamic, anatomy-based approach to address facial aging patterns, which occurs in specific superficial and deep fat compartments in the face. This technique delivers three structurally optimized fat grafts to replace and regenerate specific anatomic losses in the skin, superficial and deep fat compartments. The sizes include millifat, microfat, and nanofat. The structural differences of each fat graft make them optimal for specific areas and make the most sense from a safety perspective.DisclosuresDr. Cohen is a shareholder in the Mage Group, the owner of Lipocube and receives royalties on the Nanocube, in addition, Dr. Cohen is a shareholder in Millenium Medical Technologies, Inc. Carlsbad California. Dr. Cohen is a consultant for the Musculoskeletal Transplant Foundation, Edison NJ and receives royalties for his textbook, Regenerative Facial Surgery, QMP Medical Publishers, St. Louis Missouri. Dr. Tiryaki is an investigator for Mentor, receives book royalties from Springer, and is on the advisory board and holds equity in the Mage Group and Lipocube Ltd. Sarah Patton has nothing to declare.

References

1. Lambros V. Observations on periorbital and midface aging. Plast Reconstr Surg. 2007;120(5):1367–76.
2. Kahn DM, Shaw RB Jr. Aging of the bony orbit: a three-dimensional computed tomographic study. Aesthet Surg J. 2008;28(3):258–64.
3. Mendelson B, Wong CH. Changes in the facial skeleton with aging: implications and clinical applications in facial rejuvenation. Aesthet Plast Surg. 2012;36(4):753–60.
4. Rohrich RJ, Pessa JE. The fat compartments of the face: anatomy and clinical implications for cosmetic surgery. Plast Reconstr Surg. 2007;119(7):2219–27.
5. Rigotti G, Charles-de-Sá L, Gontijo-de-Amorim NF, et al. Expanded stem cells, stromal-vascular fraction, and platelet-rich plasma enriched fat: comparing results of different facial rejuvenation approaches in a clinical trial. Aesthet Surg J. 2016;36(3):261–70.
6. Cohen SR. Commentary on: Expanded stem cells, stromal-vascular fraction, and platelet-rich plasma enriched fat: comparing results of different facial rejuvenation approaches in a clinical trial. Aesthet Surg J. 2016;36(3):271–4.
7. Hannum G, Guinney J, Zhao L, et al. Genome-wide methylation profiles reveal quantitative views of human aging rates. Mol Cell. 2012;49(2):359–97.
8. Zuk PA, Zhu M, Mizuno H, et al. Multilineage cells from human adipose tissue: implications for cell-based therapies. Tissue Eng. 2001;7(2):211–28.
9. Cytori Therapeutics. Scleroderma treatment with Celution processed adipose derived regenerative cells (STAR). Bethesda: National Library of Medicine (US); 2015.
10. Cytori Therapeutics. Celution prepared adipose derived regenerative cells in the treatment of osteoarthritis of the knee (ACT-OA Knee). Bethesda: National Library of Medicine (US); 2014.
11. Kamakura T, Kataoka J, Maeda K, et al. Platelet-rich plasma with basic fibroblast growth factor for treatment of wrinkles and depressed areas of the skin. Plast Reconstr Surg. 2015;136(5):931–9.
12. Tonnard P, Verpaele A, Peeters G, Hamdi M, Cornelissen M, Declercq H. Nanofat grafting: basic research and clinical applications. Plast Reconstr Surg. 2013;132(4):1017–26.
13. Graziano A, Carinci F, Scolaro S, et al. Periodontal tissue generation using autologous dental ligament micro-grafts: case report with 6 months follow-up. Ann Oral Maxillofac Surg. 2013;1:20.
14. Trovato L, Monti M, Del Fante C, et al. A new medical device Rigeneracons allows to obtain viable micro-grafts from mechanical disaggregation of human tissues. J Cell Physiol. 2015;230:2299–303.
15. Cervelli V, Gentile P, Scioli MG, Grimaldi M, Casciani CU, Spagnoli LG, Orlandi A. Application of platelet-rich plasma in plastic surgery: clinical and in vitro evaluation. Tissue Eng Part C Methods. 2009;15:625.
16. Cohen SR, Tiryaki T, Womack HA, et al. Cellular optimization of nanofat: comparison of two nanofat processing devices in terms of cell count and viability. Aesthet Surg J Open Forum. 2019;1(4):ojz028.
17. Cohen SR, Goodacre AK, Womack H, et al. Topical nanofat Biocrème improves aesthetic outcomes of nonablative frac-

tional laser treatment: a preliminary report. Aesthet Surg J. 2020;40(8):892–9.
18. Zeltzer A, Tonnard P, Verpaele A. Sharp-needle intradermal fat grafting (SNIF). Aesthet Surg J. 2012;32(5):554–61.
19. Salinas HM, Broelsch GF, et al. Comparative analysis of processing methods in fat grafting. Plast Reconstr Surg. 2014;134(4):675–83.
20. Cohen SR, Hewett S, Ross L, Fischer M, Saad A, Teubel S, Delaunay F. Progressive improvement in midfacial volume 18–24 months after simultaneous fat grafting and facelift: an insight to fat graft remodeling. Aesthet Surg J. 2020;40(3):235–42. https://doi.org/10.1093/asj/sjy279.
21. Widgerow AD, Fabi SG, Palestine RF, et al. Extracellular matrix modulation: optimizing skin care and rejuvenation procedures. J Drugs Dermatol. 2016;15(4 suppl):s63–71.
22. Cohen SR, Wesson J, Willens S, et al. Standardized anatomic and regenerative fat grafting: objective photometric evaluation from 1 to 19 months after injectable tissue replacement and regeneration (ITR[2]). Aesthet Surg J. 2021;42(4):327–39.

Autologous Fat Grafting in Hand Surgery: From Osteoarthritis and Pain Management to Remodeling After Trauma and Rejuvenation

39

Elisabeth M. Haas-Lützenberger and Riccardo E. Giunta

Contents

39.1 Operative Technique: Harvesting Abdominal Fat

Autologous structural fat grafting is a well-established procedure for the treatment of volume loss, scar therapy, chronic wound therapy and regenerative and rejuvenative indications. Due to the knowledge that fat tissue from the lower abdomen consists of adipose-derived-stem cells, we recommend this area for harvesting. In our experience, adipose tissue is best extracted under sterile conditions by means of liposuction from the lower abdomen through an incision in the umbilical area with a blunt needle using a liposuction cannula with 3 mm diameter under local anesthesia. Therefore, about 50–100 mL of Tumescent solution (1000 mL of saline solution, 1 mL of 1:200,000 adrenaline, and 600 mg of lidocaine) were infiltrated at least 15 min prior to beginning harvesting fat. Depending on the amount needed, it can be either collected via a specific medical device for liposuction or with a simple Coleman cannula (Byron Medical Inc., Tucson, Ariz.) (Fig. 39.1). When fat is harvested without the assistance of a liposuction system, negative pressure can easily be generated by gently pulling back the plunger of the 10 cc Luer-Lok syringe. The hereby obtained adipose tissue is then decanted to separate the infiltrated tumescent solution from the pure fat (containing SVF and ADSCs) and thus improve the fat density. Due to different ethic regulations in countries all over the world, it is not mandatory to centrifuge the fat before grafting [1]. Simple decantation due to physical separation of fat and fluid is efficient enough to sustain the required tissue quality for fat transfer (Fig. 39.2). After harvesting fat from the abdominal area, we recommend patients to wear a corset around the abdomen to maintain pressure on the donor site for 1 week.

E. M. Haas-Lützenberger · R. E. Giunta (✉)
Department of Hand, Plastic and Aesthetic Surgery, LMU Klinikum, Munich, Germany
e-mail: R.Giunta@med.uni-muenchen.de

© The Author(s), under exclusive license to Springer Nature Switzerland AG 2023
A. Di Giuseppe et al. (eds.), *Fat Transfer in Plastic Surgery*, https://doi.org/10.1007/978-3-031-10881-5_39

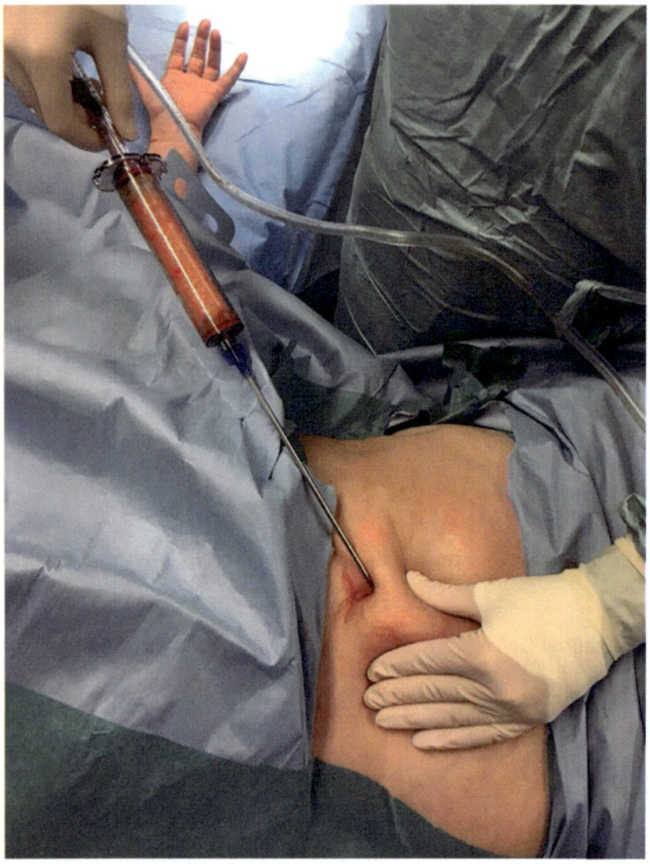

Fig. 39.1 Liposuction of peri-umbilical region after infiltration of tumescent solution

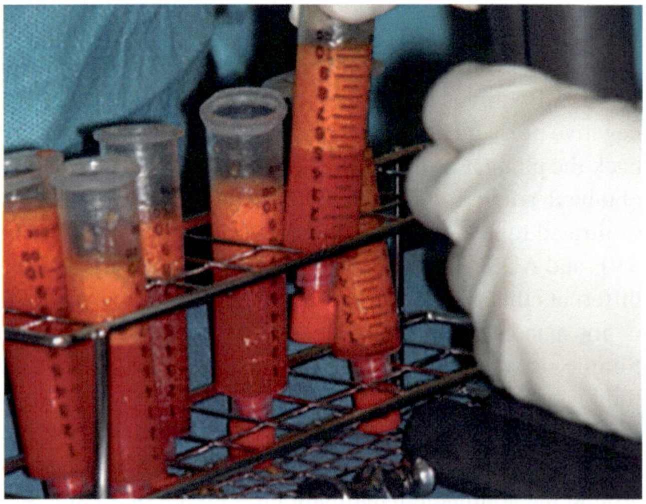

Fig. 39.2 Harvested abdominal fat tissue after decantation

39.2 Regenerative Potential of Autologous Fat for Treatment of Osteoarthritis of the Thumb

Osteoarthritis (OA) of the hand has a worldwide lifelong incidence of about 40%. The hand is beside the knee and the hip, one of the three most affected joints of the body. Suffering from OA is one of the leading causes of disability in the elderly population. The carpometacarpal joint of the thumb (CMC-1) is the most frequently affected joint, especially in the female population. Before a patient consults a hand surgeon, most conservative treatments were already performed and showed no efficacy anymore. The next therapeutical step as an injection of corticosteroid or hyaluronic acid. Unfortunately, these therapies never show a long-lasting effect. Therefore, an intermediate procedure to postpone the resection arthroplasty is widely accepted by patients. Autologous fat transfer was first applied in bigger joints such as the knee and hip. But the common scientific knowledge of the containing ADSCs and their regenerative potential led to the first clinical pilot projects in 2014 [2]. The success of one single surgery by means of pain reduction and thus improving the quality of life could have demonstrated a long-lasting effect from one up to 4 years [3, 4]. Since then, the autologous fat transfer into the CMC-1 joint has become an intermediate therapeutic option before moving to trapezectomy. The final explanation of which of the potential reasons for pain reduction are responsible for the success of the treatment and which patient criteria have to be fulfilled to respond to the therapy is yet not fully understood. The following potential hypothesis could explain its positive effect: the fat tissue acts as a filling material and produces additional "cushioning" scaring; it has power of mediation of anti-inflammatory cytokines and thus could lead to a reparative function.

39.2.1 Operative Technique

For this indication, only a very small amount of fat tissue needs to be harvested as no more than 1–2 cc fat tissue is injected into the CMC-I-joint. After harvesting the fat, it is mechanically homogenized with two syringes that were connected to a three-way valve. Subsequently, using a disposable 20-gauge needle and a 3-mL Luer-Lok (BD Medical, Franklin Lakes, N.J.) syringe, between 1 and 2 cc of lipoaspirate is injected into the thumb carpometacarpal joint under fluoroscopic radiographic guidance (Figs. 39.3 and 39.4). The operative technique used in this study has previously been described in detail by Haas et al. [3]. After wound closure and application of dressings, a plaster cast was applied for 7 days. No physiotherapy is needed, and patients are advised to use the hand without restriction after immobilization. After a simultaneous onset of pain during injection of the fat tissue due to the elongation of the joint capsule and a feeling of pressure, the pain will gradually decrease within the first 3 weeks. NSAIDs are sufficient to control the pain. The success of the therapy can be evaluated by the reduction of pain at rest and under load. The final result, whether a patient improves from the therapy or not, can be assessed after 3 months postoperatively.

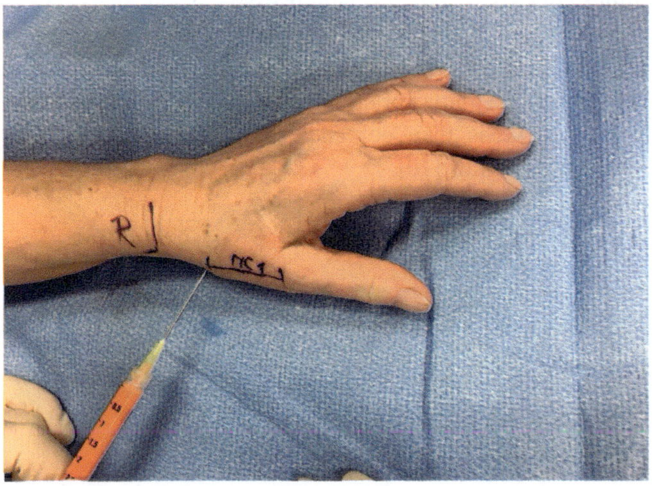

Fig. 39.3 Autologous fat transfer of 1–2 cc via a 17G-Needle under fluoroscopy guidance [4]

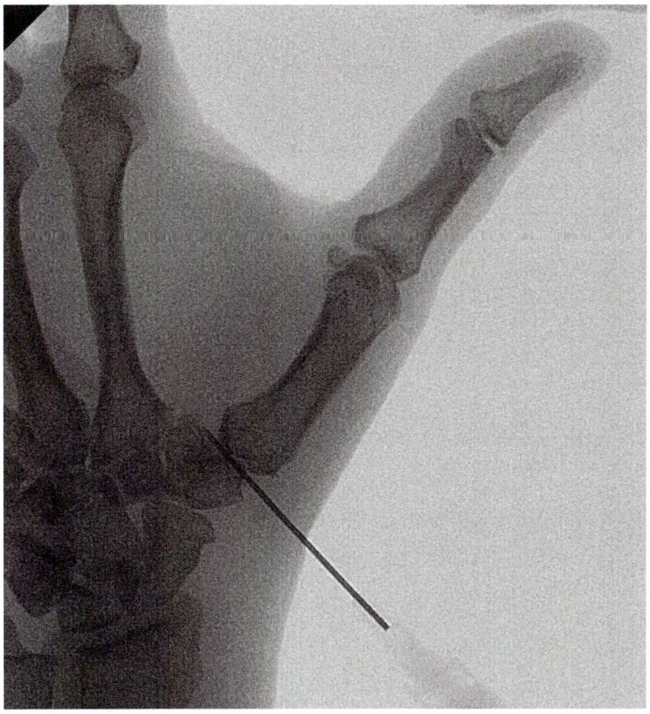

Fig. 39.4 Intraarticular infiltration of autologous fat into the CMC-I-joint

39.3 Management of Painful Neuromas

A neuroma is a benign growth of nerve tissue that may occur after a sharp or blunt injury to a nerve. There are multiple surgical treatment options for the treatment of neuroma pain. One commonly used technique is the excision of neuroma and proximal nerve stump relocation into a muscle, vein, or bone. More recent techniques respecting the physiology of nerve regeneration include end-to-side neurorraphy and targeted muscle reinnervation (TMR). Unfortunately, many of these procedures have a high rate of recurrence of pain and do have the need for concomitant professional pain management and sometimes lead to unsatisfactory results. Several clinical cases showed the positive effect of autologous fat grafting secondary to failed treatments mentioned above [5–11]. The difference of using autologous fat to the other aforementioned techniques is its mechanical and biological properties. Mechanically means the fat provides a smooth gliding layer where the nerve has free excursion and even creates a protective envelope [12]. By this means, it acts as a cushion against the surfacing skin and reduces pain caused by direct touch. It is well known that adipose fat tissue contains preadipocytes, pluripotent cells, and growth factors that may lead to positive effects of neoangiogenesis, anti-inflammatory modulation, and prevention of scar adherence. This technique was first described by Brown and Flynn in 1973 [13]. Patients usually complain about numbness, electric pain and tingling by simply touching the back of the hand. One has to differentiate between the different types of axon damage. It can lead to either a neuroma-in-continuity or an end-neuroma at the proximal stump when the neurotmesis was responsible. The Tinel's test is mostly positive. The goal of treatment is not only to reduce neuropathic pain but also to alleviate hyposensitivity in the area of the neuroma. Which can be explained that mechanical allodynia imitates hypoesthesia. Vaienti et al. named this technique "perineural fat grafting" (PFG) [10]. The indication for perineural fat grafting are (1) a condition of a terminal nerve stump or a neuroma in continuity that cannot be reconstructed, (2) neuromas that are still painful and failure of aforementioned treatment options, and (3) areas of nerve stump where no relocation technique is feasible. It can also be repeated to increase the volume of the fat depot and improve results.

39.3.1 Operative Technique

This procedure can be done as well under local anesthesia technique described above. Postoperative regional anesthesia, i.e., by means of a plexus catheter to slowly allow sensation return over multiple days, is most often very useful. The amount of fat harvested is dependent on the side and size of the neuroma. The more proximal the neuroma, the more space is warranted for the fat graft. The neuroma should be marked prior to surgery, and the incision should not be placed directly on top of it to prevent adherence and scarring of tissue (Fig. 39.5a). The harvested fat from the abdominal wall is then injected in multiple layers around the main focus of pain localized together with the patient by identifying through provoking a Hofman-Tinel sign (Fig. 39.5b, c). Additional surgical dissection by "Rigottomies" or open dissection of subcutaneous scars might be additionally helpful. The postoperative protocol includes early mobilization of the affected extremity without immobilization by means of a splint and early desensibilization training.

There are several other techniques described to ensure the nerve is wrapped with the transplanted fat. One step is to first accurately dissect the neuroma and ensure that the proximal stump consists of healthy neural tissue. Then, the nerve is "externalized" through an additional excision proximal to the first cut. The long scar is sutured, the fat is then injected via Coleman cannulas or simple 17G needles, and the externalized nerve is then cut and should become embedded within the fat graft [10].

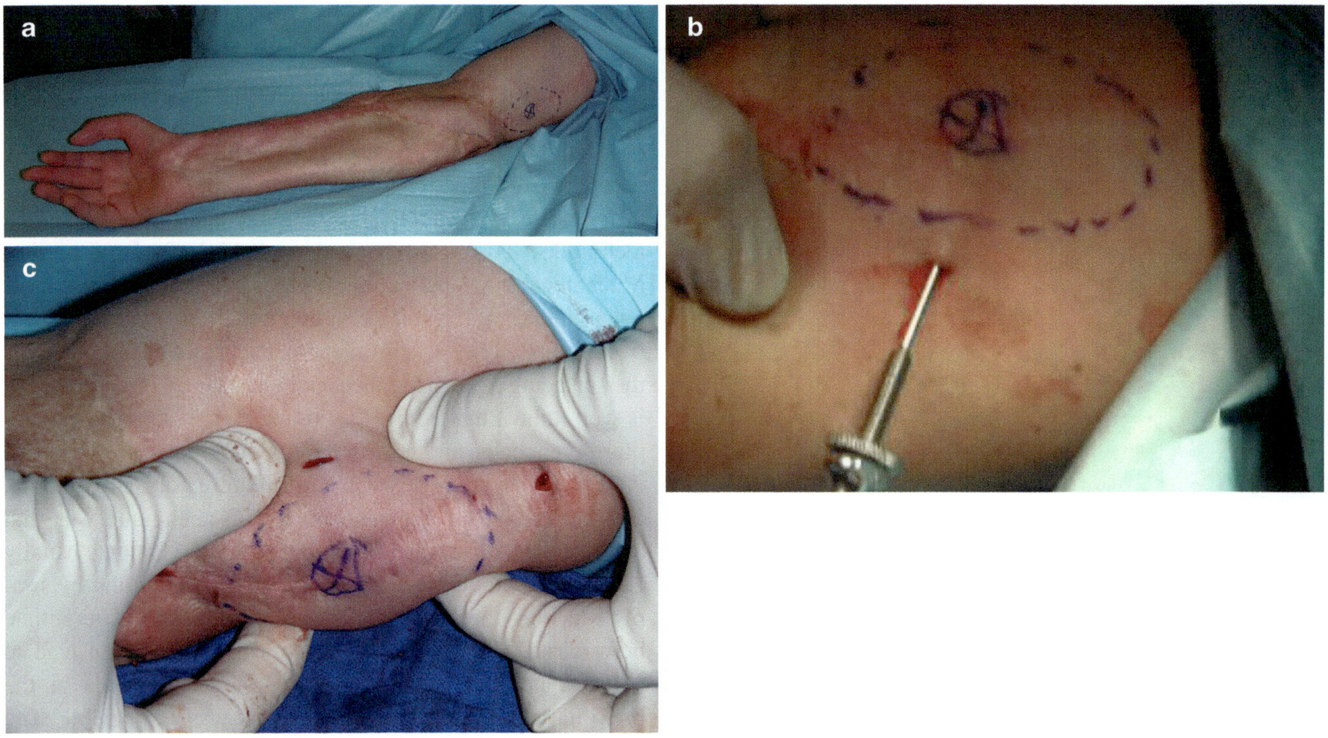

Fig. 39.5 (**a**) Patient multiple years after severe upper arm crush injury with long distance loss of the median nerve from upper arm to wrist level. In the distal upper arm stump after multiple surgical procedures residual local pain with intense Hofmann-Tinel sign (marked with an "X"). (**b**) Fan-shaped injection of fat graft to cushion between skin and scar. (**c**) Intraoperative pinch test showing increased subcutaneous fat layer

39.4 Treatment of Burn Scars and Substantial Defects

Substantial defects can be devastating, especially for children, female, and patients whose hands are the focus during their work. Different conditions result in a substantial tissue defect on the dorsum of the hand. Moreover, retractile and painful scars compromise daily mobility of the wrist and finger joints and can improve from autologous fat transfer. For a simple volume defect, fat graft can act as a filler in a mechanical way and the containing ADSCs can even improve the skin quality by taking advantage of local neoangiogenesis and potential regeneration of dermis and subcutaneous tissue. It is even discussed that the role of fat in scarred areas can lead to the release of multiple nerve entrapments so that neurogenic pain can be reduced [14, 15]. The authors describe the positive effect of fat transfer from an aesthetic and a functional point of view. These results can be detected in the early wound healing phase within the first 2 weeks after surgery. The volumetric correction also leads to an improvement of appearance which has a positive effect on patient's body image.

Burn injuries can result in hypertrophic burn scars that cause problems of aesthetic and functional concerns and moreover in one's social and psychological life. Especially altered anatomy can be limiting in everyday life. A functional impairment can develop due to hyperextension of the metacarpophalangeal joint with a boutonniere or claw deformity or contractures that are located at the first web space. Autologous fat grafting in burn injury scars as a secondary reconstruction attempt has shown that it can improve the clinical outcome as well as on the histological level [16, 17]. The histological examinations could show an improvement due to new collagen deposition, neoangiogenesis, and dermal hyperplasia, meaning tissue regeneration. The results of these studies show that even 1 year after treatment, patients report about clinical improvement. This improvement becomes noticeable by means of softness, thickness, color, and elasticity of the treated skin. There are not many cases in the literature about fat grafting for treating burn scars that are limited to the hand. But the few cases that have analyzed its effect are consistently positive regarding the changes observed in the regression of scar tissue and thus improvement of active motion, cosmetic appearance, and thus overall patient satisfaction [18]. This effect is most likely explained by the release of extrinsic and intrinsic tension that is caused by the scars.

39.4.1 Operative Technique

The literature recommends more than one single fat transfer to achieve the most effective result. Attention must be given not to "overcorrect" and produce a distorted appearance. After harvesting the necessary amount of fat as described in the technique above, a stab incision with a No. 15 blade is made adjacent to the scar contracture. A tunnel for the lipoaspirate must be produced by gently redirecting multiple passes with a blunt cannula. When addressing a relatively thick tethered scar, a release has to be performed with riggotomies or with 18G needles. The maximum amount possible to inject is achieved when fat extrusion occurs with further injections. Due to contractures and minimized spaces, the use of 1 cc aliquots was recommended by Bircoll et al. to be most beneficial [19]. Suture of the incision was done after injection to prevent further extrusion. In case of two hands affected, we recommend treating the more affected limb first.

39.5 Remodeling of the Hand After Trauma

After severe hand injury, the shape of the hand might be aesthetically disturbed by molds or troughs. This might also be a result of severe peripheral nerve injury, i.e., a complex ulnar nerve (Fig. 39.6a) injury might result in interosseous atrophy with a permanent, disturbing at first glance visible trough in the first webspace on the dorsum of the hand (Fig. 39.6b).

39.5.1 Operative Technique

Fan-shaped fat grafting remodels the subcutaneous fat layer and reconstructs the normal appearance of the first webspace (Fig. 39.6c, d).

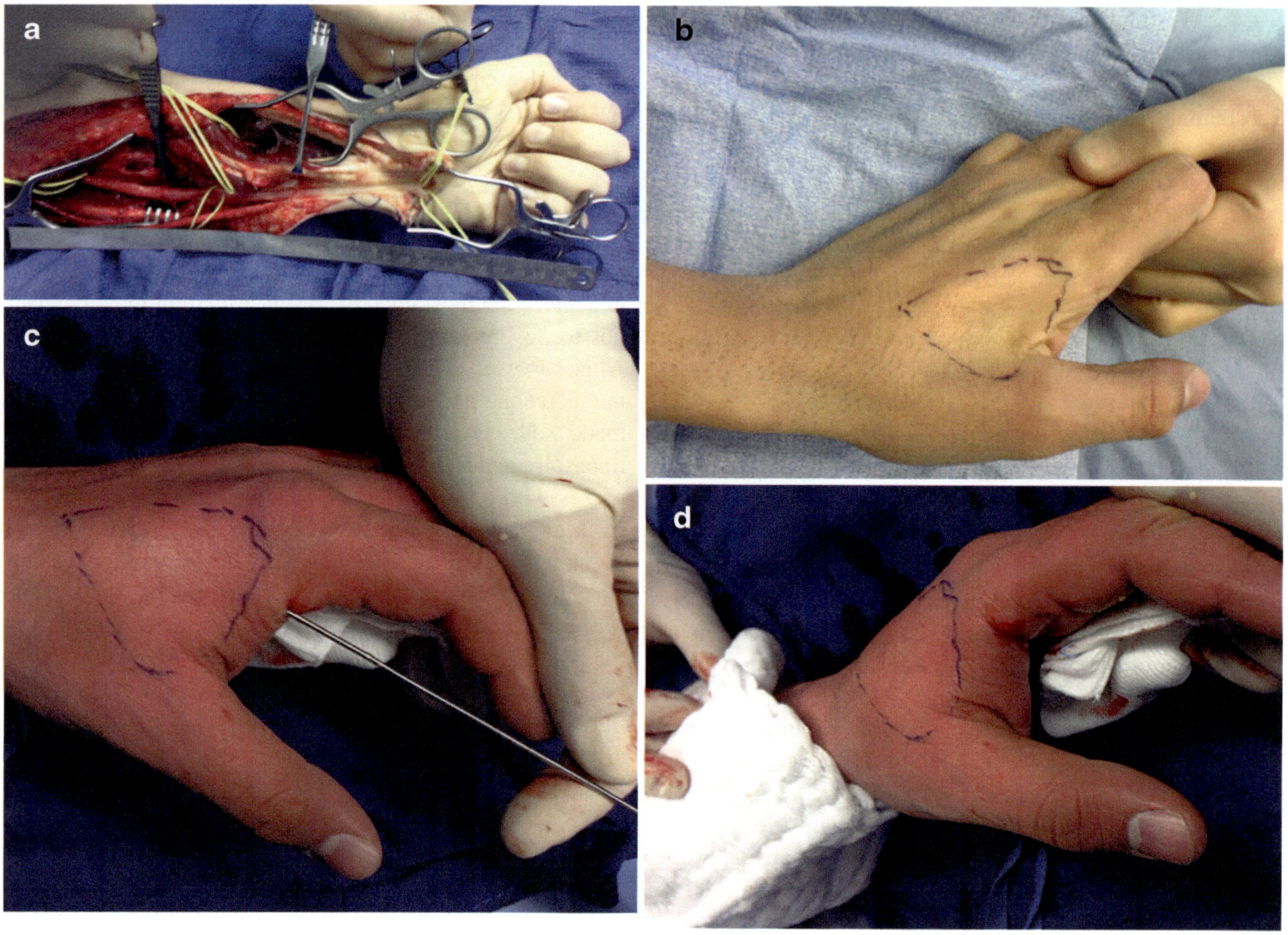

Fig. 39.6 (**a**) Complex ulnar nerve injury with need of sural nerve grafts for reconstruction. (**b**) Due to paralysis of the deep branch of the ulnar nerve and interosseous atrophy a disturbing visible trough in the first webspace results. (**c**) Fanshaped subcutaneous fat grafts fill up the trough and mimicry a functioning first dorsal interosseous muscle. (**d**) Fat grafting results in physiological appearance by enhancing the subcutaneous fat layer

39.6 Rejuvenation of the Hand ("Structural Fat Grafting")

The use of autologous fat transfer in hand for the treatment of age-related laxity was first mentioned in 1990 by Abrams et al. who described their technique of microlipoinjection [20]. Then Sidney Coleman, one of the pioneers of fat grafting, presented in 2002 his "structural fat grafting" technique at hand [21]. This term is based on the technique of the distribution of a structural fat implant throughout the dermal layers from deep to superficial. The importance of aesthetic and the desire for youth are of increasing importance in our society. The wish to improve the aesthetic appearance of the individual patient and to reduce the symptoms of aging is widespread. Apart from the face, the dorsum of the hand is permanently visible and thus an area that is of great importance. The hand is the ID of our biological age, as studies have confirmed that age can be precisely estimated by simply looking at one's hands [22]. But not only aging gives a lot of information about a person but scars, diseases, or trauma that cause deformities or malformations can be psychological stressing for an individual. The hand is like the rest of the body affected by the aging process of our bodies. This process means both atrophy of soft tissue and decline in skin quality due to sun exposure, especially of the dorsum of the hand. Additionally, other factors like mechanical stress of the hand in work and the daily direct contamination with possibly toxic substances are omnipresent. These factors can even promote the aging process, which leads to the following objectifiable changes at the dorsum of the hand.

Zhou et al. [23] analyzed the physiologic fat distribution of the hand with a B-mode ultrasound examination and developed a targeted fat grafting technique. They divided the dorsal fat compartments of the hand into three grades: (1) mild atrophy with rhytides, (2) moderate atrophy with exposed veins, and (3) serious atrophy with exposed tendons. These grades depend on the three fascial layers of the dorsum of the hand, which fuse proximal at the level of the retinaculum. Fat transfer can be securely performed into the superficial layer, because it does not contain relevant structures that can be harmed. Thus, it is recommended to perform fat grafting into the superficial and the deep layer—dependent on the grade—to address the grade-dependent volume restoration. The intermediate laminae should never be addressed because of the containing veins and nerves. Targeted fat restoration should only be applied for Grade 2 and 3 aged hands (Figs. 39.7 and 39.8).

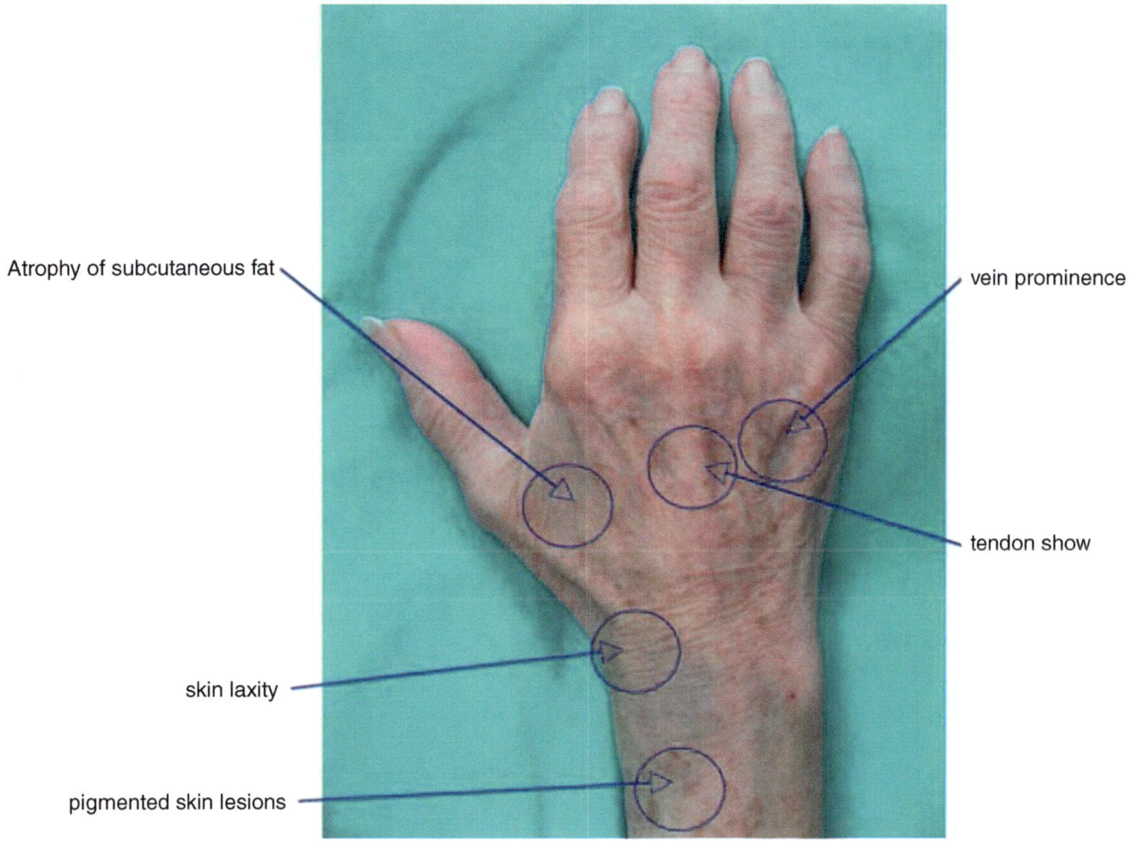

Fig. 39.7 Age-related changes of the dorsum of the hand

Fig. 39.8 Aging process of the hand. A 3-year-old girl (left), 40-year-old woman (middle), and 90-year-old woman (right) with full development of subcutaneous atrophy with resulting excess skin with wrinkling, pigmentous lesions, protrusion of tendons and vein shown. Furthermore, aesthetic impairment by means of malunion after distal radius fracture with ulna impaction [24]

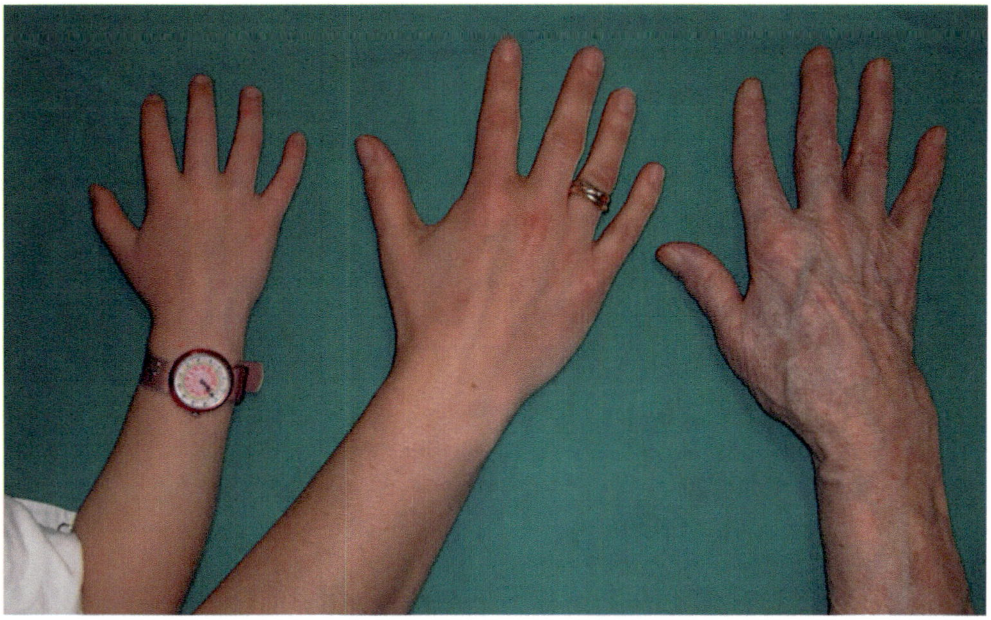

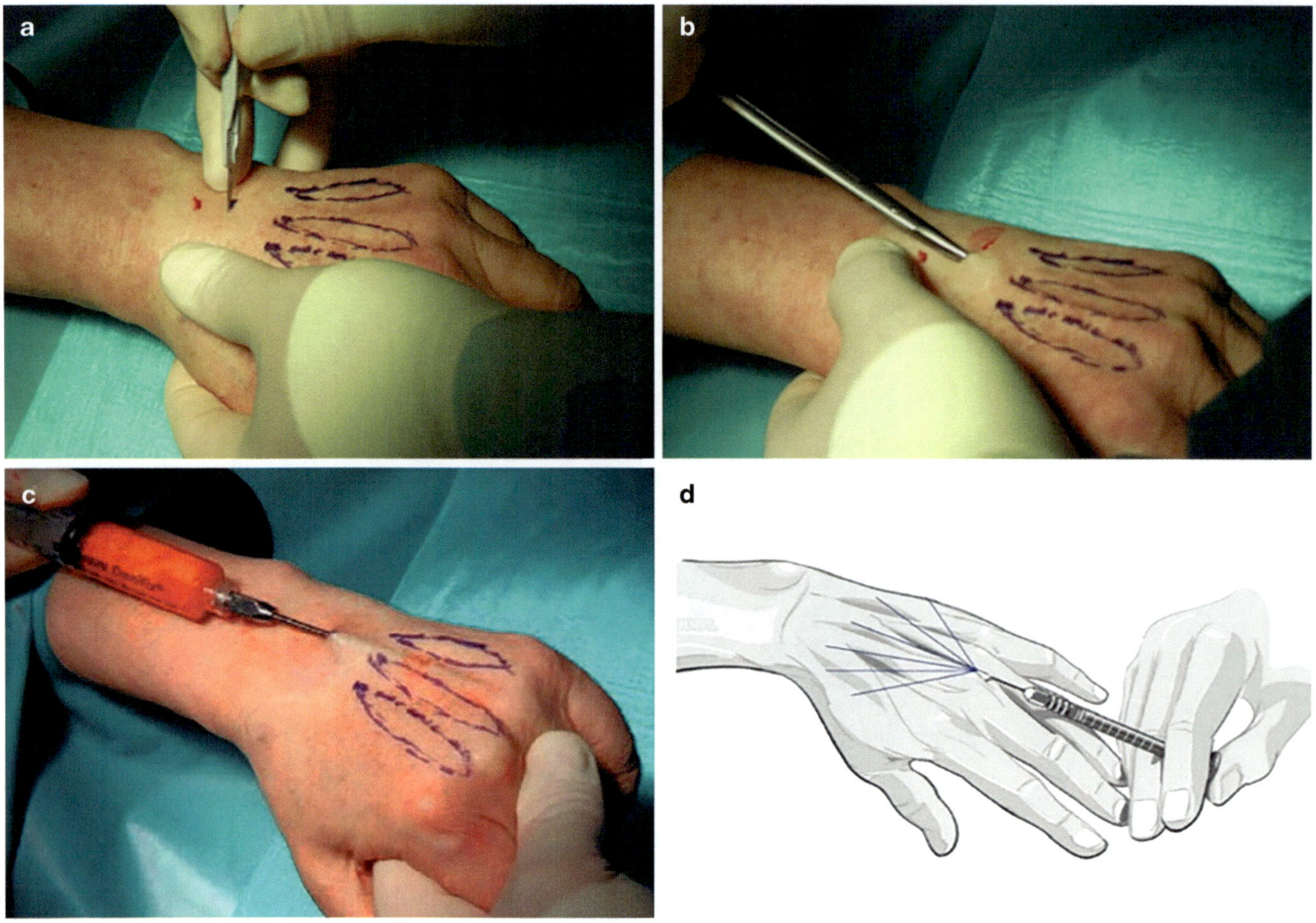

Fig. 39.9 Harvest and operative technique [24, 25]. Preoperative marking of atrophic areas between the extensor tendons on the dorsum of the hand and technique of structural fat grafting (**a**–**d**).

39.6.1 Operative Technique

At the dorsum of the hand, in our experience, the blunt injectional cannulas by Coleman are helpful. Preoperatively, volume defects are marked at the dorsum of the patient's hand (Fig. 39.9a). Then, local anesthesia with lidocaine is applied to the region by means of a hand block of the superficial branch of the radial nerve and the dorsal branch of the ulnar nerve. In most cases, five incisions allow enough access to the atrophic areas between the extensor tendons and metacarpals (Fig. 39.9a, b). One incision should be at the wrist level, and four incisions at the web spaces. Finally, the adipose tissue (Fig. 39.9c) taken from the abdomen is placed with an injection cannula where needed. This is done in multiple fan-shaped channels filling these fine channels with adipose tissue on withdrawal of the injection cannula (Fig. 39.9d) to keep the contact area to the fat graft as large as possible and thus enable a rapid incorporation; only very thin layers of adipose tissue are injected. This technique improves the survival rate of the graft. In total, at least 20 cc of fat should be transplanted to achieve an obvious volume gain. The procedure can be repeated after 6 months (Fig. 39.10). An external compression of the fat should preferably be avoided. We recommend to use solely blunt instruments for fat injection to prevent an injury to the veins and sensory nerves. Postoperatively, there is padding under the skin resulting from augmentation of the subcutaneous fat, and hence veins and tendons of the dorsum of the hand are less visual, creating the rejuvenating effect. All incisions should be sutured to prevent leakage of transplanted fat. A gauze dressing is applied for 5 days and a forearm splint for immobilization of the wrist. The most common complication is edema that dissolves within in the first weeks after surgery.

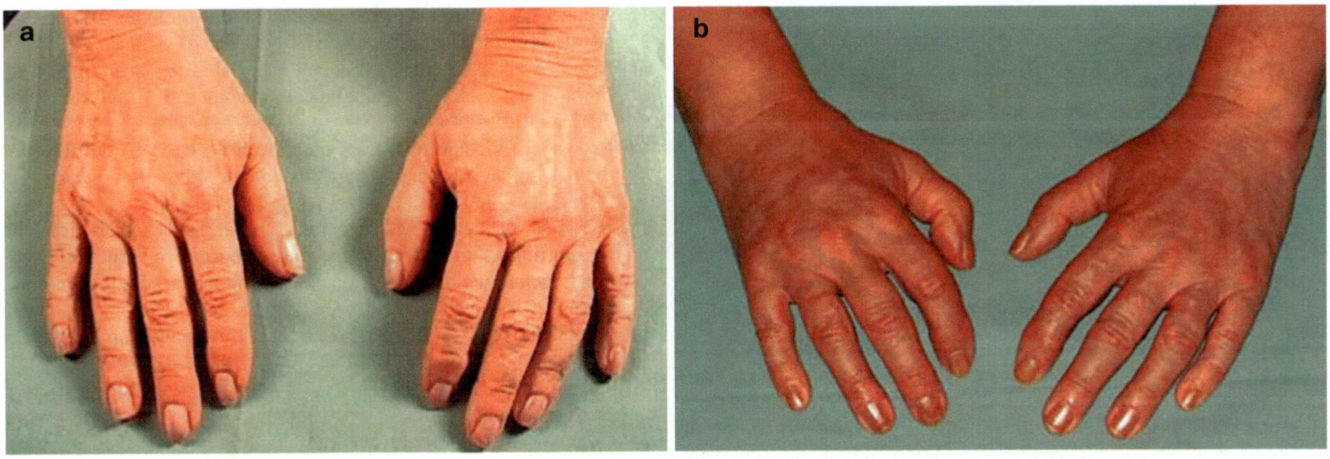

Fig. 39.10 Autologous fat transfer on the dorsum of the hand of 72-years-old woman [25]. (**a**) Preoperative situation with protruding extensor tendons und veins. (**b**) 6 months postoperatively, the subcutaneous fat tissue has significantly gained more volume and both extensor tendons as well as veins are less prominent

39.7 Discussion

The described applications of autologous fat transfer reveal the potential power of this kind of tissue transfer. In each case, individual and detailed informed consent of the opportunities and risks of the various plastic and aesthetic procedures is a compulsory prerequisite for the procedure.

The operation itself requires knowledge of the specific surgical anatomy of the hand, standard plastic surgical techniques and postoperative surgical care. Hence, surgery should only be performed by an experienced hand and plastic surgeons to prevent injuring abdominal organs during fat harvesting and not to harm the important underlying structures of the hand during transfer such as nerves, tendons, and veins. There are not many side effects known; only in rare cases, postoperative infection with atypical mycobacteria after structural fat grafting has been described, which underlines this claim [26]. Other complications were hematoma, edema, fat necrosis/oil cysts and resorption. Complications can also occur at the donor site: swelling, hematoma, paresthesia, contour disorders, and injury to deeper structures.

On the other hand, the surgical technique presented here in the hands of experienced plastic surgery is easy to understand and is a relatively safe procedure with a great effect depending on the underlying pathology. The injected fat should be as homogeneous as possible to create a smooth layer without larger nodes. After injection, no secondary modulation from outside the skin should be performed.

To evaluate the success of volume gain, objective 3D body surface imaging can be applied. It is a well-established method that allows a reliable determination of volume and surface contour analysis in the course. Its use in plastic and reconstructive surgery has grown during the last decade [27]. Therefore, it is a perfect method to document and evaluate the results of transplanted autologous fat tissue to create long time results. 3D surface imaging enables an objective analysis tool that is superior in data quality compared to a conventional two-dimensional photograph [28]. However, it does not evaluate graft vitality and is limited only to volumetric measurements.

Autologous fat transplantation is the preferred technique to address subcutaneous atrophy at the dorsum of the hand, volume deficits, burn scars and painful neuromas. Especially because autologous fat grafting appears to be currently most suitable in terms of a regenerative approach [29, 30]. In addition to the currently much-debated aspect of this method with possibly transplanted stem cells of the adipose tissue (ADSCs), the method may have particularly strong healing potential. This aspect, however, is yet unclear and is currently being investigated in numerous studies.

39.8 Conclusion

The technique of autologous fat grafting is an effective and safe surgical technique with a variety of clinical applications. It is a relatively safe procedure that can be done in an ambulant setting. Especially in the field of hand surgery, harvesting fat can be performed in local anesthesia, making it even more attractive to perform it as an outpatient-surgery. Its positive attributes, especially concerning the anti-inflammatory and regenerative effect of the containing ADSCs, make autologous fat even more interesting for new therapeutic approaches. The clinical as also scientific interest is tremendous, and we are looking with great interest and anticipation to the upcoming studies and their growing application in even more fields of hand surgery.

References

1. Rennekampff HO, Reimers K, Gabka CJ, et al. Current perspective and limitations of autologous fat transplantation—"consensus meeting" of the German Society of Plastic, Reconstructive and Aesthetic Surgeons at Hannover; September 2009. Handchir Mikrochir Plast Chir. 2010;42:137–42.
2. Herold C, Fleischer O, Allert S. Eigenfettinjektion in das Sattelgelenk zur Behandlung der Rhizarthrose—eine viel versprechende Therapieoption. Handchir Mikrochir Plast Chir. 2014;46:108–12.
3. Haas EM, Volkmer E, Giunta RE. Pilot study on the effects and benefits of autologous fat grafting in osteoarthritis of the CMC-1 joint compared to intraarticular cortisone injection: results after 3 months. Handchir Mikrochir Plast Chir. 2017;49:288–96.
4. Haas EM, Eisele A, Arnoldi A, et al. One-year outcomes of intraarticular fat transplantation for thumb carpometacarpal joint osteoarthritis: case review of 99 joints. Plast Reconstr Surg. 2020;145:151–9.
5. De Jongh F, Pouwels S, Tan LT. Autologous fat grafting for the treatment of a painful neuroma of the hand: a case report and review of literature. Cureus. 2020;12:e10381.
6. Huang SH, Wu SH, Chang KP, et al. Alleviation of neuropathic scar pain using autologous fat grafting. Ann Plast Surg. 2015;74(Suppl 2):S99–104.
7. de Gast H, Torrensma B, Fitzgerald E, et al. The treatment of chronic neuropathic pain: bio (regenerative) pain treatment through lipofilling. A short communication case series. Pain Physician. 2016;19:E495–8.
8. Fredman R, Edkins RE, Hultman CS. Fat grafting for neuropathic pain after severe burns. Ann Plast Surg. 2016;76(Suppl 4):S298–303.
9. Maione L, Vinci V, Caviggioli F, et al. Autologous fat graft in postmastectomy pain syndrome following breast conservative surgery and radiotherapy. Aesthet Plast Surg. 2014;38:528–32.
10. Vaienti L, Merle M, Battiston B, et al. Perineural fat grafting in the treatment of painful end-neuromas of the upper limb: a pilot study. J Hand Surg Eur. 2013;38:36–42.
11. Vaienti L, Gazzola R, Villani F, et al. Perineural fat grafting in the treatment of painful neuromas. Tech Hand Up Extrem Surg. 2012;16:52–5.
12. Millesi H, Hausner T, Schmidhammer R, et al. Anatomical structures to provide passive motility of peripheral nerve trunks and fascicles. Acta Neurochir Suppl. 2007;100:133–5.
13. Brown H, Flynn JE. Abdominal pedicle flap for hand neuromas and entrapped nerves. J Bone Jt Surg Am. 1973;55:575–9.
14. Mojallal A, Lequeux C, Shipkov C, et al. Improvement of skin quality after fat grafting: clinical observation and an animal study. Plast Reconstr Surg. 2009;124:765–74.
15. Klinger M, Caviggioli F, Klinger FM, et al. Autologous fat graft in scar treatment. J Craniofac Surg. 2013;24:1610–5.
16. Bruno A, Delli Santi G, Fasciani L, et al. Burn scar lipofilling: immunohistochemical and clinical outcomes. J Craniofac Surg. 2013;24:1806–14.
17. Brongo S, Nicoletti GF, La Padula S, et al. Use of lipofilling for the treatment of severe burn outcomes. Plast Reconstr Surg. 2012;130:374e–6e.
18. Byrne M, O'Donnell M, Fitzgerald L, et al. Early experience with fat grafting as an adjunct for secondary burn reconstruction in the hand: technique, hand function assessment and aesthetic outcomes. Burns. 2016;42:356–65.
19. Bircoll M, Novack BH. Autologous fat transplantation employing liposuction techniques. Ann Plast Surg. 1987;18:327–9.
20. Abrams HL, Lauber JS. Hand rejuvenation. The state of the art. Dermatol Clin. 1990;8:553–61.
21. Coleman SR. Hand rejuvenation with structural fat grafting. Plast Reconstr Surg. 2002;110:1731–44.
22. Hoang D, Orgel MI, Kulber DA. Hand rejuvenation: a comprehensive review of fat grafting. J Hand Surg Am. 2016;41:639–44.
23. Zhou J, Xie Y, Wang WJ, et al. Hand rejuvenation by targeted volume restoration of the dorsal fat compartments. Aesthet Surg J. 2017;38:92–100.
24. Giunta RE, Eder M, Müller D, et al. Structural fat grafting for rejuvenation of the hand: how much volume do we get? In: Schantz JT, Hutmacher DW, editors. A manual of current therapies in regenerative medicine. 1st ed. Aufl: World Scientific Publishing Company; 2013.
25. Giunta RE, Eder M, Machens HG, et al. Structural fat grafting for rejuvenation of the dorsum of the hand. Handchir Mikrochir Plast Chir. 2010;42:143–7.
26. Vara AD, Miki RA, Alfonso DT, et al. Hand fat grafting complicated by abscess: a case of a bilateral hand abscess from bilateral hand fat grafting. Hand (N Y). 2013;8:348–51.
27. Spanholtz TA, Leitsch S, Holzbach T, et al. 3-Dimensional imaging systems: first experience in planning and documentation of plastic surgery procedures. Handchir Mikrochir Plast Chir. 2012;44:234–9.
28. Kovacs L, Yassouridis A, Zimmermann A, et al. Optimization of 3-dimensional imaging of the breast region with 3-dimensional laser scanners. Ann Plast Surg. 2006;56:229–36.
29. Billings E Jr, May JW Jr. Historical review and present status of free fat graft autotransplantation in plastic and reconstructive surgery. Plast Reconstr Surg. 1989;83:368–81.
30. Rigotti G, Marchi A, Sbarbati A. Adipose-derived mesenchymal stem cells: past, present, and future. Aesthet Plast Surg. 2009;33:271–3.

Complications and Pitfalls

40

Chris Megapanos

Contents

40.1 Complications and Pitfalls

Liposuction, along with breast augmentation, is the most popular surgery procedure performed by plastic surgeons around the world [1]. Liposuction alone, and even when combined with fat grafting, is the surgery with the most complications and the highest mortality rate. Due to the popularity of the operation, the net number of mortality and serious complications is the largest in the specialty of plastic surgery [2]. Bibliographic analysis and research conclude that many cases with serious complications and fatalities are unreported.

40.1.1 History of Complications and Pitfalls

The first case of removing local fat from the human body was performed by Charles Dujavvier in France. He tried to remove subcutaneous fat from the calves and knees of a dancer in 1921, using Uterine Curette. The result was the injury of the femoral artery that led to the amputation of the dancer's leg. Until the 1970s, many attempts were made to remove local fat, which were accompanied by severe complications, such as hematomas and seromas.

The era of modern liposuction began with Arpad and Giorgio Fischer, who developed tools such as blunt cannulas combined with suction and crisscross tunneling techniques through multiple incisions [3]. The first good results began to appear, but with a high increase of hematomas.

Pierre Fournier was the one who adopted Fischer's technique and made it popular around the world of plastic surgery [4]. Illouz took another step further by modifying Fischer's technique (dry technique). He used a hypotonic saline solution with hyaluronidase (wet technique). He managed to reduce the number of complications further and help to achieve better aesthetic results [5]. The next step of reducing complications was taken by Hetter with the enrichment of Illouz solution with epinephrine [6] and further modification by Jeffrey Klein with the introduction of tumescent infiltration. This led to a significant reduction in the rate of complications, making the procedure very safe and slowly popularizing it in the general population [7].

In the 1980s, there was a continuous improvement of liposuction instruments and the introduction of energy-based devices with further improvement of morbidity and mortality rates [8]. In the 1990s, liposuction was introduced as one of the safest plastic surgery procedures, and several plastic surgeons around the world began to apply it. Also, it became the most popular procedure and surgeons started to combine it with fat transfer techniques [9]. While the procedure's popularity is rising, new complications such as DVT, fat embolism, skin necrosis, or visceral perforation are reported. Nowadays, the percentage of serious complications is 1.9%, with risk factors depending on the gender, as women have 2.5% more than men. Morbidity obese patients have 6.3%

C. Megapanos (✉)
GM Clinic, Athens, Greece

© The Author(s), under exclusive license to Springer Nature Switzerland AG 2023

A. Di Giuseppe et al. (eds.), *Fat Transfer in Plastic Surgery*, https://doi.org/10.1007/978-3-031-10881-5_40

higher complication rate than normal weight patients and the combination of another aesthetic procedure with liposuction raises the rate [10].

The identification of complications that accompany the fat transfer process and liposuction and the risk factors that lead to them contributes to informing plastic surgeons on better preparation, prevention, as well as thorough diagnosis and treatment.

Liposuction is associated with a variety of complications and pitfalls most of which can be avoided. These can be categorized into systemic, local harvesting sites and local grafting sites.

40.2 Systemic

Systemic complications are complications that occur intraoperatively or during the first postoperative days and have to do with systemic reactions from the cardiopulmonary system.

40.2.1 Deep Vein Thrombosis or Thromboembolism

Deep vein thrombosis is the most common systemic complication occurring to patients who undergo liposuction. According to studies, the DVT rate is 1.35% for patients undergoing plastic surgery.

DVT is manifested in the first four postoperative days.

The predisposing factors are smoking, contraceptives, hormone replacement therapy, more than two pregnancies, and age over 40 [11].

A very good evaluation tool is now the RAM caprini module which helps surgeons evaluate each patient and make decisions on prophylaxis [12].

DVT diagnosis symptoms are local pain in one leg or arm, swelling of the foot or arm, warm skin over a sore area, red or dark skin over a sore area, and an expanded superficial capillary network.

The most life-threatening complication of DVT is pulmonary embolism. It occurs when a clot that was formed in the extremity detaches and ends up in pulmonary vessels. The symptoms of PE are very acute with sudden shortness of breath, chest pain on coughing or during deep breathing, dizziness or fainting, tachycardia, tachypnea, and hemoptysis during coughing.

The final diagnosis is made with U/S or spiral CT.

Proper preoperative preparation of the patient leads to lower rates of DVT. The preparation should include meticulous record keeping of medical history based on Ram Caprini, application of anticoagulant socks, intraoperative use of a flowtron device, and admission of low molecular weight heparin.

Treatment consists of oxygen ventilatory support; anticoagulation drugs with LMWH are mandatory in managing these events. Selectively the situation may require thrombolysis or pulmonary embolectomy [13].

40.2.2 LAST: Local Anesthetic Systemic Toxicity

Following Klein's introduction of the tumescent technique, the use of local anesthetics in liposuction and fat transfer became the new standard. By using these anesthetics, complications such as allergic reaction and toxicity syndrome due to the administration of large doses began to rise.

The anaphylactic reaction to lidocaine is usually due to the preservatives that accompany the substance. The risk factor should be checked during the medical history of previous admission of the anesthetic, and, if not, a test dose should be administered subcutaneously to the patient, who should be monitored for signs of an allergic reaction.

In case of an allergic shock intraoperative, adrenaline, 100% oxygen, and intravenous fluids should be administered intravenously. If bronchospasm is diagnosed, one must proceed to secure the airway by intubation.

Local anesthetic toxicity syndrome is a complication that occurs 1–18 h after anesthesia infiltration. It can be prevented if during the medical history taking it is revealed that the patient is on cytochrome P450 inhibitors, and if lidocaine is kept below 35 mg/kg when it is used with epinephrine on infiltration. It should be noted that deaths associated with lidocaine toxicity have been observed at doses as low as 4–6 mg/kg. The symptoms observed during the onset of the syndrome are nausea, vomiting, metallic taste, numbness of the lips, tremor, light-headedness, restlessness, slurred speech, psychosis, cardiac arrhythmia, and cardiac arrest. The treatment is the administration of fat emulsion (intralipid 20%) intravenously [14, 15].

40.2.3 Pulmonary Edema

After the introduction of the super wet technique and the use of infiltration volumes over 10 L in body shaping liposuctions (total body contouring), the risk of fluid overload and pulmonary edema has risen. Cardiac and renal function assessment should be performed preoperatively and titrating the amount of IV fluid infusion. The human body can withstand up to 2000 cc intravenous fluids per hour with normal cardiac and renal function. We should also be aware of the fact that 70% of the infiltration fluid is shifted intravascularly on the first 3 days.

If we calculate the intravenous fluids taken intraoperatively and subtract the renal output, then we can easily reach the maximum volume of infiltration that should be administered without a problem [16, 17].

Pulmonary edema can be diagnosed from symptoms such as shortness of breath, nausea, coughing with frothy and bloody sputum, wheezing, anxiety, arrhythmia, and tachycardia.

The treatment of pulmonary edema is the administration of diuretics and oxygen, and therapy in severe and acute conditions is ventilatory support [18].

40.3 Harvesting Site Complications and Pitfalls

In order to collect fat for fat grafting liposuction, the surgeon must work on a harvesting area of the body. There are numerous complications that follow the harvesting site specifically. Some of them relate to the final aesthetic result, while some others are life threatening.

40.3.1 Seromas

Seroma is the collection of fluid (lymph) in a cavity of the body which is aseptic. The cavity is formed by fat harvesting. A correlation between ultrasound-assisting devices and liposuction has been found (Fig. 40.1). Also, the possibility of fluid collection is raised when liposuction is done very superficially due to damage to the lymphatic network of the skin. When an ultrasound-assisted device is used, the use of drainage and lymphatic massage is suggested.

Needle aspiration will resolve the situation. The procedure must be done under aseptic conditions. If the seroma reappears, then a drainage system must be placed under radiological guidance and prophylactic antibiotics should be administered to prevent infection. A garment must be used in order to minimize cavity formation.

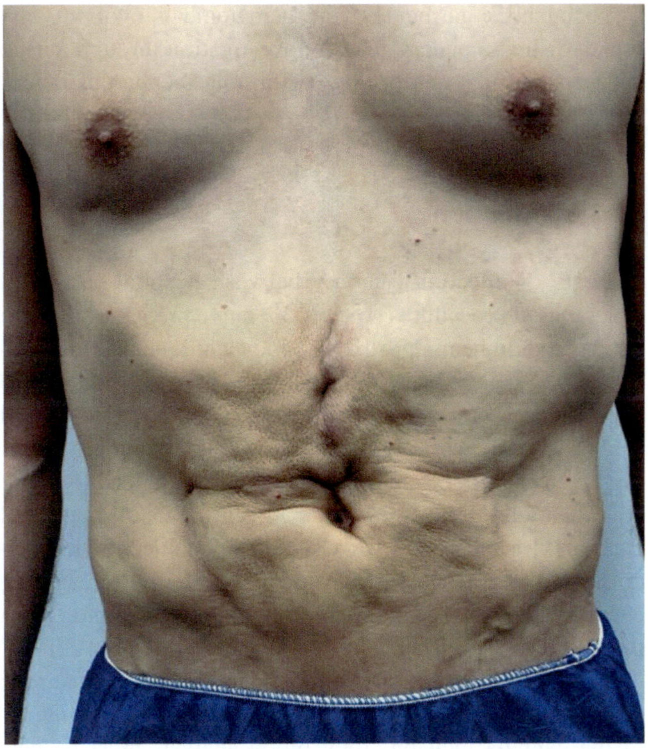

Fig. 40.1 Seroma is the collection of fluid (lymph) in a cavity of the body which is aseptic

40.3.2 Hematomas

It is a condition that can appear on the first postoperative day. The collection of blood in the cavity of the harvesting area. The use of energy-assisted devices has shown to reduce the risk of hematomas (Fig. 40.2). But the standard rule to prevent it is the use of an infiltration solution with epinephrine and pressure garments at the end of the operation. The use of anticoagulant drugs or weight loss supplements like CLA or garlic has been proven to raise the possibility of bleeding. Early diagnosis of hematomas is crucial. High-pressure bandaging must be performed over the harvesting site to stop the bleeding, except in the neck. Intravenous fluid resuscitation and monitoring of patients' Hgb should be performed to evaluate the condition. Most of the hematomas stabilize on the first day and slowly resolve only with the application of the garment. The area of the neck should be treated only with

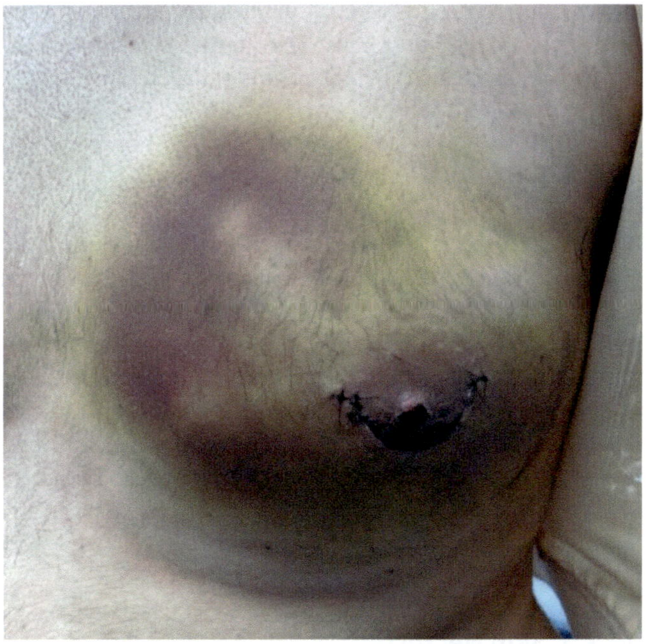

Fig. 40.2 Hematoma. The collection of blood in the cavity of the harvesting area

a special neck garment. If the hematoma is evolving and breath distress appears, then an investigation through a neck lift incision must be done by performing meticulous hemostasis with electrocautery [19].

40.3.3 Infection

The rate of infection can vary between hospital cases and office-based procedures. The preoperative preparation of the patient should be under strict aseptic technique and full sterile coverage of the body of the patient. There is a tendency to approach fat grafting and liposuction on an office-based setup as a minor surgery and doctors do not pay much attention to sterility in the surgical field. This could lead to major complications. If a surgical site infection is not diagnosed early and treated properly, it could evolve to necrotizing fasciitis or to a toxic shock syndrome which are life-threatening events with the second worst mortality rate behind pneumonic embolism. The symptoms of an undermining infection could be fever over 38 °C, erythema, warmth, blisters, or skin necrosis over the harvesting area, plus fluid running out of the cannula incision. Total blood analysis must be performed to check the WBC count and CRP. Cultures must be sent from any fluid that is correlated with the area. If the area of infection is large, then the exploration and drainage of the abscess through an incision is mandatory. Treatment with full-spectrum antibiotics must start at the beginning of the symptoms and change after the results of the culture [20, 21].

40.3.4 Skin Burn

In the last 20 years, the use of energy-based devices in fat grafting procedures has risen exponentially. With the raise in popularity of these devices, new complications appeared such as skin burns. The energy of a probe or a fiber is transited to the surrounding tissue to provoke a controlled burn or fat lysis. If the energy is too high, the area that is affected is larger. Also, if the tip of the device is near the skin for a prolonged time, then a burn could occur. These burns are always full-thickness burns and must be excised surgically leaving a scar behind [19].

40.3.5 Skin Necrosis

Skin loss is a devastating complication that leaves visible scars on the patient and prolongs the post-operative treatment time. The factors that lead to this complication is smoking, so it is recommended that the patients must stop smoking 2 weeks before the surgery. Also, the combination of liposuction with abdominoplasty (Fig. 40.3), breast reduction, thigh lift, or brachioplasty could result in skin necrosis on the area of the scar or the nipple. Compression garments are responsible for blood stasis in the superficial capillary network of the skin, if they are tighter above the venous pressure. The blood cannot flow correctly, and thrombosis could occur (Fig. 40.4). Another possible reason for skin loss is the very superficial resection of fat that could damage the superficial capillary network, especially when an energy-based device is used. The treatment of this occurrence could be loosening the garments or hyperbaric oxygen and the admission of prostaglandin E1 to increase oxygenation to the area. If the skin cannot recover after 1 week, the surgical excision must be done to clean the necrosis and close the wound [22].

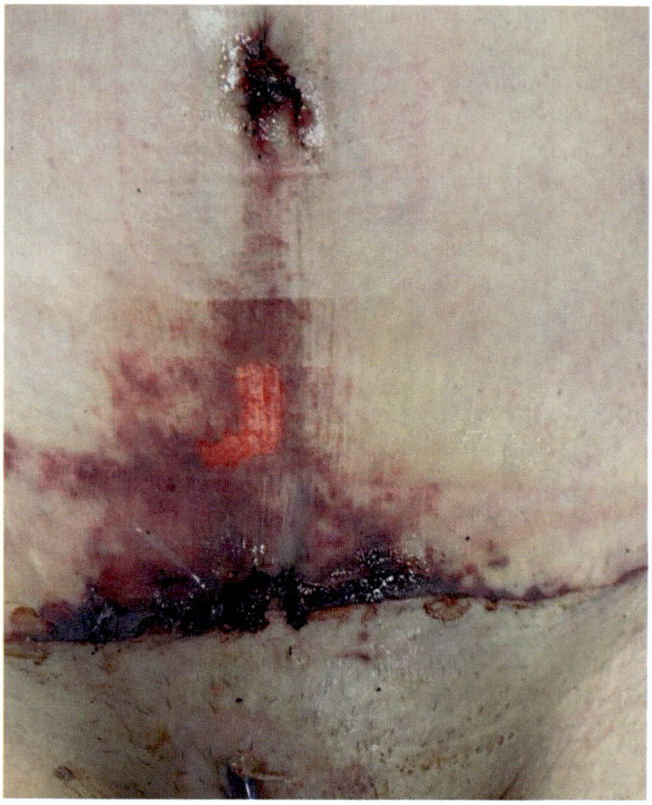

Fig. 40.3 Skin necrosis the combination of liposuction with abdominoplasty

40.3.6 Perforation of Inner Organs and Visceral

This is the rarest complication with a few cases reported around the world through retrospective questionnaires. The cases are believed to be under-reported because of the high mortality rate of this complication. It is characterized by a very difficult and late diagnosis that could lead to a fatal event of over 50%. It is responsible for 15% of the mortality of cases that have undergone liposuction. Bowel perforation

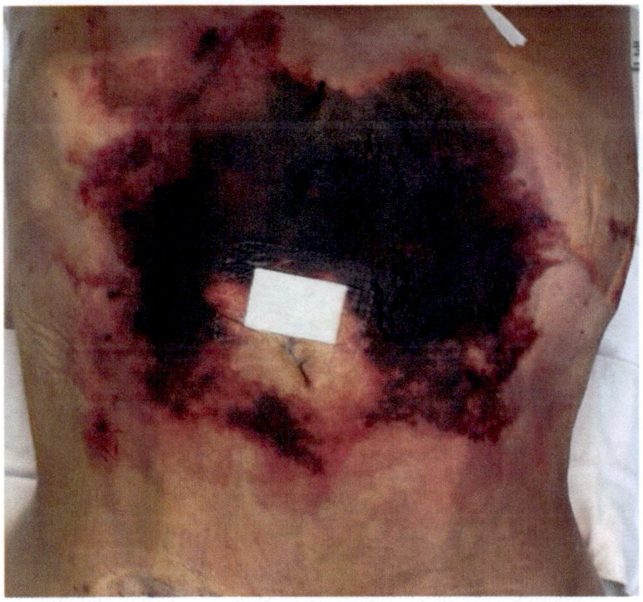

Fig. 40.4 Skin necrosis. The blood cannot flow correctly, and thrombosis could occur

is the most frequent case, and the factors vary between previous surgical scars on the abdominal wall where the cannula could deflect inside the abdominal cavity. Wall laxity in obese patients where the thickness of the oblique muscle is very low. The presence of an hernia, especially on the umbilicus, where clinical assessment is needed to be performed under Valsalva maneuver and then ultrasound confirmation of the clinical findings. The surgeon must be careful not to trust the results of the sonographic exploration of the abdominal wall due to the high rate of false negative results. Diagnosis of this disastrous complication depends on the organ being perforated. Pneumothorax, cardiac, liver, or spleen injury will occur after deflection of the cannula under the pleural arch with symptoms of dyspnea and respiratory failure after damaging the lung or hypovolemic shock because of internal bleeding due to perforation of internal organs. This event is acute and will appear at the time of the operation or during the first postoperative hours. Damage to the bowel or colon will lead to peritonitis. Acute peritonitis should be suspected when symptoms are acute abdomen with abnormal abdominal pain, abdominal muscle spasm, silence of intestinal sounds, signs of ileus such as nausea, vomiting, constipation, and gas obstruction on the first postoperative day. The final diagnosis is sealed with CT scanning or X-ray with findings of pneumothorax or signs of air or fluid inside the abdomen. After that, an emergency laparotomy or thoracotomy should be undertaken to repair the visceral or organ perforation [23–25].

40.3.7 Poor Aesthetic Result

One of the most common pitfalls of the harvesting area is a poor aesthetic result being left after fat harvesting. This happens when the surgeon over-extracts fat tissue in a small body area (Fig. 40.5). In order to minimize the surgical area, trauma could falsely lead on a road to take a lot of fat from this certain area, leaving dimples, holes or skin laxity behind. Also, the problem appears when liposuction is done superficially. To avoid this problem, the surgeon must proceed on a wider area of harvesting and on a deeper plane of fat. To correct these irregularities, it is needed first to wait at least 3 months, until the edema settles down. Then with the anterior shafting procedure and with a spatula cannula the surgeon could elevate the skin over the irregularity and make room for fat grafting. It is better to over-correct because of fat absorption. If excessive skin laxity occurs, the approach is to address the laxity with excision like an abdominoplasty or a thigh lift.

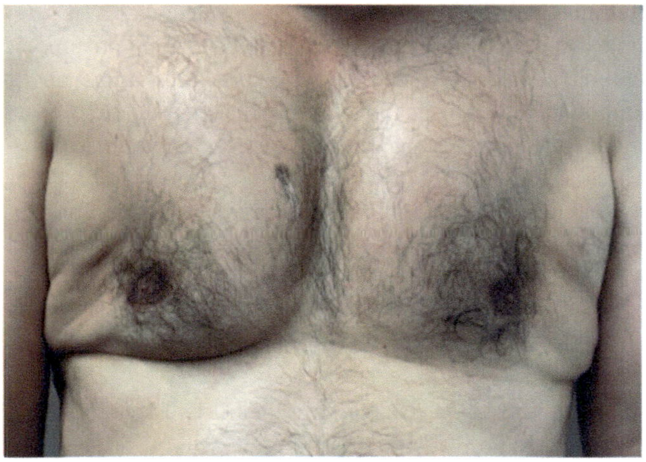

Fig. 40.5 Poor aesthetic result. This happens when the surgeon over-extracts fat tissue in a small body area

40.4 Grafting Site Complications and Pitfalls

On the grafting site, the complications and pitfalls that can occur are either severe and systemic, like fat embolism, or local to the fat being injected in a certain area.

40.4.1 Fat Embolism

The 2010s was definitely the decade of fat grafting; the changes brought to body shaping by this procedure were revolutionary. Nowadays, it is the standard procedure that accompanies liposuction. But problems arise when large amounts of fat are injected into the area of gluteus muscle for gluteus reshaping. Fat embolism could occur during fat transfer in an area with large caliber vessels like the upper and lower gluteus veins. If the intramuscular pressure is high because of a large quantity of fat and a vessel is damaged by the cannula, then fat can be moved intravenously and transferred to the lung. It must be noticed that fat embolism can occur when transferring fat also to other anatomical areas with larger caliber vessels than the gluteus with the same mechanism. The event can happen intraoperatively but also on the first postoperative day; the clinical symptoms are acute with the highest morbidity in plastic surgery.

The signs are respiratory distress, cerebral dysfunction and petechial rashes, also tachycardia, hypoxemia, pyrexia, thrombocytopenia, anemia, and hypocalcemia. It is very important to differentiate fat embolism from pneumonic embolism because the use of thrombolytics drugs on fat embolism may worsen the prognosis. There is no direct treatment but supportive measures for the respiratory and cardiovascular systems. A high dose of corticosteroids has been proven to reduce the inflammatory response and prolong the prognosis [26–28].

40.4.2 Fat Necrosis: Oily Cyst

When talking about fat grafting, then it is essential for the surgeon to know about the absorption rate of the fat or fat necrosis. Partial loss of the injected fat is inevitable. When the fat is injected into the body, a vascularization process begins. The cells in touch with the surrounding tissue of the injected cavity are the first to be revascularized and stay alive. The process completes from the periphery of the cavity to the center of the injected fat. The more distant the cell is from the periphery, the less the possibilities are for it to stay alive. To avoid fat necrosis and absorption, the injected fat must be placed in a linear arrangement into the tissue so the radius should be the smallest possible. If a bulk volume of fat is placed inside breast tissue, then the necrotic fat cells will resolve and produce an oily cyst (Fig. 40.6). In time the cyst will develop calcification and it will be palpable and radiologically detectable. The treatment is surgical excision.

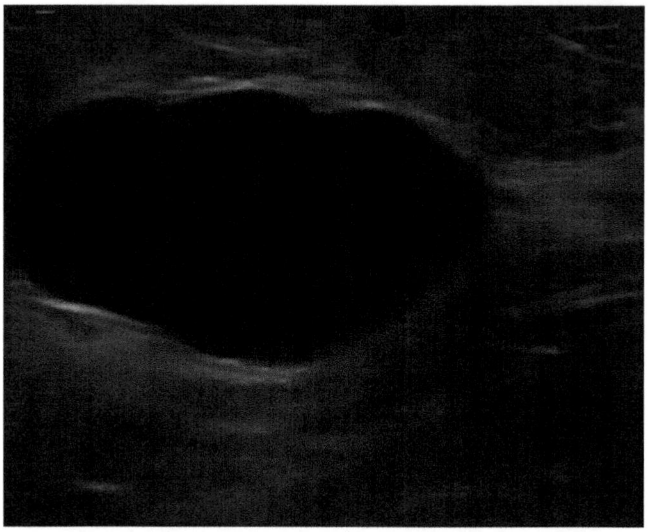

Fig. 40.6 Fat necrosis—oil cyst. If a bulk volume of fat is placed inside breast tissue, then the necrotic fat cells will resolve and produce an oily cyst

40.4.3 Poor Aesthetic Result

When injecting an amount of fat in an anatomical area to enhance it, the result cannot be predicted in an absolute manner. The absorption rate always varies from patient to patient. So, there is a possibility to end up with an asymmetry or an overcorrected anatomical area. To resolve the problem liposuction must be applied 4 months after the first procedure, waiting for the edema to settle down.

40.4.4 Hematoma

The grafting area appears to display a high rate of hematomas. It is common not to use infiltration in the grafted area. So, the chances to damage a vessel are higher. In order to prevent this, it is suggested to use a blunt lipotransfer cannula to infiltrate a small amount of the solution in the areas where vessels pass like the parasternal area for the pectoral major muscle fat grafting definition. The use of pressure garments, with high pressure especially on the first postoperative hours, is also suggested. Diagnosis of a hematoma can be made by detecting asymmetry or elevation of a grafted area. The hematomas are rarely dangerous, and they only need pressure in order to be stabilized [2].

References

1. American Society of Plastic Surgeons. ASPS National Clearinghouse of Plastic Surgery Procedural Statistics. 2018 Plastic surgery statistics report. https://www.plasticsurgery.org/documents/News/Statistics/2018/plastic-surgery-statistics-report-2018.pdf. Accessed 1 Oct 2019.
2. Cárdenas-Camarena L, et al. Strategies for reducing fatal complications in liposuction. Plast Reconstr Surg Glob Open. 2017;5(10):e1539.
3. Flynn TC, Coleman WP 2nd, Field LM, Klein JA, Hanke CW. History of liposuction. Dermatol Surg. 2000;26(6):515–20.
4. Fournier PF, Otteni FM. Lipodissection in body sculpturing: the dry procedure. Plast Reconstr Surg. 1983;72:598–609.
5. Illouz YG. Une nouvelle technique pour les lipodistrophies localisees. Rev Chir Esthet Fr. 1980;4(19):6.
6. Hetter GP. The effect of low-dose epinephrine on the hematocrit drop following lipolysis. Aesthet Plast Surg. 1984;8:19–21.
7. Klein JA. Anesthesia for liposuction in dermatologic surgery. J Dermatol Surg Oncol. 1988;14(10):1124–32.
8. Tebbetts JB. Minimizing complications of ultrasound-assisted lipoplasty: an initial experience with no related complications. Plast Reconstr Surg. 1998;102:1690–7.
9. Coleman SR. Long-term survival of fat transplants: controlled demonstrations. Aesthet Plast Surg. 1995;19(5):421–5.
10. Kaoutzanis C, Gupta V, Winocour J, Layliev J, Ramirez R, Grotting JC, Higdon K. Cosmetic liposuction: preoperative risk factors, major complication rates, and safety of combined procedures. Aesthet Surg J. 2017;37(6):680–94.
11. Murphy RX Jr, Alderman A, Gutowski K, Kerrigan C, Rosolowski K, Schechter L, Schmitz D, Wilkins E. Evidence-based practices for thromboembolism prevention: summary of the ASPS venous thromboembolism task force report. Plast Reconstr Surg. 2012;130(1):168e–75e.
12. Cronin M, Dengler N, Krauss ES, Segal A, Wei N, Daly M, Mota F, Caprini JA. Completion of the updated Caprini risk assessment model (2013 version). Clin Appl Thromb Hemost. 2019;2019:1076029619838052.
13. Martinez Licha CR, McCurdy CM, Maldonado SM, Lee LS. Current management of acute pulmonary embolism. Ann Thorac Cardiovasc Surg. 2020;26(2):65–71.
14. Shiffman MA. Lipid rescue for local anesthetic and nonlocal anesthetic drug toxicity. In: Shiffman MA, Di Giuseppe A, editors. Liposuction: principles and practice. Berlin: Springer; 2016. p. 1297–320.
15. Groban L, Butterworth J. Lipid reversal of bupivacaine toxicity: has the silver bullet been identified? Reg Anesth Pain Med. 2003;28(3):167–9.
16. Gilliland CN. Tumescent liposuction complicated by pulmonary edema. Plast Reconstr Surg. 1997;99(1):215–9.
17. Pitman GH. Tumescent liposuction complicated by pulmonary edema. Plast Reconstr Surg. 1997;100(5):1363–4.
18. Wang G, Cao WG, Zhao TL. Fluid management in extensive liposuction: a retrospective review of 83 consecutive patients. Medicine (Baltimore). 2018;97(41):e12655.
19. Toledo LS, Mauad R. Complications of body sculpture: prevention and treatment. Clin Plast Surg. 2006;33(1):1–11.
20. Rhee CA, Smith RJ, Jackson IT. Toxic shock syndrome associated with suction-assisted lipectomy. Aesthet Plast Surg. 1994;18:161–3.
21. McCormick JK, Yarwood JM, Schlievert PM. Toxic shock syndrome and bacterial superantigens: an update. Annu Rev Microbiol. 2001;55:77–104.
22. Kim YH, Cha SM, Naidu S, Hwang WJ. Analysis of postoperative complications for superficial liposuction: a review of 2398 cases. Plast Reconstr Surg. 2011;127(2):863–71.
23. Grazer FM, de Jong RH. Fatal outcomes from liposuction: census survey of cosmetic surgeons. Plast Reconstr Surg. 2000;105(1):436–46.
24. Lehnhardt M, Homann HH, Daigeler A, Hauser J, Palka P, Steinau HU. Major and lethal complications of liposuction: a review of 72 cases in Germany between 1998 and 2002. Plast Reconstr Surg. 2008;121(6):396e–403e.
25. Zakine G, Baruch J, Dardour JC, Flageul G. Perforation of viscera, a dramatic complication of liposuction: a review of 19 cases evaluated by experts in France between 2000 and 2012. Plast Reconstr Surg. 2015;135(3):743–50.
26. Wang HD, Zheng JH, Deng CL, Liu QY, Yang SL. Fat embolism syndromes following liposuction. Aesthet Plast Surg. 2008;32:731–6.
27. Gurd AR, Wilson RI. The fat embolism syndrome. J Bone Jt Surg Br. 1974;56B:408–16.
28. Shiffman MA. Fat embolism following liposuction. Am J Cosmet Surg. 2011;28:212–8.

Enhancing Body Anatomy with Fat Grafting

41

Christos Megapanos

Contents

41.1 Enhancing Body Anatomy with Fat Grafting

Achieving the ideal proportions of the human body is the main goal pursued by people seeking consultation at a plastic surgeon's practice. But which are the ideal body proportions? This is a question that troubles plastic surgeons and a fact blurred in the mind of patients. In general terms we can conclude that the "ideal" is a realistic, athletic shape of the body with visible muscular tone. But how can one define the term "realistic"?

Often, we see patients having unrealistic expectations, due to the ongoing trends and surrounding influences. A thorough consultation with the patient is the correct way to investigate and understand the personal perception one has of one's body and the expectations one has regarding the chosen procedure. A clinical examination is the means to assess the certain body's capability and probability to achieve a certain result. It is also proper, through the consultation process, to evaluate the mental and psychological condition of the patient.

The conversations can be guided by showing preoperative photos of previous patients of the same body type and the result they achieved postoperatively. It is clever to lower the expectation of a future patient, so the final result will turn out to be more satisfying. As the expectation rate rises the satisfaction rate drops [1, 2].

C. Megapanos (✉)
GM Clinic, Athens, Greece

41.2 Patient Selection

In order to achieve astonishing results, the first step of the process is patient selection.

There are different body types with different capabilities. The human body can be categorized as an ectomorph with a thin and lean body with underdeveloped muscular structure, as a mesomorph with normal shape and well-developed muscles and as an endomorph with concentration of fat localized over particular areas of the body, covering the muscles definition underneath.

BMI and fat mass percentage (FMP) are an index that helps a lot on patient assessment. A realistic approach is to select patients between 20 and 30 kg/m^2 for BMI or 20–30% for FMP, as very thin patients do not have enough fat for fat grafting, and obese patients are candidates for different operations than body shaping and enhancing [3, 4].

Lifestyle is another factor that must be assessed. In order to obtain best results, the patient must conduct a lifestyle that will prove helpful. A person with a healthy and athletic lifestyle is more likely to maintain the results over time. Skin tone and laxity, pregnancy, heavy weightlifting, use of steroids, and age are factors that affect the final result. A proper evaluation of the patient's body type, lifestyle and skin condition can lead to an appropriate operation suggestion and technique selection.

The ideal patient is the mesomorph body type with a well-developed muscular structure and proper fat deposition that covers the underneath anatomy. The ideal candidate must demonstrate a healthy and stable lifestyle, avoid the use of

© The Author(s), under exclusive license to Springer Nature Switzerland AG 2023
A. Di Giuseppe et al. (eds.), *Fat Transfer in Plastic Surgery*, https://doi.org/10.1007/978-3-031-10881-5_41

steroids, and be on a healthy, balanced, and high protein diet. In order to achieve the best results, the patient must not have a great weight fluctuation over his/her entire lifetime, that can produce severe skin laxity and lower the aesthetic result. Also it is important to check the abdominal wall with clinical examination through valsalva procedure in order to detect hernias or diastasis, of rectus abdominis [5].

41.2.1 The Athletic Body Type

So, as to achieve the athletic body type, the approach has two parts. The first one is high-definition liposuction and the second is fat grafting to enhance underline muscular anatomy.

The concept is based on the visualization of the human body and, as Alfredo Hoyos described very accurately, the transition between areas of shadows and areas with light. This visual game between shadow and light or the combination of concavities and convexities is the approach of the high-definition surgery. Following the anatomy, the surgeon must over-resect fat over the border of certain muscles and overcorrect with fat grafting above this border in order to enhance the anatomy.

Over-resection of fat must be performed on the tendinous connection of the muscles as linea alba, tendinous inscriptions of rectus abdominis and external oblique aponeurosis or on the transition border from one muscle to another, like pectoralis major with serratus, gluteus maximus with external oblique and thoracolumbar fascia and deltoid with bicep and triceps. Overcorrection with fat grafting must be done on the borders of a muscle to give proper definition, like pectoralis major and gluteus maximus, or inside the center of muscle to amplify the volume and projection [6].

41.2.2 Equipment

Regarding the equipment needed for fat harvesting, appropriate for fat grafting, there are the cannulas, canisters, and syringes. The preferred cannulas are a side hole sharp cannula to easily penetrate the fascia of a muscle for fat grafting and augmenting the center of a muscle area, and a front hole blunt cannula to work on the borders of muscles located in areas where large vessels intersect. The diameter of the cannulas must be over 3 mm to minimize the risk of damaging a vessel.

The connection of the cannula could be luer lock or toomey, which follows the preference of the surgeon, whereas a lot prefer the toomey. The length could range between 15 and 35, in order for the surgeon to be able to approach small areas with a short one, as the deltoid, or further away areas from a certain hole, as the pectoralis, the gluteus, or the gastrocnemius. The canister of choice is a plastic 500–2500 cc autoclavable multiuse canister with a luer lock adjustment tube on the bottom to drain the decanted fluids or fill the syringes. The use of a metallic mesh inside the canister is suggested to help on the separation of the fat. The syringes are 10 cc luer lock for small areas and borders, or 60 cc luer lock or toomey for large areas and volume transfer. Also, a syringe stand would constitute a good addition, to keep equipment tidy on the mayo table.

41.2.3 Preparation and Positioning

The preparation of the patient must be managed under strict aseptic technique with triclosan or povidone iodine. The non-surgical field must be completely covered with sterile covers. The preparation technique starts with the patient being in supine position. After the anesthesiologist has completed with the intubation and has secured the head, the patient is rolled on a left lateral decubitus position. The nurse prepares the back, the gluteus, the thighs, the right shoulder, and arm of the patient with the sterilization scrub. A rolled sterile surgical table cover is placed under the patient from head to feet. The patient is rolled back to supine position and the team changes places.

The patient is now to the right lateral decubitus position and the preparation continues with the rest of the back of the body. Once prep is finished the rolled surgical table cover is pulled further to cover the rest of the table. A rolled sterile towel is positioned under the gluteus area of the body. The patient returns to supine position and the towel is pulled over the edge of the table. The use of the towel is to help raise and turn the patient from supine to prone position in a sterile manner. The front part of the body is prepared with the scrub.

The sterilization area stretches from the neck and the elbows to the knees. The genitals are covered with a towel which is secured with metallic clips. A sterile sheet is placed on the thighs and secured on the table and another one is placed over the neck to the arms leaving the shoulders inside the surgical field. On this setup, after asking the anesthesiologist to raise the arms beside the head, the surgeon with the assistant could raise the patient with the gluteus towel and turn him on needed positions without risking sterilization. The nurse must assist with the feet under the sterile sheet.

The positioning of the patient depends on the part of the body that is worked on [7]. To access the upper front of the body, the patient must be in supine position with the arms sideline close, in order for the pectoralis to be relaxed and not too stretched. To access the back, the patient must be in prone position with the table bent on a 30° angle with the feet lower than the body and the gluteus up. The arms are raised on a 90° angle with the table and the elbows bent towards the anesthesiologist.

41.2.4 Harvesting Areas

The hacolour
rvesting areas depend on the amount of fat needed to be collected and the surgical planning. The best fat is collected from the lower abdomen and inner thighs. These are the areas that are suggested for harvesting for small fat grafting procedures. The surgical planning, and especially a planning of a high-definition body contouring procedure, will dictate the areas that will undergo liposuction. If a bbl procedure is planned, then the harvesting area will be the waist to achieve the transition contouring of the gluteus.

Marking the area in a certain way will guide through the stages of the operation. It is preferred to use different marking colors for different steps. The author on his practice uses blue for definition zone or over-resected parts, green for detailed definition fat grafting and red for areas that need volume and projection. Entry points are where the cannula will be inserted. It is preferred to keep the number of entry points minimum and use longer cannulas to access the grafted area [8].

41.3 Enhancing the Anatomy of Specific Areas of the Human Body

41.3.1 Pectoralis Major

A well-defined male chest is demanded by the procedure to achieve the athletic body. Talking with body builders about the definition of the male chest, it is concluded that the problem they encounter is the volume of the upper part of the pectoralis major and the sharp transition between the muscles' sternocostal origin and the sternum. Also, another area likely to be defined is the transition between the pectoralis and the deltoid and the pectoralis and the serratus.

On the other hand, the approach of the woman's chest is totally different. The demand here is volume in the breast, so the definition of the pectoralis is left intact, but volume is applied inside the breast and the muscle.

Marking of pectoralis will involve three different approaches. Marked with blue on the lateral edge of the muscle from the humeral insertion to the costal origin, this part must be over-resected to define the transition between the pectoral and serratus. Also, blue marks will be placed on a small part on the lower origin of the muscle between pectoralis major and rectus abdominis. Another blue marking will be placed over the transition of pectoralis and deltoid. With green the surgeon will mark the high-definition grafting areas, like the parasternal area, where sharp transition must be achieved and the subclavicular area in order to define clearly the angle aside the sternal notch.

Red marks will be applied on all the upper pole of pectoralis major to achieve a volumizing and projection effect. Entry points of this area are the armpit and the nipple insertion. Most of the operation can be performed from the armpit entry point. The fat grafting can be done on three plains. The first plane is under the muscle where volume can be added for projection. Two 60 cc syringes can be grafted with the fan technique.

The second plane is intramuscular. Up to two syringes must be injected for volumizing purposes. From this entry point you can access the superficial plain. First the grafting part must be done to define the transition of green zones of the edge of pectoralis, parasternal and subclavicular. After that, the part of fat removal can follow from the edge of pectoralis and the transition between pectoralis and deltoid (Figs. 41.1 and 41.2).

The entry point of the nipple can be used for high-definition superficial fat grafting on the angle aside the sternal notch and the parasternal area. Also, the same point can be used to remove fat on transition over the pectoralis and rectus abdominis of the other side. The total amount of fat that can be used is up to 300 cc. This is preferred to use the sharp cannula for

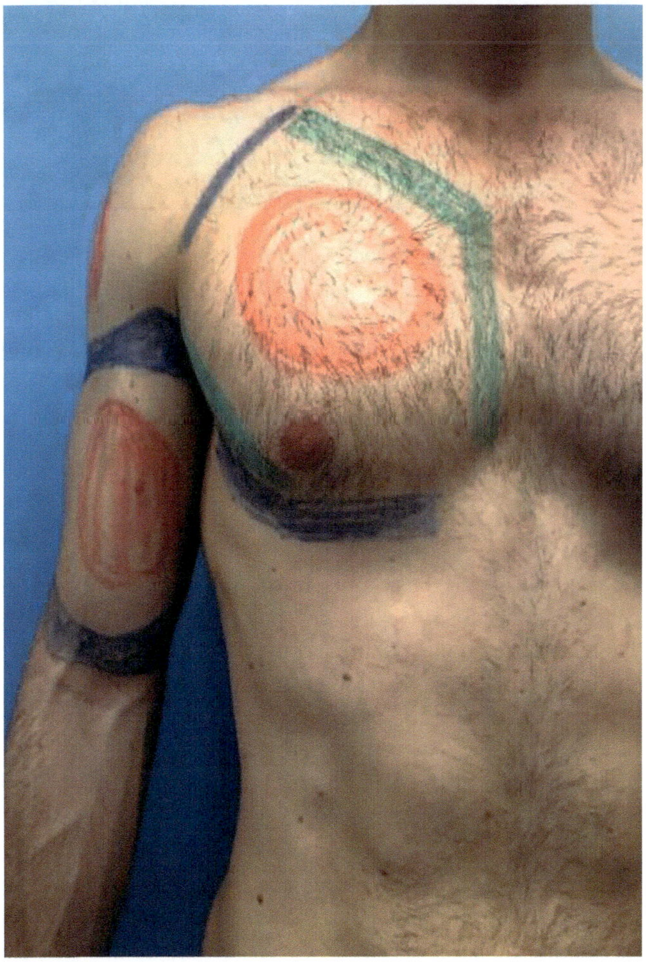

Fig. 41.1 Pectoralis and deltoid

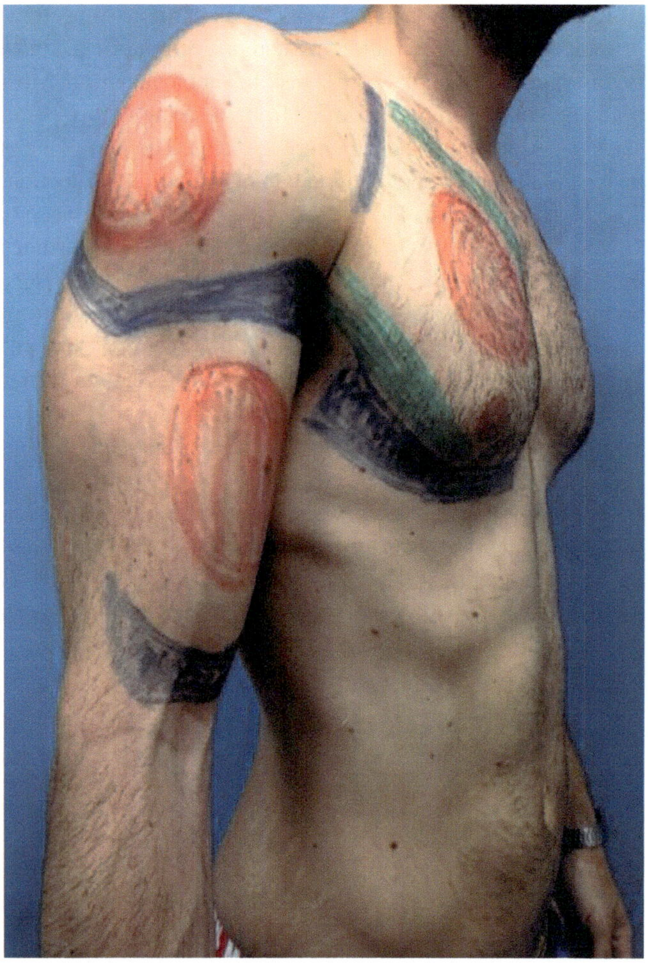

Fig. 41.2 Pectoralis and deltoid

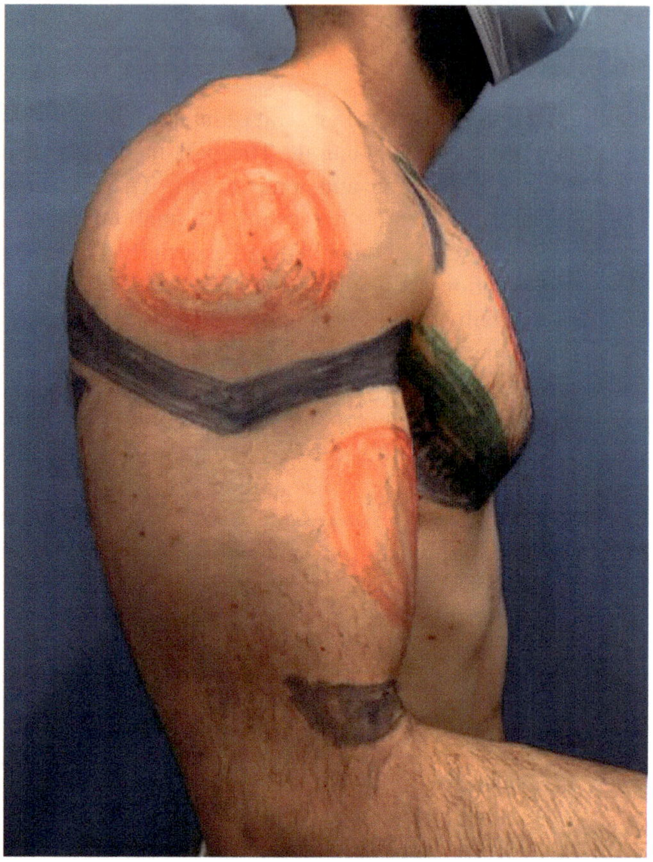

Fig. 41.3 Deltoid area

the intramuscular plain and the blunt cannula for all the others. Attention must be paid when grafting above the second and third intercostals, as hematoma can occur if the perforators of the internal mammary artery are damaged [9–11].

41.3.2 Deltoid Area

The deltoid is the basic muscle on anatomy enhancement procedure, because it is the key point of a V-shaped body. The wider the shoulders are and the slimmer the vest is, the V shape is visualized in a more prominent way. Also, definition is needed between the transition of the deltoid and the bicep and triceps. The markings of this area are blue over the transition zone of the deltoid with the bicep and triceps and red on the center of the muscle.

The entry points are the anterior below the insertion of pectoralis major to the humerus and the posterior below the insertion of latissimus dorsi to the humerus (Fig. 41.3). From this point, the surgeon can first work on the definition with superficial fat removal accessing from the supine and from the prone position. Secondly, the surgeon can inject, deep inside the muscle just above the humeral head with the sharp cannula, up to two 60 cc syringes on each side [12, 13].

41.3.3 Bicep/Tricep

These muscles are suggested to be addressed together. The volume of the fat that needs to be grafted is designated by the prominence of the deltoid to achieve a harmonious arm. The definition on the back of the arm must be achieved on a triangle between the tricep and the latissimus dorsi and on the back transition of the deltoid.

On the front definition is achieved between the deltoid and the bicep and over lacertus fibrosus and bicipital aponeurosis. The projection areas over the center of the muscles are marked with red.

The areas of high-definition fat removal are marked with blue and constitute a triangle over the origin of the long head of the tricep and can be accessed through the posterior entry point, a trapezoid area over lacertus fibrosus and bicipital aponeurosis, and can be accessed from an entry point over lateral epicondyle and the anterior and posterior definition sites of the deltoid (Figs. 41.4 and 41.5).

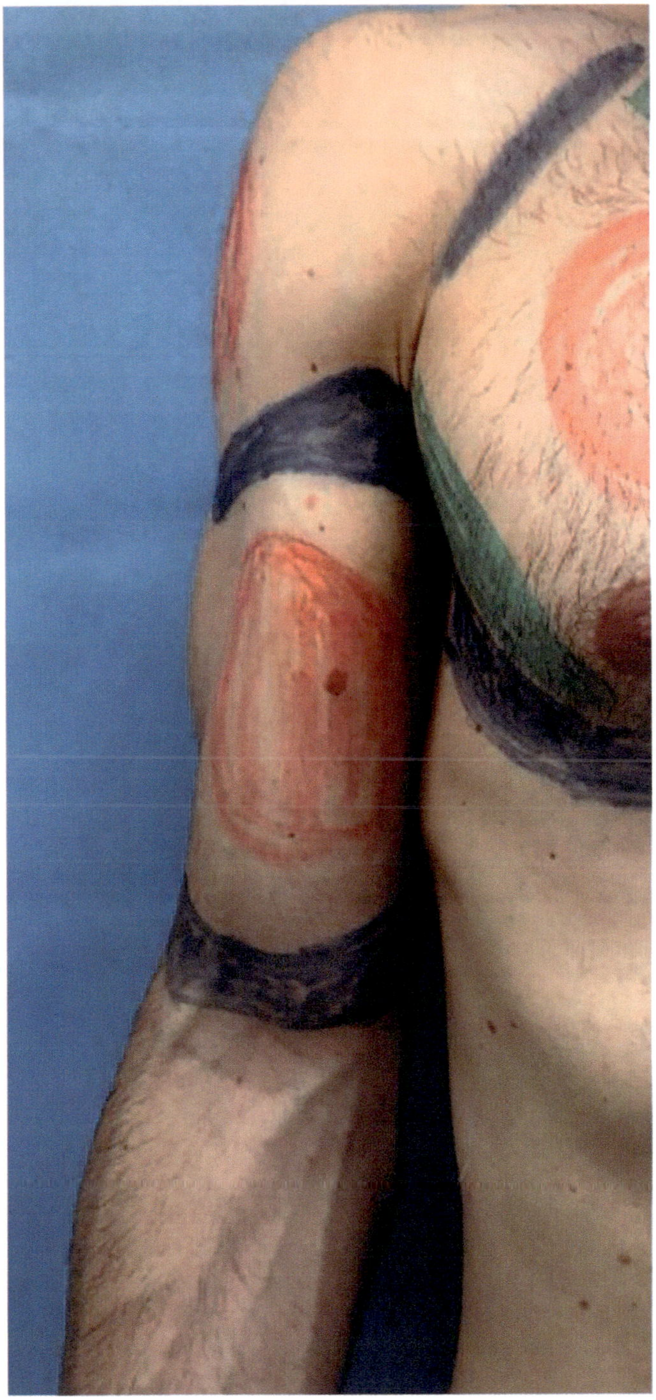

Fig. 41.4 Bicep/tricep

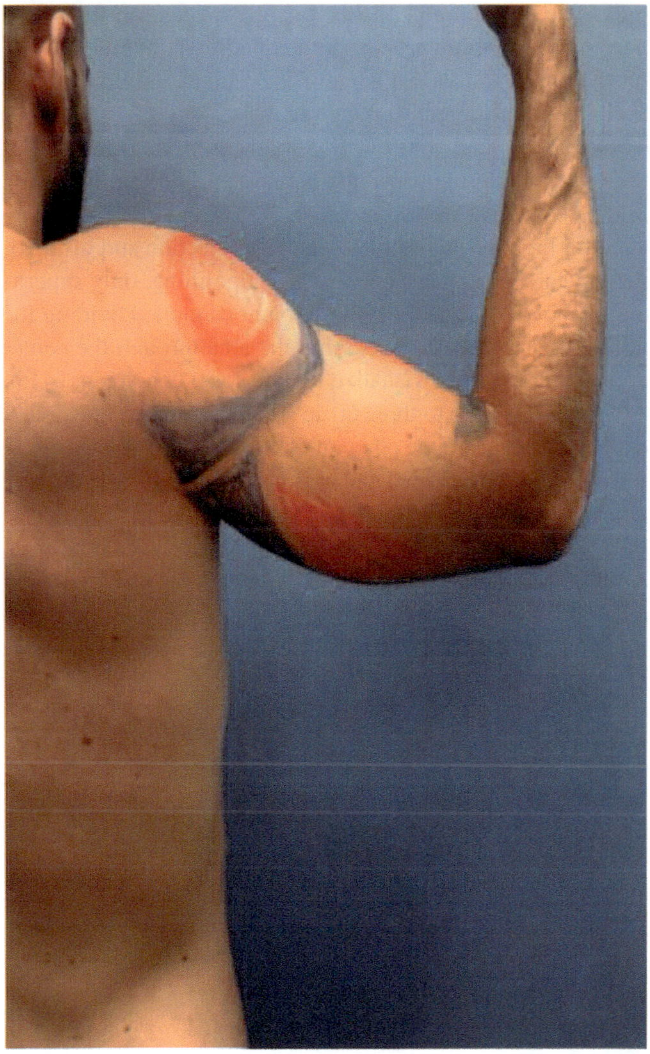

Fig. 41.5 Bicep/tricep

From the anterior entry point we can access the center of the bicep with the sharp cannula to inject up to two 60 cc syringes of fat. From the posterior access point we are allowed to enter into the middle area of the tricep to inject up to four 60 cc syringes of fat. This area demands deep knowledge of the anatomy of the arm to avoid damaging nerves and vessels. The area of the brachial crease must be avoided in order not to damage the brachial artery or the median nerve. Also, the area of bicipital aponeurosis must be respected with meticulous care not to damage radial, ulnar, and mid-basilic veins and the rest of the network of anastomoses [14, 15].

41.3.4 Gluteus Area

An oval-shaped gluteal area is requested from men and women. The approach of men and women gluteal areas changes on the definition of the transition zone of gluteus maximus and the waist. Regarding women this zone must be round and sharp, but when it comes to men it must be squarer and more undefined. Also, the area of the trochanteric notch must be grafted and rounded achieving the lateral "S" shape with an over-resect waist on women, while left untouched on men. The entry points of this area are located on the sacral part of the intergluteal fold, in the middle of subgluteal fold, on the upper and lateral edge of the gluteus maximus insertion and on the center of the gluteal dome.

Definition blue zones can be marked on the transition of the gluteus with the waist and a small part on the lower and middle part of the dome in women. Regarding men there is no need for high-definition fat resection. Projection and lifting of the gluteal area can be achieved with lipofilling, only superficially above the muscle with the blunt cannula. Dividing the dome on quartiles and accessing the area through the center entry point can help visualize the area as a cone (Figs. 41.6, 41.7 and 41.8). Up to three 60 cc syringes must be injected in each quartile. From the other entry one can perform definition fat resection or detailed fat grafting to fill the trochanteric notch and achieve the round shape of the upper and lateral pole. Injecting inside the muscle must be avoided due to fat embolism complications. Also, liposuction must be applied in a careful manner through the lower entry point of the lower middle part, avoiding damage to the rectum area [16, 17].

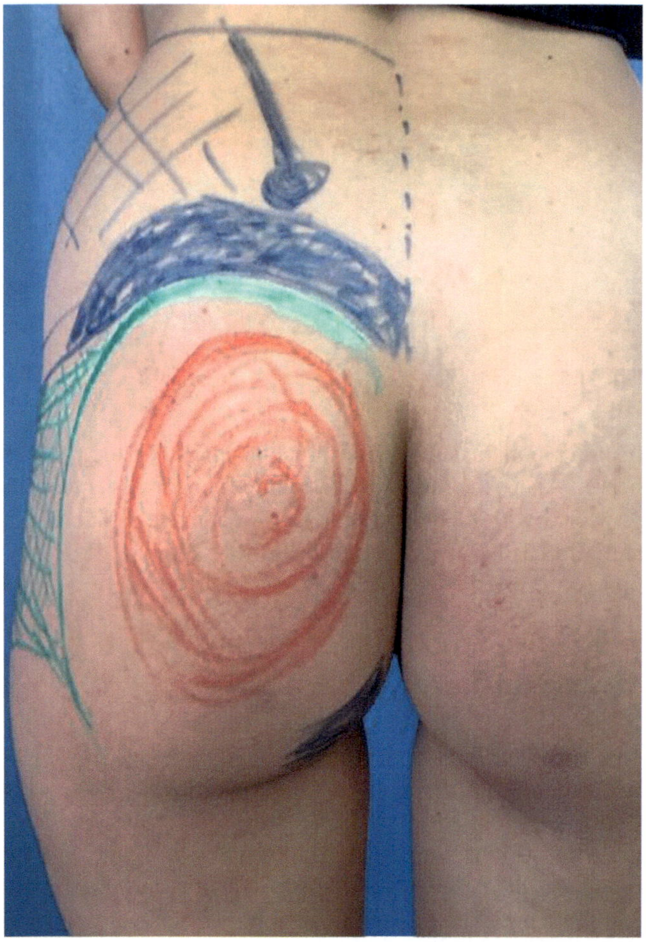

Fig. 41.6 Gluteus area. Dividing the dome on quartiles and accessing the area through the center entry point can help visualize the area as a cone

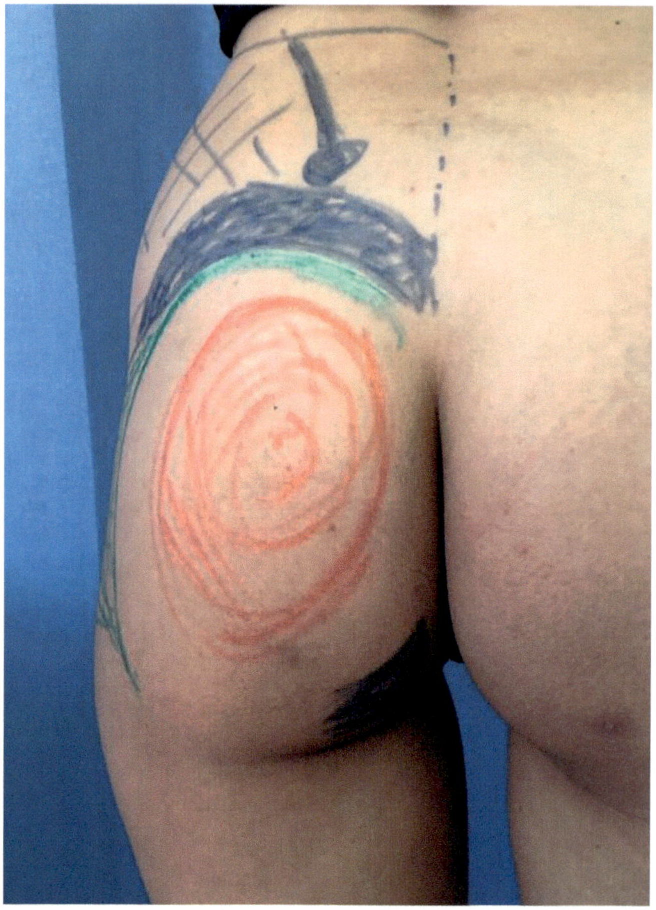

Fig. 41.7 Gluteus area

41.3.5 Gastrocnemius Area

The area of the calf defines the athletic body of women and men. Women need more definition of the muscle without additional volume and men need more volume. In order to achieve definition, approaching gastrocnemius from the back must remove fat above the origin of the lateral and medial head of the muscle, also below the heads either side of the Achilles tendon. On a frontal approach definition must be made above medial and lateral condyle of the tibial bone and along the transition between medial surface of tidal and gastrocnemius.

There are three entry points in the gastrocnemius area. The first two are located on the lateral and medial edges of popliteal fossa above the origin of the lateral and medial head of the muscle and the lower point is 10 cm above the insertion and on the medial edge of the Achilles tendon. The marking of blue zones can be drawn over the medial and lateral origin of the heads, over the medial and lateral condyle and along the transition of gastrocnemius with tibial bone. Also the posterior part below the muscle must be liposuctioned superficially.

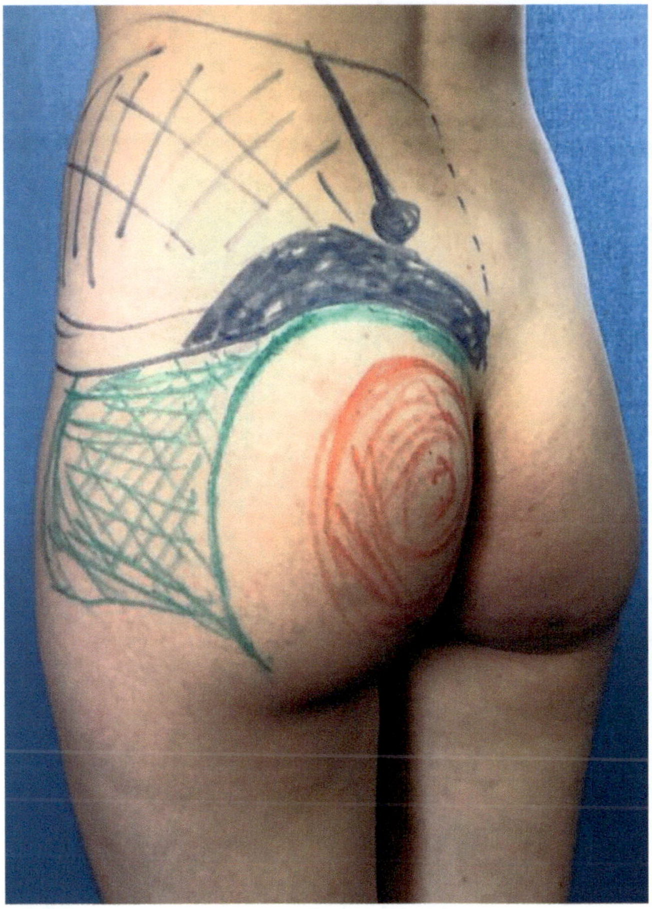

Fig. 41.8 Gluteus area

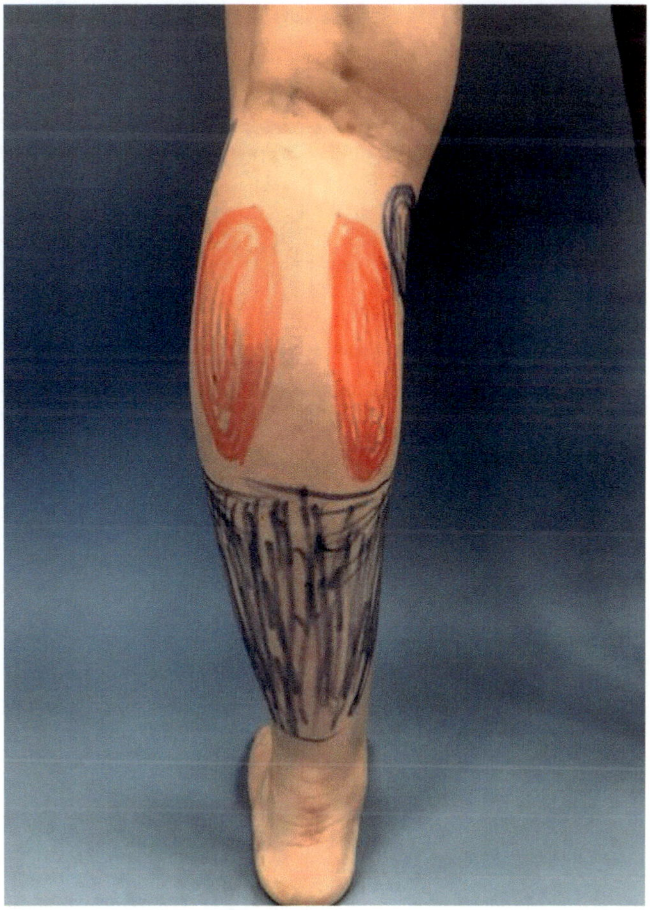

Fig. 41.9 Gastrocnemius area

The red volumizing areas are marked over the two heads of the muscle and can be accessed with a blunt cannula through the upper entry points (Figs. 41.9, 41.10 and 41.11). Injecting up to three 60 cc syringe of fat into each head will produce the volumizing effect it needs. This is the most difficult area of the body. The approach from the entry point could be tricky and lead to major complications, such as damaging saphenous, tibial, and peroneal nerves and greater saphenous vein. An excellent knowledge of the anatomy needed and very superficial fat resection is needed to be safe [18, 19].

41.3.6 Labia Majora/Vagina

As time passes the volume of labia majora is lost. Pregnancy and age also reduce the perivaginal fat that offers firmness and structure to the vaginal tube. A lot of women seek to reobtain the youthful look of the genitals that they had in the past. A small volumizing approach makes a huge difference on the beautification of the area and offers a big boost to women's confidence and satisfaction. Fat grafting above the pubic bone, alongside the hood of the clitoris and the labia majora until the height of the upper floor of the vagina, will restore the youthful look of the area. Also, fat can be transferred on the side walls of the vagina to make it firmer or on the upper floor of the tube to enhance the "g" spot for greater satisfaction [20].

The marking on enhancing the labia majora is outlined by two inverted triangles with the base over the pubic bone and the tip along the labia majora to the height of the upper floor of the vagina. The entry points are located on the lateral corners of the base of each triangle. Fat grafting must be superficial, injecting an amount of 20–30 cc of fat on each side.

Fat grafting the vagina can be performed with a blunt cannula from three entry points. On the roof of the vaginal tube and 1 cm to the inner part from the entrance a small incision can be made, and a 2 mm blunt cannula can be inserted for 2 cm. There up to 4 mL of fat can be injected on a fan tech-

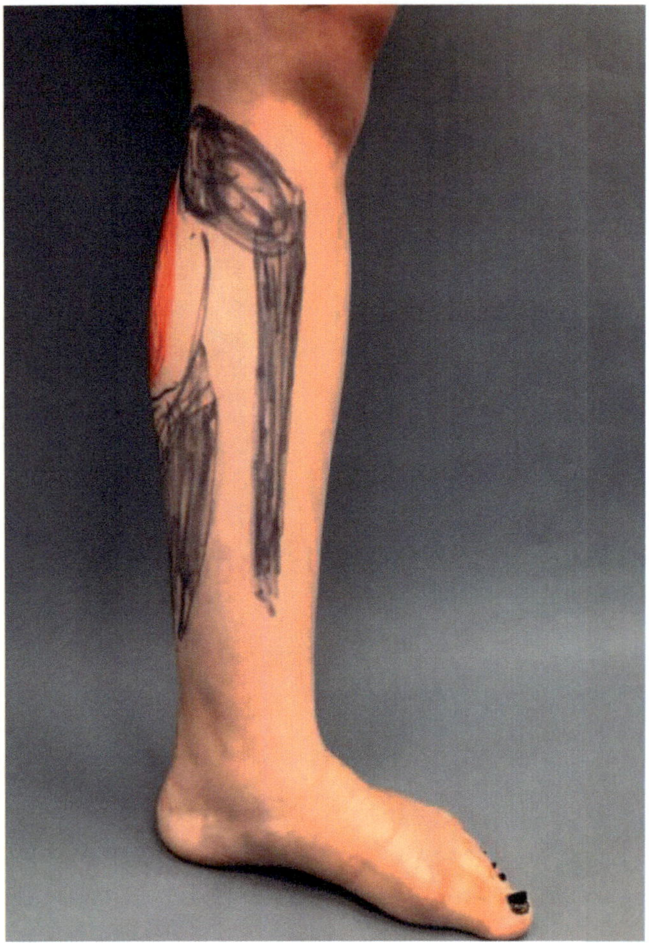

Fig. 41.10 Gastrocnemius area

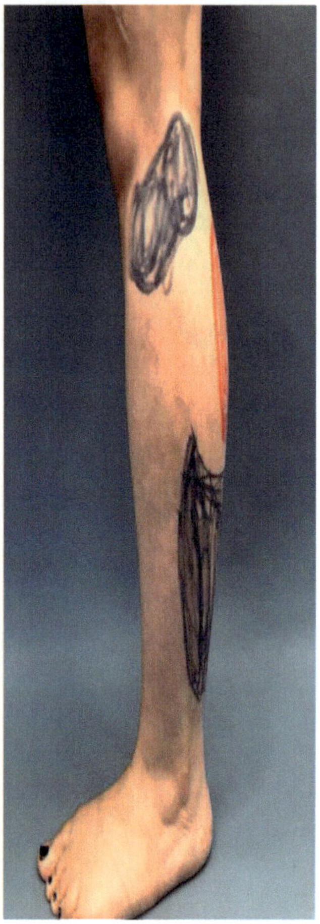

Fig. 41.11 Gastrocnemius area

nique avoiding damage to the urethra. On the sides of the vaginal wall two entry points can be made in order to inject up to 6 mL of fat on each side. The entry points are located on the medial edge of labia minora at the height of the middle of the vagina [21, 22].

41.3.7 Penis

The penis constitutes the very center of manhood, so many patients seek enhancement of this area with fat grafting. The enhancement can be performed along with a penis elongation procedure or as a standalone procedure. Patients believe that the width of the penis plays a huge role to the satisfaction of their partner. Fat grafting must be performed superficially and evenly around the body of the penis, from the base on the pubic bone and scrotum to the coronal sulcus. The entry points are located on the two lateral sides of the base of the penis above the pubic hairline. The cannula used must be a blunt 2 mm cannula. Attention must be paid to avoid the urethra and the dorsal vein [23].

References

1. Singh D. Universal allure of the hourglass figure: an evolutionary theory of female physical attractiveness. Clin Plast Surg. 2006;33(3):359–70.
2. Fisher ML, Voracek M. The shape of beauty: determinants of female physical attractiveness. J Cosmet Dermatol. 2006;5(2):190–4.
3. Perez-Gurri JA. Choosing the correct candidate. In: Garcia Jr O, editor. Ultrasound-assisted liposuction: current concepts and techniques. Berlin: Springer; 2020. p. 23–36.
4. Cornelissen PL, Toveé MJ, Bateson M. Patterns of subcutaneous fat deposition and the relationship between body mass index and waist-to-hip ratio: implications for models of physical attractiveness. J Theor Biol. 2009;256(3):343–50.
5. Hoyos AE, Guarin DE. High-definition body contouring using VASER-assisted liposuction. In: Garcia Jr O, editor. Ultrasound-assisted liposuction: current concepts and techniques. Berlin: Springer; 2020. p. 203–11.
6. Hoyos AE, Prendergast PM. The concept of human sculpting: light, shadow, and form. In: High definition body sculpting. Berlin: Springer; 2014. p. 41–8.
7. Pazmiño P. ultraBBL: Brazilian butt lift using real-time intraoperative ultrasound guidance. In: Garcia Jr O, editor. Ultrasound-assisted liposuction: current concepts and techniques. Berlin: Springer; 2020. p. 147–72.

8. Glaser DA, Kaminer MS. Body dysmorphic disorder and the liposuction patient. Dermatol Surg. 2005;31:559–61.
9. Hoyos A, Pere M. Arm dynamic definition by liposculpture and fat grafting. Aesthet Surg J. 2012;32(8):974–87.
10. Hoyos A, Perez M. Dynamic-definition male pectoral reshaping and enhancement in slim, athletic, obese, and gynecomastic patients through selective fat removal and grafting. Aesthet Plast Surg. 2012;36(5):1066–77.
11. Horn G. A new concept in male chest reshaping: anatomical pectoral implants and liposculpture. Aesthet Plast Surg. 2002;26(1):23–5.
12. Mentz HA, Ruiz-Razura A, Newall G, Patronella CK, Miniel LA. Pectoral etching: a method for augmentation, delineation, and contouring the thoracic musculature in men. Plast Reconstr Surg. 2007;120(7):2051–5.
13. Chamosa M, Murillo J, Vázquez T. Lipectomy of arms and lipograft of shoulders balance the upper body contour. Aesthet Plast Surg. 2005;29(6):567–70.
14. Hoyos A, Perez M. Arm dynamic definition by liposculpture and fat grafting. Aesthet Surg J. 2012;32(8):974–87.
15. Matlock D, Simopoulos A. Sonographic evaluation of autologous fat transfer to the biceps and triceps. Surg Technol Int. 2014;25:289–95.
16. de Pedroza LV. Fat transplantation to the buttocks and legs for aesthetic enhancement or correction of deformities: long-term results of large volumes of fat transplant. Dermatol Surg. 2000;26(12):1145–9.
17. Toledo LS. Gluteal augmentation with fat grafting: the Brazilian buttock technique: 30 years' experience. Clin Plast Surg. 2015;42(2):253–61.
18. Mundinger GS, Vogel JE. Calf augmentation and reshaping with autologous fat grafting. Aesthet Surg J. 2016;36(2):211–20.
19. Hoppmann R, Meruane M, González D, Wisnia P, Hasbún A, Villalobos B. Calf lipo-reshaping. J Plast Reconstr Aesthet Surg. 2013;66(7):956–61.
20. Goodman MP, Placik OJ, Matlock DL, Simopoulos AF, Dalton TA, Veale D, Hardwick-Smith S. Evaluation of body image and sexual satisfaction in women undergoing female genital plastic/cosmetic surgery. Aesthet Surg J. 2016;36(9):1048–57.
21. Salgado CJ, Tang JC, Desrosiers AE 3rd. Use of dermal fat graft for augmentation of the labia majora. J Plast Reconstr Aesthet Surg. 2012;65(2):267–70.
22. Triana L, Robledo AM. Refreshing labioplasty techniques for plastic surgeons. Aesthet Plast Surg. 2012;36(5):1078–86.
23. Panfilov DE. Augmentative phalloplasty. Aesthet Plast Surg. 2006;30(2):183–97.

Male Upper Arm Definition with Fat Transfer

42

Hüseyin Kandulu

Contents

42.1 Male Upper Arm Definition with Fat Transfer

42.1.1 Anatomy of Human Arm: Muscular System

42.1.1.1 Upper Arm

When the upper arm is discussed, the area between the shoulder and elbow joint should be considered. There are four muscles in this region. The biceps brachii, brachialis, and coracobrachialis muscles are located in the anterior region, while the triceps brachii muscle is located in the posterior region. The three anterior compartment muscles, including the biceps brachii, coracobrachialis, and brachialis, are all innervated by the musculocutaneous nerve. Arterial supply to the anterior region of the upper arm is provided by the muscular branches of the brachial artery. To briefly describe these muscles:

42.1.1.2 M. Biceps Brachii

The biceps brachii muscle is a large and thick muscle. It has two heads, which are located in the ventral part of the upper arm, and consists of a short head (caput breve) and a long head (caput longum). Anatomically, the short head originates from the tip of the coracoid process and the long head originates from the supraglenoid tubercle (tuberculum supraglenoidale) of the glenoid/scapula. Both heads run distally and become a confluent muscle hub before tapering along the anterior aspect of the elbow and finally entering the radial tuberosity and forearm fascia via the bicipital aponeurosis [1]. The aponeurotic tendon, called lacertus fibrosus or aponeurosis bicipitalis, blends into the forearm fascia. It is the main supinator muscle of the forearm and facilitates the flexing of the arm and forearm. In particular, it provides some flexion of the arm, flexion, and outward rotation (supination) of the forearm. While performing this function, it is connected to the N. Musculocutaneous (C5–C6) nerve. Although liposuction alone causes this muscle to become prominent,

Supplementary Information The online version contains supplementary material available at https://doi.org/10.1007/978-3-031-10881-5_42.

H. Kandulu (✉)
Huseyin Kandulu Private Clinic, Istanbul, Turkey
e-mail: huseyin@huseyinkandulu.com

© The Author(s), under exclusive license to Springer Nature Switzerland AG 2023

A. Di Giuseppe et al. (eds.), *Fat Transfer in Plastic Surgery*, https://doi.org/10.1007/978-3-031-10881-5_42

subcutaneous, sub-fascial and intramuscular fat injections can be applied during surgery. This application will increase muscle definition [1–6].

42.1.1.3 M. Triceps Brachii

The antagonist of the biceps muscle is the triceps brachii muscle, which has three heads. It is located in the posterior compartment of the upper arm. The long head originates from the infraglenoid tubercle. The medial and lateral heads originate from the humeral body. Distally, the heads converge on a tendon and enter the olecranon of the ulna. Its function is extension of the arm at the elbow and forearm extension, upper arm adduction, and shoulder stabilization. It is the strongest extensor muscle of the forearm.

42.1.1.4 The Deltoid Muscle

The deltoid muscle is a large triangular muscle located in the proximal upper extremity and is associated with the human shoulder girdle. Its base (or origin) attaches to the spine of the scapula and the lateral third of the clavicle. This U-shaped starting point reflects the entry point of the trapezius muscle. The apex (or insertion) attaches to the lateral side of the humeral body at a point known as the deltoid tuberosity.

The deltoid is divided into three different parts called anterior, lateral, and posterior. When all three parts contract at the same time, it helps the deltoid arm go past 15°. Since the pull direction of the deltoid muscle is parallel to the humeral axis, it cannot initiate abduction. During abduction, the anterior and posterior portions of the deltoid play an important role in stabilizing the arm, while the lateral head assists in raising the arm from 15° to 100°. Additional functions include ambulation. The anterior head of the deltoid works with the pectoralis major to flex the arm while walking. This is in contrast to the posterior part of the deltoid, which works with the latissimus dorsi to extend the arm during ambulation. The axillary nerve innervates the deltoid and both C5 and C6 contribute to the axillary nerve. The nerve originates from the posterior cord of the brachial plexus [7]. Beyond all these functional features, the deltoid muscle is at the starting point of masculinity. In order for the width and definition of the shoulders in the upper body to be aesthetically pleasing, fat transfer is primarily performed into the deltoid muscle.

42.2 Athletic Arm Appearance Expected in Men

The primary goal in the athletic male arm is to not make it look thinner, but to enhance the 3 main muscle groups (deltoid, biceps, and triceps) to prominence. Male patients desire a body structure having wide shoulders and prominent biceps and triceps muscles. Muscle structure, skin tone, and excess fat should be examined to evaluate whether the patient is suitable for the procedure; even if the fat structures on the arm are superficially located and skin retraction is good [8].

From an anatomical point of view, lipodystrophic arms contain a greater proportion of deeper than superficial fat in all their longitudinal sections and all their anatomical faces. It is known that the closer the sections are, the greater the amount of fat. They are regions with less thick adipose tissue on the anterior surface and distal third of the arm. In some cases, there is a specific lipodystrophic zone located in the posterior-outer region of the arm between the proximal and medial thirds. Liposuction of the lipodystrophic arms has certain limits because it is not always possible to circumferentially reduce the arms as patients may not be able to accept irregularities from retraction. We suggest that if the deltoid volume is enlarged by lipoinjection, the following effects will be achieved: restoration of the bideltoid width (making it wider than the bitrochanteric width) and great improvement in the contour of the upper limbs and optical effect of balance [9].

Arm shape is determined by the muscle structure and mass of each patient. In terms of underlying structure, three main muscles are responsible for arm shape: the deltoid, triceps, and biceps. More pronounced musculature in men is considered more athletic and healthy, but such a description can be considered a deformity in women. Slim folds and toned, smaller muscle mass are considered aesthetic for women. High-Definition Liposculpture has given both sexes an athletic appearance by shaping body fat with a multilayered and 3-dimensional shaping approach. This approach is based on creating concave and convex areas instead of flattening areas with liposuction alone. Fat is removed from some areas but grafted onto others to improve the anatomical architecture [8]. However, the general appearance of the arm is very different in men than in women. The two main muscle groups that lead to the emergence of athletic shape in male patients are the pectoral and deltoid muscle groups. The wide view of these two muscle groups causes the athletic/masculine V shape to form in the upper body. For this reason, the two main areas that should be considered before the abdominal region are the chest and arm regions. It should be kept in mind that shaping the abdominal region without shaping the upper body can cause artificial images to appear. The desired arm image to be obtained in men is to add definition to the deltoid, biceps, and triceps muscle groups using liposuction and to clarify the deltoid muscle with fat transfer.

42.3 Natural Shadows due to Muscles

Abduction, adduction, internal and external rotation, extension and flexion movements of the arm, and shadows due to muscle dynamics draw attention. In particular, the shadows of the biceps and triceps muscles at the borders of the deltoid muscle and within the triceps muscle group can be easily determined. Under bright light, the naturally protruding muscles are bright, and the fatty areas in the natural spaces between these muscles are dark, i.e., shaded. In the natural position, the deltoid is the dominant muscle in the male arm. When the arm is extended, we observe that the triceps displaces slightly upwards and creates an upper dynamic zone together with the deltoid (Fig. 42.1).

42.4 Preoperative Evaluation

Preoperative blood values should be measured and it should be checked whether they have any other problems. For this reason, hemoglobin, hematocrit, white blood cell count, creatinine, urea, and clotting time should be measured and evaluated preoperatively. If abnormalities are detected in these values, first the cause of the abnormality should be revealed and then it should be decided whether the patient is suitable

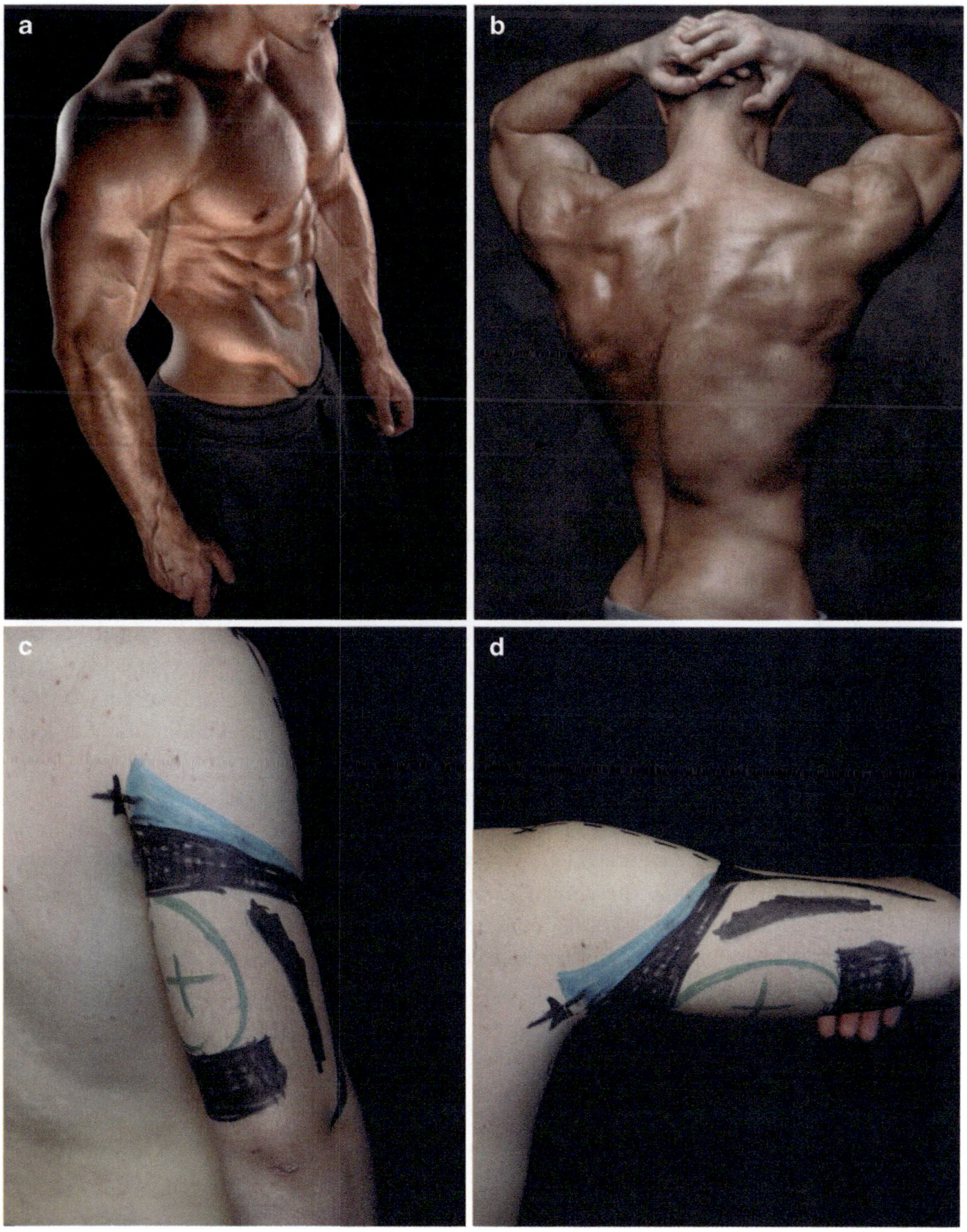

Fig. 42.1 Abduction adduction, internal and external rotation, extension and flexion movements of the arm, and shadows due to muscle dynamics (**a**). In particular, the shadows of the biceps and triceps muscles at the borders of the deltoid muscle and within the triceps muscle group can be easily determined (**b**). The area marked with light blue in the picture is the dynamic zone formed by the triceps muscle (**c**, **d**). Care should be taken when using liposuction in this area and whole fat should not be removed

for surgery. Acetylsalicylic acid evaluation should be made for each patient to be applied and consultation with anesthesiologists should be made.

In addition to such a general evaluation, the BMI, skin quality, and muscle structure should be evaluated to determine patient suitability for this operation. Since the arms and trunk are often treated together in male patients, a BMI of 32 and below is more desirable.

42.5 Preoperative Drawing Techniques and Detection of Negative Spaces

Positive areas (dynamic zones) are formed by naturally protruding muscle groups in the body. Natural spaces covered with fat tissue around them are called negative spaces. While drawing, patients should be marked in the upright position before surgery. In men, the visible areas of extra fat are marked first with the arm at adduction; these areas are mostly in the posterior region. Depending on the patient's biotype (fat, lean, athletic), the area is marked for deep extraction or more superficial liposuction to create the triceps curvature. Visible extra fat on the inner or anterior part of the arm is also marked as needed in selected patients. After that, three muscles are marked: deltoids, biceps, and triceps.

To mark the posterior sulcus (contraction mark) of the deltoid, the arm is abducted 90° and the forearm is flexed 90°. First, the deltoid boundaries are determined and the midline is marked from the shoulder to the elbow first. The negative space between the deltoid and biceps can be easily marked by backward rotation. If this negative space cannot be determined exactly, the patient is instructed to tighten and accentuate the biceps muscle. Thus, the negative space in between will show itself more comfortably. The center of the deltoid muscle and the upper 2/3 of the muscle are marked with a different color to show the area to be transferred. Approximately 100–150 cc of pure fat is injected directly into the muscle.

The anterior sulcus is then marked while the shoulder is in external rotation. Then, while the patient's upper arm is in full adduction, the patient is asked to contract the triceps muscle voluntarily and a second mark is placed on the deltoid posterior sulcus. A zone is created between the first deltoid narrowing mark and it is called the "dynamic zone." The bicipital groove is marked internally and externally. By marking the biceps muscle at 90° flexion of the elbow, a half-moon region and another dynamic region are drawn from the distal muscle tendon to the bend of the elbow [8].

In the midline, there is a tiny triangular negative area located on the midline at the lower border of the biceps triceps and deltoid. Aggressive liposuction in this area will help determine the classes of all these muscles. We know that there are medial and lateral bodies of the biceps. Superficial liposuction between the lower border of these bodies and the anterior surface of the elbow (cubital space) also supports the definition of the biceps muscle. Although this procedure is rare in my practice, I would like to share this information with my colleagues. Again, some intramuscular and superficial fat injection may be recommended to give volume to the biceps muscle. The triceps muscle consists of three trunks; when the patient is told to tighten the muscle, it is observed that the triceps moves slightly upwards.

At this stage, at the lower border of the muscle above the olecranon, a valuable negative area appears again and must be marked. Again, when the triceps muscle is contracted, the negative field shadow between the lateral and medial trunks can be identified and marked. Since the basic logic is to reveal what is naturally present, I do not recommend that any muscle body or negative area, which cannot be identified with the help of light and palpation, be relatively marked and struggled to be revealed during all these markings. It should be kept in mind that suitable candidates should be determined first for this operation (Fig. 42.2). It is easier to mark the arm in women than it is in men. Excess fat is marked when the patient's arm is in adduction. Some women may have fat deposits on the inside of the arm, which are marked with the arm on abduction. Next, with 90° arm abduction, 90° elbow flexion, and internal rotation of the shoulder, the posterior deltoid sulcus is marked and a triangular, "hollow" space is marked between the axillary posterior fold and the proximal posterior insertion of the deltoid muscle. In women, the triceps is not prominent, as the ideal female arm does not have significant muscle mass in that area. Women expect a straight and smooth arm appearance instead of convexity and prominence like that in the male arm.

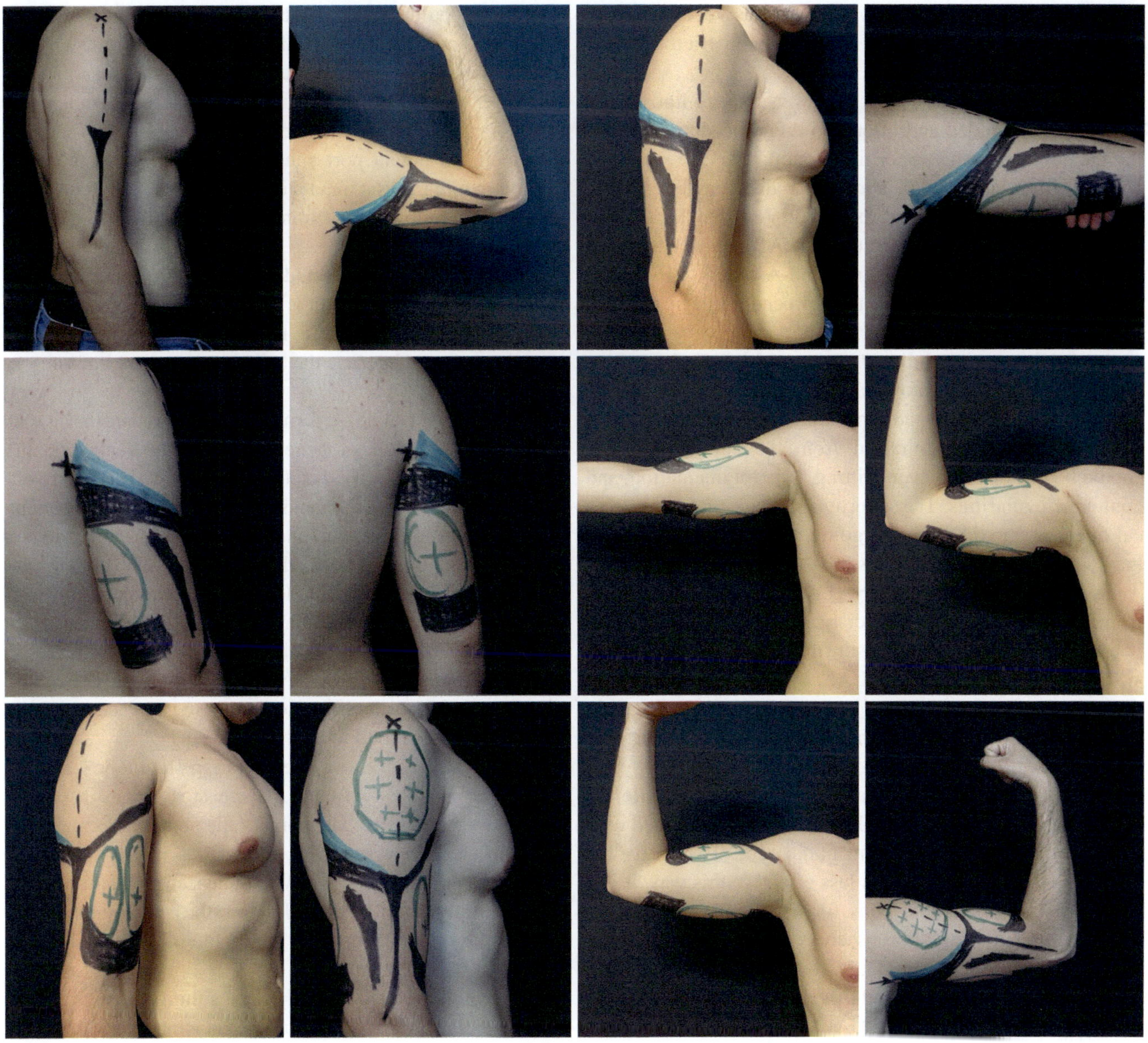

Fig. 42.2 The arm drawing stages of men shown before the operation. Black color indicates the frame, which is the negative zones. Negative spaces include the bicipital groove (inner and outer), the tricipital tendon, and the posterior margin of the triceps in the proximal and distal parts. The green color indicates the areas where fat injection can be made

42.6 Liposuction Technique

The liposuction technique has been developed over the years and has become today's modern liposuction technique. This technique was developed with inspiration from blind sharp excisional procedures. The first attempt at liposuction was made in 1921 by Dujarrier, who operated on a dancer's knees and calves. Injury to the femoral artery resulted in amputation of the leg [10]. Charles Dujarrier, a French surgeon, first experimented with the first shaping procedure of the inner thigh fat deposits in 1929, inserting a sharp curette under the skin through a small skin incision [11].

But the results were not as expected. This technique was abandoned as it resulted in excessive bleeding. The French researcher, Ives Gerard Illouz, introduced the first modern liposuction technique in 1977. In this technique he developed, he used blunt cannulas and a strong suction system. This technique was called dry liposuction because no liquid was injected into the targeted fat layers prior to suction. Further development of the wet liposuction technique introduced by Gregory Hetter of Las Vegas involved the addition of epinephrine as a vasoconstrictor to the soaking solution [12].

Recently, the development of power-assisted liposuction has further expanded and improved this procedure, increasing the popularity and use of liposuction.

Currently, we can talk about four different liposuction techniques, which include Aspiration-Assisted Lipectomy (AAL), Laser-Assisted Lipectomy (LAL), Power-Assisted Lipectomy (PAL), and Ultrasound-Assisted Lipectomy (UAL). The most commonly used procedure in liposuction surgeries is UAL (trade name: VASER®).

Aspiration-assisted lipectomy (AAL) technique uses negative pressure to remove fat from a syringe applied to a small volume, blunt-tipped aspiration cannula. While applying this technique, aspiration from the superficial layer should be avoided to prevent pitting, hyperpigmentation, and contour irregularities. The superficial fat layer contains vertical fibers. It should not be forgotten that if it is damaged, it may cause contour deformities [13–15].

In laser-assisted lipectomy (LAL) technique, it is performed with a much smaller incision compared to UAL. LAL is performed by inserting a laser fiber through a small incision [16]. Complications of LAL are very rare. Only one study reported a complication rate of 0.93%, including skin burns and local infection. From this point of view, it is a technique with very few complications.

In the power-assisted lipectomy (PAL) technique, an external power source is used. Liposuction is performed with the help of an electric vacuum pump [16]. Thanks to this power aid, the surgeon spends less effort. In this way, it is a technique that is advantageous for tissue removal in large volumes and in dense fibrous areas [15, 17]. In addition, liposuction complications should always be considered.

In the ultrasound-assisted lipectomy (UAL) technique, ultrasound energy is used. UAL transmitting ultrasound energy is used to emulsify the fat before the fat is removed. Especially with this technique, it can be advantageous in the fibrous areas, such as the back, chest, and upper side because these areas are more difficult to target using standard liposuction. It should be noted that larger incisions are required in the UAL technique. In addition, it should always be considered that there is a risk of thermal injury to the subcutaneous tissues due to the exothermic energy caused by ultrasound.

The general principle of UAL is to melt fat homogeneously by emitting sound waves specific for fat cells. It helps apply superficial liposuction and reduce complication rates, especially in areas where fat tissue is superficial (i.e., arms). General complications of liposuction can be listed in two groups as short-term complications and long-term complications. Short-term complications include wound infection, seroma hematoma, edema, ecchymosis, paresthesia, fat embolism, pulmonary embolism, and skin necrosis, while long-term complications include contour deformity, hyperpigmentation, hypertrophic scarring, and lymphedema [15]. However, complications observed, especially in arm liposuction, are noted as seroma hematoma, the friction burns at liposuction cannula port sites, and numbness in the early period, while fibrosis, inadequate definition, and very rarely cystization following fat transfer in the late period attract attention.

The patients are operated under general anesthesia. In some cases, if only the arm is to be intervened, it can also be done under local anesthesia. However, general anesthesia is mostly preferred because the technique requires intramuscular fat injection.

Liposuction can be performed under local anesthesia with or without sedation, or general anesthesia. Each modality has its advantages, disadvantages, inherent risks, and suitability for the unique demands of each operation and patient. The cannula may be a short and curved 1.5–2.5-mm-diameter infusion cannula with holes along the shaft. The oozing cannula is connected via tubing to either a syringe or a pressurized infusion pump system. The fluid is supplied by a constant fan movement, starting from the deep fat layers in the horizontal planes and advancing superficially. Because the cannula is constantly in motion, there is no need to monitor the return of blood to the tubing syringe.

When shaping the arm, the operation should be started in the probe position. Two small incisions to be made just above the elbow with the deepest point of the posterior axillary line are sufficient to work from behind. As in all liposuction applications, surgery begins with tumescent fluid infiltration. The solution used is prepared before the operation and contains 1 mg of adrenaline and 1 ampoule of 10% lidocaine in each 1000 cc RL.

After infiltration of the tumescent solution, skin protected ports are fixed with 4–0 silk to the incisions made from the elbow and posterior axillary region to prevent the friction burns at liposuction cannula port sites. The operation should be started in the prone position. Then, the fat melting phase is started with the ultrasonic system. In the Vaser® system we are using, the fat is melted by operating at 100% energy level in C mode with a five-ring 3.7-mm-thick probe. Using the probe at a deep level throughout the melting process will reduce the risk of possible complications. It is technically safe to melt all desired layers from the deep to the surface by applying passive hand pressure. As in liposuction, it would be more appropriate to start the process of melting fat from negative areas first and then move on to areas with excess fat tissue (if any). It is worth emphasizing that there is no need for liposuction in men, especially on the triceps and deltoid muscles unless there is a need to stretch for volumization. After the fat is melted by ultrasound, liposuction can be applied with cannulas of 3–5 mm in diameter and three straight or reverse triangle holes depending on the surgeon's experience. While these processes pass, the nurses prepare pure fat grafts in 50 cc luer lock injectors (these processes will be explained in detail in the next section).

Then, an average of 100–150 cc of pure fat is injected into the deltoid muscle by entering through the posterior axillary incision while still in the prone position with a 2.4-mm-diameter, three-straight hole cannula. In the application done as a bolus injection, the cannula is entered into the muscle and is slowly pulled back and the entire fat injection process is completed from the end point. Incisions used for fat grafting are always closed with 5–0 rapid vicryl. If there is not enough volume in the triceps muscle, sub-fascial and subcutaneous fat injection, ranging from 30 to 60 cc in total, can be performed from the posterior axillary incision, especially to the region of the lateral head.

If fat grafting is applied in the biceps muscle, it will be easier to reach the intramuscular and sub-fascia with this 2-mm additional incision by opening the arm medial as an additional incision instead of the anterior axillary incision in the Spin position. Here again, 30–60 cc fat can be applied, mostly at the medial head (Fig. 42.3).

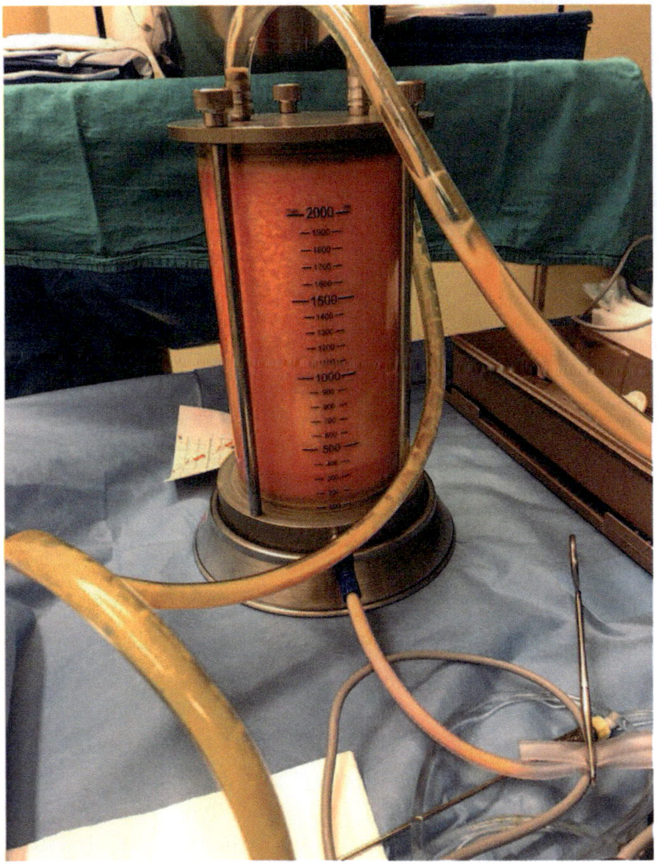

Fig. 42.3 The fat collected as a result of liposuction

42.7 Preparation of the Fat and Safe Intramuscular Injection

Fats removed from the body by liposuction are collected in closed systems (Fig. 42.3). Here, it is important to note that, when using an ultrasonic device (Vaser®) with a small bracket open, only 100% C mode energy level is need. We observe that fat cells are alive in MRI examinations of patients with a history of 1–5 years [18, 19].

The fats collected in the closed system are transferred to 50 cc injectors after separating them from the liquid part with their own weight and gravity effect without using a special system (Fig. 42.4).

Decantation is done for the fat to be grafted. Fat grafting can be done with a curved, 2.4-mm cannula. To increase the volume, 50–150 mL of fat is injected intramuscularly into the middle fascicle of the deltoid. This is done through a posterior axillary incision, by inserting a 2.4-mm curved cannula. Figure 42.5 shows preoperative (Fig. 42.5a, b, e, f) and after liposuction arm contouring and fat grafting in male patients in need of deltoid sulcus recreation (Fig. 42.5c, d, g, h). In the preoperative view, the maximum width of the scapular arch is bitrochanteric, while the postoperative maximum width is bideltoid. The more pronounced lateral curvature and triceps recess should be noted.

In the patient in the prone position, 100–150 cc fat transfer is performed using a posterior axillary incision, with a 2.4-mm-diameter and 16-cm-angled cannula, into the deltoid muscle pole. At this stage, the most important part is that the

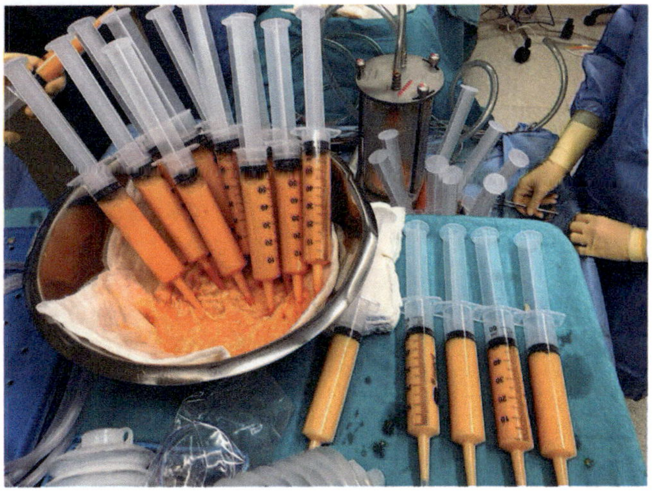

Fig. 42.4 The fats collected in the closed system transferred to 50 cc injectors after separating them from the liquid part with their own weight and gravity effect without using a special system. A 1 cc acrylic syringe is preferred for small volume fat grafting. The air bubble inside the syringe must be removed so that the amount of the injected volume can be accurately recorded. Also, 10–50 cc syringes can be used for large volume fat grafting

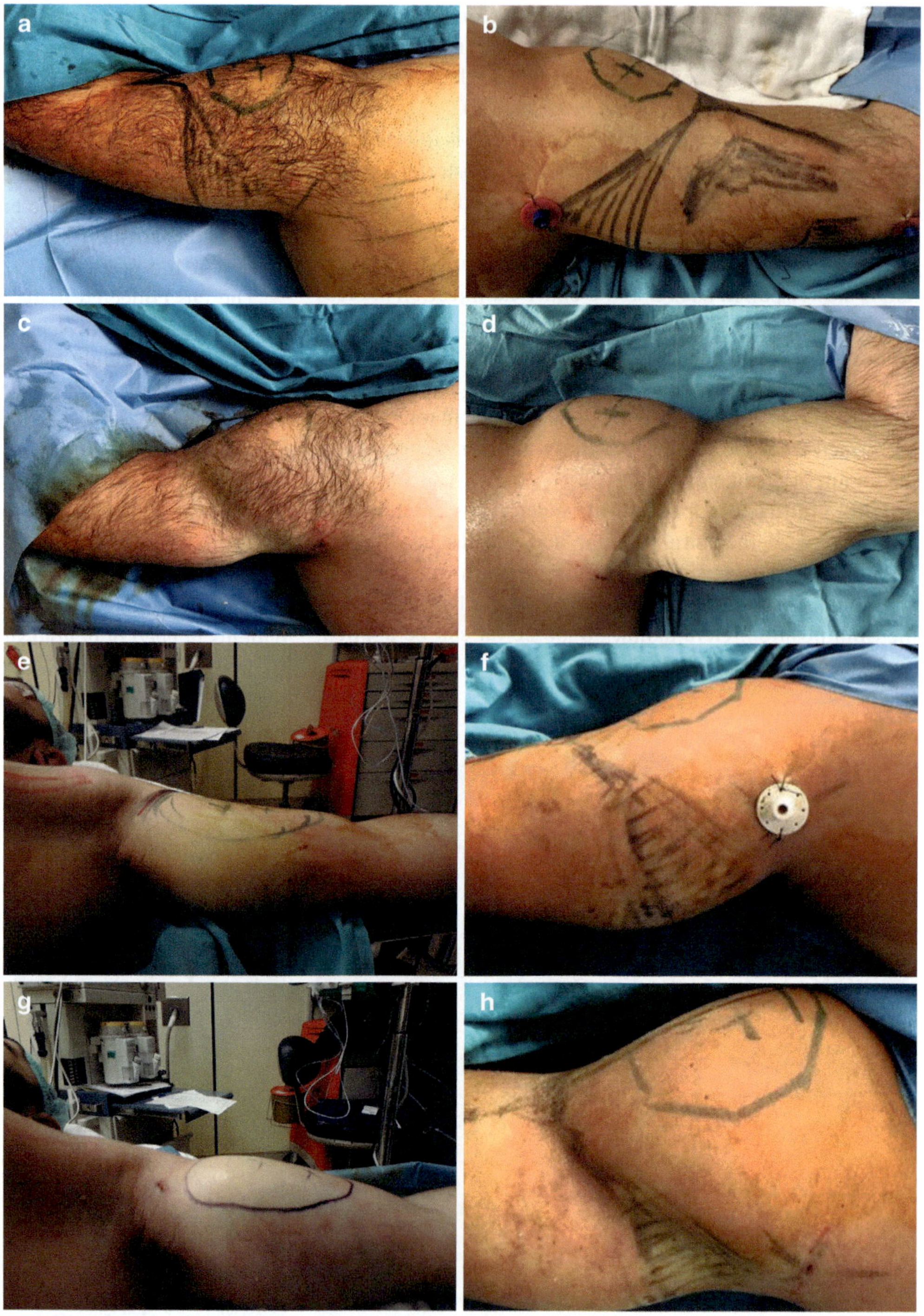

Fig. 42.5 Preoperative (**a**, **b**, **e**, and **f**) and immediate postoperative images (**c**, **d**, **g**, and **h**) of male patients. All patients underwent the arm shaping procedure described in the main text. Postoperative images show changes in shape and volume and include curved posterior areas at the proximal and distal ends

fat is injected intramuscularly (never subcutaneously) in the form of a bolus. The injection must be completed by coming back in the same direction after it enters the muscle once with the cannula and reaches the extreme point (Video 42.1). Although the injection is intramuscular, the incisions that are used for fat injection are closed with 4/0 or 5/0 rapid vicryl as a general principle.

42.8 Postoperative Care

After the liposuction application, a 10–14 days compression garment is placed on the arm that is shaped by the liposuction method, extending beyond the elbow. At the same time, an active postoperative treatment regimen, which includes infrared and local heating pads as well as a lymphatic drainage massage, should be implemented. A massage is performed to drain excess fluids from the elbow incision left open from the surgery. Patients can fully move their arms after 1 day. If necessary, physical therapy can also be applied to patients. It is recommended that patients do stretching exercises following surgery. These exercises can be given according to the patient between 1 and 6 months. In postoperative care, it is sufficient to use standard corsets with moderate pressure for 7–10 days and to use the recommended drug treatments after general liposuction procedures. While the fat-injected incision is sutured, the incision near the elbow is left open for drainage. Postoperative manual massages are an important and recommended part of this drainage. After intramuscular fat injections, I recommend patients to stay away from active sports for 6–12 weeks in which they will use the relevant muscle.

42.9 Outcomes

When the techniques described above are carefully applied by experienced surgeons for the upper arm liposuction surgery, the results are satisfactory for the patients. A few cases before and after surgery are shown in Fig. 42.6. Yet another case is shown in Fig. 42.7a, b. In order to get good results during the arm contouring process, the negative and positive zones drawn on the arm and contoured should be determined so that they have a natural appearance (Fig. 42.7c, d).

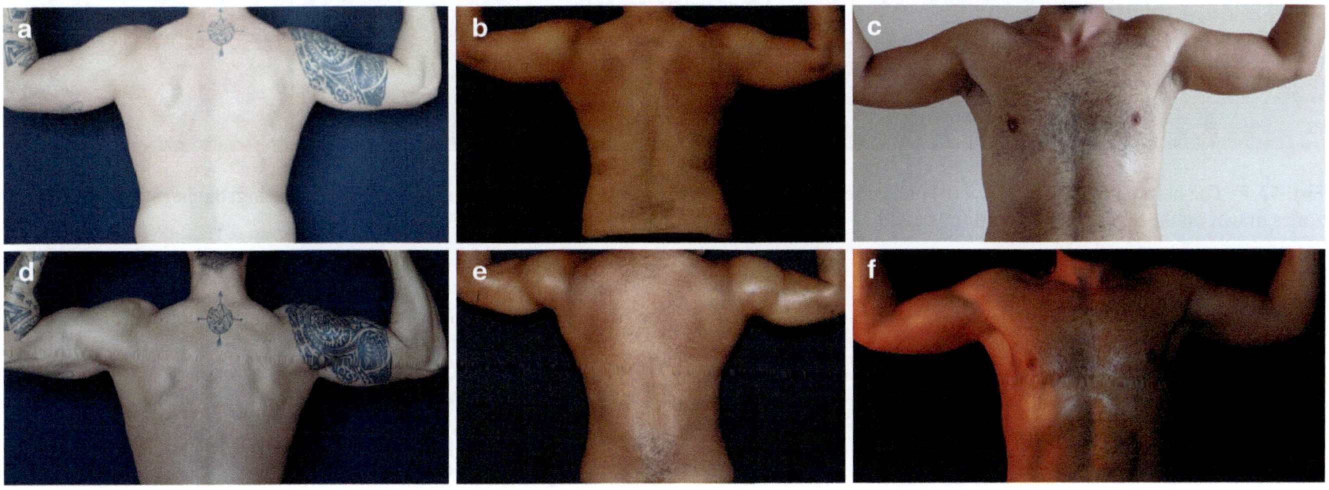

Fig. 42.6 Before photos for different patients (**a**–**c**) and after photos, respectively (**d**–**f**), for arm shaping in male patients. The deltoid definition and muscular appearance in the postoperative images are striking

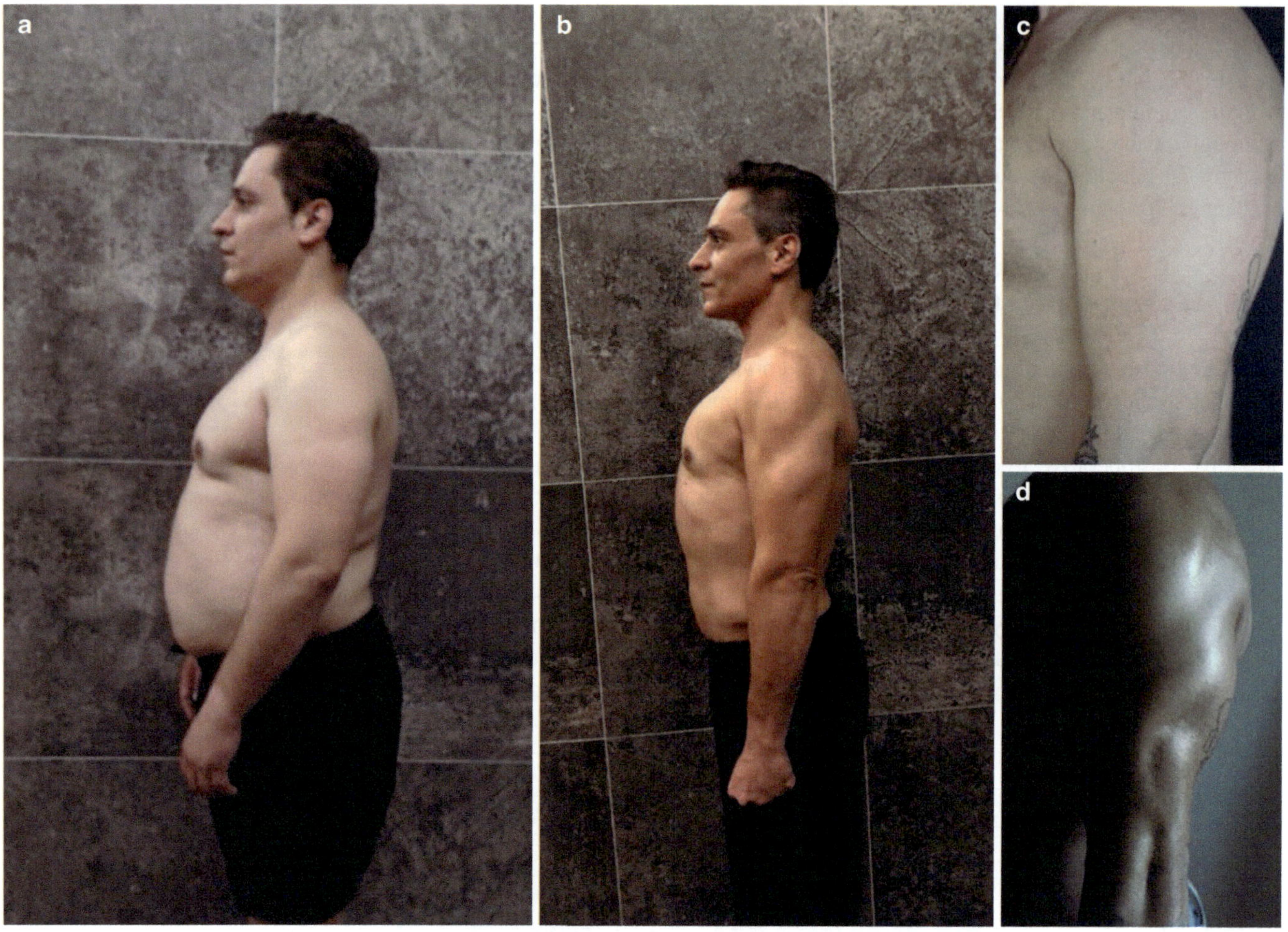

Fig. 42.7 The before (**a**) and after view (**b**) of a patient who underwent arm shaping. While planning the arm contouring, the negative and positive zones drawn on the arm and contoured (**c**) should be determined so that they have a natural appearance (**d**)

Conflict of Interest None.

Funding None.

References

1. Tiwana MS, Charlick M, Varacallo M. Anatomy, shoulder and upper limb, biceps muscle. Treasure Island (FL): StatPearls; 2021.
2. Creteur V, Madani A, Sattari A, El Kazzi W, Bianchi S. Ultrasonography of complications in surgical repair of the distal biceps brachii tendon. J Ultrasound Med. 2019;38(2):499–512.
3. El Abiad JM, Faddoul DG, Baydoun H. Case report: broad insertion of a large subscapularis tendon in association with congenital absence of the long head of the biceps tendon. Skelet Radiol. 2019;48(1):159–62.
4. Hutchinson HL, Gloystein D, Gillespie M. Distal biceps tendon insertion: an anatomic study. J Shoulder Elb Surg. 2008;17(2):342–6.
5. Moore CW, Rice CL. Rare muscular variations identified in a single cadaveric upper limb: a four-headed biceps brachii and muscular elevator of the latissimus dorsi tendon. Anat Sci Int. 2018;93(2):311–6.
6. Saluja S, Das SS, Kumar D, Goswami P. Bilateral three-headed biceps brachii muscle and its clinical implications. Int J Appl Basic Med Res. 2017;7(4):266–8.
7. Peterson SL, Rayan GM. Shoulder and upper arm muscle architecture. J Hand Surg Am. 2011;36(5):881–9.
8. Hoyos A, Perez M. Arm dynamic definition by liposculpture and fat grafting. Aesthet Surg J. 2012;32(8):974–87.
9. Chamosa M, Murillo J, Vazquez T. Lipectomy of arms and lipograft of shoulders balance the upper body contour. Aesthet Plast Surg. 2005;29(6):567–70.
10. Dolsky RL, Newman J, Fetzek JR, Anderson RW. Liposuction. History, techniques, and complications. Dermatol Clin. 1987;5(2):313–33.
11. Comiskey C. Cosmetic surgery in Paris in 1926: the case of the amputated leg. J Womens Hist. 2004;16(3):30–54.
12. Pelosi MA 3rd, Pelosi MA 2nd. Liposuction. Obstet Gynecol Clin N Am. 2010;37(4):507–19.
13. Bellini E, Grieco MP, Raposio E. A journey through liposuction and liposculpture: review. Ann Med Surg (Lond). 2017;24:53–60.
14. Husain TM, Salgado CJ, Mundra LS, Perez C, AlQattan HT, Bustillo E, et al. Abdominal etching: surgical technique and outcomes. Plast Reconstr Surg. 2019;143(4):1051–60.

15. Wu S, Coombs DM, Gurunian R. Liposuction: concepts, safety, and techniques in body-contouring surgery. Cleve Clin J Med. 2020;87(6):367–75.
16. Theodorou S, Chia C. Radiofrequency-assisted liposuction for arm contouring: technique under local anesthesia. Plast Reconstr Surg Glob Open. 2013;1(5):e37.
17. Kaoutzanis C, Gupta V, Winocour J, Layliev J, Ramirez R, Grotting JC, et al. Cosmetic liposuction: preoperative risk factors, major complication rates, and safety of combined procedures. Aesthet Surg J. 2017;37(6):680–94.
18. Doornaert M, Colle J, De Maere E, Declercq H, Blondeel P. Autologous fat grafting: latest insights. Ann Med Surg (Lond). 2019;37:47–53.
19. Shim YH, Zhang RH. Literature review to optimize the autologous fat transplantation procedure and recent technologies to improve graft viability and overall outcome: a systematic and retrospective analytic approach. Aesthet Plast Surg. 2017;41(4):815–31.

Autologous Fat Grafting Applications for Functional and Aesthetic Purposes in Gynecology

43

Pablo Gonzalez Isaza

Contents

43.1 Introduction

Subcutaneous fat tissue has been in use for many years for correcting and remodeling purposes. The first use of autologous fat transplantation was described by Neuber in 1893, by filling small soft tissue defects. Regarding female body, Czerny in 1895 reported the use of free fat graft for complete reconstruction of breast defects. The first reports of autologous fat transplantations to the genital tract were described by Kolisher and posteriorly widely accepted for different indications in gynecology which were not only for aesthetic indications but also for functional aims. It can be considered as an invaluable tool because it is readily available, easy to harvest, affordable, and biocompatible and has low morbidity.

43.2 Applications

43.2.1 Labia Majora Augmentation

There exists enough good-quality evidence regarding techniques and procedures for labia minora reduction; [1] in contrast, techniques describing labia majora augmentation are limited. While hyaluronic acid injection remains the most described technique, there is no clear agreement regarding the volume to be injected, neither the concentration nor cross linking agent for better results. However, efforts to find the best hygroscopic properties and rheology are under permanent investigation [2, 3]. Other techniques described for the augmentation of labia majora are related to flaps with or without combined labia minora reduction [4, 5].

Our experience in female genital cosmetic procedures started in 2007 in South and Central America. The injection of silicone, methacrylates, and its derivatives was common for labia majora enhancement; however, we stopped the use of these illegal substances recently due to complications, that are nowadays considered a challenging disease [6] (Fig. 43.1).

Among all the described procedures related to female genital cosmetic surgery, the option of labia majora enhancement or augmentation remains an everyday demand from our patients. Massive weight loss after bariatric surgery, chronic stretching from waxing, and just aging are the most common causes of labia majora laxity. In some occasions, it is possible to perform a combined labia majora labioplasty, when required (Fig. 43.2).

Our most common technique for labia majora enhancement is fat grafting, with the volume of injected fat varying from 15 to 20 mL, distributed in different planes per each labia majora. Fat is harvested using modified Klein solution from the inner thigh or mons pubis using a 2.0-mm cannula and 10-mL syringes with low pressure, and the fat is later separated by traditional decantation technique, for 3–5 min.

P. G. Isaza (✉)
Department of Obstetrics and Gynecology, Hospital Universitario San Jorge Pereira, Pereira, Colombia

© The Author(s), under exclusive license to Springer Nature Switzerland AG 2023
A. Di Giuseppe et al. (eds.), *Fat Transfer in Plastic Surgery*, https://doi.org/10.1007/978-3-031-10881-5_43

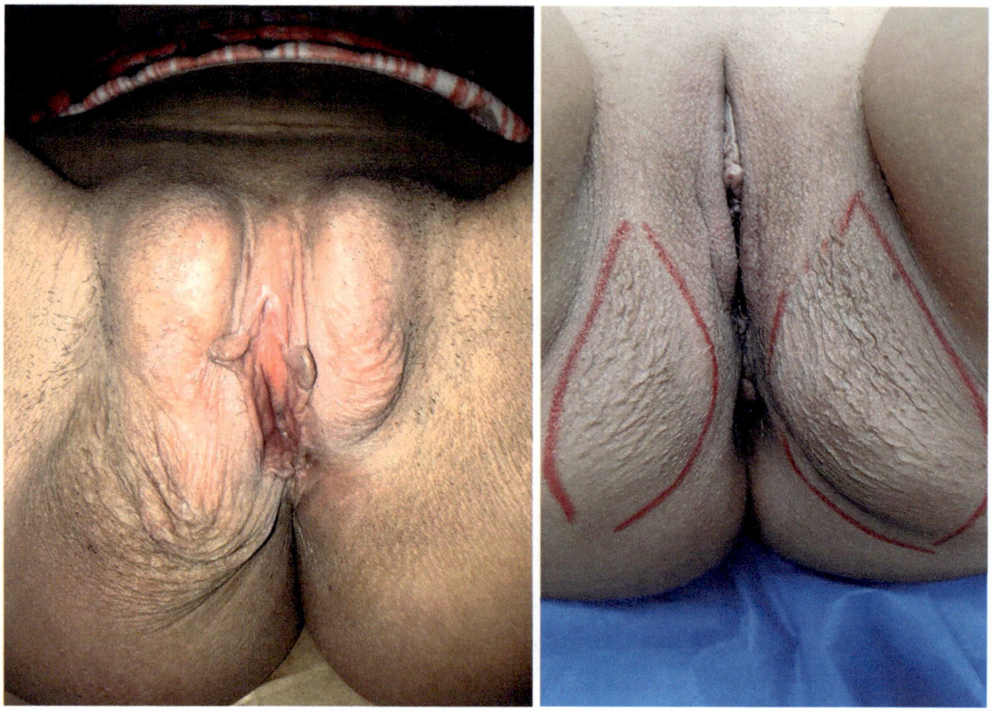

Fig. 43.1 Complications related to the use of silicone derivatives for labia majora augmentation (personal archives)

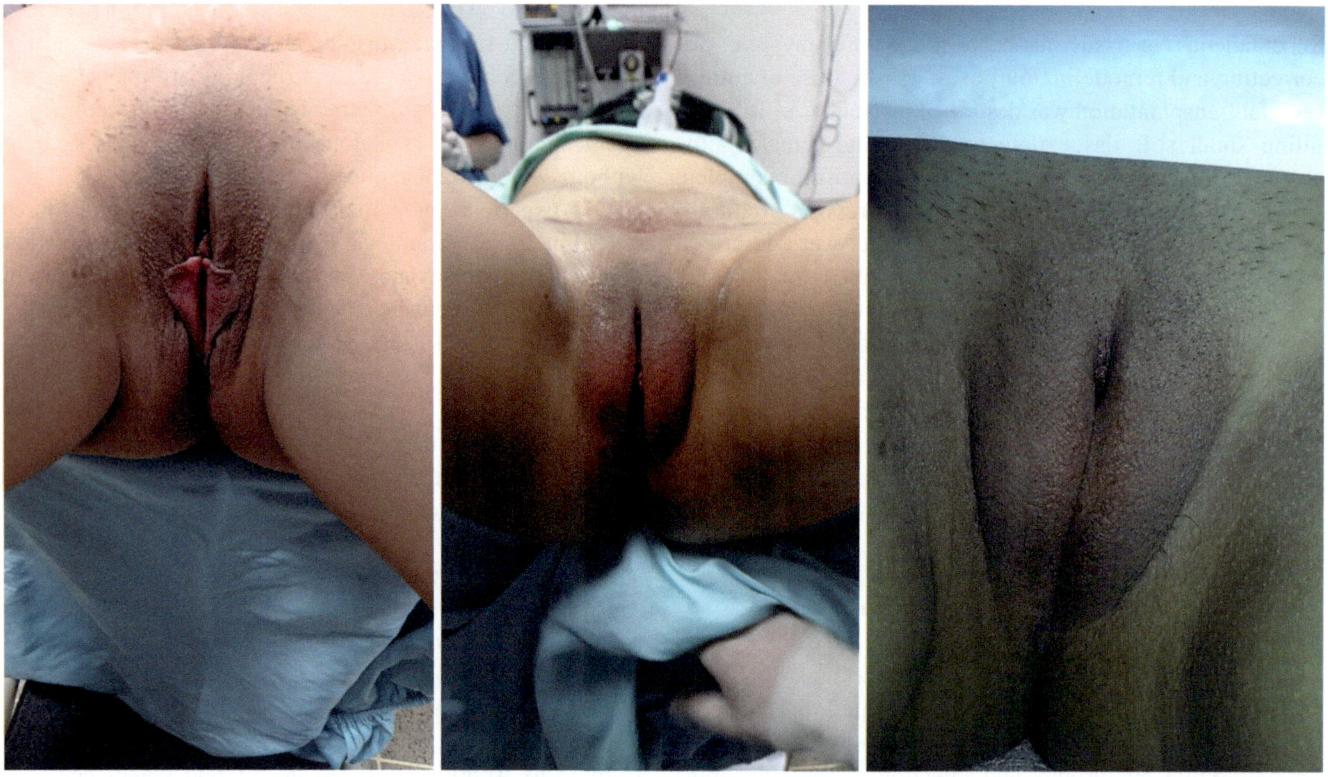

Fig. 43.2 Combined labia minora labiaplasty + autologous fat grafting

Afterward, all remnants from infiltration are removed, and purified fat is used for injection of the labia majora with a 0.8–1.0-mm reusable cannula through a bilateral suprapubic incision. The fat is injected in a retrograde technique in multiple layers to improve the survival of the implant.

Previous reports by Hersant et al. described high degree of satisfaction from patients [7].

43.2.2 Labia Majora Augmentation as a Combined Approach for the Treatment of Botched Labiaplasties

Unfortunately, due to the increasing demand for labiaplasties, there are reports in the literature about failed labiaplasties every day [8]. Sometimes there is no available remnant tissue in order to perform a reconstruction, so hiding the amputated labia by increasing the volume of labia majora is a feasible option to improve the quality of life of these patients (Figs. 43.3, 43.4, 43.5, 43.6, 43.7, and 43.8).

43.2.3 Pelvic Floor Reconstruction in Combination with Labia Majora Enhancement

Often, there is a need to combine functional and aesthetic indications to improve quality of life of patients. Urogynecological surgeries for pelvic organ prolapse, and stress urinary incontinence can be combined with aesthetic enhancement of the vulva, without adding an additional risk and with high patient satisfaction. Pardo et al. described combinations of mid-urethral slings with labia vaginal tightening procedures and labiaplasties [9].

In our experience, it is quite common to perform this combined approach and we have found high patient satisfaction and no increase in morbidity. One of our most significant cases was a 58-year-old woman suffering from chronic urinary infections, with several unsuccessful treatments. The urethral orifice was completely exposed, leaving a wide genital hiatus and short perineal distance, which favored the colonization of bacteria to the lower urinary tract. We performed a perineal repair and labia

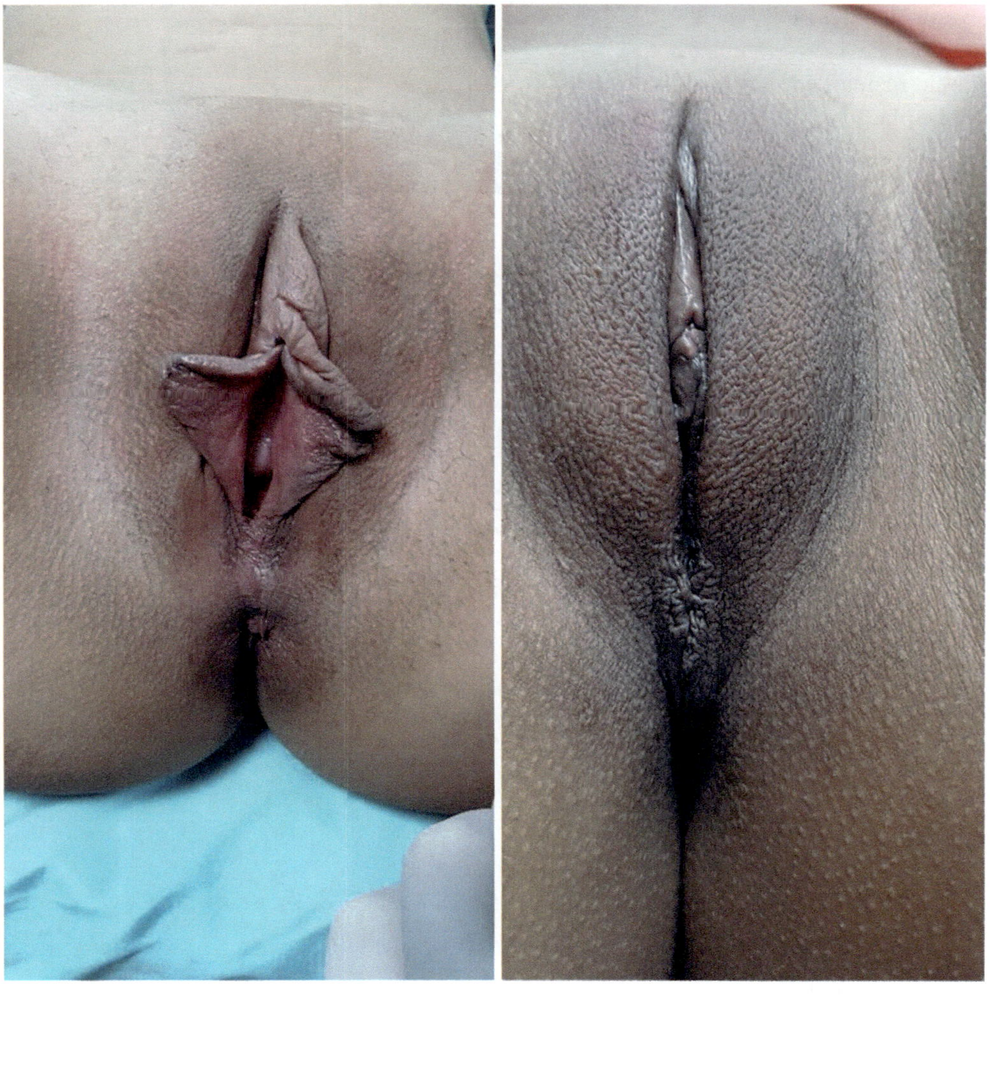

Fig. 43.3 Combined labia minora labiaplasty + autologous fat grafting

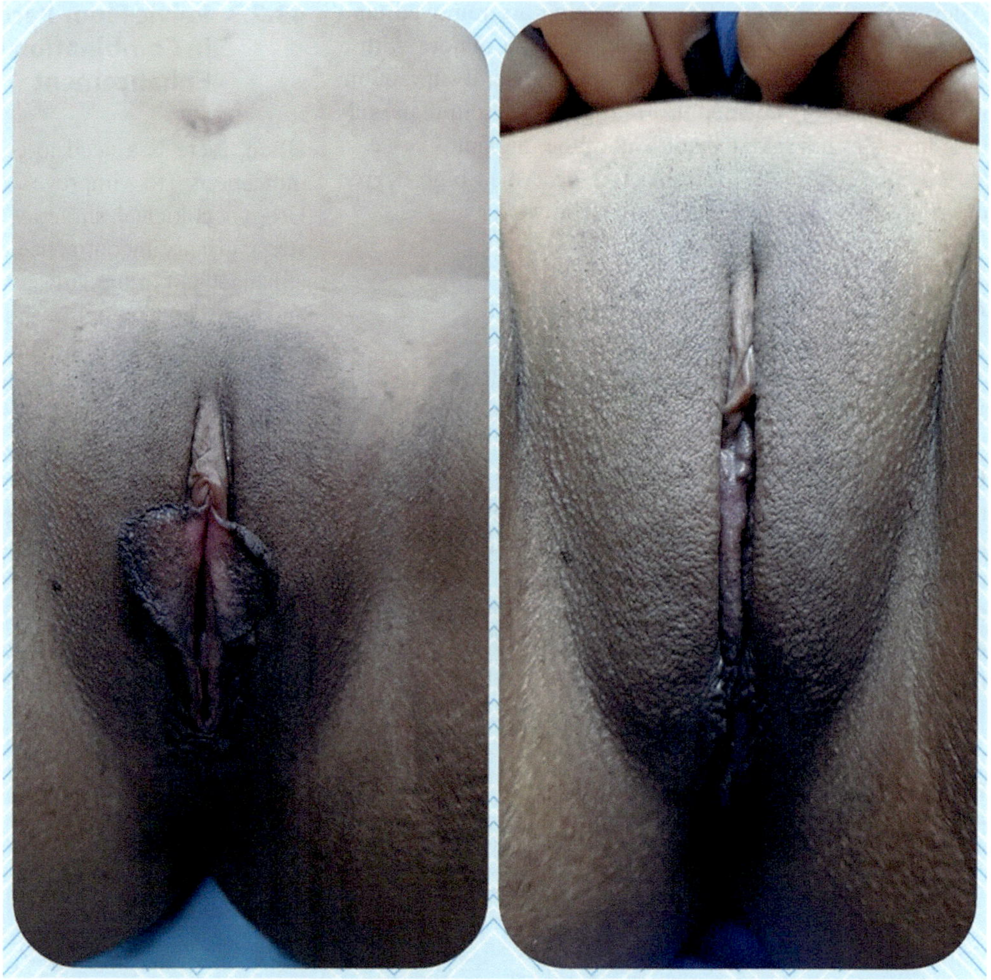

Fig. 43.4 Combined labia minora labiaplasty + autologous fat grafting

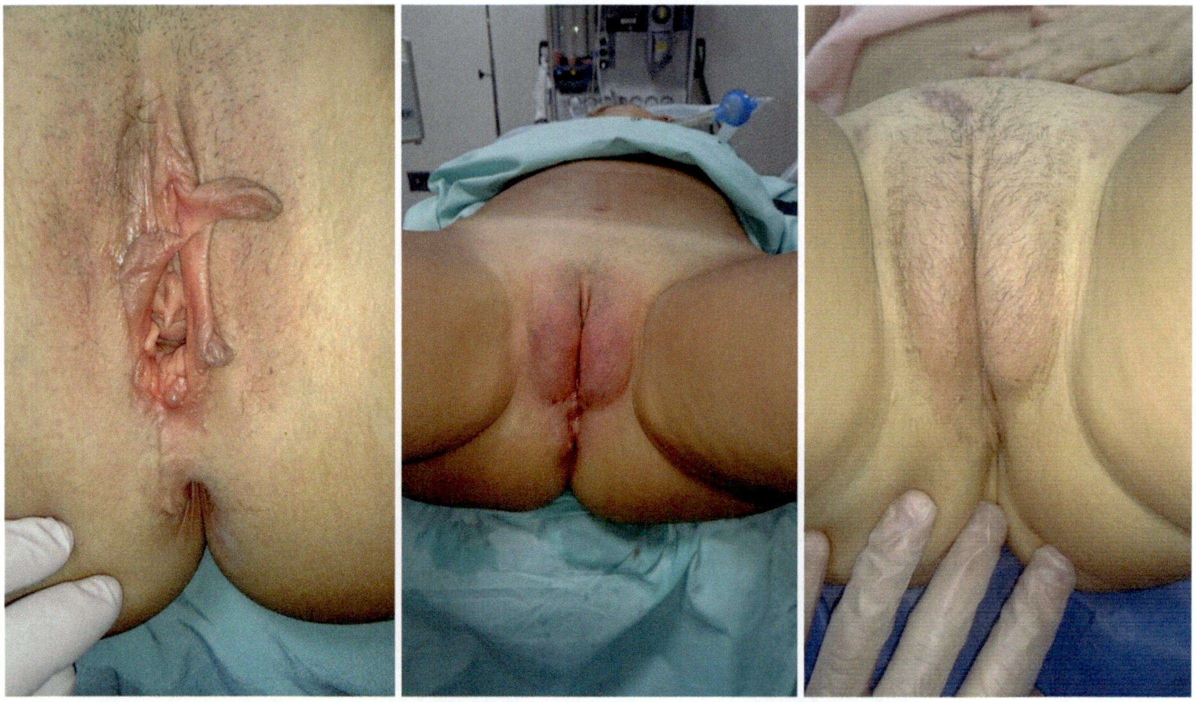

Fig. 43.5 Botched labiaplasty, labia minora reconstruction + fat grafting

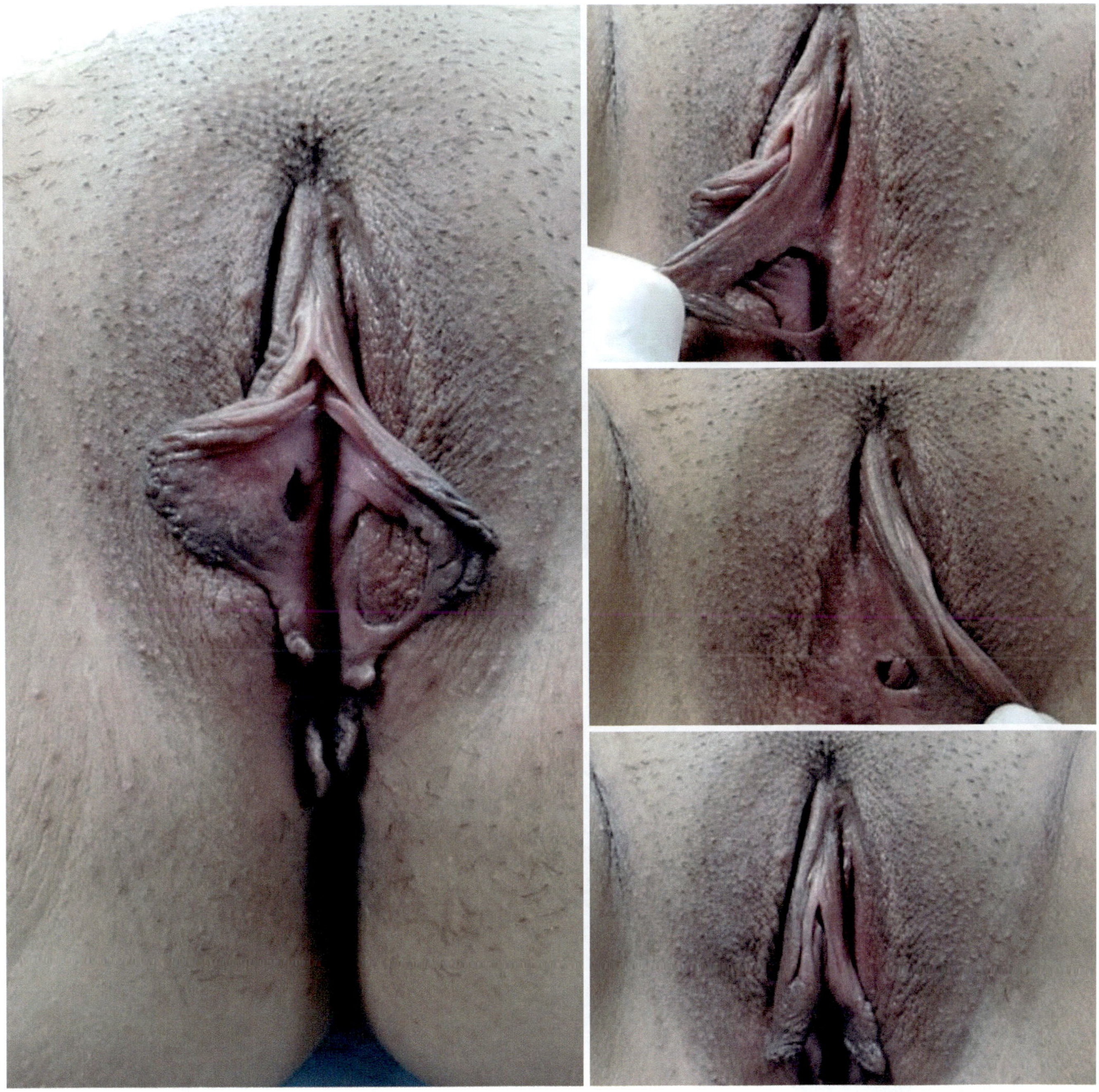

Figs. 43.6–43.7 Botched labiaplasty, combined approach clitoral hood flap reconstruction + autologous fat grafting

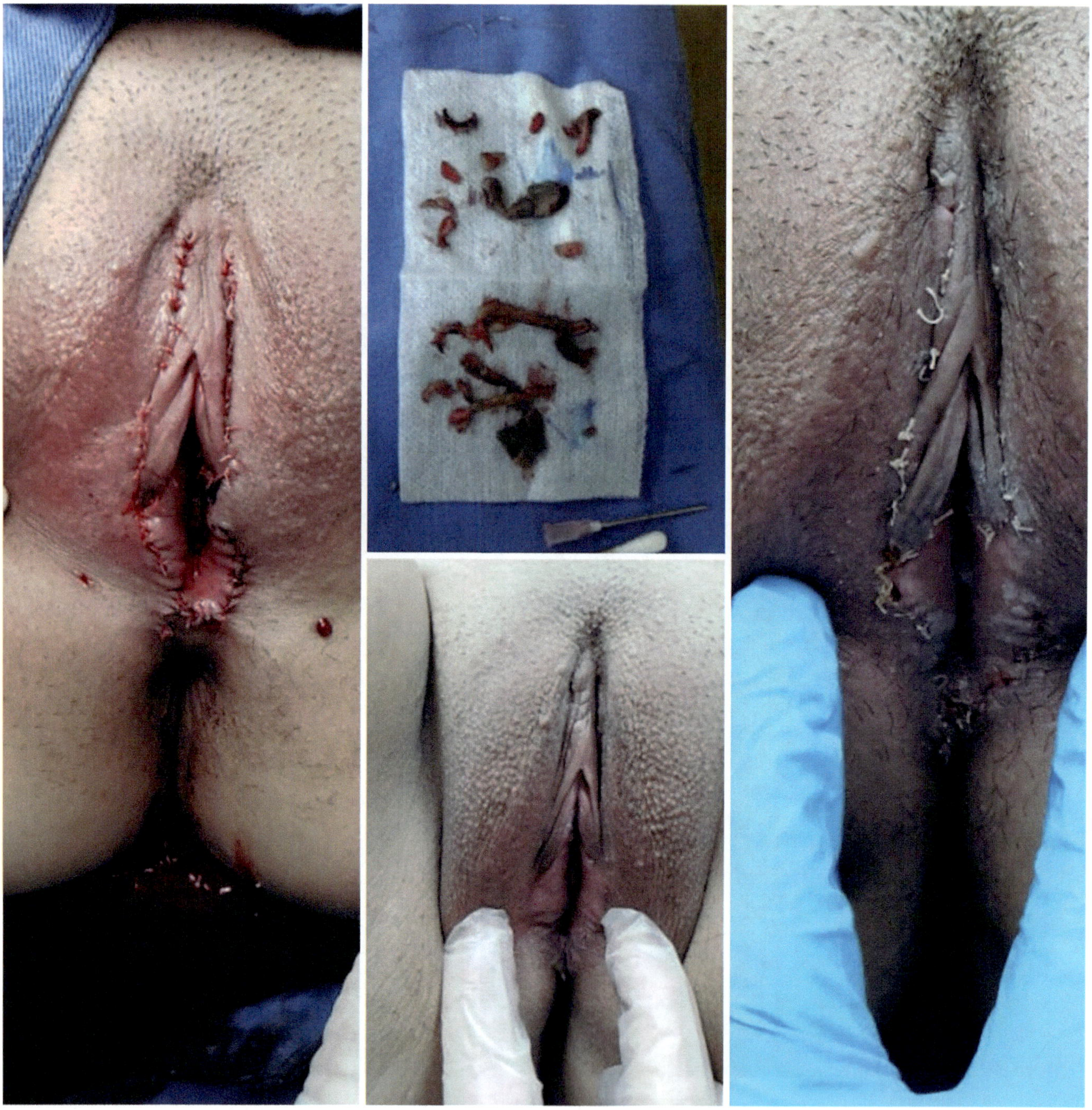

Fig. 43.6–43.7 (continued)

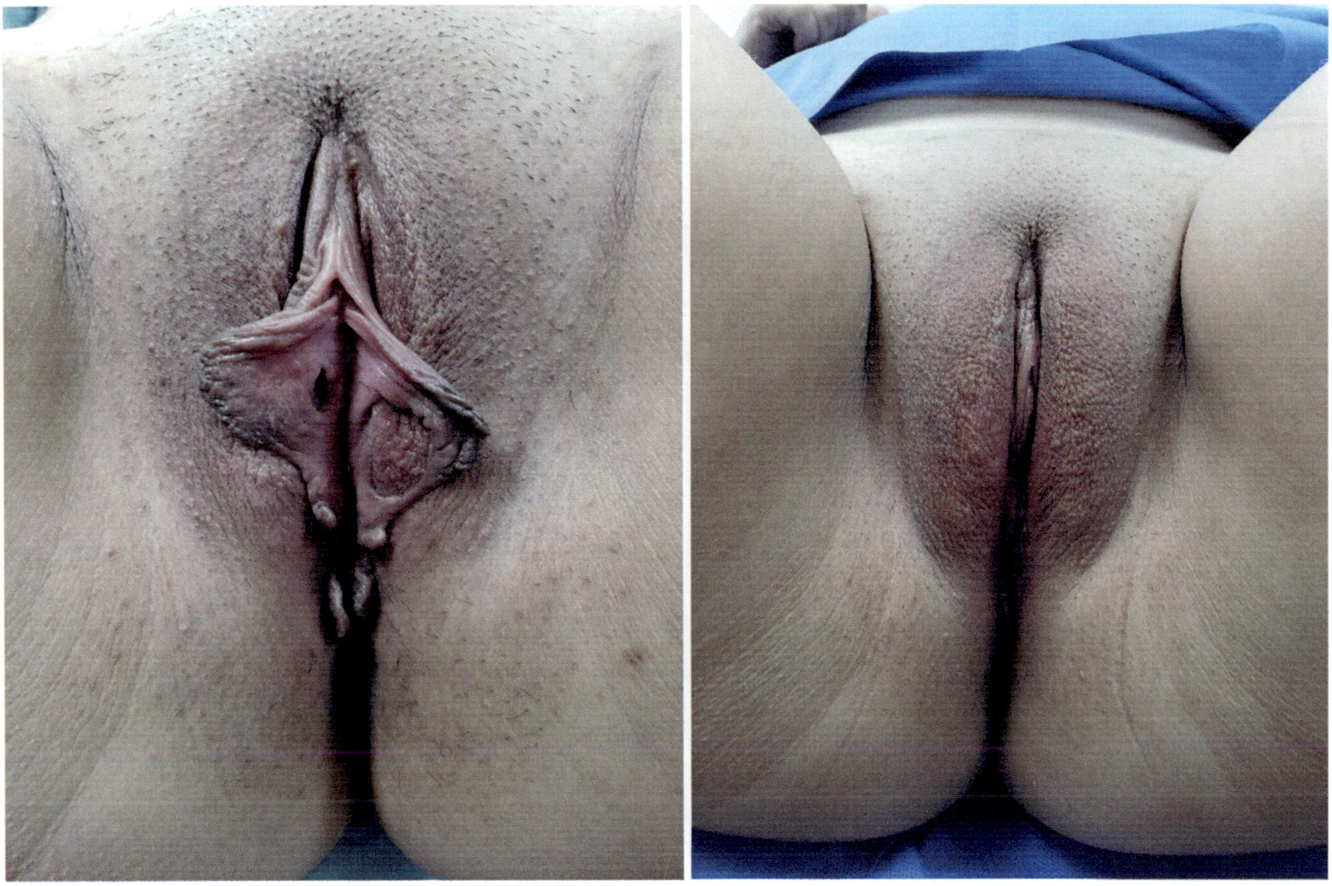

Fig. 43.8 Botched labiaplasty, combined approach clitoral hood flap reconstruction + autologous fat grafting (before 2-month postop)

majora fat grafting which turned into a total remission of her symptoms (Figs. 43.9 and 43.10).

43.2.4 Miscellaneous Applications in Gynecology

43.2.4.1 Pudendal Neuralgia

Pudendal neuralgia is a chronic multifactorial condition which has an important impact on patient's quality of life. It is one of the most common causes of chronic pelvic pain, and due to inflammation or compression of pudendal nerve, it is described by the patient, as mono- or bilateral, and can be radiating to gluteus, genitals, or thighs. Available treatments for this debilitating condition are analgesics or pain modulators, pudendal nerve block, peripheral or central neuromodulation, and surgical decompression. Unfortunately, the efficacy of these treatments is limited: the drugs have important side effects and are not always effective. Venturi et al. described a safe lipofilling technique, related to multiple transperineal injections of autologous adipose tissue with stem cells in patients with pudendal neuralgia with good results on safety and effectiveness [10].

43.2.5 Vaginal Tightening

Vaginal laxity is one of the most common symptoms of pelvic floor dysfunction. Traditional therapeutic options, such as pelvic floor physiotherapy and biofeedback, remain the gold standard strategies for this condition. Unfortunately, a vast majority of patients do not have an

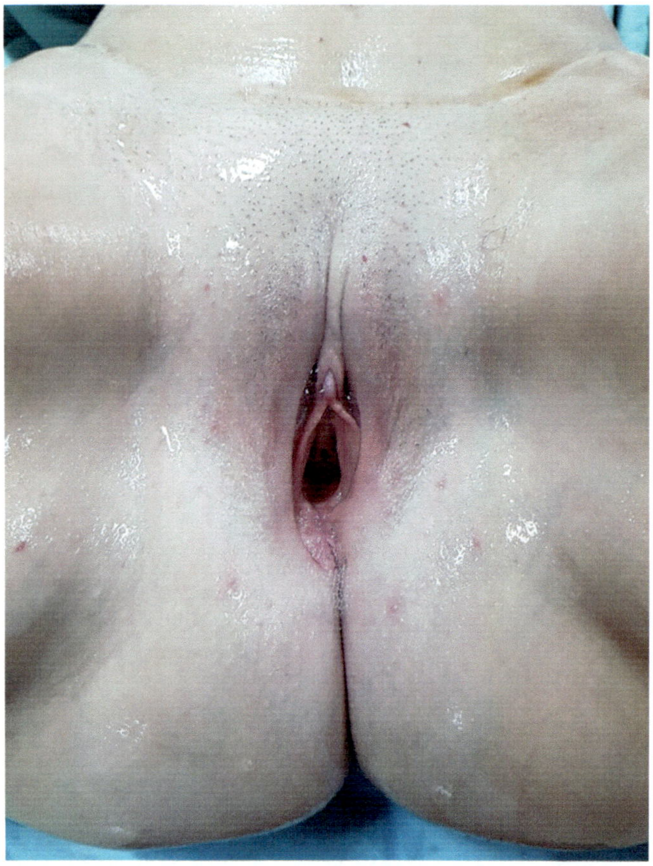

Fig. 43.9 Pre-exposed urethra short perineal length

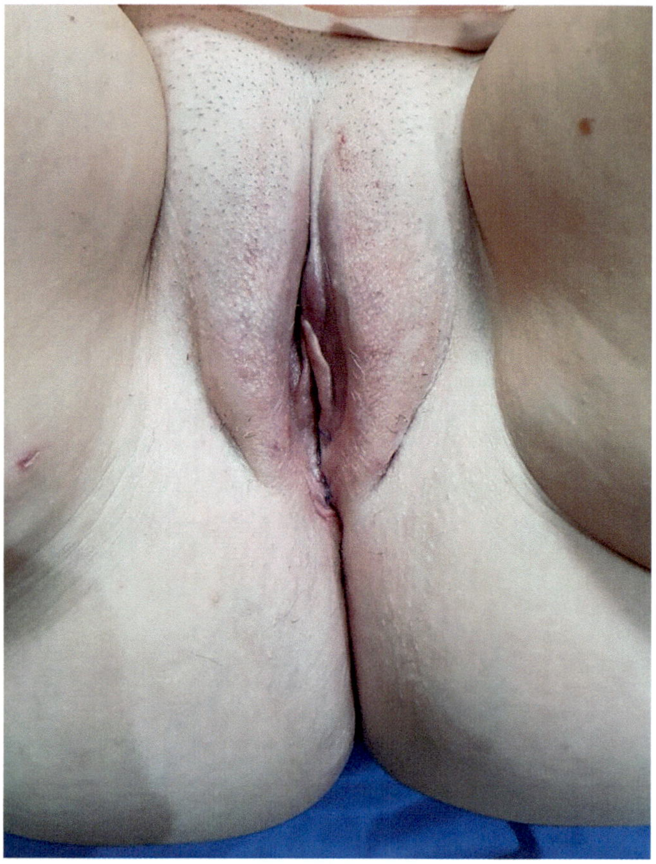

Fig. 43.10 Pop perineal repair + fat grafting

important improvement either subjectively or objectively surgical. New strategies with the use of energy-based devices appear as alternatives while surgical management is also an alternative when non-invasive options fail to offer improvement on sexual gratification. The first report by Pardo et al. described an important improvement of vaginal laxity after performing a modified colpoperineoplasty procedure and enhanced sexual gratification in the studied population [11].

A mini invasive option for vaginal tightening using autologous fat graft seems to improve sexual gratification in those patients with a history of sexual dysfunction and pelvic floor disorders [12, 13].

43.3 Complications

Autologous fat grafting for gynecological indications can be considered a safe treatment; hence, complications are extremely rare, but usually are described as the same for fat grafting, in body contouring procedures, such as granulomas, fat necrosis, oil cyst formation, infection, and temporary pain. A rare case of a woman who developed severe fat embolism after autologous fat grafting for vaginal tightening and breast augmentation surgery was described by Chenyu et al. and it was assumed to have occurred by injection of large volumes of fat directly into the vessels around the vagina. The patient survived after intensive care unit management.

43.4 Conclusion

Fat grafting in gynecology is a safe and efficient alternative highly accepted from patients. It can be done for multiples indications: aesthetic, reconstructive, or functional, alone or in combination with other surgical techniques. I consider it an invaluable tool in regenerative medicine and surgery; however, it must be performed by trained physicians in order to achieve the best results and ensure patient safety.

References

1. Oranges CM, Sisti A, Sisti G. Labia minora reduction techniques: a comprehensive literature review. Aesthet Surg J. 2015;35(4):419–31. https://doi.org/10.1093/asj/sjv023.
2. Fasola E, Gazzola R. Labia majora augmentation with hyaluronic acid filler: technique and results. Aesthet Surg J. 2016;36(10):1155–63.
3. González IP, Leibaschoff G, Esposito C, Cipolla G, Bader A, Lotti T, Tirant M, Van Thuong N, Ramo Abdulkader M, Rauso R, Serafin D, Vojvodic A, Zerbinati N. Genitourinary syndrome of menopause and the role of biostimulation with non-cross-linked injectable hyaluronic acid plus calcium hydroxyapatite. J Biol Regul Homeost Agents. 2019;33(6):1961–6. https://doi.org/10.23812/19-251-L.
4. Karabağlı Y, Kocman EA, Velipaşaoğlu M, et al. Labia majora augmentation with de-epithelialized labial rim (Minora) flaps as an auxiliary procedure for labia minorareduction. Aesthet Plast Surg. 2015;39(3):289–93.
5. El Danaf AAH. Deepithelized fasciocutaneous flap for labia majora augmentation during thigh lift. Eur J Plast Surg. 2010;33(6):373–6.
6. Guerrieri C, Hudacko R, Joseph EM. Vulvar silicone granuloma. Int J Gynecol Pathol. 2020;39(5):456–9. https://doi.org/10.1097/PGP.0000000000000632.
7. Hersant B, Jabbour S, Noel W, Benadiba L, La Padula S, SidAhmed-Mezi M, Meningaud JP. Labia majora augmentation combined with minimal labia minora resection: a safe and global approach to the external female genitalia. Ann Plast Surg. 2018;80(4):323–7. https://doi.org/10.1097/SAP.0000000000001435.
8. Smarrito S, Brambilla M, Berreni N, Paniel BJ. Nymphoplastie secondaire : étude rétrospective à propos de 44 cas. Rapport annuel de la SOFCPRE 2019 [Secondary labiaplasty: Retrospective study about 44 cases. Annual report SOFCPRE 2019]. Ann Chir Plast Esthet. 2019;64(5–6):660–6. https://doi.org/10.1016/j.anplas.2019.06.002.
9. Jack IPS, Vicente SD. Midurethral slings combined to cosmetic gynecological surgeries: our experience of 13 years. J Gynecol Womens Health. 2017;3(2):555608. https://doi.org/10.19080/JGWH.2017.03.555608.
10. Chiarioni G, Asteria C, Whitehead WE. Chronic proctalgia and chronic pelvic pain syndromes: new etiologic insights and treatment options. World J Gastroenterol. 2011;17:4447–55.
11. Pardo JS, Sola VD, Riccia PA, Guiloff EF. Colpoperineoplasty in women with a sensation of a wide vagina. Acta Obstet Gynecol Scand. 2006;85(9):1125–7.
12. Brambilla M. Intramuscular-submucosal lipostructure for the treatment of vaginal laxity. In: Paper presented at Congresso Internazionale di Medicina Estetica. Milan: Medicina Estetica; 2008.
13. Meadows LD, Avellanet YR, English J. Fat augmentation of the anterior vaginal wall: a novel use of fat augmentation in enhancing the female sexual experience. Am J Cosmet Surg. 2011;28:171–6. https://doi.org/10.1177/074880681102800312.

Breast Reconstruction with Fat Derived by Laser-Assisted Liposuction

44

Andre Ofek and Lior Heller

Contents

44.1 Introduction

Breast cancer surgery has become more prevalent in the last decade due to enhanced survival rates, the gaining popularity of prophylactic risk-reducing mastectomy, and contralateral prophylactic mastectomy. Simultaneously, breast reconstruction has become the standard of care after breast cancer surgery, shown to improve psychosocial outcomes [1].

Autologous fat grafting (AFG) is a well-known technique which gained growing popularity in the past few decades as a powerful tool for breast reconstruction. It is one of the most prevalent methods used to correct contour deformities in reconstructed breasts. This technique is used today as an adjacent procedure to implants, regional flaps or free flaps reconstruction, for contour deformities correction after lumpectomies, and even as a standalone technique for reconstruction of small breasts. It provides a biocompatible, natural, and inexpensive reconstructive solution for these deformities. This chapter will review the use of laser-assisted fat transfer for breast reconstruction.

Supplementary Information The online version contains supplementary material available at https://doi.org/10.1007/978-3-031-10881-5_44.

A. Ofek (✉) · L. Heller
Department of Plastic Surgery, Shamir Medical Center, Zerifin, Israel

Faculty of Medicine, Tel-Aviv University, Tel Aviv, Israel
e-mail: ofek@post.bgu.ac.il; liorh@shamir.gov.il

44.2 History

Neuber was the first to suggest autologous fat transfer procedure to a facial scar depression in 1893 [2].

He was followed by Hollander in 1912, Neuhof in 1921, and Josef in 1931 [3, 4].

In the 1980s, Illouz was the first to introduce modern safe liposuction procedure that preserves neurovascular structures, maintains fluid balance, and minimizes both blood loss and patient discomfort [5].

44.3 Laser Lipolysis

Lasers are used as an adjunct for liposuction as a method of disrupting the fat cells and their connective tissue. For that purpose, lasers have been used at a wide range of wavelengths from 980 to 1470 nm. Skin penetration of the different wavelengths is similar (~1.5 mm). Longer wavelengths are better absorbed by fat, making these cells more susceptible to laser lipolysis. Although fat cells are

© The Author(s), under exclusive license to Springer Nature Switzerland AG 2023

A. Di Giuseppe et al. (eds.), *Fat Transfer in Plastic Surgery*, https://doi.org/10.1007/978-3-031-10881-5_44

destroyed at all these wavelengths [6], longer wavelengths present additional advantages. For example, the long 1440 nm wavelength, which is highly absorbed in both fat and water, is more effective at reducing fat volume than shorter wavelengths [7].

In longer wavelengths, water has a higher coefficient of absorption than in shorter wavelengths, leading to the rapid heating of adipose tissue in a very localized manner. In addition, longer wavelengths translate into more energy and higher temperatures, causing skin tightening with little down time [8]. Indeed, even in the absence of liposuction, the 1440 nm laser induces long-term skin tightening, neck contouring [9], and cellulite reduction [10]. During mechanical liposuction, fat cells are sucked into the cannula at high negative pressure, affecting cell viability. Cell survival is sensitive to high vacuum negative pressure. Vacuum negative pressure is much higher in traditional vacuum-assisted than in laser-assisted liposuction [11]. Therefore, laser lipolysis is especially suited for fat grafting due to increased cell viability.

Furthermore, laser lipolysis is as effective as traditional liposuction but has fewer side effects, pain, and bruising [12]. As laser coagulates blood vessels while melting the fatty tissue, the result is less bleeding during the procedure and less bruising following it. Taken together, laser lipolysis holds several advantages over traditional liposuction, including improved cell viability, less pain and bruising, shorter recovery time, and better postoperative skin tightening [13].

44.4 Laser-Assisted Liposuction

Laser-assisted liposuction presented technological advance for liposuction and was first publicized in the early 1990s. With this technique, laser is directly pointed to the adipose tissue, liquifying the tissue, and makes it easier to suction it out [14].

Although initial results were promising [15, 16], the FDA failed to approve laser-assisted liposuction, and no further progress was made for a decade. In 2006, lasers were first used with tumescent liposculpture and "SmartLipo" (developed by Cynosure) was approved by the FDA. Clinical studies showed that a 1064 nm Nd:YAG laser was capable of reducing fat volume, as assessed using MRI imaging, without significant side effects, except mild bruising [17].

In vivo and in vitro analysis demonstrated that laser induced tissue heating acts by tissue liquefaction, adipocytes membrane disruption, and skin tightening [18].

A significant advantage of laser lipolysis over power-assisted and traditional liposuction is the shortened recovery time due to tissue liquefaction that allows physicians to use less force and lower suction, decreasing tissue trauma [19].

44.5 Autologous Fat Grafting (AFG)

Autologous fat grafting is an autologous reconstruction method.

It was described in the late nineteenth century and became popular during the twentieth century to correct cutaneous and subcutaneous defects [20].

Its advantages include the following: It is biocompatible and readily available, it is cheap, natural, easy to harvest and can even improve donor site a esthetics.

Adipose-derived stem cells can differentiate into many other cell types of mesenchymal origin, including adipocytes, chondrocytes, osteoblasts, and myoblasts.

It does not trigger an immunological response and has dynamic regenerative properties that can help regenerate damaged tissues, on top of filling volume deficiencies and correcting subcutaneous contour deformities [21, 22].

It is also a very safe procedure with very low morbidity [23].

44.5.1 Clinical Applications of Autologous Fat Grafting (AFG)

Clinical applications of AFG are diverse and include reconstructive and a esthetic breast procedures [24], correction of subcutaneous deformities resulting from diverse etiologies (i.e., facial reconstruction and rejuvenation, hand rejuvenation, Parry–Romberg syndrome, Poland syndrome, Dupuytren's, Raynaud's diseases, and pectus excavatum), and treatment of scars, ulcers, and burns [25–29].

The multipotency of the adipose stem cells allows the newly transformed fat cells to effectively replace damaged or missing cells [30].

44.6 Autologous Fat Grafting for Breast Reconstruction

Compared with implant-based reconstruction (IBR), autologous reconstructions result in a more natural shape, feel, and ptosis, requiring less maintenance, and provide a more durable reconstruction over time [31].

Autologous fat grafting for breast reconstruction was first attempted in 1895 [20] and, since then, has proven to be a powerful surgical technique after breast cancer surgical treatment.

AFG is a powerful and versatile method that can be used for different levels of deformities, including correction of breast volume, shape, and contour deformities, improves irradiated tissue vascularity and pliability, and prepares it for delayed implant reconstruction [32–35]. AFG is also used for treatment of postmastectomy pain syndrome and capsular contracture [33, 36].

AFG has been shown to improve skin quality of irradiated tissue.

It has many influences on radiated skin owing to its anti-fibrotic, proangiogenic, antioxidant, immunomodulatory, regenerative, and antiapoptotic mechanisms.

It decreases dermal thickness, decreases collagen content, and increases vascular density, thereby reversing the sequelae of radiation-induced fibrosis [37, 38].

It increases skin softness and pliability, induces volume restoration, improves hair growth in areas of alopecia, reduces pain, and improves cosmetic and functional outcomes. Fat grafting can also prepare the irradiated wound bed to better facilitate implant reconstruction [39–41]. Adipose tissue contains the highest percentage of adult stem cells. These cells can undergo multilineage differentiation and may be crucial for fat graft take [30, 42–50]. Adding fat grafting to implant-based reconstruction improves a esthetic outcomes as seen by patients and surgeons alike [35, 51]. Common locations for fat grafting are the upper breast pole and the medial aspect of the breast. These locations camouflage implant visibility and rippling [52].

Fat grafting might be even more important when the implant is located superficial to the pectoralis muscle (pre-pectoral). Fat can be grafted between the ADM and the skin, enhancing breast volume, improves flap thickness and camouflage rippling.

Fat grafting can also be used as an implant substitute when implant base reconstruction has failed d/t recurrent infections or implant exposure. The reconstructed breast might be smaller than before but will be more natural in shape and feel [53].

Fat grafting in autologous breast reconstruction can augment volume of free and pedicled flaps, can fill deformities, and can even address fat necrosis, as grafted fat can improve vascularity and tissue regeneration [54, 55].

It may be very useful to fill up missing volumes and contour deformities after breast conserving surgery (lumpectomies). It may improve asymmetry, increase volume, and improve scarring [56]. Fat grafting can even be used for primary breast reconstruction (without implant or flap), if the breasts not too big. Stark et al. have shown that primary reconstruction with fat grafting only is feasible and usually requires two–three fat grafting sessions with 3–6 months of intervals in between [35]. Fat grafting is also used for postmastectomy pain syndrome, with proposed mechanism of scar remodeling and thus preventing nerve entrapment [41, 57].

44.6.1 Fat Grafting Timing

Early fat grafting in breast reconstruction is giving faster results and might reduce the number of procedures needed. On the other hand, delaying fat grafting will allow better vascularization and will give time for contour deformities to appear. At our institution, fat grafting is usually started during exchange of tissue expander to permanent implant, later with another procedure, or as a standalone procedure [58–62].

44.6.2 Surgical Technique (Video 44.1)

Viability of the harvested fat cells is crucial for graft longevity and is influenced by every step in the process. General principles for graft survival are shortest possible fat cells ischemic time, shortest possible hypothermia, and minimal fat manipulation [34, 63].

Trials trying to identify preferred donor site regarding fat graft intake failed to do so; hence, patient and surgeon preferences should be considered when choosing the donor site. Abdomen, flanks, and lateral thighs are usually the sites of choice by patient and surgeon, supplying enough fat, and many times donor site a esthetic enhancement [64].

In general, less-traumatic methods of fat harvesting result in increased adipocyte viability and graft survival [65, 66].

Vacuum pressure should be the lowest possible (to avoid adipocytes structural damage) but high enough to be able to extract the fat cell from the body in order not to prolong surgical time/fat cell ischemia and hypothermia times [66, 67].

Optimal cannula size still lacks consensus. It should be large enough to avoid shear stress, but not too large, as to avoid donor site irregularities. Cannula with a diameter between 3 and 5 mm is good for this purpose [68–70].

Fat processing is necessary to separate adipocytes from collagen fibers, blood, and debris. These elements can cause inflammation at the recipient site, which is harmful to the grafted fat. The four commonly used methods to separate adipocytes from other elements are sedimentation, filtering, washing, and centrifugation. Filtering can be done simultaneously with liposuction, thus has the advantage of shortening fat cells time outside the body [71].

Grafts that are too large tend to undergo liquefaction, necrosis, and cyst formation. That is why one session grafted fat volume is limited, thus repeated fat grafting is necessary in many cases [72].

The BeautiFill system by Lipolife was developed for these purposes. It is a multicomponent liposuction and fat collection system developed and produced by Alma Lasers. This all-in-one platform contains all components needed for laser-assisted liposuction and fat grafting procedure, including tumescence injection device, liposuction device, sterile collection canister for collecting the fat, and a second canister to collect tumescent drainage from the fat before grafting. Fat cell survival is sensitive to high vacuum negative pressure. Vacuum negative pressure is higher in traditional vacuum-assisted than in laser-assisted liposuction [11].

The advanced low suction pressure vacuum system (450–675 mmHg) contains two in-line filtration canisters that separate fat from tumescent and serosanguinous fluid while the fat is being suctioned out of the patient's body. Since tumescent fluid and serosanguinous fluid both have detrimental effects on adipose cell viability [73, 74], serous fluids are being sucked from the fat sterile canister to a second, non-sterile canister simultaneously at the same time of liposuction. This filtration increases fat cells viability, with no OR time wasted. After aspiration is finished, collection canister is turned upside down, and syringes can be filled directly so that fat is not exposed to a non-sterile environment (Fig. 44.1). Sterility is crucial since fat cells are immediately re-injected to the patient. Preparation for reinjection is rapid, to try and keep fat cells ischemic time as short as possible.

Prepared fat can be injected through a 19–20-gauge cannula, at low pressure, with multiple passes.

The 1470 nm infrared wavelength used in this system is absorbed better by water than by fat, allowing a decrease in the effective wattage at the point of tissue interaction so that fibrous tissue is ruptured, rather than adipocytes (Fig. 44.2). The 1470 nm light has a limited depth of 1 cm (vs. 5 cm for 980 nm), improving the control of energy around the cannula tip.

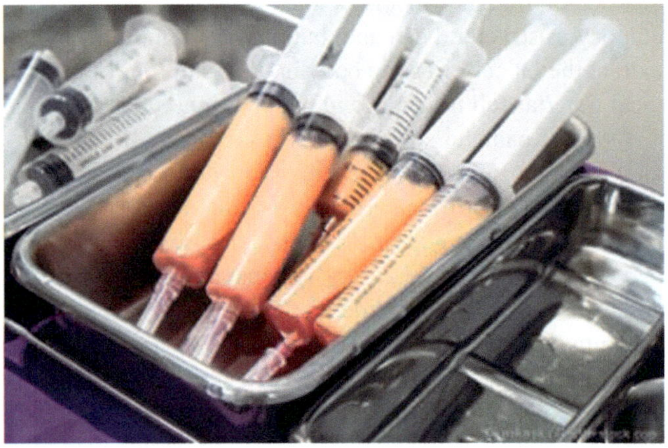

Fig. 44.1 Extracted fat (Courtesy of Alma Lasers, ©1999, All Rights Reserved)

A radial fiber diffuses laser energy in a radial manner, resulting in reduced power density compared to a bare fiber (Fig. 44.3). Bare fibers have a high-power density concentrated at the tip and are used mainly for vaporization and tissue damage in lipolysis and other surgical procedures (hysteroscopy, laparoscopy, and urology procedures). However, high-power density is less suitable for fat grafting, as the localized temperature rise damages the tissue. In contrast, a radial fiber has a much greater area energy distribution. The difference in power density between a bare and a radial fiber is substantial. The radial tip decreases energy distribution, ensuring that the fibrous components are lyzed a lot more than the adipocytes. The fiber is inserted inside the metal suction cannula (Fig. 44.4), so that fibrous components lysis and liposuction are done simultaneously.

Due to both the decreased vacuum pressure and energy distribution, this system shows greater cell survival. The viability of adipocytes ranged from an average in suction-assisted liposuction of 79.7% (50–96%) is up to 95.7% (90–99%) for BeautiFill [75].

The increased adipocyte survival rate, the ease of use, and the low risk of bruising make the system ideal for laser-assisted liposuction preceding autologous fat transfer.

44.6.3 Complications

Donor site complications are minimal and are usually temporary (bruising, swelling, hematoma formation, paresthesia, and pain). Long-term complications are usually related to liposuction technique and include hypertrophic scarring, contour irregularities, and damage to the underlying structures (i.e., intraperitoneal or intramuscular penetration) [76].

Fig. 44.2 BeautiFill—absorption coefficient vs. wavelength (Courtesy of Alma Lasers, ©1999, All Rights Reserved)

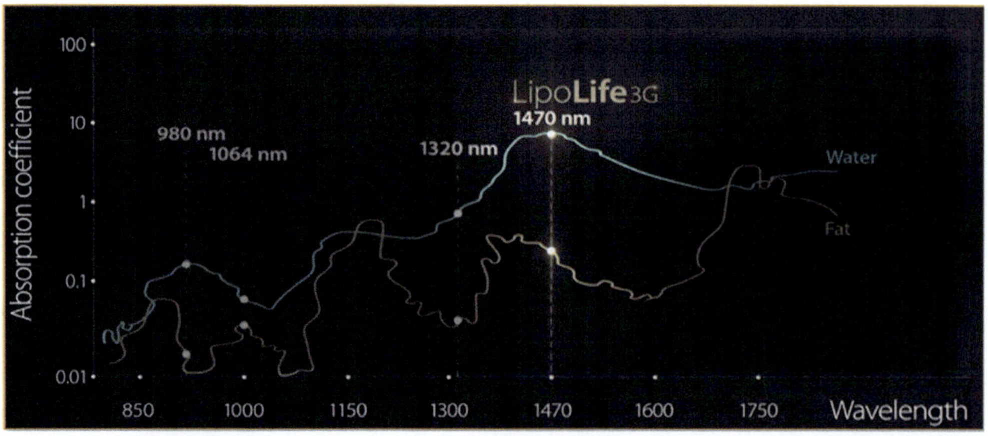

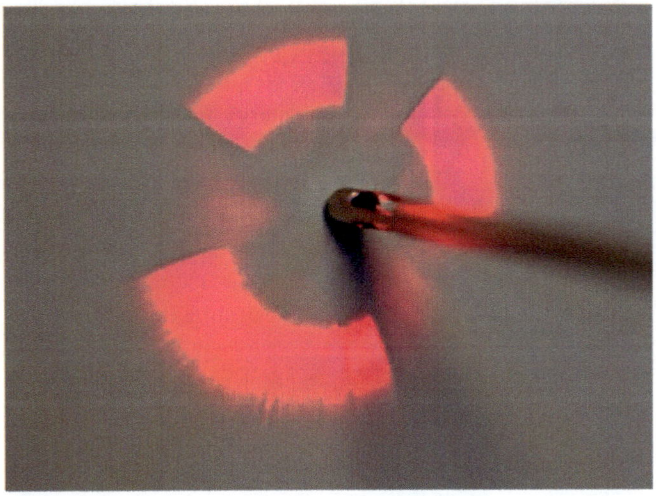

Fig. 44.3 Laser emission pattern (Courtesy of Alma Lasers, ©1999, All Rights Reserved)

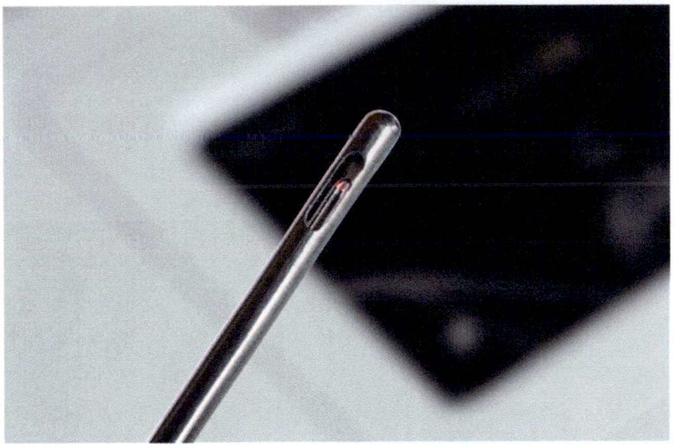

Fig. 44.4 Laser fiber inside the cannula (Courtesy of Alma Lasers, ©1999, All Rights Reserved)

Hypoperfusion to grafted fat cells can cause ischemia that might result in necrotic cysts, calcifications, fibrosis, oil cysts, and infection. These complications might also occur after other kinds of breast surgery [29, 35, 77].

44.6.4 Breast Imaging After Fat Grafting

The use of autologous fat grafting raised concerns regarding interference with breast imaging (mammography and ultrasound) and regarding its theoretical oncogenic potential [78].

Papers published by Coleman and Saboeiro showed that mammography findings after fat transfer could easily be differentiated from breast cancer mammography findings [27, 29].

Fat grafting may cause radiological changes in the breast, but these changes can be differentiated from cancer recurrence by US and mammography.

Breast cancer recurrence will usually appear as irregular mass with pleomorphic calcifications and rapid uptake of contrast, whereas fat necrosis and oil cysts will show coarse calcifications and no uptake or gradual uptake of contrast.

The consensus among authors is that fat grafting alone does not increase rates of cancer recurrence or reduce the ability to identify recurrence on imaging, but it seems more studies are needed to rule out fat grafting interference with breast imaging after breast conserving therapy and fat grafting-based reconstruction [79–81].

44.7 Conclusion

Fat grafting is an autologous method for breast reconstruction that has many advantages, including being autologous material that contain stem cells. This method is versatile and flexible.

It produces long-lasting, natural-looking, and natural-feeling results [82] (Figs. 44.5 and 44.6) and has been shown to be a safe procedure with very high patient satisfaction [83–87].

Laser lipolysis is especially suited for fat grafting due to increased cell viability.

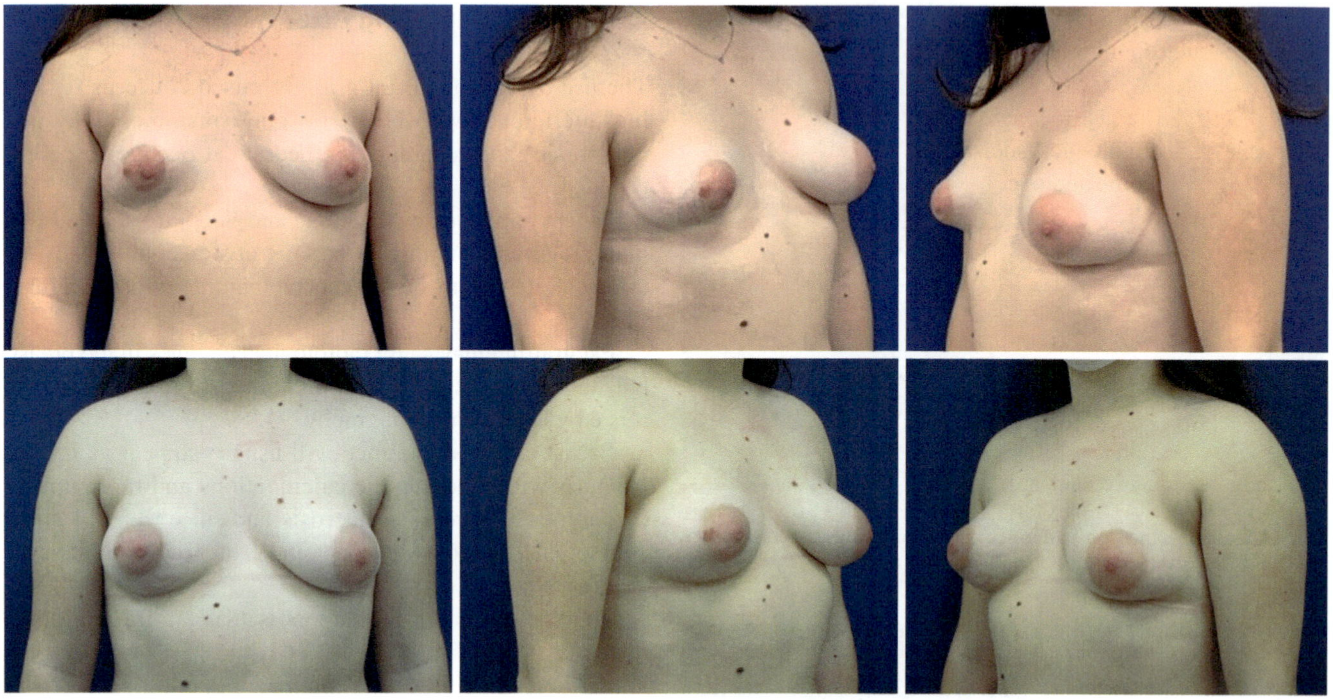

Fig. 44.5 Before (above) and 5 months after (below) one session of fat injection to the right breast for asymmetry correction (Courtesy of Dr. Andre Ofek, ©2022 YYY, All Rights Reserved)

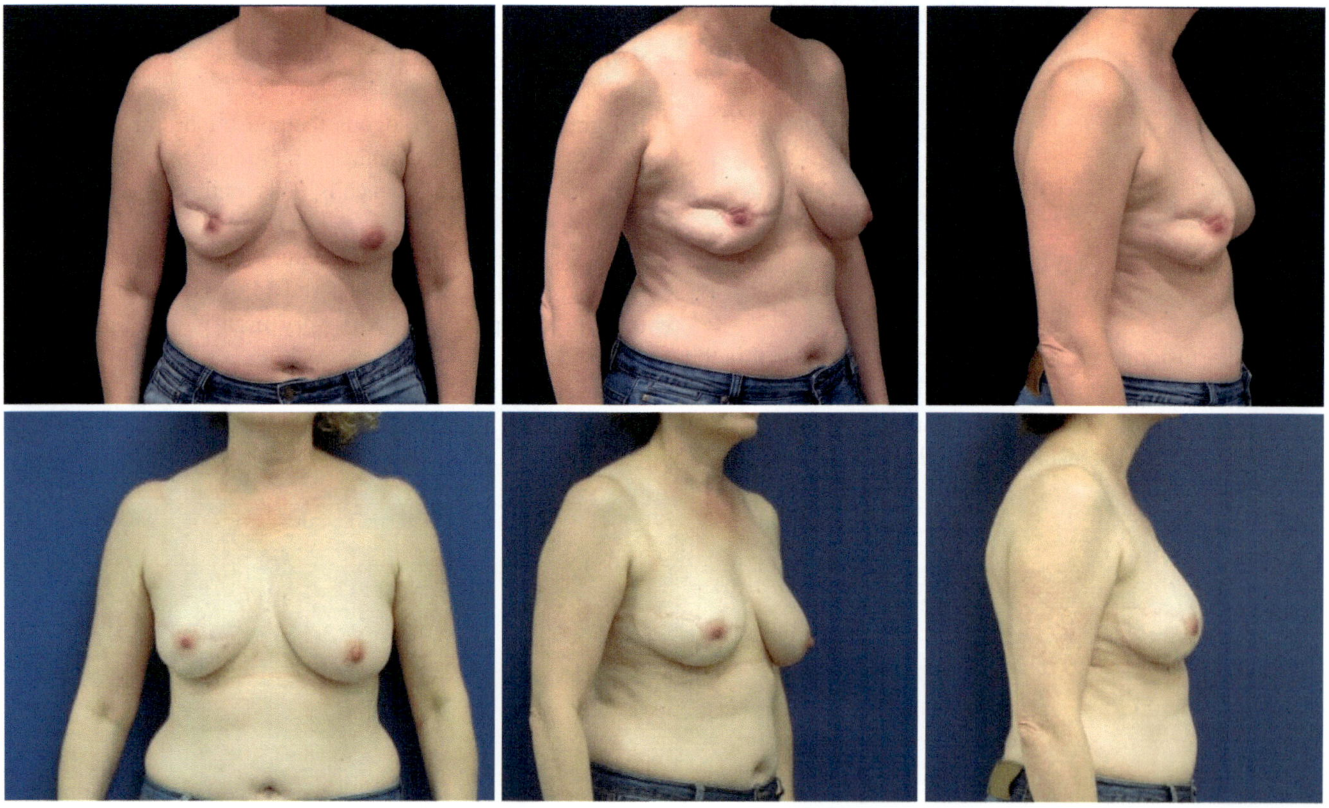

Fig. 44.6 Before (above) and 8 months after (below) one session of fat injection to a silicone implant-based reconstructed right breast (Courtesy of Dr. Andre Ofek, ©2022 YYY, All Rights Reserved)

Furthermore, laser lipolysis is as effective as traditional liposuction but has fewer side effects, pain, and bruising [12].

The future of breast reconstruction lies in adipose tissue engineering, but challenges, such as the generation of enough adipose tissue volume and oncologic safety assessment, have yet to be overcome.

References

1. Heneghan H, Prichard R, Lyons R, et al. Quality of life after immediate breast reconstruction and skin-sparing mastectomy—a comparison with patients undergoing breast conserving surgery. Eur J Surg Oncol. 2011;37:937–43.
2. Neuber F. Fettransplantation. Bericht über die Verhandlungen der deutschen Gesellscaft für Chirurgie. Zentralbl Chir. 1893;22:66.
3. Simonacci F, Bertozzi N, Grieco MP, Grignaffini E, Raposio E. Procedure, applications, and outcomes of autologous fat grafting. Ann Med Surg. 2017;20:49–60.
4. Billings E Jr, May JW Jr. Historical review and present status of free fat graft auto transplantation in plastic and reconstructive surgery. Plast Reconstr Surg. 1989;83:368–81.
5. Illouz YG. Body contouring by lipolysis: a 5-year experience with over 3000 cases. Plast Reconstr Surg. 1983;72:591–7.
6. Wassmer B, Zemmouri J, Rochon P, Mordon S. Comparative study of wavelengths for laser lipolysis. Photomed Laser Surg. 2010;28(2):185–8. https://doi.org/10.1089/PHO.2008.2480.
7. Tark K, Jung J, Song S. Superior lipolytic effect of the 1444 nm Nd:YAG laser: comparison with the 1064 nm Nd:YAG laser. Lasers Surg Med. 2009;41(10):721–7. https://doi.org/10.1002/LSM.20786.
8. Jung Y. Preliminary experience in facial and body contouring with 1444 nm micropulsed Nd:YAG laser-assisted lipolysis: a review of 24 cases. Laser Ther. 2011;20(1):39–46. https://doi.org/10.5978/ISLSM.20.39.
9. Sarnoff D. Evaluation of the safety and efficacy of a novel 1440 nm Nd:YAG laser for neck contouring and skin tightening without liposuction. J Drugs Dermatol JDD. 2013;12(12):1382–8.
10. DiBernardo BE. Treatment of cellulite using a 1440-nm pulsed laser with 1-year follow-up. Aesthet Surg J. 2011;31(3):328–41. https://doi.org/10.1177/1090820X11398353.
11. Prado A, Castillo P, Gaete F. Does vacuum pressure extraction of fat affect the infranatant cellularity of liposuction specimens? Plast Reconstr Surg. 2005;116(6):1832–3. https://doi.org/10.1097/01.prs.0000188846.80204.62.
12. Prado A, Andrades P, Danilla S, Leniz P, Castillo P, Gaete F. A prospective, randomized, double-blind, controlled clinical trial comparing laser-assisted lipoplasty with suction-assisted lipoplasty. Plast Reconstr Surg. 2006;118(4):1032–45. https://doi.org/10.1097/01.PRS.0000232428.37926.48.
13. Mordon S, Plot E. Laser lipolysis versus traditional liposuction for fat removal. Expert Rev Med Devices. 2009;6(6):677–88. https://doi.org/10.1586/erd.09.50.
14. Apfelberg D. Laser-assisted liposuction may benefit surgeons, patients. Clin Laser Mon. 1992;10(12):193–4.
15. Apfelberg DB, Rosenthal S, Hunstad JP, Achauer B, Bela FP. Progress report on multicenter study of laser-assisted liposuction. Aesthet Plast Surg. 1994;18(3):259–64. https://doi.org/10.1007/BF00449791.
16. Apfelberg DB, Rosenthal S, Hunstad JP, Achauer B, Fodor PB. Progress report on multicenter study of laser-assisted liposuction. Aesthet Plast Surg. 1994;18(3):259–64. https://doi.org/10.1007/BF00449791.
17. Kim KH, Geronemus RG. Laser lipolysis using a novel 1064 nm Nd:YAG laser: dermatologic surgery. Dermatol Surg. 2006;32(2):241–7.
18. Parlette EC, Kaminer ME. Laser-assisted liposuction: here's the skinny. Semin Cutan Med Surg. 2008;27(4):259–63. https://doi.org/10.1016/J.SDER.2008.09.002.
19. McBean JC, Katz BE. Laser lipolysis: an update. J Clin Aesthet Dermatol. 2011;4(7):25.
20. Bellini E, Grieco MP, Raposio E. The science behind autologous fat grafting. Ann Med Surg. 2017;24:65. https://doi.org/10.1016/J.AMSU.2017.11.001.
21. Coleman SR. Structural fat grafts: the ideal filler? Clin Plast Surg. 2001;28(1):111–9.
22. Coleman SR, Katzel EB. Fat grafting for facial filling and regeneration. Clin Plast Surg. 2015;42(3):289–300.
23. Maione L, Vinci V, Klinger M, Klinger FM, Caviggioli F. Autologous fat graft by needle. Ann Plast Surg. 2015;74(3):277–80.
24. di Summa PG, Osinga R, Sapino G, et al. Fat grafting versus implant-based treatment of breast asymmetry, a single surgeon experience over 13 years: a paradigm shift? Gland Surg. 2021;10(6):1920–30. https://doi.org/10.21037/gs-21-91.
25. Ergün SS, Baygöl EG, Kayan RB, Kuzu İM, Akman O. Correction of brassiere strap grooves with fat injections. Aesthet Surg J. 2015;35(5):561–4.
26. Coleman SR, Saboeiro AP. Primary breast augmentation with fat grafting. Clin Plast Surg. 2015;42(3):301–6.
27. Gutowski KA, ASPS Fat Graft Task Force. Current applications and safety of autologous fat grafts: a report of the ASPS fat graft task force. Plast Reconstr Surg. 2009;124(1):272–80.
28. Amar O, Bruant-Rodier C, Lehmann S, Bollecker V, Wilk A. Greffe de tissu adipeux: restauration du volume mammaire après traitement conservateur des cancer du sein, aspect clinique et radiologique. Ann Chir Plast Esthet. 2008;53(2):169–77.
29. Coleman SR, Saboeiro AP. Fat grafting to the breast revisited: safety and efficacy. Plast Reconstr Surg. 2007;119(3):775–85.
30. Zuk P, Zhu M, Ashjian P, et al. Human adipose tissue is a source of multipotent stem cells. Mol Biol Cell. 2002;13(12):4279–95. https://doi.org/10.1091/MBC.E02-02-0105.
31. Critchley A, Thompson A, Chan H, Reed M. Current controversies in breast cancer surgery. Clin Oncol. 2013;25:101–8.
32. Gabriel A, Champaneria MC, Maxwell GP. Fat grafting and breast reconstruction: tips for ensuring predictability. Gland Surg. 2015;4:232–43.
33. Caviggioli F, Maione L, Forcellini D, Klinger F, Klinger M. Autologous fat graft in postmastectomy pain syndrome. Plast Reconstr Surg. 2011;128:349–52.
34. Rigotti G, Marchi A, Galie M, et al. Clinical treatment of radiotherapy tissue damage by lipoaspirate transplant: a healing process mediated by adipose-derived adult stem cells. Plast Reconstr Surg. 2007;119:1409–22.
35. Stark RY, Mirzabeigi MN, Vonderhaar RJ, Bucky LP. Utilizing large volume fat grafting in breast reconstruction after nipple sparing mastectomies. Gland Surg. 2018;7(3):337–46.
36. Papadopoulos S, Vidovic G, Neid M, Abdallah A. Using fat grafting to treat breast implant capsular contracture. Plast Reconstr Surg Glob Open. 2018;6(11):1969. https://doi.org/10.1097/GOX.0000000000001969.

37. Garza RM, Paik KJ, Chung MT, et al. Studies in fat grafting: part III. Fat grafting irradiated tissue: improved skin quality and decreased fat graft retention. Plast Reconstr Surg. 2014;134:249–57.
38. Huang SH, Wu SH, Chang KP, et al. Alleviation of neuropathic scar pain using autologous fat grafting. Ann Plast Surg. 2015;74:S99–S104. https://doi.org/10.1097/SAP.0000000000000462.
39. Salgarello M, Visconti G, Farallo E. Autologous fat graft in radiated tissue prior to alloplastic reconstruction of the breast: report of two cases. Aesthet Plast Surg. 2010;34:5–10.
40. Sarfati I, Ihrai T, Kaufman G, Nos C, Clough KB. Adipose tissue grafting to the post-mastectomy irradiated chest wall: preparing the ground for implant reconstruction. J Plast Reconstr Aesthet Surg. 2011;64:1161–6.
41. Salgarello M, Visconti G, Barone-Adesi L. Fat grafting and breast reconstruction with implant: another option for irradiated breast cancer patients. Plast Reconstr Surg. 2012;129:317–29.
42. Baer PC, Geiger H. Adipose-derived mesenchymal stromal/stem cells: tissue localization, characterization, and heterogeneity. Stem Cell Int. 2012;2012:1–11.
43. Zuk PA, Zhu M, Mizuno H, et al. Multilineage cells from human adipose tissue: implications for cell-based therapies. Tissue Eng. 2001;7(2):211–28.
44. Gimble JM, Guilak F. Adipose-derived adult stem cells: isolation, characterization, and differentiation potential. Cytotherapy. 2003;5(5):362–9.
45. Strem BM, Hicok KC, Zhu M, et al. Multipotential differentiation of adipose tissue-derived stem cells. Keio J Med. 2005;54(3):132–41.
46. Fraser JK, Schreiber R, Strem B, et al. Plasticity of human adipose stem cells toward endothelial cells and cardiomyocytes. Nat Clin Pract Cardiovasc Med. 2006;3:S33–7.
47. Jack GS, Almeida FG, Zhang R, et al. Processed lipoaspirate cells for tissue engineering of the lower urinary tract: implications for the treatment of stress urinary incontinence and bladder reconstruction. J Urol. 2005;174(5):2041–5.
48. Rodríguez LV, Alfonso Z, Zhang R, et al. Clonogenic multipotent stem cells in human adipose tissue differentiate into functional smooth muscle cells. Proc Natl Acad Sci U S A. 2006;103(32):12167–72.
49. Lee J-H, Kemp DM. Human adipose-derived stem cells display myogenic potential and perturbed function in hypoxic conditions. Biochem Biophys Res Commun. 2006;341(3):882–8.
50. Huss FRM, Kratz G. Adipose tissue processed for lipoinjection shows increased cellular survival in vitro when tissue engineering principles are applied. Scand J Plast Reconstr Surg Hand Surg. 2002;36(3):166–71.
51. Qureshi AA, Odom EB, Parikh RP, Myckatyn TM, Tenenbaum MM. Patient-reported outcomes of aesthetics and satisfaction in immediate breast reconstruction after nipple-sparing mastectomy with implants and fat grafting. Aesthet Surg J. 2017;37(09):999–1008.
52. Katzel EB, Bucky LP. Fat grafting to the breast: clinical applications and outcomes for reconstructive surgery. Plast Reconstr Surg. 2017;140(5S Advances in Breast Reconstruction):69S–76S.
53. Panettiere P, Accorsi D, Marchetti L, Sgrò F, Sbarbati A. Large breast reconstruction using fat graft only after prosthetic reconstruction failure. Aesthet Plast Surg. 2011;35(05):703–8.
54. Khouri RK, Rigotti G, Khouri RK Jr, et al. Tissue-engineered breast reconstruction with Brava-assisted fat grafting: a 7-year, 488-patient, multicenter experience. Plast Reconstr Surg. 2015;135(03):643–58.
55. Khouri RK Jr, Khouri RK. Current clinical applications of fat grafting. Plast Reconstr Surg. 2017;140(03):466e–86e.
56. van Turnhout AA, Fuchs S, Lisabeth-Broné K, Vriens-Nieuwenhuis EJC, van der Sluis WB. Surgical outcome and cosmetic results of autologous fat grafting after breast conserving surgery and radiotherapy for breast cancer: a retrospective cohort study of 222 fat grafting sessions in 109 patients. Aesthet Plast Surg. 2017;41(06):1334–41.
57. Maione L, Vinci V, Caviggioli F, et al. Autologous fat graft in postmastectomy pain syndrome following breast conservative surgery and radiotherapy. Aesthet Plast Surg. 2014;38(03):528–32.
58. Hammond DC, O'Connor EA, Scheer JR. Total envelope fat grafting: a novel approach in breast reconstruction. Plast Reconstr Surg. 2015;135(03):691–4.
59. Jones G, Yoo A, King V, et al. Prepectoral immediate direct-to-implant breast reconstruction with anterior AlloDerm coverage. Plast Reconstr Surg. 2017;140(6S):31S–8S.
60. Darrach H, Kraenzlin F, Khavanin N, Chopra K, Sacks JM. The role of fat grafting in prepectoral breast reconstruction. Gland Surg. 2019;8(01):61–6.
61. Elswick SM, Harless CA, Bishop SN, et al. Prepectoral implant-based breast reconstruction with postmastectomy radiation therapy. Plast Reconstr Surg. 2018;142(01):1–12.
62. Sigalove S, Maxwell GP, Sigalove NM, et al. Prepectoral implant-based breast reconstruction: rationale, indications, and preliminary results. Plast Reconstr Surg. 2017;139(02):287–94.
63. Peltoniemi HH, Salmi A, Miettinen S, Mannerström B, Saariniemi K, Mikkonen R, et al. Stem cell enrichment does not warrant a higher graft survival in lipofilling of the breast: a prospective comparative study. J Plast Reconstr Aesthet Surg. 2013;66:1494–503.
64. Li K, Gao J, Zhang Z, et al. Selection of donor site for fat grafting and cell isolation. Aesthet Plast Surg. 2013;37(1):153–15.
65. Akagia D, Pallua N. Autologous fat grafting: in search of the optimal technique. Surg Innov. 2014;21:327–36.
66. Pu LL, Coleman SR, Cui X, Ferguson RE Jr, Vasconez HC. Autologous fat grafts harvested and refined by the Coleman technique: a comparative study. Plast Reconstr Surg. 2008;122:932–7.
67. Kakagia D, Pallua N. Autologous fat grafting. Surg Innov. 2014;21(3):327–36.
68. Ozsoy Z, Kul Z, Bilir A. The role of cannula diameter in improved adipocyte viability: a quantitative analysis. Aesthet Surg J. 2006;26(3):287–9.
69. Gause TM, Kling RE, Sivak WN, et al. Particle size in fat graft retention: a review on the impact of harvesting technique in lipofilling surgical outcomes. Adipocytes. 2014;3(4):273–9.
70. Kirkham JC, Lee JH, Medina MA, et al. The impact of liposuction cannula size on adipocyte viability. Ann Plast Surg. 2012;69(4):479–81.
71. Mojallal A, Foyatier JL. The effect of different factors on the survival of transplanted adipocytes. Ann Chir Plast Esthet. 2004;49:426–36.
72. Chan CW, McCulley SJ, Macmillan RD. Autologous fat transfer: a review of the literature with a focus on breast cancer surgery. J Plast Reconstr Aesthet Surg. 2008;61:1438–48.
73. Wang WZ, Fang XH, Williams SJ, et al. Lidocaine-induced ASC apoptosis (tumescent vs. local anesthesia). Aesthet Plast Surg. 2014;38(5):1017–23. https://doi.org/10.1007/s00266-014-0387-2.
74. Mecott GA, Gonzalez IZ, Montes de Oca R, et al. Effect of decantation time on viability and apoptosis in adipocytes after liposuction. Aesthet Plast Surg. 2019;43(1):228–32. https://doi.org/10.1007/s00266-018-1258.
75. Levenberg A, Scheinowitz M, Sharabani-Yosef O, Levenberg A, Scheinowitz M, Sharabani-Yosef O. Higher cell viability and enhanced sample quality following laser-assisted liposuction versus mechanical liposuction. J Cosmet Dermatol Sci Appl. 2015;5(3):238–45. https://doi.org/10.4236/JCDSA.2015.53029.
76. Hamza A, Lohsiriwat V, Rietjens M. Lipofilling in breast cancer surgery. Gland Surg. 2013;2:7–14.
77. Matsumoto D, Sato K, Gonda K, Takaki Y, Shigeura T, Sato T, Aiba-Kojima E, Iizuka F, Inoue K, Suga H, Yoshimura K. Cell-assisted lipotransfer: supportive use of human adipose-derived

cells for soft tissue augmentation with lipoinjection. Tissue Eng. 2006;12:3375–82.

78. ASPRS Ad-Hoc Committee on New Procedures. Report on autologous fat transplantation. Plast Surg Nurs. 1987;7(04):140–1.
79. Orbay H, Hinchcliff KM, Charvet HJ, Sahar DE. Fat graft safety after oncologic surgery: addressing the contradiction between in vitro and clinical studies. Plast Reconstr Surg. 2018;142(06):1489–99.
80. Kronowitz SJ, Mandujano CC, Liu J, et al. Lipofilling of the breast does not increase the risk of recurrence of breast cancer: a matched controlled study. Plast Reconstr Surg. 2016;137(02):385–93.
81. Juhl AA, Redsted S, Engberg DT. Autologous fat grafting after breast conserving surgery: breast imaging changes and patient-reported outcome. J Plast Reconstr Aesthet Surg. 2018;71(11):1570–6.
82. Turner A, Abu-Ghname A, Davis MJ, Winocour SJ, Hanson SE, Chu CK. Fat grafting in plastic surgery: fat grafting in breast reconstruction. Semin Plast Surg. 2020;34(1):17. https://doi.org/10.1055/S-0039-1700959.
83. Liu S, Chen S, Yang Y, Guan W. The effect and prognosis of combinative implantation by autologous-fat granule and prosthesis for breast reconstruction after radical mastectomy. Am J Transl Res. 2021;13(5):5256–63.
84. Moak T, Ebersole T, Tandon D, Tenenbaum M. Assessing clinical outcomes in autologous fat grafting: a current literature review. Aesthet Surg J. 2021;41(Suppl 1):S50–60. https://doi.org/10.1093/ASJ/SJAB148.
85. Upadhyaya SN, Bernard SL, Grobmyer SR, Yanda C, Tu C, Valente SA. Outcomes of autologous fat grafting in mastectomy patients following breast reconstruction. Ann Surg Oncol. 2018;25(10):3052–6. https://doi.org/10.1245/S10434-018-6597-0.
86. Krastev T, Alshaikh G, Hommes J, Piatkowski A, van der Hulst R. Efficacy of autologous fat transfer for the correction of contour deformities in the breast: a systematic review and meta-analysis. J Plast Reconstr Aesthet Surg JPRAS. 2018;71(10):1392–409. https://doi.org/10.1016/J.BJPS.2018.05.021.
87. Groen J, Negenborn V, Twisk D, et al. Autologous fat grafting in onco-plastic breast reconstruction: a systematic review on oncological and radiological safety, complications, volume retention and patient/surgeon satisfaction. J Plast Reconstr Aesthet Surg JPRAS. 2016;69(6):742–64. https://doi.org/10.1016/J.BJPS.2016.03.019.

45 The Use of Real-Time Ultrasound Scan Imaging in Gluteal Lipofilling as an Added Safety Measure

Omar Tillo

Contents

45.1 Introduction

The popularity of lipofilling in buttock augmentation continues to rise, making it the most popular means of volume augmentation surpassing the use of implants [1, 2]. The rapid spread of this procedure resulted in a wide variety of surgical techniques and methods that remain unsystematized and poorly understood.

Alarming reports of death caused by fat embolism [3–5] prompted several authorities to investigate this procedure and recommend the adherence to safety measures.

Some of the most important safety recommendations, if not *the* most important, are as follows: limiting the lipofilling to the subcutaneous space, avoiding injections into the deep muscle planes, and maintaining a constant awareness of the position of the injection cannula. The use of a portable ultrasound scan (USS) allows the surgeon to objectively identify and document the different planes in the gluteal area and limit the lipofilling to the safer subcutaneous plane.

Supplementary Information The online version contains supplementary material available at https://doi.org/10.1007/978-3-031-10881-5_45.

O. Tillo (✉)
Creo Clinic, London, UK

45.2 Safety First

The Brazilian butt lift (BBL) has gained a bad reputation due to the reported high rate of complications and mortalities. The rapid widespread popularity over the last few years, coupled with poor understanding of safety and efficacy, resulted in a wide variety of techniques being used by surgeons with methods that remain unsystematized.

Until recent years, fat transfer into the highly vascularized intramuscular beds has been recommended to enhance the volume maintenance and improve the adipose tissue integration [6–9].

Reports of sporadic deaths caused by fat embolism [3], then the analysis of 22 BBL fatalities in Mexico and Colombia over 10 and 15 years, respectively [4], raised the alarm bells. A survey by the Aesthetic Surgery Education

© The Author(s), under exclusive license to Springer Nature Switzerland AG 2023
A. Di Giuseppe et al. (eds.), *Fat Transfer in Plastic Surgery*, https://doi.org/10.1007/978-3-031-10881-5_45

and Research Foundation (ASERF) Task Force reported 32 fatal and 103 nonfatal fat embolisms in 198,857 procedures contributing to the risk of death from this procedure being as high as 1:2351 [5].

The case analysis by Cardenas-Camarena et al. [4] found that *intramuscular* gluteal lipofilling was associated with higher mortality due to blood vessels damage allowing the fat embolism. They recommended avoiding injections into deep muscle planes. Another systematic review by Condé-Green et al. [6] found a higher rate of fatal and nonfatal pulmonary fat embolism with deep muscle injections compared to superficial planes.

Danger zones containing the major gluteal vessels and branches were described. The authors advised avoiding "deep" intramuscular lipofilling in these areas and restricting the transfer to the subcutaneous and the "superficial" intramuscular planes [10, 11]. This concept focused on avoiding direct injury to the gluteal vessels, or the so-called "direct hit" by the injection cannula.

Del-Vecchio et al. [12] eloquently studied and demonstrated the "deep intra muscular migration" that occurs with intramuscular fat transfer. This revealed that the fat migrates into deeper planes and compartments by following the path of least resistance. Therefore, any intramuscular lipofilling even superficially would have the potential of migrating into the deep submuscular compartments. The increasing volume in the submuscular space causes the muscle to expand posteriorly excreting traction and sheer forces on the large, fixed-base and thin-walled gluteal plexus, which could tentatively tear these vessels even in the absence of a "direct hit." The low intravenous pressure would draw the fat tissue intravascularly via syphoning mechanism.

Ordenana et al. anatomical study suggested that the subcutaneous planes may provide the least risk with regard to vascular complications, as they contain the smallest caliber vessels [13]. Del-Vecchio et al. explained how the thick deep fascia, investing the superficial surface of the gluteus muscle, can act as a barricade, preventing the subcutaneously placed fat from migrating into the muscle.

The increasing evidence of the dangers of any intramuscular transfer, even into the superficial muscle planes, strengthens the importance of limiting the lipofilling to the subcutaneous space.

Subcutaneous transfer relays on the constant 3D awareness of the lipofilling cannula. This is most commonly achieved by the experienced surgeon from their anatomical knowledge and the tactile feedback received from the cannula and soft tissue. The majority of surgeons who had fatalities reported that they only injected the fat subcutaneously and did not intend to inject it intramuscularly or submuscularly. However, all the autopsy analysis demonstrated fat under the gluteal muscle and around the main gluteal vessels.

Real-time ultrasound-assisted gluteal fat grafting has been shown to be reliable in allowing the surgeon to follow the subcutaneous plane of lipofilling [14].

45.3 Anatomical Overview

The buttock consists of a layer skin, an underlying layer of subcutaneous fat and several superficial and deep gluteal muscles, in between which the important neurovascular structures are located.

The bulk of the buttock is formed by the subcutaneous fat, also called gluteal fat pads, and the superficial gluteal muscles. Cadaveric studies report the subcutaneous plane to be about 17 mm thick and the gluteal muscle to be about 25 mm thick [13]; however, these can largely vary.

The superficial gluteal muscle comprises the gluteus maximus, medius, and minus that run in an inferolateral direction in the posterior buttock area and the tensor fascia latae that runs in a vertical direction on the most lateral aspect of the buttocks and hips.

45.3.1 Gluteal Vascular Anatomy

Studies aiming to understand the vascular anatomy—to map the location and sizes of the gluteal blood vessels—demonstrated that the deeper and more medial planes of the gluteal area house the large prominent vessels, with the superior and inferior gluteal veins measuring, on average, 7.61 mm and 13.65 mm, respectively, with branches measuring around 3.83 mm. The subcutaneous vessels had the smaller diameters of 0.9 mm for arteries and 1.05 mm for veins [13].

45.3.2 Gluteal Ligamentous Anatomy

An anatomical study by Ghavami et al. [15], and others [16, 17], identified the following ligamentous and fascial structures:

- Osseocutaneous Ligaments: These include the sacrocutaneous and ischiocutaneous ligaments and two dense attachments from the lateral margins of the sacrum and the ischium, respectively, extending to the dermis.
- Fasciocutaneous Ligaments: these include the natal cleft fasciocutaneous ligament, which extends from the gluteal fascia to the dermis of the natal cleft, and the infragluteal crease ligament that defines the inferior border of the buttock. It extends from the muscular fascia in the region with loose attachments to the dermis.
- The superficial fascial system (SFS) separates the subcutaneous fat into two layers: the superficial layer

immediately under the skin and above the SFS, and the deeper layer between the SFS and the deep investing fascia of the gluteal muscle. The SFS of the buttock fuses with the gluteal crease medially, the intragluteal fold inferiorly, and the iliac crest superiorly. The SFS, in addition to surrounding and separating the fat layers, sends vertical elements through the deep fat to the investing fascia of the underlying muscle, creating a scaffold-like structure that supports the skin and adipose tissue over the muscle.

45.4 Principles of the Ultrasound Scan

The principle of USS imaging utilizes high-frequency sound waves to visualize various soft tissue structures. The full system consists of a transducer, processor, a graphics processing unit, and displaying device.

The transducer emits pulses of high-frequency sound waves. The ultrasonic waves travel through the soft tissue, penetrating the low-density structures and reflecting back against high-density tissue or objects. The transducer receives the reflected waves and sends the data to the processor. Using the sonograph principles, the processor analyses the data derived from the recaptured waves, such as the direction of travel and time delay, to map all the points of reflection. The graphics processing unit converts this map of data into a real-time two-dimentional image and sends it to be displayed on a screen.

The ultrasound wave frequency used by the current modern machines varies between 2 and 20 MHz. The low-frequency waves penetrate deeper into the tissues but generate lower-resolution images. These are often used in abdominal, obstetric, and gynecological imaging (2–5 MHz) and pelvic, vascular, and breast imaging (5–10 MHz). The high-frequency waves produce more detailed, high-resolution images, but they get absorbed easily and don't penetrate deep into the tissue. Therefore, higher frequencies are ideal to visualize superficial structures, such as the skin, subcutaneous tissue, and the musculoskeletal system (10–20 MHz).

The denser a structure is, the more reflective of the ultrasonic waves, and the brighter (hyperechoic) it appears. For example, bone reflects and absorbs all incoming waves; therefore, its surface appears bright with complete darkness or "shadow" behind it. Structures or tissues that conduct the waves well appear low in density (hypoechoic).

The skin epidermis and dermis appear as a continuous uniform hyperechoic line immediately under the transducer (the very top of the image on the screen) (Fig. 45.1). The thickness of this line varies between anatomical areas of the body, depending on the thickness of the dermis.

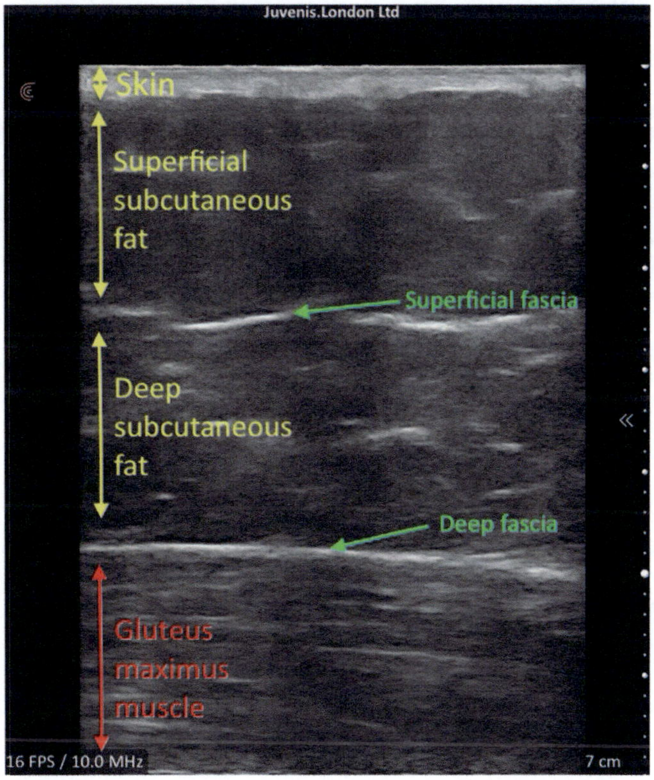

Fig. 45.1 Anatomical strictures as seem with ultrasound scan imaging

The adipose tissue appears uniformly hypoechoic, with very faint lobules and some thin septa. The superficial fascia appears as a hyperechoic thin line or lines that run horizontally in the middle of the adipose tissue, running parallel to the surface of the skin—unless this fascia is joining the dermis or the deep structures at the zones of adherence [17]. This superficial fascia separates the subcutaneous adipose tissue in the buttock to the superficial and deep compartments.

Muscles are hypoechoic; therefore, they appear relatively dark with a distinct appearance of the perimysial connective tissue. In a longitudinal plane image taken along the axis of the muscle fibers, the fascicular architecture is visible as linear structures (Fig. 45.2). In a transverse plane, perpendicular to the muscle longitudinal axis, the cross-sections of these thin fascicles give a speckled appearance (Fig. 45.3).

The deep investing fascia of the gluteal muscle appears as a hyperechoic layer separating the deep adipose tissue and the muscle.

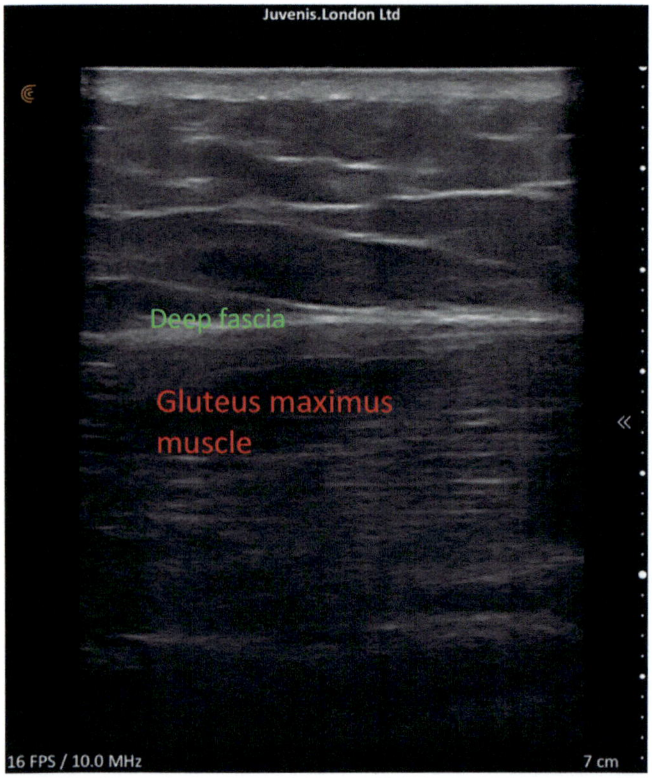

Fig. 45.2 Ultrasound scan along the axis of the muscle fibers

Metal devices and instruments reflect the ultrasound waves off their surfaces and do not allow any penetration. Therefore, the surfaces of these objects appear very brightly hyperechoic, often smooth and regular in shape, reflecting the shape of that object (Fig. 45.4). Because of their opacity, these objects leave behind them a distinct shadow that is often dark and sometimes white due to the high intensity signal interference.

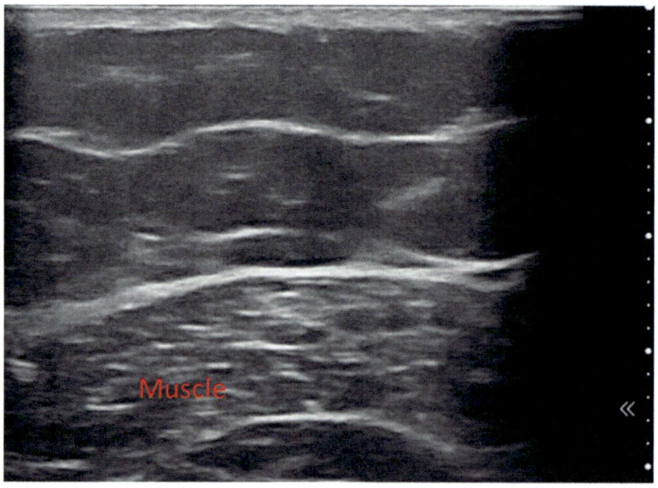

Fig. 45.3 Ultrasound scan across some muscle fibers

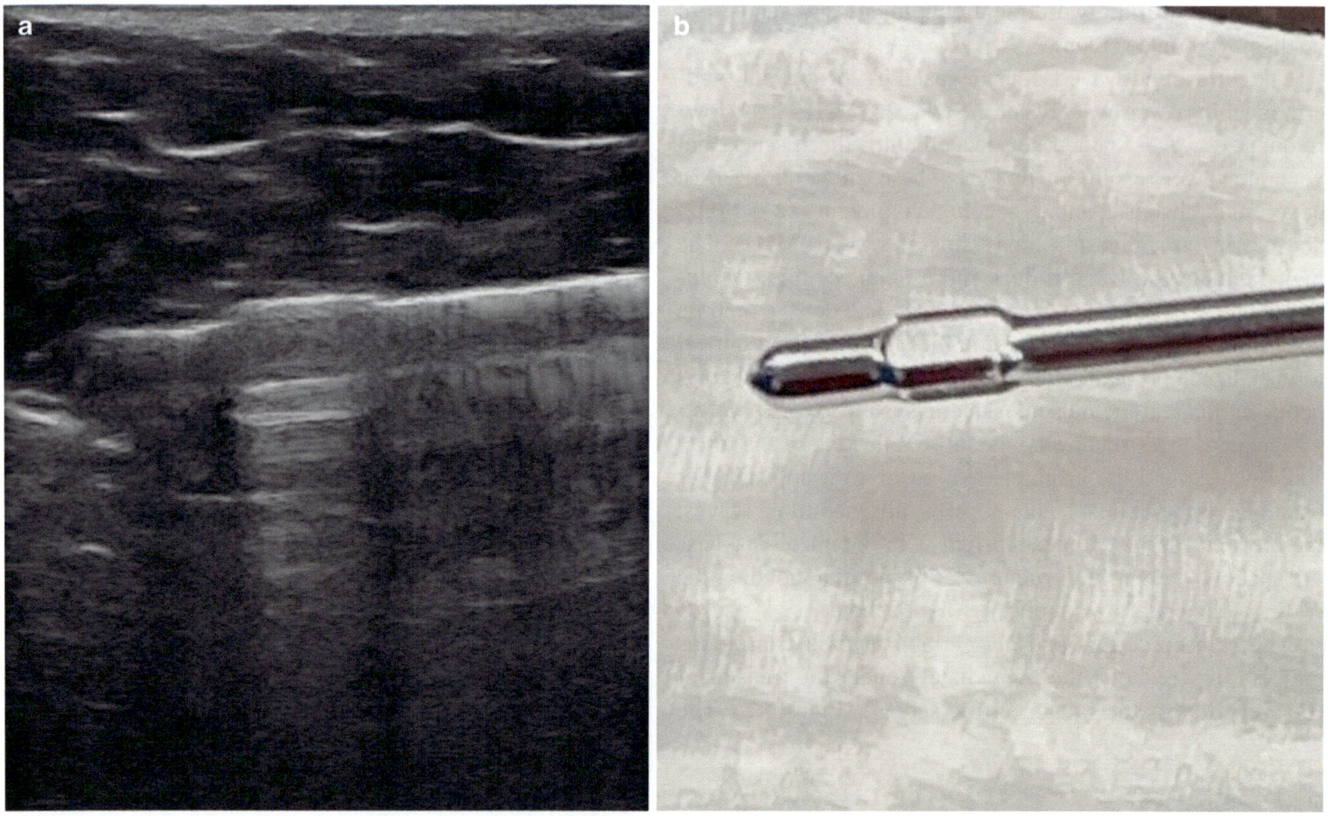

Fig. 45.4 (**a**) USS image along the axis of a flared Mercedes Liposuction cannula produces a bright hyperechoic line that reflects the shape of the tip. (**b**) The tip of a Flared Mercedes Liposuction cannula ex vivo

45.5 Preoperative Imaging

For the USS imaging, I use a Clarius L15HD device, by the Clarius Mobile Health Corporation, Austin, USA. This has a frequency of 5–15 MHz with a maximum depth of 7 cm. The device connects via WiFi to the Clarius App on a 12.9-inch iPad. Before performing the test, the connection is established and the imaging settings are set to MSK (Musculoskeletal) and a depth of 7 cm.

After marking the patient for liposuction and fat transfer, the preoperative USS imaging is performed with the patient in the standing position (Fig. 45.5). This provides controlled standardized imaging by avoiding any compression and distortion of the hip or buttocks that may result from the patient lying prone or lateral. Three measurement points are marked on each side on the hip and buttock on a horizontal line that runs across the middle of the intergluteal crease. These points are as follows:

1. The lateral trochanteric point: this is the most lateral point of the hip on the coronal line or mid-axillary line.
2. The posterolateral point: in the middle of points 1 and 3.
3. The posterior gluteal point: at the center of the buttock cheek.

Therefore, in total, six standard measurement points are marked (Fig. 45.6).

The test and its purpose are explained to the patient and their verbal consent is taken. A thin layer of gel is applied to the skin so that the ultrasound waves are transmitted from the transducer through the gel into the body. The USS transducer is applied to the skin and the image checked to identify the skin, subcutaneous adipose tissue, superficial and deep fascial layers, and the gluteal muscle. The image is frozen and the full thickness of the subcutaneous layer is measured and recorded for future comparison with postoperative studies. The test is repeated and recorded at each of the six measurement points (Fig. 45.7).

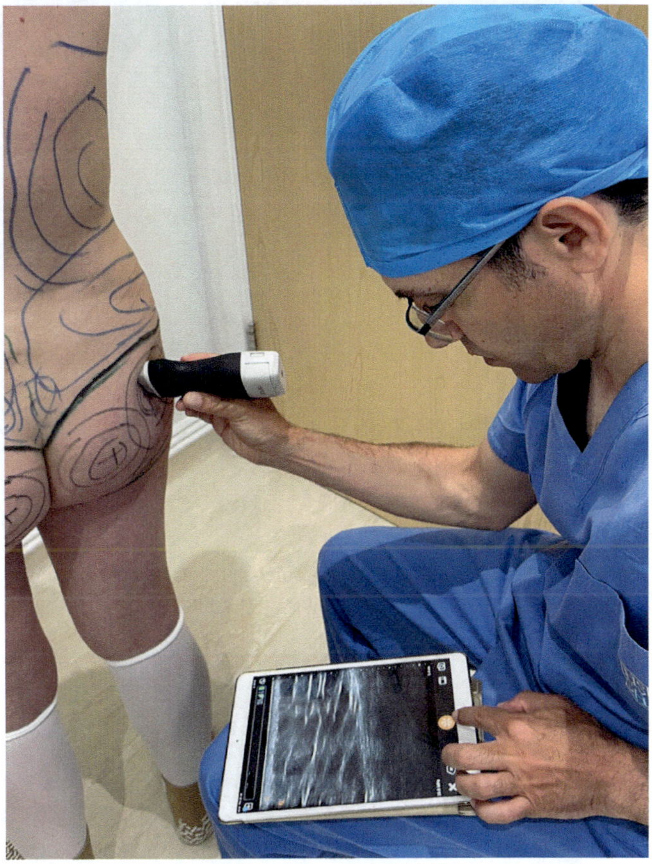

Fig. 45.5 Scanning each point with the patient in standing position

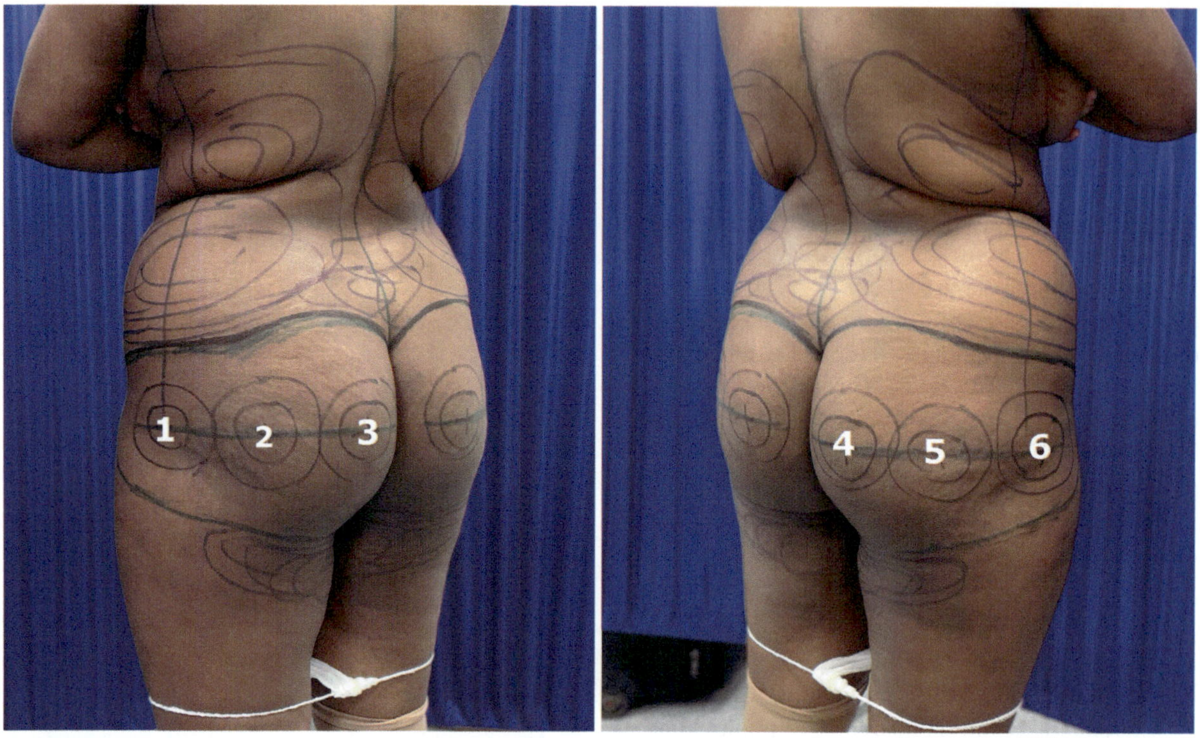

Fig. 45.6 The six standard scanning points: (1) left lateral trochanteric, (2) left posterolateral, (3) left posterior gluteal, (4) right lateral trochanteric, (5) right posterolateral, (6) right posterior gluteal

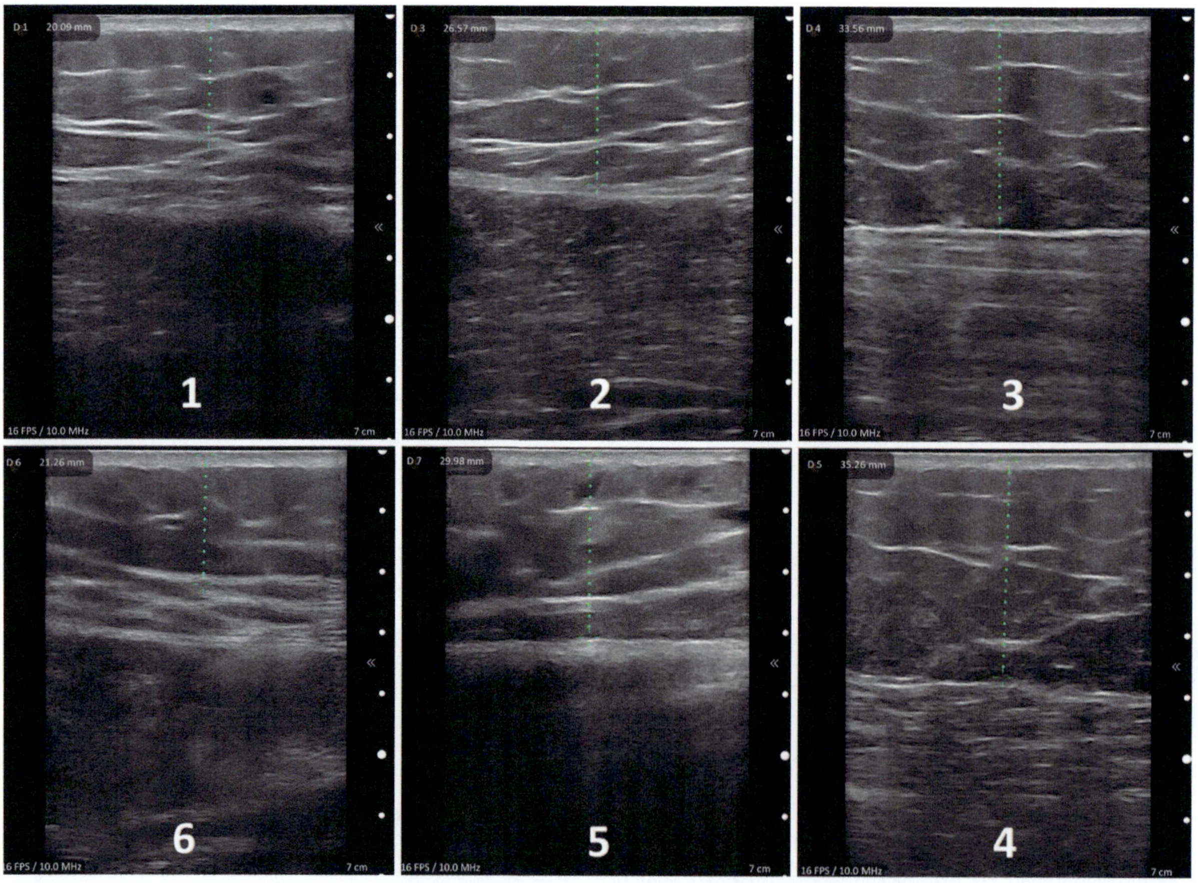

Fig. 45.7 Subcutaneous fat thickness measurements taken at six points: (1) left lateral trochanteric, (2) left posterolateral, (3) left posterior gluteal, (4) right posterior gluteal, (5) right posterolateral, (6) right lateral trochanteric

45.6 Intraoperative Imaging

After the completion of fat harvesting, the patient is secured in a prone and mild jack-knife position ready for lipofilling. The USS transducer is sealed into a sterile cover with some gel. The gluteal incisions are made for the lipofilling. A flared Mercedes, 4 or 5 mm liposuction cannula is used for the lipofilling via the Expansion Vibration Lipofilling (EVL) technique described by Del Vecchio et al. [18].

The surgeon holds the lipofilling handpiece in their dominant hand and introduces it into the subcutaneous plane, advancing it towards the first measurement point and the lateral trochanteric point. The scan transducer is held in the surgeon's non-dominant hand, applied to the skin over the first measurement point, and maneuvered to identify the distinct hyperechoic reflection of the metal cannula with the shadow behind it (Video 45.1). The scanning can be done by holding the transducer aligned either along the axis of the cannula or across it. Placing the transducer along the axes of the cannula (Fig. 45.8) produces a long bright metallic line with an intense shadow behind it. We found this useful in visualizing the full length of the end part of the cannula which makes identifying the tip easier; however, the shadow and interference behind the metal cannula obscures the depth and makes it difficult to ascertain the location of the cannula in relation to the muscle. In the cross views (Fig. 45.9), the cannula appears as a bright dot with a narrow shadow behind it that allows the visualization of deeper structures, such as fat, facia, and muscle. The tip of the cannula can be identified in cross-section imaging by following the cannula distally until the point just before it disappears (ends). The position of the cannula tip can be readjusted if needed, to be relocated in the deep subcutaneous layers between the superficial and the deep fascial system. A still image is captured and saved for documentation. The EVL lipofilling starts under continuous ultrasound scan visualization observing the grafted fat spreading over the gluteal muscle (Fig. 45.10). A video recording can be taken and stored for documentation (Video 45.2). The surgeon can stabilize the lipofilling cannula or move it gently in coordination with the ultrasound scan transducer to follow the tip and the lipofilling. Once 100–150 mL of fat is transferred under real-time imaging, the space over the muscle becomes significantly thicker allowing the surgeon to continue the lipofilling within the same plane, with constant 3D awareness, without the aid of the USS. It is impractical and counterproductive to follow the cannula with the USS through the full length of lipofilling process as the surgeon must use the non-dominant had to receive tactile feedback about the tissue turgor, fat distribution, and shape.

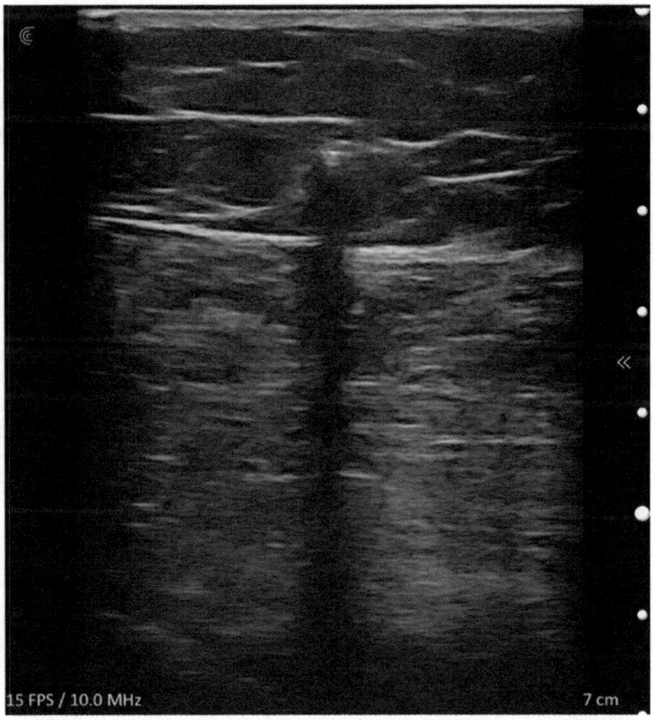

Fig. 45.9 Ultrasound scan image with the transducer across the axis of the lipofilling cannula. The cross-section of the metal cannula have the distinct appearance of very bright thin circular object with a linear narrow shadow behind it

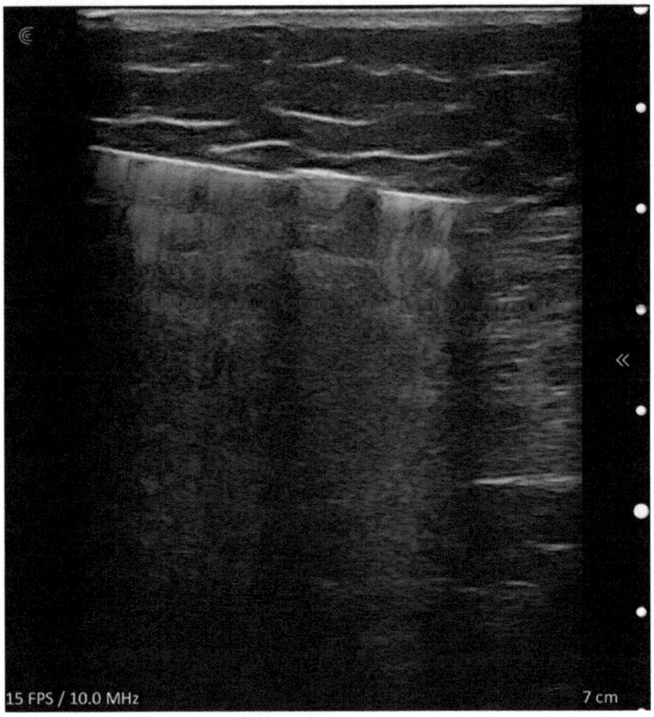

Fig. 45.8 USS image with the transducer along aligned the axis of the lipofilling cannula. The surface of the metal cannula appears as a long bright line with a broad widespread shadow behind it that often obscure the underlying tissue. Note the interruption in the line which corresponds to the flared part of the Flare-Mercedes (basket) cannula (Fig. 45.2b)

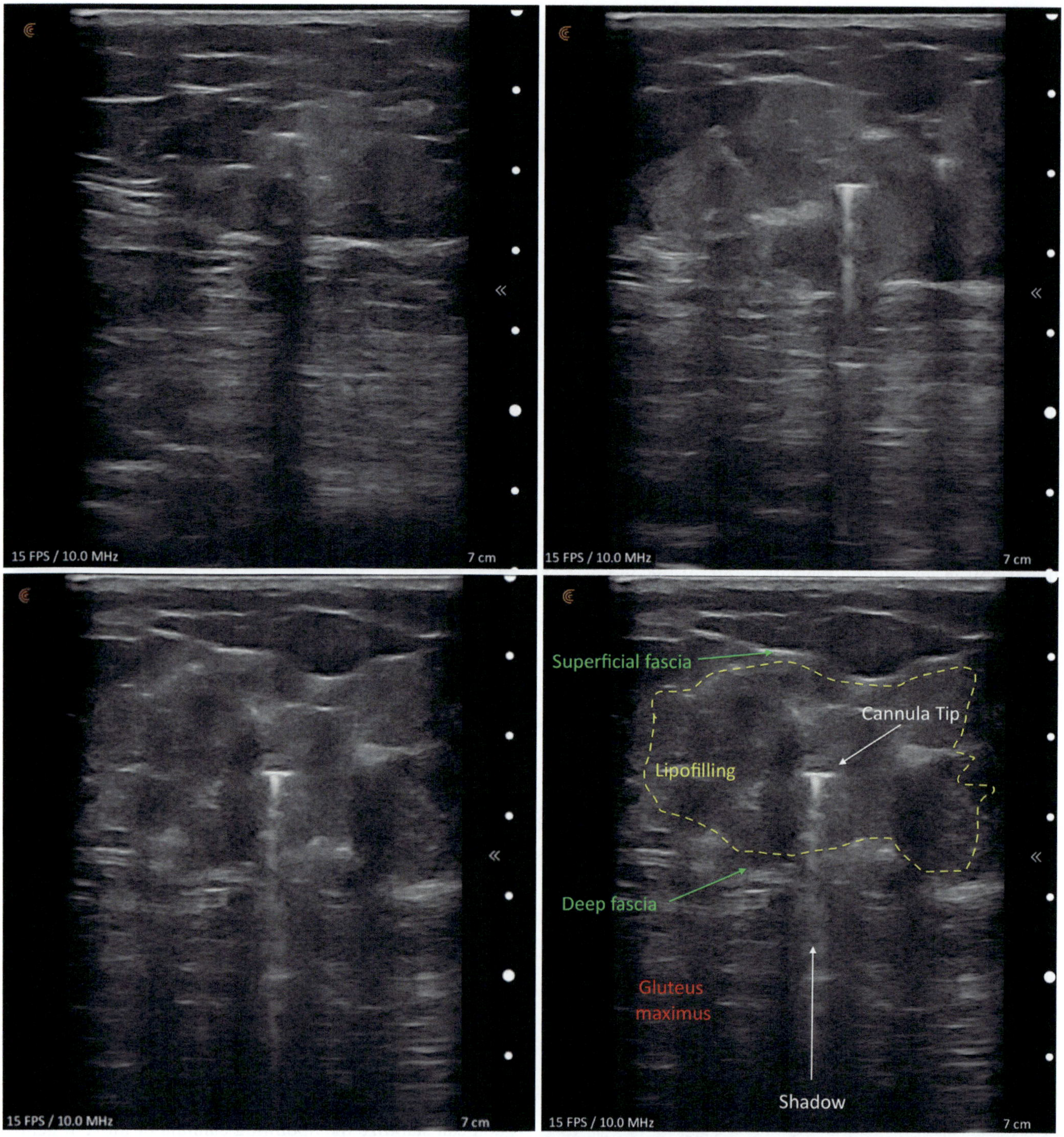

Fig. 45.10 A sequence of intraoperative USS images during EVL lipofilling demonstrating the gradual spread of the grafted fat into the deep subcutaneous space between the superficial and deep fascias

An assistant holding the transducer would not be able to follow the back-and-forth movements of the cannula. In addition, the current ultrasound scan machines have a delay in image processing which makes it impossible to produce images coming from a fast-moving transducer. The main purpose of the intraoperative USS, in our experience, is the confirmation of the accurate position of the cannula in each of the six zones at the start of lipofilling so the surgeon can maintain that depth throughout the procedure.

The USS-guided lipofilling is repeated in each of the six measurement points. In order to systemize and standardize the order of lipofilling and the image storage, the sequence of the lipofilling is performed in the following order: left lateral trochanteric, right lateral trochanteric, right posterolateral, left posterolateral, left posterior, and then finally right posterior. A blank image can be stored at the end of scanning each point to mark the transition to the following measurement point within the stockpile of images saved on the device memory. The reason for switching from side to side (left to right) is to insure equal fat distribution between the two sides. The initial fat drained from the bottom of the collection canister is the wettest and the fat at the top is the driest. If the lipofilling was fully completed on one side before switching to the other, there would be the risk of transferring graft with significantly higher content of fluid on the first side compared to the second. Switching sides between zones improves the equal distribution of wet and dry fat between the two sides.

45.7 Benefits of Ultrasound Scan Usage

45.7.1 Safety Benefits

The main advantage of intraoperative USS is the confirmation of the accurate location of the cannula tip over the gluteal muscle to avoid the inadvertent intra-muscular lipofilling. It is argued that an experienced surgeon should be able to use their knowledge and tactile perception to construct a three-dimensional awareness of the cannula. However, this ability is developed gradually, over a learning curve, and a surgeon with limited experience may make several mistakes before they master this skill. The USS can be a very useful tool in guiding their learning curve in a safe and effective way.

It is plausible that even surgeons with previous experience in liposuction may occasionally misjudge the position of the cannula in the gluteal area, due to three anatomical characteristics when this area is compared to other parts of the body:

1. The trunk's deep investing fascial systems, such as the one on the abdominal external and internal oblique muscles and rectus sheets, are thick and multilayered giving several levels of protection. In comparison, the gluteus maximus deep investing facia is a single layer of significantly thinner tissue; therefore, it is much easier to penetrate.
2. The convexity of the buttock, in comparison to the flat surface of the abdomen, makes the gluteal muscle more vulnerable to cannula penetration, especially when approached from an inferior access.
3. In prone position, the gluteus maximus muscle is at its shortest resting length, soft, unstable, with relaxed investing fascia. It is thought that the laxity of the muscle and fascia makes them more vulnerable to penetration. The mild flexion with the jack-knife position aims to provide a degree of tightness; however, this can't be maximized unless the hip was in full flexion. In comparison, the muscle and investing fascia of the trunk are relatively tighter in the body's neutral position.

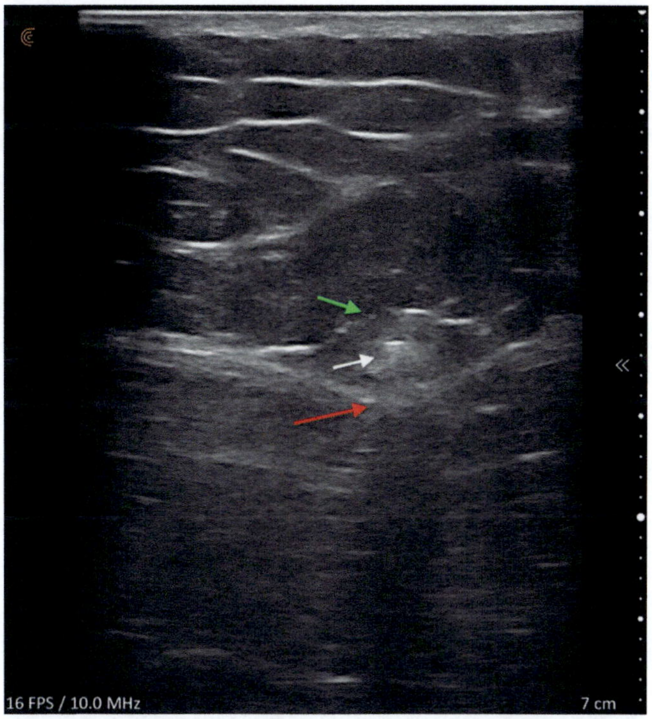

Fig. 45.11 Intraoperative ultrasound scan image showing the tip of the cannula under the deep fascia over the gluteus maximus muscle. Green arrow: deep fascia, white arrow: cannula tip, red arrow: gluteus maximus muscle fibers

When the cannula is inserted deep into the subcutaneous layers, a resistance is met against the superficial fascia, and a "pop" is felt when this is penetrated. This has always triggered a niggling "doubt" in my mind "is this the superficial fascia's pop, or did I miss a faint previous one and, therefore, this might be the deep fascia's?" This doubt did not create a big deal of anxiety at the start of my practice, as the superficial intramuscular lipofilling was still considered a safe practice. However, with the increasing calls to avoid the intramuscular lipofilling, that "doubt" started turning into "anxiety." The use of the intraoperative USS has cured this anxiety and, on two occasions, revealed unintentional intermuscular penetration confirming my doubts (Figs. 45.11 and 45.12).

45.7.2 Aesthetic Benefits

To achieve the best aesthetic outcome, the lipofilling should be concentrated into the deepest subcutaneous plane (under the superficial fascia), rather than too superficially, to achieve a maximum volume of transfer and avoid the "ice-shelf" effect (Fig. 45.13). Transferring the fat immediately under the skin above the superficial fascia results in early space saturation and tight flattening of the skin, or the so-called

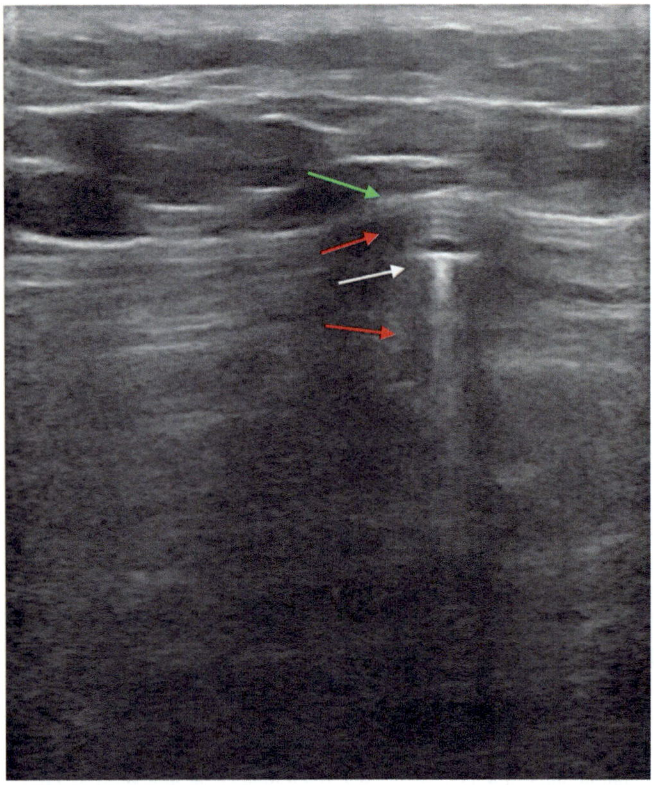

Fig. 45.12 Intraoperative ultrasound scan image showing the tip of the cannula into the superficial layers of the gluteus maximus muscle. Green arrow: deep fascia, white arrow: cannula tip, red arrows: gluteus maximus muscle fibers

"ice-shelf" effect, which prevents the desired convex expansion and limits the volume of transfer. The deep lipofilling under the superficial fascia allows the transferred fat to rest against the resistance of the muscle and therefore directs the expansion upward, causing the desired roundness of the skin rather than flattening.

45.7.3 Postoperative Monitoring

The USS is a useful tool in the early postoperative monitoring and management. It can help the diagnosis and treatment of some early postoperative complications, such as seroma, hematoma, fat necrosis and oil cysts.

Another valuable use is the measurement of outcome and fat retention. A comparative series of preoperative and postoperative measurements of the adipose tissue thickness at the standard six measurement points is a powerful objective outcome measure that can be used for auditing and research proposes.

45.8 Difficulties

45.8.1 Learning Curve

Like any new clinical skill and technology, the use of ultrasound scan comes with a learning curve. However, we found this a relatively easy-to-learn skill. It is important to attend the relevant training courses or programes on the use of the machine as well as the general principles of ultrasound scanning. Practicing the use of the machine as much as possible, in clinics and theatre, helps sharpen the surgeon's ability, speed and efficacy.

45.8.2 Added Costs

There are added costs that need to be considered before committing to the use of a new USS device. These costs include purchasing the ultrasound scan machine, its service and maintenance, and the added administrative work related to saving and filing the images and videos.

Modern technology has allowed the production of affordable portable systems that are easy to operate and provide good quality images. The multitude of their usage in plastic surgery makes them a worthwhile investment.

45.8.3 Added Operating Time

There is a presumed increase in theatre operating time to account for the imaging process. However, in a recent, yet unpublished case study, we couldn't detect any significant increase in operating time between the cases that were performed before, and those that were performed after the introduction if USS.

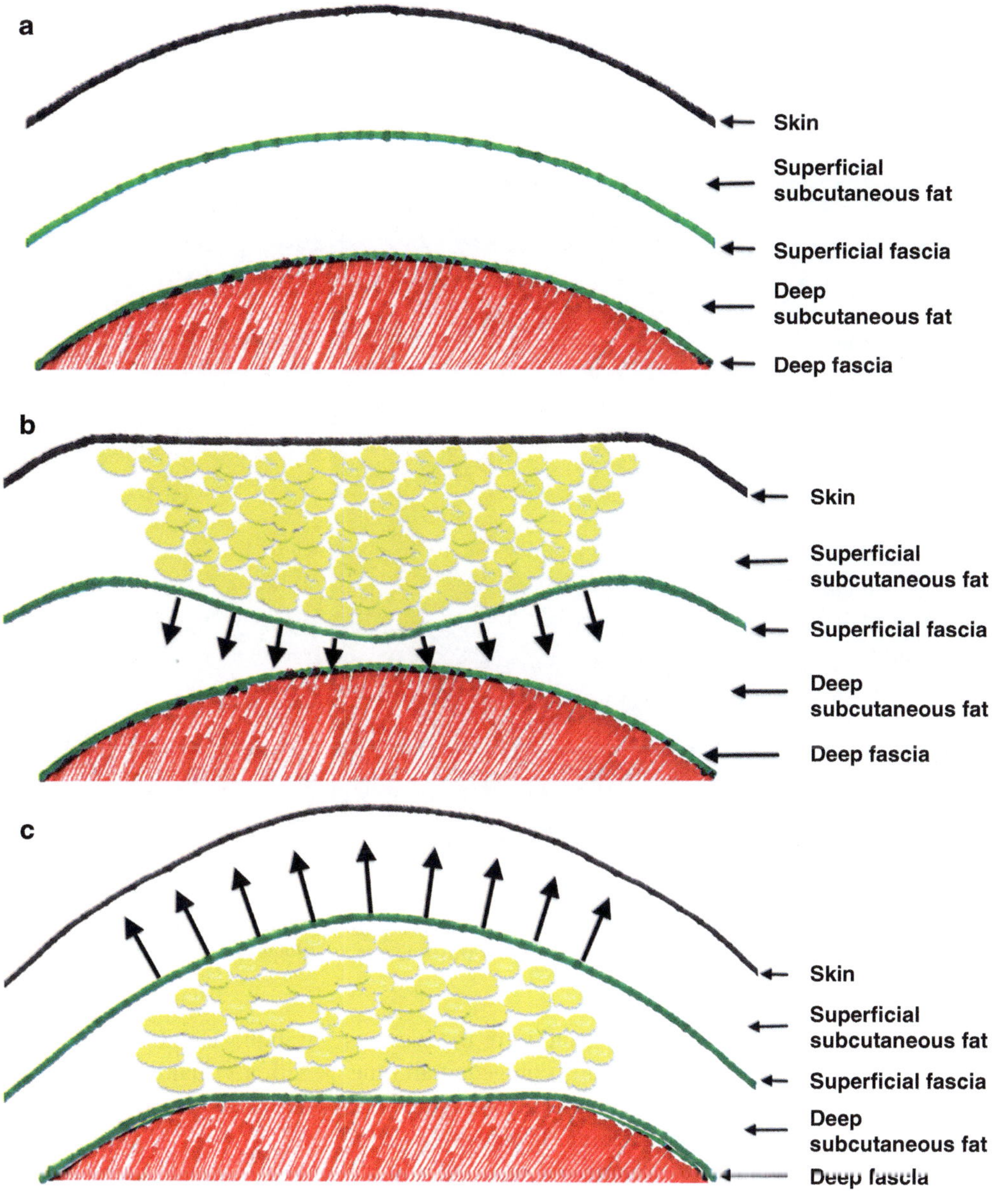

Fig. 45.13 The ice-shelf effect. (**a**) A schematic demonstration of the gluteal soft tissue layers. (**b**) The ice-shelf effect caused by lipofilling too superficial immediately subdermally causing the dermis to tighten and flatten; and the expansion progressing deeply. (**c**) Deep lipofilling against the resistance of the deep fascia directs the expansion upward causing more desired roundness

45.9 Pitfalls

1. A false extramuscolar appearance of the cannula: When scanning the cannula in cross-section views, it is extremely important to follow the shaft to the very end to confirm the position of the tip. A cross-section image taken too proximal on the shaft may show that part extramuscularly giving false reassurance, when the tip might actually be located deeper into the muscular space (Fig. 45.14).
2. Failure to visualize the lipofilling: Subcutaneous lipofilling produces the characteristic local expansion seen in Video 45.2 and Fig. 45.10. Inability to see this during the fat transfer indicates either that the fat is not progressing due to a blockage in the cannula or tubes or that the images are being taken too proximal on the cannula shaft, missing the tip. In this situation, the lipofilling should immediately be stopped, and the position of the cannula tip should be confirmed in longitudinal as well as cross-section views. If no signs of lipofilling were found around the cannula tip, then the tubes and cannula should be tested for any blockage. Figure 45.14 demonstrates an example where the cannula was thought to be extramuscular based on the first image. However, lipofilling didn't show any local expansion. Lipofilling was stopped. The cannula was followed more distally and the tip was found penetrating the superficial layers of the gluteal muscle with unintentional small localized intramuscular fat.
3. Technical difficulties: The surgeon and theatre staff should be trained to use and troubleshoot the ultrasound

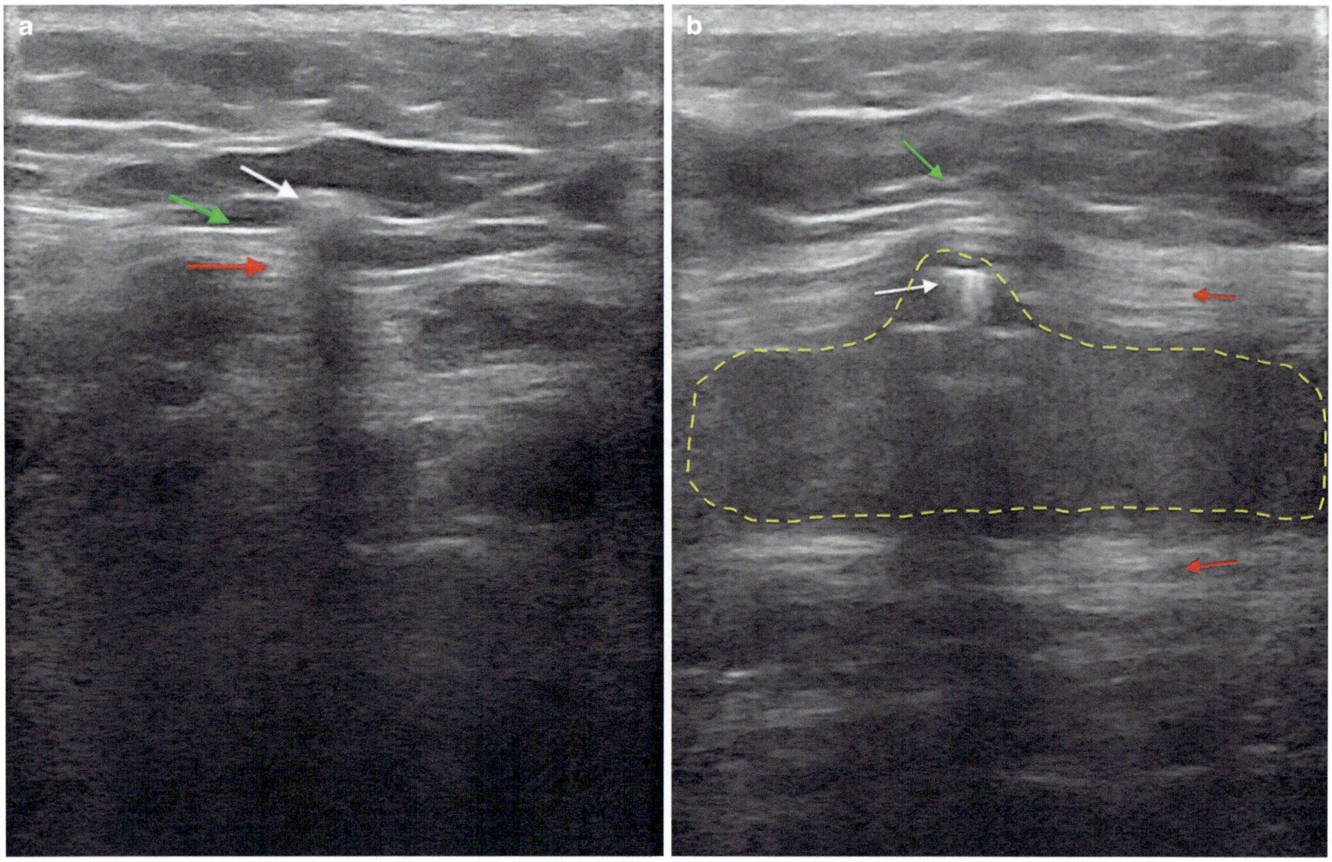

Fig. 45.14 (**a**) The cannula appears extramuscular because of the cross-section imaging was taken too proximal on the cannula shaft. (**b**) More distal visualization along the cannula shows the tip penetrated into the gluteus maximus muscle with unintentional intramuscular lipofilling. Green arrow: deep fascia, white arrow: cannula tip, red arrows: gluteus maximus muscle, yellow dotted line: intramuscular fat

scan machine. At least one spare battery for the wireless devices should be available and fully charged at all time. Any portable display device should also be charged and installed to allow easy visualization. It is recommended that the USS machine representative be present during the first few cases to help deal with any technical issues.

45.10 Conclusion

Understanding the mechanisms and causation of pulmonary fat embolisms allows us to adapt and modify our surgical techniques in gluteal lipofilling to improve the safety profile of this procedure [4, 12]. While early authors recommended intramuscular lipofilling, we now know that the transfer should be limited to the subcutaneous plane.

Fat tissue was found to be present into and under the gluteal muscle in all the reported cases of fatal fat embolism in gluteal lipofilling. However, surgeons claimed that they did not intend to inject the fat into the muscle.

The anatomical characteristics of the gluteal muscle and investing facia make them vulnerable to unintentional penetration. The USS-guided lipofilling provides the surgeon with a visual confirmation of the cannula position to avoid intramuscular lipofilling.

References

1. American Society of Plastic Surgeons. 2018 Plastic Surgery Statistics Report: 2018 Reconstructive Plastic Surgery Statistics. 2019;7–8. https://www.plasticsurgery.org/documents/News/Statistics/2018/reconstructive-procedure-trends-2018.pdf.
2. Plastic N, Statistics S. American Society of Plastic Surgeons—2020. Plastic surgery statistics report. Online. 2021;2020–2.
3. Astarita DC, Scheinin LA, Sathyavagiswaran L. Fat transfer and fatal macroembolization. J Forensic Sci. 2015;60(2):509–10.
4. Cárdenas-Camarena L, Bayter JE, Aguirre-Serrano H, Cuenca-Pardo J. Deaths caused by gluteal lipoinjection: what are we doing wrong? Plast Reconstr Surg. 2015;136(1):58–66.
5. Mofid MM, Teitelbaum S, Suissa D, Ramirez-Montañana A, Astarita DC, Mendieta C, et al. Report on mortality from gluteal fat grafting: recommendations from the ASERF task force. Aesthet Surg J. 2017;37(7):796–806.
6. Condé-Green A, Kotamarti V, Nini KT, Wey PD, Ahuja NK, Granick MS, et al. Fat grafting for gluteal augmentation: a systematic review of the literature and meta-analysis. Plast Reconstr Surg. 2016;138(3):437e–46e.

7. Oranges CM, Di Summa PG, Kalbermatten DF, Haug M, Schaefer DJ. Staying safe during gluteal fat transplantation. Plast Reconstr Surg. 2018;142(4):593E–4E.
8. Oranges CM, Tremp M, Di Summa PG, Haug M, Kalbermatten DF, Harder Y, et al. Gluteal augmentation techniques: a comprehensive literature review. Aesthet Surg J. 2017;37(5):560–9.
9. Cárdenas-Camarena L, Durán H. Improvement of the gluteal contour: modern concepts with systematized lipoinjection. Clin Plast Surg. 2018;45(2):237–47.
10. Villanueva NL, Del Vecchio DA, Afrooz PN, Carboy JA, Rohrich RJ. Staying safe during gluteal fat transplantation. Plast Reconstr Surg. 2018;141(1):79–86.
11. Rosique RG, Rosique MJF. Augmentation gluteoplasty: a Brazilian perspective. Plast Reconstr Surg. 2018;142(4):910–9.
12. Del Vecchio DA, Villanueva NL, Mohan R, Johnson B, Wan D, Venkataram A, et al. Clinical implications of gluteal fat graft migration: a dynamic anatomical study. Plast Reconstr Surg. 2018;142(5):1180–92.
13. Ordenana C, Dallapozza E, Said S, Zins JE. Objectifying the risk of vascular complications in gluteal augmentation with fat grafting: a latex casted cadaveric study. Aesthet Surg J. 2020;40(4):402–9.
14. Cansancao AL, Condé-Green A, Vidigal RA, Rodriguez RL, D'Amico RA. Real-time ultrasound-assisted gluteal fat grafting. Plast Reconstr Surg. 2018;142(2):372–6.
15. Ghavami A, Villanueva NL, Amirlak B. Gluteal ligamentous anatomy and its implication in safe buttock augmentation. Plast Reconstr Surg. 2018;142(2):363–71.
16. Illouz YG. Study of subcutaneous fat. Aesthet Plast Surg. 1990;14(1):165–77.
17. Rohrich RJ, Smith PD, Marcantonio DR, Kenkel JM. The zones of adherence: role in minimizing and preventing contour deformities in liposuction. Plast Reconstr Surg. 2001;107(6):1562–9.
18. Del Vecchio D, Wall S. Expansion vibration lipofilling: a new technique in large-volume fat transplantation. Plast Reconstr Surg. 2018;141(5):639e–49e.

Expansion Vibrating Lipofilling (EVL) in Buttocks Body Contouring: A Review of 50 Consecutive Cases

46

Omar Tillo, Alberto Di Giuseppe, and Federico Giovagnoli

Contents

46.1 Background

In November 2016, Simeon Wall published a retrospective study in the *Plastic and Reconstructive Surgery* journal on 734 cases performed with the so-called "SAFE Liposuction" [1].

The main steps of the procedure are separation, aspiration, and fat equalization:

1. Separation
   - The separation step uses an exploded-tip (flared Mercedes or basket) liposuction cannula without aspiration and takes extra care to loosen and liquefy the fat, which makes it much simpler to aspirate later, using cannulas which minimize damage to blood and lymphatic vessels in the vicinity of the treatment area.
   - This separation step, using exploded-tip cannulas, relates to the relative pressure gradient created between the wings of the cannula, which allows the targeted fat to separate from its attachments, creating a separated or emulsified fat. This pressure creates separation of the fat globules without causing an avulsion injury to the surrounding vascular structures, minimizing bleeding during the entire process.
   - As this method does not deliver thermal energy, the potential damage due to burning either to the skin or subcutaneous tissue structures is not a risk.
   - Separation of the fat allows release from the tethering structures as blood vessels and fibrous septa. Both superficial and deep fat are separated by continuous passing of the cannula.
   - Circumferential liposuction is performed to gain maximum skin retraction by the uniform undermining of superficial subcutaneous tissue while maintaining proper vascular network.
   - The surgical endpoint of the fat separation is loss of resistance in all areas to be treated.
2. Aspiration

   The clinical outcome is the main parameter to be observed when accomplished with simultaneous aspiration.
3. Fat equalization
   - After having completed aspiration, fat equalization is the final step of the technique, performed with the basket cannulas, without aspiration.
   - The aim is to smooth out irregular areas, in order to provide a uniform distribution of the subcutaneous and skin layer to the newly reduced body-contoured areas.

Supplementary Information The online version contains supplementary material available at https://doi.org/10.1007/978-3-031-10881-5_46.

O. Tillo · A. Di Giuseppe (✉)
Creo Clinic, London, UK

F. Giovagnoli
Milan, Italy

© The Author(s), under exclusive license to Springer Nature Switzerland AG 2023
A. Di Giuseppe et al. (eds.), *Fat Transfer in Plastic Surgery*, https://doi.org/10.1007/978-3-031-10881-5_46

- In case of persistent differences in the thickness of the tissue, further emulsification can be performed. Aggressive molding and equalization of the remaining fat improve body contouring, final shape, and outcome.
- A thin uniform subcutaneous tissue should be the final target to facilitate optimal skin retraction and the best contouring of the target areas. Final artistic sense of the surgeon must be able to optimize the shape and the contour of the body, recreating lines of concavity and convexity as needed.

The main advantages of SAFE liposuction are as follows:

1. Optimizing fat removal by using power-assisted liposuction especially designed "basket" cannulas, thus significantly reducing the total timing of the procedure.
2. Comprehensive body contouring of all body areas (trunk, face and neck, abdomen, arms, thighs, flanks, and calves) can undergo SAFE Liposuction.
3. Preserving vascular integrity.
4. Preventing contour deformity: the equalization step is the key to prevent contour deformity, which often follows liposuction, mostly due to fibrosis of the internal tissue, which leads to adhesions and scarring. By leaving a smooth thin subcutaneous layer of fat between the skin and the nearly contoured area, the post-op tissue adhesion is soft and uniform. Less skin leads to irregularities [2].
5. Minimizing post-op complications. SAFE Liposuction allows to harvest larger amount of fat from various areas with a low chance to having uneven, wavy, or irregular results. Other types of liposuction involving the use of thermal energy, such as laser liposuction, are more keen to lead to internal fibrosis and, thus, scarring and irregularities.

Based on steps learnt using the SAFE Liposuction, Daniel Del Vecchio [3] published an article in the *Plastic and Reconstructive Surgery* journal describing the Expansion Vibration Lipofilling (EVL) technique for fat transfer. This technique is the logical extension of the SAFE Liposuction method. The advantages here are a simultaneous disruption of recipient site connective tissue, internal expansion using exploded-tip cannulas, and backfilling of these spaces with a roller pump-propelled fat [4].

46.2 Materials and Methods

Since 2018, the authors have performed 98 cases of buttocks contouring with fat transfer using the EVL technique as described previously.

Fifty of these 98 cases were reviewed in a critical analysis of results, complications and satisfaction rate from patients. Photos were taken at 6 months post-op review.

Operating time ranged from 2 h:35 min to 4 h:23 min (mean 3 h:50 min), depending on the extension of donor areas of the case.

The average volume of fat transferred was 1785 mL (ranging from 770 to 3080 mL) to both buttocks.

Patient ages ranged from 20 to 58 years (average 33).

Patients were ASA 1 or 2.

BMI ranges between 19 and 30 (average 26).

Postoperatively, patients did wear compressive garments for 6 weeks. Complimentary MLD (Manual Lymphatic Drainage massage) twice a week for a month was advised to reduce post-op swelling.

TED stockings to prevent thrombosis and a 14 days course of subcutaneous low molecular weight heparin (Clexane 3000/4000 units depending on body weight) were prescribed to all patients.

46.3 Complications Encountered

Complications encountered were as follows: localized infection (two patients), localized fat necrosis (one patient), donor site seroma (one patient), donor skin surface irregularity (one patient), DVT, and PE (one patient). No patient had fat embolism.

The fat absorption rate was estimated to be between 30 and 40% of the total added initially.

The satisfaction rate of patients operated was high. When patients asked for their reported outcomes, 95% were satisfied with how their body looked when they were, 100% of patients were satisfied with how their clothes fit their body, and 79% were satisfied with how their body looked in swimming clothes. On a scale of 0–10 patients rated their overall satisfied from the operation as 8, and if they would recommend the procedure to a friend or family they scored 8.

Realself.com is a US-based healthcare marketplace focused on plastic surgery with 30,000 registered doctors, two million reviews and 94 million visitors annually. Based on 2450 reviews, BBL is reported as being worth undertaking by 85% of patients.

Low molecular weight heparin was given for 14 days post-op, starting the day of surgery, and early mobilization. In addition, patients wore thrombo-embolus deterrent (TED) stockings for 2 weeks and compressive garments for 6 weeks post-op.

Manual lymphatic massages were advised to all patients, once or twice weekly, for 4–6 weeks to facilitate retaining fluids absorption.

All cases were performed under total intravenous anesthesia (TIVA) as day cases and discharged from the hospital 1–2 h after completing the surgery.

46.4 Safety Issue

Ultrasound-assisted BBL is now the gold standard in authors' practice when performing BBL with EVL.

The procedure entails using a wireless ultrasound probe placed over the buttocks in order to identify the tissue layers, where the blunt basket cannula is introduced to the scanned area at the same time.

As the fat injection starts, the ultrasound images are projected wirelessly to a screen so the surgeon can follow the planes in which the cannula has been introduced.

Technology provides a precise objective method to avoid injuring the deep gluteal vessels and penetrating the muscular fascia and the deeper muscles.

It is a precise, not invasive method to monitor the cannula level and keep the injection site in a safe plane over the muscle.

Essentially, ultrasound turns a once-blind procedure into a not-blind procedure, which greatly reduces risks.

46.5 Technique

46.5.1 Devices for EVL

- Power-assisted liposuction (PAL) unit. Microaire (Microaire, Virginia, USA) (Fig. 46.1).
- Vitruvian infiltration pump by Black and Black Surgical, Georgia, USA (Fig. 46.2).
- Basket cannulas (Flared Mercedes), 3, 4, and 5 mm diameter Microaire (Microaire, Virginia, USA) (Fig. 46.3).
- Lipofilter 2500 mL Fat collection system, Microaire (Microaire, Virginia, USA) (Fig. 46.4).
- Clarius L15HD device, by the Clarius Mobile Health Corporation, Austin, USA (Fig. 46.5).
- A tumescent infiltration, rate one-to-one, of the target areas for fat removal, 1 L of normal saline with lidocaine 600 mg/L, adrenalin 1 mg/L, and 8.4% sodium bicarbonate 30 mL/L.

The position of the patient is often supine and then prone. Circumferential abdominal, torso and trunk are mainly targeted to harvest fat in the majority of cases.

Prepare the areas for liposuction by "separation" by using a 4 or 5 mm exploded-tip cannula without aspiration. Once this is completed, aspiration starts and the fat is collected in the Lipo filter canister.

Once fat aspiration is complete, the "fat equalization" is performed at this stage to correct unevenness and irregularities in remaining tissue thickness all over the areas.

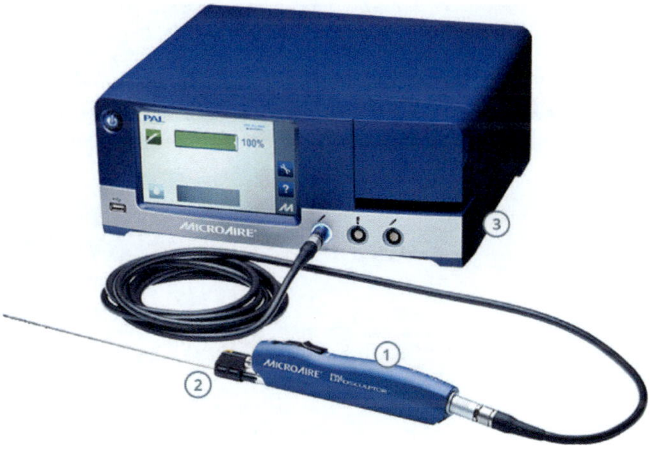

Fig. 46.1 Microaire. (1) Lightweight, autoclavable PAL handpiece. (2) Stainless steel cannula. (3) Powered by a stand-alone electric console. [Courtesy of Microaire, Virginia USA ©2022, All Rights Reserved]

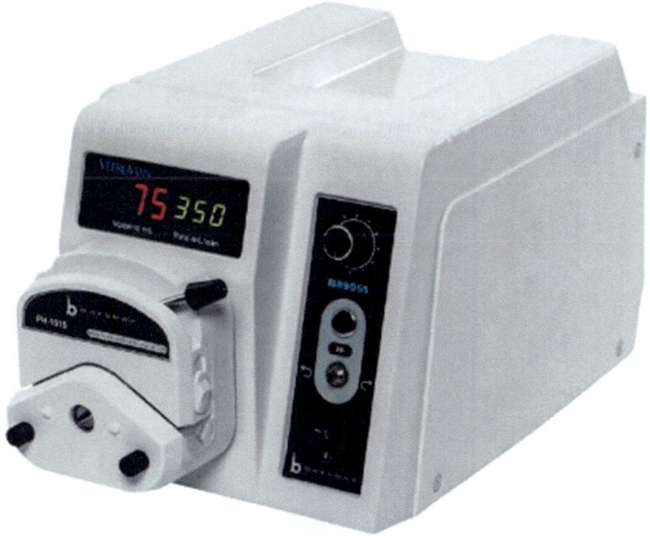

Fig. 46.2 Vitruvian infiltration pump. [Courtesy of Microaire, Virginia USA ©2022, All Rights Reserved]

Particular attention is given to shape properly the flanks in order to obtain a nice smooth contour.

46.5.2 Fat Treatment

The fat was washed while and after collection in the Lipo filter canister and left to decant for 20 min to allow the full separation of fat from fluids. Then, the fluid, blood, and oil are fully filtered and drained from the bottom of the canister until the canister is filled only by proper fat. The quantity of the fat is recorded (Fig. 46.6).

See Video 46.2.

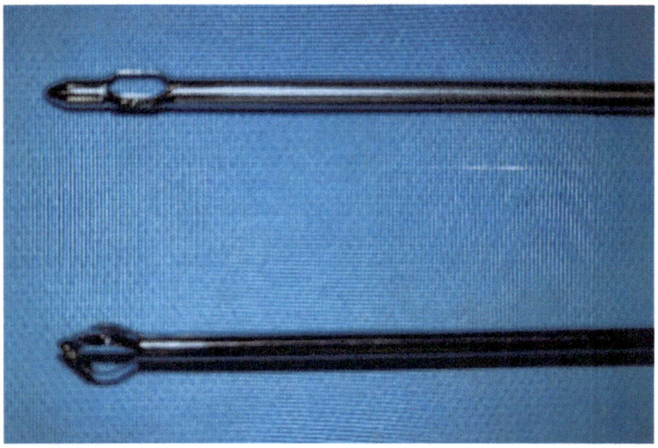

Fig. 46.3 Basket cannula with exploded tip. [Courtesy of Microaire, Virginia USA ©2022, All Rights Reserved]

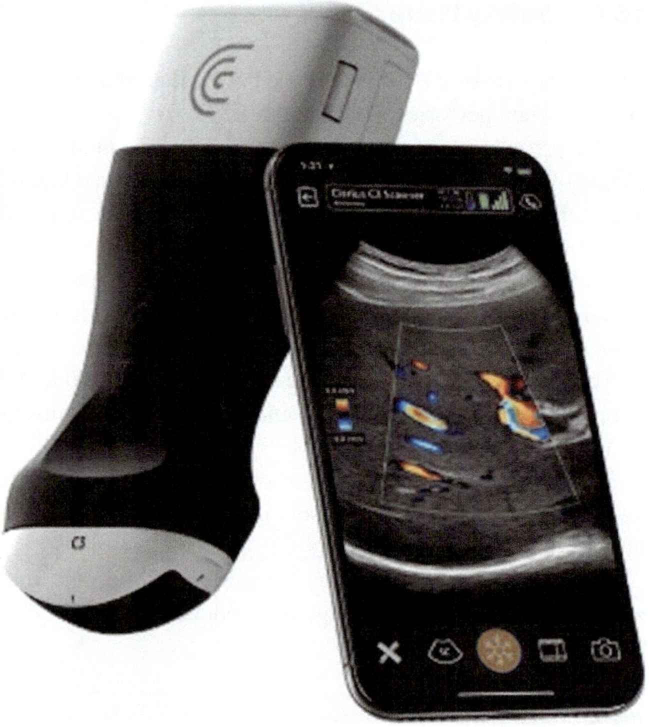

Fig. 46.5 Ultrasound. (Courtesy of Clarius Mobile Health Corporation, Austin, USA © 2022, All Rights Reserved)

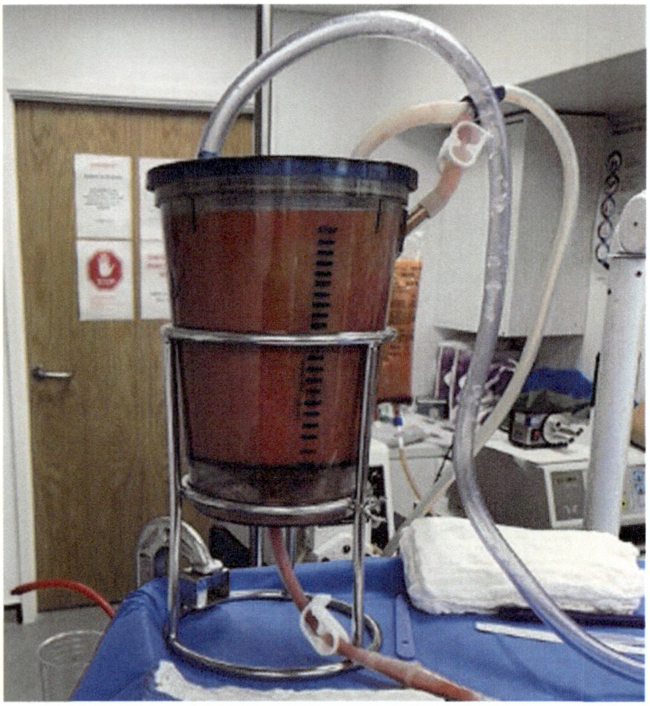

Fig. 46.4 Lipofilter. [Courtesy of Microaire, Virginia USA ©2022, All Rights Reserved]

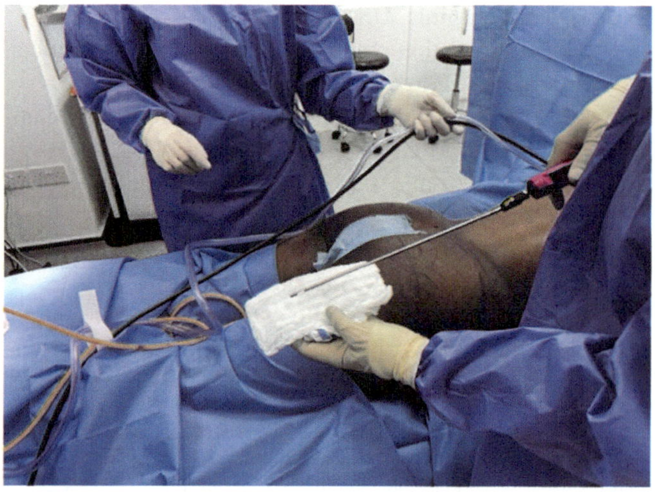

Fig. 46.6 Basket cannula ready for fat transfer

46.5.3 Fat Injection

The ultrasound device is positioned and activated to monitor the level of cannula placement in the deep subcutaneous plane. Then, the same 4 or 5 mm basket liposuction cannula is utilized to inject the fat.

Care must be taken to avoid penetrating the muscle fascia, a rather thin layer covering the gluteal muscles.

Once the surgeon confirms the subcutaneous level of the cannula, the Vitruvian infiltration pump is activated after proper connection with tubing, and the speed of infiltration is set. We do prefer a 100 mL a minute infiltration speed normally.

The power-assisted basket cannula is directed in the areas already planned. However, the expansion basket cannula does facilitate the proper separation from adhesion and septa from underlying tissue, further creating a proper space for fat positioning.

See Videos 46.1, 46.3, and 46.4.

A similar amount of fat is injected on each side if no previous asymmetry must be corrected.

The shape is progressively enhanced until the final contour is achieved.

Skin entrance sites are closed with 5.0 Vicryl rapid.

Dressing and the compressive garment are applied (Figs. 46.7, 46.8, 46.9, and 46.10).

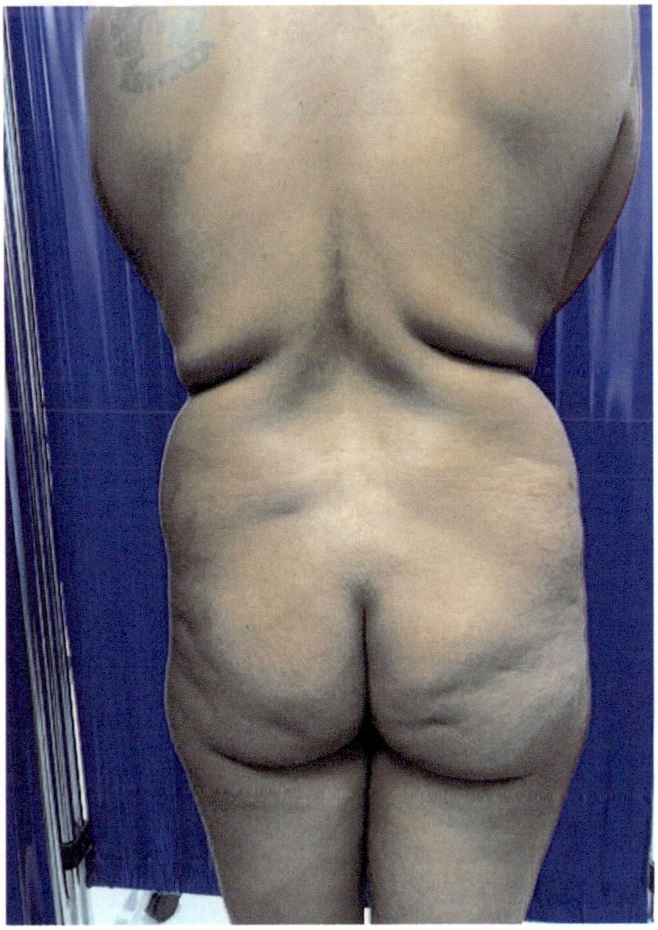

Fig. 46.7 Pre op photos of patient standing: note back rolls, loose, saggy skin at the upper and medium trunk, heavy flanks, and square buttocks

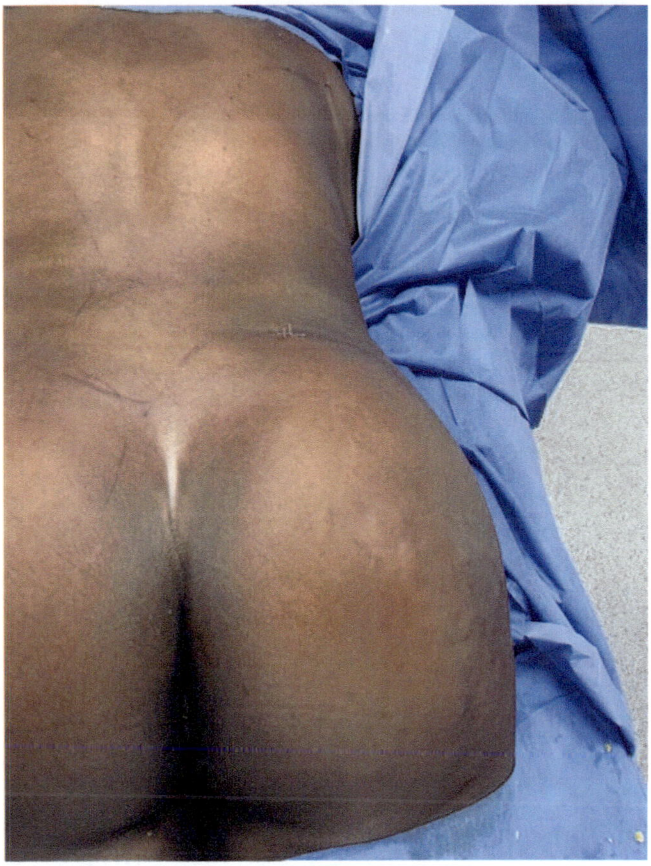

Fig. 46.8 Immediate post op result, patient prone. Back and trunk and flanks liposuction. EVL fat transfer to buttocks, 1200 mL in total

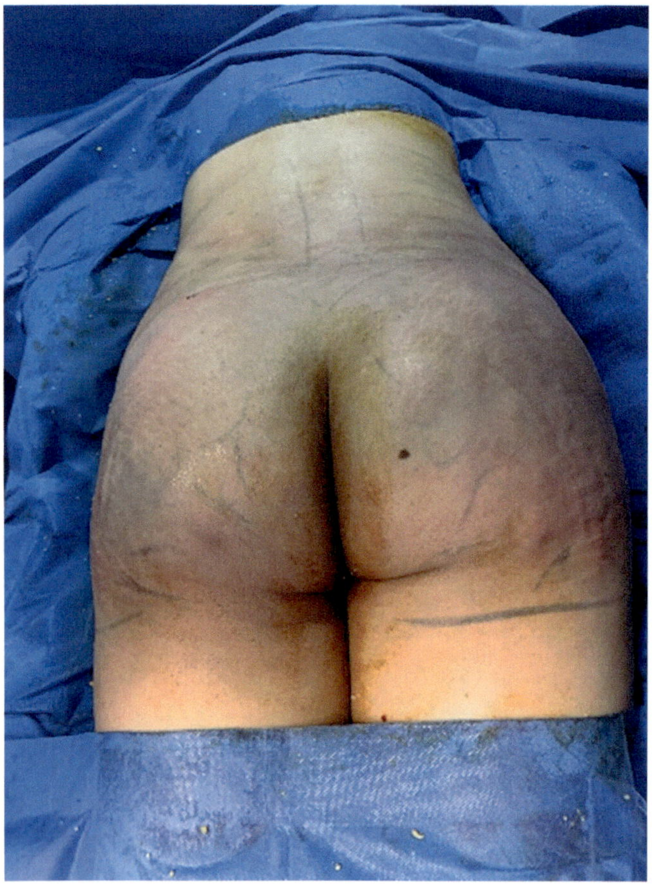

Figs. 46.9–46.10 Immediate post op, prone position. EVL to buttocks, after flanks and outer thighs debulking

46.6 Long-Term Results

The authors present here five cases at 6 months post-op follow-up.

Case 1
See Fig. 46.11.

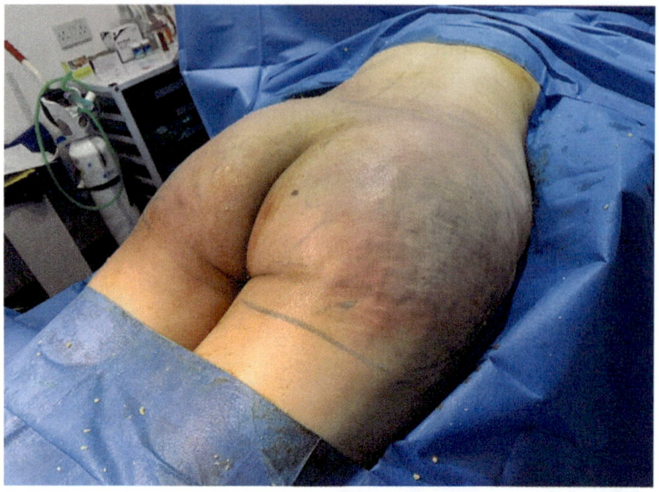

Fig. 46.11 Female, 35 years old, BMI 27. Procedure: 360 Liposuction abdomen, flanks, back and upper thighs. 2000 mL of fat removed, 570 mL transferred to each buttock (1140 mL in total). Before and 6 months after surgery. (**a**, **c**, **e**) Pre op. (**b**, **d**, **f**) Post op

Case 2
See Fig. 46.12.

Case 3
See Fig. 46.13.

Case 4
See Fig. 46.14.

Case 5
See Fig. 46.15.

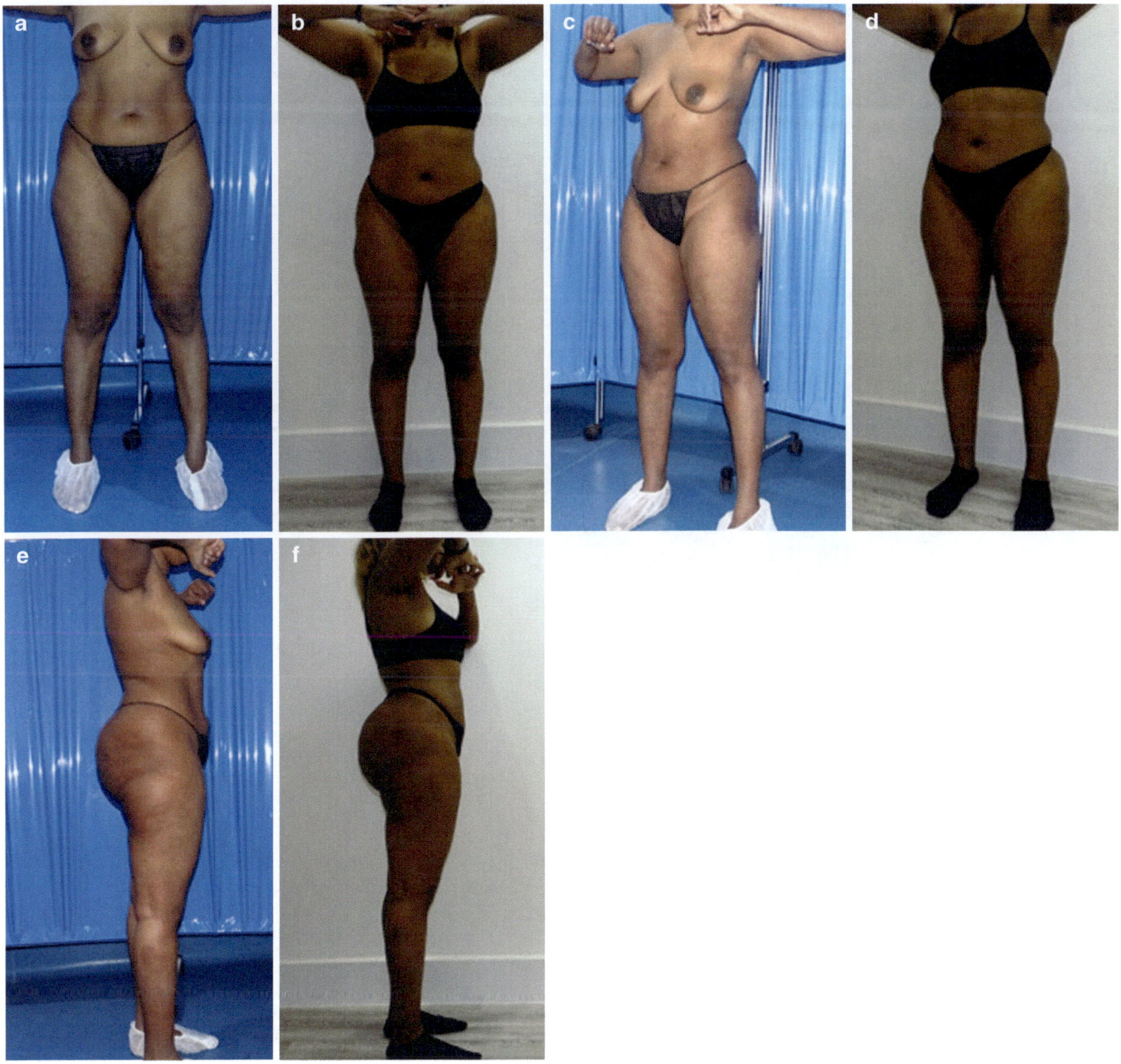

Fig. 46.12 Female, 36 years old, BMI 23. Procedure: 360 Liposuction abdomen, flanks, back, and upper outer thighs. 1400 mL of fat removed, 530 mL transferred to each buttock (1060 mL in total). Before and 6 months after surgery. (**a**, **c**, **e**, **g**) Pre op. (**b**, **d**, **f**, **h**) Post op

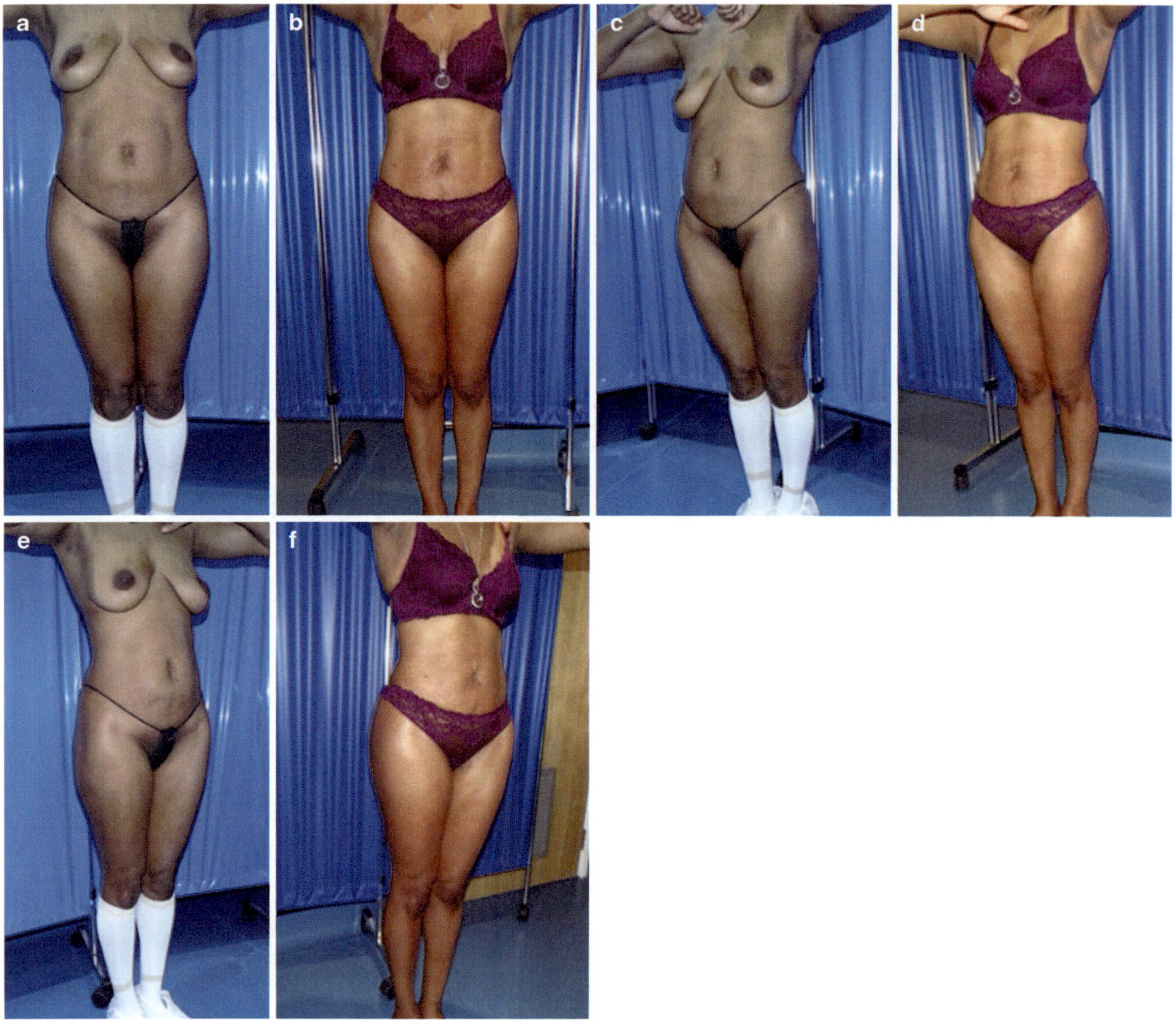

Fig. 46.13 Female, 33 years old, BMI 20. Procedure: 360 Liposuction abdomen, flanks, and back. 1300 mL of fat removed, 600 mL transferred to each buttock (1200 mL in total) using syringe-transfer technique. Before and 6 months after surgery. (**a**, **c**, **e**) Pre op. (**b**, **d**, **f**) Post op

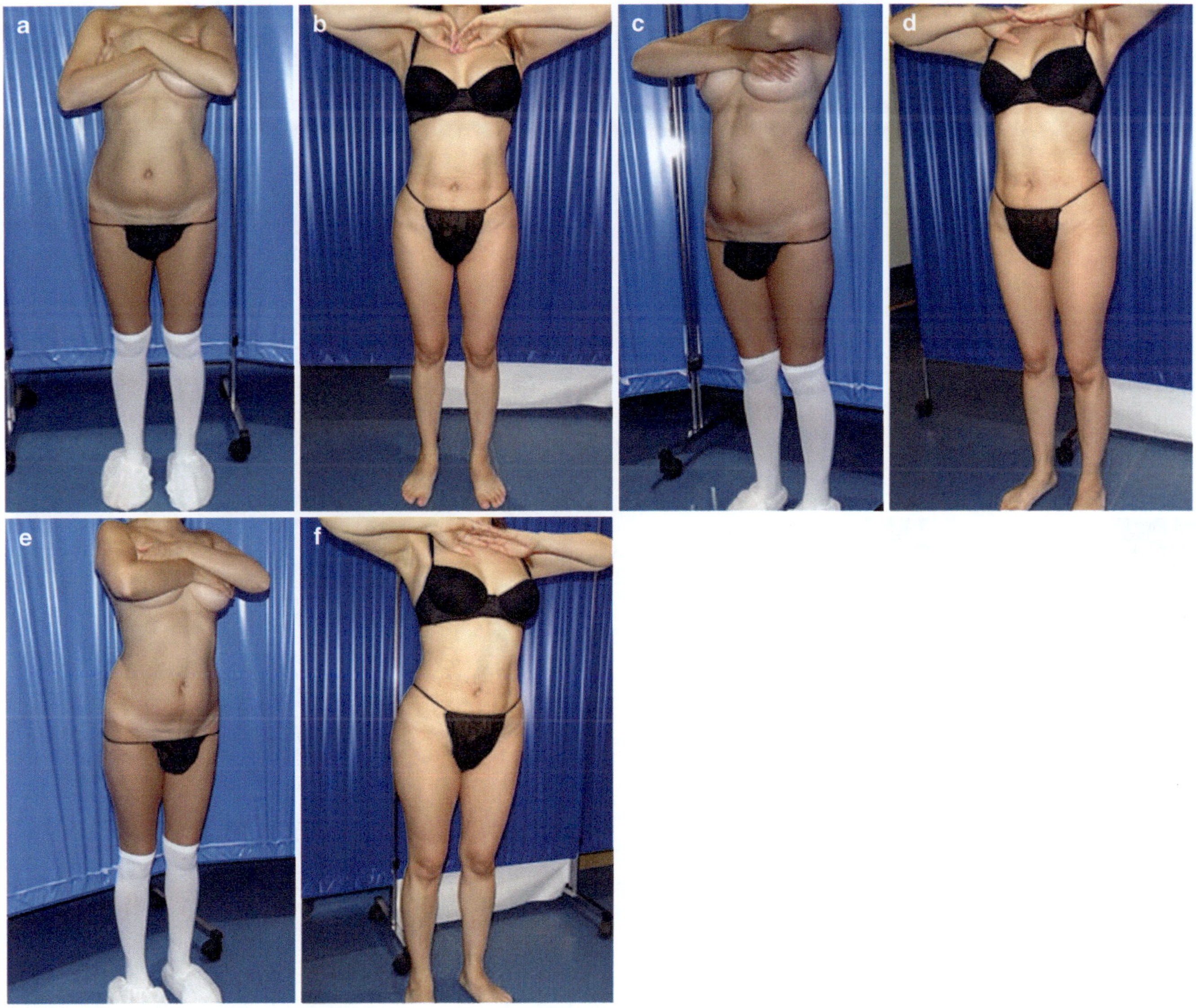

Fig. 46.14 Female, 33 years old, BMI 24. Procedure: 360 Liposuction abdomen, flanks, back and inner thighs, 2000 mL of fat removed, 750 mL transferred to each buttock (1500 mL in total) using EVL. Before and 6 months after surgery. (**a**, **c**, **e**) Pre op. (**b**, **d**, **f**) Post op

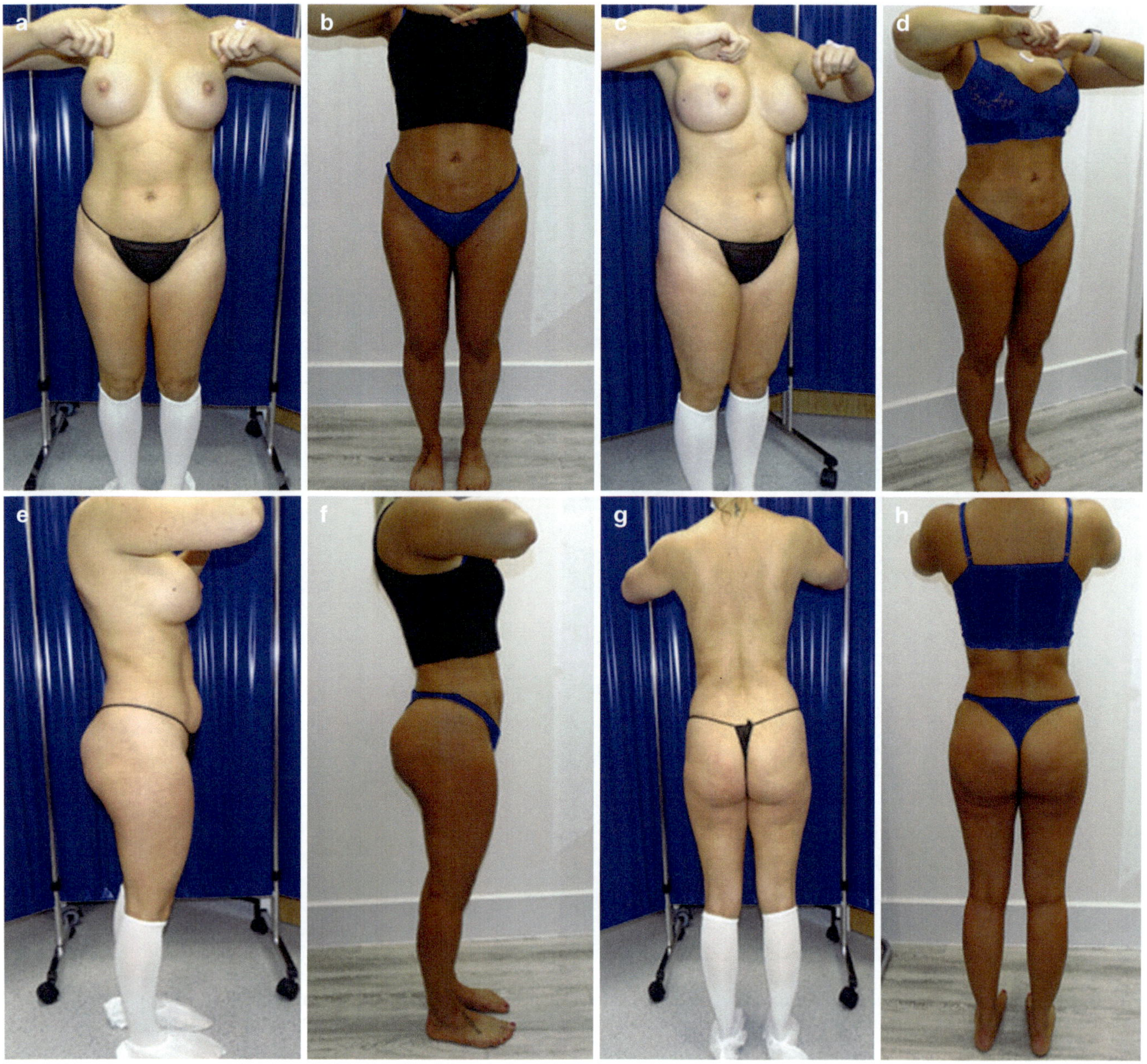

Fig. 46.15 Female, 25 years old, BMI 24. Procedure: 360 Liposuction abdomen, flanks, back and upper thighs, 2000 mL of fat removed, 900 mL transferred to each buttock (1800 mL in total) using EVL. Before and 6 months after surgery. (**a**, **c**, **e**, **g**) Pre op. (**b**, **d**, **f**, **h**) Post op

References

1. Wall S, Lee M. Separation, aspiration and fat equalization: safe liposuction concepts for comprehensive body contouring. Plast Reconstr Surg. 2016;138(6):1192–201.
2. Wall S. Safe circumferential liposuction with abdominoplasty. Clin Plast Surg. 2010;37:485501.
3. Del Vecchio D, et al. Expansion vibration lipofilling: a new technique in large-volume fat transplantation. Plast Reconstr Surg. 2018;141(5):639e–49e.
4. Khouri R, Del Vecchio D. Breast augmentation and reconstruction using pre-expansion and fat grafting. Clin Plast Surg. 2009;36:269280.

GPSR Compliance

The European Union's (EU) General Product Safety Regulation (GPSR) is a set of rules that requires consumer products to be safe and our obligations to ensure this.

If you have any concerns about our products, you can contact us on ProductSafety@springernature.com

In case Publisher is established outside the EU, the EU authorized representative is:

Springer Nature Customer Service Center GmbH
Europaplatz 3
69115 Heidelberg, Germany

Batch number: 10755964

Printed by Printforce, the Netherlands